COMMUNITY-BASED FISHERIES MANAGEMENT

COMMUNITY-BASED FISHERIES MANAGEMENT

A Global Perspective

DEVASHISH KAR

ELSEVIER

ACADEMIC PRESS
An imprint of Elsevier

Academic Press is an imprint of Elsevier
125 London Wall, London EC2Y 5AS, United Kingdom
525 B Street, Suite 1650, San Diego, CA 92101, United States
50 Hampshire Street, 5th Floor, Cambridge, MA 02139, United States
The Boulevard, Langford Lane, Kidlington, Oxford OX5 1GB, United Kingdom

Notices

Knowledge and best practice in this field are constantly changing. As new research and experience broaden our understanding, changes in research methods, professional practices, or medical treatment may become necessary.

Practitioners and researchers must always rely on their own experience and knowledge in evaluating and using any information, methods, compounds, or experiments described herein. In using such information or methods they should be mindful of their own safety and the safety of others, including parties for whom they have a professional responsibility.

To the fullest extent of the law, neither the Publisher nor the authors, contributors, or editors, assume any liability for any injury and/or damage to persons or property as a matter of products liability, negligence or otherwise, or from any use or operation of any methods, products, instructions, or ideas contained in the material herein.

Library of Congress Cataloging-in-Publication Data
A catalog record for this book is available from the Library of Congress

British Library Cataloguing-in-Publication Data
A catalogue record for this book is available from the British Library

ISBN: 978-0-12-821723-8

For information on all Academic Press publications
visit our website at https://www.elsevier.com/books-and-journals

Publisher: Charlotte Cockle
Acquisitions Editor: Patricia Osborn
Editorial Project Manager: Kelsey Connors
Production Project Manager: Omer Mukthar
Cover Designer: Christian J. Bilbow

Typeset by SPi Global, India

Dedication

This piece of work is dedicated with profound respect to the fond memory of my revered elder sister, late lamented Prof. (Mrs) Radha Rani Dev (Nee Kar), MA (English), MA (Sanskrit), who had taught me what English language is and also helped me with good English compositions on many occasions. It is also dedicated to the memory of my respected parents (late lamented Dr. Himangshu Jyoti Kar, MB (Cal.) and late lamented Mrs Sunity Kar, M.A. (Cal.) (who had made me what I am today), my respected teachers, my other family members, and to my beloved students and to my beloved fishermen.

Contents

Foreword

Nutrition and avocation are the two intricately linked aspects, which are, probably, indispensable for keeping the body and soul together. However, both these aspects are in the teeth of challenges in many parts of the world today. Incidentally, "fish" could serve as a potential entity for both nutrition and avocation, as well as could provide answers to the diminishing protein supply and shrinking employment scenario. But, for abundant availability of healthy fishes, availability of healthy water bodies is a prerequisite.

One could contemplate prolific availability of healthy and good quality of fishes. Concomitantly, in the teeth of keen competition in the today's scenario, there is also uncompromising need for efficient management of the water bodies and their coveted bioresources, notably the fishes. And, there are various *modus operandi* available today for judicious rational harvest, effective management, and conservation of resources for the posterity. And, community-based management of the resources, involving stakeholders from the grassroots level, seems to have no alternative to effective management of the resources.

I congratulate Prof. Devashish Kar for his endeavor and long-standing hard work in accomplishing this treatise on community-based fisheries management.

M.S. Swaminathan

Founder Chairman, Ex-Member of Parliament (Rajya Sabha), M.S. Swaminathan Research Foundation

Preface

"Water" is "life" and must be protected and conserved. Having its origin in water, life has evolved itself into an enthralling world of coveted and bewilderingly diverse flora and fauna. The dependence of the living world, notably the humankind, on the biological wealth of rivers, lakes, seas, and oceans cannot be overemphasized and, possibly, does not need any elucidation. Therefore, there is, perhaps, requirement to broaden and deepen our comprehension about the aquatic ecosystem with regard to its physical, chemical, and biological features and interactions. The water, as ecosystem, performs a number of significant environmental functions, notably recycling of nutrients, recharging of ground water, augmentation and maintenance of stream flow and recreation of people, to exemplify a few.

Notwithstanding the above, fresh water is one of the most important natural resources crucial for the survival of all living beings. It is even more important for humans as they depend on it for food production, industrial growth, hydropower generation, and waste disposal as well as for cultural requirements. Limnology is the science which deals with the freshwater environments, their physicochemical characteristics, their biota, and the ecosystem processes therein. Limnology is, therefore, universal in its significance.

Fishes are significant living components of both lotic and lentic systems. They constitute almost half the total number of vertebrates in the world. Of the *c.* 39,900 vertebrate species known to exist so far in the world, 21,723 are living species of fishes, of which *c.* 841 are freshwater species and the rest 11,650 are marine species. In the Indian region, alone, of the *c.* 2500 species, 930 are freshwater inhabitants and 1570 are marine species.

In view of the above, there is, probably, an urgent need for efficient management of the precious water bodies and the coveted fishes. It is estimated that the global population could reach *c.* 9.5 billion by 2050. It may not be very easy to provide food requirements for the burgeoning exponentially growing population; it rather is a matter of thought and grave concern, and a Herculean task and a Himalayan responsibility. Incidentally, projections often suggest that the current level of food production needs to be increased by *c.* 70%, which could again be a great challenge. Further, detailed projections often portray an augmentation in the production of *c.* 1 billion tons of cereals and *c.* 200 million tons of meat. Incidentally, the per capita fish consumption was believed to have increased from *c.* 6 to *c.* 19.2 kg/year in 2012, with Asia accounting for two-thirds of the global fish consumption, averaging 21.4 kg/capita/year in 2011. It is estimated that, even at the current rate of fish consumption, the world may need an additional *c.* 30–40 million tons of food fish by 2050 to take care of the expected population increase.

Concomitant to the above, it may be mentioned, in this connection, that global food fish supplies, until recently, were predominantly of a hunted origin, based mainly on wild capture fisheries. Nevertheless, like all our other staples, this predominance of wild source had changed to a farmed origin during the last few decades emanating from the sustained development progress of aquaculture to be at *c.* 6% per year over the last few decades. Incidentally, aquaculture production showed an attainment of *c.* 76,321,310 tons in 2011, in which the Asia-Pacific region contributed to > 80%, and the Peoples Republic (PR) of China is the leading nation accounting for *c.* 65.7% of the total global production. The question, consequently, thus arises whether this growth impetus in intensive aquaculture could be maintained continuously and sustainably in the days to come; especially, in the wake of increasing competition for primary resources (both physical and biological), as well as concomitantly, the need to maintain environmental integrity. It is, thus, in this context that the use of existing water bodies for the purpose of food fish production, particularly in rural areas in developing countries, assumes tremendous significance and an impelling need, mainly in order to find answer to the diminishing protein supply. Incidentally, culture-based fisheries is, generally, an environmentally friendly practice, and could also be regarded as a potential *modus operandi* for meaningful and effective use of the existing vast water resources for food fish production, in which the only external input is mainly the seed stock.

Concomitant to the above, there seems to be a compelling need for the sustainable, judicious, and rational harvest; and management, protection, and conservation of the coveted fish resources because once fish are raised, their farming is expected to grow by leaps and bounds. Otherwise, the whole stock of fish may become endangered, and ultimately exhaust and extinct vis-à-vis nothing may be left behind for our posterity for nutrition, avocation, and learning.

Under the above background, thoughts started to be generated and pouring in for finding ways and means for the management, protection, and conservation of the coveted fish resources, as well as the person or persons to whom such an honest and onerous task be entrusted with. Various methodologies and procedures have been tried; several hypotheses and theories have been advocated in different countries, in order to find the most suitable, convenient, friendly, all-pervasive *modus operandi* to successfully accomplish the task of rational and sustainable harvest; proper sharing; and management, protection, and conservation of the coveted resources in the aquatic domain. Various aspects, like sustainable development, participatory rural appraisal (PRA), comanagement, community-based fisheries management (CBFM) came into deliberations and interactions, in which it is often opined that involvement of the stakeholders from the grassroots level could be a very pragmatic approach for accomplishing the task of rational and sustainable harvest; proper sharing; and, management, protection, and conservation of the coveted resources in the aquatic domain. And, this treatise on **Community-Based Fisheries Management in A Global Perspective** is a very humble preliminary effort in this direction by the author.

Devashish Kar

Silchar, India

Acknowledgment

Today's pansy is tomorrow's hawthorn.' With the fervent prayer that this humble piece of work will benefit the people of this land, the undersigned offers his heartfelt gratitude to the numerous benevolent souls, who had never grudged him any help in his strenuous and hazardous field works spread over more than 44 years. The village folk who had kept a vigil with him in many a stormy nights could neither be named nor their help fully acknowledged.

May the undersigned utilize the privilege of this opportunity to express his profound regards and dutiful reverence to his respected late lamented parents Dr. Himangshu Jyoti Kar (Father) and Mrs. Sunity Kar (Mother), without whose blessings, this humble piece of work would not possibly have seen the light of the day.

The undersigned utilizes the privilege of this opportunity to pay his very humble profound regards to Professor M.S. Swaminathan, FRS, Ex-Director General, Indian Council of Agricultural research for his very kind Foreword.

The undersigned also records his deep sense of gratitude and profound regards to his respected teachers, notably, Dr. Subhas Chandra Dey, retired Professor of Gauhati University; late lamented Dr. Naresh Chandra Datta, retired Professor of Calcutta University; Dr. Martin Monk and Dr. R.G. Bailey of King's College, University of London for their constant encouragement, guidance, and blessings.

The undersigned also expresses his deep sense of appreciation and gratitude to his family members, notably Mr. Devajyoti Kar and Dr. Deva Prasad Kar for their help and encouragement. Last but not the least, this work would have, perhaps, been nipped in the bud, if my wife Dr. Prof. Mrs. Swarupa Kar and my children Miss Devarati Kar (daughter) and Master Dyutiman Kar (son) would not have been with me like a shadow. I must say here that I am fortunate to have the intellectual stimulation, companionship, and association with a band of excellent, intelligent, and obedient students who had been with me at different stages of preparation of this humble piece of work. In particular, Ratnabir, Ramen Barman, Romen Singh, Binku, Puinyabati, Ranibala, Gitarani, Sangita, Binky, Sulata, Papia and Uma, and Bimola, helped me in different works. But, it was, my very affectionate students, Biplab Das and Satyajit Das, who had been with me like a shadow helping me at almost every step since I had started writing the manuscript. Biplab and Satyajit had helped me substantially in the arrangement of the text and in computation works. Satyajit had enormously helped me in arranging the references. The undersigned also wishes to acknowledge the help and cooperation received from his esteemed colleagues, friends, well-wishers, and all those benevolent souls connected with this humble piece of work. Particularly, I would like to mention the name of late lamented Govinda Datta and Proprietor, M/S Jayanti Press, Silchar; and late lamented Devapriya Kar, FCA, LLB; and late lamented Prof. Mrs. Radha Rani Dev, who had not only been giving me encouragement and blessings as my brother-in-law, elder brother, and elder sister respectively; but also guided me in many aspects of preparation of this work. Further, contemporary literatures have been consulted with gratitude.

The author utilizes this opportunity to express his deep sense of gratitude to M/S Elsevier (Academic Press), USA, for them acting as a pillar of logistic support, for their excellent manuscript delivery system, and for their kind efforts in the publication of this work.

Silchar, Assam, India
The 22nd day of September, 2020 (Devashish Kar)

CHAPTER

1

Introduction

OUTLINE

Abstract

The management of natural resources by means of community-based activities is often called common pool resource management. It, of late, has become a common strategy for enhancing resource management and empowering the local stakeholders. Since the past few decades, this aspect has assumed significance based on a number of concepts, viz., comanagement, use of local knowledge, recognizing local institutions, establishing common arenas (Berkes, 1989; Ostrom, 1990; Pomeroy and Berkes, 1997), and so on. Further, there occurs, sometimes, an issue regarding what best steps could be taken to initiate such good practices, and what participatory planning methods could be used; although, there has been focus on designed principles for community management institutions (Ostrom, 1994) and identifying factors linked with sustainable common property regimes and institutions.

Keywords: CBFM, Natural resources, Fisheries management, Sustainable development, Common pool, Private management, Comanagement

1.1 Community-based fisheries management (CBFM) as an emerging concept and as an exercise

The management of natural resources by means of community-based activities is often called common pool resource management. It, of late, has become a common strategy for enhancing resource management and empowering the local stakeholders. Since the past few decades, this aspect has assumed significance based on a number of concepts, viz., comanagement, use of local knowledge, recognizing local institutions, establishing common arenas (Berkes, 1989; Ostrom, 1990; Pomeroy and Berkes, 1997), and so on. Further, there occurs, sometimes, an issue regarding what best steps could be taken to initiate such good practices, and what participatory planning methods could be used; although, there has been focus on designed principles for community management institutions (Ostrom, 1994) and identifying factors linked with sustainable common property regimes and institutions.

Attitudes toward the management of natural resources have been changing worldwide. These alteration(s) have originated mainly from concerns about the state of the resources as they come under increasing pressure to fulfill a range of demands. Food is said to be one of the most important of these demands, especially in tropical countries. It is mainly because such nations, in some form or other, use the local inhabitants to exploit the natural resources. In addition, the latter are also threatened of their sustenance, by impact from other issues like pollution, environmental degradation, etc. In fact, it is heartening to note here that the capacity of present agricultural and industrial technologies for exploitation and for causing damage has far surpassed the capacities of the societies to accumulate, grow, and control such changes. Efforts to minimize such impacts often encounter difficulties resulting from political social and economic factors, and technological and commercial factors have, possibly, little role to counter such impacts and lead to a tangible result. Concern over such trends led to convening of different kinds of meetings, conferences, workshops, symposia, meets from local, regional, national level to spreading its horizon to the global level like convening of the United Nations Conference on the Environment and Development (UNCED) in 1992 and the acceptance of its agenda 21. This has highlighted the problems and provided a model framework and guidelines for sustainable use of natural resources through governments' commitments.

Concomitant to the above, it may be relevant here to preliminarily deal with some of the pertinent terms:

(a) Fisheries management: It may be defined as the pursuit of certain objectives through the direct or indirect control of effective fishing effort or some of its components.
(b) Sustainable development: It could be defined as the development that meets the needs of the present generation without compromising the ability of the future generations to meet their own needs.
(c) Participatory rural appraisal (PRA): The challenge facing the field practitioners is to find less expensive but faster methods to acquire information through the active participation of target communities without sacrificing the scientific validity. These applied techniques in field data collection have become commonly known as participatory rural appraisal (PRA) or rapid rural appraisal (RRA).

1.1.1 Management regimes of Floodplain wetlands in general

The management regimes of the wetlands could be categorized into four broad categories, based mainly on the property rights and access to the wetland:

(a) Private management by individuals or group(s).
(b) Fishermen's cooperative society management.
(c) Community-based fisheries management (CBFM)
(d) Decentralized management, in which the government works mainly as a facilitator (generally in the open-access wetlands).

These are briefly discussed in the following.

1.1.1.1 Private management

It relates to the management of wetlands de facto by the leaseholder (lessee). The lease period which was earlier 1–5 years has been amended to 7 years in many places in order to provide incentives to the lessees for adopting stock enhancement measures. In some other places, the management of the wetland, including the fishing operations, is done by the lessee according to his choice. But, there are restriction(s) imposed on him that no wetland would be totally drained out to dry condition by the lessee. Rather, as a general rule, the lessee is required to ensure retaining sufficient amount of water in the wetland for the protection of the fish and its juveniles, as well as for providing drinking water to the cattle and the people. In the tropical countries, the lessee is also required to keep the fisheries of the wetland clear of water hyacinth and other aquatic weeds. It may further be noted here that in countries **like India and Bangladesh**, fishing by >25 persons at a time, for their own consumption or on payment made by them to the lessees, is prohibited because it is considered highly detrimental to the interest of both the lessee and the government. In the regime of private management, still further, the wetland fisheries management, including the stock enhancement measures, weed management, fishing operations in the wetlands, etc., is done by the lessee, as in wetlands of Assam in India. As a common preventive measure, access to other fishermen into the wetlands is regulated by the lessees in the **Asian** countries, either under some sharing arrangement or by paying some amount of money to the lessee.

1.1.1.2 *Cooperative management or management through cooperative society*

The lease of the fisheries in wetlands is to be given to the fishermen or fishermen cooperative societies, on priority, under the Fisheries Act in many **Asian countries**, notably **in many places in India and Bangladesh**. In the cooperative management of floodplain wetlands in these countries, the management of the fisheries operations is done by the members of the cooperative society, as per the rules prescribed by the society. Here, the ownership of the property and access to the wetland are governed by the member of the cooperative society. There are two types of sharing arrangements: (a) with the individual member(s) and (b) with the cooperative society itself. In some cooperative societies, the *modus operandi* in vogue may be that the fish catch is pooled together and sold in the market, and a part of the income is shared among the members of the community. In other types of cooperative societies, fishers are reported to give 25% of their catch to the society as a share of the revenue.

1.1.1.3 *Community-based fisheries management (CBFM)*

There has been a history of traditional management and lessee-based management approach of wetland fisheries management in many countries in **Asia**. Following this, there seems to be much interest, currently in vogue, in a novel approach among government agencies as well as among the NGOs, which is (i) the community-based approach and (ii) the comanagement approach, which involves the role of the local communities (stakeholders) in wetlands' management and their conservation.

The **community-based approach** or the **community-based fisheries management (CBFM)** approach in the case of fisheries refers to a system of management involving the stakeholder fishermen from the grassroots level in decision-making and other management procedures for the water body.

On the other hand, **comanagement (CM)** refers to a bunch of arrangements with different degrees of power sharing allowing joint decision-making by the state and the user groups (UGs) about a set of resources or an area. However, there is, perhaps, no single globally accepted definition of comanagement. Incidentally, comanagement could be referred to as a host of arrangements with different degrees of power sharing allowing joint decision-making between the state and the UGs about a set of resources or an area.

In this connection, it may be pertinent here to know the different types of comanagement. **Comanagement could be of the following types:**

Instructive type: It is the mechanism of consultation between the government and the UGs, where the decisions are taken by the government.
Consultative type: It is the mechanism of consultations between the government and the UGs, but the decisions are still taken by the government.
Cooperative type: It is the mechanism of consultations between the government and the UGs to cooperate on an equal basis in management and decision-making.
Advisory type: In this type, the UGs advise the government about the decisions to be taken, and the government usually endorses them.
Informative type: Here, the government delegates the responsibility to the UGs, who, in turn, are responsible for informing the government about their decisions.

In contrast to the above, comanagement (CM) seems to be different from community-based management (CBM) or community-based fisheries management (CBFM) in the case of fisheries: mainly with regard to the fact that the latter, i.e., the CBM or CBFM, is a self-governance system, perhaps, not involving the government in the decision-making process.

For example, the Government of Assam in India, of late, under the Assam Agricultural Competitiveness Project (AACP) has initiated community-based fisheries management (CBFM) in a number of wetlands (locally called "Beels") in Assam under the care of the Department of Fisheries of the Provincial government. These Beels have now been included under the administrative jurisdiction of the fisheries department (of the Government of Assam) by the revenue department. In these wetlands, as a *modus operandi* under CBFM, the Beel Development Committee (BDC) has been formed with the membership of fishermen and fisherwomen of adjoining village(s). The BDC is a group of 20–500 individuals, living in the adjacent village(s) of a Beel, who generally come together for effective utilization of the natural fisheries resources of the Beel; better price of the produce; and, nevertheless, for more market power for the enhancement of the livelihood of the landless, small, and marginal Beel users in a sustainable manner. The number of members (individuals) of a BDC depends on the size of the Beel and number of villages surrounding the Beel. The BDC then constitutes its executive committee (EC) for the day-to-day operation of Beel fisheries. It may be noted here that Assam is bestowed with a rich diversity of large number of wetlands with innumerable fishermen and a bewildering diversity of fishes in them. The fishery department of the Government of Assam, in fact, virtually works as a facilitator in the operationalization of CBFM. Almost similar situation could also be in vogue in other countries like Bangladesh, Myanmar, etc.

1.1.1.4 Decentralized management

In this type of management, the government works mainly as a facilitator, generally, in the open-access wetlands.

1.1.2 Little details of CBFM

1.1.2.1 What is CBFM and how it works?

CBFM could essentially be regarded as a strategy for empowering the community to manage their resources in a sustainable manner and conserve the natural wealth for posterity and prosperity. It is often said that the limitations in fisheries management in developing countries (especially in their coastal waters) have led to the formulation and implementation of CBFM. **In India, and in some of the neighboring Southeast (SE) Asian countrie**s, there had been pioneering efforts by the FAO for executing the Bay of Bengal Program for Coastal Fisheries Management (BOBP) through CBFM exercise, and this pilot program is said to have been very successful. In fact, the success of BOBP activities in **Phang Nga Bay in Thailand** is widely recognized and considered as a role model on CBFM.

Conventional fisheries management involves regulations in many **Asian and South American** countries, which aim at renewing resource abundance through the observance of different kinds of restrictions, like closed season, closed area, gear restrictions, etc. It also involves regulation of fishing effort(s) in the form of limited access, fishing quotas, taxation, etc. Monitoring and enforcement costs are generally high. It may often become quite difficult for the governments, especially in the **developing countries**, to meet such costs. Governments, who manage the resources, often may have difficulties in understanding the complexity of fisheries and the communities who go out for fishing. The fishers, concomitantly, also do not like to be always enforced unless they understand the rules and regulations. Enforcement is otherwise also expensive. Under such circumstances, the CBFM appears to be more practical and cost-effective. Further, CBFM may help in improving the efficiency, equity, and cost-effectiveness of fisheries management in the long run.

Concomitant to the above, the "fisher folk," in many countries, now seem to be gradually trying to understand and accept that "management" is not an imposition by the government but an individual and collective responsibility of all stakeholders, producers, and consumers alike. Awareness-building programs seem to have brought about such change(s) in attitude that "management" has become an integral component of sustainable exploitation and does not unnecessarily impose burden on them.

Notwithstanding the above, the basic principles of CBFM are as follows:

(a) Participation of fishermen in planning and decision-making process(es).

(b) Surveillance, evaluation, implementation, and control of management activities are important aspects. As reported, it has been a general observation, through experiences in many countries, that when fishermen themselves help to design management, a high rate of compliance could be expected. There will, perhaps, be no need for external enforcement of regulations. Consequently, the enforcement could be both situation-effective and cost-effective. Further, it is now widely recognized that in a closely knit fishing community, social sanctions are far more effective than legal sanctions.

1.1.2.2 Requirements of CBFM

CBFM is designed to manage both the fishery resources and the fishermen. Successful exercise on CBFM could happen if there is good institutional environment such as effective laws and regulations. Success is more likely if the environment is CBFM-friendly. Further, CBFM also relies on traditional knowledge and wisdom of the local community who earn their livelihood mainly through fishing. Once fishing right has been granted, they try to apply the indigenous traditional knowledge for effective management of the wetland/lake resources. While dealing with the implementation of CBFM, the issues to be looked into include the following: (i) the nature of the resources; (ii) the technologies used to exploit these resources; (iii) arrangements available for property rights associated with different kinds of exploitations; (iv) strategies for management; and (v) the effectiveness of informal and formal rules within the present management scheme. In this regard, it may be pointed out here that the success of CBFM may be questionable, without an appropriate supporting institutional framework. Further, additional costs may be needed to strengthen and build up community management capacity, where local institutional framework is weak. It could, therefore, take a long time in developing local capability and participation in fishery resource management when there is no strong local organization.

Notwithstanding the above, several prerequisites are to be fulfilled to implement CBFM. Some of the significant ones are: (i) devolution of authority of management to the local community, (ii) establishment of territorial boundaries, and (iii) incentives and motivations to be given to fishermen to set up local management systems.

When the responsibility of management of fisheries is delegated to fishermen, care must be taken to ensure a fairly equitable sharing and distribution of benefits among the fishermen. According to some workers, "fishing by rotation" could be one of the ways to achieve this objective. Fishing centers could be rotated among fishermen so that all of them get the opportunity to fish in the most productive zones. Alternatively, a "pooling system" distributes all the earnings from fishing in an area equally among the fishermen of that area. This could be another *modus operandi*. The latter method is practiced in **Japan**. Further, the community-level organizations, which are entrusted with the responsibility of fisheries management, should be economically and socially viable. Otherwise, there could be possibility that fishermen may not trust the organization. CBFM may also be integrated into the state's legal framework because community-based organizations need legal approval for their decisions to be implemented.

Notwithstanding the above, CBFM exercise may be unsuccessful in a community where the fishing resource may not be exclusive to the community. Without exclusivity at the beginning, the community may not be assured of the right over the resource. This may dampen their willingness to cooperate in adopting the CBFM mechanism. Establishment of a clear, defined fishery boundary, in the area adjacent to the community, is a basic element of CBFM mechanism. The merit of demarcating waters for the exclusive use of fishermen is that it encourages the fishermen with incentives to establish self-regulating systems. They could, thus, feel that they own a wealth-producing property. Compliance is built on trust. Control over resources by fishermen would make a management regime feasible, and this may motivate them to preserve their resources.

1.1.3 Implementation and operationalization of CBFM

Many governments, throughout the globe, today are willing to adopt the CBFM, appreciating the fact that fisheries management has not worked out as planned by the governments. It is mainly because CBFM involves more consultative and participatory mechanisms. However, the following aspects may be considered in the implementation of CBFM:

(a) To ascertain and confirm the real beneficiaries from CBFM, and the benefits given.
(b) To ascertain the capacity of the community and needs for training in CBFM, if any.
(c) To ascertain the change(s) needed at the community level with regard to the implementation of CBFM.
(d) To identify the key persons at the community level and their roles.
(e) To ascertain the roles of supportive agencies, both government and nongovernment.
(f) To organize appropriate outreach program(s) to help in empowerment and extension activities.
(g) To estimate investment(s) required in building up community capacity in effective management.

CBFM is usually implemented in two phases, as discussed in the following.

1.1.3.1 The promotional phase

This is basically an information-gathering phase. CBFM concepts are identified on the basis of a SWOT (strength, weakness, opportunity, threat) analysis. The information thus gathered and analyzed is then disseminated to relevant government officials, especially the fisheries officials, at all levels. Subsequently, seminars and workshops are organized to enhance the understanding of CBFM among stakeholders and to explain their individual roles in the implementation of CBFM.

1.1.3.2 The implementation phases

During the implementation phase, various activities are carried out. At first, the government has to chalk out the developmental programs that could promote collective activities by all or a majority of the stakeholders. This is important because success in CBFM depends on getting the stakeholders to work together, thus instilling a sense of co-ownership to that program.

In order to facilitate the integration of community groups into the fisheries management programs, step-by-step guidance on the implementation of ecosystem-based fisheries and watershed management is essential. Notably, the guidelines could provide community-based groups with a template of integrating the needs of the communities into a framework for supporting sustainable fisheries and watershed management.

Further, the guidelines may also describe how to develop a shared vision for the future. This could help identify the primary factors constraining progress toward sustainable fisheries. Moreover, there is the need to determine the indicators of progress, and to develop a focused plan of action that includes the priority activities which may contribute directly to fulfillment of the long-term vision for the future.

Notwithstanding the above, the government agencies in many Asian countries, notably in Manipur in India, e.g., the Loktak Development Authority (LDA), generally provide with support on community empowerment including supporting framework, laws and regulations, awareness generation, and financial assistance which may be required to be undertaken; all these may be negotiated and agreed upon. Local fishermen, as stakeholders, will have to be effectively involved in the decision-making process. Management plans for fisheries could be drawn up and decided by the local fishermen by following bottom-up approach. Actions and alteration(s), which may take place, may be agreeable at the community level. Effective involvement of local fishermen may further be augmented if the benefits to be received from CBFM are transparent, quick, and proportionate to their contributions. CBFM, thus, could be an answer to an effective fisheries management in Loktak Lake provided capacity is built at all levels.

What is meant by participatory wetland management?

Man is a part of the ecological system, and not detached from it. This is said to be the underlying idea, and believed to be recognized by the community-based resource management.

It is, possibly, appreciated by some quarters in the globe that today's wetlands, including those considered to be the most pristine, are the result of complex interactions among physical, chemical, biological, and anthropogenic factors over a period of time. In fact, it could be opined that almost all the wetlands of the globe have been influenced and altered by various patterns of, more or less, vigorous use by man. Nevertheless, wetland management, on a global scenario, by the local people may have a history of thousands of years.

In this connection, **participatory management** may, in general, be **defined** as follows:

It is a partnership in which government agencies, local communities, and resource users as well as, perhaps, the other stakeholders, such as NGOs, share the authority and responsibility of management of a specific area or a set of resources.

Concomitant to the above, Addun and Muzones (1997) have opined that there are five fundamental principles which are required for community-based resource management (CBRM) as follows:

1. Empowerment: The actual transfer of political and economic powers from a few individuals to the impoverished mass, and the concomitant operationalization of community management and control.
2. Equity: Benefit(s) may reach the communities as a whole, rather than reaching a few individuals only.
3. Sustainability: It relates to intergenerational equity, based on the carrying and assimilative capacity of the ecosystem.
4. Systems' orientation: It relates to the aspect that a community is a function in relation to the other communities and stakeholders. It could be similar to the situation that resources in a smaller ecosystem are ecologically linked to wider ecosystems.
5. Gender fair: It may be ensured that the women are associated with the process of regulation and management of community resources. Also, their practical and strategic needs may be addressed.

Concomitant to the above, it may sometimes be evident in some countries that the degree of community participation may vary with the local context with regard to wise use of wetlands. Such variations could be from high levels of empowerment to effective and fruitful partnerships between the government agencies and local communities. These could also extend to such situation(s) where government remains firmly in control, but the stakeholders are consulted while arriving at decisions.

In addition to the above, there appears to be a growing awareness among the people in some of the **Asian countries** that in those areas where indigenous and traditional people live, and have been living there for hundreds to thousands of years, the authority for resource and ecosystem management may, preferably, be devolved as much as possible to the local level. It may be pointed out here that the indigenous people have been demanding recognition of their rights almost **all over the globe**, and, further, they might be expecting a greater say in decisions which might be affecting their lives. Incidentally, there seems to be a growing feeling that the rich biodiversity often coincides with prolific cultural diversity. In these areas, the trend in ecosystem management is increasingly, however, toward systems of collaborative management with indigenous people (Oviedo and Brown, 1999). Nevertheless, one clear lesson, to date, is that it is unlikely that even the simplest comanagement regime could be fruitful in the long run unless a relationship of trust between communities and government agencies could be established and effectively maintained. However, in a majority of instances, in **many countries,** where the communities are in a subsistence relationship with the resources, their level of trust in the government agencies is very low or even nonexistent. Under such circumstances, long periods of dialog(s) and shared learning(s) may be required in order to build the required amount of trust needed for comanagement to work in an effective manner (Claridge and O'Callaghan, 1997a,b). In this regard, it may be kept in mind that once the communication channels go on functioning, the best way to build a solid foundation of trust, among the concerned partners, is to ensure that the plans and targets (which had been mutually agreed upon between the communities and the relevant government agencies) are implemented. Moreover, it is notable here that a sufficiently long period of analysis and problem identification, particularly during the initial phase, could ensure that resource management plans are generally formulated based not on token participation but rather on genuine mutual agreement(s) which have been derived from a clear analysis of the benefits and costs to all stakeholders, in both the short term and the long term.

Notwithstanding the above, the Ramsar guidelines (Ramsar Convention on Wetlands, 1999) for establishing and strengthening local communities' and indigenous people's participation in the management of wetlands seem to provide an extremely valuable list of indicators, which could help to measure the extent of community's involvement.

When participatory wetland management could be appropriate: Few suggestions

Experience(s) suggest the involvement of the local and indigenous people in a management partnership when:

(i) The active commitment and collaboration of stakeholders are essential for the management of a wetland (e.g., when the wetland is privately owned).
(ii) Access to the natural resources within the wetland is essential for local livelihood security and cultural heritage.
(iii) The local and indigenous people express a strong interest for their involvement in management.
(iv) It may be noted here that the case for local and indigenous people's involvement is even stronger when:
- **(a)** The local stakeholders have historically enjoyed customary/legal rights over the wetland.
- **(b)** The local interests are strongly affected by the way in which the wetland is managed.
- **(c)** Decisions to be taken are complex or controversial, e.g., different values need to be harmonized or there are disagreements on the ownership status of the land or natural resources.
- **(d)** The existing management regime has not succeeded much to make wise use.
- **(e)** The stakeholders are ready to collaborate when they are requested to do so.
- **(f)** There is sufficient time to negotiate among stakeholders before any management decisions are made.

Concomitant to the above, different platforms had been serving as sounding boards for practitioners, and had been stressing that success may be assessed from many different angles, notably the well-being of the stakeholders, satisfaction of the government agency, health of the ecosystem, and so on.

Notwithstanding the above, eight themes are usually highlighted in the effort(s) to identify and refine strategies for wise use:

(i) Supporting frameworks: These may include appropriate changes in land tenure, access to resources, right to property, and the recognition of comanagement regimes in national and local policies, legislations, development plans, and also how these could affect the local initiatives.
(ii) Local environmental knowledge: These may encompass the comanagement systems which have effectively incorporated traditional knowledge and technologies.
(iii) Economic valuations: The assessments under this category include the benefits and functions of the wetlands which are supposed to have gone beyond theory, and have been used in management planning.
(iv) Gender: These are the wetland management systems where women as well as men have a decisive influence in determining management objectives and practices.
(v) Participatory monitoring: Partnerships between the management and the local people to monitor the ecological characteristics of wetlands, and progress made toward the community's own objectives.
(vi) Effectively addressing threats to the wetlands through comanagement.
(vii) Win-win scenarios: Here, socioeconomic development(s) has/have gone hand in hand with wetland conservation.
(viii) Exit strategies: Here, comanagement initiatives have been proactively planned, and then effectively implemented, scaling down and phasing out of external assistance.

Supportive frameworks and infrastructure

It is important that a strong legal framework is almost indispensable for establishing a participatory management domain. The chances of success are, then, greatly enhanced by transparent decision-making processes and multi-sectoral planning. A policy and legislative framework which legitimize the involvement of the local people in resource management is the basic minimum input from government which will always be, possibly, required. However, concomitantly, the issues of tenure, access to resource(s), ownership of resources, comanagement, etc., are also to be addressed. In this connection, it may be mentioned here that lessons which had been learnt from involvement of people in managing marine protected areas in **Belize, Mauritania, Mozambique, and, possibly,**

also Tanzania stressed the importance of a firm legal basis to support the efforts of the local people (Wells and Gawler, 1999). Concomitantly, **in Cameroon (**Kouokam and Ngantou, 2002), a law had been enacted in 1995, which legitimized the participatory management of the forests and protected areas. In many countries, however, national legislation does not support comanagement. Furthermore, many government agencies lack training in participatory management. In many cases, legal, policies, and implementation frameworks still need to be revised and strengthened for allowing for genuine models of community empowerment.

In contrast to Cameroon, etc., **in Thailand,** the general principle of people's participation in governance, resource management, and in the formulation of developmental programs had been fully recognized in the newly revised Constitution, adopted in May 1998. Legal recognition alone, however, does not suffice. Erftemeijer and Bualuang (2002), however, pointed out that there, still, are major cultural barriers to effective grassroots level participation in the strongly stratified hierarchical Thai society.

1.1.4 Community-based fisheries management (CBFM) and its application and implication

Garrett Hardin had written an article titled, "The Tragedy of the Commons" about 34 years ago. In his article, he had argued that the eventual fate of all resources held "in common" is overexploitation because access is unrestricted and there is lack of incentive(s) among individuals toward resource protection. However, critics now assert that Hardin's thesis does not seem to properly distinguish the type of property regime(s) which are susceptible to such a process. It has been argued that it does not seem to be applicable exactly to the "common property," but seems to somehow fit into "open-access" regime(s).

However, "**common property**" could now generally be defined as "a system where the resource is held by an identifiable community of users who may exclude others and regulate the use," as opposed to "open-access" property, which is characterized by an "absence of well-defined property rights," and may lead to people "free riding" and overexploiting a resource.

A simple example of an open-access system could be a natural lake or a coastal fishery, where fishermen continue fishing as long as the returns from fishing can make up the cost. Thus, the society loses on an overall consideration, whereas it should have gained from such a resource use. The management of the resource, on the other hand, could become all the more difficult, as external factors adversely affect the fisheries, and, in turn, the community which derives its sustenance from the resource. Further, it is sometimes opined that the open-access regimes could, sometimes, be disastrous for the poor, who very much depend on the common mass for their livelihood.

Notwithstanding the above, CBFM had aimed at addressing the complicacies and challenges of developing and implementing wetland comanagement system(s) with local stakeholders by trying to adopt the following steps:

(a) Promoting the interaction and exchange of experiences, and strengthening the networks. Concomitantly, steps for identifying and refining strategies for wise use, and reviewing and to implement the best practices of participatory wetland management from around the world.
(b) Local people's involvement in wetland management could contribute significantly to maintaining or restoring ecological integrity and community well-being. Every successful comanagement initiative has the potential to stimulate positive initiatives elsewhere.
(c) Concomitantly, it also, possibly, enabled to explore the "ground-truths," and to evolve the guidelines for establishing and strengthening participatory processes to involve communities and indigenous people in wetland management. In fact, the latter aspects were under process for the 1999 Conference of Parties of the Ramsar Convention on Wetlands.
(d) However, the Ramsar guidelines summarize five major requirements for successful comanagement (as in Ramsar Convention on Wetlands, 1999):
  (i) Incentives for local and indigenous people's involvement and wise use. It is expected that everyone must benefit in the long term and long run.
  (ii) Trust among the stakeholders.
  (iii) Flexibility in operationalization.
  (iv) Knowledge exchange and capacity building as essential components.
  (v) Continuity of resources and efforts.

Indicators for measuring local and indigenous people's involvement in wetland management (in the light of Ramsar Convention on Wetlands, 1999. Resolution VII.8).

  (i) Local and indigenous people are to have achieved an economic stake or other interest(s) in the wise use of wetland resources.

(ii) The government agencies are to spell out policies supporting participatory management.
(iii) Appropriate legal and financial incentives for participatory management are to be in place.
(iv) A more equitable sharing of benefits among the stakeholders is to be the outcome from the participatory management process.
(v) Stakeholders are to express their satisfaction and happiness with their involvement in the process.
(vi) There should exist a clearly stated and widely known policy or legal document which is to make a commitment to involving local and indigenous people (stakeholders).
(vii) All key stakeholders (particularly government) are to acknowledge participatory management as legitimate and wanted.
(viii) Local and indigenous people (who are stakeholders at the grassroots level) are to be involved in making substantive amount of decisions affecting the use and management of wetland resources.
(ix) The local organizations, who take part in participatory management, are to be respected and appreciated within the community.
(x) Representatives of the local and indigenous people may be considered as their true representative and they are to be made accountable to the people.
(xi) There are to be resource use and participation rules that are appropriate to the local situation.
(xii) A management agreement is to exist among the stakeholders, which could be oral or written, formal or informal.
(xiii) The management agreement to have clearly defined boundaries and membership issues.
(xiv) The management agreement to have specifically defined stakeholders' functions, rights, and responsibilities.
(xv) The agreement for management to be approved by at least the resource-harvesting and using stakeholders and by the key decision-making groups.
(xvi) Parties to the agreement are to meet their respective commitments.
(xvii) Noncompliance with approaches, rules, rights, and responsibilities outlined in the management agreement is to be at an acceptable level, if at all there is any noncompliance.
(xviii) Any system of sanctions for infringement of rules to be agreed upon by all key parties/partners.
(xix) If there is/are evidence(s) that (indigenous traditional) measures(s) for resource management control exist(s), they are to be adequately implemented.
(xx) There is to be potential, significance and scope for collective modification of the rules relating to resource use by those affected.
(xxi) There may exist "nested" management units, i.e., there are different bodies at different levels.
(xxii) There could be evidence(s) that the local and indigenous people could influence the speed and direction of change(s) in relation to the resources with which they are concerned.
(xxiii) Facilitators/coordinators may practice adaptive management through "learning by doing."
(xxiv) There could be growing awareness among the stakeholders about new management approaches, rules, rights, and responsibilities.
(xxv) There may exist a two-way flow of information, communication, and interaction between the local and indigenous people and the relevant government agencies.
(xxvi) Information is to reach the local and indigenous people (stakeholders at the grassroots level) punctually and accurately, and in a form which is readily understandable and comprehensible.
(xxvii) Local and indigenous people are to take part in site monitoring and in the process of evaluation of the participatory process.
(xxviii) There are to be evidence(s) of regards and respects by the key government agencies toward local human systems and local ecological knowledge.
(xxix) Stakeholders could display necessary skills, capabilities, and empowerment, e.g., capacity to make decisions, monitoring skills, etc.
(xxx) Methods of measurement, which have been formulated and established by the stakeholders, could be demonstrated, as well as the degree to which local participation is intended could be quantified. These are to be actually improved and conserved as the recognized functions and values of the wetlands with regard to their wise use.
(xxxi) There could be one or more organizational structures which could facilitate local and indigenous people's involvement, e.g., a council, management body, women's group, etc., in effective wetland management.

(xxxii) A random sample of local and indigenous people could identify the community's role in wetland management. Further, the individuals who are directly involved could accurately elaborate the objective(s) of their involvement.
(xxxiii) The government agency and its staff are to have a confirmed and demonstrated commitment to participatory management, and to be in a position to accurately describe the objective(s) of the involvement of the local and indigenous people.
(xxxiv) There is to be an appropriately long-term source of funding for the ongoing participation of the local people in resource management.
(xxxv) The efforts of the local and indigenous people, who provide in-kind support (time, labor, traditional knowledge, and expertise) for effective implementation of the participatory wetland management, are to be properly recognized and appreciated.
(xxxvi) There are to be mechanisms for conflict management. There are to be also mechanisms of appeals in the case of conflicts within the management partnership.
(xxxvii) There is to be quite strong integration between local wetland management and the management of the entire catchment.

1.1.5 Conflicts and consensus

Concomitant to the above, it may be noted here that pressure on the region's fisheries have compounded mainly due to overpopulation and poverty, particularly among the fishing communities in South and SE Asia. At the same time, conflicts occur alongside competing industries and the people who are dependent on fishery resources. However, they are usually considered far more complicated conflicts than any other type. It is mainly because they are generally associated with enormous amount of socioeconomic problems, cultural differences as well as, to some extent, political power struggles; complicated concomitantly by an array of diverse interests, values, priorities, and patterns of use and exploitation by and among the resource users. As such, the social dimensions as well as the role of the resource users in streamlining conflicts could be factored-in when looking for solutions to conflicts. Further, one very important factor in resolving conflicts is the communication among the stakeholders. It may also be considered as integral in the process of understanding the conflicts which are typified in various categories, and are briefly given in the following:

(a) Type I: It is about who controls the fishery. It is quite common in **Bangladesh and Cambodia (Kampuchea)**.
(b) Type II: It is about how it is controlled, where either lack or over-enforcement is found as the primary conflict. It is prevalent in the coastal areas of **India, Bangladesh, and Cambodia**.
(c) Type III: It indicates the relationships among the users of the resource.
(d) Type IV: It portrays the relationship between the fishery and the non-fishery resource users.
(e) Type V: This type of conflict pervades when non-fishery interests and issues such as corruption among the authorities are involved, and they affect the fishery stakeholders.

All the points indicated above may be taken into account when planning for an "attitude survey" involving stakeholders from project partners in different countries, notably **India, Bangladesh, Cambodia,** and so on. It may be pointed out here that the plans and policies, which could evolve from the various project surveys, could be considerably based on attitudes and perceptions of the stakeholders, including, very significantly, the region's fishers.

Inter alia, the briefs recommend the following:

(a) Development, designing, and dissemination of appropriate information, education, and communication (IEC) materials through various means like group communication activities, print, multimedia, the folk media, and so on.
(b) Popularization of manuals like PAPD-based consensus-building tool.
(c) Establishment and institutionalization of mechanisms to sustain the participation of all the key players in the fisheries sector.
(d) Providing a sustainable participatory scheme to a qualified regional NGO by the local administration and the relevant government agencies for lending support to fisheries conflict management in the long term.
(e) Establishment of conflict management and consensus development in natural resource governance by imparting pressure on the governments to incorporate conflict management as "necessary cost of governance" and not just a "business-as-usual" activity.

1.2 Impact of CBFM on development and management of wetland fisheries: Impact of CBFM on peoples' life

CBFM is expected to give stronger security to the resources and more access to them mainly through cooperative management. Concomitantly, enhanced sustainability of the resources coupled with cooperation in different aspects are integral components of CBFM program.

Some of the important points related to the impact of CBFM exercise are briefly given in the following:

(a) The inhabitants around a wetland may not be always very expert and experienced in managing their resources in an effective manner. As such, the administration/government tries to introduce the CBFM to help them in the process.

(b) The different variables of cooperation may be highly associated with the effectiveness of the CBFM system. This could be revealed from the different types of statistical tests.

(c) Concomitantly, under certain circumstances, these patterns of cooperation may put up a negative impact on the life of the inhabitants around the wetland, which could also be ascertained through Semantic Differential Scale.

(d) In addition to the above, CBFM system could also formulate some good scope(s) for the elite classes of the society.

(e) Interviews are usually conducted based on CBFM exercises. In different instances, a substantive percentage of the respondents were believed to have opined that microcredit facility could give a boost, but it may not be available for all people. However, at the initial period of implementation of the CBFM system, organizations like the Management of Aquatic Ecosystems through Community Husbandry (MACH) and the government have provided the inhabitants different kinds of assistance, which, of late, they, probably, had discontinued, and replaced it by microcredit. However, the rural poor people may neither get any kind of assistance nor the microcredit. As such, this situation may be considered as greatly impeding them to earn their livelihoods effectively.

(f) Notwithstanding the above, another issue is also important. The migratory birds may bring in various kinds of pathogens of diseases, which could affect the fishes as well as the life of the local people.

(g) A lentic body, where fishes are stocked for short or long duration(s), is, sometimes, designated as an "aquatic storage." In **Eastern India and in Bangladesh**, there are two types of storages, viz. (i) seasonal or temporary and (ii) permanent. The former are those types of storages where fishing may be "closed" during the breeding season to enable the fishes for successful spawning. The latter, on the contrary, are those types of lentic bodies where the brood stock is protected during the dry season. However, in general, the main purpose of these storages is to protect the short-lived species in their vulnerable stage(s) of life cycle.

(h) Concomitant to the above, small, highly exploitable and rendered little-endangered trash fishes may be more available in the present context in comparison to the earlier days. It could be mainly because the water bodies, like wetlands, canals, etc., are being excavated in some places, under the CBFM and MACH Programs, to improve the fish habitat. This could improve the fish production of different kinds of fishes, which could be reflected by their occurrence in large number. But, closed wetlands might be facing problems with the increase in fish density per m^2, in the form of harboring disease due to overcrowding.

(i) In contrast to the above, the CBFM system could give increased security to aquatic resources and greater access to the resource user. At the same time, cooperative management could enhance sustainability of resources. Preliminary aquatic sanctuaries, generally in the form of aquatic storages, are set up for resource sustainability and their conservation. But, these aquatic storages may be under the stronger control of resource management organization (RMO); as a result, the common people may not get much access to the resources. Efforts are said to have been made for rectification and improvement in the situation, which could be of help to the poor fishers.

(j) Notwithstanding the above, there could be some positive changes as well with regard to the availability of aquatic resources. There may be an increase in the amount of aquatic resources in many wetlands. It could be mainly because of the fact that the RMO could plant different types of vegetations, notably the *Barringtonia acutangula*, *Pongamia glabra*, *Justicia repens*, *Nymphaea nouchali*, *Nymphoides indicum*, *Nymphoides cristatum*, *Nelumbo nucifera*, etc., which could have some role in the ecology of the wetlands. The swamp forests, further, provide habitat for fishes and other aquatic organisms. Nevertheless, the augmentation in the aquatic plant resources in a wetland is regulated by the RMO.

Concomitant to the above, on the basis of a study titled "Impact of development and management of some selected floodplain wetlands (*Beels*) in Assam (India) through community-based fisheries management," Barman et al.

(2012) have revealed the impact of CBFM on the development and management of wetlands, which is cited in the following as an example of a typical case study.

Today, community-based fisheries management (CBFM) has been gaining much importance among the different stakeholders involved in fisheries and aquaculture. Various studies have revealed that CBFM allows greater stakeholder participation in decision-making process, generates more transparent management system, enhances competence with rules and regulations, greater cooperation between stakeholders, strengthens management efficiency, reduces government cost, minimizes fish-related conflict(s), improves resource users' assessment to physical, financial, and natural capitals, and helps to promote sustainable development and management. Thus, CBFM approach could be regarded as one of the most potential and effective management approaches for sustainable development of the *Beel* fisheries in Assam. This could also be the same in other Indian provinces, as well as in the neighboring countries like Bangladesh, etc. Sitka Declaration (2005) has stated that CBFM has always promoted diverse types of fisheries, maintain habitat integrity through the use of low impact gears and practices, and so on. Further, Barman et al. (2005) indicated that due to the involvement of fisher community in the pen culture practices in the *Dek Beel* of Kamrup district of Assam, the cost of pen culture reduced to a great extent and capacity building of the *Beel* users enhanced with respect to construction and installation of pens in the *Beel*. Due to community-based management of *Beels*, in certain other Beels, viz., Ganja Khawari *Beel* (9.0 ha) in Hailakandi district, Kacharidal *Beel* (5.20 ha) in Darrang district and Barmonipur *Beel* (6.0 ha) in Marigaon district, fish production has increased substantially, community empowerment was ensured, people's participation has strengthened; all these are said have resulted in the sustainable development of these *Beels*.

The Assam Agricultural Competitiveness Project (AACP), which came into existence in the province of Assam with effect from February to March 2005, had attempted to empower the poor fish farming communities (as clusters/groups) and to enhance their socioeconomic livelihood through *Beel* fisheries development. It is a unique example of community-based fisheries management approach. Under this project, participatory development intervention has been proposed in 1000 ha of *Beels* throughout the province within 5 years by forming *Beel* Development Committee (BDC) from the surrounding *Beel* users. It is hoped that the approach will ensure (a) effective community participation, (b) environmental safeguards to the *Beels* and *Beels* users, and (c) involvement of NGO to ensure social safeguards in the entire process of development and management of *Beels*. Executive committee (EC) consisted of five community members and one NGO representative and one government official. It is responsible for the overall management of the BDC as per the project guidelines to ensure sustainable development of the candidate *Beels*. Likewise, a community-designed committee of three persons is formed as social audit committee (SAC) for those *Beels* in order to ensure effective, transparent procurement.

1.3 Present status of CBFM in the global scenario, in general

Notwithstanding the above, **multi-scale assessments** (MAs) are said to consist of significant components, notably the community-based assessments. Real-life experiences of alteration(s) in the ecosystem and human well-being are captured by them. Information about existing management systems are also given by them.

In addition, processes and patterns, which would not be otherwise evident, such as the key resources situated in fine-scaled patches and appropriate times for management interventions, could be highlighted by a fine-grained look into the MA. Further, documenting the causes and effects are as important as understanding the coevolution of knowledge, ecosystems, social institutions, and management practices; their complexities, unevenness in space and time, and non-linearity; and the feedbacks from the resource users.

Notwithstanding the above, the local communities may not be considered as mere spectators, but active managers of ecosystem's capacities to deliver services. It may be noted here that community management systems are said to be continuously evolving, although some of them may disappear, while there are others which are said to be revived or formed from scratch. Many communities, nevertheless, may possess local knowledge about the interaction(s) between human and the environment, and thus they are said to affect the ecosystem services *vis-à-vis* human well-being positively. For example, the local people in **Xinjiang, Western China**, are said to possess elaborate traditional underground water harvesting structures, which maintain both quantity and quality of the water. Contrary to the good aspects, community-based management, in itself, may not always guarantee proper management of the ecosystem. For example, the national economic incentives have led some communities to harvest the indigenous forests in the **Eastern Himalayas**. Hence, institutional frameworks, which support stewardship of ecosystem services, are required. Moreover, recognition of the role of communities as stewards of ecosystem services, and their empowerment, is essential to strengthen the capacity to manage ecosystems sustainably for the welfare of the humanity.

Concomitant to the above, diversity in ecosystem is significant in reducing community's vulnerability. In fact, most communities try to maintain a diverse range of livelihood options. Such diversities buffer the people against shocks and surprises such as climatic and economic fluctuations. For example, the local farmers in many **Indian** villages and in **Papua New Guinea** cultivate an array of wide variety of crops in order to avoid and avert the risk(s) of crop failure(s). Further, in Bajo Chirripo' in **Costa Rica**, the local communities are reported to nurture a mosaic landscape, consist of sacred places, springs, agro-ecosystems, and high mountains, thereby generating diverse livelihood options at the local level.

In addition to the above, the spiritual and cultural values of ecosystems may be considered to be as important as other services for many local communities. In this connection, it may be opined here that ecosystems may often provide a sense of habitat and identity for the local people in addition to other ecosystem services. However, these intangible values are said to include aesthetic and recreational values, and provide rationale for management. They also further precipitate management practices which enhance the ecosystem resilience through caring and custodianship. In this connection, it may be pointed out here that, in Vilcanota in **Peru**, the spiritual values and belief systems include the belief in "Pachamama" or "Mother Earth." This belief encompasses the faith that Mother Earth is a living being and she has allowed for the maintenance of a cultural identity among the Quechua people of the Southern Peruvian Andes. Moreover, in Kristianstad Wetlands in **Sweden**, the local farmers are said to have restarted cultivation in land(s), which had previously been abandoned. This is said to be not for any economic gain but for a sense of a place of living and for concomitant identity which accrues with the cultivation of this plot of land. On the contrary, these values and systems of belief are said to have eroded in many places, thereby leading to a shift in community-based management practices. As an example in this regard, it may be cited that in San Pedro de Atacama, in **Chile**, the erosion of collective indigenous identity, due to economic development, have led to the sale of land to the outsiders, and led to a consequent decline in agriculture and related traditional practices.

Notwithstanding the above, communities have been affected by large-scale processes. However, their ability to cope with and shape the alteration(s) may vary. For example, realities of the local communities may not always be reflected in the decisions taken at higher levels and scales. This could result in negative impacts at the local level. However, communities that are able to cope with these external forces are said to have learned to adapt themselves, or even taken advantage of them by establishing horizontal links with other groups. Thus, they try to form alliances with the so-called powerful actors at higher spatial scales. As a result, they could be in a position to link with the national or global processes such as policy forums, markets, and multinational agreements. On the other hand, when conditions are said to be more or less impossible to adapt to, for example, due to inflexible national policies, people may be somehow forced to migrate or encounter a reduced quality of life. For example, in Sistelo in **Portugal**, a government-sponsored afforestation program on a common-property land has diminished the locally available livelihood and coping strategies, thereby accelerating the process of rural-urban migration.

Contrary to the above, vertical and horizontal collaborations with the communities could improve ecosystem management and human welfare. In this regard, it could be opined that the key actors, the social networks, trust-building processes, and bridging organizations across scales are important vehicles in this process. In this connection, it may be pointed out that in the Kristianstad Wetlands in **Sweden**, the local managers are using an international program in order to strengthen development through the conservation of valuable local wetlands. In **Papua New Guinea**, on the other hand, trade links between islands are ensuring ecosystem diversity and enabling people to cope with the alteration(s).

In fine, it could be opined that capturing the complex and dynamic nature of the interactions between ecosystems and human beings could require complementary conceptual frameworks. It may be stated here that various community-based assessments have adopted the MA framework in order to take into account more dynamic interplays between the variables, followed by attempt(s) to capture fine-grained patterns and processes in complex systems, and allow for the integration of the various global views. Contrary to this, in Vincanota in **Peru** and in Bajo Chirripo' in **Costa Rica**, other conceptual frameworks have been used, which are reported to have incorporated both the MA principles and the local cosmologies. However, in **Southern Africa**, on the other hand, different frameworks have been used in parallel in order to offset the MA framework's shortcomings and disadvantages as a community assessment tool.

Fisheries management through peoples' participation is an important aspect in the world today. In many tropical developing countries, the mode of fisheries management is often intertwined between formal and informal or traditional systems. It is quite difficult to make clear-cut definitions of formal and informal systems mainly because of overlaps. However, a formal system is one which is legally sanctioned by the existing government or state authority. For instance, in the **Philippines**, the formal management of fisheries resources is partitioned between the national government through the Bureau of Fisheries and Aquatic Resources (IBFAR) and the local government units (LGUs), notably the municipalities (local self-governments). It may be mentioned here that the BFAR manages the commercial fisheries, while the LGUs manage municipal fishing operations or activities.

Concomitant to the above, a rights-and-rules system, collectively sanctioned by fishers or other resource users living on the bank or coasts of the water bodies, is regarded as an informal system of fisheries management. Following the customs and traditions, it might have been handed down through generations. It might exclude the nonmembers, but, concomitantly, might involve co-ownership among the fishers and stakeholders. Further, it may or may not have legal sanction of the government. The "Sasi" (closed) system in **Indonesia** could be considered as a typical example of such a system. Notwithstanding the above, a resume of some of the approaches and associated contexts are briefly discussed in the following.

1.3.1 Logic behind fast appraisal approach

To begin with, it may be stated here that a dualistic system of fisheries management exists in many fishing communities. It is often found that a formal government management system coexists with an informal management system, devised and implemented by a community of resource users. However, as the informal systems are not easily understood or observed, outsiders to the community are often found to be unaware or ignorant about the informal systems. Nevertheless, some of these systems have functioned well in meeting the management objectives and goals of the community and in achieving ecological sustainability, social equity, and economic efficiency. However, it may be borne in mind that some of the local level systems are new while others are time-tested.

In contrast to the above, information on community-based marine resource management system(s) do exist, for example, in the **Asia-Pacific region**. However, it may not be always *in vogue* mainly because of fast alteration(s) at the community, national, and international scenario, due to many reasons. In this context, it is important and pertinent to note here that the resource managers and policy makers may preferably possess up-to-date information regarding community-based management systems and their associated socioeconomic, political, and ecological contexts, if effective fisheries management efforts are to succeed. For an up-to-date status of the fisheries management system in a country, studies may preferably be contemporary, detailed, and location-specific for a comprehensive knowledge base on the range of types and functions.

In addition to the above, in order to bridge the gap(s), new management systems, different management systems, notably community-based coastal resources management (CBCRM), are said to have been introduced in various places, on top of the existing local informal ones. But, the fishers are often not very much willing to give up a system they have devised and also time-tested, due to which, the newly evolved system(s) often fail to sustain themselves. However, a management system does definitely have more chances of success if it complements or is built up on the existing informal system. Ostrom (1992) has stated that if people had been living in close association with comparatively small common-pool resources over a period of time, it could be obviously believed that they probably, evolve some system(s) to regulate and limit their use patterns. Before initiating new rules on local systems, it is imperative to launch some efforts to determine and confirm whether some of the rules and customs are already *in vogue* or not. If they do still exist, it is essential to study them carefully in order to comprehend how they could affect the use patterns over time.

Bangladesh is a small developing nation in Asia. The revenue-oriented approach in Bangladesh has not been, possibly, able to involve the poor fishers in its inland fisheries management. As such, a community-based fisheries management (CBFM) approach had been implemented as a second phase program (CBFM-2) over a period of 6 years spanning from 2001 to 2006. This was done to improve the access to fishing rights of the poor and also to improve the productivity as well as sustainability of fishery resources. The study had investigated the alteration(s) in fisher's access to livelihoods in the different kinds of water bodies, notably the various kinds of wetlands (locally called "Beels," Haors, etc., in Bangladesh) as well as in the rivers, in order to enhance their livelihood opportunities. The data for the study had been obtained from two questionnaire-based field surveys conducted by the Bangladesh CBFM project office. A baseline study has been done in 2002 followed by an impact assessment study done during mid-2006. In this study, a total of 2826 households had been randomly selected from various regions in Bangladesh comprising 1994 households at 34 (51%) CBFM project water bodies and 832 households at 10 (59%) control water bodies. This study has found that the CBFM fishers had obtained greater access to fisheries and improved livelihoods than their non-CBFM counterparts. It seemed that the fishers, of late, had changed their attitudes. They possess greater awareness about rules and regulations in fisheries, and perhaps, above all, are able to resolve conflicts in the CBFM water bodies in a much easier manner. In fine, it could be opined that long-term access rights over fishery resources could be considered as a priority for a sustainable inland fishery and livelihoods of fishers in Bangladesh.

Concomitant to the above, the Government of Bangladesh had adopted CBFM approach (as mentioned earlier) in order to maintain ecological balance in the wetlands coupled with ensuring food security. One of its major components is "co-operation." In addition to studies already mentioned, there had been studies in Bangladesh, which measured the patterns of cooperation in CBFM system and its impact on people's life. The design of the studies is often descriptive. Samples are selected through simple random sampling technique as well as through social survey technique to collect data. Patterns of cooperation and the indicators of the impacts could be measured through the Likert Scale and Semantic Differential Scale. It was found that negative cooperation may exist in the study area. Concomitantly, most of the respondents' attitude could also be negative about the effect of CBFM.

Concomitant to the above, **Fiji** is considered to represent a unique case, in view of the fact that its customary fishing-rights areas ("*qoliqoli*") are said to constitute a kind of dual ownership. This may establish a connection between indigenous owners and central government for the purpose of management. But, as reported, this kind of the so-called cooperative comanagement approach has not been uniform across Fiji. It could be mainly because of the involvement of different levels of the government and the community; as a result, the approach has become confusing with disputes. Further, in Fiji, one more issue is still unresolved, i.e., whether "ownership" of the *qoliqoli* should include both the marine resources and the sea floor; the latter being currently owned by the state. Notwithstanding the above, there are reviews which have taken critical look at the issues like "traditional conservation," origin of the prevailing *qoliqoli* system; peoples' perception regarding this; the forces driving its evolution; the impact on marine resource use; and so on. Further, it sheds some light on as well as explores the boundaries of "comanagement" in Fiji. It could also try to assess previous efforts to find a way forward. It may seem that Fiji may not be able to maintain its natural marine resources, without organizing its forces.

1.3.2 Information in this regard from Japan

As reported (Weinstein, 1998), after experimenting with western resource management methods (open-access subject to licensing) in the late 19th century, Japan rescinded the new laws, and, at the turn of the century, had returned to traditional CBFM process for all its inshore coastal fisheries. Under the mechanism, exclusive, inalienable rights had been granted to local fishing cooperatives. Membership of the cooperative societies was said to have been based on residency and a period of apprenticeship, and was not transferable. The operational rules (gear, season, fishing areas, etc.) had been decided by the cooperatives based on the experience of the fishers. The system had provided a powerful form of accountability, as management decisions had been made by the fishing groups, who shared the same fishing grounds, and error, if any, has been borne by the people who made them. According to Weinstein (1998), Japan's inshore fishery is said to be very healthy, as has been ascertained by macro-level indicators of sustainability.

Concomitant to the above, Claridge and O'Callaghan (1997a,b) noted that the government agencies are often not very prompt and fast to embrace participatory wetland management schemes, and their support for comanagement may sometimes be limited to only lip service(s). They have inferred that the development of techniques for increasing government acceptance of comanagement was one of the major challenges facing wetland conservation through CBFM process. Nevertheless, there are encouraging examples of effective supportive frameworks, although this might remain the case in some areas only.

1.3.3 Information in this regard from Senegal

As an example, it may be cited here that the Djoudj National Park in Senegal (Diouf, 2002) has been implementing a management plan framed in consultation and coordination with the neighboring communities. The plan is endowed with an institutional mechanism based on four committees, which are briefly discussed in the following.

(a) The orientation and scientific committee has provided the acumen.
(b) The park management committee included the village representatives, and had a direct role in the implementation of the management plan.
(c) The intervillage conservation committee had maximized the participation of the inhabitants of the neighboring villages. This had facilitated the exchange of opinions and decisions, and coordinated the developmental work(s) of the various technical committees on ecotourism, replanting, waterways, health, forestry/pastoralism, and so on. Enough care had been taken while designing these supporting frameworks for participatory management, which often had showed positive results. An environment of trust was built between the people and the park agents. Notably, the number of infractions within the park declined to zero when the participatory management policy had begun.

1.3.4 Information in this regard from Thailand

Lessons from Pattani Bay in Thailand (Erftemeijer and Bualuang, 2002) had provided many valuable insights into identifying and generating conditions for success, as well as the potential pitfalls and expected drawbacks, in the process of developing supporting frameworks for participatory management.

1.3.5 Information in this regard from Zambia

Til and Banda (2002) have highlighted the issues of comanagement in the Bangweulu wetland in Zambia. They have indicated the highlights of the difficulties of comanagement in a situation characterized by suspicion, mistrust, and outright hostility between the government agents and the local people, including among different ethnic groups and even within ethnic groups. They have further drawn attention to the importance of democracy in participatory management, where each and every stakeholder is free to present his or her view, and it is to be represented. Here, however, the traditional leadership of the chiefs was, probably, less helpful for the poor stakeholders. In this context, one of the biggest lessons learned from the Bangweulu is the importance of making supportive framework(s), which would effectively control the power(s) of the traditional chiefs and empower the powerless, while, concomitantly, keep the chiefs on board with the promotion of democracy and comanagement. In addition, lessons from Zambia and some other places highlighted the need to take into account the customary supportive framework(s), which were said to have been provided by the traditional kings, queens, chiefs, religious leaders, and healers. These powerful key players are to be fully involved in developing strategies for wise use.

1.3.6 Information in this regard from Bénin's coastal fisheries

Notwithstanding the above, Bénin had forged ahead to form a legal basis for participatory management of this key resource (Akambi, 2002), being faced with ever-increasing threats to the important coastal fishery. Fisheries committees have been formed by the fisheries administration in the major villages along the coast during 1993. The members of the committee were democratically elected, and entrusted with ensuring the rational use of the lagoons and their resources. A legal foundation was formed in 1996 with the adoption of a management plan for the coastal fisheries. Subsequently, in 1997, a law was passed legalizing the existence and operations of the village fisheries committees. However, 95 committees have been formed, till date, to comanage Bénin's coastal fisheries, and a functional working relationship has formed a climate of trust between the fishers and the government authorities.

1.3.7 Information in this regard from Malaysia

The Tasek Bera case study in Malaysia (Prentice et al., 2002) focused on the lives of the indigenous Semelai people reflecting their close cultural affinity to the wetland. The study had also emphasized to patronize the role of this disadvantaged group in a new nature-based ecotourism initiative. Initial results have indicated that the process of consultation was effective in raising awareness, and also in gaining trust and support.

1.3.8 Information in this regard from India, Pakistan, Kenya, and Tanzania

Pimbert and Gujja (1997) have reported on the issues of trust and equity in wetland management policies in Keoladeo National Park (NP) in India and the Ucchali wetland complex in Pakistan. They emphasized that participation must be part of a dialog process, action, analysis, conflict resolution, and change. Nzioka (2002) has presented two examples of lack of transparency in decision-making process, concerning the Tana and Rufiji deltas in Kenya and Tanzania, respectively. These developmental measure(s) were believed to have posed potential threats to the delta ecosystem(s) and to the livelihoods of local people. Thus, these examples underscore the urgency of appropriate policy frameworks, and, in particular, robust integrated coastal zone management plans. Campredon (2002), further, has emphasized the significance of a regional framework in developing a planned approach to coastal zone management.

Moreover, the case studies from the Tana and Rufiji deltas also have pointed to the importance of incorporating mechanisms for conflict management in the supporting frameworks. Still further, there might be instances where there could be a lack of trust between the local communities and the government institutions. Under such situations, it is necessary to find ways and means of overcoming the barrier(s) that may exist, as well as to build up a balanced community-ownership with the full involvement of the local government. It may be noted here that the user communities, managing the wetlands, may not be able to solve problems which originate outside the community.

Notwithstanding the above, it may be cautioned here that the current trends toward decentralization could, indeed, lead to greater reliance on comanagement structures, but if the decentralization of authority is not coordinated properly, it may result in unhealthy competition(s) and overlapping jurisdiction(s), conflicts in wetland policies, or, in the worst cases, a total surrender of the responsibility by the government agencies.

Concomitant to the above, the OECD (1996) gives useful general guidelines for building supportive frameworks for the wise use of wetlands. Moreover, promoting participatory management could require changing the attitudes, approaches, policies, etc., of the government, including legislation, etc., under a number of circumstances. The local projects and local success stories may certainly provide convincing arguments for the benefits of comanagement process, but, nevertheless, policy efforts must not be neglected in ensuring a legislative basis for collaborative resource management. It must be borne in mind, as well as appreciated, that shifting the government agencies away from rigid top-down habits and hierarchies to participatory approaches may generally require a sustained effort over a very long period of time, and hence, development agencies and the NGOs may remain prepared for a long-term commitment for bringing about novelties in the management *modus operandi*. In fact, as a corollary, it might be stated that the resource management policies may be required to be harmonized among local, provincial, regional, and national scenarios, as well as between the present and future generations. Weinstein (1998), as such, seemed to have rightly suggested the outline design and the principle(s) for successful and sustainable management of institutions based on common property resources and rights.

1.3.9 Information in this regard from Mauritania, Cameroon, Nigeria, Ethiopia, and Pakistan

The Diawling National Park in Mauritania has suffered from disruption of ecosystem functions. The local environmental knowledge could serve as the cornerstone of the restoration plan (Ba et al., 2002). The local inhabitants possess very detailed traditional knowledge of the hydrology of the river and the migratory pathways of the fish. The fishers of Takhridient have contributed their indigenous knowledge to the design of the restoration works. And, this had yielded excellent results for the fish, the shrimp, and the local economy. Similarly, many of the park regulations for the use of the resources were based on their own traditions. Moreover, the local people around the park play an active role in ensuring the respect and regard for the rules. They, probably, had also informed the park authorities of the infractions such as night fishing, using a sickle to harvest perennial grasses, and so on.

Works carried out in Waza Logoon in Cameroon (Kouokam and Ngantou, 2002) portrayed a vivid example of floodplain restoration. Here again, the project design was based on the highly detailed indigenous traditional knowledge of the local people with regard to the floodplain vegetation, hydrology, seasonal productivity of the fisheries, the cultivation of floating rice, the migratory pathways of the fishes and the wildlife, and so on. The integration of the traditional knowledge of the community into the restoration design was, doubtlessly, perhaps one of the main reasons for the success of this project.

Nigeria is a heavily populated country. Here, the demand for fish is said to far outstrip the supply (Ayeni and Mdaihli, 2002). It is often believed that the country may combine government control with local traditional method(s) which were once *in vogue*, such as the taboos on seasons, minimum size of fish to be caught, closed season(s), sacred no-fishing zones, and so on. These could be some of the essential features for considering that fisheries management could have a chance of success in the future. In this connection, it may be noted here that with the introduction of CBFM process in Kainji Lake, management decisions are now taken by representatives of the fishers and their traditional institutions. This is supposed to have brought in a quantum of changes in the attitude of the fishers from antagonists to defenders of the regulations.

Wood et al. (2002) have reported about wetland ecosystems in Southwest (SW) **Ethiopia**. Here, the development policies of the government are said to raise questions regarding sustainability of the wetlands. It is mainly because such policies favor more intensive use of the wetlands, in contrast to the less intensive uses based on tradition. More importantly, the local communities have a wealth of knowledge, which was believed to have been built up over generations, with regard to hydrology, soil dynamics, etc., and were, further, believed to have developed management practices. Such practices seemed to have permitted the long-term use of these wetlands without degradation. A strong message, in this regard, could be that the policies of development may be more sensible to the diversity of the local conditions, and the sentiment of the stakeholders. Such policies could be applicable to general wetlands, as well as to the specific wetlands and the needs of the associated communities.

In this context, exercise(s) based on participatory rural appraisal (PRA), conducted in three villages near the Ucchali wetland complex in Pakistan, had revealed important social and ecological variations among the villages. Such result(s) seemed to point to the need for integrating the site-specific knowledge and the innovation(s) of individual communities with the more general ecological principles on which management plans are usually based (Pimbert and Gujja, 1997).

It could, thus, be opined that the incorporation of indigenous local knowledge, values, and management practices may be highly required for the success of participatory management initiatives. However, such views are usually recognized in principle, but, in practice, stronger effort(s) may more often be required to incorporate local wisdom, and more importantly, the traditional management strategies, into the wetland management plans.

1.3.10 Economic valuations

It is a known fact that ecological functions rendered by the wetlands are not transacted in the markets. As such, their significance is often undervalued in the developmental decisions. But, the values and functions of the wetlands are preferably to be weighed objectively. As such, economic valuation seems to offer a powerful tool for ensuring the "wise use" of wetlands. In this regard, OECD (2002) analyzed wetland values in terms of function(s), product(s), and attribute(s) for different types of wetlands. Further, Barbier et al. (1997) have provided a good guide on economic valuations of wetlands for planners and policy makers.

Concomitant to the above, Abila (2002) has attempted to determine the replacement values for key economic activities of the local communities who depend on the wetland resources. This was supposed to have been done in response to proposals for the reclamation of the **Yala swamp** in Western **Kenya** for commercial agriculture. This study inferred that traditional sustainable uses of the wetlands by the local people are to be promoted. The *modus operandi* was based on the comparison of the replacement values for hunting, fuel wood, brick-making, papyrus, and fishing with the costs and benefits of conversion.

In addition to the above, it could, possibly, be revealed that the practice of beach seining in Lake Kainji for small freshwater sardines had resulted in substantial by-catch of juveniles of other commercially important species. In this regard, Ayeni and Mdaihli (2002) estimated the economic valuation of the total catch with and without beach seining, and these calculations have helped to raise awareness, and build consensus to eliminate beach seines among the different conflicting interest groups of fisher folk. Likewise, the community-based wetlands conservation project in the **Saiwa wetlands in Kenya** (Awer Mohamed (2002) had used the "comparative gross margin analysis" in order to help the community groups determine the relative incomes they could expect from the alternative enterprises in the wetland buffer zone (agro-forestry, fish farming, beekeeping, etc.) in comparison to maize, a more intensive land-use oriented crop.

These examples appear to illustrate the use of economic tools to inform decision-making by local stakeholders. In fact, there could be a number of examples of valuation exercises in the literature that seemed to demonstrate the economic valuation of the wetlands in the larger context of the policies of development. However, as per the feedback(s) of the participants in various workshops, notably the Dakar workshop, there seemed to be only a few studies and reports on economic valuation(s) of the wetlands, which could have impacted the policies with regard to the wetlands, especially, in developing countries. As such, lessons learned to date, from economic evaluations of wetlands, seemed to highlight:

(a) the need for developing more robust but simple indicators;
(b) the use of method(s) which are specific to different wetlands;
(c) that the technique(s) used are based on a good understanding of the human-resource interface;
(d) that data are interpreted correctly;
(e) the formulation of easily applicable techniques which could be easily used by the local communities to evaluate the probable benefits to them.

Notwithstanding the above, such methods may preferably be avoided which exaggerate the economic value of wetlands. Greatly exaggerated claims had often damaged the credibility of this potentially valuable tool for wetland conservation. In fact, greater effort(s) must be made to improve synergies between wetland economists and public relations and communications specialists. Sometimes, it may so happen that the decision makers and the general public are not aware of the economic values and ecological functions of the wetlands. This is an area which could receive more attention on priority in future, as many decisions on development are made on the basis of economic grounds.

1.3.11 Gender

Gender analysis is, probably, accepted today as a key element in successful natural resource management, although, gender roles are not always considered when wetland projects and programs are designed. However, the concept of gender is still not well understood by policy and decision makers, planners, and natural resource managers, in spite of its recognized importance. Matiza (1993) outlined the key issues which, perhaps, need to be addressed for gender to be fully integrated into wetland conservation and management. Some of these issues are as follows:

(a) To have correct gender perceptions, and increased gender awareness.
(b) To promote gender roles in research with regard to wetland conservation and management.
(c) To review traditional stereotypes in wetlands resource allocation and utilization.
(d) To try to anticipate potential conflicts between traditional culture and gender roles and empowerment.
(e) To train policy makers, planners, and wetland resource managers in the analysis of gender roles.

(f) To integrate gender issues into national wetlands policies, and also, into wetland projects and programs at the planning, monitoring, and evaluation stages.

A useful first step in facilitating women's participation could be a clearly defined user rights aspect in the society. Skill in conflict management could also be necessary. A measured approach of working with both men and women could be necessary as efforts to change unfavorable gender stereotypes. In traditional and egalitarian societies, the significance of addressing gender issues with sensitivity and patience cannot be overstated. For instance, it may be stated that in the Saiwa wetlands in Kenya, for example, men were said to be reluctant, at first, to allow women to participate in the study tours. However, their perceptions changed slowly, as they saw for themselves the benefits of involving women more actively.

Notwithstanding the above, it may be important to understand here that addressing gender issues with too much enthusiasm and too less sensitivity could backfire, and may result in the project failure. In some cases, however, deeply rooted traditional stereotypes may pose real challenges to gender-sensitive development(s). In this regard, the **Bangweulu swamps in Zambia**, probably, provide an example of strong resistance to involving women in comanagement structures, although the women were believed to play a key role in the fishing industry. It is mainly because here the comanagement is usually perceived as a political process, and hence is considered as a regime of men. Hence, there could be a possibility that the process of comanagement may be monopolized by men, and thus could have a negative impact on the position and interests of women (Til and Banda, 2002). Thus, concerted and sensible efforts are needed to ensure the representation of less powerful stakeholders in comanagement regimes in the strongly traditional societies.

Notwithstanding the above, the incorporation of women's knowledge was believed to have led to a redesigning of the artificial re-flooding scheme in the **Diawling floodplain** (Ba et al., 2002). In this regard, it may be noted here that the men had suggested releasing the flood waters early to maximize the recruitment of Tilapia. The women, on the other hand, had, however, insisted that the wetland grasses, such as the valuable *Sporobolus*, needed rain before the flooding in order to reach optimum vegetative growth. A compromise solution was tested, which produced extremely positive results for the fish, the perennial grasses, and the local community as a whole.

Concomitant to the above, successes were also recorded in many projects such as the Waza Lagoon wetland in **Cameroon** (Kouokam and Ngantou, 2002), although not yet widespread enough. In such cases, women, perhaps, have proved themselves to be ideal partners in comanagement. In this conservative context, the project had worked slowly and sensibly, negotiating with the men to permit the women to engage themselves in activities other than their domestic works at their homes. Subsequently, it may be noted here that, once resistance was overcome, and the women were allowed to participate, their contribution was much appreciated by the men. Thus, the entire community was benefited from the revisiting and changing traditional gender taboos. Consequently, the women of Waza Lagoon now have the rights to acquire land for rice cultivation (which was previously unheard of), to produce and sell goods, to manage revenues, and to participate in the mixed decision-making committees. This kind of cultural change is enormous, and may require years of patient and careful work.

The significance of engaging both men and women in wise use strategies for wetlands is widely recognized today. But, there are a good number of examples even today, where gender-sensitive strategies in wetlands conservation and development initiatives remain as an exception rather than the rule.

Notwithstanding the above, the CBFM and the PRA exercises may have many pitfalls. Thus, in order to avoid the pitfalls, wetland initiatives must ensure the following:

(a) That, gender issues are an integral component of planning, monitoring, evaluating, reporting, and dialog with the partners.
(b) That, the initiators and implementers have adequate expertise on gender issues.
(c) That, gender issues are incorporated in the Terms of Reference for all studies.
(d) Statistics are disaggregated by sex.
(e) Capacity is built with partner institutions to work with gender issues (ZBWCRUP, n.d.).

In this connection, Gawler (1999) has provided a guidance to assess whether a project is adequately addressing the gender issues in terms of project design, monitoring, and implementation.

1.3.12 Participatory monitoring and overseeing

The central theme to the strategy for sustainability is the knowledge about how systems perform, and what the effects of management efforts actually are. The science of wise use is, by necessity, an ongoing process of adaptive learning. It may be noted here that the prescriptions for one place may not be applicable for a neighboring watershed, i.e., what works in one place is not necessarily the answer for the neighboring watershed, or even the neighboring village. As such, a diversity of solutions, based on an evolving understanding of the local environmental conditions, and relevant to the local cultural scenario, is, perhaps, the need of the hour. Further, monitoring by the people themselves, to whom the solution matters most, is, perhaps, critical to the success of adaptive management.

Notwithstanding the above, Waterwatch **Australia** provides an excellent and exciting example of participatory monitoring at a nationwide scale (Chalkley et al., 2002). The "Waterwatch" is an environmental education and awareness program which promotes and supports water quality monitoring, in order to generate an ownership ethics for catchment-wide land and water management by the Australian people. Since 1993, the Waterwatch Australia had grown to include >50,000 people monitoring >4000 sites in every state and territory of the country. The Waterwatch method is based on community awareness and ownership, regional facilitators helping community groups, and strong technical support. Waterwatch is now evolving beyond monitoring, and is entering a new era of actions, pressing for solutions to waterways problems.

The key to the initial successes in Pattani Bay, **Thailand**, was the local ownership and informed local decision-making processes, where a project facilitated the community initiatives in order to restore severely degraded sites of former mangroves (Erftemeijer and Bualuang, 2002). The project approach has emphasized on the process(es) rather than the outputs, and focused on learning and strengthening of the community organization, building environmental awareness, and sharing information. The villagers themselves ascertained why their initial efforts were not effective, and concomitantly decided how the rehabilitation work could be reoriented.

The task of monitoring, when done involving the local people, is a cost-effective solution, because it draws on local knowledge. Moreover, it also has a number of potential positive spin-offs. As such, participatory monitoring provides a powerful learning element and a tool for village environmental plans. It is often believed that just having data can give people a sense of ownership. Communities involved in monitoring could see for themselves the impacts of project interventions, and may recommend corrective actions, if necessary. Baseline surveys during the planning phase or at the beginning of the intervention could generate significant level of awareness, which are said to be important, so that the communities have a reference point from which to estimate the progress. A well-designed community monitoring program is considered as one of the most effective ways to enhance skills in resource management.

Notwithstanding the above, local people may not always wish to volunteer their time for monitoring, in spite of the benefits, particularly, if they believe that the monitoring is only satisfying the scientific curiosity of the outsiders. However, in some cases, clear incentives may be needed. Further, the challenges to participatory monitoring may include the following:

(a) There is the need for robust indicators to estimate the impact of change.
(b) There is the need to secure the continuity of data collection.
(c) There is the need to ensure the relevance and quality of the data collected.

Notwithstanding the above, the capacity of community and its structural organization may not be of overambitious type. Experience(s) had portrayed that, in many cases, there could still be a requirement for outside assistance for documenting and disseminating the information collected. Nevertheless, the role of the community in monitoring may gradually increase as they build up their capacity. In fact, the monitoring technique(s) may not be highly sophisticated in many cases, and locally chosen indicators, such as "are there bigger fish catches?" or "does the water taste better?," may be more relevant and helpful. Further, Margoulis and Salafsky (1998) have provided a wealth of valuable suggestions for developing and implementing a community-based monitoring plan as well as for using the results of monitoring for learning and adaptive management.

1.3.13 Threat(s) and their effective addressing

It may be appreciated that a community may contain a diverse array of stakeholders in it. It is, thus, essential to ensure that the needs of the weaker groups within the community are addressed. In this regard, it may be noted here that threats to wetlands may often involve major issues of equity among the stakeholders, and addressing these issues could be a first effective step toward resolving the threats to biodiversity. And, comanagement could be an effective tool for conflict management and conflict resolution. One of the significant advantages of participatory management is that it opens up dialog, and increases transparency in decision-making.

Finlayson et al. (1998) have demonstrated successful strategies for addressing threats to three coastal wetlands in **Northern Australia**. The land tenure and management systems, use of resources, and management infrastructure(s) were vastly different there. However, the common denominators of these successful programs are as follows:

(i) The *modus operandi* of management was essentially cooperative, and had involved genuinely local inputs.
(ii) The mechanism of operation was based on formal arrangement(s) which acknowledged the diversity, aspirations, and economic and cultural values of the local communities.

Notwithstanding the above, the local communities, of late, had been increasingly aware and alert of the negative impacts of the large developmental projects on their resource base. As such, more and more communities are today increasingly eager to defend their rights to sustain their livelihoods as resource users. However, the local people may not necessarily have the negotiating skills needed to achieve solutions to the threats of most concern to them, particularly, when faced with a diversity of interest groups. Under such situation(s), communities may be required to be trained in order to be able to discuss their concerns with the government agencies. The extent of capacity building, which is necessary to empower communities, is to be planned and implemented seriously.

It is quite clear that management of wetlands with the involvement of the local communities and stakeholders, as the principal actors, is probably, by far, one of the most promising solutions to the ever-increasing threats to the integrity of wetlands. Moreover, it could be particularly true with the declining trend of the government resources. But, comanagement may not serve as an excuse for government inaction. To the contrary, the governments are required to set appropriate environmental standards and to be actively engaged in helping to meet these standards.

Concomitant to the above, Thampy (2002) has opined that the policy and decision makers are required to know, with more scientific certainty, how forces and parameters, such as migration, urbanization, exponential population growth, tourism, and high rates of resource consumption, etc., affect the natural ecosystems on which life depends. In this connection, it may be noted here that astute and realistic environmental policies, which are said to be compatible with human needs, and which provide incentives, may be considered as significant to achieve enduring success in maintaining ecosystem health. It is often believed that environmental issues may not be effectively tackled in isolation of the prevailing social and economic context. In this context, the Nakuru case study seemed to portray that sustainable development and ecosystem health will remain an elusive pipe dream until the issues of inequity, insecurity, and political instability are addressed effectively. Incidentally, study in this direction in Ghana was done by Olesu-Adjei, Isaac (2002).

According to De Sherbenin (1998) (as also, perhaps, many others), access to water is a human rights issue. Water is often said to be captured by powerful economic interests, to the detriment of the needs and interests of the local communities. Inappropriate (or unimplemented) national policies may often be found to underlie many of the most severe threats to wetland ecosystems, such as water diversions, agricultural policies, and pollution to name a few. However, one of the major challenges in this regard is to establish effective, integrated coordinating mechanisms for river basin, or trans-boundary wetland management (Moser et al., 1999). Likewise, contrary to this, the root cause(s) of wetland degradation may be said to be international trade or climate change. All these seem to highlight the urgency of accompanying comanagement initiatives with the required national and international policy reforms. Measures to conserve wetlands for the goods and services they provide may ultimately fail if they are not supplemented by efforts to address water allocations, human demography, pollution, climate change, etc.

1.4 Conclusions

We are often faced with the question: "What are best practices"? In fact, practically speaking, there may not be any single correct and tangible way to establish effective local community involvement in wetland resource management. This had been opined by Erftemeijer and Bualuang (2002), among many others. What could be important and useful is to share the experiences of both successes and failures with others working in the same field.

Thus, the emphasis is on the process, rather than the outputs. This could contribute to a long-lasting, sustainable change in attitudes toward resource use. Concomitantly, the process approach could be accompanied by clear and measurable indicators of progress, as defined by the target communities. Balancing an open, flexible approach based on local ownership of the process with clear prioritization of objectives could be one of the significant challenges, as mentioned by Erftemeijer and Bualuang (2002). Weighing the community's desire for immediate benefits along with the concomitant need for sustainable, long-term solutions could further add to the situation.

Notwithstanding the above, best practices in participatory management often carry the wetland managers far beyond the specific concerns of conservation and wise use. Such practices, further, try to touch upon some of the significant aspects of human life, notably democracy, equity, development, and cultural survival (Borrini-Feyerabend, 1996). Empowerment is about the people, including both men and women; taking a type of regulation over their lives; setting their own agenda; gaining skills; enhancing self-confidence; trying to solve problem(s); and making efforts to develop self-reliance. It is said to be both a process, and an outcome (**ZBWCRUP**).

It may be emphasized here that proceedings of various conferences could strengthen our collective understanding of the wise use of the wetlands by documenting some examples of best practices. There is, perhaps, no doubt that much more needs to be done at all levels, viz., local, national, and international. This could help others to learn from comanagement experiences, as well as to develop supporting policies and the capacity to implement them. As emphasized in many national and international conferences on wetlands and development, water and wetlands are the source of life. The information contained in various proceedings volumes could help in learning encouraging lessons on participatory management of wetlands in a variety of human and ecological settings. There is, perhaps, no room for complacency. In fact, before it is too late, it is increasingly urgent to develop and implement comprehensive strategies for the wise use of our source of life. Moreover, a key lesson from many of the contributions is that for a strategy of wise use, if at all to be sustainable, it cannot be based solely on a concept developed by the government or international experts, but must ensure that it incorporates the priorities and the wisdom of the local people.

1.5 Summary

(i) The management of natural resources through community-based activities is often called common pool resource management.

(ii) Attitudes toward the management of natural resources have been changing world wide.

(iii) The management regimes of the wetlands could be categorized into four broad categories, based mainly on the property rights and access to the wetland:
Private management by individuals or group(s), (b) fishermen's cooperative society management, (c) community-based fisheries management (CBFM), and decentralized management, in which the government works mainly as a facilitator (generally in the open-access wetlands).

(iv) CBFM could essentially be regarded as a strategy for empowering the community to manage their resources in a sustainable manner and conserve the natural wealth for posterity and prosperity.

(v) The "fisher folk," in many countries, now seem to be gradually trying to understand and accept that "management" is not an imposition by the government but an individual and collective responsibility of all stakeholders, producers, and consumers alike. Awareness-building programs seem to have brought about such change(s) in attitude that "management" has become an integral component of sustainable exploitation and does not unnecessarily impose burden on them.

(vi) The basic principles of CBFM are:
Participation of fishermen in planning and decision-making process(es); surveillance, evaluation, implementation, and control of management activities as important aspects.

(vii) CBFM is designed to manage both the fishery resources and the fishermen. Successful exercise on CBFM could happen, if there is good institutional environment such as effective laws and regulations.

(viii) CBFM also relies on traditional knowledge and wisdom of the local community who earn their livelihood mainly through fishing. Once fishing right has been granted, they try to apply the indigenous traditional knowledge for effective management of the wetland/lake resources.

(ix) While dealing with the implementation of CBFM, the issues to be looked into include the following: (i) the nature of the resources, (ii) the technologies used to exploit these resources, (iii) arrangements available for property rights associated with different kinds of exploitations, (iv) strategies for management, and (v) the effectiveness of informal and formal rules within the present management scheme.

(x) Several prerequisites are to be fulfilled to implement CBFM. Some of the significant ones are: (i) devolution of authority of management to the local community, (ii) establishment of territorial boundaries, and (iii) incentives and motivations to be given to the fishermen to set up local management systems.

(xi) CBFM exercise may be unsuccessful in a community where the fishing resource may not be exclusive to the community.

(xii) Many governments, throughout the globe, today are willing to adopt the CBFM, appreciating the fact that fisheries management has not worked out as planned by the governments. It is mainly because CBFM involves more consultative and participatory mechanisms.

(xiii) **Participatory management** is a partnership in which government agencies, local communities, and resource users, as well as the other stakeholders such as NGOs, share the authority and responsibility of management of a specific area or a set of resources.

(xiv) In some of the **Asian countries**, in those areas where indigenous and traditional people live, and have been living there for hundreds to thousands of years, the authority for resource and ecosystem management may, preferably, be devolved as much as possible to the local level.

(xv) The Ramsar guidelines (Ramsar Convention on Wetlands, 1999) for establishing and strengthening local communities' and indigenous people's participation in the management of wetlands seem to provide an extremely valuable list of indicators, which could help to measure the extent of community's involvement.

(xvi) "**Common property**" could now generally be defined as "a system where the resource is held by an identifiable community of users who may exclude others and regulate the use," as opposed to "open-access" property, which is characterized by an "absence of well-defined property rights," and may lead to people "free riding" and overexploiting a resource.

(xvii) Pressure on the region's fisheries has compounded mainly due to overpopulation and poverty, particularly among the fishing communities in South and SE Asia. At the same time, conflicts occur alongside competing industries and the people who are dependent on fishery resources. However, these are usually considered as far more complicated conflicts than any other type. It is mainly because they are generally associated with enormous amount of socioeconomic problems, cultural differences as well as, to some extent, political power struggles; complicated concomitantly by an array of diverse interests, values, priorities, and patterns of use and exploitation by and among the resource users.

(xviii) CBFM is expected to give stronger security to the resources and more access to them mainly through cooperative management. Concomitantly, enhanced sustainability of the resources coupled with cooperation in different aspects are said to be integral components of CBFM program.

(xix) CBFM activities like MACH Program in Bangladesh, AACP Program in Assam, India, etc., have brought about positive improvements in fisheries to improve the fish habitat. This could improve the production of different kinds of fishes, which could be reflected by their occurrence in large number. But, closed wetlands might be facing problems with the increase in fish density per m^3, in the form of harboring disease due to overcrowding.

(xx) Further, the CBFM system could give increased security to aquatic resources and greater access to the resource user. At the same time, cooperative management could enhance sustainability of resources. Preliminary aquatic sanctuaries, generally in the form of aquatic storages, are set up for resource sustainability and their conservation.

(xxi) Community-based fisheries management (CBFM) allows greater stakeholders' participation in decision-making process; generates more transparent management system; enhances competence with rules and regulations; greater cooperation between stakeholders; strengthens management efficiency; reduce government costs; minimizes fish-related conflict(s); improve resource users' assessment to physical, financial, and natural capitals; and probably, last but not the least, help to promote sustainable development and management. Thus, CBFM approach could be regarded as one of the most potential and effective management approaches for sustainable development of the lentic community-based fisheries management.

(xxii) The chalked-out programs help to form operating bodies like executive committee (EC), the Beel Development Committees (BDC), social audit committee (SAC), etc., for overseeing various community-related activities.

(xxiii) An overview of the present status of CBFM in the global scenario, in general, has been briefly presented.

(xxiv) **Multi-scale assessments** (MA) consist of significant components, notably the community-based assessments. Real-life experiences of alteration(s) in the ecosystem and human well-being are captured by them.

Suggested readings

Abila, R., 2002. Utilization and economic valuation of the Yala Swamp wetland, Kenya. In: Gawler, M. (Ed.), Strategies for Wise Use of Wetlands: Best Practices in Participatory Management. Proceedings of a Workshop held at the 2nd International Conference on Wetlands and Development (November 1998, Dakar, Senegal), pp. 116. Wetlands International IUCN, WWF Publication No. 56, Wageningen (The Netherlands), pp. 89–95.

Addun, R.P., Muzones, D.M., 1997. Community-based coastal resource management (CBCRM): Tambuyog's experience in the Philippines. In: Claridge, G.F., O'Callaghan, B. (Eds.), Community Involvement in Wetland Management: Lessons From the Field. Proceedings of the International Conference on Wetlands and Development at Kuala Lumpur. Wetlands International, Kuala Lumpur, pp. 219–230.

Akambi, L., 2002. L'approche participative pour la gestion des pêcheries dans les zones humides du sud-bénin. Programme d'Aménagement des Zones Humides, 08 BP 0837 Centre de Tri Postal, Cotonou, Benin MOTS-CLES: pêcheries, Bénin, communautés. In: Gawler, M. (Ed.), Strategies for Wise Use of Wetlands: Best Practices in Participatory Management: Proceedings of a Workshop Held at the 2nd International Conference on Wetlands and Development (November 1998, Dakar, Senegal), pp. 116. Wetlands International IUCN, WWF Publication No. 56, Wageningen (The Netherlands), pp. 43–45.

Awer Mohamed, M., 2002. Mitigating the effects of intensive agriculture on wetlands: the case of Saiwa Wetlands, Kenya. In: Gawler, M. (Ed.), Strategies for Wise Use of Wetlands: Best Practices in Participatory Management. Proceedings of a Workshop Held at the 2nd International Conference on Wetlands and Development (November 1998, Dakar, Senegal), pp. 116. Wetlands International IUCN, WWF Publication No. 56, Wageningen (The Netherlands), pp. 97–100.

Ayeni, J.S.O., Mdaihli, M., 2002. Conservation and management of Kainji lake fisheries: a community-based approach. In: Gawler, M. (Ed.), Strategies For Wise Use Of Wetlands: Best Practices In Participatory Management. Proceedings of a Workshop Held at the 2nd International Conference on Wetlands and Development (November 1998, Dakar, Senegal), pp. 116. Wetlands International IUCN, WWF Publication No. 56, Wageningen (The Netherlands), pp. 75–79.

Ba, A., Fall, O., Hamerlynck, O., 2002. Le Parc National du Diawling: expérience de co-gestion pour la restauration des plaines inondables. In: Gawler, M. (Ed.), Strategies for Wise Use of Wetlands: Best Practices in Participatory Management. Proceedings of a Workshop Held at the 2nd International Conference on Wetlands and Development (November 1998, Dakar, Senegal), pp. 116. Wetlands International IUCN, WWF Publication No. 56, Wageningen (The Netherlands), pp. 19–25.

Barbier, E.B., Acreman, M.C., Knowler, D., 1997. Economic Valuation of Wetlands: A Guide for Policy Makers and Planners. Ramsar Convention Bureau, Gland (Switzerland), p. 127.

Barman, R.C., Dana, S.S., Bhattacharya, B.K., 2005. Empowerment of women through fish culture, a case study from the West Bengal, India. In: Samanta, R.K. (Ed.), Empowering Rural Women: Issues, Opportunities and Approaches. The Women Press, Delhi, p. 255.

Barman, R.C., Kar, D., Dana, S.S., Majumdar, D., 2012. Role of availability of critical aquaculture inputs and community-managed markets on community-based floodplain wetland management. J. Appl. Nat. Sci. 4 (2), 164–166.

Berkes, F., 1989. Co-management and the James Bay agreement. In: Co-operative Management of Local Fisheries.

Borrini-Feyerabend, G., 1996. Collaborative Management of Protected Areas: Tailoring the Approach to the Context. Issues in Social Policy. IUCN, Gland (Switzerland), p. 67.

Campredon, P., 2002. Pêche artisanale et conservation du littoral de la Mauritanie à la Guinea-Bissau. In: Gawler, M. (Ed.), Strategies for Wise Use of Wetlands: Best Practices in Participatory Management. Proceedings of a Workshop Held at the 2nd International Conference on Wetlands and Development (November 1998, Dakar, Senegal), pp. 116. Wetlands International IUCN, WWF Publication No. 56, Wageningen (The Netherlands), pp. 63–67.

Chalkley, S., Edgar, B., Gowland, K., 2002. Who is watching our water?: participatory monitoring by Waterwatch Australia. In: Gawler, M. (Ed.), Strategies for Wise Use of Wetlands: Best Practices in Participatory Management. Proceedings of a Workshop Held at the 2nd International Conference on Wetlands and Development (November 1998, Dakar, Senegal), pp. 116. Wetlands International IUCN, WWF Publication No. 56, Wageningen (The Netherlands), pp. 101–105.

Claridge, G.F., O'Callaghan, B. (Eds.), 1997a. Community involvement in wetland management: lessons from the field. In: Proceedings of Workshop 3, pp. 278, Wetlands, Local People and Development, of the International Conference on Wetlands and Development held in Kuala Lumpur, Malaysia, 9–13 October 1995. Wetlands International, Kuala Lumpur, pp. 19–24.

Claridge, G., O'Callaghan, B., 1997b. Making co-management of wetland resources work. In: Claridge, G.F., O'Callaghan, B. (Eds.), Community Involvement in Wetland Management: Lessons from the Field. Wetlands International, Kuala Lumpur, pp. 25–66.

De Sherbenin, A., 1998. Water and population dynamics: local approaches to a global challenge. In: De Sherbenin, A., Dompka, V. (Eds.), Water and Population Dynamics: Case Studies and Policy Implications. American Association for the Advancement of Science, Washington, DC (USA), pp. 9–22.

Diouf, A.M., 2002. Djoudj National Park and its periphery: an experiment in wetland co-management. In: Gawler, M. (Ed.), Strategies for Wise Use of Wetlands: Best Practices in Participatory Management. Proceedings of a Workshop Held at the 2nd International Conference on Wetlands and Development (November 1998, Dakar, Senegal), pp. 116. Wetlands International IUCN, WWF Publication No. 56, Wageningen (The Netherlands), pp. 13–17.

Erftemeijer, P.L.A., Bualuang, A., 2002. Participation of local communities in mangrove forest rehabilitation in Pattani Bay, Thailand: learning from successes and failures. In: Gawler, M. (Ed.), Strategies for Wise Use of Wetlands: Best Practices in Participatory Management. Proceedings of a Workshop Held at the 2nd International Conference on Wetlands and Development (November 1998, Dakar, Senegal), pp. 116. Wetlands International IUCN, WWF Publication No. 56, Wageningen (The Netherlands), pp. 27–35.

Finlayson, C.M., Thurtell, L., Storrs, M., Applegate, R., Barrow, P., Wellings, P., 1998. Local communities and wetland management in the Australian wet–dry tropics. In: Williams, W.D. (Ed.), Wetlands in a Dry Land: Understanding for Management. Environment Australia; Biodiversity Group, Canberra, ACT (Australia), pp. 299–311.

Gawler, M., 1999. Special Emphasis on Women. NEDA-WWF (DGISWWF) Tropical Forest Portfolio Newsletter, "Is Your Project Adequately Addressing Gender?". WWF International, Gland (Switzerland), p. 4. Issue 3: Available at http://www.artemis-services.com/english/home.html.

Kouokam, R., Ngantou, D., 2002. Le processus de cogestion dans le contexe du projet Waza Logone. In: Gawler, M. (Ed.), Strategies for Wise Use of Wetlands: Best Practices in Participatory Management. Proceedings of a Workshop Held at the 2nd International Conference on Wetlands and Development (November 1998, Dakar, Senegal), pp. 116. Wetlands International IUCN, WWF Publication No. 56, Wageningen (The Netherlands), pp. 69–73.

Margoulis, R.A., Salafsky, N.N., 1998. Measures of Success: Designing, Managing, and Monitoring Conservation and Development Projects. Island Press, Washington, D.C., p. 362.

Matiza, T., 1993. Gender and wetlands management: issues and challenges in Southern Africa. In: Workshop on Gender and Water Resources Management: Lessons Learned and Strategies for the Future, Stockholm, 1–3 December 1993, pp. 1–18.

Moser, M., Silvius, M., Pirot, J.Y., Gawler, M., 1999. Wetlands in the balance… at the dawn of the third millennium. In: Wetlands—A Source of Life: Conclusions of the 2nd International Conference on Wetlands and Development, 10–14 November 1998, Dakar, Senegal. Wetlands International, Wageningen (The Netherlands), pp. 2–3.

Nzioka, P., 2002. Networking in East Africa for threatened coastal wetlands. In: Gawler, M. (Ed.), Strategies for Wise Use of Wetlands: Best Practices in Participatory Management. Proceedings of a Workshop Held at the 2nd International Conference on Wetlands and Development (November 1998, Dakar, Senegal), pp. 116. Wetlands International IUCN, WWF Publication No. 56, Wageningen (The Netherlands), pp. 57–61.

OECD, 1996. Guidelines for Aid Agencies for Improved Conservation and Sustainable Use of Tropical and Sub-Tropical Wetlands. Organization for Economic Co-operation and Development, Paris (France), p. 69.

OECD, 2002. Organization for Economic Co-operation and Development. Paris (France).

Olesu-Adjei, Isaac, 2002. Community participation in coastal resources management in Ghana. In: Gawler, M. (Ed.), Strategies for Wise Use of Wetlands: Best Practices in Participatory Management. Proceedings of a Workshop Held at the 2nd International Conference on Wetlands and Development (November 1998, Dakar, Senegal), pp. 116. Wetlands International IUCN, WWF Publication No. 56, Wageningen (The Netherlands), pp. 107–110.

Ostrom, E., 1990. Governing the Commons. The Evolution of Institutions for Collective Action. Cambridge University Press, Cambridge.

Ostrom, E., 1992. Crafting Institutions for Self-Governing Irrigation Systems. Institute for Contemporary Studies Press, San Francisco.

Ostrom, E., 1994. Neither the Market Nor State: Governance of Common-Pool Resources in the Twenty-First Century. IFRI Lecture Series. International Food Policy Research Institute, Washington DC.

Oviedo, G., Brown, J., 1999. Building alliances with indigenous peoples to establish and manage protected areas. In: Stolton, S., Dudley, N. (Eds.), Partnerships for Protection: New Strategies for Planning and Management for Protected Areas. Earthscan, London, pp. 99–108.

Pimbert, M., Gujja, B., 1997. Village voices challenging wetland management policies: experiences in participatory rural appraisal from India and Pakistan. Nat. Res. 33 (1), 34–40.

Pomeroy, R.S., Berkes, F., 1997. Two to tango: the role of government in fisheries co-management. Mar. Policy 21, 465–480.

Prentice, C., Surut, Z., Christiansen, P.C., Sinniah, P., 2002. Community development including ecotourism at Tasek Bera, Malaysia's first Ramsar site. In: Gawler, M. (Ed.), Strategies for Wise Use of Wetlands: Best Practices in Participatory Management. Proceedings of a Workshop Held at the 2nd International Conference on Wetlands and Development (November 1998, Dakar, Senegal), pp. 116. Wetlands International IUCN, WWF Publication No. 56, Wageningen (The Netherlands), pp. 47–56.

Ramsar Convention on Wetlands, 1999. Guidelines for establishing and strengthening local communities' and indigenous people's participation in the management of wetlands. In: Resolution VII.8. Resolutions and Recommendations, "People and Wetlands: The Vital Link", 7th Meeting of the Conference of Contracting Parties to the Convention on Wetlands (Ramsar, Iran, 1971), San José, Costa Rica, 10–18 May 1999. Ramsar Bureau, Gland (Switzerland). Available at http://ramsar.org/key_guide_indigenous.htm.

Thampy, R.J., 2002. Wetland conservation and development: the Lake Nakuru case study. In: Gawler, M. (Ed.), Strategies for Wise Use of Wetlands: Best Practices in Participatory Management. Proceedings of a Workshop Held at the 2nd International Conference on Wetlands and Development (November 1998, Dakar, Senegal), pp. 116. Wetlands International IUCN, WWF Publication No. 56, Wageningen (The Netherlands), pp. 111–116.

Til, J.B., Banda, M.G., 2002. Co-managing the Bangweulu fishery. In: Gawler, M. (Ed.), Strategies for Wise Use of Wetlands: Best Practices in Participatory Management. Proceedings of a Workshop Held at the 2nd International Conference on Wetlands and Development (November 1998, Dakar, Senegal), pp. 116. Wetlands International IUCN, WWF Publication No. 56, Wageningen (The Netherlands), pp. 37–42.

Weinstein, M.S., 1998. Pieces of the Puzzle: Getting to the Solution for Community-Based Coastal Zone Management in Canada. Key-note address to Coastal Zone Canada 98, Victoria, B.C.

Wells, S., Gawler, M., 1999. Involving people in marine protected areas: experiences in Central America and Africa. In: Stolton, S., Dudley, N. (Eds.), Partnerships for Protection: New Strategies for Planning and Management for Protected Areas. Earthscan, London, pp. 118–125.

Wood, A., Hailu, A., Abbot, P., Dixon, A., 2002. Sustainable management of wetlands in Ethiopia: local knowledge versus government policy. In: Gawler, M. (Ed.), Strategies for Wise Use of Wetlands: Best Practices in Participatory Management. Proceedings of a Workshop Held at the 2nd International Conference on Wetlands and Development (November 1998, Dakar, Senegal), pp. 116. Wetlands International IUCN, WWF Publication No. 56, Wageningen (The Netherlands), pp. 81–88.

ZBWCRUP. n.d., Gender integration strategy. Zambezi Basin Wetlands Conservation and Resource Utilization Project. IUCN Regional Office for Southern Africa, Harare (Zimbabwe).

Further reading

Agarwal, B., 1990. Social security and the family: coping with seasonality and calamity in rural India. J. Peasant Stud. 17 (3), 341–412.

Agarwal, A., Gibson, C.C., 1999. Enchantment and disenchantment: the role of community in natural resource conservation. World Dev. 27 (4), 629–649.

Agrawal, A., 1995. Dismantling the divide between indigenous and scientific knowledge. Dev. Chang. 26, 413–439. Crossref, Google Scholar.

Gawler, M., 2000. What are best practices? Lessons in participatory management of Inland and Coastal wetlands. In: Gawler, M. (Ed.), Strategies for Wise Use of Wetlands: Best Practices in Participatory Management. Proceedings of Workshop 1 of the 2nd International Conference on Wetlands and Development, pp. 116. Wetlands International, Wageningen, (The Netherlands), pp. 1–12. In press at http://www.wetlands.agro.nl.

McNeely, J.A., 1993. People and protected areas: partners in prosperity. In: Kemf, E. (Ed.), The Law of the Mother: Protecting Indigenous Peoples in Protected Areas. Sierra Club Books, San Francisco (USA), pp. 249–257.

CHAPTER

2

Researchers' contribution: A review

OUTLINE

Abstract

Fisheries management is a significant component of aquatic resource management. Many countries in the world have come up with policies and frameworks to develop strategies and regulations for aquatic resource management in general, and fisheries management in particular. The broad aim is to improve the economic efficiency of the fisheries. There could be various kinds of strategies for fisheries management. However, property rights-based fisheries management regimes could be considered as pragmatic and effective in the management of fisheries resources. It is mainly because they reduce or eliminate the incentives for over-capitalization in the harvesting of the fisheries resources, and thus, could contribute to economic efficiency.

Keywords: Fisheries management, Frameworks, Pragmatic approach, Biodiversity, Geodiversity, Water bodies

2.1 Introduction

Fisheries management is a significant component of aquatic resource management. Many countries in the world have come up with policies and frameworks to develop strategies and regulations for aquatic resource management in general, and fisheries management in particular. The broad aim is to improve the economic efficiency of the fisheries. There could be various kinds of strategies for fisheries management. However, property rights-based fisheries management regimes could be considered as pragmatic and effective in the management of fisheries resources. It is mainly because they reduce or eliminate the incentives for over-capitalization in the harvesting of the fisheries resources, and thus, could contribute to economic efficiency.

Community-based cooperative fisheries management is one of the community property rights-based fisheries management systems, which has received attention in recent years particularly in the developing countries. On the other hand, comanagement is a complex issue, on which there is, possibly, no consensus as yet. Interpretations and status could vary from one region to another and are influenced by a variety of regional and local parameters. Often,

Community Based Fisheries Management. https://doi.org/10.1016/B978-0-12-821723-8.00002-6

countries may find it difficult to define the term encompassing the different aspects. Concomitantly, there are efforts by certain quarters to pursue a technical agenda for the implementation of meaningful joint-management systems. As such, the matter being under process, it may not be possible to immediately offer a precise definition or policy statement on comanagement.

Fisheries are nutrition- and avocation-related interdependent ecological and social systems which call for integrated management approaches. The activities of one person or a group of users could affect the availability of the resource for others. As such, managing such common resources requires continuous and conscious efforts by a wide range of stakeholders to organize and draft rules enabling equitable and sustainable use of the resources for the welfare of all. Collective efforts are often a prerequisite for the development of community-based institutions and the devolution of authority trickling down vertically from the central stage down to the local grassroots level.

2.2 Empowering communities

Unfortunately, governments generally do not undertake comanagement as a means of empowering fishing communities and patronizing democracy. Instead, governments often consider comanagement as an instrument to achieve their administrative objectives more efficiently by involving the fishing communities in the implementation process. The difficulty could be mainly because the organizational structures of government departments have not, possibly, yet adapted to the new comanagement concept. Many fisheries departments, probably, are still staffed with scientists trained in earlier concepts and are almost exclusively aimed at resource conservation rather than the livelihoods of fishing communities. As such, community empowerment is an essential aspect of efficient resource management by the community.

2.3 A glimpse into Researchers' contributions to CBFM in different continents in the World

Review of literature enables comprehending the up-to-date status of a subject of study. This is also applicable in the case of **community-based fisheries management (CBFM). CBFM** is a potential strategy for the management of water bodies and their bioresources. Today, it is a globally acclaimed *modus operandi* mainly for the management of common property resources.

In the following, an account has been presented on the review of literature on CBFM in different countries and continents in the world.

In Asia, The continent of Asia is enormous with many countries having large number of both lentic and lotic water bodies, bewildering diversity of fishes, large number of fishermen, and a huge population of people, where "fish" forms a significant item of nutrition and avocation. There seemed to be various types of management *modus operandi* of the water bodies and their bioresources (notably, the fish resources) in Asia.

An account of the review of literature on CBFM in different countries/regions in Asia is given in the following.

2.3.1 Review of Literature in different countries/regions in Asia

Asia is an enormous continent, perhaps, the largest in the world. There are a large number of countries in the Asian continent representing a diverse array of biodiversity (both aquatic and terrestrial, consisting of fishes, amphibians, reptiles, birds, and mammals, including human beings having different types of indigenous knowledge cultural traditions, practices, etc.) and abiotic resources. There are a bewildering diversity of water bodies of different kinds. All these in Asia, perhaps, call for diverse modes of *modus operandi* of management measures.

2.3.1.1 Asia (General)

The continent of Asia is enriched with large number of small and big water bodies representing one of the largest assemblages of biodiversity in the aquatic domain of the globe. Some of the world's largest rivers and wetlands occur in this continent. But, the management of these water resources and their biodiversity is an onerous task. Nevertheless, the task, probably, has been made easy through the application of CBFM. Incidentally, there have been quite a large number of works related to CBFM.

Thilsted and Wahab (2014) have dealt with the production and conservation of nutrient-rich small fish (SIS) in ponds and wetlands for nutritional security and livelihoods in South Asia.

It is, probably, quite well known that small fish occur on their own, usually, in large number and, generally, in any kind of FW body, without much care being taken of them, in contrast to big fishes, which often require salubrious environmental conditions for proper growth, sustenance, and propagation. As such, these much easily available small fishes are to be conserved and taken care of, at least nowadays (when there is dwindling trend of fish production) for the nutritional security and livelihoods of the people, particularly in South Asia, where fish serves as one of the staple food items in nutrition and a potential source of avocation.

Efforts had been made to propagate the knowledge and technologies among the policy makers and relevant stakeholders in Bangladesh, as well as in the other regions of South and Southeast (SE) Asia, building on the successful adoption of minor carp, *Amblypharyngodon mola*. Polyculture technologies have been applied in stand-alone household ponds and ponds connected to rice fields. Such works, being disseminated, play a significant role in the promotion of further research, knowledge sharing, and refinement and adoption of production technologies to augment the production and consumption of SIS in Bangladesh, as well as throughout the South and SE Asian region.

Incidentally, Flewwelling (2001) dealt with fisheries management and their monitoring, control, and surveillance (MCS) in South Asia. The important aspect of management could, possibly, be successful, only when the status of MCS is overseen and followed up, else the exercise could go astray. Moreover, Wilson (1999) has dealt with practical implications of chaos in fisheries. Fisheries occur very much on the lap of nature and are being handled by human beings with regard to its harvest, utilization, management, etc. Contributions in this regard generally deal with the practical implications for fisheries management, if fish populations happen to behave chaotically. Such papers, perhaps, try to argue that the principal effect(s) of chaos may be manifested in terms of information, measurement, and management problems, in particular recruitment-based policies are likely to need unattainable measurement of accuracy, else could lead to unpredictable outcomes. This, perhaps, raises the questions about how sustainability could be maintained in chaotic fisheries. Further, many such communications, probably, suggest that management, based on the relatively stable ecological relationships in fishery systems, may be the most practical way to manage chaotic fisheries. Further, Pomeroy et al. (1998) dealt with fisheries comanagement with notes on key conditions and principles drawn from Asian experiences.

Management, either in the form of comanagement or community management, is a significant parameter for the sustained utilization of the resources, as well as for their preservation and conservation for the posterity. For this purpose there is, perhaps, necessity of experiences to be gained from various national and international exercises.

Incidentally, communications of this type usually deal with such aspects. As such, some papers have dealt with the presentation of the results of the research, specifically key principles and conditions, which facilitate the successful implementation of comanagement.

2.3.1.2 Asia-Pacific Region

There had been much works done on CBFM in the Asia-Pacific region. Ruddle (1993) discussed about the external forces that might cause some changes in the traditional community-based fishery management systems in the Asia-Pacific Region. Traditional *modus operandi* of community-based management measures had been based on wisdom of the people to conserve nature with nature-friendly measures.

Notwithstanding the above, in the Asia-Pacific region, as throughout the world, the traditional community-based marine resource management systems are increasingly affected by external factors that cause stresses and often lead to radical change(s) in the systems, including their demise. There may not be much novelties, but there is, probably, nothing new about this, except that the intensity of impacts and the diversity of their sources are be noteworthy. Thus, contemporary community-based marine resource management systems exist, perhaps, under environmental, social, ecological, political, and demographic circumstances that could often be quite different from those of even the recent past. In fact, of late, such systems, probably, swept up in the overall process of national modernization in the Asia-Pacific region. Further, among the principal, all-pervasive external forces are the legacy of colonialism, contemporary government policies, and the legal change(s), the replacement of traditional local authority, demographic change(s), urbanization, change(s) in education systems, modernization and economic development, commercialization and commoditization of living aquatic resources, technological change(s), change(s) in the policies of external assisting agencies, national policies for economic sectors other than fisheries, and so on. Further, such external forces, perhaps, rarely act in isolation but rather as a mutually reinforcing and potentially destructive complex. In addition, "commercialization," of late, could be considered as a somewhat more recent pressure but may not be of universal occurrence. In addition, monetization of formerly local and mainly subsistence or reciprocal exchange or barter economies, probably, now link them with external markets. This, in turn, perhaps, may lead to altered perceptions in fishing communities regarding the value of marine products, and as such to the breakdown of traditional management systems through the weakening or total collapse of traditional moral authority. Small communities may not be immune from the pressures that drive larger polities and commercial elites, and which undermine the moral imperative of local management systems from within. Moreover, the regional and global markets may also have a direct impact on them. Further, the external incentives introduce temptations for individual profit at the expense of local social equity, and may, thus, undermine systems by weakening or even destroying their moral and traditional authority.

Notwithstanding the above, the community institutions and management systems are not immutable, and change through time. Moreover, they dynamic, adapting to external as well as internal and local experiences and pressures, many of which may not be directly related to the fisheries sector. Further, the participants in community-based management systems may not be assumed *a priori* as being inherently benign resource-conservational and socially equitable actors. As such, any policy and program decision(s) about the present-day and future usefulness of local management systems are, preferably, to be based on a clearheaded and realistic evaluation of the moral authority, motives, interests, and cultural conceptions that, perhaps, underpin and drive them. Incidentally, the traditional management systems may, sometimes, decline under pressures exerted by both internal and external sources, and the latter could trigger the former, such that local phenomena may mask deep-seated problems afflicting the social institutions.

Further works related to the **Asia-Pacific region** include those by Makino and Matsuda (2011), who had dealt with ecosystem-based management in the Asia-Pacific region as part of the account on the world fisheries with a socio-ecological analysis.

The management of resources is an important aspect of a fishery. Incidentally, there are various kinds of *modus operandi* in the management of aquatic resources, and involvement of community stakeholders is an integral component of this. In this connection, it may be mentioned here that there had been efforts to ascertain various kinds of indicators of the fisheries sectors on a country-by-country basis, as well as to clarify the social and ecological conditions in the Asia-Pacific region. These are summarized as financial, industrial profile, aquatic resource diversity, food security, social security, and human resource conditions. Further, there had been efforts to propose a socially and ecologically compatible ecosystem-based management framework in the Asia-Pacific region, exemplified with a case from a UNESCO World Natural Heritage site in Japan. The local fishers were in the core of the management, under this framework. Incidentally, human capacity building and legal protection of each fishery was very important policy measures.

2.3.1.3 *Cambodia*

In Cambodia, Anon (2016), Nuov (2015), Limsong and Sitha (2015), De Silva and Song (2015), and Anon (2015a,b) worked on CBFM practices.

CBFM is an important aspect of fisheries management in Cambodia. **In Cambodia**, Anon (2016) dealt with the learning journey in Cambodia and Lao PDR in the River Mekong with the Blue Peace in the Middle East.

A learning journey, termed as the "The Mekong Learning Journey" spanning over a period of 4 days in Cambodia and Lao PDR, embarked upon, from November 17 to 20, 2014, by the members of the Blue Peace Middle East Community in order to explore and learn firsthand knowledge about cooperation on the lower Mekong River basin in SE Asia and also to gain various other kinds of experiences.

Concomitant to the above, Nuov (2015) dealt with culture-based fisheries and their management in Cambodia.

Fisheries administration, in collaboration with the regional organizations, probably, had been trying to promote an enhancement in the existing small areas of water bodies, such as the development of culture-based fisheries (CBF), including the development of community fish refuges (CFRs) pond, both of which new practices in Cambodia.

Limsong and Sitha (2015) dealt with Cambodian attempts to develop culture-based fisheries including their management.

Incidentally, one common feature in almost all the water bodies in Cambodia was, perhaps, the provision of a "conservation zone," generally, in the deeper areas of the water body, where fishing is prohibited.

De Silva and Song (2015) dealt with the challenges and constraints for developing CBF in Cambodia and possible strategies for success.

CBF seemed to an attractive development strategy, as it mobilizes paddy land farming communities (e.g., the rice farmers) to use existing water bodies for the secondary purpose of food fish production. Notwithstanding the above, support to community forestry and fisheries in Kratie and Stung Treng Provinces had been scaled up under a new Partnership Program with contribution of US$2.1 million from Switzerland for the initial phase (Anon, 2015a,b). In fact, community fisheries, community forestry, and conservation programs in Kratie and Stung Treng provinces were planned to be scaled up under joint partnership program(s) of local and international organizations, organized by the program's managing organization, the Non-Timber Forest Products Exchange Program Asia (NTFP-EP). Representatives from the Partners for Forestry and Fisheries Cambodia (PaFF) Program, together with delegates from the provincial governments of Kratie, Stung Treng, and the Forestry and Fishery Administrations and the Swiss Agency Cooperation Office in Cambodia, have declared their commitment for cooperation under the Program.

2.3.1.4 *China*

In China, Paterson and Pernetta (2015), Huang et al. (2001), and Wang et al. (2015) worked significantly on community-based management of fisheries. Fisheries management aspects are well *in vogue* in China.

In China, Paterson and Pernetta (2015) dealt with integrating fisheries and habitat management with regard to fisheries *refugia* in the South China Sea. It is, perhaps, quite a known fact that production in a fishery could be highly successful, if the habitat parameters are integrated and taken care of concomitantly along with the management aspects. Fisheries *refugia* are spatially and geographically defined marine or coastal areas in which specific management measures are administered to sustain fisheries of important species during the critical stages of their life cycle, and also to take care of their sustainable use. Concomitant to the above, Huang et al. (2001) dealt with the fish resources in the Chinese reservoirs and their utilization and management. The diversity of the landscape, aquatic resources, and climate in the reservoir catchments, probably, provide a variety of habitats for fish, resulting in a rich diversity of ichthyospecies. Concomitantly, Wang et al. (2015) dealt with culture-based fisheries in the lakes of the Yangtze River basin, China, with special reference to stocking of mandarin fish and Chinese mitten crab. In fact, the stocking of mandarin fish and mitten crab were the most successful because stock enhancement of these two species, perhaps, had been systematically conducted.

2.3.1.5 *Hong Kong*

In Hong Kong, Anon (1999a) dealt with a study on fisheries resources and fishing operations in **Hong Kong** waters under the auspices of Legislative Council Panel on Economic Services Consultancy (LCPOESC). Further, Yeung (2008) dealt with geodiversity as a new challenge for nature conservation in **Hong Kong**. Moreover, Morton (2005) reported about overfishing in **Hong Kong**, which led to the initiation of fishing crisis in the country. Further, Clarke et al. (2002) reported about the consultation with local fishers on the **Hong Kong** artificial reefs initiative.

Hong Kong is rich in aquaculture and fisheries. Anon (1999a) dealt with a study on fisheries resources and fishing operations in **Hong Kong** waters under the auspices of Legislative Council Panel on Economic Services Consultancy (LCPOESC). At the meeting of the LCPOESC, members were briefed on the major findings and recommendations of the Consultancy Study on Fisheries Resources and Fishing Operations in Hong Kong Waters commissioned by the Agriculture and Fisheries Department (AFD). Members were informed that the administration would conduct a public consultation exercise and, in the light of views received, would try to formulate a fisheries management strategy.

Concomitant to the above, Yeung (2008) dealt with geodiversity as a new challenge for nature conservation in **Hong Kong**. Many of Hong Kong's wild terrains, with their great biodiversity, of late, have attracted much attention. Indeed, also harbored within are some lesser known natural treasures of Hong Kong. They include the multifarious rock formations and geomorphological features. Incidentally, there had been contributions pertaining to the ongoing works on geodiversity and biodiversity in the fields of agriculture, fisheries, and conservation. However, geoconservation may not always rely purely on the work of the government. As such, the discussions suggest that there is, probably, a need for integrated management, knowledge building, public support, and, perhaps most importantly, a wider dissemination of knowledge. Notwithstanding the above, Morton (2005) reported about overfishing in **Hong Kong**, which finally, perhaps, led to the initiation of fishing crisis in the country. Incidentally, there had been a steady decline in fish catches within Hong Kong waters over the last *c*. 25 years. Studies have revealed that the catch had indeed fallen by c. 50%, and fish fry production had dwindled by *c*. 90% over the last decade, coupled with not much effective management measures being *in vogue*. On the other hand, it is quite a well-known fact that proper management and sustainability in harvest are prerequisites for achieving conservation of resources. The Government of Hong Kong, perhaps, had established the Committee on Sustainable Fisheries in late December 2006, probably, to tackle the overfishing problem, and also, perhaps, to study the long-term directions and goals, as well as to sort out the feasible options and strategies for the sustainable development of the local fisheries industry. Some of the salient features of the study outcome, which were submitted to the Food and Health Bureau for consideration, are as follows:

(i) The committee, after careful consideration of the situations of the international and local fisheries industries as well as the contributions of the local fisheries industry to Hong Kong's economy and society, had agreed that the local fisheries industry, perhaps, has changed in tandem with Hong Kong's economic development. However, the local fisheries industry, as an important local cultural asset with a long history, is sufficiently valuable and, probably, has the potential for sustainable development through the application of efficient *modus operandi* of community-based fisheries management (CBFM).

(ii) The goals of the sustainable development of the fisheries industry are to enable fishermen and fish farmers achieve self-reliance, effective management, help maintain their own livelihoods and cope with the changing business operating environment; provide job opportunities to the fisheries community, and ensure the supply of fresh and quality fisheries products to the people of Hong Kong. Meanwhile, the society could also benefit from a sustainable and well-managed fisheries industry, as it would help establish and maintain a salubrious aquatic (marine and FW) ecological environment, which could be abundant in resources and also could be passed on to our posterity.

(iii) In order to achieve the goals of the sustainable development of the fisheries industry, consequent upon the above, the committee has formulated two directions for promoting the goals of the fisheries industry's sustainable development with an efficient management system *in vogue*. These are: (a) to help the fishermen to develop or switchover to modernized and sustainable modes of operation and (b) to protect, conserve, and rehabilitate the marine (and also FW) ecosystem and fisheries resources.

(iv) Further, the committee, perhaps, had also proposed the following measures to promote the sustainable development of the local fisheries industry: (a) promote modernization of the fisheries industry and sustainable modes of operation, (b) regulate the fishing effort of capture fisheries, and (c) conserve and enhance the fisheries resources.

(v) The committee, probably, had found that the technology employed by the fisheries industry in Hong Kong was, perhaps, not advanced enough, resulting in high operating costs and low efficiency and hence low competitiveness. Therefore, it is recommended that the government could promote modernization and sustainable development of the fisheries industry and assist fishermen to develop or switchover to modernized and sustainable modes of operation and businesses through the adoption of the following measures: (1) enhancing fishermen's understanding of sustainable fisheries, (2) promoting fisherman cooperative enterprises, (3) providing with stable financial support to sustainable fisheries development programs, (4) reviewing fisheries loan requirements and approval procedures, (5) implementing specific support measures for sustainable fisheries operations (e.g., aquaculture), recreational fisheries, fisheries product (logistics and processing), and (6) strengthening communications with the mainland counterparts to help fishermen seek business opportunities.

(vi) It may be noted here that, of late, the catch and fishing effort of capture fisheries, probably, had far exceeded the experts' estimates of the maximum sustainable yield (MSY) and optimum fishing effort. The committee, probably, had considered that the abovementioned measures to promote modernized and sustainable operations alone may not be able to solve the problems. Moreover, measures are to be in place to control the fishing effort of local capture fisheries.

(vii) The committee, probably, has advocated for strengthening efficient fisheries management, may be through CBFM as well as law enforcement.

(viii) The committee, further, proposed certain measures to protect significant marine and fisheries ecosystems, such as fish spawning and nursery grounds, as well as to promote the rehabilitation and growth of fisheries resources in Hong Kong waters.

Notwithstanding the above, some of the other measures are as follows:

- **(a)** To strengthen the regulation of coastal and marine development projects.
- **(b)** To designate fisheries protection areas (FPAs).
- **(c)** To prohibit commercial fishing in marine parks.

(ix) The committee, probably, had believed that the excessive fishing effort could be reduced to a level, which could be below the MSY, through the implementation of the above-proposed options and measures, and significant fish spawning and nursery grounds in Hong Kong waters could be conserved by following proper measures. These proposed measures may also provide fishermen with choices, including helping the trawler fishermen to switchover to other kinds of operations. However, those who continue to engage in capture fisheries may also benefit from the improvement and recovery of marine fisheries resources. Consequently, the whole fisheries community will be better able to cope with the changing operating environment and could also achieve self-reliance.

(x) In addition to the above, the committee consults stakeholders on the proposed options and measures. The stakeholders, generally, tried to accept the long-term goals and directions proposed by the committee, as well as, usually agreed upon the positive impacts resulting from the proposed measures, although they, probably, held different opinions regarding individual measures. Thus, the committee, probably, has recommended that the government may try to conduct a more comprehensive consultation with the industry and the stakeholders while studying and formulating the implementation details of the measures, as well as strive to consider and balance the views of different parties, with a view to obtaining wide support for the proposals.

(xi) In fine, the committee, probably, had suggested that the proposed options and measures could be implemented in phases. Highest priority is given to the proposals that could be implemented using the existing resources, or, within the existing legal framework. On the other hand, the government could conduct consultations on the proposals for controlling fishing effort and enhancing fisheries resources conservation.

Concomitant to the above, the government could try to amend the relevant legislation(s) in order to enhance the conservation of aquatic ecosystems and fisheries resources, as well as to implement the proposals on maintaining an appropriate number of fishing vessels, prohibiting nonlocal fishing vessels from fishing within Hong Kong waters, prohibiting commercial fishing in the parks and sanctuaries, designating FPAs, and last but not the least, banning trawling in Hong Kong waters, after reaching consensus and formulating the appropriate measures for assisting the affected fishermen. Notwithstanding the above, Anon (2018) reported about the facts regarding agriculture and fisheries in Hong Kong. Capture fisheries and AQC contribute to nutrition and avocation in Hong Kong by maintaining a steady supply of fresh fish to local consumers, and also, probably, by giving employment to the people. Incidentally, the total capture fisheries and marine fish culture and production are equivalent to *c.* 28% of seafood consumed in Hong Kong while the pond fish farmers, perhaps, produce *c.* 3% of the FW fish eaten.

In addition to the above, Clarke et al. (2002) reported about the consultation with local fishers on the **Hong Kong** artificial reefs initiative. Consultation programs, aimed at soliciting inputs from the fishers and minimizing potential conflicts with existing fishing operations, probably, had been conducted, using local fora and information materials. The Special Administrative Region (SAR) of the Government of Hong Kong had undertaken a scheme to deploy artificial reefs (ARs) to enhance fisheries resources and conserve the marine environment as part of an ongoing effort to rehabilitate depleted local fish stocks.

2.3.1.6 Indonesia

Fisheries management is a significant aspect in Indonesian fisheries. **In Indonesia**, Crawford et al. (2004) dealt with compliance and enforcement of community-based coastal resource management regulations in North Sulawesi, Indonesia. They had reviewed the socioeconomic theory of resource management compliance and applied to community-based case examples from North Sulawesi, Indonesia. Compliance with nondestructive exploitation practices in several village sites, empirically measured by changes in live hard coral cover, was related to reef distance from a settlement area and not significantly related to visibility from a settlement.

Concomitant to the above, community-based fishery management **in Indonesia**, during in the post-conflict situation, was reported by Sidayat (2008) as a **case study** in coastal villages in North Maluku in Indonesia. It was a case study of how a local reconciliation effort and process, in the context of community-based fisheries management, could facilitate the reconciliation process between conflicting communities (Muslims and Christians) in a post-conflict situation.

Further, Satria and Matsuda (2004) reported about decentralization of fisheries management in **Indonesia**. Decentralization is very much considered as an alternative to make better fisheries management. It could be mainly because decentralization appears to be a strong strategy for increasing the efficiency and equity of developmental activities, as well as for the delivery of services. It could also play a significant role in promoting local participation and democracy. The trend of evolution of decentralization of fisheries management policy in Indonesia had revealed that the decentralization had gradually developed from deconcentration, delegation, and devolution of the parameters. After Reform Era, devolution form of decentralization was implemented due to the enactment of UU 22/1999 (the Local Autonomy Law), in which local government had gained a quantum of new authorities concerning marine-fisheries management. By such devolution, however, the community-based management system (CBMS), which routed from traditional fishing communities, has been recognized. The effectiveness of the CBMS for the sustainability of the marine resources is caused by the bottom-up planning and participatory approach that had, perhaps, led to the elevation of the local fishers' sense of stewardship over the resources. Even though this kind of decentralization practice had been dealing with several problems, this is still a better way than centralization. Many communications tried to identify some of the agenda, which are encountered both at the central and at the local level. The contributions emphasized the need for improvement in the legal framework, the capacity-building of the local government, and the revitalization of the local institution(s).

Furthermore, Radjawali (2012) examined the local measures of conservation and development with regard to live reef food fishing in Spermonde Archipelago in **Indonesia**.

Still further, Soede et al. (2001) dealt with integrated marine management concessions with regard to a new approach to an old problem. Pressures on the reef range from overexploitation and destructive fishing practices to coastal- and marine-based pollution, coastal development impacts, and global climate change.

2.3.1.7 *Japan*

Japan has a long history of fisheries management. Yamamoto (2000) opined on why community-based fishery management is well developed in Japan while dwelling upon collective fishery management development in **Japan**.

The historical development of the fishing right system in Japan, which was believed by some people to have emerged during her feudal era, probably, has led to the large-scale prevalence and practice of community-based fishery management system. However, this may not always be correct. According to history, the Japanese people had served the emperor under the military government until August 1945. Thereafter, Japan was occupied by the Allied Forces for 7 years until April 1952. During this period, the policy of the occupied forces was to make Japan a real democratic country. Within such a basic policy, the contents of Japan's fishery law was thoroughly redrafted in the pattern that fishermen would be entitled to participate in the planning process with regard to the use of fishing area and fishery resources through a fishery coordination committee in a fully democratic manner. This gave ideal circumstances to Japanese fishermen to make a collective fishery management system, which is often called a community-based fishery management system.

Concomitantly, Schmidt (2003) dealt with the multiple roles of fisheries **in Japan**. Japan has a rich potential for aquaculture and fisheries. Of late, Japan has, probably, been making efforts toward community-based fisheries management on a wider scale. According to Schmidt (2003), fisheries in Japan are of a rather particular nature.

Notwithstanding the above, as an important "side effect," the community-based fisheries management system had been of key importance for the well-being of fisheries-dependent coastal communities. Incidentally, the coastal communities play a significant resource management role in the development of the Japanese coastal fisheries management systems. In this process, various additional functions had been developed which are also of importance to the livelihood of fishers. Until the mid-1980s, it may be mentioned here that the fisheries managers in many countries had focused mainly on the biological outcome of the fisheries management systems.

Incidentally, Yamamoto (1996) dealt with community-based fisheries management in **Japan**. The **Japanese fisheries regimes** had developed, perhaps, with three fishery laws in sequence, over the past 250 years, in which, probably, a fishing right system had commonly been used. In fact, the fishing right was used mainly as a tool for the collection of fishery taxes during the feudal era, until 1867. Further, the fishing right system was established by an old fishery law (1901–1947) and was, perhaps, used mainly as a tool to lessen conflicts among various groups of coastal fishermen as well as disputes between coastal and trawl fisheries.

Concomitant to the above, the recent fishery law, enacted in 1949, refer to "Territorial use Rights in Fisheries." However, it limits coverage to sedentary resources and nonmobile gears. At the same time, the recent law(s), probably, has a pragmatic system whereby a coastal fisheries management plan might be established with the participation of fishermen. These innovations, probably, had given the fishermen a big motivation to make a community-based coastal fisheries management system (CBCFMS). Consequently, it may be noted here that the number of fisheries management organizations, formed under the initiative of fishermen, perhaps, had increased year after year and reached *c.* 1524 in 1993, since the inception of the fishery law in 1949.

2.3.1.8 *Korea*

Korean fishery has an aspect of good management. In **Korea**, Cheong (2004a,b) studied the **Korean** fishing communities in transition with a note on the limitations of community-based resource management (CBRM) by examining Korean fishing communities making the transition into tourism.

Further, Chang Ik Zhang et al. (2011) dealt with an IFRAME (Integrated Fisheries Risk Analysis Method for Ecosystems) approach for forecasting, assessing, and managing **Korean** fisheries under a changing climate.

Furthermore, Lee and Midani (2014) dealt with creative economy potentials and fisheries development in **South Korea**.

A new development paradigm is emerging in the globe which, probably, tries to link the economy and culture, embracing economic, cultural, technological, and social aspects of development, at both the macro- and microlevels. Central to the new paradigm is the fact that creativity, knowledge, and access to information are being increasingly recognized as powerful engines driving economic growth and promoting development in a globalizing world. It could be seen in many definitions that "creative economy" is, perhaps, considered as cultural and artistic products that is directly involved in the history and culture of human beings. On the other hand, in many South Asian countries, for example, Korea and Japan, fisheries and fishing have a long history, and also became part of their culture. Sometimes, the creative economy is categorized into four parts: (a) community-based fisheries management (CBFM), (b) individual transferable quota (ITQ), (c) fishing villages, and (d) stock rebuilding plans. These are the four main domains, in which, creative economy was, perhaps, used as a tool to develop the fisheries sector. Further, considering fisheries as part of the Korean culture, there have been explanations regarding how Korean government used the concept of creative economy in order to make a culture shift of this sector toward a more sustainable and efficient sector.

2.3.1.9 *Lao PDR*

Lao PDR possesses a strong base in fisheries-related activities. Baird (2000) studied integrated community-based fisheries comanagement and protected areas management in **Lao PDR** with regard to opportunities for advancement and removal of obstacles to implementation. It may be pointed out here that the comanagement of aquatic resources, including fisheries, has, perhaps, apparently succeeded in augmenting fish stocks of the villagers' fish catches in Khong District, Champasak Province, Southern Laos. However, notably, scientific evidence(s), probably, are still lacking in favor of supporting villagers' claims that aquatic resources, probably, had significantly benefited from their comanagement exercises. Concomitantly, Phounsavath (1998) reported about community-based fisheries management as an approach to participatory management of reservoir fisheries exemplified by case studies of two fishing villages at the Nam Ngum Reservoir, **Lao PDR**. The results of the studies have revealed the existence of different traditional institutional structures and organizations involved in the management of fisheries in the studied villages, as well as different fishers' behaviors and fishing practices. However, the main prerequisites for the development of a community-based fisheries management system in the study area were, probably, yet to be developed. Further, Anon (1997) dealt with the reservoir fisheries management in Savannakhet Province of **Lao PDR**. Studies have revealed that stocking alone, probably, was not sufficient to increase standing crop (stock). Access restrictions, probably, had a strong positive effect on the standing crop (stocks) of fish in small water bodies. Furthermore, Anon (2015a,b) dealt with a summary of culture-based fisheries developments in **Lao PDR**.

Many of the information are based on a number of publications, either as the primary literature or as annals and reports posted on the website of the Network of Aquaculture Centers in Asia-Pacific (NACA), and from the projects funded by the Australian Centre for International Agricultural research (ACIAR) since 1997.

2.3.1.10 *Malaysia*

Malaysia has many fisheries-related activities including the management aspects. Wong et al. (2009) reported about sustainable community-based fishery resources management with regard to the Tagal System of Sabah in **Malaysia**. They have studied the aspects of East Malaysia, as well as Peninsular Malaysia.

Sabah and Sarawak constitute East Malaysia, one of the country's two federal territories, the other being Peninsular Malaysia. East Malaysia occupies the northern third of the island of Borneo, having an area of *c*. 202,020 km^2, of which *c*. 72,500 km^2, perhaps, belong to Sabah. With a coastline of *c*. 1440 km, Sabah is surrounded by the South China Sea in the west, Sulu Sea in the northeast and Celebes Sea in the east. Sabah's fishing zones, probably, are classified into coastal [< 12 nautical miles (n mi) from the shore line] and offshore areas (12 n mi and beyond the EEZ boundary). The EEZ of Sabah is reported to be *c*. 90,000 km^2. Concomitantly, Nasuchon (2009a,b) reported about coastal management and community management in Malaysia, Vietnam, Cambodia, and Thailand, with a case study of Thai Fisheries Management. Incidentally, **Malaysia** is, probably, the only country having strong administration of coastal zone management by the federal government and is, perhaps, the strongest in surveillance and enforcement but concomitantly, probably, lacks a strong base in community-based management.

2.3.1.11 *Middle East*

In spite of being an arid region, the **Middle East** has a strong focus on fisheries. Feidi (1998) reported about fisheries development in the **Arab World**. If rationally and scientifically exploited, fisheries could play an important role in meeting the increased nutritional demand and could give a strong hand in strengthening the economy of the region (along with oil exploration) by being a renewable and self-replenishing resource.

The fisheries resources in the Arab world (consisting of a number of fossil fuel-rich countries) in the various types of aqua-regimes, including the coastal waters, the inland waters, as well as, based on the aquaculture (AQC) potential the region possesses show significant possibilities of aquaculture development in the region (perhaps, in addition to the trade with petroleum products), if properly taken care of.

Further, Anon (2012) worked on the National Fisheries Strategy (2012–2025) of the Republic of **Yemen**. Yemen's fisheries, one of the four promising sectors as identified in its Strategic Vision 2025, possess significant potential to achieve diversification in country's economy and in generating jobs for the people. This would also pave the way for the economic growth and socioeconomic development of the country.

2.3.1.12 *Myanmar*

Myanmar possesses a large number of water bodies with associated fisheries. Anon (2005a) dealt with Wetland Conservation in **Myanmar**. There are also instances of CBFM in different parts of the country, which need to be boosted.

Salient activities of different Fisheries- and Aquaculture-related organizations related to Community-based management, etc.

There have been a number of important apex organizations, like FAO, World Bank, UNEP, GEF, NACA, SEAFDEC, ACIAR, ODA, DANIDA, etc., which have significant mandates in their activities related to various aspects of fisheries and aquaculture including their promotion and management. Some of the salient activities related to the aforesaid aspects of such organizations, notably, NACA, are as follows:

De Silva et al. (2007) elaborated about NACA and its activities in various spheres including community fisheries.

NACA and its principal activities

(1) Intergovernmental organization, which promotes rural development through sustainable aquaculture.
(2) Technical cooperation among members

Some of the other activities of NACA are as follows:

(i) multi-sector participation to deal with government policy and R&D, discussion with farmer groups, traders, industries NGOs; (ii) collaborations with government, private sector, academia, and NGOs, totaling > 100 researchers in 20 countries; (iii) development of an internationally agreed document adopted/endorsed by NACA and FAO; (iv) globally accepted management principles for responsible shrimp AQC; (v) analyze and share experiences on better management of shrimp aquaculture in coastal areas; (vi) guiding principles and guiding criteria for different types of aquaculture activities; (vii) provide support for the implementation of the ongoing projects, with special reference to small-scale farmers; (vii) develop market links for responsible BMP product(s); (viii) deal with aquatic animal health issues; (ix) reduce the risk of aquatic animal disease impacting on trade, environment, and human health; (x) development of policy frameworks; (xi) implementation of practical health management strategies at farm, local, national, and regional levels; (xii) establishment of regional surveillance system(s); (xiii) provide technical support through sharing of expertise and laboratory facilities; (xiv) active participation of the primary growers; (xv) address emerging issues related to food safety, new disease (EUS, KHV, WTD, TS), etc.; (xvi) collaborate with organizations like the FAO, OIE, ACIAR, CSIRO, AusAID, European Union (EU), SEAFDEC, AAHRI, DAF, MPEDA, etc. (xvii) taking care of genetics and biodiversity issues; and (xviii) provide support to members to conserve aquatic biodiversity and genetic resources by: (a) building capacity for national aquatic resource management programs; (b) coordinate cooperative R&D programs; (c) development of broodstock management programs for economically important and newly emerging indigenous species; (d) contributing conservation plans for endangered species, e.g., Mekong giant catfish; works in this direction are at Kasetsart University, DoF Thailand, FAO, MRC, WFC, Deakin University, Malaysian Fisheries Society, etc. (e) fostering culture-based fisheries; and (f) developing extensive, community-based AQC through certain steps like the development of "best practice approaches" to culture-based fisheries, including fish and shrimp farming, etc.; effective use of small water bodies for low-cost fish production among rural communities; application of community-based management or comanagement principles to culture-based fishery and stock enhancement activities; cooperative R&D program to support the development of sustainable marine and freshwater fish farming, etc.; people-based network with institutional participation including farmers' groups, traders, industries, NGOs, etc.; multi-sector participation to deal with the government policies, R&D programs; etc.

Notwithstanding the above, the NACA members, including member countries like Australia, Bangladesh, Cambodia, China, Hong Kong SAR, India, Indonesia, IR Iran, DPR Korea, Malaysia, Myanmar, Nepal, Pakistan, Philippines, Sri Lanka, Thailand, Vietnam, etc., produce > 80% of world aquaculture production by volume.

Notwithstanding the above, Wilkinson (2007) dealt with community communication centers with regard to their establishment, applications, and constraints. Community involvement is, sometimes, a prerequisite for efficient management of fisheries. An efficient community network is, probably, essential for such an exercise. A **network** of community-based communication centers were established in Lao PDR and maintained by DoLF staff to facilitate government extension services and sharing of experience between community groups participating in culture-based fisheries (CBF) activities. The network, probably, had functioned primarily via regular (fortnightly) face-to-face meetings at district or provincial extension offices, attended by representatives from surrounding communities. These offices and their staff, probably, have served as focal points for frontline extension activities and also as relay points for technical queries requiring specialist input from head of the office or elsewhere. Incidentally, initial plans to conduct face-to-face networking via Skype and similar voice-over-IP applications through 3G mobile networks were, perhaps, not very successful, mainly due to restricted 3G coverage, poor bandwidth, and high signal latency in rural areas. Nevertheless, some discussions and information sharing were successfully done using low-bandwidth applications such as Facebook, WhatsApp, instant text messaging, etc. DoLF staff maintained logbooks of questions and advice regarding aspects of CBF activities sought by participating communities. The main subjects of queries have centered around hardships in the procurement of quality seed, nursing of fingerlings, management of the water bodies, and health/disease issues. It could also be probable that there may be scopes to utilize the communication centers to coordinate harvesting time and marketing by nearby communities.

2.3.1.13 *Nepal*

The Himalayan country of Nepal has various types of activities related to fisheries and their management. In **Nepal**, Gurung (2007) dealt with the restoration of small lakes through cooperative management, which could be a suitable strategy for poverty-laden areas in developing countries. Main threats to the lake, its aquatic life, and its very existence include infestation by excess aquatic vegetation, sedimentation, and low in-lake concentrations of dissolved oxygen (DO). Small lakes ($\leq$500ha) are relatively prone to the process of "succession and climax," although such water bodies may be of social, environmental, and ecological significance. Thus, their disappearance could spark off social chaos and disorder in areas already having to cope with other rapid environmental changes. This means that credible mechanisms for revitalizing or protecting small lakes are an important goal.

Further, Pradhan (2009) dealt with reservoir fisheries in Nepal with a note on management and conservation.

2.3.1.14 *The Philippines*

Fisheries and aquaculture are the thrust areas in the Philippines. Katon et al. (1997) have dealt with marine conservation projects for San Salvador as a case study of fisheries comanagement in the **Philippines**. Incidentally, Ahmed et al. (2004) discussed about collective action and property rights for sustainable progress in fisheries management. Fisheries are complex and interdependent ecological and social systems that require integrated management approaches. Managing such common pool resources requires conscious efforts by a broad range of stakeholders to organize and draft rules enabling equitable and sustainable use of the resources for everyone's benefit. Further, Ahmed et al. (2004) reported that, until the late 1960s, villagers on the island of San Salvador in the **Philippines** enjoyed open and unrestricted access to an abundance of coastal resources. In the early 1970s, an influx of migrants, combined with the integration of the village economy into the international market for aquarium fish and a shift to destructive fishing operations, ruined the local fishing grounds, and thus, conflicts erupted. Although, the government had claimed full control over the use and protection of marine and coastal resources, there was, perhaps, no respite in the degradation and depletion of the resource. Incidentally, the developing world seems to presents many similar examples where central government management of fisheries resources is unable to either reduce overfishing or counteract destructive fishing methods. The state often lacks the capacity to enforce property rights and regulations on resource use. The actions of one person or group of users affect the availability of the resource for others. Managing such common pool resources requires conscious efforts by a broad range of stakeholders to organize and draft rules enabling equitable and sustainable use of the resources for everyone's benefit. Collective action is often a prerequisite for the development of community-based institutions and the devolution of authority that is required from central to local authorities.

Concomitantly, Charles and Vaughan (1997) dealt with policy reforms and community-based programs to combat cyanide fishing in the **Philippines**. Experience with the CFRP over the few decades indicates that cyanide fishing could be reduced through a combination of right policies and laws, beefed-up enforcement efforts, enhanced public awareness, cyanide testing of live-fish exports, training of cyanide fishermen in cyanide-free live-fish capture techniques, and development of sustainable community-based resource management and livelihood alternatives that could transform the local fishermen into the frontline marine stewards and protectors. Concomitant to the above, in the **Philippines**, Anon (2011a,b,c) dealt with a review of the National Laws and Legislations in the Philippines in relation to Ecosystem Approach to Fisheries Management (EAFM). In fact, studies had been undertaken to review the international, national, and area-specific laws and policies in relation to EAFM in the Philippines. Many such reviews, probably, assess the policy issues and gaps, as well as the recommend areas of intervention, including policy proposals.

2.3.1.15 *Southeast (SE) Asia and the Pacific*

There have been some amount of works done related to the fisheries management aspects in the Pacific and the **Southeast (SE)** Asia in general. Pomeroy (1995) reported about community-based and comanagement institutions for sustainable coastal fisheries management in SE Asia. Moreover, it is, probably, now appreciated that resource conflicts, particularly in fisheries, could be reduced through community-based managements, and resources could now be better managed when fishers and other resource stakeholders are more involved in management and, further, if access rights could be distributed more effectively and equitably. There seemed to be an increasing commitment by the governments in SE Asia to policies and programs of decentralization and community-based management and comanagement. The planning and implementation of these management systems need the development of new legal, administrative, and institutional arrangements at both national and community levels in order to complement contemporary political, economic, social, and cultural structures. Notwithstanding the above, Komatsu (2013) dealt with a special study on sustainable fisheries management and international trade in the SE Asia and the Pacific Region. Further, there are significant contributions which tried to analyze the current status of fisheries and AQC in SE Asia and related international trade. Analyses have concluded that a policy of sustainable management for both capture fisheries and aquaculture (AQC) are of great significance but such a policy, perhaps, has never been planned or implemented with a holistic and long-term perspective. Incidentally, the current policies reflect a short-term view and the immediate needs of each nation. Therefore, capacity building of human resources and organizations, including governments, is, probably, of high significance for the formulation of holistic national policies seeking basic remedies against depletion of resources and long-term plans for the sustainable management of fisheries resources supported by intensive and extensive AQC practices. Such holistic national policies are, preferably, to include science-based management, monitoring, enforcement, coordination of capture fisheries and aquaculture, with regard to international trade policies. It may include the effects of climate change and oil price increases, as well as international market trends and regulations or barriers. Moreover, international trade is supposed to be promoted based on the sustainability of capture fisheries and AQC. ADB members and governments had been urged to provide with official development assistance for policy implementation, in particular to the private sectors that may not, otherwise, receive any small and community-related business. In such contributions, recommendations usually focus on capacity building for the long run.

2.3.1.16 *Sri Lanka (SL)*

Aquaculture and fisheries are significant areas of activities in **SL** today. In **SL**, Murray (2002a) studied the aspect(s) of failure of comanagement activities with a review of theory and lessons learnt from reservoir fisheries in the dry zone. He studied the fisheries of the Negombo estuary brush park fishery in **SL** from June 1998 to March 1999. The mean fish yield (all species) in the fishery was *c.* 12.46 mt/ha/yr. Fin fish species formed *c.* 91% of the yield and the rest was formed by penaeid shrimps and crabs. Fish yield (FY) and the twig density in the brush park exhibited a second-order polynomial relationship indicating a minimum yield at an intermediate value of twig density in the brush parks. The relationship between FY and duration of implantation of brush parks showed an optimal period of 30–40 days for high FYs. Concomitantly, Amarasinghe and Wijenayake (2015) dealt with the results of a decade of R&D efforts on culture-based fisheries in **SL**. Incidentally, **fisheries enhancement** may be defined as the technical intervention in the life cycles of fish. Culture-based fisheries (CBF) development is one of the major fisheries enhancement strategies practiced in inland reservoirs of SL. The extensive availability of inland reservoirs in the country, primarily constructed for irrigation of land crops in ancient times, probably, has favored CBF development, which is essentially a development since the late 1990s. Incidentally, it may be pointed out here that water retention period in most small village reservoirs in the country is seasonal and lasts usually for 6–9 months in a year. Thus, CBF development in these reservoirs may require fast growing fish species, notably the Chinese major carp and the Indian major carp (IMC). In fact, hormone-induced captive breeding of IMC in the government-owned hatcheries and fingerlings rearing in mini nurseries, generally maintained by the rural agricultural farmers, are the sources of seed for stocking these village reservoirs. Incidentally, the CBF in the village reservoirs of SL is a community activity, sometimes, involving agricultural farmers, who may not have any previous experience in fisheries. As such, awareness programs conducted among these farmers facilitated the establishment of CBF in small village reservoirs. However, the biological productivity of the water bodies and socioeconomic conditions of rural communities are generally found to vary from one reservoir to another.

As such, successful R&D efforts, probably, have been made for the selection of village reservoirs suitable for CBF development, based on the biological productivity-related parameters such as reservoir morphometry, allochthonous input of nutrients through livestock farming, and socioeconomic characteristics of rural communities that favor CBF. However, correct timings are necessary for CBF development in the rural reservoirs, for fingerling production to suit the monsoonal rainy season when the reservoirs get filled. Moreover, efforts were made to ascertain the **climate change** impacts, which might have resulted in a shift in peak monsoonal rainy season. These issues, perhaps, were identified together with possible resilience capacities of rural communities for sustainability of CBF. Further, dissemination of research findings through different means, such as production of documentary films, publication of a monograph, including translated versions in different regional languages, as well as holding a series of regional workshops were, probably, instrumental for CBF development at the regional level. In addition, the Asian Development Bank (ADB)-funded "Aquatic Resources Development and Quality Improvement Project or ARDQIP)," which also contained a significant component for CBF development in inland reservoirs of SL, probably, had also facilitated the R&D efforts. However, given these programs, opportunities, and facilities, developmental aspirations might not see the light of the day, unless effective management measures are in place involving the stakeholders from the grassroots level and through the application of efficient management *modus operandi* like the CBFM. Nevertheless, the recent efforts to develop CBF in SL, perhaps, included the establishment of profitable CBF with effective comanagement in selected minor perennial reservoirs and the use of *Macrobrachium rosenbergii* postlarvae for CBF in many inland reservoirs.

Further, De Silva (2001a,b) dealt with reservoir and culture-based fisheries with regard to their biology and management with great amount of emphasis on **SL** reservoirs formed by damming of rivers. This is, no doubt, detrimental to the river, as a lotic habitat is converted into a lacustrine one; although, for many Asian countries, the reservoirs, probably, serve as an expanding resource of water. Moreover, some of the reservoirs could be very diverse in terms of both size and fisheries potential. They range from small water bodies with productive culture-based fisheries to vast expanse of water with variable and often low fish production. The strategies required to achieve desirable productivity levels from these water bodies are, probably, similarly diverse. Concomitant to the above, there have been several contributions on productivity and management of reservoirs, notably, proceedings which came out of a workshop, being jointly sponsored by the Australian Centre for International Agricultural Research (ACIAR) and the Mekong River Commission Fisheries Program, which, perhaps, brought together *c.* 55 scientists from 16 countries to discuss recent progress and future challenges in the wise use and management of reservoir fisheries resources through culture-based interventions. It was noteworthy for the active participation of scientists from all major ongoing reservoir fisheries projects in Asia, variously funded by ACIAR, the Mekong River Commission (MRC), European Union (EU), Danish International Development Agency (DANIDA), as well as several national programs. Such exercises, probably, enabled the regional as well as the international specialists to share research findings and management experience; to identify gaps in existing knowledge, and from this to establish R&D priorities for the future. The major findings and recommendations of the workshop were presented to a thematic session on "Fisheries Enhancement" at the joint NACA-FAO international meeting on "Aquaculture in the Next Millennium" held in Bangkok from February 21 to 25, 2000. This had provided valuable inputs into these broader deliberations aimed at the development of global policies on aquaculture and their management.

Furthermore, De Silva (2001a,b) dealt with reservoir fisheries and their broad strategies for enhancing yields, with a note on their management and conservation. Reservoirs are significant resource of water in Asia, often having rich potential for AQC and fisheries. Incidentally, inland fisheries in the world account for *c.* 9% of total fish production, and of these, Asia, probably, accounts for *c.* 60% of the global fish production. Thus, the reservoirs and the fisheries therein are, probably, no new things to Asia. However, the inland fisheries, in particular the reservoir fishery potential, need to be exploited. They are the integral components of certain Asian cultures, dating back to *c.* 4000 years or more, and certain regulatory measures were, probably, in place with regard to their exploitation. However, an enormous expansion in reservoir acreage was witnessed in the postwar period, and the expansion was, perhaps, best exemplified in the case of mainland China, where the reservoir acreage, probably, increased to *c.* 2,47,106 ha from 1949. In this connection, it may be noted here that reservoir impoundment, particularly in developing countries, was going on almost unabated in spite of concerns of various lobby groups, often headed by environmentalists. In fact, reservoirs are, perhaps, never impounded for fisheries development *per se*. Rather, fisheries, perhaps, are gradually being recognized as a significant secondary user of reservoir water resources. Moreover, in certain instances, it is often reported that the income from the fishery could exceed that from the intended primary function of the reservoir, such as Ubolratana reservoir, Thailand. In fact, fishery aspects, so far did not and still hardly command a consideration during the planning of reservoir impoundment, perhaps, with a few exceptions in the region. However, the situation in China is, perhaps, the most significant exception where a hatchery and associated facilities (such as fry and fingerling nursery, rearing and stocking ponds, etc.) are, generally, provided for below the dam of almost every medium and large reservoirs. Moreover, the reservoir bed is prepared to facilitate harvesting, etc. and associated management aspects are also put in place. In addition, another aspect witnessed in the recent years is the incorporation of fish ladders, as seen in some recently constructed reservoirs in Thailand, such as Sirinthorn, which, in all probability, satisfies the concerns of environmentalists and conservationists and management personnel, even though there is a dearth of scientific information indicating the efficacy of fish ladders in the tropics.

Notwithstanding the above, in the light of increasing human population growth, reservoirs are becoming increasingly significant in the current millennium, as an important provider of animal protein and employment opportunities, particularly to the poorer sections of the community, which also often happens to be rural. Unlike in the past, reservoir fishery activities are considered to be a significant avenue for the rehabilitation of displaced persons, particularly exemplified in the case of Saguling and Jatinuhur reservoirs in Indonesia, and also vividly exemplified by the immigration of large number of displaced "Kaibarta" fishermen from erstwhile East Pakistan (now Bangladesh) and resettled around the wetlands in India, notably around the potential Sone Beel, Chatla Haor, etc. (Kar, 2007a,b,c, 2013). Such and similar contributions deal with fishery and fish as a food resource, as well as issues related to increasing fish yield (FY) from the reservoirs, the potential advantages of a more holistic approach to reservoir fisheries management, and so on. Moreover, attempts have also been made to highlight the associated problems linked with efforts to increase reservoir fish production and their management.

Concomitant to the above, Pushpalatha et al. (2015) dealt with the impact of introduction of culture-based fisheries on fish production in perennial reservoirs of **SL** and associated management issues. SL is blessed with a large number (> 12,000) of irrigation reservoirs. Depending on their hydrological regimes, they are broadly classified into two categories, viz. (a) perennial reservoirs and (b) seasonal reservoirs. They are secondarily used for inland fisheries. Incidentally, culture-based fisheries (CBF) in the seasonal reservoirs were, probably, initiated in the 1980s and well documented. The Government of SL, perhaps, had recognized CBF as an effective way of increasing fish supplies in the villages at affordable prices while also providing with avocation and additional income to rural farmers, thereby contributing toward alleviation of poverty. There are *c.* 2,00,000 ha of perennial reservoirs in SL. These reservoirs are categorized into three broad size classes, viz., (a) minor (< 200 ha), (b) medium (200–800 ha), and (c) major (> 800 ha). In many papers of today, the impact of introduction of CBF on fish production in minor, medium, and major perennial reservoirs has been assessed using the fish production data from a number of minor, medium, and large reservoirs. In all the three categories of reservoirs, post-CBF data revealed significant increases in fish production such as enhancement of *c.* 206% and *c.* 319% annual fish production, on the average, in minor and medium reservoirs, respectively. Available provisions under the *Fisheries and Aquatic Resources Act No. 02 (1996)* ensure the ownership of the harvested fish to the fishers' community, which is a crucial element in the success of CBF. Further, the role of fishers in community-based management measures, introduced for effective and successful CBF practices, are often also highlighted.

2.3.1.17 *Thailand*

Thailand has a rich diversity of fish resources. Henocque and Tandavanitj (2011) reported about fisheries and practice of comanagement in **Thailand**. The Coastal Habitats and Resources Management (CHARM) seemed to have developed its approach around five attributes of comanagement, viz., participation, partnership, capacity building, development of integrated approaches and methods, and learning and adaptation. Incidentally, the main rationale for the formation of the CHARM project was the recognition that overfishing of the marine waters of Thailand has negative impacts on the marine fish stocks, ecosystems, and the coastal fishermen. Incidentally, Tokrisna et al. (1997) have reviewed the fisheries and coastal community-based comanagement regime in **Thailand**. Such contributions often review successful comanagement activities where they exist. Further, attempts were made to identify the criteria of success, as well as their impacts on resource sustainability, equity, and efficacy. Such contributions often provide useful information for the successful implementation of community-based and comanagement arrangements for fisheries resources in Thailand. Further, there are works describing the community-based management regime over other coastal resources, which also included fisheries. In fact, the key factors for success include strong rural organization, NGO support, and government assistance, particularly, in providing with technical support. Further, almost equally important is that decisions are better to be made by the community and not by the government, which probably needs only to extend enabling environment with strong laws and their proper enforcement. Thus, it could be easier to emulate community-based comanagement elsewhere, once a project is successful and the living condition of the poor improved.

Notwithstanding the above, Pimoljinda and Boonraksa (1997) dealt with community-based fisheries comanagement along with a case study from Phang-Nga Bay in **Thailand**. It may be noted here that efficient "management" is indispensable for effective production and sustained harvest from a fishery. Conversely, depleted fish yield (FY) could be the outcome of disarrayed management. Consequently, the coastal ecosystem deteriorated mainly due to the degradation of fishery resources and conflicts among fishermen at various locations, including in Phang-Nga Bay, which, probably, had intensified with the fish output becoming unsatisfactory.

In addition to the above, Panjarat (2008) dealt with sustainable fisheries in the Andaman sea coast of **Thailand**. Various studies also tried to find out the ways and means for the sustainable use of the fisheries resources. Notwithstanding the above, Thailand is facing a number of problems in fisheries in spite of the fact that the country has a rich potential for AQC and fisheries. These include, among others, the depletion of fishery resources, overfishing, excessive fishing effort, conflicts among the fishers, and violation of regulations and unlawful fishing. Further, the fuel crisis, since 2001, and the tsunami event on December 26, 2004 impacted fishery activities significantly. It is anticipated that the fisheries problems may continue with significant effects on the economy of Thailand and the lives of the fishers, unless adequate measures for sustaining the marine resources are developed. However, the country is striving in that direction. There had been studies focusing on the fisheries status of Thailand and the Andaman Sea coast, as well as, addressing the fisheries problems. Such studies also tried to find out the ways and means for the sustainable use of the fisheries resources. In this connection, it may be noted here that fisheries laws are *in vogue* at global, national, and regional levels, and these also relate to the Thai fisheries. As such, such studies are also to be considered in order to find out how these principles could be effectively used in order to develop efficient instruments for sustainable fisheries management. There, probably, had also been recommendations for sustainable fisheries management in Thailand. Further, Anuchiracheeva et al. (2003) dealt with systematizing local knowledge using Geographical Information System (GIS) exemplifying fisheries management in Bang Saphan Bay, **Thailand**. The local knowledge could be converted into geospatial data on a GIS platform, and the succinct results may be used easily to guide fishery management and planning, especially, by offering directions for rights-based fisheries and comanagement. Incidentally, local knowledge is a significant source of information for fishery development. However, its general pattern of presentation, and sometimes undecipherable content(s), may make their use little difficult for the policy makers and managers. Further, there had been studies to show how these indigenous traditional local knowledge (ITLK), based on field data obtained by using participatory rural appraisal (PRA) at selected places, like Bang Saphan Bay, Prachuap Khiri Khan Province in Thailand, could be effectively systematized, analyzed, and displayed visually using a GIS for use in fisheries management. The PRA data at the fishing location, time of fishing, techniques and technology used, and species targeted were obtained from local fishers and then mapped using ArcView 3.1. Notwithstanding the above, Jutagate and Kwangkhang (2015) dealt with culture-based fishery of giant freshwater prawn with experiences from **Thailand**. The low rate of recapture of stocked *M. rosenbergii*, perhaps, made this culture-based practice not fully satisfactory, although the economic profit is generally high. In fact, the absence of proper guidelines with regard to the optimum size of the seeds to be released, as well as the appropriate time and location to be stocked, are the major difficulties, which if solved, could enhance the rate of economic returns.

Furthermore, Yakupitiyag (2016) dealt with farmer-managed systems in **Thailand** with a comment about its actual status within the purview of CBF. The *modus operandi* of AQC and classification of fisheries systems, probably, ascertains whether culture-based fisheries (CBF) are farmer-managed aquatic systems (FAMS) in Thailand or not. Incidentally, AQC may be defined as the culture of aquatic organisms in fresh, brackish, and marine waters. AQC differs from fisheries on the degree of human intervention. While AQC industry is "production"-oriented, the fisheries are "hunting"-oriented. Hence, aquaculture has control over the production process while fisheries exercise control over capturing process. Further, the degree of human intervention and control in AQC is governed by the production intensity. The extensive AQC systems, which are, probably, entirely dependent on the natural feed supply, perhaps, have a minimum human control. In AQC, mostly the humans, with a few exceptions, regulate the reproductive processes, stocking sizes and densities, use of feeds, and harvesting strategies that match market demand. In this connection, it may be noted here that the concept of FAMS, probably, captures the diversity of aquatic resource systems at the interface of AQC and wild capture fisheries. Moreover, in AQC, aquatic animals are reared usually but, not necessarily, in private ownership. Incidentally, FAMS include aquatic systems, such as rice field and farmer-owned (or rented) and farmer-managed pond systems. In CBF/CBFM, the harvesting of stocked fish done in a water body of common or community ownership, thus revealing some sort of community involvement. If this activity could be considered as extensive aquaculture, CBF could be considered as a part of FAMS, and not vice versa.

2.3.1.18 Vietnam

Fisheries and aquaculture are potential entities of nutrition and avocation in **Vietnam**. Ruddle (1998) worked and reported on the development of traditional community-based coastal marine fisheries management in **Vietnam** and their funding support. Concomitantly, there have been efforts to deal with the impact of fishery association management on the fishery community in **Vietnam**. Establishment of fishery association (FA) had brought a change in the scenario of aquatic resource management in which change(s) were, perhaps, brought about in the extent of fishers' area, fishing implements, yield, income, and fishers' awareness. Further, Nguyen (2012) dealt with community-based comanagement in Vietnamese fisheries, with emphasis on the fisheries associations in Tam Giang-Cau Hai Lagoon in **Vietnam**. Incidentally, the local FAs, probably, had played a significant role in the implementation of CBFM, and perhaps, also as a major collaborator of the government in the protection of aquatic resource and in fisheries management. Furthermore, Teh et al. (2014) dealt with the reconstruction of **Vietnam's** marine fisheries catch from 1950 to 2010. There had been works done toward reconstruction of marine fisheries catches in Vietnam from 1950 to 2010. The annual national landing statistics were incomplete and underreported. A disorganized small-scale sector and, probably, laxcity in licensing policies, perhaps, have led to unlawful and unreported catches. This reconstruction provides a more comprehensive estimate of Vietnam's marine catches by accounting for these IUU catches. Incidentally, the total reconstructed catch in Vietnam was *c.* 79 million metric tons (mmt) from 1950 to 2010, suggesting that the actual catches were, perhaps, *c.* 75% higher than *c.* 45 mmt of catches, probably, reported by FAO on behalf of Vietnam. However, this reconstruction, perhaps, did not account for the catch of foreign fishing vessels in the waters of Vietnam, which could represent another source of fishing pressure and may also be significant. Studies suggest that Vietnam's marine fisheries are on decline and urgent actions to improve the monitoring and enforcement in fisheries are needed. In addition, there have been efforts to deal with fisheries for the welfare of the global environment, and Hilborn (2008) discussed about knowledge on how to achieve sustainable fisheries. There had been reviews related to the state of the current knowledge with respect to the requirements for achieving sustainable fisheries. A range of objectives for fisheries, probably, were considered and attempts have, perhaps, been made to identify conflicting objectives as a major issue to achieving sustainability. Subsequently, efforts were made to review historical and current practices in the allocation of fish resources, as well as regulation of harvest highlighting the current knowledge. Evidences seemed to suggest that both the restriction of access and the maintenance of biological productivity are necessary for achieving biological, economic, and social sustainability. However, the tools, which could be suitable to achieve these, may differ greatly across fisheries and societies, and for both the elements of fisheries management, local solutions are needed in most cases. The studies inferred that the involvement of consumptive users through appropriate incentives could be an essential element in achieving sustainability. Concomitant to the above, Bui The Anh (2015) dealt with culture-based fisheries in Northern **Vietnam** with regard to their management status and perspective. Vietnam possesses a high potential in AQC and fisheries like many other Asian countries. In Vietnam, there are *c.* 6000 reservoirs, of which >3500 are small reservoirs. However, Northern Vietnam has a large number of such water bodies, which were considered to be suitable for development of culture-based fisheries (CBF). Fishery activities were going on in these water bodies for about 8–10 years, but on a relatively small scale. Further, these are commonly referred to as farmer-managed reservoirs, as these are generally leased out to the individuals or to groups of individuals for fishery activities and for water and fish management by the provincial authorities. Notwithstanding the above, the CBF in the northern region of Vietnam was profitable on an overall basis. In fact, this activity was, probably, a subsidiary income generator for the households, which was, perhaps, a significant contributor to rural welfare in the mountainous region of Northern Vietnam, where the total average household income is considerably less than that in the rest of the country. It was further considered that CBF could be a strategy, which could be relatively easily adoptable and adaptable for increasing the fish food supply and for generating supplementary income in rural Vietnam. The situation could, however, be made better through the application of management measures, like the CBFM. In addition to the above, Phan (2014) dealt with the review of current status of fish stock enhancement and management in **Vietnam**. Notably, most of the reservoirs in Vietnam, after impoundment, were, probably, stocked with fingerlings but only a few, perhaps, had a good recovery. The return rate from stocking, probably, ranged from *c* 2% to 10% of total stocked fingerlings.

Notwithstanding the above, AQC is implemented currently in >40% of the total reservoir area and mainly in small and medium reservoirs. Incidentally, the major species stocked, perhaps, had included silver, bighead, grass, and common carps and tilapia, as well as the Indian major carps (IMC). In fact, the growth of stocked fish depends on natural food. Incidentally, the fisheries reservoir management systems in Vietnam, of late, are categorized into three main categories: (a) state-regulated management, (b) community management, and (c) privately owned reservoirs. Concomitant to the above, the major gears used, were, probably, the gillnets, lift nets, lighted lift nets, integrated nets, cast nets, long lines, and seines. Reported yields from the reservoirs ranged from *c.* 20 to 700 kg/ha/year. Further, in Vietnam, "the national plan for reservoir fisheries toward 2020" was, probably, recently approved, and includes the stock enhancement strategy for reservoirs. Moreover, some policies for reservoir fisheries had, perhaps, been applied but the results have, probably, been less than expected. The central government of late, perhaps, has not invested much in reservoir fisheries. As such, it seemed to have remained underdeveloped. Nevertheless, some reservoirs, probably, had done successful stock enhancement exercises, including Ayun Ha (*c.* 3700 ha: in Gia Lai province), Easoup (*c.* 240 ha: in Dak Lak Province), Ea Kao (*c.* 210 ha: in Dak Lak Province), and Nui Coc (*c.* 2500 ha: in Thai Nguyen province). However, useful lessons were learnt from such practices.

2.3.1.19 CBFM in India

Studies on freshwater bodies, such as ponds, lakes, reservoirs, wetlands, rivers, and streams, have gained much significance in recent years due to their importance to human life. These water bodies harbor a wide array of aquatic organisms. Nutrition, avocation, and cultural ethos of the people living in the Indian subcontinent, since time immemorial, are seemed to be linked with the physicochemical, biological, and management aspects of the wetlands (Kar, 2013). Concomitant to the above, the lentic and lotic water bodies in India are indispensably required for the nutrition, avocation, and other social, religious, educational, scientific, economic, and cultural aspects of the innumerable people who sustain themselves on them. As such, these water bodies are to be well maintained, preserved, and conserved along with their resources. For proper conservation, efficient management *modus operandi* are to be in place for the sustenance of the people who live on them (Kar, 2007). Moreover, information on the hydrobiological conditions of any water body is of prime necessity before endeavoring to utilize it as a productive fishery, which could, in turn, sustain the life of the people and other organisms who are dependent on it (Kar, 1996). But, before dwelling upon the scientific and management aspects, an overview of the water resources of the region could give an idea regarding the planning processes of resource utilization.

India is blessed with enormous water resources in the form of innumerable rivers, streams, wetlands, lakes, etc. India has *c.* 67,429 number of wetlands covering an area of *c.* 4.1 million ha. Besides lotic territories, the lentic water bodies having 0.72×10^6 ha lake coverage in India constitute great potential of fishery resources (Kar, 2010a,b). Natural wetlands in India consist of the high-altitude Himalayan lakes, followed by wetlands situated in the flood plains of the major river systems, saline, and temporary wetlands of the arid and semiarid regions, coastal wetlands such as lagoons, backwaters, and estuaries, mangrove swamps, coral reefs and marine wetlands, and so on. In addition to the various types of natural wetlands, a large number of man-made wetlands, like reservoirs, shallow ponds, and numerous tanks, also contribute to the faunal and floral diversity and add to the country's wetland wealth (Kar, 2007).

The Northeastern (NE) region of India is a typically difficult topography with undulating terrains. Nevertheless, it provides enough potential for fish production which can supplement food requirement for the region and could provide answer to diminishing protein supply (Kar, 1990a,b). Situated between 89–97°N longitude and 20–30°E latitude, the region encompasses a vast area of *c.* 2,55,083 km^2 out of total Indian area of *c.* 3.3 million km^2. Comprising eight provinces, the region occupies a strategic position both nationally and globally. The lofty Himalayas proudly stand as sentinels to guard the northern and eastern frontiers. The Chittagong Hill tracts form a crescent on the south (Kar, 2013). The NE India, with *c.* 8% of the country's geographical area and *c.* 4% of the total population, provides a picture of mini India. Having diversities with regard to topography, water resources, biota, climate, race, language, and culture, the region is believed to harbor 100 tribes differing ethnologically, yet sharing a common destiny. The NE region is blessed with a number of lentic systems, locally called "Beel, Haor, Anua, Hola, Doloni, Jalah," etc., which alone constitute *c.* 81% of the total lentic area (0.12×10^6 ha) in Assam. These lentic systems are generally shallow and open, ranging in size from 35 to 3458.12 ha and with depth ranging from 0.25 to 3.0 m (in some, however, the maximum depth may exceed 6.0 m) at FSL. Further, in Assam, there are *c.* 1392 number of wetlands having a total of *c.* 22,896 number of fisheries of different categories, out of which the number of registered wetlands is only *c.* 394 (30.38%) covering an area of *c.* 70,000 ha. Out of this, *c.* 19,000 ha is in good condition; *c.* 15,000 ha is in semi-derelict condition and *c.* 35,000 ha is in derelict condition (Government of Assam, 2006). Incidentally, on the basis of the research findings of the present author (Kar, 2013), the largest wetland, Sone Beel, was declared as a "**Wetland of National Importance**" by the Ministry of Environment and Forests, Government of India on October 16, 2008 (*vide* Resolution No. 11 dt. 16.10.2008, of MoEF Expert Group Meeting and also Letter No. FRM 41/2008/63-A dt. 8.9.2008 from Commissioner and Secretary to the Government of Assam, Department of Forests) (Kar, 2013).

The province of Assam is a significant place in Northeast India Biodiversity hotspot. The Barak valley region is situated in the southern part of Assam. The name accrues from the river "Barak," by which the region is bathed with. The River Barak is one of the biggest rivers in Assam which originates from the Naga Hills and ends in River

Meghna on its journey toward the Bay of Bengal. With an anastomosing network of many tributaries from Mizoram, Manipur, Meghalaya, and Tripura, it forms the Barak drainage with which is associated a large number of wetlands. Notwithstanding the above, the meandering rivers, which crisscross the Northeastern Provinces, have, through the ages, played a dominant role in shaping its history. The most striking feature of the region's drainage is the magnificent flow of the mighty Brahmaputra and the Barak reckoned among the world's most majestic rivers. Originating in a Himalayan glacier, the River Tsangpo (later called the Siang, the Brahmaputra, etc.) traverses *c.* 1600 km in its primary course. Two other rivers, viz., Dibong and Lohit, join the River Brahmaputra and the main river, thus replenished, flows down the plains as the River Brahmaputra (Kar, 2013). Concomitant to the above, the river Barak originates from Japvo peak in Nagaland (*c.* 3353.65m MSL), and flows through Karang village along the Manipur-Nagaland border and drains almost the entire Manipur valley before entering Assam. It flows through the southern part of Assam (thus, forming the Barak valley), and after traversing through a stretch of *c.* 532 km from its origin, the River Barak bifurcates into two branches, viz., the Surma and the Kushiara at village Harinagar (Haritikar) at the Indo-Bangladesh border. After flowing a short spell along the Indo-Bangla border, both these rivers flow into Bangladesh to join the Meghna basin before entering the Bay of Bengal. Barak has a number of tributaries joining it on both north and south banks, viz., Jiri, Chiri, Madhura, Jatinga, (north bank tributaries) and Sonai, Rukni, Dhaleswari, Ghagra (south bank tributaries). In addition to these, River Shingla and River Longai, both originating from Mizo Hills, join River Kushiyara in Barak valley, the former flowing via Sone Beel, the biggest wetland in Assam. Also, River Baleshwar joins River Surma in Barak valley.

The province of Mizoram is drained by eight major rivers belonging to three principal drainage systems; Rivers Tuirial, Tlawng, Serlui belong to the Barak Meghna drainage system: Indo-Bangladesh; Rivers Tuivai, Mat, Kolodyne belong to the Kolodyne drainage system: Indo-Myanmar; and River Tuichong, River Karnaphuli belong to the Tuichong-Karnafuli drainage system: Indo-Bangladesh.

The province of Tripura is drained by four major rivers forming four independent basins, viz., River Manu: Manu basin; River Khowai: Khowai basin; River Gomati: Gomati basin; River Feni: Feni basin (Kar, 2007, 2013).

2.3.1.20 *Wetlands of Assam*

Assam has a total water area of *c.* 3.89 lakh ha. According to a **different estimate**, in Assam out of *c.* 1400 wetlands, *c.* 423 are registered and the remaining *c.* 969 are unregistered and under the control of both government (*c.* 505) and private ownerships (*c.* 464) (Anon, 2006a). Nevertheless, the innumerable freshwater wetlands in Assam could be classified into three broad categories, viz. (a) the perennial wetlands called "Beel," (b) the seasonal floodplain wetlands called "Haor," and (c) the river-formed oxbow wetlands called "Anua." They contain a rich diversity of ichthyofauna and also other aquatic biota, notably the coveted freshwater terrapins. Many of them are less abundant today. The wetlands in Assam cover *c.* 1.0 lakh hectare (ha). The Beels constitute *c.* 80% of Assam's lentic waters (Kar, 2007, 2013).

Beel

Beels are perennial wetlands. Wetlands are heterogeneous group of lentic bodies. They are generally situated on floodplains of major rivers and form potential habitats for rich and varied inland fishery resources (Vinci, 2002; Sugunan, 1997; Kar, 2007a,b,c, 2010a,b,c, 2013). They play a vital role in the inland fish production of the eastern and northeastern (NE) regions of India. Out of the total number of *c.* 1392 wetlands in Assam, there are *c.* 490 Beels. They cover an area of *c.* 15,494.00 ha which constitutes *c.* 0.20% of the total geographical area of the province and *c.* 15.30% of the total area under wetlands. The smallest Beel measures *c.* 2.50 ha while the largest Beel, called the Sone Beel, has a water-spread area of 3458.12 ha. Further, highest number of Beels are observed in Golaghat district (*c.* 113) followed by Dhubri (*c* 73) and Nagaon (*c.* 68) districts. However, highest area of wetlands is in Kamrup district (*c.* 15,705.00 ha) followed by Nagaon (*c.* 2,175.50 ha) and Dhubri (*c.* 1816.50 ha) districts. Some of the important Beels are Sone Beel in Karimganj district, Deepor Beel in Kamrup district, Dhir Beel in Dhubri district, Tamaranga Beel and Doloni Beel in Bongaigaon district, etc. (Dey, 1981; Kar, 2013).

Haor

These are discernible in satellite imagery by their reddish tone indicating the presence of vegetation, associated with dark blue tone indicating the presence of water. In Assam, out of 1392 wetlands, there are *c.* 610 number of Haors covering an area of *c.* 43,433.50 ha constituting *c.* 0.55% of the total geographical area of the province and *c.* 42.91% of the total area under wetlands. Their size ranges from *c.* 2.5 ha (smallest) to *c.* 1350.0 ha (largest).

Anua (river-formed oxbow wetlands)

Out of the total of 1392 wetlands in Assam, there are *c.* 290 oxbow wetlands covering an area of *c.* 5,460.60 ha which constitutes *c.* 0.10% of the total geographical area of the province and *c.* 9.27% of the total area under wetlands.

The smallest of them measures *c.* 5.0 ha while the largest measures *c.* 582.50 ha. Highest number of oxbow wetlands are observed in Golaghat district (*c* 104) followed by Nagaon district (*c.* 71) and Dhubri district (*c.* 68). Maximum area of oxbow wetlands is in Morigaon district (*c.* 2143.00 ha) followed by Nagaon (*c.* 1746.00 ha) and Golaghat (*c.* 1563.00 ha) districts. Some of the important wetlands under this category are Morikolong and Patoli Beel in Nagaon district, Mer Beel in Golaghat district, Guruajan in Morigaon district, the Anuas in Cachar district, etc.

Reservoirs

In Assam, there are *c.* 10 reservoirs covering an area of *c.* 2662.5 ha which constitutes *c.* 0.03% of the total geographical area of the province and *c.* 2.63% of the total area under wetlands. Their size range from *c.* 17.50 ha (smallest) to *c.* 930.0 ha (largest).

Tanks

There are a large number of family-owned small tanks in Assam. These are generally not reckoned in remote sensing. In Assam, *c.* 115 number of tanks have been identified from satellite data covering an area of *c.* 749.00 ha, which constitutes *c* 0.01% of the total geographical area of the province and *c.* 0.74% of the total area under wetlands. Their size varies from *c.* 2.5 to 55 ha. Most of this type of wetlands have low turbidity. Highest number of tanks are observed in Sibsagar district (*c.* 20) followed by Kamrup (*c* 18) and Sonitpur (*c.* 16). The highest area of tanks is in Sibsagar district (*c.* 267.00 ha) followed by Sonitpur (*c* 83.50 ha) and Kamrup (*c* 80.00 ha) districts. Some of the important wetlands under this category are Gaurisagar Pukhuri, Sibsagar Pukhuri, and Joysagar Pukhuri in Sibsagar district. Besides providing water to the people of the riparian zone, these tanks could also be used for rearing fishes; for raising plantation crops like coconut, areca nut, cashew nut, etc., along the banks of the ponds, etc. They could also serve as habitats for wetland birds.

Notwithstanding the above, it is an established fact that proper planning depends on the availability of reliable data. But, unfortunately, in the limnological and fisheries and in the management sector of the water bodies and their resources, there is much shortage of such useful data in the arena. As such, a concise account of the physicochemical aspects and biodiversity of the wetlands of the region could help in appreciating to acquire further information on the subject. These wetlands possess interesting features in the physicochemical characteristics of water, the range values of some of which are temperature 10–38°C, turbidity 5.0–330 NTU, pH 6.6–7.9, DO 3.2–7.1 mg/L, FCO_2 0.8–4.6 mg/L, TA 56.0–152.0 mg/L, conductivity 22.0–152.0 µmhos/cm. Further, the range of some of the soil parameters revealed as temperature 16–28°C and pH 5.2–6.6. Some of the AM species abundantly found in the wetlands of Barak valley are *Eichhornia crassipes*, *Lemna minor*, *Salvinia cucullata*, *Pistia* sp., *Azolla pinnata*, *Trapa bispinosa*, *Hydrilla verticillata*, *Vallisneria spiralis*, *Hygrorhiza aristata*, *Ipomoea aquatic*, *Nymphoides indicum*, *Nymphoides cristatum*, *Nymphaea nouchali*, *Vetiveria zizanioides*, *Euryale ferox*, *Echinochloa stagnina*, *Scirpus eriophorum*, *Eleocharis acutangula*, *Jussiaea repens*, etc. Some of the phytoplanktonic genera which are plentifully found in these wetlands are *Nostoc*, *Anabaena*, *Microcystis*, *Oscillatoria*, *Euglena*, *Phacus*, *Ankistrodesmus*, *Closterium*, *Volvox*, *Spirogyra*, *Cosmarium*, *Ceratium*, *Chlorella*, *Pinnularia*, *Synedra*, *Navicula*, *Diatoma*, *Gyrosigma*, etc. Further, some of the abundantly available zooplankton in these wetlands are *Arcella*, *Difflugia*, *Centropyxis*, *Brachionus*, *Keratella*, *Lecane*, *Trichocerca*, *Filinia*, *Simocephalus*, *Daphnia*, *Ceriodaphnia*, *Bosmina*, *Bosminopsis*, *Moina*, *Macrothrix*, *Alona*, *Pleuroxus*, *Dadaya*, *Chydorus*, *Cyclops*, *Diaptomus*, etc. (Kar, 1990a,b, 2007a,b,c, 2013; Dey and Kar, 1987, 1994).

Freshwater fishes belonging to almost all the major Orders and Families do occur in the wetlands of Barak valley region. Some of these are *Notopterus notopterus*, *Gudusia chapra*, *Salmophasia bacaila*, *Securicula gora*, *Puntius* spp., *Labeo* spp., *Catla catla*, *Cirrhinus* spp., *Rasbora* sp., *Danio* spp., *Amblypharyngodon mola*, *Botia dario*, *Lepidocephalichthys guntea*, *Mystus* spp., *Sperata* spp., *Wallago attu*, *Ompok* spp., *Eutropiichthys vacha*, *Ailia coila*, *Badis badis*, *Nandus nandus*, *Xenentodon cancila*, *Chanda nama*, *Parambassis baculis*, *Parambassis ranga*, *Trichogaster* spp., *Anabas testudineus*, *Clarias batrachus*, *Heteropneustes fossilis*, *Channa* spp., *Mastacembelus armatus*, *Macrognathus* spp., *Tetraodon cutcutia*, etc. (Kar, 1990, 2007, 2013).

Notwithstanding the above, the lentic water bodies and fisheries management scenario got a tremendous boost and facelift after the initiation of the World Bank-aided project, titled, Assam Agriculture Competitiveness Project (AACP), consequent upon the completion of the earlier Assam Rural Infrastructure and Agriculture Services Project (ARIASP). Beel Development Committees (BDC), including Social Audit Committees (SAC) were formed out of the local stakeholders who depend on the wetland(s) for their livelihoods. Under the above background, with many confusing policies in vogue, within which many fishery property rights coexist, the community-based fisheries management and development (CBFMD) scheme could go a long way in guiding and formulating consistent policies through local grassroots level initiatives, with the aim of developing greater efficiency, equity, and sustainability in inland open water fisheries. The CBFMD project also envisages to device and test models for enabling stronger organization of the fishers, with the representation of all the different interests in a fishery. Such practices could establish and ensure compliance with community-agreed fishing access rules.

Dey (1981), Kar (1984, 1990a,b, 2004a,b,c,d, 2005a,b,c, 2006a,b, 2007a,b,c, 2010a,b,c, 2012a,b,c, 2013, 2014, 2015, 2016), Dey and Kar (1987, 1989a,b,c, 1990, 1994), Kar and Dey (1982, 1986, 1991, 1993a,b), Kar et al. (2008, 2009), Deka (1999), and Bhattacharjya (2002) reviewed the earlier literature on various aspects of limnology and fisheries of the wetlands (*Beels*) of Assam in particular.

2.3.1.21 *Review of Literature: Historical Resume'*

In India, *Beels* constitute an important fishery resource in the province of Assam, West Bengal, Bihar, Manipur, Arunachal Pradesh, Tripura, and Meghalaya (Kar and Dey, 1982; Sugunan et al., 2000; Shrivastava and Bhattacharya, 2003) contributing significantly to the fish production apart from providing with livelihood to thousands of poor fishers. Among the Indian provinces, Assam has large number and areas of *Beels* (Kar, 2007, 2013) followed by West Bengal (Mukhopadhyay, 1997). The northern districts of Bihar are also bestowed with a series of natural lakes, locally called as *moans* and *chaurs*. They play a very important role in the overall fish production of the province and provide with livelihood to people living in the surrounding *Beel* area (Jha, 1997a,b). The unique *Beels* of Assam also have vast potential for fish production (World Bank, 2011).

2.3.1.22 *Limnology and fisheries of Beels*

As per the details of distribution of *Beels* in the province, Dey (1981), Kar (1990a,b, 2007a, 2010a,b,c, 2013), Sugunan and Bhattacharjya (2000), and Shrivastava and Bhattacharya (2003) mentioned that, in general, there are, in total, *c.* 1392 enlisted *Beels* in the province covering a total area of *c.* 1.0 lakh ha, out of which *c.* 423 are registered *Beels* (*c.* 30.4%) and the remaining *c.* 969 are unregistered *Beels* (*c.* 69.6%). Again, out of the total unregistered *Beels*, *c.* 505 are under the control of both government and private sectors; *c.* 464 are under semi-government/civic bodies like gram panchayats (GP), etc. The Barak Valley has *c.* 322 enlisted *Beels* covering *c.* 8000 ha of water area. The Brahmaputra Valley has *c.* 1070 *Beels* covering *c.* 92,000 ha of water area. Dey (1981) is one of the pioneers in initiating works on the limnology and fisheries of the Beels of Assam. Significant works on the limnology and fisheries of the Beels of Assam in particular have also been done by Dey and Kar (1987, 1989a,b,c, 1990, 1994), Lahon (1983), Goswami (1985), Gadgil et al. (2000), Kar (1984, 1990a,b, 1996, 1998, 1999, 2000, 2002a,b, 2003a,b, 2004a,b,c, 2005a,b,c, 2007a,b,c, 2010a,b,c, 2013, 2014, 2015, 2016), Kar and Dey (1982, 1986, 1987, 1991, 1993a,b, 1996, 2000, 2002), Kar and Sen (2007), Kar and Barbhuiya (2000a,b, 2001, 2002a,b,c, d, 2004), Kar and Barbhuiya (2011a,b), Kar et al. (1995, 1996a,b,c, 1997, 1999, 2000, 2001, 2002a,b,c, d, 2003a,b,c,d,e, 2006, b, c, 2007, 2008, 2009, 2011, 2015), Barman et al. (2008), Barbhuiya et al. (2009), Barbhuiya and Kar (2009), Das et al. (2013, 2015a,b,c,d,e,f,g,h,i), Das and Kar (2013, 2014), Kar and Kar (2014), Bimola et al. (2014), Barman and Kar (2014), and Sultan and Barman (2014). Deka (1999) and Bhattacharjya (2002) reviewed the literature on various aspects of limnology and fisheries of wetlands (*Beels*) of Assam. Further, Assam is endowed with, perhaps, the largest number (and area) of *Beels* among the Indian provinces. However, no large number of systematic studies had been made in this aspect. Moreover, various indigenous fish catching devices and, methods employed in Assam *Beels*, have been well documented (Kar, 1990a,b, 2007a,b,c, 2013). Further, Goswami and Devaraj (1993) estimated the potential yield of *Labeo rohita* from Dhir *Beel* in Southern Assam. Detailed studies on the fish catching devices of Northeast (NE) India in general, and Assam in particular, have been done by Kar (1990a,b, 2007a, 2013), Kar and Dey (1991, 1993a, 1996), and Kar et al. (1999, 2007, 2008). Further, a few researchers have concentrated on the limnology and productivity of selected *Beels* of the lower Brahmaputra Basin. Agarwala (1992) worked on the productivity indicators in *Beels*. Deka (1999) assessed the status of selected *Beel* fisheries of Assam and their impact on the fishermen community. Most of these studies were conducted in the *Beels* of Kamrup District receiving far greater scientific attention than those of the other districts. Wetlands (*Beels*, *Haors*, and *Anuas*) of Barak valley region of Assam have extensively and intensively been studied by Kar (1990a,b, 2007a,b,c, 2010a,b,c, 2013). Further, Dey (1973), Yadava et al. (1987) worked extensively in different parts of Assam and NE India. Kar and Dey (1982) and Yadava et al. (1989) reported the occurrence of *Tenualosa (Hilsa)* in certain open *Beels* like Sone, Dhir, Dora, etc. Dey and Kar (1990) reported the unique occurrence of anadromous fish *(Hilsa) Tenualosa ilisha* in the biggest wetland of Assam, viz., the Sone *Beel* while Kar and Dey (2002) reported the very rare occurrence of juveniles of *Tenualosa (Hilsa) ilisha* in Chatla Haor Wetland, which could be regarded as one of the biggest floodplain wetlands (Haor) in Assam. Sugunan and Bhattacharjya (2000) given a comprehensive overview of the ecology and fisheries of *Beels* of the province based on extensive field surveys conducted by CIFRI mainly in the Brahmaputra Valley of Assam. Deka et al. (2001) observed that fish diversity of selected *Beels* of the province had declined over the past few years due to some extrinsic and intrinsic factors with associated impact on the income of the fishermen community. Shrivastava and Bhattacharya (2003) gave a comprehensive account of fisheries of floodplain wetlands of Assam, while Bhattacharjya (2003a) outlined the fisheries enhancement measures currently practiced in *Beels* of the Province. Further, Anon (2009a) mentioned the biodiversity status and its conservation measures for *c.* 34 AACP

(Assam Agricultural Competitiveness Project) *Beels* and indicated that the total area of these *Beels* are shrinking due to encroachment and agricultural cultivation activities in some of the *Beels*. Also, the water is polluted due to various activities like cattle-bathing and other kinds of human interventions. In some *Beels*, the inlets are lost due to heavy siltation leading to reduction in auto-stocking during monsoon. This had led to decline of fish production and depletion of fish biodiversity (Kar et al., 2009).

In other regions of India, Chaudhuri and Banerjee (1965) carried out studies in Takmu *Beels* area in Manipur. Subsequently, many researchers had studied the hydrodynamics (Vass, 1992a,b), limnology and fisheries (e.g., Pathak et al., 1986; De and Mukherjee, 1992; Rana et al., 1996) of Indian *Beels*. Kumar (1985) initiated studies on the hydrobiology of a *Beel* of Bihar, which was followed by studies on their ecology and fisheries of some oxbow lakes (*Beels*/*mauns*) of the province (Chitranshi, 1992; Jaiswal and Singh, 1994; Jha, 1995, 1997a,b; Singh, 1990, 2000; Sinha and Jha, 1997). An overview of the productivity status of Indian *Beels* was given by Vass (1989), while Jha (1989) gave an overview of the *Beel* fishery resources of Bihar. Vass and Langer (1990) assessed the changes in primary production and trophic status of a Kashmir oxbow lake, consequent upon man-made modifications in the environment. Singh (1990, 2000) studied phytoplankton and zooplankton in an oxbow lake of Uttar Pradesh. Some generalized account of the community structure and seasonality of aquatic macrophytes and associated fauna in *Beels* is also available (Mitra, 1989, 1992; Mitra and Ghosh, 1992; Dey and Kar, 1989a; Kar, 1990a,b, 2007a,b,c, 2013). The removal of macrophytes in a West Bengal *Beel* resulted in > threefold increase in its fish yield. Choudhury (1989, 1992a,b), Kar, 1990a,b, 2007a,b,c, 2013), Kar and Dey, 1991, 1993a, 1996), Kar (1999, 2007b), and Yashmin et al. (2007) gave an account of the various fishing methods employed in wetlands together with their fish landings, economics, and catch composition. Parameswaran and Vass (1995) gave an overview of *Beels* in West Bengal. Sugunan and Mukhopadhyaya (1995) showed that the soil and water quality, as well as biotic communities varied considerably in these different types of *Beels*, based on the case studies in two open and closed *Beels* of West Bengal. This notion was further strengthened with the synthesis and compilation of the results of studies conducted by the Central Inland Fisheries Research Institute (CIFRI) in selected *Beels* located in all the agroclimatic subregions of West Bengal (Sugunan et al., 2000).

Published literature on aspects related to fish production potential and trophic structure of *Beel* ecosystems in India is rare. Pathak et al. (1986) showed the importance of detritus food chain in *Beel* ecosystem. There have been efforts to give a comparative account of energy dynamics of open and closed *Beels* in the Ganga and Brahmaputra basins. A brief account of the energy flow pathways generally found in open and closed *Beels* of West Bengal was given by Sugunan et al. (2000). Bhattacharjya (2002) reported some information based on the case studies in two open and closed *Beels* of the Central Brahmaputra Valley Zone of Assam.

2.3.1.23 *Socioeconomic profile of the fishers*

The socioeconomic condition of the fishers plays a significant role in the overall economic status of their family, as well as general livelihood. Socioeconomic status, therefore, is regarded as one of the most potential instruments for assessing the living standard of the fishers in the society. There have been efforts to conduct a study in community development block of Ludhiana in Punjab. Their study has revealed that farmers having higher level of education carry more favorable attitude toward scientific soil testing for enhanced crop production. There have been views stating that fishermen are mostly low in literacy and usually belong to the poorer section of the society. Bhaumik and Saha (1994), from their study on socioeconomic condition of the people of the Sundarbans in West Bengal, mentioned that *c* 84.4% of the fishers were from the scheduled caste (SC) community; the involvement of the people belonging to higher caste community in aquaculture (AQC) activities was not very significant. Bhaumik (1999) mentioned that only *c.* 21% on the fishers had education up to primary level, *c.* 7% secondary, and *c.* 2% continued studies above the secondary level. Goswami and Sathiadhas (2000) mentioned about the socioeconomic status of the fishers, and further indicated that fishers had no facility of electricity, telephone, clean water supply, and adequate health facilities along with poor literacy and low income. Khare and Punekar (2001) reported that socioeconomic status of fisherwomen such as age, caste, education, occupation, social participation, income, experience, farm size were found to be positively and significantly correlated with the attitude toward fish farming. Maiti (2003) observed that fishermen are generally the poorest of the poor, most disadvantageous with low education, income, poor basic amenities, and infrastructure. Moreover, *c.* 81.33% of the fishermen generally belong to the SC community. Also, the involvement of the people belonging to the higher caste community in fishing is very meager. Goswami (2003) mentioned that the fisherwomen belonged to the lower income group, with an average annual income of *c.* Rs. 8547.00 (as around 2007), and were represented by *c.* 59.09% of the backward classes followed by *c.* 35.45% of the SC and remaining *c.* 6% were from higher classes. Dana (1987) advocated that by giving due recognition to economic role of fisherwomen and by empowering them in different fisheries activities, especially in *Beel* fisheries, the socioeconomic condition of fishers could be substantially raised. Dey and Kar (1989a,b,c) and Kar (1990a,b, 2003a,b, 2007a,b,c, 2013) through detailed study revealed that the fishermen of certain *Beels* (notably Sone *Beel*) have been struggling against poverty since a couple of decades. The *per capita* income (PCI) of the fishermen is low and ranges from Rs. 1.30 to Rs. 8.25 (Mean: Rs. 3.83 ± 2.14) with a bigger family size, poor adult literacy (10%–20%) along with cultivable land ranging from 0.07 to 4.2 ha (Mean 0.02 ha ± 0.105) (as during 1979–1981). Babare and Sawant (2009) reported about detailed socioeconomic status of the fishermen of the Makni reservoir of Maharashtra and indicated that the size of the fishermen family ranged from *c.* 5 to 8 individuals on the average, *c.* 63% fishermen usually have pucca houses, *c* 31% have kutchha houses, and *c.* 6% live in huts. However, the level of education was not very poor with *c.* 13% studied up to the primary level, *c.* 19% up to middle level, *c.* 43% up to the high school level, *c.* 16% up to pre-degree level, and *c.* 8% up to graduate level. Further, as had been reported the fishermen might also be engaged in agriculture (*c.* 80%) followed by fisheries (*c.* 10%) and business and other services (*c.* 5% each). They basically had been found to belong to below poverty line (BPL) category with average monthly income ranging from Rs. 6000.00 to Rs. 8000.00 (as in 2007).

2.3.1.24 *Community-based Fisheries Management (CBFM)*

Contrary to the above, no large amount of works had been done on CBFM. Also, no big amount of works had been done on community-based forest management, community-based coastal resource management, and common pool resource management. Nevertheless, some works, having both direct and indirect relevance to the present sphere of research work, are reviewed in the following.

There have been views stating that community-based resource management (CBRM) is a process through which the people themselves are given the opportunity and responsibility to manage their own resources, define their needs, goals, and aspirations, and make decisions affecting their well-being, and, thus stands for peoples' empowerment for achieving productivity, accountability, equity, and sustainability in natural resource management. Sen (1997) stated that community-based management, comanagement, and cooperative management are some of many terms used to describe fisheries management systems by involving participation of both resource users and the government agencies and sharing of knowledge, mutual understanding of problems, and joint formulation of solutions that prevents overexploitation and conflicts within the fishery or between the fisheries. The salient prerequisites (Salim, 1998; Pomeroy and Williams, 1994, Pomeroy et al., 1998, Pomeroy and Rebecca, 2006) for successful fisheries based on community-based comanagement program are as follows: (a) clearly defined fishery boundary; (b) an effective information gathering mechanism through the involvement of the stakeholders from the grassroots level; (c) an effective local institutional setup to promote greater participation of all the stakeholders; (d) some forms of control mechanism to reward and punish; (e) enforcement of management rules, and (f) cooperation and strong leadership at the community level.

2.3.1.25 *Management scenario of the* Beels *in Northeast India and Assam State perspective on CBFM*

Baruah et al. (2000) mentioned about the existing settlement and management of province's *Beel* fisheries with a rider that, in today's perspective, community-based *Beel* management is imperative for their sustainable development. Considering multi-ownership and ambiguous existing method of settlement of *Beel* fisheries in the Province of Assam, the World Bank (2004, 2011) has emphasized on the preparation and adoption of a uniform guideline in the form of a comprehensive legal statute in fisheries to encourage effective community participation in settlement, development, and management of the *Beels* because this could help in the implementation of *Beel* development programs through community-based fisheries management schemes in the province of Assam (AFPRO, 2010; Anon, 2011a,b,c). Goswami and Sathiadhas (2000) stated that fish production was substantially augmented in the Lakhimpur Community Tank (possibly, due to the impact of community-based tank management programs) as compared to individual tank's production situated in the nearby localities. This also had led to the initiation of different social welfare activities among the members taking part in community management. Baruah et al. (2000) mentioned that a community-based comanagement model is quite necessary (like those applied to reservoir fisheries in Northern Brazil) for sustainable development of Beel fisheries in Assam. Bhattacharjya (2003a,b, 2008) indicated that emerging participatory approach is required for the management of the wetlands of Assam. Anon (2004a,b) demonstrated that due to the adoption of CBFM Program under the Assam Rural Infrastructure and Agriculture Service Project (ARIASP), fish productivity of farmers' ponds and community tanks enhanced by *c* 292% and *c* 175%, respectively. Further, *c* 68.2% of the project farmers, as reported, probably, had continued with the approved package of practice, either fully or partially, and on an average only *c.* 10.2% of the project farmers had, possibly, discontinued the Package of Practice. Barman (2008) indicated that there were evidences of reduction in installation cost of "pen-structures," procurement of fish as inputs, and enhancement of *Beel* users' empowerment in construction and management of the pen-structures as a result of community-based management at Deek *Beel* in Kamrup district of Assam. Barman (2009) opined for the conservation of water resources of the entire northeast (NE) region of India, through a people-led and people-centric CBFM strategy. This is, probably, very much required to meet the twin objectives of enhanced fish production, as well as for the upliftment of socioeconomic condition of the fishers. There have been advocacies for effective utilization of enormous *Beel* fishery resources in the province of Assam through community participation. In this connection, it may be pointed out here that fish production in Ganjakhauri *Beel* in Hailakandi district of Assam raised to *c.* 903 kg/ha/year mainly due to the implementation of community-based development approach. Further, the *Beel* users were involved in *Beel* supervision, including fishing activities, duck farming, and, so on. They reportedly also had shared equal distribution of income among the members leading to the overall sustainable development of the *Beel*. There have been efforts to deal with the aspects of restoration and development of Loktak Lake resources and the biodiversity of the wetland for present and future generations through participatory processes, research, and conservation activities. Anon (2011a,b,c) gave detailed modalities of management of *Beel* fisheries through community participation for sustainable development of *Beel* fishery resources in the province of Assam. Incidentally, Chandra (2011) studied the management regimes and institutional arrangements in the wetlands of Assam.

In addition to the above, some of the other significant works related to community-based fisheries management in India, are as follows.

Lobe and Berkes (2004) dealt with the **padu system** of community-based fisheries management in **South India** with a note on the change(s) and local institutional innovation(s). As a commons' institution, the padu system in Vallarpadam Island, Cochin, Kerala, tried to define the group of rights holders and resource boundaries and fishing sites. It is caste-specific, gear-specific (stake-nets) and species-specific (shrimp). As used in Vallarpadam, and other places in Kerala, Tamil Nadu, and Sri Lanka, padu is characterized by the use of lottery for rotational access. The institution functions in providing equitable access, collective social responsibility, and rule making and conflict resolution. The emergence of the institution in different places is a response to change(s) in the markets and the legislation in the 1970s. It may also be seen as a response of the fishing communities to keep their options open, i.e., to be resilient. Concomitant to the above, Gadgil et al. (2000) reported about Biodiversity Conservation Prioritization Project (BCCP). Choudhary (1979) and Kumbhar (1979) indicated that their aquatic project management system was found to be of high class with high degree of cooperation among the members, mainly due to the implementation of people-led cooperative-driven Gambhira Group Farming management program. Anon (2004a,b) indicated that due to the implementation of Community-based Integrated Watershed Development Project in **Himachal Pradesh (HP)**, there was a positive change in education and savings. Efforts had been made to tackle equity and gender issues. There had been encouragement with regard to participation of the poor, the women, and the other disadvantaged groups in planning and implementation of income generating activities leading to increase in annual income of the households, cropping intensity, irrigation facilities; enhancement of fodder, fish and milk production, and improved road connectivity in the project districts. Anon (2009a) reported that due to the implementation of cooperative wetland farming practices by the Mudialy Fishermen's Cooperative Society in West Bengal, the society was able to implement different types of social welfare schemes for the fishermen, such as weaving center, Anganwadi training center, free health checkup camps, educational loan, funeral aid, marriage aid, advances for buying TV, radio, etc., set(s), advances for starting consumer shop(s) at subsidized rates, etc., as a part of social welfare to the fisher community. Further, Various works indicated that due to the implementation of World Bank-funded Community-based South Asia Poverty Alleviation Project in Andhra Pradesh (SAPAPA), there had been improvement in the attendance of children in the schools; enhancement in the involvement of women in financial matters and in the decision-making processes; improvement in access to credit(s) and formation of community asset(s); generation of social capital in the form of trained book keepers, effective leadership, health educators, education volunteers, assets management personnel, etc. There had also appeared increased capacity building of the community members leading to enhancement of collective bargaining power of the community members, which, possibly, had enabled the people to enjoy sustainable livelihood. Notwithstanding the above, Natarajan (2005) dealt with a case study **in India** reporting about enabling better management of fisheries conflicts. The weak extension system with inadequate manpower and other resources were made the public sector incompetent to satisfy the needs and restore peace in the community. It may be mentioned here that conflicts in fisheries are the primary reasons for the initiation of the Marine Fisheries Regulations Acts (MFRAs) in India. Further, Chandra (2011) evaluated the management regimes and institutional arrangements in floodplain wetlands fisheries of **Assam**. It may be noted here that the public trust doctrine permits the government to lease, grant, and sell public resources as long as it does not unduly harm public interests. On the other hand, "privatizing" fisheries means assigning rights (property, use) to individuals, groups, cooperative(s), or communities. Incidentally, "property" could be composed of a bundle of rights (access, withdrawal, exclusion) that can be allocated to users. Incidentally, Barman et al. (2012) reported about the availability and role of critical aquaculture inputs and community-managed markets on community-based floodplain wetland management. The study revealed that community-managed markets were significantly and positively associated ($P<0.01$) with knowledge levels of Beel users on community-based fisheries management. Further, Prabhu (2011) dealt with institutional dynamics in community-based fisheries resource management for sustainable development of marine fisheries in **Kerala**. Vipinkumar et al. (2013) reported about comanagement paradigm and sociological issues in Indian Marine Fisheries Sector. Rational utilization of common property resources for sustainable development without endangering the environment could be possible through community participation. It is, perhaps, the extent of partnership between the research and the client system that, possibly, decides the fate of any technology in terms of its adoption or rejection. Incidentally, Mohamed (2004) reported about march toward ecosystem-based management of marine fisheries with an aim toward building mass balance trophic and simulation models. The utilization of sound ecological models as a tool in the exploration and evaluation of ecosystem health, probably, had been encouraged and endorsed by the leading bodies in ecosystem-based fisheries research and management. However, the potential of the available dynamic ecosystem models to make measurable and meaningful predictions about the effects of fishing on ecosystems has, probably, not been fully assessed yet.

Some of the other significant researchers on CBFM in Kerala, and may also be elsewhere, and the related aspects are as follows:

John (2009), Soumya and Shah (2005), Sreenivasan (2005), Rajasenan (2005), Scott (2001), Mccay (2001), Symes (2000a,b), Kurien (2000), Katsuyama (1999), Pitcher et al. (1998), Crane and Draper (1996), Mathew (2009), Korakandy (2008), Hari Babu and Rajan (2006), Grafton et al. (2006), Biju Kumar and Deepthi (2006), Hilborn (2005), Preikshot and Pauly (2005), Jennings et al. (2005), Owati et al. (2005), Venkatachalam (2005), D'Cruz (2004), Lodge (2004), Hannesson (2004), Damodaran (2003), Rajasenan (1999), Kurien (1991), Symes (2000a,b), Dey and Kanagaratnam (2007a,b), Thomson (2007), Santha (2007), Berkes (2006), Sonak et al. (2006), Paul (2005), Sapovadia (2004), Kurien (2000), Khakhar (2001), Baavinck (2001), Korakandy (2001), Willmann (2001), Symes (2000a,b), Mccay (2001), Kurien (2000), Wilson (1999), Crean (1999), Viswanathan (2000), Pomeroy (1995), Kallie et al. (2000), Uphoff (1986), Bal and Rao (1984), Divakarannair (2007), Ramachandran (2006), Bhatta (2003), Halfdanardottir (1993), Kurien (1991), Sathiadhas et al. (2005), Hassan and Sathiadhas (2005), Upare (2002), Upare and Dalvi (2002), Jaleela (2002), Ashaletha et al. (2001), Gracy (1998), and Nair (1998).

Notwithstanding the above, still, some of the more works done and which are related to community-based fisheries management in India are as follows.

Sampath (2005) dealt with an overview of the international and national fishery policy and related issues. The Industrial Policy Resolution, 2001 of **Government of Odisha** identified the fisheries sector as one of the three sectors wherein enterprise development could result in pro-poor economic growth. In this context, the policy, probably, emphasizes the need to assess the self-employment and entrepreneurial potential in fisheries sector, especially for the poorer sections involved in the sector. Anon (2011a,b,c) dealt with the possible approaches in community-based fisheries management (in India). A bottom-up approach could be a more pragmatic approach for effective CBFM of the aquatic resources. Notwithstanding the above, community-based management (CBM) is a significant *modus operandi* for effective management of the resources, notably the water bodies and the fish resources. There are three initiatives in community-based fisheries management (CBFM), viz., (a) management, (b) livelihoods, and (c) development. These are not well-understood everywhere. This could lead to differences in the *modus operandi* in the implementation of CBFM. Further, Anon (2004a,b) dealt with the policy briefs of good management practice in sustainable fisheries in India. The purpose of these policy briefs was to ensure effective dissemination of information collected and generated through different programs, notably, the World Bank-funded programs. Further, Anon (2006a) prepared Report on the Working Group of Fisheries for the 11th 5- year plan, 2007–2012. The Planning Commission (PC) of the Government of India (GoI) had constituted a working group (WG) on fisheries with various objectives, schemes, programs including their monitoring and implementation. The PC of the GoI had constituted a WG on fisheries with the following objectives: (a) to review the progress of schemes and programs in fisheries sector implemented during the planning period and (b) to identify the problems and constraints in the implementation of programs and schemes and to suggest measures for enhancing fish production in a sustainable manner, along with developmental programs during the plan period. However, the NITI Ayog of the GoI also have different plan and programs for the developments in fisheries sector. Incidentally, Kumar et al. (2003) reported about a profile of the people, technologies, and policies in fisheries sector in India. Growing urbanization, globalization, and rapidly changing social structures were believed to have a major impact on the fisheries structure in the country. In fact, fisheries and AQC today have emerged as important commercial activities from their traditional role in supplementary subsistence activity. Concomitantly, Datta et al. (2010) dealt with the perspectives of fisheries sector interventions for livelihood promotion. There had been significant works done on the perspectives of fisheries sector interventions for livelihood promotion in different places. Different works have used clues from recent economic theories and management tools on property rights in order to develop a perspective on intervention strategies for livelihood promotion of most intimate stakeholders, i.e., the fisherfolk, in a sustainable manner. Notwithstanding the above, Anon (2010a,b) dealt with the marine fisheries of India with regard to their issues, opportunities, and transitions. This study represents a collaborative initiative by the World Bank and the Department of Animal Husbandry, Dairying, and Fisheries, Ministry of Agriculture, Government of India, to review the marine fisheries subsector, within a broader sector that also includes aquaculture and inland fisheries.

2.3.1.26 *CBFM in Bangladesh*

A brief review of the system of management of the water bodies in Bengal (mainly during the British rule) reveals the following.

During the English rule, the British government (Govt) had given exclusive proprietary rights of management of water bodies to the "zaminders" (feudal lords). The latter, in turn, gave using rights of the water bodies to the fishers in exchange of a token sum of revenue (toll). This regulated the entry and harvest of the aquatic resources (notably the fishes) within certain limits and restrictions. Subsequently, "The East Bengal State Acquisition and Tenancy Act of 1950" had abolished the zamindari system and most of the open water bodies in nature had been reverted back to the state and they became state-owned properties. The Ministry of Lands (MoL) reserves the authority and proprietary and exclusive rights over these state-owned water bodies.

Notwithstanding the above, "**periodic leasing**" of fishing rights of the water bodies forms the integral component and basic mechanism of management of inland open water bodies in Bangladesh. Traditionally, the water bodies are leased out by the MoL for a period of 1–3 years, based on the type of water body, through the deputy commissioner (DC) of each district. In this exercise, the water bodies are leased out to the highest bidder, in which, the poverty-stricken fishers are automatically left out or eliminated on economic inability and the affluent section of the society used to be the real beneficiaries. Even when the lease was allotted to a cooperative society, the affluent section of the society used to reap the lion's share of the benefits under inequitable conditions. Moreover, under these circumstances of management of water bodies by the affluents, the water bodies were overharvested for reaping maximum benefits within the specific lease period, perhaps, without any thought of conserving the water bodies and their bioresources for future. **Under the above circumstances**, the Government of Bangladesh had launched the "New Fisheries Management Policy" (NFMP) in 1986 mainly with a view to maximize benefits to the actual stakeholders at the grassroots level and at the same time ensuring sustainability of the fish stock through the implementation of the Fish Conservation Act of 1950, which also limits exploitation to the level of maximum sustainable yield (MSY). In fact, the gradual abolition of leasing by open auction and its replacement by traditional leasing system were the main aim of the NFMP. Similar measures of management, probably, had also been undertaken by other countries in Asia like India, Sri Lanka, etc. **In due course of time**, NFMP was introduced in 10 selected water bodies in Bangladesh, which included rivers, lakes, and wetlands in different regions of the country. The responsibility of management was entrusted with the Department of Fisheries (DoF) under the Ministry of Fisheries and Livestock (MoFL). Later, the practice was extended to *c.* 264 water bodies in Bangladesh, and specific licenses for a particular fishery had been issued to genuine full-time fishers. The "Thana" (subdistrict) Fishery officer was entrusted with the responsibility of collecting the revenue and was accountable to deposit the same to the government treasury. The success of NFMP had been reflected through significant gains in certain sites. The returns to the fishers increased from 25% to 50% where the NFMP was implemented. Further, the fishery officers and the NGOs were jointly involved with the program(s) for organizing the fishers and to provide technical assistance to them. These were, probably, done under the Scheme of Improved Management of Open water Fisheries (IMOWF) Project. However, in 1995, the MoL abolished the leasing right to the lotic systems, with the result, the rivers being rendered free access to the fishers with neither revenue requirements nor any fishing restrictions and regulations but there had been a practice of leasing out of most of the "Beels." **Under the above background**, with many confusing policies in vogue, within which many fishery property rights are said to coexist, the community-based fisheries management and development (CBFMD) scheme could go a long way in guiding and formulating consistent policies through local grassroots level initiatives, with the aim of developing greater efficiency, equity, and sustainability in inland open water fisheries. The CBFMD project also envisages to device and test models for enabling stronger organization of the fishers, with the representation of all the different interests in a fishery. Such practices could establish and ensure compliance with community agreed fishing access rules.

A considerable amount of works on CBFM have been done in Bangladesh. Rab (2010) opined that fisheries are an age-old system and practice in Bangladesh being traditionally managed mostly as common property resources. There have been some amount of works done broadly on the fisheries aspects and specifically on the CBFM topics in Bangladesh (Ghani and Ahmed, 2008; Thompson et al., 2003, 2006, 2010; Hossain et al., 2010; Sultana and

Thompson, 2007; Abdullah, 2007; Sultana and Thompson, 2007). Incidentally, *Beel*, under different nomenclature, occur in many parts of the world. In Bangladesh, wetlands occur in more than half of the country's geographical area. Approximately 4.0 million ha of the wetlands form the major capture fishery habitats of the world and are the source of livelihood for rural people (Ali, 1997). Concomitantly, the fishes in the wetlands contribute to *c*. 46% of all fishes consumed (Minkin et al., 1997; Thompson et al., 1999). They support a human population of *c*. 800 people per km^2. Moreover, the wetlands are also often intensively used for agriculture and other activities, other than fisheries. Thompson et al. (2003, 2006) and Islam et al. (2006) opined that majority of rural people of Bangladesh depend on inland capture fisheries as their source of livelihoods. Further, > 70% of the households around the wetlands catch fish either for avocation or for nutrition (Minkin et al., 1997; Thompson et al., 1999).

Incidentally, Thompson et al. (1999), Sultana et al. (2002), and Mustafa and Halls (2007) indicated that there were evidences of increased participation in decision-making, greater cooperation, and empowerment among fishing communities. Ali (1997) worked on the community-based haor resources management under sustainable environment management program. The project included realistic activity packages which covered household level livelihood planning and intervention. The project components included: (a) training needs and their assessment, (b) awareness programs and institution building, (c) habitat restoration, (d) afforestation programs, (e) wise use of fish and other wetland resources, (f) establishment of sanctuaries, (g) community rehabilitation and development, (h) local level institution building, and (i) monitoring of social and biological programs. Based on lessons learnt from the community-based fisheries management (CBFM) works, of late, in Bangladesh, Islam et al. (2006), Dickson and Brooks (2007a,b) mentioned that it is little difficult to show clear benefits to communities in project areas in terms of increased incomes or clear impacts on poverty alleviation. Nevertheless, key groups, including poor fishers, were able to improve their livelihoods, and there were clear impacts on attitudes as regards issues, such as compliance to rules on water body management, increased cooperation between individuals, improved access to credit and recognition of community groups by the authorities, etc.

Concomitant to the above, Anon (2007a) dealt with scaling-up of community-based comanagement of wetlands and fisheries in Bangladesh in the context of Management of Aquatic Ecosystems through Community Husbandry (MACH). It may be noted here that MACH (2007) (Management of Aquatic Ecosystems through Community Husbandry) is a Government of Bangladesh project supported by USAID. There had been a close coordination among the project partners, viz., Winrock International, Bangladesh Center for Advanced Studies, Center for Natural Resources Studies, and Caritas Bangladesh. They all have worked in close collaboration with the Department of Fisheries since 1998. The principal objective was to buildup community-based comanagement and revive and augment sustained productivity at the ecosystem level in four water bodies in Bangladesh, viz. (a) Hail Haor in Sreemangal, (b) Turag-Bangshi River, (c) wetlands in Kaliakoir, and (d) the Kangsha-Malijhee basin in Sherpur. These water bodies cover *c*. 32,000 ha during the wet season. Further, they include > 100 distinct water bodies during the trough season. A population of *c*. 184,000 people had inhabited > 110 villages and were directly involved in the management of these water bodies.

The inland fisheries sector in Bangladesh is in the growing phase. However, the present status of the water bodies (both lentic and lotic) in Bangladesh are, perhaps, not in very good shape and are, by and large, victims of siltation and eutrophication. These have resulted in fast rate of loss of precious wetlands and rivers. Further, overexploitation coupled with encroachments is leading to the loss of coveted fish resources rendering many ichthyospecies threatened and endangered, ultimately paving the way to colossal losses in the nutrition, avocation, and economy of the nation. This is, perhaps, a common pattern in the world's fisheries but, nevertheless, Bangladesh is in the forefront of involving local institutions which could successfully manage wetlands. There have been efforts in developing successful community-based comanagement arrangements that could ensure existence of wetlands with sustainable use, productive fisheries, and meet the needs of the poor. There had been various recommendations regarding community-based comanagement arrangements. Incidentally, Mustafa and Halls (2007) reported the impact of the CBFM on sustainable use of inland fisheries in Bangladesh. The CBFM project in Bangladesh aimed to promote the sustainable use of inland capture fisheries by empowering fisher communities to manage their own aquatic resources. Works in this direction described the impact of fisheries management performance using data generated under the CBFM project, funded by the Ford Foundation and the UK Government's Department for International Development (DFID). Concomitantly, Dey and Kanagaratnam (2007a,b) reported on community-based management of small-scale fisheries in Asia, in the context of bridging the gap between fish supply and demand. Incidentally, it may be noted here that production from capture fisheries saw rapid, sustained growth from the 1950s through the 1970s but by 1990s, global capture fisheries become stagnant, warranting a sustainable approach to its exploitation. With capture fisheries stalling and demand for fish growing, many countries turned toward developing the aquaculture (AQC) sector to meet the demand-supply gap. Over the decade, the expansion of AQC, perhaps, had led to a rapid growth in fish production. However, as AQC expanded, its production, its relation to capture fisheries as food for farmed fish, probably, had increased, taking into account that, of late, nearly one-third of the world's wild fish caught is consumed as fish feed. In addition, as AQC in the developing world continues to exhibit steady growth in production, sustainability of this trend is now open to question given the rapid degradation of the capture fisheries. As such, establishing community organizations for managing fisheries is a promising means of improving the resource condition, particularly for countries with large inland and seasonal floodplains. However, this arrangement may not reduce the role of the government but emphasizes on delivering net benefits. Also, it is, probably, necessary to setup a legal framework for community-based management to ensure and sustain community participation in fisheries management.

Concomitant to the above, Dickson (2007) reported about the CBFM project with regard to proving the case for community-based and comanagement of fisheries in Bangladesh. It may be noted here that the community-based fisheries management (CBFM) project had been implemented by the Department of Fisheries (DoF) of the Government of Bangladesh (GoB) with the assistance of the World Fish Center over the period 2001–2007. There had been communication(s), which give an outline of the main principles behind the project and the activities carried out during its implementation. The project involved a wide range of partners including 11 NGOs who were responsible for the identification and development of 130 community-based organizations managing 116 water bodies including closed Beels, open Beels, rivers, and floodplains. The communications summarize the success of the project from different points of views. Some of the situations in which community management did not work under CBFM were discussed and the works had emphasized the need for further research to build up the system based on the model(s). Further, Ahmed and Dickson (2007) reported on the Institutional issues in the CBFM project in Bangladesh. This communication discusses the institutional issues in CBFM project, which are vital in achieving its long-term goal of "sustainable improvement of the livelihoods of poor people who are dependent on aquatic resources." A wide range of institutional arrangements had been practiced in this project from earlier phase (1994–1999) to next phase (2001–2007). The initial project document stated that there were three main fisheries management approaches, viz., (a) fisher-led, (b) community-led, and (c) women-led. The definition(s) of these approaches have failed to capture the full diversity of approaches adopted by 10 project partners (including DoF) during implementation. As a part of the institutionalization process, *c.* 130 community-based organizations (CBOs) had been developed and established under the project with legal entity. For better management of fisheries, efforts have been made toward linking these community-based organizations (CBOs) and local institutions in CBFM sites. In this regard, four regional CBO networking committees have been formed in four regions and a central committee also has been formed on *ad hoc* basis. This central committee is now being officially registered. The partner NGOs have been the main agents helping to develop the CBOs. It may be important to note here that Gomes (2007), founder and executive director of an NGO named "Banchte Shekha," had reported about the empowerment of women through fisheries and AQC. The NGO, based in Jessore, Southwest (SW) Bangladesh, had supported the development and empowerment of poor people, particularly women. In the CBFM project, they found that, probably, it is possible to involve women in fisheries activities, despite initial opposition from conservative groups. Banchte Shekha were responsible for organizing seven CBOs in CBFM, following their successful experiences with a single CBO in the first phase of the project. In addition to the above, Anon (2010a,b) reported about liquid assets in relation to community-based fisheries management in Bangladesh, with particular emphasis on key facts like formulation of a series of management approaches for ensuring equitable access to fisheries resources for community-based organizations (CBOs), which had resulted in significant increases in the fishing incomes of >60% of fishers in the project areas. Concomitant to the above, it may be noted here that the aim of the CBFM initiative was to provide research-based approaches so that sustainable management of inland fisheries could be handed over to CBOs. Fish sanctuaries, probably, had been established in >80 water bodies with the result that fish production increased with concomitant hike in the income of CBFM fishers and resulting in a healthier biodiversity (BD). CBFM-type approaches have, thus, been included in the new Inland Capture Fisheries Strategy and fisheries aspects of the Poverty Reduction Strategy Paper (PRSP) of the Government of Bangladesh. Further, it represents a substantive body of evidences at the regional and global levels, thereby demonstrating that co- and community-based management approaches could result in more productive and sustainable fisheries and, at the same time, could safeguard the livelihoods of poor fisher folk.

In addition to the above, the CBFM initiatives have developed a series of management approaches for ensuring equitable access to fisheries resources for community-based organizations (CBOs), which have resulted in >60% of fishers in the project areas experiencing significant increases in their fishing incomes. Incidentally, Thompson (2004) reported about lessons from CBFM in Bangladesh. According to him, inland (wetland) fisheries are one of the most important contributors to fish production in Bangladesh in particular, and in the Indian subcontinent in general. In the earlier days, they have been considered to generate government revenue without due care for sustainability of poor users. A possible solution to empowering fishing communities and improving the sustainability of management is CBFM.

In addition to the above, experiences were gained from studies in 19 rivers and Beels during the period 1996–2000 and >100 water bodies during the period 2001–2003. In this regard, some of the main lessons learnt are as follows: (i) CBFM is based on comanagement, empowering the fishing communities; (ii) development of local fisher or community-based organizations (CBOs) is essential; (iii) new institutions could be built with as much ease (or difficulty) as modifying existing ones; (iv) local government support for CBOs is, perhaps, significant for long-term sustainability; (v) establishing CBFM seems to be a slow process; (vi) strong facilitation appears to be necessary; (vii) external threats are strong limiting factors, which may not be overcome in some cases mainly due to powerful political interventions in a water body; (viii) effective well-defined partnerships between NGOs and the government may not be easy to establish but, sometimes, thought to be needed to support new community institutions for fisheries management; (ix) it could be considered as essential that communities obtain clear user rights over government fisheries; (x) in jalmohals (including rivers), the lease may be reserved for the CBOs through a supportive government agency, and the CBOs, in turn, may, preferably, need to make some lease payment; (xi) land administration at the district level may need merits of CBFM; (xii) diverse interests of stakeholders in Beels (wetlands) could be brought together where there are shared interests and concerns over declining fisheries; (xiii) coordination of CBOs and management plans in connected water bodies could be promising but needs shared trust and compliance; (xiv) non-fish aquatic resources need to be included in more integrated wetland management plans; (xv) adopting viable resource management actions, such as fish sanctuaries, probably, could help in strengthening institutional development; (xvi) CBFM appears to be little easier in small fisheries with clear boundaries; (xvii) success is generally easier in homogeneous communities; (xviii) boosting up may fast lead to growth of CBFM in adjacent water bodies, and could screen certain sites to avoid those which are dominated by strong influential interests.

Notwithstanding the above, some of the other significant works done in CBFM in Bangladesh and other related aspects are as follows:

Gazi et al. (2013) dealt with the property rights and access with regard to community-based fisheries management in Bangladesh. Chakraborty (2004) dealt with the management of Haors, Baors, and Beels in Bangladesh which could be a lesson for Lake Basin management studies. Rahman and Begum (2010) reported on the strategies for empowering the poor for wetland resources conservation in Bangladesh. Khan (2011) studied the participatory wetland resource governance in Bangladesh including an analysis in community-based experiments in Hakaluki Haor in Bangladesh. Chowdhury et al. (2012) done a sociological study on the people of Hail Haor **in Bangladesh** with regard to the patterns of cooperation in community-based fishery management. Paul et al. (2014) reported about successful practice of CBFM in seasonal waterlogged paddy lands of Begumganj in Noakhali, Bangladesh. The greater southern parts of Bangladesh, especially greater Noakhali district, are characterized by seasonal waterlogged paddy fields during rainy season where CBFM could be an option to improve socioeconomic condition of the rural people. There had been studies, which evaluated the stocking strategy in fish culture and also evaluation of the cost-benefit analysis for good CBFM practice in waterlogged area of Begumganj upazila in Noakhali District of Bangladesh. Necessary data were collected through focus group discussions and semi-structured interviews with the target group(s). Six selected sites, *viz.*, Vobanijibonpur, Kendurbag, Madhob Shing, Khwajpur, Chowmohona, and Jirtali were observed in the studies, where stocking and other management activities had been maintained from the very beginning of the culture season. Water quality parameters were closely monitored during the 6-month study period. Cow dung (2000 kg/ha), urea (50 kg/ha), and TSP (50 kg/ha) were applied as a basal dose after 7 days of liming (250 kg/ha) in order to enhance the production of natural food. In these practices of carp and prawn polyculture, the stocking density was 8–24/decameter^2 and 3–10/dm^2, respectively where prawn and carp stocking ratio was 1:3. The studies also depicted different profit margins of 36,831, 20,200, 155,700, 10,850, 10,550, and 21,600 (Bangladesh Taka) in six sites. Such studies on CBFM strategy, habitat quality, meteorological data, and field experiences denoted that a successful return of the investment, done in waterlogged areas, mostly depend on the proper management of the water bodies.

Further, Islam and Dickson (2007) dealt with the aspect of turning social capital into natural capital and concomitant changing livelihoods of fishers through CBFM. The declining trend in fish production from inland open water fisheries and limited access to resources had been crucial concerns for the livelihoods of millions of poor fishers in Bangladesh. The major objective of the CBFM project was to build local fishers' community organizations by providing training, social awareness, and credit with the aim of enhancing poor fishers' capability to access livelihood assets. Moreover, a recent impact study was conducted in sample water bodies where a total of *c.* 3000 households were interviewed to obtain information on the livelihoods of households in areas with and without CBFM. The same questionnaire format was used in the baseline survey in 2002 and in the impact survey in 2006. The study had reported that fishers in different environments with CBFM had improved their social assets which provided greater access to other livelihood assets, such as natural assets and financial assets. The results of a multivariate linear regression model had found that social capital factors had depicted a positive relationship with household income from fishing. The results indicated that social capital was a significant predictor for natural assets. Future policies could give priority to provide security of access to the fisheries resources for sustainable livelihoods of fishing communities in the wetlands of Bangladesh.

Notwithstanding the above, Gary et al. had reported about utilization of different aquatic resources for livelihoods in Asia. Among the various aspects of utilization of different aquatic resources for livelihoods in Asia, some of the significant ones are as follows: (i) Aquatic Resources as Common Property; Community-Based Fisheries Management: An Approach to Sustainable Common Property Resource Management; (ii) Economic Considerations for Community-Based Small-Scale Aquaculture in Oxbow Lakes; (iii) Considerations in CoManagement of Capture Inland Fisheries and Aquaculture in Laos; (iv) Culture-Based Fisheries in Reservoirs and Lakes in India; (v) Community Rice-Fish Farming in Bangladesh; (vi) Marine Reserves and the Role of Local Government Units in the Philippines; (vii) Local Knowledge and the Conservation and Use of Aquatic Biodiversity.

2.3.1.27 *Community-Managed Aquatic Resources: Aquatic Resources as Common Property*

A common property resource (CPR) could be broadly defined as a natural resource in which a group of people have common user rights (not necessarily ownership rights). It is common for the resources of lakes, rivers, estuaries, and marine areas to be considered as CPRs. Traditionally, CPRs are not open-access that is available to anyone, but are subject to rules and conventions, often unwritten, of local communities. Particularly, after colonization, such CPRs have come under state control. Notwithstanding the above, CPRs form an integral part of the survival mechanisms for the poor and landless. Accessibility should, thus, be ensured and the resources protected. This usually requires community management.

Why participatory management?

(i) Common property resources (CPRs) are community lifelines and almost all people have a stake in their well-being. (ii) CPRs that could once sustain local communities may no longer be able to do so, mainly because they are overstressed and mismanaged. (iii) It may not be desirable that an external agency takes control of resources that essentially belong to the community. (iv) Public awareness is to be raised and local people have to be involved in CPR management.

Oxbow lakes: a CPR

In Bangladesh, oxbow lakes have been managed by local management groups to maintain equitable access to resources.

People-managed fish stocks

In Southern Laos, local communities manage the fish stocks in the Mekong River based on conservation zones and other management regulations.

Why is CPR management difficult

The management of the CPR is difficult because of the following:

(i) Conflict between different resource users, both within and outside the community. (ii) Difficulty in excluding outsiders from resource use. (iii) Use of common resources are continually changing due to new opportunities and constraints. (iv) Lack of appropriate knowledge or skills to manage resources effectively. (v) Livelihood constraints: the local people are often dependent on the CPRs for income and have few livelihood alternatives. (vi) Individuals are often, probably, reluctant to curtail destructive activities, if other people do them. However, the benefits of individual conservation actions seemed to be shared by the entire community. (vii) Management of common resources often occur at different "scales" (local, municipal, regional, and national) and the regulations at these different levels are, sometimes, conflicting. (viii) Government responsibility for managing parts of the ecosystem often falls, mainly because of control under several different government departments (e.g., mangroves may fall under forestry and fisheries), water bodies may be under the department of land revenue or fisheries, and so on and, further, the policies and mandates of the different departments may, sometimes, be conflicting. (ix) Inequitable use and allocation of resources.

2.3.1.28 Summary

(1) Fisheries management is a significant component of aquatic resource management. Many countries in the world have come up with policies and frameworks to develop strategies and regulations for aquatic resource management in general, and fisheries management in particular.

(2) Community-based cooperative fisheries management is one of the community property rights-based fisheries management systems, which has received attention in recent years particularly in the developing countries. On the other hand, comanagement is a complex issue, on which there is, possibly, no consensus as yet. Interpretations and status could vary from one region to another, and are influenced by a variety of regional and local parameters.

(3) Community empowerment is an essential aspect of efficient resource management by the community.

(4) **Community-based fisheries management (CBFM)** is a potential strategy for the management of water bodies and their bioresources. Today, it is a globally acclaimed *modus operandi* mainly for the management of common property resources.

(5) In Asia, there have been many works covering different aspects of CBFM in different countries. For instance, **Asia** (in a **General** manner) being dealt with by Thilsted and Wahab (2014), Flewwelling (2001), and Pomeroy et al. (1998); **Asia-Pacific Region** being dealt with by Ruddle (1993) and Makino and Matsuda (2011); **Cambodia** being dealt with by Anon (2016), Nuov (2015), Limsong and Sitha (2015), De Silva and Song (2015), and Anon (2015a,b); some of the significant works covering **China** were those of Paterson and Pernetta (2015), Huang et al. (2001), and Wang et al. (2015); **Hong Kong** had notably been covered by Anon (1999a), Yeung (2008), Morton (2005), and Clarke et al. (2002); significant works in **Indonesia** were those of Crawford et al. (2004), Sidayat (2008), Satria and Matsuda (2004), Radjawali (2012), and Soede et al. (2001); in **Japan**, some of the important works are those of Yamamoto (2000), Schmidt (2003), and Yamamoto (1996); some of the works in **Korea** are those of Cheong (2004a,b), Chang Ik Zhang et al. (2011), and Lee and Midani (2014); **Lao PDR** had significantly been covered by Baird (2000), Phounsavath (1998), Anon (1997), and Anon (2015a,b); some of the important works in **Malaysia** are those of Wong et al. (2009), Nasuchon (2009a,b); in the **Middle-East**, Feidi (1998), Anon (2012) worked significantly; **Myanmar** had been worked upon by Anon (2005); **Salient activities of different fisheries and aquaculture-related organizations** related to community-based management, etc., had been elaborated by De Silva et al. (2007), Wilkinson (2007); in **Nepal**, significant works were those of Gurung (2007) and Pradhan (2009); some of the works in **The Philippines** were those of Katon et al. (1997), Ahmed et al. (2004), and Anon (2011a,b,c); **South-East (SE) Asia and the Pacific** were largely covered by Pomeroy (1995) and Komatsu (2013); in **Sri Lanka (SL)**, some of the significant works were those of Murray (2002a,b), Amarasinghe and Wijenayake (2015), De Silva (2001a,b), and Pushpalatha et al. (2015); in **Thailand**, Henocque and Tandavanitj (2011), Tokrisna et al. (1997), Pimoljinda and Boonraksa (1997), Panjarat (2008), Anuchiracheeva et al. (2003), Jutagate and Kwangkhang (2015), and Yakupitiyag (2016) emphatically worked on various aspects of CBFM; in **Vietnam**, Ruddle (1998), Nguyen (2012), Teh et al. (2014), Hilborn (2008), Bui The Anh (2015), and Phan (2014) have worked significantly on different aspects of CBFM.

(6) Concomitantly, **in India**, some of the notable works related to CBFM were those of Dey (1981), Dey and Kar (1987, 1989a,b,c, 1990, 1994), Kar and Dey (1982, 1986, 1987, 1991, 1993a,b, 1996, 2000, 2002), Kar (1984, 1990a,b, 1996, 1998, 1999, 2000, 2002a,b, 2003a,b, 2004a,b,c,d, 2005a,b,c, 2006a,b, 2007a,b,c, 2010a,b,c, 2012a,b,c, 2013, 2014, 2015, 2016), Kar and Sen (2007), Kar and Barbhuiya (2000a,b, 2001, 2002a,b,c,d, 2004), Kar and Barbhuiya (2005, 2011a,b), Gadgil et al. (2000), Lahon (1983), Goswami (1985), Anon (2006a), Vinci (2002), Sugunan (1997), Deka (1999), Bhattacharjya (2002), Sugunan et al. (2000), Shrivastava and Bhattacharya (2003), Mukhopadhyay (1997), Jha (1997a,b), World Bank (2011), Kar et al. (1995, 1996a,b,c, 1997, 1999, 2000, 2001, 2002a,b,c,d, 2003a,b,c,d,e, 2008, 2009, 2011a,b, 2012, 2015, 2018), Barman and Kar (2014), Sultan and Barman (2014), Barman et al. (2008), Barbhuiya et al. (2009), Das et al. (2013, 2015a,b,c,d,e,f,g,h,i), Das and Kar (2013, 2014), Kar and Kar (2014), Bimola et al. (2014), Chaudhuri and Banerjee (1965), Vass (1992a,b), Pathak et al. (1986), De and Mukherjee (1992), Rana et al. (1996), Kumar (1985), Chitranshi (1992), Jaiswal and Singh (1994), Jha (1989, 1995, 1997), Singh (1990, 2000), Sinha and Jha (1997), Vass (1989), Vass and Langer (1990), Singh (1990, 2000), Mitra (1989, 1992), Mitra and Ghosh (1992), Choudhury (1989, 1992a,b), Yashmin et al. (2007), Parameswaran and Vass (1995), Sugunan and Mukhopadhyaya (1995), Goswami and Devaraj (1993), Sugunan and Bhattacharjya (2000), Yadava et al. (1987), Yadava et al. (1989), Deka et al. (2001), Shrivastava and Bhattacharya (2003), Anon (2009a), Bhaumik and Saha (1994), Bhaumik (1999), Goswami and Sathiadhas (2000), Khare and Punekar (2001), Maiti (2003), Goswami (2003), Dana (1987), Babare and Sawant (2009), Sen (1997), Salim (1998), Baruah et al. (2000), NIAM (2010), World Bank (2004, 2011), AFPRO (2010), Anon (2011a,b,c), Goswami and Sathiadhas (2000), Baruah et al. (2000), Anon (2004a,b), Barman (2008, 2009), Anon (2004a,b, 2006a, 2009, 2010a,b, 2011a,b,c),

Lobe and Berkes (2004), Choudhary (1979), Kumbhar (1979), Natarajan (2005), Chandra (2011), Prabhu (2011), Vipinkumar et al. (2013), Mohamed (2004), Sampath (2005), Anon (2011a,b,c), Anon (2004a,b), Kumar et al. (2003), and Datta et al., (2010).

(7) Notwithstanding the above, some of the significant works related to CBFM in **Bangladesh** were those of Rab (2010), Ghani and Ahmed (2008), Thompson et al. (1999, 2003, 2006), Hossain et al. (2010), Sultana et al. (2002), Abdullah (2007), Sultana and Thompson (2007), Ali (1997), Minkin et al. (1997), Islam et al. (2006), Mustafa and Hall (2007), Ali (1997), Islam et al. (2006), Dickson and Brooks (2007a,b), Anon (2007a), MACH (2007), Dey and Kanagaratnam (2007a,b), Dickson (2007), Ahmed and Dickson (2007), Anon (2010a,b), Thompson (2004), Gazi et al. (2013), Chakraborty (2004), Rahman and Begum (2010), Khan (2011), Chowdhury et al. (2012), and Paul et al. (2014).

2.3.2 Review of Literature in different countries in Africa

Africa is a continent having a rich diversity of both biotic and natural abiotic resources. Thus, there seems to be a need for proper and sustained management of the resources. In fact, management of the resources through the involvement of the community from the grassroots level could be very effective. A review of literature on the community-based management of the resources in different countries/regions of the African continent is given in the following.

2.3.2.1 *Africa General*

There have been much works done on Community-Based Fisheries Management (CBFM) in **Africa** in general. Blomley et al. (2007) dealt with community conserved areas as part of a review of status and needs in selected countries of **Central and Eastern Africa**. "Community Conserved Area" (CCA) is, probably, now a commonly accepted term that refers to specific sites, resources, or species voluntarily conserved through community values, practices, rules, and institutions. Such earmarked conserved areas are, probably, essential for the domain of fisheries and aquaculture (AQC) in view of their significance as potential resources for nutrition and avocation vis-à-vis their dwindling trend in the present scenario. And involvement of the community stakeholders from the grassroots level is, perhaps, necessary for the effective implementation of such schemes. There has, probably, been growing attention paid to CCAs, over the past few years, about the potential positive role they could play to the wider objectives of biodiversity and landscape conservation, as well as meeting local objectives of sustainable management and livelihoods. However, CCAs are, perhaps, rarely considered in national policy or legal frameworks, although few legal mechanisms exist for communities to defend their resources against external forces, conversion to alternative land uses, or, in some cases, communities are being dispossessed of resources traditionally under their control as they are converted to other land uses or absorbed into the national protected area system. As such, there have been works, often commissioned by a range of concerned actors with a view to establishing the range and extent of CCAs in selected countries, in order to identify the policy drivers to take care of those earmarked areas for more effective management, often may involve the stakeholders at the community level. There could also be recommendations regarding how local conservation efforts could be supported. Such efforts could, sometimes, be seen in countries like Tanzania, Kenya, Rwanda, Cameroon, etc.

Further, Gitonga (2013) dealt with the final technical report regarding strengthening fisheries management in ACP countries, particularly in support of **East African** Industrial Fishing and Fish Processors Association (EAIFFPA) to strengthen the taskforce approach in reducing illegal fishing and trade in **Lake Victoria**. Lake Victoria is one of the largest lakes in Africa and the second largest lake in the world, covering an area of *c* 68,000 km^2 and shared by three countries, viz., Kenya, Uganda, and Tanzania. In fact, the lake is very important to the economy of the East African Community (EAC) partner nations including Rwanda and Burundi, the newest member of the community. Lake Victoria is one of the most productive FW bodies in Africa, with an estimated annual fish landing of *c.* 800,000–1,000,000 tons valued at *c.* US $ 350–400 million at the beach, before any value addition. The entire Lake Victoria fishery resources contribute to the livelihoods of *c.* 3.0 million people engaged in subsistence fishing, trade, and other related activities. The Lake Victoria, thus, probably, qualifies itself for efficient management of its resources through effective community-based fisheries management involving the stakeholders in the community in a participatory mode.

Concomitant to the above, the Nile perch is, probably, the most significant fishery contributing to the economy of the people and upon which, a lucrative processing and export industry had, probably, been built. Moreover, the production of Nile perch, probably, represents *c.* 40%–60% of the total Lake Victoria fish production, and its exports were estimated to be *c.* US$ 300 million, among the three riparian nations and generating substantial amount of foreign exchange earnings. Unfortunately there had, probably, been a continued decline in Nile perch biomass and harvests. In fact, in this context, one factor, which is, probably, not disputed is, perhaps, the tremendous increase in fishing effort, whether through legal or illegal means. The lucrative regional trade with juvenile fish, supported by high level of corruption, probably, had exacerbated the decline of the fish stocks. Under these circumstances, "sustainable management" of the Lake Victoria fisheries may, therefore, be given top priority by all stakeholders through the use of various input and output tools, such as limiting the number of fishing boats and gears (input), implementing fishing quota allocations, etc. Further, the use of key stakeholders' partnership approach for monitoring, control, and surveillance (MCS) could be envisaged by many stakeholders to be a more tenable option in the efforts to eliminate unlawful fishing and trade. In fine, community-based fisheries management (CBFM) could be a viable alternative for successful and sustained management of the Lake Victoria and its resources. Incidentally, There have been report(s) about Inland fisheries comanagement in East Africa. The importance of inland fisheries resources to the East Africa Community Partner States (countries) of Kenya, Tanzania, and Uganda had been highlighted in a number of communications. Efforts and achievements to improve management and utilization of fisheries resources for lasting social and economic benefits in the region have also been discussed, based on experiences from various other projects, viz., (a) the Integrated Lake Management project in Uganda, (b) the ongoing Implementation of a Fisheries Management Plan project on Lake Victoria, etc. Significant achievements had been perused and lessons learnt and identified for application within and beyond the region. Notwithstanding the above, Roe et al. (2009) dealt with community management of natural resources in **Africa** along with impacts, experiences, and future directions. Natural resources are coveted gifts from the nature for the welfare of the world. As such, they are to be preserved, conserved, utilized, and managed in a sustained manner for maintaining harmony in nature. And, this exercise could, perhaps, best be done by involving the community stakeholders from the grassroots level. Further, Blaikie (2006) dealt with community-based natural resource management in **Malawi and Botswana** with a comment regarding beauty of small resources. Further, Melcher et al. (2013) dealt with biodiversity (BD) and conservation in general, and further, specifically, the current questions in sustainable fisheries, water management, and higher education. The objectives were, probably, to establish an official list of ichthyospecies, a national database of information on existing biophysical characteristics of fisheries, the diversity and conservation status of fish species, and benthic invertebrates. The ecological approach had been developed, perhaps, to monitor fish and water resources which could be directly used for future assessments of aquatic ecosystems that could also be incorporated into the national database and serve as the basis for recommendations for changes to policies and practices in order to keep on updating the national BD strategy. Incidentally, the introduction of fish species in Burkina Faso had been restricted by law. But, no reliable list of species, perhaps, exists. The official and reliable species list could be the reference for the implementation of that law. The expected result could be a national database of information on biophysical characteristics of fisheries, the BD and conservation status of the fish species and benthic invertebrates, as well as the pressures on the fish populations. In compliance with the work plan, there had, probably, been attempts to ascertain the following: (i) the bio-physical characteristics of fisheries (river size, catchment size, turbidity, etc.); (ii) the diversity and conservation status of the ichthyospecies and benthic invertebrates, land use, chemical water quality, etc.; (iii) the pressures on the fish populations, e.g., overfishing, etc. In addition to the above, Anon (2008a) dealt with CBFM in **Africa** in general.

2.3.2.2 *Namibia (Africa)*

Namibia has a good stake in fisheries management. Huggins (2011) dealt with fencing fisheries in **Namibia** and beyond as part of the lessons from the developing world. Fisheries and aquaculture (AQC) are the progressive domains in the developing world, in view of their potential for nutrition and avocation. Learning from each other is a common practice in the world. And, every developing nation, perhaps, tries to develop the fisheries and AQC sectors in the country; as it also serves as a rich source of earning revenues. Namibia, probably, had *followed suit* in this direction. Thoughtful reform in the fisheries sector is essential. Different works, probably, had explained the scope of the environmental and economic benefits that could emerge from amending the structure of a regulatory regime, but logic and evidence are rarely enough to carry the day. Constructive changes in policies may often come about at the time of crisis, or when there are few special interest groups working to protect an existing structure. Namibia is, sometimes, considered backwater. It is quite a large country, almost double the size of California. However, the country, perhaps, does not have a very big population. There were few established economic interests, especially in fishing. Fortunately, a fishery policy, probably, had been adopted, which was amazingly sensible. Notwithstanding the above, fisheries around the world often suffer abuse due to the lack of property rights. The result is, probably, environmental destruction and economic waste. In a few instances, when fisheries were in collapse, developed nations, perhaps, were spurred to adopt individual quotas or some other rights-based approach that, probably, had produced better environmental and economic results. However, unique about Namibia, perhaps, was that, a catch share system was adopted in a poor nation with a population consisting of several deeply rooted tribes. In fact, this development has shown that market-based reform was, probably, not a western notion that somehow conflicts with traditional values. The lessons from Namibia and other fisheries success stories, probably, tried to illustrate that property rights and environmental protection may happen anywhere. Concomitant to the above, it could be said that efficient management system is also in place, involving the community and the stakeholders from the grassroots level, in a participatory mode for effective management of the biotic and abiotic resources.

2.3.2.3 *Ghana*

Ghana encourages collaborative fisheries Management. Brian Crawford et al. (2013) discussed about solving the fisheries crisis in **Ghana** with a fresh approach to collaborative fisheries management. Incidentally, poor governance and the open-access nature of Ghana's fisheries are the principal reasons for the decline in fish abundance.

2.3.2.4 *Lake Victoria*

Lake Victoria is one of the largest lakes in Africa and second largest in the world. Salehe (2008) reported about the capability and legitimacy of beach management units (BMUs) to improve fisher's income through the management of firsthand sales system in the **Lake Victoria** in Tanzania.

2.3.2.5 *Madagascar*

Madagascar deals with community-based programs in fisheries management. Anon (2013a,b) dealt with protection and promotion of marine biodiversity through small-scale, time-sensitive, community-based programs. The Marine Conservation Action Fund (MCAF) supports protection of locally managed marine area in **Madagascar**. MCAF had been supporting, since 1999, > 100 small-scale, high-impact projects in 30+ countries and had disbursed *c.* US $600,000 for the conservation of endangered species and habitats. Further, McClanahan et al. (2014) dealt with the perceived benefits of fisheries management restrictions in **Madagascar**. Efficient management is an essential aspect in fisheries and AQC. Madagascar has a rich potential for fisheries and AQC. Incidentally, the perceptions of the benefits of fisheries management restrictions were believed to have been evaluated in coastal Madagascar to identify restrictions that were likely to be self- and community enforced. Furthermore, Anon (2015c) dealt with community-based aquaculture as a pioneering viable alternative to fishing. It could be a model for everyone working to conserve the natural life-support systems of our troubled planet, as, probably, said by Sir David Attenborough.

Today, there seemed to be a need to diversify traditional small-scale fisheries, which also, probably, is fundamental to the livelihoods and food security of innumerable people throughout the world. Yet, the marine ecosystems and the fisheries they support are facing unprecedented pressures. There has, probably, been a crucial need to diversify the livelihoods of the people (who earn their livelihoods from the aquatic resources), mainly because of soaring demand for aquatic foods. Incidentally, in various places, the geographic isolation and the typical climates portray few economic scopes beyond fishing.

2.3.2.6 *Senegal*

Senegal is making efforts for community management of its fisheries. Dahou et al. (2001) dealt with support policies to **Senegalese** fisheries. The history of state support to Senegalese fisheries could be classified into a number of main periods with varying impacts on the fisheries' economic, social, and environmental sustainability.

2.3.2.7 *Sub-Saharan Africa*

There are preliminary efforts for fisheries management in **Sub-Saharan Africa**. Lohmeyer (2002) dealt with traditional inland fisheries management and enhancement systems in **Sub-Saharan Africa** and their potential for development.

2.3.2.8 *Tanzania*

Tanzania possesses a rich diversity of fishes. Anon (2013a,b) reported on the stock-taking exercise in **Tanzania** under the community-based natural resource management program, managed by the **Tanzania** Natural Resource Forum. The practice of stock-taking exercise considered a good practice, which is to be followed to know the status of the resources, including water resources and the fishes. Such an exercise could also help in the management of the coveted resources in a sustained manner. In any case, involvement of the stakeholders from the grassroots level in a participatory mode is a prerequisite for the success of any management operation of the natural resources. And community-based management (CBM) is one of the most efficient methods of management of natural resources. The Community-Based Natural Resource Management (CBNRM) stock-taking report is considered a part of the regional initiative under the Southern Africa CBNRM Forum (SACF).

In **fisheries** sector, the main objectives were, perhaps, to review the status of fisheries comanagement schemes such as the BMUs, to review the status of community involvement in fisheries comanagement related schemes, and to review the status of community based fisheries/water resources management for small lakes and rivers found in many parts of the country. Further, Bulayi (2001) worked on the community-based cooperative fisheries management for **Lake Victoria Fisheries in Tanzania** in Africa. Incidentally, many countries in the world have managed to develop fisheries management regimes which could improve the economic efficiency of the fisheries. Property rights-based fisheries management regimes have shown promising results in the management of fisheries resources because they reduce or eliminate the incentives for overcapitalization in the harvesting of the fisheries resources, and, thus, contribute to economic efficiency. Community-based cooperative fisheries management (CBCFM) is regarded as one of the property rights-based fisheries management systems, which has received attention in recent years, particularly, in the developing countries. In Tanzania, CBCFM seemed to be a feasible option because the current system is, probably, entirely based on common property and an open-access approach. This approach had led to increased fishing efforts, encouraged excessive fishing investment, thus declining catch trends. A CBCFM approach has been proposed for Lake Victoria fisheries in Tanzania in order to improve fisheries management. This system recognizes the sharing of management responsibilities between fishing communities through beach fisheries management units (BFMUs) and the fisheries division in the Ministry of Natural resources and Tourism. The CBCFM is likely to reduce the problem of overexploitation of fisheries resources in Lake Victoria by allocating exclusive fishing rights to the fishing communities through BFMUs in their respective villages. Legal mechanism could be developed as a basis for implementation of the CBCFM in Lake Victoria. Furthermore, Onyango (2001) studied pitfalls in building a community-based fisheries resource management regime in **Lake Victoria, Tanzania.** He discussed about reforming the fisheries management through involving local level fishery institutions (LLFI). The study was based on Tanzania's Lake Victoria fishery where LLFIs were established through the formation of local enforcement bodies. The paper takes the view that the overfishing problems that confront Tanzania's fisheries management authorities are best understood from a social perspective.

2.3.2.9 Uganda

Fish is an important item of nutrition and avocation in **Uganda**. In **Uganda**, Nyeko (2004) worked on community participation in fisheries management with regard to their prospects and challenges for comanagement.

Fisheries play a significant role in the economy of Uganda, contributing to foreign exchange, food security, and employment. In 1998, the total fish catch landed was *c.* 217,100 metric tons (mt), of which Lake Victoria contributed to *c.* 48.6%. The development of government-centered management systems seemed to have led to increasing alienation of resource users and to intentional disregard for specific regulations. The understanding of the problems faced by the current management systems had led to the recognition that user groups need to be actively involved in fisheries management if the systems are to be consistent with sustainable fisheries and to be legitimate. Community participation in fisheries management had been tried in Lake Victoria through comanagement, which was the sharing of responsibility between the government, researchers, and the resource users. A study on comanagement was conducted among the fisher communities in Lake Victoria, Uganda in July 1999 and between October and November 2000 to assess the potential for implementing a community-based comanagement approach on the fishery. Results of this study had shown that there was a general agreement between the staff and fisherfolk on the need to share management responsibilities. Fishers usually agreed that giving more responsibility for fisheries management to local fishermen could yield more positive results in terms of control of fisheries malpractices, willingness to collect and give accurate data, and compliance with the regulations. Concomitantly, Scullion (2008) worked on the inland fisheries comanagement in **East Africa**. His work highlighted the importance of inland fisheries resources to the East African Community Partner States of Kenya, Tanzania, and **Uganda**. Efforts and achievements to improve utilization and management of fisheries resources for lasting social and economic benefits in the region had been examined. Experiences and achievements from two projects, viz., the Integrated Lake Management project in Uganda and the ongoing Implementation of a Fisheries Management Plan project in Lake Victoria had been examined, and issues had been identified from the lessons learnt for applications within and beyond the region. Further, Keizire (2001) dealt with opportunities and options for financing fisheries management in **Uganda**. Fisheries and aquaculture (AQC) being profitable ventures, many countries, probably, have been investing in these sectors. Incidentally, there have been reports, which examined, from an economic perspective, arguments for introducing cost recovery in the Ugandan fisheries and the type of charges that could be appropriate. The reports also considered their likely implications for the Ugandan fishing industry, as well as their impact on the economic efficiency of both the fishery and fisheries management services. Notwithstanding the above, in the recent years, fisheries management in Uganda, perhaps, has been hampered by a severe financial crunch. The government provided only *c.* 7.8% of the estimated requirements for fisheries management services. However, a well-designed cost recovery system offers the scope for a proper financing of fisheries management in Uganda. Such studies outline a cost recovery system to finance fisheries management in Uganda. Such a system, tailored to the administrative structure of fisheries management provision in Uganda, was based on a combination of input, output, and access charges depending on the level of government providing the fisheries management services. Such studies investigate the ability of the system to recover sufficient funds for financing fisheries management, in fact management is a significant aspect of fisheries and AQC. Further, an efficient management *modus operandi* is to be in place involving the community stakeholders for effective management of the resources.

2.3.2.10 *Zambia*

Zambia is making efforts for the management of its fisheries resources involving the community. Based on **Zambian experience** on comanagement initiative, **Malasha**, Malasha (2006) stated that there was a mixed result of success and failure in achieving the comanagement goal with respect to lake fisheries management. Her works in different types of fisheries comanagement initiatives had been initiated and implemented in some of the major fisheries in Zambia, since the late 1980s. There had been diverse types of causes for instituting comanagement arrangements. These ranged from the requirement to control the influx of immigrant fishermen to the desire to encourage the use of lawful fishing gears. Since the late 1980s, various forms of fisheries comanagement initiatives had been implemented in some of the major fisheries in Zambia. The reasons for instituting comanagement arrangements had been varied and ranged from the need to control the influx of immigrant fishermen to the desire to encourage the use of legal fishing implements. This communication looked at the manner comanagement had evolved in three fisheries, viz., (a) the Lake Kariba, (b) the Lake Bangweulu, and (c) the Mweru-Luapula fisheries. It had been found that, after > 10 years of comanagement, the results were still mixed. In one Lake, there is some form of comanagement while in the other two, these initiatives have not been very successful.

2.3.2.11 *Summary*

(1) Africa is a continent having a rich diversity of both biotic and natural abiotic resources. Thus, there is need for proper and sustained management of the resources. Management of the resources through the involvement of the community from the grassroots level could be an effective one.

(2) Some of the significant works related to CBFM in Africa were those of Blomley et al. (2007), Gitonga (2013), James Scullion (2006), Roe et al. (2009), Blaikie (2006), Melcher et al. (2013), and Anon (2008a) with regard to CBFM in **Africa (General)**; Huggins (2011) with regard to **Namibia (Africa)**; Brian Crawford et al. (2013) related to **Ghana**; Salehe (2008) pertaining to **Lake Victoria**; Anon (2013a,b, 2015c) and McClanahan et al. (2014) pertaining to **Madagascar**; Dahou et al. (2001) related to **Senegal**; Lohmeyer (2002) related to **Sub-Saharan Africa**; Anon (2013a,b) with regard to **Tanzania**; Bulayi (2001) and Onyango (2001) pertaining to **Lake Victoria Fisheries in Tanzania** in Africa; Nyeko (2004), Scullion (2008), and Keizire (2001) with regard to **Uganda** and East African Community Partner States of consisting of **Kenya**, Tanzania, and **Uganda**; and Malasha (2006) related to **Zambia**.

2.3.3 Review of Literature in different countries in Australasia

The continent of Australasia possesses a rich diversity of biotic and abiotic resources. Thus, it is imperative that a good management system be in place for the effective harvest, management, utilization, and conservation of the resources. It is quite a known fact that sustained harvest and efficient management of the resources could be fully successful only through the involvement of the community from the grassroots level. An account of the review of literature on community-based management of the resources in different countries/regions in the Australasian continent is given in the following.

2.3.3.1 *Australia*

A good number of works have been done on the Community-Based fisheries Management in both freshwater and marine domains in **Australia**. Metcalf et al. (2009b) deliberated upon conceptual models for ecosystem-based fisheries management (EBFM) in Western **Australia**. In fact, the Department of Fisheries, Western Australia, was reported to be developing a framework to assess the most appropriate methodologies for the implementation of ecosystem-based fisheries management (EBFM) in Western Australia. Further, **Fetherston (2005)** dealt with sustainability certification in community-based fisheries. Community-based fisheries management (CBFM) is considered an efficient *modus operandi* for the management of fisheries in the globe. However, mismanagement of the global fisheries resources has an overwhelmingly negative impact on the ability of community-based fisheries to sustain effectively.

2.3.3.2 *Fiji*

There have been quite a good number of works done on fish and fisheries in Fiji. Annette et al. (2006) dealt with community-based marine resource management in **Fiji** with emphasis from yesterday to tomorrow. There could be various ways in which local communities could become more involved in reef management. Attempts are sometimes made to highlight comprehensively successes and failures, but with little rigorous assessment of what the conditions of long-term success are. Moreover, Fiji is considered to represent a unique case, in view of the fact that its customary fishing-rights areas ("*qoliqoli*") constitute a kind of dual ownership. This may establish a connection between indigenous owners and the central government for the purposes of management. But, as reported, this kind of so-called cooperative comanagement approach had not been uniform across Fiji. It could be mainly because of the involvement of different levels of the government and the community, with the result, that the approach became confusing with disputes. Further, in Fiji, one of the more issues is said to be still unresolved, i.e., whether "ownership" of the *qoliqoli* should include both the marine resources and the sea floor, the latter being currently owned by the state. Further, Warner (2000) dealt with conflict management in community-based natural resource management with particular emphasis on experiences from **Fiji and Papua New Guinea**. This paper discusses the problem(s) of nonviolent conflicts and disputes as a constraint to sustainable natural resource management at the community level. Further, the paper provides some background to the role of conflict in natural resource management. The report also presents a methodology designed to contribute to the removal of conflict as an obstacle to sustainability. The methodology was developed to guide a program of conflict management within NGO-sponsored community-based natural resource projects in the South Pacific. Examples of the outputs of the methodology had, probably, been drawn from conflict management activities undertaken in the Lakekamu Basin Integrated Conservation and Development Project, Papua New Guinea. The overall benefits of conflict management in community-based natural resource projects had, probably, been not much dealt with. Rather, they had, perhaps, been dealt with in relation to building social capital and sustained livelihood security. Furthermore, Anders et al. (2011) reported about mapping fish community variables by integrating field and satellite data, with emphasis on object-based image analysis and modeling in a traditional **Fijian** fisheries management area. The use of marine spatial planning for zoning multiuse areas is growing in both developed and developing countries. Comprehensive maps of marine resources, including those significant for local fisheries management and biodiversity conservation, provide with a crucial foundation of information for the planning processes. There had been attempts to use an empirical procedure, by using a combination of field and high spatial resolution satellite data, to make a bathymetric map (RMSE 1.76 m), as well as object-based image analysis to produce accurate maps of geomorphic and benthic coral reef classes (Kappa values of 0.80 and 0.63; 9 and 33 classes, respectively) covering a large (>260 km^2) traditional fisheries management area in Fiji. From these maps, attempts have been made to derive per-pixel information on habitat richness, structural complexity, coral cover and the distance from land, as well as to use these variables as input in models to predict fish species richness, diversity, and biomass. It was, probably, revealed that random forest models outperform five other model types, and that all three fish community variables could be satisfactorily predicted from the high spatial resolution satellite data. Moreover, the geomorphic zone was often shown to be the most important predictor on average, with secondary contributions from a range of other variables including benthic class, depth, distance from land, and live coral cover mapped at coarse spatial scales, suggesting that data with lower spatial resolution and lower cost may not be sufficient for spatial predictions of the three fish community variables. Concomitant to the above, Juerg M. Brunnschweiler (2009) reported about the Shark Reef Marine Reserve with emphasis on marine tourism in **Fiji** involving the local communities. The Shark Reef Marine Reserve in Fiji is an ecotourism project designed to protect a small reef patch and its fauna while preserving the livelihood of local communities. It involved the local communities by using a participatory business planning approach to marine protected area (MPA) management, generating income through diver user fees, distributed to the local villages that have exchanged their traditional fishing rights in the marine reserve for this new source of income. The Shark Reef Marine Reserve was a self-sustaining and profitable project, and was an example of a privately initiated, bottom-up approach, which included all relevant stakeholders in an area where marine rights were finely subdivided into small units.

2.3.3.3 *Oceana: Micronesia*

Rhodes et al. (2011) did interdisciplinary study of the market forces and fishery management in **Micronesia** under the Coastal Fisheries Management Program. A successful fishery with efficient management system involving the community stakeholders, probably, looks for effective markets for confirming profits in the commercial arena. Further, Anon (2014) dealt with **community-based ecosystem approach to fisheries management** (CEAFM) and **climate change adaptation** in the State of Yap, Federated States of **Micronesia** (FSM), with regard to the current status of marine resources in FSM. Furthermore, Anon (2004c) dealt with transforming the Live Reef Food Fish Trade (LRFFT) toward sustainability in the Republic of the **Marshall Islands** through community-based coral reef conservation and fisheries management. More effective assessments of the practices and studies on the impacts of the LRFFT in the Marshall Islands are, probably, required.

2.3.3.4 *Pacific Islands (1)*

Much impetus is given to the management of fish and fisheries in the **Pacific Islands**. King and Lambeth (2000) dealt with a manual on promoting the management of subsistence fisheries by **Pacific Island communities**, in connection with fisheries management by communities. It is, probably, impossible for the government fisheries agencies to properly manage subsistence fisheries without the involvement of fishing communities. In fact, most of the government fisheries agencies, probably, have limitations on budgets and staff which could make it impossible to constantly work in communities which are, generally, spread out in many different villages. But, more importantly, community involvement, probably, results in a sense of belonging and ownership of fisheries management actions and regulations. Importantly, under community ownership, fisheries management measures are enforced by the communities themselves. Incidentally, there are literature about **CBFM** in different parts of the world. They provide guidelines and suggestions on how communities could be encouraged to take a leading role in the management of their fisheries and the aquatic (freshwater and marine) environment (King and Lambeth, 2000). The Samoan model of CBFM had stood the test of time, and could also be transferable and implementable in other island countries of the Pacific, perhaps, with some alterations to suit local conditions and culture. Studies, further, revealed that *c.* 60 villages have successfully managed their own fisheries resources under CBFM for >3 years.

2.3.3.5 *Oceania: Vanuatu*

Fish and fisheries are significant entities in **Oceania: Vanuatu**. Marc et al. (2013) dealt with community-based management of near-shore fisheries in **Vanuatu**. In the Pacific region, coastal communities had compensated for chronically low capacity of governments to manage fisheries by implementing local regulations in the aquatic domain. They, probably, had looked for efficient methods of management. Thus, trajectories and factors of change in CBFM systems, since the 1990s, have been analyzed, in order to investigate the performance of CBFM in Vanuatu (Marc et al., 2013). Focal group interviews were conducted in seven villages on Efate' island in 2011 and supplemented by views of supporting literature. Results have revealed the increase in grand excessive reliance of CBFM systems on external agencies that is said to promote complex management plans. Moreover, examination of trends in CBFM systems has indicated that community and national fishing rules, which were highly acceptable to the local societies, were more likely to be enforced in the long run. In particular, the establishment of FW, coastal, and marine reserves were, perhaps, the most widely spread and best enforced community rules for the purposes of conservation, ecotourism, and fisheries. However, some of the works, probably, had tried to refute the effectiveness of CBFM in achieving sustainability of reef fisheries in Vanuatu and concomitantly, probably, highlighted the overreliance on small marine reserves as a management tool. Nevertheless, community initiatives may, preferably, be strengthened by new specific national regulations governing subsistence and commercial reef fisheries as part of a multi-scale comanagement approach.

2.3.3.6 *Summary*

(1) The continent of Australasia possesses a rich diversity of biotic and abiotic resources, with the imperative that a good management system is *in vogue* with the involvement of the community from the grassroots level for taking care of the effective harvest, management, utilization, and conservation of the resources.

(2) Significant works pertaining to CBFM in Australasia are those of Metcalf et al. (2009b), Fetherston (2005) in **Australia**; Annette et al. (2006), Anders et al. (2011), and Brunnschweiler (2009) in **Fiji and Fijian** fisheries management areas; Warner (2000) in **Fiji and Papua New Guinea**; Rhodes et al. (2011) and Anon (2014) in **Micronesia (Oceania)**; Anon (2004c) in **Marshall Islands**; King and Lambeth (2000) in the **Pacific Islands and their communities**; and Marc et al. (2013) in **Vanuatu (Oceania)**.

2.3.4 Review of Literature in different countries in Europe

Europe possesses a rich potential for aquaculture and fisheries. As such, there is, perhaps, need for efficient *modus operandi* for the effective management, utilization, and conservation of the aquatic bioresources, notably, the fishes. Community management in a participatory mode and with a bottom-up approach could be an efficient mode of effective management, sustained harvest, and proper conservation of the coveted resources. Incidentally, a good number of works had been done on the management aspects of the water bodies and their fisheries in different parts of Europe.

A review of literature on the community-based management of the resources in different countries/regions in the **European continent** is presented in the following paragraphs.

2.3.4.1 *Europe (General)*

Anon (2006b) dealt with the report of the Bergen Conference on Implementing the Ecosystem approach to Fisheries having relation to **Europe in general**. Iceland had hosted a Conference on Responsible Fisheries in the Marine Ecosystem, in 2001, organized jointly with the FAO of the UN and with the cosponsorship of Norway. It is a new type of conference, organized by the Nordic Council of Ministers and the Governments of Iceland and Norway, with technical support from FAO. It was held at Bergen in Norway from September 26 to 28, 2006 in order to address progress made toward the implementation of the ecosystem approach to fisheries. Concomitantly, Arlinghaus et al. (2002) dealt with reconciliation of traditional inland fisheries management and sustainability in industrialized countries, with emphasis on **Europe (General).** The inland fisheries sector, probably, had long been dominated by recreational fisheries, mainly in the northern industrialized countries, which normally exploit fish for leisure or subsistence, as well as, like in many other countries, provide benefits to the society. However, various factors are known to constrain the development and existence of inland fisheries, such as local user conflicts, low social priority, inadequate research, and funding, to list a few. In many cases, however, degradation of the environment and loss of aquatic habitat is the predominant concerns for the sustainability of inland fisheries. As such, there seemed to be an urgent need to little modify some of the age-old time-tested traditional community-based management practices and systems in tune with the present-day need, in order to achieve the goals of sustainable development.

Notwithstanding the above, Anon dealt with securing European fisheries, through CBFM, i.e., securing a future for the European fisheries. The continent of Europe has a rich potential for fisheries. Incidentally, it may be noted here that > 85% of the assessed fish stocks in the European waters are fished to their maximum potential, or, were overfished. No doubt, this could be a bad news for marine biodiversity, bad news for the fishermen, businesses, and communities who rely on seafood as a source of their income, and also could be a bad news for the European consumers who want to enjoy healthy, fresh, sustainable seafood. Therefore, more effective management measures for fisheries are, probably, needed in order to lessen pressure on the FW and marine environments.

2.3.4.2 *Austria*

There had been some amount of works done in Austria on various aspects related to CBFM. Schmutz et al. (2007) dealt with fish-based methods for assessing European running waters (including Austria) along with a synthesis, as well as with a brief note on their management. The European Union (EU) Water Framework Directive (WFD) requires monitoring of riverine fish fauna. When the WFD came into force in 2000 AD, most of the EU member states (MS), probably, did not have fish-based assessment methods compliant to WFD requirements. Therefore, the objectives of FAME (http://fame.boku.ac.at), a project under the Fifth R&D Framework Program of the European Commission, were, probably, to develop, evaluate, and implement a standardized fish-based method for assessing the ecological status of European running waters. Concomitantly, Zick et al. (2006) dealt with the changes in the fish species composition of all **Austrian lakes** which are > 50 ha during the last 150 years, along with associated management measures. The fish communities of all Austrian natural lakes, which are > 50 ha in surface area, were assessed; and, the historical fish communities in *c.* 1850 lakes were reconstructed. During the last 150 years, the fish communities of Austrian lakes have altered in the way that in *c.* 49% of the lakes at least one indigenous fish species, usually sensitive or small-bodied taxon, is missing. Probably, after some care, in all but one of the 43 lakes, the number of fish species later increased. In particular, certain fish species of interest to angling and commercial fisheries now occur in more lakes. Generalized linear models were used to identify variables that could explain the loss of fish species using categories of lake use (ranking score). The human population density around the lakes seemed to be one of the main causes for the loss of coveted fish species.

2.3.4.3 *Belgium*

Depestele et al. (2014) worked on the sensitivity assessment as a tool for spatial and temporal gear-based fisheries management, with reference to **Belgium**. Assessment of ecosystem health is required in the ecosystem approach to fisheries management (EAFM). Mitigation measures that address impacts exceeding the acceptable level, probably, also require assessment. The various fishing methods have different ecosystem impacts and this makes it difficult to assess them jointly. Sensitivity of the ecosystem to individual fishing method had, perhaps, been assessed previously but concurrent comparisons of different methods on the same ecosystem had, probably, not been done. There had been attempts to combine the main characteristics of established approaches to sensitivity assessment into a single roadmap, called Sensitivity Assessment of Gear Effects (SAGE). SAGE is widely applicable three-step process to assess the ecological concerns of EAFM. The methodology used in the SAGE roadmap is built on a scoring system, which, then, results in a sensitivity index of the ecosystem components of the fishing methods. The scoring system is based on a combination of expert judgment and data, both qualitative and quantitative. It allows for cross-evaluation of fishing methods and ecosystem components.

Concomitant to the above, Depestele et al. (2012) reported about an integrated impact assessment of trammel net and beam trawl fisheries. A fishery policy could be useful for everybody. Aspirations for a better protection of the marine ecosystems are formally set out in the Marine Strategy Framework Directive (MSFD), and are, perhaps, under way through the reform of the CFP. Incidentally, the Belgian marine fisheries fleet is dominated, probably, by the beam trawlers. However, the beam trawling may experience considerable amount of pressure from the latest EU legislation, as this fishery is clearly questioned for its ecological effects (e.g., discarding).

Incidentally, Margrethe et al. (2014) dealt with the aspect(s) of "what does stakeholder involvement mean for fisheries management?" The common fisheries policy (CFP), probably, sets out the guidelines for the management of European fisheries. The CFP is revised after every 10 years and the last two revisions, probably, had laid the groundwork for extending stakeholder participation in European Union (EU) fisheries management. The fishery industries and the NGOs especially are recommended to be given greater influence. However, there may not be sufficient time or fund to have everyone involved, but the challenge, probably, is, how to effectively enhance mechanisms for dialog to reach consensus.

Further, Newman (2014) dealt with structural and cohesion policies with regard to inland fisheries and the Common Fisheries Policy (CFP) in **Belgium**. The new European Maritime and Fisheries Fund, probably, has greater potential to provide increased support to inland fisheries through the strengthening of community-led local development. In many places, commercial inland fisheries are usually small-scale, labor intensive, traditional fisheries mostly using passive gears. Nevertheless, they produce high value products of local significance. However, the inland fisheries may not be much influenced by the reforms of the CFP, as they are mainly managed nationally. Diadromous species may benefit from the reforms as long as they are properly implemented. The new European Maritime and Fisheries Fund, probably, has greater potential to provide increased support to inland fisheries through the strengthening of community-led local development.

Further, Anon (2009b,c,d) analyzed the existing rights-based management (RBM) instruments in Member States (MS), and also discussed the setting up of the best practices in the EU, through various pilot projects for carrying out the Common Fisheries Policy. Rights-based approaches to fisheries management have shown potential for promoting biologically sustainable and economically viable fisheries in several parts of the world. Whilet fisheries management is a community responsibility, under the framework of the reformed EU common fisheries policy (CFP), economic management of fishing rights is a national responsibility, and, in practice, many Member States, probably, had already implemented RBM approaches in a range of fisheries across the EU. Concomitant to the above, the definition of RBM adopted for various studies is broad, including "any system of allocating fishing rights to fishermen, fishing vessels, enterprises, cooperatives, or fishing communities." Once rights had been allocated, it could become difficult for new entrants to enter a fishery. However, in most cases, if newcomers wish to enter a fishery controlled by an effort- or quota-based system, they, probably, are to buy a vessel with its associated rights (license, quota allocation, etc.). However, this could be prohibitively expensive and several countries, like, UK, Denmark, etc., have specific schemes to facilitate new entrants to the fishery, either by allowing them for starting in the small-scale sector, or by leasing out quotas. However, the most difficult type of system for newcomers to gain access seems to be TURFs (Territorial Use Rights in Fisheries), where rights are allocated, possibly, to a group of resource users only. Their establishment had often resulted in the exclusion of prior users of the resource through the conversion from an open access to a privatized regime, and, subsequently, it could be difficult for newcomers to gain membership of the association or group involved in the TURF.

2.3.4.4 *Bulgaria*

Bulgaria possesses rich resources of fisheries and associated management processes. Anon (2013c,d) dealt with the Fifth National Report (2009–2013) on Convention on Biological Diversity (CBD) of the Ministry of Environment and Water of the Republic of **Bulgaria**. "Biodiversity" being part and parcel of resource management, resource conservation, resource sustenance, and resource mobilization, the Fifth National Report on the Implementation of the Convention on Biological Diversity (CBD), prepared by the Republic of Bulgaria, was a Party to the Convention on Biological Diversity covers the period of 2009–2013. Notwithstanding the above, in accordance with the guidelines, it contains information on the following: (i) The current status of and threats to biodiversity (BD), and its socioeconomic impact. (ii) The implementation of the National Biodiversity Conservation Strategy and the Second Biodiversity Action Plan of Bulgaria (2005–2010) along with a reference to the National Program for Development in the context of Bulgaria 2020 and other sectors of the economy. (iii) Progress in the implementation of the Strategic Plan for BD, 2011–2020 and its targets, adopted by the Parties in 2010 in Nagoya. Bulgaria is one the richest BD countries in Europe. It encompasses parts of the Alpine, Black Sea, and Continental biogeographic regions. The number of Balkan and Bulgarian endemics is relatively high. As such, Bulgaria, perhaps, had established a national ecological network (NEN) (including the European ecological network Natura 2000 and protected areas) to protect its rich BD. The NEN is *c.* 34% of the national territory, which is one of the largest in Europe.

Notwithstanding the above, the fish resources are decreasing due to their overexploitation, lack of general program for regular fish stocking, and lack of regular monitoring of fish and non-fish resources. The loss of local breeds and plant varieties is, probably, mainly due to the intensification of production with limited selection of few species, introduction of alien breeds and varieties that replace the local ones.

Pollution of water, air, and soil also has impact on the loss of BD. The water quality depends on the ecological state and chemical composition of the surface waters and the quantities and chemical composition of the underground water. It may be noted here that by the beginning of the second decade of this millennium (say 2012), out of *c.* 759 surface water bodies, only *c.* 36 (*c.* 4.7%) were said to be of good quality, while out of *c.* 177 underground water bodies, *c.* 123 were in good chemical condition. Concomitant to the above, the invasive alien species might cause reduction in the habitat of other species, changes in the physicochemical characteristics of the water, possible changes in the structure and number of bacteria, phyto- and zooplankton, and changes in the trophic structure of the macrozoobenthos and fish populations.

2.3.4.5 *Climatic changes*

Bulgaria is one of the most vulnerable geographic regions to climate change. The precipitation and temperature index show that the average terrestrial temperature, probably, had risen by *c.* 1.5°C during the period 1971–2011, while the precipitation and the snow cover during the same period were relatively constant. The impact of climate change on biodiversity (BD) was demonstrated by the indicator changes in the wintering water birds in Bulgaria. This number had varied considerably during the recent past, e.g., in 2012, it was *c* 46.87% less than that in 2011. This variation, perhaps, depend mainly on the meteorological conditions in Bulgaria and northwards. Of late, shifting of the peaks in the number of some of the water birds (mainly geese, ducks, etc.), probably, had also occurred. This might reflect the need for sustainable management of the water bodies and the resources therein, on which, the habitat suitability for the migratory aquatic birds, perhaps, very much depends.

Statement of the Bulgarian fisheries administration and the Stakeholders in the fisheries and aquaculture sector in Bulgaria in Relation to the proposed European Commission Green Paper on the reform(s) in the common fisheries policy.

Anon (2013c,d) reported about the statement of the **Bulgarian** Fisheries Administration and the stakeholders in the fisheries and aquaculture sectors in Bulgaria in relation to the proposed European Commission Green Paper on the reform(s) in the Common Fisheries Policy.

Statement of the Bulgarian fisheries administration

Fisheries and aquaculture (AQC) are the integral components of nutrition and avocation for the people. As such, they, probably, qualify for sustained management by involving the stakeholders from the grassroots level. Thus, the

Green Paper, presented by the European Commission (EC), is published at a very important moment, in order to give an answer to the emerging necessity for integrating the **Common Fisheries Policy (CFP)** with the other community policies and to solve the existing structural disadvantages, which, probably, exist.

Notwithstanding the above, it, probably, had become extremely significant for Bulgaria to follow an approach for the management of the fisheries, which was supposed to had been based on the standards and the general principles adopted at EU, and on the basis of the experiences acquired after the accession to the EU and taking into account the specificity of the Black Sea. Also, the level and the specific regional conditions, perhaps, had to be considered to a maximum extent. In this connection, it may be pointed out here that the reformed CFP could accelerate the processes to find the appropriate form of collaboration in the regions, like the Black Sea region, including the management of the fish resources. The Black Sea has specific natural geographic and political features. Incidentally, probably, only two of the six coastal countries are members of the EU (Bulgaria and Romania), but their shares in the catches are very small as compared to the other countries, especially Turkey. Another specific feature was, probably, that before and after the accession of the two Member States (MS) to the EU, there was, perhaps, no form of collaboration among the six coastal countries for common management of the fish stocks. The present situation, probably, thus, poses serious difficulties in the management of the resources, which may not be easily done only by the two MS of the EU, more so, taking into account their small share in the total catches. At this stage, there is, probably, a necessity to apply a differential approach to the measures for the conservation of the fish resources, where the two basic aspects are, probably, to be pointed out, are as follows: (i) measures for the conservation of the species characterized as "local" and (ii) measures for the conservation of the species characterized as "shared" stock.

Thus, all measures for the conservation of the fish resources, such as programs for adaptation and management of the fleet, restriction of the catches, application of quotas and sustainable levels of the catches on annual or 6 monthly basis, closed zones and seasons for fishing, technical measures for restriction of the bycatches, etc., are to be adopted. Concomitantly, the discards and multi-annual plans for the management and recovery of the stocks could be efficient, only if they are applied by all the participants in the process for catches of commercially important species. In relation to the views presented previously and for the purpose of the sustainable management of the shared stocks, CFP could facilitate the establishment of regional fisheries organizations (RFO) that have to apply, in the most effective way, the community policies, taking into account the following aspects: (i) Cooperation among the Commission members, MS, and third countries (in the case of management of shared stock) to enable the RFOs to identify the necessary measures for conservation. (ii) Regulated exchange of data at regional level for the assessment of the stocks of commercially important species. (iii) Providing with opportunities to the RFO(s) to use individual approach in the management of the stocks and measures for their conservation and recovery in safe biological limits. The decisions and proposals of RFOs may be coordinated and changed by the EC wherever necessary. (iii) Exchange of information in the framework of the actions and management of the RFOs for applied fishing gears, fishing practices, fishing fleet capacity, and existing fishing opportunities. (iv) Implementation of a pilot project for sustainable implementation of CFP in the Black Sea region.

Summary of the framework for arriving at decisions on key long-term measures

Notwithstanding the above, the approach could, preferably, start from decisions, concerning the short-term development of the sector, toward development in longer period, probably, in order to solve the problem with the sustainable management of the stocks and to prevent the negative social and economic effects. Moreover, sufficient flexibility may be achieved in respect of determining the annual restrictions for the quantity of the catches and in the presence of clear proofs for their stable state or necessity of correction of total allowable catches (TACs). Notwithstanding the above, there are approaches which presume the formation of the respective conditions for the development and maintenance of the stocks in the safe biological limits during the time, thus partially satisfying the short-term expectation in the sectors and, at the same time, making conditions for general engagement of the participants in the achievement of higher levels of TAC for the following year.

Generating a practice of compliance

There is, probably, a need for forming a practice of compliance of the various common practices and regulations. As such, the following aspects may be taken into consideration: (i) Participation of the branch organizations in the process of determination and implementation of the measures laid down in the program for data collection. (ii) Participation in the stock assessment projects. (iii) Participation in the projects for the implementation of different procedures for electronic catch data reporting. (iv) Application of contemporary practices for traceability of the fish products within the whole chain.

The knowledge base for the policies and principles

Scientific knowledge and formation of data base are of great significance for the reformed CFP, because they serve as a basis for the decisions concerning the sustainable level of exploitation of the fish stocks, the impact assessment of fisheries on the marine ecosystems, and the consequences from other changes, such as the climate changes. One of the shortcomings in the process of defining our policy and the measures for the conservation and management of the resources is, probably, the lack of much clarity of the scientific advice. It is, perhaps, thus, necessary to establish a mechanism for clarity, transparency, and discussion of the scientific data and advice on an earlier stage before taking the final decision on the management of a certain stock. Efforts have been made to develop regional long-term scientific research programs with the participation of international scientists. Further, the cooperation with the branch organizations; the use of their available capacity, such as equipment, well-trained crew(s), and know-how, are of great significance for the successful and productive accomplishment of the research activities. Notwithstanding the above, the community has laid down certain guidelines in the CFP of the EU in its priority measures. In this regard, the implementation of the strategies and priorities in the operational program (OP) are supposed to ensure success in implementing the new fisheries policies. Incidentally, the OP had three main objectives: (i) to ensure competitiveness and sustainability of the fisheries sector, including management of the resources, (ii) to develop a market of fishery products and aquaculture production, and (iii) to support the sustainable development of fisheries areas and improve the quality of life in them. Achieving these goals could ensure the transition to new dimensions provided by reform of the CFP, as well as there could be integration of fisheries with other EU policies, while, concomitantly, allowing the provision of sustainable development of fisheries sector. The OP for the development of the fisheries sector is the main instrument to achieve these strategic objectives together with other financial instruments, other than the European Fisheries Fund, which were supposed to have been used by the Bulgarian authorities.

The following principles have been taken into account in EU structural policy applied also to the MS and in particular in the Republic of Bulgaria

(i) Consistency with the principles of the EU CFP, in order to achieve, in particular, a stable and durable balance between the fishing capacity and the fishing opportunities to fish stocks. (ii) Strengthening the harmonious, balanced, and sustainable development of economic activities, employment and human resources, and protection and improvement of the environment.

Notwithstanding the above, an appropriate allocation of financial resources has been achieved in accordance with these guiding principle(s) but there, probably, was also that need that these resources had to be allocated to the requirements of each objective of the strategy.

One of the main challenges in the implementation of European Union (EU) structural policy in the field of fisheries is associated with successful implementation of a "bottom-up approach," and the scope provided for the decentralization of the process of management of funds from the European Fisheries Fund (EFF) through the implementation of measures of priority in the form of "sustainable development of fisheries areas" from the EFF Operational Program. The principle of local associations of municipalities and stakeholders to participate in the decision-making process and to implement local development strategies is, probably, the actual first step in the mentioned changes in the fisheries management at community level. However, a key element in the materialization of a "bottom-up approach" is the inclusion of regional authorities and local organizations as well as individuals interested in implementing the policy.

Notwithstanding the above, with regard to the question of the advisory functions of stakeholders in the decision-making process, and also with reference to the implementation of structural policy in fisheries, it may be noted here that the programming documents (the National Strategic Plan and the Operational Program) for the absorption of funds from the EU EFF, probably, were prepared in accordance with one of the basic community principles, i.e., the principle of partnership and close cooperation with the regional administration, local economic and social partners, i.e., all the stakeholders in the sector. In addition, the organizations in the fisheries sector are involved in an advisory capacity and also in the work of the monitoring committee, i.e., the principal body for decision-making regarding the implementation of the Operational Program (OP) for the development of the fisheries sector. Concomitant to the above, public aid from the EU EEF through the OP is granted for the following priority measures, probably, related directly to the implementation of the foreseen reform(s) of the CFP activities: (i) Adaptation with regard to the efforts of the Bulgarian fishing fleet in accordance with the existing and available resources. The OP for the development of the fisheries sector provides means for increasing the selectivity of the fishing gears and sufficient financial resources for sustainable development and modernization of small-scale fishing. (ii) Investment in AQC production with good market prospects. It could be a measure of the OP, which is provided with the largest financial resources as it is a major economic activity in the fisheries sector in the long term and is likely to be a major pillar in the reformed CFP.

(iii) Establishment of producer organizations and other organizations in the sector and promotion of the exchange of experiences and best practices and other measures of common interest, such as training, investments, etc., aimed at improving the quality of fish products, certification of production, and implementing surveys, and, at the same time, encouraging cooperation between scientists and fishermen. (iv) Establishment of fisheries groups and identification of potential areas for the implementation of local development strategies. In fact, tangible results could be achieved at the regional level by applying this approach to the decentralized management of funds mainly for improving quality of life in fisheries areas, better use of fish resources in these areas, economic development of fisheries areas, and increasing the added value at the local level, and, at the same time, diversifying the economic activities in these areas.

Notwithstanding the above, the implementation of the policies and programs are supposed to ensure the success of policy both at the community and at the national level. In fact, the structural policy in the EU fisheries are usually developed in accordance with and in line with the regional policy, the employment policy, and the policy for environmental protection, and care being taken to strengthen the harmonious, balanced, and sustainable economic activities, employment and human resources, and protecting and improving the environment.

In fine, the structural policy of EU fisheries during this programming period, probably, very closely reflected the future direction of the CFP and the integration of the Common Fisheries Policy with other community policies and developmental priorities.

Views of the Bulgarian Fisheries Administration with regard to Aquaculture

Aquaculture (AQC) is an important strategy for boosting nutrition and avocation of the people. Thus, the guidelines for development are to be related to the boosting of the AQC production's share in the total volume of the fish products produced. The utilization of the EFF sources could make precondition for setting up, widening, equipping, and modernization of the production and processing facilities, as well as a potential for production of quality fish products. The investments could be directed toward fish products marketing, improvement in the hygiene and the conditions of work in the AQC farms, and also to increase the quality of the final fish product through the implementation of new species, as well as to use the production and processing innovations and production methods which decrease the negative consequences for the environment. Incidentally, the financial support for AQC could aim at achieving the following results: (i) To improve the quality of AQC fish products; market supply of the fish species, for which, there is a big demand; to make efforts to increase the share of the marine AQC products. Also, perhaps, there is the need for the adaptation of the sector to the market demand: species, prices, quality control, and products traceability. (ii) The firms, associations, and organizations in the sector could use the scientific expertise and technological potential of the scientific institutes in the field of AQC in order to optimize the activities, the methods, and the technologies for fish and water organisms' breeding and their processing for increasing the quality of the products and the economic efficiency. Meanwhile, the AQC production could be developed in the following directions: (i) Defining the AQC production as a tool for livelihood of the population that could lead to decrease in pressure on the fish stocks, and, at the same time, as a tool for the conservation and reproduction of the endangered species and recreation of the natural resources (through fish stocking of the relevant ecosystems). (ii) Increasing competitiveness in the sector. (iii) Setting up of a new strategy for AQC development, which could be applicable to all Member States (MSs). (iv) Promotion of fish and fish products as very nourishing, healthy, and healing food, as well as stimulating their consumption. (v) Setting up of conditions for fast, priority-based, and sustainable increase in volume of AQC products, fast increase in the volume of high-quality fish-stocking materials, and refining of fish-market infrastructure.

2.3.4.6 *Denmark*

Denmark has rich potential for aquaculture and fisheries. Eliasen and Sverdrup (2009) dealt with the Nordic experience of fisheries management in relation to the reform of the EU Common Fisheries Policy having relevance to **Denmark**, Nordic experience of fisheries management through Nordic cooperation is one of the world's most extensive forms of regional collaboration, involving **Denmark**, Finland, Iceland, Norway, Sweden, and three autonomous areas: the Faroe Islands, Greenland, and Åland.

2.3.4.7 *Finland*

Salonen and Mutenia (2007) dealt with alien fish species and their management in Northernmost **Finland**. An alien species, viz., peled (*Coregonus peled*), was introduced in 1972 in order to utilize the zooplankton production of the new, extensive open water areas. Although established at the beginning, this species declined subsequently, thus, probably, had caused problems for fisheries and management.

2.3.4.8 France

France possesses high potential in aquaculture and fisheries. Anon (2001) dealt with the country note on fisheries management systems with reference to France. France is an ichthyologically potential country. France's marine fisheries, of late, have produced *c.* 655,000 tons (including *c.* 24,000 tons from the overseas French departments) of fish. The catches included the crustaceans and other marine products, covering a range of >230 different species, including *c.* 150 ichthyospecies. It could be an onerous task to manage the huge aquatic biodiversity. However, the main species, in terms of volume, were the tuna, sardine, saithe, herring, mackerel, horse mackerel, and anchovy. On the other hand, the leading species, in terms of value, were tuna, anglerfish, sole, scallop, hake, and Norway lobster. Incidentally, the French fishing industry had undergone far-reaching changes since 1945. Concomitantly, Olivier and Olivier (2001) dealt with distributional issues in the operation of rights-based fisheries management systems, having relevance to **France**. Rights-based management regimes are considered by economists as an important solution to the problems of excess capacity and biological overharvesting of fisheries. In practice, adoption of such regimes, and particularly of those relying on individual quota allocations, has often met with resistance from within the fisheries concerned. A key reason for this resistance appeared to be the distributional conflicts which arise in the process of implementing the regimes. Economic analyses of the nature of the conflicts in different contexts had been done by different workers. The approach, probably, had centered around the way in which distributional conflicts could influence the operation of management systems and their impacts on fisheries, from the country to the individual firm level. There have been instances of analyses of the economic processes at the firm level based on the simulation of a fishery managed under individual transferable quotas.

2.3.4.9 Germany

Germany has a rich potential for aquaculture and fisheries. Döring et al. (2003) reported about small-scale versus trawl fisheries in the coastal waters, with reference to the case of the **German** Baltic herring fishery. The Greifswalder Bodden at the German Baltic Sea coast is the main spawning area of the western Baltic herring (*Clupea harengus membras* L) stocks. Before 1990, both trawlers and boats with static gear had been fishing there during the spawning season. After Germany's reunion, only the coastal fishermen with static gear continued fishing for herring because for bigger vessels only fishing for cod (*Gadus morhua* L) was economically feasible. Declining cods stocks and a recently built processing plant for herring on Ruegen Island changed the situation. After 10 years of absence, in 2002, trawl fishermen had resumed herring fishing leading to competition between small-scale operations and the trawl fishery. Coastal fishermen depending on local fish stocks were, probably, more aware of moderate exploitation methods. The CBFM mechanism was, probably, found to be the appropriate *modus operandi* to deal with such conflict situation.

2.3.4.10 Hungary

Hungary possesses a rich potential for aquaculture and fisheries. Anon (2008b,c,d) dealt with the fisheries operational program of the Republic of **Hungary** for the period from 2007 to 2013 and for the renewal of the fishery sector in Hungary. Hungary elaborated its own fisheries operational program (FOP) that applies to the whole territory of the country including the convergence and the non-convergence region. The FOP is coherent with the National Fisheries Strategic Plan (NFSP) and is drawn up in line with Council Regulation (EC) No. 1198/2006 (the EFF Regulation) and Commission Regulation (EC) No. 498/2007 (the Implementing Regulation). Hungary is situated in a temperate zone of the Northern Hemisphere in Central Europe, in the largest basin of the Danube valley, surrounded by the Carpathian mountains. The Danube River enters the country after draining precipitation from *c.* 131,000 km^2 relatively humid area. There are three main units of the Danube watershed: the direct local watershed of *c.* 40,000 km^2, while Tisza and Drava rivers drain water to the Danube from an area of *c.* 47,000 and 6000 km^2, respectively. The Danube and the Tisza drain waters from additional *c.* 38,000 and *c.* 110,000 km^2 of the Carpathians outside Hungary and run in part through the Hungarian country. Incidentally, there are only *c.* 4% of the surface waters, probably, borne within the country. The majority of the incoming waters are drained through the three large rivers, viz., Danube, Tisza, and Drava. Due to its basin-like property, the per capita amount of total water drained (*c.* 120 billion m^3/year) is the highest in Europe.

An analysis of national and local circumstances

In order to assess the significance of the sector in the national economy, the most commonly used indicator is the contribution to the outputs and the GDP. Incidentally, Hungarian agriculture sector, has contributed *c.* 3.0% to GDP, *c.* 6.1% to the export, *c.* 4.6% to the investments, and *c.* 5.0% to the employment in 2005. Hungarian animal husbandry, crop production, agricultural services, and the nonagricultural secondary activity had accounted, in turn, for *c.* 36.0%, 54.2%, 6.9%, and 2.9% of the gross production value of agriculture. On the other hand, of late, the Hungarian fisheries sector was said to be *c.* 2.5% of the gross production value of animal husbandry and *c.* 1% of the total Hungarian agricultural production. Incidentally, in 2006, the fishery sector has contributed *c.* 0.04% to the GDP. In fact, the significance of the fishery sector is much higher as the economic performance of several supplying and servicing sectors and also the entire recreational fishery and angling are based on this sector. However, the average income level of the employees of this sector, probably, did not reach *c.* 75% of the national average in 2005. Concomitantly, both the volume and value of the Hungarian fish and fish product import were, perhaps, significantly higher than the export. Incidentally, it may be noted here that effective management of the Hungarian fisheries by efficient *modus operandi* like the CBFM, involving the stakeholders from the grassroots level, could lead to better management of the fisheries sector. Concomitantly, Aps, Robert et al. (2004) made an overview report of the freshwater fisheries in Central and Eastern Europe (including relevance to **Hungary**) and commented upon the challenges of sustainability.

2.3.4.11 *Ireland*

Kenneth Irvine (2014) dealt with integrated water resources management (IWRM), from local to global sustainability having some relation to **Ireland**. Integrated Water Resources Management (IWRM) is a significant program, which draws its inspiration, probably, from the Dublin Principles. It is a systematic process for the sustainable development, allocation and monitoring of water-resource use with regard to its both abiotic and biotic components in the context of social, economic, and environmental objectives.

2.3.4.12 *Italy*

Anon (2004d) dealt with community-driven development and natural resource management. **Italy** is a country with high potential in aquaculture (AQC) and fisheries. However, various background information may be required for implementing successful management of natural resources, notably the water bodies and their bioresources, especially, the fisheries (Anon, 2004d).

2.3.4.13 *Mediterranean and the Black Sea*

The **Mediterranean and the Black Sea** region are rich in fisheries resources. Gwenaël Cadiou et al. (2008) reported about the artisanal fishermen in the area of Port-Cros National Park in the **Mediterranean**.

(i) **Artisanal fisheries involve local fishing units**, small amounts of capital, small vessels, and short coastal trips for small commercial or self-consumption purposes. (ii) In fact, artisanal fisheries are more sustainable than industrial fisheries in many aspects. (iii) However, the Mediterranean artisanal fishers, probably, increasingly need to adapt themselves to the new business scopes arising mainly from tourism. (iv) The artisanal fishers are to be included and properly accounted for in Marine Protected Area (MPA) designation and for management processes. (v) The MPA managers as well as the fishers share the same aim, i.e., to sustain healthy and productive seas. Further, Gutierrez (2015) dealt with the management and comanagement options for small-scale fisheries in the **Mediterranean and the Black sea**. There had also been a detailed discussion about the history of management of the fisheries. The management of natural resources, notably, the fisheries in particular, using the traditional as well as the self-designed *modus operandi* is *in vogue* since time immemorial. Time has tested many methodologies. Incidentally, community-based management and comanagement, of late, were found to be significant approaches, which have been more recently adopted globally as strong tools to counter the declining fish yield (FY). This was done, probably, in response to the perceived failure of centralized management of fisheries. (NB: The words "countries" and "states" appearing in the text refers to countries, territories, and areas without distinction.)

2.3.4.14 *Norway*

Norway lays high emphasis on aquaculture and fisheries. Hausner et al. (2012) dealt with the conditions under which community-based management (CBM) may work effectively along with an example from Sami pastoralists regarding their management of pastures sustainably. The CBM is a good exercise, and had been implemented in socio-ecological systems (SES) globally. CBM has, probably, also been the prevailing policy in Sami pastoral SES in Norway, but the outcomes tend to vary extensively among the resource groups ("siidas"). Concomitant to the above, Søreng (2008) dealt with fishing rights discourses in **Norway** with a note on indigenous versus nonindigenous voices. A pertinent question may, often arise: Who is and who should be entitled to fishing rights? This is, probably, an important issue in global fisheries management discourses. Privatization of fishing rights together with an increased emphasis on indigenous rights are among the most important factors that have put rights on the agenda.

2.3.4.15 *The Netherlands*

Fish is a significant item of nutrition and trade in the **Netherlands**. Stedinger et al. (2005) dealt with water resources systems planning and management with an introduction to methods, models, and applications. Managing water in ways that, probably, best meet all our diverse needs for water and its protection, including the needs of natural ecosystems, is absolutely essential.

2.3.4.16 *Romania*

Romania possesses a rich diversity of fishes. Anon (2008b,c,d) discussed the national sustainable development strategy of Romania for the period from 2013–2020–2030 as advocated by the Ministry of Environment and Sustainable Development of the Government of **Romania**. The Romanian **National Sustainable development Strategy** included the implementation of good practices for maintaining healthy biodiversity (BD) and their management, as well as the best available technologies for reducing pollutants, which may harm both the aquatic and the terrestrial environment.

Concomitant to the above, the Romanian **National Sustainable development Strategy** also strives for **preserving biodiversity (BD) and the natural heritage** by supporting the management of protected areas, including the implementation of the **Natura 2000 Network.** The main objective for 2008–2013 was to implement adequate management systems for the protection of the natural environment and the conservation of biological diversity, natural habitats, and the wild flora and fauna. The envisaged actions focus on the strengthening of the institutional capacity at national and local levels and involvement of public initiatives (with an important role to be played by NGOs) to comply with the relevant EU directives, particularly those concerning the fishes, birds, and habitats, in correlation with the development of Natura 2000 networks. Incidentally, the Natura 2000 sites cover *c.* 17.84% of the country's territory, including *c.* 273 sites of community importance (*c.* 13.21% of the national territory). Notwithstanding the above, the National Agency for Protected Areas and Biodiversity Conservation ensures general coordination for the preparation and implementation of the management plans for each of the areas designated for protection. The community-based methods may be applied for effective management.

Concomitantly, Mirela et al. (2005) dealt with Global Environment Facility with regard to **Romanian** Danube Delta Biodiversity Project with emphasis on case study reports of local benefits. The report had dealt with various aspects. However, what was very clear was that the various types of concessions, which had, probably, been given as facilities, have reduced the income of professional fishermen by *c.* 30%–50%, as well as, perhaps, had made subsistence fishing by the local inhabitants more difficult. On the other hand, the concessions, probably, had enhanced the amount of tax received by the government and the income of the private concession owners. In fine, the analysis of the field data revealed that while the project had, perhaps, been somewhat successful in conserving nature and its biodiversity, the livelihoods of many of the inhabitants have been negatively impacted. Further, the project had, probably, missed many scopes to benefit the local population, and therefore may not enhance the sustainability of the global environmental benefits. In contrast to this, efficient management of the water bodies and their resources by involving the stakeholders from the grassroots level in a participatory mode, through application of the *modus operandi* like the CBFM, could bring a letup in the situation. Such a measure could also strengthen the socioeconomic capacity of the inhabitants. Further, Neculita et al. (2003) dealt with **Romania's** fishing policy competitiveness in European integration context. Incidentally, fisheries in general, and aquaculture sector in particular, could be regarded as an advantage for Romania in the European competition. There had been attempts to analyze the evolution of the fisheries sector in Romania, as well as this industry's developmental stage in the European integration context. However, Romania is now subject to the Common Fisheries Policy (CFP) after its accession to the EU in 2007. In fact, the policy applied to fisheries is integrated into the Romanian agricultural policy, which, in turn, is to be integrated into the CFP. However, the main problem of the Romanian agriculture is its unsatisfactory competitiveness regarding both the domestic and European markets. Furthermore, Giger (2008) dealt with certain aspects of Romanian lentic systems and their management. Brackish biota was replaced, in general, by typical FW species. Incidentally, commercially less valuable fish, such as the giebel carp and roach became abundant while carps and pike perches drastically dwindled. In addition to the above, Anon (2009b,c,d) dealt with the operational program (OP) for fisheries in **Romania** for the period from 2007 to 2013, under the Ministry of Agriculture and Rural Development, Government of Romania. The Romanian Operational Program for Fisheries was drafted by National Agency for Fishing and Aquaculture in accordance with Article 17 (3) of Council Regulation (EC) No.

1198/2006 of July 27, 2006 of the European Fisheries Fund (EFF) following a consultation process with the socio-economic stakeholders, national and local government authorities, trade unions, NGOs and professional organizations. Concomitantly, a number of meetings were held with the stakeholders, the European Commission, and other partners, as a follow-up to the process initiated with the definition of the National Strategic Plan (NSP), in order to draft the OP, with the contribution of the thematic workgroups made for this purpose. The NSP for the period 2007–2013 had covered all the aspects of the Common Fisheries Policy (CFP) in Romania and the vision of the Romanian government for the development of the sector. The EFF had contributed, in many aspects, to the implementation of the CFP in Romania, and a specific strategy was developed as a continuation of the overall strategy adopted in the NSP. Community involvement from the grassroots level in a participatory mode could yield good results in the management of the water bodies and their resources. Notwithstanding the above, Bănaru et al. (2015) dealt with the reconstruction of the fisheries catch statistics in **Romania** from blue-fin tuna to gobies for the period from 1950 to 2010. Romania's fisheries catches were fairly well reported during the period 1950–1989. Intense trawling could have led to a strong decline in the stocks underpinning the fishery. Incidentally, Romanian fisheries consist mostly of industrial trawl vessels and artisanal fixed net fisheries known as "crawls," with a small amount of subsistence fishing for personal consumption. However, trawling had been banned since 2008. Unreported catches consisted mainly of commercial discards, subsistence catches, as well as some unreported landings. Incidentally, artisanal fisheries dominated with *c.* 66% of the catch, followed by industrial at *c.* 24%, and subsistence with *c.* 10%.

2.3.4.17 *Switzerland*

Anon (2013c,d) reviewed the key factors which support the small-scale coastal fisheries management. There have been reviews which aimed at helping to identify the role that certain foundation(s) could play important roles in managing the small-scale coastal fisheries. Incidentally, Expert Groups (EG) such as the Integrated Marine Management (IMM) are considered a leading research and development group with expertise in fisheries, AQC, and coastal zone management. Such organizations could undertake a rapid review of the key sources of knowledge (evaluations, literature reviews, and key informant interviews) and identify key factors, which could influence success in small-scale coastal fisheries management.

2.3.4.18 *United Kingdom (UK)*

Fish is an important item of food in the **UK**. Campbell et al. (2008) dealt with comanagement of sea turtle fisheries with notes on biogeography versus geopolitics. There have been instances of promotion of comanagement between local communities and the government agencies as a strategy to improve natural resource management including fisheries management. In this connection, there had been focal discussions on fisher incentives for engaging in comanagement and on the potential to scale up comanagement to a regional level. Such instances often also portray fishers' participation in socioeconomic surveys. Results of the surveys, based on three established criteria for comanagement, viz., (i) perceived crisis in stock, (ii) willingness to participate, and (iii) community cohesion have revealed that there, probably, existed fishers' support for comanagement, but the extent of support and the views of specific management interventions could vary. Further, there had been discussions pertaining to the implications of the results of comanagement in each territory; as well as, the efforts for establishing coordinated management regimes in the region.

2.3.4.19 *Macedonia (Yugoslavia)*

Macedonia possesses rich diversity of fishes. Fish forms a significant item of trade in **Macedonia**. Anon (2008b,c,d) dealt with the national report on the implementation of Convention of Biological Diversity (CBD) in the Republic of **Macedonia** for the period from 2006 to 2008. The Fourth National Report to the CBD described the development of the basic components of the biodiversity in the Republic of Macedonia; their efforts in this process; and achievement and suggestions for the future activities. Situation in many cases is not easily implemented in view of competing goals with regard to natural resources. In addition, hydrologic status of big natural lakes, notably, Ohrid Lake, Prespa Lake, and Doyran Lake, glacial mountain lakes, rivers, and other wetlands harbor significant species of aquatic flora and fauna. Concomitant to the above, the rivers, notably, the Sateska and Crni Drim, as well as the Sent Naum spring harbor important endemic species of trouts, whose population had declined. Notwithstanding the above, efficient management *modus operandi*, like the CBFM, involving the stakeholders from the grassroots level, in a participatory mode, could lead to effective management of these coveted water bodies and their resources.

2.3.4.20 *Summary*

(1) **Europe** possesses a rich potential for aquaculture and fisheries. As such, there is, perhaps, need for efficient *modus operandi* for the effective management, utilization, and conservation of the aquatic bioresources, notably, the fishes. Community management in a participatory mode and with a bottom-up approach could be an efficient mode of effective management, sustained harvest, and proper conservation of the coveted resources.

(2) Significant works pertaining to CBFM in Europe were those of Anon (2006b) and Arlinghaus et al. (2002) with regard to **Europe (General)**; Schmutz et al. (2007), and Zick et al. (2006), in **Austria**; Depestele et al. (2014), Depestele et al. (2012), Margrethe et al. (2014), Newman (2014), and Anon (2009b,c,d), in **Belgium**; Anon (2013c,d), Anon (2009b,c,d) in **Bulgaria**; Eliasen and Sverdrup (2009) in **Denmark**, Salonen and Mutenia (2007) in **Finland**; Olivier and Olivier (2001) in **France**; Döring et al. (2003) in **Germany**; Anon (2008b,c,d) and Aps, Robert et al. (2004) in **Hungary**; Kenneth Irvine (2014) in **Ireland**; Anon (2004d) in **Italy**; Gwenaël Cadiou et al. (2008), Gutierrez (2015) in the **Mediterranean and the Black Sea**; Vera H. et al. (2012), Søreng (2008) in **Norway**; Stedinger et al. (2005) in **The Netherlands**; Anon (2008b,c,d), Mirela et al. (2005), Neculita et al. (2003), Giger (2008), Anon (2009b,c,d), Bănaru et al. (2015) in **Romania**; Anon (2013c,d) in **Switzerland**; Campbell et al. (2008) in the **UK**; Anon (2008b,c,d) in **Yugoslavia (Macedonia)**.

2.3.5 Review of Literature in different countries in America

America is potentially rich in aquaculture and fisheries. Therefore, there is, possibly, a requirement of effective *modus operandi* for the sustained harvest, efficient management, proper utilization, and effective conservation of the aquatic bioresources, notably, the fishes. Participatory community management with the involvement of the stakeholders from the grassroots level could be an effective mode of efficient management, accompanied with sustained harvest and proper conservation of the coveted resources. Incidentally, quite a good amount of works had been done on the management aspects of the water bodies and their fisheries in different parts of the American continent.

A review of literature on the community-based management of the resources in different countries/regions in the **American continent** is dealt with in the following paragraphs.

2.3.5.1 *Argentina*

Argentina possesses rich potential for aquaculture and fisheries. Different workers had dealt with governance comparing fisheries management across countries and stocks. The elements of effective fisheries management are well understood. Strong management means both enacting measures to prevent overfishing and, more importantly, implementing measures to reduce fishing pressure if stocks become depleted. Incidentally, the key practices include evaluating the status of fish and shellfish stocks, designing appropriate management measures to limit fishing mortality, and enforcing these regulations to prevent or reduce negative fishing impacts.

Governance attempted predictions with upside potential Project overview

Despite awareness of what makes up strong fisheries management, there is a lack of publicly available information on the status of fisheries management globally. To address this need, Ray Hilborn and Michael Melnychuk from the University of Washington conducted an analysis to further the understanding of the effectiveness of management approaches for regulating fishing mortality. The study included an independent survey completed by experts of a country's fisheries, providing with an initial rapid assessment of fisheries management and governance systems. A total of *c.* 129 surveys across 28 countries (including, probably, Argentina) were performed, including the 20 countries that catch the most fish in the world. For each country, respondents were presented with a list of 10 species: four with the highest volume of landings, four with the greatest estimated landed value, and the remaining 2–6 species were sampled randomly in proportion to their landings and landed value. The fisheries management surveys had characterized four dimensions of fisheries governance. Within each dimension, between 8 and 14 questions were asked about the use and effectiveness of specific management strategies and tactics employed. Concomitantly, Orensanz and Seijo (2013) dealt with rights-based management in **Latin American** fisheries. For benthic fisheries, there had, probably, been a documentation of management experiences. For finfish fisheries, there was, probably, documentation of rights-based management studies for various aspects. Further, Anon (2005b,c) reported about sustainable fisheries and livelihoods in **Latin America** with reference to the imperative of recognizing the resource access rights of artisanal fishermen, coastal communities, and indigenous people. The rationale of artisanal fisheries is to provide benefits for both current and future generations, and its intrinsic characteristics make it the most effective way to exploit aquatic ecosystems on a sustainable basis. Furthermore, Schonberger and Agar

(2001) dealt with the aspects of management toward rights-based fisheries mode of management, as spearheaded by the Fisheries and Aquaculture Thematic Group in **Latin America** and the Caribbean Region. There had been regular efforts by different organizations for sustainable management of the aquatic systems and their resources. The Fisheries and Aquaculture Thematic Group was established in 1997 to improve portfolio performance by facilitating the coordination of fisheries and aquaculture activities within the fold of the Environmentally and Socially Sustainable Development Network of the Rural Development Department, of the World Bank. This was an informal publication of the Rural Development Family of the World Bank. It was intended to encourage the exchange of ideas among bank staff and all others interested in developmental issues. Incidentally, Eggert and Greaker (2005) studied the effects of global fisheries on developing countries with reference to possibilities for income and threat of depletion. For many developing countries, notably, including **Argentina**, fisheries are often characterized by poorly defined property rights, accompanied by overcapitalization, where too many vessels and fishermen catch too few fish from too small stocks. Further, trade liberalization may have an additional positive impact by promoting the development of property rights in response to increased fish exploitation. There had been studies dealing with fisheries and trade, sometimes, focusing on the developing countries. Fish is globally traded, and for many developing countries, it is an important net export good. In most of these countries, fisheries are often characterized by poorly defined property rights, accompanied by overcapitalization, where, as already stated, too many vessels and fishermen catch too few fish from too small stocks. Management is often *de facto* open access, where vessels with or without permission to fish, sometimes, try to land as much as they can catch due to limited monitoring and enforcement activities. Even in developed countries, many fisheries are poorly managed. Studies done, of late, have indicated that both freshwater (FW) and marine ecosystems are on decline globally. While trade generally is beneficial for growth and welfare, the combination of pure open access and trade liberalization may reduce both welfare and stocks for a country. It is an outcome that could be reinforced by the common use of bad subsidies. Further, trade liberalization may have an additional positive impact by promoting the development of property rights in response to increased fish exploitation. The WTO could play a role by adopting a broader classification of subsidies to help eliminate bad subsidies, such as public support of vessel construction, fuel subsidies, or fishing rights outside developing coastal countries. The WTO may also assist by distinguishing good subsidies, e.g., improving fisheries management or improving monitoring and enforcement, which are desirable targets when rich countries allocate aid resources to developing countries. Concomitant to the above, the **Environment for Development** (ED) initiative was an environmental economic program focused on international research collaboration, policy advice, and academic training, probably, often to guide in sustained utilization and management of the resources. Different centers under its support are in Central America, China, Ethiopia, Kenya, South Africa, and Tanzania, in partnership with the Environmental Economics Unit at the University of Gothenburg in Sweden and Resources for the Future in Washington, DC. Financial support for the program was provided by the Swedish International Development Cooperation Agency (SIDA). In addition to the above, Daniela et al. (2006) dealt with an estimation of compliance of the fisheries of **Argentina** with Article 7 (Fisheries Management) of the UN Code of Conduct for Responsible Fishing. The document represents the best attempt by authors at presenting a fair and objective evaluation of compliance using 44 questions derived from the Code, using a wide range of cited source materials. Concomitantly, Anon (2002) dealt with fisheries subsidies and marine resources management with emphasis on the lessons learned from studies in **Argentina** and Senegal. Such works had aimed to assess the environmental and related social and economic effects of trade liberalization and trade-related policies, as well as develop policy response packages that contribute to mitigating the negative impacts and enhancing the positive impacts identified through the assessments. Notwithstanding the above, Maria Onestini (2002) dealt with the historical background of fisheries subsidies in **Argentine fisheries** in connection with fisheries subsidies and marine resources management and world trade negotiations. In Argentina, fisheries exploitation had subsidized from both international and, to some extent, national sources, making this an intricate part of the issue.

2.3.5.2 Brazil

Brazil possesses high potential for aquaculture and fisheries. It has the large Amazon River basin. McGrath et al. (2004) reported about community fisheries and comanagement on the lower Amazon floodplain of **Brazil**. A loose regional network of community-managed lakes, probably, had proliferated throughout the Amazon floodplain system in response to the growth of Amazon commercial fisheries. This approach is widely perceived as a promising alternative for the sustainable management of floodplain fisheries. The communities, NGOs, grassroots level organizations, and IBAMA (the Brazilian environmental agency), had worked together, over the last decade, to develop a comanagement system for floodplain fisheries based on the legal recognition of community fishing agreements. There had been contributions, which had tried to examine the experience(s) of the Santarém region of the Lower Amazon, which was the major regional experiment in fisheries comanagement. Here, while considerable amount of progress had, perhaps, been made in setting up of a functional comanagement system it, however, had suffered from serious problems that had undermined its effectiveness and seemed to have threatened its long-term sustainability. Unless communities are permitted to restrict access and charged with payment of user fees, it was quite unlikely that the comanagement system could survive once funding for project implementation had terminated. There were, however, legal procedures for making the necessary changes in the design, and if such legal procedures are followed, it, probably, could increase the prospects for the long-term institutional sustainability of the system.

2.3.5.3 *Canada*

Canada has a rich resource of water bodies and associated fisheries. Wibera et al. (2004) worked on participatory research supporting community-based fishery management. They emphasized the engagement of researchers and fishers together in adopting social science approaches to the development of community-based fishers' organizations. Concomitantly, Charles (2008) reported about turning the tide toward community-based fishery management in **Canada's** Maritime. His contribution has discussed a unique nongovernmental initiative, titled, "*Turning the Tide: Communities Managing Fisheries Together*," an initiative supporting community fishery approaches in Canada's maritime provinces, which, probably, had helped to build linkages between native and non-native participants in the fishery. The initiative had been motivated by the desire of the participants for community-based fishery management in general, and in particular the recent opportunity made for the aboriginal First Nations to enter the commercial fishery on a community basis. Such paper(s) described the role played in *Turning the Tide* by three types of activities, viz. (a) workshops, networking, and capacity building, (b) exchange visits, and (c) development of resource materials. Finally, such paper(s) tried to discuss the need for grassroots initiatives to be complemented by governmental policy(ies) and logistic support in order to ensure proper implementation of community-based management.

Further, Charles (2010b) reported on combining environmental, economic, and social sustainability in fisheries with a view on policy options and cautionary notes. Furthermore, Charles (2010b) opined that a key goal of fishery policy is to balance ecological, economic, and social sustainability. However, this goal may not be met through an excessive focus on just one of these aspects, whether ecological, economic, or social. It is important to pursue policy choices which try to achieve ecological sustainability while, at the same time, trying to **actively** promote both social and economic sustainability. Incidentally, a multi-faceted bio-socioeconomic analysis is needed to assess fishery reform options properly, within an integrated sustainable development approach. Concomitantly, more attention is to be paid to economic analysis, which goes beyond the fishery sector, to include economic impacts on regions and coastal communities. Incidentally, the Rights-Based Management (RBM) provides with a set of policy choices that involve a combination of "user rights" (fishing rights relating to access, effort, and/or catch) and "management rights" (relating to comanagement and rights over involvement in decision-making). In general, the arrangement of user rights alone may not be able to solve fishery problems. In fact, they, probably, do so only in conjunction with accompanying management rights (including aspects of fishermen participation, "buy-in" and resulting improved compliance). Moreover, RBM may not be considered as a panacea for fisheries. Incidentally, some RBM choices may have negative impacts on ecological sustainability, e.g., by increasing discards in fisheries, through greater economic incentives for that practice. Further, some RBM choices may have detrimental impacts on social sustainability, particularly through a concentration of rights into fewer hands, and a consequent "withering away" of the coastal communities. Notwithstanding the above, there had been efforts for **sustainable fisheries certification** of **Canadian 4VWX Purse Seine Herring Fishery** (Gudrun et al., 2016). Acoura Marine is in the process of assembling an assessment team comprising members with experience and expertise in the areas of stock assessment, ecosystem interactions, and fishery management appropriate to fishery, in order to assist with the assessment process. The team members have experience in ecosystem interactions and fishery management. The stakeholders are often invited for comments. Concomitant to the above, Tony and Kevern (2002) dealt with the use of ecosystem models to investigate multispecies management strategies for capture fisheries. It could, probably, be used to explain trophic relationships. Ecopath models take all trophic relationships within an aquatic ecosystem into account. Ecosim is the dynamic version of Ecopath. It may look for fishing rates that may maximize ecological, social, or economic goals. Incidentally, Marcelo et al. (2002) dealt with the use of Ecosim to investigate multispecies harvesting strategies for capture fisheries of the **Newfoundland-Labrador shelf**. The model had predicted, in almost all cases, a substantial decrease in all fishing fleets, in order to maximize ecological stability.

2.3.5.4 *The Caribbean*

The Caribbean region harbors a rich diversity of fishes. Anon (2005b,c) dealt with the fishing community initiatives in community-based fisheries management and The Sitka Declaration as a proposal for Forum of Fishermen, NGOs, and the government Agencies in order to explore expansion of the use of sector's, catch share(s), and the role of Community-Based Fisheries Management in the Northeast and the **Caribbean**. The engaged and informed fishermen, probably, now believe that the communities could responsibly manage resources and regain access to diverse fisheries as those stocks were rebuild. In fact, they were, perhaps, working toward this goal with an efficient *modus operandi* or tool, CBFM, an approach where communities directly manage their own resources.

2.3.5.5 *Cuba*

Cuba is gradually progressing in its trade with fish and fisheries. Julio Abraham Baisre Hernández (2006) dealt with **Cuban** fisheries management regime with regard to its current state and future prospects. The study, probably, had reflected that the management regime, perhaps, had been successful, even though there seemed to be other factors influencing the outcome. Many reports also reflect evaluation of the current status of the most important fisheries based on the catch and effort data using various models, notably, the Gordon-Schaefer model. The results, further, seemed to have included suggestions and recommendations for future sustainable management of fisheries in Cuba, including the need for further detailed studies, etc. Further, Doyon (2007) discussed about fishing for the revolution with transformations and adaptations in **Cuban** fisheries. It tried to portray how the national fisheries had been managed since the 1959 revolution, and how the resulting processes of decentralization had occurred in the wake of the 1990s' social, economic, and political crisis. Attempts have been made to examine how large-scale fisheries had been reorganized, both ideologically and practically, as a consequence of these events. Efforts have also been made to explore how large-scale fisheries in the region are connected to an emerging small-scale fishery. Cuban fisheries demonstrate the complexities that arise from both the local context and its links to regional, national, and international processes.

2.3.5.6 *Hawaii*

Hawaii is strive toward community-based fisheries management. Tissot (2005b) reported on integral marine ecology with regard to community-based fishery management in **Hawaii**. This communication points out that successful management of fishery may often warrant a dynamic balance of disciplines to provide a fully integrated approach. There could occur conflict(s) due to multiple use of the ornamental reef-fishery in Hawaii which is community-managed *via* the implementation of a series of marine protected areas and the formation of an advisory council. As such, integral ecology had been used to analyze the conflicts due to such multiple uses. This approach had tried to illustrate how the joyful experiences of snorkelers had resulted in negative interactions with fish collectors and, thereafter, had produced social movements, political will, and ecological change(s). Nevertheless, conflicts reduced and sustainability promoted but a so-called lack of acknowledgment of divergent worldviews, including persistent native Hawaiian cultural beliefs, probably, had contributed to continued conflicts, in some form or the other.

2.3.5.7 *Mexico*

Mexico possesses rich potential for aquaculture and fisheries. Saenz-Arroyo et al. (2005) dealt with community-based marine reserve network in Northwestern **Mexico**. Marine reserves with the involvement of the community stakeholders from the grassroots level, in a participatory mode, could ensure sustained management of the resources. Concomitantly, Anon (2006c) dealt with agricultural and fisheries policies in **Mexico** with a note on the recent achievements and the efforts made in continuing with the reform agenda. There had been reports which tried to analyze the effects of Mexico's ambitious reforms to agricultural and fisheries policies since 1990. There had, probably, also been further recommendations for future reforms. The evaluation was based on the criteria for good agricultural and fisheries policy as agreed upon by the OECD countries. Such criteria, if implemented, could support economically healthy sectors that may contribute to the wider economy. They could also foster the growth of natural resources and may use inputs effectively without resorting to distorting subsidies. Notwithstanding the above, the reforms have given a new direction toward aquatic resources management, particularly, involving the stakeholders from the grassroots level in a participatory mode through the *modus operandi* of community-based management, notably, the Community-based Fisheries Management (CBFM). Incidentally, the vast stretches of community land have been restricting development but, probably, helping in resource management. Further, Enríquez and Camargo (2000) elaborated on the perspectives of rights-based fishery management in **Mexico**. The progress toward better management of fisheries in Mexico, during the earlier times, had been slow, despite increasing concerns. In fact, the principal political, social, and economic forces behind overfishing in the country, perhaps, remained largely at place, during the earlier days. However, among the major problems are the demographic pressure, open access or ill-defined property rights, excessive centralization of management decisions, and a wide array of market and regulatory disincentives for conservation. The works had tried to discuss the potential benefits and limitations of the implementation of rights-based management in order to make attempts to reverse the current overexploitation trends. Many such works, probably, have tried to conclude that property rights-based management, perhaps, has a significant potential to promote efficiency, profitability, and increased stewardship of fishery resources, particularly if implemented at the local community level. However, effective implementation may have to surmount substantial constitutional, political, economic, and cultural barriers. Furthermore, Ramade and Bourillón (2004) discussed about community-based fisheries management and eco-labeling in Baja California, **Mexico**. Moreover, McCay et al. (2014) dealt with cooperatives, concessions, and comanagement on the Pacific coast of **Mexico**. Concomitantly, Poon and Bonzon (2013) dealt with fishing cooperatives, community-based management, and territorial user rights in **Mexico**. Studies revealed that the Mexican case(s), probably, support arguments for greater community-level engagement in catch share and territorial management throughout the Pacific. The study also showed the opportunities and experiences with environmental change(s) in the development of strong and adaptive capacities for comanagement between local organizations and the government agencies. In addition to the above, Guillermo (2002) reported about community-based conservation of the *Callo dehacha* fishery by the Comcaac Indians, Sonora in **Mexico**. The fishery management scholars have suggested, of late, that there is, possibly, a need to develop a better suited small-scale fisheries management approach for developing countries, than the one offered by conventional Western fishery science. This alternative approach is based on the development of better resource access controls, community-based management, and an increased use of local traditional fishery knowledge.

2.3.5.8 *Peru*

Peru possesses high potential in fisheries. The country is also gifted with a portion of the Amazon basin. Pinedo et al. (2000b) reported about community-based natural resource management as a nonlinear process with regard to a case study in the **Peruvian Amazon** *Varzea*. Their study revealed that small communities along the waterways in the Northern Peruvian Amazonia depend on a wide variety of natural resources. Their long-term survival and development require management systems that could assure the future availability and sustainability of these resources.

2.3.5.9 *Trinidad*

Trinidad possesses a high potential in fisheries. Shing (2001b) performed a case study of the integrated coastal fisheries management project, which was a pilot project for the Gulf of Paria in **Trinidad**. The Gulf of Paria is a shallow semi-enclosed basin bounded by the west coast of Trinidad and the east coast of Venezuela. The Atlantic Ocean water enters the Gulf in the south via the constricted Serpent's Mouth and exits through the similarly constricted Dragon's Mouth at the north. The average depth is 20 m. In the Gulf, there seemed to be a clockwise circulation of water in the south. Salinity and circulation were seemed to be influenced by the outflow of rivers, such as the Amazon and the Orinoco, as well as local rivers such as the Caroni and the Oropuche. The volume of sediment brought in by these rivers had given rise to soft, muddy, sandy substrates while the nutrient-rich waters might have contributed to the high productivity of the area, supporting rich and diverse fisheries resources. The fringing mangroves and wetlands along the Gulf provide shelter to juvenile fish and support populations of shellfish such as the oysters, mussels, conch, crabs, and shrimps. In addition, the avian fauna of the wetlands is similarly rich and diverse. This has, possibly, been encouraging > 16,000 ecotourists to visit the Caroni Swamp each year.

Further, > 2500 fishers among the fishing communities operate along the coast of the Gulf of Paria. In 1991, the estimated landed catch from the Gulf of Paria was *c.* 7800 mt at an ex-vessel value of TT$70.5 million. A variety of fishing methods had been used in the marine capture fishery, which include gill nets, bottom long lines, fish pots, hand lines, and trawls. Further, there had been four components of the trawl fleet, out of which, the principal ones are as follows: (a) the semi-industrial trawlers and (b) the double-rigged industrial trawlers with fully mechanized operations. In addition, there are two categories of artisanal vessels characterized by manual operations.

2.3.5.10 *The United States (US)*

The **United States (US)** possesses high potential in aquaculture and fisheries and a strong base for trade with fishes. In the **United States**, Ayres (2009b) studied fishing for solutions in relation to a study of Community-based Fisheries Management and its implementation in the **Gulf of Maine**. His works have revealed that New England had a long-standing history of commercial fishing. However, today, many of the fishing industries are near collapse as a result of overfishing and economic shortfalls. The lobster and ground fish industries are prime examples of fishing industries in trouble, and, perhaps, form the center of discussion on many occasions. There had been works which tried to contrast the current method of fisheries management on the East coast of the **United States** with an alternative approach known as community-based fisheries management (CBFM). Concomitantly, Pablo and Carmen (2012) reported about advancing conservation issues with innovative ideas for the future generations as a community-based approach to the management of an aggregating reef fishery in the **Meso-American** reef. Conservation scientists advocate for the establishment of Marine Protected Areas (MPAs) as a global precautionary approach to the sustainability of world fisheries. **Incidentally**, Palmigiano and Crawford (2006) dealt with establishing community-based marine protected areas for cockle fishery management.

Notwithstanding the above, different workers seemed to have opined for the use of small-scale no-take zones known as community-based marine protected areas (CB-MPA) for the purpose of improved management of cockle fisheries. Cockles are important resources, especially in coastal and island populations, where they provide nutrition and avocation to thousands of families worldwide. Cockles are mainly gathered by gleaning, which is often done by women and children, probably, without mechanical assistance. In many parts of the world, expanding population and the opening of commercial markets have led to increased collection of cockles. The increased rate of collection, then, probably, led to overharvesting and depleting trend of stocks and catches. This, consequently, had also resulted in declining income for harvesters and diminishing food availability for the coastal communities.

There have been contributions, which provide information on CB-MPAs that could help prevent the decline in cockle populations and improve fisheries production. In fact, CB-MPAs had been used extensively in several regions of the world to help restore and protect wild shellfish. Various publications have tried to briefly explain the biology of important species of coral reef and mangrove cockles. Further, the publications also tried to explain the functions of the CB-MPAs, including the process of setting up, managing, monitoring, and evaluating a CB-MPA, and, finally, some case studies from countries which had set up successful CB-MPAs. Concomitant to the above, Kathy Sullivan (2014) dealt with exploration of ecosystem based fishery management in the **United States**. A concise report by the Ecosystem Sciences and Management Working Group (ESMWG), being requested by the National Oceanic and Atmospheric Administration (NOAA) Science Advisory Board (SAB), assessed the progress toward implementation of Ecosystem-Based Fishery Management (EBFM) in the United States, as well as the status and use of ecosystem science in management. Further, Anon (2007b,c) dealt with natural resources with regard to their conflicts, and conflict resolution. The World Bank has defined natural resources as "materials that occur in nature and are essential or useful to humans, such as water, air, land, forests, fish and wildlife, topsoil, and minerals, etc." Incidentally, community-based management (CBM) involving the community stakeholders from the grassroots level could be a potential method for resolving various types of conflicts with resources. Incidentally, Anon (1999b) dealt with the role of the major groups in the maintenance of sustainable oceans and seas. Incidentally, it may be mentioned here that sustainability of resources is, perhaps, an important issue in the management of resources. Moreover, it may, further, be pointed out here that the farming community is intimately associated with the resources. However, the farmers are, perhaps, not traditionally associated with the oceans, but "ocean farming" (often, popularly known as "aquaculture" or "mariculture"), is, probably, one of the most rapidly expanding industries in both the North and the South. It is an important component of sustainable fisheries management offering, at its best, increased food security in the face of declining wild fish stocks. Potential solutions to the urgent need of developing sustainable aquaculture are, probably, illustrated by case studies on traditional shrimp farms in Hong Kong and village-based clam mariculture in the Solomon Islands. Furthermore, Anon (2007b,c) dealt with the evaluation of incentive-based management with regard to sustaining **America's** fisheries and fishing communities. There had been following recommendations: (i) **To implement catch shares**: Catch shares, when well designed, are said to be one of the key components in successful fisheries management and to be implemented more widely in order to build sustainable fisheries and vibrant fishing communities. (ii) **To ensure effective design. (iii) To educate stakeholders on catch share programs and options:** Stakeholders are to plan on the basis of this detailed body of knowledge. They are also to design their strategies to tackle the frequently encountered embargos. They could also buildup successful regulations to make catch share systems that maximize benefit and minimize transition costs for their fisheries. (iv) **To improve efficiency of the designing process(es):** It may be noted here that program design(s) may, preferably, be done by small groups of representative stakeholders with clear instructions from state and federal managers as to time schedule and goal(s) for decision-making. Another critical element is, probably, the credible conflict-of-interest issue among the members of the design committee. (v) **To prioritize funding for catch share design process.** In light of their ability to help fisheries meet multiple objectives, funding pattern is to be prioritized. (vi) **To invest in future:** Some of the enhanced values made by the catch shares are to be reinvested for the welfare of the fishing communities, or could be reinvested for the advancement of any promising thrust area in the fisheries domain. Incidentally, the new revenues earned could help in continuing with running of the catch share systems, could improve data generation, or could assist in fulfilling the objective(s) related to social welfare of the fishing communities, as well as could enhance the levels of monitoring, enforcement, and research.

Incidentally, Soeftestad (1999) dealt with the report from the International **Community-Based Natural Resource Management** (CBNRM) Workshop held at Washington DC in May, 1998. This workshop was organized because of the growing interest in community-based natural resource management (CBNRM) as a third alternative to command and control market-based solutions to natural resource management (NRM). Notwithstanding the above, Sale et al. (2005) dealt with critical science gaps, which impede the use of no-take fishery reserves. No-take fishery reserves, in addition to serving as valuable biodiversity conservation roles, probably, protect a portion of the fishery stock as insurance against future overfishing. So long as there is adequate compliance by the fishing communities, it is likely that they may also sustain and even enhance fishery yields in the surrounding water bodies. In addition, Iverson (2013) dealt with the opportunities and challenges with regard to the management of **US** regional fishery. Yellow goatfish (*Mulloidichthys martinicus*) school over a coral reef in the bright clear waters of the Caribbean. Goatfish disperses at night in search of food, using their chemosensory barbels ("chin whiskers") to detect worms and other small invertebrates in the sediments. In fact, the yellow goatfish are, probably, not a target of any fishery. However, the Caribbean Fishery Management Council, probably, works to protect the reef habitat for these and other reef dwellers. Moreover, a similar species yellowfin goatfish (*Mulloidichthys vanicolensis*) is a significant commercial fish that is highly esteemed as food in the US Pacific Islands. It is known in Hawaii as weke "ula," in American Samoa as "vete," and in Guam and the Commonwealth of the Northern Mariana Islands as "satmoneti" (Chamorro language) and "wichigh" (Refaluwasch language), respectively. Incidentally, the fisheries for yellowfin goatfish are regulated by the Western Pacific Fishery Management Council as a component of the Fishery Management Plan for Coral Reef Ecosystems. Concomitantly, Donald et al. (2006) dealt with fisheries management plan for the **Minnesota** waters of Lake Superior. The Lake Superior fish community have undergone dramatic changes since the mid-1900s due to overfishing, introduction of non-native species, pollution, and land use changes in the watershed. The Lake Superior fish community, probably, has become much more complex, since the 1950s, and is now, probably, composed of both native and exotic species. The most devastating introduction to the Lake Superior community, probably, had been the sea lamprey, which virtually had eliminated the lake trout in all but a few isolated areas of Lake Superior. Incidentally, rehabilitation efforts, since the 1960s, including sea lamprey control, harvest regulations, and stocking programs, along with stricter pollution standards and best management practices for land use, had led to the restoration of healthy fish stocks throughout much of Lake Superior.

2.3.5.11 Summary

(1) **America** is a continent potentially rich in aquaculture and fisheries. Therefore, there is, possibly, a requirement for effective *modus operandi* for the sustained harvest, efficient management, proper utilization and effective conservation of the aquatic bioresources, notably, the fishes. Participatory community-based management with the involvement of the stakeholders from the grassroots level could be an effective mode of efficient management, accompanied with sustained harvest and proper conservation of the coveted resources.

(2) Significant works pertaining to CBFM in American continent include those of Anon (n.d.), Orensanz, and Seijo (2013), Anon (2005b,c), Schonberger and Agar (2001), Eggert and Greaker (2005), Daniela et al. (2006), Anon (2002), and Maria Onestini (2002) in **Argentina**; McGrath et al. (2004) in **Brazil**; Wibera et al. (2004), Charles (2008), Charles (2010a,b), Gudrun et al. (2016), Tony and Kevern (2002), and Marcelo et al. (2002) in **Canada**; Anon (2005b,c) in **The Caribbean**; Hernández (2006) and Doyon (2007) in **Cuba**; Tissot (2005b) in **Hawaii**; Saenz-Arroyo et al. (2005), Anon (2006c), Enríquez and Camargo (2000), Ramade and Bourillón (2004), McCay et al. (2014), Poon and Bonzon (2013), and Guillermo (2002) in **Mexico**; Pinedo et al. (2000b) in **Peru**; Shing (2001b) in **Trinidad**; Ayres (2009b), Pablo and Carmen (2012), Palmigiano and Crawford (2006); Sullivan (2014), Anon (1999b, 2007b,c), Soeftestad (1999), Sale et al. (2005), Iverson (2013), and Donald et al. (2006) in the **United States**.

Suggested reading

Asia

Abdullah, M.K., 2007. Fisheries co-management policy brief: findings from a worldwide study. WorldFish Center, Penang, Malaysia. 26 pp.

Action for Food Production (AFPRO), 2010. Report on Audit of the implementation of the Environmental Management Framework (EMF) and the Resettlement and Rehabilitation Policy for Implementation of Environmental Monitoring Framework. pp. 208.

Agarwala, N.K., 1992. Limnology and fish productivity of Tamranga Wetland in Bongaigaon District of Assam (India) with special reference to some productivity indicators. Unpublished Ph.D. Thesis, Gauhati University, pp. 200.

Ahmed, M.M., Dickson, M., 2007. Institutional Issues in the CBFM-2 Project, Conference Paper 02, pp. 2–12. The WorldFish Center, Bangladesh and South Asia Office, Banani, Dhaka-1213, Bangladesh; and, 1 Pathfoot Drive, Bridge of Allan, Stirling FK9 4SD, UK, pp. 1–345.

Ahmed, M., Viswanathan, K.K., Valmonte-Santos, R.A., 2004. Collective action and property rights in fisheries management. In: Meinzen-Dick, R., Di Gregorio, M. (Eds.), Collective Action and Property Rights for Sustainable Development. International Food Policy Research Institute (2020 Focus, no. 11, Brief no. 7).

Ali, A.H., 1997. Fishing gears used in the *Beel* fisheries of Brahmaputra Valley of Assam. Unpublished D.F.Sc. Dissertation, Cent. Inst. Fish. Education, Mumbai.

Amarasinghe, U.S., Wijenayake, W.M.H.K., 2015. Results of a decade of R and D efforts on culture-based fisheries in Sri Lanka. In: De Silva, S.S., Ingram, B.A., Wilkinson, S. (Eds.), Perspectives on culture-based fisheries developments in Asia. Network of Aquaculture Centres in Asia-Pacific, Bangkok, Thailand, pp. 59–71. pp. 110.

Anon, 1997. Reservoir Fisheries Management in Savannakhet Province, Lao PDR. Final Report, Fisheries Management Science Programme, Overseas Development Administration, MRAG Ltd. August 1997, pp. 1–46.

Anon, 1999a. Study on Fisheries Resources and Fishing Operations in Hong Kong Waters. Legislative Council Panel on Economic Services Consultancy, pp 1–5.

Anon, 2004a. India: Good Management Practice in Sustainable Fisheries: Policy Briefs. Published by World Bank, pp. 1–3.

Anon, 2004b. Kacharidal *Beel* Prakalpa. Leaflet. In: Published by District Administration. Darrang, Assam.

Anon, 2005a. Wetland Conservation in Myanmar. S:WorkingDocs\ Species\CentralAsianFlyway\Delhi 2005\Docs_CAF\inf_04_14_Myanmar.doc, pp. 1–4.

Anon, 2006a. Report of the Working Group of Fisheries for the Eleventh five year plan: 2007–2012. Government of India: Planning Commission, pp. 1–259.

Anon, 2007a. Scaling-up Community-based Co-Management of Wetlands and Fisheries in Bangladesh. Management of Aquatic Ecosystems through Community Husbandry, Policy brief, pp. 1–4.

Anon, 2009a. Ganjakhuwari *Beel*. A report submitted by Club Synthesis, Hailakandi to PCU, ARIAS Society, Khnapara, Guwahati-22, Assam.

Anon, 2010a. Liquid assets: community based fisheries management in Bangladesh. World Fish Center, UKAid, DFID (Department for International Development); pp. 1–5.

Anon, 2010b. India Marine Fisheries: Issues, Opportunities and Transitions for Sustainable Development. Agriculture and Rural Development Sector Unit, South Asia Region, pp., 1–124.

Anon, 2011a. Review of National Laws and Legislations in the Philippines in Relation to Ecosystem Approach to Fisheries Management (EAFM). Palawan State University, Centre for Strategic Policy and Governance, Honolulu, Hawaii, pp. 1–94. The USAID Coral Triangle Support Partnership.

Anon, 2011b. Possible approaches in Community Based Fisheries Management. In: FAO. 1–2 pp.

Anon, 2011c. Draft Final Comprehensive Legal statute in the form of Assam Panchayat (Fishery) Rule 2010. Pp. 53.

Anon, 2012. National Fisheries Strategy (2012–2025). Republic of Yemen. UNDP and IFAD, Ministry of Fish Wealth. 1–42 pp.

Anon, 2015a. Summary of culture-based fisheries developments in Lao PDR. In: Network of Aquaculture Centres in Asia-Pacific, pp. 47–51.

Anon, 2015b. PaFF Programme: Support to Community Forestry and Fisheries in Kratie and Stung Treng Provinces. Scaling up under a new Partnership Programme with Switzerland, Support to Community Forestry and Fisheries in Kratie and Stung Treng Provinces. Organised by: The Non-Timber Forest Products Exchange Programme Asia (NTFP-EP), in association with Representatives from the Partners for Forestry and Fisheries Cambodia (PaFF) Programme together with delegates from the provincial governments of Kratie, Stung Treng and of the Forestry and Fishery Administrations and the Swiss Agency Cooperation Office in Cambodia, pp. 1–4.

Anon, 2016. Blue Peace in the Middle East: Mekong learning journey in Cambodia and Lao PDR. Strategic Foresight Group and Mekong River Commission: supported by Swedish International Development Cooperation Agency (Sida), pp. 1–60.

Anuchiracheeva, S., Demaine, H., Shivakoti, G.,.P., Ruddle, K., 2003. Systematizing local knowledge using GIS: Fisheries Management in Bang Saphan Bay, Thailand. Ocean Coast. Manag. 46, 1049–1068.

Ashaletha, S., et al., 2001. Changing roles of fisherwomen of India-Issues and Perspectives. CMFRI, Kochi.

Baavinck, M., 2001. Marine Resource Management: Conflict and Regulation in the Fisheries of the Coromandel Coast. Sage Publications, India.

Babare, M.G., Sawant, P.L., 2009. Fishermen around Makni Reservoir, Mharastra. Socio-Econ. Status Fish. Chimes 29 (2), 21–22.

Baird, I., 2000. Integrating Community-Based Fisheries Co-Managementand Protected Areas Management in Lao PDR: Opportunities for Advancement and Obstacles to Implementation. IIED, Evaluating Eden Series Discussion Paper No. 14, pp. 1–22.

Bal, D.V., Rao, V.K., 1984. Fishing industry and cooperative societies, Marine Fisheries. Tata Mcgraw-Hill Publishing Compant Limited, New Delhi.

Barbhuiya, A.H., Kar, D., 2009. Status of *Tor tor* mahseer in river Barak in North-East India with a note on their morphometry and meristics. J. Natcon 21 (2), 239–246.

Barbhuiya, A.H., Das, B., Darlong, L., Tarafdar, R.G., Sharma, R., Sharma, S.S., Datta, S., Das, B., Barman, R., Deb, S., Saha, B., Kar, D., 2009. Fish Biodiversity in certain rivers of Tripura. Environ. Ecol. 27 (1), 222–227.

Barman, R.C., 2008. Fish seed raising in flood plain wet lands of Assam through Aquaculture. Fish. Chimes 28 (8), 30–31.

Barman, R.C., 2009. Community-Based Rural Aquaculture and Fisheries Management for Effective Utilization of Water Resources in North East Region of India. Study material under the course on Practical Orientation on Inland Fisheries Development Plan (19–23 October, 2009) conducted by National Institute of Rural Development, Rajendranagar, Hyderabad, India, pp. 119–132.

Barman, R.C., Kar, D., 2014. Concept, definition and approaches of Community-Based Wetland Fisheries Management. pp: 291–310, In: Kar, D. (Ed.), Research Frontiers in Wetlands, Fishery and Aquaculture, xi + 320. Dominant Publishers and Distributors Pvt. Ltd., New Delhi. ISBN: 978-93-82007-39-5, ISBN 10: 93-82007-39-3.

Barman, R.C., Dana, S.S., Bhatacharya, B.K., Kar, D., Barbhuiya, A.H., 2008. Effect of situational variables on knowledge level of Fishermen for sustainable development of Beel fisheries. Environ. Ecol. 26 (4B), 2092–2094.

Barman, R.C., Kar, D., Dana, S.S., Majumdar, D., 2012. Role of availability of critical aquaculture inputs and community-managed markets on community-based floodplain wetland management. J. Appl. Nat. Sci. 4 (2), 164–166.

Baruah, U.K., Bhagwati, A.K., Talukdar, R.K., Saharia, P.K., 2000. *Beel* fisheries of Assam: Community-based co-management imperative. NAGA, ICLARM Quart. 23 (2), 36–41.

Berkes, F., 2006. From community-based resource management to complex systems. Ecol. Soc. 11 (1), 45.

Bhatta, R., 2003. In: Kumar, A., Katiha, P.K., Joshi, P.K. (Eds.), Socio-economic Issues in Fisheries Sector in India: A Profile of People, Technologies and Policies in Fisheries Sector in India. published by National Centre for Agricultural Economics and Policy Research, New Delhi, India.

Bhattacharjya, B.K., 2002. Trophic structure and fishery potential of selected floodplain wetlands (Beels) of Brahmaputra Valley. Ph.D. Thesis, Cent. Inst. Fish. Education, Mumbai, pp. 216.

Bhattacharjya, B.K., 2003a. Enhancement tools for augmenting fish production from flood plain wet lands. In: Vinci, G.K., Jha, B.C., Bhaumik, U., Mitra, K. (Eds.), Fisheries management of floodplain wet lands in India. CIFRI, Barrackpore, pp. 34–41. Bull No. 125.

Bhattacharjya, B.K., 2003b. Bilor abestonit meen palan (Pen culture in *Beel* in Assam). Bull No. 116, CIFRI, Barrackpore, p. 33.

Bhattacharjya, B.K., 2008. Management of *Beel* fisheries in Assam. In: Souvenir. Department of Fisheries, Government of Assam, pp. 10–15.

Bhaumik, U., 1999. Role of extension for growth of riverine fisheries sector in India. In: Paper presented at the summer school (14th July to 12 August, 1999) on Ecology, fisheries and fish stock assessment in Indian rivers. CIFRI (ICAR), Barrackpore, pp. 318–325.

Bhaumik, U., Saha, S.K., 1994. Perspective on socio-economic status of fishermen engaged in fishing in the estuaries of Sundarban. Environ. Ecol. 12 (1), 181–185.

Biju Kumar, A., Deepthi, G.R., 2006. "Trawling and by-catch: Implications on marine ecosystem", Research Department of Zoology, NSS College, Pandalam 689 501, India. Curr. Sci. 90 (7), 10.

Bimola, M., Singh Shashikumar, O., Kar, D., 2014. A preliminary study on a collection of Rotifers from Thongjaorok River at Bishnupur, Manipur, India. Uttar Pradesh J. Zool. 34 (3), 65–75.

Bui The Anh, 2015. Culture-based fisheries in Northern Vietnam: Status and perspective. p. 122, In: De Silva, S.S., Ingram, B.A., Wilkinson, S. (Eds.), Perspectives on culture-based fisheries developments in Asia. NACA Monograph Series No. 3, pp: 126.

Chakraborty, T.R., 2004. Management of haors, baors, and beels in Bangladesh: Lessons for Lake Basin management. IUCN, Bangladesh, Dhaka, pp. 1–23. Center for Natural Resource Studies.

Chandra, G., 2011. Management Regimes and Institutional Arrangement in Floodplain Wetlands Fisheries of Assam: An Evaluation. Ind. J. Extens. Educ. 47 (1 and 2), 27–33.

Charles, T., 2010a. Combining Environmental, Economic and Social Sustainability in Fisheries: Policy Options and Cautionary Notes. A Report of Saint Mary's University. Halifax, Canada.

Chaudhuri, H.L., Banerjee, S.M., 1965. Report on the fisheries of Manipur with special reference to the development of Takmu *Beel* area of Loktak Lake. Misc. Contr. No. 4. Cent. Inland Fish. Res. Inst, Barrackpore (India). pp. 29.

Cheong, S.M., 2004a. Managing fishing at the local level: The role of fishing village cooperatives in Korea. Coast. Manage. 32 (2), 191–201.

Cheong, S.M., 2004b. Korean fishing communities in transition: limitations of community-based resource management. Environ. Plan. A. 37, 1277–1290.

Chitranshi, V.R., 1992. Pen culture in floodplain lakes of the Gandak basin—A case study. In: Yadava, Y.S., Sugunan, V.V. (Eds.), Compendium of the FAO sponsored training programme on Floodplain fisheries management. Central Inland Capture Fisheries Research Institute, Barrackpore (India), pp. 170–175.

Choudhary, K.M., 1979. Group farming in India—The Gambhira Experience. In: Wong, J. (Ed.), Group Farming in Asia: Experience and potentials. Singapore University Press.

Choudhury, M., 1989. Resource exploitation in *Beels*. In: Compendium of the training on management of ox-bow (*Beel*) fisheries. Central Inland Capture Fisheries Research Institute, Barrackpore, (India), pp. 102–108. Bull No. 63.

Choudhury, M., 1992a. Fishing methods in floodplain lakes. In: Yadava, Y.S., Sugunan, V.V. (Eds.), Compendium of the FAO Sponsored training programme on Floodplain fisheries management. Central Inland Capture Fisheries Research Institute, Barrackpore (India), pp. 163–166.

Choudhury, M., 1992b. Fish landings and catch composition in floodplain lakes. In: Yadava, Y.S., Sugunan, V.V. (Eds.), Compendium of the FAO Sponsored Training Programme on Floodplain Fisheries Management. Central Inland Capture Fisheries Research Institute, Barrackpore (India), pp. 160–162.

Chowdhury, I.H., Kamal, M.M., Haque, N., Islam, M.N., Akter, S., 2012. Patterns of co-operation in community based fishery management: a sociological study on the people of Hail Haor. Asian Soc. Sci. 8 (15), 8–15.

Clarke, S., Leung, A.W.-Y., Mak, Y.M., Kennish, R., Haggan, N., 2002. Consultation with local fishers on the Hong Kong artificial reefs initiative. ICES J. Marine Sci. 59, 171–177.

Crane, C., Draper, D., 1996. Socially Cooperative Choices: An approach to achieving resource sustainability in the coastal zone. In: Environmental Management.

Crawford, B.R., Siahainenia, A., Christovel, R., Asep, S., 2004. Compliance and enforcement of community-based coastal resource management regulations in North Sulawesi, Indonesia. Coast. Manage. 32 (1), 39–50.

Crean, K., 1999. Centralised and Community Based Fisheries Management Strategies, case studies from two fisheries dependent archipelagos. Mar. Policy 23, issue3.

Damodaran, A., 2003. Property Rights, Management Plans and Dynamics of Sustainability in the Coastal Kerala. In: Multidisciplinary Scientific Conference on Sustainable Coastal Zone Management, "Rights and Duties in the Coastal Zone", Stockholm, Sweden.

Dana, S.S., 1987. A study of factor effecting the attitude of livestock owners towards artificial insemination in cattle. An unpublished M. V. Sc. Thesis submitted to Indian Veterinary Research Institute, Izatnagar, Utter Pradesh, pp. 127.

Das, P., Kar, D., 2013. Studies on zooplankton diversity and physico-chemical parameters of Ramnagar annua, Assam. Int. J. Curr. Res. 5, 3058–3062.

Das, U., Kar, D., 2014. Qualitative and Quantitative Study of Planktons including Physico-chemical Properties of Water of Assam University Pond, Silchar, Assam, India. Research Frontiers in Wetlands, Fishery & Aquaculture. In: Kar, D. (Ed.), Research Frontiers in Wetlands, Fishery and Aquaculture. Dominant Publishers and Distributors Pvt. Ltd., New Delhi, pp. 247–254. xi + 320. ISBN: 978-93-82007-39-5, ISBN 10: 93-82007-39-3.

Das, B.K., Singh, N.R., Dutta, B., Kar, D., 2013. Length weight relationship of *Labeo rohita* (Hamilton-Buchanan) and *Labeo gonius* (Hamilton-Buchanan) from Sone Beel, the biggest wetland of Assam, India. J. Environ. Res. Develop. 8 (4), 6613–6618.

Das, B.K., Boruah, P., Kar, D., 2015f. Fish diversity and drainage analysis of River Siang, East Siang District of Arunachal Pradesh. Proceeding of the National Conference on Conservation of Natural Resources and Biodiversity for Sustainable Development -2014. J. Biosci. Discov. (JBSD) 6 (1-I Special), 16–20.

Das, B.K., Ghosh, A., Boruah, P., Kar, D., 2015b. Studies on the Physico-Chemical Parameters of water and Drainage analysis using RS and GIS Technology of River Siang in Arunachal Pradesh. In: Biodiversity Conservation and Sustainable Use. Print World Print and Publication, Jorhat, Assam, India, pp. 197–204.

Das, B.K., Singh, N.R., Boruah, P., Kar, D., 2015c. Fishing devices of the River Siang in Arunachal Pradesh, India. J. Fish. 3(2), 251–258.

Das, B.K., Singh, N.G., Das, S., Kar, D., 2015g. Length-weight relationship (LWR) of *Channa striata* (Bloch, 1793) of River Siang, Arunachal Pradesh. Int. J. Environ. Nat. Sci. 7, 1–6 (ISSN 2349-3763).

Das, B.K., Boruah, P., Kar, D., 2015h. Ichthyofaunal diversity of Siang River in Arunachal Pradesh, India. In: Proceedings of the Zoological Society (Springer)., https://doi.org/10.1007/s12595-015-0155-6 (ISSN: Print: 0373-5893; Electronic: 0974-6919).

Das, N., Das, B.K., Das, S., Singh, N.R., Kar, D., 2015a. Preliminary studies on Ichthyofauna diversity and Physico-Chemical Parameters of River Bhogdoi during winter and Pre-Monsoon Season in Assam. IOSR J. Agricult. Veterinary Sci. 8 (6 ver. I), 35–41.

Das, S., Das, T., Das, B.K., Kar, D., 2015i. Study of fish diversity and physico-chemical characteristics of River Singla. Int. J. Environ. Sci. 5 (6), 881–887 (ISSN 0976-4402).

Das, S., Das, B.K., Kar, D., 2015d. Length-weight relationship and relative condition factor in *Xenentodon cancila* of River Mat in Mizoram. Int. J. Biol. Biolog. Sci. 4 (2), 27–31 (ISSN 2327-3062).

Das, S., Das, T., Das, B.K., Kar, D., 2015j. Length-weight relationship and condition factor in *Channa punctata* of River Manu in Tripura. Int. J. Fish. Aquat. Stud. (IJFAS) 2 (3), 56–57.

Datta, S.K., Pal Singh, S., Chakrabarti, M., Biswas, S., Sah, B., 2010. A Perspective on Fisheries Sector Interventions for Livelihood Promotion. W.P. No. 2010-02-03, pp. 1–32.

De, D.K., Mukherjee, A., 1992. Fish breeding and recruitment in floodplain lakes. In: Yadava, Y.S., Sugunan, V.V. (Eds.), Compendium of the FAO sponsored training programme on Floodplain fisheries management. Cent Inland Capture Fisheries Research Institute, Barrackpore (India), pp. 167–169.

De Silva, S.S. (Ed.), 2001a. Reservoir and culture-based fisheries: Biology and Management. Proceedings of an International Workshop held in Bangkok, Thailand from 15–18 February 2000. ACIAR Proceedings No. 98. pp. 1–384.

De Silva, S.S., 2001b. Reservoir fisheries: broad strategies for enhancing yields. In: Reservoir and culture-based fisheries: biology and management. Proceedings of an International Workshop held in Bangkok, Thailand from 15–18 February 2000. Australian Centre for International Agricultural Research (ACIAR), pp. 7–15. GPO Box 1571, Canberra Proceedings No. 98, 390 pp.

De Silva, S.S., Song, S.L., 2015. Challenges and constraints for developing CBF in Cambodia and a possible strategy for success, pp. 51-57. In: De Silva, S.S., Ingram, B.A., Wilkinson, S. (Eds.), Perspectives on culture-based fisheries developments in Asia. NACA Monograph Series No. 3, pp. 126.

De Silva, S.S., Phillips, M.J., Mohan, C.V., 2007. A different form of dumping: The need for a precautionary approach for yet another new species for shrimp farming in Asia. In: Network of Aquaculture Centres in Asia-Pacific (2007) Aquaculture Asia. vol. 12, No. 4. Network of Aquaculture Centres in Asia-Pacific, Bangkok, Thailand, pp. 3–5 (0859-600X).

Deka, T.K., 1999. Present status of *Beel* fisheries and its impact on the socio-economic condition of fishermen community of Assam. Unpublished Ph. D. Thesis, Gauhati University, Assam, pp. 122.

Deka, T.K., Kakati, M., Goswami, M.M., 2001. Diversity of wetland fish and its impact on the income of fishermen community of Assam. J. Ind. Fish. Ass. 28, 11–22.

Dey, S.C., 1973. Studies on the Distribution and Taxonomy of the Ichthyofauna of the hill streams of Kamrup-Khasi-Garo Regions of Assam with special reference to the Functional morphology of some rheophillic Fishes, D.Sc. Thesis, xi + 299, University of Calcutta, India.

Dey, S.C., 1981. Studies on the hydrobiological conditions of some commercially important lakes (Beels) of Kamrup district of Assam and their bearing on Fish production. pp. 177, 18 Figs., 8 pls., Final Technical Report, North-Eastern Council, Govt. of India (Shillong).

Dey, M.M., Kanagaratnam, U., 2007a. Community based management of small scale fisheries in Asia: Bridging the gap between fish supply and demand. World Fish Center, Penang, Malaysia.

Dey, M.M., Kanagaratnam, U., 2007b. Community based management of small scale fisheries in Asia: Bridging the gap between fish supply and demand. WorldFish Center, PO BOX 500, GPO, 10670, Penang, Malaysia. Conference Paper 23, pp., 1–13.

Dey, S.C., Kar, D., 1987. Physico-chemical complexes of water and soil in Sone, an ichthyologically potential tectonic lake of Assam. J. Assam Sci. Soc. 30 (1), 1–11.

Dey, S.C., Kar, D., 1989a. Aquatic macrophytes of Lake Sone in Assam. Environ. Ecol. 7 (1), 253–254.

Dey, S.C., Kar, D., 1989b. Fishermen of Lake Sone in Assam: their socio-economic status. Sci. Culture 55, 395–398.

Dey, S.C., Kar, D., 1989c. An account of Hilsa ilisha (Hamilton) of Lake Sone in the Karimganj district of Assam. Bangladesh J. Zoology 17 (1), 69–73.

Dey, S.C., Kar, D., 1990. Fish Yield Trend in Sone, a tectonic lake of Assam. Matsya 15–16, 39–43.

Dey, S.C., Kar, D., 1994. Phytoplankton dynamics of Lake Sone in Assam. Bull. Life Sci. IV, 49–54.

Dickson, M., 2007. The CBFM- project: proving the case for community based and co-management of fisheries in Bangladesh. In: CBFM-2 Project. Bangladesh, The WorldFish Center Bangladesh and South Asia Office, Dhaka.

Dickson, M., Brooks, A. (Eds.), 2007a. CBFM-International Conference on Community Based Approaches to Fisheries Management. The WorldFish Center Conference Proceedings 75, Print: 38p, CD-ROM: 337p.

Dickson, M., Brooks, A., 2007b. CBFM-2 International Conference on Community-Based Approaches to Fisheries Management, Conference Proceedings 75. In: The WorldFish Center Conference Proceedings 75.

Divakarannair, N., 2007. Livelihood Assets and Survival Strategies in Coastal Communities in Kerala, India. Phd Thesis, Department of Geography, University of Victoria.

Feidi, I.H., 1998. Fisheries Development in the Arab World. Food and Agriculture Organization of the United Nations (FAO), Rome, Italy. Middle East. Nat. Environ., Bull. 103, 388–406.

Flewwelling, P., 2001. Fisheries Management and MCS in South Asia. FAO/FISHCODE Project, GCP/INT/648/NOR: Field Report C-6 (En) Rome, FAO: 56 p.
Gadgil, M., Kar, D., et al., 2000. Conserving India's Biodiversity: Let People Speak, pp. 583–589. In: Setting Biodiversity Conservation. World Wide Fund (WWF) for Nature-India, New Delhi. pp. 707.
Gazi, M., Islam, N., Yew, T.S., 2013. Property rights and access: the case of community based fisheries management in Bangladesh. J. Agric. Sci. 5 (6), 39–55.
Ghani, A., Ahmed, N., 2008. Two case studies on water bodies and Livelihoods of haor Area. Campain for sustainable Rural Livelihoods CSR Haor Group.
Gomes, A., 2007. Empowerment of women through fisheries and aquaculture. In: Dickson, M., Brooks, A. (Eds.), Proceedings of the CBFM-2 International Conference on Community Based Approaches to Fisheries Management, Dhaka, Bangladesh, 6–7 March 2007. The World Fish Center, Bangladesh Office, pp. 1–4.
Goswami, M.M., 1985. Limnological Investigations of a tectonic lake of Assam, India and their bearing on Fish production. pp. 395, Unpublished PhD Thesis, Gauhati University (Assam).
Goswami, B., 2003. Contribution of women in fishery sector. An unpublished M. F. Sc. Thesis submitted to the West Bengal University of Animal and Fishery Sciences. Mohanpur, West Bengal, pp. 58.
Goswami, P.K., Devaraj, M., 1993. Potential yield of Labeo rohita from a floodplain lake (Dhir *Beel*) in Assam, India. J. Inland Fish. Soc. India 25 (2), 12–17.
Goswami, M., Sathiadhas, R., 2000. Fish farming through community participation in India. Naga, ICLARM Quart. 23 (3), 29–32.
Government of Assam, 2006. Statistical Handbook of Assam. Directorate of Economics and Statistics, Government of Assam, Guwahati, pp. 110.
Gracy, M.M., 1998. Impact of technological advancement on socio-economic conditions of women in fisheries, Kerala. In: Hameed, M.S., Kurup, B.M. (Eds.), Technological Advancements in Fisheries. School of Industrial Fisheries, CUSAT.
Grafton, Q.R., et al., 2006. Economics for Fisheries Management, The Economics of Fishing and Fisheries Economics. Ashgate Publishing Limited, Hampshire, England.
Gurung, T.B., 2007. Restoration of small lakes through co-operative management: A suitable strategy for poverty-laden areas in developing countries. Lakes Reserv. Res. Manage. 12, 237–246.
Halfdanardottir, J., 1993. "Social Mobilization in Kerala Fishers", Priests, Unions, and Political Parties. MAST.
Hall, S., 2007. Key note paper of the CBFM International Conference. In: CBFM-2: International conference on community based app.roaches to fisheries management, 6–7 March 2007, conference proceeding -75. pp. 06.
Hannesson, R., 2004. Fishermen's Organisations and their role in Fisheries Management: Theoretical consideration and experiences from Industrialised countries. The Norwegian School of Economics, and Business Administration, Bergen, Norway.
Hari Babu, T.P., Rajan, J.B., 2006. Need for Fisheries Economic Zone (FEZ). Kerala Calling.
Henocque, Y., Tandavanitj, 2011. Fisheries and co management in practice, Thai-land. In: Climate of Coastal Cooperation, Coastal & Marine Union, EUCC, the Netherlands. pp. 102–103.
Hilborn, R., 2005. Are Sustainable Fisheries Achievable? Chapter 15, pp. 247–259, Norse and Crowder.
Hilborn, R., 2008. Knowledge on how to achieve sustainable fisheries, pp. 1-12. In: Tsukamoto, K., Kawamura, T., Takeuchi, T., Beard Jr., T.D., Kaiser, M.J. (Eds.), Fisheries for Global Welfare and Environment. 5th World Fisheries Congress 2008, pp. 45–56.
Hossain, M.I., Siwar, C., Mokhtar, M.B., Dey, M.M., Jaafar, A.H., 2010. Impact of community based fish culture on seasonal floodplain *Beel*—a comparative study. J. Scient. Res. 2 (2), 369–379. https://doi.org/10.3329/jsr.v2i2.3430. ISBN 10: 93-82007-39-3.
Huang, D., Liu, J., Hu, C., 2001. Fish Resources in Chinese Reservoirs and their utilisation. In: De Silva, S.S. (Ed.), Reservoir and culture based fisheries: Biology and management. Proceedings of an International workshop held in Bangkok, Thailand. IDRC, Ottawa, pp. 16–21.
Islam, G.M.N., Dickson, M., 2007. Turning Social capital into Natural Capital: Changing livelihoods of fishers through CBFM. In: World Fish Center, Bangladesh and South Asia Office, Dhaka, Bangladesh, Conference Paper 04. pp. 1–21.
Islam, G.N., Abdullah, N.M.R., Viswanathan, K.K., Yew, T.S., 2006. Augumenting fishers' welfare and all livelihood assets through community based management in Bangladesh. In: Paper presented at the 11th Biennial Conference of International Association for the study of common property held on June 19-23, 2006 at the Agung Rai Museum of Art (ARMA) Bali, Indonessia.
Jaiswal, V.K., Singh, U.N., 1994. Bottom fauna of an ox-bow lake of Muzzaffarpur, Bihar. Environ. Ecol. 12 (4), 888–892.
Jaleela, M., 2002. Women in Fisheries in the traditional fishing households of Trivandrum districts of Kerala. In: International Conference on Women in Fisheries, Mumbai.
Jennings, S., et al., 2005. Marine Fisheries Ecology. Blackwell Publishing House, NewYork.
Jha, B.C., 1989. *Beel* fishery resources in Bihar. In: Compendium of the Training in Management of *Beel* (Ox-bow Lake) fisheries. Cent. Inland Capture Fish. Res. Inst., Barrackpore (India), pp. 15–28. Bull. No. 63.
Jha, B.C., 1995. The floodplain fishery of the Gandak basin, Bihar. In: Howes, J.R. (Ed.), Conservation and sustainable use of floodplain wetlands. Asian Wetland Bureau, Kuala Lumpur, pp. 89–97.
Jha, B.C., 1997a. Fisheries of Muktapur oxbow lake ('man'): a case study. In: Sugunan, V.V., Sinha, M. (Eds.), Fisheries enhancement of small reservoirs and floodplain lakes in India. Central Inland Capture Fisheries Research Institute, Barrackpore (India), pp. 175–178. Bull. No. 75.
Jha, B.C., 1997b. Salient ecological feature of mans and chaurs of north Bihar and their fisheries. In: Sugunan, V.V., Sinha, M. (Eds.), Fisheries enhancement of small reservoirs and floodplains lakes in India. Central Inland Fisheries Research Institute (Indian Council of Agricultural Research), Barrackpore, West Bengal, India, p. 252.
John, A., 2009. Towards a right based system of fisheries management. Indlaw news. com.
Jutagate, T., Kwangkhang, W., 2015. Culture-based fisherry of giant freshwater prawn: Experiences from Thailand. In: De Silva, S.S., Ingram, B.A., Wilkinson, S. (Eds.), Perspectives on culture-based fisheries developments in Asia. Network of Aquaculture Centres in Asia-Pacific, Bangkok, Thailand, pp. 91–97.
Kallie, J., et al., 2000. Community-Based Fisheries Management in Samoa. In: Ministry of Agriculture, Forests, Fisheries and Meteorology. Apia, Samoa.
Kar, D., 1984. Limnology and Fisheries of Lake Sone in the Cachar district of Assam (India), viii + 201. PhD Thesis, University of Gauhati (Assam).
Kar, D., 1990a. Limnology and Fisheries of Lake Sone in the Cachar district of Assam (India). Matsya 15–16, 209–213.

Kar, D., 1990b. Limnology and Fisheries of Lake Sone in the Cachar district of Assam (India). viii + 201, PhD Thesis (Published), University of Gauhati, Assam (1984): Matsya, 15–16, pp. 209–213.
Kar, D., 1996. Biodiversity Conservation Prioritisation Project (BCPP) in India. In: Proceedings of International Project Formulation Workshop of BCPP, World Wide Fund (WWF) for Nature-India, 1 (New Delhi).
Kar, D., 1998. Biodiversity conservation prioritisation project (BCPP) in India. In: Proceedings of International Project Finalisation Symposium of BCPP, World Wide Fund (WWF) for Nature-India, 1: New Delhi.
Kar, D., 1999. Preliminary Study of Limnology and Aquatic biota of Rudra Sagar wetland in South Tripura. In: Proceedings of Regional Seminar on Biodiversity, Guwahati, Assam, 1.
Kar, D., 2000. Present status of Fish Biodiversity in South Assam and Tripura, pp. 80–82. In: Ponniah, A.G., Sarkar, U.K. (Eds.), Fish Biodiversity of North-East India. NBFGR-NATP Publication No. 2, pp. 228 (Lucknow).
Kar, D., 2002a. Present status of Biodiversity of Fishes of Barak valley region of Assam with a note on their Management and Conservation, pp. 3–10. In: Bhattacharjee, M.K., Dattachoudhury, M., Mazumder, P.B. (Eds.), Proceedings of UGC-Sponsored State-level Seminar on Biodiversity of Assam, Session Chairman's Lecture, 30 Jan 2000. Karimganj College, Assam University, Assam.
Kar, D., 2003a. Fishes of Barak drainage, Mizoram and Tripura: pp. 203–211. In: Kumar, A., Bohra, C., Singh, L.K. (Eds.), Environment, Pollution and Management. APH Publishing Corporation, New Delhi. pp. xii + 604.
Kar, D., 2004a. A Glimpse into the Fish Bioresources of North-East India with a note on their Management, Conservation and Biotechnological Potential. In: Proc. (Invited Lecture) DBT-sponsored National Symposium on Biodiversity Conservation and Sustainable Utilisation of Environmental Resources: Tripura University, 10-11 Jan., 2004: 9 p.
Kar, D., 2004d. Fish Fauna of Barak Drainage in Assam, of Mizoram and of Tripura with a note on Conservation. J. Freshwater Biol. 16 (1-4), 31–39.
Kar, D., 2005a. Fish Genetic Resources and Habitat Diversity of the Barak drainage. pp. 68–76, In: Ramachandra, T.V., Ahalya, N., Rajasekara Murthy, C. (Eds.), Aquatic Ecosystems (Conservation, Restoration and Management). Capital Publishing Company, Bangalore. pp. xiii +396.
Kar, D., 2006a. Panorama of wetlands in North-East India with special emphasis on their Eutrophication status. Sci. Soc. 4 (2), 167–180.
Kar, D., 2007a. Fundamentals of Limnology and Aquaculture Biotechnology. vi + 609, Daya Publishing House, New Delhi. ISBN: 81-7035-455-2.
Kar, D., 2007b. Sustainability issues of Inland Fish Biodiversity and Fisheries in Barak drainage (Assam), in Mizoram and Tripura, pp. 555–560. In: Kurup, M.B., Ravindran, K. (Eds.), Sustain Fish. pp. xii + 863, School of Industrial Fisheries, Cochin University of Science & Technology (CUSAT): Proceedings of International Symposium on 'Improved sustainability of Fish Production Systems and Appropriate Technologies for Utilisation', 16–18 March 2005 (Cochin).
Kar, D., 2007c. Lentic Fishery: Fishery of a Tropical Wetland (Beel) in Assam. Sci. Soc. 5 (1), 53–72.
Kar, D., 2010a. Biodiversity Conservation Prioritisation. Swastik Publications, New Delhi. pp. xi + 167. ISBN:978-93-80138-26-8.
Kar, D., 2010b. Fish diversity in rivers of Mizoram. Fish. Chimes 30 (7), 18–22.
Kar, D., 2010c. Water bodies, Fish diversity and sustainable aquaculture, pp. 168–194. In: Annual J. Womens'College (Silchar).
Kar, D., 2012a. Wetlands, Rivers, Fish diversity, Fish disease and Aquaculture in North-East India. Fish. Chimes 31 (12), 35–37.
Kar, D., 2012b. Essentials of Fish Biology. pp. vii + 244, Dominant Publishers, New Delhi. ISBN: 978-93-82007-16-6.
Kar, D., 2012c. Taxonomy, BSc, MSc, NET, SLET. New Delhi, APH Publications, p. 102. ISBN: 978-81-313-1580-4.
Kar, D., 2013. Wetlands and Lakes of the World. pp. xxx + 687, Springer, London. Print ISBN 978-81-322-1022-1; e-Book ISBN: 978-81-322-1923-8.
Kar, D., 2014. Research Frontiers in Wetlands, Fishery and Aquaculture. xi + 320, Dominant Publishers and Distributors Pvt. Ltd., New Delhi. ISBN: 978-93-82007-39-5.
Kar, D., 2015. Epizootic Ulcerative Fish Disease Syndrome. xix + 293, Elsevier, (Academic Press), USA. ISBN: 9780128025048.
Kar, D., 2016. Wetlands, Rivers, Fish, Plankton resources and Fish disease and Aquaculture in North-East India: An Overview. In: Proc. International Symposium, Lake 2016, pp. 37. Indian Institute of Science, Bengaluru, and Alva's Education Foundation, Mengaluru (India).
Kar, D., Barbhuiya, M.H., 2000a. Ichthyodiversity of Chatla Haor: A Floodplain wetland in Barak valley Region of Assam. pp. 3–6, In: Pandey, B.N., Singh, B.K. (Eds.), Advances in Zoology and Environmental Degradation and Biodiversity. Daya Publishing House, New Delhi. pp. 279.
Kar, D., Barbhuiya, M.H., 2001. Ecology of Aquatic Macrophytes of Chatla Haor, a floodplain wetland in Cachar district of Assam. Environ. Ecol. 19 (1), 231–233.
Kar, D., Barbhuiya, M.H., 2002a. Macrophytic diversity in certain Wetlands of Barak valley region of Assam. pp. 86–89, In: Ramachandra, T.V., Rajasekara Murthy, N., Ahalya, N. (Eds.), Restoration of Lakes and Wetlands. Allied Publishers (P) Ltd., Bangalore. pp. xxii + 400.
Kar, D., Barbhuiya, M.H., 2002b. Waterfowls in Chatla Haor, a seasonal floodplain in Cachar district of Assam with a note on waterfowl and wetland conservation. Environ. Ecol. 20 (3), 689–691.
Kar, D., Barbhuiya, M.H., 2002c. Aquatic macro-invertebrates of Baraknadi Salchapra Anua in Barak valley region of Assam. Environ. Ecol. 20 (2), 696–699.
Kar, D., Barbhuiya, M.H., 2002d. Status Survey of Gangetic Dolphin in Barak drainage of Assam with a note on its Management and Conservation. Environ. Ecol. 20 (1), 249–250.
Kar, D., Barbhuiya, M.H., 2004. Abundance and Diversity of Zooplankton in Chatla Haor, a floodplain wetland in Cachar district of Assam. Environ. Ecol. 22 (1), 247–248.
Kar, D., Barbhuiya, M.H., 2005. Length-weight relationship and relative condition factoring *Hilsa(Tenualosa) ilisha* (Hamiilton) of Barak drainage of Assam. Ind. J. Environ. Ecoplan. 10 (1), 265–267.
Kar, D., Barbhuiya, A.H., 2011a. Status of Chocolate Mahseer of River Barak with a note on its morphometry and meristics. J. Inland Fish. Soc. India 42 (1), 37–42.
Kar, D., Barbhuiya, A.H., 2011b. Length-weight relationship and condition factor in *Puntius amphibious* (Valenciennes) and *Puntius vittatus* Day from Barak valley region of Assam. J. Inland Fish. Soc. India 42 (1), 55–62.
Kar, D., Dey, S.C., 1982. Hilsa ilisha (Hamilton) from Lake Sone in Cachar, Assam. In: Proceedings of Indian Science Congress, pp. 69–77.
Kar, D., Dey, S.C., 1986. An account of ichthyospecies of Lake Sone in Barak valley of Assam. In: Proceedings of All-India Sem. Ichthyology, vol. 2: 3 p.
Kar, D., Dey, S.C., 1987. An account of the Fish and Fisheries of Lake Sone in the Barak valley of Assam (India). In: Proceedings of Workshop Development of Beel Fisheries in Assam, vol. 1, 13 pp.
Kar, D., Dey, S.C., 1991. Gill nets in Lake Sone of Assam with their Economics and Impact on Fishery. J. Appl. Zool. Res. 2 (2), 76–79.

Kar, D., Dey, S.C., 1993a. Variegated Encircling gears in Lake Sone of Assam. J. Appl. Zool. Res. 4 (2), 171–175.
Kar, D., Dey, S.C., 1993b. Interrelationship and Dynamics of Fish Population of Lake Sone in Assam. Environ. Ecol. 11 (3), 718–719.
Kar, D., Dey, S.C., 1996. Scooping Gears of Lake Sone in Assam. J. Appl. Zool. Res. 7 (1), 65–68.
Kar, D., Dey, S.C., 2000. Yield and Conservation of Indian major carps of Lake Sone in Assam. Environ. Ecol. 18 (4), 1036–1038.
Kar, D., Dey, S.C., 2002. On the occurrence of Advanced fry of *Hilsa (Tenualosa) ilisha* (Hamilton-Buchanan) in Chatla Haor Seasonal wetland of Assam. Proc. Zool. Soc. Calcutta 55 (2), 15–19.
Kar, S., Kar, D., 2004b. Genetics, Culture, Environment and Health among the Kuki Hill Tribe in Cachar District of Assam, India. Sci. Culture 70 (11–12), 357–359.
Kar, S., Kar, D., 2014. Ecology of Four Wetlands in South Assam: A Comparison. Research Frontiers in Wetlands, Fishery & Aquaculture. pp. 239–246. In: Kar, D. (Ed.), Research Frontiers in Wetlands, Fishery and Aquaculture. Dominant Publishers and Distributors Pvt.Ltd., New Delhi. xi + 320. ISBN: 978-93-82007-39-5, ISBN 10: 93-82007-39-3.
Kar, D., Sen, N., 2007. Systematic List and Distribution of Fish Biodiversity in Mizoram, Tripura and Barak drainage in North-East India. ZOOs' Print J. 22 (3), 2599–2607.
Kar, D., Dey, S.C., Kar, S., 1995. Fish and Fisheries of Lake Sone in Assam. J. North-Eastern Council 15 (2–3), 49–57.
Kar, D., Dey, S.C., Purkayastha, M., Kar, S., 1996a. An overview of the Impediments in Conservation of Biodiversity of Lake Sone in Assam. In: Proceedings of Seminar on Conservation of Biodiversity in Indian Aquatic Ecosystems. JawaharLal Nehru University, New Delhi.
Kar, D., Saha, D., Laskar, R., Barbhuiya, M.H., 1997. Biodiversity Conservation Prioritisation Project (BCPP) in Barak valley region of Assam. In: Proc. National Project Evaluation Workshop on BCPP, Betla Tiger Reserve and National Park, 1 (Palamu).
Kar, D., Dey, S.C., Kar, S., Ramachandra, T.V., 1999. Trawls of Lake Sone in Assam. J. Appl. Zool. Res. 10 (2), 170–172.
Kar, D., Mandal, M., Laskar, B.A., Dhar, N., Barbhuiya, M.H., 2000. Ichthyofauna of some of the oxbow lake in Barak valley region of Assam. In: Proceedings of the National Symposium on Wetlands and Fisheries Research in the New Millennium, 1: 16 pp.
Kar, D., Mandal, M., Lalsiamliana, 2001. Species composition and distribution of Fishes in the rivers in Barak valley region of Assam and in the Principal rivers of Mizoram and Tripura in relation to their habitat parameters. In: Proc. National Workshop, NATP-ICAR Project, Mid-term Review,: Central Marine Fisheries Research Institute, Cochin, 25 pp.
Kar, D., Laskar, B.A., Mandal, M., Lalsiamliana, Nath, D., 2002a. Fish diversity in Major rivers in Southern Assam, Mizoram and Tripura with a note on their Habitat parameters. Ind. J. Environ. Ecoplan. 6 (3), 473–480.
Kar, D., Laskar, B.A., Nath, D., 2003c. Length-weight relationship of *Salmostoam phulo phulo* (Hamilton-Buchanan) collected from River Karnafuli along Indo-Bangladesh border (in Mizoram). J. Appl. Zool. Res. 14 (2), 188–190.
Kar, D., Dey, S.C., Datta, N.C., 2003d. Welfare Biology in the New Millennium. xx + 97, Allied Publishers Pvt. Ltd, Bangalore.
Kar, D., Lohar, M., Ngassepam, Z., Sinom, S., Tiwary, B., Bawari, M., 2003a. Fish Bioresources in certain rivers of Assam and Manipur with a note on their assessment, management and conservation. In: Proc. Nat. Symp. Assessment and Management of Bioresources, North Bengal University and The Zoological Society, Calcutta, 28-30 May, 2003, 56 pp.
Kar, D., Dey, S.C., Roy, A., 2003e. Biodiversity of freshwater fishes of North-East India with a note on their health status. In: Proceedings of DBT-sponsored Guest Lecture Series delivered at the Dept. of Molecular Biology and Biotechnology, Tezpur (Central) University, Tezpur, 1.
Kar, D., Nagarathna, A.V., Ramachandra, T.V., Dey, S.C., 2006. Fish Diversity and Conservation aspects in an aquatic ecosystem in North-East India. ZOOS' Print J. 21 (7), 2308–2315.
Kar, D.; Laskar, B.A.; and, Nath, D. (2006c). An account of Fecundity of *Eutropiichthys vacha* (Hamilton-Buchanan): a commercially important fish in Assam. Environ. Ecol., 24 Spl (3): 726-727.
Kar, D., Barbhuiya, A.H., Arifuddidn, Barbhuiya, M.A., Chetia, P., Islam, R., Rahman, S., 2007. Traditional Riverine Fish Catching Devices of Assam. Fishery Technol. 44 (2), 137–146.
Kar, D., Barbhuiya, A.H., Thangjam, G., Devi, S.M., Deb, S., Das, B., Chanu, H., Nishima, 2008. Panorama of Fish Biodiversity in certain rivers and wetlands in Manipur. Proc. Zool. Soc. India 7 (2), 123–134.
Kar, D., Barbhuiya, A.H., Baruah, A.R., Choudhury, C., Banerjee, P., Pal, R., Bhattacharjee, A., Saikia, R., Das, B., Barman, R., Saha, B., 2009. Panorama of Fish Diversity in certain rivers, wetlands and protected areas in Assam. Geobios 36, 57–64.
Kar, D., Shomorendra, M., Singha, R., Puinyabati, H., Geetarani, B., Binky, K., Sangeeta, O., Ranibala, T., 2011. Fish diversity and Helminth fauna in the fishes of Assam and Manipur, India. In: Fishing Chimes. pp. 55-65.
Kar, D., Singha, R., Das, B.K., John, R., George, R., Mansoor, M., Selva, M., 2015. Preliminary Epidemiological Studies on Epizootic ulcerative Syndrome (EUS) in Freshwater Fishes of Assam. Int. J. Scient. Res. Sci., Eng. Technol (IJSRSET) 1 (5), 256–263.
Kar, S., Das, P., Das, U., Bimola, M., Kar, D., Aditya, G., 2018. Culture of the zooplankton as fish food: observations on three freshwater species from Assam, India. AACL Bioflux 10 (5), 1210–1220.
Katon, B., Pomeroy, R., Salamanca, A., 1997. The marine conservation project for San Salvador: a case study of fisheries co-management in the Philippines, 10–95. Fisheries Co-management Research Project Working Paper No. 23. Manila: ICLARM.
Katsuyama, K., 1999. Direction of Future Fishery Management. Japan Fisheries Agency—International Affairs Division, Tokyo, Japan.
Khakhar, K.K., 2001. Towards governance of sustainable fisheries: Role of institutions. Edited by Kanchan Chopra and Hanumantha Rao—Ecological economy for sustainable development.
Khan, S.M.M.H., 2011. Participatory wetland resource governance in Bangladesh: An analysis of community-based experiments in Hakaluki Haor. pp. xx + 297, Ph.D. Thesis University of Manitoba (Canada).
King, M., Fatasiti, U., Malipola, A., 2001. Community based management of subsistence fisheries in Samoa. In: Environmental Awareness and Community Participation, pp. 49–59.
Komatsu, M., 2013. Special Study on Sustainable Fisheries Management and International Trade in the South-East Asia and Pacific Region. Asian Development Bank Institute (ADBI) Working Paper Series No. 438, Trade Working Papers 23715, East Asian Bureau of Economic Research.
Korakandy, R., 2001. Kerala: Aquarium Reforms, Will Actual Fishermen Benefit? In: Economic and Political Weekly, 2001.
Korakandy, R., 2008. Fisheries Development in India- A Political Economy of Unsustainable Development. Kalpaz Publications, Delhi.
Kumar, K., 1985. Hydrobiological studies of freshwater *Beel* with special reference to its fish production potentialities. Unpublished Ph.D. Thesis, Bhagalpur University, Bhagalpur, Bihar.

Kumar, A., Joshi, P.K., Birthal, P.S., Kumar, A., Katiha, P.K., Joshi, P.K., 2003. Fisheries Sector in India: An Overview of Performance, Policies and Programmes. pp. 1–16, In: A Profile of People, Technologies and Policies in Fisheries Sector in India. Workshop Proceedings Series 10, National Centre for Agricultural Economics and Policy Research, Indian Council of Agricultural Research, pp. 1–171.

Kumbhar, B.L., 1979. Gambhira Cooperative Farming Society: A successful experiment in collective efforts. Ind. J. Agricult. Econ. 45 (3), 45–50.

Kurien, J., 1991. Ruining the Commons and Responses of the Commoners: Coastal Overfishing and Fishermen's Actions in Kerala State, India. Discussion Paper No. 23, May 1991, Centre for Development Studies, Trivandrum.

Kurien, J., 2000. Community Property Rights: Re-Establishing them for a Secure Future for Small-Scale Fisheries. Centre for Development Studies, Thiruvananthapuram.

Lahon, B., 1983. Limnology and Fisheries of some commercial Beels of Assam, India. Unpublished PhD Thesis, University of Gauhati (Assam).

Lee, S.-G., Midani, A.R., 2014. Creative economy potentials and fisheries development in South Korea. World J. Fish Marine Sci. 6 (1), 49–56. 2014, ISSN 2078-4589.

Limsong, S., Sitha, H., De Silva, S.S., Ingram, B.A., Wilkinson, S., 2015. Cambodian attempts to develop culture-based fisheries. 121 p. In: Perspectives on culture-based fisheries developments in Asia. NACA Monograph Series No. 3, pp. 126.

Lobe, K., Berkes, F., 2004. The padu system of community-based fisheries management: Change and local institutional innovation in south India. Mar. Policy 28, 271–281.

Lodge, M., 2004. Overcoming factors of unsustainability and over exploitation in fisheries. In: International workshop on the implementation of International fisheries instruments and factors of unsustainability and over exploitation in fisheries, FAO, Siem Reap, Cambodia.

MACH, 2007. Wetland protection and Enhancement through Sanctuaries in Bangladesh. A Technical paper.

Maiti, A., 2003. Study on the socio-economic status and life style of the fishermen in Sundarban, West Bengal, India. An unpublished M.F.Sc. Thesis submitted to the West Bengal University of Animal and Fishery Sciences, Mohanpur, West Bengal, pp. 68.

Makino, M., Matsuda, H., 2011. Ecosystem-based Management in the Asia-Pacific Region: World Fisheries: A Socio-Ecological Analysis, pp. 1–12. (Eds), Ommer, Perry, Cochrane and Cury; Blackwell Publishing Ltd., pp. 1–333.

Mathew, A., 2009. Marine Fisheries Conservation and Management in India. In: United Nations—The Nippon Foundation Fellowship Programme, Division for Ocean Affairs and the Law of the Sea (DOALOS) Office of Legal Affairs, United Nations New York, United States of America.

Mccay, B.J., 2001. Resistance to Changes in Property Rights or, why not ITQs? Rutgers University, Newa Brunswick 08901 New Jersey, USA.

Minkin, S.F., Rahman, M.M., Halder, S., Tsai, C., Ali, M.Y., 1997. Fish bio diversity, human nutrition and environmental restoration in Bangladesh. In: Open water fisheries of Bangladesh. The University Press Limited, Dhaka, pp. 183–198.

Mitra, K., 1989. Limnological features of *Beels*—Macro-vegetation dynamics. In: Compendium of the training in management of Beel (Ox-bow Lake) fisheries, Cent. Inland Capture Fish. Res. Inst., Barrackpore (India), vol. 63. pp. 29–33.

Mitra, K., 1992. Common macrophytes of floodplain lakes and their control. In: Yadava, Y.S., Sugunan, V.V. (Eds.), Compendium of the FAO sponsored training programme on floodplain fisheries management. Cent. Inland Capture Fish. Res. Inst., Barrackpore (India), pp. 55–61.

Mitra, K., Ghosh, A., 1992. Community structure, seasonality and succession of macrophytes and associated fauna in floodplain lakes. In: Compendium of the FAO sponsored training programme on Floodplain fisheries management. Cent. Inland Fish. Res. Inst., Barrackpore (India), pp. 62–65.

Mohamed, K.S., 2004. Towards Ecosystem Based Management of Marine Fisheries: Building Mass Balance Trophic and Simulation Models: Technical Notes. Winter School, Central Marine Fisheries Research Institute, Cochin (Kerala), pp. 125–129.

Morton, B., 2005. Over fishing: Hong Kong's fishing crisis finally arrives. Mar. Pollut. Bull. 50, 1031–1035.

Mukhopadhyay, M.K., 1997. Ecology of *Beels* in West Bengal. In: Sugunan, V.V., Sinha, M. (Eds.), Fisheries enhancement of small reservoirs and floodplains lakes in India. Central Inland Fisheries Research Institute (Indian Council of Agricultural Research), Barrackpore, West Bengal, India, p. 252.

Mulekom, V.L., 1998. Capacity and opp.ortunity building in a small scale fisher folk community for community-based coastal resource management: a community development approach to CB-CRM. (Draft paper).

Murray, F., 2002a. When Co-management Fails: A Review of Theory and Lessons Learned from Reservoir Fisheries in the Dry-Zone of Sri Lanka. Institute of Aquaculture, University of Stirling, Stirling FK9 4LA, UK, pp. 1–22. Conference Paper 11.

Murray, F., 2002b. When co-management fails: a review of theory and lessons learned from reservoir fisheries in the dry-zone of Sri Lanka. Fish. Manage. Ecol. 9 (1), 14–26.

Mustafa, M.G., Halls, A.S., 2007. Impact of the Community-Based Fisheries Management on sustainable use of inland fisheries in Bangladesh. In: World Fish Centre, Bangladesh and South Asia; Mekong River Commission Secretariat (MRCS), Inland Fisheries Research and Development; Institute (IFReDI), Department of Fisheries, Phnom Penh, Cambodia, Conference Paper 03, pp, 1-19; and, Proceedings of the CBFM-2 International Conference on Community Based Approaches to Fisheries Management, Dhaka, Bangladesh, p. 11.

Nair, M., 1998. Women in Fisheries-Emancipation through co-operatives. In: Hameed, M.S., Kurup, B.M. (Eds.), Technological Advancements in Fisheries. CUSAT, Kochi.

Nasuchon, N., 2009a. Coastal management and community Management in Malaysia, Vietnam, Cambodia and Thailand, with a case study of Thai Fisheries management. United Nations-Nippon Foundation Fellow. 2008–2009, pp. xx + 212.

Nasuchon, N., 2009b. Coastal management and community management in Malaysia, Vietnam, Cambodia and Thailand, with a case study of Thai fisheries management. In: Division for Ocean Affairs and the Law of the Sea Office of Legal Affairs, the United Nations, New York, pp. 1–91.

Natarajan, A., 2005. In: Enabling Better Management of Fisheries Conflicts: A Case Study in India. Paper presented at in the Regional Consolidation Workshop on Fish Fights over Fish Rights: Managing Conflicts and Exit from the Fisheries and Security Implications for South and Southeast Asia, 17-20 May 2005, International Rice Research Institute Complex (IRRI), Los Baños, Laguna, The Philippines. In: Enabling Better Management of Fisheries Conflicts; NRSP Production System, Land Water Interface. Final Technical Report (R8294), Revised Version, November 2005; Submitted by: The World Fish Center; in collaboration with research partners: The World Fish Center, Bangladesh; Fisheries Action Coalition Team, Cambodia; Mitraniketan, India; and, The University of Reading, United Kingdom, pp. 1–91.

Nguyen, T.K.A., 2012. Community-based Co-management in Vietnamese fisheries: The case of the Fisheries Associations in Tam Giang-Cau Hai Lagoon. Master's Thesis in Fisheries and Aquaculture Management and Economics FSK-3911 (30 ECTS), The, Norwegian College of Fishery Science, University of Tromso, Norway and, Nha Trang University, Vietnam, pp. 1–10.

Nuov, S., 2015. Culture-based fisheries in Cambodia. 124 p, In: De Silva, S.S., Ingram, B.A., Wilkinson, S. (Eds.), Perspectives on Culture-Based Fisheries Developments in Asia. NACA Monograph Series No. 3, pp. 126.
Onyango, P.O., 2001. Pitfalls in building a community-based fisheries resource management regime in Lake Victoria, Tanzania. pp. vii + 322, 120 Reykjavik, Iceland Final Project.
Owati, S.I., et al., 2005. Productive efficiency, property rights and sustainable renewable resource development in the mini purse seine fishery of the Java sea. Environment and Development Economics, Cambridge University Press.
Panjarat, S., 2008. Sustainable fisheries in the Andaman sea coast of Thailand. pp. 1–47. Division for Ocean Affairs and the Law of the Sea, Office of Legal Affairs, the United Nations (New York).
Parameswaran, S., Vass, K.K., 1995. An overview of floodplain wetlands in West Bengal. In: Howes, J.R. (Ed.), Conservation and Sustainable Use of Floodplain Wetlands. Asian Wetland Bureau, Kuala Lumpur, pp. 52–66.
Paterson, J., Pernetta, C., De Silva, S.S., Amarasinghe, U.S., 2015. Integrating Fisheries and Habitat Management: Fisheries Refugia in the South China Sea, pp. 113. In: Status of Reservoir Fisheries in Five Asian Countries. Network of Aquaculture Centres in Asia-Pacific, Bangkok, Thailand. pp., 113, NACA Monograph No. 2.
Pathak, V., Bhagat, M.J., Saha, S.B., 1986. Limnological significance of nannoplankton as producer in *Beel* ecosystem. J. Hydrobiol. 2 (4), 73–79.
Paul, A., 2005. Rise, fall and persistence in Kadakkodi: an enquiry into the evolution of a community institution for fishery management in Kerala, India, CDS. (Trivandrum).
Paul, P.K., Rahman, Arifur, A.F.M., Nazrul, K.M.S., 2014. Successful Practice of Community-Based Fisheries Management in seasonal water-logged paddy lands of Begumganj, Noakhali, Bangladesh. Int. J. Res. Agricult. Food Sci. 2 (2), 29–36.
Phan, P.D., 2014. Review of Current status of Fish stock enhancement in Vietnam: Consultant report. MRC Fisheries Programme, pp. 39.
Phounsavath, S., 1998. Community-Based Fisheries Management as an approach to Participatory management of Reservoir fisheries: a case study of two fishing villages at the Nam Ngum Reservoir. LAO P.D.R, pp.1-119, a thesis submitted in partial fulfillment of the requirements for the degree of Master of Science, Asian Institute of Technology, School of Environment, Resources and Development, Bangkok (Thailand).
Pimoljinda, J., Boonraksa, V., 1997. Community-based Fisheries co-management case study: Phang-Nga Bay, Thailand. Proceedings of the International Workshop on Fisheries Co-management. Andaman Sea Fisheries Development Centre, Thailand. pp 1–34.
Pitcher, T.J., et al., 1998. Reinventing fisheries management. In: Fish and Fisheries Series. vol. 23. Kluwer Academic Publications, The Netherlands.
Pomeroy, R.S., 1995. Community-based and co-management institutions for sustainable coastal fisheries management in Southeast Asia. International Center for Living Aquatic Resources Management (ICLARM), Metro Manila, The Philippines, Ocean Coast. Manag. 27 (3), 143–162.
Pomeroy, R.S., Rebecca, R.-G., 2006. Fishery Co-Management, A Practical Handbook. pp. 255, CABI Publishing, CAB International, Wallingford, Oxfordshire (UK).
Pomeroy, R.S., Williams, M., 1994. Fisheries Co-management and Small Scale Fisheries: A Policy Brief. ICLARM, Manila (The Philippines).
Pomeroy, R.S., Katon, B.M., Harkes, I., 1998. Fisheries Co-management: Key Conditions and Principles Drawn from Asian Experiences. pp. 1–23, International Centre for Living Aquatic Resources Management (ICLARM), Makati City 0718 (The Philippines).
Prabhu, R.R., 2011. Institutional dynamics in Community-based Fisheries resource management for Sustainable Development of marine fisheries in Kerala. PhD thesis, pp. 477, Department of Applied Economics Cochin University of Science and Technology (CUSAT), Kochi-22 (Kerala).
Pradhan, G.B.N., 2009. Reservoir Fisheries in Nepal. pp. 55–67, In: De Silva, S.S., Amarasinghe, U.S. (Eds.), Status of Reservoir Fisheries in Five Asian Countries. Network of Aquaculture Centres in Asia-Pacific, Bangkok (Thailand), pp. 1–119. NACA Monograph No. 2.
Preikshot, D., Pauly, D., 2005. Global Fisheries and Marine Conservation: Is Coexistence Possible? Chapter 11, pp. 185–197, in Norse and Crowder.
Pushpalatha, K.B.C., Fernando, W.A.J.R., Chandrasoma, J., 2015. Impact of introduction of culture-based fisheries on fish production in two perennial reservoirs in Sri Lanka. Int. J. Fish. Aquat. Stud. 2 (4S), 5–9.
Rab, M.A., 2010. Institutional framework for Halda fisheries: Drawing lesson from CBFM experiences. Bangladesh J. Polit. Econ. 23 (1-2), 133–148.
Radjawali, I., 2012. Examining local conservation and development: Live reef food fishing in Spermonde Archipelago, Indonesia. J. Integr. Coast. Zone Manage. 12 (4), 545–557.
Rahman, M.M., Begum, A., 2010. The strategy of empowering poor for wetland resources conservation in Bangladesh. J. Hum. Ecol. 31 (2), 87–92.
Rajasenan, D., 1999. Modernity vc Tradition and the question of Sustainability: The Changing Paradigms of the Kerala Marine Fishing Industry. Directorate of Publications and Public Relations, CUSAT (Cochin).
Rajasenan, D., 2005. In: Rajasenan, D., de Groot, G. (Eds.), Marine Fishing in Kerala: Dynamics and Challenges: Kerala Economy Trajectories, Challenges and Implications. CUSAT, Cochin. Directorate of Publications and Public Relations.
Ramachandran, M., 2006. Poverty, Inequality and Natural Resources Degradation: An investigation into Small-scale Fishery Sector of South Kerala. Ph. D Thesis, Department of Applied Economics, CUSAT, Cochin (Kerala).
Rana, G.C., Sengupta, K.K., Santra, S.C., 1996. Limnological characteristics of *Beel*s with reference to fish yield in tropics. J. Inland Fish. Soc. India 28 (1), 59–66.
Ruddle, K., 1993. External forces and change in traditional community-based fishery management systems in the Asia-Pacific Region. Maritime Anthropol. Stud. (MAST) 6 (1/2), 1–37.
Ruddle, K., 1998. Traditional community-based coastal marine fisheries management in Viet Nam. Ocean Coast. Manag. 40, 1–22.
Salim, K.H., 1998. Community-based fishery management: An alternative approach in Malaysia. BOB News, II (12), 10–14.
Sampath, V., 2005. An overview of the International and National Fishery Policy: Related Issues: A Desk Research. In: Desk Study on Fisheries Policy. pp. 1–134.
Santha, S.D., 2007. Local ecological knowledge and fisheries management: A study among riverine fishing communities in Kerala, India. Jamsedji Tata Centre for Disaster Management, Tata Institute of Social Sciences, Mumbai (India).
Sapovadia, V.K., 2004. Fisherman Cooperatives: A Tool for Socio-Economic Development. In: Indian Institute of Management, Indore, Finance and Accounting, International Institute of Fisheries and Trade Conference, July 2004.
Sathiadhas, R., et al., Ninawe, A.S., Diwan, A.D., 2005. Adoption of fishery enterprises and empowerment options for fisherwomen of India. In: Women Empowerment in Fisheries. Narendra Publishing House, New Delhi.
Satriaa, A., Matsudaa, Y., 2004. Decentralization of Fisheries Management in Indonesia. Mar. Policy 28, 437–450.

Schmidt, C.-C., 2003. Fisheries and Japan: A case of multiple roles? pp. 1–18, In: OECD (Paris), Proceedings of International Symposium on Multiple Roles and Functions of Fisheries and Fishing Communities, 13 February 2003, Aomori (Japan).
Scott, A., 2001. Introducing Property in Fishery Management. University of British Columbia, Department of Economics, Vancouver BC (Canada).
Sen, S., 1997. Building legitimacy for smart partnership in fisheries management. BOB News, II (12), 15–19.
Shrivastava, N.P., Bhattacharya, B.K., 2003. Status of fisheries of floodplain wetlands of Assam. pp. 22–29, In: Vinci, G.K., Jha, B.C., Bhaumik, U., Mitra, K. (Eds.), Fisheries Management of Floodplain Wetlands in India. CIFRI, Barrackpore. Bull. No. 125.
Sidayat, M., 2008. Community-Based Fishery Management in the Post-Conflict situation: A Case Study in Coastal Villages North Malukud. In partial fulfilment of the requirements for obtaining the degree of Masters of Arts in Development studies, Specialisation: Rural Livelihoods and Global Change (RLGC).
Singh, D.N., 1990. Studies on the phytoplankton of Mc Pherson Lake, Allahabad. Proc. Nat. Acad. Sci. India 60 (B), 237–243.
Singh, D.N., 2000. Seasonal variation of zooplankton in a tropical lake. Geobios 27, 97–1000.
Sinha, M., Jha, B.C., 1997. Ecology and fisheries of oxbow lakes (Maun) of North Bihar: A Threatened Ecosystem. Bull. No. 74, pp. 65, Central. Inland Capture Fisheries Res. Inst., Barrackpore (India).
Soede, L.P., Merk, I.A., Claussen, J., Thompson, H., Wheeles, D., 2001. Integrated Marine Management Concessions: A New Approach to an Old Problem. The Conservation and Community Investment Forum (CCIF), Sanur, Bali. pp. 1–15.
Sonak, S., et al., 2006. The problem of fit, scale and interplay and competing interest in Fisheries Management. www.Sonak&Rubinoff, J A.com.
Soumya, H.B., Shah, P.J., 2005. Breifing paper on Fisheries Policy: Resource Rights of Individuals and Communities. Centre for Civil Society, NewDelhi.
Sreenivasan, J.T., 2005. State Regulation and Collective Action in Resource Management: A case study of the Cochin Estuaries Fisheries in India. Accessible from: www.keralafisheriesmanagement.com.
Sugunan, V.V., 1997. Floodplain wetlands, small water bodies, culture-based fisheries and enhancement: conceptual framework and definitions. In: Sugunan, V.V., Sinha, M. (Eds.), Fisheries Enhancement of Small Reservoirs and Floodplains Lakes in India. Central Inland Fisheries Research Institute (Indian Council of Agricultural Research), Barrackpore (West Bengal). pp. 252.
Sugunan, V.V., Bhattacharjya, B.K., 2000. Ecology and Fisheries of *Beel*s in Assam. Bulletin No. 104, pp. 65, CIFRI, Barrackpore.
Sugunan, V.V., Mukhopadhyaya, M.K., 1995. Conservation and sustainable use of floodplain wetlands: Case studies of Bandardaha and Beloon *Beel*s. In: Howes, J.R. (Ed.), Conservation and sustainable use of floodplain wetlands. Asian Wetland Bureau, Kuala Lumpur, pp. 67–75.
Sugunan, V.V., Vinci, G.K., Bhattacharjya, B.K., Hassan, M.A., 2000. Ecology and fisheries of *Beel*s in West Bengal. Bull. No. 103, pp. 53, Central Inland Capture Fisheries Research Institute, Barrackpore.
Sultan, S., Barman, R.C., 2014. Natural Wetland Fisheries: Approaches for Judicious Utilisation, pp. 269–289. In: Kar, D. (Ed.), Research Frontiers in Wetlands, Fishery and Aquaculture. Dominant Publishers and Distributors Pvt.Ltd., New Delhi. xi + 320.
Sultana, P., Thompson, P.M., 2007. Community Fishery Management: Indication for Food security and Livelihoods. Hum. Ecol. 35 (5), 527–546.
Sultana, P., Thompson, P.M., Ahmed, M., 2002. Women-Led Fisheries Management: A Case Study from Bangladesh. The World Fish Centre, Malaysia and Bangladesh.
Symes, D., 2000a. Use Rights and Social Obligations: Questions of Responsibility and Governance. Department of Geography, University of Hull (UK).
Symes, D., 2000b. Rights based management: a European union perspective. In: Use of Property Rights in Fisheries Management. FAO, Rome, pp. 266–273. FAO Fisheries Technical Paper 404/1.
Teh, L., Zeller, D., Zylich, K., Nguyen, G., Harper, S., 2014. Reconstructing Vietnam's marine fisheries catch, 1950–2010. Fisheries Centre, The University of British Columbia, Working Paper Series, Working Paper No. 2014-17, pp. 1–10.
Thilsted, S.H., Wahab, M.A. (Eds.), 2014. Production and conservation of nutrient-rich small fish (SIS) in ponds and wetlands for nutritional security and livelihoods in South Asia. Proceedings of a World Bank/SAFANSI- Funded Regional Workshop on Small Fish and Nutrition. Dhaka, Bangladesh, 1–2 March 2014, The World Bank/South Asia Food and Nutrition Security Initiative (SAFANSI): pp 47.
Thompson, P., 2004. Lessons from Community-Based Fisheries Management in Bangladesh: Briefing Paper. World Fish Center, Dhaka, pp. 1–22.
Thompson, P.M., Sultana, P., Islam, M.N., Kabir, M.M., Hussain, M.M., Kabir, M.S., 1999. An assessment of co-management arrangements developed by the Community-Based Fisheries Management project in Bangladesh. In: Paper presented at the International Workshop on Fisheries Co-management, 23–27 August, 1999, Penag (Malaysia).
Thompson, P.M., Sultana, P., Islam, N., 2003. Lesson from community based management of flood pain fisheries in Bangladesh. J. Environ. Manage. 25, 82–92.
Thompson, P., Shelly, A.B., Rahman, A., Hossain, M., 2006. ICLARM-NGO partnership in research and community based fisheries management in Bangladesh. Available at http://www.fao.og.
Thomson, K.T., 2007. Development policies, State interventions and struggle for livelihood rights: A study of coastal communities in Kerala, India. School of Industrial Fisheries, Cochin University of Science and Technology (CUSAT), Cochin (Kerala).
D'Cruz, S.T., 2004. Artisanal Deep-Sea Fishing in Kerala: Prospects and Problems. Discussion Paper, Kerala Research Programme on Local Level Development, Centre for Development Studies, Prasanth Nagar, Ulloor, Thiruvananthapuram.
Tokrisna, R., Boonchuwong, P., Janekarnkij, P., 1997. A review on fisheries and coastal community-based co-management regime in Thailand. In: Proceedings of the International Workshop on Fisheries Co-management. pp. 1–26.
Upare, M.A., 2002. Empowerment of women for fisheries development: New initiatives for credit delivery to women. In: Singhand, M.P., Tewari, R. (Eds.), Women in Fisheries. Indian Society of Fisheries Professionals, Mumbai.
Upare, S., Dalvi, V.B., 2002. Effective fisheries management through community and development-based participation. In: Singhand, M.P., Tewari, R. (Eds.), Women in Fisheries. Indian Society of Fisheries Professionals, Mumbai.
Uphoff, N., 1986. Local institutional development: An Analytical Source Book with cases. West Hart ford.
Vass, K.K., 1989. Productivity status of *Beel*s in India. In: Compendium of the training in Management of *Beel* (oxbow lake), Fisheries. Bulletin No. 63. Central Inland Capture Fisheries Res. Inst., Barrackpore (India), pp. 57–64.
Vass, K.K., 1992a. Energy flux and nutrient dynamics in floodplain lakes, pp. 40–44. In: Yadava, Y.S., Sugunan, V.V. (Eds.), Compendium of the FAO-sponsored training programme on Floodplain fisheries management. Central Inland Fisheries Res. Inst., Barrackpore (India).
Vass, K.K., 1992b. Hydro-dynamics in floodplain lakes, pp. 23–26. In: Yadava, Y.S., Sugunan, V.V. (Eds.), Compendium of the FAO-Sponsored Training Programme on Floodplain Fisheries Management. Central. Inland Capture Fisheries. Res. Inst., Barrackpore (India).

Vass, K.K., Langer, R.K., 1990. Changes in primary production and trophic status of a Kashmir oxbow lake, consequent to man-made modifications in the environment. J. Inland Fish. Soc. India 22 (1 and 2), 43–51.
Venkatachalam, R., 2005. Sustainable Fisheries and Community Management Systems: Environment and Growth. 2nd South Asian Economics Students Meet 2005 at Lahore, Pakistan, Jan 28–30, 2005, organized by the Lahore University of Management Sciences (LUMS).
Vinci, G.K., 2002. Development of management guidelines for culture based fisheries in Beels. pp. 197, In: Sugunan, V.V., Jha, B.C., Das, M.K. (Eds.), Culture-based Fisheries for Inland Fisheries Development. Central Inland Fisheries Research Institute (Indian Council of Agricultural Research), Barrackpore, West Bengal (India).
Vipinkumar, V.P., Ramachandran, C., Shyam, S.S., 2013. Co-management Paradigm and Sociological issues in RFM in Indian Marine Fisheries Sector. Winter School on ICT-oriented Strategic Extension for Responsible Fisheries Management. Socio-Economic Evaluation and Technology Transfer Division, Central Marine Fisheries Research Institute, pp. 253–268.
Viswanathan, K.K., 2000. Community Perspectives: Exclusivity of Rights. In: Conference Vol. No, 21 (2008), pp 345–354. International Centre for Living Aquatic Resources Management (ICLARM), Metro Manila, 1229 (The Philippines).
Wang, Q., Liu, J., Li, Z., Zhang, T., 2015. Culture-based Fisheries in Lakes of the Yangtze River basin, China, with special reference to stocking of Mandarin fish and Chinese mitten crab. In: De Silva, S.S., Ingram, B.A., Wilkinson, S. (Eds.), Perspective on Culture-Based Fisheries development in Asia. Network of Aquaculture Centers in Asia-Pacific (NACA) Monograph series No. 3, pp. 126.
Wilkinson, S., 2007. Community Communication Centres: Applications, establishment and constraints. Netw. Aquacult. Centr. Asia-Pacific (NACA) XII (4), 1–52.
Willmann, R., 2001. Group and Community-based Fishing Rights: Fishery Policy and Planning Division. FAO, Rome (Italy).
Wilson, M., 1999. Facilitated Community Action process: A demand-based extension process to facilitate Community Based Management of inshore fisheries. In: Paper presented to the International Symposium on Society and Resource Management, Brisbane, Australia.
Wong, J.Z., Etoh, S., Sujang, A.B., 2009. Towards Sustainable Community-based Fishery Resources Management: The Tagal System of Sabah, Malaysia. Southeast Asian Fish. Develop. Center 7 (2), 18–23.
World Bank, 2004. Project Appraisal Document on World Bank Financed Assam Agricultural Competitiveness Project. pp. 15.
World Bank, 2011. Improved irrigation helps Assam become self sufficient in rice. World Bank India 9 (5), 1–9.
Yadava, Y.S., Singh, R.K., Choudhury, M., Kolekar, V., 1987. Limnology and productivity of Dighali *Beel* (Assam). Trop. Ecol. 28 (2), 137–146.
Yadava, Y.S., Goswami, M.M., Kar, D., Choudhuri, M., 1989. Present status of Hilsa resources in Assam. In: Das, P., Jhingran, A.G. (Eds.), Fish Genetics in India. Today and Tomorrow's Printers and Publishers, New Delhi, pp. 251–256.
Yakupitiyag, A., 2016. Farmer managed systems in Thailand: Is this CBF? Book of abstracts. In: The 3rd International Conference on Fisheries, 3 Academic Partner: Asian Institute of Technology (AIT), Thailand Strategic Partners.
Yamamoto, T., 1996. Community-based fisheries management. pp. 209–227. CBFM in Asia and the Pacific, Nihon University, 227, Tokyo (Japan).
Yamamoto, T., 2000. Collective Fishery Management Developed in Japan: Why Community-Based Fishery Management has been well-developed in Japan. Proceedings of IIFET, 2000, Japan International Fishery Research Society. Japan, pp. 1–7.
Yashmin, S., Romita, D., Pramila; Devi, B.R., Kar, D., 2007. Panorama of Fish catching devices in Manipur State of India. Sci. Soc. 5 (2), 181–188.
Yeung, K.M., 2008. Geodiversity: A new challenge for nature conservation in Hong Kong, Agriculture, Fisheries and Conservation Department, The Government of the Hong Kong Special Administrative Region. In: Proceedings of Asia-Pacific Seminar on Geological Conservation and Sustainable Development, Hong Kong, pp. 47–58.
Zhang, C.I.K., Lee, J.H., Lee, J.B., 2011. An Integrated Fisheries Risk Analysis Method for Ecosystems (IFRAME) approach for assessing, forecasting and managing Korean fisheries under a changing climate. In: Proceedings of Cook Island Workshop. pp. 1–48.

Africa

Anon, 2008a. Community-Based Natural Resource Management: How Knowledge is Managed, Disseminated and Used. International Fund for Agricultural Development (IFAD). pp. 1–51.
Anon, 2013a. Protecting and promoting marine biodiversity through small-scale, time-sensitive, community-based programmes. The MCAF (Marine Conservation Action Fund). Newsletter 2 (1), 1–4.
Anon, 2013b. Community-Based Natural Resource Management: Stock taking Exercise in Tanzania. In: Tanzania Natural Resource Forum, pp. 1–84.
Anon, 2015c. Community-Based Aquaculture: A Pioneering Viable Alternative to Fishing. Blue Ventures, London, pp. 1–9.
Blaikie, P., 2006. Is small really beautiful? Community-based natural resource management in Malawi and Botswana. World Dev. 34 (11), 1942–1957.
Blomley, R., Nelson, F., Martin, A., Martine, N., 2007. Community Conserved Areas: A review of status and needs in selected countries of Central and Eastern Africa. pp. 1–48.
Bulayi, M.E., 2001. Community-Based cooperative Fisheries Management for Lake Victoria Fisheries in Tanzania. 120 Reykjavik, Iceland Final Project 2001; Ministry of Natural Resources and Tourism; Fisheries Division; Dar-es-Salaam, Tanzania; and, University of Iceland: pp. 1–46.
Crawford, B., Olsen, S., Mills, D., Tsamenyi, M., Agbogah, K., Hutchful, G., 2013. Solving the Fisheries Crisis in Ghana: A Fresh Approach to Collaborative Fisheries Management, 2013. USAID-URI Integrated Coastal and Fisheries Governance (ICFG) Initiative, Coastal Resources Centre, Graduate School of Oceanography, University of Rhode Island. pp. 20.
Dahou, K., Déme, M., Dioum, A., 2001. Socio-economic and environmental impact of Senegalese Fishery Support Mechanisms. UNEP Fisheries Workshop; Geneva, 12 February 2001: pp. 34.
Gitonga, N., 2013. Strengthening Fisheries Management in ACP Countries. Final Technical Report. Support to East African Industrial Fishing and Fish Processors Association (EAIFFPA) to strengthen the task force approach in reducing illegal fishing and trade on Lake Victoria; Project Ref. No. EA-1.3-A3; Project Funded by the European Union; A project implemented by Fish Africa: pp. 1–56.
Keizire, B.B., 2001. Opportunities and Options for Financing Fisheries Management in Uganda: Final Project. University of Iceland, Reykjavik (Iceland), UNU-Fisheries Training Programme, pp. 1–53.
Huggins, L.E., 2011. Fencing Fisheries in Namibia and Beyond: Lessons from the Developing World. In: Meiners, R. (Ed.), PERC Policy Series No. 49. pp. 49.

Lohmeyer, U., 2002. Traditional Inland Fisheries Management and Enhancement Systems in Sub-Saharan Africa and their Potential for Development. Natural Resource Management: Back to Basics. Deutsche Gesellschaft für Technische Zusammenarbeit (GTZ) GmbH, Eschborn (Germany), pp. 1–212.

Malasha, I., 2006. Fisheries co-management: The Zambian experience. The World Fish Center, P.O. Box 51289, Lusaka, Zambia, Conference Paper No. 12: pp. 1–9.

McClanahan, T.R., Cinner, J.E., Abunge, C., Rabearisoa, A., Paubert, M., Ramahatratra, F., Andrianarivelo, N., 2014. Perceived Benefits of Fisheries Management Restrictions in Madagascar. Ecol. Soc. 19 (1), 5.

Melcher, A., Ouedraogo, R., Savadogo, M., Sendzimir, J. (Eds.), 2013. Current Questions in Sustainable Fisheries, Water Management and Higher Education: Book of Abstracts. SUSFISH Workshop and West Africa Symposium, 2013; Vienna, Austria, 23-28 of June, 2013: Sustainable Management of Water and Fish Resources in Burkina Faso, Project 56. pp. 1–39.

Nyeko, J.I., 2004. Community participation in Fisheries Management: Prospects and challenges for co-management. The East African Community EAC IRC Repository; LVBC Programmes and Projects; Fisheries Resources Department, Entebbe, Uganda; Lake Victoria Environmental Management Project, Phase I (LVEMP I), pp. 1–15.

Roe, D., Nelson, F., Sandbrook, C. (Eds.), 2009. Community management of natural resources in Africa: Impacts, experiences and future directions. Natural Resource Issues No. 18. International Institute for Environment and Development, London, UK, pp. 1–207.

Salehe, M.A., 2008. Capability and legitimacy of Beach Management Units (BMUs) to improve fisher's income through management of first-hand sales system in Lake Victoria, Tanzania. pp. 1–107, Master's thesis in International Fisheries Management, Department of Social and Marketing Studies, Norwegian College of Fishery Science, University of Tromsø (Norway).

Scullion, J., 2008. Inland fisheries co-management in East Africa. Project Management, IFMP project, Uganda, Conference Paper 13, pp. 1–22.

Australasia

Anders, K., Chris, R., Lyons, M., Phinn, S., Jupiter, S., 2011. Mapping Fish Community Variables by Integrating Field and Satellite Data, Object-Based Image Analysis and Modeling in a Traditional Fijian Fisheries Management Area. Remote Sens. (Basel) 3, 460–483.

Anon, 2004c. Transforming the Live Reef Food Fish Trade towards sustainability in the Republic of the Marshall Islands through community-based coral reef conservation and fisheries management. International Marinelife Alliance, Republic of the Marshall Islands, PHASE 1 FINAL REPORT, October 2002–March 2004, NOAA Coral Reef Conservation Grant Program FY2003; IMA, pp. 1–29.

Anon, 2014. SPC Activities: Community-based ecosystem approach to fisheries management (CEAFM) and climate change adaptation in the State of Yap, FSM: Current status of marine resources in Federated States of Micronesia (FSM). Department of Resources and Development, Government of the Federated States of Micronesia, pp. 1–2.

Annette, M.-H., Veitayaki, J., Polunin, N.V.C., Stead, S., Graham, N.A.J., 2006. Community-based marine resource management in Fiji: From yesterday to tomorrow. Conference Paper: International Coral Reef Symposium, pp. 1–8.

Brunnschweiler, J.M., 2009. The Shark Reef Marine Reserve: A marine tourism project in Fiji involving local communities. J. Sustain. Tourism, iFirst, 1–14.

Fetherston, E.H., 2005. Sustainability certification in Community-based Fisheries. In: 20th International Conference of The Coastal Society. pp. 92–96.

King, M., Lambeth, L., 2000. Fisheries Management by Communities: A Manual on promoting the management of subsistence fisheries by Pacific Island Communities. pp. 1–95, Community Fisheries Section, Secretariat of the Pacific Community, Noumea, New Caledonia, New Zealand Official Development Assistance, Australian Agency for International Development.

Marc, L., Jennifer, B., Jeremie, K., Jason, R., Caillon, S., 2013. Community-based management of near-shore fisheries in Vanuatu: What works? Mar. Policy 42, 167–176.

Metcalf, S.J., Gaughan, D.J., Shaw, J., 2009b. Conceptual models for Ecosystem-Based Fisheries Management (EBFM) in Western Australia. In: Fisheries Research Report No. 194. Fisheries Research Division, Western Australian Fisheries and Marine Research Laboratories. pp. 1–42.

Rhodes, K.L., Warren-Rhodes, K., Houk, P., Cuetos-Bueno, J., Fong, Q., Hoot, W., 2011. An Interdisciplinary Study of Market Forces and Nearshore Fisheries Management in Micronesia. pp. 1–141, The Nature Conservancy, Report No. 6/11; A Report of the Marine Programme of the Asia Pacific Conservation Region, Coastal Fisheries Management Programme, Asia Pacific Resource Centre.

Warner, M., 2000. Conflict management in community-based natural resource projects: Experiences from Fiji and Papua New Guinea. Overseas Development Institute, Working Paper 135, Portland House, Stag Place, London SW1E 5DP, UK, pp. 1–42.

Europe

Anon, 2001. Country note on fisheries management systems: France. Mar. Policy 25 (2), 103–112.

Anon, 2004d. Community-Driven Development and Natural Resource Management. In: IFAD Workshop on Community-Driven Development, Casa San Bernardo, Rome; 4–5 June, 2004. pp. 1–246.

Anon, 2006b. Report of The Bergen Conference on implementing the Ecosystem approach to Fisheries. 26–28 September 2006, Bergen, Norway: pp. 1–28.

Anon, 2008b. Fisheries operational programme of the Republic of Hungary; 2007–2013: For the renewal of the fishery sector in Hungary. Ministry of Agriculture and Rural Development of the Republic of Hungary, pp. 1–79.

Anon, 2008c. National Sustainable Development Strategy, Romania, 2013–2020–2030. Ministry of Environment and Sustainable Development, Government of Romania; United Nations Development Programme; National Centre for Sustainable Development, pp. 1–143.

Anon, 2008d. Fourth National Report of the implementation of CBD in Republic of Macedonia (2006–2008). pp. 1–32.

Anon, 2009b. Operational Programme for Fisheries Romania, 2007–2013. Ministry of Agriculture and Rural Development, Government of Romania, National Rural Development Programme, 2007–2013, Third version, pp. 1–817.

Anon, 2009c. Green Paper: Reform of the Common Fisheries Policy and A Sustainable Future for Aquaculture; 82nd plenary session; Committee of the Regions; DEVE-IV-045. European Union; Opinion of the Committee of the Regions, pp. 1–16.
Anon, 2009d. Studies and Pilot Projects for carrying out the Common Fisheries Policy No FISH/2007/03. European Commission: An analysis of existing Rights Based Management (RBM) instruments in Member States and on setting-up the best practices in the EU; Final report, Part I, pp. 1–135.
Anon, 2013c. Convention on Biological Diversity. 5th National Report, 2009–2013, Ministry of Environment and Water, Republic of Bulgaria, pp. 1–129.
Anon, 2013d. Key Factors Supporting Small-Scale Coastal Fisheries Management: Synthesis review. Evaluation Office, Rockefeller Foundation, IMM Ltd.: pp. 1–122.
Arlinghaus, R., Mehner, T., Cowx, I.G., 2002. Reconciling traditional inland fisheries management and sustainability in industrialized countries, with emphasis on Europe. Fish Fish. 3 (4), 261–316.
Bănaru, D., Le Manach, F., Färber, L., Zylich, K., Pauly, D., 2015. From bluefin tuna to gobies: a reconstruction of the fisheries catch statistics in Romania, 1950–2010. Fisheries Centre, The University of British Columbia, Working Paper Series, Working Paper, 2015 – 48, pp. 1–11.
Cadiou, G., Boudouresque, C.F., Bonhomme, P., Le Diréach, L., 2008. The management of artisanal fishing within the Marine Protected Area of the Port-Cros National Park (North-West Mediterranean Sea): A success story? ICES J. Marine Sci. 66 (1), 41–49.
Campbell, L.M., Silver, J.J., Gray, N.J., Ranger, S., Broderick, A., Fisher, T., Godfrey, M.H., Gore, S., Jeffers, J., Martin, C., Andrew, M., Richardson, P., Sasso, C., Slade, L., Godley, B., 2008. Co-management of sea turtle fisheries: Biogeography *versus* geopolitics. Mar. Policy, 1–9.
Depestele, J., Courtens, W., Degraer, S., Haelters, J., Hostens, K., Houziaux, J.-S., Merckx, B., Polet, H., Rabaut, M., Stienen, E.W.M., Vandendriessche, S., Verfaillie, E., Vincx, M., 2012. An integrated impact assessment of trammel net and beam trawl fisheries. Science for a Sustainable Development, SD/NS/08A, p. 233, Belgian Science Policy Office (SSD), Brussels (Belgium).
Depestele, J., Wouter, C., Steven, D., Jan, H., Kris, H., Mardik, L., Eunice, P., Bea, M., Hans, P., Marijn, R., Henning, R., Eric, S., Sofie, V., Filip, A.M.V., Magda, V., 2014. Sensitivity assessment as a tool for spatial and temporal gear-based fisheries management. Ocean Coast. Manag. 102, 149–160.
Döring, R., Hart, P., Johnson, M., 2003. Small scale *versus* trawl fisheries in coastal waters: The case of the German Baltic Herring fishery. Proceedings of the Workshop 'Who Owns the Sea', Tjärno, Sweden, pp. 113–140.
Eliasen, S., Sverdrup, S., 2009. Nordic experience of fisheries management: Seen in relation to the reform of the EU Common Fisheries Policy. TemaNord 2009: 579, pp. 1–93.
Giger, W. (Ed.), 2008. Proceedings of the Swiss-Romanian Research Programme on Environmental Science and Technology (ESTROM) 61. National Institute of Marine Geology and Geo-ecology, Geo-Eco-Marina 14, pp. 1–95.
Gutierrez, Nicolas L. (2015). Management and Co-management options for small-scale fisheries in the Mediterranean and the Black sea; Updating National Biodiversity Strategies and Action Plans. In book: First Regional Symposium on Sustainable Small-Scale Fisheries in the Mediterranean and Black Sea. Publisher: Food and Agricultural Organization of the United Nations. Editors: Abdellah Srour, Nicola Ferri, Dominique Bourdenet, Davide Fezzardi and Aurora Nastasi: pp.147-184.
Hausner, V.H., Fauchald, P., Jernsletten, J.-L., 2012. Community-based management: under what conditions do sami pastoralists manage pastures sustainably? PLoS One 7 (12), 1–9.
Irvine, K., 2014. Integrated Water Resources Management (IWRM): From local to Global sustainability: Towards Integrated Water Management. TIMe project 2014-W-DS-20, pp. 1–57.
Margrethe, A., Claire, W.A., Helen, J.B., Christine, R., 2014. What does stakeholder involvement mean for fisheries management? Ecol. Soc. 19 (4), 35.
Mirela, A., Michael, M.C., Alexandra, C., Nicoleta, D., Dirk, R.F., Veronica, M., Emil, P., Cosima, R., 2005. Romania Danube Delta Biodiversity project: Local benefits Case Study Report. Stockholm Environment Institute, Global Environment Facility, Romania DDBP Local Benefits Case Study Report, pp. 1–91.
Neculita, M., Sarpe, D.A., Afanase, C., Neculita, V., 2003. Romania's fishing policy competitiveness in European integration context. Development, Energy, Environment, Economics, pp. 1–4.
Newman, S., 2014. Structural and Cohesion Policies: Inland fisheries and the Common Fisheries Policy. Institute for European Environmental Policy European Union, pp. 1–46.
Olivier, G., Olivier, T., 2001. Distributional issues in the operation of rights-based fisheries management systems. Mar. Policy 25 (2), 103–112.
Robert, A., Robin, S., Tamara, K., 2004. Freshwater Fisheries in Central and Eastern Europe: The Challenge of Sustainability. Overview Report. IUCN, Warsaw, Poland. English-language edition by: Robin Sharp; Project Coordinators: Zenon Tederko Tamara Kutonova, pp. 1–94.
Salonen, E., Mutenia, A., 2007. Alien fish species in northernmost Finland. Finnish Game and Fisheries Research Institute, Helsinki, TUTKIMUKSIA, 2 /2007, Finland, pp. 1–17.
Schmutz, S., BOKU, Cowx, I.G., Haidvogl, G., Pont, D., 2007. Fish-based methods for assessing European running waters: A synthesis. Fish. Manage. Ecol. 14, 369–380.
Søreng, S.U., 2008. Fishing rights discourses in Norway: Indigenous *Versus* Non-indigenous Voices. MAST 6 (2), 77–99.
Stedinger, J.R., Jozef, P.M., Monique, D., Villars, T., 2005. Water Resources Systems Planning and Management: An Introduction to Methods, Models and Applications. In: Loucks, D.P., van Beek, E. (Eds.), Studies and Reports in Hydrology. UNESCO Publishing, pp. 1–690.
Zick, D., Gassner, H., Filzmoser, P., 2006. Changes in the fish species composition of all Austrian lakes > 50 ha during the last 150 years. Fish. Manage. Ecol. 13, 103–111.

America

Anon, 1999b. The Role of Major Groups in Sustainable Oceans and Seas: A series of case studies prepared jointly by WWF International and the Division for Sustainable Development in collaboration with Equilibrium Consultants. Background paper prepared for the Commission on Sustainable Development, Seventh Session, New York, April 1999. United Nations Department of Economic and Social Affairs, Division for Sustainable Development and World Wide Fund for Nature (WWF).

Anon, 2002. Fisheries Subsidies and Marine Resources Management: Lessons learned from Studies in Argentina and Senegal. In: United Nations Environment Programme, pp. 1–64.

Anon, 2005b. Sustainable Fisheries and Livelihoods in Latin America: The Imperative of recognising the Resource Access Rights of Artisanal Fishermen, Coastal Communities and Indigenous Peoples, Santa Clara, Argentina, 1–4 March. In: Proceedings of Workshop, pp. 1–4.

Anon, 2005c. Fishing Community initiatives in Community Based Fisheries Management and The Sitka Declaration: Proposal for Forum of Fishermen, NGO's, and the Government Agencies to explore expansion of the use of Sectors, Catch Share, and Community Based Fisheries Management in the Northeast and Caribbean. pp. 1–2.

Anon, 2006c. Agricultural and Fisheries Policies in Mexico: Recent achievements and continuing with the reform agenda. OECD-Multi-lingual Summaries, pp. 1–6.

Anon, 2007b. A Study Guide Series on Peace and Conflict for Independent Learners and Classroom Instructors Natural Resources, Conflict, and Conflict Resolution. United States Institute of Peace, Washington, DC. pp. 1–36.

Anon, 2007c. Sustaining America's Fisheries and Fishing Communities: An evaluation of incentive-based management. In: Environmental Defense. pp. 1–36.

Anon, n.d., Governance and Marine Fisheries: Comparing across countries and stocks, pp. 1–2.

Ayres, S., 2009b. Fishing for Solutions: A Study of Community Based Fisheries Management and its Implementation in the Gulf of Maine. Reflection Papers for the Class of 2009; The Goodwin-Niering Center for Conservation Biology and Environmental Studies, The Gulf of Maine Research Institute, Portland, Maine, pp. 1–19.

Charles, A., 2008. Turning the Tide: Toward Community-Based Fishery Management in Canada's Maritimes. vol. 49 American Fisheries Society Symposium, pp. 569–573.

Charles, T., 2010b. Combining Environmental, Economic and Social Sustainability in Fisheries: Policy Options and Cautionary Notes. Presentation Notes. Policy Day—Fisheries Policy Reform in the European Union, Montpellier, France, pp. 1–56.

Daniela, K., Marcelo, V., Pitcher, T.J., 2006. An estimation of compliance of the fisheries of Argentina with Article 7 (Fisheries Management) of the UN Code of Conduct for Responsible Fishing. pp. 1–19.

Donald, R.S., Ostazeski, J.J., Theodore, N.H., Steven, A., 2006. Fisheries Management Plan for the Minnesota Waters of Lake Superior. Fisheries Management Section, Division of Fish and Wildlife, Minnesota Department of Natural Resources in cooperation with the Lake Superior Advisory Group, Lake Superior Fishery Area, Special Publication, 163: pp. 1–100.

Doyon, S., 2007. Fishing for the Revolution: Transformations and Adaptations in Cuban Fisheries. MAST 6 (1), 83–108.

Eggert, H., Greaker, M., 2005. Effects of Global Fisheries on Developing Countries: Possibilities for income and threat of depletion. Int. Rev. Hydrobiol. 90 (4), 422–432.

Enríquez, R., Camargo, G.B., 2000. Perspectives for Rights-Based Fishery Management in Mexico: Microbehavior and Macroresults. In: Proceedings of the Tenth Biennial Conference of the International Institute of Fisheries Economics and Trade Presentations. pp. 1–27.

Pablo, G.-D., Carmen, G.M., 2012. A community-based approach to the management of an aggregating reef fishery in the Mesoamerican Reef. Appl. Biodivers. Sci. Persp., 1–11.

Gudrun, G., Andy, B., Bob, A., 2016. Canadian 4VWX Purse Seine Herring Fishery, Sustainable Fisheries Certification. Final Report, Marine Stewardship Council (MSC), Certification Body: Food Certification International Ltd (Trading Acoura Marine), pp. 1–288.

Guillermo, X.B., 2002. Community-based conservation of the *callo dehacha* fishery by the Comcaac Indians, Sonora, Mexico, pp. 1–187. A Thesis submitted to the Faculty of the School of Renewable Natural Resources, in partial fulfillment of the requirements for the Degree of Master of Science with a major in Renewable Natural Resources Studies in the Graduate College, the University of Arizona (USA).

Hernández, J.A.B., 2006. Cuban fisheries management regime: Current state and future prospects. Final Project, Institute of Economic Studies, University of Iceland and Dirección de Regulaciones Pesqueras, Ministerio de la Industria Pesquera, 5ª Ave. y 246, Barlovento. Playa, Ciudad de La Habana, Cuba, pp. 1–33.

Iverson, C.P., 2013. US Regional Fishery Management Councils: Opportunities and Challenges. pp. 1–40.

Marcelo, V., Johanna, H., Alida, B., 2002. The use of Ecosim to investigate multispecies harvesting strategies for capture fisheries of the Newfoundland-Labrador shelf, pp. 68–72. In: Pitcher, T., Cochrane, K. (Eds.), The Use of Ecosystem Models to Investigate Multispecies Management Strategies for Capture Fisheries. Labrador Shelf, pp. 1–159. A report co-published as an FAO Fisheries Report: Using Ecosim for Fisheries Management: FAO/Fisheries Centre Workshop, Page 3; A Research Report from The UBC Fisheries Centre and the Food and Agriculture Organisation of the United Nations: Fisheries Centre, University of British Columbia, Canada, Fisheries Centre Research Reports, 10 (2): 1–159.

McCay, B.J., Micheli, F., Ponce-Díaz, G., Murray, G., Shester, G., Ramirez-Sanchez, S., Weisman, W., 2014. Cooperatives, concessions, and co-management on the Pacific coast of Mexico. Mar. Policy 44, 49–59.

McGrath, D.G., Cardoso, A., Sá, E.P., 2004. Community fisheries and Co-management on the lower Amazon floodplain of Brazil. Food and Agricultural Organisation of the United Nations, pp. 1–17. In: Proceedings of the second international symposium on the management of large rivers. FAO, Phnom Penh, pp. 207–221.

Onestini, M., 2002. Subsidies in Argentine Fisheries, pp. 11–24. In: Anon (Ed.), Fisheries Subsidies and Marine Resources Management. Lessons learned from Studies in Argentina and Senegal. United Nations Environment Programme, pp. 1–64.

Orensanz, J.M., Seijo, J.C., 2013. Rights-based management in Latin American fisheries. FAO (Rome), Fisheries and Aquaculture Technical Paper No. 582, pp. 1–155.

Palmigiano, K., Crawford, B., 2006. Establishing Community-based Marine Protected Areas for Cockle Fishery Management. Western Indian Ocean Marine Science Association, University of Dar es Salaam, University of Hawaii, Hilo and the Coastal Resources Centre, University of Rhode Island, pp. xix + 22.

Pinedo, D., Summers, P.M., Smith, R.C., Saavedra, J., Zumaeta, R., Almeyda, A.M., 2000b. Community-based natural resource management as a non-linear process: A case study in the Peruvian Amazon *varzea*. Eighth IASCP Conference, Bloomington, Indiana 30 May–4 June, 2000, pp. 1–28.

Poon, S.E., Bonzon, K., 2013. Territorial use rights for fishing: cooperative, catch share design manual. In: Environmental Defense Fund, Catch Shares. vol. 3, pp. 1–161.

Ramade, M., Bourillón, L., 2004. Community-based fisheries management and eco-labelling in Baja California. Mexico, pp. 1–17.

Saenz-Arroyo, A., Torre, J., Bourillon, L., Kleiberg, M., 2005. A community-based marine reserve network in North-western Mexico. In: Proceedings of Symposium and Workshop of the North American Marine Protected Areas Network. North American Commission for Environmental Co-operation, pp. 19.

Sale, P.F., Cowen, R.K., Danilowicz, B.S., Jones, G.P., Kritzer, J.P., Lindeman, K.C., Planes, S., Polunin, N.V.C., Russ, G.R., Sadovy, Y.J., Steneck, R.S., 2005. Critical science gaps impede use of no-take fishery reserves. Trends Ecol. Evol. 20 (2), 74–80.

Schonberger, S.N., Agar, J.J., 2001. Towards Rights-Based Fisheries Management: Fisheries and Aquaculture Thematic Group, Latin America and Caribbean region: Argentina: The World Bank. Rural Development Family: Working Paper, 22816, pp. 1–120.

Shing, C.C.A., 2001b. Case Study of the Integrated Coastal Fisheries Management Project: A Pilot Project for the Gulf of Paria, Trinidad. CANARI Technical Report No. 280, Division of Fisheries, Caribbean Natural Resources Institute (Canari), Trinidad and Tobago, pp. 1–13.

Soeftestad, L.T., 1999. The International Workshop on Community-Based Natural Resource Management (CBNRM), Washington DC, 10–14 May 1998. Workshop Report, Washington, DC, pp. 1–36.

Sullivan, K., 2014. Exploration of Ecosystem Based Fishery Management in the United States. A Report from the NOAA Science Advisory Board, pp. 1–25.

Tissot, B.N., 2005b. Integral Marine Ecology: Community-based Fishery Management in Hawaii; World Futures. J. Gen. Evol. 61, 79–96.

Tony, P., Kevern, C. (Eds.), 2002. The Use of Ecosystem Models to Investigate Multispecies Management Strategies for Capture Fisheries. Fisheries Centre Research Reports, University of British Columbia, Vancouver and FAO Fisheries Report, 10 (2): 156.

Wibera, M., Berkesb, F., Anthony, C., John, K., 2004. Participatory research supporting community-based fishery management. Mar. Policy 28, 459–468.

Further reading

Acharya, S., Barman, R., Bhattacharjee, P.C., Botre, S., Chhatre, A., Gandhi, K., Ganguli, A., Ghate, U., Kar, D., Mishra, M., Panigrahi, R., Pramod, P., Seshagiri Rao, P.R., Srivastava, D.S., Vardhan, M., 1998. Conservation. pp. 107–137, In: Gadgil, M. (Ed.), The Hindu Survey of the Environment. Kasturi and Sons, Chennai. pp. 216.

Agardy, T., 2000. Effects of fisheries on marine ecosystems: A conservationist's perspective. ICES J. Marine Sci. 57 (3), 761–765.

Allison, E.H., Ratner, B.D., Bjorn, A., Willman, R., Pomeroy, R., Kurien, J., 2012. Rights-based fisheries governance: from fishing rights to human rights. Fish Fish. 13, 14–29.

Ayres, S., 2009a. Fishing for Solutions: A Study of Community Based Fisheries Management and its Implementation in the Gulf of Maine. Senior Integrative Project Abstracts for the Class of 2009. pp. 117.

Barber, C.V., Pratt, V.R., 1997. Policy Reform and Community-Based Programmes to Combat Cyanide Fishing in Philippines. World Resources Institute and International Marinelife Alliance, The Philippines. SPC Live Reef Fish Information Bulletin #3, pp. 26–35.

Barbhuiya, A.H., Kar, D., 2012. Treatise on Mahseer Fishes. pp. vii + 293, Dominant Publishers, New Delhi). ISBN: 978-93-82007-17-3.

Barman, R.C., Kar, D., 2015. Role of Education and Social Participation on community-based Floodplain Wetland Management. ESTIA (Multidiscip. Ann. Res. J.) 1, 16–22. ISSN: 2454-1451.

Baticados, D., Agbayani, R., 1998. Case study of institutional arrangement in the fisheries co-management of Malalison. South East Asian Fisheries Development Center, Island, Central Philippines, Hoilo.

Baylon, C.C., 2001. Micro-enterprise development in selected fishing communities in the province of Iloilo, Philippines. The Project Report of The Food and Agriculture Organization (FAO) of the United Nations (UN). pp. xix + 313.

Bhattacharjya, B.K., 2009. Conservation and sustainable use of *Beel* Fisheries In Assam. In: Souvenir. Department of Fisheries, BTC, Government of Assam, pp. 60–65.

Binky, K., Kar, D., 2010. Intensity of Parasitisation among Fishes of Karbhala wetland in Cachar district of Assam. Life Sci. Bull. 7 (2), 233–234.

Binky, K., Kar, D., 2011. Ichthyospecies Diversity of Karbhala Wetland in Cachar district of Assam. Environ. Ecol. 29 (1), 17–19.

Binky, K., Shomorendra, M., Kar, D., 2011. Nematode Parasites of Karbhala Wetland in Silchar Assam. Biol. Forum (Int. J.) 3 (2), 18–21.

Binky, K., Shomorendra, M., Kar, D., 2012. Redescription of two known species of genus *Metaclinostomum* Pandey and Baugh, 1965 (Clinostomatidae) and one new species of genus *Allocreadium* Looss, 1900 (Allocreadiidae). In: Biospectra. pp. 1–6.

Binky, K., Shomorendra, M., Kar, D., 2015. A Treatise on the Helminth Fauna of Fishes. pp. xxvi + 262, Asiatic Publishing House, New Delhi.

Chakraborty, T.R., 2006. Management of haors, baors, and beels in Bangladesh: Lesson for Lake Basin management. In: Natural Resource Studies. pp. 1–14.

Charles, A., 2006. Turning the Tide: Toward Community-Based Fishery Management in Canada's Maritime. American Fisheries Society Symposium.

Choudhury, M., 1987. Some aspects of biostatistical studies on catch, growth and ecological parameters of freshwater Indian Major Carps in the lower reach of the River Brahmaputra, Assam (India). Unpublished Ph. D. Thesis, Gauhati University, Assam, pp. 202.

Cunningham, S., Neiland, A.E., 2009. Wealth-based fisheries management: using fisheries wealth to orchestrate sound fisheries policy in practice. Mar. Resour. Econ. 24, 271–287.

Das, B., Kar, D., 2011a. Essentials of Limnology and Fishery Science. p. 190, Mangalam Publishers, New Delhi.

Das, B.K., Kar, D., 2011b. Habitat mapping, spatial analysis of fish diversity of River Subansiri during winter season in Assam and Arunachal Pradesh (India). Environ. Ecol. 29 (4A), 1948–1951.

Das, B.K., Kar, D., 2015a. Physico-Chemical Parameters and Drainage Types of River Siang in Arunachal Pradesh, India. In: Mishra, G.C. (Ed.), Conceptual Framework and Innovations in Agroecology and Food Sciences. Krishi Sanskriti Publications, New Delhi, pp. 53–56. pp iii + 82.

Das, B.K., Kar, D., 2015b. Drainage Pattern and their Geological Significance of River Siang in Arunachal Pradesh, India. ESTIA (Multidiscip. Ann. Res. J.) 1, 32–38. ISSN: 2454-1451.

Das B.K. and Kar D. (2015c). Habitat Inventory Parameters and Drainage Analysis of River Siang in Arunachal Pradesh. Int. J. Marine Sci. Vol. 5, No.48: 1–7, http://ijms.biopublisher.ca. (Impact Factor: 1.543) (Canada).

Das, P., Kar, D., 2016d. Composition, abundance and diversity of zooplankton population from three different wetlands from Barak valley, Assam. Int. J. Appl. Nat. Sci. 5 (5), 41–46 (NAAS: 2.74; Impact factor: 3.2816)(Accepted).

Das, B.K., Boruah, P., Kar, D., 2014a. Icthyo-Faunal diversity and conservation strategies of Siang River in Arunachal Pradesh. In: Singha, D.N.G.C. (Ed.), Natural Resources: Utilizing and Conserving for Human Welfare and Sustenance. Global Publishing House, Visakhapatnam, India, pp. 139–151.

Das, B.K., Boruah, P., Kar, D., 2014b. Physico-chemical parameters of River Siang in Arunachal Pradesh, India. In: Banerjee, P., Rajguru, U., Sarkar, C.R. (Eds.), Proceedings of the National Seminar on Advances in Physical and Biological Sciences. Jagiroad College, Morigaon, Assam, India, pp. 78–88.

Das, B.K., Boruah, P., Kar, D., 2014c. Fish diversity and Habitat Mapping of River Siang in Arunachal Pradesh using Remote Sensing and GIS. In: Mishra, G.C. (Ed.), Innovative Energy Technology Systems and Environmental Concerns: A sustainable approach. Research India Publication, New Delhi, pp. 13–20.

Das, B.K., Boruah, P., Kar, D., 2014d. Potential Application of Geographic Information System (GIS) and Remote Sensing (RS) in River Analysis. Int. J. Recent Scient. Res. 5 (4), 828–830.

Das, S., Das, T., Das, B.K., Kar, D., 2014. Length-weight relationship and condition factor in *Barilius sacra* of River Torsa in West Bengal. Int. J. Scient. Res. Rev. 3 (3), 65–69.

Das, B.K., Boruah, P., Kar, D., 2016b. Habitat mapping and fish diversity of River Siang in Arunachal Pradesh. Global J. Environ. Sci. Res. 3 (1), 9–18. Taiwan.

Das, S., Purkayastha, D., Kar, D., 2016a. Length-weight relationship and Relative Condition factor in *Puntius ticto* of River Singla in the Karimganj district of Assam, India. Int. J. Appl. Res. 16, 32–38.

Datta, B., Kar, D., 2011. Aquaculture in Dibrugarh district of Assam. Environ. Ecol. 27 (2), 1520–1523.

Devi, Maibang, B., Das, P., Das, U., Kar, S., Aditya, G., Kar, D., 2016. Rotifer Species Assemblages in Three Freshwater Habitats of Manipur, India. In: Proceedings of the Zoological Society, Springer., https://doi.org/10.1007/s12595-016-0201-z. pp. 1–9.

Dhar, N., Goswami, M.M., Kar, D., 2004a. Aquatic macrophytes in Baskandi Lake, an Oxbow Lake of Barak basin, Assam. Environ. Ecol. 22, 606–608. Spl-4.

Dhar, N., Goswami, M.M., Kar, D., 2004b. Co-operative Movement: Need of the hour in Capture Fishery Sector. In: Goswami, P.J., Acharjee, P.R. (Eds.), Cooperative Movement in Assam: Problems and Prospects. GC College Credit and Thrift Co-operative Society Ltd., Silchar (Assam), pp. 113–116. pp. xiv + 124.

Dutta, B., Biswas, S.P., Kar, D., 2013. Hydrobiological parameters of ponds of NR Hatchery at Lahowal at Dibrugarh in Assam. Environ. Ecol. 31 (2B), 844–846.

Dutta, B., Boruah, P., Kar, D., 2015b. Fishes of the Towkak River: with special emphasis on the habitat requirements. Int. J. Manage. Social Sci. (IJMSS), 3 (3), 38–44.

Dutta, B., Baruah, L., Baruah, P., Kar, D., 2015a. Studies on the potential Habitat Paraameters for the fishes of the Towkak River in Assam and Nagaland, India. In: Biodiversity Conservation and Sustainable Use. Print World Print and Publication, Jorhat, Assam, India. pp. 52–57.

Dutta, B., Das, N., Kar, D., 2016. The fish catching devices with their efficacy and cost-benefit analysis in the Towkak River in Assam and Nagaland, India. Fish Aquac J 7, 156. https://doi.org/10.4172/2150-3508.1000156. (USA), Fisheries and Aquaculture Journal, Dutta et al., Fish Aquac J 2016, 7:1.

Edwin, L., Hridayanathan, C., 2004. Overfishing—A study with reference to the ring seine fishery. Ind. J. Fisher. 51 (3). CMFRI.

Foltz, C., Pomeroy, R.S., Barber, C., 1996. Proceeding of the Visayas-wide conference on community-based coastal resource management and co-management. 1–115, Fisheries Co-management Research Project Working Paper No. 4, Manila, ICLARM Forum J. 3, 56–63.

Frid, C.L.J., Clark, R.A., Hall, J.A., 1999. Long-term changes in the benthos on a heavily fished ground off the NE coast of England. Mar. Ecol. Prog. Ser. 188, 13–20.

Gawler, M., 2000. What are best practices? Lessons in participatory management of inland and coastal wetlands. In: Gawler, M. (Ed.), Strategies for wise use of wetlands: Best practices in participatory management. Proceedings of Workshop 1 of the 2nd International Conference on Wetlands and Development. Wetlands International, Wageningen, The Netherlands. in press at http://www.wetlands.agro.nl.

Gazi Nurul, I., Barr, J., 2006. Edited by D. Malcolm, Livelihoods Impact, Policy Brief, 3. the WorldFish Center, Bangladesh and South Asia Office.

Geetarani, B., Shomorendra, M., Vishweshwari, A., Kar, D., Jha, A.N., 2010. Helminthic infection in relation to the host length and sex in some freshwater fishes, *Anabas testudineus* (Bl.) and *Channa gachua* (Ham.). Biospectra 5 (2), 233–236.

Geetarani, B., Shomorendra, M., Kar, D., 2011. Helminth-fauna of Air-breathing Fishes of Utra Lake in Manipur. Environ. Ecol. 29 (1), 176–178.

Graham, J., Charles, A., Bull, A., 1971. Community Fisheries Management Handbook. Gorsebrook Research Institute, Saint Mary's University. 2006. pp. xiii + 437.

Halls, A.S., Mustafa, M.G., 2006. Final Assessment of the Impact of the CBFM Project on Community—Managed Fisheries in Bangladesh Report to the World Fish Center, Bangladesh, September 2006. 83pp.

Hassan, F., Sathiadhas, R., 2005. Fisherwomen of Coastal Kerala: A Socio-economic perspective. In: The Seventh Indian Fisheries Forum Proceedings, AFSIB, Mangalore.

ICES, 1998. Report of the Working Group on the Ecosystem Effects of Fishing Activities ICES, Copenhagen. ICES CM 1998/ACFM/ACME: 01 Ref.: E, 263pp.

ICES, 2000. Ecosystem effects of fishing. In: ICES Marine Science Symposia. 210. ICES Journal of Marine Science 57. 791pp.

Islam, N.M., Dickson, 2006. Livelihoods Impact of the CBFM - 2 Project. WorldFish Center, Bangladesh Office. Working Paper 2006/02.

Joshi, H.D., Ramteke, K.K., 1991. Floodplain wetland-an important aquatic resource for enhancement: aquafind. (aquatic fish database, estd. 1991).

Kalita, M.N., Kar, D., Singh, H.T., 1993. Studies on toxicity of certain piscicidal plants of Assam to fishes. J. Assam Sci. Soc. 35 (2), 125–129.

Kar, D., 2002b. Waterbodies and Fishes: Resources and Development. Conserv. Forum J. 1, 15.

Kar, D., 2003b. Peoples' Perspective on Fish Conservation in the Water bodies of South Assam, Mizoram and Tripura: pp. 325–328. In: Mahanta, P.C., Tyagi, L.K. (Eds.), Participatory Approach for Fish Biodiversity Conservation in North-East India. National Bureau of Fish Genetic Resources (ICAR), Lucknow. v + 412.

Kar, D., 2004b. Inventorying of Fish Biodiversity in North-East India with a note on their Conservation. In: Proc. National Conference on Fish and their Environment, 9-11 Feb 2004, B.A. Marathwada University, Aurangabad. p. 1.

Kar, D., 2004c. Conservation and Ichthyodiversity in the River Systems of Barak drainage, Mizoram and Tripura. In: Proceedings of National Symposium Management of Aquatic Resources for Biodiversity Maintenance and Conservation, 2-4 Feb 2004, J.N. Vyas University, Jodhpur, p. 1.
Kar, D., 2005b. Fish Biodiversity and Habitat Parameters of rivers in Barak drainage (Assam), in Mizoram and in Tripura. Himalayan J. Environ. Zool. 19 (1), 41–45.
Kar, D., 2005c. Fish Diversity in the Major Rivers in Southern Assam, Mizoram and Tripura, pp. 679–691. In: Nishida, T., Kailola, P.J., Hollingworth, C.E. (Eds.), Proc. 2nd International Symposium on GIS and Spatial Analyses in Fisheries and Aquatic Sciences, 2–6 Sep 2002, University of Sussex, Brighton (UK). Vol. 2. Fisheries and Aquatic GIS Research Group, Kawagoe, Saitama (Japan).
Kar, D., 2006b. Study of Fish Diversity in a Wetland. In: Activity Guide `Biodiversity'. National Childrens' Science Congress, NCSTC Network and Dept. of Science & Technology, Govt. of India, pp. 37–38.
Kar, D., Barbhuiya, M.H., 2000b. Length-Weight Relationship and Condition Factor in *Gudusia chapra* (Hamilton-Buchanan) and *Botia dario* (Hamilton-Buchanan) in Chatla Haor (Floodplain Wetland) in Cachar district of Assam. Environ. Ecol. 18 (1), 227–229.
Kar, D., Barbhuiya, A.H., 2013. Frontiers of Wetlands, Fishers and Aquaculture Research. pp. 263, Mangalam Publications, New Delhi. ISBN: 978-93-81142-99-8.
Kar, S., Kar, D., 1989. Food habit and malaria among the Ao Nagas of Nagaland: a brief report. Soc. Sci. 2 (2), 129–132.
Kar, S., Kar, D., 1994. Health and Disease pattern among the Ao Nagas. Social Sci. VI (7), 417–422.
Kar, S., Kar, D., 2004a. Osmotic fragility and malaria among the Ao Nagas of Nagaland. Environ. Ecol. 22 (1), 243–244.
Kar, S., Kar, D., 2016a. Zooplankton diversity of a freshwater pond in Cachar district of Assam, India. Int. J. Life Sci. 4 (1), 125–128.
Kar, S., Kar, D., 2016b. Zooplankton diversity of a freshwater perennial pond in Silchar city of Assam, India. Am. Int. J. Res. Formal, Appl. Nat. Sci. 14 (1), 10–14.
Kar, D.; Dey, S.C.; Kar, S.; Michael, R.G.; and, Gadgil, M. (1996b). Ichthyoecology, Management and Conservation Fish Resources of Lake Sone in Assam (India). Tiger Paper (FAO, UN), XXIII (3): 27-32.
Kar, D., Purkayastha, M., Kar, S., 1996c. Biodiversity Conservation Prioritisation Project: A case study from Sone Beel in Assam. In: Proc. National Workshop on Biodiversity Conservation Prioritisation Project (BCPP), World Wide Fund (WWF) for Nature-India and Centre for Ecological Sciences, Indian Institute of Science (Bangalore).
Kar, D., Laskar, B.A., Nath, D., Mandal, M., Lalsiamliana, 2002c. *Tor progenius* (McClelland) under threat in River Jatinga, Assam. Sci. Culture 68 (7–8), 211.
Kar, D., Laskar, B.A., Nath, D., Mandal, M., Lalsiamliana, 2002d. *Tor progenies* (McClelland) under threat in River Jatinga, Assam. Sci. Culture 68 (7–8), 211.
Kar, D., Laskar, B.A., Nath, D., 2002e. *Tor* species (Mahseer fish) in River Mat in Mizoram. Aquacult (Nature Conserv.) 3 (2), 229–234.
Kar, S., Kar, D., Arjun, J., 2002b. Genetic Survey among the Kuki Tribal Populations in Loharbond area of Cachar, Assam. Environ. Ecol. 20 (1), 82–85.
Kar, D., Baishya, A.K., Mandal, M., Arjun, J., Laskar, B.A., 2003b. A preliminary account of the Vegetations in Dargakona locality of Silchar in Assam. Environ. Ecol. 21 (2), 254–258.
Kar, D., Laskar, B.A., Nath, D., 2006b. Length-weight relationship of *Neolissochilus hexagonolepis* (McClelland) and *Garra lissorhynchus* (McClelland) from River Jatinga in Cachar district of Assam. Indian J. Fish 52 (4), 495–496.
Kar, D., Barbhuiya, A.H., Das, B., Purkayastha, M., Barbhuiya, M.H., Saha, D., Laskar, R.A., Arjun, J., 2010b. Medicinal plants in certain regions of Assam, India. Environ. Ecol. 28 (3B), 1965–1968.
Kar, D., Barbhuiya, A.H., Saha, B., Devi, S., Medeline, G.M., Das, B., 2010a. Protection of fish diversity in Meghalaya needs conservation measures: in the context of fishing devices used. Fish. Chimes 30 (1), 124–127.
Khan, M.S., Apu, N.A., 1998. Fisheries Co-management in Oxbow lakes of Bangladesh. Chittagong University, Chittagong, pp. 1–61.
Khare, N.K., Punekar, S., 2001. Socio-economic status of fisherwoman in relation to trapa-cum-fish farming. In: Paper presented at International Conference (December, 12-14) on Women in fisheries, CIFE, Mumbai.
Krishnan, M., Birthal, P.S., Ponnusamy, K., Kumaran, M., Singh, H., 2000. Aquaculture Development in India: Problems and Prospects. In: Workshop Proceedings held at National Centre for Agricultural Economics and Policy Research, New Delhi, September 6–7, 1999.
Kurien, J., 1978. Towards an understanding of the Fish Economy of Kerala State. WP: 68, CDS, TVM.
Kurien, J., 1995. The Kerala model: its central tendency and the outlier. Soc. Sci. 23, 70–80.
Kurien, J., Achari, T., 1998. Overfishing the Coastal Commons; Causes and Consequences. In: Guha, R. (Ed.), Social Ecology. Oxford University Press, New Delhi, p. 389.
Laskar, B.A., Das, S., Nath, D., Kar, D., 2002. Ecological studies in Puneer Haor Wetland in Cachar district of Assam with special emphasis on Aquatic macrophytes, ichthyofauna and wetland birds. Conserv. Forum J. 1, 25.
Malasha, I., 2006–2007. Fisheries Co-management, the Zambian experience. In: CBFM-2 International conference on community based app. roaches to Fisheries management. 6–7 March 2007, Dhaka, Bangladesh. The World Fish Center, Zambia. vol. 4: pp. 12–22.
Malcolm, D., Muzaffar, A., 2006. Capturing the Benefits, Policy Brief 1. The WorldFish Center, Bangladesh and South Asia Office.
Masae, A., 1998. An analysis of fisheries co-management arrangement: the case of Ban Laem Makham, Sikao District Trang Province, South Thailand. In: Paper presented at the Regional Workshop on Fisheries Co-management, Phuket, Thailand.
McGrath, D.G., Cardoso, A., Sá, E.P., 2005. Community Fisheries And Co-Management On The Lower Amazon Floodplain of Brazil. Pesquisa Ambiental da Amazônia (IPAM), Santarém.
McGrath, D.G., Cardoso, A., Sá, E.P., 2007. In response to the growth of Amazon commercial fisheries, a loose regional network of community-managed lakes. J. Ecoplan. 66, 56–75.
Metcalf, S.J., Gaughan, D.J., Shaw, J., 2009a. Conceptual models for Ecosystem Based Fisheries Management (EBFM) in Western Australia. Fisheries Research Report No. 194, Department of Fisheries, Western Australia, 42 p.
Mustafa, M.G., Halls, A.H., 2006. In: Malcolm, D., Brooks, A. (Eds.), Fisheries Yields and Sustainability, Policy Brief 2. the WorldFish Center, Bangladesh and South Asia Office.
Narzary, A., Das, S., Das, B.K., Singh, N.R., Kar, S., Das, P., Butta, B., Kar, D., 2015. A preliminary study on zooplankton diversity of Ramnagar Anua, Srikona Beel and Tapang Hoar of Cachar district, Assam, India: A Project Report. J. Chem, Biol. Phys. Sci. 5 (3), 2809–2817.

Nath, M., Singh, N.R., Das, B.K., Dutta, B., Das, U., Das, P., Kar, S., Kar, D., 2015. A preliminary study on Fish Diversity of Kakri and Deo rivers around Dharma nagar in Tripura. Int. J. Theoret. Appl. Sci. 7 (2), 6–13.

NRC, 1999. Sustaining marine fisheries. National Research Council, National Academy Press, Washington, DC. 164 pp.

Nyeko, J.I., 2002. Community participation in fisheries management: prospects and challenges for co-management. J. Aquat. Sci. 23 (4), 32–45.

Parvin, J., 2006. Social capital and household well-being in fisheries community: A survey of Community-based Fisheries Management (CBFM) project in Rural Bangladesh, pp. 38. Unpublished Master's thesis, Lund University, Centre for East and South-East Asian studies, South Asian Studies (Sweden).

Patgiri, A., Goswami, M.M., Kar, D., Barbhuiya, M.H., 2001. Comparative study of length-weight relationship and relative condition factors in major carps in ponds in Guwahati. Ind. J. Environ. Ecoplan. 5 (1), 179–180.

Patgiri, A., Goswami, M.M., Kar, D., Barbhuiya, M.H., 2002. Quality assessment of biological feed in four carp species of Assam. Environ. Ecol. 20 (1), 86–88.

Pemsl, D.E., Seidel, L.J.L., White, Ahmed, M.M., 2008. Community-based fisheries management project in Bangladesh. In: CGIAR Science Council, Impact Assessment of Policy-Oriented Research in the CGIAR:Evidence and Insights from Case Studies. A study commissioned by the Science Council Standing Panel on Impact Assessment. CGIAR Science Council Secretariat, Rome, (Italy).

Pinedo, D., Summers, P.M., Smith, R.C., Saavedr, J., Zumaeta, R., Almeyda, A.M., 2000a. Community-based natural resource management as a non-linear process: a case study in the Peruvian Amazon Varzea. In: Eighth IASCP Conference, Bloomington, Indiana 30 May–4 June, 2000, Instituto del Bien Común. 5: pp. 65–112.

Puinyabati, H., Shomorendra, M., Kar, D., 2010a. Helminth parasites of fishes of Chatla Haor, Silchar, Assam. Environ. Ecol. 28 (3A), 1852–1854.

Puinyabati, H., Shomorendra, M., Kar, D., 2010b. Infection of Helminth parasites in relation to the length of fish host. Natl. J. Life Sci. 7 (1), 49–50.

Puinyabati, H., Shomorendra, M., Kar, D., 2012. Trematode parasites of some freshwater fishes of Manipur and Assam. Biospectra 7 (4), 61–68.

Puinyabati, H., Shomorendra, M., Kar, D., 2013a. *Senga silcharensis* n. sp., a new cestode from the intestine of the fresh water fish *Channa punctatus* (Bloch) from Chatla Haor, Silchar, Assam. Sci. Culture 79 (5–6), 245–247.

Puinyabati, H., Shomorendra, M., Kar, D., 2013b. Basic Parasitology. pp. xxi + 201, Asiatic Publishing House, New Delhi.

Ranibala, T., Vishweshwari, A., Kar, D., Jha, A.N., Puiyabati, H., Shomorendra, M., 2010. Seasonal occurrence and infestation intensity of a digenetic trematode and two cestodes in fishes of Awangsoi fishery, Bishnupur, Manipur. Biospectra 5 (2), 229–232.

Ranibala, T., Shomorendra, M., Kar, D., 2011a. Prevalence and intensity of Nematode parasites infecting the fish *Anabas testudineus* in Loktak Lake. Environ. Ecol. 29 (3B), 1611–1612.

Ranibala, T., Shomorendra, M., Kar, D., 2011b. Intensity of Helminth infection with special reference to Cestodes in the fish *Clarias batrachus* (Linn.) of Loktak lake, Manipur. Conserv. Forum J. 3, 20–25.

Ranibala, T., Singh, M.S., Kar, D., 2015. Essentials of Fish Helminthology. pp. xx + 207, Asiatic Publishing House, New Delhi.

Ranjeev, B., Chakrapani, B.K., Kar, D., Nagarathna, A.V., Ramachandra, T.V., 1996. Fish Mortality in Bangalore lakes, India. Electron. Green J. (USA) 6, 5.

Rao, G.S., 2011. Marine Fisheries in India: The path ahead. In: 9th Indian Fisheries Forum. Central Institute of Brackish Water Aquaculture, Chennai. 19–23 December 2011.

RuhAfza, R., Vickers, T., Kanagaratnam, U., 2006. Fishing for a Future: Women in Community-based Fisheries Management. Booklet 2, The World Fish Center, Bangladesh and South Asia Office (Dhaka).

Sangeeta, O., Shomorendra, M., Kar, D., 2010. Study of seasonal occurrence of fish nematodes of Oinam Lake in Bishnupur District of Manipur. Life Sci. Bull. 7 (2), 259–261.

Sathiadas, R., 2005. Policy issues for Marine Fisheries Management in India. In: The Proceedings of 7th Indian Fisheries Forum, AFS,IB, Mangalore.

Sathiadhas, R., Ramachandran, C., Vipinkumar, V.P., Prathap, S.K., 2007. Status of fisheries economics and extension research in India: opportunities and challenges. p. 404, In: Modayil, M.J., Pillai, N.G.K. (Eds.), Status and Perspectives in Marine Fisheries Research in India. Central Marine Fisheries Research Institute, Cochin.

Shing, C.C.A., 2001a. Caribbean Natural Resources Institute (CANARI) Case study of the Integrated Coastal Fisheries Management Project: A pilot project for the Gulf of Paria, Trinidad. CANARI Technical Report No 280, pp. 226.

Singh, N.R., Kar, D., 2014. Abundance and distribution of Helminth parasites in the Fishes of Sone Beel, the biggest wetland in Assam. Int. J. Scient. Res. 3 (12), 67–69.

Singh, M.S., Singh, N.R., Devi, I.N., Devi, B.D., Ranibala, T., Kar, D., 2015. Check list of fish species of Loktak Lake, Bishnupur District. Biol. Forum: Int. J. 7 (1), 171–179.

Singh, N.R., Singh, M.S., Kar, D., 2015b. A check list of Fish species diversity of Sone Beel, the biggest wetland in Assam. NeBIO (Int. J. Environ. Biodivers.) 6 (1), 21–26.

Singh, N.R., Singh, M.S., Kar, D., 2015c. Role of Fishing gears and their significance in Fishery technology of Pumlen Lake. Life Sci. Leaflet 65, 13–25.

Singha, R., Puinyabati, H., Shomorendra, M., Kar, D., 2010. Helminth Parasites in relation to length of three Freshwater Fishes of Dolu Lake of Silchar, Assam, India. Environ. Ecol. 28 (4A), 2506–2508.

Singha, R., Singh, M.S., Kar, D., 2014. Parasite Fauna in the Wetland Fishes of India. Today and Tomorrow's Printers and Publishers, New Delhi, p. 209.

Sonowal, M., Kar, S., Das, B.K., Das, S., Singh, N.R., Das, P., Dutta, B., Kar, D., 2015. Preliminary Studies on Zooplankton diversity of Certain Wetlands of Southern Assam. Life Sci. Leaflets 67, 68–78.

Soucy, R.H., 2005. Management of Natural Resources in the Southern Highlands Project (MARENASS: 1997–2005). Project to Peasant Communities in the Highlands (1993–1999), pp. 422.

Sultana, P., Thompson, P.M., 2004. Method of consensus building for community-based fisheries management in Bangladesh and Mekong Delta. Agr. Syst. 82, 327–353.

Tasker, M.L., Knapman, P.A., Laffoley, D., 2000. Effects of fishing on non-target species and habitats: Identifying key nature conservation issues. pp. 281–289, In: Kaiser, M.J., de Groot, S.J. (Eds.), Effects of Fishing on non-target Species and Habitats. Blackwell Scientific Publishers, Oxford.

Tissot, B.N., 2005a. Integral Marine Ecology: Community-Based Fishery Management in Hawai, I. World Futures. J. Gen. Evol. 61, 79–96.

Umi, N.D., Singh, N.R., Kar, D., 2015. Studies on Fish affected with Epizootic Ulcerative Syndrome with special emphasis on Parasitic infestation: A Project Report. Biol. Forum: Int. J. 7 (1), 155–164.

Wiber, M., Berkes, F., Charles, A., 2007. Participatory Research Supporting community-based fishery management. American Fisheries Society Symposium 49.

Wong, J.Z., Etoh, S., Sujang, A.B., 2005. Towards Sustainable Community-based Fishery Resources Management: The Tagal System of Sabah, Malaysia. Southeast Asian Fisheries Development Centre, pp. 12–31.

Yamamoto, T., 1995. Development of a Community-based fishery management system in Japan. Mar. Resour. Econ. 10, 21–34.

Yongyut, T., 2006. Community-based Wetland Management in Northern Thailand. Int. J. Environ. Cult. Econ. Soc. Sustain. 2 (1), 49–62.

Onyango, P.O., Jentoft, S., 2001. Embedding co-management: Community-based Fisheries Regimes in Lake Victoria, Tanzania. Conference Paper, 14, pp. 1–21.

CHAPTER

3

Materials and methods

OUTLINE

Abstract

There are standard methodologies for conducting community-based research works in fisheries management. However, implementation of methodologies is to be preceded by systematic literature search in peer-reviewed documents in order to identify the correct, adaptive, and replicable procedures. The literature search could also portray qualitative and quantitative information and evidences of the impacts of various fisheries management practices with particular emphasis on community-based fisheries management (CBFM). Efforts could also be directed toward gathering information regarding other associated management practices, notably comanagement, etc., around the globe.

Keywords: Study site, Community management meeting, Fisheries, Data collection, Peer review, Water bodies

3.1 Introduction

There are standard methodologies for conducting community-based research works in fisheries management. However, implementation of methodologies is to be preceded by systematic literature search in peer-reviewed documents in order to identify the correct, adaptive, and replicable procedures. The literature search could also portray qualitative and quantitative information and evidences of the impacts of various fisheries management practices with particular emphasis on community-based fisheries management (CBFM). Efforts could also be directed toward gathering information regarding other associated management practices, notably comanagement, etc., around the globe. In addition to laying emphasis on CBFM, attempts are also to be made in defining and distinguishing other community management approaches, notably the community-based comanagement, etc., in order to cover the whole spectrum of management scenario. Incidentally, comanagement is a formal consultation mechanisms between the government and users for governance. The presence of a well-established local CBFM organization, with power of decision-making in fisheries management (having been entrusted with the community management of water body), could be considered an ideal case for studies related to CBFM. On the contrary, fisheries without sufficient or consistent information on CBFM regime and *modus operandi* during the pre-implementation phase could be excluded from further study and analyses. Efforts are also to be made to group the contextual variables in the CBFM exercise during different phases of the management regime, including HDI, Corruption Perception Index (if any), Governance Index, ecosystem, fishing sector, and resource types, as well as a number of CBFM attributes. In addition, aggregated social, economic, and ecological binary outcomes may be used to indicate success in CBFM.

Moreover, a regression tree model could be constructed which could graphically depict quantitative relationships between predictor attributes and CBFM success. Incidentally, efforts could be made to fill-in the gap values by using surrogate splits inside the regression model. Further, there could be a thought of using a random forest model of a large number of trees (say, *c*. 10,000) in order to estimate the relative importance of selected attributes in determining CBFM success. Concomitantly, the significance of contextual variables (for instance, fishing sector) could also be investigated by grouping them in the random forest models, and by running independent models for each category, e.g., artisanal, industrial, etc. Incidentally, model accuracy for trees and random forests could be quantified using standard matrices.

Notwithstanding the above, a typical procedure for conducting CBFM studies is given in the following.

3.2 A typical procedure for conducting CBFM study

3.2.1 Locations of the study site

Study site(s), having not much difficult access, could be selected for typical CBFM studies. Concomitantly, one of the sites could be kept as it is, without any CBFM intervention, which would serve as a "control," in order to ascertain the impact of CBFM exercises on other sites (Islam et al., 2008; Chapagain, 2007; Barman and Kar, 2014; Kar, 2003, 2007, 2010, 2013).

3.2.2 Land and the people

Basic information is to be collected regarding the following aspects:

- **(i)** Geographical position of the study site.
- **(ii)** The water bodies (both lentic and lotic).
- **(iii)** The physiography of the terrain including catchment, limnology, and the ecological history.
- **(iv)** The lifescape of the site(s), including the people (the stakeholders from the grassroots level), their socioeconomic scenario, their management options and scenario, development aspirations, conflicts and consensus, conservation efforts, and so on (FAO, 1962, 1974, 1979; Kar, 2007, 2013).

3.2.3 Methods of data collection

The basic socioeconomic data could be collected through participatory rural appraisal (PRA) method by applying PRA tools, viz., community meeting and small focus group discussion, direct observation, participatory mapping, wealth ranking with the help of key informants/knowledgeable individuals (KI) (Townsley, 1996; Talukdar and Barua, 1999; Kumar, 2002; Kar, 2003; Barman et al., 2006; Chapagain, 2007). Semi-structured interviews could also be conducted with *Beel* users (stakeholders) using different techniques. The schedule could be pretested with non-sample respondents of the area. Data could be collected in local language and the responses may be recorded both in English and in the local language (Barman, 2004; Kar et al., 2009).

PRA team may consist of technical officers of fisheries department, representatives of NGO(s), local knowledgeable individuals (KIs), as well as others, as and when required. In addition, supplementary data could also be collected from the offices of the concerned district nodal NGOs, district fisheries development officers (DFDO), directorate of fisheries (DoF) of the province, as well as from different ongoing national and international projects.

3.2.3.1 *Community meeting and small focus group discussion*

Community meeting and small focus group discussion are the two basic important techniques for gathering community information and their dissemination. It is a cooperative approach to decision-making through discussion (Pomeroy and Rivera-Guieb, 2006). In fact, there may be the need for a number of such meetings to be held with the stakeholders. The KIs could collect firsthand information on the involvement of community members including women in all the stages of management and development of the water bodies through CBFM. Some of the aspects of collection of information are as follows:

3.2.3.2 *Direct observation*

Direct observations could be done on the activities of the management committee (say, water body development committee of WBDC) related to management and development of the water bodies; distribution of incomes of the users of the water bodies, etc. (Ray, 1999).

3.2.3.3 *Interviews with knowledgeable individuals (KIs)*

The KIs are considered as local experts because of their knowledge about the water bodies and their resources. Further, the KIs could be interviewed to know about the various facilities and difficulties of the community at large (Townsley, 1996; McCaslin and Tibzinda, 1997; Pomeroy and Rivera-Guieb, 2006).

Incidentally, the local fishery demonstrators, fishery extension officers of the government departments, the NGO coordinators, village headmen (locally called "*Sarpanch*" in India), and local school teachers could be considered as KIs for the purpose.

3.2.4 Wealth ranking

It is a participatory analytical exercise carried out by the KIs based on certain socioeconomic parameters, viz., income, land holding, employment, etc., in order to reveal the socioeconomic status of the users of the water bodies, as well as classify them into various wealth classes, e.g., Rank A (rich), Rank B (medium), Rank C (poor), and Rank D (very poor) (Chapagain, 2007). A brief account of this economic categorization is given in the following.

3.2.4.1 *Rich (Rank A)*

People who possess land measuring >2.0 ha, whose annual income is > Rs. 1,10,000.00, and who have government/nongovernment job.

3.2.4.2 *Medium rich (Rank B)*

People who possess land measuring >1.0 ha and up to 2.0 ha, whose annual income ranges from Rs. 55,001.00 to Rs. 1,10,000.00, and who have no government/nongovernment job.

3.2.4.3 *Poor (Rank C)*

People who possess land measuring >0.40 to 1.0 ha, whose annual income ranges from Rs. 22,001.00 to Rs. 55,000.00, and who generally work as daily wage laborers.

3.2.4.4 *Very poor (Rank D)*

People who possess land measuring up to 0.40 ha, whose annual income is up to Rs. 22,000.00, and who work as daily wage laborers.

3.2.5 Procedure for the selection of respondents

A substantive number of households of water body users are to be selected, preferably, by statistical random sampling (Zar, 2009) out of the total number of households surrounding the water bodies. Both men and women, among the inhabitants, are to be included. They could be considered as the respondents for the study.

3.2.6 Application of statistical methodology

Independent variables: These include the following aspects.

Personal characteristics:

These include the following parameters:
(a) age, (b) education, and (c) family size.

Socioeconomic characteristics:

These include the following aspects:
(a) type of the dwelling house, (b) caste, (c) land under possession, (d) occupation, (e) income (annual, pre-CBFM, and post-CBFM), (f) social participation, (g) community-based physical asset, (h) sociopolitical empowerment, (i) community-based fishery organization(s), (j) involvement of NGOs, etc.

Psychological characteristics:

These include the following matters:
(a) Participation of the users of the water bodies in development program(s), (b) decision-making ability of the users of the water bodies, (c) women empowerment, (d) management capabilities of the members of the committees for water body development, and so on.

Situational characteristics:

These include the following aspects:
(a) Availability of critical aquaculture inputs, (b) community-managed markets.

Dependent variables:

These include the following parameters:
Knowledge level of the users of the water bodies in CBFM, (b) sustainable development and management of fisheries of the water bodies; (c) livelihood of the community for sustenance; etc.

3.2.7 Operationalization of the variables and their measurement

Independent variables:
These include the personal characteristics.
Personal characteristics:
These include the following parameters.
Age:
It refers to the chronological age of the users of the water bodies, and consists of the following three categories (Goswami, 2003; Barman et al., 2006):
Young users of the water bodies: This relate to those users of the water bodies whose age is up to 30 years.
Middle-aged users of the water bodies: It refers to those users of the water bodies whose age ranges from 31 to 50 years.
Elderly users of the water bodies: It includes those users of the water bodies whose age is > 50 years.
Education:
It refers to the academic qualification acquired by the users of the water bodies from formal institution(s) (Dana, 1987; Goswami, 2003; Barman et al., 2006; Kar, 2007, 2010, 2013). The formal education levels used in such kinds of studies are as follows:
Illiterate: This includes those users of the water bodies who are not educated at all.
Primary level: This level includes those users of the water bodies whose education is up to the primary level (below "O" level), and could read and write in the local vernacular language.
Middle level: It includes those users of the water bodies whose education is up to the middle school level (class V–VIII, i.e., also below "O" level). Here, English could be one of the subjects of study.
High school level (including higher and senior secondary): This level includes those users of the water bodies whose education is up to high and higher (senior) secondary (HS/SS) level(s), where English is almost certainly one of the subjects of study in the curriculum (i.e., "O" level, "A" level).
Graduation level and above: It includes those users of the water bodies whose education is up to the graduation level or above (university level).
The education level of the respondents could be determined with the help of a classification, and scoring could be done as follows:

(a) primary: 1;
(b) middle: 2;
(c) high school (including higher or senior secondary): 3; and
(d) graduation: 4.

Family size:
It refers to the total number of members in the families of the community.
Nuclear family:
It refers to total number of persons in the family of a user of the water body, which includes his wife, sons, daughters, etc., irrespective of their age.
Joint family:
It refers to the number of family members in a nuclear family, as well as the parents (father, mother), sister(s), brother(s), and also the grandparent(s) of the user of the water body. The total family size could be categorized in the following manner:

(a) small-sized family (1–4 members);
(b) medium-sized family (5–7 members); and,
(c) large-sized family (>7 members).

Socioeconomic characteristics:
Type of constructed house:
It refers to the condition of the house(s) of the user of the water body, regardless of whether they are of concrete, semi-concrete, or thatched houses. Pattern of the houses, probably, indicate the socioeconomic condition of the users of the water bodies (Dey and Kar, 1989; Kar, 1990, 2000, 2003, 2007, 2013; Maiti, 2003; Goswami, 2003; Barman, 2004).

The variable was categorized with the scoring of:

(a) concrete: 3;
(b) semi-concrete: 2; and
(c) thatched: 1.

Caste:

It refers to the caste of the users of the water bodies. Various categories of castes could be considered, notably (a) scheduled caste (SC), (b) scheduled tribe (ST), (c) general caste (GC), and so on, depending on the societal and administrative categorization.

Land possession:

It refers to the total bottom land area (culture or not cultured) of the water bodies in possession of the each user. Incidentally, under this aspect, the users of the water bodies could be categorized into four categories, viz.,

(a) very small area of land under possession (up to 0.40 ha);
(b) small area of land under possession (ranging from 0.40 and to 1.0 ha);
(c) medium amount of area of land under possession (ranging from 1.0 to 2.0 ha); and
(d) large area of land under possession (> 2.0 ha).

Occupation:

This variable could be classified into five categories with the scores allotted as follows:

(a) service: 5;
(b) agriculture: 4;
(c) business: 3;
(d) day labor: 2; and
(e) fishing: 1.

Income:

It refers to the total annual income of the family of a user of the water body earned through fishing, aquaculture, and agriculture. Some of the non-fishery users of the water bodies could also earn through jobs, trade, day labor, etc. Incidentally, the users of the water bodies could be categorized into the following income groups based on annual income (Anon, 2004, 2008):

(a) (a) very poor (< Rs.22,000.00 per annum);
(b) poor (from Rs. 22,000.00 to < Rs. 55,000.00 per annum);
(c) medium rich (from Rs. 55,000.00 up to Rs. 110,000.00 per annum); and
(d) rich (> Rs. 110,000.00 per annum).

(Incidentally. 1 US $ is generally = Indian Rupees 70.00 approximately.)

Social participation:

It refers to the degree to which the users of the water bodies are associated with different organizations, like village panchayat (gram panchayat), panchayat samiti, fish farmers club, rural development club, zilla parishad (ZP), fishermens' cooperative societies, etc. This variable could be classified into four categories and the following scores could be allotted:

(a) member of one organization: 1;
(b) member of > 1 organization: 2;
(c) office bearer: 3;
(d) office bearer-cum-member of an another organization: 4; and
(e) not a member of any organization: 0 (zero).

Community-based physical asset(s):

It refers to the common infrastructure available to users of the water bodies. It may include the following:

(i) community fish pond(s) (in which, pisciculture could be done as a common pool exercise);
(ii) common model fish farm with or without eco-hatchery;
(iii) common crafts and gears used in harvesting;
(iv) common institutes like schools, libraries, etc.

This variable could be quantified by allotting the following scores:

(a) strong infrastructure (5–6 out of 6);
(b) medium infrastructure (3–4 out of 6);

(c) poor infrastructure (2 out of 6); and
(d) very poor infrastructure (<2 out of 6).

Sociopolitical empowerment:
It refers to the simple average (Hiremath and Raju, 2008) of the following areas, viz.,

(i) Control/influence over resource(s).
(ii) Participation in policy making.
(iii) Participation in law/policy implementation and enforcement and it could be applied to measure the extent of the participation of the users of the water bodies in the aforesaid arenas. The variable could be quantified after adding the total scores (in a scale of 1–50) and could be categorized as follows:
  (a) high sociopolitical empowerment (> 40 out of 50);
  (b) medium sociopolitical empowerment (30–40 out of 50); and
  (c) low sociopolitical empowerment (<30 out of 50).

A brief description of the subareas under the sociopolitical empowerment is given in the following.

(a) *Control/influence over resources*:
It refers to pattern of using the resources of the water bodies. It also includes how the users of the water bodies may obtain lease and their management norms for conservation, rational harvest of the aquatic resources, their management, etc.
(b) *Participation in policy making*:
It refers to the extent to which the users of the water bodies are involved in making policy decision(s) for effective development and management of the water bodies.
Responses from the respondents could be taken against the above two areas after framing a dichotomous type of questions. Respondents could be awarded a score of "**1**" or "**0**" based on their response of "**yes**" or "**no**," respectively.
(c) *Participation in law/policy implementation and enforcement*:
This reveals how the users of the water bodies are involved in the implementation and enforcement of the laws/policies of the water body management committees (WBMC) for effective development and management of the water bodies.
(d) *Conflict resolution/management*:
Conflict(s) may arise on different issues and for various kinds of causes, viz. sharing of the benefit(s) of the project(s), choices in the allocation of the resources or the involvement of cost(s) at different levels, and so on. All these, probably, need certain kinds of approach to management. It may be noted here that such types of conflicts may usually be the focus in most of the conflict management issues (Pomeroy and Rivera-Guieb, 2006).

Responses from the respondents could be taken against the above two areas, i.e., (a) "participation in law/policy implementation and enforcement" and (b) "conflict resolution/management," after framing a dichotomous type of questionnaire. Respondents may be awarded a score of "3," "2," "1," or "0" based on their response of "Fully agreed," "Agreed," "Disagreed," or "No decision," respectively.
Psychological characteristics:
These refer to a number of independent psychological variables, viz.,

(a) participation of the users of the water bodies in the development programs of the water bodies;
(b) decision-making ability of the users of the water bodies;
(c) women empowerment in the water body management committees (WBMC);
(d) community-based organization(s); and
(e) management capabilities of the water body management committees (WBMC), etc.

These could be measured by framing schedules, and also could be awarded with a score of "**1**" or "**0**" based on their response of "**yes**" or "**no**," respectively (Barman et al., 2006; Barman and Dana, 2009).
Then, the respondents could be grouped into three categories, viz.,

(a) "low";
(b) "medium"; or.
(c) "high".

for each of the aforesaid variables. Later, the scores for each item could be added in order to find the total scores against each variable. A short description of the variables is given in the following.

(a) *Participation of the users of the water bodies in the development programs of the water bodies under WBMC*:

It refers to the degree to which the users of the water bodies are associated with different stages of the development program(s) of the water bodies.

(b) *Decision-making ability of the users of the water bodies*:
Decision-making is the process of consciously choosing courses of action from available alternative(s) and integrating them for the purpose of achieving desired results through planned activities (Ray, 1999).

(c) *Women empowerment under WBMC*:
The concept of empowerment could be related to the concept of freedom (liberty) (Srinath and Tangamani, 1993). Empowering is said to equip oneself with facilities in order to improve his/her living condition(s) (Devadas et al., 1988). Here, the concept of women empowerment is, probably, related to the involvement of women users of the water bodies in each stage of the development program(s) of the water bodies (Barman et al., 2007).

(d) *Community-based Organization(s)*:
Here, community-based organization(s) usually refer to *Water Body Management Committees (WBMC)*, which could be a group of 20–500 individuals living in the riparian villages around a water body (say, a wetland, which is locally called *"Beel"* in India and Bangladesh). The users of the water body generally come together to work hand in hand for effective utilization of the water body; thereby ensuring the enhancement of the avocation (livelihood) of the users of the water body in a sustainable manner (Anon, 2004, 2011).

(e) *Management capabilities of the Water Body Management Committees (WBMC)*:

It portrays the efficiency of a WBMC with regard to the following aspects:

(a) management of the WBMC;
(b) effective record keeping by the members of the WBMC;
(c) mobilization and effective utilization of the funds of the WBMC; and
(d) capacity building exercise of the members of the WBMC.

Success or failure of a community-based water body development depends on how efficiently the management of the water body is carried out by the WBMC. Depending on the management efficiency of a WBMC, the productivity of the water body may be enhanced to a desirable level.

Situational characteristics:
Few (say, two) independent (situational) variables could be considered, viz.,

(a) availability of critical aquaculture inputs and
(b) community managed market(s)

These could be measured by developing schedules. Accordingly, responses from the respondents may be taken (Dana, 1987; Goswami, 2003; Barman, 2008; Barman et al., 2008; Kar, 2007, 2013). Respondents could be awarded a score of "3," "2," "1," or "0" based on their responses of "Fully agreed," "Agreed," "Disagreed," or "No decision," respectively. A brief description of the aforesaid variables is given in the following.

(a) *Availability of critical aquaculture inputs*:
This relates to the decisive inputs for sustainable development of fisheries, which include availability of quality fish seeds, application of lime, organic and inorganic fertilizers, other different infrastructures for aquaculture, fishing, and harvesting implements, etc. (Kar, 2007, 2013; Barman et al., 2008).

(b) *Community managed markets*:
This involves the trade (selling and buying) with aquaculture and related aspects by the users of the water bodies. As a general norm, the marketing facility may be well developed in the event of good management practices (Kar, 1990, 2007, 2013; Barman et al., 2008).

Later, the respondents could be grouped into three categories as follows:

(a) "low,"
(b) "medium," and
(c) "high."

The scores of each item could be added to find the total scores against each variable.

Dependent variable:

(a) *Knowledge level of the users of the water bodies in community-based fisheries management (CBFM)*:
Here, "knowledge" refers to the information and understanding of the users of the water bodies with regard to CBFM. In such studies, the knowledge level of the users could be measured with the help of "knowledge test" scale developed by Dana (1987), Sheikh et al. (1993), Barman (2004), and Barman et al. (2007). However, little modification(s) could be done as per requirement of the study.

Regarding enumeration, for each of the correct answer, a score "X" may be given; and for the incorrect answer, "O" could be assigned. Then, such scores could be finally added to yield a knowledge score out of 100.

Later, the respondents could be grouped into three categories, viz.,

(a) low knowledge score: <75 out of 100;
(b) medium knowledge score: 75–85 out of 100; and
(c) high knowledge score: >85 for each of the aforesaid variable(s).

The scores for each item could be added to calculate the total scores against each variable.

A suggested procedure for developing knowledge test:

At first, the major items and subitems of knowledge with their definite correct answers may be collected after detailed review of literature and discussion with the fishery officers, scientists, and other persons of the provincial government departments.

The selected items and subitems of "*knowledge test*" could be given to different scientists and fishery officers (say, 20 in number) who may act as judges. Each judge could be requested to distribute the marks (weightages) to each of the major items and the subitems, considering their relative importance to CBFM so that the total mark becomes 100. At last, final score for each of the items and subitems are to be finalized by calculating arithmetic means of the scores given to the items and subitems by the experts (Dana, 1987; Shaikh et al., 1993; Barman, 2004; Barman et al., 2007).

(b) *Sustainable development and management of fisheries of the water bodies*:

Sustainable development and management of the fisheries of the water bodies may not be simply a concept only, but it could also be considered as a process of enhancing and ensuring good fish production in a systematic and sustained manner without lessening the existing fish production potentials of the water bodies so that the good level of fish production could be continued and enjoyed by the users of the water bodies (Barman et al., 2006). In this connection, reference could also be drawn to the activities and achievement(s) related to the areas of (a) profit sharing plan, (b) adoption and replication of the package of practice(s), and (c) transparency/accountability in the fisheries development in the water bodies.

This variable could be analyzed by taking responses from the respondents on a prepared schedule. Respondents could be awarded a score of "**1**" or "**0**" based on their response of "**yes**" or "**no**," respectively. The variable could be quantified after adding the total scores (in a scale of 1–15) and categorized as follows:

(a) high sustainability (12–15 out of 15);
(b) medium sustainability (9–11 out of 15);
(c) low sustainability (6–8 out of 15); and
(d) very low sustainability (<6 out of 15).

(c) *Community's livelihood for sustenance*:

It may be considered as a fact that household sustainable livelihood could be ensured when the inhabitant(s) may be in a position to cope up with and recover from stresses and strains, as well as could maintain their capabilities and assets without undermining the natural base (Hiremath and Raju, 2008).

This variable may be analyzed by taking responses from the respondents on questionnaire schedule.

Respondents may be awarded a score of "1" or "0" based on their response of "yes" or "no," respectively. The variable could be quantified after adding the total scores in a scale ranging from 1 to 15, and could be categorized as follows:

(a) highly sustainable livelihood (5–6 out of 6);
(b) medium sustainable livelihood (3–4 out of 6); and
(c) low sustainable livelihood (< 3 out of 6).

3.2.8 Statistical analyses of data

The collected data are to be checked and put in proper format for coding in IBM sheets. For making simple comparisons, the frequency tables may be constructed and percentages calculated. The data could also be fed into the computer and the following statistical techniques may be used to analyze the data using the Software Statistical Package for Social Sciences (SPSS) or other types of statistical software.

3.2.8.1 *Descriptive statistics*

Descriptive statistics could be used for various statistical analyses.

(i) *Comparison of mean*:
This analysis could be done to see the effect of different independent variables on other independent and dependent variables, notably the effect of education; family size; type of constructed house; occupation; social participation; community-based physical assets; participation of the users of the water bodies in development program(s); decision-making ability; women empowerment; community-based fisheries organization(s); involvement of NGOs; availability of critical aquaculture inputs; mean scores of dependent variables, like land holding, income, sociopolitical empowerment, and management capabilities of the water body management committees (WBMC); community-managed marketing; pre-CBFM income; post-CBFM income; knowledge scores of *Beel* users on CBFM; sustainable development and management of fisheries of the water bodies and community's livelihood for sustenance, etc., could be calculated and tested by nonparametric tests like Kruskal-Wallis test, Chi square test, etc.,

(ii) *Correlation coefficient*:
A correlation is a simple number that describes the degree of relationship between dependent and independent variables (source: http://www.socialreserchmethods.net/kb/ststdesc.php). The significance of the estimated correlation coefficient may be tested at 5% and 1% level of significance by using the statistical table for biological, agricultural, and medical research (Fisher and Yates, 1975).

(iii) *Regression analysis*:
Regression analysis portrays the effect or influence of independent variables on dependent variables. Variables could be selected on the basis of "F" values and R^2 values, following stepwise regression analysis. Such stepwise multiple regression analysis could be done to study the predictor variables for selecting the best predictor (Ray and Mondal, 2000).

(iv) *Factor analysis*:
This technique may be used to make a group of variables into a single factor (based on correlation between variables), which, however, can convey all the essential information of the original set of variables. Factor analysis could be extracted by Equamax with Kaiser normalization method of principal component analysis (PCA) on the basis of *Eigen* values > 1 with rotation converged in 18 iterations.

The main applications of factor analysis techniques are as follows:

(a) to reduce the number of variables, and

(b) to detect structure in the relationships between variables, i.e., to classify variables. Therefore, factor analysis may be applied as a data reduction or structure detection method. Thus, basic idea of factor analysis is to reduce the data into the group that could describe the relationship between the variables (Kothari, 1996; Ray and Mondal, 2000; http://www.stasoft.com/textbook).

(v) *Path analysis*:
This technique could be used to get a clear picture of the direct and indirect effects of the independent (causal) variables on dependent (effect) variable. Basically, the main purpose of doing path analysis is to get a clear picture of the direct and indirect effect of the independent variables on the dependent variables (Ray and Mondal, 2000).

Similarly, Chandra (2011) worked on the Management Regimes and Institutional Arrangement in Floodplain Wetlands Fisheries through participatory rural appraisal (PRA) techniques. Some of the aspects on which the study was based include the access to fishing in floodplain wetlands, formation of the ownership and control rights, domains of management in wetlands and role of institutions as controlling agencies in the allocation of the fishing rights, etc. In addition, Rahman and Begum (2010) have dealt with the strategies for empowering the poor for the conservation of wetland resources in Bangladesh. Further, aspects like resources user group (UG) formation, capacity building, alternative livelihoods and microcredits, institution building, etc., were also dealt with. Similar study has also been done by Chowdhury et al. (2012).

3.3 Summary

(i) There are standard methodologies for conducting community-based research works in fisheries management.

(ii) Typical procedure for conducting CBFM study has been discussed, covering aspects like locations of the study site, land and the people, methods of data collection, community meeting and small focus group discussion, direct observation, interviews with knowledgeable individual (KI), wealth ranking, rich (Rank A), medium rich (Rank B), poor (Rank C), very poor (Rank D), procedure of selection of respondents, application of statistical methodology, and so on.

(iii) Further, aspects like operationalization of the variables and their measurement, as well as application of statistical methodologies have also been emphasized.

(iv) However, implementation of the methodologies is to be preceded by systematic literature search in peer-reviewed documents in order to identify the correct, adaptive, and replicable procedures.

(v) Comanagement is a formal consultation mechanism between the government and users for governance. The presence of a well-established local CBFM organization, with decision power in fisheries management (having been entrusted with the community management of water bodies), could be considered as an ideal case for studies related to CBFM.

(vi) On the other hand, fisheries without sufficient or consistent information on CBFM regime and *modus operandi* during the pre-implementation phase could be excluded from further study and analyses.

(vii) The education level of the respondents could be measured using standard procedures.

(viii) In addition to above, the following aspects have also been mentioned as part of the studies:
- **(a)** participation of the users of the water bodies in the development programs of the water bodies under water body management committees (WBMC);
- **(b)** decision-making ability of the users of the water bodies;
- **(c)** women empowerment under WBMC;
- **(d)** management capabilities of the WBMC;
- **(e)** situational characteristics;
- **(f)** availability of critical aquaculture inputs;
- **(g)** community managed markets;
- **(h)** knowledge level of the users of the water bodies in community-based fisheries management (CBFM);
- **(i)** suggested procedure for developing knowledge test has also been dealt with;
- **(j)** in addition, sustainable development and management of fisheries of the water bodies and community's livelihood for sustenance have been mentioned; and
- **(k)** Chandra (2011) worked on the Management Regimes and Institutional Arrangement in Floodplain Wetlands Fisheries through participatory rural appraisal (PRA) techniques.

Suggested reading

Anon, 2004. Project Paper on World Bank Funded Assam Agricultural Competitiveness Project (Fisheries Component). Directorate of Fisheries, Assam, p. 96.

Anon, 2008. Souvenir, Matshya Mahotsav, 08 and Seminar Cum Workshop on Development of Fisheries in Assam. Department of Fisheries. Government of Assam.

Anon (2011). Draft Final Comprehensive Legal statute in the form of Assam Panchayat (Fishery) Rule 2010. pp. 53.

Barman, R.C., 2004. Community based fisheries management with peoples participation for sustainable development of *Beel* fisheries. M.F.Sc. thesis (unpublished) West Bengal University of Animal and Fisheries Sciences, Mohanpur, West Bengal, Nadia.

Barman, R.C., 2008. Fish seed raising in flood plain wet lands of Assam through aquaculture. Fishing Chimes 28 (8), 30–31.

Barman, R.C., Dana, S.S., 2009. Impact of participation and involvement in decision making on knowledge level of *Beel* users for sustainable development of *Beel* fisheries. Indian J. Anim. Sci. 79 (9), 91–95.

Barman, R.C., Kar, D., 2014. Concept definition and approaches of community based Wetlands Fishers and management. In: Kar, D. (Ed.), Research Frontiers in Wetlands, Fishery and Aquaculture. Dominant Publishers and Distributors Pvt. Ltd., New Delhi, pp. 291–310. xi + 320. ISBN: 978-93-82007-39-5, ISBN 10: 93-82007-39-3.

Barman, R.C., Dana, S.S., Bhattacharjya B. K., Majumdar D. (2006) Predictive factors of knowledge levels of *Beel* users on sustainable development of *Beel* fisheries. MANAGE Extension Research Review, MANAGE, Hyderabad, India 7 (2): 61–69.

Barman, R.C., Panigrahi, A.K., Dana, S.S., 2007. Empowering the women through *Beel* fisheries: a new thought for sustainable development of *Beel* fisheries. In: Proceedings of All India Seminar on Sustainable Aquaculture for augmentation of export with special reference to environment, engineering and value addition, Kolkata. pp. 149–152.

Barman, R.C., Dana, S.S., Bhattacharjya, B.K., Kar, D., 2008. Effect of situational variables on knowledge level of fishermen for sustainable development of *Beel* fisheries. Environ. Ecol. 26 (4B), 2092–2094.

Chandra, G., 2011. Management regimes and institutional arrangement in floodplain wetlands fisheries of Assam: an evaluation. Ind. J. Extens. Educ. 47 (1 and 2), 27–33.

Chapagain, B.P., 2007. Impact of community forestry on livelihood improvement of rural people (A case study of Shir Khola CFUG, Maisthan, Mahottari). An unpublished thesis submitted for partial fulfillment of the requirement for the degree of Bachelor of Science in Forestry. Pp. 85.

Chowdhury, I.H., Kamal, M.M., Haque, N., Islam, M.N., Akter, S., 2012. Patterns of co-operation in community based fishery management: a sociological study on the people of hail Haor. Asian J. Soc. Sci. 8 (15), 8–15.

Dana, S.S., 1987. A study of factor effecting the attitude of livestock owners towards artificial insemination in cattle. An unpublished M. V. Sc. Thesis submitted to Indian Veterinary Research Institute, Izatnagar, Utter Pradesh, pp. 127.

Devadas, R.P., Lakhshmi, S.R., Ramathilagan, A., 1988. Home Science Education and women's Empowerment. Paper Presented for Association of Women in Development Conference. Global Empowerment of Women, Washington.

Dey, S.C., Kar, D., 1989. Fishermen of Lake Sone in Assam: their socio-economic status. Sci. Cult. 55, 395–398.

FAO (1962). Rapport au Government de la Re' publique du Niger sur la situation et' evolution de la pe^che au Niger: Base' sur le travail de M. Jaques M.A. Daget', Rapp. FAO/PEAT, 1525: 27 pp.

FAO, 1974. Design of fisheries statistical surveys: inland waters. In: Food and Agricultural Organisation. Fisheries Tech. Pap. vol. 133, pp. 1–122.

FAO, 1979. China Mobilization of Rural Communities for Reforestation. FAO, Rome.

Fisher, R.A., Yates, F., 1975. Statistical Tables for Biological, Agricultural and Medical Research. (Oliver and Boyd). Edinburgh and London, p. 146.

Goswami, B., 2003. Contribution of Women in Fishery Sector. An unpublished M.F.Sc Thesis submitted to the West Bengal University of Animal and Fishery Sciences. Mohanpur, West Bengal, pp. 58.

Hiremath, B.N., Raju, K.V., 2008. Rural livelihood system from food security to livelihood security. In: MDP on Collective Enterprise for Livelihoods Enhancement, July 14–18, 2008, IRMA, Anand, pp. 1–6.

Islam, G.N., Abdullah, N.M.R., Viswanathan, K.K., Yew, T.S., 2008. Augumenting fishers' welfare and all livelihood assets through community based management in Bangladesh. In: Paper presented at the 11th biennial conference of international association for the study of common property held on June 19-23, 2006 at the Agung Rai Museum of Art (ARMA) Bali, Indonessia.

Kar, D., 1990. Limnology and Fisheries of Lake Sone in the Cachar district of Assam (India). viii + 201, PhD Thesis (Published), University of Gauhati, Assam (1984): Matsya, 15–16: pp. 209–213.

Kar, D., 2000. Socio-economic development of the fisherwomen through aquaculture with emphasis on integrated farming in the villages around Chatla Haor wetland in Silchar, Assam. In: Seminar presented on DBT-sponsored Awarness Workshop on Biotechnology-based Programmes for Women and Rural Development, NEHU, Shillong. vol. 1, pp. 13–14.

Kar, D. (2003). Peoples' perspective on fish conservation in the water bodies of South Assam, Mizoram and Tripura, pp. 325-328. In: Participatory Approach for Fish Biodiversity Conservation in North-East India (Eds.) Mahanta, P.C. and Tyagi, L.K., v + 412, National Bureau of fish genetic resources (ICAR) (Lucknow).

Kar, D., 2007. Fundamentals of Limnology and Aquaculture Biotechnology. vi + 609, Daya Publishing House, New Delhi, ISBN: 81-7035-455-2.

Kar, D., 2010. Biodiversity Conservation Prioritisation. pp. xi + 167, Swastik Publications, New Delhi, ISBN: 978-93-80138-26-8.

Kar, D., 2013. Wetlands and Lakes of the World. pp. xxx + 687, Springer, London. Print ISBN 978-81-322-1022-1; e-Book ISBN: 978-81-322-1923-8.

Kar, D., Barbhuiya, A.H., Thangjam, G., Devi, M.S., Deb, S., Das, B., Chanu, H., Nishima, 2009. Panorama of fish biodiversity in certain rivers and wetlands in Manipur. Proc. Zool. Soc. India 7 (2), 123–134.

Kothari, C.R., 1996. Research Methodology: Method and Techniques. Wishwa Prakashan, New Delhi.

Kumar, S., 2002. Methods for Community Participation: A Complete Guide for Practitioner. Vistaar Publication, New Delhi, p. 327.

Maiti, A., 2003. Study on the Socio-Economic Status and Life Style of the Fishermen in Sundarban, West Bengal, India. An unpublished M.F.Sc. Thesis submitted to the West Bengal University of Animal and Fishery Sciences, Mohanpur, West Bengal, pp. 68.

McCaslin, N.L., Tibzinda, J.P., Swanson, B.E., Bentz, R.P., Sofranko, A.J., 1997. Assessing Target Group Needs, in Improving Agricultural Extension: A Reference Manual. FAO, Rome.

Pomeroy, R.S., Rivera-Guieb, R., 2006. Fishery Co-Management, A Practical Handbook. CABI Publishing, CAB International, Wallingford, Oxfordshire OX10 8DE, UK. pp. 255.

Rahman, M.M., Begum, A., 2010. The strategy of empowering poor for wetland resources conservation in Bangladesh. J. Hum. Ecol. 31 (2), 87–92.

Ray, G.L., 1999. Extension-Communication and Management, fourth revised ed. Naya Prokash, 206 Bidhan Sarani, Calcutta, India. pp. 19.

Ray, G.L., Mondal, S., 2000. Research Methods in Social Sciences and Extension Education. Naya Prokash, Calcutta. pp. 324.

Sheikh, A.A., Arya, H.P.S., Nikam, T.R., 1993. Knowledge level of farmers on poultry production technology. Ind. J. Extens. Educ. 29 (3 and 4), 92–93.

Srinath, K., Tangamani, K., 1993. Empowering women through extension. Ind. J. Extens. Educ. 29 (3 and 4), 33–39.

Talukdar, R.K., Barua, U.K., 1999. Participatory aquaculture development: Element and process. In: Paper presented at the summer short course (July 19th to August 28, 1999) on Participatory Rural App.raisal and Rural Aquaculture. Prospect and Perspective Dept. of Extension Education (Faculty of Agriculture) and Fisheries Research Centre, AAU, Jorhat, Assam, pp. 35–37.

Townsley, P., 1996. Rapid Rural Appraisal, Participatory Rural Appraisal and Aquaculture. FAO Fisheries Technical Paper 358, FAO, Rome. pp. 10.

Zar, H.S., 2009. Biostatistical Analysis. vii + 663, Pearson Education, Dorling Kindersley (India).

Further reading

Barman, R.C., Dana, S.S., Panigrahi, A.K., 2004. Information sources utilization pattern of *Beel* users for sustainable development of *Beel* fisheries. J. Environ. Sociobiol. 1 (1-2), 1–94.

Halls, A.S., Arthur, R., Bartley, D., Felsing, M., Grainger, R., Hartmann, W., Lamberts, D., Purvis, J., Sultana, P., Thompson, P., Walmsley, S., 2005. Guidelines for designing data collection and sharing systems for co-Managed Fisheries. Part I: a practical guide. In: FAO Fish. Tech. Pap. 494/1 and 2.

Lobe, K., Berkes, F., 2004. The padu system of community-based fisheries management: change and local institutional innovation in South India. Mar. Policy 28, 271–281.

MACH (2007). MACH achievement. A completion Report, 1–2.

Pido, M.D., Pomeroy, R.S., Carlos, M.B., Garces, L.R., 1996. A handbook for rapid appraisal of fisheries management systems (version 1). ICLARM Educ. Ser. 16. 85 p.

World Bank, 2004. Project Appraisal Document on World Bank Financed Assam Agricultural Competitiveness Project. pp. 15.

CHAPTER

4

Aspects of participatory management of water bodies

OUTLINE

Abstract

The water bodies, notably the wetlands, and their bioresources are coveted wealth of nature. They need to be taken care of through appropriate and efficient *modus operandi* of management. But, it is an onerous and Herculean task for a single person or one single community to accomplish. Studies have portrayed that the involvement of local stakeholders from the grassroots level in wetlands' management contribute significantly to maintaining or restoring ecological integrity and community well-being.

Keywords: Water bodies, Frameworks, Wetland management, Fisheries, Bioresources, Community

4.1 Introduction

The water bodies, notably the wetlands, and their bioresources are coveted wealth of nature. They need to be taken care of through appropriate and efficient *modus operandi* of management. But, it is an onerous and Herculean task for a single person or one single community to accomplish.

Studies have portrayed that the involvement of local stakeholders from the grassroots level in wetlands' management contribute significantly to maintaining or restoring ecological integrity and community well-being. There have been appreciations that successful management initiatives at certain locations have the potential to stimulate positive initiatives elsewhere. The deliberations and resolutions adopted at various meets seem to encourage people to focus on solutions, as well as on honest assessments of practical experiences in participatory wetland management in addition to providing opportunity for "ground-truthing" the guidelines for establishing and strengthening participatory processes by involving the communities and indigenous stakeholders in wetland management.

Notwithstanding the above, there have been efforts to address the complexities and challenges of developing and implementing wetland management systems with the help of local stakeholders. Some of the significant strategies in this regard are as follows:

(i) Promotion of exchange of knowledge, expertise, and experiences through networks.
(ii) Identification and refinement of strategies for wise use of wetlands through best practices of participatory wetland management from around the world.

In this regard, it may be pointed out here that various strategies for wetland management have been developed through different national and international meets, notably the International Conference on Wetlands and Development held in 1995; International Conference on Wetlands and Development held in 1998; Current Trends in Wetlands and Fisheries Research in the New Millennium held in 2000 in Assam, India; Wetlands 2012 International Symposium held in 2012 in Assam, India; International Conference on Wetlands and Fisheries Research held in 2012; and so on.

Such conferences and workshops seem to function as sounding boards for practitioners, policy makers, academicians, voluntary organizations, etc. The deliberations often seem to stress that success must be judged from many different angles, such as welfare of the stakeholders, satisfaction of the government agencies, ecosystem health, etc.

Incidentally, the following themes have been emphasized in the efforts toward the identification and refinement of strategies for wise use of wetlands.

(i) *Assisting frameworks in wetland management*:
Suitable and appropriate changes may be brought about in land tenure, resource access, property rights, and the appreciation and recognition of management (notably, the community-based fisheries management or CBFM) regimes in international, national, and local policies, legislations, development plans, etc., and their impact on the local initiatives.

(ii) *Local environmental knowledge in wetland management*:
Indigenous local traditional knowledge and technologies could be effective in wetland management.

(iii) *Economic valuations in wetland management*:
This includes assessments of the benefits and functions of the wetlands that go beyond theory, and have been used and time-tested in management planning of the wetlands.

(iv) *Gender issue in wetland management*:
Not only the "men" but also the womenfolk could play a significant role in the management of the wetlands. In fact, collaborative joint management systems, involving both men and the womenfolk, could lead to one of the superb management systems for the wetlands. There could be a decisive influence on determining management objectives and practices.

(v) *Participatory monitoring in wetland management*:
It is an important aspect of wetland management involving participation of many people. There could be partnerships between the management body and the local people in order to monitor the ecological characters of the wetlands as well as the community's progress in this direction.

(vi) *Effectively addressing threats in wetland management*:
Threats posed to wetlands could be addressed by involving the stakeholders in wetland management.

(vii) *Win-win scenarios in wetland management*:
It relates to the socioeconomic development going hand in hand with wetland conservation.

(viii) *Exit strategies in wetland management*:
It relates to the community management initiatives being pro-actively planned for, and then effectively implemented followed by scaling down and phasing out of external assistance.

4.2 What is participatory wetland management?

The fundamental inner idea of community-based resource management (CBRM) is said to be the acceptance that human beings are certainly part and parcel of the ecosystem, and neither could part with it nor could be disintegrated from it. The wetlands of today, along with those believed to be prehistoric, are said to be the consequences of complex interactions among physical, biological, and human forces spread over time. Incidentally, almost all the wetlands of the Earth have been influenced and altered by patterns of, more or less, intense human use.

In addition to these, there have, probably, been gradual evolutionary changes in the wetlands through abiotic successional processes. Nevertheless, wetland management by the local people is believed to have a history of millennia.

Notwithstanding the above, participatory management could be defined as "a partnership in which the government agencies, the local indigenous communities, the resource users, and also, perhaps, other stakeholders such as the NGOs, etc., share the authority and responsibility for management of a specific area or set of resources."

Concomitant to the above, the following basic principles are usually required for effective community-based resource management.

(i) *Equipping with power*:
It relates to the real transfer of economic and political powers from a few persons to the impoverished many, and the subsequent control and operationalization by community management.
(ii) *Equity issues*:
It relates to the benefits for the communities as a whole, rather than a few persons only.
(iii) *Sustainability aspects*:
It pertains to the intergenerational equity aspects, said to be based on the carrying and assimilative capacities of the ecosystem.
(iv) *Systems' orientation*:
The community is said to function in the context of other communities and stakeholders, just as resources are ecologically linked to wider ecosystems.
(v) *Gender fair (with special reference to women)*:
Women are fairly involved in the regulation and management of community resources, and their practical and strategic requirements are to be addressed.

4.3 Supporting frameworks

The degree of community participation in the wise use of wetlands seems to vary with the local context, viz., from high levels of empowerment to effective partnerships between the government authorities and the local communities. There seemed to be an acceptable means of settlement (way out) that the authority for resource and ecosystem management are justifiably to be devolved to the maximum extent possible to the local stakeholders, as the indigenous and traditional local people have been living in the locality since millennia. In fact, of late, the indigenous people, almost throughout the whole world, are, probably, trying to seek recognition of their rights, and a greater say in decisions affecting their lives.

Notwithstanding the above, most developing countries depend heavily on the exploitation of its natural resources, especially the biological component. Concomitantly, most of these resources are believed to be found among the less affluent rural communities whose livelihood depends mainly on the harvest and utilization of these resources. In fact, it may be mentioned here that sustainable harvest, efficient management, and effective conservation, *vis-à-vis* development of the locality, probably, depend largely on strengthening the capacity of the local individuals and communities to implement several aspects including conservation initiatives.

4.3.1 An example of community-based fisheries

An interesting success story of community-based fisheries (based on participatory management) could be obtained from Japan (Weinstein, 1998).

It may be mentioned here that Japan rescinded the new laws, after experimenting with western resource management methods (open access subject to licensing) in the late 19th century, and returned to traditional community-based management for all inshore coastal fisheries at the turn of the century, Incidentally, exclusive, inviolable rights have been granted to the local fishing cooperatives. Further, a period of apprenticeship was required to become a member of the cooperative(s). Moreover, the membership was nontransferable. Further, the rules for operation (gear, season, fishing areas, etc.) were decided by the cooperatives, based on the experience of the fishers. Thus, the system seems to provide a robust accountability. It was, probably, because the management decisions were made by fishing groups sharing the same fishing grounds. In addition, any mistake(s), if done, are believed to be borne by the people who make them. Incidentally, Japan's inshore fishery was said to be very healthy, as measured by macro-level indicators of sustainability (Weinstein, *op cit*). In addition to the above, other case studies from different workshops provide encouraging examples of effective supportive frameworks.

4.4 When is participatory wetland management advisable?

The poor conservation results that followed decades of intrusive resource management strategies and the so-called planned development strategies have forced the later policy makers and planners to reconsider the role of community in resource management, harvest, and conservation. Consequently, under such a scenario, as a deviation from earlier views and practice; and, in a path-breaking exercise from the previous works on societal development and management; communities have, perhaps, deterred progressive social change(s). Current champion writers on fish management aspects are, perhaps, making every efforts to emphasize the role of community in bringing about decentralization, meaningful participation, cultural autonomy, sustainable harvest, and conservation of resources. Thus, it, probably, became the right time to advocate and propagate participatory management of resources, including wetland management, as wetlands are said to potential hubs of coveted aquatic resources (Chambers and McBeth, 1992; Chitere, 1994; Etzioni, 1996).

4.5 Indicators for measuring local and indigenous people's involvement in wetland management

The so-called present-day concept of community rarely receives serious attention from all corners; and it, probably, needs support from those concerned with resource use and management, despite its recent popularity. As such, attempts are, perhaps, going on to redress the deficiencies (if any) by investigating and exploring the conceptual origins of "community" concerning resource harvest, management, and conservation. Incidentally, it may be mentioned here that three aspects of community (including its indigenous and other inhabitants) are very important to those who advocate a positive role for communities in resource management. These are as follows:

(i) community as a small spatial unit;
(ii) community as a homogenous social structure; and
(iii) community as shared norms.

Notwithstanding the above, there have been suggestions from some quarters for a more political approach. It has also been argued that community is to be examined in the context of conservation by focusing on the multiple interests and actors within communities, as well as on how these actors influence the decision-making process. Also, there is focus on the internal and external institutions, which shape the decision-making process. In fact, a focus on institutions rather than "community" is likely to be more fruitful for those interested in community-based natural resource management. Incidentally, research and policy, sometimes, try to move away from universalist claims, either for or against the community. Instead, community-based conservation initiatives may, preferably, be formulated on the basis of the ability of the community to tackle their internal differences and processes, and their relations with the external actors.

4.6 Local environmental knowledge

"Community" as shared norms also has an independent positive effect on resource use and conservation. In fact, it may be mentioned here that shared community level norms may promote conservation in two different ways:

(i) Firstly, norms may specifically prohibit some actions. For instance, in many rural areas in the semiarid western Rajasthan, the existing norms hinder the villagers from cutting khejri trees (*Prosopis cineraria*), especially, when these trees are present in the local area, which could be a common area set aside for grazing, and also, often dedicated to a religious deity. In the same region, the people of the Bishnoi community have strong resentments and norms against the killing of wild animal species such as the black buck deer. Cook (1996) tried to detail how the Amung-me in Irian Jaya protect certain groves of trees as sacred, and a marsupial (amat) that plays a role in the propagation of the Pandanus trees. Attempts have been made to explain why women belonging to Juang and Saora tribal communities in Odisha follow strong norms about the timing(s) and season(s) for collecting non-timber forest products (NTFP). There could also be other examples of "conservationist" norms.
(ii) Secondly, the existence of community norms could promote cooperative decision-making processes within the community. If members of a community believe in shared identities and common experiences, they also might be willing to cooperate over more formal decisions to manage and conserve resources. In fact, the presence of community-level norms may facilitate resource management by preventing certain unacceptable behaviors, and, at the same time, could encourage other law-abiding persons (Coleman, 1990).

4.7 Economic valuation

There seems to be a need for an understanding of at least some history of the participatory concept's use in order to comprehend the current widespread involvement of "community" in management exercises. Such a history, perhaps, portrays the ways in which "community," probably, had moved in and out of different scenarios. Further, it, probably, also makes us cautious in accepting community as a panacea to problems concerning the effective conservation of natural resources. Incidentally, current perceptions of community appear to be strongly linked to the patterns of involvement of the researchers of the 19th and early 20th centuries. Such patterns, perhaps, had impacted the economic aspects of industrialization, monetization, and production. Researchers of those days, probably, had found that the world was moving from relationships based on status, kin networks, and joint property to one based on contract, territory, and individual rights. Such underlying images of societal evolution influenced various kinds of formulations (including economic ones) in the community and the society.

4.8 Participatory monitoring

The basic elements of some of the earlier policies and the information available in some of the earlier publications about the local communities seem to have considered "people" as "hindrances" to the ancient and "rational" utilization of resources. A kind of logic, probably, had undergirded the belief that the goals of conservation and the interests of local communities were diametrically opposite. Conservation required protection of threatened resources like the forests, wildlife, pastures, water bodies, fishes, etc. But, the humankind rely on these coveted resources for their food, fodder, shelter, avocation, etc., and thus, perhaps, had got the opportunity to exploit them without restraint, probably, when the harvesters and consumers of these resources were less in number. Studies by various researchers have depicted these by schematic representation(s), and explained several theoretical metaphors that, probably, had served to guide policies, but, perhaps, in an unsatisfactory way. Nevertheless, they were believed to have at least provided a persuasive explanation for resource degradation and depletion. The population of many villages in the tropical countries has grown by leaps and bounds, even with migration to cities.

While many of these beliefs, probably, still persist, most of the present-day thoughts about the role of the community in conservation, perhaps, have changed radically. Incidentally, communities, today, with their significant role in participatory monitoring, are, probably, considered as the loci of conservationists' thinking. Incidentally, many international agencies, such as the World Bank, IDRC, SIDA, CIDA, WWF for Nature, Conservation International, The Nature Conservancy, The Ford Foundation, The MacArthur Foundation, the USAID, etc., probably, have all "found" community to be a strong platform for the implementation of sustained management operations. All these organizations are believed to direct a handsome amount of money and efforts toward the implementation of policies and programs related to community-based sustained harvest, management, and conservation of resources. Moreover, a good number of scholarly papers and policy-centric reports also feature community-based sustained harvest, management, and conservation of resources (notably, Arnold, 1990; Clugston and Rogers, 1995; Dei, 1992; Douglass, 1992; Perry and Dixon, 1986; Raju et al., 1993; Robinson, 1995). A collection of essays, of late, on community-based conservation was said to have exemplified the swing toward community, and tried to opine that "Communities down the millennia, probably, had developed elaborate rituals and practices to limit intake levels of resources; restrict access to critical resources; and, look into proper distribution of harvests" (Western and Wright, 1994). Further, a host of other more specific factors were believed to have helped the advocates of community-based conservationists.

4.9 Effectively addressing threats

Certain views of the community as an organic whole, probably, have continued to influence the present-day perceptions to a significant degree, as well as, perhaps, have accounted for some of the attractions, which the community is believed to have held for millennia. In addition, many of the researchers of the so-called social change(s) were said to have highlighted the disappearance of strong role of community, and its subsequent replacement by other forms of social organization(s). On the contrary, some researchers believe that analytical categories representing discontinuous social states, probably, have overshadowed the real processes of historical change, while the so-called researchers of social change(s), perhaps, have generally accepted the ongoing nature and irreversibility of changes. Nevertheless, a strong correlation seem to exist between those who view progress positively and community negatively.

4.9.1 Community and conservation

Concomitant to above, the history of community in conservation is also a history of revisionism. A comparison of the probable status of prehistoric ecosystem and the life of the primitive people could not, probably, yield much scientific information. However, a rejuvenating attempt, on behalf of the indigenous and the local communities was said to had been made by some quarters; which had, probably, been debated by later anthropological and historical researchers. All these, probably, had much practical and policy implications for future workers.

Incidentally, it is often argued that, demographic growth could only increase harvesting and consumption pressures. In addition, penetration by market forces, could link local systems of resource use to a larger network of demand. This, probably, had further increased pressure on the harvest of natural resources. Concomitantly, the poorly articulated and enforced property rights arrangements, probably, had provided with disincentives for individuals to protect resources. Harmonious manner of resource management might had occurred in the past; but, later, effective management and conservation of resources, probably, had occurred through heavy hand of the State; or, through equally heavy, although less visible, hand of the market forces as well as the stringent regulations of private property rights. Such, ideas, probably, had excluded the locals in the management and conservation exercise of the resources; which, was, probably, not very helpful in the long run.

4.10 Win-win scenarios

4.10.1 Community as common interests and shared norms

The idea of community as shared norms and common interests depends strongly on the perceptions of its members. In this regard, all communities are said to be only "imagined communities," and this imagined sense of community attracts researchers of the conservation to the "community." It is this notion of community that was believed to have grown out of common location, small size, homogeneous composition, and/or shared characteristics. As some researchers have opined, a community exists among individuals who share "common interests and common identity, and said to be growing out of shared characteristics." In fact, "common" and "shared," rather than, individual and selfish are, probably, what makes successful resource management more likely. In a community, there could be instances, where "individuals are said to give up some of their individuality to behave as a single entity in order to accomplish the goal(s)" (Kiss, 1990).

Incidentally, internalized norms of behavior among the members of a community could guide resource management outcomes in desired directions. Community as shared norms is itself an outcome of interactions and processes that take place within communities.

4.11 Exit strategies

Concomitant to the above, it may be mentioned here that conservationists' norms may not always be equated with particular identities, such as "woman" or "the indigenous." Sometimes, although may be remotely, norms may become a significant part of the problem to a conservationist, if a norm promotes exploitation, thus posing an enormous embargo for those interested in community-based conservation. For instance, as a result of land laws in the early colonial periods of many countries, say, in Latin America, there was a strong norm that land is only useful when cleared of trees and used for agriculture. Concomitantly, wildlife is, sometimes, considered as a threat to crops and human lives in many parts of Africa, and hence, it may not be considered as a resource to be conserved (Marks, 1984; Naughton-Treves, 1997). Further, norms may not always be considered as a set of beliefs that communities must always hold, and never give up. They, usually, come into being in relation to particular contextual factors, but, at the same time, may not remain static and rigid, even when codified and written. Incidentally, those who conceptualize community as shared norms, may, sometimes, fail to recognize the difficulties this position poses for conservation efforts. Incidentally, conservationists' norms may not always be easily introduced into a community by external actors. Indeed, it is, probably, not easily known which strategies successfully bring about the change(s) in the norms, which people hold about conservation, especially when the resources in question are a critical part of the family income.

4.12 Conclusions

4.12.1 Community as a homogeneous social structure

Indeed, one of the characteristic features of the community, receiving a good amount of attention in its construction as a social artefact, is, probably, its homogeneous composition. Typically, observers assume "communities" to be groups of similarly endowed (in terms of assets and incomes), relatively homogeneous households which possess common characteristics in relation to ethnicity, religion, caste, or language. Incidentally, such homogeneity is assumed to foster cooperative solutions, reduce hierarchical and inimical interactions; and could also promote better resource harvest, management, and conservation. The idea that a community is homogeneous could match well with the beliefs about its spatial boundaries. The village sites of the less affluent countries are believed to be the commoner sites, where most of the advocates of community-based resource management usually locate their projects and analyses. People living within the same location(s) may indeed have similar occupations, depend on the same resources, speak similar kind of language(s), and usually belong to the same ethnic or religious group. These uniformities may facilitate regular interactions among the group members.

Notwithstanding the above, it is, perhaps, not very clear, at what point, the label "homogeneous" could be applied; even if members of a group are similar in several aspects. In fact, the word "homogeneous" refers to "exactly of same kind." At the same time, it is, perhaps, not very clear, how these shared characteristics are critical to conservation. Further, it is, perhaps, important to analyze the degree of homogeneity and those dimensions of it that are important to resource conservation; mainly, because, all human groups are stratified to some extent or the other. In this regard, it may be mentioned here that, only few studies have been done on what social homogeneity might be. In fact, some of the studies tried to portray that, within the same group (e.g., Masai or pastoralist, fishermen or cattle farmer, or women), multiple axes of differentiation might exist. It may be so required that, on the basis of certain studies with resource use at the local level, it might become necessary not to vividly focus on intra-community conflicts (Agrawal, 1994a; Gibson and Marks, 1995; Ilahaine, 1995; Madzudzo and Dzingirai, 1995; Moore, 1996a,b). On the other hand, it may so appear from the field studies that, even highly differentiated communities may be able to take steps to use local resources sustainably (e.g., Agrawal, 1994b). As such, it could be inferred that there is no easy linear relationship between social homogeneity and sustainable resource use.

4.13 Summary

(1) The water bodies, notably, the wetlands and their bioresources are coveted wealth of nature. They need to be taken care of through appropriate and efficient *modus operandi* of management.

(2) Studies seem to portray that the involvement of local stakeholders from the grassroots level in wetland management could contribute significantly to maintaining or restoring ecological integrity and community well-being.

(3) Participatory wetland management strengthens efficient management (through a participatory process) by involving the communities and indigenous stakeholders in wetland management.

(4) There have been efforts to address the complexities and challenges of developing and implementing wetland management systems with the help of local stakeholders.

(5) Various strategies of wetland management have been developed through different national and international meets.

(6) Such conferences and workshops seem to function as sounding boards for practitioners, policy makers, academicians, voluntary organizations, etc.

(7) Success may be judged from a number of different angles, such as welfare of the stakeholders, satisfaction of the government agencies, ecosystem health, etc.

(8) The fundamental inner idea of community-based resource management (CBRM) is the acceptance that human beings are certainly part and parcel of the ecosystem.

(9) Almost all the wetlands on the Earth have been influenced and altered by patterns of, more or less, intense human use. In addition to these, there have, probably, been gradual evolutionary changes in the wetlands through abiotic successional processes. Nevertheless, wetland management by the local people is believed to have a history of millennia.

(10) Participatory management could be, usually, defined as "a partnership, in which, the government agencies, the local indigenous communities, the resource users, and also, perhaps, other stakeholders, such as the NGOs, etc., share the authority and responsibility for management of a specific area or set of resources."
(11) Some of the basic principles usually required for effective community-based resource management are as follows:

Equipping with power, equity issues, sustainability aspects, systems' orientation, and gender fair (with special reference to women).

(12) The degree of community participation in the wise use of wetlands varies with the local context.
(13) Japan's inshore fishery was very healthy, as measured by macro-level indicators of sustainability.
(14) Three aspects of the community (including its indigenous and other inhabitants) are very important to those who advocate a positive role for communities in resource management. They are: (i) community as a small spatial unit; (ii) community as a homogenous social structure; and (iii) community as shared norms.
(15) There have been suggestions from some quarters for a more political approach. It has also been argued that community is to be examined in the context of conservation by focusing on the multiple interests and actors within communities, as well as on how these actors influence decision-making.
(16) "Community" as shared norms is said to also have an independent positive effect on resource use and conservation.
(17) There seems to be a need for an understanding of at least some history of the participatory concept's use in order to comprehend the current widespread preoccupation with community.
(18) Most of the present-day thoughts about the role of the community in conservation, perhaps, have changed radically. Incidentally, communities, today, with their significant role in participatory monitoring, are, perhaps, considered as the loci of conservationists' thinking.
(19) The history of community in conservation is also a history of revisionism.
(20) The idea of community as shared norms and common interests depends strongly on the perceptions of its members.

Suggested reading

Agrawal, A., 1994a. Rules, rule-making and rulebreaking: Examining the ®t between rule systems and resource use. In: Ostrom, E., Gardner, R., Walker, J. (Eds.), Rules, Games and Common-Pool Resources. University of Michigan Press, Ann Arbor, pp. 267–282.
Agrawal, A., 1994b. I don't need it but you can't have it: Politics on the commons. Pastor. Develop. Network 36 (July), 36–55.
Arnold, J.E.M., 1990. Social Forestry and Communal Management in India. In: Social Forestry Network Paper 11b. Overseas Development Institute, London.
Chambers, R.E., McBeth, M.K., 1992. Community encouragement: returning to the basis for community development. J. Commun. Develop. Soc. 23 (2), 20–38.
Chitere, O.P. (Ed.), 1994. Community Development: Its Conceptions and Practice with Emphasis on Africa. Gideon S. Were Press, Nairobi.
Clugston, R.M., Rogers, T.J., 1995. Sustainable livelihoods in North America. Development 3, 60–63. Sept.
Coleman, J.S., 1990. Foundations of Social Theory. Harvard University Press, Cambridge, MA.
Cook, C., 1996. The divided island of New Guinea: People, development and deforestation. In: Sponsel, L.E., Headland, T.N., Bailey, R.C. (Eds.), Tropical Deforestation: The Human Dimension. Columbia University Press, New York, pp. 253–271.
Dei, G.J.S., 1992. A forest beyond the trees: tree cutting in rural Ghana. Hum. Ecol. 20 (1), 57–88.
Douglass, M., 1992. The political economy of urban poverty and environmental management in Asia: access, empowerment and community based alternatives. Environ. Urban. 4 (2), 9–32.
Etzioni, A., 1996. Positive aspects of community and the dangers of fragmentation. Dev. Chang. 27, 301–314.
Gibson, C., Marks, S., 1995. Transforming rural hunters into conservationists: an assessment of community-based wildlife management programs in Africa. World Dev. 23, 941–957.
Ilahaine, H., 1995. Common Property, Ethnicity, and Social Exploitation in the Ziz Villey, southeast Morocco. In: Paper presented at the IASCP conference.
Kiss, A.e., 1990. Living with Wildlife: Wildlife Resource Management with Local Participation in Africa. The World Bank, Washington DC.
Madzudzo, E., Dzingirai, Y., 1995. A Comparative Study of the Implications of Ethnicity on CAMPFIRE in Bulilimamangwe and Binga. Centre for Applied Social Sciences Working Paper University of Zimbabwe, Harare, Zimbabwe.
Marks, S., 1984. The Imperial Lion: Human Dimensions of Wildlife Management in Central Africa. Westview Press, Boulder, CO.
Moore, D., 1996a. In: University of Zimbabwe, and Program for Land and Agrarian Studies, University of Western Cape, South Africa (Eds.), A River Runs through it: Environmental History and the Politics of Community in Zimbabwe's Eastern Highlands. Center for Applied Social Sciences. Working Paper Series.
Moore, D., 1996b. Marxism, culture, and political ecology: Environmental struggles in Zimbabwe's eastern highlands. In: Peet, R., Watts, M. (Eds.), Liberation Ecologies: Environment, Development, Social Movements. Routledge, New York.
Naughton-Treves, L., 1997. Wildlife versus farmers: vulnerable places and people around Kibale National Park, Uganda. Geogr. Rev. 87 (1), 462–488.

Perry, J.A., Dixon, R.K., 1986. An interdisciplinary approach to community resource management: preliminary field test in Thailand. J. Dev. Areas 21 (1), 31–47.
Raju, G., Vaghela, R., Raju, M.S., 1993. Development of People's Institutions for Management of Forests. Viksat, Nehru Foundation for Development, Ahmedabad, India.
Robinson, M., 1995. Towards a new paradigm of community development. Commun. Develop. J. 30 (1), 21–30.
Weinstein, M.S., 1998. Pieces of the Puzzle: Getting to the Solution for Community-Based Coastal Zone Management in Canada. Key-Note Address to Coastal Zone Canada 98, Victoria, BC.
Western, D., Wright, M.R. (Eds.), 1994. Natural Connections: Perspectives in Community-Based Conservation. Island Press, Washington, DC.

Further reading

Conklin, H., 1957. Hanunoo Agriculture. United Nations, Rome.

CHAPTER

5

Community-based fisheries management in different continents and countries across the world

OUTLINE

Community Based Fisheries Management. https://doi.org/10.1016/B978-0-12-821723-8.00005-1

Abstract

Management is a significant aspect of harvest and conservation of resources. Effective management *modus operandi* is a prerequisite for sustained harvest, utilization, and conservation of resources. The following is a detailed account of the management aspects of 'fish' resources in different continents and countries of the world.

Keywords: Fisheries management, Continents, Countries, Global scenario

Management is a significant aspect of harvest and conservation of resources. Effective management *modus operandi* is a prerequisite for sustained harvest, utilization, and conservation of resources. The following is a detailed account of the management aspects of 'fish' resources in different continents and countries of the world.

5.1 Asia

Introduction

Fisheries management is a significant component of aquatic resources management. Many countries in the world have come up with policies and frameworks to develop strategies and regulations for aquatic resource management, in general, and fisheries management, in particular. The broad aim is to improve the economic efficiency of the fisheries. There could be various kinds of strategies for fisheries management. However, property rights-based fisheries management regimes could be considered as pragmatic and effective in the management of fisheries resources. It is mainly because they reduce or eliminate the incentives for overcapitalization in harvesting the fisheries resources, and thus, could contribute to economic efficiency.

Community-based cooperative fisheries management is one of the community property rights-based fisheries management systems, which has received attention in recent years, particularly in the developing countries. On the other hand, comanagement is a complex issue, on which there is no consensus as yet. Interpretations and status could vary from one region to another, and are influenced by a variety of regional and local parameters. Often, countries may find it difficult to define the term encompassing the different aspects. Concomitantly, there are efforts by certain quarters to pursue a technical agenda for the implementation of meaningful joint-management systems. As such, it may not be possible immediately to offer a precise definition or policy statement on comanagement.

Fisheries are nutrition and avocation-related, interdependent ecological and social systems. They usually call for integrated management approaches. It may be noted that the activities of one person or a group of users may affect the availability of the resource for others. As such, managing such common resources may require continuous and conscious efforts by a wide range of stakeholders. Such efforts may be done to organize and draft rules enabling equitable and sustainable use of the resources for the welfare of all. In fact, collective efforts are often a prerequisite for the development of community-based institutions and the devolution of authority which generally trickles down vertically from the central stage to the local grassroots level.

Communities empowered

However, government(s) generally may not undertake comanagement activities as a means of empowering fishing communities and patronising democracy. Instead, government(s) often consider comanagement as an instrument to achieve their administrative objectives more efficiently by involving the fishing communities in the implementation process. The difficulties could be mainly because the organizational structures of government departments have not yet adapted themselves to the new comanagement concept. Further, most fisheries departments are still staffed with scientists trained in earlier concepts. They are almost exclusively aimed at resource conservation rather than fishing communities' livelihoods.

Notwithstanding the above, a key goal of fishery policy is to balance ecological, economic, and social sustainability. This goal cannot be met through an excessive focus on just one of these aspects, whether ecological, economic, or social. It is important to pursue policy choices that achieve ecological sustainability. Concomitantly, social and economic sustainability **could be promoted actively**. On the other hand, a multifaceted bio-socioeconomic analysis is needed, in order to assess fishery reform options properly within an integrated sustainable development approach. At the same time, more attention could be paid to economic analysis that goes beyond the boundaries of fishery sector for assessing the economic impacts on regions and coastal communities. However, rights-based management (RBM) provide a set of policy choices which involve a combination of "use rights" (mainly fishing rights relating to access, effort and/or catch) and "management rights" (relating to comanagement and rights over involvement in decision-making). In general, use rights arrangements alone may not be able to solve fishery problems. They do so only in conjunction with accompanying management rights (including aspects of fishermen participation, "buy-in" and resulting improved compliance). However, RBM may not be a panacea for fisheries. It could be because: (a) some RBM choices could have negative impacts on ecological sustainability, e.g., by increasing discarding in fisheries, through greater economic incentives for that practice and (b) some RBM choices could have detrimental impact(s) on social sustainability, particularly through a concentration of rights into fewer hands, and a consequent "withering away" of coastal communities.

In the last two decades, policy prescriptions for addressing the global crisis in fisheries had been centering around strengthening fisheries governance through clarifying exclusive individual or community rights of access to fishery resources. However, keeping in view a focus on the small-scale fisheries in developing-countries, in particular, it may appear that the assumed economic incentives for resource stewardship could be found to be insufficient when we think of proper reform of fishery governance. This may be in view of the fact that there could be other sources of insecurity in people's lives which may be unrelated to the state of fishery resources.

In this connection, it could be said that more secured and less vulnerable fishers may become more effective and motivated fishery managers in the context of participatory or rights-based fisheries governance. Further, insecurity among the poor fishers could be most effectively addressed by social and political development that invokes the existing legal framework supporting the Universal Declaration of Human Rights. This perspective goes well beyond the widely advocated notion of "rights-based fishing" and aligns with what fishery sector analysts call the "rights-based approach" with the same terminology used in the context of international development. Moreover, there could be enhancement in the chances of achieving both human development and resource sustainability in small-scale fisheries of developing countries by embedding the fisheries governance aspect within a broader perspective of human rights. Incidentally, it may be stated that the traditional or local community-based management (CBM) systems have a long history of existence in **Southeast Asia.** However, a majority of these systems have weakened or disappeared due partly to the institutional restructuring under colonial administrations, technological modernization, socioeconomic stratification, and unequal concentration of power and wealth within coastal communities.

Fisheries management is a significant component of aquatic resources management. Many countries in the world have come up with policies and frameworks to develop strategies and regulations for aquatic resource management, in general, and fisheries management, in particular. The broad aim is to improve the economic efficiency of the fisheries. There could be various kinds of strategies for fisheries management. However, property rights-based fisheries management regimes could be considered as pragmatic and effective in the management of fisheries resources. It is mainly because they reduce or eliminate the incentives for over-capitalization in harvesting the fisheries resources and, thus, could contribute to economic efficiency.

The **Asian continent** is gifted with rich varieties of large number of water bodies and a bewildering diversity of BD including fishes. As such, there have been different kinds of *modus operandi* for the management of the water bodies and their bioresources. Different workers have been working on the various types of management *modus operandi* including **Community-Based Fisheries Management (CBFM)** in Asia.

5.1.1 Asia general

Various workers have worked on Management *modus operandi* in **Asia in general**.

Thilsted and Wahab (2014) had dealt with production and conservation of nutrient-rich small indigenous fish species (SIS) in ponds and wetlands for nutritional security and livelihoods in South Asia. It is quite well known that small fish occur on their own, usually, in large number and, generally, in any kind of FW (freshwater) body, without much care being taken of in contrast to big fishes which, often, require salubrious environmental conditions for proper growth, sustenance and propagation. As such, these, much easily available small fishes are to be conserved and taken care of, at least now a days (when there is dwindling trend of fish production) for the nutritional security and livelihoods of the people, particularly in South Asia, where, fish serves as one of the staple food items in nutrition and a potential source of avocation.

Concomitant to the above, fish is an important entity in the life and culture of the people of Asia in general and the Indian subcontinent in particular, being a source of rich nutrition and avocation for the people. The abundantly distributed FW bodies, notably the rivers, wetlands, ponds, etc., are the habitats and homes for a large number of FW fishes. In fact, *c.* 50% of 260 FW species are small fishes, popularly known as "SIS." Incidentally, researches carried out by the Bangladesh Agricultural University (BAU), Bangladesh Fisheries Research Institute (BFRI), and University of Copenhagen (UoC), Denmark over the last few decades have established that some SIS are highly nutritious, containing large amount of bioavailable micronutrients, notably, vitamin A, iron, zinc, and calcium. As such, World Fish Bangladesh promoted the development and nationwide dissemination of SIS production technologies, in collaboration with the Department of Fisheries and NGOs, as a measure in its continued endeavor to address hunger and malnutrition. Efforts had been made to propagate the knowledge and technologies among the policy makers and relevant stakeholders in Bangladesh as well as in the entire South and SouthPolycultureeast (SE) Asian region, building on the successful adoption of minor carp, *Amblypharyngodon mola.* Polyculture technologies in stand-alone household ponds and ponds connected to rice fields could be considered with emphasis. Such works, being disseminated, play a significant role in the promotion of further research, knowledge sharing, and refinement and adoption of production technologies to augment the production and consumption of SIS in Bangladesh as well as throughout the South and SE Asian region.

Notwithstanding the above, SIS are the potential sources of essential macro- and micronutrients, which play a significant role in the elimination of malnutrition and promotion of micronutrients availability as well as a potential source of avocation among the populations of many South and SE Asian nations. Incidentally, it may be noted that of the *c.* 260 FW fish species in Bangladesh, > 140 are classified as SIS. In fact, SIS are an integral component and the staple food of the Asian diet (particularly, India and Bangladesh). The SIS contains high amounts of vitamins and minerals, including calcium, iron, and zinc, and many SIS are eaten whole, with organs and bones. Further, some SIS, notably, *A. mola*, are also rich in vitamin A. Furthermore, the SIS are often cooked with vegetables and a little oil, and thus, they contribute to the food diversity of the consumers. The Department of Fisheries (DoF), Government of Bangladesh, researchers from Bangladesh Fisheries Research Institute (BFRI) and the universities as well as the other extension agencies and NGOs have shown interest of participation in the promotion of SIS in different parts of Bangladesh as well as in the neighboring countries. In addition, the SIS are often considered not only as a major source of animal-protein, contributing to improved food and nutritional security, but are also recognized as a potential source of earning livelihoods of the people in South and SE Asia. Incidentally, the Bangladesh Agricultural University (BAU) had been working with partners from Denmark, over more than a decade, and had developed production technologies for pond culture and stock enhancement in the wetlands. Notably, the WorldFish had been implementing these technologies in collaboration with national institutions and NGOs. Nevertheless, such efforts of large-scale SIS production could be more successful with the application of efficient *modus operandi,* like the CBFM involving all the stakeholders from the grassroots level. Thus, one of the principal aims of many such efforts is to bring all the stakeholders, policy makers, extension workers, researchers, NGOs, and development partners together to share knowledge of the different people as well as for developing production technologies, and policy briefs for formulation of strategies for wider dissemination of pond culture and wetland-based production and conservation technologies. Moreover, different workshops are expected to generate ideas, update information for further research and development of sustainable technologies for production, management, and conservation of SIS for the all-round success in this direction, and for the benefit of the people of the region at large.

Concomitant to the above, Flewwelling (2001) have dealt with fisheries management and their monitoring, control, and surveillance (MCS) in South Asia. The important aspect(s) of management could be successful only when the status of MCS is followed up, else the exercise could go astray.

Incidentally, there have been contributions depicting the results of the country review and assessment missions to 11 Asian countries under the Norwegian funded FISHCODE Project GCP/INT/648/NOR) undertaken during January 1999 and October 2000. It has focused on the MCS component of fisheries in Bangladesh, Cambodia, India, Indonesia, Malaysia, Maldives, Myanmar, Philippines, Sri Lanka (SL), Thailand, and Vietnam. The objectives of the missions were threefold:

(i) To assess fisheries management and MCS capability.
(ii) To assist, wherever possible, with advice, for immediate MCS steps to enhance current systems.
(iii) To ascertain potential regional or subregional mechanisms for cooperation in fisheries management, including the MCS component.

In fact, a number of such reports have provided a comparative analysis of the findings of the various efforts as well as presented a synopsis of the potential regional cooperation(s), and then provided more detailed information and suggestions to enhance MCS on a regional basis.

Concomitantly, Wilson and Kleban (1992) have dealt with practical implications of chaos in fisheries. Fisheries occur very much on the lap of nature and are being handled by human beings with regard to its harvest, utilization, management, etc.

There have been paper(s) dealing with the practical implications for fisheries management, if fish populations happen to behave chaotically. Many such papers try to argue that the principal effect(s) of chaos may be manifested in terms of information, measurement, and management problems; in particular, recruitment-based policies are likely to need unattainable measurement accuracy, else could lead to unpredictable outcomes. This raises the questions about how sustainability could be maintained in chaotic fisheries. Further, many such communications suggest that management, based on the relatively stable ecological relationships in fishery systems, may be the most practical way to manage chaotic fisheries. Different works also try to hypothesize that successful cases of traditional (community) resource management are likely to be based on qualitative ecological approaches rather than the direct quantitative manipulation (e.g., quotas) of exploited populations. Incidentally, Estellie Smith (1991) and Chris (1991), in a recent article and comment, discussed the implications of chaos theory for fisheries management. Both Smith and Finlayson have described "chaos theory" as a competing paradigm for the conventional "linear paradigm" of fisheries management with only minor quibbles. Nevertheless, their views are accepted with minor disagreements with regard to the practical implications of chaos in fisheries. By chaotic fishery, briefly, it could be meant as one in which the time path of individual species may not have an equilibrium tendency, and may vary unpredictably within certain limits. This contrasts with standard theory (including its stochastic aspects) that assumes that population abundance tends toward some predictable equilibrium.

Nevertheless, the principal inference of many such works is that the best hope for successful management of chaotic fisheries lies in "ecologically adapted" management system.

Notwithstanding the above, Pomeroy et al. (1998) have dealt with fisheries comanagement with notes on key conditions and principles drawn from Asian experiences. Management, either in the form of comanagement or community management, is a significant parameter for the sustained utilization of the resources as well as for their preservation and conservation for the posterity. For this purpose, there is a necessity of experiences to be gained from various national and international exercises.

Concomitant to the above, the International Center for Living Aquatic Resources Management (ICLARM) in Manila, the Philippines, and the Institute for Fisheries Management (IFM) at the North Sea Center, Hirtshals, Denmark, in 1994, in collaboration with National Research Partners (NARS) in several Asian and African countries had initiated the fisheries comanagement research project. Incidentally, the collaboration between ICLARM, IFM, and NARS was based on a mutual interest to gain practical experience in research in fisheries comanagement, to demonstrate its applicability as a sustainable, equitable, and efficient management strategy as well as to develop models for use and adoption by governments, fisheries communities, NGOs, and others. The first phase of the project had ended in 1998 and a second phase was also planned. The project strategy was to conduct research in a variety of aquatic resource systems and countries around the world, with the aim to determine if fisheries comanagement could be a viable management strategy under varying conditions (political, social, cultural, economic, biophysical, technological, etc.). In fact, the overall purpose of the project was to ascertain the prospects for successful implementation of fisheries comanagement strategies. The project might not have much advocated or promoted fisheries comanagement, but systematically and comparatively have documented and assessed models and processes of comanagement at national government and community levels as well as their results and impacts, which could be emulated. General principles and conditions which facilitate successful implementation of fisheries comanagement were also identified.

Incidentally, there had been communications dealing with such aspects. As such, some papers have dealt with presentation of the results of the research, especially, key principles and conditions, which facilitate the successful implementation of comanagement as identified through the various projects and research activities in Asia. Such papers usually began with a discussion of the strategy and data sources used in the research. Incidentally, the projects have used key conditions for successful common pool resources institutions as identified by Ostrom (1990, 1992) as a foundation for the research. There had been revisits, discussions, and assessments in light of their applicability for fisheries comanagement in Asia. The contributions usually infer with policy implications for fisheries comanagement in Asia and worldwide.

Notwithstanding the above, the activities of the fisheries comanagement research are conducted usually through three components as follows:

(i) comparative case studies of fisheries comanagement,
(ii) researches on the subject in different countries, and
(iii) exchange of information.

In this respect, the first component is to make use of secondary data sources, such as the project reports, research reports, NGO reports, scientific journal articles and other published materials, to gain insights into approaches, processes, performance, results, and the impacts of comanagement at both national government and community levels.

The second component is to make use of a variety of research activities, including historical reviews of comanagement experiences, case study analysis, impact evaluations of comanagement arrangements; hypothesis testing of advantages or benefits of comanagement; legal, institutional, and policy analysis; pilot sites; and workshops, etc., in order to evaluate and document the approaches, institutional arrangements, performance, and legal and policy factors affecting implementation of fisheries comanagement.

The third component is networking and training activities among the research partners. The research works make use of comparative analytical approach, relying on a common research strategy and research framework for use in each partner country and resource system, in order to integrate and improve the understanding and implementation of comanagement strategies. In addition, the institutional analysis research framework provides a structured approach to examine and document the origin, current status, operation, and performance of fisheries comanagement arrangements.

Notwithstanding the above, researchers have undertaken works in six Asian countries to ascertain implementation of fisheries comanagement to varying degrees (Pomeroy et al., 1996; Ahmed et al., 1997). Of the six countries, the Philippines was the most experienced with comanagement and CBM of coastal resources. Further, the Philippines also had the strongest policies and laws supporting comanagement in the Local Government Code of 1991, the new Fisheries Code, and the national development plan.

In addition, the new **Thailand** constitution and national development plan strongly support CBM initiatives. The master plan for fisheries development up to 2010 in Vietnam and the policies of the Ministry of Fisheries support participatory approaches to resource management and local user rights.

The **Malaysian** Department of Fisheries have been developing a new fisheries policy, which would endorse comanagement arrangements.

In **Bangladesh**, the government and the NGOs are jointly promoting sustainable use of open-water fisheries resources through the active participation of users in management. The program had sought to empower fishing communities to become comanagers of these fisheries. Further, in **Bangladesh**, the Oxbow Lakes were of a size that could be easily managed and monitored by the fishers. Furthermore, in **Bangladesh**, membership in the lake fisheries teams of the Oxbow Lakes was clearly defined to include those fishers living around the lake (Khan and Apu, 1998). Still further, in the Oxbow Lakes of **Bangladesh**, Muslim and Hindu fishers work together on the lake fisheries teams (Khan and Apu, 1998).

In **Indonesia**, the national development plan had endorsed more active participation of the fishers in economic development. In addition, there was increasing support by local governments and NGOs to revitalize traditional resource management systems through comanagement arrangements.

Incidentally, the researchers used the key conditions for successful common pool resource institutions developed by Elinor Ostrom (1990, 1992) as working hypotheses for the research. In fact, the analysis of comanagement was included in the area of common property theory (Pomeroy and Berkes, 1997). Further, 11 key conditions, described by Pomeroy and Williams (1994), have served as the starting point for analyzing the emergence and institutional sustainability of comanagement arrangements. The researchers found that the boundaries were of high importance to successful implementation of comanagement.

In **San Salvador** Island and **Malalison Island, the Philippines,** the marine sanctuary had boundaries identified with buoys to inform outsiders of its existence and to allow fisher organization members to more easily monitor the area (Katon et al., 1997; Baticados and Agbayani, 1998). The researchers, further, found that clearly defined membership was of high importance to successful implementation of comanagement.

5.1.1.1 Cohesion in the group

The fisher group(s) or organization(s), who permanently reside near the area are to be managed. There was a high degree of homogeneity in terms of kinship, ethnicity, religion, or fishing gear type, among the groups. The researchers found that group cohesion, especially in terms of group homogeneity, was of some importance to successful implementation of comanagement. In the **Philippines,** successful comanagement activities have occurred in both socioeconomically and culturally homogeneous and heterogeneous communities (Pomeroy et al., 1996). However, this is not to downplay the importance of this condition, for there were many communities in **Vietnam, Thailand, Indonesia, and the Philippines,** where successful implementation of comanagement was dependent on the high level of socioeconomic and cultural homogeneity of the community.

5.1.1.2 *Organizations existing*

Many fishers have some amount of prior experience with traditional community-based systems and with organizations, where they were representative(s) of all resource users and stakeholders interested in fisheries management. The researchers have found that this condition was of medium importance to successful implementation of comanagement. While it was useful for fishers to have had some amount of prior experience with traditional resource management systems and with organizing the collective actions, it was not a prerequisite for success. There were many cases in the **Philippines, Thailand, and Bangladesh,** where comanagement was successful despite the fact that fishers had no previous experience with organizing resource management (Katon et al., 1997, 1998; Masae, 1998; Khan and Apu, 1998).

5.1.1.3 *Costs less than the benefits*

Individuals might have an expectation that the benefits to be derived from participation in and compliance with CBM could exceed the costs of investments in such activities. This condition was found to be of high importance for success of comanagement. In the **Philippines,** NGOs spend a great deal of time in preparation for society's activities, e.g., educating the fishers about the benefits and costs of comanagement and the economic implications of choosing among different management and development strategies (Foltz et al., 1996; van Mulekom, 1998).

5.1.1.4 *Participation by the affected people*

Most individuals who are affected by the management arrangements are included in the group that makes and can change the arrangements. The same people, who collect information on the fisheries, also take decisions about management arrangements. This condition was found to be of high importance for the successful implementation of comanagement. In the Oxbow Lakes of **Bangladesh**, the lakes fisheries teams allowed all members to have equal voting rights in making management decisions (Khan and Apu, 1998). In **San Salvador Island and Malalison Island, in the Philippines**, all members of the fisher organization were involved in making and changing the rules (Katon et al., 1997; Baticados and Agbayani, 1998).

5.1.1.5 *Enforcement of management rules*

The simple management rules, if enforced, could be monitored by the fishers, having duties, perhaps, being shared. The researchers had found that enforcement of management rules was of high importance for success in comanagement. In **San Salvador Island, in the Philippines**, the fishers shared responsibility for guarding the marine sanctuary (Katon et al., 1997).

5.1.1.6 *Legal rights for organization*

The fisher groups or organizations have the legal right(s) to organize and make arrangements related to its needs. Moreover, they had been enabling legislation(s) from the government defining and clarifying local responsibilities and authorities. This condition was found to be of medium importance for successful implementation of comanagement. In the **Philippines**, for example, NGOs helped in organizing fishers to take responsibility for resource management before there was any formal legislation from the government. Moreover, the responsibility and authority of the local fisher organizations for resource management in the **Philippines** had been formally clarified under the Local Government Code of 1991. In **Thailand**, a similar situation existed, and the fishers had been organized for comanagement without any legal right from the government. However, the new **Thailand constitution** supports the right for fishers to organize.

5.1.1.7 *Leadership and cooperation at the community level*

There was an incentive and willingness on the part of fishers to actively participate, with time, effort, and money, in fisheries management. Moreover, there is an individual or core group who takes responsibility of the management process. Researchers found that this condition was of high importance for the successful implementation of comanagement. In the **Philippines**, research workers, evaluating the impacts of CBM have revealed that communities where fishers had positive levels of cultural values and attitudes toward collective actions were consistently related to perceptions of positive change and were more successful (Pomeroy et al., 1996). It was, further, found that, strong local leadership was critical for success.

5.1.1.8 *Delegation of authority through decentralization*

The government established formal policy and/or laws for the decentralization of administrative functions and delegation of management responsibility and/or authority to local government and local group organization levels. This condition was found to be of medium and low importance for successful implementation of comanagement.

Comanagement has been successfully implemented, throughout the **Asian** region, without the formal policy support of government in many nations. Only recently, have the governments of the **Philippines and Thailand** developed and implemented policies for decentralization and delegation of authority to local fishers' organizations for resource management. It may be pointed out here that the existence of formal policies and laws for decentralization does increase the chances of success for comanagement (Katon et al., 1997, 1998; Baticados and Agbayani, 1998).

5.1.1.9 *Collaboration and coordination between government and community*

A coordinating body was established, external to the local group or organization and with representation from the fisher groups or organizations and the government, to monitor the local management arrangements, resolve conflicts, and reinforce local rule (s) *in vogue*. This condition was found to be of medium and low importance for the successful implementation of comanagement. In some cases, such a formal coordinating body does exist, such as the San Miguel Bay Management Authority in **San Miguel Bay in the Philippines** (Pomeroy and Pido, 1995), but it is not common. In other cases, coordination between the government and the community is, perhaps, informal. This is done, primarily, through dialogs, meetings, and consultations. A reexamination of Ostrom's 11 key conditions for successful common pool resource institutions, based on knowledge and experience gained on fisheries comanagement in Asia, revealed that 6 of the 11 conditions were of high importance for the successful implementation of fisheries comanagement.

5.1.2 Asia-Pacific region

Notwithstanding the above, traditional systems of CBFM are time-tested. They, probably, need protection and boosting. An attempt was made in the following discussion (**Ruddle, 1993**) to analyze and exemplify the historical and contemporary external forces, which drive change(s) in traditional community-based aquatic resource management systems in the **Asia-Pacific Region.**

Ruddle (1993) discussed about the external forces that might have caused some changes in the traditional community-based fisheries management systems in the Asia-Pacific Region. Traditional *modus operandi* of CBM measures was based on wisdom of the people to conserve nature with nature-friendly measures.

However, various external forces act from different directions in order to bring about changes in the traditional community-based fisheries management systems in the **Asia-Pacific** region. In fact, effective management of the natural resources could be achieved through the application of efficient management measures like the CBM of the natural resources, including the fisheries management by involving the stakeholders from the grassroots level in a participatory mode.

Concomitant to the above, there have been contributions pertaining to the historical and contemporary external forces that had driven changes in traditional community-based aquatic resource management systems in the Asia-Pacific region, and in a contribution of similar type, Ruddle (1993) tried to analyze, exemplify, and summarize **the** policy alternatives regarding the future of such systems.

5.1.2.1 *Development of model fisherman villages*

Some of the salient objectives and plans in this direction are as follows:

(i) Providing with basic civic amenities like housing, drinking water, and construction of community hall for the fisher folk.
(ii) The villages are to be provided with tube wells at the rate of one tube well for every 20 houses.
(iii) It is notable that a fisherman village with at least 75 houses is eligible for availing financial assistance for the construction of a community hall for recreation and common working place.

5.1.2.2 *Group accident insurance program for active fishermen*

It operates in the following manner:

(i) INR 50,000.00 against accidental death or permanent total disability and.
(ii) INR 25,000.00 for permanent partial disability.

These could be revised as per the prevailing market rates.

5.1.2.3 *National cooperatives development corporation (NCDC)*

It works in the following manner:

(i) Strengthening of the existing cooperatives by providing assistance on liberal terms for the expansion of their activities.

(ii) Development of potential areas by organizing functional cooperatives in these sectors.
(iii) The Apex Management organization formulated specific schemes and patterns of assistance for enabling the fishery cooperatives to take up activities relating to production, processing, storage, marketing of fishes, etc.

5.1.2.4 *Employment status*

The estimated statistics in this regard are as follows:

5.2. *c*. 9.0 lakh full-time and *c*. 11.0 lakh part-time fishermen in marine fisheries-related activities.
5.3. *c*. 4.0 lakh in fish marketing *c*. 3.0 lakh in repairs of fisheries requisites, *c*. 50,000 in fish processing and *c*. 4.0 lakh in other ancillary activities.
5.4. In addition, *c*. 31.5 lakh people are engaged in fishing and farming operations directly or indirectly.

The above figures could change with time.

5.1.2.5 *Difficulties in the operations of cooperatives at optimum levels*

Some of the difficulties are as follows:

(i) Weak collaboration/coordination among the various institutions involved in the management and operation of cooperatives at primary, district, and provincial levels.
(ii) Rivalry among the individuals/groups.
(iii) Lack of adequate infrastructure including lack of trained and skilled man power and deficiencies in financial resources.

Nevertheless, FISHCOPFED may play a significant role in strengthening the linkage and coordination among the cooperatives at various levels.

5.1.2.6 *Human rights and issues related to environment*

There is a common notion that "Rights entail Duties." Therefore, it could be opined that the humankind is duty bound to protect the environment with regard to both biotic and abiotic resources. Further, the 1972 Stockholm Declaration had proclaimed that mankind's natural and human-made environment are both essential to his/her well-being ad to the enjoyment of basic human rights, even the right to life itself. In 1986, the United Nations General Assembly recognized the relationship between the quality of human environment and the enjoyment of basic human rights [UNGA resolution 2398 (XXII), 1986]. Further, the 1992 Rio Declaration had emphasized the sustainable development and environmental protection. Moreover, the Agenda 21 had called for the fulfillment of basic needs, improved living standards for all, better protected and managed ecosystems, and a safer and more prosperous future.

5.1.2.7 *Issues of equity and access to fisheries*

Most fisheries regimes may be considered as rights-based. Individual transferable quota (ITQ) regimes have been theoretically constructed and specified according to neoclassical economic prescriptions. In such regimes, maximization of net benefits in terms of narrowly defined economic values is paramount. Equity considerations are notably absent while important collateral economic values bearing on the welfare of fishery-dependent communities are routinely ignored. With respect to equity impacts, ITQs are especially noted for large windfall gains from public resources to small groups and for corporate and geographical concentration of fisheries access rights. This result relates to the way in which ITQs, typically, were implemented.

5.1.2.8 *Categorization of fisheries management systems*

Type characteristics

(a) Private ownership with right to transfer

Ownership rights are held by individual fishermen as in the case of fixed engines like stake nets and China nets.

(b) Ownership by the State

Ownership right of certain reservoirs and wetlands are, generally, retained with the government and access for fishing is banned for the fishermen or the public.

(c) Sole Ownership

All fishing activities dictated by the government as the sole owner, as in the case of certain reservoirs, and the fishing right is auctioned to fishermen groups or societies.

(d) Limited Access Schemes

Entry and withdrawal rights are restricted by the government as in paddy fields during periods of no cultivation.

(e) Closed season

This measure was adopted for trawlers in coastal waters and for giant prawn fishing in certain inland water bodies during breeding season. Moreover, bottom trawling in coastal waters is banned generally from July 15 to August 31 every year by a law. Also, there exist regulation(s), in general, regarding "no fishing" during the breeding season of any fish.

5.1.2.9 *Common ownership (unofficial: without any legal or governmental support)*

Resource is owned and managed jointly by a small group of self-governed fishers. In such practices, there seemed to be no legal or governmental support for this kind of arrangement. Each participant in the fishery then depends on the arrangements or rules worked out in the fishermen societies or groups.

5.1.2.10 *Sri Subrahmanyeshwara Aqua Club in West Godavari district of Andhra Pradesh*

Cluster-based shrimp farms, aqua clubs, etc., had been setup in Andhra Pradesh (AP) in order to promote cooperative approach in the management of shrimp farming. A total of 128 aqua clubs with a membership of *c.* 3367 farmers are existing in the province. However, one of the best examples of initiatives of this kind is the Sri Subrahmanyeshwara (SS) aqua club in Mogalthur Village in West Godavari district of AP. The club comprises 58 farmers with 108 ponds spread over 58 ha and the farmers involved were mostly small-scale, practicing improved traditional farming with low investments. Such efforts by the shrimp farmers in Tamil Nadu (TN) and AP could also to be replicated in the other coastal provinces, where shrimp is fast developing on commercial scales.

5.1.2.11 *Suggestions for India*

The following suggestions of fisheries management are proposed for India, keeping in view the incentive structure of ITQs, and the organization of the Japanese cooperatives:

(i) To calculate the maximum sustainable yield (MSY) along a coastline.
(ii) The share (in weight) of each fishing family could be obtained by dividing the MSY by the number of families.
(iii) Thus, the share in percentage of each family is determined, which could remain constant for ever.
(iv) The actual amount of fish entitlement could depend on MSY as calculated periodically.
(v) The quotas could still bind people together, regardless of caste/clan.
(vi) This may enable the formulation of norms and rules easily. Enforcement could also become easier in such close-knit environment.

5.1.2.12 *The role of public-private partnership in the management of estuarine fisheries: Learning from the Kerala model of comanagement*

Cherai poyil

Cherai poyil is a brackish water pond with an area of *c.* 210 ha located in the Kerala province of Indian peninsula at the extreme Northwest (NW) opening of the Cochin estuary, which opens into the Arabian Sea at Azhikode. Cherai poyil is a typical example of how a local organization (gram panchayat or GP) and communities together can share the responsibility to allocate fisheries to various fishing groups and private entrepreneurs in the northern side of the Cochin estuary. The fishing territory managed jointly by local fishing communities and the GP is known as "kappu."

5.1.2.13 *Rights and responsibilities of the panchayats in managing poyil fisheries*

Today, kappu is the property of Pallipuram GP and its management is believed to be coordinated by a subcommittee which included the elected councillors, fisher representatives, and leaders of local political parties. The GP learnt from its past experience(s) that direct mundane supervisory functions were expensive and community cooperation was highly essential to ensure better governance.

In fine, cooperative resource governance is a unique arrangement by which local governments accepted the role of communities in resource management and provided more space for their activities and initiatives in the management of the natural resources. The timely restructuring and flexibility of the system in tune with the changing forces of modernization and external pressures could be considered as a welcome step for its success.

Notwithstanding the above, in the **Asia-Pacific region**, the traditional community-based aquatic resource management systems are increasingly affected by external factors which cause stresses and often lead to radical change(s) in the systems, including their extinction. Such a trend is *in vogue* also in many parts of the world. There is nothing new about this, except that the intensity of impacts and diversity of their sources have increased in recent decades. Thus, contemporary community-based aquatic resource management systems exist under various environmental, social, ecological, political, and demographic circumstances that could often be very different from those of even the recent past. However, nowadays, such systems are swept up in the overall process of national modernization in the Asia-Pacific region. Among the principal, all-pervasive external forces are the legacy of colonialism, contemporary government policy and legal change(s), the replacement of traditional local authority, demographic change(s), urbanization, changes in education systems, modernization and economic development, commercialization and commoditization of living aquatic resources, technological change(s), the policies of external assisting agencies, and national policies for economic sectors other than fisheries. Such external forces may not work in isolation but rather sometimes work as a mutually reinforcing and potentially destructive complex.

Concomitant to the above, somewhat more recent pressures (which may not be universally so) are the commercialization and monetization of formerly local and mainly subsistence or reciprocal exchange or barter economies, which now link them with external markets. In addition, certain altered situations could lead to changed perceptions in fishing communities regarding the value of aquatic products. Such situations could lead to the breakdown of traditional management systems through the weakening or total collapse of traditional moral authority. In fact, small communities may not be immune to such effects.

Notwithstanding the above, deliberate legislative action(s) may do away with traditional community-based systems and these were believed to be also widespread. Incidentally, the **Philippines** provides an example. According to early Spanish chroniclers, systems of community-based coastal and riverine fisheries management existed in the Philippines in pre-Hispanic and early-Colonial times, based on the independent villages ***(barangay)***, around Manila and in the Tagalog Region of Luzon Island (Blair and Robertson, 1903-09). But, as Spanish colonial rule intensified, the ***barangay*** was eliminated as an administrative unit and the village sea territory disappeared with it (Lopez, 1985). Today, the coastal or "municipal fisheries" of the Philippines operate under an open-access regime, perhaps in both inland and marine "municipal waters"; the latter within 3 NM of the municipal coastline (Santos, 1980). This situation had its origins in the colonial periods of both Spain and the United States.

Concomitant to the above, the case of the island nation, **Fiji**, is highly interesting, in that it provides a documented example of a blatant attempt by local colonial officials to destroy a traditional management system in favor of expatriate entrepreneurs, and in defiance of the expressed wishes of the British Crown and the unambiguous orders of the metropolitan government.

On the other hand, it may be noted that traditional fishing rights still regulate activities in most traditional coastal communities of **Papua New Guinea.** Further, it may be mentioned that many fishermen lacked access to alternative resources from which to make a living. Their livelihood and incomes were dependent on the continued recognition of their traditional fishing rights. Further, in Papua New Guinea, the attitude of the colonial authorities toward community-based aquatic resource management was quite unlike the case of Fiji or most other colonies in the **Asia-Pacific Region.** In Papua New Guinea, the customary fishing and aquatic resources rights of indigenous peoples were recognized or protected by the various colonial administrations. The indigenous people such as the Trobriand Islanders were allowed to exploit these resources commercially (Tom'tavala, 1990). The indigenous peoples' fishing rights in waters adjacent to a landholding were explicitly recognized in 1952.

5.1.2.14 Policy of the government

The situation of traditional community-based aquatic (mainly marine) resource management systems have changed a little with national independence. Incidentally, for most parts of the **Asia-Pacific region**, the present unsystematic and ad hoc statutory framework relating to community-based aquatic resource management is the result of the absence of appropriate policy, which, in turn, is partly a legacy of colonial neglect. Moreover, it could partly be also a reflection of a preoccupation with other development priorities and, partly, because of devolution of power to the local level, as in the case of **Indonesia**. It could be an anathema in a vast, sprawling archipelagic nation.

About Papua New Guinea, Tom'tavala (1990) observed that in early 1990, neither the national government nor the provinces had a policy for traditional marine fishing rights, despite the official admission that conflicts over them are among the most prevalent contemporary disputes. This actually or potentially had been impeding economic development, thus leading to social and political instability.

Notwithstanding the above, in the **Asia-Pacific region**, the government concept regarding traditional community-based aquatic resource management could range from confusion, as in **Kiribati and Fiji**, to the incorporation of traditional systems within modern management regimes, as in much of **Melanesia**. Further, it could also be somewhat like the deliberate undermining of traditional systems and their replacement with centralized systems of administration, as in **Indonesia**.

Notwithstanding the above, the government policy had often been to replace traditional CBM systems with open-access systems. But some nations and communities have rejected this approach. Such an initiative and response occurred in 1976 in Palau (Johannes, 1988), and in **Japan** during the Meiji era (1867–1912), and again under the post-war US occupation in Japan (Ruddle, 1985, 1990). Indeed, Johannes (1988) observed in the case of the **Pacific Islands** that the level of resentment aroused in fishing communities by official attempts to eradicate community-based resource management (CBRM) systems was spontaneous. As such, such attempts could be impossible in a democracy. Apart from near urban centers, almost every disappearance of a traditional system in **Oceania** had resulted from legislation and other actions under undemocratic colonial regimes. Such problems could only become more complex as the demographic change(s), urbanization, commercialization, and other pressures become more pronounced.

5.1.2.15 Support of tradition

In many **Pacific Island nations**, and especially those in **Melanesia and also in parts of Polynesia and Micronesia,** despite an absence of clear-cut policy and the requisite statutes, the contemporary value and role of traditional community-based aquatic resource management systems are recognized. There has also been scattered and somewhat ad hoc attempts to implement this recognition.

Notwithstanding the above, in **Fiji**, the Native Land Trust Board had been attempting to increase integration of the traditional CBFM system with state law(s), by seeking more formal Fijian ownership of proposed marine parks. This planned devolution of management responsibility has received ministerial support, of late (Adams, 1993). A similar process was underway in parts of **Polynesia** and **Micronesia.**

Further, the policies of many governments, particularly in Melanesia, as exemplified by **Vanuatu** (Amos, 1993) recognized that traditional systems are an integral part of a matrix that regulate social and political relationships and define cultural identities and ways of life, rather than being concerned with just fishing rights and the organization of economic activities.

Furthermore, in some countries, it is a tenet of policy that customary law(s) may empower CBM as well as could form a basis for comanagement of local marine resources. In the context of economic and social change(s), rights to resources could become stronger and some groups may attempt to obtain codification of their customary rights. This has occurred in Papua New Guinea (Wright 1985, 1990). However, it may be noted that formal acceptance by legal and political institutions may provide communities with a legally codified basis for control over resources whereas the state-supervised codification may destroy much of the adaptive flexibilities. However, Tom'tavala (1990) concluded that this interpretation, except for principles of common law and equity, could be overridden by the provisions of any other law which prescribes anything to the contrary.

Notwithstanding the above, fishing is also an integral component of many traditional inland and coastal societies. For the **Trobriand Islanders**, for example, the sea and marine life have immense significance. Similarly, for the Indian (particularly, Northeast Indian) traditional fisher folk, for example, the rivers and wetlands and the biota therein great cultural significance, embracing the whole fabric of their beliefs about their origin, their prowess, their relationships, and their ceremonies. Incidentally, the stories of origin try to speak about the maritime origin of certain lineages, their relationships with certain fish species, the acquisition of fishing rights by others, and so on. The sea was also fundamentally important in the ceremonial, interisland *kula* trade. Similarly, the rivers and wetlands are basically significant and related to the cultural ethos of many East Indian fisher folk communities, and fish happened to be an important symbolic and nutritional trading commodity with inland and extra-inland Indian villages (Kar, 1990, 2007a,b,c, 2013). Thus, the recognition of traditional fishing rights is inseparable from the protection of indigenous culture, which is part of the fifth "National Goal and Directive Principle," inscribed in the Preamble to the Constitution (1975) of many countries (Tom'tavala, 1990). However, even the discussion for a legislation to register traditional claims could itself lead to changes in the systems. For example, in **Solomon Islands**, in some cases, exaggerated claims are made because it was anticipated that at some future time a community's exclusive fishing areas could be registered, hence it would be wise to stake a claim to an enlarged exclusive area. This had occurred in Western Province.

Although, in some cases, this may just be an opportunism, in other cases it may be seeking to protect both the inshore exclusive area and an offshore open-access area, from which a community's subsistence is habitually derived.

5.1.2.16 Centralization

The national policy of **Indonesia**, with regard to natural resources, is that all marine waters are state property, to be managed centrally, through the provincial, regency, and village offices of the Directorate General of Fisheries, for the benefit of the entire nation. This may be assumed to be a legally sanctioned open-access system nationwide, regulated theoretically, since it is practically unenforceable throughout the vast archipelago only by certain technical considerations regarding gear type, vessel size, and mesh limitations. But the centralization philosophy is an integral part of the nation-building effort in Indonesia (Bailey and Zerner, 1992). Probably, the control of management and rents extracted from marine resources is one of the ways of achieving these goals. Thus, the traditional community-based aquatic (particularly, marine) resource management systems are little undermined. Further, it may be noted that in many areas of Indonesia, local fisheries officials uphold the statutory law of free access. This had resulted in occasional violence (Zerner, 1991) or disenchantment. For example, on Ambo Island in the Balabalakang Islands of the Makassar Strait, an indigenous royalty system was *in vogue* for the last few decades to regulate harvesting. Although having worked well until the mid-1980s, the system, of late, was apparently not working. It was believed to be due to the blatant disobeying of the rules by the outsiders. In addition, the government policy of open access in fisheries had further worsened the situation.

5.1.2.17 External replacement of traditional local authority

The decline of traditional aquatic (notably marine) resource management systems may often be traced to the decline of traditional and indigenous political systems and their replacement by Western political systems during the colonial era. For example, in the **Cook Islands**, regular contacts with Westerners changed the traditional systems. Conversely, it had led to the initiation of British legal and administrative code. In 1901, the Cook Islands have been annexed to New Zealand. The early Cook Islands Parliament became a legislative assembly in 1946, and the elected island councils were established later. These had little undermined the traditional power holders ("**ariki**"), and also the basic component(s) of the traditional resource management system (Sims, 1990).

Notwithstanding the above, **in Palau,** the traditional social organization was undermined by modernization. The greatest blow to the integrity of the kin group had been the vesting of property ownership on the individual. As a consequence, authority of the chief had become minimal, and the economic and political functions of the chiefs had become vestigial (McCutcheon, 1980).

On the other hand, in **Tokelau,** as is typical throughout the region, monetization was believed to have eroded the respect, and hence the authority of the Council of Elders because they are now salaried and have ventured into nontraditional areas like budgeting. It had also enabled islanders to pay relatively painless cash penalties for transgressing fishing regulations (Toloa et al., 1991).

5.1.2.18 Changes in demography

Demography depicts the population structure of a region. Demographic change(s), whether through natural growth, immigration or population decline from various causes, is an all-pervasive factor throughout the Asia-Pacific region. In those countries and regions where human populations are exponentially growing and expanding geographically, there is naturally an increased demand for fish and aquatic products. This combined with the prevalence of a cash economy creates new opportunities to sell surpluses. However, the impact of this phenomenon on fishing communities is exacerbated by the adoption of new technologies in order to increase harvests to provide a regularly marketable surplus. This had also brought about a host of social changes.

Concomitant to the above, immigration, in particular, exerts a major impact on traditional community-based systems. Likewise, excessive emigration or going out of a population results in population decline. In some locations, where there are no longer enough people to merit maintenance of a system, neighboring communities may decide to merge their territories, as had occurred in Okinawa Prefecture, **Japan** (Akimichi and Ruddle, 1984).

Notwithstanding the above, in the **Cook Islands**, localized population decline had contributed to the decline in traditional management systems (Sims, 1990). There demographic dislocation caused by diseases had decimated the local populations. Further, the population of Rarotonga, for example, had declined from *c.* 6000 or 7000 due to mortalities. The pattern was also similar on the outer islands. On **Penrhyn Island**, at one time, only *c.* 88 people remained, resulting in abandonment of two villages at the southern end of the lagoon (Maude, 1982; Crocombe, 1964). Such drastic changes ensured the decline in traditional society, institutions, and practices.

5.1.2.19 Urbanization and alterations in the education systems

It may be pertinent to note that the heavy influx of in-migrants, such as those occur under urbanization, is a major factor causing the breakdown of systems, and against which it is impracticable to defend them. Kiribati provides an example. The government had not sustained the customary owners, arguing that since the lagoon and its resources belong to the state, and every citizen was entitled to harvest the resources (Teiwaki, 1988).

Notwithstanding the above, the breakdown of CBM systems in the vicinity of urban centers is quite usual. However, it is desirable not to interfere with the rational development of the coastal zone, as in **Japan**.

Concomitant to the above, it may be mentioned that fishermen may also earn money by selling their catch to hotels, restaurants, foreign project administrators, and foreign contract workers (McCutcheon. 1980). Further, the traditional fishing skills have declined with economic and educational changes. In Tokelau, for example, this had been manifested through increased pressure on easy-to-catch species, like parrotfish (Scaridae), and a concomitant reduction in pressure on stocks such as giant maori wrasse or eel, the catching of which requires either special knowledge or intense physical effort (Toloa et al., 1991).

In addition to the above, the tendency to disparage local knowledge may not be new, although the causes for disparagement have changed. For example, in **New Caledonia**, the process of discrediting the "superstitions of primitive peoples" had been going on for generations among European colonists, administrators, educators, and missionaries (DaN, 1989; Warren, 1989; Slkkerveen, 1989).

Notwithstanding the above, there have been changes in the *modus operandi* of various parameters including resource harvesting (Teiwaki, 1988).

5.1.2.20 The Tuna Bait fisheries: The impact of the industrial fisheries sector

In the South Pacific, the single most important and widespread inshore fisheries problem, involving the traditional CBFM systems, occurred when tenured waters were also used by industrial fisheries to obtain live bait for the pole-and-line tuna industry. Incidentally, tuna is undoubtedly one of the most significant renewable natural resources in the region and an essential source of export earnings. As such, this has become a high level policy issue.

5.1.2.21 Development and related management projects

The efforts of international assistance agencies to promote fisheries development may, sometimes, be misdirected, despite laudable intellectual efforts, considerable staff commitment, and tremendous amount of financial expenditure. For example, efforts to help small-scale fishermen may infrequently be based on capital-intensive approaches, such as purse seining, trawling, etc., which may distort local economies. They often, therefore, may have a negative impact on the livelihood and welfare of most fishing families and members of fishing communities.

Notwithstanding the above, three groups of donor participants, with widely differing values and perceptions, may usually be involved in fisheries development and management:

(a) International assistance agencies, whose programs, perhaps, have often emphasized upon capital-intensive, production-oriented technological solutions than the national and local problems.
(b) National policy makers and planners, who are frequently Western-trained and share the same "technology fix" values, reinforced by national economic imperatives to boost export earnings and encumbered by the need to cater to the wealthy local vested interests.
(c) Members of local fishing communities, the "target populations," who, for the most part neither trust the members of the first two groups nor share many of their values.

In fact, the small-scale fisheries development characteristically had been a top-down process characterized by attempts to impose external values on fishing communities.

Concomitant to the above, the development planners often fail to recognize that small-scale fisheries constitute complex cultural systems that had evolved from a long-standing and complex interplay of local resources, physical environments, social organization, value system, and information (Cycon, 1986; Panayotou, 1980). Thus, individual fishery systems are complex adaptive strategies based on a depth of traditional social behavior and knowledge concerning local natural resources. In fact, the small-scale fisheries developmental projects clearly begin with a clean slate. However, a cursory examination of the plans for most development projects reveal that the planners may require much time to implement (Emmerson, 1975).

Notwithstanding the above, the new technologies and management systems, promoted by persons working in fisheries development and management, could also usually be freighted with alien values, especially those concerning preferred social organization, distribution of benefits, and division of labor. As such, when a new technology is introduced, economic concerns, perhaps inevitably, tend to displace other social values among the fishermen, as they now require more cash than *hitherto* in order to repay credit for expensive items such as nylon nets, engines, fuel, and new boats, etc. All innovations are, of course, not inherently bad, since existing fishery systems represent the temporary end result of this ongoing process. However, problems arise when innovation is attempted without reference to existing systems operated by target populations. In fact, traditional management systems and associated regulations are, generally, disrupted by the introduction of expensive implements including gear types, such as nylon nets, which are, usually, beyond the financial capacity of small-scale fishermen. Consequently, they often become low-paid laborers for businessmen, who could afford to buy the new type of gear(s). However, the small-scale fishermen often eventually may lose their resource rights, mainly because of their inability to save from their low salaries to purchase nets. In fact, ill-planned development programs for fisheries may have ripple effects that extend beyond the immediate fishing community. In **Java,** Indonesia, for instance, the fishermen need more financial resources to meet their increased financial commitments to pay for new gears and boats. As such, any management system is to be affordable to all the stakeholders.

5.1.3 Bangladesh (B'dsh)

The CBFM has emerged as one of the most viable options for managing fisheries resources in many developing countries including **Bangladesh**. The fisheries of Bangladesh are known to support the livelihoods of millions of poor people. However, capture fisheries are declining as a result of high rates of exploitation and habitat degradation. Incidentally, it is believed that >70% of the households fish in the floodplains, and the fish are used either for avocation or for food (Minkin et al., 1997; Thompson et al., 1999).

In Bangladesh, there have been studies related to the impact of the community-based fisheries management project (CBFM-2) on household welfare by examining how various types of livelihood assets contribute to household income (Islam and Dickson, 2007). The CBFM projects had been implemented since 1995, through a partnership with the Department of Fisheries, 11 NGOs, and the World Fish Center (WFC), working in 116 water bodies with >23,000 households living around the project sites. The major objective of the CBFM was to build local fishery community organizations by providing training, building social awareness, and giving access to credit facilities, with the aim of enhancing poor fishers' capability to access livelihood assets. The studies have attempted to report on the results from an impact survey of *c.* 2826 households from *c.* 40 water bodies in four different regions of Bangladesh. It was observed that CBFM households had significantly improved their social capital and had got better access to land and fishing grounds compared to households in non-CBFM control sites. The results of regression analysis have shown that the contributions of social capital and natural capital factors were important in improving household incomes. However, future policy options are needed to be considered on a priority basis to invest more in human, physical, and natural capital assets.

Many of the fisheries resources are state properties, and consequently, management control very often falls into the hands of rich and influential lessees. In this connection, it may be noted that the leaseholders tend to allow fishing by as many fishers as are willing to pay user fees to ensure that they make a profit (Ullah, 1985; Naqi, 1989; McGregor, 1995).

Concomitant to the above, access to resources had been a constant debate in rural economies and the livelihoods of rural households. Increased access to resources depends on which assets are more relevant to the types of livelihood. Incidentally, the indicators of livelihood security could be grouped into five types of capitals, viz. (a) social capital, (b) natural capital, (c) financial capital, (d) physical capital, and (e) human capital (DFID, 1999).

(a) "Social capital" relates to the social resources (networks, membership of groups, and relationships of trust, access to wider institutions of society, etc.) upon which people draw in pursuit of livelihoods.
(b) "Natural capital" points to the natural resource stocks from which resource flows, which are useful for livelihoods, are derived, e.g., land, water, wildlife, BD, and wider environmental resources.
(c) "Financial capital" represents the financial resources which are available to people (whether savings, supplies of credit or regular remittances, or pensions) and which provide them with different types of livelihood options.
(d) "Physical capital" is the basic infrastructure (transport, shelter, water, energy, and communication), the production equipment, and means that enable people to pursue their livelihoods.
(e) "Human capital" is the skills, knowledge, labor, and good health, which are important to the ability to pursue different livelihood strategies.

Concomitant to the above, Sen (1997) argued that capabilities enhance people's ability to be agents of change. However, sustainable rural livelihoods could be conceptualized in terms of recent debates on access to resources (Berry, 1989; Blaikie, 1989), asset vulnerability (Moser, 1998), and entitlements (Sen, 1981). Incidentally, the fishers, in general, have limited access to livelihood assets. They are mostly illiterate, landless, have poor housing condition, lack of employment, poor capital assets, and lack of funds. Thus, security of access to fisheries resources is vital for the livelihood of poor fishers. It is argued that inequality in livelihood assets among the user groups (UGs) might be associated with different degrees of control and access to the fisheries resources.

5.1.3.1 *Community-based fisheries management (CBFM)*

The CBFM project, funded by the Ford Foundation and the UK Government's Department for International Development (DFID), aimed to promote the sustainable use of, and equitable distribution of benefits from, inland fisheries resources by empowering communities to manage their own resources. The project was implemented in two phases, viz. (i) 1994–99 (CBFM-1) and (ii) 2002–2005 (CBFM-2) by the WFC and the Department of Fisheries (DoF) of the Government of Bangladesh (GoB) with the support of 11 nongovernmental organizations (NGOs). Incidentally, it may be noted that by 2005 the project had facilitated the establishment of *c.* 130 community-based organizatons (CBOs) in different types of water bodies located throughout Bangladesh representing >23,000 poor fishing households.

The water bodies under the project were diverse, comprising mostly of rivers, closed Beels, open Beels, and also significant areas of floodplains (largely private land) as well as small Beels (<8 ha). It may be noted that each CBO was responsible for the management of a defined area of fish habitat of different types of water bodies. In fact, the CBOs were encouraged to implement various types of management interventions to help manage their fishery resources in a sustained manner. In closed and open Beels, CBOs had to take over a commitment to pay the lease fees in return for which they secured control over management of the water body. This involved a clear change in tenure and access, as in most cases, fishers in the newly established CBFM community groups (CBOs) earlier had no access to fishing in those water bodies before the project because the lease was held by a single person or a "fisherman's cooperative" controlled by a few rich and influential individuals. In floodplains, the land was privately owned before the project, and there was no effective change in access or tenure because no lease was required. Incidentally, the community groups operating in these areas were encouraged to implement measures to improve the state of the fish stocks in the floodplain, in particular, by excavating dry season refuges for fish. Concomitantly, the situation in the rivers was similar because leasing was abolished in 1995. This led to a free-for-all fishing, which tended to favor the most powerful, who could afford to install and maintain fish aggregating areas known as "kathas" or "kheo"s.

Incidentally, the main objective of the project had been to test models for sustainable management of the fisheries. Further, it had also tried to encourage the fishers and others living in project areas to develop alternative livelihoods through training and credit support. Incidentally, there had been academic works which provided information on whether poor fishers have benefited in terms of increasing income by efficient and equitable access to livelihood assets.

5.1.3.2 *Methodology*

Sampling and collection of data

The main tool for assessing livelihood impacts was a pair of questionnaire-based field surveys, which could be regarded as a baseline study carried out in 2002 shortly after the start of CBFM-2, and an impact study carried out in mid-2006. The survey covered *c.* 1994 households (including both project beneficiaries and others) at 34 project water bodies plus *c.* 832 households in 6 "control" water bodies. The questionnaire used in the impact survey was based on the baseline survey format which separated households into five categories based on their poverty and fishing profiles.

The baseline and impact questionnaires have covered a wide range of socioeconomic and livelihood parameters, details of aquatic resource use, fishing involvement, access, compliance, existing NGO support, and scales to measure more subjective indicators including social capital. Moreover, survey results were analyzed using descriptive statistics to show differences in the key livelihoods indicators according to water body type (closed Beels, open Beels, floodplains, rivers). Comparisons were also made against households from "control" water bodies.

Factor analysis

It may be noted that social capital cannot be measured by a single variable. A set of variables were used to measure the social impact. Factor analysis was used to construct the indices of social capital. Principal component analysis (PCA) was widely used to find the important principal components as un-rotated factor based on the criteria of eigenvalues >1. PCA extracts a maximum amount of variance to compute the factor scores calculated only from highly loaded factors. The factor scores are weighted according to the factor loadings. The Bartlett's test of sphericity (Bartlett, 1954) and Kaiser-Meyer-Olkin (KMO) test were used to ensure the correlations between the factors. The Bartlett's test of sphericity was significant ($p<0.05$) and considered appropriate. The KMO index ranged from 0 to 1, with 0.6 suggested as minimum value for a good factor analysis (Tabachnick and Fidell, 1996).

Regression analysis

The factors that contribute to household income were analyzed using a regression model.

5.1.3.3 *Impact of CBFM on household income*

Descriptive statistics showed that there were large rises in the annual incomes of households over a couple of years between baseline and impact. Average household incomes were believed to have increased.

(i) in project areas by *c*. 31% ($P < 0.01$);
(ii) in all types of water bodies by *c*. 21%;
(iii) in closed Beels by *c*. 24%;
(iv) in open Beels by *c*. 37%;
(v) in floodplain Beels and rivers by *c*. 57%.

However, these rises matched similar increases (average, *c*. 37%) in the household incomes of people living in the "control" sites:

(i) *c*. 22% in the closed Beels;
(ii) *c*. 42% in the open Beels;
(iii) *c*. **33% in the floodplains; and.**
(iv) *c*. 30% in the rivers.

This could mean that the overall income gains in the project areas may not be directly attributed to the project. There was substantial rises in average income levels in general; a clear trend that had been recognized in other recent livelihood studies in Bangladesh. It was expected that the project activities could result in increased fish production in project water bodies.

Notwithstanding the above, household income sources/income diversification data in the impact survey had shown that project fishers' incomes from farming and remittances had increased significantly over the project period while their earnings from wage laboring had shown a significant decline. Although fishers generally have few landholdings, they are now getting access to land through leasing or share cropping. There had been suggestions that the agricultural sector had remained the major sector of the economy and the better off households were in the best position to capitalize on the shift to high-value production. Agricultural growth is playing a significant role in rural poverty alleviation. Concomitantly, the CBFM study had found that fishers had shifted their employment from wage laboring to self-employment activities in the agriculture sector (farming and fishing), However, in the "control" sites, the fishers still tend to rely on the wage laboring for their livelihoods. The availability of micro-credit and fishing access may be important factors for increasing local employment opportunities. The increased income from remittances due to out-migration was found to be significant in project sites compared to the "control" sites. For poor households, migration was stress-driven. They usually migrate to other regions or districts for seasonal employment in paddy harvesting, road and building construction, rickshaw (tricycle) pulling and other non-fishing type labor-oriented activities, which, ultimately, could lead to loss of precious fishermen "gene." Incidentally, it was found that household members were increasingly residing temporarily away from their village homes to find better work.

5.1.3.4 *Access to fishing*

In most of the closed Beels, open Beels, and floodplains, fishers are now able to participate in fishing in the water bodies nearest to their homes compared to the situation before, when the access to water bodies was highly restricted. In the rivers, however, the number of fishers increased. Incidentally, the overall number of fishers who participated in fishing in the closed Beels and floodplains in the project sites was less in 2006 as compared to 2002. In the "control" sites, the fisher's participation in fishing had not been concentrated in their own water bodies, except in the floodplains. It, probably, depicted that CBFM fishers have increased their awareness levels and had more of an incentive to use their own fisheries efficiently.

5.1.3.5 *Land and assets of household*

The baseline and impact surveys have recorded land ownership as well as rented or sharecropped land holdings. Land is regarded as one of the most significant assets for poor people. The results have shown that fishers in both project and control sites had increased their land holdings, mainly by renting or sharecropping and also by purchasing land. The increasing land holdings of fishers correlates well with the finding that agriculture has become more important for many fisher households. This indicates that the poor people had improved their financial situation allowing them to invest in land over the project period. These self-employment opportunities at their farms and related activities might have enabled them to resist frequent migration.

5.1.3.6 *Microcredit*

Households in both CBFM project sites and control sites had increased their access to credit from a range of credit sources. NGOs were an important source of credit, and both beneficiary and nonbeneficiary households had more than doubled the amount of credit taken over the last few years of the CBFM project. Beneficiary households had a 273% increase in the amount of credit from "interest-free" sources, which might have been used for mitigating short-term crises while the average amount of credit received from this source by non-beneficiaries in project sites decreased. Incidentally, the non-beneficiary households in "control" sites have increased their access to this type of credit by only *c.* 50%. This indicates that CBO members might have become more trusted in lending and borrowing at the project village level. Beneficiary households had also become less reliant than other households on money lenders ("mohajans"). This is the most exploitative type of lending as they charge very high rates of interests. On the other hand, households in the control sites had become much more reliant (*c.* 173% increase) on money lenders as in the case of non-beneficiary households (*c.* 163% increase) in project sites. However, some of the beneficiary households have increased their borrowings from the "mohajans" by *c.* 34% over the project period. Moreover, all types of households greatly increased their overall exposure to credit, meaning that the potential negative impact of "mohajan' credit did not have much impact on the household incomes on an overall basis.

5.1.3.7 *Social capital*

An index was constructed using PCA in order to measure changes in social capital. Six variables with high loadings were aggregated to form the "social capital index."

The scores of four separate items were aggregated to form the social capital index. Perhaps, the **most important and dominating factor** found in social capital was household's membership in organizations. The majority of the poor households have affiliation with production-oriented organizations, such as NGOs, various cooperative societies, and local credit management societies. The poor fishers were not involved in the social, religious, and cultural organization due to their poverty and difficulty in tiding over trying situation(s). Their primary concern was to look for work for their daily food and other necessities. Good leadership did not exist much among the community in the project sites. In both Chapandaha and Hamil Beel, the CBFM participants elect their executives for Beel management committee through voting. The organized fishers participate in stocking and other production-related activities, such as protecting and harvesting fish. Incidentally, poaching was a common threat for stocked water bodies. This was controlled through surveillance provided by the fishers in the project sites. It had been observed that there was improvement in fishers' confidence in using fish culture technologies. The **second major social factor** is that the participants actively participate in the decision-making process of fisheries management rules. Concomitantly, the fisheries management rules, introduced by the management committee, were aimed to enhance fish production. Some of these rules were fishing restriction in the fish sanctuary, restriction on destructive gear used, and 3-month "closed season." The fishers seemed to have strictly obeyed the first rule, but sometimes violated the other two rules when they could not find any income-generating work.

Notwithstanding the above, the **level of knowledge** is important in social capital variable. Low level of formal education is, generally, observed among the fishing communities. Partner NGOs conducted awareness campaigns and training programs on leadership, accounting management, productive activities, and fisheries management to improve fishers' level of knowledge. The **last social capital variable** is the fishers' influence on resource use. The fishers had exclusive access to use the fisheries resources as well as could resist outside threat. They could decide who, when, and where to fish, and thus they were able to control overfishing on their own.

5.1.3.8 *Capital index: Physical*

It may be noted that, generally, the poor fishers are landless or functionally landless. They usually have small amount of productive assets. Moreover, the physical assets endowments were good indicators of income, welfare, and livelihood. Various studies have found that the organized fishers have improved their assets over the last couple of years, such as using better construction materials for their house, and so on. The level of income has also increased, and they have got some other productive assets to be used in making additional income, such as rickshaw (tricycle) pulling, petty trade (grocery, tea stall, fish selling, etc.), and crop cultivation. Women were seemed to be largely engaged in cattle and poultry rearing using credits from the NGOs. In fact, the significant variables were the house materials, area of homestead land, sanitary latrine and productive assets, etc., in constructing the Physical Capital Index using the PCA. Incidentally, these four variables load highly on a single common factor.

The scores of four separate items were **aggregated to form the Physical Capital Index.** Incidentally, land is a very scarce resource in Bangladesh. The average area of homestead is *c.* 0.05 ha in the study area in Bangladesh.

However, most of the fisher households do not own any cultivable land. And, many of them cultivate land owned by others, either through sharecropping or by means of mortgage arrangements. More and more poor fishers in all survey areas had been engaged in high-yielding variety (HYV) "Boro" paddy cultivation. Almost a similar situation existed also in the wetlands in Barak valley region of India (Kar, 1990, 2007a,b,c, 2013; Dey and Kar, 1989a,b,c; Kar et al., 1993).

Notwithstanding the above, improvement in housing was a significant factor for the rural households. The **first important variable** was the value of the house. The area of homestead land was the **second dominating variable,** followed by sanitary condition. The CBFM participants were seemed to have greater awareness in health and sanitation. Further, the NGOs had provided their group members sanitary latrines at low cost to. The **last significant variable** was the household assets (including the productive ones), such as livestock, rickshaw/vans, shallow tube well, bicycle, watch, radio and television sets, etc. These assets play an important role as alternative sources of incomes, during unemployment and incidence(s) of natural crises, such as floods, cyclones, etc., which usually result in loss of fish and other crops.

5.1.3.9 *Human, economic, and natural capital assets*

The poor fishers are generally illiterate, access to formal education is very limited. Human capital includes age of household head, attendance in community meetings, participation in trainings, and access to information. Generally, only the variable indicating the level of formal education is significant in the PCA. The role of financial capital is very important for explaining the livelihood of poor fishers. Fishers, generally, have very limited access to credit. They are usually not able to pay lease money for fisheries resources and cannot, generally, invest in productive sector(s) to generate income. Financial capital includes the variables: amount of credit received, value of household assets, value of fishing implements, and income from asset sales. In fact, usually, only the variable indicating the amount of credit received was said to be significant. The land and fishing ground are, generally, considered as natural assets for the fishers. Fishers are generally landless, but they have traditional access to fishing in the floodplains owned by the private landowners or open-access common property. Access to fishing to such fishing grounds depends on the extent of social linkages among the community in that particular location. One variable in each of the three types of assets was found to be significant in the PCA method. These variables are: (i) level of formal education for human capital; (ii) amount of credit received for financial capital; and (iii) area of fishing for natural capital.

The mean difference between the project and the "control" for the social capital index (SCI) was highly positive in all types of water bodies. The comparisons between the project and the control households show that the net increment for the social capital score was 1.228 and for financial capital was 2.209. In Bangladesh, poor fishermen are, generally, deprived of opportunities in making fisheries management decisions. The "jalmahals" (water areas) are generally controlled by the rural elites and the maximum share of benefits from fishing usually goes to them. On the other hand, the fishermen today are, usually, able to participate in making fisheries management decisions in the CBFM areas. Each of the participants were believed to have received 2000 Taka more in the CBFM sites compared to the "control" sites, which indicates that they have the potential to increase their employment and income opportunities due to the project interventions. The value of household physical assets for the project participants was higher by *c.* 500 Taka. Concomitantly, the fishing area of the participants was higher by 0.827 ha. However, the level of education did not show any significant difference between the project and the control areas.

Notwithstanding the above, the government had transferred the user rights of water bodies under the CBFM project, and had provided administrative support to the fishers. The NGOs usually have full time staffs at the village level, which facilitate coordination between the government and the fishers. The organized fisher's participation in making decisions has increased and contributed to better management of fisheries and improved access to livelihood assets. Moreover, the CBFM participants elect their executives for Beel Management Committee (BMC) through voting. Further, the fishers had improved social linkages, which enhanced their ability to gain economic power and livelihood security in the project areas.

5.1.3.10 *Regression analysis*

Multiple regression analysis is generally used to examine the link between the household income and various livelihood asset variables. The standard equation is estimated separately for the project and control sites by the ordinary least-squares technique.

The results of regression seemed to indicate that the social capital, employment, and the area of fishing were significant predictors of household income in the project area. On the other hand, in the control areas, education, household size, and age variables were said to be significant. In fact, the contribution of social capital factor was an important variable in determining household income.

Notwithstanding the above, the social factor had played a very significant role in poverty alleviation in Bangladesh.

5.1.3.11 *The impact of human development training conducted by CBFM partner NGOs*

This had helped the organized fishermen in gaining sociopolitical knowledge. Associational involvement had helped them to participate actively in decision-making on livelihood issues. There is/are also evidence(s) that they have the capability to influence their access to livelihood assets. The coefficient for the social capital factor was significant for the project area, as expected from community-based management. The regression model reported positive coefficient for social capital. The fishers in the project area got easier access to credit due to their institutional identity. However, access to financial capital was not only from CBFM project NGOs but other organizations were also providing credits. Incidentally, **greater access to credit was a spill-over effect due to high social capital than human capital in Bolivia.** Further, the coefficient of fishing area was negatively linked but was moderately significant. The implication of negative relationship of fishing area was that capture fisheries had been restricted by the land owners. The land owners had introduced either fish farms called "gher" or had opted for cultivating rice crops. The organized community had, further, established fish conservation strategies, such as setting up of fish sanctuaries and imposed "closed season" ban on fishing for 2–3 months during fish breeding periods. In the control sites, the fishing area coefficient was positive, but was not a significant determinant of income. Incidentally, land is a scarce resource in Bangladesh. The coefficients are negative and not significant in the project as well as in the control sites. The poor fishermen in the project areas involved in farming mainly in sharecropped land but their crops were subject to natural calamities. Poor people have limited access to financial assets due to their lack of ownership of other assets. The very poor people, further, were not eligible even for formal credit from banks and NGOs. The coefficient for credit in both project and control areas were positive but were not significant predictors for household income. Although credit is very important for the participants, it was not a significant contributor to household income. The possible reason could be that the poor participants had used their credit for nonproductive activities, such as household consumption, health care, and festivals. The physical asset coefficient of both project and control areas were negative and were not significant predictors of household income. The important implication of this relationship was that the poor people cannot retain their assets during crisis periods. Flooding and other natural hazards occurred almost every year, causing people to sell or mortgage the assets to meet their basic needs such as food, house repair, and health care. It had been reported that almost **a similar scenario had existed in Zimbabwe.** A severe natural shock could wipe out the productive assets, which could result in increased livelihood vulnerability and reduced productivity.

5.1.3.12 *Inferences and policy implications*

A number of contributions seemed to have depicted the impact of CBFM on household welfare, which are usually examined by investigating how the various types of assets contribute to household income. Comparisons are generally made between the sample households from project and control areas. Fisher's income from fishing and non-fishing activities may portray an increase. But, the rise in income was said to be higher in non-leased water bodies (floodplains and rivers) compared to leased water bodies (open and closed Beels). Moreover, the fishers may diversify their livelihood options and may have increased access to land and fisheries in the project areas. Factor analysis show that social capital contributes significantly to household livelihood assets in the project area compared to control area. The project households might receive higher amounts of credit from multiple sources compared to the control households, and they could utilize credit for more productive activities. Beneficiary households of the project, notably, say, CBFM-2 might have access to nonexploitative sources of credit and may become less dependent on money lenders compared to non-project sites.

In view of the fact that the regression results often indicate that the social capital, employment, and area of fishing are significant predictors of household income in the project area, the important policy implications of certain studies could be that the user groups (UGs) of community-based organizations, who primarily depend on fisheries for their livelihood, may often need strong facilitation by the NGOs and the government to establish access to the fisheries. As such, posting of experienced staff of DoF and NGOs could be vital for the success of CBFM. In addition, the fisher households may need assets for their security during the periods of crisis. As such, there would be a strong need for establishing a social safety net so that the poor fishers may feel secure to use their physical assets as investments.

Notwithstanding the above, provision of public works at critical times might be a good option for generating employment opportunities. Concomitantly, the health services are very poor in Bangladesh and the poor people, thus, have to spend a good portion of their income for health care. The provision of free and effective primary health-care facilities at the village level are to be given priority. Furthermore, the security of access to the fisheries resources need to be taken as a priority in future policy formulation in natural resources management, particularly since the poorest fishers rely on fishing both for their nutrition and avocation. Incidentally, the community-based approach is tested as an alternative to the current revenue-orientated leasing model. Future policy makers may need to consider the successes shown by this new model of community-based institutional development for the management of the vast inland fisheries of Bangladesh.

Concomitant to the above, the popular press seem to regularly carry articles on the crisis in global fisheries (O'Riordan, 2013). Relentless overfishing, unlawful and unreported catches, environmentally damaging fishing methods, and unwanted bycatches all make the headlines. Incidentally, people usually think about the global fisheries crisis from the perspective of large-scale fisheries, mainly because those stories are usually about trawlers and high seas fleets. But small-scale fisheries such as those in Bangladesh deserve to be taken more seriously not least because of the number of people who depend on them but mainly because of their importance to developing countries. It is worth bearing in mind, for example, that >95% of the *c.* 200 million people involved in fishing are small-scale fishers, processors, and traders, and further that *c.* 90% of these are said to be from the developing countries. Incidentally, it may be noted that the developing countries produce and consume *c.* 70% of the world's fish and that small-scale fishers account for *c.* 70% of this production. So, small-scale fisheries are to be given due importance. They are important because they can act as an economic driver for rural development. Even in remote places, small-scale fishers are linked to the cash economy, thus providing with a model for other sectors that might help drive economic growth in rural areas. They are important because a large number of people depend on them for nutrition and avocation.

Concomitant to the above, in Asia, *c.* 30% of dietary protein, on an average, is obtained from fish (*c.* 51% in Bangladesh, *c.* 58% in Indonesia, and *c.* 75% in Cambodia). On the other hand, in Africa, *c.* 200 million people obtain *c.* 22%–70% of their dietary animal protein from fish. They are important because they provide a social safety net during difficult times. For many, fishing is what you turn to when crops fail or other types of crises hit. But small-scale fisheries are vulnerable due, among others, to overfishing, competition with industrial fleets, and environmental degradation. Pollution of inland and coastal waters has become a major problem as has been the destruction of key fish habitats and fishing grounds. In the hill streams, it is mainly due to the collection of rocks, etc., from the river beds, thus devastating the potential habitats of the coveted rheophilic fishes. Further, the small-scale fishers may face conflicts with other water-resource users which could become more intense as we experience more and more climate change. As such, there is urgent need to find approaches that could reduce the vulnerability of fisheries to such threats and make them more resilient. Nevertheless, there are certainly hopes for the future, despite the problems. Incidentally, there are good examples from around the world, where sustainable small-scale fisheries are making a difference for rural households and national economies.

Notwithstanding the above, a superb illustration of how investment in fisheries could pay off is the CBFM project in Bangladesh. Here, community management had yielded fish production of *c.* 1230 tons/year worth *c.* US$ 0.9 million. Incidentally, fish abundance increased at *c.* 17% per year whereas control sites had significant declines. Moreover, fish BD also had improved in *c.* 70% of the monitored sites. Further, CBFM fishers had increased their earnings from fishing by *c.* 21% compared to *c.* 15% for fishers in control areas. These could be considered as impressive results. Thus, the poor people could benefit from improvements in small-scale fisheries, something that governments are increasingly recognizing. They are now trying to incorporate the small-scale fisheries sector into poverty alleviation strategies mainly by promoting CBM and comanagement arrangements as part of governance reforms. Yet, despite these success stories, it could be said that supporting pro-poor small-scale fisheries is a complex, multi-faceted problem, with many interdependencies and entry points. Incidentally, the failure to embrace this complexity could lead to piecemeal solutions as well as may generate faith in magic bullets. In fact, success may not be achieved from simplistic single initiative approaches, be they CBM, protected areas (PAs), or any of the other popular tools in the manager's box. In fact, to build on success, therefore, there is the need to adopt a much more structured and systematic approach to assessing and managing fisheries. In particular, we may need to think beyond the fish resource itself and consider the needs and wishes of all the people who have a stake in the fishery. As a general framework for moving forward, a three-step approach is usually required as follows:

(i) We need to thoroughly evaluate the fishery and the context in which it operates.
(ii) There is need to reconsider who needs to be involved in management.
(iii) We are to manage in a way that allows us to learn and adapt.

Concomitant to the above, the first step in this process, which is generally regarded as a "diagnosis," is crucial. For any small-scale fishery, there is need to ask where it fits into the wider economy and identify the socioeconomic drivers within the fisheries sector as well as those external to it, which have profound impact on the fishery. In doing so, concomitantly, there is need to identify the key relationships, dependencies, and assumptions made by us, and the pathways to impact. Only then the best management strategy could be identified with important focus for management. The result of this process could be an impact framework for resilient small-scale fisheries. Incidentally, at the WFC, there had been efforts to develop an impact web approach in which one identifies a set of interlinked impact pathways, with each pathway showing the steps to a solution. It is believed that with this more structured analytical approach to support management, one could expect better decisions. In fact, work in Bangladesh on CBFM had modeled the way for this approach in many respects. Moreover, the project did consider the bigger picture, particularly the issues such as water allocation, access to credit, the role of women, interaction with agriculture, and so on. In addition, the identification of the most effective management strategies had also been at the core of the CBFM approach. However, integrating community managed approaches into official policies for water body and wetland management was critical. Further, the project has also taken a long-term structured experimental approach and is a platform for further improvement.

Notwithstanding the above, one of the main lessons learnt from the CBFM project was that one may need to take a long-term perspective. In the studies done in Bangladesh, the first 3 years were taken up for social mobilization, group formation, consensus, and confidence building leaving the remaining years to demonstrate the effectiveness of the local institutions (CBOs) and strengthen them in readiness for project phase out. Subsequently, setting up of the project up to learning was crucial. However, the CBFM project leaves a legacy for other small-scale fisheries interventions to follow. It is a classic illustration of how investment in fisheries could pay off. In particular, under CBFM management, there is hope for success in fisheries and spiral of decline could be avoidable. It has succeeded by encouraging diversity and participation and by reacting to circumstances rather than applying a fixed approach. If there is one lesson above all others, it is that there is no magic cookie cutter solution to these challenges. Incidentally, context may not be everything. Instead, we may need structured but adaptive and flexible approaches, if investment in fisheries is to secure their multiple benefits.

In fine,Ithe **CBFM** project, phase II was implemented by the Department of Fisheries (DoF) of the Government of Bangladesh (GoB) with the assistance of the WFC over the period 2001–2007. There have been communications which tried to give an outline of the main principles behind the project and the activities carried out during its implementation. The project involved a wide range of partners including 11 NGOs who were responsible for the identification and development of 130 community-based organizations managing 116 water bodies including closed Beels, open Beels, rivers, and floodplains. Such papers also tried to summarize the success of the project from different viewpoints. The papers, often, had attempted to discuss some of the situations where community management did not work very successfully under CBFM, and thus, had emphasized that there is a need for further research to build in this direction based on the situation *in vogue*.

Further, there have been discussions regarding the institutional issues in CBFM project, which were vital in achieving its long-term goal of "sustainable improvement of the livelihoods of poor people dependent on aquatic resources" (Ahmed and Dickson, 2007). A wide range of institutional arrangements had been practiced in this project from phase I (1994–1999) to phase II (2001–2007).

Concomitant to the above, the initial project document stated that there were three main fisheries management approaches, viz., (i) fisher-led, (ii) community-led, and (iii) women-led.

The definition of these approaches fail to capture the full diversity of approaches adopted by 10 project partners (including DoF) during implementation. As a part of the institutionalization process, 130 CBOs had been developed and established under the project with legal entity. Efforts were made toward linking these CBOs and local institutions in CBFM sites for better management of fisheries. In this regard, four regional CBO networking committees were formed in four regions and a central committee also had been established on ad hoc basis. This central committee is now being officially registered. The partner NGOs were the main agents helping to develop the CBOs for making them sustainable. Further, exit or phase-out plans and post-project sustainability plans for the CBOs have been finalized and are being implemented for better management of the fisheries.

Another important process in CBFM is the cluster management. Overall, it could be concluded that as a measure to manage and preserve natural fisheries, establishment of CBOs as local level institutions, using cluster management and linking with local/regional/central level CBO networks, may be essential for social, biological, and political reasons, as it includes people from all sectors. Therefore, people's participation as well as cluster management and CBO networking system are very important in order to manage people's common pool resources.

Notwithstanding the above, the CBFM project in Bangladesh aimed at promoting the sustainable use of inland capture fisheries by empowering fisher's communities to manage their own aquatic resources. There have been communications which tried to describe the impact of fisheries management performance using data generated under the CBFM project, funded by the Ford Foundation and the UK Government's Department for International Development. Various research works, using quantitative indicators of catch per unit area (CPUA), catch per unit effort (CPUE), catch per fisher's day (CPD), BD index (H'), fishing intensity (DPUA), and destructive fishing ratio

(DFER) at up to 86 project water bodies across the country, representing a variety of different habitats were compared with the research results of those water bodies operating under the existing fisheries management practices (which could serve as control water bodies). Estimates of the slope coefficients for each performance indicator were compared among habitat types and between CBFM and control water bodies using ANOVA (GLM). Fish production was found to have increased significantly through time at CBFM water bodies. Trends in fish production through time were upwards at *c.* 77% of the 64 project water bodies that were monitored for at least 3 years without data gap. Trends in fish abundance, indicated by annual average and daily catch rates by fishers, also had depicted an upward trend at *c.* 72% of monitored water bodies. Changes in BD index (H′) with time were found to be positive and significantly greater than in control water bodies. Species assemblages were richer and more abundant in CBFM compared to control water bodies. Trends in BD were also upwards at *c.* 70% of monitored water bodies. Considering all management approaches together, the score-based performance indicators suggested that a fisher-managed approach ensured maximum benefits, and followed by community-managed and women-managed approaches. In fine, it could be opined that CBFM appeared to perform significantly better than the existing management system in Bangladesh. Future works might choose to lay greater emphasis on identifying habitat-specific interventions and arrangement to meet precise management objectives. Existing information sharing networks could support experimentation and learning for future initiatives.

Concomitant to the above, Mustafa and Halls (2007) reported on the impact of the CBFM on sustainable use of inland fisheries in Bangladesh. The CBFM project in Bangladesh aimed to promote the sustainable use of inland capture fisheries by empowering fisher's communities to manage their own aquatic resources. Works in this direction described the impact of fisheries management performances using data generated under the CBFM project, funded by the Ford Foundation and the UK Government's Department for International Development (DFID).

Notwithstanding the above, the Bangladesh Environmental Lawyers' Association (BELA) provided legal support to the CBFM project, in particular to the 130 Community-Based Organizations established under the project (Kabir and Hassan, 2007). The project was implemented against an uncertain legal background due to many changes in the way that wetlands and fisheries in Bangladesh have been managed over recent decades. Many of the key interventions such as sanctuaries have yet to receive legal recognition. Many currently accepted norms and practices have come about through gazette notifications or individual decisions, rather than being supported by Acts or clear policies. Incidentally, the papers in this regard tried to outline the legal background for community managed fisheries in Bangladesh and the challenges faced during the implementation of the project. The paper(s) also tried to suggest what needs to be done, if community managed fisheries are to become more widespread in Bangladesh.

In addition to the above, an NGO based in Jessore in Southwest Bangladesh named "Banchte Shekha" has supported the development and empowerment of poor people, particularly women (Gomes, 2007). In the CBFM project, they found that it was possible to involve women in fisheries activities, despite initial opposition from conservative groups. "Banchte Shekha" was responsible for organizing seven CBOs in CBFM works, following their successful experiences with a single CBO in the first phase of the project.

Incidentally, among the reasons for the loss and declining productivity of the wetlands of Bangladesh is a lack of understanding of the underlying causes of wetland resource depletion (Islam and Rahman, 2007). Other constraints to sustainable management include a general lack vis-à-vis of a need for building consensus among stakeholders and their inability to take up appropriate measures. Of late, it had been recognized that local resource users, managers, and controllers could play key roles in wetland management. Incidentally, efforts had been made to involve the local user communities in wetland and fisheries management in various capacities since the mid-1990s. However, different organizations and projects adopted different modes of community participation in project planning and implementation. Incidentally, there had been papers, which tried to describe cluster-based management systems, formation processes, legal status, scope of works, and community responses to the approaches in five different sites managed by CNRS under the CBFM project. The papers tried to present some key achievements of cluster–/apex-based approaches relevant to wider area issues, such as restricting harmful fishing practices, reclaiming "khas" (government) lands from illegal occupation, fish friendly operation of sluice gates, basin-wise integrated management, stopping the sale of banned gears, and conflict resolution in upstream-downstream areas, which may have been impossible for single water body-based CBOs. A cluster approach also seemed to encourage individual CBOs managing their respective water bodies within the wider area through periodic sharing of lessons, assisting each other's needs, and realizing services from local agencies which will contribute to developing capacity and sustainability for future.

Concomitant to the above, since 1998, the Management of Aquatic Ecosystems through Community Husbandry (MACH) project had established what is best described as community based comanagement of three large wetland systems covering in total *c.* 32,000 ha at the full storage level (FSL) during the rainy season and *c.* 4600 ha of water area at the dead storage level (DSL) during the dry season (Paul and Chowdhury, 2007). The project was supported by USAID and the Government of Bangladesh, and was implemented by Winrock International, CNRS, Caritas, and BCAS working closely with the Department of Fisheries and Ministry of Fisheries and Livestock of the Government of Bangladesh (GoB). The key objectives of the MACH approach was to establish community organizations and then embedding within them, the institutions for the purpose of sustainable and wise use of the wetland resources. At the same time, there were efforts to formally link these with the existing local government system. Moreover, there were efforts to restore wetland productivity and improve the livelihoods of the poor people through these interventions. The organizations involved comprised 16 resource management organizations representing all local people with interests in wetlands and fisheries, 13 federations of resource user groups comprising mainly poor fishers and other poor wetland users, 25 union parishads, and the administrations of 5 upazilas. Incidentally, comanagement was formalized through Upazila Fisheries Committees where representatives of all bodies usually sit to coordinate and oversee management of the systems. These organizations were said to observe "closed seasons," did excavation of *c.* 46 ha of Beels, and *c.* 30 km of canals to enhance the water-holding area during the dry season, etc. They also established *c.* 56 sanctuaries of *c.* 173 ha area, and had planted *c.* 605,000 trees. The results of their efforts had portrayed augmentation of fish catches to the extent of 2–5 times over 1999 baselines of *c.* 58–171 kg/ha, and later reaching *c.* 316–388 kg/ha across the whole wetland systems in 2004–2005. There had also been increase in fish consumption to the extent of *c.* 45% over the same period which had benefited the landless as much as large landowners. Substantive amount of revolving loan funds had been transferred to community organizations along with training. This had helped *c.* 5200 poor households in the form of increasing their supplementary incomes by *c.* 50% while also reducing their dependence on fishing by about two-thirds.

For sustainability issues, the Upazila Fisheries Committees had also been endowed with quite a large amount of funds, the interest from which had been primarily used for continued restoration of the wetland habitats by the resource management organizations. The wetland ecosystem and its catchment being vital, efforts had been made for plantation of trees as well as for the promotion of contour cultivation on hills in order to reduce soil erosion and siltation of the wetlands. Moreover, the ability to address threats was enhanced. It may be mentioned that in **Kaliakoir**, the number of textile-related industries increased from *c.* 20 in 2003 to *c.* 166 in late 2005, and the surface water was said to be unfit for use as it was below the national standards. In view of this, the communities had to possess their own water quality monitoring program. Thus, the Upazila Fisheries Committee and Department of Environment agreed to sign an MoU to cooperate to enforce the existing anti-pollution laws as a priority.

Notwithstanding the above, MACH had also supported the Department of Fisheries (DoF) to take up similar activities in some of the other or future fisheries project sites and to assist the new inland capture fisheries team of the department. Incidentally, the MACH approach had already been taken up at the policy level. The Inland Capture Fisheries Strategy of the DoF had incorporated establishing Upazila Fisheries Committees nationally, as a key element to incorporate and work with an expanding network of community-based organizations. There were also efforts to devise mechanisms to restore and sustain major wetlands as high priority issues.

Concomitant to the above, it is quite well known that Sunamganj, in Bangladesh, is a land with abundant resources of floodplains with seasonally flooded depressions locally called "Haors" and smaller water bodies called "Beels" (Mohsin, 2007). However, this seems to be not in conformity with the definition of "Beel" and "Haor" prevalent in **India**, particularly, in **Northeast India**, where, a "**Beel**" is a perennial wetland, which retains water throughout the whole year, while a "**Haor**" is a seasonal floodplain wetland, which retains water in it only during part of a year (Kar, 2007a,b,c, 2013). Notwithstanding the above, in Bangladesh, people's livelihoods and culture are largely influenced by the Haor economy. Concomitantly, the Beel fisheries also play a very critical role. Although the Beels have a wide range of resources, the people at large had not been able to benefit from them. Incidentally, a small number of people have been exploiting the resources by their bad power and influence, overriding all the codes of resource management and maintenance. The research works tried to implement a process which had been derived largely from a CBRM approach. It is a demand-driven approach set by a comprehensive discussion with a wide diversity of stakeholders. Incidentally, access to Beel could be achieved through two steps:

(i) Handing over the Beel to genuine fishers on a long-term basis; and.
(ii) initiating Beel resource development activities.

The projects under such kinds of research works have been entrusted with "Beels" (both < and > 20 acres in size) through MoUs from the concerned ministries, and later handed over to the Beel user groups (BUG) that comprised mainly the fishers. A set of criteria are usually followed to select the BUG members and BUG functions under a bylaw that contains some core values to form an independent institution of Beel users.

Concomitant to the above, in all aspects, from Beel selection to formation of the BUG and initiating Beel resource development, a participatory approach is generally followed, where local people's knowledge and experts' views are duly consulted at all stages of the work. In fact, the BUG members are trained to raise their skill levels on group management and Beel resource development. The Beel resource development activities comprise demarcation, re-excavation, swamp tree plantation, establishing sanctuaries, and introducing conservation measures. Surveys and studies on fish catch and consumption, fish processing, and marketing are also included in Beel fisheries development under the project.

Notwithstanding the above, one visible strength of the project's approach is its partnership with other departments and institutions. An apex body was implementing the Beel fisheries activities at the field level with cooperation from the local administration, DoF, and the WFC.

Incidentally, the project had a target of accessing *c.* 300 Beels in Sunamganj district by June 2014. However, *c.* 55 Beels of different sizes had, of late, been accessed and handed over to BUGs. Incidentally, out of *c.* 55, *c.* 50 had been harvested and *c.* 10 had been brought under developmental activities. In fact, the project was at an early stage at that time. It was said to be advancing through learning by doing. Meanwhile, a good number of lessons had already been learnt. Clearly, sustainable Beel resource management is largely a social issue rather than a technical issue, and people's views are to be utilized with full potential for such natural resource management (NRM).

Incidentally, a number of floodplain AQC projects had sprung up, of late, in the Daudkandi area of Comilla District (Gregory et al., 2007). A number of unique organizational and financing arrangements were key to this development, which had facilitated the development of necessary infrastructure through issuing shares to farmers who possess plot(s) of land in the targeted floodplain area. Incidentally, a short review was carried out, in February 2007, to better understand how floodplain AQC was affecting a range of local social, economic, and environmental issues. The production and economic performance of floodplain AQC projects was impressive, and as such they obviously have a positive effect on local economies, general security, people's nutritional status, and employment and service opportunities. However, this approach, productive though had the ability to effectively change a seasonal, open, common property water resource into a closed, privately owned one. As a result, the opportunity for some traditional livelihood foraging activities had been reduced or lost. In general, the poor people in the area were unable to benefit directly from the project share. Incidentally, the review had concluded that the newly formed embankments were restricting the movement of wild fish stocks onto the floodplain, and the internal refuges formed were not having a significant impact on the dry season survival of the wild fish stocks, except, in cases, where rarer species had been deliberately targeted for induced propagation. The review had, further, recommended that the government funds could be best spent on carrying out a rehabilitation drive for both the fishes and the fishermen.

Notwithstanding the above, in Bangladesh, experiences from good practices for a common property resources (CPR) had identified that it could often be necessary to choose CPR members from the resource users with clearly defined rights to use the resource with defined physical boundaries (Nathan and Apu, 2007). In fact, the long-term security of tenure is said to be a precondition for the establishment of common property resources (CPRs) in the water bodies by the users (mainly fishers and adjoining agriculturists of the water body) themselves. In fact, the sustainability of such CPRs depends on the equity in sharing expenses and income, monitoring by the users themselves, graduated sanctions for violations of CPR rules, and development of local forums for resolving conflicts.

Concomitant to the above, it may be noted that **Bangladesh** has had comprehensive experience of community-based management for inland capture fisheries from several projects (revenue and externally funded) over the last few decades (Arne Andreasson, 2007). The lessons were extensively used for the elaboration of various strategies and programs, which are supposed to seek for consolidation of gains in the matter as well as expansion of community-based managements linked to institutional and legal reforms and a recognition and strengthening of the roles of civil society and the private sector.

Incidentally, the Ministry of Fisheries and Livestock had adopted the National Fisheries Strategy (NFS) in January 2006. It has been developed by the Department of Fisheries over an extended period through a consultative process involving all relevant stakeholders. The strategy was to guide the implementation of the National Fisheries Policy, the PRSP, and the Road Map for PRSP. The strategy had covered eight areas termed as "sub-strategies," viz., inland capture fisheries, marine capture fisheries, aquaculture, aquaculture extension, shrimp culture, quality control, human resources development, and monitoring and evaluation.

The strategy has a pro-poor focus, and the guiding principles are people's participation for community-based resource management, decentralization of planning and decision-making, sustainability, and conservation.

Incidentally, production from capture fisheries saw rapid, sustained growth from the 1950s through the 1970s, but by 1990s global capture fisheries became stagnant, warranting a sustained approach to its exploitation (Dey and Kanagaratnam, 2007). Thus, with capture fisheries stalling and demand for fish growing, many countries turned toward developing the AQC sector to meet the supply gap. Over the decades, the expansion of AQC had led to a rapid growth in fish production. However, as AQC expanded its production, its use from the capture fisheries as food for farmed fish has also increased, taking into account that, of late, nearly one-third of the world's wild caught fish is consumed as fish feed. Incidentally, as AQC in the developing world had continued to exhibit steady growth in production, sustainability of this trend is now open to question given the rapid degradation of the capture fisheries. Incidentally, establishing community organizations for managing fisheries had been a promising means for improving the resource condition, particularly for countries with large inland and seasonal floodplains. However, as several papers had outlined, this arrangement is not to reduce the role of the government, but there was emphasis on delivering net benefits. Also, it is necessary to setup legal framework for community-based management so as to ensure and sustain community participation in fisheries management, enhancement of fish stocks, protection against habitat destruction, pollution, and degradation of biodiversity. The strategies had emphasized the need for reform of access rights (lease systems) and institutional development for decentralization and effective public sector service delivery.

5.1.3.13 *Recommendations for CBFM in Bangladesh from the CBFM international conference*

The lessons, experiences, and achievements presented at the conference had supplemented and corroborated the existing knowledge base over the past few decades. This had enabled the scientist to draft, as a post conference activity, brief recommendations to chalk out the way forward, specifically for CBFM in Bangladesh. However, many of the key issues and recommendations proposed could be relevant also to other parts of the world adopting a community comanaged approach to managing common property aquatic resources.

- **(i)** Appropriate mechanisms needed to be designed and developed in advance to assess the impacts of the interventions on the wider community. This could clearly demonstrate impacts of community-based management interventions.
- **(ii)** The CBOs are to be given official recognition at all levels as legal entities, and the relevant laws and various other aspects like sanctuary establishment and their maintenance, etc., are to be formulated and enacted for implementation as soon as possible.
- **(iii)** A policy paper needed to be prepared and reviewed in consultation with the Ministry of Land for fixing a nominal lease fee for sustaining new approaches to community management of wetlands in Bangladesh. The policy is to recognize that the lease fee could, preferably, be used as a means of regulating access rather than revenue generation.
- **(iv)** Watershed management policies implemented by the Department of Fisheries (DoF) is to reflect the need for networking.
- **(v)** The newly established Association of Fisheries CBOs (Central Committee of the CBO Networking Committee in CBFM exercise) needs to be institutionalized and strengthened through registration by the appropriate authorities, including DoF.
- **(vi)** Local government control over fisheries is to be strengthened through forming Upazila Fisheries Committees (UFCs). It is important to establish strong linkages between CBOs and UFCs and to encourage NGOs and other agencies to facilitate community management of fisheries. Local government officials need to be involved in policy development recommendations at all levels.
- **(vii)** Local government officials need to be made more aware of the processes and approaches used in community management of fisheries. This could be achieved through the provision of information and training (nontechnical).
- **(viii)** Management of interconnected water bodies can be improved if they are managed as large, watershed-based areas. This requires the formulation and implementation of cluster and apex management policies and guidelines as well as coordination between neighboring Upazila Fisheries Committees.
- **(ix)** CBOs are to be empowered so that key representatives (local trainers or legal aides) understand fisheries-related legal issues with regard to land and water usage, and develop training modules on paralegal issues related to fisheries management. The involvement of specialist lawyers, environmental associations, and rights-based NGOs are to be facilitated to promote legal issues for CBOs and establish strong linkages between the Association of Fisheries CBOs and paralegal workers/environmental lawyers.
- **(x)** Long-term tenure for leaseholders could be crucial to ensure sustained and healthy development of the water bodies. Likewise, support by the external agencies could, preferably, be for a long period in order to build capacity of CBOs. As such, the future community-managed fisheries programs may continue for long periods involving the CBOs, and, further, all programs and projects are to be at least of 10-years duration.
- **(xi)** The potential development partners are to be made more aware of the Inland Capture Fisheries Strategies of DoF. The ICF Development Program(s) outline the other related development projects proposals so that they could coordinate easily.

- **(xii)** The future community-managed fisheries interventions are to stress upon the linkages between public administrative levels, and that these are to be between officially recognized institutions and also to be clearly understood by all stakeholders.
- **(xiii)** New projects and programs may try to consider ways and means to improve intergovernmental policies which address wetland comanagement objectives.
- **(xiv)** It is important to complement future fisheries interventions with entrepreneurship development in order to diversify fishers' livelihood opportunities, especially, during the periods of "closed seasons" or bans.
- **(xv)** Prior to the implementation of future community-based interventions, a full understanding of inter- and intra-community social dimensions is required. The designs for future CBFM projects are to include sociocultural and economy-related issues and the processes of empowerment are to be fully participative, and, as far as possible, to be woven into local culture and traditions.
- **(xvi)** More emphasis is to be laid, preferably, on understanding the local/traditional indigenous knowledge and information for better management of the wetlands and their resources, e.g., edible and medicinal plants, and other valuable flora and fauna. Better management of these diverse aquatic resources by the community members are to be stressed upon.
- **(xvii)** Greater attention is to be paid to broaden the range of products, other than fish, that could be grown from the large amount of aquatic resources, and could be considered for value addition in order to increase the economic value of these products. This may be linked to alternative income-generating activities, which may not be water body-based but could certainly represent a diversification of livelihood options, and thus, could reduce vulnerability of the stocks of other resources.
- **(xviii)** Physical mapping, resource identification, area demarcation (where appropriate), and distribution plans are to be developed and implemented at the initial stage in the CBFM-type interventions. In fact, the boundaries of community management activities, if any, are to be explicitly mentioned in the policy document(s).
- **(xix)** It is recommended that the government could carry out a comprehensive zoning exercise on floodplain AQC in Bangladesh by initiating a detailed study. This could enable effective, planning, monitoring, and regulation of floodplain AQC as well as help in avoiding potential conflicts between the traditional fisher communities and the floodplain AQC farmers.
- **(xx)** The Bangladesh Department of Fisheries and its partners have developed an Inland Capture Fisheries Strategy, which supports expansion of community comanaged approaches. This is to be implemented as soon as possible in order to improve the sustainability of fisheries, thereby helping the communities dependent on fisheries.
- **(xxi)** Public sector agencies supported by development partners may, preferably, continue with the existing projects, and scale up to a larger number of water bodies so that data generated through their involvement and works in quite a large number of water bodies could generate sufficient knowledge within the sector for validating their success. To this end, the Department of Fisheries are to gear up and strengthen their nascent wings in order to promote and coordinate further programs in this direction.
- **(xxii)** The various plans, programs, strategies, and *modus operandi* for CBFM are to be regularly reviewed, renewed, and updated in order to make them adaptable to the altered situations.

Concomitant to the above, the fisheries sector, of late, is playing an increasingly significant role in the economic development of the people of **Bangladesh**. Incidentally, the inland fisheries comprise the fisheries in the rivers, ponds, *Beels*, floodplains, *Haors*, *Baors*, etc. There are *c.* 260 fish and *c.* 24 prawn species in the inland freshwaters (FW) of Bangladesh. Incidentally, fish production from AQC increased to a great extent but open wild water fish production is said to be in slow progress as compared to AQC. In fact, it may be noted that only *c.* 34% of the total fish production comes from the inland open waters in the wild (DoF, 2012).

Concomitant to the above, the greater southern parts of **Bangladesh**, especially the greater **Noakhali** district, is characterized by seasonal water-logged paddy fields during rainy season where CBFM could be an option to improve socioeconomic condition (Paul et al., 2014). There had been studies **to evaluate the culture stocking strategy and cost-benefit analysis for good CBFM activities in waterlogged areas of Begumganj upazila of Noakhali District in Bangladesh.** Required data were collected through focus group discussions and semi-structured interviews with the target group(s). In a study, six selected dogis, viz., Vobanijibonpur, Kendurbag, Madhob Shing, Khyajpur, Chowmohona, and Jirtali were observed in this study, where stocking and other management activities were maintained from the very beginning of the culture season. Water quality parameters were closely monitored during the 6months of the study period. To increase the production of natural fish food, cow dung (2000kg/ha), urea (50 kg/ha), and TSP (50 kg/ha) were applied as a basal dose after 7 days of liming (250 kg/ha). In these carp and prawn polyculture farms, the stocking density was 8–24/decimal and 3–10/decimal, respectively, where prawn and carp stocking ratio was 1:3. The study also depicted different profit margins of 36,831, 20,200, 155,700, 10,850, 10,550, and 21,600 taka in six dogis. The work, based on CBFM strategy, involving studies on habitat quality, meteorological data, and field experiences, had portrayed that a successful return from the investment done in the waterlogged areas mostly depends on proper management of these factors.

Notwithstanding the above, the seasonal waterlogged paddy land fish culture is a recently introduced cultural system which ensures the best use of these unused land as an additional income avenue. Incidentally, in Noakhali district, most of the paddy fields remain water logged during the rainy season. Water is generally retained for several months, and thus, render the land unavailable for crop production from May to October each year. Incidentally, Sarker (2005), Das and Hossain (2005), Das et al. (2009), Hossain (2009) had identified *c.* 22,186 ha of seasonal waterlogged paddy land (locally called *dogi*) in Begumganj upazila of Noakhali District. So, proper AQC practices through CBFM are most suitable in these waterlogged paddy fields ("dogis"). This could boost the economic upliftment of the people of this region. Moreover, culture in these waterlogged areas is cost-effective, especially for the rural poor people, as they provide the essential source of nutrition and avocation as well as support the livelihoods of large number of people in Bangladesh. Ownership regimes of the floodplains in Bangladesh are diverse and complex with (i) some floodplains being completely under public ownership; (ii) some under public but surrounded by privately owned lands; and (iii) some under completely private ownership. It is generally agreed that any significant increase in fish yield from inland capture fisheries can be achieved from fisheries enhancement activities. Incidentally, it may be noted that fisheries enhancement through stocking is often synonymously used as "**culture-based fisheries,**" and is said to be one of the most commonly practiced management interventions in lakes, reservoirs, and floodplains in different parts of the world in order to augment fish production. In culture-based fisheries (CBF), the young fish are regularly stocked in open or semi-closed waters and recaptured at a larger size without aiming to establish a self-recruiting population. The stocking size of the fish may generally vary depending on the type of the water bodies to be stocked and the fish species being raised, while the harvesting size could be greatly influenced by the market demand and the consumption pattern of fish in the locality.

Notwithstanding the above, the specific objectives of the fisheries exercise, as referred to the above, are said to be as follows:

(i) To assess the suitability of community based fisheries (*dogi*) management.
(ii) To assess the cost-effectiveness through comparing stocking and water quality management in the aforesaid areas.

Pre-stocking management aims at proper preparation of ponds to remove the causes of poor survival, unsatisfactory growth, etc., and also ensures the availability of natural food in sufficient quantity and quality for the spawn, fry, and fingerlings to be stocked. The stony lime was applied at the rate of 1 kg/decimal. Sumithion was applied in 1 decimal at the rate of 1.5–2 mL for 1-ft water depth to control predatory aquatic insects, as the fry and fingerlings are usually vulnerable to predation by certain aquatic insects. Moreover, the fry were usually conditioned before releasing into the nursery ponds. The bags with fry were kept in pond water for at least half an hour to make temperature of the water within the bag equal to the temperature of the water outside the bag, i.e., the pond water. Dried bushy bamboo, dried branches of trees, etc., were placed in the water body in order to provide shelter to the juvenile fishes.

It may be noted that the CBFM initiative had developed a series of management approaches for ensuring equitable access to fisheries resources for CBOs, which had resulted in > 60% of the fishers in project areas experiencing significant increases in their fishing incomes in **Bangladesh** (Anon, 2010a,b).

The aim of the CBFM initiative was to provide research-based approaches so that sustainable management of inland fisheries could be handed over to CBOs. In this connection, fish sanctuaries have been established in > 80 water bodies with the result that production of fish has risen, the income of CBFM fishers increased, and BD has improved. Incidentally, CBFM-type approaches have been included in the Government of Bangladesh's new Inland Capture Fisheries Strategy, and also in the fisheries aspects of the Poverty Reduction Strategy Paper (PRSP). Moreover, at regional and global levels, it represented a substantive body of evidences demonstrating that comanagement and community-managed management approaches yield more productive and sustainable fisheries. Concomitantly, such an exercise safeguards the livelihoods of the poor fishing households.

Notwithstanding the above, pilot projects had begun in the mid-1980s, but in order to scale-up benefits, the approach of the major institutions in charge of managing inland water fisheries were required to be changed, as it was felt. As such, the overall goal of CBFM projects was, therefore, to make efforts to improve inland fisheries management policy and stimulate more sustainable, equitable, and participatory management of the resources.

It is quite a known fact that fish from Bangladesh's vast inland waters are vital to the livelihoods of millions of poor people. But catches and species diversity in the water bodies have been declining due to problems, such as habitat degradation (siltation and conversion of land into agriculture), increased fishing pressure, destructive fishing methods, and acute shrinkages of wetland water spread area at DSL during the dry season. The situation was exacerbated by fisheries policies in *c.* 12,000 government-owned water bodies ("jalmohals") being based on short-term, revenue orientated, leasing. This, excluded the poor fishers, and encouraged the leaseholders to overexploit the fisheries. On the other hand, the CBFM initiative had developed a series of fisheries management approaches for ensuring equitable access to fisheries resources for CBOs. Incidentally a total of *c.* 164 fish sanctuaries have been established so far in >80 water bodies under the CBFM approach. The project involved *c.* 14,000 CBO group members and a further *c.* 9000 direct beneficiaries. When interviewed in 2007, as part of an impact study, the majority of CBOs (*c.* 74%) reported increases in fish production, despite an increase (*c.* 30%) in the number of fishers accessing the water body. Further, fish catch monitoring had shown increased fish abundance and diversity. Incidentally, *c.*123 CBOs were said to be active and practicing CBFM, indicating the potential longevity of the intervention. In fact, the CBFM project is said to be unique because of the length of time over which community-managed fisheries approaches had been implemented (*c.* 11 years), the depth to which they have been studied, and the number and diversity of sites in which approaches were tested. Further, the CBFM type approaches have been included in the Government of Bangladesh's new Inland Capture Fisheries Strategy. At the regional and global levels such mechanisms represent a substantive body of evidence demonstrating that comanaged and community-managed approaches could result in more productive and sustainable fisheries, at the same time safeguarding the livelihoods of poor fishing households. Led by the WFC, and supported by DFID, the project had involved a number of Bangladeshi partners including the Department of Fisheries and a number of NGOs.

With help from these partners, as stated earlier, *c.* 130 CBOs, comprising mainly of poor fisher folk, were formed to manage the water bodies, particularly with regard to sustainable management of inland aquatic resources and for equitable and sustained distribution of resources. As a result of the initiative, securing fisheries access for poor fishers had improved the nutrition and health for their family, including their social status and education. The poor households could improve their standard of housing and sanitation, with an increase in the number of houses with toilets. Spending on health, savings, land, and festivals was also noted to have increased, whilet fish consumption was significantly higher in CBFM sites, as compared to sites where the approach was not implemented. Moreover, level of awareness had also improved. Overall, CBFM approaches in Bangladesh improved fishers' access to open-water fisheries, and enabled poor fishers to catch diverse types of small fishes for their consumption and profession. Further, bimonthly household fish consumption was significantly higher at CBFM sites compared to the control sites.

It is difficult, for cultural reasons, to involve Bangladeshi women in open-water fisheries activities. However, using female-managed CBOs in Narail and Jessore demonstrated that it was possible for women to manage and operate floodplain fisheries just as good as the male-managed fisheries. Training and information dissemination activities had contributed to increased awareness and understanding of community people, including women, about the importance of BD and fish conservation. Such activities had an impact on national and regional fisheries policy. For example, there is explicit mention of CBFM as the preferred management approach in the Fisheries Subsector PRSP Road Map of 2006 and the 2007 Inland Capture Fisheries Strategy.

Concomitant to the above, the impact study demonstrated changes in opinion and awareness of CBFM among policymakers from government institutions and others working in community-based management or inland fisheries management. Incidentally, interviewing *c.* 26 experts had revealed a clear idea of the basic concepts of CBFM and a positive opinion of this approach, considering it as a way forward for inland fisheries management in Bangladesh. They had seen these activities through field visits and in the documentation of the project, and had recognized that the approach works well. The majority had felt that more participation and involvement of the resource users (i.e., the small-scale fishers) in resource management is important and desirable. This is an important policy shift, especially among the government bodies. However, it is still too early to tell whether this will result in a large-scale change in the management of inland water bodies in Bangladesh. The main objective was to enhance production through the conservation of fisheries resources and BD, and to give priority to the fisher folk for the leasing of water bodies. The World Bank was also said to be, of late, finalizing plans to invest in substantial scaling-up of the comanaged and community-managed approaches for more effective and sustainable aquatic resource management essential to the livelihoods of the innumerable poor rural people in Bangladesh and also in many other places in the world.

5.1.4 Cambodia

Cambodia possesses rich fisheries potential. In **Cambodia** and Lao PDR, Anon (2016) dealt with the learning journey in River Mekong with the Blue Peace in the Middle East.

A learning journey, termed as the "The Mekong Learning Journey" spanning over a period of 4 days in Cambodia and Lao PDR, embarked upon, November 17–20, 2014, by the members of the Blue Peace Middle East Community in order to explore, learn, and gain firsthand knowledge about community cooperation on the Lower Mekong River Basin in SE Asia and various other experiences. Incidentally, the delegation from the Middle East consisted of Members of Parliament, senior policy makers, academic and technical experts, and leading members of the Blue Peace Media Network.

The participants had the opportunity to learn and gain experiences directly from the members working in the Mekong River Commission as well as to derive key knowledge for the Middle East. The Learning Journey had opened doors to new experiences and insights regarding the cooperation and efficient water management achieved by the developing countries in the Lower Mekong River Basin. The Mekong Learning Journey was organized by Strategic Foresight Group in coordination with the Mekong River Commission. Following the Rhine Learning Mission, which was held in September 2013, the Mekong Learning Journey was a continuation of the "exchange of experiences" in community resource management and activities under the Blue Peace Initiative. It was sponsored by the Swedish International Development Cooperation Agency (IDCA). Incidentally, the Mekong River Commission (MRC) along with its riparian members have built a platform for open dialog, information exchange, and joint development regarding the Lower Mekong River Basin with the motto of "Meeting the Needs, Keeping the Balance." In fact, drawing lessons from the Integrated Water Resources Management (IWRM) model, etc., the member countries have jointly developed comprehensive data collection systems and basin wide models to study climate change, flooding and navigation potential, These models have been used to develop future strategies for agricultural activities, emergency response systems, forecasting, and other similar plans. The riparian countries in the Lower Mekong River Basin had slowly moved from humanitarian cooperation to encompass wider economic cooperation by starting with small technical projects such as joint hydrological stations, fisheries, irrigations projects, etc. The participants of the Learning Journey had the rare and unique opportunity to learn about these and many other areas of cooperation in the Mekong River Basin. They had interacted with the staff at the MRC Secretariat in Cambodia, the International Cooperation and the Communication Section and Environment Division in Lao PDR, the National Mekong Committees of Cambodia and Lao PDR, the Phnom Penh Autonomous Port and Department of Meteorology and Hydrology (DMH), and a hydro-Met station at Vientiane. The topics covered were Knowledge and Information, Flood Management, Navigation, Fisheries, Basin Development, Climate Change, and Sustainable Hydropower. Each session highlighted the role and functions of MRC in bringing about cooperation between the member states in these areas. It was especially useful for the participants from the Middle East to learn about the principles which formed the basis of cooperation among the riparian states, with regard to "accommodation, adaptation, and adjustment in a peaceful manner," a remarkable and vital feature which went to the core of the success of MRC.

Concomitant to the above, Nuov (2015) dealt with CBF in **Cambodia**. Cambodia's inland fisheries resources play a significant role in national economic, food, and nutritional security. Fisheries administration, in collaboration with the regional organizations had been trying to promote an enhancement of existing small areas of water bodies, such as the CBF development, including the development of community fish refuges (CFRs) pond, both of which are new practices in Cambodia.

Notwithstanding the above, the CFRs pond is a form of natural stock enhancement or a culture-based system in the rice fields, and could promote rice field fisheries production during the wet season by a community management approach. The site selection, the pathway connection, fish species-composition, timing of the stocking, management mechanisms, etc., if managed properly, could serve as the key success stories of the activities. The stocking density of indigenous species may vary from one site to another from 750 to 10,000 seed/ha, or from 20 to 320 kg brood fishes/ha. Such exercises could enhance the rice field fish production, and the local households could catch fish *c.* 348 kg/ha/year, which could be *c.* 6 times more than the normal catch (*c.* 60 kg/ha/year); *c* 800 ponds could be promoted through the CFRs pond development scheme in Cambodia.

Concomitant to the above, the CBF is not a new concept, but a new practice as well as the CFRs pond stocking exercise of fish in existing small water bodies (reservoirs) with a view to enhance fish production for sustaining the fisheries. CBF normally depends on species and composition stocked, size of seed and timings of release; active community and local authority involvement; management structure; common plan(s); agreement(s), such as prohibition of fishing in the no-fishing sanctuary zones, etc. Establishment of such zones is the major key for the success of CBF. The application of CBF by Fisheries Administration, supported by NACA, had been implemented in *c.* 16 reservoirs in 4 provinces in 2012, with a total stock of *c.* 1,518,000 seed fishes of selected ichthyospecies and prawns. After a period of 6 months, when the study ended, the total fish production in the reservoirs has increased considerably. The best results in CBF were obtained when: (a) The natural productivity of the reservoirs are relatively high, (b) the predatory aquatic organisms were eliminated or minimized, (c) suitable species and composition are utilized while stocking, (d) seed stock is of a good quality and of appropriate size, (e) the water level (WL) recedes at a suitable rate, (f) poaching is curtailed, (g) fish sanctuaries are established in all culture areas, and (h) "closed season" (no fishing) is observed for a 5- or 6-month period.

Such measures, if adopted on a sustained basis and through CBFM *modus operandi*, could lead to highly productive fisheries.

Meanwhile, Cambodian people are among the prolific fish eaters in the world. Concomitantly, Limsong and Sitha (2015) dealt with **Cambodian** attempts to develop CBF. There are indicators which show a reflection of the relative availability, particularly of FW fish and other aquatic animals, at an affordable price. Fish is one of the staple foods and an important source of animal protein for human consumption in Cambodia. On the average, fish buildup > 75% of the animal protein and in some areas of the country, the aquatic resources makeup *c.* 80% of the available animal protein. According to Ahmed et al. (1998), on an average, the aquatic organisms makeup > 80% of the animal protein consumed in the country; and *c.* 90% people are dependent on fishing in provinces, such as Siem Reap. However, the fish consumption in Cambodia had been estimated to be *c.* 63 kg/person/year on an overall scenario. This could be many times greater than the global average, thus probably, reflecting the significance of the fisheries sector to the diet and culture of Cambodian people. Incidentally, it may be noted that the application of CBF in Cambodian waters had commenced with the initiation of a project under the auspices of the Australian Centre for International Agricultural Research (ACIAR Project FIS/2011/013), coordinated by the NACA. For initial trial *c.* 16 small reservoirs located in four different provinces were selected. These reservoirs differed from each other in surface area, mean depth, and the catchment land use features, the latter evaluated using GIS software. However, for choosing the reservoirs, initial consultations were held with the village communities, who were responsible for water regime management, and their agreement was obtained for monitoring and cooperating through the trial period. Incidentally, one common feature in all the reservoirs selected and, for that matter, in all the water bodies in Cambodia was the provision of a "conservation zone," generally, in the deeper areas of the water body, where fishing is prohibited. Incidentally, it may be noted that CBF are a form of AQC that utilize small water bodies, both perennial and seasonal, which cannot support a fishery through natural recruitment processes, for food fish production through a stock-recapture strategy. CBF are also environmentally friendly as the only external input is seed stock. In addition, it also applies a comanagement approach utilizing the downstream farming communities in most instances (who are already organized into functional entities for part-time agricultural farming) as the principal beneficiaries. Moreover, the CBF are an attractive development strategy as it mobilize the part-time agricultural farming communities (e.g., the rice farmers) to use the existing water bodies for the secondary purpose of food fish production. Furthermore, the strategies to optimize benefits from the **CBF**, however, may vary in detail from one country to another as well as across the agroclimatic regimes. Incidentally, CBF activities had been conducted for > 2 months growth cycles, and in all instances the fish production had been augmented above the levels that were obtained prior to the implementation of CBF.

Notwithstanding the above, it is believed, however, that the yields could be enhanced by implementation of the conservation measures, and by the application of efficient management *modus operandi*, like the CBFM.

Incidentally, There have been efforts to deal with the challenges and constraints for developing community-based fisheries (CBF) in **Cambodia** and a possible strategy for success.

Initial consultations with the village communities responsible for water regime management were held, in choosing the reservoirs, and their agreement obtained for monitoring and cooperating through the trial period. Incidentally, one common feature in all the reservoirs selected was the provision for a "conservation zone," generally, in the deeper regions of the water body, where fishing is not permitted.

In fact, CBF are a form of AQC that uses both perennial and seasonal small water bodies for pisciculture. In fact, such fisheries may not be able to support fisheries through natural recruitment processes. Moreover, CBF fisheries are said to be environmentally friendly as the only external input is seed stock. It also tries to extend benefits to the downstream farming communities in most occasions through a comanagement approach; who are already organized into functional entities for dry land agriculture.

Further, the principal challenges and constraints are the EUS fish disease (Kar, 2015) and adverse impacts of climate change, while the main strategies against the challenges could be hygienic management and the implementation of CBFM *modus operandi* involving the community stakeholders from the grassroots level.

Notwithstanding the above, the support to Community Forestry and Fisheries in Kratie and Stung Treng Provinces in **Cambodia** had been scaled up under a new partnership program with contribution from Switzerland of US$2.1 million for the initial phase (Anon, 2015).

The PaFF (Partners for Forestry and Fisheries Cambodia) program aimed to improve, increase, and scale-up support to Community Forestry and Fisheries in Kratie and Stung Treng provinces in order to enable communities (in particular, ethnic communities and women) secure access to natural resources as a way to improve food security and increase their income. The PaFF program was supposed to be implemented with four local and international organizations, viz., the Non-Timber Forest Products Exchange Program Asia (NTFP-EP), RECOFTC (The Center for People and Forests), World Wide Fund (WWF) for Nature, and Culture and Environment Preservation Association (CEPA) as well as with the in-kind contribution of the Royal Government of Cambodia (the Forestry Administration and the Fishery Administration at subnational levels). In fact, the partnership program interventions had combined the rich experience of these multi-sector partners to enable the forestry and fishery-dependent communities respond better to the challenges and tensions created by fast economic growth coupled with environmental degradation, while improving nutrition and income levels through diversified and sustainable income strategies from natural resources. This program was a welcome move with its strong focus on strengthening livelihoods of poor people combined with the sustainable use of natural resources. In fact, the partnership program brings important and experienced organizations together to work closely with the authorities. PaFF, however, need continuous and consistent political support from all the stakeholders so that it could contribute positively toward addressing the critical challenges **Cambodia** had been facing in sustainable management of its natural resources, and alleviate poverty by enhancing equitable and sustainable development for all Cambodians, especially women and vulnerable groups. The program was appreciated by different quarters noting that the partnership program emphasized a strong integrative and collaborative approach. Livelihood and NRM, technical support, and capacity building activities were planned in the program around key potential products, notably, processed fish, bamboo, forest honey, mushrooms, rattan, cassava, including ecotourism activities, and so on. These activities were managed by Community Forestry and Fishery groups, and it had been expected that support from the authorities and market actors could be enhanced through the PaFF Program. Incidentally, NTFP-EP Asia is a regional organization active in supporting capacity building of community-based organizations in the aspects of sustainable harvest, value addition, and marketing of non-timber forest products (NTFPs). NTFP-EP Cambodia country program has been supporting various community-based organizations in their NTFP development initiatives and projects in six (6) provinces including Kratie and Stung Treng, since 2007.

In continuation of the above, WWF had been running a landscape conservation program, since few decades, in Cambodia (particularly in Kratie and Stung Treng provinces) and in Lao PDR (in the Siphandone islands). Moreover, WWF was a leading player in a regional campaign to stop the development of hydropower dams in the Mekong mainstream, such as the Don Sahong Dam in Lao PDR which could block fish migration and natural flow of sediments critical to food availability, nutrition, and agricultural production and other livelihood activities of the communities living along the Mekong River and its surroundings.

Concomitant to the above, the program was also appreciated by the local stakeholder organizations, like the CEPA, which had asserted that PaFF could provide an important platform to bring NGO partners together to help the communities more effectively and efficiently in order to reach the critical goal of securing community rights to use and manage their natural resource and improve their livelihood. In particular, PaFF could scale-up and continue with the support to CF through further livelihood and sustainable forest and fisheries management interventions. In fact, the PaFF Program aimed to support up to 75 communities in the implementation of their community forestry and fishery management plans, and in the overall development and equitable sharing of livelihood and natural resource protection benefits out of community managed forests and fisheries in the two provinces. In particular, the program could work around key BD hotspots of the Mekong Flooded Forest and Ramsar Site in Kratie and Stung Treng provinces, home to the critically endangered Irrawaddy Dolphin and to a significant population of poorest resource-dependent communities in the country. In this first phase, the PaFF Program included 30 communities in eight districts in Kratie and Stung Treng provinces, representing *c.* 4000 households, over an area of *c.* 18,000 ha of community forestry and *c.* 10,000 ha of community fisheries.

5.1.5 China

China has a long history of reservoir fisheries production and management activities. The earliest stocking activity of fish seed collected from rivers took place in Dong Qian Hu reservoir, which was impounded *c.* 1000 years ago in Zhejiang province (Shi, 1996). Many reservoirs have been constructed in China since the 1950s for various purposes and are also used as potential habitats for fisheries. Currently, reservoir fisheries have become an important component of the fisheries industry in China, and are making a substantial contribution to people by providing with fish as nutritional supplies and rural livelihoods (Liu et al., 2008).

Notwithstanding the above, China is currently in a new stage of development. Environmental conservation and wise use of resources are of major concern in developmental activities. Nevertheless, there is a great potential for reservoir fisheries development in China, but nevertheless, challenges do also exist (Miao Weimin, 2008).

Concomitant to the above, AQC has been in practice in China since over 3000 years, but reservoir fisheries are a new sector in China, which have been growing with the large-scale construction of water projects. Incidentally, >86,000 reservoirs have been built in China for irrigation, flood control, power generation, navigation, fisheries, etc., in a little over 5 decades, as a result 2 million ha of reservoir area have become suitably available for fish production (Wu, 1998).

China is known to be densely populated. The construction of reservoir projects has inundated a large area of agricultural land, resulting in the displacement of large number of people; as a compensation for which, the Chinese government emphasized the development of reservoir fisheries to provide alternative livelihoods to the displaced people.

Concomitant to the above, China made considerable amount of progress in the arena of reservoir fisheries in the past *c.* 60 years, mainly with regard to their development and management. At the same time, the development of reservoir fisheries in China had also encountered some challenges.

5.1.5.1 *Reservoir resources and their status*

Nature of reservoirs

Reservoirs were constructed by the ancient Chinese, mainly for the purpose of harvesting and regulating rain water for crop irrigation. The earliest reservoir in China, Dong Qian Hu, was constructed *c.* 1000 years ago in Zhejiang Province. However, there were only *c.* 20 reservoirs in the whole of China until 1949, when the People's Republic of China (PRC) was founded. Since the 1950s, innumerable reservoirs of different sizes had been constructed in China, mainly for the purposes of flood control, irrigation, and hydroelectric (HE) power generation.

Notwithstanding the above, in China, reservoirs are usually classified into three categories, according to the geographic characteristics of the locations, viz. (i) valley type, (ii) branch type, and (iii) plain type.

Incidentally, the reservoirs in China are, generally, classified into three types based on the surface area and according to their storage capacity as follows:

(i) Large (surface area > 667 ha), > 400 in number.
(ii) Medium (surface area between 67 and 667 ha), *c.* 3500 in number.
(iii) Small (surface area < 67 ha), 8200 in number.

Numerically, small reservoirs dominate the total reservoir resource in the country, but the highest storage capacity is registered for large reservoirs.

5.1.5.2 *Potential of reservoir fisheries development*

China lies across a range of climatic zones and reservoirs are also located in areas of different climatic, geographic, and topographic conditions. As such, the potential of reservoirs for fisheries development varies between locations and individual reservoirs.

Reservoirs in the NE China have the highest fertility due to high organic content in the soil of the watershed area. However, due to the long winter with very low temperature, in this region, the growth of fish and other aquatic animals is slow. Nevertheless, the productivity is still higher in the reservoirs of this region than in the reservoirs of most other parts of China. Fish production in many reservoirs in the region could reach 230 kg/ha/year (Shi, 1996). In addition, the reservoirs located in the lower reaches of Yangtze River are also said to be highly productive. For instance, the average annual FY of all reservoirs (except Xin'anjiang Reservoir) in Zhejiang province have reached >2700 kg/ha/year from natural and AQC-based fisheries.

Notwithstanding the above, comparatively, the productivity of reservoirs located in Loess Plateau and the western part of China is much lower than those in other areas due to various reasons, such as high water turbidity, low water temperature in high altitudes, and high salinity. On the other hand, the capture fisheries production of reservoirs in Southern China is not very high. High temperature and long growing period of fish greatly facilitate cage culture in these reservoirs. Concomitantly, the valley-type reservoirs have very limited potential for fisheries development, mainly due to the high current velocity, high turbidity, and short water retention period. Branch and plain type reservoirs are usually built in the middle and lower reaches of rivers at low altitudes and the environmental conditions of these reservoirs are said to be more or less stable. Also, they are biologically productive. As such, these reservoirs are suitable for fisheries activities.

5.1.5.3 *Biological characteristics of reservoirs*

Natural food organisms

The availability of natural fish food organisms in reservoirs is one of the most important basis for fisheries production. Plankton are said to be the most dominant food organisms in all reservoirs. In many reservoirs, phytoplankton (PP) abundance may range between 5 and 10 mg/L. In Fengman reservoir (a mega reservoir in Jilin province), plankton biomass was as high as 10.86 mg/L (Zhang, 1986). In smaller reservoirs, the PP biomass could be much higher. For example, mean PP biomass in Kunlong Reservoir (*c.* 188.7 ha) in Shandong province is *c.* 20.57 mg/L (Xing et al., 2007). On the other hand, zooplankton (ZP) biomass may usually be lower, and in Fengman reservoir, it was 6.63 mg/L (Zhang, 1986).

5.1.5.4 *Reservoir fish fauna*

The information on the fish fauna of Chinese reservoirs is relatively limited. Generally, reservoir fish fauna are less complex than other inland water bodies, particularly lakes. This could be due to specific hydrological characteristics in reservoirs. Since reservoirs in China are located in different geographic and climatic zones, fish fauna in different reservoirs also vary greatly. The fish fauna in a reservoir in Liaoning province of NE China is reported to consist of 29 species belonging to nine families (Liu et al., 2003); out of which, 21 species belonged to family Cyprinidae.

5.1.5.5 *Introduced fish species*

It may be noted that introduction of fish species had been extensively carried out in Chinese reservoirs since the 1950s, as an effort to improve fish faunal composition in reservoirs for better utilization of natural productivity. Various translocated species, such as grass carp, silver carp, bighead carp, *Protosalanx hyalocranius* (Abbott), *Megalobrama amblycephala* (Yih), *Hypomesus olidus* (Pallas), and *Plagiognathops microlepis* (Bleeker), were believed to have significantly improved fish production.

Concomitant to the above, it may be said that the introduced fish species play a very important role in reservoir fish production in China. Silver and bighead carps account for the major share of fish production in most reservoirs in China.

5.1.5.6 *Fisheries*

Utilization of fisheries resources in reservoirs of China

Management *modus operandi,* notably, **CBFM**, play a significant role in management of the fisheries.

Notwithstanding the above, reservoir area utilized for CBF in China, during 1995–2006 had shown a steady growing trend. Approximately 83% of the total reservoir area in China was said to be used for CBF.

5.1.5.7 *Production levels and trends*

Environmental conditions of reservoirs vary greatly in China, and there is a great difference in FY across the reservoirs. Concomitantly, **fisheries management** practices and their effectiveness also differ from one reservoir to another. All these result in differences in production level, and as such, it could be rather difficult to have a clear understanding on the production level of different types of reservoirs. Some well-managed reservoirs of moderate size with conducive environmental conditions may produce as high as 2 tons/ha of fish. For example, FY of Huayuan reservoir (*c.* 246 ha, water depth *c.* 19.8 m on average) has reached *c.* 2032.52 kg/ha in 2004 (Gao et al., 2006). However, fish production is said to be < 100 kg/ha in some of the large reservoirs in China.

5.1.5.8 *Jurisdiction of reservoir water and fishery resources*

Main fisheries strategies

The Main fish culture methods and strategies are as follows: (i) Capture fisheries, (ii) CBF (extensive culture), and (iii) intensive farming.

The fisheries strategies vary according to these three types.

Capture fisheries are usually operated by individual fishers, using small traditional fishing gears, such as gill nets, traps of different types, lift nets (sometimes assisted with light attraction), and floating dragnets. Individual fishers need to be licensed by the Reservoir Fisheries Management Authority. These fishers are permitted to fish only naturally recruited species using specified mesh sizes of fishing gear in specified areas.

CBF are usually operated by companies/community groups or individual households depending on the size of the reservoir. They are responsible for looking after all production links from stocking, management to harvesting. The fishing operation is usually run on large scale with combined gears, such as gill nets, blocking nets, driving nets, and set-bag nets. **The CBFM mode of management could play a significant role in the effective management of community-based fisheries.**

In contrast to the above, **intensive culture** in reservoirs include mainly the pen and cage culture. Pen culture is practiced mainly in shallow reservoirs. It is usually operated by individual households. On the other hand, cage culture is generally practiced in relatively deep water, which takes advantage of good water quality. However, cage culture operation may vary in scales. It could be operated either by commercial companies or by individual households.

5.1.5.9 *Achievements of reservoir fisheries in China*

One of the most significant changes, during the last few decades, when most of the progress was made in reservoir fisheries development in China, was the transition from **extensive culture** to **semi-intensive** and **intensive culture**. The capture of naturally occurring fish populations was said to be the main component of reservoir fisheries during the 1950s. This is still the case in some large-sized reservoirs. In fact, some strategies, such as introducing fry into reservoirs through diversion of water from rivers, establishing PAs for the reproduction of fishes, observing closed area and closed season, prohibiting fishing in the breeding areas and during the peak recruitment seasons and regulating the size at capture had been adopted in order to rationally exploit the natural fish resources in reservoirs during the early phase of development (Liu and He, 1992).

Further, stocking of fish into small and medium-sized reservoirs became a very important fisheries practice with the success of artificial propagation of the four domestic carps (silver carp, *Hypophthalmichthys molitrix;* bighead carp, *Aristichthys nobilis;* grass carp *Ctenopharyngodon idella;* and, black carp, *Mylopharyngodon piceus*).

This not only brought about significant economic efficiency but also helped to employ a lot of displaced people. For instance, the FY from Fuqiaohe Reservoir in Hubei Province (*c.* 1340 ha) have increased from *c.* 30 tons prior to stocking to *c.* 300 tons after stocking, and > 300 displaced people were employed as fish farmers or fisheries management officials (Liu and Huang, 1998).

Notwithstanding the above, some of the principal achievements of Chinese reservoir fisheries are as follows:

(a) Rapid increase in fish production in reservoirs.
(b) Establishment of stocking and fisheries management systems, notably, **CBFM**.
(c) Fish productivity assessment in reservoirs.
(d) Improvement in capture fisheries techniques.
(e) Improvement in fish escaping avoidance techniques.
(f) Standardization of fisheries extension techniques.

5.1.5.10 *Production*

Incidentally, it may be noted that the reservoir fisheries are under the jurisdiction of the Chinese Ministry of Water Resources (Liu, et al.,2001). However, all other fisheries, including marine fisheries and fisheries from ponds, lakes, rivers, and rice fields, are under the administration of the Chinese Ministry of Agriculture (Wu, 1998). Further, in most of the medium and large-sized reservoirs, an administrative agency is, generally, set up to manage the stocking, capture, sale, and availability of sufficient water in the reservoir. Usually, a hatchery is built below the dam in medium and large reservoirs to supply fingerlings for stocking. For new reservoirs, building a fish hatchery is required by law.

5.1.5.11 *Fish productivity assessment in reservoirs*

A, more or less, countrywide survey of reservoir limnology had been conducted in the 1980s. Different types of reservoirs were identified based on morphological characteristics, principal nutrient concentrations, and biomass of aquatic organisms (Dai and Xu, 1996). The methodology was developed to predict silver carp and bighead carp production. Based on the analysis of 15 factors affecting fish yield (FY) in 32 reservoirs, Xiong (1996) has shown that mean depth, catchment area, water temperature, precipitation, dissolved oxygen (DO) content, total phosphorus (TP), phytoplankton (PP) biomass, and the number of stocked fingerlings are the major factors which are said to influence FY. These factors could be used as an ecological index for estimating FY from reservoirs.

5.1.5.12 *Rapid increase in fish production in reservoirs*

Stocking large-sized fingerlings in reservoirs is also practiced in order to ensure fast growth, high recapture rate, and rich fish production. Generally, silver carp and bighead carp of > 15 cm in length are stocked. The stocking density is calculated as $d = f/ws$, where d is the annual stocking density (fish/ha), f is the annual fish yield (kg/ha), w

is average harvesting size (cm/kg), and *s* (%) is the recapture rate (Li and Xu, 1995). Incidentally, FY from reservoir fisheries was only *c.* 54,000 tons in 1949 when the People's Republic of China (PRC) was founded. However, the consumption patterns of people have changed as a result of increased buying power, with the liberalization of the economy in China in 1978. Consequently, the FY from the reservoirs have increased from *c.* 112,000 tons in 1978 to >2.23 MT in 2005, currently averaging *c.* 1200 kg/ha. Still, the reservoir fisheries sector is considered to have even more potential for further development. Incidentally, the reservoir areas suitable for fish production represent, *c.* 30.1% of the total FW area in China. However, the fish production from reservoirs represent only *c.* 11.7% of the total FW production (FBMA, 2006). Nevertheless, the reservoir fisheries could play a more significant role in the future.

5.1.5.13 *Improvement in fish capture techniques*

Efficient fish capture from large and medium-sized reservoirs was a difficult task mainly due to greater depth of water. As such, a combination of capture techniques, called "United Fishing Method" UFM), is used to capture fishes, which use driving nets, bar nets, gill nets, stake nets, etc. At the same time, the blocking and driving operations are also conducted simultaneously with capture so that all fish are forced to aggregate in a certain location which could facilitate capture.

5.1.5.14 *Improvement in fish escape preventing techniques*

Fish escapes from reservoir spillway(s) and turbine(s) are a common problem, causing great loss to the fish farmers. A new electric facility is said to have been developed to prevent fish from escaping from the reservoirs. This facility can both drive fish away from turbine and spillway area and prevent the fish from escaping.

5.1.5.15 *Fisher organizations*

CBF are a common practice in many reservoirs of China, and are managed by companies or community groups. Of late, the government has promoted the establishment of fishers' associations in order to assist individual fishers in marketing and to facilitate management of fishing activities. Of these two types of organizations, companies and community groups are said to be very effective to develop reservoir fisheries, particularly culture-based fisheries. On the other hand, the fishers' associations are still in the process of development. They are expected to play a vital role in reservoir fisheries management in China in future.

5.1.5.16 *Marketing*

Marketing of the products is an essential aspect of commerce. Fisheries products from reservoirs are marketed in different ways according to the practices and scope of fisheries activities. For fisheries operated by the community groups and companies, there are well-established marketing channels. They usually have access to distant markets which may bring them better benefits. On the other hand, community groups and companies often may have their own cold storage and processing facilities, which could increase the value of the products. Tourism and restaurant business also play significant roles in augmenting the value of fisheries products.

In contrast to the above, individual fishers in small-scale fisheries may not have access to distant markets, where fisheries products could fetch higher prices. The fisheries products in small-scale fisheries are, therefore, sold to local and nearby consumers. Thus, establishment of fishers' organizations may help small-scale fishers to get better market access.

5.1.5.17 *Socioeconomics of reservoir fisheries*

It may be well known that fisheries production may, usually, generate relatively higher incomes for rural households as compared to agricultural crop farming and forestry. Therefore, average income of the fishers is a little higher than the income from agriculture. On the other hand, fishing and AQC in reservoirs are relatively tough occupations compared to other economic activities, such as agriculture, and as such they are less attractive to the younger generations.

5.1.5.18 *Biological and environmental constraints*

Eutrophication is a common problem in many inland water bodies in China. Such environmental degradation is generally caused by many factors associated with rapid economic development. Protection and restoration of the aquatic environment is of greater concern to the government and the public. Fisheries and particularly AQC have become easy targets to be blamed for environmental degradation. More and more strict regulations are imposed on the development of AQC in inland water bodies including reservoirs. Cage culture has been discontinued in some reservoirs, which are mainly used for water supply in urban areas. Eutrophication of reservoirs also affects the reservoir fisheries in a number of ways. Serious eutrophication is said to alter the species composition of fish communities in the reservoirs. Population densities of fish species of high economic value have declined significantly in reservoirs. Degradation of water quality is also known to affect the growth of fish and feed conversion in culture-based fisheries and AQC.

5.1.5.19 Socioeconomic constraints

Currently, CBF are the main approach to augment fish production in reservoirs. Silver carp and bighead carp are the major constituent species in CBF production in reservoirs in China. These two species are, however, generally considered as low-quality fish. High production of these fish species often may drastically reduce the market price, which could result in low economic efficiency.

Notwithstanding the above, in China, quite a large number of small and medium-sized reservoirs are given on contract basis to companies or individual households for, so-called, better fisheries management. Incidentally, conflicts may arise between the community groups conducting CBF practices and the traditional fishers. In fact, unlawful fishing and poaching still exist in some reservoirs.

5.1.5.20 Policy constraints

Reservoir fisheries are a relatively small contributor to the national fisheries Industry. Hence, it may receive limited attention from the government. In general, reservoirs are located in remote areas where it is relatively difficult to access support systems, such as technical extension services and marketing because related infrastructure is usually weak. Moreover, the fishers and AQC operators are more vulnerable to natural disasters.

5.1.5.21 Prevalence of the sudden and virulent fish disease

For a long time, reservoirs have been famous for good water quality and the prevalence of fish diseases had not been a problem. In the recent years, however, the FW fishes of China, India, and the neighboring and contiguous countries, have been attacked by a *hitherto* unknown virulent enigmatic fish disease, variously called **Epizootic Ulcerative Fish Disease Syndrome (EUS)**, which has been sweeping almost semi-globally today (said to have initiated from Japan and Australia, and sweeping all the way through Far East, SE, and South Asia, since the 1970s; and now said to have reached United States) causing large-scale mortality among the FW fishes, rendering many of them endangered, and *vis-à-vis* causing nutritional and avocational difficulties to the people (Kar, 2007a,b,c, 2013, 2015).

In addition, other types of sudden and violent fish diseases may be prevalent in the Chinese reservoirs. These could also be due to intensive culture, such as cage culture, cove culture, and applications of fertilizers, which cause big losses to fish farmers (Gao et al., 1997). It is predicted that, with the development of culture scale and culture intensity, the diseases in AQC could even be more harmful.

5.1.5.22 Establishment of stocking and management systems

Bighead carp and silver carp are two of the four major Chinese carps in China, which have been traditionally cultured since many years in ponds and lakes of China. They are said to possess a number of favorable biological characteristics, such as rapid growth rate, easy capture, strong disease resistance, and easy availability of fry and fingerlings. They are, generally, used as the dominant stocked species in reservoirs for intensive and semi-intensive AQC. They, probably, account for up to *c.* 70% of the total catch in almost all reservoirs. In most reservoirs, the ratio of bighead carp stocked to silver carp could be 2:1 or 3:1. Nevertheless, in reservoirs, where fertilizer is applied, the ratio of bighead carp is about half that of silver carp. For example, in Qingyunhe Reservoir in Liaoning Province (*c.* 92 ha), which is fertilized with the waste water from a chemical fertilizer factory, stocking of silver carp has resulted in good economic returns.

5.1.5.23 Future strategies

Reservoirs are an important natural resource with multiple uses. Incidentally, in China, the per capita availability of land is low, it being a country with very high population density. As a result, land-based economic activities, such as agriculture, have little scope for contributing to furthering the rural economy. Nevertheless, economically and ecologically sound fish production is possible from reservoir fisheries through effective management. As such, reservoir fisheries could play a significant role in improving fish supply and generating income for the rural population. Nevertheless, the following approaches are important for the long-term sustainability of reservoir fisheries of China.

5.1.5.24 Development of eco-friendly fisheries in reservoirs

In the past, reservoir fisheries were mostly fish production-centered, when fish supply was inadequate to meet the demand in China. Currently, fish production meets the demand, and environmental concerns prevalent in reservoir fisheries management are taken care of in order to produce fish without adverse impacts on the environment.

Incidentally, "water conserving fisheries" is said to have become a concept accepted by more and more people for long-term sustainability of fisheries. Furthermore, good reservoir fisheries practices could assist in coping with degradation of the aquatic environment, particularly eutrophication problems.

5.1.5.25 Development of leisure fisheries

Reservoirs are generally located in areas with beautiful natural landscapes so that there is a high potential for the development of sport fishing and tourism. Leisure fisheries may be considered as an important approach to improve the economic performance of fisheries, where it is applicable (Hu et al., 2005).

5.1.5.26 Improvement of fish species composition in reservoirs

Reservoir fisheries production, at present, is said to be mainly dependent on silver and bighead carps. These two ichthyospecies could, undoubtedly, still remain as important species in reservoir fisheries due to their efficiency in utilizing plankton. In addition, other indigenous species could be introduced into the reservoirs, after careful study to improve the species structure and fisheries efficiency.

Notwithstanding the above, the challenges are highlighted and development strategies are discussed with a view to identify strategies for the sustainable development of reservoir fisheries in China.

Notwithstanding the above, stocking sturgeons into reservoirs is a rational choice for making better use of reservoir resources and improving economic situations (Liu and Hu, 2000). The organic detritus is an important food resource of fishes (Chang, 1995). In Heilongtan Reservoir, an eutrophic reservoir of *c.* 1693 ha in Sichuan Province, the annual fish production potential based on organic detritus was estimated at *c.* 473 tons (Chang, 1995; Hu et al., 1998). However, large fish populations that feed on organic detritus are rarely found in reservoirs. *Xenocypris davidi* and *Xenocypris microlepis* feed mainly on organic detritus and are commonly found in the Yangtze River and the associated water bodies (IHHP, 1976). Stocking of these two species into reservoirs or strengthening the protection of their reproductive populations may increase fish production and economic returns significantly. Piscivorous fish have a very important role in aquatic ecosystems. Some large-sized pelagic piscivorous fish, like *Elopichthys bambusa*, could harm the stocked fingerlings of silver carp and bighead carp. As such, population sizes of this piscivorous species needs to be strictly controlled. However, some small or medium-sized piscivorous fishes like mandarin fish, *Siniperca chuatsi*; Chinese snakehead, *Channa argus*; and, large icefish, *P. hyalocranius* may not harm stocked fingerlings. Instead, they can convert the low valued trash fish into high valued reservoir fisheries in China. Furthermore, piscivorous fishes command a high market price, and a lower stocking rate may produce a higher economic return, decreasing pressure of fisheries (Liu et al., 1998). Studies in North America and Europe showed that stocking of piscivorous fishes into lakes could improve water quality by trophic cascading effects (Carpenter and Kitchell, 1988; McQueen, 1990). However, determination of appropriate stocking number of piscivorous fish in a water body is important. Bioenergetics models may be used to predict the growth and food consumption of fish (Hewett and Johnson, 1992; Brandt and Hartman, 1993). Studies on the productivity of trash fish in a reservoir and the bioenergetics model of the piscivorous fish could help in determining the appropriate stocking rates of piscivorous fish based on predator-prey relationships.

Concomitant to the above, catching demersal fish efficiently in reservoirs is somewhat problematic. Hence, studies on fishing gear technology for capture of demersal fish are necessary. Cage culture practices also need to adopt changes, such as in cage structure and location of cages in a water body, which may increase the economic efficiency, decrease pollution, and control incidence of diseases (Peng et al., 2004). A strategy to develop intensive culture of high-valued fish through the use of the discharged water below reservoir dams needs to be explored. In Miyun Reservoir, Beijing, several different cold-water fish species, such as salmonids and sturgeons, are cultured, resulting in good economic returns. Through these culture practices, water pollution can be efficiently controlled. Concomitantly, carrying capacity of reservoirs for AQC is a main concern in recent years because of the deteriorated water quality in reservoirs. Standards for determining the carrying capacity of reservoirs with different functions are to be established. A standard method for assessing the carrying capacity of water body for cage culture was developed (Peng et al., 2004), but needs to be validated. Carrying capacity of reservoirs for other AQC practices are also to be developed.

5.1.5.27 Strengthening the support for the management of reservoir fisheries

The support system may be strengthened to fully ensure the sustainability of reservoir fisheries in China, and for effective fisheries management of reservoirs. There is a big need to set up new institutional arrangements to coordinate reservoir fisheries development with other concerned partners. Infrastructures such as marketing and transportation need to be improved. In-depth studies on the relation between different fisheries production modes, environment, and new techniques for improving ecological and economic benefits of reservoir fisheries should be strengthened.

In addition to the above, improvement in fisheries management mechanisms in reservoir fisheries may, preferably, include functional zoning of the water bodies, licensing systems for different kinds of fishery-related activities, and better coordination between the interests of different stakeholders. Participation of fishers' and farmers' organizations in the decision-making processes and application of the technique of **CBFM** could be vital in the improved management strategies.

5.1.5.28 *Future strategies and management*

Some strategies and management plans are needed to be considered for sustainable development of reservoir fisheries in China. The renewal of stocked species is an effective strategy, instead of the traditionally stocked planktivorous fish species, silver carp, and bighead carp, which are said to have low market demand. Research priorities could be directed toward the selection of high-valued planktivorous fish species which can partially replace the traditionally stocked species. During the early 1990s, the icefish (*Neosalanx taihuensis*), a planktivorous fish mainly feeding on zooplankton, was stocked into some reservoirs in Central and SW China. Many reservoirs have benefited from this new species. For example, Zhanghe Reservoir, an oligotrophic reservoir of *c.* 7000 ha, in Hubei Province, without any other stocked species, was stocked with *c.* 200,000 fertilized new icefish eggs during 1992 (Liu and Hu, 2000). From 1996 to 1998, the annual FY of the new icefish was said to be > 100 tons with a market value of > 1.5 million Chinese yuan (6.8 Yuan = 1 US$). This amount was said to be equivalent to the market value of *c.* 300 tons of all the fish products in Fuqiaohe Reservoir, an eutrophic reservoir of *c.* 1500 ha in Hubei Province. Incidentally, the latter reservoir is known for the early development of many reservoir fisheries techniques and also for the research base of the Institute of Reservoir Fisheries, the Chinese Ministry of Water Resources and the Chinese Academy of Sciences (Liu and Huang, 1998). In Geheyan Reservoir, an oligotrophic new reservoir of *c.* 7000 ha in Hubei Province, the new icefish was stocked in 1996, which produced > 300 tons in 1998. The new icefish may spawn naturally in reservoirs, so that regular stocking is not required. They can be easily harvested by trawl nets so that large-scale capture with the United Fishing Method (Li and Xu, 1995) is not necessary. Furthermore, fishers in reservoirs with the new icefish are also benefited by selling fertilized eggs. In 1997, fishers in Zhanghe Reservoir sold *c.* 54.45 million icefish eggs and have earned *c.* 1 million Chinese yuan.

Concomitant to the above, the American paddlefish, *Polyodon spathula*, a planktivorous fish, mainly feeding on zooplankton is another alternative for partial replacement of bighead carp. This species is known for its high-quality meat and high priced eggs in North America (Liu and Yu, 1990). Since the early 1990s, China has been introducing the American paddlefish every year from the United States. It was said to have shown high growth rates, both in the open water and in cages. This fish could be easily caught with gill nets as well as it may spawn and hatch naturally in some large reservoirs (Liu and Hu, 2000). Also, currently, the artificial propagation of the American paddlefish is said to be successful in China so the import of fish from the United States is no longer a necessity. The only concern could be that the American paddle fish may escape from stocked reservoirs into the Yangtze River and crossbreed with the endangered Chinese paddlefish, *Psephurus gladius*. Further, the silver carp and the bighead carp use only the trophic spatial niches in the top layer in the reservoirs. As such, there appear to be a provision to stock demersal fishes into reservoirs in order to enable full use of the food resources and space in a reservoir ecosystem. Concomitantly, the common carp (*Cyprinus carpio*), a benthos predator, is commonly found in Chinese lakes, but large populations of the fish are rarely found in Chinese reservoirs mainly because of the lack of aquatic macrophytes (AM) as their spawning substrates. In all these exercises, proper management of the water body with regard to its physical, chemical, and biological characteristics as well as its productivity and sustainable and rational harvest are prerequisites. And, CBFM, involving the community is the best strategic tool for achieving the goal of efficient management of reservoirs and to yield effective results. Moreover, the carrying capacity of reservoirs for AQC is a main concern in recent years because of the deteriorated water quality in reservoirs. A CBFM approach could find a tangible answer for this too.

Notwithstanding the above, Paterson and Pernetta (2015) dealt with Integrating Fisheries and Habitat Management with regard to fisheries *refugia* in the **South China** Sea. It is quite a known fact that production in a fishery could be highly successful if the habitat parameters are integrated and taken care of concomitantly along with the management aspects. Incidentally, fisheries *refugia* are spatially and geographically defined marine or coastal areas in which specific management measures are administered to sustain important species of fisheries resources during critical stages of their life cycle for their sustainable use. Concomitantly, Huang, et al. (2001) dealt with the fish resources in the **Chinese** reservoirs and their utilization. The diversity of the landscape, aquatic resources, and climate in the reservoir catchments provide a variety of habitats for fish, resulting in a rich diversity of ichthyospecies. Further, Wang et al. (2015)

dealt with CBFin the lakes of the Yangtze River basin in **China**, with special reference to stocking of mandarin fish and Chinese mitten crab. In fact, the stocking of mandarin fish and mitten crab are the most successful because stock enhancement of these two species had been systematically conducted.

5.1.6 Hong Kong

Rapid industrialization and large-scale infrastructural projects in **Hong Kong** and the Pearl River Delta in recent years have placed an extraordinarily heavy burden on the coastal environments. The magnitude of the problem could increase as sewage discharge, ocean dredging/disposal, and land reclamation continue to expand in the next few years. Basics of aquatic environmental impact assessment are based on a good understanding of BD that permit us to monitor environmental changes. However, there is no baseline detailed and exhaustive database on the FW, marine, and estuarine fauna which could permit the evaluation of the damage from various large-scale infrastructural projects to aquatic environments. As such, a comprehensive database on local fish fauna is, therefore, urgently necessary for the planning and implementation of management and conservation strategies of the natural resources, notably, the water bodies and the fishes, in Hong Kong (Ni and Kwok, 1999). Incidentally, although not too many, there had been some contributions on the systematic report on fish fauna (notably, marine) in Hong Kong waters. From years of fishing surveys and a thorough literature review, a total of 25 orders, 124 families, 390 genera, and 834 fish species in Hong Kong waters have been identified. Incidentally, 97 fish species had been listed as new records to the area. In view of the above, a complete record on the local ichthyofauna is very much needed for the planning and materialization of conservation efforts and fisheries management in Hong Kong waters. Incidentally, the Hong Kong Government's policy on fisheries and BD is essentially one of free enterprises, with minimal government intervention. Efforts are said to have begun, only recently, to establish inventories of marine flora and fauna for the Hong Kong waters (Morton and Tseng, 1982; Morton, 1990, 1992, 1997). Incidentally, the main policy of the Hong Kong Agriculture and Fisheries Department is "to facilitate sustained production and enhancement of the local fishery industry through improvement of productivity." However, this policy for improving fish productivity is often sacrificed in the interest of economic development. For example, there is no baseline data for the **management of fisheries** resources and conservation of fish fauna.

With reference to the above, a survey of the literature reveals that only a few studies have been published on parts of Hong Kong's fish fauna. The earliest studies have portrayed that Hong Kong had been a fishing port for centuries. Currently, Hong Kong is said to have >6.5 million people residing in an area of *c.* 1064 km ^{2} with *c.* 800 km coastline. Moreover, Hong Kong is said to be one of the world's most vibrant success stories with a flourishing economy built on free enterprise and free trade. These economic activities has had a great impact on the aquatic environmental protection and **fisheries resources management**. It is not surprising that most of Hong Kong's waters are said to be intensely polluted. Until recently, *c* 1.5 million m ^{3} of untreated sewage and industrial waste water, equivalent to *c.* 1000 Olympic-sized swimming pools, flow straight into its waters every day, thus generating a stink and a bad appearance (Hong Kong Environmental Protection Department, 1996). The magnitude of the problem has increased as ocean dredging/disposal and land reclamation continue to expand in addition to rapid industrialization in the Pearl River Delta in recent years. Now, the query is, "what could be the effect of these developments on the aquatic domain, and how can we proceed with the issues of 'conservation' without knowing what is to be done."

Incidentally, there had been attempts to design, collect, and, to compile published and unpublished baseline information on the ichthyofauna, notably the marine fish species, in Hong Kong waters. Efforts had been made to collect information on the marine fish fauna in Hong Kong waters through a thorough review of all possible available sources including survey results and verification with contributors whose lists were not officially published. Such a list of ichthyofauna serve as a first complete list of the BD of marine fishes in Hong Kong and as baseline data for conservation and **fisheries management** programs for Hong Kong waters.

Some of the works on Hong Kong waters, in this regard are as follows: A synopsis of the fishes of China, Part I-X' (1930–62) and "Studies of Hong Kong Fishes" (Fowler, 1930a,b,c,d,e, 1931a,b,c,d, 1938a,b). Subsequently, 116 fish species were included in "A list of fishes from Hong Kong" (Shen, 1964). The latest one is "The marine fishes of Hong Kong Part I" (Chan, 1968), which was published *c.* 30 years ago. This dealt with 9 families including 71 species of perch-like commercial fishes. In addition, there had been a few unpublished fish lists about Hong Kong waters (Gaiger, 1974; Chan, 1984; Chan and Wong, 1989) that were intended for internal use in the Hong Kong Agriculture and Fisheries Department. Also, there had been sporadic fishing surveys (Thompson and Horikoshi, 1982; Thompson et al., 1982; Wong, 1982; Chong, 1984; Leung, 1991, 1994, 1997a,b; Ni, 1994a,b, 1995a, 1997; Lee, 1997; Hong Kong Agriculture and Fisheries Department, 1998). Further, there had been ecological studies of fish communities (Chan and Tseng, 1982; Wu, 1984; Richards and Wu, 1985; Lam, 1990c) in parts of Hong Kong waters.

There were also systematic studies on (a) flatfish (Shen, 1967a,b); (b) new species (Chan, 1965, 1966a,b,c, 1970, Lam, 1986, Hodgkiss, 1996); (c) halfbeaks (Collette and Su, 1986); (d) gobies (Lam, 1990a,b), butterfish (Pan and Ni, 1996); and (e) Trichiuridae (Wang et al., 1992, 1993, 1994; Kwok and Ni, 1999) in the adjacent waters around Hong Kong. Further, there had also been Zoological Studies 38(2): 130–152 (1999), either in Hong Kong's territorial waters or in adjacent waters. In the taxonomic studies, all the family names were verified and reclassified (wherever necessary) according to Nelson (1994) and genera was followed after Eschmeyer (1990). The scientific names were mainly followed after Froese and Pauly (1997) and Eschmeyer (1998), while the synonyms were clarified based on Shen et al. (1993), Froese and Pauly (1997), and Eschmeyer (1998). However, the localities of the species, as determined in previous studies, were the main concern. Those not found in the related region (based on other references, e.g., Zhang, 1955; Zhu et al., 1963; South China Sea Fisheries Institute, 1979; Shen, 1984a,b; Chen and Zheng, 1987; Institute of Oceanology Academia Sinica, 1991, 1995; Huang, 1994), and also those whose sampling was questionable had been excluded. Furthermore, five species, viz., *Strongylura anastromella*, *Strongylura leiura*, *Strongylura strongylura*, *Tylosurus crocodilus crocodilus*, and *Tylosurus acus melanotus* (Belonidae), were added to the checklist.

All results of the studies were summarized with standard historical information pertaining to basic taxonomic research, like names of Class and Order, Family, Genus, Species (including those found in Hong Kong territorial waters, Hong Kong adjacent waters); contributors (researcher who had identified and/or published the fish species); keys for columns "Cited in," and so on.

Further, in addition to the surveys conducted by Ni (1994a,b, 1995a,b,c, 1997), a thorough literature review had been undertaken to perfect the checklist in terms of species coverage and accuracy. In addition, the reviewed literature also included published fish books (Chan, 1968; Hong Kong Agriculture and Fisheries Department, 1968, 1972); published fish lists (Fowler, 1930a,b,c,d,e, 1931a,b,c,d, 1932a,b,c, 1933, 1934a,b,c,d, 1935a,b,c, 1936a,b,c, 1937a,b, 1938a,b, 1939a,b, 1940a,b, 1941, 1949, 1953, 1954, 1956, 1958, 1959, 1960, 1961a,b, 1962; Shen, 1964), unpublished fish lists (Gaiger, 1974; Chan, 1984; Leung, 1994), unpublished fish identification guides/diagnostic keys (Chan and Wong, 1989), unpublished survey reports (Chong, 1984; Leung, 1994; Lee, 1997; Leung et al., 1997; Hong Kong Agriculture and Fisheries Department, 1998), and published papers (Chan, 1965, 1966a,b,c, 1970; Shen. 1967a,b; Chan and Tseng, 1982; Thompson and Horikoshi, 1982; Thompson et al., 1982; Wong, 1982; Wu, 1984; Richards and Wu, 1985; Collette and Su, 1986; Lam, 1986; Lam, 1990a,b,c; Leung, 1991, 1997a,b).

Some of the other significant studies include the following:

Chan (1966a), Chan (1966b), Chan (1966c), Chan (1968), Chan (1970), Chong (1984), Chan (1984), Hodgkiss (1988), Chan and Wong (1989), Lam (1990a), Lam (1990b), Lam (1990c), Leung (1991), Leung (1994), Hodgkiss (1996), Leung (1997a,b), Lee (1997), Leung et al. (1997), World Wide Fund—Mai Po Reserve publications, etc.

Notwithstanding the bewildering ichthyodiversity in Hong Kong waters, and supported by enormous number of publications, the issue which follows is the *modus operandi* for their efficient management. And, in this direction, there had been thoughts and efforts for an effective management of the resources. And, in this pursuance, **CBFM** in a **participatory** mode was adjudged as one of the most effective *modus operandi* for efficient management of the water bodies and their bioresources.

Hong Kong is a tropical country with rich resource of BD. In spite of enormous amount of database on ichthyodiversity exploration, it could be contemplated that more fish species could be found in Hong Kong waters because of the unique habitat and its hydrology. Incidentally, Hong Kong waters are a mixture of three coastal currents (the Chinese Coastal Current, the South China Sea Drift, and the Kuroshio), Pearl River runoff, and upwelling and downwelling. In addition, they are said to vary seasonally and geographically (Williamson, 1970; Morton and Wu, 1974; Tang and Ni, 1996).

Notwithstanding the above, the major objective of constructing the inventory of ichthyofauna (FW, Marine, and Estuarine) is to provide baseline data to stimulate other kinds of studies on Hong Kong's fish systematics, fisheries biology, environmental impact assessments, management, and, so on. There is no doubt that a number of fish species are still missing in the list, although they were believed to be present in Hong Kong's waters, and obviously could find a place in the Inventory of Fishes, for instance, the species inhabiting coral reefs (e.g., Families Labridae, Pomacanthidae, Pomacentridae, and Scaridae). These fishes are less likely to be collected with the help of conventional commercial fishing/sampling methods. Nevertheless, the checklist could, at least preliminarily, serve as an urgently needed baseline for the planning and implementation of management and conservation strategies for the fishes in Hong Kong.

In continuation of the above, the major difficulty in validating species from most of the references was that there were no preserved specimens to check. Thus, in order to construct a robust database of Hong Kong's marine fauna, one may have to establish a museum of Ichthyofauna (with proper storage) of Hong Kong's fishes (FW, marine, or estuarine) in order to validate and construct a robust database of Hong Kong's Fish BD. Nevertheless, the Hong Kong Agriculture and Fisheries Department had maintained a considerable library of specimens, in which all the scientific names of the specimens are said to be based on FishBase (Froese and Pauly, 1997) and Catalog of Fishes (Eschmeyer, 1998). Further, the genus names are based on Catalog of the Genera of Recent Fishes (Eschmeyer, 1990), and the family names are based on Fishes of the World (Nelson, 1994). Moreover the fish species are designated as those found in Hong Kong territorial waters, or Hong Kong adjacent waters. Fish species listed as new records to the area are noted.

Notwithstanding the above, fisheries are said to be potential resources for nutrition and avocation of the people of **Hong Kong**. There could be bone of contentions and conflicts with this coveted resource. This could often call for legislative issues and legal counseling (Anon, 1999). Moreover, contentious issues are quite common in economic affairs. Consequently, there could be many consultancy services in dealing with such apple of discords.

With reference to the above, in Hong Kong, there are said to be meetings held of the Legislative Council Panel on Economic Services. In one such meeting, held on September 28, 1998, members had been briefed on the major findings and recommendations of the Consultancy Study on Fisheries Resources and Fishing Operations in Hong Kong Waters (hereafter referred to as "the study") commissioned by the Agriculture and Fisheries Department (AFD). Members were informed that the administration would try to conduct a public consultation exercise, and, in the light of views received, could formulate a **fisheries management** strategy. The administration undertook report progress to members. In this connection, it may be noted that there were contributions which tried to summarize the views received during the consultation exercise, and also, make efforts to setout the approach, which AFD proposed to take in working out the way forward.

Concomitant to the above, there have been studies which indicate that fisheries resources in Hong Kong have declined. However, it may be noted that depletion in fisheries resources is an almost worldwide affair. In view of this, efforts have been made in different countries to regulate fishing operations in order to avoid overfishing and to re-stock fisheries resources as a measure to conserve these precious resources and to promote the overall sustainable development of the fishing industry.

Notwithstanding the above, in view of different types of difficulties faced with regard to social, economic, legal, etc., aspects in fisheries and AQC, as stated earlier, there had been attempts for finding an efficient method of **management and conservation** of the water bodies and their resources. And, in striving to find a way out in this aspect, **CBFM** in a **participatory** mode have been found to be an effective tool for the efficient management of the water bodies and their bioresources. Under the above background, the consultants had examined and prioritized 13 management options on the basis of ease of implementation, cost-effectiveness, and scientific robustness in order to conserve fisheries resources in Hong Kong waters, and with reference to the circumstances in Hong Kong. Of the 13 options, the following six **management** measures were considered to be of high priority and recommended to be implemented, at first: (a) Establishing a fishing license program, (b) limiting entrants, (c) setting up of nursery and spawning ground protection areas, (d) habitat enhancement, (e) habitat restoration, and (f) restocking.

Conversely, the medium and low priority **management** options were recommended to be considered later.

Notwithstanding the above, a consultation exercise had been carried out between September end and December end, 1998 to solicit views of the public, including major fishermen associations, major green groups, academics, relevant advisory committees, and all Provisional District Boards (PDBs), in order to help formulate a **fisheries management strategy** suitable for Hong Kong, as certain fisheries management measures were likely to be controversial.

5.1.6.1 Feedbacks obtained in general

Concomitant to the above, majority of the submissions (including those from some fishermen associations) were found, generally, to support the implementation of **management** measures in order to help to rebuild fish stocks in Hong Kong waters. Further, they also agreed with the consultants that the six high priority **management options** could, preferably, be implemented, at first. However, some fishermen associations had raised objection(s), raising apprehension(s) that the proposed fishing license program may limit the development of the fishing industry, and that the other fisheries management measures, notably, protection of nursery and spawning grounds, habitat enhancement, etc., may not be cost-effective, and may not achieve the desired economic value in the case of Hong Kong.

Concomitant to the above, there also exist divergent views on the extent and pace of implementation. In fact, the fishermen's main concern is the likely impact of the management measures on their livelihood. Moreover, some of them had also suggested that *ex-gratia* payment may preferably be made to the affected fishermen to compensate them for the possible loss of fishing grounds resulting from some of the **proposed management** measures.

On the contrary, green groups have pressed for a faster pace of implementation of fisheries management measures, and had suggested that additional measures be adopted to rectify the present less helpful situation.

Apart from the high-priority options, other options were also suggested in the submissions including (i) the control of the use of nonselective fishing methods (such as trawl and trammel nets) in certain areas, and (ii) the promotion of recreational fishing as an alternative employment for the fishermen.

5.1.6.2 Options with high priority

Establishing a fishing license program and limiting new entrants

The workers in the field had come across a number of aberrations with regard to operationalization of the **management *modus operandi* of fisheries** and AQC. As such, the consultants had recommended that the fishing license program may, preferably, be implemented as a first priority. The program is supposed to enable the government to take a more proactive role in managing fishing activities. It could also furnish with valuable information about fishing operations, such as gear type, catch, and area fished, which are said to be important for stock assessment and **fisheries management** purposes. In addition, the consultants had also proposed that fishing in Hong Kong waters could certainly be restricted to existing vessels with fishing licenses, upon the establishment of the fishing license program; and no license may be issued to new entrants. This may have to be done with a view to contain the number of fishing vessels at the present level. However, according to their opinion, this measure may have to be kept under regular review in the light of the prevailing fish stock levels.

In continuation of the above, it may be noted that majority of the submissions seemed to have supported the fishing license program *vis-à-vis* limiting new entrants. However, some fishermen organizations were said to have suggested that the licensing program may, preferably, be introduced in some selected areas only. Conversely, others had raised objection(s) to these two options, as they had apprehended that these could limit their fishing operations in Hong Kong waters, and thus, could affect their livelihood.

5.1.6.3 Setting up of nursery and spawning ground protection areas

Although sounds little harsh, it could be said that atrocities in the aquatic world are quite common today, which sometimes extends beyond the threshold and tolerable limits. Hence, comes into picture the issue of restricted localities and concomitant declaration of PAs or sanctuaries, notably the "aquatic sanctuaries." This option involves either restricting or prohibiting all forms of fishing in areas, where commercially important ichthyospecies spawn, or where there is an abundance of juvenile fishes. Incidentally, the consultants have identified most of the eastern and southern waters (said to be amounting to about one-third of Hong Kong waters) as spawning and nursery grounds, and also had recommended for their protection. As such, there is need to make amendments to the Fisheries Protection Ordinance, in order to make room for legal backup for the implementation of the proposal. This could affect efficient management of the water bodies and their bioresources.

Some fishermen organizations and PDB members had suggested that the establishment of nursery and spawning protection grounds may, at first, be introduced in some selected areas only, while the other fishermen organizations were said to be little sceptical about such an option, as they were anxious about its impact on the livelihood of fishermen.

5.1.6.4 Habitat restoration and Habitat enhancement

Habitats, by and large, are degraded today, and hence need enhancement. With regard to habitat enhancement, the consultants were said to have recommended implementation through the deployment and management of the artificial reefs (ARs), which could effectively rebuild fish stocks in Hong Kong waters. Concomitantly, on habitat restoration, the consultants had recommended the earliest opportunities to restore disturbed habitats whenever and wherever possible. Moreover, Environmental Impact Assessments (EIA) could form part and parcel of habitat inventory studies. Although largely supportive, there had been questions regarding the cost-effectiveness and the economic value of these measures in enhancing fish stocks in the case of Hong Kong.

5.1.6.5 Re-stocking exercise

It had been reported that restocking, with proper **management**, could help in rebuilding the depleted fish stocks (resources), and hence could be said to be a useful tool for resource enhancement. However, this option could be considered as complementary to other high priority options, such as habitat enhancement or protection of sensitive areas. Restocking is now employed in places like Japan, Taiwan, etc. Juveniles of commercial species grown in hatcheries are released into the sea for later capture as adult fish. In view of the mixed results observed with programs of this sort in Japan, the program was not much encouraged.

Notwithstanding the above, there were said to be divergent views on this option. Some fishermen organizations were said to believe that restocking is the most effective means of rehabilitation of the depleted resources, and its implementation could provide good incentives for the fishermen to cooperate and support the **fisheries management** strategies. On the other hand, some people believed it to be cost-intensive and had questioned its cost-effectiveness and economic value. Moreover, the local green groups had cautioned that restocking be attempted only in a well-managed fishery.

Notwithstanding the above, the consultants had recommended for a trial to be conducted before embarking upon such a so-called expensive program. The trial could examine the best time, location, and species for restocking and the return rate of the released fish.

5.1.6.6 *Medium priority and low-priority options*

"Options" in any arena are said to be "open." They could be selected/accepted according to priority or need. In this regard, it could be said that the medium priority options show promise, but may have obstacles, which are to be overcome before they could be implemented. Incidentally, the options, which could be ranked as medium priority, include the operationalization of a demersal trawl ban, introduction of fishing quotas, mesh size regulation, establishment of pelagic fisheries protection areas, and so on. The consultants seemed to have agreed with the above viewpoints. However, the consultants do not seemed to have recommended the implementation of low-priority options, including restricting age/size at first capture, banning unselective gears, sea ranching, etc., mainly due to drawbacks including cost implications, lack of sufficient scientific information, etc. On the contrary, the consultants, in some of the discussions, seemed to have emphasized on a faster pace of control over fishing activities, and with concomitant implementation of some of the medium priority measures at the beginning.

5.1.6.7 *Way forward*

With reference to the above, given the divergent views received during the consultation period, AFD was said to have set up a working group (WG) on **fisheries management** to advise on matters relating to the formulation and implementation of fisheries management strategy in Hong Kong waters. Such matters had included, among others, the case for implementing the six **high priority management measures** recommended by the consultants as well as the extent and pace of implementation. The WG comprised representatives from the fishing community, green groups, academic and relevant government departments/bureaux, etc. The WG had also the mandate to formulate strategies for **effective management** of the fisheries.

Notwithstanding the above, there had been attempts for finding an efficient method of management and conservation of the water bodies and their resources. And, in striving to find a way out in this aspect, **CBFM** in a **participatory** mode had been suggested as an effective tool for efficient management of the water bodies and their bioresources.

Notwithstanding the above, AQC and fisheries are said to be highly potential in providing with rich nutrition and substantive avocation to the people. In **Hong Kong** also, culture and capture fisheries are said to make a significant contribution by maintaining a steady supply of fresh fish to local consumers (Anon, 2018). The total capture fisheries and production through marine fish culture amount to *c.* 28% of the total seafood consumed in Hong Kong. On the other hand, pond fish farmers produce *c.* 3% of the total freshwater (FW) fish eaten.

5.1.6.8 *Capture fisheries*

Hong Kong's fishing fleet comprises *c.* 4000 vessels, and almost all of these are mechanized. Approximately 33% of the vessels are said to be > 15 m in length. They are mainly trawlers, liners, and gill-netters, and are believed to be engaged in fishing mainly outside Hong Kong waters along the northern continental shelf of the South China Sea. Incidentally, the estimated production was *c.* 170,000 tons in 2013 valued at *c.* US $2.3 billion. Major species of fish caught were hairtail, mackerel, scad, big-eye, pomfret, and croaker.

5.1.6.9 *Culture fisheries and aquaculture*

In Hong Kong, total area of fish ponds in use is *c.* 1150 ha, and are located mainly in the Northwest (NW) New Territories. In fact, most of the fish farms are engaged in carp polyculture (bighead carp, silver carp, common carp, and grass carp) in combination with tilapia or grayy mullet. However, the other cultured species include the jade perch, sea breams, and spotted scat. Total fish production from pond was said to have amounted to *c.* 2187 tons, valued at *c.* US $44 million, in 2013. Further, *c.* 990 licensee units were engaged in marine fish culture in the *c.* 26 fish culture zones designated under the Marine Fish Culture Ordinance. The common species cultured include the green grouper, brown-spotted grouper, Russell's snapper, mangrove snapper, **cobia**, and pompano. Incidentally, the total marine fish culture production was said to have amounted to *c.* 1005 tons, valued at *c.* US $94 million, in 2013.

Hong Kong, geographically, consists largely of steep, hillsides. Further, *c.* 7 km^2 of land are actively farmed. Farms are usually small in size and produce mainly leafy vegetables, pigs, and poultry.

5.1.6.10 Consumption

Incidentally, the daily fresh food consumption by Hong Kong's population of >7.0 million, of late, is *c.* 913 tons of rice, *c.* 2270 tons of vegetables, *c.* 4580 head of pigs, *c.* 53 head of cattle, and *c.* 34 tons of poultry. In fact, much of this is imported. However, Hong Kong's primary producers help to satisfy some of the demands.

Approximately 1150 ha of fish ponds, located mainly in the NW New Territories, are engaged in AQC. Incidentally, the total pond fish production in 2013 amounted to *c.* 2187 tons, valued at *c.* US $ 44 million.

The total capture fisheries and marine fish culture production were equivalent to *c.* 28% of seafood consumed in Hong Kong, while the pond fish farmers produce *c.* 3% of the FW fish eaten.

5.1.6.11 Development and services

It is believed that AFD tried to pursue a number of fisheries conservation and **management** measures in order to promote sustainable development of the fishing industry and to conserve fisheries resources in Hong Kong waters. Moreover, with a view to enhance the fisheries resources, the department has deployed 668 units of artificial reefs with a total volume of 179,000 m^3 in Hong Kong waters, apart from strengthening enforcement against destructive fishing practices. **Further, results** of the underwater monitoring survey had revealed that the deployed artificial reefs had supported a higher diversity and abundance of fish species, as compared with the nearby natural habitats. Incidentally, >220 species of fish, including many high-valued species, such as groupers, breams, snappers, and sweetlips, use reefs for feeding, nesting, and as spawning and nursery areas. In addition, continuous emphasis is said to be laid on helping the fishermen to switch over to sustainable fisheries and other related operations through provision of technical advice, training, liaison services, grants, and credit facilities. Incidentally, the AFD was said to have organized 14 free-of-charge training courses for fishermen to enhance their knowledge of sustainable fisheries development and **management,** and had also provided training allowance to eligible participants.

Concomitant to the above, the AFD is said to have been continuing with the implementation of the voluntary **Accredited Fish Farm Scheme** in a bid to enhance the environmental hygiene standards of local fish farms and the quality of cultured fish. Moreover, AFD had conducted AQC studies and provided technical services related to the development of efficient pisciculture methods and husbandry techniques. In addition, AFD is believed to continue with the promotion and development of local fish hatchery and nursery. Further, AFD is said to organize training programs for fish farmers on various aspects of fish hatchery and management.

In addition to the above, AFD has put in place a fish health management program, in order to help the fish farmers in preventing and managing fish diseases. Notwithstanding the above, "Good Aquaculture Practices Program" aims to promote modern fish farm management. Under the program, regular farm visits were arranged to collect water and fish samples for analysis. Further, the department also introduced modern farm management techniques and practices to fish farmers through workshops and seminars. Perhaps, last but not the least, AFD has been continuing to administer various loan funds for fishermen and fish farmers for production and development purposes.

Concomitant to the above, the department was said to administer the Fish Marketing Organization(s) which is a nongovernment, nonprofit-making corporation vested with the Director of Marketing (currently the Director of Agriculture, Fisheries and Conservation) to provide orderly fish marketing services to both the fishermen and the fish traders.

Notwithstanding the above, overfishing seems to be a common phenomenon anywhere in the world today. Declining species diversity, depletion of coveted bioresources and dwindling of yield are *in vogue* in almost every nook and corner of the globe today. In Australia, it is said to be Patagonian toothfish, abalone, and the resources of the Great Barrier Reef; in Asia, shark finning; in the Caribbean, Garoupas; in the Pacific, Tuna; and in the Atlantic, Cod, to name but a few. We have come to regard over fishing as a fact of life, but in a few places, it has become a major social problem.

In continuation of the above, there was a steady decline in fish catches from **Hong Kong** waters over the last 25 years approximately (Morton, 2005). In 1998, the Agriculture and Fisheries Department of the Hong Kong Government has instituted a commission of consultancy and study to assess the situation and identify ways to remedy any identified problem. The study has found that the catch had indeed fallen by *c.* 50%, and that fish fry production had decreased by *c.* 90% over a few years. Moreover, 12 of the 17 commercial fish species assessed were said to be "highly overexploited" and the remaining 5 "fully exploited." The local fishery was further shown to have shifted from one dominated by large, slow growing, high value species to one comprising small, faster growing, and low value species. To understand how this situation had come about, one may have to delve into history.

Prior to the World War II, Hong Kong was said to be a remote colonial outpost, far from the gems, such as India and Malaysia, under the imperial crown of the British Empire. Thus, artisanal fishing could easily provide the few people then resident in the infant colony with nutrition and avocation needed through fish and fishery. The scenario changed when, in 1928, one of the most important persons in the history of colonial Hong Kong, G.A.C. Herklots, was appointed as a Reader in Biology at the University of Hong Kong. Herklots had started developing the subject of Biology in Hong Kong by inviting to it various marine and other scientists and founding, in 1930, The Hong Kong Naturalist. In 1937, Herklots was given a small grant by the Hong Kong Government to set up a Fisheries Research Unit within the University of Hong Kong. This was mainly because the prewar, local fishing vessels were unable to compete with an increasingly aggressive and mechanized Japanese fleet. Although the Fisheries Research Station was not formally established within the newly founded Department of Biology of the University until 1940, subsequently, Herklots, in cooperation with S.Y. Lin, the Superintendent of Fisheries Research, had begun, in 1937, to salt dry the fish in response to the Japanese invasion of China, and for other impelling reasons. It was called "siege fish"; and the government encouraged its stockpiling. Herklots also produced a fisheries research publication, named the Journal of the Hong Kong Fisheries Research Station. Subsequently, the University's Fisheries Research Station had reappeared as a division of the Department of Agriculture, Fisheries and Forestry, *albeit* with laboratories within the University, after the war and with the liberation of China and Hong Kong in 1946. Such a situation could not and did not last long, and in 1952, a Fisheries Research Unit was inaugurated as a subdepartment of the Department of Zoology of the University. A research vessel, the r.v. Alister Hardy was launched in 1953. Between 1952 and 1960, fisheries research results were published in the Hong Kong University Fisheries Journal.

Incidences of "fish-kills" or "fish mortalities" are said to be of quite a common happening today, sometimes could be on an enormous magnitude and exponential scale. The possible reasons could be multifarious, like pollution, toxication, asphyxiation, or, could be due to *hitherto* unknown virulent, enigmatic **Epizootic Ulcerative Fish Disease Syndrome (EUS),** which has been sweeping the freshwater (FW) fishes in an epidemic dimension, unhindered, unimpeded, and unabated, almost semi-globally today, causing large-scale mortality among them, rendering many of them endangered; causing fear psychosis among the fish-eating people; causing untold misery to the fishermen, fish farmers; and devastating the economy of the nation (Kar, 2007a,b,c, 2013, 2015). In addition, with the opening up of China following the end of the Cultural Revolution, Hong Kong, at first, the only door into this new commercial arena, has literally boomed as real estate prices and investment bases have risen and broadened. Subsequently, with the fast development of Hong Kong, the need for a more modern infrastructure had arisen, and this was accommodated on reclaimed land.

Concomitant to the above, the fishermen, capture and mariculture alike, have blamed pollution, dumping, dredging, reclamation, etc., for the loss of their inshore incomes, and such intensive activities in such a small body of territorial water (1650 km ^{2}) certainly had been a contributing factor to the decline in fish catches. But, it is not the main reason. Further, the recreational fishermen are also to be advised to consider about the declining ichthyodiversity as well as the yield trend, while going for recreational fishing on a large-scale.

5.1.6.12 *Fisher folk, their livelihood, their struggle for existence*

The hard truth is that the fishermen, like the farmers, generally have to live a difficult life in a world that is continually changing. The price of fish goes up and down, sometimes by leaps and bounds. In the world of AQC and fisheries, there could be glut of cultured prawns; the capture fishery, which often collapses; one government may subsidize its fleet, while another may not; limits of territorial water may get bigger; sea areas may be closed, and may be opened again; the weather may be salubrious or inclement; the price of fuel may go up and down; the economy may grow or subside; crews are hard to enlist, even, the crew may be sick. Under these circumstances, the life for a fisherman is not just hard, it is dominated by an array of uncertainties and vagaries, unlike ours (non-fishermen), which is said to be relatively ordered and organized, and, as a consequence, the human being simply does what he feels best for that moment. He may fish anywhere and everywhere he can in the hope that he can maintain a living from his catch. He argues against all legislations to control him because any such restrictions lessen his spectrum of fishing options. Thus, if he can, he may violate the law, may be by working under the cover of darkness, falsifying his catch records, buying unlawful fuel, hiring illegal crews; in fact, may do anything to make a living, sometimes, in a shrinking marketplace. However, fishermen may not certainly be regarded as "criminals"; their actions could simply reflect their need in today's world to be so-called opportunistic and as frugal as possible in the face of hardship, even though the fishing masters might have brought it on themselves. It may be mentioned that in Hong Kong, it is still the capture fishermen, who put *c.* 95% of the fish on our plate, often under conditions of great hardship and danger, generally, at a price we could afford, and thus we are to offer our respects and regards to him. Especially, in many parts of the world, the fishermen have gone from being ignored, lionized, derided, and vilified.

Notwithstanding the above, Hong Kong could learn from the European experience because in reality, its fishing problems are simpler, in comparison to the multinational complexities of the European Union's fishery policies. The global experience has, moreover, collectively identified a suite of economic and political measures and innovative and practical strategies, which could be used to manage any fishery. However, whether they would be successful or not is largely a result of one simple human characteristic, i.e., goodwill. Incidentally, the conservationists often have a very poor understanding of a fisherman's quite difficult life; never, for example, had probably experienced; neither having been on a vessel nor talked to a fishing family. Conversely, the fishermen, because of their adherence to tradition among a tightly knit group of people with almost similar backgrounds, often believed to have a poor understanding of the fast changing world. Many, even today, are illiterate; and mainly because of a lack of understanding of the issues and concerns expressed by others, simply try to cling to the past and what they know worked for their fathers.

In addition to the above, the fishermen are resistant to change, especially where it affects their perceived right to free access of what they regard as their. But, they are required to change themselves under the present scenario, and what is needed are new initiatives that could help them to understand and appreciate that a new order is inevitable, and that it is better to work with, rather than against, the other components of an increasingly complex society, who have other suggestions for the use of the world's aquatic domain.

On the one hand, fishermen fail to recognize that fisheries resources are in fact the common property of every Hong Kong citizen while the green groups fail to understand the long history and pride of local fishermen who exploit the resource, admittedly for their own livelihood, and concomitantly, the residents of this city have fresh fish in their markets every day. One cannot simply remove the income of innumerable fishermen and their dependents. At the same time, the so-called benefits, which conservation and sustainability may bring consequential socioeconomic problems.

Notwithstanding the above, as already stated, AQC and fisheries are important and potential arenas for nutrition and avocation of the people of Hong Kong. At the same time, the devastating damage to the water bodies and wanton destruction of the aquatic bioresources are posing threats to the biosphere. Under this debilitating scenario, production, management, conservation, and sustainability are some of the keystone words, which are indispensable for enriched BD and production of bioresources in the globe. Many governments (Govts) have set up different committees on sustainable fisheries since long time. Concomitantly, the Government of Hong Kong had decided to study the long-term directions and goals as well as feasible options and strategies for the sustainable development of the local fisheries industry in **Hong Kong** (Anon, 2010a,b).

The committee, after careful consideration of the situations of the international and local fisheries industries as well as the contributions of the local fisheries industry to Hong Kong's economy and society, had agreed that the local fisheries industry had altered in tandem with Hong Kong's economic development. However, the local fisheries industry, as an important local cultural asset with a long history is sufficiently valuable and has potential for sustainable development.

Notwithstanding the above, the principal goal of the sustainable development and management of the fisheries industry are to enable the fishermen and fish farmers to achieve self-reliance, help them to maintain their own livelihoods, provide job opportunities to the fisheries community, cope with the changing business operating environment, and probably, last but not the least, to ensure the supply of fresh and quality fisheries products to the people of Hong Kong. Moreover, the human society could certainly benefit from a sustainable fisheries industry, as it is supposed to assist in establishing and maintaining an overall healthy aquatic environment that could be abundant in resources, and that could be passed on to our posterity.

Concomitant to the above, the committee has formulated two directions for promoting the industry's sustainable development in order to achieve the goals of the sustainable development of the fisheries industry as follows:

(i) To assist the fishermen to develop or switchover to modernized and sustainable modes of operation, and (ii) to protect, conserve, and rehabilitate the coveted aquatic ecosystem and fisheries resources.

Further, the committee has proposed the following measures in order to promote the sustainable development and management of the local fisheries industry:

(i) To promote modernization of the fisheries industry and sustainable modes of operation.
(ii) To control fishing effort of capture fisheries.
(iii) To conserve and enhance the fisheries resources.

Concomitant to the above, the committee finds the technology employed by the fisheries industry in Hong Kong as not advanced enough, resulting in high operating costs *vis-a-vis* low efficiency, and hence low competitiveness. Therefore, it recommended that the government should promote modernization coupled with sustainable development of the fisheries industry and assist fishermen to develop or switch over to modernized and sustainable modes of operation, management and businesses by adopting the following measures:

(a) Enhancing fishermen's understanding of sustainable fisheries management.
(b) Promoting fishermen cooperative enterprises for cooperative or community management.
(c) Providing with stable financial support to sustainable fisheries development and management programs.

(d) Reviewing fisheries loans requirements and approval procedures.
(e) Implementing specific categorical support measures for sustainable fisheries operations and management, for different categories of fisheries programs, e.g., AQC, recreational fisheries, fisheries product logistics and processing, etc.
(f) Strengthening communications with the mainland counterparts to help fishermen seek business opportunities.

With reference to the above, the catch and fishing effort of capture fisheries have far exceeded, at present, the experts' estimates of the MSY and optimum fishing effort OFE). As such, the committee feels that the abovementioned measures to promote modernized and sustainable operations and management alone cannot solve the problems. And hence further measures are to be in place to control the fishing effort of local capture fisheries. Thus, the committee proposes the following measures to regulate the fishing efforts to a sustained level:

(i) Maintain an appropriate number of fishing vessels, (ii) prohibit fishing by nonlocal vessels in Hong Kong waters, (iii) ban trawling in Hong Kong waters, (iv) help the trawler fishermen to switch to other operations, and (v) strengthen fisheries management and law enforcement.

In addition to the above, the committee also proposes certain measures to protect important marine and freshwater (FW) fisheries ecosystems, such as fish spawning and nursery grounds, and to promote the rehabilitation, growth, and efficient management of fisheries resources in Hong Kong waters. The principal measures include the following:

(i) Strengthening the regulation for coastal and marine development projects, (ii) designating fisheries protection areas (FPAs), and (iii) prohibiting commercial fishing in marine parks.

In continuation of the above, the committee considered that the current excessive fishing effort could be reduced to a level below the MSY through the above-proposed options and measures, and thus, the important fish spawning and nursery grounds in Hong Kong waters could be conserved. In addition, such proposed measures could also provide fishermen with choices, including assisting trawler fishermen for switching over to other kinds of operations. However, those who continue to engage in capture fisheries may, however, benefit from the improvement and recovery of marine/FW fisheries resources. As a result, the whole fisheries community may be in a better position to cope with the changing operating environment and achieve self-reliance.

Notwithstanding the above, the committee had consulted the stakeholders on the proposed options and measures. Although, the stakeholders may hold different views regarding the individual measures, they, generally, may accept the long-term goals and directions, and also may agree with the positive impacts resulting from the proposed measures. Concomitantly, the committee had recommended that the government could conduct a more comprehensive consultation with the industry and stakeholders, while studying and formulating the implementation details of the measures as well as could strive to consider and balance the views of different parties, with the aim of obtaining wide support for the proposals.

In fine, the committee had suggested that the proposed options and measures could be implemented in phases. Highest priority may be given to the proposals that could be implemented using existing resources or within the existing legal framework. Conversely, the government could conduct consultations on the proposals for regulating the fishing efforts and for enhancing the management strategies and conservation of fisheries resources. Further, the government may amend the relevant legislation(s) in order to enhance the management scenario and conservation of marine and FW ecosystems and fisheries resources as well as to implement the proposals on maintaining an appropriate number of fishing vessels, prohibiting nonlocal fishing vessels from fishing within Hong Kong waters, prohibiting commercial fishing in marine and FW parks, designating FPAs and banning trawling in Hong Kong waters, after reaching consensus and formulating the appropriate measures for assisting the affected fishermen.

Concomitant to the above, in order to tackle this problem, the World Trade Organization (WTO) had established a negotiating group, with the aim of initiating negotiations to abolish subsidies not conducive to conservation of fisheries resources and sustainable exploitation. The WTO also considered that, in the absence of a **fisheries management system** among its member countries, all subsidies granted to the fisheries industry would be regarded as unfavorable to the sustainable fisheries development, and in future will be subject to constraints.

Notwithstanding the above, there had been recommendations for the establishment of "Sustainable Fisheries Funding Scheme" for the purpose of providing financial assistance for projects and researches conducive to the sustainable development of the fisheries industry. Such works may be entrusted with certain locally registered NGO-type nonprofit making fisherman organizations, academic institutions, fisherman cooperative enterprises, trustworthy private organizations, and so on. Moreover, if so required, the government might even consider allowing fishermen or fisherman's organizations to manage fishing grounds, while the government and academic institutes could provide technical advice. In the long run, the fisheries industry could be properly managed and the industry and the fishermen's culture could be sustained. Concomitant to the above, AQC products are said to be available quite on a large scale in the market now a days. But, product's safety is of utmost importance today, mainly because of increasing public concerns over food quality and food safety. Further, the competitiveness of local fisheries products could be enhanced, mainly because of the rising demands for good quality local fisheries products. Thus, the development of local AQC could further be boosted through improvement in the AQC management, culture techniques, product quality, and food safety. The public may also enjoy quality, healthy, safe, diversified, and stable supply of local fisheries products. Moreover, the future development of the AQC industry may, preferably, be based on local species, with less dependence on wild fish fry and fish feed. A stable supply of quality fisheries products, which are farmed using environmentally friendly, low contamination, high-technology and high-quality culture techniques, could be available in the market. Through branding strategies and stringent hygienic standards, the local quality products could be supplied to local markets, the mainland, and even international markets. This could enhance AQC's operating environment. Further, a relook is necessary for regulation of sport and recreational fishery with an eye on management and conservation of fishes and water bodies.

With reference to the foregoing account, it appears that there is a need for an effective strategy for the efficient management of the water bodies and their bioresources, notably, the fishes. In this direction, efforts made had led to the thinking of implementation of efficient *modus operandi* of management, notably the **CBFM** involving the stakeholders from the grassroots level in a **participatory** mode.

Concomitant to the above, many of **Hong Kong**'s wild terrains, with their great BD, seemed to have attracted much attention in recent years. Indeed, also harboured within are some lesser known natural treasures of Hong Kong: multifarious rock formations and geomorphological features (Yeung, 2008). There have been contributions which discussed the ongoing works on bio- and geodiversity by the Agriculture, Fisheries and Conservation Department(s). Significantly, geo-conservation alone, cannot rely purely on the work of the government, and there had been suggestions for integrated management, knowledge building, public support, and, most importantly, a wider dissemination of knowledge. In continuation of the above, it may be stated that most people's impression of Hong Kong is that it is a crowded concrete jungle with little else to offer besides tower blocks. This belies the fact that Hong Kong also has sylvan environments of stunning natural beauty. Incidentally, >50% of the land is said to be wild mountains and ravines in this small territory of *c.* 1100 km^2. Moreover, no one visiting Hong Kong's countryside could fail to be impressed by its bewildering diversity. Probably, every turn of a trail brings new and fascinating vistas; rugged uplands, deep ravines, coasts and islands; each of them said to be magnificent in its own way. Thus, we come to realize the greatness of nature's power relative to our own, after relishing the beauty of our countryside.

Concomitant to the above, geodiversity is said to be an essential element of our stunning scenery on the coast and inland. Concomitantly, Hong Kong is said to be rich in natural resources and endowed with bewildering BD. Hong Kong, being located at the edge of the tropical zone, possesses indented coastlines together with a mountainous topography, which provide a wide range of habitats for an impressive number of different species of animals and plants. Incidentally, >40% of the total land area in Hong Kong is designated as country parks or other special conservation areas. These are said to be managed for the purpose of nature conservation, education, recreation, nature-based activities, and scientific studies. Various types of habitats, such as woodland, shrubland, plantations, reservoirs, rivers, streams, and mangroves are found within these PAs. All these habitats are said to be important to the local flora and fauna. Nearly all animal species recorded in Hong Kong could be found in the PAs, including: mammals (*c.* 100%), birds (*c.* 100%), reptiles (*c.* 95.7%), amphibians (*c.* 100%), freshwater (FW) fish (*c.* 100%), butterflies (*c.* 98.2%), and so on. Incidentally, phenomena like soil erosion, soil pollution, habitat destruction, habitat alteration, etc., could have cascading deleterious effects on the ecosystem and the BD (Marafa and Chau, 2006a,b). Moreover, ecologically, data on changes in habitats and species populations, the numbers of keystone species, etc., could provide with indices for showing whether nature-based activities had affected the natural environment. Concomitantly, flagship conservation measures, like habitat restoration, habitat enhancement, declaration of protected areas (PA), etc., particularly, through community management measures could help in finding an answer to formulation of effective conservation strategies. At the same time, conservation of the BD, notably, the ichthyodiversity, enshrined in the ecosystem, could also be taken care of with regard to its management and conservation through efficient management *modus operandi*, like, the **CBFM**, involving the stakeholders from the grassroots level in a **participatory** mode. In fact, both biodiveristy (BD) and geodiversity (GD) could effectively be managed through integrated management. And, it is said to be essential in view of the fact that the bedrocks, boulders, cobbles, gravels, and fines, which constitute the substratum of a lotic system, particularly, in its rheophilic upstream region, are essential and prerequisites for the sustenance of the coveted hill-stream fishes (Kar, 2007a,b,c, 2013; Kar and Sen, 2007).

Notwithstanding the above, it is quite known that there has been depletion of fish diversity and fish stocks in different parts of the globe. **Hong Kong** is not free from this. As such, the Government of the Hong Kong (GoHK)

Special Administrative Region (SAR) has undertaken to deploy artificial reefs (ARs) in order to enhance the fisheries resources and conserve the marine environment, as part of an ongoing effort to rehabilitate depleted local fish stocks (Clarke et al., 2002).

A two-stage consultation program had been conducted, which aimed at soliciting inputs from fishers and minimizing potential conflicts with existing fishing operations using local fora and information materials. Salient inferences of the consultation had included recognition of the need for management and enforcement to improve fisheries resources, any difference(s) in support for artificial reefs between trawlers and small-scale fishers may be minimized and resolved through mutual dialogs, etc. The consultation program, further, has provided a model for future **community involvement** initiatives; and this could be said to be a glimpse of introduction to **CBFM**, which could be regarded as a welcome beginning in both FW and marine sectors.

Hong Kong is said to be rich in AQC and Fisheries. The country has transformed from a small fishing and trading port into one of Asia's largest and most powerful economic and transport hubs over the past century. Nevertheless, fishing had traditionally been, and continues to be a significant industry and a way of life for many inhabitants. In fact, it may be mentioned that a large proportion of *c.* 19,000 fishers based in Hong Kong are believed to be descendants of the original "Tanka" and "Hoklo" fishing communities, and still maintain strong and localized community networks. As already indicated earlier, "fish" happens to be a potential source of nutrition and avocation of the people. In Hong Kong, the significance of fisheries products to the population's diet is illustrated by the high consumption rate of fish (*c.* 58 kg/person/year), which is 2–3 times higher than that of most of the western countries (FAO, 1999).

Notwithstanding the above, many aspects of the fishing industry have changed radically in the past few decades. Many of the small-scale fishers, who fish from small sampans and often use artisanal gears, are unable to exploit offshore waters, and have, consequently, turned to raft-based mariculture to supplement or replace their income from capture fisheries. Concomitant to the changes in the fishing industry, fish stocks in Hong Kong waters had been subjected to a variety of pressures and impacts. Incidentally, the government had been pursuing a policy of large-scale land reclamations since the late 1980s. This, consequently had resulted in the loss of feeding, spawning, and nursery grounds of the demersal and pelagic species of fishes. In addition, the coastal water quality worsened during the same period, owing to accelerated construction activities and increased urban and industrial sewage and waste disposals (Yung et al., 1999; Wu et al., 1998). Concomitantly, the available information on the species composition of local fish stocks, despite limited historical fisheries records, seem to indicate a trend of declining number of commercially valuable inshore demersal species, and a reduction in size of those species still persist (Pitcher et al., 2000). In other words, decline could be said to be due to urban development and pollution coupled with fishing pressure.

Concomitant to the above, the seabed is subjected to heavy trawling pressure (Valente et al., 1996; Leung, 1997a,b; Selby and Evans, 1997). Such a view was also supported by a recent stock assessment, which also had concluded that the current poor status of fish stocks is primarily be due to less recruitment and growing trend of overfishing, and to a lesser degree due to development-related effects (ERM, 1998). Nevertheless, the fishing community, regardless of the reason for the decline, has grown increasingly vocal in its complaints about pollution and other development-related effects. As a result, the government had instituted a system of ex gratia allowances, which had been granted to the fishers to offset the impacts of marine/FW development. In view of the above, an artificial reefs (AR) deployment program was initiated by the Hong Kong SAR Government in 1996 in response to a growing awareness of the significance of both enhancing fisheries resources and conserving the marine environment, with probable concomitant initiatives in the FW sector as well. Concomitant to the above, obviously, the key stakeholders were the local fishers, who not only could be the most affected but also could have the greatest influence on the ultimate success of the program. Incidentally, international and regional experiences have demonstrated that if fishing effort cannot be managed, AR may serve only to aggregate the remaining fish and the benefits of stock enhancement may not be realized (Pitcher et al., 2000; Polovina, 1991; Morton, 1996). Thus, one of the goals of the consultation was, therefore, to investigate the interest in, and opportunities for, **community involvement** in management and enforcement. In addition, there had been more consultation(s) before the final recommendations for AR deployment and further formulations (Kennish et al., 2002). In this connection, it may be opined that no-take reefs could be preferred to fishable reefs from a stock-enhancement perspective. However, limiting the size of totally PAs may be counterproductive, as had been opined by some workers (Sadovy, 1999; Pitcher et al., 2000).

Notwithstanding the above, involvement of the **community** could be regarded as a beginning of the concept of **CBFM** involving the stakeholders from the grassroots level in a **participatory** mode.

5.1.7 India

India has extensive wetlands of *c.* 2.02 lakh ha. These are low-lying areas and are generally situated along the major rivers basins like the Ganga, Brahmaputra, Barak, Godavari, Cauvery, and Krishna river basins. They are one of the most important fishery resources of the provinces of Assam, West Bengal, Bihar, Manipur, Arunachal Pradesh, Meghalaya, etc. They provide livelihood support to innumerable poor people, which include numerous fishermen. Incidentally, Assam has >1.0 lakh (1 lakh = 100,000) ha of wetland area, which include the perennial, seasonal, and river-formed oxbow wetlands, which are locally called "*Beel, Haor*, and *Anua*" respectively (Kar, 2007a,b,c, 2010a,b,c, 2013; Shrivastava and Bhattacharya, 2003; Sugunan and Bhattacharjya, 2000; Vass, 1997; Barman et al., 2008).

Incidentally, India has a long coastline of *c.* 8119 km; a continental shelf of *c.* 0.5 million km^2; an extensive EEZ of *c.* 2.02 million km^2; *c.* 1.24 million ha of brackish water area; *c.* 1,91,024 km length of rivers and canals; *c.* 3.15 million ha of reservoirs; *c.* 2.25 million ha of ponds and tanks; *c.* 0.82 million ha of Beels, oxbow lakes, and derelict water bodies; *c.* 0.24 million ha of flood plain wetlands; *c.* 0.29 million ha of estuaries; *c.* 1.65 million ha of mangroves, swamps, lagoons, etc. (Ayyappan and Diwan, 2004).

Concomitantly, the Northeast (NE) India, in general, and the province of Assam, in particular, is bestowed with myriads of aquatic wealth with the total area of *c.* 3.89 lakh ha. The Beels (perennial wetlands), Haors (seasonal floodplain wetlands), Anuas (river-formed oxbow wetlands) constitute the most potential wetlands of the province covering >1.0 lakh ha of water-spread area. Incidentally, there are *c.* 1392 wetlands in Assam (Dey, 1981), of which *c.* 423 are registered and the rest *c.* 969 are unregistered and under the control of both government (*c.* 505) and private ownerships (*c.* 464). These water bodies serve as significant fish habitats providing nutrition and avocation not only to the people of the NE India and the province of Assam but also to a big part of the nation as a whole. Moreover, these pristine lentic systems also serve as sources of water for drinking as well as for industrial purposes. The wetlands are considered to be one of the most productive ecosystems owing to their characteristic interactions between terrestrial and aquatic systems.

Notwithstanding the above, as stated above, the province of Assam is gifted with a large number of diverse types of wetlands. Before embarking upon the management aspects of the wetlands of India, an account of some of the typical wetlands of the region are described in the following in order to project an overview of the physicochemical and biological characteristics of the wetlands (Kar, 1990, 2007a,b,c, 2013).

5.1.7.1 Sone Beel wetland

Sone Beel (Plate 1, Fig. 1) is situated at an altitude of 23 m MSL, between 92°24′ 50″–92°28′25″E and 24°36′ 40″–24°44′30″N within Karimganj district of Assam and falls in a valley geologically called syncline.

PLATE 1 The Sone Beel, the biggest wetland in Assam.

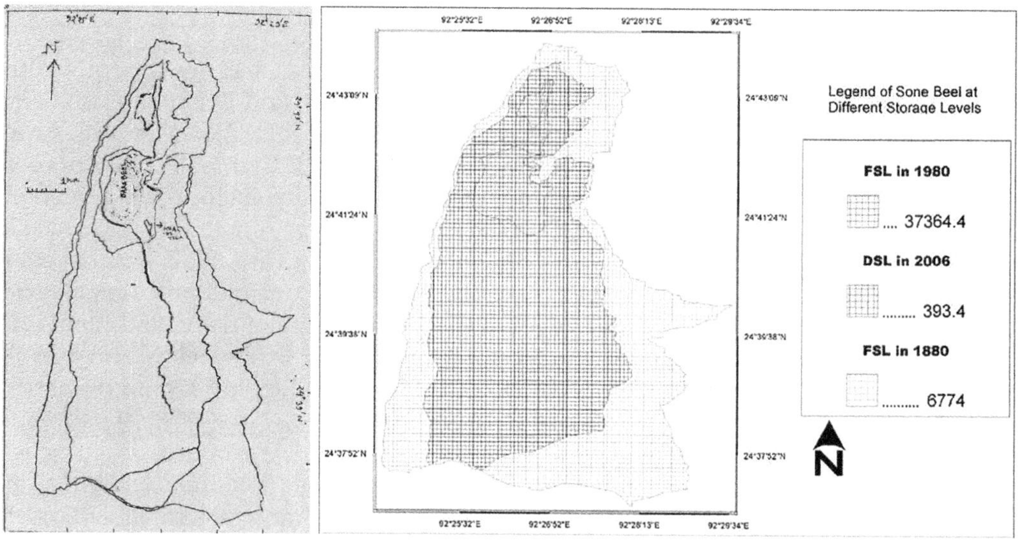

FIG. 1 Maps of Sone Beel.

The physiography of the district consists of small hillocks intervened by wide low valleys. The hillocks have NE-SW and NE-SSW trend near the Barail range and N–S trend toward south away from the Barail range. Notably, Sone Beel, the biggest "Beel" (wetland) in Assam, is situated between two hill ranges, viz., the Badarpur-Saraspur range and the Chowkirmukh-Dohalia range. In the east, the neighboring structure is the Badarpur line of folding, while there is the Chargola anticline toward the west. A typical geomorphological feature is the tightfoldedness of the anticlines represented by hillocks having very high dips of the sedimentary beds. Information obtained from ONGC and GSI (Personal communication) revealed that Cachar represents a type area of Surma sediments exhibiting only tertiary deposits (70 million years old). Investigations into the rock samples of this wetland revealed that the hillocks around the wetland were formed after Tipam sedimentation. Most of the wetlands in this region, including the mighty Sone Beel, might have been originated after the Dupitila sedimentation during the Mio-Pliocene period.

Around Sone Beel, the soil in the catchment of the plains is generally loamy, but occasionally sandy or gravelly admixed with quartz. Conversely, hilly portion of the catchment generally consists of fine grain sand stones bearing many angiosperms, and thus, form the evergreen forest. Further, the catchment of the wetland also includes the reserve forests (RF) of the province, notably, the Singla RF. During monsoon, the wetland receives some humic as well as inorganic and organic nutrients from the hillocks and cultivable areas, particularly around the swollen tail end of the wetland.

The principal feeder of Sone Beel is the major inflow, River Singla, which originates as "Thing Tlawng Lui" at a height of *c.* 365.21 m MSL in Mizo Hills, from where, after traversing a meander course of 62.75 km, it enters Sone Beel. River Singla drains a total catchment area of 46,105 ha. The wetland also receives water from 12 minor inlets and many other canals flowing from both hills and plains, all of which together drain a total of 18,941.9 ha of the catchment area of the wetland. Out of this, 11,003.9 ha lies in the plains, while 7938.0 ha falls in the hilly portion of the wetland. These, otherwise, form, respectively, 58.09% and 41.91% of the total catchment area of the wetland. Moreover, 12 minor inlets occur in different parts of the wetland.

The major outflow (there being no minor outflow) of the wetland, River Kachua originates from the northern-most end of the wetland. It drains out the wetland water into the mighty river Kushiyara after covering a length of 19.30 km. Although River Kachua was blocked by a blind dam constructed by the Government of Assam in 1950–51, the dam was replaced by a lock gate in 1964 after experiencing navigational and fishery problems. The maximum outflow of the wetland was found to be 87.03 $m^3 s^{-1}$ in August 2008, while the maximum inflow was recorded to be 33.91 $m^3 s^{-1}$ in July 2008. Concomitantly, the minimum inflow and outflow of the wetland were recorded as 0.0027 and 0.087 $m^3 s^{-1}$, respectively.

The maximum length (L) and breadth (B) of the wetland at full storage level (FSL) were measured as 12.5 and 3.9 km, respectively. Interestingly, these values were found to shrink (reduce) to 4.07 and 2.22 km, respectively, at its Dead Storage Level (DSL). The area of Sone Beel at FSL was measured as 3458.12 ha, while at DSL, the area diminished to only 409.37 ha. The length of the shoreline was measured as 35.4 km while the shore and volume developments were recorded as 1.69 and 0.15, respectively, with mean depth of 0.29 m. The gross volume of the wetland was found to be $101.54 \times 10^6 m^3$. Although the wetland did exhibit variable WL ranging from 0.07 to 5.69 m at FSL (June–September) and 0.02 to 2.08 m at DSL (November–April), the average depth of the wetland was found to vary from 0.18 m (February 2007) to 3.34 m (September 2007) in 2007 and 0.16 m (January 2008) to 3.38 m (June 2008) during 2008. The wetland basin tended to become deeper from south to north. The contours in the west were found to be almost parallel and closer than their counterparts in the east. The wetland, as a norm, was found to exhibit a decreasing trend of its storage capacity from post-monsoon period to winter and to early summer; after which, it regains its storage capacity with the onset of monsoon.

Silt islands (SI) were recorded on the northern and southern parts of the wetland. Of these, the Gopikanagar SI (area 3.74 ha and 25 MSL) and Khagdi tila SI (area 3.31 ha and 21 MSL) were noteworthy. Interestingly, the wetland surface itself is situated 23 m above the mean sea level (MSL).

5.1.7.2 *Physicochemical characteristics of water*

The turbidity ranged from 20.56 to 185.54 TU while the overall thermal trend of the wetland depicted a state of isothermy with a bimodal peak occurring during summer and monsoon. The water temperature varied from 18.7°C to 32.3°C and it followed the air temperature fairly closely. The water of the Beel was circumneutral with pH varying from 6.0 to 7.9; DO ranged from 2.6 to 5.9 mg/L, FCO_2 fluctuated from 0.9 to 14.5 mg/L, and TA between 25.0 and 76.0 mg/L. The solubility of DO ranged from 9.4 to 11.9 mg/L while its percentage saturation varied from 32.3% to 74.4%. As an index of dissolved salts, the conductivity of Sone water was low. The lake soil was quite unproductive with pH ranging from 4.9 to 5.0, OC 0.25%–1.74%, available phosphorus ranged from 0.15 to 1.93 mg/100 g of soil, and available potassium ranged from 1.6 to 24.8 mg/100 g.

5.1.7.3 *Aquatic macrophytes (AM)*

AM was found to exhibit a heterogeneous assemblage of 23 species in Sone Beel. It may be noted that association of different species of AM, forming phyto-social units, are generally encountered in the wetlands. In Sone Beel, among the floating varieties, *Salvinia cucullata* and *Azolla pinnata*, *Nymphoides cristatum*, and *Nymphoides indicum* were found to form distinct phyto-social units. Among the emergent varieties, *Eleocharis acutangula* and *Sagittaria trifolia*, and *Scirpus eriophorum* and *Echinochloa stagnina* were found to be associated with each other. Among the submerged types, *Hydrilla verticillata*, *Vallisneria spiralis*, and *Nechamandra alternifolia* were found to make an association with each other. Interestingly, *Trapa bispinosa*, the floating form, also exhibited association with this group. Further, *Euryale ferox* was found to remain solitary and thus formed monospecific unit.

AM biomass was found to vary from 0.58 to 21.90 kg/m^2 (average 2.48 ± 0.82) having the maximum in December and the minimum in May. In addition to WL and amplitude of flooding (Welcomme, 1979), the growth and distribution of the AM was found to be influenced by water quality of the wetland.

Eichhornia crassipes was the sole perennial species in the wetland, followed by *H. verticillata* and *T. bispinosa*, occurring during most of the months of the year. During dry season, the emergent varieties (like *E. stagnina*, *E. acutangula*, *S. eriophorum*, *Oryza sativa*, *S. trifolia*, and *Polygonum flaccidum*) and the submerged types (like *H. verticillata and V. spiralis*) generally succeed, flourish, and show abundance at less WL when the wetland exhibits a decreasing trend in its depth. It indicates an indirect relation of AM biomass with WL ($r = -0.130 \pm 0.442$. $P > 0.05$) during this period. Higher ranges of conductivity during this period was found to sustain (Pearsall, 1938) a rich biomass of *T. bispinosa*.

The wetland water during dry season, in general, portray high DO produced by photosynthesis at rich insolation, in which FCO_2 is consumed and shows a fall. Concomitantly, a direct relationship of AM biomass with DO ($r = 0.500 \pm 0.340$, $P < 0.05$) and an inverse relationship with FCO_2 ($r = -0.780 \pm 0.178$, $P < 0.05$) were recorded in the study. The pH and TA, which were found to depict a rise, portrayed their direct relationship ($r = 0.850 \pm 0.126$, $P < 0.05$; $r = 0.022 \pm 0.454$, $P < 0.05$) with AM biomass. *H. verticillata* and *V. spiralis*, as indicated earlier, were found to be closely ($P < 0.05$) associated during this period (χ^2, 34.67). Also, a close ($P < 0.05$) association between floating *T. bispinosa* and submerged *V. spiralis* (χ^2, 24.15) and between submerged *H. verticillata* and emergent *S. eriophorum* (χ^2, 31.89) was discernible during this period.

With the onset of monsoon, the floating varieties, viz., *Nymphaea nouchali*, *Nymphoides cristatum*, and *N. indicum* generally occur in their flowering stage. Most of the emergent varieties encountered during winter, spring, and summer get submerged and decayed during monsoon. However, the littoral and sublittoral zones of the Beel, during monsoon, are generally found to be moderately infested with emergent varieties like *Cynodon dactylon* and *Vetiveria zizanioides*. *H. verticillata*, among the submerged varieties, do occur. Onset of monsoon, as stated earlier, cause decay of AM, thus rendering poor AM biomass with corresponding low DO, pH, and TA but high FCO_2 during the periods. Concomitantly, a direct relation of AM with DO ($r = 0.940 \pm 0.053$, $P < 0.05$), pH ($r = 0.160 \pm 0.443$, $P > 0.05$), and

TA ($r = 0.530 \pm 0.320$, $P < 0.05$) were recorded. Significant ($P < 0.05$) phyto-social association between *C. dactylon* and *E. crassipes* (χ^2, 12.44), *C. dactylon* and *N. indicum* (χ^2, 10.21), *Hygrorhiza aristata* and *E. crassipes* (χ^2, 13.87), *H. aristata* and *N. cristatum* (χ^2, 18.96) were recorded during this season. However, none of the AM species formed significant phyto-social relation with *E. ferox* due to its thorny body, and the latter, thus, formed a monospecific unit.

Almost all the monsoon varieties, as stated above, also occurred during the autumn, but with profuse flowering restricted mainly to *N. cristatum* and *N. indicum*. With the reappearance of dry winter, most of the emergent varieties gradually start reappearing with the decrease in WL of the wetland. *E. ferox* was seen to occur only for a limited tenure during the period from the beginning of Spring to mid-Summer. Since the plant is generally considered by the local fishermen as an impediment to fish movement and hindrance in fishing, they are usually removed from the Beel at the end of Summer.

High species diversity among the AM species were evident in this wetland. And, the level was found to be high (biassed estimate of $H' = 2.015$; expected value, $E(H') = 2.014$; variance of $H' = 1.431\ E\text{-}3$) in Sone Beel (Dey and Kar, 1989a).

5.1.7.4 Plankton

"Plankton," the living fraction of material, found in water and moved passively by wind or current, is composed of bewildering varieties of microscopic organisms. The "phytoplankton," which are predominantly autotrophic, are the primary producers of organic matters in the aquatic habitats. The nutritionally dependent animal component constitutes the "zooplankton."

5.1.7.5 An account of limnoplankton of Sone Beel

About 47 different forms of phytoplankton belonging to five groups, as indicated above, have been recorded till date in Sone Beel. Of these, the Chrysophyta included the maximum number and Pyrrophyta, the least. The phytoplankton density in the Beel varied from 48 to 5308 (average 1027) units/lit. The minimum population is generally recorded during the rainy season and maximum during the dry season. The cyanophytes and the euglenophytes generally exhibit prolific growth during the Spring and Summer with occasional abundance during monsoon. The chlorophytes usually show higher occurrence intensity during summer and early monsoon with infrequent richness during Winter and Spring. The chrysophytes are generally recorded throughout the year. The pyrrophytes were found to register their presence mainly during the Autumn.

About 19 different forms of zooplankton, belonging to five groups, have been recorded in Sone Beel. The zooplankton density varied from 6 to 380 (average 49) units/L. Low density is generally recorded during February–March and high density during November–December. In Sone Beel, a considerable portion of zooplankton population is considered to succumb to the day-in and day-out fishing operations, thus leading to their poor population in the Beel. Nevertheless, the copepods have been found to occur throughout the year. On the other hand, the rotifers were quite abundant mainly during early monsoon and autumn. However, the cladocerans generally exhibited abundance during Autumn and Winter (Dey and Kar, 1994).

Concomitant to the above, there is a bewildering diversity of fishes in the lentic systems of this region.

5.1.7.6 Fish diversity in Sone Beel

About 70 species of fishes belonging to 49 genera under 24 families and 11 orders have been recorded in Sone Beel, the biggest wetland in Assam.

1. *Pisodonophis boro*	32. *Mystus corsula*	63. *Glossogobius giuris*
2. *Gudusia chapra*	33. *Mystus tengara*	64. *Anabas testudineus*
3. *Hilsa (Tenualosa)ilisha*	34. *M. vittatus*	66. *Trichogaster fasciata*
4. *Chitala chitala*	35. *Sperata seenghala*	67. *Macrognathus aral*
5. *Notopterus notopterus*	36. *Rita rita*	68. *M. pancalus*
6. *Amblypharyngodon mola*	37. *Ompok bimaculatus*	69. *Mastacembelus armatus*
7. *Aspidoparia morar*	38. *Wallago attu*	70. *Tetraodon cutcutia*
8. *Barilius bendelisis*	39. *Ailia coila*	
9. *Catla catla*	40. *Clupisoma atherinoides*	
10. *Cirrhinus cirrhosus*	41. *C. garua*	

11. *C. reba*	42. *Eutropiichthys vacha*
12. *Cyprinus carpio carpio*	43. *Silonia silondia*
13. *Devario devario*	44. *Pangasius pangasius*
14. *Esomus danricus*	45. *Nangra nangra*
15. *Labeo bata*	46. *Glyptothorax telchitta*
16. *L. calbasu*	47. *Clarias batrachus*
17. *L. gonius*	48. *Heteropneustes fossilis*
18. *L. nandina*	49. *Chaca chaca*
19. *L. rohita*	50. *Xenentodon cancila*
20. *Puntius chola*	51. *Aplocheilus panchax*
21. *P. conchonius*	52. *Channa gachua*
22. *P. sarana*	53. *C. marulius*
23. *P. ticto*	54. *C. punctata*
24. *Rasbora daniconius*	55. *C. striata*
25. *Bengana elanga*	56. *Monopterus cuchia*

5.1.7.7 Fish yield (FY) of Sone Beel

Of the 70 ichthyospecies of Sone Beel, 59 species under 39 genera belong to the primary FW group while 11 species under 10 genera belong to the category of peripheral FW group (Nichols, 1928; Darlington, 1957). On the other hand, on the basis of Indian and Extra-Indian Territorial Distribution (Motwani et al., 1962; Kar, 1990), 28 ichthyospecies of Sone could significantly be placed under two groups, viz. (a) widely distributed species and (b) species of Northern India. Further, among the other species, one species, viz., *Glyptothorax telchitta* was found to be a true hill stream form; while five species, viz., *Botia dario, Lepidocephalichthys guntea, Acanthocobitis botia, Somileptes gongota,* and *Nangra nangra* were recorded as semi-torrential forms (Dey, 1973). Nearly 39 fish species were found to belong to the Plainwater group (Dey and Kar, 1990).

A conservative survey conducted by Northeastern Council (NEC) (Government of India) resulted in the following information:

c. 172 fish species belonging to 73 genera and 27 families have so far been recorded from NE Hill region. Of these, 33 representatives are quite common to Indo-Gangetic fish stock, 19 representatives are common with the fauna of the Eastern Himalayas, and 11 representatives are common with the Myanmarese fauna. Of these 23 fish species are said to be endemic in this region (NEC, 1991).

Of the 70 ichthyospecies, 84.2% belong to the primary FW group while the rest are of peripheral class. The annual fish yield from the Beel was found to be 358.21 mt during 1979–80 and 312.16 mt during 1980–81 having an annual average yield of 335.18 mt. Per hectare fish yield was determined as 103.5 and 90.26 kg, respectively, during the investigating years (Kar, 1990). Subsequently, further studies revealed that the annual fish yield from the Beel was 308.45 mt during 2009–10 and 301.55 mt during 2010–11. The yield trend appeared to show a dwindling trend (Kar, 2012a,b,c). Inconsistent trend of natural recruitment in such water bodies (Dey, 1981) coupled with higher catches in the turbid water might be the causative factor for higher yield during the first year. The forage fish containing mainly the cyprinids and chandids which constitute the bulk landing, could have gained ascendance over the predatory catfish to add further reasons to such propositions. The fish yield from Sone Beel appeared to be low as compared to 250 kg/ha obtained by Vaas and Schurman (1949) in some tropical FWs of Java. Further, Jhingran (1991) have recorded the range of fish yield from the Indian reservoirs to be as low as 6–7 kg/ha/year.

The constant deposition of "silt" expose the fishes in the shallower areas to shortage of living space (Welcomme, 1979). This has impact on the production through an eventual diminution of the water spread area (Kar, 1990). Moreover, the lake soil with pH 5.0–5.9, organic carbon 0.25%–1.74%, available phosphorus 0.15%–1.93%, water with pH 6.0–7.9 and DO 2.4–5.9 mg/L were not very suitable for fish production (Banerjea, 1967; Michael, 1969). Also, dense strands of aquatic macrophytes, often known to bind up nutrient materials, thus could result in the reduction in natural food availability for the fish and lead to poor fish yield (Bennet, 1962). Further, as an attempt to increase fish yield (Fernando and Furtado, 1975) by introduction of *C. carpio* into the Beel has resulted in an adverse effect on the indigenous ichthyodenizens of the Beel which could well account for overall poor yield from Sone Beel (Dey and Kar, 1990).

Of the 70 fish species, only *Puntius chola, Parambassis* sp., *Chanda nama,* and *Gudusia chapra* reflected Annual Average Yield (ARY) > 10%. Among them, *P. chola,* the minnow, contributed to the bulk of the fish landing from the Beel registering an average ARY of 21.39% and 27.82% during the study periods, thus portraying its significant contribution to the overall fish yield from the Beel. It is significant to note that the Indian major carps (IMC) of the Beel consists of five distinct species depicting an average ARY (as a group) of 21.39% and 27.82% during study periods. Of them, *Labeo rohita* alone portrayed an average ARY of 0.23% and 0.24% during the study period, which, incidentally, was the highest among the other species of the group. Non-judicious and irrational harvest of the juvenile IMCs in greater proportions have caused imbalance in the IMC fishery of the wetland culminating in overall poor contribution of this group in the total fish yield of the Beel (wetland). Concomitant to IMCs, the Indian shad, *Hilsa(Tenualosa) ilisha,* with an average ARY of 0.04% (which is admittedly low) revealed a single run in Sone Beel during the monsoon against two runs in other water bodies of India. Hilsa is generally more prevalent in the NE, SE, and SW zones of the Beel due to fluviatile nature of these zones. Indeed, the day-in and day-out fishing operations could spare very little scope for the recruited shads to make their return journey. *C. carpio* was the only exotic carp recorded in the Beel with an average ARY of 0.155%. Among the other carps and minnows, *Puntius sarana, Rasbora daniconius,* and *Salophasia bacaila* depicted average ARY of 0.085%, 0.63%, and 0.93%, respectively. Among the cobitids, which significantly includes the "ornamental fishes," *Botia dario* accounted for 0.485% as average ARY, while *Lepidocephalichthys guntea* dominated the cobitids with an average ARY of 0.97%.

The "C"-class fish (Swingle, 1950) of the Beel usually contain average ARY of 1.75% and 4.83% collectively. However, of all the predatory fishes, *Wallago attu* alone contributed to an average ARY of 2.94% as revealed from the study (Kar, 1990). Nevertheless, *Mystus vittatus* dominated the predatory fishes with an average ARY of 7.75%. Fishes like *Sperata seenghala, Mystus cavasius, Ompok bimaculatus,* and *Eutropiichthys vacha* portrayed high commercial value and depicted average ARY of 0.075%, 0.03%,0.77%, and 0.09%, respectively. The live catfishes include *Clarias batrachus* and *Heteropneustes fossilis* which exhibited average ARY of 0.11% and 1.81%, respectively. The bottom-dwelling murrels, which usually contribute significantly to the total fish yield of the lake, contain species like *Channa gachua, C. punctata,* and *C. striata* with average ARY of 0.27%, 7.0%, and 0.24%, respectively. *Channa marulius* is generally a rare species in Sone Beel with an average ARY of only 0.015%.

5.1.7.8 *Fish catching devices in Sone Beel*

In all, 26 varieties of fishing gears have been recorded in Sone Beel. However, *Mahajal, Daljal,* and *Chatjal* were the monsoon varieties, while the hooks and lines, the cages and large-meshed cast net were operated chiefly during the winter. The gill net and the triangular scooping net does not seem to have any seasonality. The CPGH (Catch/Person/Gear/Hour) (Dey, 1981; Kar, 2007a,b,c, 2013) of Mahajal was the highest (1.28–1.69 kg) followed by Chinese dip net (0.53–0.93 kg), long line (0.52–0.68 kg), cage (0.21–0.56 kg), and gill net (0.14–0.21 kg).

On the basis of the research findings of the present author, Sone Beel was declared as a "**Wetland of National Importance**" (Kar, 2013) by the Ministry of Environment and Forests and Climate Change, Government of India on October 16, 2008 (*vide* Resolution No. 11 dt. 16.10.2008, of MOEF Expert Group Meeting; and, also, Letter No. FRM 41/2008/63-A dt. 8.9.2008 from Commissioner and Secretary to the Govt. of Assam, Department of Forests).

5.1.7.9 *Management and conservation aspects Sone Beel as a case study*

In view of "fish" being the principal aquatic animal resource and staple food of the region, and the burgeoning demand for good quality and higher quantities of fish for the health-conscious people, efforts are being made toward sustaining and conserving these enormous freshwater habitats, such as the Sone Beel, and their bioresources. However, there are certain forms of environmental pressures which adversely affect fish, such as habitat loss, water contamination, drought, competition, predation, fishing pressure and disease, which must be tackled to achieve sustainable growth of fish, which is a prerequisite for formulating effective management strategies and conservation (Kar et al., 1996c).

5.1.7.10 *Impact of habitat loss and topographical changes*

The construction of embankments and roads along Sone Beel often destroys shoreline habitats, resulting in sedimentation which spoils large quantities of fish eggs and also increases turbidity. Under such circumstances, molluscs, such as mussels, can neither feed nor easily settle at the sediment-clogged wetland bottom, and thus, could contribute less to the food chain. Furthermore, the IMCs and the Indian shad [*Hilsa (Tenualosa)ilisha*] have been losing habitats due to the construction of the "blind dam" in the River Kachua and diversification of the waterways, thus, registering a sharp decline in their population in the Beel (Kar, 1990). Prior to 1951, River Kachua used to flow wild and free from the Sone Beel before emptying itself into River Kushiyara. Both *Hilsa* (Dey and Kar, 1989b) and the IMCs (Kar and Dey, 2000) used to migrate long distances from Bangladesh to enter directly into the breeding grounds in Sone Beel. The blind dam in Kachua (constructed in 1951) was replaced by a "lockgate" in 1964, and the main outflow had since been diverted through River Khagra. In addition, diversification of both the major inflow and the outflow through a network of channels within the lake has recently become a common practice in Sone Beel, mainly for agricultural purposes. Consequently, unless fish ladders, fish elevators, etc., are provided in the dams and lockgates, the long-term future of the depleting fish species is not optimistic because the migratory paths have been seriously disrupted. The impact of offsite erosion on the Beel fisheries is also very serious (LaRoe, 1986).

5.1.7.11 Contamination of the Beel water

Sone Beel receives a large amount of organic load in the form of sewage disposal which occasionally results in the depletion of the DO in some sections of the Beel (Dey and Kar, 1987).

Although pesticides are applied to the paddy cultivation at the draw-down level of the Beel, there is no potential threat of industrial pollution just now, in view of the absence of any big industry around the Beel. As such, incidents of "direct fish kills" due to drastic reduction of DO level shifts in fish species (Wheeler, 1979), and the occurrence of fish with cancers that have been recorded in freshwater bodies elsewhere (Klee, 1991) have not been seen in Sone Beel (Kar and Dey, 1987).

5.1.7.12 Effects of drought

Sometimes Sone faces prolonged periods of drought. In early 1979, the WL of the Beel was so low that the spawning ground of the large-sized fishes were practically nonexistent. Also, occasional high water temperatures could reduce the production of minor carps (Dey and Kar, 1990).

5.1.7.13 Competition and predation

Introduction of exotic fish may be done only after examining the possible impact they might have on the autochthonous fish species of the Beel. The introduction of *C. carpio* in Sone Beel has resulted in a substantive depletion of the population of endemic species by competition, predation, or hybridization (Moyle, 1976), although such a practice has been suggested to be useful elsewhere.

5.1.7.14 Fishing pressure

Overfishing has been exerting heavy pressure on the Beel's resources, with fishing operations carried out throughout the year. Both the quality of the fishing gears and the frequency of fishing operations are to be regulated.

5.1.7.15 Remedial measures

Sone Beel and its channels and associated rivers historically have been used for recreational and commercial fishing, navigation, mass transport, colonization, agricultural irrigation, water supply, and waste disposal (Kar, 1990). Therefore, in order to ensure a balanced development of the biota and maintain the maximum sustained yield, (MSY) the Beel habitat might be altered to increase the competitive advantage of desirable species and, more importantly, to increase the total carrying capacity for maximum fish biomass production. In order to achieve positive effects through habitat manipulation, possible measures include alteration of land, water and vegetation; habitat construction, and biotic manipulation (Owen, 1985).

5.1.7.16 Alteration of land

This procedure covers many aspects. Primarily, the Beel shore and stream bank erosion could be stabilized through afforestation. Rock deflectors could be built to direct stream currents. Artificial spawning sites, which have been found to be useful elsewhere (Cooper, 1980), could be made with sand, gravel, or nylon mats in those portions of the Beel which are heavily mud-laden and do not serve as suitable natural spawning grounds.

5.1.7.17 Alteration of Beel water

This is a difficult task for an enormous Beel like Sone Beel. However, attempts could be made in this direction by controlling the WL, turbidity, aeration, and fertility. Manipulation of the WL would be useful in controlling vegetation and spawning, and would, thus, affect predation. It could also be an effective way for eliminating undesirable species by lowering the WL to expose and isolate their eggs (Noble, 1980). Furthermore, when the prey species become overabundant, lowering the WL could bring them out from the aquatic vegetation which serves as their cover, thus making them more susceptible to predators.

5.1.7.18 *Turbidity*

Turbidity is a menacing problem in this region and Sone Beel is subject to prolonged turbidity, both during monsoon (Dey and Kar, 1987) and dry seasons (Kar et al., 1994). The latter is due to vigorous fishing at low WLs. Reasonably clear water is a prerequisite for the healthy growth of biota, notably the fishes. Methods to reduce turbidity include, among others, the following: (i) directly adding green or dry organic matter to the Beel, which, after decomposing, causes the particles in the water to precipitate; (ii) establishing herbs on the Beel-bottom at DSL; and (iii) adding alum or gypsum to the Beel water.

5.1.7.19 *Nutrients*

Nutrients are an important factor in fish health, but Sone Beel is oligotrophic (Hutchinson, 1967) with a low nutrient status (Kar, 1990), which could be due to the characteristics of the surrounding watershed (Klee, 1991). The deficit could be improved by directly adding organic fertilizers to the Beel water, thereby boosting phytoplankton production. Inorganic agricultural fertilizers may also be used, but with caution against as heavy doses can cause fish kills by phytoplankton bloom. Such measures have only met with limited success elsewhere (Klee, 1991).

5.1.7.20 *Vegetation management*

Sone Beel has a rich flora of AM (Dey and Kar, 1989a) that provide a higher biomass during the dry season than during the monsoon. The AM are a source of oxygen, provide cover for the young fish from predators, and serve as a surface on which food organisms could attach themselves and grow. However, excessive vegetational growth in the Beel could be harmful to the fishery, if it upsets the predator-prey relationship, and often leads to the stunning of fish growth due to overpopulation (Kar, 2007a,b,c, 2013).

Concomitantly, a low phytoplankton biomass in the Beel could also be due to less nutrients being made available to them due to excessive vegetational growth (Dey and Kar, 1994). Elimination of excess of AM is essential and could be done mechanically by manipulating the WL and by the use of various types of cutting machines. The harvested vegetation could be used as a soil additive in the form of compost (Henderson and Markland, 1987). Spraying or dusting of algicides, herbicides, and pesticides on the Beel water could be tested to chemically control the growth of AM, but repeated doses may prove disastrous by resulting in fish kills due to low DO and copper toxicity. Biological control of AM growth in Sone could be attempted through the introduction of herbivorous fishes which would control a variety of weeds without interfering with other fish species. The ideal species that are closest in meeting these requirements are the grass carps and the tilapia.

5.1.7.21 *Biotic manipulation*

Manipulating the fish population in the wetland could be done by controlling predators, competitors, and through artificial propagation. Sone Beel harbors a rich population of cormorants, herons, and cranes which help keep the fish population from exceeding the carrying capacity of the Beel. Thinning their population would likely to increase the population of both desirable and undesirable species of fish.

Nongame fish species which are often considered useless (e.g., *Chanda baculis, C. nama, Badis badis, Lepidocephalichthys guntea*, and *Danio devario*) may be removed by large-scale seine net operations and the Beel restocked with more desirable species. However, seining may endanger rare native species, thus reducing the BD and overall richness of the Beel ecosystem.

Artificial propagation involves the stocking, AQC, and breeding of pollution-resistant fish. Sone Beel has eight prominent fishing centers where intensive AQC could be practiced through polyculture and other techniques for raising fry and fingerlings of desired species in hatcheries for future stocking of the Beel.

5.1.7.22 *Habitat construction*

In addition to improving the existing habitat of the Beel, ponds and small reservoirs could be constructed adjacent to the Beel, conforming to standard specifications with proper management procedures for maintaining the fish stocks (Owen, 1985). This could release fishing pressure in the Beel.

5.1.7.23 *Aquaculture and agriculture*

In order to reduce capital investment, attempts are now being made to combine AQC with agriculture. An example of this being practiced in Sone Beel and its environs, particularly, at DSL, is the paddy-cum-fish culture, which results in high yields of both fish and rice.

5.1.7.24 *Fish disease*

Despite the high biotic potential of the fishes of Sone Beel, there have been heavy mortalities, since 1988, affecting about 70% of the fish population. The cause is a hitherto unknown fish epidemic called the EUS (Kar, 2007a,b,c, 2013, 2015).

5.1.7.25 *Conservation*

A hardline approach to the management and conservation of the aquatic resources of Sone Beel is to impose bans or restrictions on angling activities, such as closing a specific area to fishing, forbidding fishing during a particular season, restricting the number or size of fishes caught, restricting the types of fishing gears, promoting aquacultural practices, and supporting state and national regulations protecting a particular endangered species.

While imposing restrictions, it is essential to be adaptable and take into consideration the full spectrum of the users of the resource. Concomitant to administrative restrictions, encouraging results could be obtained through mass education of the Beel users (particularly, the fisher folk) by the NGOs regarding rational harvest and conservation and management of the Beel resources.

The potential health hazard posed to the Beel fishes by EUS is to be tackled through regular monitoring of the Beel environment and fish health, and by adopting quarantine measures.

5.1.7.26 *Epilogue*

It is now gradually being realized that many of the management techniques would not be required if fishery habitats are properly protected, and that priority be given to finding ways to prevent the problems from occurring in the first place, rather than relying on quick-fix solutions. Beel and stream improvement programmes could now begin to focus on the entire human-land relationship within a watershed. Financial incentives could be offered which would encourage farmers to practice good soil and water resource management. These are some of the notable problems which need to be addressed in order to find a lasting solution.

In brief, with regard to **Management and Conservation of Sone Beel Wetland in Assam**, it could be narrated as follows:

Profitable harnessing of fish resources from these natural "Beels" depend today on systematic management of these water bodies.

At DSL, Sone Beel has eight deeper fishing centers locally called "Bundhs" (November–April: 5–200 ha approx.): 1–1.75 m.

Now the "Bundhs" witness only "capture fishing."

Conversely, scientific pisciculture in these "Bundhs" could increase FY and generate employment and income.

Concomitantly, certain other measures forming part of the management strategies (as well as may be applicable to other similar "Beels") are as follows:

1. Regulation of siltation at the upstream region of the inlet(s) coupled with construction of check dams, wherever necessary, along with concomitant allotment of "Patta" (land-ownership right) to the local inhabitants to prevent further encroachment into the Beel mainly for paddy cultivation.
2. Leasing out of the Sone Beel to people's organization like Sone Beel Fishermens' Cooperative Society (SBFCS) instead of individuals with complete responsibility of its management to SFFCS as well as imposition and realization of its proper share. Government may do developmental works through the earnings.
3. SEFCS may further earn through collection of token tolls from the fishermen as well as the fish traders on the basis of fish caught and purchased, respectively.
4. The generation of employment and income through proper management would prevent avaricious exploitation of the Beel, check exodus of fishermen from the Beel to the cities, and thus could help conserving fishermen as fishermen.
5. Proper steps should be taken by the Board of Management to control fish disease like the EUS in order to prevent large-scale fish mortality.
6. Similar management strategies could also be thought of in the other lentic and lotic systems.

5.1.7.27 *Chatla Haor*

Chatla Haor (Plate 2, Fig. 2) is situated between 92°46′11.8″N and 24°42′38″ E in the Cachar district of Assam (Kar and Barbhuiya, 2000a,b; Kar, 2007a,b,c, 2013).

PLATE 2 The Chatla Haor, one of the biggest seasonal floodplain wetlands in Assam.

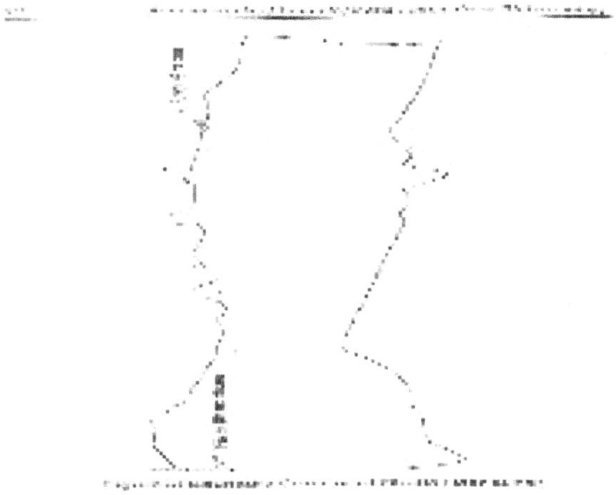

FIG. 2 Map of Chatla Haor at full storage level (FSL).

It was said to be a "Beel" (perennial wetland) some decades ago having its water spread area reaching Silchar town. Due to gradual siltation and eutrophication occurring naturally in the successional process, accelerated by man-made interference, today it has become a "Haor" (seasonal wetland) and retains water for approximately 6 months in a year having practically no DSL. So, it is almost completely dry during the winter. Having a water spread area of 1600 ha (approx.) at the full storage level (FSL), Chatla is considered to be one of the biggest "Haor" in Assam.

Around Chatla, the soil in the catchment is generally sandy loam, but shore vegetation is thin. The Haor is drained by a number of small inlets (viz., Jalengachhara, Baluchhara, Salganga) and an outlet (viz., River Ghagra) which drains itself into River Barak. The catchment of the Haor includes a small portion of the Innerline Reserve Forest (IRF). During monsoon, the Chatla, like other similar wetlands, receives some humic as well as inorganic and organic nutrients from the hillocks and surrounding cultivable lands.

The maximum length (L), breadth (B), depth (D), and water spread area (A) of the wetland at FSL have been measured to be 10 km, 2.5 km, 5.5 m, and 1600 ha, respectively. Prominent silt islands (SI), viz., Bairagitila and Harintila, have been found to occur toward the eastern shore of the Haor. Other small SIs, viz., Haltia, Diblia, and Barshangan occur toward the SW side of the Haor.

Among the inlets, River Salgonga originates from the foot hills of the Mizo Hill range while the Jalengachhara and Baluchhara inlets, which are mostly rheophilic in nature, flow down into the Haor from the IRF. The only major outlet, River Ghagra, drains the water of the Haor directly into River Barak traversing a tortuous course of *c.* 14 km from the northern boundary of the Haor.

5.1.7.28 *Phytoplankton*

Studies revealed the occurrence of the following genera/species of phytoplankton in Chatla Haor: *Nostoc, Anabaena, Oscillatoria, Spirogyra, Ankistrodesmus, Closterium, Euglena, Phacus, Ceratium, Cosmarium, Volvox, Pinnularia, Navicula, Diatoma, Gyrosigma,* etc. *Oscillatoria* sp. occasionally forms blooms during spring. *Euglena* and *Ceratium* occasionally shows luxuriant population during rainy season. *Pinnularia* and *Navcula* often bloom during autumn. The surface water is often red during dry season, possibly due to the red pigment in cyanobacteria.

5.1.7.29 Zooplankton

Studies revealed the occurrence of 18 species of zooplankton consisting of two species each of Protozoa and Copepoda, six species of Rotifera, and eight species of Cladocera. Occurrence of *Arcella* sp. among the protozoans and *Brachionus calyciflorus* among the rotifers, indicate eutrophy of the wetland. Two protozoans, viz., *Arcella* sp. and *Paramoecium* sp., represented *c.* 11.11% of the total zooplankton taxa in the wetland. Works reported *Arcella* sp. in eutrophic waters. Six rotifers contributed to *c.* 33.33% of the total zooplankton taxa; of which, *Filinia* sp. and *Lecane* sp. were found to be abundant. The identified Copepods included the *Cyclops* sp. and the *Diaptomus* sp. and they represented *c.* 11.11% of the total zooplankton taxa. The total zooplankton count was found to be 68 ± 45 units/L (Kar and Barbhuiya, 2004).

5.1.7.30 Aquatic macrophytes

A total of 23 species of AM have been recorded in this wetland. These could be classified as follows:

Five free floating, 4 rooted floating, 2 submerged, and 12 emergent. Of these, 6 AM species could be found throughout the year. These are *Azolla pinnata, Eichhornia crassipes, Salvinia cucullata, Trapa bispinosa, Justicia repens,* and *Cynodon dactylon.*

J. repens showed flowering during March–May, while *Nymphaea nouchali* bloomed during June–August followed by *Nymphoides cristatum* and *N. indicum,* which exhibited profuse flowering during September–October. *Ipomoea aquatica* depicted significant growth of population during July–August. *H. verticillata, V. spiralis* among the submerged varieties, and *Alternanthera sessilis, Cyperus platystylis, E. stagnina, E. acutangula, Enhydra fluctuans, Scirpus eriophorum, Sagittaria trifolia* among the emergent varieties succeeded at a lesser WL during the dry season. The floating varieties, viz., *A. pinnata, E. crassipes, S. cucullata* are recorded throughout the year and found to be associated with each other. Further, among the submerged varieties, *H. verticillata* and *V. spiralis* were found to be associated with each other.

Significant phyto-social association were also observed among *Nymphaea* sp. and *Nymphoides* sp. and among *E. acutangula, S. eriophorum,* and *E. stagnina.*

5.1.7.31 Fish diversity

About 57 species of fishes, belonging to 28 genera, 17 families, and 9 orders have been recorded in Chatla Haor (Kar, 2007a,b,c, 2013):

1. *G. chapra*	31. *O. bimaculatus*
2. *Hilsa(Tenualosa) ilisha*	32. *W. attu*
3. *Chitala chitala*	33.*Ailia coila*
4. *N. notopterus*	34. *C. atherinoides*
5.*A. mola*	35.*C. garua*
6. *Catla catla*	36.*E. vacha*
7. *Cirrhinus cirrhosus*	37. *E. murius*
8. *C. reba*	38.*C. batrachus*
9. *Cyprinus carpio carpio*	39.*H. fossilis*
10. *Devario devario*	40. *X. cancila*
11. *Esomus danricus*	41. *Aplocheilus panchax*
12. *Ctenopharyngodon idella*	42. *C. gachua*
13. *Hypophthalmichthys molitrix*	43. *C. punctata*
14. *Labeo calbasu*	44.*C. striata*
15. *L. gonius*	45.*M. cuchia*
16. *L. rohita*	46.*Parambassis* sp.
17. *P. chola*	47.*P. ranga*
18. *P. conchonius*	48.*Chanda nama*
19. *P. ticto*	49.*Badis badis*

Continued

20. *P. sarana*	50.*Nandus nandus*
21. *R. daniconius*	51.*Glossogobius giuris*
22. *Salmostoma bacaila*	52.*A. testudineus*
23. *Botia dario*	53.*T. fasciata*
24. *Lepidocephalichthys guntea*	54.*T. lalius*
25. *Acanthocobitis botia*	55.*M. armatus*
26. *M. cavasius*	56.*Macrognathus pancalus*
27. *M. corsula*	57.*Macrognathus aral*
28. *M.tengara*	
29. *M. vittatus*	30.*S. seenghala*

Zoogeographically, the ichthyospecies of Chatla Haor contains 79.62% of primary FW fish while the rest (20.38%) belong to the peripheral class (Nichols, 1928; Darlington, 1957; Kar, 1990). Further, on the basis of Indian and Extra-Indian territorial fish distributional pattern (Motwani et al., 1962), ichthyospecies of Chatla Haor did contain fishes belonging to the groups called "widely distributed species" (notably, *Puntius, Ompok, Channa, Anabas*) and species of Northern India (notably, *B. dario, Lepidocephalichthys guntea,* etc.). Ecomorphologically (Dey, 1973), fish species of Chatla Haor contained the "semi-torrential" form (*B. dario)* and the "plainwater" forms (notably. *A. mola, C. catla, C. carpio, Puntius spp., Mystus spp).*

Species diversity by Shannon-Weaver Index revealed high diversity of the fish species of this wetland (H = – 6.15) (Kar and Barbhuiya, 2000a,b; Kar, 2007a,b,c, 2013).

5.1.7.32 *Fish yield*

Of the 57 species, as revealed from our studies, Cypriniformes was found to constitute *c.* 32.3% of the total fish population followed by Channiformes (*c.* 22.8%), Siluriformes (*c.* 14.02%), Clupeiformes (*c.* 10.52%), and Osteoglossiformes (*c.* 0.00029%). Family Notopteridae showed the least abundance (*c.* 0.00064%). Interestingly, among the clupeids, occurrence of *Hilsa (Tenualosa) ilisha* (*c.* 0.000041%) in Chatla is a remarkable feature. Further, occurrence of advanced fry stages of hilsa (45.5–128.0 mm in length) in Chatla is an exceptional feature in the zoogeography and biology of the fish (Kar and Dey, 2002). On the other hand, yield of *G. chapra* revealed an alarming trend of decline in this wetland. Furthermore, occurrence of juveniles of Indian major carps (IMC) in Chatla indicate the possibility of this wetland serving as a breeding ground for these large growing fishes (Kar, 2007a,b,c, 2013).

5.1.7.33 *Satkarakandi Anua*

Satkarakandi Anua (Plate 3, Fig. 3) lies between 92°52′41.6″E and 24°45′8.9″N (altitude 20.73 m MSL) in the Sonai Revenue Circle in the Cachar district of Assam. The wetland is situated at a distance of *c.* 15.5 km away from Silchar city. River Barak flows toward the northern side of the wetland. The villages Satkarakandi and Dhanehary are situated on the eastern shore of the Anua. The villages Barahali and Satkarakandirpar are situated toward the south and west of the wetland. The NE and western shores of the Anua are protected by earthen embankments about 4 m in height as a preventive measure against flood.

PLATE 3

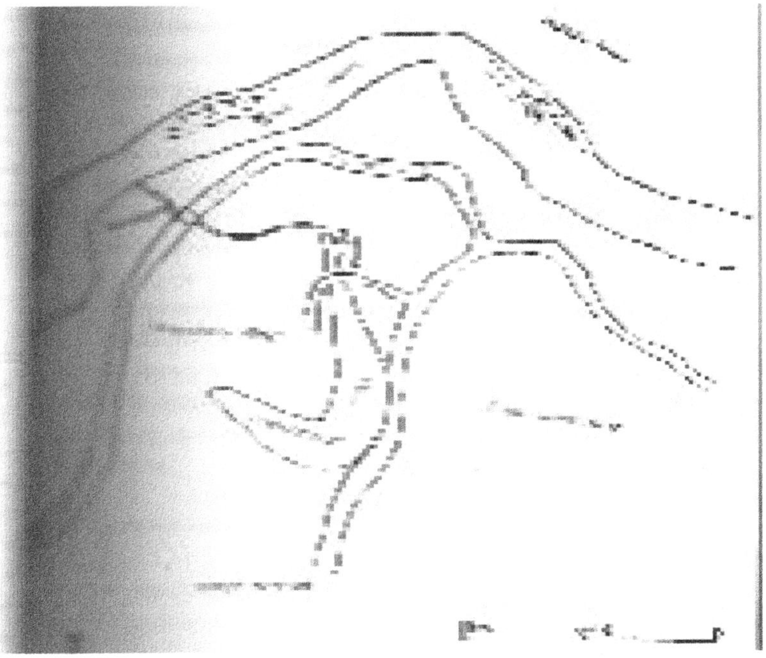

FIG. 3

The maximum L and B of the wetland were found to be *c.* 1.75 and 0.3 km, respectively. The Anua was found to have an average depth of *c.* 3.0 m at FSL. During monsoon, the Anua establishes connection with River Barak through a channel on the northern side of the wetland which is, however, guarded by a sluice gate as a flood-mitigation measure.

The wetland was found to exhibit interesting trends in its physicochemical features of water. The average values of physicochemical parameters of water were found to be: water temperature 26°C, turbidity 25.5. NTU, pH 5.7, DO 4.95 mg/L, FCO_2 45.0 mg/L, TA 101 mg/L, conductivity 49 μmhos/cm, PO_4 0.406 mg/L, and NO_3 0.32 mg/L.

5.1.7.34 Aquatic macrophytes (AM)

This wetland is a victim of acute eutrophication being infested by luxuriant population of *E. crassipes*. Other forms of AMs include *Lemna minor, E. acutangula, S. eriophorum, S. cucullata, A. pinnata, I. aquatica*, etc.

5.1.7.35 Plankton

The following genera/species of phytoplankton generally occur in this wetland, *Nostoc, Oscillatoria, Anabaena, Microscysis, Ankistrodesmus, Chlorella, Closterium, Euglena, Ceratium, Cosmarium, Pinnularia, Navicula*, etc.

Studies revealed the occurrence of following genera/species of zooplankton in Satkarakandi Anua: *Arcella* sp., *Brachionus* sp., *Lecane* sp., *Keratella* sp., *Filinia* sp., *Trichocerca* sp., *Daphnia* sp., *Moina* sp., *Alona* sp., *Chydorus* sp., *Dadaya* sp., *Cyclops* sp., *Diaptomus* sp., etc.

5.1.7.36 Fish diversity

In Satkarakandi Anua, 32 icthyospecies belonging to 16 families and 8 orders have been recorded. Some of the fish species are *N. notopterus, G. chapra, Salmophasia bacaila, A. mola, Osteobrama cotio cotio, P. chola, Puntius conchonius, Puntius ticto, B. dario, Mystus cavasius, M. vittatus, O. bimaculatus, W. attu, C. batrachus, H. fossilis, Anabas testudineus, Xenentodon cancila, Channa punctatus, Trichogaster fasciatus, C. nama, Mastacembelus armatus, Macrognatus pancalus, Macrognatus aculeatus*, etc.

5.1.7.37 Fish yield

Total landing of fish ranges from 140 to 425 kg/ month. The total annual fish landing was recorded as 2299 kg, while the average annual landing was 255.44 kg. *G. chapra* showed maximum yield (78%) followed by *Puntius ticto* (12%).

The management aspects of Satkarakandi wetland revealed that the wetland suffers from various problems like overgrowth of aquatic weeds, closure of feeding channels, unsustained fishing, addition of domestic sewage, etc. These could be overcome by certain steps like deweeding, opening of the feeding inlets, sustained fishing, etc.

5.1.7.38 *Baskandi Anua*

Baskandi Anua (Plate 4) oxbow wetland is situated between 24°48′36″N and 92°55′22.7″ E in the Lakhipur subdivision of Cachar district in Assam. It is said to have been formed due to change in the course of River Barak. Baskandi Anua borders villages Chandrapur Part II and Nijabaskandi in the north; villages Ratangram and Hazarigram in the southwest, and village Chandrapur Part III in the south.

The Anua is situated near the Manipur range of hills. Toward the east lies the Manipur valley while the Silchar subdivision is situated in the west. The North Cachar Hills is situated toward the north and the province of Mizoram lies toward the south. The catchment soil is found to be mainly sandy loam. Rain is the main source of water for the Anua. The wetland also receives water from the surrounding catchment having human habitation. The catchment vegetation includes herbs, shrubs, and trees including a lot of bamboos.

Baskandi Anua is situated about 20 km away from Silchar city along the NH 53 which connects Silchar with Imphal. The Silchar-Jiribam railway line passes near the wetland. The Rural Development Block HQ is at Baskandi.

The length (L), breadth (B), and area (A) of Baskandi Anua have been found to be, respectively, 2.230 km, 205 m, and 39.2 ha at FSL and 2.090 km, 190 m, and 36.7 ha at DSL. The wetland basin tends to be deeper towars the southern side as compared to the northern side. The Anua exhibits variable WL ranging from 0.25 to 5.85 m at FSL (June–September) and from 0.14 to 4.12 m at DSL (November–April).

5.1.7.39 *Aquatic macrophytes (AM)*

About 16 species of AM have been recorded in Baskandi Anua. They belong to six free-floating (*A. pinnata, E. crassipes, S. cucullata, L. minor, Pistia stratiotes, Wolfia* sp); two rooted submerged (*H. verticillata, V. spiralis*); six rooted with floating leaves (*N. nouchali, Nymphoides indicum, N. cristatum, T. bispinosa, Euryale ferox, Nelumbo nucifera*); and two rooted emergent (*Jussiaea repens, Muradania nudiflora*). Of these, 6 AM species were found to occur throughout the year. These are *A. pinnata, E. crassipes, S. cucullata, T. bispinosa* and *J. repens*. (Dhar et al., 2004a,b; Kar, 2013).

Studies conducted during the period 1998–2000 revealed wet AM biomass to be ranging from 4.4 to 11.4 kg/m^2. Wet biomass was observed to be higher during monsoon and post-monsoon reaching trough value during winter with a concomitant increasing trend during summer. Significant positive correlation of AM biomass with water temperature ($r = 0.1820$) has been recorded during the study period.

5.1.7.40 *Plankton*

Studies revealed the occurrence of 49 forms of phytoplankton belonging to 48 genera under 26 families, 11 orders, and 6 classes. The sequential analysis of abundance of phytoplanktonic groups generally revealed Chlorophyceae (43.51%) > Myxophyceae (34.78%) > Bacillariophyceae (15.58%) > Dinophyceae (4.47%) > Euglenophyceae (1.14%) > Chrysophyceae (0.52%). Some of the abundant phytoplanktonic forms are *Microcystis aeruginosa, Anabaena, Notoc, Oscillatoria, Chlorella, Volvox, Ankistrodesmus, Closterium, Cosmarium, Ceratium, Euglena, Phacus, Spirogyra, Diatoma, Fragillaria, Synedra, Gyrosigma, Navicula, Pinnularia*, etc.

Studies revealed the occurrence of 33 forms of zooplankton belonging to 22 genera under 17 families, 7 orders, and 4 classes. The sequential analysis of abundance of zooplanktonic groups generally revealed Rotifera (64.41%) > Copepoda (25.14%) > Protozoa (7.34%) > Cladocera (3.11%). Some of the abundant zoooplanktonic forms are *Arcella, Centropyxis, Brachionus, Keratella, Lecane, Trichocerca, Filinia, Daphnia, Ceriodaphnia, Simocephalus, Moina, Bosmina, Bosminopsis, Macrothrix, Alona, Pleuroxus, Chydorus, Dadaya, Cyclops, Diaptomus*, etc.

5.1.7.41 *Fish diversity*

About 13 species of fishes belonging to 10 genera, 6 families, and 4 orders have been recorded in Baskandi Anua which are as follows (Kar, 2013; Kar et al., 2000d,b,c,d; Dhar et al., 2004a,b):

PLATE 4 Baskandi Anua.

1. *C. striata*, 2. *Glossogobius giuris*, 3. *C. punctata*, 4. *G. chapra*, 5. *Macrognathus aral*, 6. *M. pancalus*, 7. *S. bacaila*, 8. *Badis badis*,9. *Chanda ranga*, 10. *P. ticto*, 11. *R. daniconius*, 12. *A. mola*, 13. *P. Conchonius*.

5.1.7.42 Fish yield

G. chapra showed the highest yield (*c.* 75%) followed by *Puntius, Mystus, Amblypharyngodon,* etc. The yield has since declined during the decade beginning 2000 CE with *G. chapra* showing depleting trend.

Concomitant to the above, in addition to data on Sone Beel, Chatla Haor, Satkarakandi Anua, and Baskandi Anua, our further study had portrayed the occurrence of different species of fishes in different other wetlands, like 25 species in Dolu lake, 40 species in Karbhala lake, 31 species each in Awangsoi lake and Oinam lake, 33 species in Loktak lake, and 32 species in Utra lake. Concomitantly, it was revealed that 22.72% of the fishes of Dolu lake, 5.76% of Chatla Haor, 19.16% of Karbhala lake, 23.83% of Awangsoi lake, 33.86% of Oinam lake, 49.75% of Loktak lake, and 48.8% of Utra lake were found to be parasitized (Puinyabati et al., 2010a,b; Singha et al., 2010; Binky and Kar, 2010; Ranibala et al., 2011a,b; Geetarani et al., 2011; Sangeeta et al., 2010a,b; Kar et al., 2011).

Management of the wetlands is a very important aspect today and is said to be part and parcel of the efforts toward their sustainable resource utilization for the nutrition, avocation, and other related aspects of the people, for whom these coveted water bodies are indispensable in their everyday life. Besides effective management by efficient *modus operandi* like the CBFM could be regarded as a prerequisite for achieving the goal of sustainable management and conservation.

Among the various recommendations, the significant ones include management of the Beels, specifically by CBFM, involving the stakeholders from the grassroot level for a healthy social life and balanced development; management of Eutrophication of the Beels; management of the silt load; management of *hitherto* unknown EUS fish disease; rehabilitation of the fishermen; eco-friendly tourism; management of the aquatic microhabitats and the substrata; establishment of fisheries training institute to take care of scientific methods of fishing and AQC (pisciculture); helping the fisherfolk in landing, stocking, transportation, and marketing of the fishes; raising the awareness among the riparian stakeholders for education, economy and sustainable management; all these leading to socioeconomic upliftment and development of the fisherfolk. Further, promotion of practices, like induced breeding, integrated fish farming (IFF) in localized areas to increase the fish stock in the wetlands; determination and analysis of basic physicochemical, biological, geographical, and climatological features using a Geographic Information System (GIS) platform could go a long way in sustainable management of the wetlands.

Further, the *hitherto* unknown dreadful, virulent, enigmatic is the **EUS**, which has been sweeping the FW fishes in an epidemic dimension, unhindered, unimpeded, and unabated, almost semi-globally today, causing large-scale mortality among them, since 1988, rendering many of them endangered; causing fear psychosis among the fish-eating people; causing untold misery to the fishermen, fish farmers; and devastating the economy of the nation. Our studies, involving aspects like limnology, chemistry, physics, bacteriology, mycology, and virology involving tissue culture and electron microscopy, revealed interesting findings including isolation of a virus. There is an urgent need for managing this unimpeded disease (Kar, 2007a,b,c, 2013, 2015).

Incidentally, the Beels of Assam are said to constitute > 70% of the total lentic water area of the province. The average existing fish production from the Assam Beels is said to be *c.* 173 kg/ha/year against its production potential of *c.* 1000–1500 kg/ha/year (Barman, 2004; Kar, 2010a,b,c). This reveals that the full potential of the Beels have not been completely utilized, thus leading to poor fish production of the province. Barman (2004) and Barman et al. (2008a, b) have indicated that situational variables like availability of critical inputs and transportation and marketing facilities were positively and significantly associated with the knowledge level of Beel users for sustainable development of Beel fisheries. Thus, in this regard, the CBFM may be a viable alternative in this regard. The existing knowledge level of the Beel users seemed to play a significant role on successful implementation of the CBFM.

Notwithstanding the above, a study (Barman et al., 2012a, b) was conducted in three Beels of the Assam to see the role of two situational independent variables, viz., availability of critical AQC inputs (X1) and community-managed markets (X2) on dependent variable, viz., knowledge levels of Beel users for community-based Beel fisheries management (Y). The study had portrayed that community-managed markets were significantly and positively associated ($P < 0.01$) with knowledge levels of the Beel users on community-based Beel fisheries management. Also, between these two variables, community-managed market (X2) was found to be the most significant contributing variable ('t'-value—2.91**) on knowledge levels of Beel users for community-based Beel fisheries management (Y).

The study was undertaken in three Beels, viz., Kutuha Bar Beel (16.00 ha) in Dibrugharh district, Amuguri Basa Pathar Beel (49.80 ha) in Golaghat district, and Talu Malu Beel (22 ha) in Dhubri district of the province of Assam. Further, 200 households, represented by both males and females, were randomly selected from the surrounding three Beels and were taken as the sample size of the study. Data were collected during the period 2007–2009 with the help of personal interviews through structured interview schedule. Initially, data were collected through local language and replies were recorded simultaneously in English. Correlation matrix was prepared. Afterwards, correlation coefficient, regression analysis was done to analyze the data using various softwares and statistical packages.

For measurement of dependent variable (Y), scale developed by Dana (1987), Sheikh et al., 1993), Barman (2004), and Barman et al. (2008a, b) were taken into consideration with necessary modifications. Respondents were grouped into three categories, viz., low, medium, and high based on mean SD and their knowledge score on CBFM. Also, frequency tables were constructed and percentages were calculated for making simple comparison.

The study revealed that majority, i.e., 146 (*c.* 73%) of the Beel users obtained medium availability of critical input scores, 36 (*c* 18%) low availability critical input scores, and 18 (*c.* 9%) obtained high availability critical input scores. Similarly, majority of the Beel users, i.e., 181 (*c.* 90.5%) obtained medium scores, 11 (*c.* 5.5%) high score, and 8 (*c.* 4%) obtained low score on community-managed marketing. Further, majority, i.e., 100 (*c.* 50%) obtained medium knowledge scores, 52 (*c.* 26%) low knowledge scores, and 48 (*c.* 24%) obtained high knowledge scores with regard to Beel users' knowledge level in CBFM.

Concomitant to the above, regression analysis revealed that community-managed marketing was the most significant contributing variable that could explain the maximum variability of Beel users' knowledge level on CBFM ('t' values 2.91**, significant at 1% ($P < 0.01$) level of significance on Beel users knowledge level). Further, availability of AQC inputs and community-managed marketing are the essential elements for effective development of fishery and AQC. Incidentally, the Beel users are free to procure their AQC inputs by themselves as per their choice, need, and satisfaction and even at low cost under the community-managed marketing system. This is because there is no involvement of middlemen in overall procurement and marketing. These inputs may be quality fish seeds, different organic and inorganic fertilizers to produce quality fish seeds and table fish, prophylactic and therapeutic drugs, fishing equipments, etc. Community-managed marketing may also bring "ownership status" and "community feelings" with respect to management of Beel fisheries through direct involvement of Beel users in each stage of procurement and marketing. The study also clearly indicated that community-managed marketing is a very important situational variable for successful community-based fisheries. Almost similar findings had also been reported by Dana (1987), Barman (2004), and Barman et al. (2008a, b).

From the study outlined above, it was concluded that community-managed markets were significantly and positively associated with knowledge levels of Beel users on community-based Beel fisheries management. Also, between these two variables, community-managed market was found to be the most significant contributing variable on knowledge levels of Beel users for community-based Beel fisheries management. It had revealed that this variable is one of the important factors for sustainable development and management of Beel fisheries through community-based Beel fisheries management program.

Concomitant to the above, wetlands are almost universally acclaimed as among the most productive life support systems in the world and are of immense socioeconomic importance in view of their providing with food, fodder, and water for domestic, irrigation, and industrial purposes. A large population of human beings, living in and around these wetlands, depend on them for their sustenance. Moreover, the wetlands are also the CPRs and are also said to be under different management regimes (Ganesh Chandra, 2011). These wetlands are used for multiple purposes and by multiple user groups and systems. A large number of stakeholders participate in the Beels, both within the fisheries and across other sectors like forestry, animal husbandry, agriculture, flood control, etc. Multiple institutions play their role through ownership and control over the fisheries of wetlands. A large number of stakeholders are associated directly and indirectly with the wetlands, which include, among others, the fishers, the lessees, various provincial government departments including the Assam Fisheries Development Corporation (AFDC), financial institutions, research institutes, NGOs, etc. Thus, the different stakeholders and user groups (UGs) constitute a diverse set of interests, goals, and priorities for the fisheries. Moreover, the livelihood of the fisherfolk and their families has been dependent on the water bodies and their fish resources for their nutrition and avocation since time immemorial (Dey, 1981; Kar, 1990, 2007a,b,c, 2013; CIFRI, 2000; Sugunan and Sinha, 2001; Ramsar Convention Secretariat, 2004).

Incidentally, it may be noted that the adoption of conservation measures and sustainable production practices are embedded in the Assam Fishery Rules 1953 (with amendments in 2005), which controls the wetlands fishery.

Notwithstanding the above, there had been attempts, through various types of communications, to describe management regimes and institutional arrangements in the frame work of stakeholders' mechanism of access to utilize various wetland resources. In addition, there have been various types of objectives of study, notably:

(a) to review the institutional arrangements;
(b) to identify the major issues, challenges, concerns, and choices for sustainable management of fisheries, and so on.

There were works portraying data collected from the wetlands located in different parts of Assam under various schemes through participatory rural appraisal (PRA) exercises; formal and informal discussions with the fishermen, lessees, village elders, local heads, and various other stakeholders; formal and informal communications with the provincial governments, and so on. The framework of this evaluation study was based on the access to fishing in wetlands, formation of the ownership and control rights, domains of management in wetlands and the role of institutions as controlling agencies in the allocation of the fishing rights, etc.

A fishery system is a plexus of subsystems. It is said to be also a part of broader natural and human-made artificial system, and is affected by the environment, economy, and society within which it exists. Nevertheless, at the wetland level, fisheries may be described at several axes, notably:

(a) Resource scale: small wetland to large size wetlands.
(b) Fish scale: small local stock to exotic fish stock.
(c) Fishery scale: small boat with small and simple gear(s) to motorized boats and gill nets, beach seine, etc.
(d) Administrative scale: local, district, and provincial levels; small and large fisheries agencies; field level staff; cooperatives; etc.

5.1.7.43 *Property rights scenario in wetlands*

It may be pointed out that the public trust doctrine permits the government to lease, grant, and sell public resources as long as it does not unduly harm public interests. On the other hand, "privatizing" fisheries means assigning rights (property and use) to individuals, groups, cooperative(s), or communities. Incidentally, "property" could be said to be composed of a bundle of rights (access, withdrawl, exclusion) that can be allocated to users.

Concomitant to the above, it may be noted that in the past, an individual fisherman had held only one exclusive right, viz., the right to own the fish he caught. Other rights were held in common. As fish became less abundant and, at the same time, competition and conflict increased, there had been a need to enact legislation(s) in order to regulate different people; unfriendly practices, such as gear restriction, "closed fishing season," etc. Further, later it, had become more clear that regulatory techniques were not much effective; thus, the idea of restricting access to the fishery was introduced. In this regard, there had been a number of ways to restrict access, including limited entry (licensing), effort rights, harvesting quotas, etc. However, these are limited rights, defined by law, whose ultimate ownership remains with the public. Group ownership of fishing rights by communities, cooperatives, or corporations is also said to be possible (Petruny-Parker et al., 2004).

Notwithstanding the above, establishing property rights to fisheries could generate stable expectations among the users and managers. It also provide the owners with an incentive for long-term sustainability. However, the government is said to retain the responsibility to conserve fishery resources for the public, who are involved through the management processes.

Concomitant to the above, the wetlands are CPRs. The property rights are said to be the fundamental rules of allocation and access. These are wide and may vary from the highly controlled exclusive private properties to open-access wetlands (Beels, etc.) with almost no control. Various factors like the size of the wetland, traditional and customary rights, physiographic dimensions, accessibility and river connection(s), etc., are the determining variables for the nature of property rights. Incidentally, the wetlands are managed for various objectives like economic benefit(s), livelihood security, sustainability, equity, conservation of BD, maintenance of the ecosystem functions, etc.

Further, the customary rights of the tribal and other indigenous ethnic groups are usually safeguarded lawfully in most of the wetlands (Phukan, 2006). These rights are for choices of species, gears, and could be for other specific purpose(s). However, it may be noted that the use of small gears and other domestic gears are usually free from much strict regulations. However, these rights are generally limited to self-consumption only. Moreover, the marginal areas of wetlands usually become open after the harvesting season and women fishers usually fish in these areas. Nevertheless, the access to the fisheries is predominantly governed by the lease holders; but, at the same time, it may be mentioned that the traditional rights of access are also been integrated in the property rights. Generally, fishing for the large-growing fishes are determined by the lease holders, while catching of the small fishes could be open or may be accessible only for domestic consumption.

5.1.7.44 *Stakeholders in the wetlands*

There are a large number of stakeholders which are associated, directly or indirectly, with the wetlands. These stakeholders include the following:

(a) Local user communities, i.e., those people who live in and around the wetlands and directly use the resources, and who are partly in a subsistence relationship with the resources and partly in a market relationship, i.e., the fishers, lessees, etc.

(b) Local communities having an indirect interest in the management of the wetland resources, e.g., the local communities who rely on some function(s) of the wetland but do not directly use the resources, e.g., villagers in the riparian villages and areas.
(c) Remote user communities, i.e., who come from some distance to use the resources and who may be in competition with the local users, may have a long-standing arrangement with the local communities, or may use a different component of the resources, i.e., share fishers and laborers.
(d) Users of wetland resources (individuals or groups) who have a purely commercial relationship with the resources, i.e., lessees, fish traders, middlemen, etc.
(e) Suppliers and marketers associated with the wetland resource users, who could be a diverse group, including middlemen for wetland products, suppliers of other different kinds of inputs, such as fuel and equipment(s), providers of credit, etc.
(f) Government agencies (public sector) with the duty of management of some aspect(s) of wetland resources. This might include a range of agencies with sectoral responsibilities for different resources, e.g., Department of fisheries, Assam Fisheries Development Corporation (AFDC) in Assam, Revenue Department, gram panchayat(s), Department of Forest, etc.
(g) Research organization(s), notably scientific organizations and their laboratories, educational Institutions, etc.
(h) Various NGOs, and so on.

Notwithstanding the above, each of the stakeholders are said to operate at different levels of management. Therefore, a number of management domains exist with different components and outputs. Thus, interventions in the management process which focus solely on a particular user community, or even on a user community and the relevant government agency, are often undermined and unaccepted by a bigger portion of the community, mainly because their views and priorities have not been included in the project design considerations. In this connection, it may be mentioned that among the different stakeholders the fishers constitute one of the most significant stakeholders as their life and livelihoods are dependent on the wetland resources.

5.1.7.45 *Institutional arrangement with regard to management regime*

Institutional arrangement

The wetlands in Assam may be categorized in two types on the basis of ownership:

(a) Government fisheries.
(b) private fisheries.

Concomitant to the above, the ownership of the Assam wetlands fall under different government departments, notably:

(i) Department of Fisheries.
(ii) Department of Revenue.
(iii) Department of Forest.
(iv) Assam Fisheries Development Corporation (AFDC)
(v) Gram Panchayats, etc.

However, the formation and the transfer of fishing rights are determined either through tendering methods or by direct transfer based on the criteria by different departments of the government.

5.1.7.46 *The aim of ownership and regulation rights of the government*

It may be mentioned that the government departments generally take care of the resources for enabling better utilization of the wetland resources, principally for the fishing population, conservation of resources, revenue collection, etc.

Concomitant to the above, the management rights in the registered perennial wetlands (locally called "Beel") are usually given, on lease, to the highest bidder (either individual or fishermen cooperative society). The lease of the registered wetland(s) is, generally, given only to the fishermen. Incidentally, the term "fisherman" in Assam are said to include the persons belonging to the schedule caste (SC) and other backward caste(s) in the community and said to have been defined under government circular (number RGF/62/57/103, dated February 15, 1960 and subsequent amendments thereafter), and engage themselves in any of the following kinds of profession(s) in which.

(a) the persons undertake fishing by themselves or in a fishing group.
(b) the persons directly undertake fish trade, such as marketing of mainly fresh fishes, preserved fishes, fishing implements, etc.
(c) members of the fishermen cooperative societies undertake fishing or fish trade, etc.

In addition to the above, the system(s) followed by the different government departments for limiting access to the fisheries in wetlands are as follows:

(a) Revenue Department: Open tender system in which price is generally fixed on the basis of FY in 1 year.
(b) Fishery Department of the provincial Government: It is usually done by the provincial government. However, the Director of Fisheries is usually said to lease to the highest individual auction bidder among the fishermen communities or the fishermen cooperative society.
(c) AFDC: Lease is usually given through tenders to the fishermen community or to the fishermen cooperative society.
(d) Panchayat: They generally go for "direct transfer" or through limited tenders to the villagers' group(s) or to the individuals on the basis of criteria ascertained by them.
(e) Department of Environment and Forest (DoEF): They are said to usually follow the system of permit(s) and license(s), mainly in their notified areas like the wetlands situated in the sanctuaries and National Parks (NPs). However, the wetlands are not given on lease and fishing is restricted only to the local inhabitants principally for their sustenance.

5.1.7.47 *Management scenario of the wetlands*

The management regimes of the wetlands in Assam could be classified into four types, mainly based on the property rights and their access as follows:

(a) private management (by individuals and groups)
(b) fishermens' cooperative society management.
(c) CBFM (decentralized management in which the government usually works only as a facilitator)
(d) open access with no management.

5.1.7.48 *Private management*

It mainly relates to the management of the Beels by the lessee. The lease period which was earlier 1–5 years have been amended to 7 years now, for providing with incentives to the lessee for adopting stock enhancement measures. The management of the Beel, including the fishing operations, is done by the lessee according to his choice subject to the following restrictions:

(a) No wetland should be drained dry by the lessee, who shall be required to leave sufficient water for the protection of the fish juveniles as well as for drinking water purposes of the human and cattle population in and around the wetland.
(b) The lessee is usually required to keep the water body (Beel) devoid of water hyacinth and other choking aquatic weeds.
(c) Fishing operations by >25 persons at a time even for their own consumptions, also on payment to the lessees, is to be largely prohibited in as much as it is highly detrimental to the interest of both the lessees and the government.

Concomitantly, the Beel fisheries management, including stock enhancement measures, weed management, fishing in the Beel, etc., are said to be done by the lessee. Further, access of other fishermen to operate in the Beel on payment of some amount of money to the lessee or under some kind of sharing arrangement where a share of the fish catch is taken by the lessee as fee are generally managed by the lessee and his team.

5.1.7.49 *Cooperative management*

In cooperative management of wetlands, the management of fisheries operations is usually done by the members of the cooperative society as per the rules prescribed by the society. Here, the access to the property is said to be governed by the membership of the cooperative society. In some societies, the fish catch is pooled together and sold in the market and a part of the income shared among the members of the communities. In other societies, fishers give *c.* 25% of their catch to the society as share of revenue.

It may be mentioned that the lease of the fisheries in wetlands are preferably given to the fishermen or fishermen cooperative societies on priority basis, under Assam Fisheries Act.

5.1.7.50 *Community-based fisheries management*

In Assam, there seemed to be an upsurge of interests among the government agencies and the NGOs in community-based management and comanagement approaches of the Beel fisheries following the history of traditional management and lessee-based management approaches, which involves local communities in Beel management and conservation. In this connection, it may be noted that the Government of Assam (GoA), under the Assam Agricultural Competitiveness Project (AACP) has initiated management practices in several Beels of Assam through CBFM under the aegis of Department of fisheries. These Beels have been put under the administrative control of fisheries department of the Government of Assam (GoA) after diverting them from the control of the revenue department of the Government. In these wetlands, Beel Development Committees (BDC) has been formed with the membership of fishermen and women of the riparian villages. The BDC is a group of 20–500 individuals living in the adjacent villages of a Beel coming together for efficient utilization of the natural fisheries resources of the Beel, for better price of the fishes caught, for more market power of the small and marginal Beel users, and for the enhancement of their livelihood in a sustainable manner. The number of individuals in a BDC depends on the size of the Beel and the number of inhabitants in the surrounding villages. The BDC, in turn, is said to constitute their executive committee (EC) for day-to-day management-related works in the Beel fisheries. Incidentally, it may be mentioned that the fisheries department of the GoA acts mainly as a "facilitator" in CBFM practices (Barman et al., 2012a, b).

5.1.7.51 *Open access*

In Assam, many of the unregistered Beels are "open access" in nature, i.e., there is almost no control or regulation of access to these Beels and their resources, although it is believed that the access to these categories of Beels as well as to their resources is based on the local rights, i.e., the fisherman operating in these Beels, preferably from the adjoining localities or from the same community or tribe, etc. However, only capture fisheries are practised in these Beels which are said to be enriched by **auto-stocking** of fishes during the rainy season. Further, in some of the open access Beels, even different community types of fishing, notably, "katal," "Jheng," "Kheo," etc. (as are locally called), are sometimes practised by various families in and around the Beel.

5.1.7.52 *Challenges, concerns, and choices for sustainable management of fisheries in wetlands*

The Beel management, incidentally, could face with some basic queries. In fact, moving from a sectoral approach to an integrated one could involve balancing alternatives. All the challenges faced by wetland fisheries, in turn, could, concomitantly, raise the concerns with regard to ecosystem health, social justice, livelihoods, and employment as well as food security and safety. Alteration(s) in livelihood outcome resulting from particular (indigenous) mechanism(s) of access may offer enhanced access to existing resources, access to new resources, and the development of new mechanism(s), e.g., through the application of new technologies (Franks and Cleaver, 2007). A full recognition of the complex nature of fishery systems could lead to the recognition that fisheries issues, like other large-scale environmental and social issues, may not be merely ecological and science-based issues, but, rather, could be social, economic, institutional, and political issues, and thus could require strong processes that could necessarily involve social values and issues of social justice and equity (Garcia and Charles, 2007).

In fine, wetlands in Assam are dynamic ecosystem resources, where a large number of stakeholders operate for gaining access and livelihood. Understanding various stakeholders and their level of involvement in the management is significant for bringing them into effective governance. The inclusion of the entire active group and the presence of active linkages within and among them could strengthen the process of governance. All understanding of management regimes and institutional arrangements in the Beels could lead to a deeper appreciation of how the riparian society puts its management exercises in a more or less orderly fashion with regard to sustained harvest of this key natural resource. Beel fisheries, as one of the multiple use and multi-user systems, are said to operate in an environment of multiple objectives. The real objective is governed by the institutional arrangement and management regimes which decide the priorities in resource use.

In addition to the above, some of the significant works done in the aspect of wetland management in other parts of India are as follows:

Datta et al. (2010) have worked on the perspectives of fisheries sector interventions for livelihood promotion in different places.

Incidentally, the main objective of some of the works were to help the workers in the field to speedily share their research findings with professional colleagues and to test their research findings at the prepublication stage.

The largely typical types of stylized features and the so-called lacklustre performance of this sector, said to be *in vogue*, vis-à-vis the distinctive features of the fisheries resources, fishers, and their geographic contexts had called for specialized and sustained efforts to promote livelihood of usually poor, backward, and unorganized fisher communities, which are, nevertheless, the most intimate stakeholders of this sector and its underlying resources. In fact, there had been communications which use clues from recent economic theories and management tools on property rights in order to develop a perspective on intervention strategies for livelihood promotion of most intimate stakeholders, i.e., the fisherfolk, in a sustained manner. Incidentally, stakeholder cooperation and public-private-community partnership are said to be efforts toward resolving the multidimensional problems of this sector. It, probably, stratifies and brings out the *pros* and *cons* of the existing fishing efforts into **four categories of models**, viz.,

(i) the traditional marketing model;
(ii) state-led models of livelihood promotion and fisheries development (including cases of para-statal cooperatives);
(iii) entrepreneur or leader-driven models; and.
(iv) technology-driven models, through selected illustrations from different parts of the country and covering both marine and inland (including brackish water) sectors of fisheries.

Further, there had been works, which, after identifying the major ingredients for sustainable livelihood development around fisheries, finally had tried to articulate Dr. APJ Kalam's concept of PURA (providing urban amenities to rural areas) to recommend a rural entrepreneur-led hybrid model of fisheries development in order to solicit sustainable and growth-oriented cooperation among the suppliers (i.e., stakeholders to fishery resources), labor (including fishers), and capital (including professionals). The final goal of many such works was to derive inspiration from **Coase Theorem** and the **Japanese model of Keiretsu** in order to empower the producers and suppliers of fish, viz., the fisherfolk and to place them at the center stage of control of rural entrepreneur-led private organizations wherein the fisher community will not be deemed as mere consumers or vendors of fish but may also enter as dignified coproducer partners with significant shares in residual claim and residual control in those organizations.

As already stated, India is endowed with enormous number of FW and marine water bodies and fish resources. The vast coastal line as well as the inland areas harbor a rich marine and nonmarine living resources. Innumerable people are dependent on these resources for sustenance through commercial exploitation (Natarajan, 2005).

India has *c.* 3638 marine fishing villages, *c.* 2251 traditional landing centers, *c.* 1400 small-fish landing centers, six major fishing harbors, and *c.* 41 minor fishing harbors. Based on the infrastructure and facilities, the different fishing crafts land their catches. The country has *c.* 138 fish landing centers with modern facilities established with support from the Union and provincial governments. The fisher population of India, of late, is said to be *c.* 5.96 million, which includes full-time (professional), part-time, and occasional fishers. The marine fishing sector may be classified into the following categories:

(i) nonmotorized artisanal sector using country craft with traditional gears;
(ii) motorized sector;
(iii) mechanized sector using inboard engines of 50–120 HP, and.
(iv) deep-sea fishing with bigger boats (25 m and above) and engines of 120 HP and above.

In this connection, it may be noted that in 1996, India had a total fishing fleet of *c.* 238,125 units comprising *c.* 160,000 traditional crafts, *c.* 31,726 motorized crafts (converted from traditional), and *c.* 46,918 mechanized vessels operating with different gear combinations. Incidentally, it has been inferred that the major problem in Indian marine fisheries is inadequate fisheries management system. Considering the country's diverse and vast coastline, efforts are to be specific to the fisheries' situation in each coastal zone. Though the sector has several problems, management of resources is considered as one of the most important aspects. Several reports seemed to have informed that the inshore fishery is stagnant, and there is no further scope for exploitation. It, probably, appeared that the weak extension system with inadequate man power and other kinds of resources were said to have made public sector incompetent to satisfy the needs and maintain peace in the community.

Some of the pertinent issues in this aspect are as follows:

Policies

Natarajan (2005) dealt with a case study in India reporting about enabling better management of fisheries conflicts.

Fisheries conflicts were seemed to be one of the principal reasons for the formulation of the Marine Fisheries Regulations Acts (MFRAs) in India (Natarajan, 2005). The GoI issued guidelines to all maritime provinces to formulate rules and regulations to be passed by the respective provincial legislatures in order to manage the marine fisheries. It was believed that these guidelines are intended mainly to avoid confrontation between the mechanized and artisanal sectors rather than as suitable regulatory measures for the sustainable exploitation of the fisheries resources. The guidelines were said to have been first issued in 1978 followed by its amendment in 1980. TN and Odisha were among the maritime provinces, bordering the Bay of Bengal, to have passed Marine Fishing Regulation Acts. Other provinces have followed ad hoc measures to prevent or tackle conflicts between artisanal and mechanized sectors. These acts are said to entail the following:

(a) Registration of all fishing vessels, including non-mechanized country crafts at their respective base ports.
(b) Licensing of fishing vessels for fishing in specified areas.

(c) Regulation, restriction, or prohibition of fishing in any specific area by such class or classes of fishing vessels that are supposed to fish in their own specified area(s) only.
(d) Regulation, restriction, or prohibition of catching in any specified area of such species of fish and in such periods as may be specified.

These acts are said to have, thus, equipped the provincial governments with the authority to regulate and control fishing activities in their respective provinces according to specific local needs. The area of operation of mechanized vessels in different provinces are said to range from 10 to 23 km.

Conflicts arise within and among these sectors mainly due to the inequalities existing among and between them. Demarcation of the respective fishing areas (Niche width) for the three groups is not very clearly defined and done. Conflicts arise within the sector mainly because of fishing in areas assigned to the other groups. Other conflicts arose due to social and economic factors. Incidentally, acceptance and implementation of fisheries management ideas are said to be a slow and gradual process. In fact, it may not be realistic to be too ambitious and optimistic. Nevertheless, a socioeconomic approach coupled with a bio-economic view, handled with understanding, tact, and foresight and farsight may ensure proper *modus operandi* of good management practices as well as sustainability of the resources. However, it is important to note that the communities themselves have been in the forefront to conserve their valuable resources with social methodologies in many parts of India, including coastal and inland. It may be further significant to peruse that prominent fisheries management techniques, such as comanagement (CM) and community-based Fisheries Management (CBFM) practices do exist in different parts of the country, but may be in different stages of development and implementation.

Institutions and their governance

It seems to be a common matter that the stakeholders of the fishing industry, almost always, differ in their views about the management failures with respect to resource sharing and conflicts. This could be due to the weaknesses of the institutional designs and approaches. Institutions working on the development of the sector seem to opine that resource and conflict management are of two types, viz. (a) formal and (b) informal.

Incidentally, it may be mentioned that the institutions that take care of the fisheries sector are usually and mostly vested under the government in most of the Southeast (SE) Asian nations, and India is one of them. Moreover, many countries are said to be in the process of drafting policies by which governance could be decentralized at the community level. Apparently, informal institutions that work effectively on fisheries management generally support the involvement of the stakeholders from the grassroots level in the management process. Examples seemed to support the statements that fisheries resources often managed by the community-based fisheries organizations (CBFO) supersede the government initiatives. There are said to be cases in India where often the public institutions seek the help of informal institutions to work on fisheries management. With the existing weak infrastructure, manpower, and economics, many countries may have to plan in using the informal institutions and CBFOs in managing fisheries resources and other issues pertaining to the sector as discussed previously. However, the colossal loss on the extinction of many traditional institutions formed a great vacuum in the system. In various countries, the policy makers are in the process of rethinking about reviving the traditional institutions for effective management of resources and conflicts. Incidentally, the stakeholders' initiatives, including those by the NGOs, trade unions (TUs), etc., have come up against weak implementation of rules by governments (Kurien, 1978, 1995a,b,c; Kurien and Achari, 1990; Shajahan, 1996).

Objectives

The project envisioned promotion of institutions and practices toward resolution of conflicts that are disadvantageous to poor fishers by enabling better management of fisheries conflicts. The project also intended to promote conflict assessment and resolution tools as well as consensus-building methods, targeting the key stakeholders. Specifically, the project aimed at the following aspects:

(i) Determining the best ways of communicating good practices in managing conflicts.
(ii) Promoting key lessons and practices from earlier projects on conflict resolution and consensus building, including Participatory Institutional Survey Conflict Evaluation Exercise (PISCES) and Participatory Action Plan Development (PAPD) developed by the Bangladesh-based Center for Natural Resources Study.
(iii) Adoption, adaption, and demonstration of the results to three key countries with large number of poor people dependent on fisheries.

5.1.7.53 *Studies in India*

In India three sites were selected. They were the following villages:

(1) Pedajalaripetta and
(2) Bheemunipatinam in Visakhapatinam District in Andhra Pradesh province.
(3) Sakthikulangara Village, Kollam District in Kerala province.

Incidentally, Sakthikulangara was the site identified by the project for the PAPD field trial.

The fishery profile and other details of the three sites are as follows:

Site 1. Pedajalaripeta village in Visakhapatinam

More than 90% of the Pedajalaripetta villagers depend on fisheries resources for their livelihood, and these fishery resources are mainly marine. The village has a total of *c.* 2136 households and a total population of *c.* 8128, which include *c.* 3184 males, *c.* 2691 females, and *c.* 2253 children. Of this number, *c.* 6459 were fishers. Fishing crafts used are of the following types:

(i) nonmotorized,
(ii) motorized fiber boats, and.
(iii) wooden crafts.

All these types describe the village's traditional and motorized types of fishing, operating in inshore and offshore fishing areas. Fishers often use gill nets, trammel nets, hook and line, and shore seines. The important fish species include the tuna, seer fish, shark, and sailfish. The village has good infrastructure facilities. There are social welfare organizations in the village. Socioeconomic problems range from social backwardness, economic stagnation, low catch/income to no gainful subsidiary occupation mainly due to lack of access to institutional finance.

Site 2. Bheemunipatinam village in Visakhapatinam

Bheemunipatinam is also a marine fishing community near Visakhapatinam Municipality. The village has *c.* 8763 households with a total population of *c.* 42,000. The main occupation of the people in the area is related to fishing, animal rearing, fruit and vegetable vending, and rural artisanal work. The landing center at Bheemunipatinam has traditional boats, mainly the stitched type. Incidentally, Bheemunipatinam's literacy level is *c.* 29.6% and the average income is *c.* Rs 1205.00 per month (2005). The village has one primary health center, 15 primary schools, one under primary school, three upper primary schools, one higher secondary school, one junior college, one girls polytechnic, and one teacher training center. Traditional and motorized boats operate in the area. The catches include seer fishes, shark, and sail fish.

Site 3. Sakthikulangara

Sakthikulangara is an important coastal fishing village in Kollam District of Kerala. The fishers comprise the major population of the village. The introduction of mechanized boats has brought about major changes in its fisheries sector, particularly in the development of infrastructures. All types of fishermen operate from this landing center. A study, however, was said to had indicated that *c.* 64% of *c.* 1209 families in Sakthikulangara are in debt. The total debt incurred by Sakthikulangara's fisher families amounted to Rs. 229.2 lakhs, with an average debt of Rs. 29,766.00 per household (1 US $ = Rs. 46.00 during that period).

5.1.7.54 *Application of PISCES*

The project made use of PISCES methodology and had employed timelines, institutional wheel, and semi-structured interview schedule to collect information useful for the evaluation of fisheries conflicts.

Timelines

Timelines are used to get a clear idea of what events in the past are considered important and how they occurred in sequence. In the study sites, conflicts that occurred during the British period to the present were recalled. Such conflicts were drawn over the timeline. The inferences on the timelines indicated that the conflicts in fisheries resources date back to 1970 and were mostly between the traditional and mechanized fishers. Conflicts involved burning of boats and the rights of traditional fishers. The inferences were used in the later part of the studies, specifically for analyzing conflicts with different parameters.

Institutional wheels

Institutional wheels are used to identify the relationships among different stakeholders within the community. Such information are used to collect further information on attitudes related to fisheries conflicts. Later, the inter-relations among institutions working in the area are identified. Key stakeholders identified were the government, local government bodies, voluntary action groups, NGOs, community-based fisheries organizations, fishermen's associations, etc.

Semi-structured interview schedule

In the study, the semi-structured interview schedule was administered to collect information regarding the fisheries conflicts in the sites. The tool helped the team in identifying these conflicts and in cross-checking the identified conflicts with other social methodologies of the project.

Stakeholder interactions through country-planning workshop

The study was said to have designed communication planning matrices, through stakeholder consultations in a national workshop, for the important conflicts identified in the various sites.

Nature, categories, and reasons of conflicts

These are significant aspects to be evaluated. The nature of fisheries conflicts were identified through the country planning workshop held in Visakhapatnam in AP. In fact, some of the major conflicts identified in the study area are briefly as follows:

(a) Conflicts due to resource sharing and indiscriminate fishing practices among certain groups of fishers. Specifically, the conflicts were due to the use of small mesh-size nets, trawling in breeding grounds, etc.
(b) Those due to weak marketing structure;
(c) Those due to the intrusion of mechanized boats into the traditional fishers' area. This happened to be one of the most common conflicts in the study sites.
(d) Those due to pollution due to effluents and oil spills from different types of industries. These naturally caused conflicts.
(e) Conflicts between promoters of tourism and traditional fishers also prevailed in the study sites. The latter argued that tourism led to displacement of fisheries from the coastal areas.
(f) Conflicts occurred when there were prohibitions on juvenile fishing, catching brooders, buying wild brooders, and on restricting mechanized boats to fish in inshore waters.
(g) Loss of living aquatic resources and decrease in catches due to oil spills and discharges of effluents led to conflicts.

Notwithstanding the above, there had been discussions on the methods to avoid these conflicts. Some of these are as follows:

(a) One such method, as suggested by the stakeholders, was to submit written complaints to the boat operators' association as well as to inform the state fisheries department.
(b) Another method suggested by the stakeholders was to restore peace through community gathering. Moreover, the traditional fishers and the mechanized groups should exert efforts together to avoid brooder catches. Incidentally, the stakeholders seemed to have suggested that these conflicts could be prevented through community meeting, by stopping the transport of live brooders, observing "closed seasons," etc.
(c) The fishers are to be self-motivated, as far as possible, and also laws are to be enforced properly.
(d) Convening community meetings using mass media and conducting researches were considered as important in instituting preventive measures to avoid conflicts.
(e) On the other hand, the stakeholders considered mass rallies, written complaints, and imposing penalties as some of viable methods to help avoid conflicts.
(f) There could be strict monitoring of the discharge of effluents.

Concomitant to the above, the workshop consolidated the fisheries conflicts based on the following typologies:

Type I Access to designated fishing zones.
Type II Poor enforcement.
Type III Gears and advanced technology and encroachment of fishing grounds of the traditional fishers by the commercial fishers.

Type IV Traditional fishers and prawn brood stock fishers.
Type V Lack of proper management and enforcement by the authorities.

5.1.7.55 *Attempt and attitudes toward conflict resolution*

Attitudes toward fisheries conflicts were identified according to the types of stakeholders, such as

1. primary stakeholders (who are directly related to fisheries sector, exploiting the resources).
2. fisheries managers (who have the stake in the fisheries sector and responsible in managing the fisheries conflicts).

Further, the attitude statements were selected based on different frames of references. Nevertheless, pilot testing was done to test the interview schedule for its validity and reliability. Moreover, the tool was modified based on the conclusions of the pilot survey. The data on the attitude of the fisheries conflicts were analyzed with simple percent analysis.

In addition to the above, the team selected two important topics for interventions based on the project objectives after conducting attitude survey. Incidentally, two community workshops were held at the sites, covering topics such as Marine Fisheries Regulation Acts (MFRA) and the FAO Code for Responsible Fisheries (CCRF). In fact, the MFRA of AP had important information that required dissemination. However, drawbacks in this aspect was found to be another cause for the incidence of fisheries conflicts. The contents of the MFRA discussed are as follows:

(i) zoning of fishing areas for different groups of fishermen;
(ii) banning different illegal fishing practices;
(iii) registration of the boats;
(iv) utilization of institutional and other facilities for restoration of peace.

The second intervention was with the CCRF. The information dealt with in detail included the following:

(i) overcapacity and overfishing issues.
(ii) ban on different unlawful fishing practices.
(iii) significance of the Marine Protection Areas.
(iv) industrial pollution and its impact on fisheries.
(v) destruction of mangroves and other breeding grounds.
(vi) use of chemicals and antibiotics in AQC.
(vii) impact of tourism on fisheries.
(viii) use of research data on fisheries.
(ix) save sea for your future generation.

5.1.7.56 *Plan for communication*

Concomitant to the above, identification of the best communication strategy for resolving conflicts in the project sites was analyzed through community workshops. The best communication strategy for resolving conflicts was identified using the communication channel identified by the stakeholders themselves with the developed communication planning matrix.

Some of the significant communication channels identified in the national workshop are as follows:

(a) face-to-face meetings at the village level;
(b) information through circulars, mass media, training programmes, workshops, written complaints, community assemblies, etc.

Further, the various communication channels through which the conflicts were to be resolved were discussed and analyzed through a participatory mode. The ranking method was used to collect information related to the best communication channels to be used, and these are given in the following **Ranking of communication channels**.

(i) Face-to-face meeting: 3; (ii) Circulars: 4; (iii) Mass media: 5; (iv) Training programmes: 2.
(v) Workshops: 2; (vi) Written complaints: 1; (vii) Community gathering: 3.

Notwithstanding the above, the communication channels suggested by the stakeholders were ranked based on their preferences, as shown in the following:

(a) Written complaints in the top rank. The reason for suggesting this channel could be their earlier successful attempts in combating conflicts by writing complaints to the relevant authorities.

(b) The second most prioritized communication channel was training programmes and workshops. The experiences of the stakeholders in acquiring knowledge and skills through training and workshops would have been due to their participation in these activities.

(c) The third best was community gathering. The other channels perceived by the stakeholders in the process were media and circulars.

Concomitant to the above, coastal fisheries are said to constitute a significant component of the fisheries sector in India. They have characteristics of common pool resources and are also said to be subject to the problem of common pool resource dilemmas (Prabhu, 2011). Incidentally, the **common property or common pool resource** is a form of resource management, in which a well-delineated group of competing users participate in extraction or use of a jointly held resource according to explicitly or implicitly understood rules about who may take how much of the resource (Stevenson, 1991). Incidentally, common pool resource management literature emphasizes the inevitability of institutions, both formal and informal, for resolving such dilemmas. The fishing industry, both traditional and mechanized, and the general public are said to have a vital interest in safeguarding and sustaining the economic and nutritional role of fisheries. Thus, it could be said that there is an urgent need to strengthen and put in place efficient measures to limit fishing effort and rehabilitate coastal areas and aquatic resources, if sustainability and viability are to be ensured. Incidentally, it may be mentioned that the fisheries sector in India is associated with the poor, illiterate, and undernourished population belonging to one of the most economically weakest sections of the society, although the sector immensely caters to the country's protein requirements and registered highest export earnings growth rate among agriculture commodities. Most of the fish stocks in inland and marine sectors have either been overexploited or reached their MSYs. For the stocks to recover and continue to provide production at an optimal level, collaboration between fishers and public sector institutions is needed in observing closed seasons, shifting from open-access system to the user right regime, reducing fishing effort and capacity, establishing marine and FW PAs/sanctuaries, participatory management, etc. (Prabhu, 2011). These could take the shape of CBFM and co-management (CM).

Concomitant to the above, it may be stated that CBRM had emerged as a way to involve the resource users and to utilize the indigenous institutional arrangements and knowledge in fisheries management. Moreover, communities of fishers could regulate access and enforce rules, under certain conditions, through community institutions and social practices in order to ensure fisheries resource sustainability. It may be pointed out that community-based fisheries resource management refers to the community level efforts, which seek to elucidate how ecological and social dynamics influence the fisheries resource management activities of diverse groups of people and how these activities, in turn, help to produce and shape particular kinds of environment. Available literature seemed to suggest the existence of community institutions organized along caste, kinship, or religious lines, which in some cases have played some role in regulating resource use, resolving conflicts, ensuring equitable access to resources, and in providing some form of social insurance.

Concomitant to the above, the FAO of the UN partnered with the South Indian Federation of Fishermen Societies (SIFFS) and attempted for pilot trials on CBFM as part of the fisheries livelihoods activities under the UN tsunami recovery support program (Anon, 2011a,b). Incidentally, the SIFFS had identified the Nagapatinam coast in TN and the Alapad Village along the Kerala coast for these trials. In these two areas, SIFFS were said to have been already working with traditional fishers' panchayats as well as with the elected panchayat leaders under the Panchyanti Raj system and had established traditional fisher marketing cooperatives in Nagaptinam.

5.1.7.57 Nagai experiment

Over the years, there had been a growing realization among fisher folks that things were going wrong. They learnt it in a hard way when dwindling catches and declining catch quality started affecting their livelihoods. However, the tsunami disaster in 2004 had brought to light some of the pertinent resources issues that were *hitherto* neglected and had formed an environment, which could be said to be non-conducive to addressing the problems, and which had been pulling the sector down. As many other nations, India too is challenged to improve fisheries management in spite of having relevant policies. Indeed, most of the policies, regulations, and management measures, such as zone regulations, mesh regulations, seasonal regulations, etc., are often formulated in a top-down fashion and with limited stakeholder participation. As a result, adherence to the regulations had been poor.

Notwithstanding the above, during in-depth consultations with the village leaders, SIFFS realized that fisheries management and related issues needed to be addressed collectively, and consensus was reached on the establishment of a 21-member Nagai Karai Fisheries Management Council. The council selected 12 villages to be represented in a management committee which were requested to undertake fisheries management debates, advocacy for fishers' welfare, and arriving at consensus decisions.

5.1.7.58 *Alapad experiment*

The 17-km coast of Alapad village was plied by too many high investment ring seiners (60 ft. and above with powerful outboard or inboard engines), each operated by 40–60 fishers. This was an unsustainable and uneconomic situation, and SIFFS found the elected GP to be an effective institution for bringing about a change. The GP president, along with nine traditional institutions, discussed the issue and decided to constitute a democratic forum. As a result, the Fisheries Management Council Alapad (FIMACA) was inaugurated in December 2009. The council took a series of bold decisions, some of which are as follows:

(i) no more increase in the number of ring seine boats operating from the village;
(ii) the ring seiners were to be replaced only by smaller boats;
(iii) a ban on night fishing, and.
(iv) a reduction in the number of trips by ring seiners to one per day.

5.1.7.59 *Results*

The Nagai and Alapad interventions have resulted in microlevel initiatives in CBFM. Incidentally, the new institutions continued with their functions and maintained democratic processes with added encouragement, and also with horizontal and vertical initiatives to implement decisions. From that time onwards, there were efforts to propagate and spread CBFM further across to other villages and districts as well as allow building up of district and provincial level councils and federations by fully involving the government in the process. These efforts could finally buildup the true fisheries comanagement.

In fact, this process had provided enough evidences regarding the importance of addressing local problems and working through local institutions suited to the fishers. Incidentally, the follow-up Fisheries Management for Sustainable Livelihoods (FIMSUL) project, initiated by the government with FAO support, is expected to buildup further based on the previous achievement(s) in many places, notably, in TN and Pondicherry.

Notwithstanding the above, different types of management systems are said to be prevalent in various parts of India. The "**Padu system**" in Vallarpadam Island of Cochin in Kerala, as a commons institution had tried to define various aspects of the group of rights holders and resource boundaries as well as fishing sites (Lobe and Berkes, 2003). It is said to be caste-specific, gear-specific (stake nets), and species-specific (shrimp). As used in Vallarpadam, and elsewhere in Kerala, TN, and SL, the padu is characterized by the **use of lottery for rotational access**.

Incidentally, the institutions are said to function in providing equitable access, collective social responsibility, rule-making, and conflict management. The emergence of the institution in the various study areas is said to be a response to change(s) in the markets as well as due to the legislation in the 1970s. It may also be seen as a response of the fishing communities to keep their options open, i.e., to be resilient.

Small-scale fisheries are often characterized by a diversity of gears, which often may also reflect the diversity of species and ecosystems in which they operate. The fisheries and the communities which depend on them for their livelihoods are also characterized by an assortment of local management institutions (Kurien, 1992; Johnson, 2001). Incidentally, in south India and SL, one of these management systems had survived and evolved for the approximately three centuries. In fact, the padu system of fishing site rotation is said to manifest itself in at least three distinct geographic regions, where the fishers generally utilize stake nets to fish for shrimps in the lagoons and estuarine ecosystems. Panini (2002) described padu as a traditional system of granting entitlements to eligible members of a particular community to undertake specific fishing activities in certain designated fishing grounds of the lagoon during specified seasons. Mathew (1991) and Amarasinghe et al. (1997) have documented two occurrences of the padu system, one in the Indian province of TN and the other in SL. In fact, we need to go into the basics of common property theory in order to put the padu system and use of coastal commons in perspective, Incidentally, the Common property (common pool) resources are said to share two characteristics:

(a) exclusion or the control of access of potential users is difficult, and
(b) each user is capable of subtracting from the welfare of all other users (Berkes, 1989; Feeny et al., 1990).

These two universal characteristics of commons are referred to as the **exclusion problem** and the **subtractability problem**, respectively. Thus, Ostrom et al. (1999) defined the common-pool (or common-property) resources as "those in which

(i) exclusion of beneficiaries through physical and institutional means is especially costly, and
(ii) exploitation by one user reduces resource availability for others."

Many systems of CBRM are said address exclusion and subtractability problems by devising collective actions (e.g., Kurien, 1995a,b,c; Johnson, 2001). The key is said to be the ability of a community using a common resource to limit the access of outsiders and self-regulate its own harvest. Incidentally, common property works through incentives. If members of a group are assured that future harvests would be theirs by right and not end up being harvested by another group, they have the incentive to self-regulate. We expand on the two concepts. **Exclusion** refers to the ability to exclude people other than the members of a defined group. In fact, several marine tenure systems are designed to deal with exclusion but stresses of population growth, technology change, and economic transformation may contribute to the breakdown of mechanisms for exclusion (Berkes, 1989). However, one of the significant conclusions of the literature is that the legal recognition of community resource rights is the key to success, as in Japanese coastal fisheries (Ruddle and Akimichi, 1984). Incidentally, conflicts among the competing groups could be inevitable without legal protection and, consequently, the local resource use rights may become fragile (Sen and Nielsen, 1996; Pomeroy and Berkes, 1997). Further, problems of exclusion may become especially important in an increasingly interconnected world in which local resource rights could be under pressure. Hence, attention may shift to cross-scale institutional interactions to deal with impacts at a multiplicity of levels (Berkes, 2002).

Notwithstanding the above, **subtractibility** refers to the ability of social groups to design a variety of mechanisms to regulate resource use among the members. In many CBM systems, users have devised self-governing rules, monitoring mechanisms, and sanctions. Incidentally, much of the common property literature addresses the subtractibility issue and the ability of groups to make rules-in-use. Ostrom (1990) listed eight design principles for effective community-based institutions, but there may be as many as 40 design principles or critical enabling conditions important for the success of commons institutions (Agrawal, 2002).

Concomitant to the above, common property analysis focuses on institutions, looks at access and self-regulation, and asks questions about rules and who has rights and control over a resource. However, these rights and relationships are rarely static. Rather they tend to change over time in response to various influences. Hence, some researchers started to study the dynamics and resilience of common property systems, instead of describing community-based institutions as if they were fixed in time. For example, Seixas and Berkes (2003) focused on the cycles of resource collapse and institutional changes in a Brazilian lagoon fishery in an attempt to understand governance systems as an adaptive process. The significance of this approach relates to the ability of a community to deal with change(s). Incidentally, a resilient system has the ability to absorb shocks and stresses and still maintain the functioning of the society and the integrity of the ecosystem on which the society depends (Adger, 2000; Gunderson and Holling, 2002).

Concomitant to the above, the objectives of the paper dealing with examination of the padu system in the stake-net fishing community on Vallarpadam in central Kerala in India, are as follows:

(i) To analyze the local property rights system, with emphasis on its origins, evolution, functions, and adaptability.
(ii) To investigate the wider context of padu systems by comparing the Kerala case to similar system(s), which is said to be *in vogue*, in SL.

Incidentally, the observations of the present study are based on field research from the stake-net fishery on Vallarpadam Island, with a brief background of the study area and its people.

Notwithstanding the above, comparison of some of the key characteristics of the padu systems had been illustrated by three case studies. It had portrayed that the three examples are similar in form and function. This comparison of the three cases not only provides an understanding of the institutional design principles for small-scale fisheries and systems of fishing site allocation but also allows some generalizations about the ability of community-based systems to address exclusion and subtractibility problems in coastal marine commons. Incidentally, it could be concluded with a discussion of the emergence of the common property institutions in terms of resilience. However, it could be argued that institutional change(s) may be seen in the context of dealing with impacts and building resilience to absorb shocks and stresses.

5.1.7.60 Methods of study

The community selected for study was the community of Vivekananda Chandrika Dheevara Sabha. This was selected after visits to several locations in the area. Although the entire community was of the fishermens' caste and also engaged primarily in stake-net fishing, we focused our study on a particular subset of the community consisting mainly of the unlicensed stake-net fishers because of the CBM institutions through which they regulated access to their fishing grounds and made rules among themselves. The research methods utilized were said to be primarily qualitative, mainly those characterized as **rapid rural appraisal** (Pido, 1996; Chambers, 1997). Incidentally, details of the stake-net fishery were gathered through a variety of methods, including participant observation and semi-structured interviews. Focus groups and key informant interviews were used for in-depth knowledge associated with the fishery, viz., seasonal cycles and species details. In fact, interviews with faculties at the local university, caste, organization leaders, and fisheries department officials provided some of the historical and administrative contextual information about the fishery. Interviews were the primary method used. A total of 58 semi-structured interviews were conducted, the majority of them with the unlicensed stake-net fishers in the selected community for study. Incidentally, about two-thirds of the target community was interviewed. Key informants in this group included the leaders of the fishers' organizations. Each was interviewed at least 3 times through the course of the research. Further, visits were organized in the homes of the unlicensed fishers in order to explore livelihood details and to obtain further information on the functioning of the fisher organization. Also, the officials of the provincial Fisheries Department, Dheevara caste organization, and Cochin Port Trust were interviewed. Toward the end of the research period, discussions have been held with the three focus groups with 48 unlicensed stake-net fishers. Participation was said to be good, with > 50% of the unlicensed stake-net fishers present. It could be mainly because the meetings were coordinated to coincide with the scheduled meetings of the three organizations.

Incidentally, there were two main goals of the focus groups:

(i) To verify and cross-check details on seasonality, species harvested, and gears used by fishers.
(ii) To conduct a brief survey of the participants to obtain details on demography, occupational diversity, and marketing.

Incidentally, all the interviews were done in Malayalam language involving interpreter(s), who were familiar with the overall project and its objectives, and who operated as a project partner. Quantitative data regarding fishery statistics/records and maps of the area were obtained from the appropriate government offices, like the Marine Products Export Development Authority (MPEDA); District Fisheries Office, Ernakulam; State Fisheries Department, Trivandrum; the local Universities, notably, the Centre for Development Studies in Trivandrum and the Cochin University of Science and Technology; and so on.

Concomitant to the above, the people in the study area and the people of South India, in general, possess a rich setting for the study of community-based marine resource management systems (Kurien, 1992; Johnson, 2002; Gadgil et al., 2000). The province of Kerala, with its traditions of social organization, high standards of social services, and high rates of literacy, is where the citizen science movement in India had started (Gadgil et al., 2000).

A piece of research was conducted in the Cochin Estuary. It is a 242,600 ha brackish water ecosystem (locally called "Cochin backwaters"), which supports a multitude of livelihoods and a wealth of BD. The estuary and lagoon system is also home to one of the country's largest military and commercial ports and the city of Ernakulam. One of the traditional livelihoods supported by this ecosystem is the inshore small-scale fishery. It is characterized by various types of fishing gears. The inshore fishery produces > 42,000 metric tons (mt) of fish per year, representing > 20 different species. The largest component of the harvest is shrimp, which makes up more than one-fourth of the catch (Government of Kerala, 1993). Vallarpadam Island is located close to the mouth of the Cochin Estuary, and is a part of the Mulavukad GP. It is accessible only by boat, with ferry services linking the residents with neighboring Vypin Island to the west and to the city of Ernakulam. The study was done with the people of Vivekananda Chandrika Dheevara Sabha (VCDS), a caste community of Hindu fishers living on the northern end of Vallarpadam Island. The island is home to various other Hindu caste groups as well as a large number of Latin Catholic Christians. The life of each community centered around the respective temple or church. The residents of VCDS are members of the Vala fishing caste, which is said to be a subcaste of the Dheevara. The 302 Vala families are represented collectively by a caste organization that serves to maintain the temple, organize festivals, and provide leadership during weddings and funeral ceremonies. Further, VCDS is linked with district and provincial level Dheevara Sabhas that serve as umbrella organizations and perform more of a political function, lobbying the provincial government on behalf of its members. The Vala caste represents the principal group of small-scale fishers on Vallarpadam. It may be mentioned that out of 302 Vala families (*c.* 1380 persons) in VCDS, *c.* 206 families gain an income from fishing, while *c.* 117 families depend on fishing as their sole means of livelihood (Vivekananda Chandrika Dheevara Sabha Household Survey, 2001). Incidentally, the fishers from this community predominantly utilize a single type of fishing gear, the "stake net," or "Ooni Vala" (in Malayalam language). They usually operate on seven distinct fishing grounds (called "Ooni padu") surrounding the northern tip of Vallarpadam Island. However, in three Ooni padu, the fishers operate without state-sanctioned licenses.

5.1.7.61 Stake-net fishery and the "padu system"

The dominant gear used in the small-scale inshore fishery is locally called "stake net" ("Ooni Vala") and is used by small-scale fishers to harvest shrimps, as they migrate from the estuary back out to sea. This gear is said to account for *c.* 57% of the total catch in the estuary and is the prevalent gear used in Kerala's backwaters in which there are *c.* 12,900 nets (Kurup et al., 1993). The stake net is a bag-shaped trap net, which is attached to permanent stakes that are fixed to the bottom. A typical stake net in the study area is a conical bag, *c.* 7–15 m long with a circumference of 16 m or more at the mouth. It is a passive gear that filters the out-flowing tide. The stakes are generally setup in long rows, called "Ooni padu," in areas of strong tidal currents. Each of the seven stake-net fishing padu grounds at the northern tip of Vallarpadam Island has a specific name and a specified number of stake-net fishing sites. There are a total of 288 stake nets operating in these seven padu grounds, and 126 fishers have been utilizing them, all of them being members of the Vala fishing caste, and all were said to be residents of VCDS. These 126 fishers could be further classified into two distinct groups, viz., those who had state-sanctioned licenses (48 fishers) and those who did not (78 fishers). Incidentally, four of the padu grounds (consisting of *c.* 144 stake nets) were used solely by *c.* 48 licensed fishers. The sites in these four padu grounds were allocated to specific fishers and were never changed, i.e., the license was specific to the site in the padu grounds. Licenses for this group of fishers were renewed yearly by the Kerala State Fisheries Department for a nominal fee. The remaining three padu grounds (consisting of *c.* 144 fishing sites) were used by the group of 78 unlicensed stake-net fishers.

Notwithstanding the above, this study focuses mainly on the unlicensed fishers. These three groups of fishers, on three padu grounds used a system of rotational access. The three groups were represented by "sanghams" (in Malayalam for "association" or "society"). The three sanghams shared a common system for fisheries management, and fished in the same waters with the same gear(s). Incidentally, they were not linked to each other. Rather, each sangham operated independently to assign fishing rights for the padu grounds in which their members fish. The three sanghams are not formally recognized by the provincial fisheries department, but, nevertheless, they are registered with the Provincial Registrar's office at the High Court of Kerala. It is correct to say that sanghams are recognized by the Kerala Provincial government, but, that their actual functions for management of the fishery are not recognized. Incidentally, the main functions of the sanghams are briefly discussed, following an examination of the evolution of the padu system and the sanghams.

5.1.7.62 Evolution of the padu system and the sanghams

The system, which is said to be *in vogue* among the unlicensed stake-net fishers on Vallarpadam, is not associated with the local GP (the lowest form of government organization), or not linked with the caste organization. This in itself is interesting, given Kerala's emphasis on decentralization of decision-making authority to the GP level. The fishers in the studied community have been using a CBM system, and formed an independent institution, the "sangham" in order to facilitate the padu system of fishing site allocation for each fishing ground. However, the stake-net fishers from Vallarpadam did not refer to their management system as a padu system. However, since all of the locations for stake-net rows in the vicinity of the island were named with "padu" as the suffix (which in Malayalam means "fishing site" or "spot"), and since there are other padu systems in South Asia that is the term used. The Kerala Fisheries Department became autonomous in 1967 and attempted to institute licensing arrangements for the fishery, replacing an older system of land and fishing site holdings, called "pattayam." Incidentally, the State legislation required State-sanctioned licenses for all kinds of fishings since 1974. In the stake-net fishery, this replaced the pattayam system, and thus was said to be one of the factors that precipitated the situation in which the resource seemed to have reverted to an open-access condition. Incidentally, the fisheries department was in fund crunch as well as lacked staff, money, and equipment to enforce the legislation, resulting in two decades of conflicts over access to the productive fishing grounds.

Notwithstanding the above, the period from 1973 to 1980 appeared to be a critical time of change for the stake-net fishery of Vallarpadam, as it was marked by conflicts over access to the stake-net fishery and the padu fishing grounds, which were traditionally used by the residents of VCDS. Subsequently, in 1974, corresponding to the boom in global shrimp markets (Kurien, 1992) and the new fisheries legislation, a group of 21 Vala families from VCDS installed a row of unlicensed stake nets, *c.* 300 m north of Panumbukadpadu, one of the traditional pattayam rows, which had received license from the province. The fishers from Panumbukadpadu complained to the Kerala Provincial Fisheries Department, asking them to remove the stakes. The 21 families who had formed this row ("Unionpadu") took the matter to the municipal courts and received a stay order based on the grounds that they were fishermen by profession (caste), and that this was their livelihood and they had the right as members of the Vala caste. This process finally concluded in late 1979, when official licenses were granted to the fishers from Unionpadu. The precedent for challenging the state-sanctioned property rights regime for the shrimp fishery had been set and other challenges could soon follow. Subsequently, three more rows of unlicensed nets appeared in the early 1980s, challenging the state for the right to fish on traditional caste-based occupation as fishers. Incidentally, the establishment of these latter three rows had marked the culmination of collective action on the part of the fishers who wanted access to the stake-net fishery. The challenges proposed by these three groups differed from the earlier challenges, in that the fishers had organized themselves in order to present their collective claim to the stake-net fishery as their caste-based right.

Notwithstanding the above, the first of the three sanghams to emerge in the early 1980s was made up of *c.* 21 families from VCDS who had joined together to form the stake-net row of Kodipadu, a row with *c.* 43 unlicensed stake-net locations. After a period of conflict, the Kodipadu sangham filed a petition to the Hon'ble High Court of Kerala in 1987 arguing for their caste-based right to the stake-net fishery. As a result of their action, a stay order was issued by the Hon'ble High Court, which prohibited the fisheries department from removing their so-called "illicit" stakes from the backwaters, but licenses were not granted. Following the issuance of this stay order, two other sanghams formed to claim similar rights, forming the unlicensed rows of Ayapenpadu (*c.* 31 families) and Muruganpadu (*c.* 26 families). According to interviews with sangham leaders, they did so based on the precedence set by the members of Kodipadu. Considering their "illicit" status (the term, probably, locally used by the fishers themselves), sanghams were well organized. Incidentally, the padu systems governed by the sanghams have three basic functions, which are as follows: (i) facilitating equitable access, (ii) providing collective social responsibility, and (iii) providing mechanisms for conflict resolution and rule-making.

5.1.7.63 *Facilitating equitable access to fishing grounds*

The stake nets are said to be fished with the stakes planted in long rows. The location of the row in the channel is an important factor in determining the amount of catch, but equally important is the specific location of a net in the row. For example, the secretary of the sangham for Ayapenpadu (a row with *c.* 62 nets) identified stakes 15–21 and 45–52 as the best locations in the row. Further, the president of Muruganpadu sangham stated that those fishing in the best locations could expect double the income as compared to those at other locations on the same row.

One of the key elements of the padu system was the attempt to redistribute the catch fairly among the fishers by rotating access to fishing locations. All of the sanghams in the study area had instituted a lottery system that rotates access to fishing locations to ensure equal opportunity to prime fishing locations. Meetings are called once a year by the sangham for all members to draw lots. Participation is generally very high, as the lottery is important in determining a fisher's livelihood for the coming year. The name of each fisher in the sangham is written on a small piece of paper, folded, and placed into a brass pot. For example, there are 21 fishers in Kodipadu, so there will be 21 names in the pot. A second pot is filled with 21-folded pieces of paper, with the numbers 1–21. Members will take turns randomly drawing a name from the first pot and a number from the second pot to determine the location of one of the two nets. The second net location is determined by the same draw, so that if fisher "A" had drawn the number "1," he will fish at location 1 and 22 on the row. Fisher "B," who had drawn number "2," will fish at location "2" and 23, and so on. The drawing of lots is repeated each year, so that the chance at the prime fishing locations is randomized from year to year. This system is in place in all of the three sanghams. An interesting variation to this system on one of the padu sites illustrated the adaptability of the system. Until 1996, there were 52 nets on Muruganpadu and 26 fishers in the sangham. Their lottery system operated as above with two nets per fisher. However, due to increased sedimentation of the channel, in which their nets are located, it was decided collectively that the 13 net sites most affected by the increase in sedimentation are to be removed from the row. This left the fishers with only 39 locations to share among the 26 members, and thus required a modification of the lottery process. It was decided that locations 1–26 would be selected using the usual lottery arrangement, but each of the nets from 27 to 39 would be rotated between two fishers who would share access to the location. The two fishers who were allocated locations 1 and 2 on the row would share location 27, those allocated 3 and 4 would share location 28, and so on. On any given day, fisher "A" would tie his net at the shared location, and on the following day, fisher "B" would use this site. The decrease in the number of net sites had meant less income, but the decision was arrived at collectively and allowed all members of the sangham to maintain equitable access to the remaining productive locations.

5.1.7.64 *Providing collective social responsibility*

Each of the three sanghams had a measure of collective responsibility incorporated into its structure and function. These measures included financial support for families in the event of the death of a family member or on the occasion of a marriage. Sanghams made collective donations to the temple during important festivals. Sanghams implemented shared responsibility for maintenance of the stakes on the row. These measures were supported in one of the two ways. In Kodipadu and Muruganpadu, the funds required for the collective well-being of their members were funded by auctioning a sangham net. In Ayapenpadu, the funds were collected as the need arose.

In addition to the collective responsibilities mentioned previously, sangham members had also borne the collective responsibility for maintenance of the row. In all of the sanghams, a committee was appointed for row maintenance and stake replacement. At the time of year when locations are assigned, the sangham determined the state of repair of the stakes. Fishers are asked which stakes are in need of repair and then a day is set for the repair or replacement of the stakes.

5.1.7.65 Mechanisms for conflict resolution and making of rules

Another function served by the sanghams was resolving internal conflicts among the members, and making rules to facilitate the operation of the sangham. Conflicts within the sangham are dealt with during meetings called by the sangham officials. Issues are presented orally or in writing, and often center around the allocation of fishing sites, more specifically conflicts arise usually on the right to use another fisher's location should he be absent at the time when the nets are set. When disputes between two parties could not be resolved informally, they are taken to the sangham for resolution, with the elected leaders serving as arbitrators. It may be noted that conflict resolution mechanisms apply only to the fishers of the same row, as there are rarely instances in which fishers will set their nets on a different row. As such, there is no mechanism to deal with such a conflict, where no formal linkages exist between the three sanghams.

Notwithstanding the above, Ayapenpadu dealt with the problem of fishing rights to locations in the absence of right holders, by allowing the boat partner of the absentee fisher to set his nets at the location and keep the catch. In this case, the rules were informal and there were no sanctions in place to enforce them. If a member was sick or physically unable to set his nets, other members would set and collect his nets giving his family the full amount from the catch. A written application was required from the sick member to the sangham in order to facilitate this process and avoid misunderstandings and potential conflicts.

Concomitantly, Kodipadu adopted a different approach to deal with the allocation of fishing rights to an absentee member's location. At one of the sangham meetings that took place during the study period, a rule was instituted stating that when a fisher was absent, the fisher at the neighboring location (higher number) had the right to fish in his location. However, the rule came about as a result of a specific conflict between the two fishers that had been addressed in writing to the sangham officials. During the meeting, there was a heated discussion, after which consensus was reached, a decision was made and a sanction was put in place to ensure compliance.

Notwithstanding the above, while the padu system attempted to ensure equitable access to the fishing grounds, it also prohibited fishing during the incoming tide when shrimp are migrating in from the sea to the backwaters. Fishers were keenly aware of the importance of allowing the shrimp to enter the backwaters to breed. Sanctions to enforce this rule were different for each sangham and ranged from written warnings to penalties of a substantive amount of money. In the case of repeated infarctions, offenders faced expulsion from the sangham.

5.1.7.66 Regional context of padu systems

A number of community-based fisheries had devised elaborate management systems that served to address questions of equity, distribute access rights, and mediate conflicts. Rotational access systems based on lottery are said to be not unique to South Asia only. Berkes (1986) provided a case study of a small-scale fishery in Alanya, Turkey, in which the local fishers devised a lottery system to allocate fishing rights to productive trammel net fishing sites in an equitable manner with the aim to reduce conflict. Incidentally, the classical example of systems of rotational access is the study by Alexander (1995) of a beach-seining community in Southern SL. In a fishing community with only two good sites available for seining, fishers developed an elaborate rotational system of access for the equitable use of the limited fishing grounds. They did so by limiting the access to the fishery, numbering all of the nets used, and using a lottery to allocate sites. Alexander (1995) used the term padu to refer to fishing grounds, but there were efforts also to focus on two other padu systems, which use stake nets and provide a direct comparison with the current case.

Notwithstanding the above, the word "padu" refers to "site" or "location" in SL (Alexander, 1995) as well as in TN and Kerala. More generally, the term refers to the system of rotational access in stake-net fisheries (Mathew, 1991). According to the fishers of Pulicat Lake in TN, padu has been practiced since "time immemorial" (Mathew, 1991). The term "Padu" refers to fishing sites in three distinct geographic regions, viz., Pulicat Lake in the State of TN, the Negombo Estuary in SL (Amarasinghe, 1997), and the present case in Kerala. However, there are a number of characteristics that had allowed for comparison of these three padu systems despite the diversity in the approaches to allocate access.

The first of these categories identifies and describes the role of the padu system in defining a group of rights holders, its caste-specific nature, and the role of the institutions in the functioning of the system.

The second category identifies and examines how the padu system determines the boundaries for fishing grounds and further defines the actual fishing sites shared by the members.

The third category is about the operation of the padu system, which is characterized by the use of specific gears to catch specific species and the use of a lottery system to allocate access to the fishing grounds.

In fine, it could be said that the padu system had emerged as a general system with locally adapted institutions that address critical management issues in the community.

5.1.7.67 *Discussion and conclusions*

The comparison of the three padu systems is said to be useful because it permits generalizations about the factors which help in the successful resource management.

In the Negombo fishery, the shrimp fishers were able to resolve both the exclusion and the subtractibility problems of commons' management. They solve the exclusion problem through the limitation of membership in the four Rural Fisheries Societies (RFSs). All shrimp fishers were members of the RFSs, and all RFS members could harvest shrimp with stake nets. They were subject to the rules made by each of the RFSs at the local level, coordinated across the four RFSs at the regional level, and backed up by the government. Hence, rules were nested across three levels of organization. Incidentally, the Negombo fishery was said to have lived through cycles of conflict and crisis by being able to reorganize itself and adapt to new conditions. It appeared to be biologically, economically, and socially successful under these arrangements (Amarasinghe, 1997; Atapattu, 1987), not because these arrangements were perfect but because they provide a measure of resilience for the local common property institutions.

Concomitant to the above, the **Pulicat Lake fishery in TN** shares some of the characteristics of the SL case. The three villages carried out their lottery activities at the local level, and this was further facilitated by the caste organization. The entry into the fishery at both the village and lagoon level were dealt with but the nagging exclusion problem remained. There was pressure on the fishery from the nonlocals. Moreover, conflicts had increased steadily since the boom in shrimp markets in the 1970s. Lack of state level involvement in the padu system, specifically the lack of legal recognition of its management functions meant that exclusion remained problematic. This, in turn interfered with the ability of fishers to enforce their own rules (Mathew, 1991).

Incidentally, the **Vallarpadam fishery** was even more problematic than the TN case, probably, because of the exclusion problem even in the immediate fishing area. There was no mechanism for the three groups of "illicit" fishers to coordinate with each other. Incidentally, each Sangham was able to limit its own membership, but there was no mechanism for the overall control of the number of fishers in the area as a whole. This was because the government did not recognize the management role of the sanghams and also did not even license them. However, it did not (or legally cannot) exclude them from the fishery, either. However, each sangham dealt with subtractibility through elaborate and well-articulated rules to provide for equity, social responsibility, and conflict management among its members. But the lack of legal recognition and cross-scale coordination had meant that subtractibility had remained as a problem at the regional level, and may eventually result in an unsustainable fishery.

Notwithstanding the above, **the SL** case showed full cross-scale governance from the local to the national level, and there was no provincial level government in SL. The importance of such cross-scale governance in matching the scale of management institutions to lagoon ecosystems has been documented elsewhere (Kalikoski et al., 2002). Incidentally, the **TN** case portrayed incomplete cross-scale governance and is intermediate between the SL and Kerala cases. Incidentally, the **Kerala** case seemed to have no cross-scale governance, no active intermediate-level institutions, and no government-community comanagement (Pomeroy and Berkes, 1997). However, this may not be considered as a failure of the local governance in the three Vallarpadam Island sanghams in Kerala. The sanghams did have a sophisticated set of locally adapted rules but fell short on cross-scale linkages.

Notwithstanding the above, institutional development and linkages in the SL case had a record going back at least to the 18th century. By contrast, the **Vallarpadam** fishery had a relatively short history of institutional development. The three sanghams only emerged in the 1980s, as a response to changes in provincial governance and the condition of the shrimp markets in the 1960s and 1970s. However, chances are that the **Vallarpadam** area padu system was not a local invention or a novelty to emerge in the 1970s. Rather, it was likely to had been borrowed from ancient traditions of resource use to emerge as a post-1970 folk system. In fact there were evidences to propose continuity. Incidentally, according to some fishers on **Vallarpadam,** the basic idea of a rotational, territorial use-rights system is age-old. Indeed, elsewhere in the Cochin area, there seem to be other padu systems, some of which might reach back earlier than the 1970s. However, these padu systems may be seen as part of the dynamic diversity of locally adapted commons institutions for coastal resource management in **South Asia.** In fact, these institutions try to respond to various external drivers and seem to be changing all the time.

The formation of sanghams on Vallarpadam and the reinstitution of padu may be seen as a response to change in markets and legislation. It is part of the attempt by fisher communities to improve their livelihoods and solve conflicts, and, in general, to deal with the world around them (Jentoft, 2000). It may, in part, also be a response to compensate for the loss of resilience in the fishery. Incidentally, there are evidences in the Vallarpadam fishery, and in Kerala and South India, in general (Kurien and Paul, 2000), of a loss of gear diversity and species diversity in the catch. Such a reduction in options, with a resultant loss of resilience makes the fishery vulnerable to economic and ecological fluctuations. This is offset by the emergence of diversity in common property institutions (i.e., the sanghams) which help communities to deal with change, as in the case of Ayapenpadu and Muruganpadu sanghams. Diversity in local commons institutions may, in turn, help increase social resilience (Adger, 2000), and hence the resilience of the fishery as a whole. The challenge to resource management in the small-scale coastal fishery, then, is not to maintain stability, but to maintain diversity and flexibility.

5.1.7.68 *Summary*

(i) Padu defines the group of rights holders.
(ii) Three sanghams (society or association) allocate access to the fishing grounds.
(iii) Padu is caste-specific. All fishers in a group are generally of the same caste.
(iv) Padu specifies institutional basis of rights holders.
(v) Sharing of the fishing grounds among the three villages is organized by the caste organization.
(vi) Individual villages operate the lottery independently.
(vii) Each fishing ground operates independently; no coordinating institution.
(viii) Lottery system run by individual sanghams.
(ix) Padu defines the geographical boundaries, fishing grounds, and sites.
(x) Fishing takes place close to the opening of the estuary into the sea.
(xi) Fishing grounds are located close to the estuary opening into the sea.
(xii) Two main fishing grounds are divided into 22 fishing sites that can accommodate 65–68 stake nets.
(xiii) Three main fishing grounds are further divided into 25 sites that can accommodate 56 nets.
(xiv) In addition three fishing grounds are divided into a total of 78 sites that can accommodate 78 stake nets.
(xv) The Padu is gear-specific stake-seine net, Suthu Valai (stake net), Ooni Vala (stake net).
(xvi) Padu is species-specific.
(xvii) Padu uses a lottery for site allocation.
(xviii) The four RFS rotate access to fishing sites on a daily basis using a 7-day cycle and a yearly lottery to assign starting points.
(xix) The three villages rotate access to the fishing sites on a daily basis within a monthly cycle of assigned days.
(xx) The three sanghams organize lotteries for the allocation of access.
(xxi) Lottery takes place once in a year at the annual meeting.

Notwithstanding the above, there remain significant differences among the three cases, despite the similarities which were apparent. There were two areas: cross-scale institutions (or nestedness) and legal recognition. Both are critical elements of Ostrom's (1990) design principles for effective institutions, and both seem to play key roles in facilitating the management of resources at the local level.

Notwithstanding the above, in the Negombo Estuary, institutions were nested (organized hierarchically) from the village to the lagoon level. Amarasinghe et al. (1997) identified the facilitating and organizing role of the Roman Catholic Church with the four individual RFSs as one of the reasons for the effectiveness of the padu system. In the case of the Pulicat Lake fishery, the padu system links across several scales. Incidentally, The GP facilitated the padu system for three villages. The elaborate system of sharing was said to allocate rights among the three villages, but was coordinated at the level of the fishing caste organization. This linkage addresses, and in most cases resolves the competition for the limited number of fishing locations. In contrast to this, in the case of the padu system on Vallarpadam, each sangham operated independently. They were not linked to one another or to any province or caste organization. However, legal recognition of traditional management represented a crucial step in empowering the fishers to manage their resources.

Notwithstanding the above, in the Negombo fishery, the four RFS were said to have been given legal status by the Negombo (Kattudel) Fishing Regulations, and resource sharing practices were reinforced by government regulations (Amarasinghe et al., 1997). In TN, legal protection was said to be weaker. The fishers had the right to organize but had no territorial use rights on the padu grounds. The CBM system was not recognized by TN State law initially (Mathew, 1991). Finally, the Kerala case was said to be still weaker.

Notwithstanding the above, it may be noted that fishermen villages are to be of ideal type having different facilities for their life (Ramachandran, 2005). Incidentally, a model fisherman village could have the following facilities:

(1) For providing basic civic amenities there should be facilities, such as housing, drinking water, sanitation, and construction of community hall for fisherman, etc.

(2) The villages are to be provided with tube wells at the rate of one tube well for every 20 houses.
(3) A fisherman village with at least 75 houses is said to be eligible for availing financial assistance for the construction of a community hall for recreation and common working place.

5.1.7.69 *Group accident insurance scheme for active fishermen*

These could operate in the following manner:

(i) Rs. 50,000.00 against accidental death or permanent total disability.
(ii) Rs. 25,000 for permanent partial disability. These values might have, of late, been revised.

5.1.7.70 *National cooperatives development corporation (NCDC)*

The cooperatives are strong peoples' organizations, which play a significant role in the socioeconomic life of the fisherfolk. There have been many schemes related to the fishermen's cooperatives. Some of the schemes aim at different purposes. Some of the significant points in this regard are as follows:

(a) Strengthening of existing cooperatives by providing with assistance on liberal pattern for expansion of their activities.
(b) The NCDC have formulated specific schemes and patterns of assistance for enabling the fishery cooperatives to take up activities relating to production, processing, storage, marketing, etc.

5.1.7.71 *Employment*

There were *c.* 900,000 full-time and *c.* 110,000,000 part-time fishermen in marine fisheries-related activities. Of these, the probable number of fishermen in different types of activities is as follows: (i) Four lakh in fish marketing, (ii) three lakh in repairs of fisheries requisites, (iii) *c.* 50,000 in fish processing, and (iv) four lakh in other ancillary activities.

Incidentally, *c.* 31.5 lakh persons are engaged in fishing and AQC farming activities, directly or indirectly.

5.1.7.72 *Development of model fisherman villages*

The fisherfolk generally live in the villages around the water body. They too need basic civic amenities. The following points may be considered:

(i) Provision of basic civic amenities, such as housing, drinking water, sanitation, and construction of community hall for fishermen.
(ii) Provision of tube wells at the rate of one tube well for every 20 houses in the villages.
(iii) A fisherman village with at least 75 houses is said to be eligible for availing financial assistance for the construction of a community hall, mainly for recreation and common working place.

Nevertheless, the cooperatives often face difficulties in their operations. There could be many problems. However, some of the problems in the operations of cooperatives at the optimum levels are as follows:

(i) Weak collaboration/coordination among various institutions involved in management and operation of cooperatives at primary, district, and provincial levels.
(ii) Rivalry among individuals/groups, lack of adequate infrastructure, lack of human and financial resources, lack of skills among the management workers and fishers, and so on.

As such, the "FISHCOPFED" play a significant role in strengthening the linkage and coordination among the cooperative at various levels.

5.1.7.73 *Human rights and environment*

The 1972 Stockholm Declaration proclaimed that man's natural and human-made environment are essential to his well-being and to the enjoyment of basic human rights, even the right to life itself. In 1986, the United Nations General Assembly (UNGA) recognized the relationship between the quality of human environment and the enjoyment of basic human rights **[UNGA resolution 2398 (XXII) 1986]**. The 1992 Rio Declaration emphasized sustainable development and environmental protection. Moreover, Agenda 21 called for the fulfillment of basic needs, improved living standards for all, better protected and managed ecosystem, and a safer and more prosperous future.

5.1.7.74 *Equity and fisheries access*

It may be noted that most of the fisheries regimes may be considered to be rights-based. However, ITQ regimes had been theoretically constructed and specified according to neoclassical economic prescriptions. In such regimes, maximization of net benefits in terms of narrowly defined economic values was said to be paramount. Incidentally, equity considerations are said to be notably absent. On the other hand, important collateral economic values bearing on the welfare of fishery-dependent communities are routinely ignored.

Concomitant to the above, with regard to equity impacts, ITQs are especially noted for large windfall gains from public resources to small groups, and for corporate and geographical concentration of fisheries access rights. This result relates to the way in which ITQs, typically, have been implemented.

5.1.7.75 *Classifications of fisheries management systems*

Type characteristics

(a) Private ownership with transferable right

Ownership rights are held by individual fishermen as in the case of fixed engines like stake nets and Chinese nets.

(a) State ownership

Ownership right of certain reservoirs and lakes are usually maintained with the government and access to fishing is banned for the fishermen or the public.

(b) Sole ownership

All fishing activities dictated by government as the sole owner as in the case of certain reservoirs, the fishing right is auctioned to fishermen groups or societies.

(c) Limited entry schemes

Access and withdrawal of rights restricted by the government as in paddy fields during periods of noncultivation.

(d) Closed season

This was said to have been adopted mainly for the trawlers in coastal waters, and for giant prawn fishing in certain inland water bodies during the breeding season. Bottom trawling in coastal waters is said to be banned from July 15 to August 31 every year by law.

5.1.7.76 *Common ownership (unofficial: without any legal support or governmental support)*

It is a resource owned and managed jointly by a small group of self-governing fishers. Unlike in other countries, there is no legal or governmental support for this arrangement. Each participant in the fishery, thus, depends on the arrangements or rules worked out in the fishermen societies or groups.

5.1.7.77 *Type characteristics*

Sri Subrahmanyeshwara Aqua Club in West Godavari District, AP

To promote cooperative approach in management of shrimp farming, cluster-based shrimp farms, aqua clubs, etc., had been set up in AP. A total of *c.* 128 Aqua Clubs with a membership of *c.* 3367 farmers is now said to be existing in the province. However, one of the best initiatives of this kind is the S·S Aqua club in Mogalthur Village is West Godavari district of AP. The club comprises *c.* 58 farmers with *c.* 108 ponds spread over *c.* 58 ha and the farmers involved were mostly small-scale, practising improved traditional farming with low investments. The above two initiatives by the shrimp farmers in the TN and AP need to be replicated in the other coastal provinces also where shrimp is fast developing on commercial scales.

5.1.7.78 *Models for India*

Every country requires a model tailored to its need. So does India.

Keeping the incentive structure of ITQs in mind, and the organization of the Japanese cooperatives, the following models of fishery management have been proposed for India:

(i) To calculate the MSY along a coastline (of sea or FW wetland).
(ii) The share (in weight) of each fishing family would be arrived at by dividing the MSY by the number of families.
(iii) Thus, the share in percentage of each family is determined, which may remain almost constant forever unless there is change in the population structure.
(iv) The actual amount of fish entitlement would depend on MSY as calculated periodically.

(v) Whether of a particular caste/clan or not, the quotas would still bind people together.
(vi) This enables the formulation of norms and rules easily. Enforcement also becomes easier in such closely knit environment.

5.1.7.79 *The role of public-private cooperation in the management of estuarine fisheries: Learning from the Kerala model of comanagement in Cherai poyil*

Cherai poyil was said to be a brackish water pond with an area of *c*. 210ha located in the Kerala State of Indian peninsula at the extreme northwest opening of the Cochin estuary, which opens to the Arabian Sea at Azheekode. Cherai poyil is a typical example of how a local self-government (GP) and communities together share the responsibility to allocate fisheries to various fishing groups and private entrepreneurs in the northern side of the Cochin estuary.

The fishing territory managed jointly by local fishing communities and the GP is known as "kappu."

5.1.7.80 *Rights and responsibilities of the GP in managing poyil fisheries*

Today, kappu is the property of Pallipuram GP, and its management is coordinated by a subcommittee that include elected councillor, fisher representatives, and leaders of local political parties. The GP has learnd from its past experience that direct mundane supervisory functions were expensive and community cooperation was highly essential to ensure better governance.

Concomitantly, both the communities and local GP recognized that CBM might not be very much feasible for want of modern skills to manage socioecological complexities of the modern markets. Thus, it was quite natural that both the organizations looked for viable management alternatives as direct management by any one agency appeared to be economically and politically expensive. Thus, the GP decided to lease out the primary fishing rights, as the leaseholder possessed resources and market information. And, then, onwards, the present practice of auctioning system had got stabilized. An important factor was "choice," which assisted in the stable growth in the volume of trade from the village to international markets.

Notwithstanding the above, increase in the demand for prawns had motivated a number of private entrepreneurs to lease out the kappu fisheries and manage it according to locally laid down norms of the GP. The local traders and the merchants were said to have been motivated to take over the fishing rights over kappu. Incidentally, the top most authority of kappu management is said to be the GP, which participate in this joint management exercise mainly because it receives money from this contract. It leases out the water body to the contractor who in turn allocates fishing rights to various gear groups including women. This transfer, however, was subject to a set of rules and regulations regarding the use of different fishing practices, type of gears, mesh size regulation, etc.

5.1.7.81 *Roles of various partners in cooperative fisheries management*

The partners are said to have specific significant roles in management in a cooperative management system. The partnership arrangement between the local GP and the leaseholder and their relationship with local communities is said to be mutually beneficial to the participating agents in a number of ways. The GP is said to participate in these joint management efforts, as this partnership yields cash income and monitory returns. For the leaseholder, there is clarity in the ownership and tenure, which restraints the rest of the communities to encroach upon his property. Communities may accept the procedure as an ideal alternative that delivers the required management services to kappu fisheries by reducing their risks and uncertainties.

5.1.7.82 *Local state: Upholding formal rules and regulations of central and provincial governments*

The GP, being the owner of poyil, has to make sure that various gear groups and the leaseholder(s) comply with the legal codes and regulations passed by both the central and the provincial government from time to time. These are the rules that are said to impinge upon the economic activities directly or indirectly (fisheries, AQC, public works, irrigation, agriculture, tourism, mining, etc.) undertaken by the local communities on the ecosystems.

5.1.7.83 *Failure to regulate externalities*

Kappu communities had also failed to prevent destructive activities undertaken by other users and the costs associated with these externalities. For instance, local AQC farms pollute the kappu by constantly altering the water while all the sediments and pollutants from these farms flushed into the kappu. During harvests, they pollute the farms to collect all prawns, and the water released from these farms pollutes kappu.

5.1.7.84 Conclusion

Cooperative resource governance is a unique arrangement by which local governments accept the role of communities in resource management and provide more space for their activities and initiatives in the management of the natural resources. The timely restructuring and flexibility of the system in tune with the changing forces of modernization and external pressures is needed for success in management.

Concomitant to the above, in India, the reservoirs contribute considerably to the inland fish production (Vass and Sugunan, 2009), which had been estimated at *c.* 93,000 tons (Anon, 2006a,b,c). In spite of this fact, reservoir fish production had been treated as a by-product, relegating it as a less important fish production system. For this reason, and could be for other reason(s), reservoir fisheries had not been able to make significant progress in India. As such, India may not be in a very strong position to contribute to inland fish production of the country to the extent it could have been. Occasional improper management is also said to be a cause for less amount of success in reservoir fisheries. CBFM may, preferably, be applied for effective management of the reservoirs.

However, subsequently, the managers of a majority of the water and power projects in the country realized that fisheries need to be given its due importance in overall reservoir management including initiation of CBFM. It was also felt that reservoirs in India offer ample scope for FY optimization through effective management (Vass, 2007a), preferably, through the application of CBFM. Further, from a social perspective, it is believed that the benefit of increased FY and income generated by the fishing communities of reservoirs to be more or less equitably distributed if CBFM is implemented. However, it may need further ground works. Incidentally, efforts may be directed to achieve the goals related to the objectives of reservoir study with regard to yield optimization, sustainable harvest and management, and environmental conservation.

5.1.7.85 Reservoirs

GoI was said to have **defined reservoirs** as man-made impoundments formed by obstructing surface flow, by erecting a dam of any description on a river, stream, or any flowing water course. However, water bodies <100 ha in area had been excluded from this definition. The Ministry of Agriculture (MoA), GoI had classified reservoirs, for the purpose of fisheries management, as (a) small: <1000 ha; (b) medium:1000–5000 ha, and (c) large: >5000 ha.

Notwithstanding the above, various types of classifications may be *in vogue* in different provinces. The estimated total areas are *c.* 1,485,557 ha, *c.* 507,298 ha, and *c.* 1,160,511 ha for small, medium, and large reservoirs, respectively (Sugunan, 1995). It may be mentioned that in the absence of effective management system, for example, through CBFM, the whole exercise of damming a river and conversion of a lotic system into a lacustrine system could be futile.

Incidentally, India has *c.* 19,370 reservoirs spread over *c.* 15 provinces with *c.* 3.15 million ha surface area at full storage level, and this is expected to increase due to execution of various water projects in the country. Incidentally, the province of TN is said to have the highest reservoir area of small reservoirs (Sreenivasan, 1989) followed by the provinces of Karnataka and AP, Madhya Pradesh (MP), while AP, Gujarat, and Rajasthan also have a higher area of medium reservoirs. Incidentally, Karnataka is said to have the most large reservoirs.

5.1.7.86 Morphometric and hydrological features

Morphometric factors (basically area, mean depth, irregularity of shoreline) are said to have significant bearing on productivity in reservoir ecosystems. Mean depth (volume/area) in reservoirs is considered to be one of the most important morphometric parameters indicative of the extent of the euphotic-littoral zone (Rawson 1952, 1955). A relationship between FY and MEI has been worked out in some African reservoirs by Henderson and Welcomme (1974) and Henderson et al. (1973). Concomitantly, Ramakrishniah (1990) is said to have worked out the MEI of a few Indian reservoirs, but no statistical relationship was established. However, in the Indian reservoirs, depth is not always a constraint to productivity. Many of the small and medium reservoirs are of low mean depth (*c.* 4–7 m) and depict high fisheries potential.

Community involvement for proper application and utilization of the morphometric and hydrological features is essential through proper consultations and interventions.

In contrast to the above, the Gangetic basin is more fertile than any other riverine basin, notably, the Indus, etc., because the gangetic alluvial soil boosts productivity of the reservoirs and the basin as a whole (CIFRI, 1986). On the other hand, the upper Indus flows through glacial terrains before emerging onto the plains in the lower reaches.

Concomitantly, in the Gangetic basin, the predominant soil formation is an illitic clay-dominated alluvium zone with high potassium content and moderate nitrogen and phosphorus content, while the Indus basin soils are poor nutrients mainly in the upper reaches, with low-to-moderate nitrogen and phosphorus content in Punjab, Haryana, and Himachal Pradesh (CIFRI, 2006).

Notwithstanding the above, in the shallow reservoirs, the reservoir bed may serve as a source of autochthonous nutrient loading (De Silva, 1988), while, by contrast, in the Indian reservoirs, it has been observed that the catchment determines the water quality to a larger extent than the basin soil (Natarajan, 1976, 1977, 1979; Jhingran, 1986). The extent of catchment has a positive impact on the reservoir productivity, provided the catchment is moderately fertile. **The catchment (C) to reservoir area (A) at FRL is considered to be an index of allochthonous input**. The ratio could vary in a narrow range in the case of small and medium reservoirs in India. Again, effective **management** of the reservoirs through CBFM could make proper utlization of the productivity of the reservoir.

Concomitant to the above, flushing rate (inflow/storage capacity) is also significant as it regulates both the degree and regime of nutrient loading.

Notwithstanding the above, drastic fluctuations in WL have been observed in the majority of Indian reservoirs to impact plankton, benthos, and periphyton pulses (Unni, 1993). Proper regulation and **management** of the abiotic factors, notably the WL fluctuations, could lead to better results.

5.1.7.87 *Physicochemical features of reservoirs*

Coarse and fine sand are said to predominate the basin soil in the majority of the downstream reservoirs of Rajasthan, Haryana, UP, and Punjab followed by MP, HP, Jharkhand, and WB. Silt deposition is said to have been observed mainly along the river courses, and in majority of the reservoirs in the Gangetic plains, the sediment is mostly sandy-clay type. Proper **management** of the substrate type could lead to higher productivity.

5.1.7.88 *Chemical characteristics*

Soil reaction

Soil is, generally, neutral to moderately alkaline in reservoirs, thus reflecting a moderately productive trend. However, moderately acidic basin soils were observed in some reservoirs in Bihar and Jharkhand due to the forest-covered catchment. It may be mentioned that the release of phosphorus is basically dependent on soil pH. Bottom sediment containing a high level of organic matter with slow decomposition rate develops acidity due to humic and short-chain fatty acids leading to less productivity. Incidentally, the favorable pH range for biological productivity is said to be 6.5–7.5. Proper education and **awareness at the community** level is essential for effective soil management.

5.1.7.89 *Nutrients*

Available nitrogen

Nitrogen is the fundamental and primary constituent of protein and is basically present in soil in organic form getting mineralized gradually to ammonium (NH_4^+) and nitrate (NO_3^-) forms, thereby making their availability to fish food organisms. Reservoirs in Punjab, HP, MP, UP, and WB have moderate amount of available nitrogen content but moderate-to-high productivity status. Proper **management** at the stakeholders' level is said to be essential for the management of nutrients in the soil.

Available phosphorus

Soil available phosphorus (SAP) (mg/100g soil) may, preferably, be in the range of 4.7–6.2 for high biological productivity but it is said to be in low profile in the Indian reservoirs. Its range (mg/100g) in reservoirs, especially in Punjab, Haryana, HP, and Jammu and Kashmir (J and K), is said to be of very high order, while it was low in other reservoirs. Proper **community** interventions could enhance the productivity.

5.1.7.90 *Physical characteristics of water*

The transparency of water, generally, decreases during the monsoon due to the inflow of water being loaded with dissolved and suspended organic and inorganic particles, which are subsequently said to be stabilized during the post-monsoon period. Water transparency has a significant role in light attenuation with depth, affecting productivity. Reservoirs with rocky and gravelly catchment, such as those in Jammu and Kashmir, HP, etc., are said to have deeper euphotic zones of >2m. In these reservoirs, the temporal variation in transparency is very wide unlike reservoirs of MP (Unni, 1985). The **reduced transparency** in these reservoirs is due to suspended inorganic particles, and hence **phytoplankton production is much less**. Proper management at the community level is essential for maintaining good physical characteristics of water.

Thermal stratification

In tropical Indian reservoirs thermal stratification is not stable and prolonged. The degree of stratification is low in comparison to temperate reservoirs. Even though the period of thermal stratification is short, it has a significant influence on biological productivity (Pal, 1979).

5.1.7.91 *Chemical features of water*

Dissolved oxygen (DO)

In Indian reservoirs, DO is, generally, higher during the pre- and post-monsoon periods as compared to the monsoon days, mainly due to the downpour. An oxycline is a basic feature in any productive reservoir. Deep tropical and subtropical reservoirs are said to depict a stable clinograde (i.e., decreasing DO levels with depth) with sharp fall in oxygen concentration at the hypolimnion, which mostly occurs during the pre-monsoon period. Incidentally, **highly productive water could possess a DO concentration of > 5 mg/L.** However, a very high concentration of DO leading to supersaturation may become lethal to fish fry. Management of the reservoir with regard to maintenance of DO level is to be done **involving the community**, with proper interventions, wherever necessary.

Free carbon dioxide (FCO_2)

Some reservoirs may be free from aquatic macrophytic (AM) vegetation, or AM may be confined and limited to littoral zone only. FCO_2 may be in traces or absent at surface waters. Conversely, a substantial amount of FCO_2 could be found in surface waters of a number of reservoirs in India (Unni, 1993; Singh et al., 1980). **Community-involved** management could also manage the fluctuating harmful level of FCO_2 through consultations and actions in this regard.

pH

Indian reservoirs seldom show drastic pH fluctuations mainly due to natural buffering capacity of water. However, in general, water pH is low during the monsoon mainly due to dilution of alkaline substances or dissolution of atmospheric CO_2 with a resultant increase during the post-monsoon and pre-monsoon periods. However, majority of the reservoirs have moderately alkaline pH due to alkaline soil reaction. **In community-managed practices**, the community members could monitor the pH through the application of lime from time to time or otherwise.

Specific conductivity (SC)

SC is an index of the amount of water-soluble salts present in the water. In normal FW, chloride content could lie within the range of 10–30 mg/L. If the value is somewhat more than this, it is said to indicate local pollution. The average chloride content (mg/L) in the reservoirs was in the range 7–30 with higher values observed, generally, during the pre-monsoon months in some reservoirs of Rajasthan and Haryana.

In FW, electrical conductivity (EC) is also said to be considered as an index of productivity when the ionic composition is dominated by CO_3 and HCO_3 system because conductivity is known to be related to total dissolved solids (Henderson and Welcomme, 1974). **Community, trained appropriately**, could take care of the SC level of the water bodies.

Total alkalinity (TA)

It may be interesting to note that many of the Indian reservoirs had portrayed moderate **TA within the range of 40–205 mg/L, thus representing moderate productivity** trends. Incidentally, **the low values of TA in Jharkhand and Bihar reservoirs could be due to the acidic basin soil** of lateritic type originating from dry forest areas.

In general, total alkalinity (TA) is high during pre-monsoon period with a substantial decrease in the monsoon period mainly due to dilution. Optimum level of TA may be maintained by **community management** of the water bodies through proper consultations and intervention measures.

5.1.7.92 *Nutrient status*

In Indian reservoirs, available N (NO_3+NO_2) in water is said to be typically very low throughout most of the year, except in monsoon and, to some extent, for a shorter period during the pre-monsoon period. Incidentally, during the monsoon, its availability is said to be due to runoff from fertile catchments, and in summer it is said to be due to a faster rate of decomposition of bottom organic load, rendering nitrogen availability to standing water phase.

Concomitant to the above, in Indian reservoirs, except for a shorter period during the monsoon, the availability of phosphate is of a very low order and rarely exceeds 0.1 mg/L. **CBM** could also take care of proper management of nutrient inputs.

5.1.7.93 *Biological features of reservoirs*

Plankton

Plankton assumes the principal role in maintaining trophic structure, as neither the AM nor the periphyton get a strong hold due to effects of drawdown and high flushing rates in the Indian reservoirs.

Notwithstanding the above, in most of the reservoirs, green algae and the blue-green algae (BGA) form the bulk of the phytoplankton community followed by diatoms. Incidentally, the occurrence of *Microcystis* is said to be significant. Among the zooplankton (ZP), the copepods are important followed by cladocerans, rotifers, and protozoans. Moreover, altitude of a water body also plays an important role in the distribution of *Microcystis*.

Macrobenthos

Shallow reservoirs, such as Bachhra, Baghla, Kerwa, and Dahod, were found to be particularly rich in benthic fauna due to favorable substratum, rich organic matter, and minimal WL fluctuations (Katiha et al., 2007). The deeper reservoirs usually have poor benthic macrofauna, while the reservoirs with a rocky and gravelly bottom have no macro-benthic communities. Concomitntly, dipterans were said to be the highly significant in the reservoirs followed by gastropods, bivalves, and different types of annelids including the oligochaetes.

The significance of bottom fauna in relation to FY in reservoirs have been highlighted by Hayes (1957).

Aquatic macrophytes (AM)

AM are potential biotic entities in a lentic body, notably a reservoir. Some reservoirs have a rich growth of AM due to low flushing rates, e.g., Ramgarh, Jaisamand (Rajasthan), Bachhra, Baghla, Pahuj (UP), Halali and Dahod (MP), Sone Beel (Assam), etc. The biomass varies from 0.2 to 5.2 kg/ m^2 wet weight. The peak is virtually observed in summer and winter, and decreases during the monsoon season.

AM are mostly submerged, emergent, and floating types and were mostly abundant in the littoral areas. AMs provide substrata to many insects, molluscs, and other invertebrates, which serve as fish food organisms. They accumulate large quantities of inorganic nutrients, thus affecting the PP growth adversely. Moreover, the submerged weeds give shelter to the minnows, which compete with the IMCs for food and living space. Further, luxuriant growth of AM prevents sunlight penetration, and could cause fish mortality due to DO depletion.

Some of the common AMs found in Indian reservoirs are as follows:

Hydrilla, Vallisneria, Lemna, Pistia, Hygrorrhiza, Trapa, Eichhornia, Nymphaea, Nymhoides, Nelumbo, Salvinia, Azolla, Eleocharia, Echinochloa, Scirpus, Cynodon, Euryale, Ceratophyllum, Najas, etc. However, AM like *E. ferox* may also cause severe difficulties in the operation of fishing gears. On the other hand, some types of AM may help in the oxygenation of the water, and also facilitate the collection and anchoring of fish eggs.

Incidentally, *Potamogeton* was said to be predominant while *Hydrilla, Vallisneria, Ceratophyllum,* and *Najas* were also recorded among the AM. The AMs, in general, and the *E. ferox,* in particular, are generally regarded as impediments for fishing operations, but, concomitantly, these submerged AM facilitate the collection and anchoring of fish eggs also.

Notwithstanding the above, the reservoirs along with their biotic components (notably the fishes) could be harmed through infestation by luxuriant growth of AM, leading to **eutrophication.**

In view of the above, AMs could effectively be managed by involving the **community stakeholders through CBFM** mechanisms, which could take care of not only the fisheries management but also the management of eutrophication issues.

5.1.7.94 *Fish and fisheries in the reservoirs*

Species composition

Fisheries are the potential entity of nutrition and avocation of the people. As such, maximum care is to be taken in its effective management through the application of efficient management techniques like **CBFM.**

The ichthyospecies contributing to the fisheries of the reservoirs located in the upper reaches of the rivers of the Ganges system are said to be mahseer, viz., *Tor putitora, Tor mosal, Tor. tor, Neolissochilus hexagonolepis*; the snow trout, *Schizothorax plagiostomus*; medium carps, *Labeo dero* and L. *pangusia*; and the goonch (*Bagarius bagarius*).

The reservoirs located in the middle and lower reaches of the rivers harbor mainly the IMCs, notably, *L. rohita, Labeo gonius, Labeo calbasu, Cirrhinus mrigala,* and *Catla catla*, and other carps such as *Labeo bata, Labeo boga, Labeo boggut, P. sarana, Chagunius chagunio*, etc.

Incidentally, the major catfish species contributing to the fisheries of the reservoirs of the Ganges river system are *W. attu, Silonia silondia, Pangasius pangasius, Rita rita, Sperata aor, S. seenghala*, etc. Concomitantly, the smaller cat fishes found in these reservoirs are *Clupisoma garua*, *E. vacha*, *M. cavasius*, *O. bimaculatus*, etc. The formation of reservoirs has adversely affected the indigenous fish stocks of the mahseer, snow trout, *L. dero*, and L. *dyocheilus* of the Himalayan streams.

Status of fisheries

Any fishery that is effectively managed by efficient techniques, like CBFM, could yield good results.

In HP, the two large reservoirs, Gobindsagar (*c.* 16,867 ha) on River Sutlej and Pong (*c.* 24,629 ha) on River Beas, and a newly constructed small reservoir Chamera (*c.* 900 ha) on River Ravi are highly productive. The annual FY of Gobindsagar reservoir increased from *c.* 17 kg/ha (1975) to *c.* 96.4 kg/ha (1992–93), which is very high when compared to the national yield rate of large reservoirs (*c.* 11.43 kg/ha) in India.

However, stock enhancement through silver carp is preferred, as the fish is of low economic value, highly perishable, and of low consumer preference. As such, in such a situation, an aware community (through CBFM) may exercise its role in selecting the most preferred fish for stocking. In the present situation, it has been proposed that the fishery dominated by higher valued IMC be reestablished. CBFM has been followed in this case to take community's opinion.

Concomitant to the above, in UP the Rihand (*c.* 46,538 ha), one of the important large reservoirs popularly known as "catla mine," had a dominance of *C. catla* (*c.* 73%–99%) from 1971 to 72 to 1980–81. However, due to failure of catla breeding, the catla fishery collapsed. Moreover, **CBFM technique** of fishery management was not followed in this case.

Notwithstanding the above, studies conducted by CIFRI in small reservoirs have documented a number of success stories. Small reservoirs, viz., Gulariya, Bachhra, and Baghla in UP have depicted a threefold increase in FY ranging from 50 to 150 kg/ha (Jhingran, 1986).

The culture-based capture fishery management measures coupled with **CBFM measures** were instrumental for remarkable enhancement in fishery of these reservoirs.

Concomitant to the above, the FY potential of Gulariya reservoir was reported to be *c.* 234 kg/ha which resulted from a stocking rate of *c.* 500 fingerlings/ha. The production increased from *c.* 8 to 22 tons (*c.* 150 kg/ha) after the application of the new **management** package **CBFM.**

Studies in this direction have been done by other researchers in different reservoirs (Dubey and Chatterjee, 1977). Incidentally, the recapture rate of the IMC in relation to stocking was very poor in some reservoirs, to the extent of only 2%, indicating that the reservoir was not under CBFM mode of management. Further, the reservoirs having hilly catchment area is favorable for breeding and propagation of mahseer. The rocky bed, submerged aquatic plants, molluscs, and insects are also suited to mahseer. Incidentally, the Government of MP has reserved the Kerwa reservoir (482 ha) for the development of a sport fishery for mahseer.

Concomitantly, Pal (1979) reported that Kulgarhi reservoir (193 ha) was the first reservoir in India where the exotic silver carp (*H. molitrix*) was introduced on an experimental basis. The silver carp has adversely affected the growth of catla in the reservoir as both the fishes compete for the same food resources. Moreover, the reservoir was not under CBFM mode of management.

5.1.7.95 Fishing practices

Gears

Fishing gears are the essential implements for fish catching. They vary in type, *modus operandi*, and catch record in different water bodies. In Sone Beel, the biggest wetland (3458.12 ha at FSL) in Assam, there are 26 different varieties of fishing gears and devices, and there is gear specificity to season type, fish type, and fishermen type (Kar, 1990, 2007a,b,c, 2013; Kar and Dey, 1991, 1993a,b; Kar et al., 1999a,b,c, 2006a,b,c,d, 2007a). The management practices, notably, through CBFM could play a profound role in regulating the fishing gears and devices for sustainable harvest of the fish resources. It may be noted that among the major fishing gear in the reservoirs, the surface gill nets of different mesh sizes (20–150 mm bar size) are quite common. The standard size of gill net is 50 × 2 m with a head rope and floats and footrope with or without sinkers. The net is made of nylon twine. Currently, the plastic twine (mono-filament) nets are also commonly used in the reservoir fisheries; *c.* 77% of the catch is entangled. Dragnets (shore seine) are also used in some reservoirs, especially for catching catfishes and weed fishes. In UP, MP, Assam, etc., the dragnets are called "Mahajal." The other fishing gears, such as cast nets, scoop nets, long lines, and traps, are also operated, but the catches are not much significant. The presence of underwater obstacles restricts the operation of active gear(s) in reservoirs and the choice is often limited to passive gear(s) such as gill nets. The most common among them is the Rangoon net, an entangling type of gill net without a foot rope. "Uduvalai" is another entangling type of net used in reservoirs, which has a reduced fishing height and is usually operated in shallow marginal areas to catch small fishes.

Shore seines of different dimensions and mesh sizes are operated in many reservoirs. Although a number of other fish catching devices, such as the long lines, hand lines, pole and line, cast nets, dip nets, are in use, their contribution to the total catch may not be always very significant.

It may be emphatically mentioned that operation of fish catching devices are regulated by the community. Besides, CBFM could play a significant role in judicious use of the fishing gears for sustainable harvest of the fishes.

5.1.7.96 *Fishing crafts*

Fishing crafts are significant components associated with fishing gears. Coracle, a saucer-shaped country craft is the major fishing craft. It is profoundly used in the reservoirs of peninsular India. It is made of a split bamboo frame, covered with buffalo skin. In addition to being simple and inexpensive, coracle is durable and has very good manoeuvrability in waters. It is also a versatile craft used for laying and lifting of nets, besides navigation and transport of fish and other materials. Coracles of Krishnarajasagar are prepared by wrapping HDPP over the bamboo frame with the help of coal tar as an external covering in place of buffalo skin. This modified version of coracle is cheaper and more durable (Parameswaran and Murugesan, 1984).

In Sone Beel wetland in Assam, two different types of boats are usually used, viz., (a) flat-bottom boat and (b) round-bottom boat. The former is generally used for operating big encircling gears, notably, the Mahajal, Chatjal, Ghuranjal, etc., while the latter is usually used in operating small, domestic gears like the cast net, Chhat jal, etc. In Chatla Haor of Assam, almost a similar situation is *in vogue* (Kar, 1990, 2007a,b,c, 2013; Kar et al., 2006a,b,c,d). In addition, in Gobindsagar, all the fishers possess their boats. But, the reservoir fishers are too poor to own boats. In many reservoirs like Vallabhsagar and Hirakud, the fishers could get their boats through a subsidy and other financial assistance from the government or funding agencies. In Vallabhsagar, boats aredistributed by the state among the fishers at a subsidy of 50% from various government-sponsored schemes. Wooden boats are used for fishing in a number of reservoirs, especially in North India. Flat-bottom, locally fabricated boats ranging in length from 3 to 7 m are, generally, used in Kyrdemkulai, Hirakud, Malampuzha, Gobindsagar, Gandhisagar, and Rihand reservoirs. Further, a plank-built, flat-bottom canoe of 2–3 m in length is the most popular fishing craft of Gandhisagar. Mechanized boats are not used in reservoir fishing to any appreciable extent. A 9.1-m long wooden, mechanized boat has been introduced by the CIFT (Central Institute of Fisheries Technology) in Hirakud reservoir but these are too expensive for the fishers. It is significant to note that no motorized crafts, either for fishing or for fish transport, are used in large water bodies like Nagarjunasagar, Tungabhadra, and Krishnarajasagar in the earlier days. Dugout canoes, carved out of palm trees, are used in Yerrakalava reservoir. The fishers rely on improvised materials in most of the reservoirs in the country.

5.1.7.97 *Fish production trends*

In India, the reservoirs are underexploited with an average annual FY of *c.* 20.15 kg/ha, with mean production in small, medium, and large reservoirs being 48.05 kg/ha/year, 12.39 kg/ha and 11.31 kg/ha, respectively.

Incidentally, the overall FY from the reservoirs remained low. This could be due to poor management as a result of the lack of understanding of reservoir ecology, trophic dynamics, inadequate stocking, wrong selection of species for stocking, low size of stocking materials, and irrational harvest. Nevertheless, the existing so-called wide gap between the potential productivity and the actual yield could be bridged by adopting the available package of practices, developing the necessary infrastructure facilities, and providing with conducive socioeconomic environment.

Based on hydrological conditions and primary productivity, incidentally, the large reservoirs have a production potential of *c.* 65–190 kg/ha, the medium reservoirs 145–215 kg/ha, and the small reservoirs 285–545 kg/ha per year. There is a wide gap between potential and actual yield, which could be mitigated, and *c.* 70%–80% of the potential yield could be achieved through proper scientific management, notably, CBFM; by developing the necessary infrastructure facilities; and by providing with conducive socioeconomic environment.

Notwithsatnding the above, the FY from Sone Beel (3458.12 ha at FSL) was 302.12 mt per year on the average (Kar, 1990, 2007a,b,c, 2013). Adequate management measures are a prerequisite in achieving rich FY. Also, efficient management practices, like the CBFM, could be an effective tool in achieving the target of high FY per unit area of the lentic area.

5.1.7.98 *Efforts in yield enhancement*

It is quite well known that various efforts have been made in India and in other countries to scientifically manage the reservoir fisheries. The ICAR and other organizations had made efforts through linkages with the provincial government that regulate the reservoirs to a big extent.

However, unfortunately no horizontal expansion could take place due to various constraints faced by the provincial departments.

It could be possible to achieve the potential target production especially from small reservoirs in India. Therefore, scientific management, if continued on a long-term sustainable basis, by the reservoir owners of the provincial departments, could significantly increase the overall reservoir fish production. This could help improve the livelihoods of large number of people in comparison to benefiting only a few in AQC activities, which, in the long run, may have environmental implications too.

5.1.7.99 *Socioeconomic and institutional settings*

Socioeconomic and institutional settings are important aspects of lentic water study. However, in India, very few studies have been performed on the socioeconomic status and institutional framework of reservoirs, particularly with reference to fishing activities. It may be mentioned that fisheries had been developed as a secondary user in most reservoirs, so the fisheries component was not included with due significance during the planning phase of the reservoir construction. Consequently, there were no fisheries-related studies during the pre-impoundment phase. In most reservoirs, thus, impediments to fishing, such as submerged tree stumps, rocks, etc., are said to be present, and their removal after the construction of reservoir is said to be much cost intensive. Thus, there could be representations from the Fisheries Authorities in the Reservoir Management Boards for better management of the fisheries in the reservoirs. These water bodies, if managed properly for fisheries operations, could be very effective in enhancing the income of rural communities in addition to providing employment opportunities as in the case of Gobindsagar and Pong Dam reservoirs in Himachal Pradesh (Katiha, 1994). Also, management of a lentic body, notably a reservoir, could be done in a very effective way by involving the stakeholders in the community from the grassroots level through the application of efficient *modus operandi* like the CBFM in a participatory mode for reaching the desired management goal.

Notwithstanding the above, reservoir fisheries are considered as a secondary activity compared to the irrigation and hydropower generation. The department of fisheries or other fisheries agencies may obtain the fishery management rights from the owners of the reservoirs, either by paying some royalty or by paying a nominal amount of tax as in Rihand in UP and Kansbati in West Bengal, or free as in Gobindsagar and Pong Dam in HP. In some situations, the fisheries department, in turn, may transfer the fishing rights to some other government agencies, cooperative societies, or to private agencies; and the department, in turn, could receive the royalty from them.

Concomitant to the above, one of the common institutional arrangements for the exploitation of fishery resources in reservoirs is a leasing system. It, however, depends on trends of fish production, income and expenditure of the department, socioeconomic conditions of the fishers, government policy toward cooperatives, and status of fisheries resources of the province. Incidentally, past experiences reveal that reservoirs entrusted to state agencies for the management of fish stocks through stocking and control of harvesting had consistently maintained a good production level, while those under the control of private contractors have not been encouraging. On the other hand, the results have been moderate, where the fisheries management was vested with the cooperative societies. In addition, it may be mentioned that, when reviewing the fisheries management practices for various-sized reservoirs, the appropriate management strategies for large and medium reservoirs may involve stock enhancement aspects, while for small reservoirs, culture-based management could be considered as appropriate. Further, the less availability of reliable catch statistics and FY estimates in reservoirs are a major setback for defining scientific management strategies for the Indian reservoirs.

Notwithstanding the above, there is a need to merge the twin objectives of conservation and yield optimization in reservoir fisheries management. The fishers and fish vendors are known to strive to increase the production to maximize profit. It is thus the responsibility of the state to ensure that economic expediency of development does not mar ecological reasoning. In this connection, it may be mentioned that the virtually open-access nature of fisheries exploitation, as practiced in AP, is counterproductive to resource conservation and management. There may be fair possibilities of linking reservoir fisheries development with poverty alleviation programs, but the progress made so far in this direction is, perhaps, not very much encouraging. The possibilities of generating additional employment openings may not be much good in the near future in majority of the Indian reservoirs. On the contrary, many reservoirs have surplus manpower, which could be diverted to other reservoirs, enabling them to continue fishing without adverse effects on the income level of the existing fishers (Paul and Sugunan, 1990).

5.1.7.100 *Management practices*

Management relates mainly to proper handling of the lentic body for (a) its own sustenance, (b) sustenance of the people living around, and (c) for the sustenance of the biota living within.

In India, "fisheries" are said to be under individual provincial control, and there is a great deal of variation in the management practices followed by different provinces ranging from outright auctioning to almost free-fishing access. Cooperative societies and the provincial-level fisheries development corporations are also involved in the management of fishing and marketing operations. The nature of their involvement and their role in the fishery and market interventions may vary from one reservoir to another within the same province. Incidentally, the reservoirs in India, with very few exceptions, are public water bodies owned by the government departments, such as district administration, departments of fisheries, irrigation and power generation, and so on. In many cases, management of their fisheries is handed over to the fisheries department. The management options for developing fishery in lakes and wetlands in different river basins in India had been highlighted by Vass (2007b,c). Those with suitable modifications could be tested in small reservoirs on a pilot scale.

Notwithstanding the above, the fishers are required to obtain a license from the local office of the provincial fisheries department. The authorities of the provincial department determine the number of fishers and fishing units to be operated in a reservoir. The directorate at the provincial department usually determines such regulations, for example, the mesh regulations and the decisions regarding closed seasons, which are policy matters. Incidentally, in some provinces, such as MP and HP, the fisheries development corporations are said to be responsible for managing reservoir fisheries. These corporations act as overseeing bodies for numerous cooperative societies operating in the reservoirs, and undertake marketing functions to ensure that fishers get the correct price for their catch. However, their market interventions have not markedly benefited the fishers (Sinha and Katiha, 2002).

Notwithstanding the above, crop sharing is a very popular mode of exploitation in TN, where the fishers are provided with all fishing implements, in exchange for a royalty payment from the landings (sometimes up to 50% of the catch) to the provincial government. In many cases, licenses are issued either free of charge or for a very small nominal fee. Reservoir fisheries development is also often linked to many social welfare and relief activities under village and district administrations. In such cases, immediate social goals are pursued overlooking the long-term effects on the fish stocks.

In fine, it could be stated that management of the water body through the involvement of local stakeholders is the best *modus operandi* in the management of the lentic body. Also, CBFM could play a significant role in the sustainable management of the water bodies.

5.1.7.101 *Future action plan*

FY from the reservoirs are to be augmented. Some of the measures to increase FY from the reservoirs are as follows:

(i) Extensive transfer of technology (TOT) from the State Department of Fisheries to the grassroots level of fishers and fisher organizations. The government officials might also require training.
(ii) Reappraisal of the current state policies for the ownership, fishing rights and fisheries management, and leasing (lessee, lease tenure, and amount) to encourage the reservoir fisheries development activities.
(iii) Enforcing appropriate guidelines for stocking and other management protocols, strict monitoring of stocking and harvesting schedules, and so on.
(iv) Issuance of notification for "closed season" or fishing holiday during the breeding season.
(v) Declaration of the lotic sector of reservoir as a highly sensitive and protected area during the breeding season.
(vi) Effective enforcement of size and mesh regulations to conserve the breeding stocks.
(vii) Stringent prohibitory regulation(s) against introduction of exotic fish species. Rather, encouragement of region-specific introduction and establishment of coveted indigenous local fish species.
(viii) A partnership approach may be encouraged by involving various user groups, particularly the stakeholders from the grassroots level, mainly for CBFM.
(ix) To involve sister organizations like the irrigation, agriculture, and environment departments with the fisheries department to help maintain sustainable fisheries.

5.1.7.102 *Community fisheries in India*

There have been extensive studies related to fisheries in the developing world. South Asian countries possess rich fisheries potential. Incidentally, South Asia's five coastal countries, viz., Bangladesh, India, Maldives, Pakistan, and SL, account for <2% of the world's total coastline. Yet, these coastal zones are rich in marine resources, which help and support *c.* 135 million people (Paul and Nishida, 2008). India alone have *c.* 4000 fishing villages, which support *c.* 15 million residents and provide food security for the nation (FAO, 2011). Further, in India, as in many regions of the world, coastal and marine ecosystems had been governed in a piecemeal and erratic fashion, which had resulted in unsustainable pressures on natural resources. A study conducted by the PEW Charitable Trusts and the "Sea

Around Us" Project at the University of British Columbia Fisheries Centre predicts that over the next 50 years, India could lose up to 40% of its fish stocks. This problem was, however, exacerbated by the institutional regulatory framework, which spreads the management of coastal and marine resources across multiple agencies without adequate coordination, technical capacity, or enforcement (Paul and Nishida, 2008). However, this was not always the case. The coast of India has a long history of fishing villages practicing an array of traditional CBFM systems. Incidentally, Edgar Thurston was one of the first to chronicle the fishing castes of India in 1909. Works indicated the existence of vibrant and sustainable use patterns and resource management within the fishing communities. However, the traditional caste system for fisheries management, according to the Indian Director of Integrated Coastal Management, provided informations such as (i) assertion of rights over fishing areas, (ii) establishment of rules of access, (iii) a balance of fishing activities with resource capacity, and (iv) systems of governance, which helped to maintain the social integrity of the villages.

However, these local management systems had come under increasing stress, primarily due to central government's initiatives during the postindependence era. Customary arrangements were pushed aside to make room for a new emphasis on achieving the socialist ideal through planning, science, and technology. In the short term, this path accelerated growth in the fisheries sector but, concomitantly, it also augmented the damage to local fisheries.

Notwithstanding the above, the government focused on excluding both traditional and modern fishing practices with seasonal bans to remedy this situation. The bans included, notably, bans on fishing in specific areas and restrictions on the operation of certain fishing gear(s). But the bans were not much effective. Incidentally, the government's disregard for the local systems, probably paved the way for the fact that, the knowledge and understanding accumulated over generations were not taken advantage of while designing the new systems. Today, the GoI with a nudge from the World Bank is coming full circle to accept that there is much to learn from customary fishing systems. In fact, the World Bank and the GoI are supporting a number of regional initiatives through the Integrated Coastal Zone Management program(s) to promote the sustainability of coastal and marine ecosystems. Nevertheless, one of the main objectives of this program was to improve national marine management through capacity building at the state and local levels. Another objective was to introduce and enforce zones along the coastline to help establish property rights and monitor the health of the fisheries. Nonetheless, the traditional systems were built upon centuries of experience and understanding of the fisheries and could certainly show a way forward.

Notwithstanding the above, given the largely informal nature of the fishing economy, many of the management regimes, such as the ITQ, would be difficult to establish in the Indian context. However, new approaches from fishing villages across India are surfacing. Works portrayed the existence of detailed local systems governing access to the fishing grounds among the sea communities on the Coromandel Coast. Also, in the province of AP, there have been efforts to summarize the success of the CBM approach as follows:

(i) Management systems had varied between villages and built on specific conditions of the natural and social environment in the given area.
(ii) The assertion of rights of access and withdrawal and their regular practice, probably, had provided with a sense of security of tenure to the fishers and the fishing grounds that could be equated in farming terms to acquisition of land tenure for cultivation.
(iii) Participatory decision-making is a key ingredient of CBM. The community-based system allows people to initiate and practice a fishing regulation because the community had decided on the rule.

Notwithstanding the above, one of the major components to the pilot programs (primarily in the provinces of Gujarat, Orissa, and West Bengal) was the management of fisheries resources by those who depend on them as well as by those who have the greatest stake in protecting them. It could be opined that an important lesson learnt from the fisheries projects was the importance of beneficiary involvement in project preparation and implementation.

Notwithstanding the above, more research and case studies are required on community run fisheries. But the local institutions are shedding light on marine management strategies in India. Incidentally, "the traditional systems," were built upon centuries of experience and understanding of the fisheries, and thus could certainly show a way forward. In fact, devolving control of fisheries at the local levels was said to be a step in the right direction. In addition, "the historic rights of the traditional fishers to fisheries resources are to be secured" (Kar, 2003c).

In addition to the above, **some of the other significant studies on** CBFM **in India** are as follows:

Vipinkumar et al. (2013) reported about comanagement paradigm and sociological issues in RFM in Indian marine fisheries sector.

The open-access regime existing in the harvesting of marine fishery resources in our country warrants stronger emphasis on invoking technological innovations as well as management paradigms that reconcile livelihood issues with concerns on resource conservation. In fact, innovations do not happen in a sociopolitical vacuum. It is the extent of partnership between the research and the client system that decides the fate of any technology in terms of its adoption or rejection. Rational utilization of CPRs for sustainable development without endangering the environment could be possible through community participation. Incidentally, fisheries are a source of nutrition and avocation for >6.0 million fishers and fish farmers in India. In fact, the fisheries sector had been recording faster growth as compared to the agricultural sector throughout a number of decades, and is also contributing in a significant way to the economic growth of the nation.

Notwithstanding the above, marine fishing has been considered a primary livelihood option since time immemorial for the people of the coastal belt in India. The harvestable marine fishery resources in India support the so-called deprived coastal community with sufficient nutritional security which is otherwise unreachable for such segments.

Concomitant to the above, the marine fisheries sector produces *c.* 2.6 MT of fish per annum (2003) and *c.* 3.0 million people are employed in the primary, secondary, and tertiary sectors of the marine fisheries, which provide livelihood security to *c.* 18–20 million people (Sathiadhas et al., 2007).

Notwithstanding the above, fisheries development is a provincial subject in India, but, nevertheless, the Union government at the Center promotes fisheries development through provincial level programs, **planning and implementation units** (PIU). The development plans for the fisheries sector aimed at improving fish production, promotion of export, and efficient management.

But, fisheries production is showing a depleting trend which is adversely affecting the livelihoods of fishers as well as making a large population vulnerable. Moreover, stock assessment and irreplenishable nature of abundance in the stock vis-à-vis conflicts of various types have become the part and parcel of the fisheries system in the country, fisheries being an open-access regime. Consequently, the government had to introduce regulatory mechanisms, such as gear selectivity, seasonal area closures, and regulations that control the fishing effort and catching, mainly to address the livelihood issue. In fact, this could be considered as a top-down government-driven management approach. However, the government-managed models of management had proved to be unsuccessful as indicated by poor compliance of action and regulations, thus resulting in crisis and adverse effects on the livelihood of fishers.

Concomitant to the above, Mohamed (2004) reported about march toward ecosystem-based management of marine fisheries with an aim at building mass balance trophic and simulation models.

Much attention was recently paid at assessing the impacts of fisheries on the marine ecosystems as a whole. This had in part been driven by the need to ensure conservation of biological diversity and sustainable use of resources of the biosphere, which are key provisions of the convention agreed at the UN Rio Summit. The utilization of sound ecological models as a tool in the exploration and evaluation of ecosystem health and state had been encouraged and endorsed by the leading bodies in ecosystem-based fisheries research and management (NRC, 1999). However, the potential of the available dynamic ecosystem models to make measurable and meaningful predictions about the effects of fishing on ecosystems has not been fully assessed yet.

Incidentally, Sampath (2005) dealt with an overview of the International and National Fishery Policy and related issues.

Policies show path to development and a correct policy could lead to meaningful development. Incidentally, the Industrial Policy Resolution, 2001 of **Government of Odisha** has identified the fisheries sector as one of the three sectors wherein enterprise development could result in pro-poor economic growth. In this context, the policy emphasizes the need to assess the self-employment and entrepreneurial potential in fisheries sector, especially for the poorer sections involved in the sector.

Incidentally, Odisha is one of the major maritime provinces offering vast scope for development of FW (inland), brackish water, and marine fisheries. The long coastline of the province, measuring *c.* 480 km and covering an area of *c.* 24,000 km^2, provides huge potential for marine fisheries development. Concomitantly, the FW resources of the province include *c.* 1.16 lakh ha area under tanks and ponds, *c.* 2.56 lakh ha reservoir area, *c.* 1.80 lakh ha area under swamps and tanks ("jheels"), and *c.* 1.55 lakh ha water area of rivers and canals. On the other hand, the brackish water resources are *c.* 2.98 lakh ha of estuaries, *c.* 32,587 ha of cultivable brackish water area, *c.* 8100 ha of backwater area, and *c.* 9000 ha of Chilika lake (Kar, 2013).

Concomitant to the above, a conservative estimate of fish production in the province is *c.* 3.23 lakh tons as against consumption requirement of 11 kg/capita/annum (as per WHO). However, of late, the level of fish production in the Odisha province was only *c.* 2.595 lakh tons, thus, leaving a net deficit of *c.* 1.635 lakh tons.

A reform in Odisha depicts the average per capita annual income of those involved in the fisheries sector of *c.* Rs.6787.00 in contrast to the all-India average of *c.* Rs.10,204.00. Enhancement of the livelihood and enterprise opportunities in this sector, therefore, need to focus on the poor workers involved in fisheries and the small/medium entrepreneurs. Inadequate infrastructure facilities, information and communication channels, expertise, etc., in the province coupled with not much efficient *modus operandi* of management not much involving the stake holders from the grassroots level, were identified as the limiting factors in promoting cost-effective and environmentally sustainable methods of production, harvesting, and management of the resources.

As a consequence, Odisha is left with a big untapped water area for fish production covering varied ecosystems and open-water resources like lakes, ponds, reservoirs, etc. Further, the existing policy of the province lacks a number of aspects, which are needed for a well-coordinated and sustainable fisheries development and management, not only for meeting the nutritional requirement of the people of Odisha but also for ensuring livelihood security and entrepreneurial development.

In view of the above, a number of desk research/studies had been undertaken. One such desk research/study was on the fishery policy-related issues, undertaken by the UNDP Project Team during July–September 2005. This report has provided the necessary inputs on the fishery policy *in vogue*, both at the national and international levels.

Furthermore, Anon (2004) dealt with the policy briefs of good management practice in sustainable fisheries in India.

The purpose of these policy briefs was to ensure effective dissemination of information collected and generated through different programmes, notably, the World Bank-funded programs. Study of Good Management Practice in Sustainable Fisheries was undertaken under ACP Fish II Feasibility Study (EC) program and a Workshop on Fiscal Reform in Fisheries (DFID and GTZ), World Bank Study.

The project, "Study of Good Management Practice(s) in Sustainable Fisheries" was undertaken during 2003 by SIFAR with funding from the World Bank. This had resulted in an initial (brief) report followed by the substantive report which contributed to a recent World Bank internal process aimed at justifying future investments in fisheries sector development.

Notwithstanding the above, the FAO undertook a feasibility study on behalf of the European Commission (European Aid Cooperation Office—AIDCO). This comprised an extensive consultation process with fisheries sector participants from >60 ACP countries, together with the preparation of a range of major project proposals covering capacity building for more effective fisheries management in Africa, the Caribbean, and the Pacific.

5.1.7.103 *Dealing with the complexity and change in fisheries management*

The case of small-scale fisheries in Andhra Pradesh, India

Overview The fisheries sector is often located within complex societal systems with particular social, economic, and institutional arrangements. These systems are also increasingly subject to the impact of change, which may have major effects on the livelihoods of the people. As such, fisheries management planning is to take this wider context into consideration and to design appropriate management systems which could adapt itself to the change(s). However, in many parts of the world, there is still a need to gain a better understanding of the underlying relationships involved at all levels and in all policy domains (e.g., community, sectoral, and society).

Case study(1) The province of AP has 900 km of coastline on the east coast of India and *c.* 870,000 fishers live in fishing communities. However, the communities vary widely with distinctive fishing systems, fish disposal and marketing systems, and social and political organizations.

Two villages, notably, Uppada and Boddu Chinna Venkataya Palem (BCV Palem), among many, provide interesting examples of how traditional community-based fisheries management systems (TCBMS) could regulate the fishing practices successfully. Uppada is a marine fishing community of *c.* 13,000 inhabitants, and the community had undergone considerable changes over time. In fact, technological advances had influenced the types of boats used and the variety of fishing gears, although coastal erosion had seen the village move several times. However, the increasing globalization of the fishing industry, particularly, the advent of shrimp farming impacted Uppada in the form of affecting the catch prices. Moreover, alternative community resources were not abundant in Uppada.

In contrast to the above, the BCV Palem was a riverine fishing community of just 4000 people. Technology had little impact here and the levels of rent generated by the fishery are much smaller mainly because of the nature of creek fishing in shallow waters. However, the community had the access to considerable sources of alternative community resources, e.g., extensive tracts of mangrove.

Nevertheless, the fisheries in both the communities are managed through TCBMS, which are rarely sector specific to fishing, but provide a framework for managing livelihoods and the community as a whole. Hence, they generally encompass the sociocultural, religious, political, administrative, and economic aspects of the community organization, and, within this context, fishing activity was also said to be managed. Notably, in both Uppada and BCV Palem, the well-developed TCBMS had operated through the traditional Indian village management system, called the "panchayats." The "panchayat" literally refers to the system of village-level organization with five elders ("pancha" meaning five in Sanskrit), and they are often formed around castes. The institution of caste provides the basis for codifying and ensuring effective enforcement of the customary laws in the form of "caste codes." However, the specific nature of TCBMS varies between fisheries. In BCV Palem, fishing activities are carried out by a number of fishing systems confined to creeks and backwaters. Also, the traditional systems of management and control related to fisheries and fishing are elaborate and have an impact on the different systems.

Notwithstanding the above, the Five-Year Plans are drafted for an overall development and welfare of the different sectors including fisheries and AQC. Incidentally, Anon (2006a,b) prepared the Report of the Working Group (WG) of Fisheries for the 11th Five-Year Plan, 2007–2012. **The Planning Commission (PC)** of the GoI constituted a WG on Fisheries with various objectives, schemes, and programs including their monitoring and implementation. The PC of the GoI, in the planning process had emphatically stressed upon the sustainable harvest and utilization of the water bodies and their bioresources through efficient CBM *modus operandi* involving the stakeholders from the grassroots level. Also, as such effective measures are to be in place for all-round development of these poverty-stricken stakeholders, particularly, the fisher folk. In fact, the Five-Year Plan provides a road map for the overall development of the fisheries sector along with the socioeconomic emancipation of the fisher folk.

The main objectives of the WG on Fisheries of The PC of the GoI are as follows:

(a) To review the progress of schemes and programs in fisheries sector implemented during the previous plan.
(b) To identify the problems and constraints in the implementation of programs and schemes, and to suggest measures for enhancing fish production in a sustainable manner, along with developmental programs for the subsequent plan.

India is one of major fish producing countries in the world with third position in fisheries and second in AQC. The sector has high potential for rural development, domestic nutritional security, employment generation, gender mainstreaming, and export earnings.

Indian fisheries sector has been witnessing a steady growth since First Five Year Plan. The annual fish production increased to > 6.3 MT during 2004–2005 from *c.* 0.75 million metric tons (mmt) in 1950–51. Conversely, the marine fish production had gone up from 0.53 million tons in 1950–51 to a maximum of 2.99 million tons in 2002–2003.

Concomitantly, the contribution of inland sector increased at a higher rate from *c.* 0.218 mmt during 1950–51 to *c.* 3.52 mmt during 2004–05. However, fisheries and AQC contribute, of late, to *c.* 1.04% of the national GDP and to *c.* 5.34% of agriculture and allied activities. Indian exports and its share in the global trade had depicted a steadily increasing trend over the years.

Notwithstanding the above, according to an estimate, India is endowed with enormous fisheries resources in a coastline of 7516.6 km and *c.* 2.02 million km^2 of exclusive economic zone (EEZ) including *c.* 0.530 million km^2 of continental shelf. The inland abiotic resources connected to fisheries include the rivers and canals (1.95 lakh km), reservoirs (*c.* 3.15 million ha), wetlands (*c.* 0.35 million ha), estuaries (*c.* 0.26 million ha), other types of FW bodies (*c.* 2.41 million ha), and brackish water bodies (*c.* 1.24 million ha). These resources are one of the main sources of livelihood for the rural poor, particularly the fisher community. The sector could provide livelihood for > 90 lakh persons at subsistence level of annual income. In all, *c.* 14 million people are engaged in fishing, AQC, and ancillary activities.

Demand for fish and fishery products is increasing considerably, both at domestic and export fronts. The projected demand for fish in the country, by 2012, was *c.* 9.74 mmt, which was met by the projected supply of fish of *c.* 9.60 mmt during 2012, with major share of *c.* 5.34 mmt from inland AQC followed by *c.* 3.10 mmt from the marine fisheries. Further, the fisheries of the island systems, viz., the Andamans and Lakshadweep as well as deep-sea resources, including the tunas, were less exploited. Concomitantly, the socioeconomic welfare of the fishing communities should also be boosted in order to encourage them for better performance.

Concomitant to the above, there was a need for the processing in both export and domestic markets and improvement in fishing harbors. Effective marketing system in the identified areas was a key need for the development of this sector.

Notwithstanding the above, the fisheries of the natural waters, including coastal and inland open, are under pressure due to high fishing intensities, pollution, open access, human-made modifications, water abstraction, etc., and this could lead to problems in maintaining sustainable fisheries. However, sustainable exploitation of fish stocks, in these waters, could be achieved through **community participation, comanagement, and CBM**.

In the case of AQC, scopes exist for bringing more candidate fish species under the umbrella of AQC, both in fresh and brackish waters with a focus on food fish, high value species, ornamental species, and those with potentials for sport and tourism. Ready availability of inputs like seed, feed, fertilizer, medicines, fisheries requisites, etc., are believed to be preconditions for development. Further, establishment of "aqua-shops," as a single window facility, for the purpose, could bring major change(s) in the sector.

In continuation of the above, for all round development of the fisheries sector, it is high time to.

(i) implement the code of conduct for responsible fisheries;
(ii) utilize the strong traditional wisdom and know-how by active community participation;
(iii) address gender issues;
(iv) strengthen the capacities of the fishers and the aqua-farmers regarding the latest technological practices, administrative skills, for disaster management, etc.;
(v) intensify efforts for treating AQC *at par* with agriculture;
(vi) explore areas of public-private partnerships (PPP) mode;
(vii) strengthen the institutional credit support and public investments;
(viii) harmonize fisheries policies in concurrence with the recent developments, both at national and international levels;
(ix) strengthen the inter- and intralinkages between fisheries sector and other concerned departments.

Incidentally, the nomenclature of the Department of Animal Husbandry and Dairying (DAH and D) was changed to Department of Animal Husbandry, Dairying, and Fisheries during the 10th Five Year Plan. In fact, the department had implemented schemes under two broad heads, which are as follows:

(a) "Development of Inland Fisheries and AQC".
(b) "Development of marine fisheries, infrastructure, and postharvest operations" during the plan period.

These were implemented along with some ongoing schemes from the IX plan, modified to some extent. The seven schemes are as follows:

Development of marine fisheries, development of inland fisheries and AQC, marine infrastructure and postharvest operations, strengthening of database and IT scheme, welfare of fishermen and training and extension, establishment of fisheries institutes, and National Fisheries Development Board.

Concomitant to the above, it is relevant to look at the prevailing concerns that need to be addressed in order to design intervention points during the subsequent plans. The concerns pertain to water availability and allocation, BD loss and depletion of fish stocks, excess coastal fishing, enhancing fish productivity, oceanic and deep-sea fisheries, impact of climate change on fisheries, trans-boundary fisheries issues, inland and coastal pollution, extensive sedimentation of the rivers, estuaries and the lakes/wetlands, effective compliance of code of conduct for responsible fisheries, increasing input costs of water and power, high marine fishing costs and low profitability, mechanization in fisheries and AQC, cold and hygienic fish handling, quality assurance issues in exports, overseas market fluctuations, disaster management, credit and insurance, inadequate database, and poor linkage in domestic marketing. The programs in the forthcoming plan would need to address these aspects to build up a greater resilience and sustainability.

In continuation of the above, the areas with promise that need to be supported in order to achieve diversification in terms of nonfood fisheries are as follows:

Ornamental fisheries, seaweed, pearl culture, aqua-tourism, sport fisheries, and molluscan fisheries for ornamental purposes. Further, the strategies for realizing the potentials of the sector are as follows:

(i) Ensuring adoption of responsible and sustainable fishery practices and enhancing fish productivity in all cultivable waters.
(ii) Establishing agro-aqua farms, aqua-shops, and fishery estates to incorporate activities from production to consumption.
(iii) Spreading fish quality literacy among fishers and aqua-farmers.
(iv) Improving facilities for fish landing and handling at harvest and postharvest stages.
(v) Developing social marketing techniques.
(vi) Introducing aquarian reforms with regard to leasing and management of waters.
(vii) Ownership and community management and training in different aspects of fisheries and AQC.

Notwithstanding the above, the main objectives of the GoI with regard to development programs in fisheries and AQC during the 11th Five Year Plan as follows:

(i) Enhancing the production of fish from Indian waters on an environmentally sustainable and socially equitable basis.
(ii) Addressing the *hitherto* unexplored potentials of Indian fisheries, e.g., island fisheries and nonfood fisheries.
(iii) Conservation of aquatic resources and genetic diversity.
(iv) Preservation of health of ecosystems and the health of the fishes.
(v) Increasing the profitability of the fishers and the aqua-farmers through an integrated approach from production to consumption.
(vi) Promoting fish as a healthy food and meeting the changing requirements of both domestic and export markets.
(vii) Strengthening of infrastructure in harvest, postharvest, value addition, and marketing and upliftment of fishers and aqua-farming communities with gainful employment opportunities and capacity strengthening.

Concomitant to the above, the ongoing components of capture fisheries and AQC, which were planned during the previous plan had been recommended for continuation during the next plan period under the scheme for Development of Inland Fisheries and AQC. The suggestions made in the subsequent plans did not differ much from the suggestions in the previous plans. Nevertheless, the new initiatives suggested under the scheme are as follows:

(i) Survey, assessment, and evaluation of inland aquatic resources.
(ii) Improving the productivity of inland open-water fisheries including reservoirs.
(iii) Magur breeding, hatchery establishment, and seed production.
(iv) Advanced and upgraded protocols for carp breeding, seed certification, and certified brood banks for food and ornamental fishes under FW AQC.
(v) Hatchery establishment and seed production of sea bass and pearl spot under brackish water AQC.
(vi) Establishment of mahseer seed production facility.
(vii) Construction of community raceways for trout farming in hilly regions under cold-water fisheries.
(viii) Establishment of centers/facilities in the provinces for surveillance, monitoring, and reporting of fish diseases, notably the *hitherto* unknown, virulent **EUS** (Kar, 2007a,b,c, 2013, 2015).
(ix) Establishment of fish quarantine centers under conservation of fish stock and diversity and general activities including establishment of aqua-shops.
(x) Provincial level soil-water testing and disease diagnostics laboratories and construction of aqua-tech park.

Notwithstanding the above, under the scheme "Welfare programmes, Governance, TOT and Capacity building", the following were suggested:

(i) Continuation of the ongoing programs on the development of model fishers' villages.
(ii) Group accident scheme for active fishers and saving-cum-relief under welfare programs.
(iii) Establishment of awareness centers, training programs for fishers, and extension literature and video films. Trainers' training and refresher courses under capacity building and awareness development is recommended.

However, the new initiatives suggested are as follows:

(i) Hut insurance scheme, group insurance of fishers and farmers, Mediclaim policy.
(ii) Incentives to fisherwomen groups (SHG).
(iii) Safety at sea and assistance to FISHCOPFED under welfare programs.
(iv) Formulation of National Fisheries Development Policy under policy issues.
(v) Comanagement and CBFM.
(vi) CCRF and certification of boatyards and hatcheries under management and governance.
(vii) Capacity building at the provincial level.
(viii) Use of IT in ToT.
(ix) Programmes for EDUSAT, vocational education in fisheries schools.
(x) Capacity building of fisheries cooperatives, associations, NGOs, SHGs, and revitalization of cooperatives.
(xi) Strengthening of service delivery system under human resource development (HRD).
(xii) Capacity building/training and demand generation for fish under cooperatives.
(xiii) Public-private partnerships.

(xiv) Location-specific field trials and demonstration units.

(xv) Trickle down extension and involving:

  (a) NGOs in training of trainers, under demonstration and exhibitions.

  (b) Fisheries Technology in Dissemination Centers (FTDC) and ToT units at ICAR institutes/SAUs under coordination and linkages.

(xvi) The scheme on "Strengthening of Database and Information Networking" include the activities of sample survey for the estimation of inland fishery resources, their potential and fish production; census on marine fisheries; catch assessment survey for inland and marine fisheries; development of GIS for inland and marine fisheries and delineation of brackish water areas, and assessment of productivity and production.

Concomitant to the above, the National Fisheries Development Board (NFDB) had been established, which could give a major fillip to the Indian fisheries sector.

Concomitant to the above, Kumar et al. (2003) reported on a profile of the people, technologies, and policies in fisheries sector in India.

Fisheries is said to be a potential sector in India for the nutrition and avocation of people. Its role in nutritional security, employment generation, and in augmenting the economy of the country (particularly in earning foreign exchange) has been significant. Growing urbanization, globalization, and rapidly changing social structures have a major impact on the fisheries structure in the country. Moreover, fisheries and AQC emerged as important commercial activities from their traditional role in supplementary subsistence activity. There had been reports on the aspects like, "strategies and options for increasing and sustaining fisheries and AQC production to benefit poor households in Asia," as Proceedings of the Workshop organized by the National Centre for Agricultural Economics and Policy Research, New Delhi in association with ADB.

Concomitant to the above, the fisheries sector in India had witnessed an impressive growth from a subsistence traditional activity to a well-developed commercial and diversified enterprise. Fisheries issues were said to have emerged from being merely an obscure sectoral concern to an important growth sector, over the last few decades, with an expanding role in economic development and food security. Further, fisheries sector has emerged, of late, as a vibrant sector, and is considered a strategic subsector for promoting agricultural diversification. However, information regarding the development process of this important sector is quite fragmentary and not very well documented.

Fishing, as an occupation, had been *in vogue* since time immemorial. Till recently, it was reckoned to be a supplementary enterprise practiced by the fishermen community at subsistence level with little external inputs (Krishnan et al., 2000). But, with the changing consumption scenario, emerging market forces, and technological developments, fisheries sector in India is undergoing a transformation.

5.1.7.104 *India's share in World fish production*

Fish production in India, as reported, reached 5.6 mmt during the period 1999–2000. It was, however, only a mere 0.75 mmt during the period 1950–51. On the other hand, the global fish production during the same period has risen from *c.* 23.5 mmt to *c.* 120 mmt during 1999–2000. The growth of fish production in India had been at a faster rate, as compared to growth in global fish production, mainly due to increasing contribution from inland fish production.

5.1.7.105 *Growth of fisheries sector in India*

Fisheries sector plays a significant role in the Indian economy. Incidentally, the share of agriculture and allied activities in the GDP is not very promising. But, the contribution of fisheries sector to the GDP had gone up from 0.46% in 1950–51 to *c.* 1.16% during 1999–2000 (at the prevailing prices). Moreover, the contribution of fisheries to agricultural GDP (Ag-GDP) increased more impressively during the period from 1950 to 51 to 1999–2000 from mere *c.* 0.84% to *c.* 4.19%. This was believed to be largely due to sustained annual growth rate of well over *c.* 4% in the fisheries GDP during the last few decades. The fisheries sector had recorded faster growth, as compared to the agricultural sector, in all the decades. This was mainly due to hard labor put-in by those involved in the fisheries sector, including the poverty-stricken fishing communities, who are said to be the stakeholders at the grassroots level. As such, it is imperative that appropriate and adequate measures are to be in place for the socioeconomic upliftment of the fishing communities and for the overall development of the fisheries sector through application of efficient *modus operandi* like the CBFM.

Incidentally, it may be noted that > 6.0 million fishermen and fish farmers are fully dependent on fisheries for their livelihood in India. According to the concerned department of the GOI, the marine subsector accounts for *c.* 39% of the total national fish production of 7.60 mmt. On the other hand, inland fishing accounts for *c.* 61% of the total fish production, largely through FW AQC from large inland agriculture water storage areas (tanks), smaller farm ponds, and reservoirs. India is the world's third largest fish producing nation and second in inland AQC.

Incidentally, Anon (2010a,b) dealt with the marine fisheries of India with regard to their issues, opportunities, and transitions. This study represents a collaborative initiative by the World Bank and the Department of Animal Husbandry, Dairying, and Fisheries, MOA, GOI to review the marine fisheries subsector, within a broader sector that also includes AQC and inland fisheries.

5.1.8 Indonesia

Indonesia is a BD-rich country with high fisheries potential. Community-Based Fishery Management (CBFM) in the post-conflict situation was reported by Sidayat (2008) as a case study in coastal villages of North Maluku in **Indonesia**.

This research work was reported to be based on personal experiences of the researcher(s). They had witnessed the social conflict erupted in North Maluku in Indonesia. It was a case study of how a local reconciliation effort and process, in the context of CBFM, could facilitate the reconciliation process between conflicting communities (Muslims and Christians) in a post-conflict situation. In fact, many researchers have opined that small groups, shared rules, institutional arrangements, and external factors could all be crucial for the success of a community-based system, although little, if any, attention had been paid to its potential for post-conflict reconciliations. Nevertheless, a substantive number of scholars of the reconciliation processes had tended to look at the reconciliation process, but had been less inclined to go into in-depth discussions on how to achieve it successfully through natural day-to-day processes. In fact, every community has a different way of living, and reconciliation is understood and takes place differently according to its context. However, the difficult part of this study is the process of linking theories on community management with those on reconciliation because they usually remain very discrete.

The study found that the use of CBFM in the post-conflict situation in North Maluku involve a number of complementary but important factors. Some of these are as follows:

(a) Fishery resources potency, (b) cultural values and traditions, (c) economic livelihood activities, and (d) developmental activities (which are generally introduced and exogenous).

Notwithstanding the above, joint fishery management provide with an everyday forum, in which traditional values play their role(s) in important ways. However, two factors were believed to be significant in facilitating the reconciliation process: (a) Moral commitment to a shared set of rules and (b) the mediating/dispute-settling role of the community leaders.

Shared rules enhances the cooperative behavior, minimizes the desire for competition, and enhances solidarity. They encourage people to behave for the common good, and to consider other people's needs and interests as well as their own. Community leaders helped to bridge the gap between the more powerful and less powerful, or between two communities, e.g., between Muslims and Christians. In fact, such mediations are very much required when conflict(s) arise over access to share of resources. These values/norms/practices were essential in implementing the process of reconciliation naturally; and, CBFM could be considered as a relaxed, everyday process. Concomitantly, in spite of its advantages, fishing, when done in excess, may have the potential to weaken the same social cohesiveness of these communities. In this connection, it may be noted that if the commercial demand for fish continues to increase by leaps and bounds vis-à-vis incomes from farming continue to fluctuate or even fall; in that case, larger commercial markets could undermine the moral authority of traditional leaders, and, at that point, CBFM may no longer help to keep an increasingly differentiated and heterogeneous community together. Unequal power relations and reduced power of the weak have the potential for more conflict(s). However, this study is about reconciliation in the wake of past violence, and, in that context, its contribution has been invaluable.

Notwithstanding the above, Soede et al. (2001) dealt with Integrated Marine Management Concessions with regard to a new approach to an old problem.

Indonesia, a coastal country, is said to have a rich marine resource, and, hence, there is priority with regard to their conservation. In fact, the Indonesian archipelago sits at the heart of the biological bull's eye for coral species diversity and development. It is clear to many observers that the reefs of Indonesia are in trouble. Pressures on the reef range from overexploitation and destructive fishing practices to coastal and marine-based pollution, coastal development impacts, and global climate change. Many have called for quick and decisive solutions to reverse this trend.

Notwithstanding the above, the Conservation and Community Investment Forum (CCIF) team advocated an approach proposing an entirely new, systematic approach to conserving marine resources in Indonesia: the development of an integrated marine management concession. The approach combines the traditional solutions and activities of the site-based conservation and enterprise development with the immediate and sustainable formation of a conservation concession funding mechanism. Efforts could be directed toward tying the existing traditional approaches to site-based community development and enterprise development together under an integrated conservation concession approach.

5.1.9 Japan

Japan has a significant contribution to fisheries. Fisheries in Japan is rather of a particular nature (Schmidt, 2003) because:

(i) The importance of fish in the national diet and hence a strong demand for fish is particularly pronounced in Japan.
(ii) Japan has a long tradition of coastal CBFM that had been used to ensure that the resource base is harvested in a sustainable manner.

Notwithstanding the above, as an important "side effect," the CBFM system had been of key significance for the well-being of fisheries-dependent coastal communities. In fact, it may be mentioned that in the development of the Japanese coastal fisheries management systems, the coastal communities play a key resource management role. In this process, a number of additional functions had been developed, which were also of importance for the livelihood of the fishers.

Notwithstanding the above, it may be mentioned that, until the mid-1980s, the fisheries managers in most countries have focussed mainly on the **biological outcome** of the fisheries management systems. Often the fisheries had been managed exclusively in the domain of marine biologists who saw it as their job to ensure that fish stocks were harvested at the **MSYs**. However, in the beginning of the 1980s, the focus changed to include the **economic aspects** as well; and the concept of **maximum economic yield (MEY)** become a management objective in some countries.

Nevertheless, the fisheries managers, of late, have begun to focus more on the **social aspects** and functions of fisheries. This has largely developed in response to the fact that many fisheries are overexploited, and that fishing profits have been registering a fall. Moreover, in fisheries, where fisheries remuneration is largely based on the so-called "share basis," the effects of fewer fish may have immediate income consequences for the skippers and crew alike, and hence impact the coastal communities that are dependent on fish as a source of nutrition and avocation. Furthermore, the researchers try to point out that the viability of the fisheries management systems and the fishers' adherence to the rules to depend largely on social and cultural aspects, and to a lesser degree on the biological and economic aspects of fishing.

Notwithstanding the above, the development and implementation of a particular fisheries management system had significant repercussions in terms of biological, economic, and social outcomes for fishers and fishing communities alike. There had been efforts to discuss some of the important outcomes that one could expect from the fisheries comanagement system in Japan. Contribution in this direction also provided a brief introduction to the Japanese fisheries sector as a background to understand the different management systems that are applied. **In fine**, such contributions discuss the multiple roles of fisheries and whether the concept of multi-functionality, which is usually discussed threadbare in the agriculture sector, would also be a useful analytical framework to fisheries.

5.1.9.1 The Japanese fisheries sector: Some basic facts and data

Catch/harvest

The Japanese fisheries sector could be classified into three main categories as follows:

(i) Distant-water fisheries, (ii) off shore fisheries, and (iii) coastal fisheries.

In fact, these three categories are quite different in terms of resource management and fishing techniques used. The relative significance of the three types of fisheries has changed considerably over the recent decades as, with the introduction of 200 mile EEZ, Japan lost access to traditional long-distance fishing grounds in many places around the globe. Furthermore, overexploitation and fluctuating stocks of some resources have had impacts on the relative importance of the three types of fisheries. Incidentally, it may be mentioned that, in 2000, Japan harvested a total of 6,250,000 tons of fish, amounting to a total value of *c.* JPY 1.76 trillion from marine fisheries and AQC (corresponding to *c.* US $11 billion). By quantity, this made Japan the second largest producer in the world with only China producing more. Nevertheless, at 6.25 MT, this could be a comparatively meagre harvest as the post-war peak production reached *c.* 11.83 MT (12.82 MT if FW and AQC are included) in 1984. However, more detailed statistics on the catch composition revealed that the decrease in production had been caused largely by a drop in the long distance and offshore catches while catches in coastal waters experienced less dramatic changes. However, the value of the coastal fisheries was high, and also had contributed most of the three types of fisheries to the overall value of Japan's catches. Further, while distant water and offshore fisheries are characterized by volumes, the coastal fisheries target high-valued species with ready market demand. Japanese consumers were willing to pay high prices for freshness. Hence, and despite falling quantities, the value of the coastal fisheries was high and had contributed most of the three types of fisheries to the overall value of Japan's catches.

Notwithstanding the above, over the past 250 years, **Japanese** fisheries regimes have developed three fishery laws in sequence, in which fishing right system was commonly used (Yamamoto, 1996).

During the feudal era, until 1867, the fishing right was used mainly as a tool for the collection of fishery tax. The fishing right system, which was an old fishery law (1901–47), was used mainly as a tool to reduce conflicts among different groups of coastal fishermen and disputes between coastal and trawl fisheries.

The current fishery law, enacted in 1949, refers to "TURF." It had limited coverage to sedentary resources and non-mobile gears.

Concomitantly, the current law had made a beginning, where by a coastal fisheries management plan might be established with the participation of fishermen. These innovations had given the fishermen a great motive to form a community-based coastal fisheries management system. As a result, since the inception of the current fishery law in 1949, the number of fisheries management organizations, formed under the initiative of fishermen, increased year after year and reached 1524 in 1993.

Concomitant to the above, Japan has both inland and marine fisheries, the latter being insignificant in quantity and value. Further, the marine fishery is administratively classified into (a) coastal, (b) offshore, and (c) distant water fisheries. Moreover, the coastal fishery is classified into (a) coastal capture fishery and (b) coastal AQC.

Incidentally, the coastal and offshore fisheries operate in Japan's waters, i.e., her territorial waters, and usually within her 200 mile EEZ. The distant water fishery operates in the high seas and, may be, in the EEZ of foreign countries. In fact, the coastal fishery is the mainstay of Japanese marine fishery, and accounts for 94% of the total number of fishing establishments. All of them are fishing households.

Notwithstanding the above, in a number of CBFM in Phang Nga Bay, in terms of quantity, the coastal and offshore fisheries produce 30% and 56% of the total production, respectively. Moreover, in terms of value, the coastal fishery produces 55% of the total value.

It may be mentioned that according to Japanese researchers, there is no internationally agreed definition for "CBFM." However, in Japan, it is understood that CBFM is a system of fisheries management run under the initiative of fishermen. Its activities usually cover the management of fisheries resources, fishing efforts, and fishing grounds. However, management covers not only conservation of fisheries resources, such as setting catch limits, but also propagation of fisheries resources through aquatic (FW and marine) ranching.

Concomitant to the above, CBFM in Japan has been developed mainly for the coastal fishery and partly for the offshore fishery. Further, owing to the complexity of Japan's coastal and offshore fisheries, the quantity of fish caught and fishing gears employed could be of varied types. Also, a variety of CBFM approaches could be in operation.

5.1.9.2 *Change in fisheries management regime*

Over the past 250 years, the Japanese fishery has been administered by regulatory regimes such as (a) the "Ura" Law (1743–1867), (b) Old Fishery Law (1901–1948), and (c) the Current Fishery Law (1949). These three laws have used the concept of fishing right for different purposes.

5.1.9.3 *The "Ura" law (1743–1867)*

Decreed by the feudal government in 1743, "Ura" Law was the first fishery law in Japan.

Its objective was to ensure tax collection from villages by granting them fishing rights. A similar tradition existed before the "Ura" Law was enacted.

Under the "Ura" Law, all villages along the coast were classified as either fishing or farm villages. The fishing villages were granted an "Osumi-tsuki," a fishing right that allowed the villagers to fish in their sea area. Those living in farm villages were allowed to collect only seaweeds for use as fertilizer. An Osumi-tsuki was sometimes awarded to an individual when he made a special contribution to a Samurai Lord. Such occasions, however, were the exception.

5.1.9.4 *The "Blank period" (1868–1900)*

The feudal era ended in 1867, and the Ura law became invalid. The policy of the new government was to modernize every aspect of Japanese administration. A special fishery mission sent to Europe took note of fishery laws in France, Germany, and England, but none of them suited conditions in Japan. It took 32 years to get a new law in place. This period was characterized by resource disputes among the fishermen. Certain intermediate measures were also taken but without much success.

5.1.9.5 Old fishery law (1901–1948)

The "Old Law" was formulated by the new government and approved by the Parliament in 1901. The "Old Law" was Japan's first own modern-day legislation, as all other laws were the modifications of European laws.

According to the Old Law, fishing rights were classified into the following categories: (a) exclusive fishing right, (b) set net fishing right, (c) specific fishing right, and (d) AQC right.

It stipulated that exclusive fishing rights were to be granted only to a Fishery Society (FS). As a result, all fishermen in a fishing village had to organize their own FS. Otherwise, they were not allowed to fish. While the Old Law was being drafted, fishermen made it known that they wanted traditions and rights established by the Samurai Lords to continue. Consequently, all Osumi-tsuki granted by feudal lords were converted into a coastal fishing right known as an Exclusive Fishing Right (traditional). However, the Old Law also had a category covering resources not covered by any of the exclusive fishing right (traditional). It was known as an Exclusive Fishing Right (new).

Owing to their importance, the central government was responsible for issuing the exclusive fishing rights. They were for 20 years and could be renewed. When the traditional right was renewed, its geographic area of coverage could not be expanded. Conversely, when the new right was renewed, in response to a request from a FS; the geographic area of the right was expanded seaward so as to cover migratory resources such as sardines, mackerels, squids, etc. Such an expansion of the new right had taken place in accordance with the progress of small boat mechanization. The new right had no restriction in terms of type of species and fishing gear to be covered. The prefectural governments took charge of granting: (a) set net fishing right, (b)specific fishing right for beach and boat seines, and (c) AQC right valid for 5 years.

These rights were granted to individuals who were capable of carrying out fisheries or AQC, but the fishermen's society was not given this right. With the introduction of medium-size mechanized boats, fishing gears such as otter trawl, pair trawl, and Danish seine began to operate in near-shore waters. This resulted in severe conflicts with coastal fishermen. This was an occasion when offshore fishery appeared in Japan. In response, the government introduced a "restricted fishing license system" within the framework of the Old Law. Moreover, closed areas for the trawl fishery were also established. Enforcing the new regulations required large expenditures by both the central and prefectural governments on patrol boats and inspectors.

In fine, for the success of CBFM, fishermen are, preferably, to regard the resources as their own. When the fishermen consider the fish stocks as their own property, they would adopt a more positive attitude toward conservation and management of the resources (FAO, 1983). Incidentally, the 1983 FAO Expert Consultation on the Regulation of Fishing Effort suggested that property rights could take many forms including individual catch quota, "TURFs," etc. Incidentally, the 1949 revision of Japanese fishery law has led to alteration of the characterization of Japanese fishing rights and brought it close to TURFS by limiting its coverage to sedentary resources and nonmobile gears. This had led to a more positive involvement of the fishermen with CBFM. In many FCAs (Fishermens' Cooperative Association), marine ranching is being intensified at the cost of FA or by sharing the cost among fishermen. Marine ranching had also given the fishermen a perception that fish released are their own, and increased the chances to initiate a CBFM system. For the formation of the CBFM, both (a) a fishery law which is the legal framework for the award of fishing rights and issue of licenses, and (b) fishermen's organizations were said to be indispensable.

While new fishing rights were being granted by the Current Law, new FCAs were already in existence with good solidarity among the fishermen. This had facilitated the formation of a community-based coastal fisheries management system. For the formation of a new CBFM system, a mutual agreement among fishermen is indispensable.

Such an opportunity may best occur at the fish market hall of the FCA, where daily sale of the catch of fishermen takes place. This might enhance the chances of the fishermen to create an idea of CBFM. Such a fish marketing system with auction could strengthen the financial stability of the FCA, as commission charged to each catch sale could be the constant income of FCA. It may, however, be argued that the absence in the size of sustainable yield in many FMOs (Fishery Marketing Organization) i a weak point in Japan's CBFM system. In recent years, however, fishermen have been trying to have MSY/TAC with the help of the prefectural fisheries experimental station. Incidentally, some people attribute the success of CBFM development in Japan to a long history of fishing rights. In a country without such a history, fishermen may not accept a fishing rights system, and hence there will be no chance to develop CBFM.

Concomitant to the above, it had been reported that in Thailand (1993), the fishermen believed in the concept of ownership of resources, although, the fishing rights were not legally endorsed. There is, therefore, a possibility of developing a CBFM system in Thailand.

5.1.10 Korea

There are various types of mechanisms for fisheries development and management in **Korea**. An account of the Integrated Fisheries Risk Analysis Method for Ecosystems (IFRAME) approach is briefly given in the following (Chang Ik Zhang, et al., (2011).

5.1.10.1 Indirect control devices in Korea

TAC-based management for 11 species

(i) Resource rebuilding activities by artificial reefs, releasing fry and juveniles, seaweed beds, marine ranches, and MPAs.
(ii) Buyback program to reduce fishing vessels operated separately, not systematically.
(iii) To develop tools for managing fisheries resources by environmentally sound techniques and in a sustainable way.

5.1.10.2 Ecosystem-based resource management

Why ecosystem-based fisheries management?
Some of the salient points are as follows:
FAO (2003) stressed implementation of ecosystem approach to fisheries (EAF).

(i) Further, WSSD (2002) encouraged the application of the ecosystem-based approach to fishery.
 (a) Shortcomings of a single species management led to overfishing in many areas.
 (b) However, *c.* 15% were said to be underexploited or moderately exploited in Korea; FAO (2010).
(ii) Limited management only, possibly *in vogue*, on sustainability issues in some regions.
(iii) Ignoring habitat quality, BD, and socioeconomic benefits occasionally could be detrimental.
(iv) Numerous studies on ecosystem indicators were carried out (Jennings, 2005).
(v) However, only a few approaches had synthesized indicators to obtain an integrated assessment (e.g., ERAEF by Australia, MSC's FAM, EBFA by ROK).
 (a) IFRAME (Integrated Fisheries Risk Analysis Method for Ecosystems)

IFRAME is based on the following assumptions:

(i) NEMURO model provides an adequate first approximation of the lower trophic level (LTL) response to shifts in temperature, solar radiation, and various layer depths.
(ii) The response of higher trophic level (HTL) species to climate change could be derived as a function of LTL production and species interactions.
(iii) Therefore, Ecosim could be used with scenarios derived from NEMURO to predict fish production.
(iv) Direct fishing impacts and harvest strategies could be roughly approximated by manipulating fishing mortality.

IFRAME could be little detailed as follows:

(i) A framework for evaluating the performance of management strategies.
(ii) An extension of EBFA approach (Zhang et al., 2010), which is made up of three components, viz. (a) assessment, (b) forecast, and (c) management.

Impacts of climate change on fish BD: Christopherm et al. (2006).

Notwithstanding the above, different researchers have worked on different aspects of fisheries, notably:

Productivity of fish populations: Hollowed et al. (2009).
Distribution of fish populations.
Spawning biomass of fish populations.

But, in spite of this, there is limited knowledge and poor understanding on the relevant mechanisms of key ecological processes.
Fishing with current TAC level could cause ecological overfishing, suggesting to reduce the catch level.
Notwithstanding the above, suggested policies and measures were responding to climate changes in Korea.

5.1.10.3 Current management through ecosystem-based fisheries management (EBFM)

Goal

Managing and rebuilding species: Managing and protecting species, fisheries, and their ecosystems.

Objective

(i) Sustainability of target species.
(ii) Sustainability of target and bycatch species, habitat quality, BD, and socioeconomic benefits, considering climate changes.
(iii) **Information:** Scientific research and collection of fishery data.
(iv) Nonscientific knowledge and information from fishers and other stakeholders.

Bodies

Central and local governments, and all relevant stakeholders are to establish fishery management councils.

Flexibility

Restricted, flexible, adaptive way, responding to climate changes.

Range of areas

Areas within one nation's EEZ.
Cooperating with neighboring nations by establishing a RFMO.

Period

Short term, midterm, and long term.

Further research

(i) Ecological process studies on the indicators and reference points under changing climate, such as ecological interactions with physical factors, are essential.
(ii) Studies on the ecological effects of biogeochemical changes, such as acidification and reduction in oxygen content.
(iii) Studies on climate change impacts on habitat quality and socioeconomic attributes.

Concomitant to the above, there seemed to be enough wetlands with quite good fishery management practices in **Korea**. Incidentally, the coastal landscapes in Southwestern Korea include a diverse array of tidal wetlands and salt marshes (Hong et al., 2010). These coastal zones link the ecological functions of marine tidal wetlands and FW ecosystems with terrestrial ecosystems. They are rich in biological diversity, and also play a significant role in sustaining ecological health and processing environmental pollutants. Moreover, the Korean tidal wetlands are particularly important as nurseries for economically important fishes and habitats for migratory birds. Diking, draining, tourism, and disposal of agricultural and urban wastes had adversely affected the Korean tidal wetlands. Further, recent large development projects had contributed to further losses. Environmental impact assessments conducted for projects affecting tidal wetlands and their surrounding landscapes are to be customized for application to these special settings. However, adequate environmental impact assessments may include classification of hydro-geomorphic units and consideration of their responses to biological and environmental stressors. Incidentally, the Korean laws and regulations are changing to be more favorable to the conservation and protection of tidal wetlands, as is true worldwide. In addition, more public education needs may be required at the local level to build support for tidal wetland conservation. Incidentally, some of the key public educational aspects include the role of tidal wetlands in maintaining healthy fish populations and reducing impacts of nonpoint source pollution. There is also a need to develop procedures for integrating economic and environmental objectives within the overall context of sustainable management and land uses.

Incidentally, Wilson, et al. (2006) reviewed the research conducted in Asian countries during the second phase of the Worldwide Collaborative Research Project on Fisheries Comanagement. Building on the results of the first phase, the research tried to focus on the stakeholder conflicts at social and geographical scales. Several conclusions emerged from common patterns. Community motivations for comanagement are often related more to the protection of fisheries resources from outsiders than to conservation. Access rights are important but exclusion from food resources in a context of widespread poverty is to be approached carefully. Cross-scale institutional linkages make adaptive management possible by bringing together groups with broad local foci and ones with narrow trans-local mandates. The role of the government is to balance the interactions between the various groups. However, this may not be considered as a role that is compatible with top-down management.

Notwithstanding the above, a new development paradigm is emerging that links the economy and culture, embracing economic, cultural, technological, and social aspects of development at both the macro- and microlevels

(Lee and Rahimi, 2014). Central to the new paradigm is the view that innovation, knowledge, and access to information are increasingly recognized as powerful engines driving economic growth and promoting development in a globalizing world.

In many definitions, in many instances, the so-called innovative economy is considered as cultural and art products that are directly involved in the history and culture of human beings. On the other hand, in many South Asian countries, e.g., in Korea and Japan, fisheries and fishing have long history and became part of their culture. Incidentally, there had been efforts to categorize the innovative (so-called creative) economy into four parts: (i) community-based fisheries management, (ii) individual transferable quota, (iii) fishing villages, and (iv) stock rebuilding plans. These are used as tools to develop the fisheries sector. Considering fisheries as part of the Korean culture, further attempts had been made to reveal the *modus operandi* of the Korean government in using the concept of so-called innovative (creative) economy in order to shift this sector toward a more sustainable and efficient sector.

In the contemporary world, there seemed to be development and emergence of a new initiative(s), which may exacerbate the paradigm shift of economy and cultural links. As such, under such a background, the embracing economic, cultural, technological, and social innovative or the so-called creative economy becomes a topical issue of different aspects of development at both the macro- and micro-international economic and development agenda.

Concomitantly, in the center stage of the new paradigm is the fact that the access to information, sense of innovativeness, and the so-called creativity are prerequisites in framing policies, particularly in the developing countries. However, there seemed to be no simple definition of the powerful engines driving the so-called "creativity" that encompasses all the various dimensions of economic growth and promotion of development in different forms in the globalizing world. The so-called "creativity" in this context could refer to some form of innovativeness. Incidentally, the so-called creative industries had developed rapidly, after the 1980s, with the formulation of new ideas; and through the application of these, they are becoming an important part of knowledge-based ideas to produce original works in the modern economy (Wikipedia, 2011). We often see the so-called creative products, functional creations, scientific inventions, and industries in our daily life. They, generally, refer to many technological innovations. There are, thus, economic kinds of activities of our life, such as fashion, movie, aspect to innovations(s) or so-called creativity, etc., which could be observable in the way they contribute to design, art, and, so on. Also they are often viewed as entrepreneurship, fostering innovation, etc. Moreover, such aspects could be product(s) of productivity and promotion of economic growth. Incidentally, knowledge economy is related to the development of people; and, in fact, the innovative or the so-called "creative economy" is a holistic concept of knowledge economy. So, in many countries, the innovative or the so-called "creative economy" deals with complex interactions between culture and industries, and are often referred to as the "cultural industries."

Notwithstanding the above, the so-called term "creative industries" is a relatively new term, and is based on the licensing system of fishing vessels and government policies thereof (UNCTAD, Creative Economy Report 2008, 2009). It mainly refers to technical measures, such as closed season, closed area, mesh size regulation, etc. Incidentally, the report is the first cultural policy in the history of size regulation (of fish catching devices, etc.) as well as input control to manage the importance of the creative fish stock and the fishing industry for the last century in Austria. It also enhanced the knowledge economy of Korea with regard to fish stock assessment (Reis, 2008). Then, the concept was adopted and developed further in the United Kingdom (UK). The cultural aspect of fishing industry given a tremendous boost to the so-called creative industries and creative economy in the fisheries sector, and all these had a deeper implication in the industry. It was more evident in South Korea.

In continuation of the above, the so-called term "creative economy" had appeared in 2001, and was quite *in vogue* in Korea and other countries which are rich in fishery. Such an aspect may assist in ranking and could reveal how they combine to generate extraordinary value to different agricultural commodities.

Notwithstanding the above, these seemed to be two kinds of creative economy in Korea, viz. one relates to people's fulfillment with regard to the growing importance of recreational fishing, and that the other generates a product, which has contribution to economics. The first one has a universal characteristic of humanity, although its benefits are often difficult to assess; and is also insufficiently found in all societies and cultures. On the other hand, the second is recognized as a more generalized aspect.

Concomitant to the above, there have been some attempts to portray the positive impacts of stronger industrial societies, such as to assign a higher value of innovative (creative) economics in South Korea with regard to the expected novelties on science and technological innovations. Incidentally, The Fisheries Act of Korea (1953) is said to be a significant basis for on intellectual property rights (IPRs) issues. The legal framework for Korea's fisheries contains the concepts and provisions on the management and regulation of various kinds of restrictions, including significant queries as to how much one could regulate the different aspects and limitations of fisheries, including licenses, property rights, rights-based fishing, etc.

Notwithstanding the above, in many authorizations, notifications, enforcements, etc., there have been efforts to consider the so-called innovative (creative) economy as major legal instruments for fisheries also involving the cultural and artistic products that might have been included under a "Fishery Resources Management Act," which had brought in the historical and cultural aspects as well. On the other hand, in some of the South Asian countries, for example, in Korea and Japan, there had been a long history of governance, management, conservation, utilization, and development of the fisheries resources; and they might have become part of the so-called "Fisheries Act" for the culture and management of the fisheries including AQC in inland waters, the "Distant Water Fisheries Development Act" for the management of high seas, issues arising out of increasing number of fishers' groups joining the fisheries, the promotion of international fisheries, gradual emergence of CBFM, and so on.

In 2011, the "Korea Fisheries Resources Agency (FIRA) Aqua Farm Management Act" had come into force for the effective and efficient operation of AQC and pollution mitigation; fisheries stock enhancement; management of artificial reefs, fry release, etc. There are other acts and regulations to take care of various other aspects like the Marine and Plant Disease Control Act, the Agricultural and forests Act, The Fishery Products Quality Management Act, and so on.

The relative dispersion of coastal small-scale fisheries in the international waters, where overseas fishing is carried out, is considered to add to maintaining economically viable rural communities. It also tried to balance the trend toward growing coastal total allowable catch (TAC) system. Incidentally, the **TAC system** was introduced in 1999 to manage and control the harvest of the aquatic resources. In fact, according to history, fisheries had often been the basis for the EEZ of Korea, to be consistent with the UN human settlements and coastal development as well as with the Convention on the Law of the Sea, which came into effect in rural and urban environments. Artisanal fisheries were also taken care of. As of 2011, Korea's TAC system covered 11 species and also assisted in the generation of the capital needed by fishers and fish farmers to invest in the promotion of fishes like the mackerels, Jack mackerels, squids, red snow agriculture, crab and blue crab AQC, etc., with a total TAC of *c.* 425,000 tons. However, in well-managed fisheries, high resource rents could also be achieved. Further, individual vessel quotas were generated under the system, and used to finance investments within or allocated by respective fishers' cooperatives. Subsequently, in 2001, **CBFM**, having considerable aesthetic value, was introduced. In addition to its direct contribution, the already-established rules and regulations in the fisheries sector are often responsible for significant direct or indirect impact on the sustainability of the local fisheries as well as multiplier effects on the economic developments. It may be mentioned that under the CBFM, fishers' groups take intra-sectorial interactions and measures, e.g., voluntary management measures on their own fisheries; ancillary activities, such as net making; actively participating in decision-making processes in capture fisheries and AQC through management and settlement of conflicts and disputes; income generation; fishing ground and resource management; management of fish farming; and so on. In addition, other economic activities, like stock enhancement, small trades, etc. are also undertaken.

Concomitant to the above, these multiple economic occupations not only bridge the gap caused due to depletion of resources but also take care of the often-great seasonality in the abundance of fishery resources. In fact, this has a long cultural history in South Korea.

Further, these complementary pursuits (by the Korean government) compartmentalize the fisheries sector into four main parts, notably (i) the dynamics; for example, the supply of capital and labor of the fishing activity, which may evolve in close relation to fisherman communities; (ii) fisheries management and the agricultural activities undertaken by the household; (iii) the ITQ system; and (iv) the infrastructure developed for fisheries, e.g., villages roads, landing sites and coastal havens, water-retaining stock rebuilding plans (like ponds), etc. All these could tend to trigger further economic developments in other sectors such as tourism, agriculture, etc. Such strategies have made significant contribution to the Korean fisheries employment opportunities, especially in the remote and marginal areas. In addition, all these strategies help in other ancillary activities in Korea, like boat building and maintenance; workshops management, including engines and gear handling and maintenance; net making, repair and maintenance; handling, processing, packing, and transport of fish and fish products, and so on. All these measures have led to effective management of the fisheries by the fishermen communities in Korea. Further, the Korean government has introduced a new self-landing site-based activity. Furthermore, as part of the management system, fishermen voluntarily earn their livelihoods from fisheries, including those engaged in production, harvesting, etc. They also take active part in decisions on the management, etc. Many receive support from the government for improvement in their fisheries to earn better livelihood. The rapid development of resource management to administer a sustainable AQC for the local and export markets; and, the gradual grant of ownership and independence in fisheries; the gradual development of awareness in fisheries; various kinds of transformation in many areas related to fisheries, etc., are the other aspects of upcoming management scenario of the aquatic resources in Korea. Still further, a fishermen-oriented and community-based semi-industrial activity had also been contributing substantially to the CBFM in Korea.

In CBFM, the fishing community is responsible for alerting the relevant authorities in fisheries management as well as various adjustments in the fisheries sector, like management of various kinds of hazards in pollution, taking care of various types of difficulties faced by the fishers and fish farmers, and, so on. In fact, promotion of CBFM could help in resolving disputes between communities, such as disputes related to major changes in the ecosystems, disputes related to fish processing industries, etc. Various types of conferences under the aegis of CBFM could resolve several types of conflicts across the board, if necessary, by taking the help of various kinds of literature like the FAO Code of Conduct for Responsible Fisheries, fisheries discussions, etc. Moreover, a public fishing village guidance service could significantly increase their future contribution to each participating community for the sustainable development and management of the aquatic resources.

Notwithstanding the above, the **Development of South Korean Innovative (Creative) Economy in** communities in *c.* 160 areas have given supports to their **fisheries,** like many of the fisheries around the world.

It may be mentioned that the Korean government had initiated the **ITQ** system in 2011 as part of "fisheries resource and oriented comanagement system for more effective management plan" for the implementation of responsible fisheries (Arnason, 2000). The ITQ system allows individuals to trade with fisheries under the catch quota system among the group of fishermen in fishing villages set among themselves. The system also allows organization(s) of fishermen to go for such similar kinds of activities. Still further, such as a fishery system also allows the implementation of self-regulation according to relevant fishery-related ITQ system, in general; and could be categorized as follows:

(i) Laws and regulations with the endorsement of the local government so that fishery could be controlled.

(ii) Increased fishing efficiency: Fishermen-oriented comanagement system is designed to decrease the level of uncertainty, enhance a sense of responsibility of the fishermen profit maximization, prevent illegal fishing, etc.

(iii) Another characteristic of Korean deduction of administrative costs in ecosystem-based fisheries management (EBFM) is that it is premised fisherman's voluntary participation by connecting with community as well as by generating a "Resource Zone"; and concomitant sustainability-based management of fisheries. In fact, self-management could prevent overfishing as well as could systematize the traditional management of fishing activities. Such a strategy could also promote fisheries management by the fishing communities by focusing significantly on fishing villages, thus guaranteeing fishermen's and fishing community's fishing rights. In this system, fishermen communities try to issue orders to prevent illegal fishing through voluntary decisions. This was believed to be one of the main reasons for the implementation of ITQ system in Korea (Kim, 1999; Hwang et al., 1998).

In addition to the above, under the traditional fisheries management system, a complete settlement of EEZ agreement had been accomplished among three countries, viz., Korea, China, and Japan based on conservation-dependent approaches. Further, South Korea might have had supplementary approaches, as well. The Korean fisheries sector is ready to take new gain(s) from applying TAC system. In fact, the Korean Government had considered this as a business-type fisheries-based comanagement, involving innovative (creative) economy in order to shift to the solo ownership of allowable biological catch (ABC) based on scientific framework (Fleischer and Pizam, 1997; Hjalager, 1996; UNCTAD, 2010).

Notwithstanding the above, the locational advantages in the form of nearness could lead to increased foreign exchange inputs through national and international tourism, and further could account for up to *c.* 40% of GDP (UNCTAD, Creative Economy Report 2010, 2013).

Fisheries management programs, which had been proven through experiences over millennia, could be considered as effective and efficient management *modus operandi.* Also, this became a part of the Korean culture and history. The fishery and AQC industry has been significantly playing a holistic major role in Korean economy. Korea is following an ecosystem-based approach to fisheries management, which is unique and may not be easily compared with other nations of the world.

5.1.11 Lao PDR

Lao PDR is an ichthyologically potential country. **In Lao PDR,** Baird (2000) studied Integrated Community-Based Fisheries CoManagement and Protected Areas Management with regard to opportunities for advancement and obstacles to implementation.

The Lao People's Democratic Republic (Lao PDR or Laos) is a small country in SE Asia with an ethnically diverse and largely rural population of *c.* 4.5 million (Claridge et al., 1997). Laos had established 18 National Biodiversity Conservation Areas (NBCAs) by Prime Minister's Office decree (PMD 164) in October 1993. An additional two areas were added to the National Protected Areas (NPA) system in 1995 (IUCN, 1999), summing the total number of NBCAs to 20, covering *c.* 12% of the country. A number of international organizations (IOs), nongovernmental organizations (NGOs), and bilateral and multilateral donors had given financial and technical support to help establish Laos' NPA system. These efforts had been led by IUCN, the World Conservation Union, Wildlife Conservation Society (WCS), WWF for Nature, Swedish International Development Agency (SIDA), The Netherlands Government, and the World Bank. Lao government's coordination at the central level had mainly been confined to the responsibility of the Center for Protected Areas and Watershed Management (CPAWM), being entrusted with the Department of Forestry, Ministry of Agriculture and Forestry of Laos.

Fisheries and AQC have been providing nutrition (notably the subsistence protein) and avocation to the people living in and around the protected areas (PA) in Central and Southern Laos. The significance of fish and fisheries with regard to PA management was often been neglected due to the focus on biological surveys and management initiatives on large mammals and birds.

In fact, the comanagement of natural resources by villagers is one of the government policies in Laos. Also, there is considerable amount of support for making cooperative management systems with regard to natural resources, including the aquatic resources. It may be pointed out that the comanagement of aquatic resources including fisheries has apparently succeeded in augmenting fish stocks, and villagers' fish catches, in Khong District, Champasak Province, Southern Laos. However, it may be noted that scientific evidence(s) are still lacking in favor of supporting villagers' claims that aquatic resources had significantly benefited from their comanagement exercises. Nevertheless, villagers from some communities with aquatic resource comanagement systems, such as, Kokpadek and Chan villages had reported that increased fish catches had unexpected spin-off benefits in terms of management and conservation measures adopted. There were a number of advantages to initially emphasizing fisheries comanagement in the PAs.

Notwithstanding the above, large-scale participation of villagers in fisheries comanagement was more feasible than with large mammals and rare birds because all villagers come in contact with fish on a daily basis, while only a small portion regularly encounter large mammals and rare birds. Moreover, fish generally have a high reproductive capacity, resulting in a relatively high potential for quick and visible positive results to management initiatives. Further, it is easier to define and demarcate PAs for fish due to environmental limitations, compared to many species of mammals and birds, especially those that have large ranges. In addition, villagers also have more indigenous ecological knowledge (IEK) about fish, as compared to large mammals and rare birds.

Therefore, the sustainability of fisheries is crucial to villagers' food security, fish being one of the staple food items of the people of Lao PDR. If fish availability is less, villagers might not have much options other than increasing hunting pressure on vulnerable populations of birds and mammals.

Notwithstanding the above, fisheries comanagement could act as a good entry point for conducting community-based collaborative management activities regarding a wide range of natural resources, other than fisheries, including large mammals, rare birds, and terrestrial forests. In fact, many non-timber forest products (NTFPs) may have a high potential for initiating comanagement activities in PAs as well. However, there remained a number of hindrances to implementing living aquatic resource management in the PAs. These included mainly the following:

(i) Problems related to mutual trust and respect between the villagers and the staff of the PAs.
(ii) Institutional constraints.
(iii) Lack of documentation in Lao and Khmer language with regard to indigenous fish and wild-capture fisheries, biological constraints related to migratory fish species, village to village conflicts, etc.

Notwithstanding the above, community-based living aquatic resources comanagement may not always be the most appropriate first step for cooperating with communities in the Pas; but nevertheless, it serves as an important opportunity worthy of increased and careful consideration.

Notwithstanding the above, Phounsavath (1998) reported about CBFM as an approach to participatory management of reservoir fisheries exemplified by a case study of two fishing villages at the Nam Ngum Reservoir, **Lao PDR**.

There had been works with regard to the assessment of the present management system of the fisheries in the Nam Ngum reservoir in Lao PDR to investigate the existing potentials of the local communities to be involved and to participate in the development of a CBFM system. The study has primarily focused on two major fishing communities of the Nam Ngum reservoir, viz., the Phonsavath and Xai Oudom villages, in order to investigate their existing institutional structures and types of organizations that had directly affected the management of fisheries. Different attitudes of the local fishers toward the establishment of a CBFM system were also investigated. A total of 150 fishers' households were interviewed using structured questionnaires. The assessment of local institutions and organizations was conducted through the method of Rapid Appraisal of Fisheries Management Systems or RAFMS (Pido et al., 1996).

The results had revealed the existence of different traditional institutional structures and organizations involved in the management of fisheries in the two villages as well as different fishers' behaviors and fishing practices. The majority of all respondents from both the villages (*c.* 79%) agreed toward the establishment and legislation of village management zone(s), however, with the preference for a district-level partitioning of the reservoir fishing boundaries. Many of the respondents (*c.* 93%) were well aware of the current fishing regulations. Moreover, a significant proportion of the respondents (*c.* 85%) were in agreement to comply with the establishment of a fishing license system, provided that the regulation could receive unanimous support from the majority of the fishers. Further, almost all the respondents agreed to comply with the regulation of "closed" fishing season without any exception. Furthermore, the majority of the respondents (*c.* 59%) had a positive attitude in relation to the establishment of a fishers' association. However, only a small number of the respondents (*c.* 43%) were interested to participate in the activities of such an organization, other than to remain as members only. Nevertheless, the most favored structure of fishers' association was suggested to be established at the village level, and appropriately divided into fishing units under equal conditions for all types of fishers irrespective of the types of fishing gears. The main prerequisites for the development of a CBFM system in the study area still were not yet developed, such as interests and willingness for the establishment of fishers' management units under the fishers' associations. Also, many of them had not shown much interest in voluntary and responsible compliance with fishing regulations, mainly because means to facilitate such development were either lacking or not very clearly defined. However, the establishment of management zones and fishermen associations could be promoted at the village level. Nevertheless, further works could be required for possible stratification of reservoir fishing boundaries at the district level.

Concomitant to the above, Anon (1997) dealt with the reservoir fisheries management in Savannakhet Province of **Lao PDR**.

The principal objectives of the works were to devise management strategies to increase the individual and community income from reservoir capture and culture fisheries, while maintaining the role of reservoirs in providing subsistence for the fishermen and dry season habitat for natural fish populations.

The main outputs of the works are as follows:

Reservoir fisheries in Savannakhet were analyzed in ecological/technical, socioeconomic, and institutional terms. Particular emphasis was laid on the enhancement of fisheries in the small water bodies. Stocking enhancement and community management of the fisheries of small water bodies have been actively promoted by the provincial government, and this effort was spreading rapidly. Incidentally, the fisheries of the selected and studied small water bodies are found to be of four types, as follows:

(a) open-access type without stocking, (b) open-access type with stocking, (c) restricted access type without stocking but rented out to individuals, and (d) community management with restricted access and stocking.

Notwithstanding the above, access restrictions has a strong positive effect on the standing crop (stocks) of fish in small water bodies. Stocking alone was not sufficient to increase standing crop (stock). On the other hand, stocked water bodies with restricted access usually portrayed the highest standing stocks of wild fish.

However, there appeared to be no significant effect of stocking on fish species diversity, although further analyses may be required in this regard. Nevertheless, preliminary analyses had suggested that yields from community fisheries could be lower or may be equal to the yields achieved from open-access water bodies. In fact, Catch Per Unit of Effort (CPUE) in community fisheries was substantially higher than that in open-access fisheries. The given yield could, therefore, be achieved with much lower effort. Meanwhile, the community fisheries and the water bodies, which were rented out, seemed to have generated substantive income for the rural people. Villages with community fisheries seemed to have a higher income than other villages, with the fisheries often contributing to $>50\%$ of the total income. In addition, renting out also, usually, had generated significant income. Incidentally, the village income was used mainly to improve the schools, religious institutions, etc., or for other infrastructure development like road, electricity, etc. However, the contribution of the community fisheries to household fish consumption was marginal ($<10\%$) with $>80\%$ of the household consumption based on fishing in open-access water bodies. However, the village income derived from the fishery had led to a reduction in household contributions to village activities, and thus had provided an indirect benefit to the individual households. Further, it may be noted that a significant proportion (often $>50\%$) of the yield from the community fisheries was exported from the village, thus reflecting a low in-village demand to buy fish. Further, it may be noted that all the community ponds were located close to the villages, and were, generally, fished by most of the households before the initiation of access restriction period. Villagers were believed to fish in the alternative water bodies, except in one village where no alternative water bodies had existed nearby; and, individual fishing with small gears were still allowed in the community fishery. Furthermore, it may be noted that in all the villages surveyed, the major part of the household fish catches had been obtained from the inundated rice fields and from the rivers, and not from small water bodies, which suggest that the loss of individual access to community fisheries had only a little effect on the household consumption. Nevertheless, the villagers generally regarded the community fishery as beneficial, with community income being the largest perceived benefit.

In fine, enhancement, as currently practiced in community fisheries, could raise the community income significantly, but concomitantly, it could have a negative and marginal effect on individual fish consumption and income from fishing and, further, such a practice could protect the stocks of wild fish in the enhanced water bodies. Moreover, substantial improvements in overall yield and community benefits are also technically feasible, but may require improved marketing. In addition, improvements in individual benefits may also be possible and desirable; probably, at the same time, not least to ensure the institutional sustainability of enhanced fisheries. However, this may require more complex institutional arrangements.

Notwithstanding the above, the results suggest various ways of improving the management scenario, particularly, more active involvement of the stakeholders from the grassroots level in large number in a participatory mode and through more effective implementation of the CBFM. But, substantial amount of uncertainties may still remain, which could best be resolved through an adaptive approach to further development efforts aimed at small water bodies. In addition, the close integration of technical, socioeconomic, and institutional analysis in various projects could lead to significant advances in the understanding of enhancement in the systems. Incidentally, various contributions had summarized the integrated approach, with particular reference to the predictability of enhancement outcomes. A number of contributions to such fields have been presented at various symposia, notably, at the FAO/DFID Expert Consultation on Inland Fisheries Enhancements in Dhaka, Bangladesh, and so on.

Notwithstanding the above, **the main contributions of the outputs to the goals of the project** are as follows:

The patterns of work, as outlined above, are to fulfil the objectives related to the technical aspects, socioeconomic benefits, expenditure and returns, etc., as well as the alternative arrangements for developmental aspects, with emphasis on enhancement of community income, conservation of wild stocks, and increased returns to labor. However, low effective demand appears to be the main factor limiting yields from community fisheries, and thus improvements in marketing and/or access arrangements are crucial to increasing yields. Further, the communities are required to experiment actively with management regimes in order to identify improvements; and this process may be facilitated by appropriate support from the provincial government. The reduction in fishing effort and increase in wild stocks, which are, sometimes, found in community fisheries, show how enhancement, by precipitating institutional change(s), may contribute to a more sustainable exploitation of aquatic resources.

Such efforts, as indicated above, may serve, on an overall basis, as significant steps toward an integrated understanding of enhanced fisheries systems.

5.1.12 Malaysia

Malaysia is a significant country in SE Asia with a high fisheries potential. Nasuchon (2009) reported about coastal management and community management in **Malaysia**, Vietnam, Cambodia, and Thailand, with a case study of Thai Fisheries Management.

This report had reviewed approaches to coastal management and community management in Malaysia, Vietnam, Cambodia, and Thailand. It also had included the case study on fisheries management in Thailand. The data had been collected from secondary sources, like articles in journals, technical papers, website(s), etc. Malaysia is the only country having strong administration of coastal zone management by the federal government and is the strongest in surveillance and enforcement, but concomitantly, lacks CBM. On the other hand, community management in Vietnam was established from community needs that they want to protect their resources, and also has community traditions to support resource management. Conversely, Cambodia is very young in coastal management and almost the entire policy for aquatic bodies is focused on FW, mainly because *c.* 75% of the aquatic protein comes from the Tonle Sap and Makhong rivers. Concomitantly, Thailand has decentralized fiscal responsibility to the local level for the collection of taxes and administration of funds. Fisheries communities enjoy the right to manage their resources, an approach which was promoted by the Department of Fisheries. With respect to fisheries management, Thailand had faced challenges of overexploitation of the resources and a lack of real data on the number of fishing gears; both of which are to be addressed so as to implement effective fisheries management.

The Gulf of Thailand is surrounded by four countries, viz., Malaysia, Thailand, Cambodia, and Vietnam (southern part). This area has highly exploited resources base as they respond to market demand. The FAO had reported that, in 1950, AQC production was *c.* 603,941 MT, which was augmented to *c.* 26.592 mmt in 1996, then rapidly enhanced to *c.* 45.924 mmt in 2004. In fact, four countries in SE Asia Indonesia, Thailand, Vietnam, and the Philippines are the top 10 exporters of AQC products.

Notwithstanding the above, there were overexploitation of the resources mainly because of its demand in the market. In addition, coastal AQC is highly developed, and as such Thailand lost *c.* 64.3% of its mangrove areas. Concomitantly, Vietnam had rapidly increased its shrimp farms from 3000ha in 1980 to 60,000ha in 1992. Cambodia had also increased its mangrove areas for AQC in the past few years.

Concomitant to the above, the Integrated Coastal Zone Management (ICZM) was started in Australia and Sweden in the early 1970s, but was not formally called as such. In fact, the United States was the first to introduce legislation in the form of Coastal Management Act > 20 years ago.

Notwithstanding the above, the World Commission on Environment and Development started looking at the future. The commission had recognized the sustainability of living resources of the oceans. Subsequently, they had agreed to management in five zones, viz., inland areas (river catchments), coastal lands, coastal waters, offshore waters, and high sea. Moreover, the UN also held a conference in 1992, at Rio de Janeiro, during which, the states elaborated Agenda 21, within which Chapter 17 addressed the protection of the oceans, coastal zones, use, and development of living resources, and so on.

Incidentally, Thailand and other neighboring countries started coastal management in 1947, but Thailand was the first country to establish management of fisheries resources, although not as ICZM. Incidentally, ICZM was started in Vietnam in 1994. Then, Thailand initiated a concept of fishing rights around 1995 based on the Japanese system.

Notwithstanding the above, **Malaysia** implemented Integrated Coastal Zone Management (ICZM) in its peninsular region in 1996, followed by Cambodia in 1997. Later, the concept of CBM was significant and implemented in this region with the support of international agencies. However, it was not very easy to achieve successful results for these programs in poverty-stricken countries. People need food and money for their livelihood and day to day survival, and thus conservation of natural resources for the future becomes less important than the everyday needs. Nevertheless, these regions attempted to encourage the community to appreciate conservation and management of their resources through CBM. Further, this region has also examined new techniques, so as to understand their applicability with respect to their needs. Thus, the project, "Partnership in Environment Management for the Seas of East Asia (PEMSEA)" was initiated in 1994. The pilot projects were demonstrated in Batangas (The Philippines) and Xiamen (China). The experiences and lessons learned through this Integrated Coastal Management (ICM) project were then applied in six more countries, viz., Cambodia, DPR Korea, Indonesia, Malaysia, Thailand, and Vietnam.

Concomitant to the above, all the countries involved in CBM policies had implementation techniques that differ in their approach, often resulting in successful projects in some countries and project failures in others. Incidentally, the experiences gained from many projects have provided fruitful lessons for exposing the weaknesses and also in recognizing strengths.

5.1.13 Middle East

Arab fisheries resources, including the coastal waters of the Arabian countries, inland waters, and AQC potential, are indeed significant sectors for the development of the **Arab countries** and their people (Izzat, 1998). In fact, it could be opined that fisheries sector may play an important role in meeting increased demand for food, and in activating the economy of the region, if rationally and scientifically exploited. These sectors could surpass economic return from oil exploitation. In fact, the stock of fossil fuel is said to be depleting.

Seas are said to border the Arab countries on all sides: the Gulf of Arabia on the east, the Atlantic on the west, the Mediterranean on the north, and the Indian Ocean on the south. In addition, the Red Sea and various gulfs and rivers, notably, the Nile, the Tigris, and the Euphrates; and the natural and man-made lakes, which are said to constitute the inland water resources, provide the Arab countries with high potential for fisheries and AQC. Further, The EEZ expands the Arab marine waters to rich international fishing grounds. Concomitantly, engaging in AQC activities in marine and inland waters, as well as, better exploitation of the available fisheries resources could lead to a more extensive and beneficial fishery industry. It could become one of the main pillars of Arab economic activities, comparable, more or less, to the fishery industry in the major fishing nations, such as Japan and Norway. In this connection, it may be noted that before the discovery of oil in some Arab countries, fish had traditionally been a potential source of animal protein, particularly along the coastal areas. Subsequently, increasing incomes from oil, which provided higher purchasing power and made other protein sources available, thus causing diversification of protein intake; and concomitantly had reduced fish consumption. Nevertheless, of late, due to different economic and social reasons, demand for fish and its consumption in fresh or processed form have risen again. In fact, consumers have begun to turn to cheaper foods, often consuming traditionally unmarketable fish products, thus developing new uses of fish, by either diversifying methods of preparation or by accepting imported products, which were not locally known earlier.

Notwithstanding the above, fish marketing opportunities are opening up increasingly in the Arab nations. Such an increase in the demand for fish products, thus, may be met either from better exploitation of fisheries resources or through the import of quantities that may not be obtained from local sources.

5.1.13.1 *The issue of fisheries globally*

In the past decades, the fast growth of the exploitation of fisheries resources had taken its toll during the 1950s and 1960s, when the global catch from commercial fishing grew three times faster than the world population. However, the production rate was slower afterwards. Currently, the global fishery production (catch and AQC), averages *c.* 100 MT per year (*c.* 113 MT in 1995). However, per capita supplies are said to be on the decline. As such, these trends had an obvious impact on the prices, thus rendered a traditionally inexpensive source of protein much less accessible to the poor. However, the depletion in aquatic resources was due to two factors:

(a) The development of excess fishing capacity with respect to existing fish stocks.
(b) The deteriorating condition of fish stocks due to the pollution of sensitive aquatic systems, in both FW and marine domains.

Notwithstanding the above, the small-scale fisheries in developing countries have generally played a less significant role beyond their communities, although the pressure of growing local demand had occasionally contributed to the overexploitation of coastal waters.

Concomitant to the above, the development of fishing capacity, fishing efficiency, and income levels had occurred mostly in the large-scale (industrial) and medium-scale fishing subsectors. Concomitantly, the large number of small-scale fishermen had no access to that help, thus finding it increasingly difficult to survive in an overexploited environment. Thus, the crisis of fisheries was a crisis of livelihoods, from this perspective. Moreover, the implications have gone far beyond dwindling fish stocks, which were available to small-scale fishermen. Incidentally, fishing is known to provide principal source of income for *c.* 100 million fishermen and their families. Thus, fishing and fisheries contribute more than any other animal production activity. Fishery, as already had been said, helps in protein intake in most of the developing regions of the world, including most of the Arab nations, and is important for the food security of many coastal populations. In fact, many wild marine and FW resources are on the decline by all accounts, and majority of them are among the world's poorest. There are one billion people, nearly a fifth of the global population, for whom fish is the main source of nutrition and avocation. The scarcity of fisheries resources, thus, often leads to conflicts among various groups.

Notwithstanding the above, *c.* 85% of the world's fish harvest is still caught in the wild, and this practice has reached the limits of sustainability in many areas. This, in part, could be due to the fact that the means of exploitation have become very efficient. For instance, fishing fleets of today use sonar, radar, aircraft, and satellites to track shoals of fishes. Perhaps, the winches and motors and drift nets typically trap >18 tons of fish. This enables the trawlers to boost not only their catches but also the **bycatches, i.e., the species that are netted but are unwanted and consequently discarded.** The issue of discards is a very disturbing one, since it results in a very large wastage of fisheries resources. FAO has estimated that *c.* 18–39 mt of fish per year may be discarded at sea to catch about 50 mt of fish suitable for human consumption.

5.1.13.2 *Fisheries of exclusive economic zone*

Every country in the world has a permissible zone for exploiting the resources of the marine habitat. The introduction of EEZs in the mid-1970s and the adoption of the United Nations Convention of the Law of the Sea provided a new framework for the management of marine resources. The new legal regime of the oceans has given the coastal states the rights and responsibilities for the management and use of fishery resources within their EEZs. However, many coastal countries including various Arab nationscontinued to face serious challenges, lacking experience and financial and physical resources to achieve greater benefits from the fisheries within their EEZs. Subsequently, the FAO Committee on Fisheries (COFI), at its 19th Session in 1991, called for the development of new concepts which could lead to responsible and sustainable fisheries. Consequently, the International Conference on Responsible Fishing (ICRF), held in 1992 in Mexico, had requested the FAO to prepare an international Code of Conduct to address these issues. The outcome of this conference was an important contribution to the 1992 United Nations Conference on Environment and Development (UNCED), particularly Agenda 21. Later, the United Nations Conference on Straddling Fish Stocks and Highly Migratory Fish Stocks was convened, to which the FAO provided important technical backup. Subsequently, the Agreement to Promote Compliance with International Conservation and Management Measures by Fishing Vessels on the High Seas was adopted in November 1993.

As a result of these developments, the Code of Conduct for Responsible Fisheries, adopted in October 1995 by the FAO Conference, had provided a necessary framework for national and international efforts to ensure sustainable exploitation of aquatic living resources in harmony with environmental management activities. Further, it had also brought about a significant change in the conditions under which fisheries are carried out on a worldwide basis. However, the extent of national EEZs and the nature of jurisdiction claimed differed from one country to another. However, the zone had extended to 200 nautical miles offshore in the majority of cases; and in all cases, jurisdiction over fishery resources was included.

Concomitant to the above, all the Arab countries, *albeit* in ranging degrees, have access to fisheries resources in their EEZs.

5.1.13.3 *Small-scale fisheries*

This subgroup is characterized by a fleet of small motorized boats, 5–6 m in length, of small catching capacities, equipped with various kinds of traditional fishing gears. Fishermen composing this group are scattered all along the coastlines of Arab nations. Their numbers are quite large and much of their fishing is subsistence fishing for local markets, families, and clans. However, the official statistics of their catches are not recorded systematically and, therefore, lack in reliability. However, the species they target are mostly demersal, near-shore species, crustaceans (mainly shrimps), etc. Incidentally, the average share of coastal fisheries in the Arab world fish landings was estimated at *c.* 80%–85%.

5.1.13.4 *Inland fisheries*

Inland fisheries resources are those that are landed from the various inland water bodies. These are estimated to cover *c.* 1.5 million km^2. The major nations, where these are available, are Egypt, Sudan, Iraq, and Syria. The inland abiotic resources include the lakes, rivers, marsh lands, swamps, reservoirs, etc. Incidentally, the lakes in the Egyptian Delta region are the main fish production water areas, in addition to Lake Nasser in the south, Lake Qarun, and the River Nile. Concomitantly, in Sudan, the main inland fisheries are in the Blue and White Niles, in addition to the main stem of the Nile. On the other hand, Iraq's main fisheries are located on the Tigris and Euphrates rivers as well as some manmade reservoirs. Other rivers with almost similar or little smaller size fisheries occur in Syria, Lebanon, Jordan, and Mauritania.

Notwithstanding the above, most of the inland fisheries in the Arab countries are characterized by subsistence fishing in order to meet immediate food supply needs of the population living in the vicinity of the water bodies. However, any excess to their food needs is sold to the local markets. However, the major problems facing these rivers are as follows:

(a) reduced volumes from upstream diversions and flood control; and
(b) compromised water quality from pollution.

5.1.13.5 *Aquaculture*

AQC is a potential source of nutrition and avocation for the people. Its global importance indeed grows each year, mainly because of population growth and the fact that the natural fisheries resources have reached their MSY. The Arab world is no exception. Many countries with FW resources, such as Egypt, Sudan, Iraq, and Syria were practicing AQC since many years. Other countries where FW is scarce are engaging themselves in mariculture activities, farming, fish along their coastlines, such as in Tunisia, Morocco, Saudi Arabia, Kuwait, and Bahrain. Moreover, the UAE and Oman also established research centers with the aim of developing mariculture in their waters. Incidentally, it may be mentioned that in 1995, fish from AQC reached a production level of *c.* 68,000 mt, which was *c.* 4% of the total fish catches in the Arab region, valued at *c.* US $300 million. On the other hand, production in 1984 was only *c.* 22,000 mt valued at *c.* US $59 million. Thus, the 1995 level represented an increase of *c.* 67% in quantity and *c.* 95% in value.

Incidentally, the main species farmed are those indigenous species that are popularly accepted in the individual Arab countries. In Egypt and Sudan, Tilapia is farmed, and is also being introduced for farming in Saudi Arabia, where Egyptian and Sudanese minorities live. Likewise, groupers and rabbit fish are farmed in the Gulf Cooperation Council States (Bahrain, Kuwait, Oman, Qatar, Saudi Arabia, and UAE). On the other hand, shrimp farming was attempted along the Egyptian coast of the Red Sea. Further, there were also experimental projects aimed at developing the commercial cage farming of sea bass and sea bream in Egypt, which had begun in 1990 in Port Said in the east and Mersa Matrouh in the Mediterranean coast in the west. Subsequently, when research studies on fish farming become feasible and as the technologies and know-how become readily available, fish from AQC production is expected to expand substantially.

Notwithstanding the above, the Arab region, with a coastline stretching to *c.* 23,000 km, offer considerable possibilities for fish farming both in inland and marine waters. It, however, remained to be seen whether AQC production of fish could substantially increase fish supply in the region and significantly reduce either the supply-demand gap or dependence on the imports.

Notwithstanding the above, subsistence fishing is practiced, to a considerable extent, in the region, especially in the countries bordering the Arabian Peninsula and along the shores of inland lakes and rivers, particularly in Iraq and Egypt. Therefore, due to the lack of extensive and large-scale culture and marketing system, it can be assumed that a portion of the total production is consumed by the fishermen and their families.

Concomitant to the above, in spite of the possibilities, there still remains some gap in the management system of the water bodies and their resources. There is an efficient management system in place, particularly involving the stakeholders from the grassroots level, and CBFM *modus operandi* could be tried for efficient management.

It remains to be seen, however, whether AQC production of fish could substantially increase fish supply in the region, and whether it could significantly reduce either the demand-supply gap or dependence on imports.

5.1.13.6 Management and conservation

There is enough resources of water bodies and aquatic biota in the Arab countries but not much thought and works had been put and done on their management. Nonetheless, management and conservation are integral aspects for proper sustenance of the abiotic and biotic resources. It could be opined that efficient methods of management like the **CBFM** in a **participatory** mode could serve as effective *modus opernadi* for the efficient management and conservation of the water bodies and their bioresources by involving the stakeholders from the grassroots level.

Notwithstanding the above, Anon (2012) had worked on the National Fisheries Strategy (2012–2025) of the Republic of **Yemen**.

Fisheries account for a major portion of the global economy and a major source of nutrition and avocation for the people in both the developed and the developing world. Also, fish happened to be a denizen of one of the largest domains in the world, i.e., aquatic regime. In fact, a close attention to this issue is mandatory in order to sustain and maintain healthy and productive oceans going forward.

Nevertheless, the global fisheries, of late, are undergoing a dramatic change, due to changing climates, environmental concerns, food security needs, and technological advancements. Adapting to these alterations, while ensuring the resource sustainability has become the primary need and focus of today's fisheries. Yemen's fisheries, one of the four promising sectors as identified in its Strategic Vision (2025), has significant potential to achieve diversification in country's economy and in realizing the job-generating pro-poor economic growth for the country's socioeconomic development. Incidentally, the country's long coastline of *c.* 2520 km is endowed with diverse coastal habitats and harbors a rich marine resource of wealth. However, persistent challenges impede the development in the arena, which is further exacerbated by the prevailing socioeconomic situations. These had emphasized the need for a long-term strategy with a coherent approach integrating the resource, environment, livelihoods, economic, and research concerns as well as, concomitantly, identifying the priority needs to generate synergistic benefits for enhancing fisheries sector utilization.

Accordingly, the NFS formulation was the subject matter of an intensive, interactive, and complex policy development and planning process, using participatory and consultative approaches involving all the stakeholders in its planning, formulation, and validation. Thus, the NFS establishes a living dynamic document that directs the country's fisheries sector development in the medium to long term and specifies the key areas for development, thus providing the essential path through which fisheries could become a frontline player in the national development process. Also, the strategy comprises a dynamic action and a working plan along with a monitoring-evaluation framework to support its implementation and to review, examine, and update its progress from time to time.

The overall "goal" of the NFS was "to enhance the fisheries sector contribution to the national economic growth while ensuring the environmental and resource sustainability, improving community livelihoods, strengthening the food security needs, and job-generating economic empowerment opportunities with a special focus on youth and women."

5.1.14 Myanmar

Myanmar is bestowed with rich resources of both lotic and lentic water bodies. Anon (2005) dealt with Wetland Conservation in **Myanmar.**

Myanmar with an area of *c.* 667,553 km ^{2} is gifted with vast areas of natural forests, potential aquatic resources like the wetlands, rivers, etc. The forests range form beach and dune forests, mangrove forests in coastal and delta areas in the south to varied terrestrial vegetation of dry and moist deciduous forests and evergreen forest in the central and coniferous-dominated subtemperate forest types on the high mountain parts of far north. The wetlands and rivers, likewise, are also of various types ranging from small ponds and streams to vast wetlands and big rivers. There are also instances of CBFM in different parts of the country, which need to be boosted.

5.1.15 Nepal

Nepal is a relatively small, landlocked, and hilly country with an area of *c.* 147,181 km^2. There have been contributions related to water bodies, fisheries, and associated management aspects in Nepal (Gurung, 2007; Pradhan, 2009).

The country is endowed with abundant inland FW resources collectively covering *c.* 0.82 million ha or *c.* 3% of the country (DOFD, 2006). Approximately 500,000 ha of water surface is available for fish production, of which *c.* 100,000 ha consists of lakes, reservoirs, and fish ponds (NARC, 1991). Notwithstanding the above, different studies revealed that the inland water resources in **Nepal** are broadly classified into rivers, lakes, reservoirs, ponds, swamps, and irrigated farm lands.

5.1.15.1 Inland water resources: Nature and distribution

Rivers

The major river systems in Nepal consist of the following: (a) the Koshi in the east, (b) the Gandaki in the central region, (c) the Karnali in the west, and (d) the Mahakali in the far west.

These river systems are fed by > 6000 rivulets and streams originating in the high glacial Himalayan, Mahabharata, and Siwalik Mountain regions, and cover an area of *c.* 395,000 ha or *c.* 48% of total inland water resources of the country (DOFD, 2006). The rivers of Nepal form the part of the headwaters of the Ganges River Basin, one of the largest rivers in the world. Discharge from Nepal contributes to *c.* 40% of the total annual flow of the Ganges System and *c.* 70% of flow in the dry season (Water Resources Strategy of Nepal, 2002; Pradhan, 2009).

Lakes

Lakes of various sizes are said to be scattered all over the country covering an area of *c.* 5000 ha or *c.* 0.6% of natural waters. Lakes occur from southern low plains (*c.* 60 m MSL) to > 5000 m MSL (Gurung and Wagle, 2002). The lakes situated at an altitude of > 3300 m MSL are not usually affected by human activities, while those at the foot hills are generally affected to varying degrees by different human activities.

Concomitantly, there are usually three types of lakes in Nepal: (a) the glacial lakes, (b) the tectonic lakes, and (c) the oxbow lakes.

There are 44 major glacial lakes in the northern Himalayan region located > 4000 m MSL. On the other hand, the tectonic lakes occur in the hilly regions. These lakes, when drained, are replaced by flat basins such as the Kathmandu valley, Banepa, Panchkhal, Mariphant (Palpa), Dang, and Surkhet. Some of the famous lakes of tectonic origin in the mid-hills are Phewa (*c.* 523 ha), Begnas (*c.* 328 ha), and Rupa (*c.* 115 ha). The age and origin of these lakes are not known. Conversely, > 2 dozen oxbow lakes are present in Nepal, and most of them are located in Chitwan National Park, Nawalparasi, Bardiya, and Kailali (Sharma, 1997) in the low altitude plains toward the southern part of the country. Lakes are also classified as major and minor types. Incidentally, some of the major lakes are Rara Lake, Phewa Lake, Begnas Lake, and Rupa Lake or Rupakot Lake. On the other hand, the minor lakes include Khaptad Lake, Baragon Lake, Tilicho Lake, Phoksundo Lake, Dudhpokhari Lake, Jageswar Lake, and Panch Pokhari.

Reservoirs

It is quite well known that reservoirs are built to collect water to generate hydropower and/or for irrigation and may be for navigation. In Nepal, there are two types of hydropower projects:

(a) Run-of-the-river type and (b) traditional reservoir type.

The former type projects are small compared to reservoir-type projects. As such, many small and large reservoirs are built in different parts of Nepal. Currently, the total extent of reservoirs comprises *c.* 0.2% of the total water area in the country (Pradhan, 2009).

Incidentally, some of the significant reservoirs are as follows:

(a) The Indrasarobar reservoir (*c.* 220 ha), which was impounded by damming the Kulekhani River in the mid-hills to produce hydroelectric power.
(b) The Jagdishpur reservoir (*c.* 155 ha), which was impounded for irrigation by damming the Budhi Ganga River in the plains area.

Notwithstanding the above, other small reservoirs built by runoff scheme for irrigation and hydropower generation are as follows:

Trisuli (*c.* 16 ha), Marsyangdi (*c.* 62 ha), Panauti (*c.* 2 ha), Sunkoshi (*c.* 4 ha), and various others.

Incidentally, there is high potential for expansion of reservoir area in Nepal. A feasibility study on various river basins and systems indicated that construction of an additional *c.* 92,400 ha of FW reservoir could be possible, which include the following:

Gandaki basin (45,000 ha) (Prosser, 1978), Bagmati river system (9000 ha) (Shrestha et al., 1979), and Karnali river system (38,400 ha) (Himalayan Power Consultant, 1889). These could be added upon their completion to the country's reservoir inventory.

In addition to the above, a considerable area of > 6500 ha of fish ponds are situated in various parts of the country. There is also *c.* 12,500 ha of marginal swamps, while irrigated paddy fields cover *c.* 398,000 ha in Nepal.

No doubt, the lotic and lentic systems could be well managed by involving the stakeholders from the grassroots level, among which the fishermen form a significant entity. CBFM could be an effective tool for the efficient management of the lotic and lentic water bodies.

5.1.15.2 Biological aspects

Nepal is said to be endowed with a bewildering diversity of vast abiotic and biotic resources. However, only a small portion of this is exploited for fisheries purpose, in spite of the vast availability of Nepal's water resources and their great zoogeographical significance. Incidentally, the earliest record of fish and fisheries of Nepal goes back to the 18th century (Kirkpatrick, 1793). The first scientific report on the ichthyofauna of Koshi and Rapti rivers were by Francis Buchanan (Hamilton, 1822). Similarly, the first study of the fish fauna of Nepalese lakes focused on Phewa, Rupa, and Begnas lakes of Pokhara Valley.

Concomitantly, the first report of limnological work from Nepal was a study on some aquatic fauna in Kalipokhari from Eastern Nepal (Brehm, 1953). Subsequently, the first comprehensive limnological investigation in Nepal was initiated by Loffler (1969) in high altitude mountain lakes of Khumbu Himal. Later, Hickel (1973) investigated the lakes of Pokhara valley during the research scheme of Nepal Himalayan Expedition. Ferrow and Swar (1978) conducted a survey on the biological and limnological conditions of lakes and natural water bodies in Pokhara Valley with reference to the existing fish populations, their feeding habits, and biology. Of late, Pradhan (1982) conducted a preliminary study of Syarpu Daha, a mid-hill lake of the western region located at *c.* 1372 m MSL, and had reported a single species of *Schizothorax*. Terashima (1984) reported three new cyprinids of the genus, *Schizothoraichthys* from Lake Rara in NW Nepal. These species are endemic to Lake Rara. Pradhan and Swar (1988) reported *Acrossocheilus (Neolissochilus) hexagonolepis* (katli), *Puntius chelynoides* (karange), *Tor tor*, and *Schizothorax* (asala) from Indrasarobar reservoir, and outlined the potential for expansion of fish culture in the reservoir. Concomitantly, Aizaki (1986) studied the trophic status and water quality of high altitude lakes in Mount Annapurna region, and Yadav (1994) worked on the water quality and benthic fauna of Palung, Thudo, and Chitlang rivers. All these rivers drain into the Indrasarobar reservoir. Further, Gautam (2003) studied the fish diversity and aquatic living resource of Lake Rupa, Kaski, and recorded 23 species of fish belonging to six families and 18 genera under five orders. Further, indigenous fishes, such as *Barilius barna, Barilius bendelisis, Barilius vagra, Neolissocheilus hexagonolepis, T. putitora, C. chagunio, Noemacheilus (Acanthocobitis) botia, Garra annandale, Puntius sophore, P. conchonius, Esomus danricus, Danio rerio, Danio dangila, D. devario, M. pancalus, X. cancila, C. batrachus, H. fossilis, S. plagiostomus, C. gachua* **(Scientific names to be confirmed)**, etc., had been reported from Harpan Khola, Phewa Lake (Prajoo, 2007).

Since long time, the indigenous ichthyofauna of Nepal had not been largely impacted by exotic species, except some stray cases here and there. However, a decline of *c.* 42% in FY of native species (*Mystus* spp. and *Puntius* spp.) had been observed after the introduction of bighead carp (*A. nobilis*), silver carp (*Hypopthalmicthys molitrix),* and grass carp (*C. idella)* in Begnas Lake, Pokhara valley (Swar and Gurung, 1988). The plankton feeding fish species are eco-friendly, as they are net removers of nutrients. As such, they exert beneficial impact on the water bodies, which are often subjected to eutrophication. Pradhan and Pantha (1995) reported that fish production removed *c.* 272 kg of phosphorus (P) and *c.* 2048 kg of nitrogen (N) from Lake Phewa, Pokhara. However, considering the nutrient loading in the lakes, removal of nutrients using plankton feeding fish is one of the alternatives to keep the lake environment healthy and balanced. Plankton feeding exotic fish introduced in the lakes of Pokhara Valley do not breed naturally. As such, they do not exert any negative impact on the environment in terms of mass propagation affecting the populations of indigenous fish species. In addition, the effect of proliferation of the exotic African cichlid species, *Oreochromis mossambicus,* in Nepalese water bodies needs to be assessed. Meanwhile, farming of African catfish, *Clarias gariepinus* and rainbow trout, *Onchorhyncus mykiss* is becoming popular in Nepal. Both these species being carnivorous, a strong preventive protocol is needed to prevent their escape into the natural waters.

5.1.15.3 *Environmental aspects of reservoirs*

In Nepal, lakes and reservoirs in the mid-hills are used for multiple purposes, such as recreation, irrigation, hydropower generation, navigation, fisheries, and drinking water supply. At the same time, the reservoirs trap agricultural chemicals and wastes from catchment areas, thus adding nutrients into water. Turbidity due to suspended solids and siltation as a result of erosion from uplands is also a parameter, which is seriously affecting the fish communities. As a result, some fish species gradually disappear. Incidentally, **river impoundment upsets the long established balances of temperature, gaseous distribution, and energy flows of the lotic system.** In fact, the most noticeable changes are the stratification of water temperature into epilimnion, thermocline, and hypolimnion. In Indrasarobar reservoir, the temperature varies from 12°C to 17°C during winter and from 17°C to 27°C during summer. Also during summer, the reservoir is stratified with a thermocline at 13–14 m and DO becomes very low in deeper layers (Pradhan and Swar, 1988).

In addition to the above, the dams affect fish spawning, migration, and the species composition of the rivers. Hill stream fishes, such as *Pseudocheneis sulcatus*, *Glyptothorax* spp., *Noemacheilus* (*Acanthocobitis*) spp., *Barilius* spp., *Garra* spp., etc., found in Kulekhani River were threatened in the post-impoundment situation, and their population had declined. Concomitantly, the stream fish populations, of late, are confined to *Acrossocheilus* (*Neolissochilus*) *hexagonolepis*, *P. chelynoides*, *Schizothorax richardsonii,* and *Tor* species. Incidentally, the population of the previously dominant *S. richardsonii* had dwindled and *N. hexagonolepis* had become dominant among the indigenous fishes. On the other hand, the new biota adapted to the lacustrine habitat had established with abrupt and drastic changes in species composition but there seemed had been increased the growth of phytoplankton, zooplankton, and zoobenthos, such as nymphs, chironomids, ceratopoganid larva, etc. Concomitantly, AM growth is an ecological problem in the shallow regions of reservoirs.

5.1.15.4 *Fisheries*

Fishing in the lakes and reservoirs is a traditional practice in Nepal but is not well organized. Fishers use traditional gears for **subsistence** fishing. The capture fisheries in these waters are of **artisanal** nature and have no industrial fisheries potential. The fisheries development potential in lakes and reservoirs seemed to be low, from a national perspective. However, the potential is high at the local level since it is relatively easy to manage and develop because of the nature of these water bodies. In fact, the potential FY per unit area from lakes and reservoirs is much higher than that of rivers. As such, the lake and reservoir fisheries can contribute significantly to local nutrition and livelihood enhancement. In view of the potential, the government, of late, seemed to have accorded due priority toward fisheries development in lakes and reservoirs through fish conservation, fisheries management, fish culture in cages and enclosures, open-water stocking management, etc., for enhanced production. Concomitantly, there seemed to be efforts directed toward **effective management** of the water bodies and their fish resources through efficient management measures like CBFM. The policy is to utilize natural productivity of these water bodies in fish production limiting to extensive culture practices. By doing so, the ecological balance could be properly maintained for sustainable utilization of the resources.

Concomitant to the above, cage fish culture in lakes and reservoirs with Chinese carps (mainly silver carp and bighead carp) was initiated with the support of FAO/UNDP during the late 1970s in Phewa, Begnas, and Rupa lakes of Pokhara valley and later extended to Indrasarovar reservoir, Kulekhani during early 1980s with support from the International Development Research Centre (IDRC), Canada. Incidentally, in Nepal, bighead and silver carps are popular for cage culture. These ichthyospecies are stocked at low stocking densities (*c.* 10–12 fingerlings/m^3). *L. rohita* is also stocked in small numbers as a biological cleaning agent in cage fish culture (Sharma, 1979). The stocking ratio of silver carp and bighead carp are usually manipulated depending on the abundance of phytoplankton and zooplankton.

Concomitant to the above, the cage fish culture system had been quite successful in terms of utilizing natural productivity in fish production. However, studies are being undertaken to examine the viability of intensive culture of common carp and rainbow trout in cages with supplementary feeding.

Notwithstanding the above, polyculture of planktivorous carps in lake enclosures in Pokhara valley was developed as a popular AQC activity. It has, further, underlined the potential role of the lakes in increasing fish production. The activity, at the moment, is managed by community federations with positive outcomes, thus indicating application of **community management**.

Notwithstanding the above, open-water stocking could be effective, and, of late, it is said to be practiced and managed by cooperative societies in Rupa, Kaski. On the other hand, in Jagdishpur reservoir, located in Kapilvastu district, open-water stocking is being managed by a group of private enterprises. However, the viability of these activities has not yet been assessed.

Concomitant to the above, the fish production from cage fish culture during 1981–82 had been reported to be *c.* 10 tons from 3110 m^3 of cages with an average productivity of 3.2 kg m^3. The fish production from cage fish culture reached *c.* 180 tons from 30,000 m^3 cages during 2001–2002. This increase in cage volume by almost 10 times and fish production by *c.* 18-fold in 20 years indicates the growing popularity of the activity.

In continuation of the above, in Nepal, FY from capture fisheries in lakes and reservoirs are reported to be *c.* 160 and *c.* 242 kg/ha/year, respectively. This is low as compared to reservoir FY in other countries. The average FY in Chinese reservoirs is *c.* 370 kg/ha/(Lorenzen, 1967). Incidentally, India is the second largest fish producer in reservoir fishery in Asia, producing > 10,000 mt of fish/year (Srivastava, 1985). However, FY from Indian reservoirs in terms of kg/ha/year is much lower (Sugunan, 1995). Incidentally, in Nepal, fish production from lakes and reservoirs reached 1163 tons during 2005–2006, which was *c.* 6% of the total fish production of 20,016 tons from capture fisheries in the country. Incidentally, the total fish production in Nepal had been recorded as 45,425 tons during 2005–2006, and contributed 2.75% of the agriculture gross domestic production. Further, the per capita fish production in Nepal during 2005–2006 was 1755 g. Concomitantly, the fisheries and AQC development activities in Nepal employed *c.* 511,000 individuals of 139,000 families as well as benefited *c.* 758,000 people, which was *c.* 3% of the total population during 2005–2006. Further, it has been estimated that capture fisheries had employed > 83% individuals in the fisheries sector, of which women employment consisted of *c.* 60% in 2005–2006. A preliminary analysis on employment and income generation has portrayed that AQC sector had employed fewer people but had higher income potential compared to capture fisheries.

5.1.15.5 Fisher organizations

Fisher organizations are formed mainly to take care of various issues related to fisher folk. Often the fisher organizations, which are said to be the stakeholders, play a significant role in the effective **management** of the water bodies through efficient systems, like the CBFM. In Nepal, the cage owners in Phewa, Begnas Rupa lakes, and Indrasarovar reservoir have developed a group/community concept in managing the enterprise. The organizational modality developed does not follow the prototype concept, since it is entirely dependent on local conditions. However, each organization seems to be functioning effectively. The fisher communities have been organized in an "**Association/ Federation Concept**" in Lakes Phewa and Begnas and Indrasarovar reservoirs. The "**cooperative concept**" has been adopted by the fisher community of Rupa Lake (as mentioned earlier) for better management of the fisheries resources. This lake is small and isolated. It is basically utilized by the local people for irrigation purposes. The cooperative organization seemed to have plans to manage the resources in a holistic and integrated manner by inducting cross-sectoral enterprises, such as ecotourism, aquasports, etc. Also, aspects like forestry conservation and navigation could be inducted for better economic returns. Open-water stocking and associated fish production and concomitant management activities have been initiated, of late, at Jagdishpur reservoir which had been built primarily for irrigation in Kapilvastu, Western Tarai district in Nepal. However, in some water bodies, the fish production and associated management activities are managed by a group of private entrepreneurs.

5.1.15.6 Economic viability, marketing aspects, profitability, and value addition

Extensive management of cage fish culture in the lentic bodies is considered to be easier and eco-friendly because fish in cages utilizes natural food for growth. The system of extensive cage culture of planktivorous fish in the lentic bodies could be beneficial with quick return on investment compared to other agricultural practices. For this purpose, one fisher family may need at least five cages of *c.* 250 m^3 to make it profitable and sustainable (Swar and Pradhan, 1992). Further, the cage culture of carps in the lakes of Pokhara valley had been viable with a net benefit of Rs. 31.2/m^3 (Sharma, 1990). This technology is one of the most economic methods of fish rearing in Nepal. Incidentally, extensive cage fish culture is based on natural productivity and no supplementary feeding is practiced, which, consequently, lessens the cost of production considerably. As such, this AQC system is affordable to poor and low-income farmers. The future expansion of these practices could make an important contribution to improve the economic status of the communities in the mid-hills.

Concomitant to the above, bulk of the production is marketed locally. In addition, there seems to be a substantial demand of dry fish in the local markets. Incidentally, a solar drier plant was installed in Indrasorovar reservoir for use by the local fisher communities to make dried fish and to introduce the concept of value addition to the product for higher net returns. Moreover, a fish processing plant had already been installed in Pokhara and is being operated with success.

5.1.15.7 Constraints to reservoir fishery development

Policy/legislation

It may be noted that the hydropower and irrigation projects have been accorded high national priorities for economic development in Nepal. However, there appears a basic disparity in national policy regulations with regard to judicious utilization of natural water resources. The damming of rivers for reservoir schemes for hydropower and irrigation projects overlooked the biological and socioeconomic aspects in many cases. In fact, these cross-sectoral issues may be accorded equal significance by incorporating adequate and accurate mitigation measures through holistic management approaches to maintain the biological, environmental, and socioeconomic aspects of the impacted area in a sustainable manner. The importance of fisheries resource as well as the need for conservation were realized quite early in the country. The legislation, "The *Aquatic Life Conservation Act 1961"—"Jal char Sanrakshan Ain—2017"* was adopted, but could not be implemented for some time. In 1999, the Act was amended. The changes included important aspects such as its scope and the definition of different terms specified in the Act. The other aspects included in the Aquatic Life Conservation Act are restrictions on killing and capture methods, punishments, citizen's obligations, role, and the responsibility of local authority and technical authority. However, the Act was not in operation till recently because laws and regulations were yet to be approved by the government. The said Act could assist the community in the effective management of the water bodies.

5.1.15.8 Ecological impacts

The damming of a river for hydropower and irrigation development converts the lotic habitat of the river into a lacustrine one. Consequently, some native fish species are threatened and gradually disappear due to the loss of spawning grounds, migratory routes, and imbalance of physiochemical factors and energy flow mechanisms of flowing river habitat. Nevertheless, the limnophilic species thrive and flourish. In addition, many native fish species are threatened by various reasons, notably, they are driven into turbines of hydropower plants. This issue is, generally, not considered in the basic engineering and design of the projects, so mitigation measures might not have been adopted. As such, appropriate mitigation mechanisms at the policy level are urgently required in order to safeguard the loss of native fish species and other aquatic fauna and minimize the BD losses.

5.1.15.9 Socioeconomic aspects

Technological development in civic facilities could, generally, be considered as useful to the society but utilization of natural water resources for development interventions may cause serious socioeconomic problems. Such development activities may, occasionally, displace human settlements, thereby affecting livelihoods, life styles, and values. Fisher communities are often considered as highly vulnerable and deprived, and mostly belong to poor classes, and hence the most seriously impacted group of people due to any human intervention with nature. Rehabilitation of these utterly poverty-stricken people is of urgent necessity, even if it is cost intensive. The issue could be addressed by providing appropriate alternative livelihoods and settlements to the affected communities through appropriate mitigation measures and efficient management measures, notably, CBFM.

Concomitant to the above, there had been a **case study** on Lake Rupa with regard to its fisheries and management. **Lake Rupa** is a small, subtropical, shallow lake with a surface area of *c.* 100 ha situated 600 m a.s.l. in Central Himalaya, Nepal. The Lake Rupa watershed is located between 28°08′N to 28°10′N and 84°06′E to 84°07′E, at 600 m a.s.l. in Central Nepal. The lake has an estimated surface area of 100 ha. The lake's total catchment area is *c.* 30 km^2. The lake is narrow and elongated in shape, extending in a north-to-south direction, being surrounded by hills on the north, east, and west sides. Rice fields are located in the watershed between the hills and the lake in the north, with the lake's inlet stream originating from hilly terrain. The outlet drains into the rice fields at the southern end of the lake. The main threat to the lake is its dangerously low WL and its heavy aquatic vegetation, suggesting a transitional stage in the process of lake succession.

This degraded lake was studied between 2000 and 2006, with the goal of determining whether or not it could be restored by a community-based cooperative of local people living in its catchment.

Main threats to the lake, its aquatic life, and its very existence include infestation by excess aquatic vegetation, sedimentation, and low in-lake concentrations of DO. Small lakes ($\leq$500 ha) are relatively prone to the process of "succession and climax," although they may also be of social, environmental, and ecological significance (Gurung, 2007). Thus, their disappearance could spark off social chaos and disorder in areas already having to cope with other rapid environmental changes. This means that credible mechanisms for revitalizing or protecting small lakes are an important goal. Accordingly, a lake cooperative of *c.* 329 households living in close vicinity to Lake Rupa was formed in 2002, by prioritizing traditional fishers, women, and other deprived community members with the goal of respecting the citizenry and equity. Following the cooperative's campaigns directed at weed removal and fish stocking, its fisheries has improved. In addition to the cooperative's restoration activities, several water quality parameters including water transparency and DO and chlorophyll-*a* concentrations were monitored on a monthly basis. The measured water transparency was inconsistent, exhibiting large variations between 2000 and 2003. Relatively low, but consistent, values were measured after 2004. The results implied that after the removal of weeds, sufficient nutrients and solar radiation are subsequently available to the lake's phytoplankton communities. Supporting this notion is that the water transparency became consistent and the DO concentration increased to >3.8 mg/L during the critical months (March–May) after 2004. These water quality indicators indicated improvement in the ecological conditions of the (degraded) Lake Rupa, suggesting that the establishment of cooperatives, such as that highlighted in this study, could be a powerful and sustainable mechanism for restoring degraded lakes in similar socioeconomic settings by maintaining equity, by connecting communities with their resources, and by facilitating integrity, equity, citizenry, and social justice.

Human social dynamics and their involvement in understanding environmental degradation and restoration are central to the studies of sustainability (Janssen and Carpenter, 1999; Cairn Jr, 2003; Leigh, 2005). Incidentally, **restoration may be defined** as **the act of bringing back to a former state or position** (Bradshaw, 1996). The restoration of any degraded landscape, especially lakes, could be an expensive, time consuming, and labor-intensive process. Thus, several modalities of lake restoration had been applied in the past, focusing primarily on habitat, water quality, improvement of BD, and other related aspects (Richter, 1986; Shapiro, 1990; Perrow et al., 1997). Small lakes (≤500 ha) are commonly perceived to disappear over time as a result of "succession and climax" (Niering, 1987). Smaller lakes in Asia, however, are generally used intensively for multiple purposes (Gopal, 2005), suggesting that they represent a multitude of social and ecological benefits (Bhandari, 1998). Thus, the disappearance of such lakes could lead to social chaos and financial deprivation for the inhabitants of communities that depend on them (Gurung and Bista, 2003; Gurung et al., 2005). This factor could be critically important to communities with low literacy and economic activities as many deprived poor communities subsist on natural lakes for food, fishing, and boating (IUCN, 2004; Bhandari, 2006). In contrast, affluent people mostly use such water systems for irrigation and other corporate needs (Costa-Pierce, 1998), providing the potential for conflicts where the water resource is limited. Nepal is a small mountainous country well known for its BD, culture, ethnicity, and community-based forestry and also aquatic activities (Gautam and Shivakoti, 2002; Bhandari, 2006). Community-based natural resource management (CBNRM) may lead to social conflicts (Leal, 1996), notably where social discrimination exists in a heterogeneous society, with people belonging to different ethnic castes (Chakraborty, 1998), and with an economy of inequitable sharing (Upreti and Upreti, 2002). Any community desiring social justice might be able to reduce social tension emerging from a skewed distribution of wealth generated from common resources, even if less income is generated (McConney, 1998). This implies that extensive fisheries could be encouraged for small lakes, as this activity is compatible with the traditional practice of sharing the wealth and resources of nature among many individuals. Consistent with this perspective, fisheries in Lake Rupa had a flourishing, plankton food-based cage fish culture, which had begun in 1975 (Pillai and Swallows, 1980). The fish production increased to *c.* 50 mt per annum in 1990, involving >150 families in the surrounding communities. Gradual deposition of silt in the lake, resulting from landslides, floods, and nutrient runoff had caused a low WL in the lake, followed by eutrophication and heavy infestation with aquatic plants (IUCN, 1996; Rai, 2000a, b). A seasonal decomposition of aquatic plants and the resulting DO depletion had caused the collapse of the cage fishery, and destroyed the livelihoods of fishers dependent on this industry. The possibility of lake restoration in such instances is often viewed from a biophysical perspective. More crucial and central to a balance aimed at sustainability, however, are social values, which are often neglected in such activities (Klessig, 2001; Klessig et al., 2004). Prior to this study, a number of obstacles to this goal had existed, including inadequate finances as well as lack of mutual trust and respect among villagers. Thus, the present study had tried to examine an approach to restoration of a degraded lake that involves the cooperation of local households. **A cooperative (International Cooperative Alliance, ICA) is an autonomous association of villagers who unite voluntarily to meet their common economic, social, and cultural aspirations through a democratically controlled enterprise** (Kimberly and Robert, 2004). To implement this cooperative approach, which is a central theme of the lake restoration effort, it was assumed that establishment of a win-win situation is essential in attempting to manage a common resource in a heterogeneous society without causing social conflict, and in which the outcome could benefit all the stakeholders. This approach was possible because of the ability to develop criteria that judiciously prioritize citizenry, traditional knowledge, occupations, skills and the economic status of the cooperative members. This understanding and perseverance was the real strength behind this approach for sustainable management of this common resource in a society comprising differing ethnicity and financial status. It also is pointed out that **the concept of sustainable development had been defined as a process that stimulates a sustainable coevolution of human activities and environmental changes** (Janssen and Carpenter, 1999). A recent seminar work by Leigh (2005) described community-based restoration as the best instrument for recovering from ecological crisis. This approach could also be helpful in the peaceful and rapid coevolution of human activities and environmental changes in societies characterized by weak economic support systems. Moreover, it delineates a bottom-up social mechanism for the restoration of natural resources and their management for sustainable development. As there are limited studies on the restoration of small lakes through the participation of local community-based cooperatives, this report attempted to describe how equity and social justice-based cooperation in a heterogeneous society could play a major role in the restoration of small lake ecosystems in a sustainable manner. The main objective of the present study, therefore, was to examine whether or not a degraded lake could be restored by stimulating and involving local communities.

There are *c.* 15,000 individuals, representing different ethnic communities, in the human settlements in the catchment area of the lake. These communities could be categorized broadly as follows:

(i) Those dependent entirely on fishing for their livelihood, as per their tradition (*pode*).
(ii) Those financially weaker, and generally involved in fishing for their livelihoods (*damai, kami, sarki, gurung, magar),* and.
(iii) Those generally with a sound financial status, although the daily income of some might not exceed $US2 (*brahmin, chettri, newar*).

Concomitant to the above, in order to resolve the impasse of establishing a cooperative, a number of draft regulations were discussed, with active participation of people living toward the upper side and the lower (down) side of the lake. Several workshops and public meetings, involving fishers, village chiefs, and government officials, were convened. The result of these activities was the establishment of a cooperative that incorporated 329 local households located within the vicinity of the lake. The establishment of the cooperative was said to be in the line with National Wetland Policy of the Government of Nepal (Ministry of Forest and Soil Conservation, 2003). Its major goal was lake conservation through the enhancement of lake fisheries management, to facilitate continuing benefits to society and the environment. The immediate objectives of the cooperative are as follows:

(i) To prevent human encroachment in the dry and shallow parts of the lake.
(ii) To increase the financial income derived from the lake through the management of recapture fisheries.
(iii) To implement lake restoration measures from funds generated by fish production activities.

Notwithstanding the above, for initial establishment of the cooperative, each cooperative member invested 5000 NRs (*c.* US $71) for lake restoration, purchase of fish nets, and other activities. However, the poor fishers and other members who could not afford to contribute to the investment cost, contributed through engaging in activities such as weed removal, mending of nets, and transportation of fish seed for the cooperative. A meshed net of *c.* 530 m in length and *c.* 7 m in width was placed across the outlet of the lake to prevent the stocked fish from escaping. Fish seeds weighing *c.* 0.5–2.0 g (average individual body weight), which were available from the nearby Fisheries Research Station, were stocked from the beginning of 2002. Grass carp (*C. idella*) were expected to control submerged vegetation by their grazing activities, while the common carp (*C. carpio*) uprooted submerged plants by their browsing food-searching habits. On the other hand, silver carp (*H. molitrix*) and bighead carp (*A. nobilis*) were stocked to control plankton blooms. The cooperative employed 10 fisher and financially deprived members as the permanent staff since 2002 for fishing, marketing, data collection, and financial book-keeping activities. FY data collected during 2001 were maintained for inspection, whenever required.

In addition to the above, in order to monitor the water quality changes, water samples from a fixed site in the middle of the lake were collected on a monthly basis at 0, 1, and 2.5 m depths from 2000 to 2006. A plastic 2-Lvan Dorn sampler was used where the depth fluctuated between 3 and 4 m during the study period. The measured parameters were water temperature, water transparency, chlorophyll-*a*, and DO concentrations. The water transparency was measured with a 20-cm diameter Secchi disk with alternative white and black bands. Chlorophyll-*a* samples were obtained by filtering lake water through Whatman GF/C glass fiber filters (Whatman International Limited, Maidhdone, USA) (<1.2 μm pore size). The chlorophyll-*a* concentrations were determined spectrophotometerically, based on the methods of Lorenzen (1967) and UNESCO (1969). The DO concentration was measured using the Alsterberg Azide modification of the Winkler method (Kar, 1990, 2007a,b,c, 2013).

5.1.15.10 Results

Formation of the cooperatives

An executive committee of 11 members was formed to resolve local lake issues and everyday problems. The term of each executive committee member was fixed at 2 years, with the first committee being formed in 2002. The structure of the cooperative, and the composition of the executive committee, was accepted on the basis of the accepted structure. It was designed to maintain harmony, equal opportunities, and respect for the watershed citizenry. The chairperson was to be elected according to democratic norms. Two vice-president posts were reserved for women and traditional fishers, which were to be nominated among the cooperative members. The posts of general secretary and treasurer were to be elected. Out of the four members, two were to be elected; while, the remaining two were to be nominated by traditional fishers and the women members, respectively.

5.1.15.11 *Awareness and fisheries-based employment of cooperative members*

The total fish production was *c.* 18.5 mt in 2004, with indigenous fish comprising *c.* 70% of the total yield. The contribution of indigenous fish to the total yield, including rohu (*L. rohita*), mrigala (*C. mrigala*) and bhakur (*C. catla*) as well as other native species were higher than the exotic, bighead (*A. nobilis*), silver (*H. molitrix*), and grass carp (*C. idella*). In addition to the income, capture fisheries provided jobs and employment to *c.* 80% of the cooperative member fishers. Incidentally, after the establishment of the cooperative, the number of direct beneficiaries increased from a few to *c.* 329 households. Although, only a few fishers initially had jobs related to the lake, a considerable number of people are subsequently involved in weed removal, fish harvest, value addition and marketing activities, etc.

5.1.15.12 *Discussion*

The probable major causes of the degradation of Lake Rupa are as follows:

(i) sedimentation, (ii) excess growth of rooted aquatic vegetation, and (iii) the absence of authoritative local institutions to take care of the lake.

Notwithstanding the above as a result of the decomposition of dead aquatic weeds (Rai, 2000a,b), there was drastic depletion of DO in the lake in April and May. The present findings, however, had tried to illustrate that the establishment of the cooperative, as an authoritative institution of local households had not only resulted in the curtailment of the infestation by the aquatic vegetation but also helped in the clearing of excess MA growth. Moreover, the drastic DO deficiency, especially during March to May, had gradually improved since 2003, being largely overcome since 2004. This decreasing anoxia in the lake is a significant indicator of environmental improvement, particularly, with regard to the sustainability of aquatic organisms in the lake. Similarly, the measured water transparency was inconsistent during the period 2000–2003, with the lake exhibiting large seasonal increases and decreases in the values. After fish stocking and the manual removal of aquatic weeds from the lake, the water transparency became relatively consistent after 2004. Earlier trends in water transparency could be the result of abundant AM in the lake, which dominated the in-lake phytoplankton populations, as macrophytes uptake nutrients from the lake bottom sediments (Takamura et al., 2003; Mironga, 2005). Incidentally, several studies have illustrated the occurrence of low water transparency after fish stocking in lakes (Jeppesen et al., 1997; Rowe, 2007). However, the lowest transparency in Lake Rupa in July 2002 and 2003 was caused by heavy silt-containing runoff from the lake catchment during the monsoon season.

It may be mentioned here that one of the nuisance weeds in the lake was water hyacinth (*Eichhornia* sp.). As such, members of the cooperative had effectively cleared them. As part of the relevant campaign, the cooperative has announced a reward of NRs 75 (*c.* US $1.05) for every kilogram of water hyacinth collected from the lake. This strong incentive also could help prevent further recolonization of the lake by water hyacinth. A strategy of fish stocking was applied earlier to Lake Rupa, which exhibited extensive growth of aquatic vegetation, to the extent that the lake was beginning to be transformed into swampy land. It is expected that the fish stocking could control the excess aquatic weeds in the lake to some extent, as several earlier studies demonstrated that grass carp (*C. idella*) can effectively control submerged and floating macrophytes (Mitchell, 1980). *C. carpio*, a bottom browsing fish, can be effective in uprooting aquatic plants and causing the release of nutrients from the lake bottom sediments. Thus, after this well thought-out activity, only *c.* 10% of Lake Rupa was, later, estimated to be occupied by AM. The annual FY has tripled in 2004 to *c.* 18.5 mt, which was more than that in 2002 after restocking. However, the present FY from the lake is low (*c.* 185 kg/ha), compared to other managed lakes for which the yield was *c.* 540 kg/ha (Liu, 1984). However, higher FY could be achieved in the future, if more efficient lake management efforts were undertaken (Shapiro, 1990; Jeppesen et al., 1997; Heerdt and Hootsmans, 2007). In fact, activities such as the establishment of tree plantations, improvement in catchments, removal of silt from the lake bottom, etc., are planned by the cooperative, utilizing financial resources generated through capture fisheries. Moreover, combining fish species that exhibit diverse feeding habits in lakes may cause complex ecological interactions (Shapiro, 1990; Padisák, 2002). Further, the indiscriminate stocking of fish species into Lake Rupa, for example, might cause interspecific food competition. In order to avoid the adverse impacts, major attention should be given to enhancing the native fishery (Swar and Gurung, 1988). Although it is known that restocked exotic fish might not have yet been established in Lake Rupa, the cooperative is to take measures to protect the lake's endangered and vulnerable fish species. It is also important to educate fishers who use gill nets, as these nets are not environmentally friendly with regard to birds, amphibians, and reptiles. This implies that with the current restoration model for Lake Rupa, care is to be taken to recognize and consider the needs of the specific aquatic communities in the lake.

Notwithstanding the above, inclusion of traditional poor fishers, low-income people, and women in the cooperative's executive committee reflects the respect due to the citizenry, indigenous knowledge, and gender base development. As the cooperative was designed to offer a win-win situation to all its members, their link to the cooperative could be expected to enhance its harmony and stability. Based on the success of the present Lake Rupa restoration approach, it is expected that communities living next to or near other lakes, wetlands, and swamps will also be encouraged to undertake such restoration efforts. Arriving at this conclusion, it is also noted that it may not be necessary to establish a fishery-based cooperative for all degraded lakes. Instead, this activity could be based on the consensus of local community and the specific BD conservation needs of individual lakes.

Another item meriting mention is that Lake Rupa's catchment area previously experienced indiscriminate killing of birds, fish, and wild animals. Such activities in the lake and its surrounding catchments, however, have since been regulated by the cooperative, implying that such an approach could also be helpful in restoring the BD of the lake and its catchment as a whole.

5.1.15.13 Lessons that had been learned

Notwithstanding the above, some of the most significant difficulties facing many small lakes in Nepal are as follows: (i) sedimentation associated with heavy monsoon rainfall followed by landslides, felling of trees, and floods; (ii) occasional unsustainable traditional agricultural practices in the catchments; (iii) eutrophication; (iv) water abstractions, (v) deforestation, (vi) weed infestation, and (vii) occasional encroachment of the wetland area by the people.

In fact, many Asian lakes have to cope with the impacts of human activities in their catchment areas (Groombridge and Jenkins, 1998). As a result of recent increases in population and poverty, the sustainability of small lakes situated near human settlements might be questioned. The growing trend of converting these lakes and their catchments into intensive AQC or agriculture systems are to be approached cautiously because of their potential environmental impacts, and the socioeconomic problems they may buildup. Intensive AQC in such settings might be essentially a capital-intensive activity undertaken by corporations aiming to make a quick profit that is often shared by only a very small portion of the watershed population (Costa-Pierce, 1992). Based on the experiences highlighted in this study, it is recognized, however, that community-run fisheries may be able to avoid the "tragedy of the commons" syndrome, whereby a common resource is everyone's resource to exploit, but no one's responsibility to sustain (Leal, 1996; Buyali, 2001).

Fish and wildlife resources associated with Lake Rupa correspond to a mosaic of interconnected aquatic and terrestrial habitats that have important roles as the main intersection for migratory birds between the Himalayan Mountains and the southern plains of Nepal during the spring and fall. Many other birds and waterfowl also use the lake and its surrounding forests as feeding and breeding habitats. In recognition of these environmental realities, one of the main objectives of the cooperative was the enhancement of fisheries and wildlife by restoring Lake Rupa. Thus, the answer to the question posed in this study, i.e., "Restoration of small lakes through **cooperative management: A suitable strategy for poverty-laden areas in developing countries?"** is "Yes." It was believed that this approach could save hundreds of small lakes around the world from disappearing, especially in developing countries, for which the paucity of financial and other resources can have dire consequences. This approach of establishing legally constituted cooperatives to manage small lakes and their resources illustrates that it is possible to make an equity-based, win-win perspective for poverty reduction, even in conflict-prone societies, by utilizing human social dynamics.

5.1.16 The Philippines

The Philippines is a coastal country with rich fisheries and AQC potential. In **the Philippines**, Baticados and Agbayani (1998) observed that in San Salvador Island and Malalison Island, the marine sanctuaries have boundaries identified with buoys to inform outsiders of its existence to **more easily monitor the area by making and changing community's rules as well as by sharing responsibility among the community members for guarding the marine sanctuary with high level of enforcement of rules.**

Further, Foltz et al. (1996) mentioned that in the **Philippines**, NGOs usually spend a great deal of time in social preparation, i.e., educating the fishers about the importance, benefits, and economic implication of fisheries comanagement in comparison to different fisheries management and development strategies.

There have been efforts to deal with capacity and opportunity building in a small-scale fisherfolk community for community-based coastal resource management (CB-CRM) as a community development approach to CB-CRM. Pomeroy et al. (1996) worked on the impact evaluation of CB-CRM projects in the **Philippines**. Pomeroy et al. (1998) further dealt with fisheries comanagement as key conditions and principles drawn from Asian experiences. Katon et al. (1998) worked on the Mangrove rehabilitation and coastal resource management of Mabini-Candijay as a case study of fisheries comanagement arrangements in Cogtong Bay, the **Philippines**. Interestingly, Masae (1998) made an analysis of fisheries comanagement arrangements in the case of Ban Laem Makham, Sikao District, Trang Province, South **Thailand**. Incidentally, Khan and Apu (1998) dealt with fisheries comanagement in the Oxbow lakes of **Bangladesh**. Baylon (2001) further studied the microenterprise development in selected fishing communities in the province of Iloilo, the **Philippines**. He worked on the development of microenterprises in Banate Bay, Iloilo and Southern Iloilo. This project was implemented by the University of the Philippines in the Visaas in coordination with the Banate Bay Resource Management Council, Inc. and the Southern Iloilo Coastal Resource Management Council. Notwithstanding the above, There had been works, which dealt with policy reforms and community-based programs to combat cyanide fishing in the **Philippines**.

The exact number of cyanide fishermen operating in SE Asia and neighboring countries is unknown. Consumers have a significant role to play in pressuring the aquarium fish industry to take action on imports of cyanide-caught fish. In fact, publicity and the consumer pressure in Europe and North America have had some impact on aquarium fish importers, and had led to efforts such as the Marine Aquarium Fish Council in the United States, etc.

The Philippines has a rich potential in fisheries and AQC. The country possesses a rich market for aquarium fishes. In this connection, it may be noted that, > 1.0 million kg of deadly sodium cyanide had been squirted onto coral reefs in the Philippines, since the 1960s, in order to stun and capture ornamental aquarium fish destined for the pet shops and aquariums in Europe and North America. More recently, a growing demand for larger reef food fish had vastly increased the incidence and spread of cyanide fishing. In fact, the combined demand for aquarium and live food fish had spread cyanide fishing throughout the Philippines and also Indonesia and into neighboring countries such as Papua New Guinea, Vietnam, the Maldives, and Fiji.

Notwithstanding the above, cyanide kills corals and reef invertebrates along with many nontarget fish. Large number of fish that are captured live die in transit, mainly due to their poison-weakened state. The spread of cyanide fishing is particularly tragic in the countries of the Indo-Pacific, being deadly in the marine environment. The region may justifiably be called "the Amazon of the Oceans," as the global center of marine BD for corals, fish, molluscs, and reef invertebrates. In addition, cyanide fishing also threatens the livelihood of poor coastal people in the region, where dependence on fish protein is very high, mainly for both nutrition and avocation.

Notwithstanding the above, the Philippines, the birthplace of cyanide fishing, is also the only country with a program in place to eradicate the practice. The Bureau of Fisheries and Aquatic Resources and the International Marine life Alliance (IMA), the Philippines, an NGO jointly developed and implemented the Cyanide Fishing Reform Program (CFRP) since the early 1990s. Experience with the CFRP over the few decades indicates that cyanide fishing could be reduced through a combination of the right policies and laws, beefed-up enforcement efforts, enhanced public awareness, cyanide testing of live-fish exports, training of cyanide fishermen in cyanide-free live-fish capture techniques, development of sustainable CBRM, and livelihood alternatives that could transform the local fishermen into the frontline of marine stewards and protectors.

Concomitant to the above, one of the key aspects of the CFRP's initial success was the initiation of policy reforms in both source and consumer countries to form anti-cyanide fishing incentives and enforcement mechanisms. In addition, a second significant element had been the development of effective partnerships with the fishing communities themselves, focusing on transfer of nondestructive technology and improvement of local livelihoods.

Notwithstanding the above, there had been contributions, which tried to identify the key actors in the live-fish trade as well as analyzed their roles and interests, including elaborating the policy reforms of the governments of both exporting and importing countries. Further efforts are also to be directed toward establishing incentives for a cyanide-free live-fish trade as well as to examine the community-based strategies that lie at the heart of the CFRP's efforts to counter this ominous threat to the planet's marine BD.

Concomitant to the above, the live reef fish trade in SE Asia has an estimated annual retail value of *c.* US $1.2 billion, the live food fish trade worth *c.* US $1 billion (mostly with Hong Kong), and exports of aquarium fish to Europe and North America worth *c.* US $200 million.

5.1.16.1 *Cyanide fishermen*

The exact number of cyanide fishermen operating in SE Asia and neighboring countries is unknown. In fact, the number of hard-core cyanide fishermen throughout the Indo-Pacific region does not exceed 20,000. This was based on the estimates in the Philippines, where there had been *c.* 4000 cyanide fishermen. Incidentally, cyanide fishing is not a ubiquitous problem like the slash-and-burn farming, practised by innumerable poor farmers. Nevertheless, poverty could be one of the causes of cyanide fishing, as many cyanide fishermen are certainly very poor. Rather, cyanide fishermen are a fairly small and discrete group, who respond to very specific incentives: a new technology, a ready market for the product, and the lack of viable livelihood alternatives. Experience in the Philippines suggests that when cyanide fishermen were introduced to cyanide-free techniques for live-fish capture as well as ensured a fair price for their catch, they were willing and often eager to give up using the poison and to talk about ways to ensure the long-term sustainability of their local reefs and fisheries, which could certainly provide them with efficient management of the water bodies and their resources, often involving the community through CBFM.

5.1.16.2 *Live reef fish export*

The number of business farms involved in live-fish export business in SE Asia is also unknown but it appears to be expanding rapidly. For instance, there were only three companies exporting aquarium fish from the Philippines in the early 1960s and export of live food fish did not exist. By the 1990s, there were *c.* 45 aquarium fish exporters in the country, and *c.* 8 business farms exporting live food fish. Further, at least 10 companies run holding tanks for live food fish in Bali, Indonesia, which is a major transhipment point. Conservative estimates of the annual volume of Asian trade in live food fish alone range between 20,000 and 25,000 tons.

5.1.16.3 *People involved in the live reef fish trade*

Quite a large number of people are involved in live reef trade. They are usually equipped with boats and fishing gears. People just buy fish from them. The production side is left to them.

5.1.16.4 *Live reef fish consumers*

Consumers have a significant role to play in pressuring the aquarium fish industry to take action on imports of cyanide-caught fish. In fact, publicity and consumer pressure in Europe and North America has had some impact on aquarium fish importers, and led to efforts such as the Marine Aquarium Fish Council in the United States, etc.

Notwithstanding the above, Anon (2011a,b) dealt with a review of the National Laws and Legislations in the **Philippines** in relation to Ecosystem Approach to Fisheries Management (EAFM). EAFM was emanated from the realization that the world's largest food basket is not infinite. Therefore, proper management of the fisheries, if possible, by involving the stakeholders from the grassroots level through CBFM is to be undertaken to ensure that the posterity will continue to benefit from these resources.

The **Philippines** possess a rich potential in AQC and fisheries. However, its marine ecosystems were portraying signs of overfishing, which could be detrimental, despite the Philippines' abundant fisheries resources (FAO, 2011). In fact, the country is experiencing rapid depletion of its aquatic resources, with *c.* 75% of its coral reefs degraded and its mangrove forest and its other fisheries ecosystems steadily dwindling since the recent past (Green et al., 2003). However, this problem is not exclusive to the Philippines only, with the world's marine resources facing similar depletion over the years mainly due to overexploitation and destructive fishing methods (that devastate the aquatic ecosystem) as well as due to poor fisheries management.

Notwithstanding the above, the concept of EAF is also sometimes used interchangeably with other systems. In fact, the EAF to fisheries strives to balance diverse societal objectives by taking into consideration the knowledge and uncertainties about biotic, abiotic, and human components of ecosystems and their interactions, and applying an integrated approach to fisheries within ecologically meaningful boundaries. Incidentally, it may be noted that only the FAO of the UN attempted to come out with specific guidelines to implement EAFM, although several international agreements have adopted the EAFM concept. In fact, the FAO had issued a Technical Guideline for Responsible Fishing, in 2003, which set the standard for fisheries management in implementing EAFM (FAO, 2003). However, adherence to the guideline is voluntary.

Notwithstanding the above, studies had been undertaken to review the international, national, and area-specific laws and policies in relation to EAFM in the Philippines. Many such reviews assess the policy issues and gaps as well as recommend areas of intervention, including policy proposals.

Concomitant to the above, the **Philippines** had a number of overseas-funded projects on fisheries management and/or coastal resources management (CRM) that had adopted some of the EAFM principles and concepts in both project design and project implementation. Further, the Philippines has also a relatively broad and comprehensive legal framework on fisheries and coastal-resources management. Moreover, the Philippines is a signatory to several international agreements that relate to general environmental conservation, protection of the aquatic (notably, marine) environment, and fisheries management. As such the Philippine government has passed special laws to implement some of these international covenants in keeping with its international commitments.

Notwithstanding the above, the government of the Philippines adopted the Integrated Coastal Management (ICM) policy as a national strategy to implement the sustainable development of the country's coastal and marine resources in order to achieve food security, sustainable livelihood, poverty alleviation, and reduction of vulnerability due to natural hazards. The Philippines Development Plan of the government explicitly stated that it wishes to follow the ecosystem approach to the management of fisheries (EAMF/EAFM) as well as other aquatic (notably, marine) resources, with an effort to achieve complete implementation of its strategy of enhancing coastal and marine resources management.

With reference to the above, it may be stated that the Fisheries Code, NIPAS, and several other special laws of the Constitution of the Philippines (1987) provided platforms for developing Integrated Fisheries Management (IFM) and CRM approaches, which had facilitated the adoption and implementation of EAFM concepts. Incidentally, NIPAS is primarily concerned with the establishment and management of protected areas (PAs). In fact, the Fisheries Code encourages the adoption of a holistic approach to fisheries management and contains provisions that address EAFM concepts and embody EAFM principles. Among others, it provides for the establishment of Aquatic (marine and FW) sanctuaries within municipal waters. This was complemented by the Local Government Code of 1991 (RA 7160), which had devolved a number of environment and natural resources functions from the national agencies to the local government units (LGUs).

Studies performed in various places, notably, in the province of Palaw as an illustrative case, depicted its unique BD and its distinct legal framework on sustainable development and management of the resources. Moreover, Palawan's unique character is highlighted by its special law, the Strategic Environmental Plan for Palawan (1992) (SEP law, Republic Act No. 7611), which provides a specific management scheme and governance mechanism for sustained development and management of resources. Further, other activities include undertaking research and surveys of fishery areas, zonation of specific water bodies and fishery areas, establishing aquatic (notably, marine) protected areas, capacitating LGUs and community stakeholders as well as developing local ordinances and programs and supporting livelihood efforts. Furthermore, the study revealed that, although there is no EAFM-specific legislation yet, there are policies and programs that had pursued specific management strategies, and also had designated conservation areas in the coastal zone, where the EAFM approach could be operationalized.

Notwithstanding the above, in spite of delving deep into the aspect, there are gaps in the policy matters, which could stall efforts toward integrated fisheries management and EAFM.

5.1.16.5 *Agriculture and fisheries modernization*

It may not be quite uncommon that there may be overlapping mandates of the various institutions involved in fisheries management. This concern was repeatedly pointed out in the earlier policy studies and development plan(s) of the government of the Philippines. Ideally, EAFM is to work within a simpler and streamlined institutional arrangement.

Notwithstanding the above, various studies recommended the following intervention areas in order to move toward EAFM:

(i) Enhancement of information and educational activities on EAFM among national and local agencies as well as concerned civil society groups and community stakeholders.
(ii) Providing policy support to municipalities or areas that could be the critical habitats to fishery resources (both FW and marine), and, further, those areas could be considered as vulnerable spots in fishery crimes.
(iii) Initiation of legislative proposals to include specific provisions on EAFM in the ongoing review of the Philippine Fisheries Code as well as in pending bills on CRM and coastal fisheries resource management and also in aquatic resource protection and management.
(iv) Development of appropriate institutional arrangements for scaling up of fisheries management as an organizational modality toward achieving EAFM as reflected in the earlier proposals of Pomeroy et al. (2010).
(v) Enhancement of multiagency engagements so that institutions, like DENR, PCSDS, LGU leagues, NGOs, etc., or bodies working on CRM and fisheries management could collectively develop strategies to address the policy gaps.

In fact, some of the initiatives on MPAs, fishery reserves, and sanctuaries could serve as the foundation for EAFM approaches and strategies.

(vi) There is a need for the Review of National Laws and Legislations in the Philippines in relation to EAFM.

5.1.16.6 *Background and relevant context: The aims and policy questions*

The Philippines has rich potential in AQC and fisheries and various laws, policies, and programs related to fisheries and the aquatic environment. Among the many objectives, a significant one is the measures taken to address the policy gaps, taking into consideration the various fisheries and coastal resource management initiatives that have been undertaken.

5.1.16.7 *Some of the limitations*

The studies, sometimes, are limited in view of having specific legal framework on sustainable development, its distinctive BD status, and the range of initiatives on coastal resources management and fisheries management.

5.1.16.8 *Framework and procedure(s) of policy review*

There were policy studies involving existing legislations and literature for relevant information. Peer reviewers from the government were involved to enhance the study. The study team reviewed laws and policies relevant to EAFM, including discussions with environmental lawyers and policy analysts. Further, the international instruments and agreements, national government laws, policies, and programs were also evaluated.

5.1.16.9 *EAFM and its relationship with sustainable development*

5.1.16.9.1 Evolution and beginning of EAFM

The Philippines is, generally, recognized among the top fish-producing countries in the world with a fish production of *c.* 2.63 MT of fish, crustaceans, molluscs, and aquatic plants (FAO, 2011). However, the country's marine ecosystem is depicting signs of overfishing, which could be detrimental, although the fisheries resources were said to be abundant.

The realization that the world's largest food basket is not infinite has brought about the concept of EAF or EAFM, and that proper management measures are to be undertaken to ensure that future generations will continue to benefit from these resources. In addition, the EAFM strives to balance diverse societal objectives by considering the knowledge and uncertainties about biotic, abiotic, and human components of ecosystems and their interactions, and applying an integrated approach to fisheries within ecologically meaningful boundaries (FAO, 2003). In fact, the EAFM concept and approach is for sustainable fisheries development and seeks to address the multiple needs and desires of societies without jeopardizing the options for future generations to benefit from the full range of goods and services provided by marine ecosystems (FAO, 2003). The EAFM takes into account the aquatic ecosystem including the marine environment, which includes the coral reefs and mangroves, alongside human activities, including coastal development, tourism, and fisheries management (USAID, 2011). Moreover, the EAFM was applied formally to the fisheries management through the Reykjavik Conference on Sustainable Fisheries in the Marine Ecosystem in 2001. Probably, equally noteworthy is the World Summit on Sustainable Development (WSSD) in 2002, where countries made commitments to introduce an ecosystems approach to marine resource assessment and management by 2010. Moreover, the concepts embodied in EAFM are not new, and had been contained in previous international agreements. Further, the United Nations Convention on the Law of the Sea (UNCLOS) provides protection and conservation of marine habitats. This was followed by the United Nations Fish Stocks Agreement (UNFSA). Furthermore, a parallel initiative in the development of the EAFM concept was made by the United Nations Conference on the Human Environment (1972) and, later, by the United Nations Conference on Environment and Development (UNCED) (1992). UNCED was significant, as it paved the way for the Rio Declaration, the Convention on Biological Diversity (CBD), and Agenda 21. These initiatives were followed by the drafting of the Malawi principles and the World Summit on Sustainable Development (WSSD) in 2002. EAFM was also included in the UN Food and Agriculture Organization's (FAO) Cancun Declaration, which was subsequently followed by the 1995 Code of Conduct.

Review of National Laws and Legislations in the Philippines in Relation to EAFM for Responsible Fisheries (CCRF) and the Reykjavik Declaration eventually promoted and introduced EAF.

The FAO, in 2003, issued a Technical Guideline for Responsible Fishing, which had set the standard for fisheries Management. This was related mostly to the implementation of EAFM (FAO, 2003). Incidentally, the Philippines and the other ASEAN countries adhered to this concept, and later developed the regional guidelines for the proper implementation of EAFM.

5.1.16.10 *Efforts toward EAFM in the Philippines*

The Philippines introduced and applied EAFM concepts and principles through various government and nongovernmental programs, most notable of which are the following:

5.1.16.11 *Fisheries improvement for sustainable harvest (FISH) project*

The FISH Program was funded by the US Agency for International Development (USAID). This project was built upon the foundation of a coastal resource management to achieve the next crucial benchmark in managing fisheries and coastal resources in the Philippines. This benchmark called for the development of ecosystem-based approaches to fisheries management driven by informed, disciplined, and cooperative stakeholders at national and local levels of engagement. The FISH project was implemented in the Calamianes Islands (Palawan), Danajon Bank (Bohol, Southern Leyte, Leyte, and Cebu), Surigao del Sur, and Tawi-Tawi.

5.1.16.12 *Coral triangle initiative (CTI)*

The CTI had aimed to integrate EAFM in its action plans concerning coral reefs, fisheries, and food security. CTI is a multi-government partnership that aims to safeguard the marine and coastal resources of the Coral Triangle, the world's richest marine ecosystem (USAID, 2011). It is centered on political commitments and proactive implementation by the governments of the Coral Triangle area (Indonesia, The Philippines, Malaysia, Timor Leste, Papua New Guinea, and Solomon Islands). CTI is supported and carried forward by the private sector, international agencies, civil societies, and NGO partners.

5.1.16.13 *Ginintuang masaganang ani for fisheries program (GMAFP) (2002–2004)*

The GMAFP (2002–2004) is a government program designed to provide national directions and framework to develop and manage the country's fisheries resources for food security and the socioeconomic upliftment of subsistence fisher folk.

5.1.16.14 *Integrated coastal resources management project (ICRMP)*

The ICRMP, supported by the Asian Development Bank (ADB) aimed to sustainably manage coastal resources and uplift the socioeconomic conditions of people living in the coastal areas.

5.1.16.15 *Fishery sector programme (FSP) and fisheries resources management project (FRMP)*

The FSP and the FRMP were funded by the ADB. The FSP project undertook fisheries management at different levels, focused on governance reform, and utilized a comanagement approach (ADB, 1999). FRMP promoted sustainable use of resources and was implemented from 1998 to 2005 (ADB, 2007).

5.1.16.16 *World wildlife fund (WWF), The Philippines*

The WWF-Philippines, under the Coral Triangle Support Project (CTSP), worked on research projects regarding the live reef fish trade (LRFT) and its effects on the province of Palawan. The project reviewed the LRFT supply chain, the relevant policies on LRFT, and its implementation.

5.1.16.17 *The Philippine-Canada local government support program (TPCLGSP)*

The TPCLGSP is a joint undertaking of the governments of the Philippines and Canada, with the goal of helping the government in realizing its objectives of equitable growth and poverty reduction. The Integrated Coastal Resource Management is one of the development projects under this program (LGSP, 2011).

5.1.16.18 *Small grants program-global environment facility (SGP-GEF)*

The SGP-GEF, under the United Nations Development Program (UNDP), funded several projects involving coastal and fisheries resource management in the province of Palawan, notably, "Review of National Laws and Legislations in the Philippines in relation to EAFM," and so on.

5.1.16.19 *Honda Bay and Snake Island fisheries management and mangrove conservation project (HBSIFMMCP)*

The HBSIFMMCP (November 2002 to November 2006) was a project, which involved community-based resource conservation and formulation on a comprehensive resource-management plan for Honda Bay and Snake Island in Puerto Princesa.

5.1.16.20 *Community-based marine sanctuary management and livelihood support project in Caramay, Roxas, Palawan (CBMSMALSPICRP)*

The CBMSMALSPICRP (February 2004 to January 2006) sought to address the problem of declining FY/fish catch of marginalized fishermen in Caramay, Roxas as a result of the destruction of marine resources due to illegal fishing activities, overfishing, and intrusion of commercial fishers in prohibited zones. The project also had an ecotourism component that served to monitor illegal activities, reinforce advocacy on resource protection, and increase members' income.

5.1.16.21 *Coastal resources management*

The primary objective of this project was to review the National Laws and Legislations in the Philippines in relation to EAFM as well as to generate a model for community-based coastal resource management using a multi-sectoral approach (1998–2000).

5.1.16.22 *Strengthening the fisheries and aquatic resources management councils (FARMCs)*

The FARMCs is a project (2008–2011) aimed at strengthening the FARMCs for comanagement of fisheries resources. FARMCs were given capacity building trainings, which include protected area management, adaptation to climate change, law enforcement training, and the introduction of the Fish Catch Monitoring and Database System (or NEMO as nickname).

5.1.16.23 *Strengthening governance and sustainability of small-scale fisheries management in the Philippines: An ecosystem-based fisheries management approach in the Philippines (SGASOSSFMITP:AEBMAITP)*

(Phase 1/2)

The SGASOSSFMITP:AEBMAITP is an Ecosystem-based Fisheries Management Approach (EBFMA) (Phase 1/2), and was funded by the Bureau of Agricultural Research (BAR) and being spearheaded by the World Fish Center. Its goal is to strengthen the governance and sustainability of small-scale fisheries management in the Philippines. Fisheries management in, general, and EAFM, in particular, is one of the key concerns that has been developed and pursued globally. There were several international instruments and agreements, which contained EAFM concepts and principles, notably, the UNCED and the Agenda 21 of the Biodiversity Convention.

Sustainable development was the theme of the UNCED summit in June 1992. It resulted in nonbinding statements containing a framework on convention on climate change, a CBD, a Review of National Laws and Legislations in the Philippines in Relation to EAFM, declaration of sustainable development of forests, etc., virtually a very wide ranging environmental action for the next century.

The Rio Declaration of 1992 was a nonbinding statement that contained 27 broad guiding principles on environmental policy. It emphasized, among others, the right of human beings to ecological security and environmental protection as part of economic development. Agenda 21 is a program that was approved during the Rio Summit and contained 40 action plans designed to promote sustainable development [(CBD), Rio de Janeiro, Brazil, 1992].

As a significant outcome of the Rio Summit in 1992, this covenant requires contracting parties to develop national strategies, plans, or programs for the conservation and sustainable use of biological diversity. Such measures include the establishment of a system of PAs, rehabilitation of degraded habitats/ecosystems, and developing legislation for the protection of threatened species. It provides a mechanism that would give financial and technical assistance to developing countries for assessing and conserving BD.

5.1.16.24 *Cancun declaration*

This ICRF (organized by the government of Mexico, in close consultation with the FAO) led to the adoption of the Declaration of Cancún, which include a request to FAO to draft an International Code of Conduct for Responsible Fishing. The principles embodied in the Declaration included the following:

(a) Cooperation by States on bilateral, regional, and multilateral levels to establish, reinforce, and implement effective means and mechanisms to ensure responsible fishing in the high seas.
(b) That the States fishing on the high seas would better cooperate with other States to ensure conservation and rational management of the living resources.

5.1.16.25 *United Nations conference on straddling fish stocks and highly migratory fish stocks (UNFSA)*

The UNFSA aims at ensuring the long-term conservation and sustainable use of straddling and highly migratory fish stocks, and includes general principles for their conservation and management. Incidentally, its provisions include the application of precautionary approach, compatibility of conservation and management measures, cooperation for conservation and management, regional fisheries management organizations (RFMOs), collection and provision of information and cooperation in scientific research even, if possible, with the nonmembers of RFMOs, and spelling out of the responsibilities of the States. Similarly, it provides for international, subregional, and regional cooperation in enforcement; procedures for inspection; measures taken by poor States in various aspects of management of resources; special requirements, if any, and forms of cooperation with developing countries; and settlement of disputes, if they arise.

Notwithstanding the above, specific subjects covered include mainly those on fisheries management, fishing operations, AQC developments, integration of fisheries into coastal area management, postharvest practice and trade, and fisheries research. The code, further, laid emphasis on the role of regional and subregional fisheries organizations in its implementation (FAO, 2011).

The Reykjavik Conference on Responsible Fisheries in the Marine Ecosystem, University Cinema and Conference Center in Reykjavik, Iceland, October 2001.

The outcome of the conference resulted in the approval of Reykjavik Declaration on Responsible Fisheries in the Marine Ecosystem. It states that each nation could, individually and collectively, incorporate ecosystems consideration in the management of their fisheries, in order to achieve sustainable and responsible fisheries in aquatic (notably, marine) ecosystems. The declaration, further, required that scientific advice on fisheries management shall be based on an "Ecosystem Approach" (IISD, 2001). The efforts for the incorporation of ecosystem-based management considerations of the fisheries in the Philippines were benefited from the outcomes of this conference.

5.1.16.26 *The Malawi principles, Lilongwe, Malawi, January 26–28, 1998*

Notwithstanding the above, during a workshop on EAF, the international group had identified the following 12 principles and characteristics of EAF:

(i) Management objectives are a matter of societal choice.
(ii) Management could be decentralized to the lowest appropriate level.
(iii) Ecosystem managers are to consider the effects of their activities on adjacent and other ecosystems.
(iv) It is necessary to recognize potential gains from management. As such, there is a need to understand the ecosystem in an economic context, e.g., mitigating market distortions, aligning incentives to promote sustainable use, internalizing costs and benefits, etc.
(v) A key feature of the ecosystem approach is the conservation of ecosystem structure and functions.
(vi) (vi) Ecosystems are to be managed within the limits of their functioning.
(vii) The ecosystem approach is to be undertaken at the appropriate scale.
(viii) The ecosystem approach could seek appropriate balance between conservation and use of BD.
(ix) The ecosystem approach is to consider all forms of relevant information, including scientific and indigenous local knowledge, innovations, and practices.
(x) The ecosystem approach could involve all relevant sectors of society and scientific disciplines.

5.1.16.27 *World summit on sustainable development (WSSD), Johannesburg, South Africa, August 26 to September 4, 2002*

Concomitant to the above, the WSSD adopted a Political Declaration and a Plan of Implementation in relation to capture fisheries. In the declaration, the Heads of States agreed to develop and facilitate the use of diverse approaches and tools, including the **ecosystem approach**, the elimination of destructive practices, the establishment of MPAs, and the integration of marine and coastal areas into key sectors. Concomitantly, the destructive practices in fisheries include the use of noxious substances and inappropriate fishing gears, among others.

5.1.17 Sri Lanka (SL)

SL is a small island country having high potential in AQC and fisheries. Incidentally, the reservoirs in SL play a significant role in fish production and their management. There have been works dealing with the reservoir fisheries in **SL** and related management aspects.

SL is an island in monsoon-fed Asia with rich water resources. There are *c.* 103 perennial rivers in SL, most of which drain from the central highlands into the western, southern, and eastern coasts. In addition, a multitude of reservoirs on the island support rich aquatic BD (Fernando and Weerawardhena, 2002). The reservoirs in SL consist of.

(a) ancient reservoirs, which have been restored and rehabilitated and
(b) new reservoirs constructed primarily for irrigation and/or generation of hydroelectricity.

5.1.17.1 *Reservoirs in Sri Lanka*

The total extent of reservoirs of the country is *c.* 170,000 ha (Costa and de Silva, 1995). As such, reservoir density in SL is *c.* 2.7 ha per every km^2 of the country, which is the highest density of reservoirs in the world (De Silva, 1988). There are > 10,000 reservoirs in SL, but most of them are < 100 ha in surface area. These small rural reservoirs, which are distributed across the undulating landscape of the dry zone, are not randomly located but also occur in the form of distinct cascades, which are positioned within well-defined small watersheds or meso-catchment basins (Panabokke, 2001). According to a statistics, there is one reservoir of *c.* 6000 ha; about six reservoirs of *c.* 3000 ha, and 40 reservoirs of *c.* 500 ha in surface area (Fernando, 2000).

Notwithstanding the above, the inland fishery of SL is essentially a capture fishery in major and medium-sized perennial reservoirs of the country (Upali S. Amarasinghe and D.E.M. Weerakoon, 2015).

Recently, attempts have been made to develop CBF in small rural reservoirs in SL. In most of these small village reservoirs, water is retained from the inter-monsoonal rains in November–December to a peak dry season in August–October. It had been shown that these seasonal reservoirs and other small-sized perennial reservoirs, where virtually no capture fisheries are practiced, could be used to develop CBF through stocking of hatchery-reared Chinese and IMCs and subsequent recapture after growth period of 7–9 months (Thayaparan, 1982; De Silva, 1988).

5.1.17.2 *Ichthyofauna in the reservoirs of Sri Lanka*

In Sri Lanka freshwaters (SL FWs), *c.* 85 native fish species belonging to 19 families were reported to occur (Newrkla and Duncan, 1984; Silva and Davies, 1988; Pethiyagoda, 2006; Weliange et al., 2008). In addition, *c.* 24 exotic species were introduced into SL FWs since the late 19th century and three of them (*Chitala ornatus*, *C. batrachus*, and *Pterygoplichthys multiradiatus*) were accidentally introduced as exotic fish species (Amarasinghe et al., 2006; Shirantha and Amarasinghe, 2006). Since the introduction of the two exotics, viz., *Salmo trutta* and *Onchorynchus mykiss* during the late 19th century for purposes of establishment of sport fishing, several exotic species were introduced, either deliberately or accidentally, into SL FWs. Most exotic species were introduced during the 20th century to develop inland fisheries in SL (Amarasinghe, 1998a).

5.1.17.3 *A concise history and recent trends of inland fisheries of Sri Lanka*

Siriweera (1986, 1994), based on the evidence(s) found in the stone inscriptions, stated that inland fish were exploited from reservoirs since time immemorial. Laws were in place at that time to control fisheries and establish fishing rights (Siriweera, 1986, 1994; Ulluwishewa, 1995). However, prior to the 1950s, FYs from SL reservoirs were very low (Fernando, 1984, 2000). The reason for these low yields was the inability of indigenous riverine fish species to sustain themselves in high population densities in the lacustrine habitats of the reservoirs (Fernando and Holcik, 1991). However, a productive fishery was established in the reservoirs of SL after the introduction of the exotic cichlid species, *O. mossambicus* into the country's FWs in 1952 (De Silva, 1988).

Notwithstanding the above, the capture fisheries in the major reservoirs of SL are usually characterized by the following:

(a) the use of non-mechanized fiberglass canoes, *c.* 6 m in length, manned by two persons;
(b) the gear used comprises exclusively gill nets (*c.* 6–20 net pieces per craft), ranging in mesh size from 7.5 to 12.7 cm, with each net piece having a standard length and height of 60 and 1.5 m, respectively; and.
(c) the predominant catch is exotic cichlid species, *O. mossambicus* and *Oreochromis niloticus*.

5.1.17.4 *Status of capture fisheries in reservoirs of Sri Lanka*

The much increase in inland fish production resulted in an elevation of reservoir FY from negligible levels before 1952 to a very high level of *c.* 283 kg/ha/year in 1980s (Fernando, 1984; De Silva, 1988).

It may be noted that reproductive performance of *O. mossambicus* is high due to the extensive availability of shallow littoral zones in Sri Lankan reservoirs (Fernando and Indrasena, 1969). Further, the high yields of the exotic cichlid species in SL reservoirs could be due to their strong ability to colonize the lacustrine habitats, which the indigenous fish fauna may be lacking (Fernando and Holcik, 1991). Further, the lacustrine species may, sometimes, be composed of riverine and marsh-dwelling fish species, which may not be able to sustain dense populations in reservoir habitats. Moreover, *O. mossambicus* is able to digest the abundant phytoplankton group (particularly the BGA), which is not much preferred by the indigenous fish species (Costa and Abayasiri, 1978; De Silva, 1985). Moreover due to its ability to change its dietary habits from season to season depending on the availability of food, *O. mossambicus* has successfully colonized in reservoirs of SL. Chandrasoma (1986) and Amarasinghe (1997) also reported similar trends in some other reservoirs of SL.

5.1.17.5 *Fishery resource evaluation in reservoirs of Sri Lanka*

The reservoir fishery of SL may be treated as "common pool resource" or "open access resource" in terms of exploitation pattern. The fisheries of individual reservoirs are stabilized at the level of zero economic revenue, i.e., where the revenue from the fishery is balanced with the cost of fishing. The relationship between the FY (kg/ ha/ year) and fishing intensity (boat-days/ha/year) of individual reservoirs could, therefore, be used to determine the number of boats to be employed in a given reservoir, provided an estimate of fish production potential is available. Empirical approaches for stock assessment are particularly useful for situations where reliable data for stock assessment using standard methods are not available as is the case with most of the SL reservoirs.

Notwithstanding the above, Amarasinghe et al. (2002b) have shown by comparing actual FY in various SL reservoirs with the FYs predicted by these models that the predictive power of these models is very high. Further, GIS could be used as an effective means for data gathering and processing for a wide range of planning and management procedures (Meaden and Kapetsky, 1991).

5.1.17.6 *Aspects of management of capture fisheries in perennial reservoirs*

During the period 1979–89, it may be mentioned that the government of SL worked to develop the capture fisheries in its reservoirs, by providing the fishers with fiberglass canoes and gill nets under a subsidy scheme. Thus, the fishing effort in the reservoirs increased considerably as a result of this subsidy scheme (De Silva, 1988). Concomitantly, the fisheries authorities also imposed fisheries regulations to control fishing effort and size of the fish landed. Use of mechanized boats and any kind of shore seine nets were forbidden in perennial reservoirs and the minimum permissible mesh size for the gillnet fishery was 8.4 cm. However, gill nets of mesh sizes <8.4 cm and beach seines, which are also not unlawful, were operated in some reservoirs sporadically (Amarasinghe and De Silva, 1992b). However, some fishers have tried to increase efficiency of gillnetting by beating water with wooden poles or weighted ropes to drive the fishes toward gill nets (Amarasinghe and De Silva, 1992b).

During the early 1980s, fisheries cooperative societies (FCS) were functioning effectively, probably, because the fishers in inland reservoirs had to become members of the FCS in order to be eligible for receiving boats and gill nets under the state-sponsored subsidy scheme. As FCS were well functioning, fishers tended to arrive at collective agreements regarding resource exploitation, such as complete curtailing of beach seining and increasing the minimum mesh size of gill nets up to 10.2 cm. This resulted in a considerable increase in inland fish production in the country with a reported peak of 39,300 mt in 1989 (Amarasinghe and De Silva, 1999). On the other hand, due to discontinuation of state support for inland fisheries development during the period 1990–94, inland fish production declined markedly (Amarasinghe and De Silva, 1999) as a result of interruption of government funding for monitoring and stocking programs. Consequently, fishers began to use smaller meshed (<7.5 cm) gill nets that resulted in increase of overfishing of the cichlid species, which were dominant in the fishery. However, inland fisheries had nearly fully recovered since the state restored its support in the mid-1990s.

Notwithstanding the above, the available evidence(s) seem to suggest that in the absence of state-sponsored monitoring procedures or centralized management systems, reservoir fisheries could not be effectively managed under the prevailing socioeconomic milieu. Incidentally, Amarasinghe and De Silva (1999) reported management systems in two reservoirs, viz., Maduru Oya and Muthukandiya reservoirs, where fishing communities were also actively involved in making decisions for management of the fisheries. In both the reservoirs, there is well-functioning FCS. Fishers arrive at collective decisions on fisheries management and environmental protection through the FCS. This comanagement procedure, in which the centralized administrative authority (i.e., the Ministry of Fisheries) and the fishing community share the responsibilities of making decisions for fishery resource management (Berkes, 1994; Pomeroy, 1995; Sen and Raajaer-Nielsen, 1996). This had been useful for preventing overexploitation of fishery resources even during the period of nonexistence of state-sponsored monitoring procedures from 1990 to 1994 (Amarasinghe and De Silva, 1999). Such a system of management involving the stakeholder fishermen from the grassroot level in decision-making and other management procedures for the water body, could be regarded as **Community-based Fisheries Management (CBFM) an,** could be one of the most effective and efficient methods of management of the water bodies.

5.1.17.7 *Aspects of management of CBF in the village reservoirs*

In CBF, artificially propagated seed stocks are, usually, released into water bodies, which are not primarily and traditionally used for fish production and the fishes are recaptured upon reaching a desirable size (Lorenzen, 2001). Incidentally, it may be mentioned that, generally, CBF could involve ownership, either singly, as in the case of Vietnam, where farmers lease-out small reservoirs (Nguyen et al., 2001); or, collectively, in the form of a cooperative, such as, the case in oxbow lakes in Bangladesh (Middendrop and Balarin, 1999).

Notwithstanding the above, the seasonal reservoirs are highly productive due to the cattle manure accumulated in the drawdown areas from cattle grazing during the trough season and remnants of terrestrial vegetation. However, the small puddles in most seasonal reservoirs may not dry up completely, so that they could shelter some of the indigenous live fish species, notably, the *Channa* spp., *Anabas testudineus*, *Clarias* spp., etc. These native species, although harvested in some reservoirs, may not usually portray high FY. In fact, the potential of small village reservoirs for the development of CBF was indicated first by Mendis (1965). Accordingly, attempts were made to utilize these seasonal reservoirs for the development of CBF during the 1960s (Indrasena, 1965). He had revealed that, in the 1960s, some seasonal reservoirs were stocked with *O. mossambicus* fingerlings of *c.* 7.5–10.0 cm size and that during 8–9 months of water retention period, they grew up to *c.* 25–30 cm in size (Fernando and Ellepola, 1969). The community could come into the picture for proper management of such fisheries by the application of CBFM.

Concomitant to the above, a program had been implemented to formulate a suitable strategy for the development of CBF in seasonal reservoirs of SL, based on the recommendations by Rosenthal (1979) and Oglesby (1981). In addition, works of Thayaparan (1982), De Silva (1988), Chakrabarty and Samaranayake (1983), and FAO/UNDP (1980) (often funded by organizations, like FAO/UNDP) had portrayed promising prospects of CBF in seasonal reservoirs, of course, if efficiently managed through effective systems, like CBFM. Incidentally, Chandrasoma and Kumarasiri (1986) reported that in 15 seasonal reservoirs, FY ranged from 220 to 2300 kg/ha within a single growing season.

5.1.17.8 Species cultured

It may be noted that the species suitable for stocking in seasonal reservoirs could reach the marketable size in *c.* 6–8 months, as the culture period in seasonal reservoirs is 7–9 months. Concomitantly, these species could also utilize the natural food resources available in the reservoirs. It may be noted that seasonal reservoirs in SL exclusively rely on common carp, Chinese carps, and the IMCs, as these species were artificially spawned successfully in SL (Weerakoon, 1979; Balasuriya et al., 1983; De Silva, 1988).

5.1.17.9 Aspects of success of culture-based fisheries in Sri Lanka

CBF in seasonal reservoirs have many success stories. The main secret behind the success could be the involvement of the stakeholders from the grassroots level in many cases through CBFM. It had been suggested that at a stocking density of *c.* 2000 fingerlings/ha, an average yield of *c.* 750–1000 kg/ha could be achieved (Chakrabarty and Samaranayake, 1983, Chandrasoma and Kumarasiri, 1986; De Silva, 1988). However, fingerling availability, selection of suitable seasonal reservoirs, post-stocking management and, last but not the least, proper community management through CBFM are essential to achieve these high yields.

Notwithstanding the above, Pushpalatha (2001) and Pushpalatha and Amarasinghe (2007) reported case studies of rural AQC in SL for the production of fish fingerlings in ponds and cages.

With the stocking density of *c.* 5000 fry per cage, high survival rates (*c.* 55%–92%) were achieved within the culture period of 58–80 days. This case study, in addition, indicates that in SL, it may be possible to introduce a strategy for rearing of fish fry to fingerling size in net cages and earthen ponds. However, low-cost feed for fry rearing, subsidy schemes to cover the initial cost of cage and pond construction, CBFM-type management through the involvement of the community, etc. are needed for the sustainability of this strategy.

Further, production of bigger fingerlings (*c.* 10–14 cm) might help to minimize mortality rates of juveniles, as practized in China (Li, 1988).

5.1.17.10 Regulatory aspects and community management

In SL, culture-based fishery activities are community dependant. Also community management through CBFM plays a key role in efficient management of the water bodies. However, the final yield from individual reservoirs may depend not only on biological criteria such as the natural productivity of the water bodies and the hydrological regime but also on the socioeconomic factors including the degree of participation and cooperation within the community and the relative easiness of marketing the produce (De Silva et al., 2005).

Incidentally, it may be noted that the small village reservoirs come under the jurisdiction of the Department of Agrarian Services, which establishes a village farmers' organization(s) in each reservoir. These farmers' organization(s) is responsible for day-to-day water management. However, with the improvement in liaison between the Department of Agrarian Services and the Ministry of Fisheries during the 1980s, a well-structured institutional link has developed between the National Aquaculture Development Authority (NADA) of SL and the Department of Agrarian Services, through which the field-level coordination was considerably facilitated. In fact, it may be pointed out that the **Agrarian Development Act No. 46** of 2000 provides legal provision for the establishment of the major rural institution, like the farmers' organizations, the mandate of which constitutes the fisheries and AQC development whenever the farmer community decides to adopt it. In fact, in most seasonal reservoirs, farmers' organization(s) have setup "AQC committees." These committees are responsible for the management of CBF. Moreover, it may be pointed out that a proportion of the profit from the CBFis utilized for improvement of the reservoir, such as strengthening of the earthen embankment ("bund") for effective fisheries development, so that there is a strong relationship between the AQC committee(s) and the farmers' organization(s). Further, the members of the AQC committee are actively involved in the labor-intensive activities, like preparation of the reservoir beds in most seasonal reservoirs, prior to stocking; removal of impediments to fishing, such as submerged decaying tree stumps; AMs; etc. in order to facilitate operation of seine nets for harvesting the stocked fish. On the contrary, in some reservoirs, farmers may not remove AMs until the onset of harvesting because the presence of AM in the reservoir prevents poaching of stocked fish. However, community decision in this matter could play a key role.

Notwithstanding the above, the NADA of SL is responsible for extension activities pertaining to inland fisheries and AQC. Fisheries extension staff are usually appointed in place of the former fisheries inspectors, whose major responsibility is to promote "technology transfer," rather than enforcement of regulation. The enabling policy, legislative environment, and mechanisms to empower stakeholder communities from the grassroots level are the strong positive aspects for realizing the potential of CBF development in the rural reservoirs, and this could be the biggest application of CBFM. There could be many applications of CBFM, notably, in induced breeding of IMC and rearing of fry-fingerling through community participation, issuing aquaculture licenses for seed supply, sharing of technical know-how of farmers' organizations, aquaculture committees for proper AQC management, and so on.

5.1.17.11 *Present trend of research on culture-based fisheries in Sri Lanka*

The optimal stocking density for the CBF in seasonal reservoirs in SL is *c.* 2700 fingerlings per ha. Here, the reservoir area is, generally, considered as the effective area (i.e., *c.* 50% of the area at full storage level). Geographical information systems (GIS) and remote sensing (RS) methodologies could be used to determine the reservoir area accurately (Wijenayake et al., 2005). This is of particular importance due to the reasons that determination of the reservoir area with much precision is crucial for proper planning of CBF and that the available information on reservoir area in the databases is usually and largely underestimated because they had been, generally, based on the maps generated from the aerial photographs.

Notwithstanding the above, there is a need to rank the suitability of the reservoirs in SL, for the development of CBF, in view of the large number of seasonal reservoirs in the country. Thus, in order to develop a suitable ranking system or scale, aspects such as the physicochemical, biological, catchment, and hydrological characteristics of the water bodies as well as the socioeconomic aspects need to be taken into consideration. Thus, De Silva et al. (2005) performed a preliminary analysis to develop a ranking system to determine the suitability of 14 non-perennial reservoirs in Southern SL. The ranking was achieved using GIS and analytic hierarchy process. The latter permits a set of heterogeneous factors to be equated to a common denominator (Saaty, 1977). Thematic layers of water quality, catchment characteristics, market-related aspects, and socioeconomic factors, corresponding to each of the seasonal reservoirs studied were evaluated using the scores that were obtained according to the **weighted linear combination(s)**. The final weighing for each of the 14 reservoirs were done and the results indicate that none of the reservoirs could be considered as "poor"; rather, one reservoir was considered as "excellent," six reservoirs were considered as "fair," and the remaining seven reservoirs were considered as "good." The results indicate that it is possible to develop such a ranking system.

Notwithstanding the above, over the recent decades, **comanagement** has become an increasingly popular form of management in many developing countries. Viewed as a means of promoting sustainable and equitable management of natural resources, it has seen wide application in small-scale inland fisheries. However because of its worthy credentials, there has been insufficient critical assessment of the results. However, there was a failure of the comanagement mechanism for a brief period in SL between 2001 and 2002. Also, there had been communications that often commence with a review of the underlying theory, which explore the reasons for failure of a comanagement initiative in the Sri Lankan reservoir fisheries during the said period.

Incidentally, the comanagement thinking arose from an earlier view of common property theory which had advocated for private or public ownership of the commons as the only rational response to increasing resource pressure. This view subsequently gave way to the comanagement view, which stated that, under certain conditions, management was best left to local users, with state, civil society, or market sector support wherever necessary. Critics of comanagement's rationalistic underpinnings point to a lack of cultural or political embeddedness. This limits the understanding of power and accountability distribution and how it shapes collective behavior. Others suggest that the contentiousness of the common property theory/comanagement debate had resulted in neglect of key empirical features of fisheries commons: most critically the complex causes and consequences of change(s) in fishing effort. In SL emphasis is on understanding the environmental, technical, and socioeconomic drivers of effort change. The comanagement initiatives took place in two small and medium-sized reservoirs (*c.* 788 and *c.* 1546 ha) in Northwest province. These are believed to support artisanal gillnet fisheries, predominantly for exotic tilapias marketed on a highly local basis. The comanagement strategy involved local fishery officers and fishing cooperatives, centered on fishing restrictions designed to sustain yields and increase mean catch size. Despite initial enthusiasm, the restrictions collapsed after 4–5 months with the onset of opportunistic free riding. This was due to lack of enforcement capacity, disproportionate hardship faced by the least economically mobile fisher groups, and political patronage relations. The other reason(s) could be that, as compliant partners, the cooperatives had weak leverage in the decision-making process. They were also not much in a position to do comprehensive zoning exercises to enable the effective, planning, monitoring, and regulation of floodplain AQC; and, the cooperative also could not avoid potential conflicting situations between traditional fisher communities and floodplain AQC farmers.

Despite the short effective duration of restrictions, simultaneous analysis of local market trends demonstrated a remarkable rebound in yields pointing to the resilient regenerative capacity of the tilapia fishery. Furthermore, historic commodity price data revealed long-term tilapia price stability and a close match between demand and supply even as fishing pressure increased. Despite negative impacts on indigenous BD, these features suggest that fears of overexploitation are unfounded in a number of instances, and question the underlying premise for the comanagement interventions.

Finally, results were used to highlight contextual differences with floodplain fisheries in Bangladesh, where declining wild harvests and the emergence of a vibrant AQC industry present very different challenges. Results caution against overreliance on comanagement as a blue print approach. There is instead a need for detailed contextual analyses which move beyond a view of fisheries as closed systems to incorporate broader livelihood-based perspectives. Such a situation also exists in SL and in other parts of Asia.

In fine, the reservoir fishery in SL is very productive and provides a cheap source of animal protein for rural communities, provided it is efficiently managed through effective means of CBM involving the community stakeholders. Further, the stock assessment procedures used for assessing the fisheries of SL reservoirs enabled identification of optimal fishing strategies. In fact, potential fish production could be estimated using catchment land-use characteristics, which may be quantified by GIS methodologies. As the FY and the fishing intensity are linearly related in the reservoir fishery of SL, the number of boats to be employed in each reservoir could be determined using this relationship on the basis of estimated potential fish production. Moreover, **length-based fish stock assessment methodologies** portrayed the optimal size of first capture of various fish species. This corresponds to the stretched gillnet mesh size of 8.5 cm as determined from gillnet selectivity studies.

Notwithstanding the above, small cyprinids are abundant in perennial reservoirs of SL. They are not much exploited mainly due to poor consumer preference. However, due to high fishery potential of the minor cyprinids, a subsidiary gillnet fishery could be introduced to exploit small cyprinids in the reservoirs. Incidentally, in Udawalawe reservoir, the IMCs (mainly *C. catla*) and indigenous *Labeo dussumieri* form a significant portion of the landings, especially, during the rainy seasons. According to the information gathered from fishers, egg masses of IMCs were often found attached to submerged AMs in the inflow areas of the reservoir suggesting that self-reproducing populations of IMCs have been established in the Walawe river basin. Thus, studies on the reproductive biology of IMCs in the SL inland waters are necessary for defining reservoir fisheries management strategies. Further, the much availability (>39,000 ha) of non-perennial reservoirs in SL, which usually dry up completely during July–September and again fill during inter-monsoonal rains during December–January, permit development of CBF. In fact, the CBF in seasonal reservoirs provide a means of increasing food supply in the rural areas of SL due to their high yields. As such they could benefit the poor sectors of the rural communities significantly. In addition, the CBF in seasonal reservoirs of SL have little impact on the environment because they are dependent on the existing water bodies and do not involve supplementary feeding. On the contrary, it is a fact that there are environmental concerns with regard to intensification of AQC. In this connection, Naylor et al. (2000) highlighted that the use of fishmeal in aquafeeds would bring about detrimental effects on capture fisheries.

Notwithstanding the above, there seem to be an increasing emphasis on the need of fisheries management in an ecosystem context because of the fact that the fisheries have a direct impact on the ecosystem, which is also impacted by other human activities (Garcia et al., 2003). On the other hand, the CBF in seasonal reservoirs of SL, however, involve minimal external inputs, especially due to the reason that supplementary feeds are not used. Moreover, the seasonal reservoirs do not harbor rich population of indigenous fish communities; in view of the fact that most of these water bodies completely dry up during some months of the year.

Conversely, net cage culture of fish fry to fingerling in perennial reservoirs is promoted under the inland fisheries development plan in SL. It is contemplated that this AQC strategy could produce part of the fingerling requirement for furthering the development of CBF in seasonal reservoirs.

Concomitant to the above, it is imperative that the stakeholder fishing communities from the grassroots level be made active partners in the decision-making processes for the efficient management of reservoir fishery of SL, often through CBFM strategies. Moreover, scientific information from different parts of the world may be collected through the active participation of the fishing communities, as reported by Ticheler et al. (1998) and Neis et al. (1999), for comparable fisheries in other parts of the world. Nevertheless, a *modus operandi* of comanagement and CBFM may be expected, in which there is involvement of the centralized administrative authority (i.e., Ministry of Fisheries) and the fishing community for making decisions for fishery resource management (Berkes, 1994; Pomeroy, 1995; Sen and Raajaer-Nielsen, 1996). Such a *modus operandi* could be introduced for the effective management of the reservoir fishery of SL.

5.1.18 Thailand

Thailand is a country with rich fisheries and aquacultural potential. Incidentally, the reservoirs in Thailand play a significant role in fish culture and their management. There have been works which dealt with the reservoir fisheries in **Thailand** and relevant **management aspects**.

In **Thailand**, the inland fisheries have long been recognized and operated in major rivers, floodplains, canals, swamps, wetlands, lakes, and reservoirs (Tuantong, 2015). There are seven major river basins in the country according to zoogeographical classification, viz., Chao Phraya Basin, Mekong Basin, Eastern Basin, Southern Basin, Salween Basin, Mekhlong Basin, and Tenasserim Basin (Sataporwanit and Wongrat, 2000). Incidentally, the natural lakes in Thailand include Bung Borapet in the central region (*c*. 13,000 ha), Nong Harn in the NE region (*c*. 12,500 ha), Kwan Phyao in the northern region (*c*. 2100 ha), and Songkla Lake (*c*. 96,000 ha) in the south (Pawaputanon, 1992). However, most of the Thai reservoirs were impounded during the second half of the 20th century.

Notwithstanding the above, the FW fish fauna in Thailand is very diverse with *c*. 573 indigenous species and *c*. 16 introduced species (Vidthayanon et al., 1997). In fact, the fisheries resources in Thailand are regarded as a public wealth (Bhukaswan, 1987) and fishing is a long-standing tradition in the country and is said to be considered as an integral part of the heritage and culture, particularly in rural areas of Thailand (Coates, 2002). Incidentally, FY from the inland fisheries contribute *c*. 7.3% to the country's overall fish landings and provides *c*. 3 kg/*capita*/year. Notably, inland fisheries production increased from *c*. 200,000 tons in 1995 (Virapat and Mattson, 2001) to *c*. 203,000 tons in 2004 (Department of Fisheries, 2007). However, the proportion of the total production from various habitats is as follows: (i) rivers, streams, and canals: *c*. 34.2%; (ii) swamps/marsh/floodplain: *c*. 31.3%; (iii) rice-fields: *c*. 22.2%; (iv) reservoirs: *c* 8.3%; (v) lakes: *c*. 2.1%; and (vi) AQC ponds: *c*. 2.0% (Coates, 2002).

However, the inland fisheries are essentially artisanal and are based mainly on the indigenous species (*c*. 80%–90%) such as the carps, minnows, snakeheads, and catfishes (Virapat and Mattson, 2001). The rest are thought to constitute the introduced species, such as the tilapias as well as IMCs and the exotic carps. When the country is divided into four regions (i.e., north, NE, south, and central), the highest production in inland capture fisheries was reported from the NE region (*c*. 40.8%) and the lowest production (*c*. 10.4%) was in the south (Sataporwanit and Wongrat, 2000).

5.1.18.1 *Reservoir fisheries in Thailand*

Reservoirs in Thailand are ubiquitous countrywide, especially, in the north and NE regions. They had been impounded for either hydropower or irrigation purposes, and fisheries are considered a secondary benefit from the impoundments. However, there are at least two Thai reservoirs that were impounded and used primarily for fisheries (Virapat and Mattson, 2001). Virapat et al. (2000) reported an inventory of *c*. 28,956 existing reservoirs in Thailand ranging from 0.01 ha to >1000 ha. However, according to Virapat and Mattson (2001), there are *c*. 27,779 Thai reservoirs. Virapat et al. (2000) also estimated fish production from these reservoirs to be between 122,314 and 318,909 tons per year.

5.1.18.2 *General aspects*

Most of the Thai reservoirs possess relatively small catchment areas (<10,000 km ^{2}). The water storage volume ranges from 108.5 million m^3 in Mae Chang to 17,745 million m^3 in Srinagarind (Bernacsek, 1997). On the other hand, the surface area of major reservoirs in Thailand ranges from 1200 to 41,000 ha (Verdegem, 1999) and the cumulative area of Thai reservoirs is *c*. 432,176 ha (Virapat and Mattson, 2001), which is significantly higher as compared to the approximation of *c*. 3×10^5 ha (De Silva and Amarasinghe, 1996; Bernacsek, 1997).

In Ubolratana reservoir, one of the largest reservoirs in the country, the pelagic fish catch was significantly related to the surface area variation while the littoral catch was related to the total reservoir area and to the difference between the minimum and maximum water spread of the reservoir (Moreau et al., 2004). Further, **the morpho-edaphic index** (MEI; the ratio of conductivity to the mean depth) portrayed a very highly skewed distribution with no relationship to FY (Bernacsek, 1997).

5.1.18.3 *Fish communities*

The structure of fish assemblages is more or less of similar magnitude in all the Thai reservoirs. The reservoir fish communities consist of cyprinids, which form > 50%, followed by catfishes, murrels, and miscellaneous species. The exotic species are the Chinese carps and the IMCs as well as the Nile tilapia. There is also evidence that changes of fish species composition occurred after impoundment (Srichareondham et al., 2000).

Further, in reservoir fish communities, there appear marked differences in distribution of species in the inshore and offshore areas. The small-sized species, such as Thai river sprat (*Clupeichthys aesarnensis*), silver knife barb (*Paraluebuca typus*), etc., are usually caught in the offshore area, while the medium- to large-sized species, such as spotted wolf barb (*Hampala dispar*), carp (*Dangila lineartus*), etc., are generally found in higher abundance in inshore area (Sihapitukgiat et al., 2002). Prchalová et al. (2003) investigated the vertical distribution of fish in Ubolratana reservoir based on hydroacoustic studies. During daytime, most fish species followed a general tendency of forming shoals or staying near the bottom and during early dusk, shoals disintegrated and single fish appeared mostly in deeper layers. At night, the entire water column was filled with organisms as most fish moved gradually to the upper part of the water column (0–5 m) and stayed there throughout the night.

5.1.18.4 *Fisheries*

It may be important to note that the majority of the fishers in the Thai reservoirs are subsistence fishers. They use different types of gears based on season, WL, and fishing ground. The fish landings are mostly of indigenous species, which form > 80% of the production. In general, most fishers use gill nets, long lines, cast nets, scoop nets, and big lift nets (Jutagate and Mattson, 2003). Incidentally, there are two types of lift nets, viz. (i) fixed lift nets and (ii) movable lift nets. These are used for catching small clupeids (Sihapitukgiat et al., 2002). The reservoir fishers take all species regardless of size variation (Bhukaswan and Chookajorn, 1988). The average FY in Thai reservoirs is *c.* 48 kg/ha/year (Moreau and De Silva, 1991). The empirical model between catch per unit effort and fishing effort in Thai reservoirs showed that the estimated MSY and optimum fishing effort were *c.* 93 kg/ha/year and 10 fishers km^2, respectively (De Silva et al., 1991).

5.1.18.5 *Stock enhancement programs*

Stock enhancement in Thai reservoirs had been carried out since the early 1950s (Pawaputanon, 1992). Moreover, in Thailand, stocking programs were carried out not only in reservoirs but also in the rice fields, rivers, and swamps. Stocking is generally considered as a means of providing benefits to the poor communities (Coates, 2002). Some of the species stocked are the grass carp (*C. idella*), silver carp (*H. molitrix)*, common carp (*C. carpio*) (Pholprasit and Waiwuttho, 1970; Benchakarn, 1972; Duangswasd and Janesirisak, 1973), mrigal (*C. mrigala*), and rohu (*L. rohita*) (Benchakarn and Nookuor, 1988; Pawaputanon, 1988). Interestingly, the IMCs had performed better than the Chinese carps mainly because the IMCs showed better survival and FYs (Virapat, 1993). Since the early 1990s, stocking programs had been shifted from exotics to indigenous species. The popular species to be stocked are silver barb (*Barbonymus gonionotus*), seven-line barb (*Probarbus jullieni*), the critically endangered species, the Mekong giant catfish (*Pangasianodon gigas*) (Hogan, 2004), etc. The performance of the Mekong giant catfish in the reservoir environment was investigated with regard to various aspects, such as feeding habits, which comprised both phytoplankton and zooplankton (Yamagichi et al., 2005b,a), and its preferred habitat during both day and night, where it stayed in shallower region at night, deeper during the day and avoided hypoxia areas of the reservoir (Mitaura et al., 2005). It possesses peculiar reproductive biology in the reservoir habitats (Manosroi et al., 2003).

5.1.18.6 *Cage culture*

The development of cage and pen culture, especially in Asia, has substantially increased reservoir fish production (Petr, 2007). In Thailand, cage culture is recognized as one of the important activities in lakes and reservoirs, and may be traced back to 1950 (Thanasomwang et al., 1983). However, cage culture in reservoirs brings about competition with other reservoir uses, particularly fishing, navigation space, and water. The cage culture practices in Thailand are semi-intensive to intensive with the use of artificial feed (Lin and Kaewpaitoon, 2000). Popular species, such as Nile tilapia, red fin bagrus (*Hemibagrus wyckioides*), etc., are usually cultured in cages to marketable sizes. Also, there had been attempt(s) to do cage culture of broodstock of African catfish (*C. gariepinus*) and hybrid catfish (Chabjinda and Teanchareon, 1991). Incidentally, Sihapitukgiat et al. (2000) have shown that growth performance, production, and economic return of the cage culture in the river systems were better than that in reservoirs. However, cage culture in reservoirs is still profitable compared to other agricultural activities. Many of the studies on cage culture in reservoirs were carried out to determine optimum stocking densities of various species (e.g.,

Chabjinda, 1996; Sriwatana and Kasisuwan, 1996); appropriate feed formulae for cultured fish (e.g., Thawornnan and Sorasit, 1997; Sirichaipun et al., 2001), and impacts of wastes from cage culture on reservoir water quality (Jiwyam and Chareontesprasit, 2001).

5.1.18.7 Socioeconomics and management aspects

It was demonstrated that the financial income generated by the fishery in Ubolratana reservoir might be higher than that of electric power generation from the reservoir, which was the primary objective of the impounding exercise (Bhukaswan, 1985). In Thai reservoirs, the fishery (fishing) activities are exclusively the domain of males, whereas females have almost exclusive responsibility of cleaning and processing of fish at landing sites and elsewhere. The catches are mostly sold in local markets and smaller catches are always taken for household consumption or for processing to yield fermented fish.

Notwithstanding the above, Sihapitukgiat et al. (2002) mentioned that a subsistence-oriented fisher commonly has two or three income sources, e.g., rice cultivation and/or wage labor in rice fields, or even nonagricultural work. Generally, poor fishers with multiple income sources usually fish only seasonally but some of them, who could spare time, may fish all the year round, for example, setting their gill nets after work and lifting them before work next morning.

Concomitant to the above, in terms of income, a recent study showed that at the equilibrium stage, high profits were derived from fishing activities as 42.2, 39.6, and 27.3 Baht/kg from the trap, long lines, and gillnet fisheries, respectively (Saowakoon et al., 2008) as compared to 2.8 Baht/kg from the small clupeid lift net fishery (Jutagate et al., 2001a).

5.1.18.8 Conflicts in reservoir fisheries

In general, there are no obvious conflicts in dam construction and operation in Thailand, except, for the cage culture in the reservoirs. However, one exception is Pak Mun Reservoir. It was constructed by damming Mun River 6 km upstream from the confluence with the River Mekong. A fish ladder was incorporated, although this was the first fish ladder in SE Asia. In fact, many studies have shown that the "**combined pool and weir-type" fish ladder** has been unsuccessful and has also performed poorly in terms of helping the fishes to move upstream (Schouten et al., 2000; Roberts, 2001). Consequently, fish production in the upstream areas declined, as the reservoir dam acted as a barrier for upstream migration of the fishes. It was evident that *c.* 37 fish species had disappeared and the fish catch declined by *c.* 60%–80% from the upstream areas since the construction of the dam (Jutagate et al., 2001). Therefore, the riparian fishers in the area, known as "Assembly of the Poor" had requested the government to resolve the problem. In fact, based on a scientific study, since 2003, the Government has implemented the opening of all sluice gates annually for 4 months from May to August, in order to allow migratory fishes to move upstream so that they can complete their life cycles (Jutagate et al., 2003b; Jutagate et al., 2005a,b).

5.1.18.9 Regulations

In Thailand, the *Fisheries Law 1947* had been enacted on the basis of FW fisheries, which was the leading sector at that time. Section 32 of *Fisheries Law 1947* allowed Minister/Governor to issue decrees on fishery regulations. Most of the current regulations were issued under this section. Other relevant sections are Sections 6 and 7. According to Section 6, fishing grounds are divided into four types, viz. (i) sanctuary, (ii) auction, (iii) permission, and (iv) public areas. Section 7, under approval of the Minister, is said to grant authority to provincial committees to announce specific fishing measures in their provinces (Sihapitukgiat et al., 2002). So far, the most effective measure in the inland fishery regulation has been the "closed fishing" season, which is set from May 16 to September 15 countrywide except in some specific areas.

Concomitant to the above, Henocque and Tandavanitj (2011) dealt with **fisheries and comanagement in practice in Thailand.**

Thailand possesses a rich potential in AQC and fisheries in FW, marine, and estuarine domains. There is thus a need for proper management of the resources in different habitats (Henocque and Tandavanitj, 2011). The Coastal Habitats and Resources Management (CHARM) developed its approach around five attributes of comanagement, which are as follows: (i) participation, (ii) partnership, (iii) capacity building, (iv) development of integrated approaches and methods, and (v) learning and adaptation.

In fact, comanagement is a dynamic partnership between the resource users and the government. It works at the interface between the ecosystem and the social system and seeks changes in the state of the former by bringing about changes in the behavior of the latter.

Concomitant to the above, the principal cause for the building up of the CHARM scheme was the recognition that overfishing in the seas of Thailand had caused negative impacts on the marine fish stocks, ecosystems, and the coastal fishermen. In fact, the total marine annual catch is double the estimated MSY. It may be noted that the Thai government responded, in partnership with the European Commission (EC), by carrying out the large-scale CHARM project between 2002 and 2007.

Notwithstanding the above, the achievements of the fisheries institutions observed in Southern Thailand had highlighted the validity of comanagement in addressing conflicts and gaps in fishery management. In fact, comanagement and rights-based fisheries management offer a working decentralized system, sustainable at a national level as well as also include some management rights entrusted to the local fishery institutions. This establishes them as partners of the government in management, with user rights, which aim to change their behavior toward becoming more responsible for local marine resources.

Concomitant to the above, CHARM supported >200 local fishery projects and had portrayed that the future of coastal resources comanagement could be better handled by this mechanism for better coastal governance in Thailand. However, success of such a mechanism lies on the one hand with skilled self-organized community-based organizations, and on the other hand with strong, committed, and enlightened local government. An example of the outcome of this comanagement approach is provided in the improvement of fishery status in the Phang-Nga Bay during 1997–2005.

Notwithstanding the above, it may be noted that self-organization and collective action do not always come easily to fishermen, which often follow the local traditions. For this reason, capacity building by the government agencies is a prerequisite for ensuring the involvement of most communities in local fisheries management, including conflict resolution. Incidentally, institutional capacity building begins at the national level, and subsequently includes the provincial management bodies and the fishing communities. Further, the future fields of engagement may also be broadened in order to enhance community-based activities.

Further, **Pimoljinda and Boonraksa** dealt with **CBF comanagement** along with a case study from Phang-Nga bay in **Thailand**.

Efficient "management" is indispensable for effective production and sustained harvest from a fishery. Conversely, depleted FY could be the outcome of disarrayed management. Consequently, the coastal ecosystem deteriorated mainly due to the degradation of fishery resources and conflicts among fishermen at various locations, including in Phang-Nga Bay, where conflicts intensified with the fish output becoming unsatisfactory. Various factors have aggravated this situation as follows:

(i) Inappropriate methods, particularly the farm management measures, which had been established by the government.
(ii) Lack of participation and poor cooperation among the members of the fisher community.

However, the CBFM was appropriate to solve this problem, the essence of which is said to be that the fishermen/fishermen's organizations/communities and the government sector could share their responsibilities in fishery management. Such a system had been initiated in two pilot coastal fishing villages of Phang-Nga Bay, viz., Ban Bang Chan and Ban Haad Sai Pleug Hoy in 1995. Incidentally, the result of CBFM in the Bay was successful in the short term. The fishery management, through the CBFM approach had formed better overall outcomes than those of the former system(s), and fishers were convinced that the CBFM may produce good overall outcomes in the future than the present time.

Notwithstanding the above, some of the key conditions for the success of CBFM are as follows:

(i) socioeconomic status of the community, such as community homogeneity and people's dependence on fisheries including active participation;
(ii) cooperation and coordination among the various stakeholders;
(iii) collaboration, support, and advice from outside agencies; and.
(iv) the follow-up and evaluation to be carried out closely and continuously by responsible agencies.

Concomitant to the above, the process of CBFM in the Bay had begun with the organization of a workshop to get the acceptance, commitment, and consensus of key stakeholders on the CBFM work plan components before selection of the appropriate target villages, followed by organization of the meeting and selection of a village committee. Subsequently, the stakeholders were familiarized with the CBFM objectives and work plan. Later, the problems faced as well as the needs of the villages were taken into consideration in the work plan. Incidentally, it may be noted that the implementation of all the activities were conducted mainly by the fisher folk themselves under the supervision and support of Andaman Fisheries Development Center (AFDEC) and other organizations, such as the Bay of

Bengal Program (BOBP), Phuket Fisheries Association, etc. Lastly, close and continuous follow-up and evaluation of the implementation were done during the bimonthly meetings of the representatives from the target villages. In fact, the venue of the meetings rotate from one village to the other in order to achieve the effective output, and these had given them the scope to exchange experiences, both positive and negative, in order to improve, adjust, and to develop the current activities for the success of CBFM approach.

5.1.18.10 Proceedings of the international workshop on fisheries comanagement

General situation of Thai marine fisheries

Fisheries, notably, the marine fisheries in Thailand operated in two distinct fishing areas, viz.,

(i) The Gulf of Thailand on the South China Sea, which has a coastline of *c.* 1825 km.
(ii) The West Coast of Thailand on the Andaman Sea, with a coastline of *c.* 740 km.

Concomitant to the above, Thai marine fisheries developed rapidly since the 1960s following the introduction of new fishing gears and the extension of new fishing technologies, the venture of fishing fleets into new fishing grounds, improvement in fishing vessels, and the development of supporting infrastructure. Further, the significant development of marine fisheries, particularly commercial fisheries was due to the successful introduction of trawl fisheries, the development of purse seines, and the development of light-luring techniques. Further, commercial fisheries development had been accompanied by the parallel growth in the number of boats and fishing efficiency of the small-scale fisheries. These developments resulted in a spectacular increase in marine resources in both demersal and pelagic sectors, which augmented FY from *c.* 0.3 mmt in 1962 to > 1.0 and 2.0 mmt in 1968 and 1977, respectively. Furthermore, the landing of marine capture fisheries reached *c.* 2.8 mmt, of which *c.* 1.9 mmt was from the Gulf of Thailand and *c.* 0.9 mmt from the Andaman Sea Coast. Still further, the contribution of national landings from the Andaman Sea Coast increased from 5% in 1967 to 32% in 1996, with the Andaman Sea fisheries playing a much more significant role than in the past.

Notwithstanding the above, the significant development of marine capture fisheries had resulted not only in a considerable increase in landings but also in an equally considerable level of degradation of both marine fishery resources and marine environments, particularly in coastal areas. This is due to unrestrained fishery operations conducted in the absence of a comprehensive management framework and lack of consequent conservation and rehabilitation of coastal fishery resources.

Concomitant to the above, the development has also not provided with any benefit to the coastal communities of artisanal fishermen, who could only add to the progressive erosion of the fishery resources base in their traditional fishing grounds. These fishermen had also responded to the increasing pressure with an equal increase in fishing efficiency and fishing efforts in the immediate inshore areas. Incidentally, the artisanal landings represented < 10% of the total marine fishery landings but the economic stability of the network of coastal small-scale villages remained a problem of major social significance mainly due to the lack of proper implementation of efficient management measures, like the CBFM.

5.1.19 Vietnam

Vietnam is a country with rich fisheries and aquacultural potential. **In Vietnam**, there have been works dealing with the impact of Fishery association management on the fishery community.

Tam Giang is a potential lagoon system and has a length of *c.* 68 km along the coast of central Vietnam and an area of *c.* 22,100 ha. Animals were abundant, including many kinds of shrimps, fish, and crab. Vinh Giang is a commune, which belonged to the Tam Giang lagoon system. It is a place where fishing activities started long time ago and developed a strong fishery culture. There are many fishing groups, who used this resource, including fixed gears fishing, mobile gears fishing, and AQC. The number of fishers were, evidently, increasing but, on the other hand, the resource under harvest depleted. Accordingly, conflict was almost a routine affair among the household groups in connection with aquatic resource(s) exploitation and use. Conversely, they used various methods to increase their efficiency, which also included destructive fishing methods. Such a situation led to the overexploitation and depletion of resources. The local government tried to take some measures in resource management but the impact was not significant. Specifically, they had given emphasis on reform in the administrative management mechanism, and slight motivation to reduce injudicious exploitation of the resources, in general, and fisheries resources, in particular. In fact, the centralized management mechanism did not seem to be very much suitable in common pool resource management (Tuyen, 2008). Thus, very importantly, it is necessary to renew the *modus operandi* of the management system in AQC and fisheries in the lagoon. The implementation of an improved and comprehensive fisheries management system could be done through the application of modern fishery technology in fishing methods as well as in management regulations in order to reach a goal that is not harmful to environment and BD of aquatic resources. However, the collaborations, were not very close enough to bring about a radical change in the system, although the local government and other organizations had coordinated in management of fisheries and AQC there. In fact, planning for the areas of fishing and AQC were not very clear. In addition, the government as well as the other organization(s) did not have enough of man power and funds for efficient resource management in large areas. Moreover, the fishers had not participated actively in the management, protection, and conservation of aquatic resources. Thus, overexploitation had continued to grow from time to time. These situations had affected proper management of aquatic resources as well as fishers' livelihood.

To solve this problem, in view of the above, the community had to eliminate property rights regime to reinforce **aquatic resource management** (Tuyen, 2010). Thus, the community-based fishery management model involving the community was initiated and, subsequently, established in many communes. Moreover, Vinh Giang commune was the first commune to allocate right in resource management and harvest in the aquatic domain. Thus, the nature of the CBFM model was decentralized with the participation of the fishers and the local government.

Incidentally fishery association could be said to be a socio-professional organization. In fact, the participation of the fishers in the association is voluntary. They, generally, participate in activities such as fishing, AQC, processing as well as in other logistic services in fisheries. In fact, the establishment of fishery associations brought about a change in the scenario of aquatic resource management in which change(s) were brought about in the extent of fishers' area, fishing implements, yield, income, and fishers' awareness.

Concomitant to the above, such researches could depict the advantages and disadvantages in the *modus operandi* of aquatic resource management process as well as some amount of information could also be obtained with regard to the changes and improvements in the livelihood of fishers through support by the fishery associations, and so on.

Concomitant to the above, Nguyen (2012) dealt with community-based comanagement in **Vietnamese** fisheries, with emphasis on the fisheries associations (FA) in Tam Giang-Cau Hai Lagoon.

Fisheries and AQC are potential means of nutrition and avocation. As such there is high demand and keen competition in the field of AQC and fisheries. Such a situation in the teeth of demand and competition, coupled with often open-access regime could lead to the degradation of aquatic (FW and marine) resources, occasionally, associated with conflicts in sharing them. Such a situation could be *in vogue* in many parts of the world, and there had been attempts to study them. A typical case is the fishing grounds in Tam Giang Cau Hai lagoon in Thua.

CBFM, particularly through FA, was introduced and established in Thien Hue, Vietnam in 2003. As of 2011, there were 62 local FAs in the region.

Incidentally, the local FAs played a significant role in the implementation of CBFM, and also as a major collaborator of the government in the protection of aquatic resource and in fisheries management.

In fact, various such research works were based on primary data collected through a questionnaire survey followed by interviews with the key informants, accompanied by efforts to highlight the demands of implementation, level of awareness and participation, effectiveness and weakness and methods for measuring the transaction costs, etc. In addition, there are collections of secondary data and literature, etc., on comanagement and fisheries management with different types of *modus operandi.*

Certain such studies often present the existing legal framework, which often is quite clear and also innovative for supporting CBFM development and implementation. However, some constraints still prevail, such as poor level of education, lack of budget and a low rate of women participation, as found in Giang Xuan FA, although there was an overall level of satisfaction with enhanced income of the fishers. However, many of the Giang Xuan FA members were little reluctant and passive in developing action plans and providing with ideas. In fact, their roles were limited to the terms of providing suggestions and recommendations in monitoring and in evaluation of the management process. Moreover, such studies also tried to demonstrate the "**transaction cost concept**" and approach for assessing such costs in fisheries comanagement systems. Incidentally, transaction costs in fisheries comanagement are the total cost of three items that consist of:

(a) information collection; (b) collective fisheries decision-making; and (c) collective operation.

Notwithstanding the above, there are three stages in incurring transaction costs in fisheries comanagement as follows:

(a) Stage I: Mainly focuses on meetings/discussions associated with recognition of new management regime.
(b) Stage II: Mainly focuses on improved education and amendment of regulations.
(c) Stage III: Concentrates on monitoring and patrolling.

Incidentally, the transaction costs in Stage III are lower than those of Stages I and II in Giang Xuan FA. Many such studies usually try to conclude with certain recommendations for improving local FA systems, and *modus operandi* of implementation of CBFM.

5.1.20 Asia: Summary

(1) Management is a significant aspect of harvest and conservation of resources. Effective management *modus operandi* is a prerequisite for sustained harvest, utilization, and conservation of resources. Many countries in the world have come up with policies and frameworks to develop strategies and regulations for aquatic resource management, in general, and fisheries management, in particular.

(2) In **Asia**, fisheries management is a significant component of aquatic resources management. The broad aim is to improve the economic efficiency of the fisheries. There could be various kinds of strategies for fisheries management.

(3) Fisheries are nutrition and avocation-related, interdependent ecological and social systems. They usually call for integrated management approaches. Managing common resources may require continuous and conscious efforts by a wide range of stakeholders. Collective efforts are often a prerequisite for the development of community-based institutions and the devolution of authority which generally trickles down vertically from the central stage to the local grassroots level.

(4) The **Asian continent** is gifted with rich varieties large number of water bodies and a bewildering diversity of BD including fishes. As such, there have been different kinds of *modus operandi* for the management of the water bodies and their bioresources. Different researchers have been working on the various types of management *modus operandi* including **CBFM** in Asia.

(5) Flewwelling (2001), Wilson and Kleban (1992), Estellie Smith (1991), Chris (1991), Pomeroy, et al. (1998), Ostrom (1990, 1992), Pomeroy et al. (1996), Ahmed et al. (1997), Katon et al. (1997), Baticados and Agbayani (1998), Katon et al. (1997, 1998), Masae (1998), Khan and Apu (1998), Katon et al. (1997), Pomeroy and Pido (1995), etc., dealt with various aspects of CBFM in **Asia in General.**

(6) Ruddle (1993), Blair and Robertson (1903-09), Lopez (1985), Santos (1980), Tom'tavala (1990), Ruddle (1985, 1990), Johannes (1988), Adams (1993), Amos (1993), Wright (1985, 1990), Kar (1990, 2007a,b,c, 2013), Bailey and Zerner (1992), Zerner (1991), Sims (1990), McCutcheon (1980), Toloa et al. (1991), Akimichi and Ruddle (1984), Teiwaki (1988), McCutcheon (1980), Toloa et al. (1991), DaN (1989), Warren (1989), Slkkerveen (1989), etc., dealt with various aspects of CBFM in **Asia-Pacific Region**.

(7) Minkin et al. (1997), Thompson et al. (1999), Islam and Dickson (2007), Ullah (1985), Naqi (1989), McGregor (1995), Sen (1997), Berry (1989), Blaikie (1989), Moser (1998), Sen (1981), O'Riordan (2013), Ahmed and Dickson (2007), Mustafa and Halls (2007), Kabir and Hassan (2007), Gomes (2007), Islam and Rahman (2007), Paul and Chowdhury (2007), Rick Gregory (2007), Paul and Chowdhury (2007), Rick Gregory (2007), Nathan and Apu (2007), Arne Andreasson (2007), Dey and Kanagaratnam (2007), etc., dealt with various aspects of CBFM in **Bangladesh** including, notably, popular press featuring articles on the crisis in global fisheries; institutional issues in CBFM project, which were vital in achieving its long-term goal of sustainable improvement of the livelihoods of poor people dependent on aquatic resources; the Bangladesh Environmental Lawyers' Association (BELA) providing legal support to the CBFM projects; the NGO, Banchte Shekha' supporting the development and empowerment of poor people, particularly the women, depicting a lack of understanding of the underlying causes of wetland resource depletion including their productivity; management of Aquatic Ecosystems through Community Husbandry (MACH); various floodplain AQC projects springing up in the Comilla District; choosing common property resources (CPR) members from the resource users with clearly defined rights to use the resource with defined physical boundaries; various symposia held in Bangladesh including Recommendations for CBFM in Bangladesh from the CBFM International Conference; improve socioeconomic condition of the people through use of seasonal water logged paddy fields during rainy season for short-period AQC including evaluation of the culture stocking strategy and cost-benefit analysis for good CBFM activities in waterlogged areas of Begumganj upazila, Noakhali District indicating $>60\%$ of the fishers in the project areas experiencing significant increases in their fishing incomes in **Bangladesh (B'dsh)**. Further, Khan and Apu (1998) dealt with fisheries comanagement in the Oxbow lakes of **Bangladesh**.

(8) Anon (2016), Nuov (2015), Limsong and Sitha (2015), De Silva (2003), Silva and Song (2015), Anon (2015), etc., worked on different aspects of CBFM in **Cambodia** including, notably, the learning journey in River Mekong with the Blue Peace in the Middle East; CBF; challenges and constraints for developing community-based fisheries and a possible strategy for success; support program for scaling up community forestry and fisheries under a new partnership program with contribution from Switzerland; and so on.

(9) In **China**, Shi (1996), Liu et al. (2008), Weimin (2008), Liu and He (1992), Liu and Huang (1998), Liu et al. (2001), Wu (1998), Dai and Xu (1996), Xiong (1996), Gao et al. (1997), Hu et al. (2005), Hewett and Johnson, 1992, Brandt and Hartman (1993), Peng et al. (2004), Paterson and Pernetta (2015), Huang et al. (2001), and

Wang et al. (2015) worked on various aspects of CBFM and related aspects; notably, information about earliest stocking activity of fish seed collected from rivers; information providing fish as nutritional supplies and rural livelihoods; potential for reservoir fisheries development in China; integrating fisheries and habitat management; information regarding fish resources in the **Chinese** reservoirs and their management and utilization; CBF in lakes of the Yangtze River basin; establishing PAs for the reproduction of fishes; employment of displaced people, thus enhancing economic efficiency; tackling of Virulent Epizootic Ulcerative Fish Disease (EUS) problem (Kar, 2015); and so on.

(10) Ni and Kwok (1999), Morton and Tseng (1982), Morton (1990, 1992, 1997), Fowler (1930a,b,c,d,e, 1931a,b,c,d, 1938a,b), Shen (1964), Chan (1968), Gaiger (1974), Chan (1984), Chan and Wong (1989), Thompson and Horikoshi (1982), Thompson et al. (1982), Wong (1982), Chong (1984), Leung (1991, 1994, 1997a,b), Ni (1994a,b, 1995a, 1997), Lee (1997), Hong Kong Agriculture and Fisheries Department (1998), Chan and Tseng (1982), Wu (1984), Richards and Wu (1985), Lam (1990c), Wang et al. (1992, 1993, 1994), Kwok and Ni (1999), Nelson (1994), Eschmeyer (1998), Chan (1966a), Chan (1966b), Chan (1966c), Chan (1968), Chan (1970), Chong (1984), Chan (1984), Hodgkiss (1988), Chan and Wong (1989), Lam (1990a), Lam (1990b), Lam (1990c), Leung (1991), Leung (1994), Hodgkiss (1996), Leung (1997a, b), Lee (1997), Leung et al. (1997), Anon (1999) worked on various aspects of CBFM and related aspects in **Hong Kong,** notably, comprehensive database on local fish fauna for the planning and implementation of management and conservation strategies of the natural resources, notably, the water bodies and the fishes in Hong Kong; studies of Hong Kong Fishes; fish lists and sporadic fishing surveys; legislative issues and legal counseling; formulation of a **fisheries management** strategy through public consultation exercise including **CBFM** involving the stakeholders from the grassroot level in a participatory mode; settingup of nursery and spawning ground protection areas and aquatic sanctuaries; efforts toward habitat restoration and habitat enhancement; restocking exercise to help in rebuilding the depleted fish stocks; setting up of WG on fisheries management by the Agriculture and Fisheries Department (AFD) including high priority management measures; maintenance of a steady supply of fresh fish to local consumers through culture and capture fisheries; EUS fish disease problems (Kar, 2015); programs for livelihood of the **fisher folk**; dealing with heavy trawling pressure; **community involvement** in management and enforcement; and so on.

(11) **In India**, Michael (1969), Banerjea (1967), Jhingran (1991), Dey (1973, 1981), and Dey and Kar (1989a, b, c, 1990, 1994) are some of the earliest researchers to study the limnology, fisheries, and management aspects of the Wetlands and "Beels" (as called locally in Eastern India).

Incidentally, **enormous amount of works** have been done on the limnology, fisheries, including fishing gears, socioeconomics of fisher folk, and management aspects of the wetlands (Beels) and rivers in Northeast (NE) India by Kar (1984, 1990, 1996, 1998, 1999a,b, 2000a,b, 2001, 2002a,b, 2003a,b,c, 2004a,b,c,d, 2005a,b,c,d, 2006a,b, 2007a,b,c, 2010a,b,c, 2012a,b,c, 2013, 2014, 2015, 2016).

Incidentally, it needs a special mention that on the basis of the research findings of the present author, Sone Beel (the biggest wetland in Assam) was declared as a **"Wetland of National Importance"** (Kar, 2013) by the Ministry of Environment and Forests and Climate Change, GOI on October 16, 2008 (*vide* Resolution No. 11 dt. 16.10.2008, of MOEF Expert Group Meeting; and, also, Letter No. FRM 41/2008/63-A dt. 8.9.2008 from Commissioner and Secretary to the Govt. of Assam, Department of Forests).

Kar and his associates did further works on the limnology, fisheries, fish parasites, zooplankton, and their culture; and management aspects of the wetlands and rivers in India, in general, and in NE India, in particular (Kar and Dey, 1982a,b, 1986, 1987, 1988a,b,c,d, 1991, 1993a,b, 1996, 2000, 2002; Kar and Barbhuiya, 2000a,b, 2001, 2002a,b,c,d, 2004, 2005, 2009, 2011a,b, 2013; Kar and Mazumdar, 2004; Kar and Sen, 2007; Kar and Kar, 2016a,b; Kar et al., 2018; Kar et al., 1996a,b,c,d; Kar et al., 1997; Kar et al., 1999c; Kar et al., 2000a; Kar et al., 2001b; Kar et al., 2001a; Kar et al., 2002a; Kar et al., 2002d; Kar et al., 2002g; Kar et al., 2002c; Kar et al., 2002b; Kar et al., 2003e; Kar et al., 2003g; Kar et al. (2003b); Kar et al., 2006c; Kar et al., 2006a; Kar et al., 2003d; Kar et al., 2007a; Yashmin et al., 2007; Kar et al., 2008; Kar et al., 2009; Kar et al., 2010a,b; Kar et al., 2006d; Kar et al., 2011; Puinyabati et al., 2010a,b; Singha et al., 2010; Binky and Kar, 2010; Ranibala et al., 2011a; Geetarani et al., 2011; Sangeeta et al., 2010a,b; Kar and Kar, 1989; Kar and Kar, 1994; Kar et al., 2002g; Kar and Kar, 2004a,b; Kar et al., 2003a).

Concomitant to the above, enormous amount of works 2002 on the *hitherto* unknown **EUS** have been performed by **Kar, one of the pioneersong worker in the world, along with his associates,** and almost all of them have been mentioned in an earlier publication of a book by Elsevier, USA written by the present author (Kar, 2015).

Kar, notably, Kar and Dey (1988b), Kar and Dey (1990a,b,c), Kar and Das (2004a,b), Kar and Upadhyaya (1998), Kar et al. (1990a,b, 1993, 1994, 1995a,b, 1999a,b, 2000a,b,c,d, 2003a,b, c, d, e, f, g, h, i, j, k, 2004a,b,c, 2006a,b,c,d, 2007a,b,c), Kar et al. (1996d), Kar et al. (1998a,b,c,d), Kar and Upadhyaya (1998), Kar et al. (1999a), Kar et al. (2000a,b,c), Kar et al. (2002e,h,i,j), Kar et al. (2003c,f), Kar et al. (2004a,b,c), Kar et al. (2006b), Kar et al. (2007b,c), Kar et al. (2015), Riji et al. (2016).

Concomitant to the above, some of the works done specifically on CBFM **in India** are, briefly, those of Barman (2004), Dana (1987), Sheikh et al. (1993), Shrivastava and Bhattacharya (2003), Sugunan and Bhattacharjya (2000), Barman et al. (2008, 2012), CIFRI (2000), Barman and Kar (2014), Sugunan and Sinha (2001), Kurien (1978, 1995), Kurien and Achari (1990), Shajahan (1996), and so on.

Incidentally, Chandra (2011) worked on an evaluation of the management regimes and institutional arrangement in floodplain wetlands fisheries of Assam. Datta et al. (2010) worked on the perspectives of fisheries sector interventions for livelihood promotion in different places. Natarajan (2005) dealt with a case study in India reporting about enabling better management of fisheries conflicts. Prabhu (2011) dealt with institutional dynamics in community-based fisheries resource management for sustainable development of marine fisheries in Kerala. The FAO of the UN, partnered with the SIFFS, attempted for pilot trials on CBFM as part of the fisheries livelihoods activities under the UN tsunami recovery support program (Anon, 2011a,b). Lobe and Berkes (2003) dealt with the padu system of CBFM with regard to changes and local institutional innovation(s) in South India. It is said to be caste-specific, gear-specific (stake nets), and species-specific (shrimp). As used in Vallarpadam and elsewhere in Kerala, TN, and SL, the padu is characterized by the use of lottery for rotational access to those system(s) and characterized as rapid rural appraisal. Incidentally, the **Pulicat Lake fishery in TN** shared some of the characteristics of the SL case. The three villages carried out their lottery activities at the local level. Mathew (1991) and Ramachandran (2005) dealt with development of model fisherman villages. Vass and Sugunan (2009) dealt with status of reservoir fisheries in India, in general, with brief notes on their management aspects. Katiha (1994) dealt with natural resource management as a case study of reservoir fisheries in Himachal Pradesh. Vipinkumar et al. (2013) reported about comanagement paradigm and sociological issues in RFM in Indian marine fisheries sector. Concomitant to the above, Mohamed (2004) reported about march toward ecosystem-based management of marine fisheries with an aim toward building mass balance trophic and simulation models. Sampath (2005) dealt with an overview of the International and National Fishery Policy and related issues. Anon (2004) dealt with the policy briefs of good management practice in sustainable fisheries in India. Anon (2006a,b) prepared the report of the WG of Fisheries for the 11th Five year plan, 2007–2012. **The PC** of the GoI had constituted a WG on Fisheries to report on the various objectives, schemes, and programs including their monitoring and implementation. The PC of the GoI, in the planning process had emphatically stressed upon the sustainable harvest and utilization of the water bodies and their bioresources through efficient CBM *modus operandi* involving the stakeholders from the grassroots level. As such, effective measures are to be in place for all-round development of these poverty-stricken stakeholders, particularly, the fisher folk. In fact, sustainable exploitation of fish stocks, in these waters, could be achieved through community **participation, comanagement, and CBM**. Incidentally, the National Fisheries Development Board (NFDB) had been established, which could give a major fillip to the Indian fisheries sector. Kumar et al. (2003) reported about a profile of the people, technologies, and policies in fisheries sector in India. Anon (2010a,b) dealt with the marine fisheries of India with regard to their issues, opportunities, and transitions.

Concomitant to the above, some of the other workers who have worked on various aspects of limnology, fisheries, and management aspects of the water bodies, and whose works are related to India and elsewhere are as follows:

Ayyappan and Diwan (2004), Nichols (1928), Darlington (1957), NEC (1991), Vaas and Schurman (1949), Welcomme (1979), Bennet, (1962), Swingle (1950), Vass (1997), Stevenson (1991), Pido (1996), Chambers (1997), Kurien (1992), Johnson (2002), Gadgil et al. (2000), Government of Kerala (1993), Kurup et al. (1993), Kurien (1992), Berkes (1986), Alexander's (1995), Ostrom's (1990), Mathew (1991), Amarasinghe et al. (1997) (Sri Lanka), Pal (1979), Parameswaran and Murugesan (1984), Paul and Sugunan (1990), Dubey and Chatterjee (1977), and so on.

Incidentally, the fisheries management authorities in **India**, and also in some other countries started to recognize that a fishery may not always be managed effectively without the cooperation of fishing communities.

Notwithstanding the above, lessons are also said to be emerging from **India's** traditional community-based fishing. In fact, community-based cooperative fisheries management is a viable option in some of the artisanal fisheries in developing countries. Nevertheless, a key component emerging from the pilot programmes in

India is the bottom-up management of fisheries resources by those who depend on them, those who have the most knowledge about them, and those who have the greatest stake in protecting them. This bottom-up approach coupled with the need of the GoI to introduce and enforce more fishing zones along the coastline as part the new National Fishing Policy. This had showed more promise for stronger property rights and healthier fisheries.

(12) In **Indonesia**, some of the works on CBFM in the post-conflict situation was reported by Sidayat (2008) as a case study in coastal villages of North Malukud in **Indonesia**.

Further, Soede et al. (2001) dealt with integrated marine management concessions with regard to a new approach to an old problem.

(13) In **Japan**, Schmidt (2003) dealt with fisheries and Japan as a case of multiple roles.

Yamamoto (1996) worked on CBFM in Japan and stated that, over the past 250 years approximately, **Japanese** fisheries regimes have developed three fishery laws in sequence in which a fishing right system was commonly used.

(14) In **Korea**, an account of the Integrated Fisheries Risk Analysis Method for Ecosystems (IFRAME) approach was briefly followed by Chang et al. (2011). Christopherm et al. (2006) worked on the impacts of climate changes on fish BD. Hong et al. (2010) dealt with land use in Korean tidal wetlands with a note on their impacts and management strategies. Douglas Clyde Wilson et al. (2006) had worked on the cross-scale linkages and adaptive management and had reviewed the researches done in Asian countries during the second phase of the Worldwide Collaborative Research Project on Fisheries Comanagement. Arnason (2000) initiated works on **ITQ** system in Korea. This system was done as part of "Fisheries resource and oriented comanagement system for more effective management plan" for the implementation of responsible fisheries.

(15) In **Lao PDR**, Baird (2000) studied integrated community-based fisheries Comanagement and PAs management with regard to opportunities for advancement and obstacles to implementation. Phounsavath (1998) reported about community-based fisheries management as an approach to participatory management of reservoir fisheries exemplified by a case study of two fishing villages at the Nam Ngum Reservoir in Lao PDR. Anon (1997) dealt with the reservoir fisheries management in Savannakhet Province of **Lao PDR**.

(16) In **Malaysia**, Nasuchon (2009) reported about coastal management and community management.

(17) In **Middle-East**, Izzat (1998) dealt with fisheries development in the Arab World including the coastal waters of the Arab States, inland waters, and AQC potential, and opined that they are significant sectors for the development of the **Arab countries** and their people.

(18) Anon (2012) worked on the National Fisheries Strategy (2012–2025) of the Republic of **Yemen**.

(19) In **Myanmar**, Anon (2005) dealt with wetland conservation.

(20) In **Nepal**, Gurung (2007) worked on the restoration of small lakes through cooperative management as a suitable strategy for poverty-laden areas in developing countries. Pradhan (2009) reported about the reservoir fisheries in Nepal.

(21) In **The Philippines**, Baticados and Agbayani (1998) observed that in San Salvador Island and Malalison Island, the marine sanctuaries had boundaries identified with buoys to inform outsiders of its existence to more easily monitor the area by making and changing community's rules as well as by sharing responsibility among the community members for guarding the marine sanctuary with high level of enforcement of rules. Further, Foltz et al. (1996) mentioned that, in the **Philippines**, NGOs usually spend a great deal of time in social preparation, i.e., educating the fishers about the importance, benefits, and economic implication of fisheries comanagement in comparison to different fisheries management and development strategies. van Mulekom (1998) dealt with capacity and opportunity building in a small-scale fisher folk community for community-based coastal resources management as a community development approach to CB-CRM. Pomeroy et al. (1998) dealt with fisheries comanagement as key conditions and principles drawn from Asian experiences. Katon et al. (1998) worked on the mangrove rehabilitation and coastal resource management of Mabini-Candijay as a case study of fisheries comanagement arrangements in Cogtong Bay, the Philippines. Pomeroy et al. (1996) worked on the impact evaluation of community-based coastal resource management projects in the **Philippines**. Baylon (2001) further studied the microenterprise development in selected fishing communities in the province of Iloilo, the **Philippines**. He worked on the funding support for the development of microenterprises in Banate Bay, Iloilo and Southern Iloilo. This project was implemented by the University of the Philippines in the Visaas in coordination with the Banate Bay Resource Management Council, Inc. and the Southern Iloilo Coastal Resource Management Council. There have been works dealing with the policy reforms and community-based programs to combat cyanide fishing in the **Philippines**. Anon (2011a,b) dealt with a review of the National Laws and Legislations in the **Philippines** in relation to EAFM.

(22) In **SL**, there have been works dealing with the reservoir fisheries in SL and related management aspects. Upali S. Amarasinghe and D.E.M. Weerakoon (2015) opined that the inland fishery of SL is essentially a capture fishery in major and medium-sized perennial reservoirs of the country.

(23) In **Thailand**, there have been works dealing with the reservoir fisheries and relevant aspects. Masae (1998) made an analysis of fisheries comanagement arrangements in the case of Ban Laem Makham, Sikao District, Trang Province, South Thailand. Henocque and Tandavanitj (2011) dealt with fisheries and comanagement in practice in Thailand. Pimoljinda and Boonraksa dealt with community-based fisheries comanagement along with a case study from Phang-Nga bay in Thailand.

(24) In **Vietnam**, there have been works dealing with the impact of fishery association management on the fishery community. Nguyen (2012) dealt with community-based comanagement in **Vietnamese** fisheries, with emphasis on the FA in Tam Giang-Cau Hai Lagoon.

5.2 Africa

The **African continent** is said to be rich in water bodies and aquatic resources having myriads of wetlands, lakes, and rivers. This calls for efficient sustainable harvest and management of the water bodies and their bioresources, notably, the fisheries. Africa also possesses a rich AQC potential. However, thoughtful reform was not very common. Different management measures had explained the scope of the environmental and economic benefits that could emerge from amending the structure of a regulatory regime from time to time.

5.2.1 Africa general

"Fish" is a significant resource for nutrition, avocation, and recreation of human kind. Fish is also an item of national and international trade in both live and dead forms. Thus, animal welfare regulations may also be applicable for trade with live fish, along with an urgent need for efficient management *modus operandi* for effective management of the coveted resources. But, logic and evidence(s) are rarely enough to carry the day. As such, there could be a need for proper monitoring of the efficient management measures by some organization(s), and suggest ways and means for their proper and effective implementation. Moreover, constructive changes in policies may often come about at the time of crisis or when there are few special interest groups working to protect an existing management structure.

The Conference of the Parties (COP) to CITES recommends in Resolution Conf. 10.21 (Rev. CoP16) that the Live Animals Regulations (LAR) of the International Air Transport Association (IATA/LAR) and the *CITES* guidelines for the nonair transport of live wild animals and plants suggested to meet CITES transport requirements for live specimens. IATA/LAR regulations include specifications for containers used to transport live animals applicable to a list of >1000 species, including fish.

5.2.1.1 Sustainable management of wild-capture fisheries

Anon (2009) identified global marine capture fisheries as an underperforming global asset. The report concluded that *c.* 75% of the world's fisheries had underperformed in 2004, mainly due to overexploitation. The lost potential revenue was estimated at *c.* US $50 billion annually. However, covering a span of *c.* 30 years, this had amounted to a loss of *c.* US $2 trillion.

Notwithstanding the above, fisheries sector could become a basis for economic growth and a potential resource generation for alternative livelihoods in many countries, including Senegal, through comprehensive reform(s). Concomitantly, a nation's natural capital or bioresource in the form of fish stocks could be greatly increased as well as the negative impacts of the fisheries on the aquatic (both FW and marine) environment could be reduced by adopting fisher-friendly management operations.

In addition to the above, a report by Anon (2012) also illustrated the potential economic gains to be expected from a transition toward sustainably managed fish stocks. A number of developing countries (including Senegal) had attempted to study certain case studies dealing with economic benefits at a local fishery level. For example, one study on Eastern Pacific tuna stocks has demonstrated that the annual revenue of the tuna fleet could be augmented by *c.* US $93 million under better management practices. Another study concluded that the potential economic gains from restoring the stock of the Hilsa Shad fishery, one of the largest single species fisheries in India and Bangladesh, could be to the extent of US $260 million annually. But, unhappily, much economic benefit could not be achieved with the Hilsa fishery in Sone Beel, one of the largest (area 3458.12 ha at FSL) wetlands in Assam and India in the business-as-usual (BAU) scenario (Kar, 1990, 1996, 1997, 1998, 2000, 2002, 2003, 2004, 2007a, b, 2012, 2013; Kar et al., 2008;

Kar and Barbhuiya, 2011a, b; Kar, 2005a, b, c; Kar and Dey, 1982, 1986, 1987; Dey and Kar, 1990; Yadava et al., 1989). Incidentally, there had been a *hitherto* rare occurrence of juveniles (advanced stage of fry) of *Hilsa (Tenualosa) ilisha* in Chatla Haor seasonal floodplain wetland of Assam (Kar and Dey, 2002). Further, the case of the city of Kayar, in Senegal, is particularly illustrative with regard to the potential economic, social, and environmental benefits arising from the sustainable management of fish resources.

5.2.1.2 *Current questions in sustainable fisheries, water management, and higher education*

Notwithstanding the above, Melcher et al. (2013) dealt with the current questions in sustainable fisheries, water management, and higher education.

5.2.1.3 *Fish diversity and assessment of aquatic ecosystems in Burkina Faso BD and conservation*

The objectives were to establish an official list of ichthyospecies, a national database of information on the existing biophysical characteristics of fisheries, the diversity and conservation status of fish species, and benthic invertebrates. The ecological approach was developed to monitor fish and water resources, which could be directly used for future assessments of aquatic ecosystems that could also be incorporated in the national database and serve as the basis for recommendations for changes to policy and practice in order to achieve the national BD strategy.

Incidentally, the introduction of fish species in Burkina Faso was restricted by law. But, no reliable list of species exists. The official and reliable species list that would be made could be the reference for the implementation of that law. The expected result could be a national database of information on biophysical characteristics of fisheries, the BD and conservation status of the fish species and benthic invertebrates, and the pressures on the fish populations. In compliance with the work plan, there had been attempts to ascertain the following aspects:

(i) The biophysical characteristics of fisheries (river size, catchment size, turbidity, etc.).
(ii) The diversity and conservation status of the ichthyospecies and benthic invertebrates, land use, chemical water quality, etc.
(iii) The pressures on the fish populations, e.g., land use, chemicals, water quality, overfishing, etc.

Concomitant to the above, the total estimated number of described fish species in **Burkina Faso** was 131. The first estimations indicated *c.* 114 described species in the Volta River catchment and the most recent results indicated *c.* 131 species in the Pendjari River, which belongs to the Volta basin. Probably, no evaluation had been done using the criteria of the IUCN Red List. Nevertheless, there were some indications of *c.* 10 threatened species that had not been observed since 1968. Moreover, *c.* 8 species are considered as endemic (Lévêque et al., 1990).

A tentative list of 54 species belonging to 19 families was prepared. Some of the abundant species are as follows: *Barbus macrops, Brycinus nurse, Clarias anguillaris, Ctenopoma kingsleyae, Distichodus rostratus, Gymnarchus niloticus, Hemichromis bimaculatus, Hemichromis fasciatus, Heterotis niloticus, Lates niloticus, Marcusenius cyprinoides, Marcusenius senegalensis, Oreochromis niloticus, Parachanna obscura, Paradistichodus dimidiatus, Polypterus senegalus, Sarotherodon galilaeus, Schilbe intermedius, Siluranodon auritus, Synodontis punctifer, Synodontis schall, Tilapia zillii, Mormyrus rume*. Some taxa such as *B. nurse* sp., *Hemichromis* sp., *Sarotherodon* sp., *Tilapia* sp., *Clarias* sp., *Barbus* sp., *Labeo* sp., *Synodontis* sp., etc.

5.2.1.4 *Studies on the impacts of policies, legislations, and institutions in the management of fish and water resources*

As a landlocked and Sahelian country, **Burkina Faso** does not have much surface water. The original water bodies and water courses are intermittent. For that reason, the Burkinabe commercial fisheries were not developed until the 1970s. This was favored by the development of artificial lakes following the drought of 1973–74. Then, various policies, laws, and institutions tried to ensure a sustainable management of fish and water resources. Nowadays, it is necessary to assess their interactions and impacts at macro-, meso-, and microlevels. The methodology consisted of literature search at the international and national (macro), regional (meso), and local (micro) levels. There was the use of sustainable livelihoods approach due to its ability to attract attention on the five major components, viz., the context of vulnerability, capital assets, policies, institutions and processes, and strategies and results of livelihoods. Preliminary results indicated that macrolevel policies and legislations were not known at regional and local levels. The national organization in charge of fisheries was unknown because no tangible activity was undertaken in the field or to target the direct stakeholders. The field police fisheries officers and foresters were not inclined to work on fisheries in the areas of monitoring, surveillance, and control. As a result, prohibited fishing methods were more and more used. Actually the methods of management differed depending on the ministry charged to manage fisheries. The occasional nomadism of fisheries institutions (from one ministry to another) was sometimes highly criticized by all the stakeholders. The insufficiencies in the direct stakeholders' organizations and their illiteracy do not allow the regulatory texts to be well known. However, it is hoped that the capacity of the four most important fisheries (Bagre, Kompienga, Sourou, and Ziga reservoirs) will improve.

Notwithstanding the above, the implementation of "closed fishing season" could improve the recruitment of fishes. Four other smaller fisheries are under the concession schemes which give a private body, e.g., the local fishermens' association, the exclusive right of access to the resource. Further, the fisheries stakeholders were often motivated to cooperate for the development of their activities.

5.2.1.5 Gender and fisheries perspectives in the SUSFISH project

Women were involved in the SUSFISH project's formulation and would be involved in its implementation under the direction of a gender and fish expert position from the GDRF. Incidentally, the GDRF took advantage of the Sustainable Fisheries Livelihoods Program in West Africa to increase its capacity in gender.

Women were also targeted in the social part of the research. They had an important stake in the appropriation and the management of water and fish resources. Sustainable management of the water resources and fisheries became an imperative and could not succeed without an appropriate approach that had integrated the basic principles of the gender. The overall objective was to favor the participation of the stakeholders, particularly vulnerable groups, in the process of research in order to guarantee an appropriation and a valuation of the results. In SUSFISH, the gender approach aimed at integrating gender issues at all stages of research. It had concerned integration of gender vision in the protocol and the amendment of the tools to allow the participation and the expression of the needs for all stakeholders, particularly, the vulnerable groups. There were workshop(s) on gender internalization, focusing on how women and young people in work teams could develop the sensibilities and the building of good conditions for women participation in field works. Research on sustainability inspired the team to schedule raising awareness in a frame that encompassed the fisheries sector. Further, the gender profiles presented the disparities of access to the resources between the women (*c.* 20%) and men (*c.* 80%). Further, in certain fisheries used by both migrants and autochthons, vulnerable groups, particularly women had relatively low power of control and decision-making in the use of water and fish resources. This trend had correlated with the increasing imbalance in well-being between men and women, rich and poor, etc. Incidentally, in Burkina Faso, there were *c.* 32,699 fishermen, among whom *c.* 14% were women; *c.* 2983 transformers, among whom *c.* 82% were women; *c.* 3375 wholesale fish merchants, among whom *c.* 54% were women. While men and women complement each other in fisheries activities, many obstacles limit women's activities in fishing. Women are charged with many different tasks which require competence and time, and also on the ground, e.g., production and repairing of nets; processing and selling of fish, etc. As such, traditional roles limit and restrict the activities of women. Further, they are also limited in their time and mobility through their responsibilities for raising the children and for the household activities. Further, the assignment of traditional roles, perceptions, and stereotypes about women make barriers and perpetuate beliefs that affect the participation of women in the fisheries. These trends only recently became evident because research had been for a long time insensitive to gender issues. In fact, the specific needs and the strategic interests of each category of stakeholders were rarely taken into account in many strategies of development. Thus, the principal objectives of many such working teams of today, perhaps, are:

(a) To strengthen the contribution of the vulnerable groups, particularly the woman in the governance of fisheries.
(b) To integrate research better for integration of vulnerable groups and particularly women's and young peoples' roles in sustainable management and conservation of the resources, often involving the community stakeholders from the grassroots level in a participatory mode through the application of efficient *modus operandi* like the CBM.

5.2.1.6 Fisheries health and food security

Different research results often provide a scientific and technical basis to develop methods to assess health risks for fisher folk communities by using the ecological status of waters and fish. Health risks associated with fishery activities are to be taken care of. The main objective of such efforts is "Food supply and health care illustrated in appropriate case studies."

The related factors are to be identified and described. These results are to be incorporated in academic training curricula, and also to be communicated by the team engaged with health policy formulation and implementation. Expected results could lead to critical analysis of connections between fisheries, environment, health, and food security; some scientific and technical basis to monitor the ecology of waters; the factors of risks of riverside communities' health; etc.

Further, poverty leads to poor health status and poor health also contributes to monetary poverty; and all these are related to fishery activities. In many fishery-related places, like **Burkina Faso**, poverty is more rural and *peri*-urban, and are often characterized by recurrent droughts and irregular rainfall. Poverty profile manifests itself in various ways: food insecurity, poor access to clean water, endemic malnutrition, and so on.

Notwithstanding the above, fish could improve the nutritional status. FAO statistics indicated that fish is a source of *c.* 22%–50% of protein in **sub-Saharan Africa**, and may contribute to calorific intake, which may raise *c.* 80 cal daily. There were observations that in 360 households in Burkina Faso, fish intake contributed to improving food and nutritional security. In fact, *c.* 96% of the surveyed households consumed fish and for *c.* 67% of households, fish is often part of the constituents of their daily diet as source of animal protein.

Concomitant to the above, fish is an affordable source of animal protein for the rural and poor population, in some places, although it may not be true for every place. Moreover, assessment of the contribution of fish in fishermen economy showed that in many places fish sale gain is usually used to improve the household food and nutritional security (in *c.* 80% in all study sites). Furthermore, there were also works related to the analysis of the microbiology and quality of the fish consumed in many places. Moreover, there had been assessments of the fishing health risks, water-related diseases, morbidity, etc., among the fishermen communities. Many such parameters of study and research are often included in the curricula of academic programs at various levels. However, further works should be done on the role of women in the arena of fisheries, particularly, in fish management (particularly, in fish processing and fish sale, etc.) as well as in households food security and improving the nutritional status.

Notwithstanding the above, some staff of the Fisheries Department, Ministry of Agriculture/Animal and Fish Resources could also use the research results in their lectures in the national agricultural schools.

The schools could also use the gained knowledge to improve the fisheries course. For this purpose, they could be involved in the assessment of the technical abilities of field level fisheries staff. Some staff of the Fisheries Department, Ministry of Animal and Fish Resources could teach in the three governmental agricultural schools that train field level fisheries staff. Further, there could be fisheries-related technical advisor to the Minister of Secondary and Higher Education of the country.

Therefore, this is an important asset to be used in educational arena. In addition, the university professors could involve certain Master's and Doctoral students to help and monitor the scientific activities in the schools and similar institutions. Further, the students, who seek for field supervision of their research at the fisheries department, could also be trained for further training to others.

5.2.1.7 *Systems analysis of environmental, economic, sociopolitical/cultural factors that influence sustainable fisheries*

The process of participatory activities could have immediate impact on the policy of managing aquatic ecosystems in Burkina Faso. Such activities could continue over a longer period, both in education and in future policy formulation and implementation. Moreover, methods of systems analysis and the research findings could identify leverage points for policy interventions to make fisheries more sustainable, and, moreover, could be incorporated in academic curricula at the university level. Further, future policy processes in Burkina Faso could employ the methods of scenario development in order to determine challenges and scopes in establishing sustainable fisheries management.

Concomitant to the above, a deeper understanding of the complex underlying processes and correlations in socioeconomic and "socio-ecologic" systems could be achieved through the involvement of local experts. The expected results could help in the development of the following:

(a) Plurilingual material (biographic narratives), which could be used for further terminology work on fisheries and ecologic management in Burkina Faso and might be useful for the analysis of the historical dimension in ecological development processes.
(b) Input for innovative approaches in the dissemination strategies.
(c) It could also help in the involvement of various groups of the local society in the research processes.
(d) It could facilitate a fostered dialog between researchers, practitioners, and local community on sustainable fisheries policy.

5.2.1.8 *Expected results*

(a) There could be generation of data and opinions on factors critical to sustainable fisheries management.
(b) There could also be interactions and group exercises with stakeholders on systems analysis of critical factors identifying barriers and bridges to establish sustainable fisheries in Burkina Faso.
(c) These could help in the preparation of report(s) on policy recommendations for establishing sustainable fisheries in Burkina Faso.

Concomitant to the above, Anon (2008) further dealt with CBFM in **Africa, in general**.

Among other African countries, **in Tanzania**, there is participatory approach to irrigation development, which is extended to fisheries and AQC development. Revenues could be generated through the sale and export of wild-capture fisheries. Moreover, enhanced export revenues may arise from the export in an eco-friendly sustainable manner.

Concomitant to the above, Roe et al. (2009) dealt with community management of natural resources (including fisheries) in **Africa** along with impacts, experiences, and future directions.

Natural resources are coveted gifts from the nature for the welfare of the world. As such, they are to be preserved, conserved, utilized, and managed in a sustained manner for maintaining harmony in nature. Also, this exercise could be best done by involving the community stakeholders from the grassroots level.

Notwithstanding the above, no information could be found for **Senegal** in this regard. However, the 1993 Forestry code (Code Forestier) allow participatory management of the natural forests (including fisheries) including ownership of the trees planted, and management of the forests by local Groups (GIEs). Fees, taxes, and fines over forest and fisheries products were to be returned to the local management committees. According to the new code, Rural Councils are allowed to sell rights to cut trees on plots of national domain forest. In 1997, a decentralization policy was adopted providing for local communities to make associations (for different community works including fisheries) (GIEs). Further, the National Environmental Action Plan (NEAP) included CBRNM mechanisms at the Communities' Rural (CR) level (Christopherson et al., 1998).

Notwithstanding the above, Land Policy (2005) allowed for ownership of land by chiefs and families for community activities like fisheries, etc. In addition, not much information could be found for **Somalia**. In **South Africa**, Community Property Association Act of 1996 allow communities to establish legal common property institutions (including fisheries), which could be meant for the development of fisheries and AQC.

Notwithstanding the above, the **Sudan** Comprehensive Peace Agreement (2005) has called for recognition of customary land rights including community lands for fisheries and AQC.

In Kenya, the fishing communities used the traditional ecological knowledge (TEK) and social norms to define the rules for resource management and governance since the 1900.

Not much information could be found for **Swaziland**.

In **Tanzania**, the Fisheries Act (2003) allowed for the establishment of Beach Management Units (BMU) along with FW coastlines. Further, the National Strategy for Growth and Reduction of Poverty, 2005, aimed at improving sustainable use through CBNRM including fisheries. The strategy had placed a strong emphasis on improving local land and resource tenure, and increasing local incomes from natural resources, such as forests, fisheries, and wildlife (United Republic of Tanzania, 2005).

Concomitant to the above, not much information is available on details of CBM of fisheries, although the matter is of high importance today in view of participatory role of the stakeholders for effective management of the coveted aquatic resources, notably, the fisheries.

In FW lakes, wetlands and aquifers, tanks, and irrigation channels, fish, livestock, water supply, and irrigation are all to be taken care of for domestic and commercial use and for all-round efficient management (Adams et al., 2000).

5.2.2 Botswana, Malawi

The CBNRM is an established policy goal of rural development, in various forms, especially in Africa. It is also a simple and attractive one that communities, defined by their rigid spatial boundaries of jurisdiction and responsibilities, by their distinct and integrated social structure and common interests, may be able to manage their natural resources (including fisheries) in an efficient, equitable, and sustained manner. The natural resources in question are usually, though not exclusively, common pool resources.

Incidentally, Piers (2006) dealt with the community-based natural resource (including fish and fisheries) management in **Malawi and Botswana**.

In Southern Africa, these resources are typically forests, open woodland, or grasslands for livestock grazing; wood supply, medicines, farm land for gleaning; crop residues after harvest for grazing; wildlife for game meat and safari incomes; fish in FW lakes, aquifers, tanks, ponds, lakes, streams, irrigation channels, etc., for various livelihood needs, like nutrition and avocation, and also for domestic, commercial, irrigation, and for water supply to people and livestock (Adams et al., 2000).

There have been contributions reflecting results of in-country research on CBNRM in two contrasting African nations, viz., **Malawi** and **Botswana**. In **Malawi**, the rural people endured decades of sustained dispossession by a neopatrimonial despot, and of late, face serious food insecurities and extreme poverty. Incidentally, > 60% of the population live below the poverty line and > 85% of the rural population live on customary land. Illiteracy is *c.* 50%, and *c.* 30% of Malawi's households are female headed. The government, of late, had pursued a program of progressive legislation for forests removing restrictions to the access and use of woodland, and has specifically targeted women as key resource users.

There was a decentralization policy since 1998, a Strategic Plan for CBNRM approved in 2001, and CBNRM was also implemented in forestry and artisanal fisheries.

Concomitant to the above, **Botswana**, a comparatively wealthy African nation, is generally designated as a middle income country with a GDP per capita of *c.* $9500. It was able to provide education, health, and social security, and a minimum level of welfare to its population. However, unemployment and rural poverty are high (*c.* 40%). Botswana has low population-land resource ratios and its government has taken the devolution of powers seriously in order to manage natural resources (including fish and fisheries) since the mid-1980s. This involves CBNRM initiatives since 1998, following assistance from USAID. Incidentally, **Malawi and Botswana** have had very different histories of governance but many rural inhabitants of both the countries had witnessed the growing interference into, and resulting dissolution of, local bodies combined with territorial incursions by the state and private parties to establish plantations and state forests in Malawi.

Notwithstanding the above, the aspect of CBNRM was not largely in use until the 1980s. However, there was a notion that communities could satisfactorily manage their own resources (including and fish and fisheries) according to their local custom, knowledge, and technologies. This has a long history. Incidentally, the ideas of community have constantly been shaped and reshaped by different outsiders from time to time from colonial Governor-Generals, political advisors, European settlers, and more recently rural development consultants and academic writers. Thus, the idea of CBNRM had evolved through time, and had been specific to particular countries. However, over the past 15 years, there had been a convergence of various strands of meanings in the international development literature and in the practice of international funding institutions (IFIs). Recently, social and community forestry and also fisheries in India and Nepal and most countries of Southeast Asia as well as Natural Resource Management Committees in Malawi were found to have some kind of close similarities at a general level. These had resulted from similar strategic policy designs from IFIs. Still, at the level of the details of administrative, legal, and financial structures and of policy implementation, the aspect of CBNRM may mean widely different things to different people.

Nevertheless, it remains powerful today even in the minds of many government officials the significant role of the stakeholders from the grassroots level in the implementation of CBNRM programs (Taylor, 2001, 2002).

5.2.3 Ghana

Ghana in Africa possesses a rich potential in fisheries. Crawford et al. (2013) discussed about solving the fisheries crisis in **Ghana** with a fresh approach to collaborative fisheries management.

Fisheries and AQC, being lucrative vistas, often are faced by disputes, crisis, etc., in both FW and marine domains. Such disadvantageous situations are to be tackled by effective management *modus operandi*, like the efficient CBFM involving the community stakeholders, or through other collaborative methods.

Incidentally, most of Ghana's fishing fleet consists of traditional canoes.

5.2.3.1 *The present status of Ghana's marine fisheries*

Notwithstanding the above, Ghana's marine fisheries capture sector consists of three main types of fishing fleets, viz., the artisanal canoes (mainly, but not all, motorized), semi-industrial boats (wooden-planked vessels including the "china-china" boats), and industrial vessels (large-scale trawlers and tuna boats). However, the canoes employ the most people (*c.* 92% of direct employment) out of these fleets, and accounted for the majority of the total annual catch in 2007 (*c.* 73% of total landings), but decreased to *c.* 28% in 2012 due to the decline of the small pelagic landings.

5.2.3.2 *The importance of Ghana's marine fisheries*

The fisheries sector is significant because it generates > US $1.0 billion as revenue each year, as well as also accounted for **at least *c.* 4.5% of Ghana's gross domestic product (GDP)**. Approximately 210,000 people work directly in Ghana's fisheries sector, and the sector employs, directly or indirectly, *c.* 2.2 million people or *c.* **20% of the population** (Atta-Mills et al., 2004). Incidentally, *Sardinella* sp. and other small fish harvested near the surface (pelagic fish) are of critical concern because these are the staples of the canoe fleet, which employs the largest number of people. The small pelagics make up *c.* 85% of the canoe catch and are the dominant target of the semi-industrial fleet.

Notwithstanding the above, fish is also important as an inexpensive, high-quality protein source. More than **60% of the animal protein in the Ghananian diet comes from fish**, and buying fish accounts for more than one-quarter of expenditures among the poor households. Moreover, Ghana's difficulties in sourcing high-quality, low-cost food protein will be exacerbated by population growth as a net import of *c.* 40% of the fish is needed to feed its people. Thus, if local supplies continue to decline due to ineffective fisheries governance and management, the cost to the nation of outsourcing quality animal protein could increase dramatically.

Concomitant to the above, AQC, while showing promise in proving significant increases in fish supply, produced only 19,092 mt in 2011, which was *c.* 5% of the national supply. Moreover, it is clearly unrealistic to propose that increases from AQC will equal or replace the potential of a well-managed capture fishery any time soon. Looking beyond food security, the poor governance and management also leads to overfishing, which in turn paves the way for a significant loss of wealth. Moreover, the World Bank estimated that with better management an additional *c.* US $50 million annual net economic gain could be generated from Ghana's fisheries (The World Bank, 2009).

5.2.3.3 *The depletion in Ghana's fish landings*

Ghana's marine fisheries was once in crisis. Landings (fish caught and retained) of all stocks declined dramatically over the last decade. In fact, the official national statistics indicated a 30% decline from a high of 492,776 mt in 1999 to 333,524 mt in 2011. However, the reality in coastal fishing communities was much worse than the figures indicated. Incidentally, the shrinking harvest was particularly dramatic for the pelagic, small, sardine-like *Sardinella* sp., which was, by far, the most critical for coastal livelihoods and food security. Catches declined by *c.* 66% from a high of 252,112 mt in 1996 to 84,980 mt in 2011. In fact, this, probably, had occurred at the time of increase in fishing efforts at a time by all fleets.

Notwithstanding the above, worse still, the *Sardinella* catch for the all-important canoe fleet is now *c.* 20,000 mt, down from a peak of *c.* 140,000 mt in 1992. Incidentally, the small pelagic fisheries have typically depicted large swings in annual landings because reproductive cycles are strongly linked to environmental factors associated with highly variable upwelling marine ecosystems, such as in the Gulf of Guinea. Manifestation of this could be seen over many years of fluctuations in the landings data for Ghana. However, the precipitous decline, which is, now seen is well beyond this normal variability and is undoubtedly due to overfishing.

5.2.3.4 *Aquatic desert*

Notwithstanding the above, when the recruitment is poor and the size of the stock (biomass) is low, excessive fishing pressure could result in the collapse of the stock, from which, recovery was unlikely without drastic changes in the management, such as a total moratorium on the fishery.

5.2.3.5 *The reason(s) for the decline in marine fisheries*

There are several reasons for the decline in marine fisheries in Ghana. Any strategy to revert this decline is to be integrated to address this multifaceted and complex situation. **Poor governance and the open-access nature of Ghana's fisheries are the principal reasons for the decline in fish abundance.** These factors have led to a large increase in the number of fishing vessels including the semi-industrial and canoe fleets that primarily had targeted the all-important pelagic species. There was no control on total fishing effort or the number of vessels allowed to fish. Moreover, overfishing occurred as more and more vessels had entered the fishery, until all profitability had dissipated. For the canoe fleet, this situation was exacerbated by the premix fuel subsidy. While fishing subsidies offered an immediate benefit to fishers by lowering operating costs and thereby increasing profits, this benefit was quickly lost under open-access conditions, leaving all operators worse off in the long run. In fact, the subsidies had encouraged more people to turn to fishing and increase effort, thus allowing fishers to keep harvesting the dwindling stocks well beyond a point or threshold limit, whereas in the absence of subsidies, fishing would become unprofitable. This policy had aggravated the overfishing problem. Incidentally, the fast rise in the number of canoes clearly indicate this increased effort.

5.2.4 Lake Victoria

Lake Victoria is one of the biggest lakes in Africa and in the world. Notwithstanding the above, Salehe (2008) reported about the capability and legitimacy of BMUs to improve fisher's income through management of first-hand sales system in **Lake Victoria** in **Tanzania**.

It is a common belief that different management tasks may be suitable for different forms of comanagement arrangements, depending on the particular institutional and organizational setup. There had been studies related to the examinations of the possibility of implementing the marketing-oriented fisheries comanagement in Lake Victoria (Tanzanian side) for Nile perch fisheries. The community-based organization, the BMU), was selected as fishers' representative organization to investigate how could the fishers' marketing performance be improved vis-à-vis reducing poverty in the fishing community.

The results indicated that the globalization of Nile perch fishery had attracted a number of actors to join the fishery, and among them were the middlemen, the persons of the processing plants, and the fishers themselves. In fact, the recent first-hand sales system has greater contribution on poverty alleviation of many fishers; although a few players benefitted from the resources, especially, the investors of the fish processing plants and the fish agents. The so-called deliberate attempts (perhaps) to not identify the actual problems, needs, and scopes, which had existed in the fishing communities, could be better resolved under the comanagement arrangement system, else the nonparticipatory mode of management could make the fishers' community less motivated in the participation of resource management activities. Fight for better fish prices and issuing of credits to fishers were the highly prioritized first-hand sales initiatives that were proposed by the fishers' community to be undertaken by the BMU organization. Capacity building for the BMU is highly needed in order to make the organization undertake its responsibilities efficiently and effectively. Theories used in such studies include, among others, the poverty theory, comanagement, SLA, and middlemen theory.

5.2.4.1 *Beach management units (BMU) in East Africa*

Notwithstanding the above, **Lake Victoria** is one of the biggest lakes in the world. BMUs for Lake Victoria in East Africa emerged as an effective fisheries management strategy. BMUs are organizations that bring together those involved in a fishery, including boat owners, processors, traders, and boat and net repairers, to work with the government in managing fisheries resources. Each BMU has at least 30 boats, an assembly and committee, and legal authority over BMU activities. Incidentally, there were > 1000 BMUs for Lake Victoria (Lake Victoria Fisheries Organization, 2010).

Concomitant to the above, the fishing in **Lake Victoria**, which covers parts of Tanzania, Uganda, and Kenya, was previously "managed" by the national governments, which viewed fisheries resources as common property. Anyone could gain access to the resource. This had led to increasing fishing effort (in terms of number of fishers and vessels) and declining catches (Bulayi, 2001). In addition to overfishing, introduction of exotic species, poor land use practices, and pollution from various sources, all contributed to the depletion of fish populations (Lynch, 1996). Because of the enormous ecological and social cost of this loss, BMUs had become a more attractive option to policy makers.

Notwithstanding the above, the BMUs emerged in **Uganda** in 2003. The duties of the BMUs in Uganda cover the daily management of the local fishery, issuing fishing permits, registering fishing gears, and working with the government fisheries department to enforce regulations against unlawful fishing practices. Further, the BMUs also collected fishing data to help guide their management decisions. In this connection, Waldman pointed out that the local committees were allowed to keep *c.* 25% of the money generated from licenses and landing fees to fund their operations [**World Resources 2005** (WRI, 2005, 65–68)]. These local institutions demonstrate that decentralization could be ecosystem-friendly and could serve the interests of the nation's fishers.

Notwithstanding the above, the BMUs had declared three nonfishing zones, designed to protect known nursery areas, in order to help maintain the fish stock. Additionally, the fishing statistics that BMUs in Uganda had collected show the need for greater local awareness of the people to reduce fishing pressure and, therefore, to control unlawful fishing. However, it is too early to assess, if these improvements in management could translate into more income for local fishers. Incidentally, there were anecdotal reports of higher daily catches (Lake Victoria Fisheries Organization, 2010).

Notwithstanding the above the most significant benefit of devolution of resource management to local levels was that it had given the people a voice and a direct incentive to manage resources, whereas they previously had none. In Uganda, for example, the local culture discouraged the women from joining fishing crews and, more broadly, owning property. But, under BMUs, some women started fishing from the shores of Lake Victoria. Moreover, a few had become boat owners and a handful are now the official members of a BMU (WRI, 2005). This was a significant step, as the absence of rights to land and resources, particularly, affected the poor women, who missed opportunities for economic and environmental development. Incidentally, the World Bank's Agenda for Development pointed out that agricultural output in many **sub-Saharan African countries** could have increased by up to *c.* 20%, if women had equal access to agricultural inputs (Saito et al., 2005). However the most significant benefit of devolution of resource management to local levels was that it had given people a voice and a direct incentive to manage resources, where they previously had none.

Notwithstanding the above, the BMUs are said to be just one example of CBNRM. Incidentally, the fisheries management authorities in **India** and also in some other countries have started to recognize that a fishery may not always be managed effectively without the cooperation of fishing communities.

Notwithstanding the above, the property aspect came in the form of **BMU around Lake Victoria**. In this connection, it could be opined that if a community of fishers exhibit a high degree of social, cultural, and economic homogeneity, then they would be well positioned to successfully manage the resources (Nobel Laureate Ostrom, 1990). This appeared to be the case for the BMUs of Lake Victoria. These arrangements indicate that decentralization could be ecosystem-friendly and could serve the interests of the local fishing communities. The BMUs had managed to empower the locals successfully, fence off critical breeding grounds to anglers, and generate income from licenses and landing fees to fund their operations. Further, effective decentralization requires, at a minimum, that local institutions, irrespective of whether they are official bodies such as village councils or more informal arrangements such as cooperatives, are formed on principles of representation and are accountable to their constituents. Even more important was that the state could grant these local institutions actual authority over resources, something that was far from common.

5.2.5 Madagascar

Madagascar possesses expertise in local management of its fisheries resources. Anon (2013a, b, c) dealt with protection and promotion of marine biodiversity through small-scale, time-sensitive, community-based programs in **Madagascar.**

MCAF supports protection of locally managed marine area in **Madagascar**. The Marine Conservation Action Fund (MACF) had been supporting, since 1999, > 100 small-scale, high-impact projects in 30+ countries, and had disbursed *c.* $600,000 for the conservation of endangered species and habitats. MCAF is made possible through generous support from the Oak Foundation, the Curtis and Edith Munson Foundation, and individual donors.

The "Vezo" or the "people of the sea" head out for daily fishing trips in the handmade boats, known as "pirogues." In fact, preservation of the traditional culture of the Vezo was a prerequisite for the protection of the marine resources of the Velondriake region.

Notwithstanding the above, the waters of the Velondriake region off the SW coast of Madagascar encompass a diversity of habitats including seagrass beds, coral reefs, and mangroves. These ecosystems provide food and livelihoods for the Vezo people, who inhabit the region. Concerns arose almost a decade ago that increasing pressures from a growing population as well as overfishing threaten to deplete these coveted resources. The local villagers, given these concerns, have begun experimenting with seasonal closures of selected fishing grounds. Also, significantly, the closures were successful, ultimately generating higher yields for the fishermen. The early success of these closures coupled with the cooperative management among the villagers had led to the formation of the Velondriake Association, composed of representatives from 25 villages who had been overseeing the management of their shared marine resources. Thus, in addition to continuing with the seasonal closures, the association, being supported by the community, established six permanent marine protected areas (MPAs), which, based on scientific surveys, were deemed to be ecologically significant. Once the PA was established, it was important that the boundaries of the permanent no-take areas be clearly marked to allow for effective enforcement. Initial attempts to demarcate these areas were hampered by the loss of the buoys, which were not well suited for the sea conditions.

Notwithstanding the above, the MCAF issued grant in support of FishNET, a new tool to combat illegal fishing. In fact, **illegal, unreported, and unregulated** (IUU) fishing is a worldwide problem facing our oceans today. IUU fishing could occur in many forms, such as failing to report catch, exceeding quotas and bycatch limits, fishing in PAs, using prohibited gears, etc. In fact, it presents a grave threat to sustainability of fisheries, healthy oceans, and the livelihoods of the many fishers, who comply with the law. However, there are a number of efforts underway to combat IUU fishing, including an innovative, technological approach called FishNET developed by Shah Selbe, an engineer with Boeing Space and Intelligence Systems.

As Mr. Selbe noted, "the foundation of the FishNET effort is an integrated system that completely revamps the way we handle the information about our oceans." One could imagine a system that would harness the power of a networked society to allow users to report ocean threats via the Internet, SMS text messaging, phone calls, etc.

Concomitant to the above, it may be noted that innumerable marine species are known to inhabit the waters of the Velondriake region. Many of them are threatened or endangered, and a few others are endemics. Also, CBFM could be an efficient *modus operandi* for the successful management and conservation of such coveted ichthyospecies. The FishNET concept could be an innovative and comprehensive technological tool for mitigating the threat of illegal fishing. It could also have the immediate device of alerting local and migrant fishers of the protected area's boundaries, and could allow for increased control of poaching. Moreover, in the long run, securing the PA could contribute to sustainable fisheries and a healthier and more productive ecosystem.

Notwithstanding the above, in the teeth of injudicious overfishing as well as due to high demand for shark fins, the scalloped hammerhead shark (*Sphyrna lewini*), which is now included in Appendix III of CITES (Convention on International Trade in Endangered Species) has been threatened. Also, only community-based *modus operandi* could protect such threatened species through the involvement of the community participation from the grassroots level. Incidentally, the "Misión Tiburón" is making some efforts to improve protections for the scalloped hammerhead shark in Costa Rica.

Concomitant to the above, McClanahan et al. (2014) dealt with the perceived benefits of fisheries management restrictions in **Madagascar**.

Efficient management is an essential aspect in fisheries and AQC. Madagascar has a rich potential in fisheries and AQC. Incidentally, perceptions of the benefits of fisheries management restrictions had been evaluated in coastal Madagascar to identify restrictions that were likely to be self- and community enforced.

There had been surveys focused on 24 Malagasy fishing villages adjacent to coral reefs. Studies revealed that resource users' perceptions of the benefits of restrictions were generally high and widespread but some less positive perceptions were found in three villages located near MPAs. Perceptions of the benefits of gear restrictions had widespread support. However, closed areas, closed seasons, and minimum sizes of fish were less common, and thus restrictions on species were supported only infrequently. It is, therefore, often advocated for a management implementation approach that uses these scales of perceived benefits and prioritizes support for the most widely accepted restrictions most broadly, with the less accepted restrictions matched to specific supportive locations. However, the socioeconomic and wealth variables were not very clearly associated with perceived benefits at the village level, which result from a stronger influence of village history than the socioeconomic conditions. At the individual fisher level, however, there are evidence(s) that experienced people involved in decision-making, having livelihood alternatives, and having permanent housing had other opinions and also frequently were more supportive of management restrictions. As such, incorporation of such information into forums and management plans may not always increase the rate of adoption and compliance with needed fisheries restrictions.

Incidentally, Anon (2015) dealt with community-based AQC as a pioneering viable alternative to fishing.

"**Community-based AQC: a pioneering viable alternative to fishing**": "a model for everyone working to conserve the natural life-support systems of our troubled planet" as had been said by Sir David Attenborough.

Today, there is a need to diversify traditional small-scale fisheries; and this also seems to be fundamental to the livelihoods and food security of innumerable people throughout the world. Yet, the marine ecosystems and the fisheries they support are said to be facing unprecedented pressures. In addition to capture and culture fisheries, there had been a crucial need to diversify the livelihoods of the people earning their livelihoods from the aquatic resources to other types of trades in the aquatic domain, notably, the diversified AQC practices, mainly because of soaring demand for aquatic foods. This could also reduce pressure on some of the common type of aquatic resources, notably, the fishes. Incidentally, in various places, the geographic isolation and the typical climates portray few economic scopes beyond fishing.

Notwithstanding the above, *c*. 90% of global fish stocks are either overfished or fully fished. Incidentally, the demand for fish as a source of protein is expected to rise by >20% by 2030. The AQC now accounts for *c*. 50% of the total fish supply for human consumption.

5.2.5.1 *Farming the sea*

Notwithstanding the above, "Blue Ventures" is working with coastal communities and private sector AQC trades in Madagascar to develop viable livelihood opportunities, which could be appropriate to the local environment and culture. In addition, the community-based AQC initiative(s) has been providing coastal people in SW Madagascar with new sources of income, allowing improved access to food and education, while at the same time mitigating pressure on marine biodiversity (BD). Further, Blue Ventures connects isolated coastal communities with lucrative international markets for seaweeds and sea cucumbers, through collaboration with the University of Toliara's Marine Science Institute (IHSM), local seafood exporter Copefrito, and AQC company in the Indian Ocean Trepang (IOT), thus enabling families to develop their own AQC businesses. In addition, the AQC specialists trained >700 people to become farmers of sea cucumbers (*Holothuria scabra*) and red "cottonii" seaweed (*Kappaphycus alvarezii*). In fact, the sea cucumbers (called "trepang" after processing) are in high demand in Asian markets, where they are considered as a delicacy, health food, and aphrodisiac, while the red seaweed is widely used in food and cosmetics industries as a texturing agent. In this connection, it may be noted that *c*. 700 people were trained to farm sea cucumbers and seaweeds. Some of them belong to the "Savings Women Associations" from villages and farming associations.

5.2.5.2 *Why seaweed and sea cucumbers?*

Some of the reasons are as follows:

(i) Well-suited to Madagascar's extensive shallow coastal lagoons.
(ii) High demand from lucrative international markets.
(iii) Farms may operate with low running costs.
(iv) Production methods are not much complicated, requiring minimal initial training and producing less adverse environmental impacts.
(v) An established network of business and research partners provide hatchery technology and supply, assured access to markets, and technical expertise to maximize the benefits to communities.
(vi) Efforts to catalyze local entrepreneurship.

Notwithstanding the above, there had been commitments to develop models for community-based AQC in which farms are fully owned and operated by communities themselves. The trained AQC team(s) had been providing materials and technical guidance and also helped with start-up costs. There had also been efforts to assist and nurture small trade development with training programs, in partnership with CITE, a Malagasy NGO supporting local socioeconomic and entrepreneurial development. This could buildup the technical, financial, managerial, and organizational skills needed by the fishermen and fisherwomen to manage their own AQC business. There could be involvement of the community through CBFM *modus operandi.* This support is basic for improving revenues and also for working toward the long-term sustainability of AQC trade in these isolated communities.

Concomitant to the above, there had been trends of new farmers and villages joining the AQC program every year, there by increasing the number of people benefiting from alternative livelihoods. Incidentally, it may be mentioned that more than half of the farmers supported by Blue Ventures were women, who often use this new income to help pay for children's school fees and supplement their family's diet.

5.2.5.3 *Scaling success*

Initiating innovations in AQC may not be a simple task. It may require strong technical partnerships and practical experiences. In fact, commercial nature of many AQC trades may not always publicize the results for obvious reasons. Moreover, the experiences gained in overcoming technical, logistic, and financial challenges in production are not much shared. However, the Blue Ventures is ready to promote and share the best practices in community-based AQC, thus maximizing the number of people who could benefit from these livelihood activities. There had been efforts to improve AQC initiatives along with others in the field, thus developing a new Indian Ocean network for practitioners and at the same time sharing the experiences from Madagascar.

5.2.5.4 *Partners for innovation*

The AQC program(s) are often supported by Norges Vel, an international NGO with expertise in tropical mariculture development. Moreover, there were collaborations with a range of local partners. Notwithstanding the above, the AQC practitioners attend seminars, symposia, and workshops in different places on community-based AQC in the Western Indian Ocean, being hosted by NGOs like the Blue Ventures, WIOMSA, etc. Further, it may be noted that new farmers and villages join various AQC programs every year, thus, increasing the number of people benefiting from the alternative livelihoods.

5.2.6 Namibia

Namibia is something of a backwater. It is a big country, but its population is only 2 million. It had long been an appendage of South Africa and gained independence only in 1990. The slate was, thus, clean, as there were few established economic interests, especially in the fisheries sector. Fortunately, the result was the adoption of a fisheries policy, which was amazingly sensible.

Notwithstanding the above, fisheries around the world suffer from abuse(s) mainly due to the lack of property rights. As such, the result was environmental destruction and economic waste. In a few instances, when the fisheries were in collapse, developed nations were spurred to adopt individual quotas or some other rights-based approaches that produced better environmental and economic results.

Unique about **Namibia** was that a **catch share system** was adopted in a poor nation with a population consisting of several deeply rooted tribes (Laura, 2011). Incidentally, market-based reform is not a western notion that somehow conflicts with traditional values. The lessons from Namibia and other fisheries success stories had illustrated that property rights and environmental protection could happen anywhere. "Fencing Fisheries in Namibia and Beyond" is a part of the PERC Policy Series of papers.

Namibia is famous for its wildlife safaris, sand surfing, adventurous anglers, and shark fishing. The indigenous Namibians, who were disenfranchized and discriminated against under the colonial rule, were finding ways to participate in the nation's new fisheries regime. While developing skills and strengthening the local economy, Namibians are rebuilding fish stocks and becoming a leader in marine management. However, there is still much progress to be made toward a true free-market democracy, but Namibia's advancement in instituting rights-based management adds to the lessons learnt in natural resource reform in emerging nations.

Concomitant to the above, the clean, cold waters off the coast of the Namib Desert are home to some of the richest fishing grounds in the world, with the potential for sustainable yields of *c.* 1.5 mmt per year (Namibia, 2010). Commercial fishing and fish processing are a significant and growing sector of the Namibian economy, contributing to *c.* 5.7% of the GDP and accounting for *c.* 18% of Namibia's foreign exchange earnings (MFMR, 2007). Incidentally, the principal species found off the coast of Namibia are hakes, sardines, anchovy, and horse mackerel. When Namibia gained independence in 1990, several fish stocks were on the verge of collapse.

Notwithstanding the above, prior to independence, "management" of Namibia's fisheries consisted of constant jockeying for control. The fishery was an open-access one, over which neither South Africa, as the de facto authority over Namibia nor the United Nations (UN) as the *de jure* authority were able to exercise jurisdiction (Manning, 2005). Also, the fishing grounds became an international free-for-all, with no rule of law that could be enforced. However, mid-water and bottom-trawl vessels were operating off the coast of Namibia during the years prior to independence (Beaudry et al., 1993). According to an account by the African Economic Digest (1993), the USSR had a 32% market share in the sale of the country's fish, followed by Spain with *c.* 26%. In fact, unrestricted access to Namibia's fishing grounds had devastating effects on the fishery. There were simply too many fishers catching too many fish. The hake catch, for example, grew from *c.* 47,600 tons in 1964 to *c.* 815,000 tons by around a decade, which was considered a 1700% increase. Notwithstanding the above, the pilchard or sardine fishery had been one of the most important fisheries in the region, in addition to the hake fishery. Incidentally, in 1968, the total landings for the Namibian sardine were recorded as *c.* 1.4 MT which dropped steadily to a modest *c.* 100,000 tons in the following decade (Jurgen, 1998; Panduleni, 2004). With the collapse of the Namibian sardine, came a growing concern that the hake fishery would be next.

Notwithstanding the above, Namibia's fisheries are recovering now and the trend toward collapse had reversed since independence to the point that Namibian fisheries management is now considered as a "model." Also, the few fishing ports that were once stark desert coast ghost towns are today thriving hubs, where Namibians flock for jobs with the processing plants and fishing fleets. The question thus arises how did pastoral Namibians with no real experience as a nation state, let alone fisheries management, could accomplish this feat? Two words, viz., "**property rights**." Namibians instituted a system that enforced an exclusive fishing zone (EFZ), and had individual rights to fish stocks. In essence, they had **erected fences for their fisheries**. As several scholars pointed out "property rights" had provided the foundation for economic and environmental prosperity (Demsetz, 1967; North, 1990; De Soto, 2000; Anderson and Leal, 2001).

5.2.6.1 An overview

Concomitant to the above, a recent joint report from the World Bank and the FAO of the UN revealed the extent to which marine fisheries could be considered an under-performing global asset (Ragnar et al., 2009). In fact, the difference between the potential and actual net economic benefits from marine fisheries, according to the report, was in the order of *c.* US $50 billion per year. These benefits, however, could be much higher, if institutions are given more incentives for greater investment(s) in the marine assets. The payoffs from the fisheries reform(s) could enhance economic growth in the developing countries as well as could protect fisheries for future generations (Leal, 2010). Unfortunately, the pre-1990 state of Namibia's marine resources was, sometimes, the norm for most of the world's ocean fisheries. According to the World Bank (2011a, b), *c.* 25% of the world's marine fish stocks are considered as overexploited and an additional *c.* 50% are fully exploited. Incidentally, this dismal economic and ecological scene was most visible in Africa. In fact, the marine resources around the continent provide livelihood and employment to millions of fishers as well as foreign exchange for trade and food security. Yet, the state of World Fisheries and Aquaculture Report (FAO, 2008) explained that all FAO-designated marine fishing areas around Africa are overexploited (Cunningham and Neiland, 2010).

Notwithstanding the above, continued mismanagement of African fisheries means losing the potential to realize an additional *c.* US $2–5 billion in revenues to coastal nations per year, according to Tim Bastook, fisheries advisor to the UK Department for International Development (FOROYAA Newspaper, 2010). Furthermore, as the latest PEW Ocean Science Series had pointed out Africa had the highest marine life losses relative to its actual catch (Srinvasan et al., 2010).

The people of the Republic of Namibia range from subsistence farmers (> 70%) to an urban population consisting of traders and civil servants (BBC, 2010). They belong to different cultural and ethnic groups including the Owambo,

Herero, Damara, Kavango, Himba, Nama, and, at least, six other smaller tribes. Namibia is a low-middle income country with an estimated annual GDP per capita of *c.* US $4310 (World Bank, 2011a, b). The country had portrayed extreme inequalities in income distribution and standard of living, with two in five Namibians living on <US $1.25 per day (Economist, 2011). Average life expectancy was *c.* 52 years for both men and women, and infant mortality was *c.* 46 per 1000 live births (CIA, 2011).

Notwithstanding the above, Namibia was ranked in the Freedom of the World Index with a low score of *c.* 2.69 in the "Property Rights and Legal System" category. However, later on, it, probably, had begun to implement a limited access regime to its fisheries (Gwartney et al., 1990).

In contrast to the above, in the past, the nations that had successfully implemented **catch share systems** were developed nations that scored closer to 10 in a mark-awarding system. However, it was not a place where one would expect to find a rights-based fishing sector, given Namibia's low score. Yet, it had emerged. In fact, a combination of unusual historic, geographic, and sociopolitical factors contributed to the success of Namibia's fisheries management (Deacon, 2010). In this connection, it may be noted that after the World War II, the economy of Southwest (SW) Africa (now Namibia) experienced a colonial boom, reaching a peak of *c.* US $20,000 per capita for Europeans and *c.* US $150 for black Namibians during the late 1970s (Encyclopedia Britannica, 2010). Incidentally, the economy was based on minerals, cattle, diamonds, and fish (largely for fish meal and canned sardines). Further, the exploitation of West Africa's marine fisheries by the European, Asian, and Russian fleets increased sixfold between the 1960s and 1990s, and, evidently, there was little benefit to the Namibians. Further, the so-called lack of restraint on fishery exploitation had set the stage for stock depletions. Consequently, two-thirds of black Namibians were in severe poverty by the late 1980s, and the nation as a whole was in an economic mess. Moreover, adding to the misery was a 16-year-old war in Angola-Namibia's troublesome northern region.

Notwithstanding the above, South Africa had clung to its control over Namibia, while Angola also attempted to capture the colony and establish it as a communist nation. Nevertheless, poverty, war, and the desire for human rights had served as the catalyst in the rise of the SW Africa People's Organization (SWAPO), a Marxist liberation movement that had fought a decade-long campaign for Namibia's independence against South Africa and apartheid. Namibia became an independent nation on March 21, 1990. Incidentally, at that time Namibia inherited a fisheries sector, which was on the verge of its collapse (Manning, 2005). Nevertheless, commercial fishing, which was represented mostly by processing, was Namibia's third largest sector of the economy, despite this condition. However, the local people and the new government had an incentive to revitalize the remaining stocks (Van Zyl, 2001). Subsequently, Namibia's fishery exploitation by the European, Asian, and Russian fleets increased sixfold between the 1960s and 1990s. However, the region was located in one of the few major ocean upwelling systems in the world, which meant that water from the deep-sea traveled up to the surface, making it highly productive for marine life. Yet, the hyperarid Namib Desert is said to stretch the length of Namibia's long coastline, and reach up to *c.* 150 km inland. Incidentally, the sand dunes made road and rail construction along the coast nearly impossible. Consequently, the only harbors in Namibia were Walvis Bay and Luderitz. Meanwhile, the existence of just two ports had simplified the job of monitoring the landed catches (Sumailia et al., 2005). Further, a lack of a historical association with the seameant that despite the richness of Namibia's marine fisheries, few Namibian people virtually had traditions of ocean fishing. Moreover, there was no customary claim to the resource (Manning, 2005). In one sense, this might have been beneficial. In fact, the absence of an established domestic fishing industry meant that organized user groups did not exert undue influence on the formation of fisheries policy. Namibia had begun with a "clean slate" (Deacon, 2010; Sumailia, et al., 2005).

5.2.6.2 *Beginning from the scratch*

The inexperienced Namibian government, being faced with the task of rebuilding a fishery, which had long been exploited by the foreigners, contacted the fisheries experts primarily from Norway to assess and monitor the state of its various fish stocks (Manning, 2005). Meanwhile, the fisheries scientists, with international experience conducted research at a level comparable to that usually found in developed nations (Van Zyl, 2001). Meanwhile, the new government had learnt enough about the state of their marine life, and was thinking of ways and means to **establish a system of property rights** to their fishery. Firstly, Namibia declared jurisdiction of a 200-mile EEZ, which was established under the Law of the Sea Treaty (United Nations, 2011). This was a crucial step to close Namibia's open-access waters. Secondly, the government had ordered unlicensed foreign vessels to exit the zone. Incidentally, on the day the law became effective, it was estimated that >100 foreign vessels were fishing unlawfully in the Namibian waters (Nichols, 2005). Meanwhile, the fledgling government had no aircraft to patrol *c.* 500,000 km^2 of the ocean, but nevertheless made a strong effort by hiring a private helicopter to fly out volunteers from the Namibian Defence Force to arrest Spanish trawlers at gunpoint. The vessels were ordered-in, impounded, and the skippers and officers were put on trial in the relatively stable court system inherited from South Africa. This had sent a clear signal to the poachers that Namibia was serious about enforcing rights to its waters.

Concomitant to the above, the next step toward reducing the fishing pressure was to set the **TAC** for individual species. A dramatic reduction in the TAC was necessary to enhance the chance(s) of recovering the fisheries. This was especially significant for the hake fishery, which represented the majority of total fishery earnings (Manning, 2005). Meanwhile, the government had set the hake TAC at a remarkably conservative *c*. 60,000 mt in 1991 to encourage the fishery to grow. This was the lowest amount caught since 1964, when foreign trawlers had begun to arrive. Further, the hake fishery was also regulated through the size of the nets and the area of restrictions (Van Zyl, 2001).

Notwithstanding the above, the TACs were divided into fishing quotas for eight species, notably, hake, horse mackerel, orange roughy, alfonsino, sardines, red crab, rock lobster, and monk. Once the TACs were set, a combination of harvesting rights, individual quotas, enforcement policies, fees, and a monitoring system was established.

5.2.6.3 *Renewable harvesting rights form the core of the Namibian fisheries management system and limited permission of access to the fishery*

It may be noted that "rights" were issued to the bidders for a period of 4–20 years, depending on various factors like investment in vessels, fishing experience, and level of Namibian ownership (Ithindi, 2003).

Notwithstanding the above, **individual quotas** were issued for harvest by the rights holders. However, preference was given to firms owned by the Namibians and those which employ the Namibian citizens. Furthermore, vessels were not given a license to fish, unless they were registered in Namibia, or are at least *c*. 51% Namibian owned, and have a crew that is at least *c*. 85% Namibian. This structure was a result of the "Namibianization" process adopted by the fledgling government to help recompensate those who suffered under apartheid rule (Manning, 2005).

Notwithstanding the above, Namibianization was aimed to shape the specifics of the nation's quota system in various ways. The non-Namibian fishers were tired due to increased tax which was levied from them, as compared to lesser amount paid by the Namibian-owned vessels. Nevertheless, this practice helped to explain why Namibia had not granted fixed rights to the non-Namibians and had granted them only for a period of 3 years. It is probably also debatable that why fishing quotas were technically nontransferable. However, according to fishery economist Robert Deacon, this was "due to the deliberate policy to employ Namibian citizens." While, this no doubt caused some amount of "proximate inefficiency," Deacon commented that the broad distribution of the benefits from fisheries across a large number of Namibians seemed to have protected the efficiency of the Namibian **fisheries** in the long run (Deacon, 2010).

Concomitant to the above, the democratization process of Namibia arguably played a significant role in the initial success of fishery management, although the process was later found to be messy and full of frictions. The question thus arises, what, then, is a fishing right in Namibia? However, the government had overcome the common property problem by renting out a limited number of rights to a resource. Incidentally, the Namibians chalked out decisive ways to lease and buy shares; although, the fisheries are still a public resource (individual quotas are not transferable, other than with permission of the Ministry of Fisheries). However, the rights were leased out under the guise of chartering vessels, and the services of processing and marketing the catch were also included in the contract. The operating company pays the fee, and the quota holder (i.e., small holding company) has nothing to do with the quota for the rest of the year. However, the "chartering" of a vessel in such cases was no different than "leasing" the quota, as commented by Manning (2005). Nevertheless, this arrangement had formed a web of joint ventures with large operating companies at the center. In fact, the small quota holders gained by having a legal way into the market that could be exploited personally or by renting the quota. However, the operating companies had gained by acquiring more quotas and efficiency in scale of operation. However, there were still some road bumps with Namibia's arrangement, despite these gains. In this connection, Manning had commented that it is usually hard for the "new comers" to enter the operations side and hence they often collect a smaller portion of the rent. More broadly, the process of privatization was incomplete. As such, it will be difficult for Namibia to realize the full potential of its fishery until individual quotas become perpetual and transferable.

5.2.6.4 *Enforcement and efforts for implementation*

It is essential for the rights holders to have confidence that their rights will be enforced. Compliance with limitations on fishing activities, however, depend on fishers' recognizing that limitations are necessary. As stated in a Policy Brief by the World Bank, "regulation with the consent of the regulated, leads to greater compliance and lower costs of enforcement" (2004). In fact, evidences show that fishers with rights to fish often try to monitor and enforce fishing rights themselves (Leal et al., 2008). It may be recalled that licensed vessels offered detailed intelligence reports to the Namibian government to help enforce their rights. They even paid for a helicopter to nab the intruders (Manning, 2005). However, Namibia remained vigilant in its efforts to discourage unlawful fishing (Van Zyl, 2001).

Notwithstanding the above, the fishing rights holders paid levies to finance the Fisheries Observer Agency. In fact, evidence indicated that this monitoring system was effective as violations per inspection declined drastically since the early 1990s (Bergh and Davies, 2005). Further, Namibia's fisheries department installed a vessel monitoring system (VMS) (FAO, 2010).

5.2.6.5 *Significant snags*

Some fishery experts believed that the Namibianization of the industry has not properly defined the altered situation, i.e., who the beneficiaries are as much as one might think. However, Manning (2005) suggested that the foreign interests and wealthy national companies have found ways to protect their interests in Namibia's fisheries sector through competent attorneys and political handouts. Spain, for example, has given millions of dollars as aid to Namibia and some of it directly to the fishing industry (iWatch News, 2011). In contrast to this, Melber (2003) criticized Namibia's fishing sector, claiming that it had promoted the interests of a privileged black minority but had failed to correct "the grossly unegalitarian social structure and distribution of wealth." Concomitantly, Deacon (2010) argued that as the Namibian fisheries policy had deliberately broadened the set of stakeholders, it was difficult for the elite groups to appropriate rents for themselves.

Notwithstanding the above, there were extensive evidences that transaction costs often block beneficial reallocations of a resource (Libecap, 1989). Indeed, evidence(s) indicating excess capacity in the vessels and processing for a few companies in Namibia had signified some wastefulness in the system and a tendency to focus on short-term operating costs (Manning, 2005).

In addition, various scholars argued that the so-called altered catch shares made in the nature of fishing jobs—the shift from a larger, part-time, more seasonal work force to a smaller, full-time, higher paid and safer work force—was both economically and ecologically positive (Costello et al., 2008; Heal and Wolfram, 2008). However, it could be difficult to comprehend the full potential of the fishery sector, without longer term quota rights and transferability (Scott, 2010).

Notwithstanding the above, from "Failing to Flourishing" Namibia's postindependence fisheries policies offered fresh hopes and change in tune with other developing nations. The fishery stopped dwindling. Further, the asset value of fish stocks increased by *c.* 40% of the hake stock and, probably, there was a general increase in the economic value of the total stock (Lange, 2003). Moreover, there were enormous scopes to improve the chances for successful fisheries reform in the developing world. In addition, during this same time, the contribution of the fisheries sector to GDP had risen from *c.* 4% at independence to *c.* 10% during the late 1990s, *c.* 6% from the fish processing sector, and *c.* 4% for the harvesting subsector (Manning, 2005). As of 2007, the fishing sector in Namibia contributed *c.* 5.7% to the GDP and accounted for *c.* 18% of Namibia's total foreign exchange earnings (MFMR, 2007).

Further, recognizing the recovery of the hake fishery, the TAC for hake was increased by *c.* 5000 tons for the 2010–11 fishing season, meaning *c.* 140,000 tons of hake could be landed in 2010–11 (Hartman, 2010). However even more significant for Namibia was that the hake industry as a whole earned the country *c.* N $1.4 billion per year; the exchange rate was *c.* N $7 to US $1 (Windhoek Observer (Namibia), 2011). Furthermore, the Namibian government and fisheries sector deserved credit for operating without subsidies (Nichols, 2005). This was, perhaps, striking given the global pattern of government subsidies for the fishing industry. Further, the objective of generating jobs for the Namibians was also generally viewed as a success by the public. The domestic fishing industry that emerged had employed *c.* 14,000 locals out of *c.* 2 million Namibians, a substantive number in a country, where unemployment was *c.* 51% (Heita, 2010).

Notwithstanding the above, there were, sometimes, revenue losses on different exercises, for example, revenue lost in a process in 1999 was > N $128 million or *c.* N $51,000 per job, which was often regarded as an expensive way of making new jobs (Eide et al., 2003). Nevertheless, Namibia's catch share system could be commended despite such shortfalls in view of its social, economic, and environmental impacts, which are superior to what occurs in most coastal nations.

Notwithstanding the above, fishery management regimes are multidimensional, meaning there is a broad range of pathways toward better management. Nevertheless, two fundamental ingredients for success in Namibia are as follows:

- **(i)** A clear assignment of rights.
- **(ii)** The ability to develop rights-based reform pathways compatible with the political economy circumstances of the nation (Deacon, 2010; Cunningham and Neiland, 2010).

In continuation of the above, the Namibians were able to **fence their fishery by establishing rights to an** EFZ **and then enforcing those rights**. They, further, **defined rights by establishing catch shares for fish stocks**. An incentive structure quickly emerged to alter the desire for fishers to race for fish, which reversed the process of fishery collapse, rather raised the total value of the fishery. Further, in addition to forming rights, a key component to this system's success was establishing the principle that natural capital may not be free and that management of the resource could be a part of the cost. However, Namibia introduced payments for fishing quotas early, even when the resource was depleted (Manning, 2005).

Concomitant to the above, there are huge opportunities to improve the chances for successful fisheries reform in the developing world, as the case of the Namibian fisheries demonstrates. However, some scholars have opined that a catch share system could be impractical in countries where there is little domestic capacity to manage resources sustainably or where there is little will to protect property rights held by nonelites (Ragnar et al., 2009). In such cases, another pathway such as devolving responsibilities to lower levels of government or to the fishing communities could hold promise (Leal, 2010).

In continuation of the above, community fishing rights, where defined groups are given exclusive rights, is attractive where other rights-based approaches may not be applicable for sociopolitical or enforcement reasons (Ragnar, 2001). Concomitantly, among the advantages of community fishing rights were that they could be socially acceptable and facilitate effective law enforcement on the basis of physical proximity and traditional institutions (Ostrom, 1990; Wilson, 2001; Ragnar, 2001). Nevertheless, well-functioning community management arrangements displayed benefits in both household income and environmental conditions (Shyamsundar et al., 2005).

Concomitant to the above, a brief overview of the fisheries management and other related aspects in certain other countries is given in the following.

5.2.7 Status in Mozambique

Not much good fisheries management could be seen in **Mozambique**. There the fisheries sector contributed *c.* 4% to the GDP and *c.* 28% to the country's foreign exchange earnings, with prawns being Mozambique's largest single export (Omar, 2006). However, the main domestic fisheries sector consisted of *c.* 90,000 subsistence and small-scale fishers, according to Maria Omar, a member of Mozambique's Ministry of Fisheries. But the fishery was under constant threat from outsiders with larger and better equipped vessels. Like many coastal countries, the property rights to the marine resources were not clearly defined or enforced in Mozambique, meaning that the country was a target for uncontrolled, unlawful, and unregulated fishing (Lopes and Pinto, 2006). Meanwhile, an Institute for Security Studies Report (2007) estimated that unlawful fishing of shrimp and tuna in Mozambique amounted to *c.* US $38 million as lost revenue. Moreover, the local economies suffered as small-scale fishers lost potential revenue from shallow coastal waters that were overfished by the outsiders.

5.2.7.1 *Feeding frenzy for fish: A view*

Notwithstanding the above, the problem in **Mozambique and in most other coastal countries** is a failure to escape the freedom-of-the-seas mindset and more importantly a **failure of the governments to establish property rights to fishers** in a way that could maximize both economic benefits and long-term survival. Incidentally, it may be mentioned that all the FAO-designated marine fishing areas around Africa were overexploited particularly with regard to fish. Therefore, in an attempt to combat the collapse of fish stocks, the UN in conjunction with many coastal nations framed the Law of the Sea Treaty (LOST) during the late 1970s. This treaty has defined specific jurisdictional limits on the ocean area which the countries might claim including a 200-mile EEZ (United Nations, 2011). Moreover, the treaty also called for technology and wealth transfers from developed to undeveloped nations as well as required nations to adopt specific regulations regarding the marine environment. Such provisions were among the reasons the United States had not ratified the treaty. Incidentally, defining the EEZ could be a positive step toward fencing fisheries. Further, LOST relied on the UN's bureaucracy focusing too much on regulations and too little on economic incentives (Meese, 2005). Concomitantly, under the prevailing "management" approach, it is for the governments to impose regulations restricting vessels, gears, seasons, and catch characteristics. However, such regimes fail to instil, in each fisher, a regard for the future value of the resource. Incidentally, it may be noted that in a regulated commons, shares of the catch are usually up for grabs, leading to a destructive "race for fish." In fact, attempting to prevent overfishing through regulations is not only expensive, rather it often may not work. Perhaps, in order to counter this, the fishers may become more adept at catching larger amounts of fish in shorter period of time, and could be also in fewer trips. They simply use bigger boats, bigger nets, and bigger crews (Leal, 2002). As such, this style of fisheries control could lead to perilous outcomes because fishers may not be able to save fish for the future. If they restrain their catch to leave enough fish to reproduce for the following seasons, the fish may be caught and taken away by someone else. Notably, each fisher in common gets all the benefits of catching more fish now, while facing only a fraction of the future cost of stock depletion. The evidence may not be overwhelming that this approach fails to deliver desirable economic, social, or biological results.

5.2.8 A brief account in other regions (in a little comparative manner)

Incidentally, it may be noted that only a few countries led by **New Zealand and Iceland**had begun experimenting with a new approach to fishery management, while the rest of the world was busy focusing on strict fishing regulations. This approach, known as "**catch shares**," was designed to secure a legal right to the individual communities or fishery associations to fish (Deacon, 2009).

Notwithstanding the above, the catch share system combines two features to help overcome the destructive race for fish as follows:

(i) A "**TAC**" in which the size is determined based on scientific criteria about sustainable stock levels.
(ii) Then, the fishing entities (individuals or defined groups) are assigned shares of the TAC. These shares could be used, sold, or leased to others but no one is allowed to harvest more than the amount specified in their quotas.

In addition, the most common form of **catch share management** is the **ITQ**, which allocates to its owners the right to harvest a specified quantity of fish every year. In essence, the fishers gain "**property rights**" to shares of the catch, much as they have rights in their boats and gears. Collectively, the "rights owners" have the incentives to protect and maintain the quality of the fishery because the value of their shares hinges on their sustainability.

Concomitant to the above, the "catch shares" help in protecting the fisheries by providing with an economic rationale for conserving the resources. In fact, the fishers in the catch share system want the fishery to remain productive, just as the shareholders of a company want the firm to excel in order to enable the shares to gain value (Environmental Defense Fund, 2010).

Case studies of fisheries in **British Columbia, Iceland, New Zealand, the United States**, and in certain other areas try to demonstrate that this system could have the potential to dramatically improve the economic and biological health of the fisheries. In this connection, it may be mentioned that when catch shares were adopted in the Alaskan halibut fishery in 1995, the fishing season lengthened from 48h to >8months a year, meaning the boom-and-bust market cycles had disappeared (Leal, 2005). The fishers had received higher prices because they could deliver fresh halibut the year round. Moreover, safety problems had been reduced, as there was no need to fish in hazardous conditions. The costly technological arms race switched from maximizing vessel size and catch to maximizing the productivity of the fishery and the value formed by offering a superior product over a longer duration of the year. The result was said to be that the actual harvests no longer exceeded the total allowable catch each season. Moreover, the fishers gained property rights to shares of the catch, much as they had rights in their boats and gears.

Concomitant to the above, the catch share system might also help improve the biological health of the fisheries. Some benefits had arisen from slowing the race for fish. There might have been less dumping of unintended catches, for example, and less discarding of fishing gears that were cut loose when anglers were in a hurry.

In addition to the above, in the **Gulf of Mexico**, the catch shares for red snapper cut the accidental killing of fish by *c.* 70% and led to a 60% increase in red snapper population (Environmental Defense Fund, 2010). Other ecological benefits might also arise when secured access to a catch gives incentives to fishers for reducing overfishing and avoid sensitive spawning areas to promote stock recovery (Deacon, 2009). Concomitantly, damage to wreckfish habitat, off the South Atlantic coast, for example, decreased with the introduction of catch shares because there were fewer weights dragging across corals and also because the slower pace of fishing allowed fishers to be more selective. They were also able to comply with a ban on bottom long-lining, which is a mainline with baited hooks attached at intervals by branch lines (Griffith, 2008).

Concomitant to the above, a highly cited report in "Science" by Costello et al. (2008)' takes the study of catch shares to a new level. The authors compared *c.* 121 fisheries, where ITQs and other catch share systems were implemented, and >11,000 fisheries around the world, where such systems were not implemented or absent. They found that once catch shares were established, the process of collapse stopped and in many cases, fish stocks had rebound. The authors estimated that had catch shares been adopted in all fisheries beginning in 1970, the incidence of fisheries collapse might have been reduced by two-thirds. Rather, in the failed fisheries, which we see today, fish stocks could be growing while still supporting fishers and nourishing consumers.

Notwithstanding the above, the catch share systems were expanding around the world with *c.* 150 operating today (Costello et al., 2010). Incidentally, in the United States there are 15 catch share programs (CSPs), which according to the Environmental Defence could flourish. The system produced an 80% increase in revenues for 5years after implementation.

In continuation of the above, obstacles to instituting property rights in world fisheries remain formidable, despite these positive advances. Less than 2% of the world's fisheries had adopted rights-based strategies (Leal, 2010). Namibia is part of that small number and is making efforts in proving to the rest of Africa and beyond that under the right conditions, the use of local knowledge and a system of property rights compatible with political and economic circumstances, fisheries could be improved.

5.2.8.1 Inference

It may be pertinent to note that about half of the world's population lives within 200 km of a coastline (Creel, 2003). However, many coastal regions are experiencing a decline in marine resources and a loss of fish stocks. The problem was, however, particularly acute in the developing countries, where fish play a significant role in the nutritional status and livelihood of innumerable people (Cunningham and Neiland, 2010).

Notwithstanding the above, the reasons for environmental decline in coastal habitats are complex. But, a lack of property rights plays a significant role. Incidentally, the property rights-based fisheries management regimes demonstrated success in the management of fisheries resources in developed countries. Common property problems are usually eliminated by establishing property rights over the fish stock, which concomitantly reduced the incentives for overcapitalization of the resource and, consequently, also contributes to economic efficiency.

In contrast to the above, in the developing world, catch share systems are rare because the legal, managerial, and fiscal frameworks are often absent (Leal, 2010). Namibia, however, demonstrated that a catch share structure could be built from the ground-up and could work beyond the developed world. However, Namibia still suffers from a high rate of poverty and inequality, even after 20 years of independence; few would have imagined that a new nation in an area dominated by strife would have a relatively stable democratic government, a free press, and an economy that has grown at an average rate of *c.* 4.2% a year since independence in 1990 (Economist, 2011). As evidence of good governance, the 2010 Economic Freedom of the World Report allotted Namibia with a score of 7.58 out of 10 in the "Property Rights and the Legal System" category, up from 2.69 in 1990 (Gwartney et al., 2010). Further, Namibia secured high marks for both political rights and civil liberties in the most recent survey by Freedom House (Puddington, 2011). Furthermore, in the latest index of good governance in Africa, published by the Mo Ibrahim Foundation, Namibia secured sixth position out of 53 countries (Economist, 2011). Nevertheless, the Namibians are trying to prove that tools for effective resource management could be universal and there is no one exclusively "African solution." In fact, any nation and its resources could benefit from pursuing **market-based economic principles** and establishing secured **property rights. Namibia did this by instituting the principle that a right to benefit from the natural capital of a country had real value and a payment should be made for it. This right and payment came in the form of fishing quotas under a catch share system**. This system was then enforced, which helped to reassure fishing companies that their rights were secure. Yet, the **catch share system** implemented in Namibia represents only one way of **fencing a fishery**. It could be opined that "depending on the nature of a fishery, other methods may work better. These might divide up and sell lobster pots, numbers of fish, numbers of boats, bits of the ocean, or even individual reefs. The best choice depends on the value and underlying biology of each fishery" (Economist, 2008).

Notwithstanding the above, lessons are also emerging from **India's** traditional community-based fishing. In fact, community-based cooperative fisheries management is a viable option in some of the artisanal fisheries in developing countries. Nevertheless, a key component emerging from the pilot programs in **India** is the bottom-up management of fisheries resources by those who depend on them, have the most knowledge about them, and have the greatest stake in protecting them. This bottom-up approach coupled with the need of the GoI to introduce and enforce more fishing zones along the coastline as part the new National Fishing Policy shows more promise for stronger property rights and healthier fisheries.

Concomitantly, there is growing evidence from the developing countries around the world that insecured property rights could limit economic development and resource conservation. In this connection, it may be noted that during the middle of the 19th century, the development of **American** agriculture was similarly limited when farmers were unable to protect frontier lands from encroachment by livestock. As such, fencing techniques, such as barbed wire could be erected in order to solve this problem. This incidence enabled individuals to establish property rights and secure their resources. **Such fences** are obviously not possible in the deep sea, but property rights could still be erected for fisheries.

Notwithstanding the above, **fences for fish** vary among countries to the extent that they approach full property rights but overall securing rights to marine resources saved fisheries in the developed world. Similar strategies could also work in the developing world. Traditional systems of participatory governance mixed with market approaches to resource management, such as **catch shares, devolution of control to local levels, and CBM for fisheries** could help to improve both environmental quality and quality of life in Africa and beyond.

5.2.9 Senegal

Fishing is said to be one of the key sectors of the Senegalese economy. Incidentally, Anon (2013a, b, c) discussed about the concerted management of the sea bream fishery in Kayar, **Senegal**.

It provided *c.* 200 billion FCFA (*c.* US $400 million) of annual receipts for nearly one-third of total export revenues. Moreover, fishing engages *c.* 17% of the working population (*c.* 600,000 people) and provides *c.* 75% of the animal protein requirement of the population. Further, the economic and social performance of the Senegalese fishing industry is directly linked to the dynamics of the artisanal subsector, which ensured > 80% of national production (out of a total of *c.* 450,000 tons/year) and as much as *c.* 70% of the exports. However, this performance is threatened by environmental constraints such as the overexploitation of major demersal fish stocks.

Notwithstanding the above, sanitary certification of fish exports from Senegal had begun with the European Commission sanitary regulation on fish products during the early 1990s. The delivering of catch certification became more complex since January 1, 2011, due to the obligation for export countries to provide certificates which state the origin of the fish. For this cause, a number of units within the Ministry of Fishery and Marine Affairs are involved in the export certification scheme. The Division of Inspection and Control is the authority responsible for verifying that products have followed sanitary and hygienic standards equal to those in Europe. Moreover, the Directorate of Maritime Fishery is responsible for delivering sanitary and origin certification, while the Directorate of Protection and Surveillance of Fisheries provides the catch certificate.

In 1995, the National Collective of the Artisanal Fishermen of **Senegal** (CNPS) implemented a policy of daily quotas on sea bream in Kayar, which could be regarded as the first **comanagement** initiative in **West Africa**. Such a measure, which limits fishing to three boxes of *c.* 13 kg of fish each/canoe, was respected by the fishermen. The sanctions were highly dissuasive in the cases of nonrespect of the quota limit. This included a penalty of *c.* EUR 50 confiscation of the catch beyond the limit and the prohibition of fishing in case of recidivism. The immediate commercial effect reflected through an increase in the price of fish and consequently of fisher's incomes.

5.2.9.1 Extension of the experiences of Kayar to other Senegalese fisheries

The Kayar **comanagement** fishery approach had been extended to other fisheries along the coast of Senegal, using the concept of territorial fishing rights and shared responsibility between the states and fishing communities. Several projects, initiated by the Ministry of Fisheries and funded by the World Bank, had developed a **comanagement system** at the national level. For example, the project, of late, 2005–12, titled, "Integrated Management of the Marine and Coastal Resources in Senegal" had recommended an immediate delegation of powers to **stakeholders** and the promotion of local initiatives regarding **fish stock management**. The concept of traditional centralization of fishery management is, thus, under review. A new legal framework is being developed, which recognizes the authority of stakeholders to take decisions and actions. These recent alteration(s) in the governance of fisheries were implemented in four pilot sites along the Senegalese territory.

Concomitant to the above, there were other notable initiatives like the project for "Sustainable Management of Marine Resources," funded by Japanese International Cooperation Agency (JICA) (2009–13). It adopted the same principles of fish stock management by communities inhabiting areas near fishing grounds. Local initiatives had been encouraged with involvement of all stakeholders. Overall, both the projects aimed to reinforce management capabilities of both fishermen organizations and the state administrations. Subsequently, pursuing the process of comanagement, the Ministry of Fisheries, in partnership with fishermen organizations and NGOs, had also developed a national strategy for the immersion of artificial reefs and the generation of a MPA network. In fact, these fishery management tools had been integrated into the regional planning policy and the general framework of fishing territorial unit rights. Nevertheless, the Kayar experience reveals that it is possible to reverse downward trends of fish stocks. Moreover, it also portrayed that **involving fishermen organizations is a key factor to success in implementing any fishery management measures.** As such, the Senegalese government had been taking steps to propagate the Kayar experience along the coast of Senegal. On a more global level, the certification of fish exports, contributed to raising the sanitary conditions of all the fish landed in Senegal, with a possible positive effect for local consumers. Parallel to upgradation of the export fish industry, significant efforts had also been made by the state authorities to improve the distribution of fish products in the domestic markets.

5.2.10 Tanzania

The **Tanzania Natural Resource Forum (TNRF)** embarked upon a process to establish a series of dialogs on **CBNRM.** The aim of this initiative was to provide a platform for multi-stakeholder exchange of information, sharing experiences, and agreeing upon a common course of action for the success of CBNRM in the country. The platform disseminates information on CBNRM successes, challenges, and the areas which need more focus in the natural resources sector, thereby promoting learning and providing opportunities to influence CBNRM practices and policies in ways that could lead to success. It is against this background that TNRF undertook a national stocktaking exercise on CBNRM. This initiative was said to be also part of the regional Southern Africa CBNRM Forum (SACF), of which TNRF is a focal point for Tanzania.

Anon (2013a, b, c) dealt with CBNRM stock taking exercise in Tanzania.

5.2.10.1 *About the Tanzania natural resources forum (TNRF)*

The TNRF is a network organization registered in 2006 as an NGO. It aims to achieve improved natural resource governance for sustainable rural livelihoods and better conservation outcomes. TNRF bridges the gap between people's local NRM needs and practices, and national NRM priorities, policies, laws and programs (of the authorities). As a network organization, it brings together members to change policies and practices in the natural resource sector for better results.

5.2.10.2 *Background on community-based natural resources management (CBNRM)*

CBNRM with a conservation and rural development approach emerged during the early 1990s in **Tanzania and in most of sub-Saharan African countries** following a conservation paradigm shift from state-centric top-down conservation approaches to democratic decentralization and participatory bottom-up conservation and rural development approaches. Many factors contributed to the conservation paradigm shift and emergence of CBNRM approaches including the following:

(i) A clear failure by the poor newly independent African countries, including Tanzania, to halt rampant wildlife poaching and massive deforestation as well as loss of water resources during the 1970s.
(ii) Increased pressure by the national and international NGOs such as the IUCN, WWF, and AWF to participate in natural resources management claiming to be better equipped with resources (man power, finances, and technical capacity) than the prevailing system in the newly independent African countries.
(iii) CBNRM aimed to confer natural resources management rights and powers to the local rural communities in managing resources found in their areas with the assumption that these communities have long-term and vested interests in the sustainable utilization of these resources as well as they could be time-tried and time-tested institutions for their management. The underlying premise was that the local communities are likely to be better managers of the resources, since their livelihoods are almost entirely dependent on these resources.

Notwithstanding the above, CBNRM had begun as an approach in the wildlife sector and soon spread to other sectors including forestry, water resources, rangelands, and fisheries. In Tanzania, CBNRM models included the Community Conservation and Outreach services (CCS) under TANAPA; the community-based conservation (CBC) (later Wildlife Management Areas or WMAs) approach under the Wildlife Division (WD), the participatory forest management (PFM) approaches, which was categorized into Joint Forest Management (JFM) and Community-Based Forest Management models; fisheries comanagement approaches such as BMUs; water user associations (WUAs) in the water sector including wetlands; communal rangelands; village land management through village land use plans, among others.

Notwithstanding the above, in general, CBNRM had fast expanded into a number of geographical areas over the last few decades. A number of important policies and legislation reforms occurred with the aim of making an enabling policy and legislation environment for effective CBNRM implementation at the national level. Public, private, and civil society stakeholders were actively working in close collaboration with the local communities in CBNRM implementation in various parts of the country covering all the relevant natural resources subsectors. However, the natural resources sector had at one time perhaps, plagued with loss of resources, uncontrolled exploitation, lack of accountability, corruption, and poor management. It appeared that government lacked the capacity needed to control collusion between the interests exploiting the resources and the officers charged with oversight and control of the resources. The question remains: Is the CBNRM approach really operational in Tanzania? If so, why it was not working as expected?

5.2.10.3 *Background need for a Tanzania CBNRM forum*

The Tanzania Natural Resources Forum (TNRF) dwelt upon an exercise of generating a series of deliberations on CBNRM with the aim of providing a basis for multi-stakeholder interchange of information, sharing experiences, and unanimity on a common course for the success of CBNRM in the country. The **platform** was supposed to disseminate information on CBNRM success stories; and, thereby fostering a greater appreciation and understanding of the actions and enabling environment on CBNRM that could pave the way to success.

Concurrently, TNRF joined a regional civil society network on CBNRM issues called Southern Africa CBRM Forum (SACF), whose secretariat seats at Lusaka in Zambia. The ongoing transformation of conservation paradigm from more project-oriented toward high level of thinking for good governance entailed the need for dialog within Tanzania and collaboration with other regional initiatives in order to address natural resources management. However, for the meaningful dialog among and between stakeholders to happen, it was thought to be important to understand and analyze the current situation of CBNRM implementation in the country. It is against this background that TNRF had undertaken the CBNRM stocktaking exercise. This exercise had been carried out in partnership with the WWF for Nature: Coastal East Africa Initiative (WWF-CEAI) and with WWF-Namibia.

5.2.10.4 *Objectives, scope, and approach of stocktaking exercise*

Objectives of the stocktaking exercise

The stocktaking exercise intended to establish the current situation of CBNRM implementation in Tanzania, taking stock of the legislative and policy environment; champions of the CBNRM projects on the ground, benefits attained, challenges faced, how they are addressed, and important lessons learnt so far.

Scope of the stocktaking exercise

The following sections try to provide the scope of the stocktaking exercise:

(i) Sectors to be covered: all CBNRM models/programs/projects under the forestry, wildlife, and fisheries.
(ii) Review of the current policy and legislation environment and its evolution under each of the above sectors.
(iii) To identify specific provisions in the policies and legislative documents that mention community involvement in forest, fisheries, and other natural resources' management.
(iv) To analyze critically the extent to which the policy and legislative system(s) support or prohibit community involvement in natural resources (forestry, fisheries, etc.) management.
(v) To identify areas in need of revision with the aim of improving modalities of community involvement.
(vi) To explore and report on whether the customary community level institutions that had traditionally been managing resources over time are recognized and given priority or, otherwise, considered in the official arrangements for managing the resource, e.g., in fisheries, forests, wildlife, rangelands, etc.
(vii) To review the linkages between CBNRM and broader economic development and poverty reduction strategies, plans, policies, and programs at the national level.
(viii) To review the origin and current status of CBNRM models in Tanzania including:

Forestry: To review the status of implementation of PFM (JFM and CBFM), etc.
Fisheries: To review the status of fisheries comanagement schemes, such as BMUs, etc.

(ix) To review the status of community involvement in fisheries comanagement-related schemes.
(x) To review status of community-based fisheries/water resources management for small lakes and rivers found in many parts of the country.

5.2.10.5 *Fisheries subsector*

CBNRM in the fisheries sector: Policy and legislative environment which support CBNRM in the fisheries sector

In Tanzania, community participation in fisheries management was supported by three policy and legislative documents.

CBNRM models in fisheries sector

Community involvement in the fisheries resources management in Tanzania was mainly through collaborative management arrangements, unlike community involvement in forestry and wildlife managements. These are known as BMUs organized spatially either according to a beach landing site or according to coastal villages.

Challenges with regard to implementation of fisheries comanagement

Notwithstanding the above, there had been some factors that lead to weak capacity to enforce fisheries legislation, including insufficient financial and human resources. These are as follows:

(i) Inadequate promotion of value-added fish and fishery products to increase income.

(ii) Weak mechanism for data collection, processing, analysis, and storage, thus affecting monitoring in fisheries management.
(iii) Lack of research-based information on performance of fisheries comanagement in both marine and FWs.
(iv) Poor mechanism for dissemination of fisheries data and information.

The legislation and policy environment on community fisheries management

Three policy and legislative documents provide guidelines on CFM in Tanzania as follows:

(a) National Fisheries Sector Policy and Strategy Statement of 1997.
(b) The Fisheries Act No. 22 of 2003.
(c) The Fisheries Regulations on Comanagement of 2005.

Concomitantly, various user guides had been developed to interpret and implement the policy provisions.

The National Fisheries Sector Policy and strategy statement

It may be noted that the goal of the fisheries policy is to promote conservation, development, and sustainable management of the fishery resources for the benefit of present and future generations. The policy was believed to provide with guidance for promoting sustainable exploitation, utilization, and marketing of fishery resources. The policy, further, aimed to achieve its goal toward sustainable management and development through ensuring provision of food, income, employment, foreign exchange earnings, effective protection of aquatic flora and fauna, and environment. The fisheries policy initiated a directional change from state-centered control to collaborative initiatives. The Fisheries Legislation of 2003 provided the legal backup to the National Fisheries Policy.

Fisheries act no. 22 of 2003 and fisheries regulations of 2005

The Act and Regulations had stipulations relating to fisheries comanagement including:

(i) Sustainable development, protection, conservation, AQC development, regulation and control of fish and fishery products, aquatic flora and their products, and related matters.
(ii) Section 18 of the Fisheries Act provided for the establishment of BMUs (Community-Based Collaborative Management Units).

Fisheries Regulations (2005) on comanagement

Notwithstanding the above, specific regulations on fisheries comanagement include the Regulation 104 on comanagement under Part VIII (General Provisions), and sub-Regulations (1–4).

In addition, the Fisheries Regulations (2005) was amended and the amended regulations strengthen collaborative management activities in the fisheries sector.

5.2.10.6 Collaborative fisheries management in Tanzania

Fisheries managers had increasingly recognized that a fishery could not be managed effectively without the cooperation and participation of fishers in order to make laws and regulations to work (Pomeroy and Berkes, 1997). Comanagement systems were said to have emerged as a partnership arrangement using the capacities and interests of local fishers and communities, complemented by the ability of the government to provide with enabling legislation(s), enforcement and conflict resolution, and other assistance. Comanagement could be considered as a middle course between state level concerns for the efficient and equitable management of fisheries and local level concerns for self-governance, self-regulation, and active participation (Pomeroy, 1998). Such systems, thus, need to be sufficiently dynamic, if they are to cope with the changing socioeconomic environment which the communities usually face.

Notwithstanding the above, the history of CFM dates back to 1990s, when the government realized the need to involve the local communities in fisheries management. These efforts had been followed by the 1997 Fisheries policy that included a clause on comanagement under Objective (8) and in 1998, formal collaborative fish management units called the BMUs were formed. The BMUs in turn were made up of various local stakeholders surrounding the water bodies. In this regard, it could be said that the Lake Victoria Fisheries Organization had pioneered the comanagement initiatives around Lake Victoria as implemented by the Lake Victoria Environment Management Program (LVEMP) in Phases I and II.

Further, the Fisheries Act [2003, part 5, section 18 (1)] also catered to the need for the establishment of BMUs in all coastal villages in order to supplement implementation of CFM schemes. A number of initiatives were made throughout Tanzania and one of them was the Rufiji-Mafia-Kilwa (RUMAKI) Seascape Program by WWF. RUMAKI was developed as a strategy to empower the coastal communities to take advantage of provisions of the Fisheries Act (2003) in order to decentralize decision-making, planning, and management of fisheries resources, in collaboration with the government.

Concomitant to the above, the comanagement initiatives had been clearly elaborated in the Fisheries Regulations of 2009, which included setting up of mechanisms for its implementation. Thus, collaborative management in Tanzania had been recognized as an agreement between BMUs, district authorities, and the fisheries division. These regulations under section 133 (10) provided for the BMUs to associate with others to form high-level comanagement structures known as CFM areas for the purpose of planning, management, and development.

5.2.8.7 *Opportunities in the fisheries sector*

The enforcement of fisheries legislation enhanced sustainable conservation, protection, development, management, exploitation, utilization, and marketing of fishery resources in order to provide food, income, employment, foreign exchange earnings, effective protection of aquatic flora and fauna as well as the environment for the benefit of present and future generations.

5.2.10.8 *Constraints in the implementation of fisheries policy*

Notwithstanding the above, a number of constraints hampered successful comanagement of fisheries in Tanzania, as follows:

(i) Un-harmonized sector-related policies linked to conservation, development, management, and utilization of natural resources including fishery resources.
(ii) The existing administrative setup on the management of fishery resources (central government and local authorities) could not be very efficient or very effective.
(iii) Weak capacity for the enforcement of fisheries legislation including insufficient financial and human resources.
(iv) Inadequate promotion of value-added fish and fishery products to increase income.
(v) Not very strong mechanism for data collection, processing, analysis, and storage, hence affecting monitoring in fisheries management.
(vi) Presence of not much reliable data, poor mechanism for dissemination of fisheries data, and information.

5.2.10.9 *Recommendation by stakeholders to improve fisheries comanagement*

The Fisheries CoManagement Guidelines, 2009, contain a list of solutions to the challenges for ensuring sustainable fisheries management through fisheries comanagement. Stakeholders consulted in this exercise have underscored these solutions and stressed that these provide additional strategies for multi-stakeholder engagement in improving fisheries management in the country. Some of the points in this regard are as follows:

(i) Continuous conservation, development, management, and utilization of fishery resources in a sustained manner with support from donors.
(ii) Strengthening the collaboration among the key stakeholders on the sustainable utilization, development, management, and conservation of fisheries resources for the benefit of both the present and the future generation.
(iii) Establishment of an effective mechanism to lessen unlawful fishing practices, control/regulate fishing capacity, and to promote alternative livelihood for the coastal communities.
(iv) Continuous training to fisheries staff and fishing communities on natural resources management and conservation, entrepreneurship skills, good leadership and governance, conflict management, and cooperation.
(v) Promoting and strengthening community-based collaborative management of natural resources in marine waters.
(vi) Reviewing and updating fisheries policy and legislation from time to time.

5.2.10.10 *Lessons and scopes in the fisheries co-management*

There are a number of scopes and lessons, which the stakeholders could tap for contributing toward improved fisheries comanagement. Some of these are as follows:

(i) Political will to support sustainable fisheries management.
(ii) Availability of key stakeholders dealing with sustainable fishery resources.

(iii) The presence of development partners and the private sectors, notably the NGOs, which among others, support sustainable conservation and development of natural resources.
(iv) Continuous training to fisheries staff and fishing communities on natural resources conservation and management, entrepreneurship skills, good leadership and governance, conflict management, and cooperation.
(v) Promoting and strengthening community-based collaborative management of natural resources in marine waters.
(vi) Reviewing and updating the fisheries policy and legislation from time to time.
(vii) Establishment of national CBNRM forum

One of the objectives for the stocktaking exercises was to ascertain whether there is a need for establishing a national CBNRM forum. Apart from individual interviews during this exercise, discussions and recommendations were also collected during the validation workshop of the stocktaking report.

Notwithstanding the above, stakeholders agreed, in general, that TNRF could continue to coordinate and convene the national CBNRM forum. However, the following recommendations were provided from various stakeholders throughout the consultation process.

(i) To buildup on existing forum
TNRF is a membership-based organization with representatives from civil society organizations, academia, private sector, and ordinary citizens. It had already built a platform for collaboration and sharing of information.
(ii) To make it clear about distinction of roles
In situations, where it is necessary to establish the proposed forum, there is a need to clearly distinguish the roles of TNRF and that of the proposed CBNRM forum. What different type(s) of things could be proposed for CBNRM forum, which are already being done by the TNRF. It was an issue that need to be discussed and determined.
(iii) To try to have continual financing
The proposed CBNRM forum could be more helpful if financial mechanisms were included that could help its continuity.
(iv) To have a strong coordination
The coordination could, preferably, be at a central point to link all interested parties, which include a range of players, notably, the government, private sectors, civil society organizations, academia, local communities, etc., all involved in a single forum.
(v) To try for a conclusion on the status of CBNRM in Tanzania

The following could be drawn as conclusions on the status of CBNRM and its approaches in forestry, fisheries, and wildlife sectors in Tanzania:

(i) There seemed to be weak governance, especially, at the local level in terms of capacity to respond to challenges of management structures, such as WMAs, BMUs, etc.
(ii) There seemed to be inadequate recognition on the contribution of CBNRM to national development planning and policy by the government.
(iii) There seemed to be no effective devolution of power for all models, notably, BMU, JFM, and CCS models. However, there seemed to be rhetoric power devolution in models, especially in CBFM and WMA models.

Nevertheless, there were some success stories, although scattered, where benefit sharing mechanisms had been clear, and collaborative governance being enhanced.

Notwithstanding the above, CBFM appeared to be more progressive than JFM in terms of decision-making and rights given to local communities to manage forests and fisheries resources.

5.2.11 Summary

(1) The **African continent** is rich in water bodies and aquatic resources having myriads of wetlands, lakes, and rivers. This calls for efficient sustainable harvest and management of the water bodies and their bioresources, notably, the fisheries.
(2) There have been a number of works pertaining to community-based resource management in **Africa (in General).** Anon (2009) identified global marine capture fisheries as an underperforming global asset,

including those of Africa. A report by Anon (2012) also illustrated the potential economic gains to be expected from a transition toward sustainably managed fish stocks including those of Africa. According to a report, the potential economic gains from restoring the stock of the Hilsa Shad fishery, one of the largest single species fisheries in India and Bangladesh, could be to the extent of *c.* US $260 million annually. But, unhappily, it may be emphatically mentioned that much economic benefit could not be achieved with the Hilsa shad fishery in Sone Beel, one of the largest (area 3458.12 ha at FSL) wetlands in Assam (India) in the BAU scenario (Kar, 1990, 1996, 1997, 1998, 2000, 2002, 2003, 2004, 2007a, b, 2012, 2013; Kar et al., 2008; Kar and Barbhuiya, 2011a, b; Kar, 2005a, b, c; Kar and Dey, 1982, 1986, 1987; Dey and Kar, 1990; Yadava et al., 1989). Incidentally, rather, there had been a *hitherto* rare occurrence of juveniles (advanced stage of fry) of *Hilsa (Tenualosa) ilisha* in Chatla Haor seasonal floodplain wetland of Assam (Kar and Dey, 2002). Incidentally, Melcher et al. (2013) dealt with the current questions in sustainable fisheries, water management, and higher education with regard to Burkina Faso. Notwithstanding the above, fish could improve the nutritional status. FAO statistics indicate that fish is a source of *c.* 22%–50% of protein in **sub-Saharan Africa.** Further, Anon (2008) dealt with CBFM in **Africa, in general**. Roe et al. (2009) dealt with community management of natural resources (including fisheries) in **Africa (in general)** along with impacts, experiences, and future directions.

(3) In **Botswana, Malawi**, Piers (2006) dealt with community-based natural resource (including fish and fisheries) management in **Malawi and Botswana**.

(4) In **Ghana**, Crawford et al. (2013) discussed about solving the fisheries crisis in **Ghana** with a fresh approach to CFM.

(5) **Lake Victoria** is one of the biggest lakes in Africa and in the world. In **Lake Victoria**, Salehe (2008) reported about the capability and legitimacy of BMUs to improve fisher's income through the management of firsthand sales system in **Lake Victoria** in **Tanzania**.

(6) In **Madagascar**, Anon (2013a, b, c) dealt with protection and promotion of marine BD through small-scale, time-sensitive, community-based programs. McClanahan et al. (2014) dealt with the perceived benefits of fisheries management restrictions in **Madagascar**. Incidentally, Anon (2015) dealt with community-based AQC as a pioneering viable alternative to fishing in Madagascar.

(7) Unique about **Namibia** was that a **catch share system** was adopted in a poor nation with a population consisting of several deeply rooted tribes (Laura, 2011).

(8) In **South Africa**, Community Property Association Act of 1996 allowed communities to establish legal common property institutions (including fisheries), which could be meant for the development of fisheries and AQC.

(9) In **Senegal**, Anon (2013a, b, c) discussed about the concerted management of the sea bream fishery in Kayar, **Senegal**.

(10) In **Somalia**, Land Policy (2005) allowed for ownership of land by chiefs and families for community activities like fisheries, etc. (Unruh and Turray, 2006).

(11) In **Sudan**, comprehensive Peace Agreement (2005) called for the recognition of customary land rights including community lands for fisheries and AQC.

(12) In **Tanzania**, Anon (2013a, b, c) dealt with CBNRM stocktaking exercise. The **TNRF** embarked upon a process to establish a series of dialogs on **CBNRM**. The aim of this initiative was to provide a platform for multi-stakeholder exchange of information, sharing experiences, and agreeing upon a common course of action for the success of CBNRM in the country. Notwithstanding the above, among other African countries, **in Tanzania**, there is participatory approach to irrigation development, which is extended to fisheries and AQC development. Revenues could be generated through the sale and export of wild-capture fisheries. Moreover, enhanced export revenues may arise from the export in an eco-friendly sustainable manner. Further, in **Tanzania**, the Fisheries Act (2003) allowed for the establishment of BMU along FW coastlines (TNRF, 2009a, b). Further, the National Strategy for Growth and Reduction of Poverty (2005) aimed at improving sustainable use through CBNRM including fisheries. The strategy placed a strong emphasis on improving local land and resource tenure and increasing local incomes from natural resources such as, forests, fisheries, and wildlife.

(13) In **Uganda**, the BMUs emerged in 2003. The duties of the BMUs in Uganda covered the daily management of the local fishery, issuing fishing permits, registering fishing gears, and working with the government fisheries department to enforce regulations against unlawful fishing practices.

(14) Some of the other researchers who had contributed to CBFM in Africa and may be elsewhere also are Adams et al. (2000), Manning (2005), Beaudry et al. (1993), Demsetz (1967), North (1990), Anderson and Leal (2001), and Eide et al. (2003).

5.3 Australasia

The continent of **Australasia** possesses a high potential in fisheries and AQC. As such there is a need for efficient methods of management and conservation. There are various *modus operandi* for the management and conservation of **fishery** resources in Australasia.

5.3.1 Australia

Australia has rich fisheries potential. There have been much works done on various aspects of fisheries in different parts of the country. In **Western Australia**, Metcalf et al. (2009) deliberated upon conceptual models for EBFM.

The Department of Fisheries, Western Australia was reported to be developing a framework to assess the most appropriate methodologies for the implementation of EBFM in Western Australia. The implementation of EBFM could be considered as an operational extension of ecologically sustainable development (ESD), the principles of which were adopted by Western Australia in 2002, for application in the management of fisheries. The EBFM aims at assessing and managing ecological impacts as well as social and economic outcomes related to fish and fisheries at a regional level. The degree to which EBFM could be implemented in terms of activities or processes are yet to be ascertained. This report forms component of a project designed to address the question as to what benefits EBFM could generate if applied in Western Australia. Fisheries management in Western Australia traditionally focused on maintaining the sustainability of targeted stocks. As such it has comprehensive data sets on the catches, biology, and ecology of many exploited stocks. Incidentally, quantitative data regarding the ecosystem, social and economic impacts of fisheries, and formal consideration of the roles of external drivers and governance are limited. Beyond these limitations, there were also the challenges to identify linkages and important relationships both within the ecosystems (including exploited stocks), and across social, economic, and governance facets of fisheries management.

Qualitative modeling (also known as loop analysis) uses feedbacks to investigate the impacts of perturbation on system stability and produce predictions of change in ecological, social, and economic aspects of systems. This technique may be used when quantitative data are lacking, as it requires only the signs of interactions between the model variables (i.e., positive, negative, or zero). Further, there may be qualitative information based on expert opinion(s) suggesting that an increase (perturbation) in the number of iconic fish in an area could have a positive effect on the general community through the maintenance of cultural and heritage values. Qualitative modeling may also be used to identify the critical thresholds for change in a system and may also be useful to highlight gaps in data. It could also assist in identifying significant relationships. Such information, along with the capacity to investigate system stability, may allow the test of the efficacy of alternative management strategies and could be used to indicate likely methods of helping species recovery following perturbation. In addition, qualitative modeling could be used to investigate five separate systems within the West Coast Bioregion. Ecosystem dynamics and the importance of social and economic links could also be examined using different scenarios for each of the five systems. The systems included the following:

(i) The role of direct stakeholders in influencing government decisions.
(ii) Management needs in Cockburn Sound.
(iii) The impacts of the market on the western rock lobster fishery.
(iv) Alterations to recreational fisher behavior following hypothetical changes in management.

Notwithstanding the above, investigations involved the examination of system stability and the inclusion or removal of links between particular variables of interest to compare predictions between alternative models. In addition, model predictions following a positive perturbation, i.e., positive disturbance or increase could also be done to examine various scenarios.

Concomitantly, Elizabeth (2005) dealt with sustainability certification in community-based fisheries.

CBFM is an efficient *modus operandi* for the management of fisheries in the globe. However, some quarters believe that management of fisheries at the global scale through the application of CBFM *modus operandi* may not always be very easy.

Although world fish production reached *c.* 120 mmt in 1997, *c.* 200 million people around the globe, who depend on fishing for their livelihood find themselves squeezed out of this way of life. Survival of these artisanal fishermen is inextricably linked to sustainability in marine resources, and the intrinsic characteristics of artisanal fisheries are usually small vessels, low waste, seasonal diversity, and fishing within a small distance from the shore. Incidentally, proper monitoring may tend to make overharvest a near impossibility. Moreover, the regulations and subsidies that favor industrial fisheries were making perverse incentives for communities and governments alike to harvest unsustainably.

5.3.1.1 *The marine stewardship council (MSC)*

International failure to protect fish stocks prompted conservation organizations to examine alternatives to traditional management in order to ensure both fish stocks and fishing communities to continue with rich fishing in the future. One such organization, the Marine Stewardship Council (MSC), had developed a strategy for providing market-based economic incentives to fisheries that operate in a sustainable manner via an eco-label certification program. The MSC is a nonprofit, NGO that seeks to promote the long-term sustainable use of the world's marine resources. The initial mission of the MSC was "to work for sustainable marine fisheries by promoting responsible, environmentally appropriate, socially beneficial, and economically viable fisheries practices, while maintaining the BD, productivity and ecological processes of the marine environment." Fisheries that voluntarily conform to MSC standards of sustainability could request assessment of their fishery. A fishery is to have all aspects of their fishery certified as sustainable in order to use the product logo of the MSC to gain access to high-dollar specialty or "green" markets. This so-called "chain of custody" takes the fish product from the point of capture, i.e., the fishery, to the point of sale, the market, and ensures that no other, unsustainably harvested products have been mixed-in.

Rigid inspection of the chain of custody is essential to the credibility of the MSC, and guarantees that the product purchased by the consumer is, in fact, from a sustainable source (MSC, 2001). The MSC hopes to use this process to provide an economic benefit to fisheries operating in a sustainable manner and to provide an economic incentive for other fisheries to change their unsustainable practices. Ideally, fishers operating within an MSC certified fishery could exert a higher price for their products from socially and environmentally conscious consumers (Charles et al., 2003). Since the first MSC fishery certification in 2000, the overwhelming majority of fisheries that achieve certification follow a pattern: originating in developed nations with excellent data collection, and exclusive access to their resource. This may not be the case for community-based fisheries, which are usually characterized by a lack of sufficient data, lack of exclusivity, and a lack of organization (McGoodwin, 1990). However, in contrast to many industrial fisheries, the MSC certified only one community-based fishery as sustainable, begging the question: can this model effectively appraise the sustainability of a community-based fishery? Both scientists and conservation organizations expressed concern that the MSC certification program may not be able to measure and assess community-based fisheries, especially those located in developing nations and in areas where management is not based on Western-style scientific information (Schärer, 2001). In order to evaluate the success of the MSC certification model on community-based fisheries, there were works which examined the failed certification of one such community-Prainha do Canto Verde in Northeast Brazil. Efforts are said to be in progress to implement such *modus operandi* on a wider scale.

5.3.2 Fiji

Fiji is a country with high fishery potential. Dealing with fisheries may involve different types of conflicting situations. The word "conflict" generally carries negative connotations. It is often viewed as the opposite of cooperation and peace. This idea of conflict as negative may not always be helpful. Incidentally, Michael Warner (2000) dealt with conflict management in community-based natural resource projects with a note on experiences from **Fiji and Papua New Guinea**.

Notwithstanding the above, in nonviolent settings, it could often be seen as a force for positive social change. Its occurrence could be a visible demonstration of society adapting to a new political, economic, or physical environment. Incidentally, the management of renewable natural resources is an area of international development and, of late, is subjected to a wide variety of fast changing developmental pressures. In fact, the pressure is exerted on individuals and groups in a number of ways, including the inception of new technologies, commercialization of CPRs, more involvement of rural communities in conservation measures in their avocation, and privatization of rural public services. Other pressures result from growing consumerism, government policies that may not be much supportive of CBNRM, and a general decline in the terms of trade for agricultural produce. Incidentally, the conflicts and disputes which arise from these factors are not something that could be easily avoided, suppressed, or eliminated. As such, development models are required, which consider conflict as a potential obstacle to sustainable development.

Notwithstanding the above, less common, but of increasing relevance, is the design and implementation of "smart" community development projects, which could be particularly in the area of renewable natural resources impacting the life and livelihood of fisheries development and management (Ndelu, 1998). This approach capitalizes on the stakeholder cooperation. The design of community projects to overtly contribute to resolving conflicts and building peace is part of the new strategic thinking of a number of donor agencies, including the Swedish International Development Co-operation Agency (SIDA), the Canadian International Development Agency (CIDA), the UK Department for International Development (DFID), and the World Bank. The approach could also be related to the new "Framework for Cooperation" between the World Bank and the United Nations High Commissioner for

Refugees (UNHCR), which argues for conflict resolution strategies that "bridge humanitarian aid and sustainable development" (World Bank, 1998).

5.3.2.1 *Poverty alleviation*

The ongoing debate to define **sustainable development** seemed to suggest that conflicts over the utilization of renewable natural resources could be avoided or reduced through greater stakeholder participation during project planning and management (World Bank, 1995; UNCED, 1992; Winterbottom, 1992). This emphasis on participation is particularly relevant to "the poor."

In fact, it is by no means certain that promoting NRM that make poorer groups economically more secure could reduce intergroup tensions.

5.3.2.2 *Biodiversity conservation*

Growing conflicts between wildlife conservation interests and local communities over the utilization of natural resources are well documented (Jusoff and Majid, 1995; Wells and Brandon, 1992, 1993; Machlis and Tichnell, 1985). The dominant response to these disputes has been the schemes that raise the value of conservation to local people through the distribution of revenues from tourism, etc. or through community development designed to compensate for loss of access to conservation-worthy resources. However, the effectiveness and reach of these schemes have been limited. Further, in many conservation and PAs, conflicts over resources persist (Metcalf, 1995; IIED, 1994).

In fact, they persist for a number of reasons as follows:

(i) The continuing dominance of conservation goals over the livelihood needs of local people.
(ii) An emphasis on reducing the dependency of local people on resources of conservation value, rather than increasing their stake in sustainable resource management.
(iii) Introspective community participation planning techniques that omit consideration of external constraints.

5.3.2.3 *Types of conflicts in natural resource management*

There could be different types of conflicts in NRM in terms of their occurrence, as follows:

(i) Those occurring at the micro-micro level.
(ii) Those occurring at the micro-macro level.

The former could occur among community groups. On the other hand, the latter could occur between the community groups and the external agencies like the government, private, civil society organizations, etc. (Grimble and Wellard, 1997). Moreover, the micro-micro conflicts could further be categorized into the following types:

(i) Those taking place either within the group directly involved in a particular resource management regime (e.g., a forest user group or ecotourism association).
(ii) Those occurring between this group and those not directly involved (e.g., between the user group and women entering the forest to collect fuelwood) (Conroy et al., 1998).

5.3.2.4 *Types of conflicts arising in NRM intra-micro-micro conflicts*

These could be of the following types:

(i) Disputes over land and resource ownership, e.g., between private and community land owners.
(ii) Disputes over land boundaries between individuals or groups.
(iii) Latent family and relationship disputes.
(iv) Disputes due to natural resource projects being captured by élites and/or those who happen to own resources of a higher quality.
(v) Breaking of CPR constitutional or operational rules, such as protection agreements for grazing areas, fish net and mesh sizes, forests, etc.

5.3.2.5 *Inter-micro-micro conflicts*

These could be of the following types:

(i) Conflict between land owners and resource users.
(ii) Conflict between indigenous CPR groups and more recent settlers.
(iii) Lack of cooperation between different community groups.

5.3.2.6 *Micro-macro conflicts*

These could be of the following types:

(i) Contradictory natural resource needs and values, e.g., between wildlife habitat protection and local livelihood security.
(ii) Cultural conflicts between community groups and outsiders.
(iii) Disputes over project management between community groups and outside project sponsors.

5.3.2.7 *Causes of nonviolent conflict*

The causes of nonviolent conflicts in CBNRM may be classified into four principal categories as follows:

(i) demographic change.
(ii) natural resources competition.
(iii) developmental pressures.
(iv) structural injustices.

The combination of demographic change and the limits to sustainable harvesting of renewable natural resources (forests, water bodies, fisheries, grazing areas, marine resources, wildlife, and agricultural land) are often cited as the underlying cause of conflict over natural resources, both among community groups and between community groups and outside public and private organizations.

In fine, the disputes and conflicts over CBRNM could be viewed in the context of a complex web of demographic change, sensitive natural environments, new development pressures, structural economic and legal inequalities, personal and ethnic differences, and the multiple interests of different individuals, groups, and organizations from both inside and outside rural communities.

Further, in **Fiji**, Anders et al. (2011) dealt with mapping fish community variables by integrating field and satellite data, object-based image analysis and modeling in a traditional **Fijian fisheries management area**. The use of marine spatial planning for zoning multiuse areas is growing in both developed and developing countries. Comprehensive maps of marine resources, including those important for local fisheries management and BD conservation, provide a crucial foundation of information for the planning process. Using a combination of field and high spatial resolution satellite data, an empirical procedure could be used to create a bathymetric map (RMSE 1.76 m) and object-based image analysis to produce accurate maps of geomorphic and benthic coral reef classes (Kappa values of 0.80 and 0.63; 9 and 33 classes, respectively) covering a large (> 260 km^2) traditional fisheries management area in Fiji. From these maps, per-pixel information on habitat richness, structural complexity, coral cover and the distance from land could be derived and used as input in models to predict fish species richness, diversity, and biomass. Studies show that random forest models outperform five other model types, and that all three fish community variables could be satisfactorily predicted from the high spatial resolution satellite data. Studies also revealed the geomorphic zone to be the most important predictor on the average, with secondary contributions from a range of other variables including benthic class, depth, distance from land, and live coral cover mapped at coarse spatial scales, suggesting that data with lower spatial resolution may be sufficient for spatial predictions of the three fish community variables.

Many of the people worldwide live within 100 km of a coral reef. They depend on small-scale reef fisheries to sustain their livelihoods (Bryant et al., 1998; Andrew et al., 2007). Incidentally, *c.* 60% of reefs globally are imperiled by human activities (Bryant et al., 1998). As such, protection of reef fish populations requires urgent and active management. However, tools that have long been promoted by fisheries scientists for managing single species fisheries (e.g., MSY) do not capture the ecosystem complexities of multispecies coral reef fisheries (Gaichas, 2008). Concomitantly, newer advanced models applied to coral reef fisheries, e.g., Ecopath with Ecosim (Ainsworth et al., 2008), require large amounts of data that are often resource intensive to acquire and usually do not include spatially explicit outputs. Meanwhile, the MPAs have emerged as an alternative, widespread tool for coral reef fisheries management. Its application resulted in increases in the assessment of the abundance and biomass of targeted species where successfully implemented and enforced (Lester et al., 2009). The coral reef managers could greatly benefit from comprehensive fish resource maps, which are optimally designed networks for individual MPAs. Maps of fish biomass could be used to guide placement of MPAs toward naturally productive areas, a known driver of MPA success (Jupiter and Egli, 2011). Further, in the absence of accurate information on single-species distributions, maps of species richness and diversity may be used to ensure coverage of local BD hotspots. Incidentally, ecological field research produced a large body of literature describing functional and statistical relationships between variables, which quantify the aspects of the reef fish community and the local habitat. For example, the positive effect of structural complexity on reef fish species richness and abundance had been demonstrated experimentally (Syms and Jones, 2000), and areas of high structural complexity had been shown to contain more species (Gratwicke and Speight, 2005) and greater total biomass (Friedlander and Parrish, 1998) compared to areas with less structure. Other factors, alsohad shown similar correlations with one or more aspects of reef fish communities, such as water depth (Lara and González, 1998), history of disturbances (Connell, 1978), coral cover (Jones et al., 2004), location on the reef (Friedlander and Parrish, 1998) and shelf (Christensen et al., 2003), presence of sea grass beds and mangroves (Dorenbosch et al., 2005; Mumby et al., 2004a,b), benthic composition and patch structure (Pittman et al., 2004, 2007; Grober-Dunsmore et al., 2008), etc. Such environmental variables and the patch mosaic they make interact with and influence the local fish communities at a range of spatial scales. Methods have been developed for mapping several of these key environmental variables using remote sensing. The spectral reflectance characteristics of an area could be used to characterize the dominant benthos (Mumby et al., 1997) and estimate both live coral cover (Mumby et al., 2004a,b; Newman et al., 2007; Isoun et al., 2003) and water depth (Lyzenga, 1978; Maritorena et al., 1994; Stumpf et al., 2003). The location on the reef may both contribute to classification of the benthos and be used to map geomorphic reef zones (Roelfsema et al., 2010). Spatial analysis of modeled depth estimates may be used to quantify structural complexity. Moreover, the maps of the benthos could be used to calculate indices of patch diversity, size, and shape as well as distances to other marine and adjacent terrestrial and FW habitats (Mellin et al., 2010).

Concomitant to the above, the ability to map these environmental variables with RS has in turn made it possible to employ statistical models to produce spatially explicit predictions of fish community variables (Knudby et al., 2010a,b). Only a handful of studies (Anders et al., 2011) have been published, of late, using this approach.

Notwithstanding the above, ecotourism is a commercial domain where community could also be involved. The Shark Reef Marine Reserve in **Fiji** is an ecotourism project designed to protect a small reef patch and its fauna while preserving the livelihood of local communities (Brunnschweiler, 2009).

It involves local communities by using a participatory business planning approach to MPA management, generating income through diver user fees, distributed to the local villages that had exchanged their traditional fishing rights in the marine reserve for this new source of income.

Notwithstanding the above, the Shark Reef Marine Reserve is a self-sustaining and profitable project, and is an example of a privately initiated, bottom-up approach, which includes all relevant stakeholders in an area where marine rights are finely subdivided into small units (Brunnschweiler, 2009).

Artisanal inshore reef fisheries provide a vital source of food and income for many coastal communities, and living standards in such communities are directly or indirectly linked to small-scale fisheries (Jiménez-Badillo, 2008; Kronen, 2004; Souter and Linden, 2000). Concomitantly, the artisanal fisheries have the potential to transform reefs in ways that compromise their ecological and economic values (Bascompte et al., 2005; Carr et al., 2002; Dulvy et al., 2004; Hawkins and Roberts, 2004). Further, many coral reef fisheries are poorly managed and have negative effects on the structure of the reef fish community (Cinner and McClanahan, 2006; Mangi and Roberts, 2007; Zeller et al., 2006). Degraded coral reef ecosystems, in turn, have socioeconomic impacts on communities with strong reliance on coral reefs (Ochiewo, 2004; Turner et al., 2007). MPAs that permanently or partially close off fishing grounds have been widely adopted as a leading fishery management tool for marine conservation.

However, many marine reserves are ineffective or may remain as "paper parks" and critical gaps in the biological knowledge such as connectivity and the anticipated fish recruitment may impede the use of no-take zones and preclude implementing them with confidence that they could sustain surrounding fisheries (Christie and White, 2007; Grafton et al., 2005; Lester and Halpern, 2008; Mora et al., 2006; Palumbi, 2004; Sale et al., 2005). While the "where" and "why" questions about marine park establishment are relatively easily addressed, it is often the "how" that is the most difficult question to accomplish successfully (Aswani, 2005; Kelleher and Kenchington, 1992; Souter and Linden, 2000; Wescott, 2006). Incidentally, in poor, developing countries, no-take MPAs are, generally, proposed as a versatile marine fishery management tool. However, they are often difficult to establish or they may remain ineffective due to their incompatibility with traditional customary marine tenure systems or the lack of funds to manage them (Cinner and Aswani, 2007; Faasen and Watts, 2007; Foale and Manele, 2004; Reid-Grant and Bhat, 2009). Traditional marine parks that conserve resources through strict regulation of access may not work because of the dependence of local communities on these fishing resources for their economic and physical well-being (Cinner, 2007; Jiménez-Badillo, 2008).

A holistic approach to these issues tried to focus on the local people, who are most affected by the implementation of the marine reserve(s). Such holistic approach is to recognize and appreciate the needs, perceptions, and attitudes of local stakeholders toward the environment as well as conservation measures are to be planned before implementing new management policies. Also, traditional systems of resource management are to be taken into consideration (Aswani, 2005; Hoffmann, 2002). Incidentally, a **hybrid management system**, merging customary and modern management, could be an effective way to improve reef ecosystem health and provide significant conservation benefits for reef fish stocks while at the same time meeting utilitarian community goals and taking into account the customary practices or traditional knowledge (Adams, 1998; Aswani, 2005; Caillaud et al., 2004; Cinner, 2007; Cinner and Aswani, 2007; Jiménez-Badillo, 2008; McClanahan et al., 2006; Rodríguez-Martínez, 2008).

Concomitant to the above, the introduction of management measures, such as "closed areas," may need to be coupled with projects to develop alternative income strategies (Jiménez-Badillo, 2008; Souter and Linden, 2000). Further, instead of the traditional approach of protecting reefs, it could be better to return them to a more productive state and then deciding how best to use the conservation gains to benefit people in a sustainable manner. Incidentally, an alternative approach is to make incentives during the planning and implementation phases to protect the resources that provide livelihoods (Bell et al., 2006; Sale et al., 2005; Weiant and Aswani, 2006). Coupling marine tourism, which is a component of the wider ecotourism sector and which is growing rapidly in terms of both its volume and value (Garrod and Wilson, 2004), with marine park management offers an income stream that additionally may have a positive influence on conservation awareness and support (Diedrich, 2007; Johannes, 2002; di Sciara et al., 2009).

Concomitant to the above, the marine megafauna is attracting increasing attention in ocean conservation planning and threatened predators such as sharks. They are often used politically to promote reserve declaration (Hooker and Gerber, 2004). Further, there could be various ways in which local communities are, and could become more, involved in reef management. Annette et al. (2006) dealt with community-based marine resource management in **Fiji** in the context of yesterday to tomorrow. Attempts are sometimes made to highlight comprehensively successes and failures but with little rigorous assessment of what the conditions of long-term success are.

Fiji is considered to represent a unique case, in view of the fact that its customary fishing-rights areas ("*qoliqoli*") constitute a kind of dual ownership. This may establish a connection between indigenous owners and central government for the purposes of management. But, as reported, this kind of so-called cooperative comanagement approach had not been uniform across Fiji. It could be mainly because of the involvement of different levels of the government and the community; with the result that the approach becomes confusing with disputes. Further, in Fiji, one of the more issues is said to be still unresolved, i.e., whether "ownership" of the *qoliqoli* should include both the marine resources and the sea floor, the latter being currently owned by the state.

Notwithstanding the above, there are reviews which may take critical look at the issues like "traditional conservation," origin of the prevailing *qoliqoli* system, peoples' perception regarding this, the forces driving its evolution, the impact on marine resource use, and so on. Further, it shed some amount of light on as well as explores the boundaries of "comanagement" in Fiji. It could also assess previous efforts to find a way forward.

5.3.3 Micronesia (Oceania)

Micronesia is a fishery-rich country with community management aspects *in vogue*. In **Micronesia**, Rhodes et al. (2011) conducted interdisciplinary study of the market forces and fishery management under the Coastal Fisheries Management Program.

A successful fishery with efficient management system involving the community stakeholders looks for effective markets for confirming profits in the commercial arena. Notwithstanding the above, there had been contributions toward interdisciplinary assessment and analysis of the market forces, both internal and external, that are driving the increasing demand for coastal fish resources within Micronesia. There had also been attempts to investigate the potential scopes for reinforcing community and traditional rights-based coastal fisheries management with contemporary rights-based and market-based fisheries management approaches that are increasingly being applied internationally to manage fisheries. At a time when Micronesia's coastal fisheries were recognized by its leaders as being increasingly overharvested, there was also a so-called transformation toward traditional rights-based coastal fisheries management. This had been occurring at a time when internationally there is mounting recognition of the potential of rights-based and market-based fisheries management approaches.

Such studies reflect the following aspects:

(i) To identify, review, and analyze the internal and external market forces and their impacts on coastal fisheries sustainability in Micronesia.
(ii) To review the range of existing community and traditional rights-based marine resource management practices in Micronesia and to assess them in terms of modern rights-based fisheries management approaches and the EAFM.
(iii) To evaluate and propose possible approaches to incorporate modern rights-based fisheries management with community and traditional Micronesian fisheries management systems to reinforce local management of coastal fisheries, preferably, within an ecosystem approach context.

(iv) To identify appropriate and innovative sustainable financing options for coastal fisheries management within Micronesia, including market-based approaches and alternative livelihood options.

Notwithstanding the above, various studies contributed to the discussions around a shift in Micronesia's coastal fisheries policies to support and reinforce community and traditional rights-based fisheries management systems and the adoption of an EAFM. The protection of vital coastal fisheries for communities and the development of innovative approaches to coastal fisheries management and financial sustainability had also been dealt with within the context of the Micronesia Challenge. The initial focus was on applying these approaches within Micronesia. However, they could also be more broadly relevant within the Pacific, including Hawaii.

5.3.3.1 *Contextual aspect*

The diverse natural resources of Micronesia are natural capital for local people. The leaders took action by launching the Micronesia Challenge by recognizing the growing local and global threats to their natural resources and the Micronesian way of life. This could be considered as an unprecedented commitment to protect the habitats and resources that sustain their people. However, the Micronesia Challenge jurisdictions are facing significant threats to the management of their coastal fisheries due, in part, to internal and external market forces. It thus places increasing demands on the coastal fisheries resources at a time of weakening the traditional resource management systems. Incidentally, all the five jurisdictions had identified the overharvesting of reef fisheries as a priority threat that needs prompt attention. But, the measures could be adopted in a manner appropriate to the existing governance systems, both traditional and legislative. In fact, Micronesia offers a wide spectrum of fisheries governance, from strong traditional management practices (that reinforce tenure and access rights, such as those in Yap State, FSM) to government-managed, open-access fisheries (like Guam, which are driven primarily by market forces). As a result, the region provides the scope for innovative coastal fisheries management approaches by developing culturally appropriate fisheries management systems. Some of these were built on the region's long traditions of sustainability, while, concomitantly, incorporating modern market-based responses.

Concomitant to the above, the significance of coastal fisheries to the economics of Micronesia reflected in the increasing attention received from the value of the coastal fisheries to the national economy. Further, the coastal fishery is significant to the local markets (Rhodes et al., 2011). However, internationally, there had been a shift to market-based and rights-based approaches to fisheries management to increase the ecological and economic performance of fisheries. At the same time, there were attempts to improve the sustainability of coastal communities (Costello et al., 2008). Similarly, there was a recognition within the Pacific region, including Micronesia, regarding the need to adopt and implement an EAFM.

Concomitant to the above, the purpose of certain other studies was to undertake an interdisciplinary (fisheries, economics, and social science) assessment and analyses of the internal and external market forces that were driving the increasing demand for coastal fish resources within Micronesia. Incidentally, there were also attempts to investigate the potential scopes to reinforce community and traditional rights-based coastal fisheries management with contemporary rights-based and market-based fisheries management approaches. These are increasingly being applied internationally to more effectively manage fisheries. The long-term goal ensures sustainable coastal fisheries management practices that address the three contemporary challenges of overharvesting, escalating market forces, and a continuing shift away from the traditional management of fishing.

Concomitant to the above, Anon (2014) dealt with community-based ecosystem approach to fisheries management (CEAFM) and climate change adaptation in the State of Yap, Federated States of **Micronesia** (FSM), with regard to the current status of marine resources in FSM.

Federated States of Micronesia (FSM) possess a rich potential in fisheries and AQC. In FSM, recent surveys indicated that the catches of fish and shellfish are declining in lagoons and inshore reefs. However, the main reasons for this decline are as follows:

(i) Overexploitation due to growing human population.
(ii) Use of efficient and destructive fish catching devices and practices.
(iii) Other land-based activities, such as near-shore infrastructure development, affecting marine habitats.
(iv) Impacts of climate change, which are expected to add to the already existing local threats to mangroves, coral reefs, sea grasses, and intertidal flats, resulting in the decline in the quality and area of all habitats.

Notwithstanding the above, projections had depicted progressive decline in productivity of all components of coastal fisheries. Incidentally, national and provincial governments, municipalities, and island communities implemented a CEAFM program with technical assistance from the Secretariat of the Pacific Community (SPC) and in partnership with NGOs in order to address this problem.

Incidentally, the CEAFM program represents a combination of three different perspectives: (i) fisheries management; (ii) ecosystem management; and (iii) community-based management.

Notwithstanding the above, CEAFM embraces the management of fisheries, within an ecosystem context, by local communities supported by governments and other partners, notably, SPC/GIZ CCCPIR's support.

Since mid-2012, the regional SPC/GIZ program Coping with Climate Change in the Pacific Island Region (CCCPIR) had been supporting CEAFM with the aim of improving community awareness on climate change impacts and assisting communities in maximizing scopes to adapt to social, economic, and environmental effects of climate change. This could strengthen the capacity of local communities to respond to climate change impacts through the application of integrated coastal resource management, conservation, and adaptation. Thus, the resilience of marine ecosystems could be improved as well as the livelihoods of people depending on them.

Incidentally, the GIZ is a federally owned enterprise that supports the German government in the field of international development cooperation. Notably, GIZ had been cooperating with the Pacific Island partners for >30 years in strengthening the capacity of people and institutions to improve the livelihoods of communities of the present generation and also of the posterity. In fact, GIZ is an implementing agency providing support through technical cooperation to balance economic, social, and ecological interests by means of multi-stakeholder dialog, participation, and collaboration and community consultations in the Yap State.

5.3.3.2 *SPC activities*

The State of Yap was chosen to begin the activities in the pilot sites under the supervision of the FSM national government.

5.3.3.3 *Activities*

Coastal fisheries resources management

Community consultations were undertaken and community fisheries management plans were developed.

The management activities included the following:

- **(i)** Strengthening of MPAs.
- **(ii)** Control of destructive fish catching devices and practices.
- **(iii)** Control of land-based activities that may impact the marine environment.
- **(iv)** Rehabilitation of the critical habitats.
- **(v)** Strengthening of community enforcement and monitoring.
- **(vi)** Identification and development of alternative livelihoods [deployment of fish aggregating devices (FADs) and postharvest activities] with the aim of diverting fishing pressure from coastal fisheries to offshore areas and facilitating climate change adaptation.
- **(vii)** Strengthening of community leadership and governance.

Alternative livelihoods and adaptation: Fish aggregating devices (FADs)

FADs improve access for coastal subsistence and artisanal fisheries to skipjack, yellow fin tuna, and other pelagic fish as coral reef production registered a decline because of climate change. However, the FADs may be used for management purposes, as they alleviate fishing pressure on inshore areas by facilitating access to offshore fisheries. They also provide alternative livelihoods to local communities.

However, the activities included the following:

- **(i)** community training on FAD construction, management, deployment, and maintenance;
- **(ii)** ongoing data collection and monitoring activities to evaluate social and economic benefits of FADs;
- **(iii)** community training on FAD fishing gear and technology; and.
- **(iv)** stockpiling materials for repair, maintenance, and new deployments.

Capacity building

Capacity building is an important task mainly to increase efficiency of the workers. This was undertaken through training attachments for state counterparts at SPC HQs on resource management planning processes and monitoring.

Next steps for CEAFM and climate change adaptation program are as follows:

(i) State-wide awareness campaigns/climate change open day.
(ii) Official launch by national and provincial governments, partners, and communities.
(iii) Capacity building for FSM national researchers to conduct resource monitoring and scientific observations in all provinces.
(iv) Exploration of postharvest and preservation options for value addition.
(v) Vulnerability and adaptation assessments for selected communities.

Furthermore, Anon (2004) dealt with transforming the Live Reef Food Fish trade (LRFFT) toward sustainability in the Republic of the Marshall Islands (RMI) through community-based coral reef conservation and fisheries management. In fact, further assessments of the practices in and impacts of the LRFFT on the Marshall Islands could lead to better understandings.

The Marshall Islands are located in the Central Pacific. In fact, the scenario of fisheries may not be called as very bright and promising in the globe today, including those in the Marshall Islands. As such, more effective assessments of the practices in and impacts of the LRFFT on the Marshall Islands are required with the following objectives that could help to build management capacity and community awareness.

Participatory and scientific assessments

To collaborate with Marshallese communities, the College of the Marshall Islands (CMI), and the government agencies in the assessment of the LRFFT, its socioeconomic and environmental impacts, degree of management and linkages to subsistence fisheries, producing a series of participatory management recommendations for building local capacity to eliminate the destructive aspects of this trade, sustain reef fisheries and protect RMI's coral reef ecosystems, etc.

Awareness

To raise the awareness of the Marshallese communities about the destructive fishing practices. Also, to raise awareness about the LRFFT and coral reef conservation in general. Further, to show to the communities and governments in islands, through demonstration, about more informed decision-making mechanism(s) for sustainable use and conservation of their coral reef ecosystems.

Capacity building and implementation

Capacity building is essential to help the Marshallese build and maintain community and government capacity for marine conservation through the development and implementation of local and national management plans and initiatives for the LRFFT. In this connection, two major stakeholder groups had been identified: (a) the government and (b) the communities.

These were the focal points for awareness and capacity building programs.

5.3.4 Pacific Islands

The Pacific Islands are rich in AQC and fisheries. King and Lambeth (2000) dealt with fisheries management by communities in the form of a manual on promoting the management of subsistence fisheries by **Pacific Island communities**. It is impossible for the government fisheries agencies to properly manage subsistence fisheries without the involvement of fishing communities. In fact, most of the government fisheries agencies have limitations on budgets and staff which make it impossible to constantly work in communities that are, generally, spread out in many different villages. But, more importantly, community involvement results in a sense of belonging and ownership of fisheries management actions and regulations. Importantly, under community ownership, fisheries management measures are enforced by the communities themselves. Incidentally, there are literature about **CBFM**. They provide guidelines and suggestions on how communities could be encouraged to take a leading role in the management of their fisheries and the aquatic (FW and marine) environment. The Samoan model of CBFM stood the test of time and is also transferable and implementable in other island countries of the Pacific with some alterations to suit local conditions and culture. Studies, further, revealed that *c.* 60 villages are successfully managing their own fisheries resources under CBFM for > 3 years.

There could be an appeal to the government, the management bodies, and the various agencies in the world to consider the benefits of CBFM and to develop a suitable program for the participation of all the stakeholders including the village stakeholders from the grassroot level, so that efficient management of the precious abiotic and the coveted biotic resources are ensured with cut to the bottom guarantee.

Notwithstanding the above, there are some thoughtful views in this regard:

(a) There seemed to be a common concern that catches of seafood in inshore lagoons and reefs are declining in the Pacific Islands.
(b) Also, that there seemed to be a growing recognition that fishing communities themselves hold the key to prevent this decline.

Concomitant to the above, fishing and the aquatic (FW and marine) environment had always been playing a significant role in the lifestyle and culture of the people in the world, in general, and in the Australasian Continent and Pacific Island countries, in particular. Now, after many years of rapidly changing lifestyles, people are being encouraged to eat more local foods, i.e., to eat plants and meat that are traditional and healthy. Incidentally, one of the most traditional and appropriate foods for people living in the islands is, of course, seafood. But it is of no use encouraging people to eat more seafood, if it is becoming increasingly difficult to get and more expensive to buy. However, catches of the most accessible seafood, notably the fish, seaweed, and shellfish of the lagoons and reefs, have been declining in many island countries over several years. Probable reasons for the decline in inshore catches include overexploitation, the use of highly mechanized, efficient and destructive fishing methods (including the use of explosives, chemicals, and traditional plant-derived toxins), and environmental disturbances. The latter are said to result not only from natural events, such as cyclones and storms, but also from human activities and interferences. Even though inshore catches of seafood are decreasing in the Pacific Islands, **subsistence fishing** (the catching of fish to eat rather than to sell) often still results in a total catch that could be several times larger than that from commercial fishing. In spite of the obvious importance of subsistence fishing, government authorities usually emphasize on conservation and fisheries management efforts on commercial fish stocks. Nevertheless, the governments may be concerned about the falling fish catches in the villages, but their ability to do something about the problem is limited. Incidentally, it may be noted that most fishery agencies have limited number of staff and small budgets. In this connection, it may be noted that although most Pacific Islands have enacted national laws to protect fish stocks, but enforcing such laws in widespread urban and rural areas is often impossible without community involvement and support.

In continuation of the above, fishing communities are often repositories of valuable indigenous traditional knowledge (ITK) concerning fish stocks and have a high level of awareness of the aquatic (FW and marine) environment. In addition, many villages have some amount of control (either legal or traditionally assumed) over their adjacent waters. Together, these factors provide an ideal basis on which communities could be encouraged and motivated to manage their own aquatic (FW/marine) resources. So, literature were made available in response to growing regional interest from fisheries and environmental agencies as well as from NGOs, particularly, about the *modus operandi* of management operations. Some of the pertinent and principal queries on which responses are usually sought are as follows:

(a) How could villagers be encouraged to manage their own fisheries?
(b) What could be done to support village communities in their efforts to conserve fish stocks and protect their aquatic environment?
(c) What particular effective conservation measures could the village communities take?
(d) What could be the responsibility of the communities and the government(s)?

"Management" and "conservation" are the essential aspects in the aquatic (FW, Marine) domain. The pertinent literature in the field are, generally, based on the following premise. The responsible and effective management of the fisheries resources could be achieved only when fishing communities themselves see it as their responsibility rather than that of the government, regardless of national legislation and enforcement. There are literature designed to provide some technical background on fisheries and the aquatic (FW/Marine) environment and to be a guide on promoting and encouraging their management. Also, manuals are available on the subject pertaining to specific sites, e.g., with regard to the Pacific Island communities. Such management efforts are community-based. However, the community-based methods presented in the concerned manuals may not be the only ones possible and in any case could be adopted after adapting to the local situation(s) (culture, customs, etc.) to a greater or lesser extent.

5.3.4.1 Fisheries

In its broadest sense, a fishery includes the aquatic (FW/Marine) environment, a resource or target species of aquatic food, and the people involved in fishing and in handling the catch. Incidentally, both the FW and the marine environments consist of the living and nonliving surroundings of the respective organisms. Thus, both corals and seawater are parts of the environment of a reef fish while FW fish, FW terrapin, and FW are parts of the FW environment. Incidentally, a fisheries resource is a population of fish, or other FW/marine animal, which is exploited.

Concomitant to the above, a detailed account on the pertinent subject related to the **Pacific Islands** is given in the following lines.

A fishing operation may be a simple one, such as the collection of shellfish on a reef, or may be a much more complicated one, such as the catching of tuna by a large fishing vessel. Handling the catch may range from the simple storage of fish on ice to the technologically sophisticated procedure of canning.

Notwithstanding the above, "fisheries" are often classified into two sectors: (a) **noncommercial** sector and (b) **commercial** sector.

The **noncommercial or subsistence sector** involves the catching of fish to eat rather than to sell. The **commercial** sector may, further, be classified into the following:

(i) Artisanal type: usually small-scale fishing to supply local markets.
(ii) Industrial type: involves large-scale fishing for canneries and export.

The present example is concerned with subsistence fishing among the coastal communities in the Pacific (Anon, 2000).

5.3.4.2 Fishing gears and methods

Fishing gears and methods used depend on the species fished. Fishing techniques in the Pacific may vary from very simple, such as hand grabbing or gleaning of shoreline invertebrates, to complex and expensive commercial operations, such as purse seining for tuna. The roles of men and women in using different fishing gears and methods are largely dictated by whether they require a boat or a canoe, and whether they involve the fisher being away for extended periods. Fishing methods that can be used from the shore, on the reef or within the lagoon, and close to the village are usually practiced by the women. For example, women are often involved in handline fishing from the edge of the reef, netting in the lagoon, and gathering invertebrates from the inshore areas. Although it is becoming more common in some areas for women to accompany men on boat trips, fishing with boats is traditionally and generally a male activity. Incidentally, a large range of fishing gears is used by commercial and artisanal fishers, and some of the principal basic types relevant to subsistence fishing in the Pacific are briefly discussed in the following.

5.3.4.3 Reef gleaning

The collection of marine animals and seaweeds in the lagoons or on the reef flat at low tide is a common activity usually practiced by the women and the children. A variety of species are collected in this way, including sea cucumbers, sea urchins, crabs, shellfish, seaweed, eels, small fish, worms, jellyfish, octopus, etc. Collection could be done by hand grabbing; by digging in the sand or mud with the feet for bivalves; by overturning or breaking corals and rocks; and by using sticks and metal hooks to draw octopus, crabs, or fish from holes in the reef. However, the quantity of food collected by one person in this manner may be quite small, but the impact on the reef and marine life could be substantial.

5.3.4.4 Traps

Traps are devices said to be designed to allure the entry of particular types of animals, which are then prevented from escaping either by particular aspects of their behavior or by the design of the trap itself. Portable traps are regularly used in Micronesia and parts of **French Polynesia**. However, cane, bamboo, and mangrove wood traps were in the past throughout the South Pacific. Portable traps may be either baited or unbaited. Incidentally, the principle of baited traps is that the animals attracted to the bait enter the trap through tapered openings from where it is difficult to escape. Baited traps or pots are, generally, used to catch various carnivorous species of crustaceans, molluscs, and fish. Incidentally, only two species of lobsters, viz., *Panulirus penicillatus* and *Panulirus longipes* enter the traps. Concomitantly, traditional traps, generally made of cane and baited with sea urchins or chitons, were used to catch lobsters in Samoa, Vanuatu, and Tonga, but currently most of them are caught by spear or hand.

Incidentally, the barrier and fence traps represent one of the oldest ways of community fishing. It may be noted that the simplest traditional traps use the ebbing tide to strand fish in hollowed out areas on reefs and sandbanks, and are contained by v-shaped or semicircular walls of stone or coral. Fence traps include a wall built at right angles from shorelines and reefs to guide migrating coastal fish, such as the mullets, into a large retaining area.

5.3.4.5 Hooks and lines

Hook and line gears are used in a wide range of configurations, the simplest being handheld lines with one or more baited hooks.

In many commercial fisheries, however, the hooks are more circular in shape, rather than J shaped. Circle hooks are almost similar in design to the bone or shell hooks which have been used since prehistoric times in the Pacific Islands. When a fish strikes a circle hook, the point rotates around the jawbone, ensuring that the fish remains caught without the fisher having to maintain pressure on the line. Steel circle hooks are now used to catch tuna, sharks, and deepwater snappers. Hand-lining gear consists of one or more baited hooks attached to a line, which is weighted at the bottom in the case of demersal target species. If more than one hook is used, these are connected to the main fishing line by short sidelines or snoods.

Concomitantly, handline fishing, in shallow coastal waters, is a popular subsistence and recreational fishing method and is commonly practiced by women and children. Incidentally, handlines (or droplines) are, generally, used to catch demersal fish in deep water. Mechanical means may be used to haul in the line.

5.3.4.6 *Longlines*

A longline consists of a mainline with hooks set on short sidelines or snoods. A horizontal longline may be set near the surface for pelagic fish such as tuna or on the seafloor for demersal species such as the sharks. A vertical longline with hooks set from snoods along its lower section could be set perpendicularly in the water column in areas where the sea floor is steep or rugged. Incidentally, horizontal longlines set by fishermen in the Samoan tuna fishery are generally 8–40 km in length. This rapidly expanding commercial fishery, generally, targets albacore, yellow-fin, and big-eye tuna.

5.3.4.7 *Trolling and lures*

Natural or artificial lures attached to lines may be towed (or trolled) behind the boats to catch pelagic species, such as Spanish mackerel, dolphinfish, and tuna. In general, lures are designed to allure fish by having one or more of the following characteristics:

(a) An erratic movement when towed through the water (to resemble an injured prey).

(b) A bright or reflective surface, and fluttering appendages of feather, plastic, rubber, or cloth. Instead of artificial lures, small, silver fish such as garfish and flying fish, or pieces of larger fish, may be threaded onto one or a series of hooks (Preston et al., 1987). Traditionally, pole-and-line fishing was a significant community fishing method using live bait.

Notwithstanding the above, the fishery in French Polynesia is now a highly developed and competitive local industry, supplying fresh tuna to the local markets.

5.3.4.8 *Gill nets, tangle nets, and barrier nets*

Gill nets are panels of netting held vertically in the water column by a series of floats attached to their upper edge (the float-line or corkline), and weights attached to their lower edge (the footrope or lead line). These nets may be either anchored in shallow water, or set to drift in the open ocean. As passive gear, their catching ability depends on the movement or migration of fish through the area where the nets are set. Gill nets are, generally, used in shallow water to catch species such as mullets and mackerels. In deeper water, they may be set on the seafloor for demersal species such as the sharks or near the surface for pelagic fish such as tuna. The main advantages of gill nets are that they are highly selective, i.e., they usually have a mesh size designed to catch fish of a specific size range, and very small and very large fish are not caught. Incidentally, the principal determinant of the range of lengths of fish caught by gill nets is the "**hanging coefficient**" or "**hanging ratio**," which may be defined as the ratio of the length of the headline to the length of the stretched net. If the hanging ratio is low, say <0.5, the net is believed to hang slack in the water rather than being tight.

Notwithstanding the above, the **Barrier nets** could be set across the reef passages and channels to trap fish as they try to return to deeper water on a falling tide.

5.3.4.9 *Seine nets and drive-in netting*

A **beach seine**, in its simplest configuration, consists of a long panel of netting which is dragged around shoreline with schools of fish. The net is weighted to keep the lower side of the panel in contact with the seafloor, and has floats to keep its upper side at the sea surface. Some beach seines have a central panel of loose netting which forms a bunt or cod end to retain fish. However, means of employing beach seines vary, although in many cases one end of the net is anchored on the shore and a boat is used to drag the other end seawards in a large arc and back to the shore before hauling.

Concomitant to the above, **drive-in net fishing** is a group activity, which often involves the whole community.

5.3.4.10 *Cast nets and scoop nets*

Cast nets are, generally, common types of gears, which are used in shallow water by men and women to catch schooling species such as mullets, rabbitfishes, and scads. Fishes are stocked in shallow water and a circular, weighted net is thrown, which spreads horizontally over the school like a parachute, entangling the fish in the net. Flying fish are caught in **Polynesia** by fishermen at night, using handheld **scoop nets** and lanterns or torches. Incidentally, the Flying fish are located using canoes or dinghies and caught in the scoop nets while lying on the surface, being dazed from the light.

5.3.4.11 *Spears*

Spears, as fish catching devices, are used by men, women, and children in a variety of ways, from both above and below the water. The spear may be used from land, from a boat, or by diving beneath the water with sling-type spears and spear guns. Men and women often use torches and spears at night to catch fish at low tide. Incidentally, the use of modern, underwater flashlights has had a large impact on inshore marine life.

5.3.4.12 *Fish aggregating devices (FADs)*

Many species of fish, which inhabit the open sea, could be lured to floating objects. For example, some tunas congregate around drifting logs. This behavior has been used in the deployment ofFADs, e.g., floating rafts anchored offshore to attract pelagic fish. Incidentally, a range of materials, e.g., coconut logs, bamboo, aluminum pontoons, etc., are used in the construction of rafts. The benefits of FADs are that they reduce the search time involved in a fishing trip and, therefore, may reduce fuel costs as well as may increase fish catches. However, the costs of building and setting FADs are high and, moreover, their average life span is <2 years, mainly because of storms, wear and tear, etc.

5.3.4.13 *Fisheries management*

"Management" of resources, notably, the bioresources (including "fish"), is an essential aspect of Fishery Science and AQC. Incidentally, a group of fish of the same type, or species, living in one area is referred to as a **fish stock**. Incidentally, there are forces acting on a fish stock for controlling its numbers. It is a fact that the population of fish increases through reproduction of adult fish, which eventually results in small fish being added or recruited into the stock. On the other hand, the weight, or biomass, of the fish stock increases with the growth of individuals. Concurrent with these increases, the stock is being reduced in numbers and biomass by natural mortality and by fishing mortality as well. Fishing mortality refers to fish caught by fishers and natural mortality refers to the death of fish by other means, most commonly by predation. If a fish stock is unexploited or is fished at a low level, losses due to mortality are balanced, on the average, by gains through the recruitment of young fish or juveniles. The number of fish in the stock could, therefore, fluctuate around an average level as long as too many fish are not caught. It is for this reason that fish are referred to as a renewable resource, i.e., fish can continue to be caught and used as food forever, as long as the numbers caught are replaced by more young fish being produced. If exploitation is high, however, the number of adult fish may be reduced to a level where reproduction could not to replace the numbers lost. In such cases, the number of fish in the stock could decrease. In severe cases of overfishing, the stock may even disappear altogether. In Pacific Islands, many fish stocks have been overfished. Some species of giant clams are now extinct in Micronesia and one species has disappeared from Vanuatu, Fiji, and Tonga since the 1970s. Stocks of mullet, which have been caught in large numbers in fence traps, are very low in some islands. The renewability of fisheries resources depend on our ability to ensure that too many fish are not caught. This implies that fisheries, and particularly the amount or types of fishing, have to be controlled or managed. Concomitantly, the stock may also be reduced by natural mortality, and in exploited species by fishing mortality as well (King, 1995). Incidentally, fisheries scientists in various Pacific Islands and in regional organizations, such as SPC and the University of the South Pacific are studying the basic biology and distribution of some of the commercially important ichthyospecies. Many had attempted to determine how much of a particular species could be caught on a sustainable basis and this quantity is sometimes referred to as **the MSY**. In order to manage fisheries, that is, to ensure that fishing is done on a sustainable basis, it is necessary to apply one or more regulations. National governments in the Pacific Islands has imposed a variety of regulations that restrict fishing (input controls), restrict the catch (output controls), or protect the aquatic environment.

5.3.4.14 *The role of subsistence fishing*

Fishing has always been playing a significant role among the Pacific Island communities, for cultural, nutritional, and, more recently, economic reasons. In fact, many coastal communities have developed a strong cultural identity based on fishing and the marine environment because of their dependence on the sea and its resources. Moreover, for many countries with limited productive soil and a lack of potential for agriculture, the sea is providing the major source of protein. Subsistence activities and informal employment are recognized as their significant contributions to the economic and social well-being of Pacific island communities.

5.3.4.15 *Traditional culture and the division of labor*

Division of labor is an integral component of the traditional culture of many communities. In most countries, fishing offshore with boats has been the domain of men, while the women, usually, had concentrated their activities on the inshore areas, collecting a wide variety of species from the reef and inshore areas. Different fishing roles and areas for men and women reflect community and family obligations of both. Incidentally, the women and the children, usually, collect species from the inshore areas for obvious reasons, often as a means of supplementing their diet, when the weather is too inclement for the men to go out in their boats. Thus, the women and the children could stay close to the village by gleaning the reef, and this enables them not to neglect other necessary tasks such as gardening, cooking, etc. The men, on the other hand, often fish from the boats or canoes and are away from the village for much longer time periods. Incidentally, **community fishing** methods such as fish drives and palm frond sweeps may involve all the men, women, and children. Further, the women rarely collect or fish alone. They are often accompanied by friends, relatives, and children. Reef gleaning could be a way the family and friends may enjoy as a community activity. However, the social values of such enjoyments are difficult to quantify.

5.3.4.16 *The significance of subsistence fishing*

The relationship between the coastal communities and the sea is intimate since time immemorial. In fact, a number of customs and traditions were developed as a result of the close relationship of coastal communities with the sea both to protect the resource and to ensure harmony within the village. In fact, in many cultures throughout the Pacific, the distribution of excess catch throughout the community is based on customary obligation toward certain family and community members. Women are often the ones responsible for this type of customary distribution of the catch. In addition, some of the fishing cultures developed specialized roles guiding exactly the form of fishing that a fisherman was entitled to pursue given his social standing. Thus, in Palau, offshore fishing for several species of shark was practised only by a few prestigious specialists, with the prestige being related to the danger of the fishing method, rather than an appreciation of the shark as food. Further, in some societies, there existed a professional class of men whose occupation in the community is only fishing. Furthermore, other cultures have placed restrictions on who could catch, cook, or consume certain species, e.g., turtles in some parts of Polynesia. In this connection, it may be mentioned that religious taboos and magic were often associated with fishing before the advent of Christianity to the Pacific. Incidentally, in Kiribati, the significance and rituals associated with fishing tied the fishermen to the sea spiritually and gave them a feeling of respect for the marine environment (Mollica, 1999). It may be noted that many customs, traditions, taboos, and rituals had been eroded among the fishing communities of the Pacific Island but the significance of fishing to the cultural identity of the community remained very strong.

5.3.4.17 *Informal employment*

Informal and subsistence employment traditionally forms the basis of the Pacific Island economic and social structure. Incidentally, the informal sector included the production, home consumption, trade, and marketing of surplus agriculture and marine products, handicrafts, etc. In fact, a significant proportion of people's livelihood in the Pacific is still derived from the informal production. This is especially valid in the case in islands and villages away from the main population centers, where the cash economy and formal job opportunities are limited. Incidentally, the low levels of unemployment and poverty in the Pacific Island countries are largely due to the subsistence sector, which provides a livelihood to the people who might, otherwise, be destitute. In fact, the Pacific Island governments recognize the importance of supporting the subsistence sector and enhancing their productivity, as it supports the majority of the Pacific people. However, measuring the nonmarket output of subsistence activities is very difficult. Moreover, collection of catch data from innumerable and widely distributed fishing communities is time consuming and often could be impracticable for the generally understaffed fisheries agencies. Dalzell et al. (1996) estimated that *c.* 80% of the catch from inshore fisheries in the South Pacific, whether from reefs, estuaries, or FW, is taken for subsistence purposes, with the remainder going to the commercial markets.

5.3.4.18 *Lack of much information on inshore fisheries*

There is dearth of information on different aspects of inshore fisheries. Most Pacific countries have some amount of estimation of commercial fish production but there is usually very little information collected on the subsistence fisheries production. The development of the commercial fishing industry and the collection of information to support commercial fisheries management had been the main priority of both national governments and the outside funding agencies. In the early stages of the development of a fishery, whether it be subsistence, artisanal, or commercial, each increase in the amount of fishing effort produces a corresponding increase in the annual catch or yield. At this stage, catch rates (catch per unit of fishing effort, or CPUE) could be high, encouraging more people to enter the fishery. However, as the amount of fishing effort grows, the resulting increase in yield may not be as great and mean catch rates could go down. On the other hand, if fishing effort continues to increase, then the adult stock may be reduced to the extent that insufficient young fish are produced to maintain the fish stock. In fact, continued excessive fishing could cause FYs to decrease, on the average, year after year. **Reduction in fish stocks is, therefore, indicated by lower catches per unit of effort. In fact, catch-and-effort information are the basic data requirements for monitoring the "health" of a fishery.** Other indications of overfishing include changes in the size structure of catches, changes in the catch composition, and a shift of fishing effort to more distant grounds. In most of the fisheries of the developed countries in the temperate zone, mathematical models are used by a centralized administrative authorities to make decisions about management strategies and the regulation of fishing pressure. The collection and use of this type of information could be very difficult and also expensive for the Pacific Island countries and territories with limited budgets and fisheries personnels. This is compounded by the difficulty in dealing with a fishery that can involve a wide range of fish stocks and species, a variety of different harvest methods, many fishers per unit of catch, and a large number of distribution channels. If government fisheries agencies recognize that the key to the successful **management** of inshore marine resources lies in the hands of the **fishing communities**, who harvest those resources, they could take the first step of encouraging and motivating those communities. A parallel benefit of the fisheries personnels working closely with village fishing communities is that the collection of scientific data on subsistence fisheries could be greatly facilitated by community involvement. Fishing communities could also provide much amount of local or indigenous knowledge to supplement the scientific information.

Notwithstanding the above, a host of information, including estimates of sustainable yield by area, could be obtained from extensive surveys on subsistence fisheries. Moreover, where data are collected from different areas with similar ecological characteristics, it is possible to apply a surplus yield model (over area rather than time) not only to provide an approximate estimate of the average sustainable catch but also to indicate villages where resources are currently under stress (King, 1995).

5.3.4.19 *Problems with fisheries*

Declining fish catches

Declining trend in FY could be a common issue in many places of the world. Despite the lack of data in most Pacific Island countries, it could be said that the coastal inshore and reef areas are heavily exploited, and in many cases overexploited. This is especially applicable in the case of the near main urban communities. Incidentally, surveys conducted in some Pacific countries and territories indicated a reduction in landings of inshore species (Horsman and Mulipola, 1995; Saucerman and Kinsolving, 1995; Dalzell et al., 1996). Incidentally, reductions in total inshore fish landings in a country may be caused by few people are going for fishing or there being less fish to catch. Moreover, lifestyle changes could mean that, in some areas, less people are going for fishing due to loss of traditional culture, increased involvement in the formal employment sector, and the availability of cheap, convenient sources of protein, such as tinned fish and mutton flaps. Nevertheless, most fisheries agencies and fishing communities acknowledge that catch rates of fish and shellfish from the lagoons and inshore reefs of many areas had been declining for a number of years. Reasons for the decline in fish stocks include overexploitation, the use of destructive fishing methods, and environmental disturbances, to name a few. Overexploitation has resulted from a combination of factors, including growing population sizes, the use of highly mechanized, and destructive fishing methods. Further, the use of modern materials, such as chicken wire for fence traps and monofilament nylon for gill nets have made fishing efforts more effective. Moreover, in some cases, modest developments, such as the availability of underwater torches, which allow the spearing of fish resting under corals at night, have resulted in a dramatic increase in fishing efficiency. Nevertheless, some fishing methods are considered to be destructive mainly because they damage the environment or are nonselective in what they kill. Damaging the environment in which a fish or animal lives reduces the quality of the said environment, and could make it uninhabitable for further life. Moreover, the destructive fishing methods also tend to kill everything, including coral polyps, very small fish, and shellfish as well as the food of the target species. Further, the destructive fishing methods include the use of explosives and chemicals, such as the bleaching agents, and traditional plant and animal toxins. In addition, fish drives and some collecting activities may cause damage to the corals either directly as a result of breaking or overturning coral to catch sheltering fish or indirectly through the impact of many people moving over the reef. Moreover, some traditional destructive fishing practices are problematic as a result of growing population sizes. It may be said that, in the past, the aquatic

(FW and marine) environment was able to sustain occasional, localized damage(s) mainly because the frequency of the activities were low and fewer people were involved.

Notwithstanding the above, in some countries, the use of explosives and poisons to disable and capture fish could present a serious threat to the aquatic (FW/Marine) ecosystems and the long-term viability of the fisheries. These destructive fishing methods are include the use of toxic plants, commercially available poisons such as bleaches (Sodium hypochlorite), pesticides, explosives, etc. Toxic plant materials may be derived from the roots of the climbing Vine (*Derris elliptica*), and the nut of the coastal tree (*Barringtonia asiatica*), which are ground into a paste and wrapped in small parcels made of leaves. The fishers drive fish into the shelter of a preselected coral head, where two or three parcels of toxic material are placed. More seriously, commercial poisons including bleaches are poured into pools, which may be isolated at low tide to capture small coral fish. Explosives are either thrown from a canoe into a school of fish, such as mullets, or set on a coral, where fish may be lured to gather by setting bait. Explosives and severe toxins are believed to be many times more damaging to small animals such as the fish larvae and coral polyps, than they are to large fishes. Incidentally, destroyed coral reefs could result in low fish production, and may not recover for >20 years.

Concomitant to the above, environmental disturbances could result not only from natural events such as cyclones and storms but also from human activities. These activities could include the destruction of nursery areas (including mangrove areas), road construction, and land reclamation. Moreover, corals are collected for sale as souvenirs and coral blocks are used for building. Further, harbor dredging and coastal building projects may often release silt into the water, and this may block off the sunlight or may smoother the coral. Poor land management practices have resulted in erosion and the siltation of lagoons. Environmental disturbances and habitat destruction are linked to an increased incidence of **ciguatera**, a form of poisoning caused by eating affected fish.

The need for awareness

"Awareness" happened to be an important parameter for successful implementation of management and conservation. Incidentally, the knowledge of the island and coastal people regarding the marine environment are, sometimes, underestimated. Most coastal communities have good awareness of, and concern for, their marine environment. However, it may not be exactly the same case with the FW environment. Incidentally, the Pacific Island fisheries are managed within a framework of traditional knowledge accumulated over centuries. In fact, many communities have had, and in some cases still have, traditional rules and regulations to protect the fish stocks. These may include taboos on destructive fishing such as catching small fish and fish at particular times of the year when they spawn. Many communities have traditional rights to fish in adjacent coastal areas under customary marine tenure. Fishers were small in number and were mindful of customary methods of conserving fish stocks. Further, communities which cared for their fish stocks were highly regarded. In parts of Fiji, for example, the word "kaiwai" is used to describe coastal or sea people who keep and use the marine environment wisely. On the other hand, in some islands, however, customary rights eroded with growing local population sizes and a trend toward a money-based economy and commercial fishing are sometimes found. In places with large population, however, commercial fishing to supply local markets provides an important source of employment. Further, in some cases, fishermen travel long distances from their homes to catch fish in areas which were previously exploited and controlled by local communities. Younger people in villages may not appreciate the same conservation values as their parents and grandparents. Thus, there are literature available that encourage communities to take actions necessary to manage their marine resources properly and make fisheries sustainable. A prerequisite to these actions is a community awareness of the aquatic (FW/marine) environment and a concern for the existing problems. Incidentally, the process of community involvement may be illustrated by the following sequence: awareness > concern > action.

In fine, it could be said that fisheries have to be managed, and that in order to make fisheries sustainable, regulations may have to be imposed. If the majority of fishers are aware of the need for conservation, the compliance for strict fisheries regulations is likely to be high and fisheries management is likely to be more effective.

Environmental education in schools

A school is the ideal place to instill a noble concept among the children. Thus, education of the young is the best way to ensure future protection of the aquatic (FW and marine) environment and the sustainability of the aquatic resources, notably, the fisheries resources. It is the "education" that provide an increased awareness of environmental issues and produces future citizens who could be more environmentally responsible individuals. Thus, introducing students to environment-related subjects at an early tender age is very important as young people could be particularly more receptive to learning environmental values. In addition, information and values communicated to the young could be a strong means of raising the awareness of their parents and the community, in general.

5.3.4.20 *Raising community awareness*

Community awareness and environment education are the prerequisites for success in effective management and conservation of resources by applying efficient *modus operandi*, like the **CBFM** in a **participatory mode** by involving the stakeholders from the grassroots level. Such people with awareness could recognize and appreciate the importance of the aquatic (FW and marine) environment, possess an awareness of environmental issues, and discuss technical and social solutions to problems.

Such environmentally aware and educated people could be in a better position to sort out the difficulties and problems than the governments and scientists as commented by some quarters. As such, aspects of environment education could be incorporated into the school curricula and all relevant government and nongovernmental organizations could take initiatives in this direction to develop budding citizens with environment awareness for effective management, conservation, and sustainability of the resources.

As suggested there are two ways to introduce studies on the marine environment into high school curricula:

(a) a multidisciplinary approach.
(b) an interdisciplinary (single subject) approach.

(a) Multidisciplinary approach

In this approach, aspects of marine environmental studies could be introduced into subjects that the students are currently studying, such as general science, social science, ecology, biology, and geography. This multidisciplinary approach may be the most convenient way to provide students with exposure to environmental topics in the short term.

(b) Interdisciplinary (single subject) approach

In this approach, marine environmental studies are included in high school curricula as a separate subject. This could be the better long-term choice, as it provides environmental education with the status that it deserves.

Nevertheless, whichever approach is used to introduce environmental education, it eventually aim to take students beyond a study of ecology, to include relationships between ecological concepts and environmental issues as well as solving of environmental problems. These sequential stages are briefly listed in the following (King and King, 1995):

(i) **Ecological concepts**: Students may be provided a background knowledge of ecological concepts based on biological, chemical, and physical properties of the environment.
(ii) **Awareness of environmental issues**: Students may be exposed to broader issues such as sociology, economics, and human behavior including personal, political, and cultural activities and how these interrelate.
(iii) **Problem solving in relation to environmental issues**: Students may be encouraged to discuss community-focused solutions to actual local problems. The balance between the need for development and protection of the aquatic environment, for example, is of real concern in many Pacific Islands.

Notwithstanding the above, similar *modus operandi* could also be made available for FW sector.

Radio and television

Radio is one of the easiest and effective mass media for infusing environment-related lessons and messages to far-flung masses for effective management and conservation of the resources. Concomitantly, television could add pictorial effect to this transmission.

Printed material

Notwithstanding the above, the print media can also reach the people in the remote places, with the lessons on environment and messages for its proper management, sustenance, and conservation.

Direct contact

Among all the tools and means of communication, direct (face-to-face) contact is the most effective and impactful way of imparting lessons in environment education and in conveying the messages for sustained harvest, effective management, and conservation of the resources.

Extension work

"Extension" could be broadly defined as working with people to provide, or to build on, skills and knowledge, or to promote community action, to achieve particular goals. Concomitantly, fisheries extension, therefore, could mean working to provide communities with the skills necessary to increase aquatic (FW/sea) food production.

Concomitantly, it could mean working to provide communities with the means and motivation to conserve fish stocks and protect the aquatic (FW/marine) environment. One type of extension involves an agency deciding what kind of skills and knowledge would be required to achieve a particular goal, and then providing these to the community. This method is, sometimes, referred to as "top-down," i.e., one in which an agency (say a Fisheries Department) believes that it has the knowledge to achieve a goal, and proceeds to either provide instruction and skills to, or to impose regulations on, the community. An example of this is the setting of minimum size limits for fish by government agencies. Fish size limits are usually set without reference to the community (they are usually set by fisheries biologists) and rules are imposed on fishers by the government agencies on the understanding that it is for the public good.

Concomitant to the above, another type of extension involves an agency working to encourage and assist communities to achieve their own particular goals. In this case, the community is encouraged to define its needs and propose solutions to problems. As the agenda are set and actions are taken by communities with government agencies playing a supportive role, this type of extension is, sometimes, referred to as "bottom-up." In this case, the community would set its own conservation rules, and it (rather than the government) has ownership of the rules and a responsibility to enforce them. Because communities play the key role, this type of management is referred to as **CBRM**, which if applied to management of "**fisheries**" could be called **CBFM.** Such managements can be defined as "arrangements under which a community takes prime responsibility for managing resources."

Notwithstanding the above, there are many other arrangements under which resources could be managed. One which could be regarded as falling somewhere between the above two methods is "**comanagement**," in which government representatives, communities, and other user groups manage fisheries resources on a cooperative or shared basis. Concomitantly, in **CBFM,** the community is the major partner and player.

5.3.4.21 *Government attitude toward fisheries management by communities*

In this connection, it could be said that the government would feel relieved if the community shoulders the responsibility of managing the environment and its resources in a sustainable manner. But, sometimes, the government shows reluctance initially, when the community is embarking on a community-based program, particularly when enough sense of responsibility is not displayed. However, it may be essential to overcome this initial government reluctance.

Notwithstanding the above, fisheries agencies, as repositories of technical and scientific expertise, sometimes, have historically assumed responsibility for taking actions and setting national laws to protect fish stocks and the aquatic (FW and marine) environment. So, under such circumstances, the people in the government may not so easily delegate the responsibility to the community, as the former may feel a loss of power or that it is avoiding its own responsibility by handing over the responsibility of aquatic (FW and Marine) conservation in the hands of fishing communities.

5.3.4.22 *Involving communities in marine resource management*

However, experience(s) revealed that a government agency promoting community management could gain in many ways.

5.3.4.23 *Extension staff skills*

Historically, extension staff are required to have appropriate technical knowledge and skill and an ability to train people and transfer skills. However, if CBM is the goal, facilitating skills are more important. The term "facilitate" means "to make easy." In other words, the extension officer, as facilitator, could "make things easier" for a community to achieve its own objectives by its own actions. In this role, the extension officer is more a listener than a teacher. Facilitation refers to the process of encouraging other people to offer their views and to take their own actions. Ideally, a community-based extension officer may have a balance of both basic scientific knowledge and community, facilitating and motivating skills. An understanding of biology, ecology, conservation, fishing techniques, aquatic-food handling, and resource management practices are, undoubtedly, valuable. However, most government agencies have scientific personnel available to provide with more specialized technical advice to communities. In the process of encouraging communities to take conservation actions, extension staff are required to facilitate meetings. A facilitator needs to unobtrusively encourage groups to define problems and to propose their own solutions, without imposing his or her own views on the proceedings. Previous experience with problem-solving techniques could be an advantage. One particular problem-solving technique, which is easily learned and suitable for working with groups, involves the use of a **problem/solution tree.** Some of the most successful community-based extension officers, both men and women, are extroverted people, who like people and are good at encouraging all individuals in a group to give their opinions. This may (depending on local custom) involve joking and friendly dealings. However, what is not required is an egotistical bully who is likely to embarrass people and impose his or her own opinions on the group.

Notwithstanding the above, many younger people may require training in cultural skills in order to address community groups which could include the village elders. However, efficient, experienced, and skilled staff are to be used for successful CBFM.

Notwithstanding the above, it may be mentioned that having female fisheries extension staff often could make it easier, or culturally more acceptable, to facilitate community meetings involving women. Women not only makeup an important and influential part of the community but are also very much involved in subsistence fisheries. In Fiji, for example, >50% of the rural subsistence catch is believed to be taken by women (Rawlinson et al., 1995). In the Pacific, women play a significantly large role in the capture, postharvest, and marketing sectors but still fail to be well represented in most fisheries agencies. Fisheries agencies, thus, could aim to increase the number of active female staff members. Currently, many of the women employed in government fisheries agencies have clerical or secretarial positions but they might be more profitably employed in the field.

5.3.4.24 *Community ownership*

Community ownership is an essential feature for developing a sense of belonging and a feeling of responsibility among the stakeholders for efficient management of the resources. Community ownership would be optimized if as many people as possible are allowed to contribute to the process of developing the Village Fisheries Management Plan (VFMP). This may require involving all groups including women and the elderly persons in order to ensure the widest community participation and eventual ownership of the plan. The length of the extension process in each village may have to be sufficiently extended in order to allow the community time to establish ownership of their plan(s) and undertakings. Ownership by the community may require sufficient time for the people to consider their own problems and causes, and think of their own solutions. It may take many months of facilitated discussions by community groups before the plan could be regarded as owned by the community.

Further, it may be unreasonable to expect communities to readily take conservation measures as many subsistence fishers require aquatic food (notably, seafood) for their families; sometimes, on a daily basis. This could reduce the present stock of seafood further. However, certain restrictions, like ban on destructive fishing methods and imposing fish size limits, may cause a short-term decrease in catches. This may, in turn, have positive impacts on fish stock. Accordingly, a community-based extension program which may not provide support by way of promoting alternative means of obtaining seafood is unlikely to be sustainable.

5.3.4.25 *Extension process*

The extension process may be based on laid-down effective procedures. In order to enable the communities to manage their fisheries resources, they must have:

(a) An awareness of problems with the aquatic (FW/marine) environment and fisheries resources.
(b) A concern for these problems and their effects on the community.
(c) A willingness to take actions to address these problems and difficulties.
(d) Some traditional, legal, or assumed control over their adjacent fishing areas.
(e) The traditional power or organization to make and enforce their own regulations.

Nevertheless, the level of community awareness must be assessed before embarking on a CBFM program. Conversely, if community awareness is felt as low, greater efforts might be needed to raise it. However, care may be taken not to underestimate community knowledge. The use of damaging fishing methods, such as dynamiting, may not necessarily indicate community ignorance rather it may mean that such methods are used and are condoned for economic reasons. The use of dynamite may result in large quantity of fish catches in a short time, even if there could be long-term destruction of precious coral reefs and associated coveted fish stocks.

Notwithstanding the above, it may be decided that public awareness-raising activities could be part of the fisheries extension program, but it may be that the most important need in communities is not for education, but for motivation and support. The key extension tasks, therefore, are to provide this motivation and technical advice and most importantly to enable the communities to realise that they and not the government have the primary responsibility to manage their fisheries and aquatic (FW/marine) environment. Nevertheless, preparation for a fisheries extension program consists of designing a culturally appropriate extension process and training extension staff in order to facilitate the process effectively. The extension process have to be designed specifically to encourage communities to discuss problems and propose solutions relating to fisheries and the aquatic (FW/marine) environment. In many countries, the extension process involve recognizing village leaders as the prime instruments of change, while still allowing ample opportunities for other community groups to participate. In the early stages of the program, it would be necessary to contact communities to explain the program. Later, and if the program is successful, this will become less necessary as village leaders will approach the agency to express interest in joining the program.

5.3.4.26 *Fisheries advisory committee*

The Fishery advisory committee does the following:

(a) Prepares village fisheries management plan.
(b) Presents plan to the village leaders' meeting, which may either approve or reject the plan.

Concomitant to the above, the **Village Fisheries Management Committee (FMC)** does the following:

(a) oversees community activities.
(b) takes actions listed in management plan.
(c) probably, does joint assessment of community management.
(d) provide agency support for community.

5.3.4.27 *Community-based fisheries management*

The committee related to this does the following:

(a) assessment of community awareness of problems.
(b) concern for problems.
(c) willingness to take actions.
(d) village leaders' meeting.
(e) discussion of extension process.
(f) group meetings.
(g) community groups determining key problems; their causes and solutions.
(h) election of members from each group for management:

(i) assessment of community awareness, concern, and willingness to act for sustainable harvest, management, and conservation of resources.

Following an initial expression of interest by the members of a community, extension officers could assess whether the community as a whole is ready to commit to the process. The three key components are as follows:

(a) An awareness of problems with the fisheries and the aquatic (FW/marine) environment.
(b) A concern for these problems.
(c) Willingness to take actions to solve these problems.

If the abovementioned assessment is positive, it is usually culturally appropriate to arrange for a meeting with the community leaders. If the meeting decides to accept the process, it may have to agree to arrange for separate meetings of several different village groups. This meeting could be necessary to allow particular sectors of the community to express opinions which they, otherwise, may not do in large groups dominated by community leaders. The village may also be encouraged to nominate its own groups (e.g., fisher groups, church groups, etc.), but extension staff may need to ask that women and, if applicable, untitled men have their own groups. Each of the nominated groups could then arrange to hold village group meetings (VGMs) at some time in the year.

5.3.4.28 *Community group meetings (CGMs): Problem/solution trees*

Each group meeting requires a facilitator and, if possible, a second person to act as a recorder of the discussions. Incidentally, there may be three or more different community groups in a particular village. Thus, it could be very convenient to arrange for the individual group meetings to be held on the same day.

5.3.4.29 *The community fisheries management committee (FMC)*

The FMC is appointed by the village leaders in order to administer the conservation undertakings of the community. Members of the previous Fisheries Advisory Committee (FAC) are most likely to be appointed to the FMC but this remains a village decision. Once the VFMP is formally agreed to, the agency has to agree to make regular contact with the FMC and provides the technical support agreed to under the plan.

5.3.4.30 *Ensuring continuing community commitment*

Once a community has prepared its VFMP, it is responsible for taking actions and enforcing regulations that are listed in their plan. However, it is unlikely that a community would be able to sustain these actions without some support, at least in the short-to-medium term. In particular, post-management plan activities could include regular contact with villages and support for the village FMCs.

5.3.4.31 *Review of fisheries management in participating communities*

Review could be said to be an important exercise in monitoring progress. In this case, the extension staff may need to assist the communities in reviewing their conservation and fisheries management efforts after a regular interval of time.

5.3.4.32 *Things which a community extension officer may not do*

(a) Preferably not give the name and opinions or comments of the extension officer, when he is trying to encourage people to give the best.
(b) Preferably not prompt the community with opinions of the extension officer.

5.3.4.33 *Community versus national fisheries regulations*

It could be said as a fact that fisheries have to be managed and that in order for fisheries to be sustainable, regulations that control fishing and fish catches have to be imposed and enforced, wherever necessary.

However, in the case of CBFM, communities themselves try to enforce their own fisheries regulations.

5.3.4.34 *Community conservation measures*

A responsible community is believed to chalk out its own conservation measures for its resources. In many respects, conservation measures and regulations that a village may impose on its own community would reflect and support those imposed by the government. The difference is that the measures and regulations are owned, imposed, and enforced by the community. Communities are to be allowed to take actions and impose regulations that they have developed by themselves (as long as these do not contradict national laws).

Notwithstanding the above, many communities have tradition-based controls on fishing.

5.3.4.35 *Community-based management measures*

A responsible community, generally, formulates its own management measures. Incidentally, some village communities in **Samoa** have limited the number of fishers permitted to use fence traps. Concomitantly, the number of fishers in the past were also regulated, in effect by restrictions in access to a community fishing area. Trespassers, who fished without permission in an area not controlled by their own group, would be punished by the clan leaders. However, some communities with strong traditional control over the marine resources, are still able to practice this but public ownership of the sea up to the high tide mark had made it legally unenforceable under the law of the national government. The increasing mobility and range of fishers, sometimes, had also made it difficult to maintain village regulation and control over the fishers who fish in the high seas.

5.3.4.36 *Limiting the efficiency and types of fishing gears*

The fishing gears are the implements used to harvest fish resources from the water body. Incidentally, the highly mechanized fishing gears may overharvest the fish resources much beyond the sustained limit. As such, the use of some highly mechanized fishing gears may be restricted or banned in the interest of conserving fish stocks and allowing more people to use the resource. Limitations on gear types may include banning a specific fishing method in particular areas, particular season(s), or on a particular species. For example, the use of gill nets may be banned in lagoons or the use of scuba diving to catch lobsters may be banned. These regulations are equally applicable to subsistence fisheries and the commercial fisheries, particularly, with regard to endangered ones.

In **Samoa**, for example, some communities have restricted the use of underwater torches for spearfishing at night.

Further, commercial gillnetting has been banned by communities in parts of Fiji, and was supported by the government.

Concomitant to the above, there are certain restrictions and regulatory measures, which, as part of associated effective management measures, are administered and imposed for efficient and sustainable management of resources and for their conservation. These are briefly discussed in the following.

5.3.4.37 *Certain regulatory measures to help in management process*

(a) Banning destructive fishing

There seemed to be a number of destructive methods of fishing in different parts of the world, which devastate the fisheries. Such methods involve the use of chemicals, bleaches, phyto-toxins (e.g., *Derris*) explosives, etc. They are unlawful in most of the Pacific Island countries. They are deplorable. In some places, such dangerous methods (which devastate not only the biota but also the abiotic components of the water body, notably, the habitat) are generally practiced in the villages away from the notice of many law-enforcing authorities. Nevertheless, the village communities may, by themselves, strictly impose and enforce the law against such nefarious practices. They could also add further stringent measures of the land.

Moreover, some communities in **Samoa** have banned the traditional smashing of corals to catch small sheltering fish. Local clans of Marovo lagoon in the **Solomon Islands** have enforced prohibitions on the use of dynamite and plant poisons.

(b) Closed areas and Closed seasons

An area in the water body where fishing is strictly restricted or banned in order to protect the juveniles and/or brooders is, generally, regarded as a "closed area." Concomitantly, the season(s) during which fish catching (by any means) is strictly restricted or totally prohibited, in view of the breeding season of the fishes occurring during that/those period(s), is, usually, termed as "closed season(s)." Concomitantly, a highly restricted area where endangered less abundant fishes are stocked for the purpose of all-round protection of their population as well as to help them to grow in population size, being protected, could be regarded as an "Aquatic sanctuary."

Notwithstanding the above, both "closed area" and "closed season" could serve as effective measures to ensure sustainability of population, their effective management, and conservation. As examples, shallow water mangrove habitats are known to be the nursery areas for many species and are permanently closed to fishing in some coastal areas. Moreover, in some countries, known breeding areas for trochus are permanently closed to fishing.

Incidentally, if the spawning season of a particular species is known from ITK, for instance, a "closed season" at the time of spawning may allow adults to breed without much interference. Closures may also be used to prevent stocks from being overfished. Incidentally, villages in **Vanuatu** periodically close their fishing areas for the collection of trochus and green snails. Following a radio advertisement by a government fisheries biologist, offering to provide advice to villages interested in managing their trochus stocks, a number of villages in Vanuatu implemented no harvesting during the closed periods (Johannes, 1998). Incidentally, the closures could be considered similar to customary taboos in design and enforcement, but were also based on basic biological principles explained to them by the government fisheries team.

Notwithstanding the above, the lagoon is open to trochus diving during the closed year (Gillet and Lam, 1999). Concomitantly, in **Samoa**, a large number of villagers have chosen to establish small areas adjacent to the "closed area" as their stop-gap fishing ground. In terms of total fisheries production, a small reserve is unlikely to be as effective as a large one. Larger reserves are more likely to provide suitable breeding areas for small inshore pelagic fish such as mullets and scads. However, even the small reserves may be beneficial for nonmigratory species (King and Faasili, 1998a).

5.3.4.38 *Minimum mesh sizes*

Concomitant to "closed season," strict administration of "minimum mesh sizes" in nets, and "escape gaps" in traps in many fisheries allow the juvenile fishes to escape, and this allows them to grow to a size and age at which they **may reproduce at least once before capture**. In many island countries, governments have imposed mesh size regulations and rules set by the local fishing communities who could support and enforce these regulations. On the other hand, some communities may set their own larger mesh sizes, to further reduce the catch of small fish.

5.3.4.39 *Size limits (minimum legal lengths)*

"Size limit" is the imposition of minimum size of a fish which could be caught. A fish that is less than the prescribed and imposed size limit cannot be caught and has to be returned to the water body. Imposition of size limit also helps in managing and maintaining an ecologically and biologically sustained population of the fish species. Further, it may be mentioned that, **traditionally, size limits have been imposed to allow individual fish to spawn at least once before capture.** Minimum legal size limits have been applied by national governments in Pacific Islands to many ichthyospecies as well as invertebrates like sea cucumbers, trochus, pearl oysters, giant clams, spiny lobsters, mangrove crabs, etc.

Concomitant to the above, the village communities may decide to support and enforce national regulations on minimum sizes. Incidentally, it may be noted that, some villages in **Samoa** have set their own minimum size limits, which are larger than those set under national regulations.

5.3.4.40 *Rejection of females, or spawning females*

Female fishes, particularly the gravid ones, if caught are not to be retained but to be released back to the water body. Otherwise, it would be an unlawful act. However, such a regulation, not to catch the female fishes, could only be applied and administered sensibly to such ichthyospecies in which the sexes could be distinguished easily by fishers in the field itself as well as where the catching method(s) may not cause any harm to the individual fish caught. However, the sex of most fish cannot be easily determined in the field itself only by external examination. Nevertheless, the regulation may have some amount of application in certain cases of subsistence fisheries.

5.3.4.41 *Catch quotas*

"Catch quotas" are also important to maintain a sustained stock of fishes. Fisheries agencies may determine that total catches are not to exceed a certain amount called a "quota," in order to protect the fish stocks. As an example, it could be said that in the trochus fishery in the Cook Islands, fisheries scientists have estimated that fishermen could be allowed to catch *c.* 30% (or *c.* 40 mt) of the total trochus stock each year. The fishery would be closed, once this quota has been reached. On the other hand, in subsistence fisheries, however, catch quotas have little application. Nevertheless, applying a daily quota, or bag limit, for particularly desirable species could be a possibility.

5.3.4.42 *Community compliance and enforcement of regulations*

CBFM is a people-oriented and people-centric work. Under CBFM, the rules and regulations are enforced by the village communities. Moreover, fisheries rules and regulations are more likely to be effective under such CBM, as they are enforced by communities with a direct interest in their continuation and success.

5.3.4.43 *Supporting community-based fisheries management*

CBFM, as already mentioned, is a people-oriented and people-centric activity.

CBFM depends on the availability of professional technical support for the communities involved. Nevertheless, scientific input is required to help the communities with finding alternative sources of aquatic food (say, seafood) and avocation, and to advise on and monitor community actions. Whether community-based or not, most fisheries conservation measures, including the prevention of destructive fishing and the imposition of fish size limits, are supposed to cause a short-term decrease in catches. The same is true also for the marine reserves as they reduce the area available for fishing. As most subsistence fishers require aquatic food (seafood in the case of marine fishers and FW food in the case of FW fishers) for their families on a daily basis, it is unreasonable to expect fishing communities to adopt full conservation measures which could initially reduce present catches of aqua-food even further without offering alternatives. Accordingly, an extension program should, preferably, include the promotion and development of alternative sources of aqua-food. Alternatives should include the introduction of medium-sized, low-cost boats in order to divert fishing pressure to areas immediately beyond the vulnerable areas, the promotion of village-level AQC, and, the restocking of depleted species of less abundant biota in village areas. Support should also include government assistance with non-fisheries development, particularly in livestock and agriculture. This requires the support of government departments other than those involved with fishing and the aquatic (FW/marine) environment.

The overall aim is to compensate communities for the short-term decrease in availability of aqua-food which inevitably accompanies the imposition of fisheries regulations, whether community-based or not. A community-based extension program that does not promote alternative means of obtaining aqua-food is unlikely to be sustainable. In this process, the community agrees to certain responsibilities, while the government agrees to reciprocal actions necessary to support the community's undertakings. These may involve the national fisheries agency providing technical advice on how to care for the aquatic (FW and marine) environment and on the development of alternative sources of aqua-food. Incidentally, the NGOs may assist both the government and the community in this work, both in the area of information and training.

5.3.5 Vanuatu (Oceania)

In the Pacific region, coastal communities have compensated for chronically low capacity of governments to manage fisheries by implementing local regulations in the aquatic domain. They have looked for efficient methods of **management**. Thus, trajectories and factors of change in CBFM systems, since the 1990s were analyzed to investigate the performance of CBFM in Vanuatu in connection with CBM of near-shore fisheries in **Vanuatu** (Marc et al., 2013). Focal group interviews were conducted in seven villages on Efate´ Island in 2011 and supplemented by views of supporting literature. Results revealed the increase in grand excessive reliance of CBFM systems on external agencies that promoted complex management plans. Moreover, examination of trends in CBFM systems indicated that community and national fishing rules, which were highly acceptable by the local societies, were more likely to be enforced in the long run. In particular, the establishment of FW, coastal, and marine reserves were the most widely spread and best enforced community rules for the purposes of conservation, ecotourism, and fisheries. However, some of the works tried to refute the effectiveness of CBFM in achieving sustainability of reef fisheries in Vanuatu, and, concomitantly, highlighted the overreliance on small marine reserves as a management tool. Nevertheless, community initiatives could, preferably, be strengthened by new specific national regulations governing subsistence and commercial reef fisheries as part of a multi-scale comanagement approach.

CBFM is an effective tool for efficient management of water bodies and their bioresources. In this connection, it may be stated that market-based factors have encouraged the users of goods, such as common pool marine resources, to employ extraction strategies, which led to the general overexploitation of worldwide fisheries. Moreover, the decline of a large number of jointly used fisheries that were managed by top-down government-based fisheries management (GBFM) strategies further suggests that this governance strategy may not be particularly effective (Hunt, 2003) and, obviously, calls for a different approach, which emerged in the late 1980s, based on the observation that under certain conditions, users could self-organize to achieve successful governance without individual appropriation of resources (Ostrom, 1990; Ostrom et al., 1999; Berkes, 1989). It is now well-established that the knowledge of local stakeholders could aid in making decisions about the **management** and the conservation of resources in the aquatic regime (Neis et al., 1999; Berkes et al., 2000; Webb et al., 2004).

With reference to the above, an upsurge of interest in community-based natural resources management attempts to address this desirable change in governance, and, consequently, has produced considerable scientific debate (Shackleton et al., 2010). Incidentally, CBFM spread quite rapidly in the developing countries and particularly in the Pacific region, where this system has grown upon a fertile cultural acceptability to reduce the general decline of coral reef resources in addition to other bioresources (Ruddle et al., 1992; Johannes, 2002).

Concomitant to the above, CBFM regimes, like GBFM tools, may include different regulatory measures such as species and gear restrictions, closed seasons, access rights, and reserves in the aquatic domain. Incidentally, a number of small-scale case studies have established that CBFM may improve the use of local social capital in terms of governance, communication, knowledge, enforcement, cooperation, reciprocity, and adaptive capacity (McClanahan et al., 1997; Alcala, 1998; Ostrom and Walker, 2003; Pretty, 2003; Lobe and Berkes, 2004; Berkes, 2008). Moreover, external governmental and/or nongovernmental organizations (NGOs) have also increasingly been associated with CBFM initiatives through collaborative or **comanagement** arrangements (Adams, 1998), although the latter operates within the prescribed implementation periods of external projects. Therefore, CBFM may appear as a powerful, integrated approach to maintain natural ecosystems, benefits to fishers and community livelihoods. However, biological evidence(s) of the effectiveness of CBFM regulations on fisheries remains scarce. Moreover, the outcomes of CBFM are context-dependent, and, importantly, inconclusive findings are not welcome in the *modus operandi* of the system.

5.3.5.1 CBFM in the long term

CBFM could be considered as an important strategy for the effective management of resources. A CBNRM program could be considered as successful only if there are certain positive outcomes in issues like signs of BD conservation; fulfilling some of the local socioeconomic needs to some extent; signs of local livelihood improvement, indications of local capacity building like social cohesion, leadership, enhanced financial resources, successful linkage with supportive government policies; etc. Moreover, the dynamics of the local sociopolitical relationships, fishing activities, and various environmental factors could also be significant in this regard. Concomitantly, time could also be a major determinant toward implementation of various issues related to CBFM. Notwithstanding the above, CBFM is highly sensitive to temporal change(s) in the factors mentioned above and could, therefore, be perceived as a site-specific adaptive process (Dressler et al., 2010; Blaikie, 2006; Maliao et al., 2009; Ruddle, 1998; Agrawal and Gibson, 1999; Aswani, 2005).

Concomitant to the above, the historical and sociopolitical context in Vanuatu (South Pacific) provides the relevant conditions to analyze the functioning of CBFM, since the latter was fully recognized by national legal and political frameworks. Moreover, the communities have been assisted by the Vanuatu Fisheries Department (VFD), since the country's independence in 1980, which had provided support to CBFM through programs initially aimed to enhance trochus (*Trochus niloticus*) fisheries. Further, the VFD largely relied on communities and the sea tenure owing to its limited man power and financial capacities in order to frame and to enforce coastal fisheries management.

Concomitantly, the National Fisheries Regulations have been established in1982, 2005, and 2009, which mainly affected the commercial sector, i.e., high-value, export, and world's flagship species (i.e., marine turtles and mammals). In fact, the communities are said to be the legal owners of near-shore resources and generally manage their aquatic (FW and marine) tenure at the community level using locally specific regimes. Further, subsistence of fisheries extend over *c.* 3100 km ^{2} in waters from 0 to100 m depth across the archipelago, and play a key role in this nation's protein supply. In addition to the above, the considerable amount of financial resources invested by the national government through fisheries development programs have had relatively little impact. Fishing for subsistence continues to dominate in terms of number of fishers and aquatic product consumption, and engages *c.* 50% of Vanuatu's rural population. In addition, as in other Pacific islands, coastal and urban population growth, technological innovations, and improvements in commercial networks have raised near-shore resource sustainability issues, and also new challenges for CBFM (Ruddle, 1994; Hickey, 2007; Cillaurren et al., 2001; David and Cillaurren, 1992; Mael, 2007; Johannes and Hickey, 2001).

Notwithstanding the above, there were contributions, which, 31 years after Vanuatu's independence, aimed to investigate the effectiveness of CBFM as well as tried to suggest practical management regulations based on community and national governance structures in the light of challenges in the fisheries domain. Further, trajectories in CBFM Systems, since the early 1990s, were considered, and the factors that had contributed to their dynamics, were analyzed using empirical case studies and a review of the supporting literature spanning for two decades. In addition, there were assessments on the knowledge level of the community members about fishery management and the influence of external agencies on local fishery management.

5.3.5.2 *Local fisheries management systems*

In many countries, fisheries are managed at the community level under the authority of one or several chiefs depending on the village. The tenure in the aquatic domain was open to all fishers in each village. However, 12 fishing rules were in force in at least one village in 2011, including two national rules (found in *c.* 100% of villages) and 10 community rules (found in *c.* 14%–86% of villages). Finfish fishing was believed to be unregulated in >77%–100% of the water bodies in many nations. On the other hand, the commercial fishing of a number of fish species were effectively regulated in many other countries, and even, sometimes, by restrictive community rules. Reserves and sanctuaries in the aquatic regime (Freshwater Reserves and Marine Reserves) could be considered as the strict widespread community rule (in *c.* 86% of villages); while, there could be temporary closure of fishing for many other species. This could be regarded as the second-most common community rule(*c* 43% of villages). The expected benefits varied widely among villages and among stakeholder groups within villages. In particular, the conservation of aquatic ecosystems, attractiveness to tourists, and benefits to fisheries were the most widely shared objectives, in *c.* 60%, 40%, and 20% of villages, respectively.

5.3.5.3 *Issues and suggestions for fisheries management*

Efficient fisheries management is an important aspect of management activity, "fish" being a potential item of nutrition and avocation. Incidentally, the stakeholders' perceptions showed that *c.* 54 groups (*c.* 92% of all groups) have shown concern about the evolution of the state of their fisheries, which had been subjected to different types of challenges across villages. Incidentally, the most commonly perceived management issue was poaching from within or neighboring communities due to social disputes or household needs (*c.* 43% of villages), excessive fishing effort (*c.* 14%), excessive use of gill nets (*c.* 14%), within- and between-village disputes (*c.* 14%), and environmental impacts on aquatic habitats (*c.* 14%). The stakeholders have systematically discussed courses of action to improve the current management arrangements to address issues pertaining to fisheries. The most common suggestion concerned the restrictions of gill nets (in *c.* 43% of the villages), the development of marine reserves (*c.* 29%), the establishment of other fishing rules (*c.* 29%), and the development of alternative sources of income (*c.* 14%). In all villages (*c.* 25%–88% of groups), there was expression of willingness to strengthen cooperation with outside agencies to improve the local management system. Points on which cooperation was sought included the nature of the relevant fishing restrictions, the formulation of management plans, and the general raising of awareness. The results also displayed that 76% and 95% of the fisheries issues and suggestions for management, respectively, were shared by fewer than half of the stakeholder groups from the same village.

Concomitant to the above, **multivariate classifications** confirmed that the local perceptions were heterogeneous within villages, and that the stakeholder groups did not generally share a similar vision. In particular, divergences were frequently observed within villages between leaders and management committees as well as between these groups and most of the fisher groups, that were grouped in very different classes. Moreover, the classes derived from management issues and suggestions, respectively, joined together very different groups of stakeholders, suggesting that the groups that shared similar views on current fisheries issues expressed different suggestions for management.

5.3.5.4 *Discussion*

Intervention from unrelated quarters were not much welcomed with regard to sustainability of the fishing restrictions. CBFM of fisheries in Vanuatu combined community and national rules, and has followed a trend similar to that observed across the Pacific, since the 1990s. The growth of community restrictions of fishing activities was interpreted as a sign of adaptive capacities of the communities of Vanuatuto facing the increasing exploitation of resources. There had been data spanning for few decades to reveal that most community rules were short-lived, and that CBFM systems relied on the support of external governmental or nongovernmental agencies. Moreover, the apparent dynamism of community management in the 2000s was partly exogenous, according to some quarters, and maintained by the intervention of agencies mobilized by global conservation and sustainable development goals. These external interventions greatly influenced the nature and complexity of the local fisheries regulatory frameworks over the last couple of years through the participatory implementation of management plans. The reasons for the rapid decline in the number of community regulations since around 2007–10 may be further investigated. In particular, most fishing regulations had been written into the village-based management plans. These were given up following the departure of external agencies, despite their ceremonial launching.

Notwithstanding the above, the evolution of CBFM systems since the 1990s suggest that in a context of weak government capacity, the sustainability of local fishing regulations may be related to their social acceptability at the community level (i.e., including all social groups). However, acceptability could be linked to local sociopolitical and economic mechanisms acting on the costs, the incentives, the social and cultural foundations of fishing rules, etc. Moreover, those rules that benefit from a social consensus at the community were easier to maintain in the absence of external intervention.

Notwithstanding the above, according to some quarters, there could be difficulties in the operation of CBFM in view of the suspicion that only long-term rules are likely to improve fishery resource status in a sustainable way, and without a sustainable status of the resources effective implementation of CBFM may not be possible.

5.3.5.5 *Strengthening community initiatives through specific national fisheries regulations*

It could be opined that community-based aquatic sanctuaries could be promising option(s) for management of fisheries, but the results of certain studies highlighted the overreliance of CBFM on a number of related and often also unrelated management tools. Nevertheless, combining the respective advantages of CBFM and GBFM could increase the capacity of each of these management scales (Hviding, 2003; White, 2007; Hicks and McClanahan, 2012; Lane and McDonald, 2005).

In fine, there are studies suggest the need to move from an idealized concept of CBFM to an operational stage that includes evaluation of the performance of CBFM regimes. Also there is the need for the assessment of the types of external support that could be appropriate. In particular, providing external assistance for tools of high social acceptability and supporting local needs could very likely prove more effective in achieving sustainability of fisheries than short-lived generic approaches derived from conservation, **ecosystem management, and participatory models**. In addition, comanagement innovations between the government and the local communities could still be needed in order to optimize the design and the implementation of effective fishing regulations in the country over the long run (Wilson et al., 2013).

5.3.6 Summary

(1) The continent of **Australasia** possesses a high potential in fisheries and AQC. As such, there is a need for efficient methods of management and conservation.

(2) In **Australia,** much amount of works has been done in **Western Australia**, where Metcalf et al. (2009) deliberated upon conceptual models for EBFM. Elizabeth (2005) dealt with sustainability certification in community-based fisheries.

(3) In **Fiji**, Michael Warner (2000) dealt with conflict management in community-based natural resource projects with a note on experiences from **Fiji and Papua New Guinea**. Anders et al. (2011) dealt with mapping fish community variables by Integrating Field and Satellite Data, Object-Based Image Analysis and Modeling in a traditional **Fijian fisheries management area**. Concomitant to the above, the ability to map these environmental variables with RS has, in turn, made it possible to employ statistical models to produce spatially explicit predictions of fish community variables (Knudby et al., 2010a,b). Only a handful of studies (Anders et al., 2011) have been published, of late, using this approach. Notwithstanding the above, ecotourism is a commercial domain where community could also be involved. The Shark Reef Marine Reserve in **Fiji** is an ecotourism project designed to protect a small reef patch and its fauna while preserving the livelihood of local communities

(Brunnschweiler, 2009). Notwithstanding the above, the Shark Reef Marine Reserve is a self-sustaining and profitable project, and is an example of a privately initiated, bottom-up approach, which includes all relevant stakeholders in an area where marine rights are finely subdivided into small units (Brunnschweiler, 2009). Further, Annette et al. (2006) dealt with community-based marine resource management in **Fiji** in the context of yesterday to tomorrow. Fiji is considered to represent a unique case, in view of the fact that its customary fishing-rights areas ("*qoliqoli*") constitute a kind of dual ownership. Incidentally, in Fiji, for example, > 50% of the rural subsistence catch is taken by women (Rawlinson et al., 1995).

(4) In **Micronesia (Oceania)**, Rhodes et al. (2011) performed interdisciplinary study of the market forces and fishery management under the Coastal Fisheries Management Program. Anon (2014) dealt with CEAFM and climate change adaptation in the **State of Yap**, Federated States of **Micronesia** (FSM), with regard to the current status of marine resources in FSM. Furthermore, Anon (2004) dealt with transforming the LRFFT toward sustainability in the RMI through community-based coral reef conservation and fisheries management.

(5) In **Pacific Islands**, King and Lambeth (2000) dealt with fisheries management by communities in the form of a manual on promoting the management of subsistence fisheries by **Pacific Island communities**. Concomitant to the above, there have been detailed account on the pertinent subject related to the **Pacific Islands**. There are examples concerned with subsistence fishing among the coastal communities in the Pacific.

(6) In **Samoa**, for example, some communities have restricted the use of underwater torches for spearfishing at night. Moreover, some communities in **Samoa** have banned the traditional smashing of corals to catch small sheltering fish. Local clans of Marovo lagoon in the **Solomon Islands** have enforced prohibitions on the use of dynamite and plant poisons. Notwithstanding the above, the lagoon is open to trochus diving during the closed year (Gillet and Lam, 1999). Concomitantly, in **Samoa**, a large number of villagers have chosen to establish small areas adjacent to the "closed area" as their stop-gap fishing ground. Incidentally, **CBFM** is a people-oriented and people-centric activity.

(7) In **Vanuatu (Oceania)**, Marc et al. (2013) dealt with CBM of near-shore fisheries in order to investigate the performance of CBFM in Vanuatu. Incidentally, closures of fishing areas may be used to prevent stocks being overfished. Incidentally, villages in **Vanuatu** periodically close their fishing areas for the collection of trochus and green snails. Following a radio advertisement by a government fisheries biologist, offering to provide advice to villages interested in managing their trochus stocks, a number of villages in Vanuatu implemented no harvesting during the closed periods (Johannes, 1998).

(8) Some of the other researchers contributed to CBFM in Australia and may be elsewhere also are Conroy et al. (1998), Bryant et al. (1998), Andrew et al. (2007), Gaichas (2008), Ainsworth et al. (2008), Lester et al. (2009), Jupiter and Egli (2011), Syms and Jones (2000), Gratwicke and Speight (2005), Friedlander and Parrish (1998), Lester et al. (2009), Syms and Jones (2000), Syms and Jones (2000), Gratwicke and Speight, 2005, Lara and González (1998), Connell (1978), Jones et al. (2004), Friedlander and Parrish (1998), Christensen et al. (2003), Dorenbosch et al. (2005), Mumby et al. (2004a,b), Pittman et al. (2004), Pittman et al. (2007), Grober-Dunsmore et al. (2008), Gratwicke and Speight (2005), Friedlander and Parrish (1998), Lara and González (1998), Connell (1978), Jones et al. (2004), Friedlander and Parrish (1998), Christensen et al. (2003), Dorenbosch et al. (2005), Mumby et al. (2004a,b), Pittman et al. (2004), Pittman et al. (2007), Grober-Dunsmore et al. (2008), Gratwicke and Speight (2005), Friedlander and Parrish (1998), Lara and González (1998), Connell (1978), Jones et al. (2004), Friedlander and Parrish (1998), Christensen et al. (2003), Dorenbosch et al. (2005), Mumby et al. (2004a,b), Pittman et al. (2004), Pittman et al. (2007), Grober-Dunsmore et al. (2008), Hooker and Gerber (2004), Rhodes et al. (2011), King (1995), King and King (1995), Hunt (2003), Ostrom (1990), Ostrom et al. (1999), Berkes (1989), Neis et al. (1999), Berkes et al. (2000), Webb et al. (2004), Shackleton et al. (2010), McClanahan et al. (1997), Alcala (1998), Ostrom and Walker (2003), Pretty (2003), Lobe and Berkes (2004), Berkes (2008), Dressler et al. (2010), Blaikie (2006), Maliao et al. (2009), Ruddle (1998), Agrawal and Gibson (1999), Aswani (2005), Ruddle (1994), Hickey (2007), Cillaurren et al. (2001), David and Cillaurren (1992), Mael (2007), Hviding (2003), White (2007), Hicks and McClanahan (2012), Lane and McDonald (2005), Wilson et al. (2013), and so on.

5.4 Europe

Europe, in the temperate zone, has different types of water bodies in different parts with diverse types of aqua-fauna. There is a need for efficient *modus operandi* for the management of the water bodies and their bioresources in different parts of the continent.

The deliberations begin with an account of the situation in **Europe (in general)**.

5.4.1 Europe (in general)

There have been many works on Fisheries management in Europe (in general). Anon (2006) dealt with the report of The Bergen Conference on implementing the EAF.

Iceland has hosted a conference on responsible fisheries in the marine ecosystem, in 2006, organized jointly with the FAO of the UN and with the cosponsorship of Norway. It was a new type of conference organized by the Nordic Council of Ministers and the Governments of Iceland and Norway with technical support from FAO. It was held at Bergen in Norway from September 26 to 28, 2006 in order to address progress made toward implementing the EAF.

An ecosystem approach for sustainable use was a central theme of the implementation of international agreements, such as the CBD of the UN. Its principles are also embodied in the FAO Code of Conduct for Responsible Fisheries and binding in law such as the UN Fish Stocks Agreement. Incidentally, a political commitment to implementing the ecosystem approach by incorporating ecosystem considerations into fisheries management has resulted from the Reykjavik Conference in 2001. The requirement for incorporating the ecosystem approach in responsible fisheries was reaffirmed and consolidated at the World Summit on Sustainable Development in Johannesburg (2002), where a target year of 2010 was set for its achievement.

Concomitantly, many nations and regional and international governmental organizations and NGOs have embraced the ecosystem approach (consistent with the above commitments) and the members were progressing toward its actual implementation. Incidentally, experience had been gained at various scales, including the local, regional, and national (e.g., the Large Marine Ecosystems) scales in both developing and developed parts of the world. However, there are still widespread uncertainties despite some amount of progress made as regards the scientific and governance requirements, and the possible social and economic consequences of implementing an ecosystem approach. Thus, the aims of the Bergen Conference were mainly to review the following:

(i) concepts and address of the implementation issues related to applying the EAF;
(ii) to exchange experiences made and constraints encountered so far and
(iii) to identify strategies and best practices that could facilitate further implementation in practical fisheries management.

Concomitant to the above, Arlinghaus et al. (2002) dealt with reconciliation of traditional inland fisheries management and sustainability in industrialized countries with emphasis on **Europe** (in general).

The inland fisheries sector had long been dominated by recreational fisheries in northern industrialized countries, which normally exploit fish for leisure or subsistence as well as like in many other countries provide benefits to the society. However, various factors are known to constrain the development and existence of inland fisheries such as local user conflicts, low social priority, inadequate research, and funding, to list a few. In many cases, however, degradation of the environment and loss of aquatic habitat are the predominant concerns for the sustainability of inland fisheries. As such, there seemed to be an urgent need to little modify some of the age old time-tested traditional CBM practices and systems in tune with the present day need, in order to achieve the goals of sustainable development.

In fact, the need for concerted efforts to prevent and reduce environmental degradation as well as conservation of FW fish and fisheries as renewable common pool resources was the greatest challenge facing sustainable development of inland waters and fisheries. Incidentally, in inland fisheries management, the declining quality of the aquatic environment associated with long-term inadequate and often inappropriate fisheries management has led to an emphasis on enhancement practices, such as stocking, to mitigate anthropogenic stress. However, this may not be always the most appropriate management approach. As such, there seemed to be an urgent need to alter many traditional inland fisheries management practices and systems to focus on their sustainable development. In this regard, it may be said that there have been communications reviewing the literature regarding the inputs needed for sustainability of inland fisheries in industrialized countries. Hence, there were problems facing sustainable inland fisheries management, the inland fisheries environment, its benefits, negative impacts, and constraints as well as historical management and paradigms. Incidentally, trends and current practices were often dealt with in detail in order to understand the issue better. In various contributions, the major philosophical shifts, challenges, and promising integrated management approaches are often discussed in a holistic framework.

Notwithstanding the above, the following aspects are considered as the key elements for sustainable development of inland fisheries: communication, information dissemination, education, institutional restructuring, marketing outreach, management plans, decision analysis, socioeconomic evaluation, and research extending into the human dimension, in addition to traditional biological and ecological sciences. If these inputs are integrated with traditional fisheries management practices, the prospects of sustainability in the inland fisheries sector may certainly be enhanced.

Notwithstanding the above, **Anon** dealt with securing European fisheries, through CBFM, i.e., securing a future for the European fisheries (in general).

The continent of Europe has a rich potential in fisheries. Incidentally, it may be noted that >85% of assessed fish stocks in European waters were fished to their maximum potential or were overfished. No doubt, this could be a bad news for marine BD, bad news for the fishermen, businesses, and communities who rely on seafood as a source of earning their income, and also could be a bad news for the European consumers who want to enjoy healthy, fresh, sustainable seafood. In fact, more effective management measures for fisheries are needed in order to lessen the pressure on the FW and marine environments. Moreover, reform(s) of the CFP needs to deliver better strategy, enhanced stakeholder engagement, and appropriate scope. However, some of the important alterations, which are significantly required, are as follows.

5.4.1.1 *Long-term management plans (LTMPs)*

LTMPs for all EU fisheries are to be in place. These plans are to meet clear minimum standards setout in the new regulation, and also aim to achieve centrally agreed targets. They are to also assess the capacity and environmental impact because good fisheries management decisions will be based on these assessments. These ecosystem-based plans will move Europe away from the politically motivated annual quota negotiations and set fisheries on a more stable track toward rapid recovery and development.

5.4.1.2 *Effective regionalization*

Stakeholders are to be at the heart of the decision-making process. This could be accomplished with a new management regime which has regional or member state stakeholder development of the LTMPs and comanagement of the fisheries, once the plans are in place. We shall, probably, be able to design workable and effective management strategies by involving stakeholders more directly, to ensure that the EU meets its environmental commitments.

5.4.1.3 *Scope*

For effective management the CFP principles are to be applied to all fisheries in EU waters, including the Mediterranean, and to the European vessels, wherever they fish in the world's oceans.

Incidentally, FAO estimated that Europeans' average per capita consumption of seafood could rise from *c.* 22 kg/year in 1998 to *c.* 24 kg/year in 2030. Thus, better management of fish stocks is essential to restore marine ecosystems.

5.4.1.4 *Improved governance*

The present CFP is not very sustainable as it failed to make adequate incentives for long-term conservation. A new governance system is, thus, required to tackle the global fisheries crisis and deliver long-term health for the oceans and security for all who are dependent on them for their avocation and nutrition.

Notwithstanding the above, mandatory LTMPs with fishermen and other key stakeholders engaged in the decision-making process are the ways forward for all European fisheries. In fact, the CFP may set high level objectives but the plans could have the flexibility for each fishery to tailor the way in which they meet them including how to limit access and assure compliance to measurable objectives and indicators. The regulation needs to set out clear standards that each LTMP are to address, such as environmental impact and capacity assessments of the fishery, which could guide measures to address overcapacity and ecosystem impacts. The plans are to cover the whole fishery rather than a particular species and also to provide a framework for the most appropriate management tools for each fishery, which may be accomplished through an ecosystems approach.

5.4.1.5 *LTMPs need to be ecosystem based*

Effective management tools to address wider ecosystem impacts, including time and/or area closures and catch selectivity improvements, are to be identified and addressed.

5.4.1.6 *Paving the way for regionalization*

The CFP needs to set clear high-level objectives while providing greater freedom on how to meet them: **where, when, and how to fish**. This could better reflect the local and regional circumstances. Moreover, effective CFP regionalization means adopting LTMPs at the level of the fisheries. In fact, unproductive EU **micromanagement** could be avoided by placing more responsibility in the hands of those who know the fishery best.

5.4.1.7 *Address overcapacity and improve profit margins*

It may be noted that overfishing of key stocks has resulted in dwindling catches. Many fisheries are today running close to or at a loss. LTMPs are to legally oblige each fishery to assess capacity and put in place a reduction strategy when overcapacity is identified. Incidentally, solutions could include prioritizing the removal of more destructive fishing methods or the retention of higher employment generating vessels, depending on the priorities of the fishery and its community.

5.4.1.8 *Delivering a system: A real reform of the CFP*

The Scottish Conservation Credits Scheme was set up in 2008. The scheme credits fishermen for adopting conservation measures which are most suitable to the fisheries covered by the scheme (mixed whitefish and Nephrops trawl fleets) and to its objectives (cod conservation). It had brought together decision makers, scientists, industry, and NGO representatives for monthly steering group meetings. In fact, the scheme was a prime example of comanagement that could easily be replicated. The better understanding and high compliance could generate the sense of value and a degree of ownership among the stakeholders in the management process.

5.4.1.9 *Embrace rights-based management (RBM)*

The adoption of effective RBM systems could provide incentives, which could be successful in addressing overcapacity, rebuilding fish stocks and improving the overall environmental performance of fisheries. Moreover, RBM could also help in strengthening the economic efficiency of fisheries and phase out the dependence on subsidies. Further, when designed correctly, RBM could do improved stewardship. However, any system of RBM could, preferably, be individually tailored and carried out within the framework of the LTMP. Nevertheless, well-designed ones could preserve important principles, such as community linkage or accessibility to new entrants, etc.

5.4.1.10 *To place the environment at the heart of fisheries management*

The CFP currently assigns equal importance to ecological, economic, and social considerations. However, a thriving and sustainable fishing industry may not be able to exist without a healthy aquatic ecosystem. Hence, the ecological pillar needs to be given top priority in order to enable the recovery of aquatic species and habitats. Further, the new CFP also needs to establish a means of delivering commitments of MSs under the Marine Strategy Framework Directive (MSFD) in order to achieve Good Environmental Status for fish stocks and marine habitats by 2020. For this to happen, management systems are to be built into LTMPs that could identify and prevent significant threats both to aquatic species and to habitats.

5.4.2 Austria

Austria possesses a high potential in AQC and fisheries. The historic European lake types were based solely on commercially important fish species, and, as such, the formation and selection of the sentinel and type specific fish species may not be always very convincing. Incidentally, types were described by different authors (e.g., Haempel, 1930; Bauch, 1963; Spindler, 1997). The proposed lake typology could provide a basis for the **effective management** of the fisheries of the Austrian lakes (Gassner et al., 2005).

Notwithstanding the above, of late, new lake classifications based on fish assemblages were reported for Finnish (Tammi et al., 2002), French (Argillier et al., 2002a, b), and German (Mehner et al., 2005) lakes. In contrast to most European classification systems, well-defined lake classification systems based on fish communities were developed for North American lakes. For example, Tonn and Magnusson (1982) categorized Wisconsin lakes according to the number of fish taxa present. Johnson et al. (1977) classified 2500 Ontario lakes based on the combination of four major recreational fishes present in each lake and found that seven limnological parameters had predicted the lake type in which each species combination occurred. For Texas reservoirs, Dolman (1990) used cluster analysis, thereby revealing five different groups based on species associations. In Minnesota, Schupp (1992) classified lakes starting with limnological variables continuing with fish community variables associated with the limnological classification. The proposed lake typology is said to play a significant role in the **effective management** of the fisheries of the Austrian lakes.

Concomitant to the above, the interest in the assessment and classification methods for European FW ecosystems has increased significantly due to the establishment of the Water Framework Directive (WFD) of the European Union (EU) (Ruoppa and Karttunen, 2002). The WFD requires hydro-morphological, physicochemical, and biological elements to be used for the assessment of the ecological status of water bodies. The biological communities representative of the referenced conditions have to be derived with respect to the water body type, and, preferably, to reflect an anthropogenically least disturbed or undisturbed status as the best (EU, 2000).

However, large (>50 ha) alpine and prealpine lakes, hosting an undisturbed, natural fish community are lacking in Europe. The fish communities of nearly all alpine and prealpine European lakes are more or less degraded by human activities, such as eutrophication, shoreline degradation, barriers to migration and fisheries management activities, etc., to list a few. Existing classification systems are based on the fish communities currently found, and thus may not reflect natural fish communities of European, and in particular of Austrian prealpine or alpine lakes, mainly because of their degradation by human activities. This represents a tremendous scientific deficit, given the enormous importance of fish in lake and reservoir ecosystems.

Incidentally, there had been studies with the main aim of developing a lake typology based on fish assemblages for natural Austrian lakes, which are >50 ha in area. More specifically, the classification procedure could be based on the reconstructed historical fish communities of these lakes, which are termed as "near-natural." Other aims of the studies were to find fish species having some indicator value for the presumed classes of lakes in analogy to fish zones of European running waters. In order to find such indicator species, there had been efforts to develop an index, which is supposed to enhance the objectivity of such selections. Other aims could be to compare basic abiotic variables between the lake types based on fish communities. A classification of lakes based on their fish community could be helpful also for **fisheries management**. Stocking programs, harvest regulations, or fishing methods could be coordinated with respect to the type of lake based on the specific fish community.

5.4.2.1 *Methods*

Data bases

Data from 43 natural Austrian lakes, which were >50 ha in surface area, were used for this study, ranging from lowland lakes of the eastern part of Austria to prealpine lakes in the northern and southern parts, and also to high altitude alpine lakes. The recent trophic status of these natural Austrian lakes ranges from oligotrophic (*c.* 28 lakes, *c.* 65.1%) to oligo-meso-trophic (*c.* 7 lakes, *c.* 16.3%) to mesotrophic (*c.* 7 lakes, *c.* 16.3%), and to one eutrophic lake (*c.* 2.3%) (BMLFUW, 2002). All the lakes were more or less influenced by different recreational activities, commercial fisheries, and angling. The historical fish communities of the lakes were reconstructed hierarchically based on historical documents, museum specimens, inquiries to the fisheries managers, and if the data set about a fish species was too small, they were reconstructed based on expert opinion. In fact, expert opinion was used to assess historical presence or absence of certain species in 12 lakes, 2 fish species in 2 lakes, and 3 fish species in 1 lake. Moreover, all the historical data were usually checked for synonyms and fitted to the current nomenclature. For each lake, all the data were summarized in a preliminary list of fish species. To this list, certain criteria were usually applied, e.g., alien species and typical species of lotic habitats, etc. Further, there had been exhaustive information on the reconstruction procedure and the historical fish communities of the Austrian lakes (Gassner et al., 2003a, c; Zick et al., 2004). The reconstructed fish communities (presence-absence data) represent the situation of about 1850–1900, i.e., the time before intensive human influence took place and was, therefore, considered "near-natural."

Classification procedure

Based on these reconstructed historical fish communities (presence-absence data), the lakes have been classified into different groups. Hierarchical cluster analysis methods were used for the purpose of classification (Multi Variate Statistical Package, MVSP 3.12). In the first step, the program calculates the similarity among fish communities using Jaccard's Coefficient. In the second step, the centroids were compared in an average linkage algorithm resulting in a dendrogram (Krebs, 1989). The centroid function computes the distance between two clusters as the difference between centroids. However, untransformed binary data sets (0/1) were used for all these calculations in order to distinguish between presence (1) and absence (0) of a fish species in a certain lake. The dendrogram represents different lake groups with highly similar fish species assemblages. Incidentally, type specific fish species and sentinel species (key species) of the different lake groups were defined by a newly developed "lake-type specific fish community-index" (LFI). It was based on the consideration that the ideal indicator species for a given group of lakes occurs in all lakes of that group and in no lake of another group. The LFI is based on the occurrence probability of each fish species within one lake type and the occurrence of each fish species among other lake types.

Mathematically,

$LFI = (\Sigma\, pi)^{*}$ range
pi = occurrence probability of a fish species in one lake group
range = total number of lake groups in which a given species occurs (in our cases 1–4).

Contrary to the above, the **sentinel fish species were defined** by an occurrence probability of 1.00 within a lake group and an LFI < 8.0. In fact, **type specific fish species were defined** by an occurrence probability of 0.80. In addition, the dissimilarity of the fish communities as well as the contribution of each species to the total similarity/dissimilarity between the groups of lakes were calculated using the software package Primer v5/SIMPER (Clarke and Warwick, 1997; Clarke and Gorley, 2001). These analyses were based on the Bray-Curtis similarity/dissimilarity calculations (Krebs, 1989). The results of the LFI calculations and the similarity analysis had been compared. Furthermore, the different lake groups were characterized by a set of biotic and abiotic factors, i.e., lake area, maximum depth, altitude, volume, catchment area, and fish species number. Lake Constance was excluded from the analyses as it represents an outlier with regard to lake area, maximum depth, volume, and catchment area. The means of these factors of each lake group were compared using ANOVA (Sigma Stat 2.03) and bootstrap analyses (SIMSTAT).

Results

There had been a decision to discriminate four groups of natural lakes, based on the results of the cluster analysis and plausibility testing of different possible groupings and also based on the similarity of their fish communities. The four groups could be separated based on centroid distances < zero.

(i) Group 1

Lakes ($n = 16$, centroid distances −0.27) could be clearly separated from the other three groups and were characterized by high altitude (*c.* 941 m MSL), small surface area (*c.* 152.2 ha), and low fish species number (*c.* 5.2).

(ii) Group 2

Lakes of group 2 ($n = 10$, centroid distances of −0.21) were characterized by a mean altitude of *c.* 634.4 m MSL, relatively large mean surface area (*c.* 1209.3 ha), and high maximum water depth (94.3 m).

(iii) Group 3

The third group ($n = 14$, centroid distances of −0.09) displayed a relatively low maximum water depth (*c.* 45.4 m) and a long retention time (*c.* 4.1 years).

(iv) Group 4

Incidentally, the lakes of the eastern part of Austria represent group 4 ($n = 3$, centroid distances −0.09) and could be characterized by low altitude (*c.* 115 m MSL) and very low maximum water depth (0.9 m).

Concomitant to the above, pike, roach, perch, tench, chub, and stone loach were found in lakes of all four lake groups. More specifically, pike could be found in all lakes of groups 2, 3, and 4, whereas roach and perch occurred in all lakes of groups 2 and 3. The calculated LFI of pike, roach, perch, tench, and chub is > 8.0; and thus, these fish species were unsuitable as sentinel species. Pike-perch was found as an ideal sentinel fish species for group 4 lakes because this species was exclusively found in the historical data sets of all lakes of this group but was lacking in all other groups (occurrence probability = 1.0; LFI = 1). However, bleak could be categorized as sentinel fish species for group 3 lakes. Bleak was found in all the lakes of group 3 (occurrence probability = 1.0), additionally in one lake of group 4 and in no lake of the two other groups (LFI = 2.67).

Notwithstanding the above, minnow proved to be a sentinel fish species for group 2 lakes (occurrence probability = 1.0) and was also found in numerous lakes of group 1 and in a few lakes of group 3, resulting in an LFI of 6.24. For group 1 lakes, arctic char could be classified as sentinel fish species. Arctic char was found in all lakes of group 1 (occurrence probability = 1.0) and also in a few lakes of groups 2 and 3, respectively.

The criterion for the designation of type specific fish species was an occurrence probability of ≥ 0.80 within a specific group of lakes. In the char lake(s), a total number of 15 fish taxa were found, with bullhead, minnow, and European lake trout being "type specific fish species." In the minnow lake type (= group 2), on the other hand, a total number of 16 fish taxa were counted. Further, type specific fish species are bullhead, arctic char, European lake trout, stone loach, chub, roach, perch, bream, and pike. The lake type "bleak lake" (= group 3) encompassed a total number of 25 fish taxa. Type specific fish species are chub, pike, roach, perch, bream, rudd, tench, and wels. Concomitantly, in the pike-perch lake type (= group 4), also a total number of 25 fish taxa were found, with tench, pike, and carp as "type specific fish species." The highest species contribution to the differences between groups of lakes was found for arctic char (group 1), minnow (group 2), bleak (group 3), and pike-perch (group 4) in the analysis using SIMPER (Clarke and Warwick, 1997; Clarke and Gorley, 2001). Further, fish community comparisons between the different lake types revealed that minnow lakes and bleak lakes had a lower dissimilarity compared to arctic char lakes and pike-perch lakes which had a high dissimilarity (Gassner et al., 2005).

A brief discussion

In consequence to the above, based on the fish communities, four different types of lakes could be defined for the Austrian lakes >50ha in surface area. In contrast to numerous published lake classifications (e.g., Haempel, 1930; Bauch, 1963; Johnson et al., 1977; Tonn and Magnusson, 1982; Dolman, 1990; Schupp, 1992; Spindler, 1997; Argillier et al., 2002a, b; Tammi et al., 2002; Mehner et al., 2005), which were based on the recent fish communities, attempts had been made in the many recent exercises (Gassner et al., 2005) to use the reconstructed historical fish communities for the classification procedure. These so-called "near-natural" fish communities are believed to represent the situation from around 1850 to the preindustrial period before intensive human influence took place. Nevertheless, historical information could be very helpful for the reconstruction of earlier conditions as shown by Steedman et al. (1996) and Kelso et al. (1996) for the Great Lakes region. Incidentally, historical records on fisheries in Austria, used in some of the studies, dated back to the 15th century. Moreover, many of these old records were hand written and also time- and cost-intensive to evaluate. Furthermore, these writings were often biased toward commercially important fish species, whereas commercially unimportant species, like the minnow, bullhead, and stone loach, were often documented imprecisely. Nevertheless, the huge amount of data and specimens deposited in the museums and the accomplished reconstruction procedure enabled a satisfactory reconstruction of the historical fish species composition.

Concomitant to the above, the application of historical data to the reconstruction of native fish communities always incorporate an undefined uncertainty, based mainly on the different quality and quantity of historical data. Conversely, it may be noted that if historical data were used carefully, they may give important and often the only information about the native fish species composition of lakes (Steedman et al., 1996).

Notwithstanding the above, the applied clustering method is an exploratory data analysis tool that allows sorting of different objects into groups in a way that the degree of association between two objects is maximal if they belong to the same group and minimum if otherwise. However, cluster analyses refer to similarity analyses and clustering algorithms are not related to statistical significance testing. In fact, the end results of classification may be evaluated by the ecological and environmental requirements (Krebs, 1989). These methods are mostly used when no a priori hypothesis exists, in this case the typology of Austrian lakes. Therefore, there had been efforts to cluster the Austrian lakes without any presumption of the possible number of fish-based lake types, at first. In the second step, there had been efforts to analyze the ecology of the fish communities, and the morphometric and geographic requirements of the different clustered groups, ranging from two to six. There had been highest plausibility of certain parameters, which were significantly different among the groups of lakes (ANOVA Student-Newman-Keuls test; Dunn's Method; Bootstrap confidence interval 95%) and fish community comparisons between the different lake types were based on average dissimilarity (numbers) calculations (Gassner et al., 2005).

Concomitant to the above, different possible groupings for the four groups were represented in some of the **recent works** (Gassner et al., 2005). The four groups were well distinguishable in the ecology of their fish community, in their abiotic factors, and also in their geographical distribution; thus, represent discrete types of Austrian lakes.

Notwithstanding the above, one of the significant reasons for the development of this lake classification based on historic fish communities was the implementation of the WFD of the EU. In addition to the formulation of an abiotic classification of the European water bodies, the WFD requires that the reference condition (s) and the assessment procedure(s) be specific for the types of water bodies. As such, the fish-based classification of lake(s) is a significant tool to fulfil the requirement(s) of a biotic classification of Austrian lakes. Further, reconstructed historical fish communities provide also a basic element for defining the reference conditions in order to assess the alteration(s) in the fish species composition. Further, the species composition of a fish community, in addition to other factors, is dependent on the abundance and age structure, which are significant quality element(s) of the WFD (EU, 2000).

Notwithstanding the above, **another important study performed** (Gassner et al., 2005) was the generation of basic information for the **management of fisheries**. Incidentally, the regulation of the **fisheries management** in the Austrian lakes is executed by the government of the nine Austrian provinces. Generally, the stocking programs during the last decades were not adapted to the ecological requirements of the lakes. Hence, compared to the historical situation the number of fish species increased significantly because of intensive stocking activities with species which were not indigenous members of the respective fish community. The proposed lake typology, with the development of sentinel and type specific fish species, possesses the potential to provide significant guidelines for the **effective management** of the fisheries of the Austrian lakes. Further, application of efficient fisheries management *modus operandi*, notably, the **CBFM**, involving the community stakeholders from the grassroots level could lead to effective management of the water bodies and their fish resources. Further, the historical fish communities of the Austrian lakes are significantly influenced by the catchment area, altitude, habitat requirement(s), and colonization possibilities after the last ice age. Incidentally, the species diversity was, generally, higher in the larger lakes of the minnow-, bleak-, and pikeperch lake types and smaller in the arctic char lake type, thus portraying the well-known relationship between fish species number and lake area (Eckmann, 1995). A cold-water-adapted fish community in the high altitude arctic char lake type and a warmwater-adapted community in the lowland pike-perch lake type are clearly distinguishable, and similar to lakes in Germany (Mehner et al., 2005). While comparing minnow and bleak lake types, it became apparent that the majority of the minnow lakes is found the north of the Alps whereas most of the bleak lakes are found south of the Alps. This correlates with the notion that the minnow lake type is inhabited by more cold-water-adapted species compared to the bleak lake type. Moreover, the spawning guilds proposed by Balon (1975) reflect the four different fish-based lake types, ranging from exclusively lithophil spawners in the arctic char lake type to a combination of mostly lithophil and few phytolithophil spawners in the minnow lake type. The sentinel and type specific fish species of the bleak lake type consist mostly of phytophil and phytolithophil spawners whereas the pike-perch lake type consisted of phytophil spawners only.

Further, Schmutz et al. (2007) dealt with fish-based methods for assessing European running waters along with a synthesis, and a brief note on management.

The EU, WFD requires monitoring of riverine fish fauna. When the WFD came into force in 2000, most of the EU MSs did not have fish-based assessment methods compliant to WFD requirements. Therefore, the objectives of FAME, a project under the fifth R&D Framework Program of the EC, were to develop, evaluate, and implement a standardized fish-based method for assessing the ecological status of European running waters.

There have been contributions which attempted to synthesize the outputs of FAME and also tried to define future research needs. Two different methodologies were used:

(i) The so-called spatially based modeling.
(ii) The site-specific modeling;

the latter leads to the European Fish Index (EFI).

Incidentally, the advantage of the EFI is that, despite being a single index, it is applicable to a wide range of environmental conditions across Europe precluding the need for intercalibration. The EFI could support the WFD toward harmonized/standardized assessment and management of running waters in Europe, thus enabling comparative analyses of the ecological status of running waters across Europe.

5.4.3 Belgium

Belgium possesses a strong base for fisheries. **In Belgium**, Depestele et al. (2014) worked on the sensitivity assessment as a tool for spatial and temporal gear-based fisheries management.

5.4.3.1 Lessons from the case studies

Assessment of ecosystem health is required in the EAFM. Mitigation measures that address impacts exceeding the acceptable level also require assessment. The various fishing methods have different ecosystem impacts and this makes it difficult to assess them jointly. Sensitivity of the ecosystem to individual fishing method had been assessed previously, but concurrent comparisons of different methods on the same ecosystem had, probably, not been done.

There had been attempts to combine the main characteristics of established approaches to sensitivity assessment into a single roadmap called Sensitivity Assessment of Gear Effects (SAGE). SAGE is a widely applicable three-step process to assess the ecological concerns of EAFM. The methodology used in the SAGE roadmap is built on a scoring system, which results in a sensitivity index of the ecosystem components of the fishing methods. The scoring system is based on a combination of expert judgment and data, both qualitative and quantitative. It allows for cross-evaluation of fishing methods and ecosystem components. Sensitivity maps are usually made by using the spatial and temporal distribution patterns of the ecosystem components in the index. The uncertainty of the sensitivity scores and maps are estimated through a pedigree index. The index is based on proxy representation, empirical basis, methodological rigor, theoretical understanding, and degree of validation, all of which measure the strength of the research results. Various methodologies could be illustrated using case studies that compare the ecosystem effects of different gears, notably, the beam trawl and trammel net fisheries. However, the selected examples may not result in unexpected outcomes, but rather could be chosen to evaluate the applicability of the methodology. The examples try to illustrate how a semiquantitative frame work, which includes the uncertainties associated with scientific assessments, could deliver holistic advice to fisheries managers in a fully transparent manner.

Notwithstanding the above, sensitivity could be calculated in several different ways in scientific literature because its value(s) depends strongly on the selected factors and how they are combined (Garthe and Hüppop, 2004). Incidentally, there had been proposals to combine two factors, viz., intolerance and recoverability. In turn, these factors consist of various partitions or attributes such as susceptibility to being smothered (Tyler-Walters et al., 2009). This is in contrast to other approaches, where all attributes are positioned at the same level (Stelzenmüller et al., 2010). Incidentally, it may be mentioned that, the **intolerance-recoverability approach** was better and hence preferred because recoverability attributes are intrinsic to the ecosystem component and intolerance clearly relates to the fishing aspects.

In fact, the overall framework could be considered as transparent and repeatable (Stelzenmüller et al., 2010) and comply with essential requirements of sensitivity assessments (Hiddink et al., 2007).
Concomitant to the above, Depestele et al. (2012) **reported about an integrated impact assessment of trammel net and beam trawl fisheries.**

A fishery policy is to be useful for everybody. There is a wide spread agreement on the critiques to the CFP and fishing impacts on the marine ecosystem, e.g., low fish stocks, loss of BD, high levels of discarding, and so on. Aspirations for a better protection of the marine ecosystems are now believed to be formally setout in the Marine Strategy Framework Directive (MSFD) and are, perhaps, under way through the reform of the CFP. Incidentally, the Belgian marine fisheries fleet is dominated by the beam trawlers. However, the beam trawling may experience considerable amount of pressure from the latest EU legislation, as this fishery is clearly questioned for its ecological effects (e.g., discarding).

One example could be the need for fisheries measures in Natura 2000 sites in the Netherlands, which is likely to impact the Dutch beam trawlers. Similar measures are likely to be discussed and implemented in the short term for the BPNS to fulfil the requirements from the habitats. Passive fisheries, on the other hand, such as the trammel net fishery, may be prompted as an alternative gear for the Belgian flatfish-directed beam trawl fleet. In this connection, the WAKO-I project explored the state-of-the-art knowledge on beam trawl and trammel net effects in the Belgian Part of the North Sea. The project served as a stepping stone in the exploration of multidisciplinary research and compiling expertise from biological-ecological and fisheries science. Incidentally, the WAKO-II had been built upon this base and it is supposed to had bridged projects with fisheries projects that were exclusively focused on the fundamental ecosystem research (such as those financed through the EU European Fisheries Fund). From this basis, WAKO-II aimed at making progress in developing an integrated assessment tool, which could scientifically underpin the policies that had reconciled the interests of both the marine environment and the fisheries. To be effective in the CFP, such a tool should be at the level of the fisheries management unit.

Notwithstanding the above, the tool also needed to be set in the context of risk analysis, in order to take ecosystem-based considerations effectively onboard. The WAKO-II project, therefore, had focused on the development of a risk/sensitivity assessment of the effects of beam trawl and trammel net fishery. The WAKO-I inferred that our scientific knowledge especially on trammel netting is limited. Nevertheless, part I of the WAKO-II project focused on filling parts of the knowledge gaps. Part II was set up around the development of an appropriate assessment tool, given the need for ecosystem-based management decisions despite incomplete knowledge. The development of the tool had been demonstrated with selected examples from the divergent effects of both the fisheries. These examples as well as the link with the actual fishing efforts at the BPNS explored the application of risk/sensitivity assessment as a tool for spatial and temporal planning and evaluation of fisheries.

Further, in **Belgium**, Margrethe et al. (2014) dealt with the aspect(s) of "what does stakeholder involvement mean for fisheries management?"

The CFP sets out the guidelines for the management of European fisheries. The CFP is revised after every 10 years, and the last two revisions had laid the groundwork for extending stakeholder participation in EU fisheries management. The fishery industries and the NGOs especially are recommended to be given greater influence. However, there may not be sufficient time or fund to have everyone involved, but the challenge is how to effectively enhance mechanisms for dialog to reach consensus.

Various papers reported results from an international survey on fisheries stakeholders' preferences for the three pillars of sustainable fisheries activities as defined in the CFP, viz.: (a) ecological, (b) economic, and (c) social.

Notwithstanding the above, results of the various surveys have shown that industry member preferences were significantly different from the preferences of authorities, scientists, and NGOs. However, the preferences of the three latter groups did not differ significantly across stakeholder groups. This raises the question that what consequences the stakeholder involvement in the CFP might have.

Notwithstanding the above, if the recommendations of the RACs, which usually consist of two-thirds industry, are ratified by the EU and formally incorporated into management, there could be different priorities for fisheries management than the ones which exist, if management were developed solely by national and EU authorities. Moreover, industry dominance of the RACs, coupled with the significant heterogeneity in the so-called preferences of the NGOs, often may exist because of the breadth of interests represented, from fishers' welfare organizations to bird watchers. Further, it could mean that nonindustry views might not be well represented in the recommendations put forward to the EU.

Concomitant to the above, results of various studies may, however, have implications for stakeholder involvement in fisheries management in terms of the selection of representatives for the different stakeholder groups as well as the inclusion of new or more stakeholders and/or stakeholder groups in the future.

In addition to the above, the various results have also demonstrated that the selection of representatives to be involved in fisheries management based only on our predefined stakeholder groups may not be very appropriate. It could be because of the fact that preferences may not be very much group-specific, but rather individual. This could pose challenge for representation for all those involved in fisheries management as well as also raise questions about whether representatives may be selected individually, e.g., through a referendum-like process, through ballot, or through random sampling, rather than as members of a particular group. But, in spite of this, such a *modus operandi* could come with its own challenges, for instance, who are to assess the individuals, and what parameters and preferences are to be the basis of the assessment? However, the priority is to clarify what features and preferences, group and/or individual, are to be considered within fisheries management. However, there may not be sufficient time or fund to have everyone involved but the challenge is how to effectively enhance mechanisms for dialog to reach consensus, taking into consideration the preferences and targets for sustainable fisheries. At the same time, an ecosystem-based approach to fisheries management is an important parameter, which needs consideration on a larger context.

Notwithstanding the above, Newman (2014) dealt with structural and cohesion policies with regard to inland fisheries and the CFP in **Belgium**.

The new European Maritime and Fisheries Fund has greater potential to provide increased support to inland fisheries through the strengthening of community-led local development.

Commercial inland fisheries are usually small-scale, labor intensive, traditional fisheries mostly using passive gears. Nevertheless, they produce high value products of local significance. However, the inland fisheries may not be much influenced by the reforms of the CFP, as they are mainly managed nationally. Diadromous species may benefit from the reforms as long as they are properly implemented. The new European Maritime and Fisheries Fund has greater potential to provide increased support to inland fisheries through the strengthening of community-led local development.

Incidentally, Bartch et al. (2009) dealt with a Handbook reporting, Voices from the Fisheries for preserving local Fisheries knowledge, which help in linking generations and in improving environmental literacy linking to sustainable fisheries management involving the stakeholders from the grassroots level through the *modus operandi* of CBFM.

Concomitantly, NOAA's National Marine Fisheries Services,' Local Fisheries Knowledge (LFK) Project, piloted in Maine had been a learning journey for everyone involved in designing an educational program, where knowledge and participation intersect to engage young people as vital community members. They try to address issues that could impact the future vitality and sustainability of their small fishing communities. Undoubtedly, this is a monumental task. Nevertheless, some of the outcomes that sustained the workers through the many challenges of implementing the LFK project are as follows.

(i) Receptivity from the community

Formulating partnerships and common agendas between the diverse stakeholders, notably, the regulatory agencies, schools, students, and local community members, etc., was a challenging task. However, this has been one of the greatest strengths of this exercise.

5.4.3.2 *Level of commitment by students*

The students, in general, have realized the value of their work to their community and national institutions, like the NOAA's National Marine Fisheries Service.

5.4.3.3 *Learning that transpired for both students and adults*

Teachers had enhanced their knowledge of the relationships between their communities and the marine environment. They also enhanced their understanding of the communities in which their students live in. They also tried to develop new relationships with community partners, including NOAA.

Notwithstanding the above, there had been attempts to share strategies, lessons learnt, and sample curricula for use in a variety of formal and informal educational settings interested in attempting similar worthwhile endeavors of documenting and preserving LFK.

Notwithstanding the above, Anon (2009a, b) analyzed the existing RBM instruments in MSs and on setting up the best practices in the EU through various pilot projects for carrying out the CFP.

There had been reports presenting a review and analysis of the RBM systems in place in EU coastal MSs. Rights-based approaches to fisheries management have shown potential for promoting biologically sustainable and economically viable fisheries in several parts of the world. While fisheries management is a community responsibility, under the framework of the reformed EU CFP, economic management of fishing rights is a national responsibility and, in practice, many MSs had already implemented RBM approaches in a range of fisheries across the EU.

Various studies satisfy the need for a review of these existing RBM practices, as identified by the commission. The studies tried to analyze the attributes and effects of RBM systems and also investigated their degree of success in contributing to achievement of the CFP objectives of sustainability in the exploitation of stocks, matching fleet size with available fish resources, and economic viability of the fishing industry.

Concomitant to the above, the definition of RBM adopted for this study was broad including "any system of allocating fishing rights to fishermen, fishing vessels, enterprises, cooperatives, or fishing communities." As such, the main types of RBM systems covered are as follows:

(i) limited nontransferable licensing type (LL);
(ii) limited transferable licensing (LTL) type;
(iii) community catch quotas (CQ);
(iv) individual nontransferable effort quotas (IE);
(v) individual transferable effort quotas (ITE);
(vi) individual nontransferable catch quotas (IQ);
(vii) vessel catch limits (VC);
(viii) individual transferable quotas (ITQ); and
(ix) territorial use rights in fisheries (TURF).

Once rights had been allocated, it could become difficult for new entries to gain access to the fishery. However, in most cases, if newcomers wish to enter a fishery controlled by an effort- or quota-based system, they are to buy a vessel with its associated rights (license, quota allocation, etc.). However, this could be prohibitively expensive and several countries like United Kingdom, Denmark, etc., have specific schemes to facilitate new entrants to the fishery, either by allowing them to start in the small-scale sector or by leasing out quotas. However, the most difficult type of system for newcomers to gain access seems to be TURFs, where rights are allocated to a group of resource users only. Their establishment had often resulted in the exclusion of prior users of the resource through the conversion from an open access to a privatized regime and, subsequently, it could be difficult for newcomers to gain membership of the association or group involved in the TURF. In addition, rights may have varying validity. Incidentally, TURFs tend to have the longest period of validity (often in perpetuity). However, in some cases, validity of the right could be short (e.g., 1 year), but rights are renewable, thus effectively making the validity period much longer. Further, in some cases, RBM systems had evolved over time, without the nature or duration of the right being specifically defined. Furthermore, a number of different "rights" may be required to be able to fish, e.g., a license, quota allocation, days at sea, etc.

Notwithstanding the above, markets have evolved around the sale, transfer, and lease of rights even in the case of "nontransferable" rights. It could be because vessels with rights may typically be sold together with those rights. However, vessels with attached licences may often be sold for more than the pure asset value of the vessel itself, thus, implying a value to the license or fishing right. Markets also exist for individual nontransferable catch quotas (IQ) and individual nontransferable effort quotas (IE) because it is possible to buy companies that have quota and/or to sell or rent vessels with unused quota.

5.4.4 Bulgaria

The natural resources in a country are to be properly managed in order to avoid unsustainable harvest and overexploitation. **Bulgaria** had made every effort in this direction for sustainable harvest and conservation of the natural resources. Meanwhile, there had been a Green paper presented by the EC, which was published at a very important moment, in order to give an answer to the emerging necessity for integrating the CFP with the other community policies and to solve the existing structural and also operational disadvantages.

Incidentally, Anon (2009a, b) discussed the statement of the **Bulgarian** Fisheries Administration and the stakeholders in the fisheries and AQC sector in **Bulgaria** in relation to the proposed EC Green Paper on the reform in the CFP.

Fish and fisheries are potential entities of nutrition and avocation for the people in a country. As such "fisheries" sector deserves tremendous amount of boosting and patronage for making its role effective.

The Green paper, presented by the commission, has been published at a very important moment in order to give answer to the emerging necessity for integrating the CFP with the other community policies and to try to solve the existing structural disadvantages, after taking into consideration the specificity and the characteristics of the geographical regions and the fisheries sectors of the different countries.

5.4.4.1 *Taking into account the specificity and the characteristics of the geographical regions and the fisheries sectors of the different countries*

It is very important for Bulgaria to follow an approach for the management of the fisheries, which is based on the standards and the general principles adopted at EU, and also take into account the specific regional conditions to maximum extent. Further, the experiences gained after the accession of the country to the EU could also be considered. Also, the specificity of the Black Sea could be an item of consideration. In this regard, the reformed CFP could accelerate the processes to find the appropriate form of collaboration in the regions including the management of the fish resources.

Notwithstanding the above, on the basis of the experience gained after the accession to the EU and taking into account the specificity of the Black Sea, it seemed to be very important for Bulgaria to follow an approach for the **management of the fisheries**, which is based on the standards and the general principles adopted at EU level as well as take into account the specific regional conditions to the maximum extent. In this regard, the reformed CFP would accelerate the processes in order to find the appropriate form of collaboration in the regions, like the Black Sea region including the management of the fish resources. In fact, the Black Sea possesses specific natural geographic and political features. It may be noted that only two of the six coastal countries are members of the EU (Bulgaria and Romania) and their share in the catches is very small compared to the other countries, especially Turkey. Another specific feature is that before and after the accession of the two MSs to the EU, there seemed to be no form of collaboration among the six coastal countries for common management of the fish stocks. The present situation leads to serious difficulties in the management of the resources, which cannot be done only by the two MS of the EU and, moreover, taking into consideration their small share in the total catches. At this stage, there seemed to be some amount of necessity to apply a differential approach to the measures for the conservation of the fish resources, where the two basic aspects could be pointed out:

(a) Measures for the conservation of the species characterized as "local."
(b) Measures for the conservation of the species characterized as "Shared" stock.

Concomitant to the above, all measures for the conservation of the fish resources, such as programs for adaptation and management of the fleet; restriction of the catches, application of quotas and sustainable levels of the catches on annual or 6 monthly basis, closed zones and seasons for fishing, technical measures for the restriction of the by-catches and the discards, multi-annual plans for the management and recovery of the stocks, etc. All these measures would be efficient only if they are applied by all the participants. Thus, for the purpose of the sustainable management of the shared stocks, CFP could facilitate the establishment of Regional Fisheries Organizations (RFO) that have to apply in the most effective way the community policies, taking into account the following:

(a) Cooperation among the commission, MS, and third countries (in the case of management of shared stock) to enable the RFOs for the identification of the necessary measures for sustainable harvest and conservation.
(b) Regulated exchange of data at regional level for the assessment of the stocks of commercially important species.
(c) Providing opportunities to the RFO(s) to use individual approach in the management of the stocks.
(d) Measures for their conservation and recovery in safe biological limits. The decisions and proposals of the RFOs may be coordinated and altered by the EC wherever necessary.
(e) Exchange of information in the framework of the actions and management of the RFOs for the applied fishing gears, fishing practices, fishing fleet capacity, and existing fishing opportunities.
(f) Implementation of a pilot project for sustainable implementation of CFP in the Black Sea region.

CBM involving the stakeholders from the grassroots level would lead to efficient management of the fisheries (CBFM).

5.4.4.2 *Overcoming the five structural disadvantages of the policy*

Concentration on the objectives of the policy

It would rather be a subjective issue, at this stage, to take into account the social and economic effect(s) of the measures, envisaged in the CFP, as pointed out in the proposal of the EC. The problem could assume dangerous dimensions under the conditions of financial and energy crises. Therefore, a preliminary SWOT analysis of all the impacts, effects, and interactions should be done before planning any restrictive measures.

5.4.4.3 *Concentration on the framework for making decisions on key long-term principles*

Statement of the fish product producers organisation "big fish"

The approach could start from the decisions concerning the short-term development of the sector and, gradually, advance toward development in the longer period in order to solve the problem with the sustainable management of the stocks and prevent the negative social and economic effects.

The common decisions to be taken by the council at the highest political level practically could lead to concentration on short-term considerations at the expense of the long-term sustainability of the European fishery in terms of ecological, economic, and social plan. The governments and the sector may be allowed to adopt the implemented policy to their needs and to find the best solutions in technical and economic plans.

5.4.4.4 *Encouraging the industry to take more responsibility in implementing the CFP*

Statement of the Bulgarian fisheries administration

The Bulgarian branch organizations in their entity, being direct participants in the entire process of the sector, may have to take part at each stage of the decision-making process on the conservation and management of the resources. The decisions proposed by them have to be in compliance with the policy on conservation of the resources, in each one of the possible and applied legislative restrictive measures. Moreover, they could submit adequate, current information about the market demands and trends. Further, they could share their views concerning the quantities and regional allocations of the catches, taking into consideration the recent biological state of the stocks.

Indicators for the assessment of the state, at the moment, could be formulated for the purpose of ascertaining the methods of risk analysis.

5.4.4.5 *Statement of the fish product producers organisation ("big fish")*

The representatives of the association have proposed the establishment of the commercial fishing licensing regime on the basis of certain references, issued to the candidate by acknowledged branch organizations, authorized with representative functions for a given subsector or region.

The common practice decisions taken/to be taken by the council at the highest political level, practically, could lead to concentration on short-term considerations at the expense of the long-term sustainability of the European fishery in ecological, economic, and social plan. The governments and the sector could be allowed to adopt the implemented policy to their needs and find the best solutions in technical and economic plan. Regarding encouraging the industry to bear more responsibilities for the implementation of the CFP, there should be proposal to work with bottom-up initiatives. Rights may be attributed to the organizations (the sector) for proportional allocation of the quotas on the basis of documentary evidence for the catches during previous period. Sanctions could be specified for persons violating the rules.

5.4.4.6 *Statement of the Bulgarian fisheries administration*

The responsibilities and delegated power may be oriented in the direction of financing different surveys for the assessment of the stocks, studying the import/export fish markets, implementation of different practices connected with the origin traceability of the fish products, and so on.

Concomitant to the above, in order to improve the data collection systems in short- and long-term plans and to ensure coordinated information for the system, it would be necessary to take into consideration the experience in the traceability of the raw materials and products according to the requirements in the regulations. It is indispensable to have mandatory labeling of the trade packages in accordance with the existing requirements, containing the number of the approved producer, fisherman, or retailer. Marking of the bigger size fish species should ensure possibility of identification for the controlling bodies within the whole market chain.

5.4.4.7 *Maximum sustainable yields (MSY)*

The mixed fisheries is facing the risk of taking onboard nontargeted by catches, especially in the cases of fishing ban for given species, because of quota exhaustion. The fishermen, in that case, have no other option but to discard the catches into the sea. This idea of discard ban in the reformed CFP has received considerable support.

5.4.4.8 *Developing a culture of compliance*

Some of the steps are as follows:

(a) Participation of the branch organizations in the process of determination and implementation of measures laid down in the program for data collection.

(b) Participation in stock assessment projects.
(c) Participation in projects for the implementation of different procedures for electronic catch data reporting.
(d) Application of contemporary practices for traceability of the fish products within the whole chain.

5.4.4.9 *Trade and markets: From catch to consumer*

Statement of the fish product producers organisation

Fisheries and AQC sectors earn quite a substantive amount of income from the small part of the sum paid by the consumers as price for the fish and fish products. So there is every reason to take steps for further increasing the income of the fishermen, the fish farmers, the fish traders, and the aquaculturists. The community, at large, may be involved to make headway in this regard through **cooperative management** or through **CBFM**. At the same time, efforts could be made to improve the quality of the fishery products including proper and genuine certification, which would help them to earn confidence of the buyers and the consumers. In one sense, it could be made compulsory to turn the quality of the products into one of our main priorities in the future CFP.

5.4.4.10 *The knowledge base for the policy*

Statement of the Bulgarian fisheries administration

Concomitant to the above, scientific knowledge and building up of a data base are of immense significance for the reformed CFP mainly because they serve as a basis for the decisions concerning the sustainable level of exploitation of the fish stocks. Moreover, it is very much necessary to establish a mechanism for transparency and interaction of the scientific data with the stakeholders and other related community. Also, advice could be sought from them at an earlier stage before taking the final decision on the **management** of a certain stock. We may chalk out long-term regional scientific research programs with the participation of the national and international scientists and more importantly the stakeholders.

5.4.4.11 *Structural policy and public financial assistance*

Statement of the Bulgarian fisheries administration

The structural policy and, in particular, absorption of funds from the EU European Fisheries Fund reflect the new principles for the development of EU "Fisheries and Maritime affairs" policy. In fact, the European legislation in the fields of structural measures, the priority axes, measures and mechanisms for allocating the funds, as laid down in Council's Regulation as well as the programming documents are a transitional instrument and means of building up long-term sustainability in the development of fisheries sector. Further, the Operational Program for the Development of the Fisheries Sector of the Republic of Bulgaria largely reflect the objectives outlined in the Green Paper on the reform of the community laid down in the CFP of the EU. Further, **management measures through CBFM** could play a significant role.

5.4.4.12 *The following principles are taken into account in EU structural policy with particular reference to the Republic of Bulgaria*

(a) Consistency with the principles of the EU CFP in order to achieve a stable and durable balance between the fishing capacity and the fishing opportunities to fish stocks.
(b) Strengthening the harmonious, balanced, and sustainable development of economic activities related to fisheries, human resources and employment generation, and protection and improvement of the environment, in general.

Further, the Operational Program for the Development of the Fisheries Sector provide means for increasing the selectivity of fishing gears as well as sufficient financial resources for sustainable development and modernization of small-scale fishing sector. Their implementation will ensure the success of policy at the **community level with the involvement of CBFM at the local, regional, and national levels**. Incidentally, the structural policy in the EU fisheries, generally, is developed in accordance with, and in line with, the regional policy, the employment policy, and the policy for environmental protection. Also, care is usually taken to strengthen the harmonious, balanced, and sustainable economic activities, employment generation for the human resources, and protection and improvement of the environment.

5.4.4.13 *Aquaculture*

Statement of the Bulgarian fisheries administration

The guidelines of development have to be related to the enhancement of the AQC production share in the total volume of both the FW and marine fish products produced. The utilization of the funds (e.g., the European Fishery Fund sources) could buildup precondition(s) for setting up, widening, equipping, and modernization of the production and processing facilities as well as a potential for production of quality fish products. The investments could generally be directed to the fish products marketing, improvement of the hygiene and the conditions of work in the AQC farms as well as to increase the quality of the final fish products.

Notwithstanding the above, the financial support for AQC is aimed at achieving the following results:

(a) Improving AQC fish products with regard to quality and supply in the market, for which there is usually a high demand.
(b) Adaptation of the sector to the market demand(s), species, prices, quality control, and so on.
(c) The firms, associations, and organizations in the sector may use the scientific expertise and technological potential of the scientific institutes in the field of AQC.

Nevertheless, the AQC production may be developed in the following directions:

(a) Defining the AQC production as a tool for livelihood of the population that could decrease the pressure on the wild fish stocks.
(b) Concomitantly, AQC could be an effective tool for conservation and reproduction of the endangered species and reclamation of the natural resources through fish stocking of the relevant ecosystems.
(c) Promotion of fish and fish products as most nourishing, healthy, dietetic, and healing food as well as stimulating their consumption.
(d) Setting up of conditions for the fast and sustainable increase of the quantity of AQC products, fast augmentation of the volume of high-quality fish-stocking material, and refining of fish-market infrastructure.

In **inference**, it could be said that the structural policy of EU fisheries closely reflect the future direction of the CFP and the integration of the CFP with other community policies and development priorities, often involving mechanisms of CBFM.

The production from AQC may be an integral part of the future agreements for cooperation and exchange of good practices. It could be also integrated in the Common Fishery Policy with special objectives and tools.

Notwithstanding the above, it is often considered that the role of the production from AQC in the CFP has to be defined by each MS individually, so as to provide development on a national basis. The instruments for the integration of the AQC with CFP have to be implemented on a national level in close cooperation with the producers of AQC. It may be noted that the fish farmers are usually in a position to carry out quantitative planning of their production. The sales are rhythmic throughout the year without considerable price fluctuations. The products in the market may be of high quality with distinct information for the consumers.

Notwithstanding the above, the fish offered for fattening has to be supplied by licensed companies with a guaranteed quality. The quantities may be planned by the producers, due to very unstable rates of the sales. The farms for breeding may have a vast variety of genetic material with great possibilities for selection with respect to flexibility to the demands of the market.

Concomitant to the above, BD happens to be an important aspect in the life and livelihood of the people of Bulgaria. The Fifth National Report on the Implementation of the CBD prepared by Republic of Bulgaria as a Party to the CBD includes important aspects of conservation and **management** of natural resources.

Further, Anon (2013a, b) dealt with the Fifth National Report of the CBD of the Republic of Bulgaria with some aspects of **management** and conservation of the resources.

In accordance with the guidelines, it is supposed to contain information on the following aspects:

(a) the current status of and threats to BD and its socioeconomic impact;
(b) implementation of the National Biodiversity Conservation Strategy (NBCS) and the Second Biodiversity Action Plan (SBAP) of Bulgaria along with a reference to the National Program for Development: Bulgaria 2020 and other sectors of the economy;
(c) progress in the implementation of the strategic plan for BD and its Aichi targets adopted by the parties in 2010 in Nagoya.

Incidentally, Bulgaria is one of the richest countries in BD among many countries in Europe. It encompasses parts of the Alpine, Black Sea, and continental biogeographical regions. The number of Balkan and Bulgarian endemics is relatively high. To protect this BD, Bulgaria has established a national ecological network [including the European ecological network Natura (2000) and PAs], which is *c.* 34% of the national territory and is one of the largest in Europe.

5.4.4.14 Status of biodiversity

Bulgaria is one of the BD rich countries in Europe. More than 3100 algae from *c.* 580 genera, 15 classes, and 7 divisions were reported. The Bryo-flora of Bulgaria contains *c.* 754 species, of which *c.* 34% are of conservation importance (CR:18, EN: 25, VU:124, and NT: 20). In addition, Pteridophytes and Phanerogams (including neophytes) consist of *c.* 4100 species, of which *c.* 270 are Balkan and *c.* 174 are Bulgarian endemics. Incidentally, there have been listing of *c.* 801 red species, of which 1: Ex, 12: RE, 208: CR, 297: EN, 204: VU, and 79: NT, which constituted *c.* 19.5% of the national flora. Further, *c.* 340 neophytes have been registered in Bulgaria, of which *c.* 16 were reported only during the period 2009–13. Further, plant communities are represented by > 1250 associations and > 300 formations.

Concomitant to the above, the forest communities account for *c.* 34% of the total BD in the country. On the other hand, in Bulgaria, animals amount to *c.* 30,359 species from 28 types and 75 classes, which is only 50% of the estimated total BD in the country. Incidentally, the Bulgarian herpetofauna ranks third in Europe with regard to species richness after Spain and Greece. Moreover, *c.* 1200 animal species are endemic (*c.* 790 Bulgarian and *c.* 410 Balkan endemics). Further, > 780 cave species occur in Bulgaria, of which *c.* 110 species are typical troglobionts. The new edition of the Red Data Book of Bulgaria had portrayed *c.* 442 species, which depicts EX: 30, CR: 87, EN: 107, VU: 137, NT:14, LC: 42, and DD: 25.

However, the number of invasive alien species has increased during the past few years. In fact, the most vulnerable to invasions are the aquatic ecosystems including both natural and man-made habitats. Nevertheless, the trends in the population of some species deserve attention in view of their rate of BD loss (e.g., the common birds' index, etc.). In fact, the general trend seemed to portray that the number of birds in agricultural fields had dropped by *c.* 12% during the period 2005–11.

Notwithstanding the above, *c.* 977 habitat types from all hierarchical types occur in Bulgaria, according to the Palaearctic classification. The number of habitats at elementary level is *c.* 542, of which *c.* 96 are restricted only to Bulgaria. Further, the Red Data Book (vol. 3, "Habitats," 2012) contain *c.* 166 habitats, classified as follows: CR: 28, EN: 71, VU: 47, and NT: 20. Furthermore, the total area of the network is *c.* 4,053,941.44 ha, of which *c.* 56.47% are forests, *c.* 32.35% agricultural land, and the rest is other types of land according to the National Priority Actions Framework for Natura 2000. Incidentally, cultivated land, pastures, and meadows are the prevailing types of agricultural land in Natura 2000 network.

Notwithstanding the above, the following groups of biological resources (sensu CBD) are utilized in Bulgaria:

(a) forests: timber and nontimber resources,
(b) game species (*c.* 16 mammal and *c.* 26 avian species),
(c) more than 40 species of Black Sea and FW fish, molluscs, crabs, and snails,

In addition, Bulgaria has a rich pool of local genetic resources.

Concomitant to the above, the total AQC production during the period 2009–12 amounted to an average of *c.* 8427.4 tons per year with its peak in 2010. Of this, the biggest quantities were produced from rainbow trout, followed by different types of carp like the bighead carp, grass carp, silver carp, followed by catfish. However, the main species for marine AQC is the blue mussel.

Notwithstanding the above, the main threats to BD could be classified as follows:

(a) Deterioration, fragmentation, and loss of habitats. Main driving forces are the various kinds of human activities.
(b) Overexploitation of the biological resources and loss of genetic resources.

On the other hand, the main threats to the game species are the alteration(s) of the habitats, change of structure and ways of use of agricultural crops, and poaching. The **fish resources** are dwindling mainly due to their overexploitation, lack of general program for regular fish stocking, and lack of regular monitoring of fish- and nonfish resources.

(c) The loss of local breeds of plant and animal varieties is mainly due to the intensification of production, limited selection to few species, and introduction of alien breeds and varieties that replace the local ones. Pollution of air, soil, and water are also the causes of BD loss. Further, the water quality seems to depend on the ecological state and chemical composition of the surface waters and the quantities and chemical composition of the underground water. In the near future, of all the **759 surface water bodies**, *c.* 36 (*c.* 4.7%) would be of very good quality, while of all the 177 underground water bodies, *c.* 123 would be in good chemical conditions.
(d) Introduction of invasive alien species to be prevented. Approximately 60 angiosperms were invasive or potentially invasive species that compete with the local flora. They may also bring about alteration(s) in the composition and structure of indigenous plant communities. Moreover, of all the 347 species of aerial and terrestrial arthropods that occur in Bulgaria, 52 species are widely distributed and have negative impact on

forestry, agriculture, AQC, horticulture, etc. Of all the 29 alien marine invertebrates in the Black Sea, 9 would be invasive and they can be predators, compete with local species, or change habitats. There are three FW shells, viz., *Corbicula fluminea*, *Anodonta woodiana*, and *Dreissena rostriformis bugensis*. Another form, viz., *Dreissena polymorpha* also occurs, which were believed to be aggressive alien species in Europe. They might cause reduction in the habitat of other species as well as may play a significant role in altering the physicochemical characteristics of the water. They could also change the structure and number of bacteria, phytoplankton, and ZP, and also bring about changes in the trophic structure of the macrozoobenthos and fish populations.

Concomitantly, it may be mentioned that the status of BD has a bearing on the life and livelihoods of the fishermen and also on the **management** and conservation of the water bodies and their bioresources.

(e) Climate change

Bulgaria is one of the most vulnerable places to be prone to the effect(s) of climate change. The precipitation and temperature index showed that the average terrestrial temperature in Bulgaria has risen by *c.* 1.5°C during the period 1971–2011. In fact, the impact of climate change on BD in Bulgaria is, today, demonstrated by the changes portrayed by the indicator biota, which, in turn, is reflected by change in the number of wintering water birds in Bulgaria.

Notwithstanding the above, the rich BD of any country is to be sustained and maintained. For taking care of this aspect, the stakeholders from the grassroots level serve as the most appropriate platform. The stakeholders from the grassroots level may aggregate into communities to take care of the natural resources (both biotic and abiotic), by sharing various kinds of responsibilities.

Concomitant to the above, "fish" happens to be one of the potential entities to ensure both nutrition and avocation of the people. Also, preservation and conservation of the ichthyodiversity and their habitat are a prerequisite and indispensable aspect in order to ensure cut-to-the bottom guarantee for the humanity for both their nutrition and avocation. Also, CBFM could be a strategically important tool for the effective management of the fish resources.

Further, the major strategic document that outlines the socioeconomic development, particularly, of the fisherfolk of Bulgaria is the National Program for Development: Bulgaria 2020. Among the national priorities are the environmental protection (including BD), sustainable use of biological resources, and development of agriculture and pisciculture sectors. In fact, BD conservation had been among the priority goals of the National Plan for Development of the Republic of Bulgaria. This is included in the National Program for Reforms of Bulgaria within the EU. Further, the National Environmental Strategy, 2009–18 and the related action plan envisage actions for limitation and halting of BD loss through its sustainable management. Further, Bulgaria, as a party, outlined a set of national priorities to implement the Aichi targets.

Further, the Rural Development Program supports the agroecological measures and maintenance of "Natura 2000" network in rural areas.

Notwithstanding the above, there are program(s) which also support activities related to the restoration of the forest potential and implementation of preventive measures to prevent the loss of Water Framework plans in Bulgaria. The Operational Program (OP) for the development of the fishery sector corresponds to the European Fishery Fund. There were had been measures suggested for the maintenance of the natural fish resources.

Notwithstanding the above, there were plans for the development of the sectors like River Basins Management Plans; National Program and Plan for Fishery and AQC; National Program for Support of the Sustainable Development of Fish Resources; National Plan for the Conservation of the most Important wetlands in Bulgaria; Quantitative status of surface waters; species composition, abundance, and distribution of the ichthyofauna in the river basins; improvement of hydrological regime and reduction of the eutrophication in wetlands; and so on. Concomitantly, there were also plans for taking care of the technical and environmental measures for the artificial barriers to fish migration, construction of fish ladders, restoration of river basins affected by aggregate extraction and others, to assess the implementation of the objectives of sustainable management of fish stocks, quantity of catches of fish and other aquatic organisms and share of economically valuable species in the catches, etc.

Incidentally, only a limited number of fish-stocking exercises were conducted to support the sustainable development of fish resources. Moreover, bans on fishing had been introduced year-round or, at least during the spawning season, e.g., there had been a 4-year period of ban on sturgeon fishing. In the implementation of the measure "annual survey and assessments of population status of resource species of fish and other aquatic animals" were conducted. There could be several measures in the National Plan for BD. Moreover, trawl photos had been made for stock assessment of turbot as well as the method of acoustic stock assessment of pelagic fish species in the Black Sea had been tested. However, the stocks of most commercially valuable fish species in the Danube River and Black Sea had not been evaluated. Moreover, a project entitled, "Implementation of activities of the National Action Plan for sturgeon fish in order to improve the status and conservation of sturgeons in Bulgaria" (2012–15) was initiated.

Further, the Fishery Operating Program (FOP) financed a total of 134 projects, of which 92 were totally paid with the total value of €22,670,549.44. Most of the measures and activities financed by this Program had direct or indirect impact on BD. The achievement of the goals for sustainable management of water enterprises and forests in the rural areas and the outcome of the achievement of BD conservation are measured by the following indicators, viz., forest area, composition of tree species, forest restoration, defoliation, extent of deterioration of forests, index of ordinary birds in Bulgaria, number of the populations of the brown bear, progress in AQC production, etc.

Further, the activities were financed by different components of the Fishery Operating Program like AQC, fishery in the inland waters, processing and marketing of products from fishing and AQC, and so on. The activities are supported by an international, EU-funded project implemented in the Balkans. The study and preservation of traditional knowledge is part of the university curricula and research programs of a number of academic institutions. It is also incorporated in the management plans of the National and Nature Parks. Moreover, BD-related research is conducted in a large number of academic institutions and university departments. Some of these are as follows:

(a) Agriculture and forestry: Agricultural Academy with 27 institutes, 19 experimental stations, and 2 experimental plots; professional high schools for agriculture, forestry, and food technology; Agrarian University—Plovdiv; Thracian University, Stara Zagora; Forestry University, etc.;
(b) BD and biological resources: Bulgarian Academy of Sciences. It is a group of institutes dedicated to BD, bioresources, and ecology, such as Institute of BD and Ecosystem Research; Forestry Institute; Institute of Plant Physiology and Genetics; and National Natural History Museum and Botanic Garden;
(c) Sofia University "St. Kliment Ohridski" with its Faculties of Biology and Geologogeography;
(d) Plovdiv University: Faculty of Biology;
(e) Southwest University with its Faculty of Nature and Mathematics.

Concomitant to the above, Bulgaria was implementing consistent policy, during the period 2010–13, to increase the funding of projects related to sustainable use of BD. A total sum of €70,656,514 from the state budget had been spent for such projects. Additionally Bulgarian NGOs and private donors have invested €538,395 in total.

In fine, the implementation of the CBD in Bulgaria in 2009–13 could be assessed as positive due to the following facts:

(a) The BD issues had been integrated into the relevant national legislation and program documents which constitute the legal and operational framework of the country as an EU MS.
(b) Measures from the past (2007–13) and forthcoming (2014–20) periods of the OPs functioning in Bulgaria support BD conservation and its sustainable use.
(c) The National Ecological Network (including PAs and Natura 2000 sites) covered *c.* 34% of the country's territory, thus ensuring effective in situ conservation of BD. However, the management plans of many of the PAs and Natura 2000 sites may have to be elaborated or updated.

5.4.4.15 *Updated information on the status, trends, and threats to biodiversity and consequences for human well-being*

Importance of biodiversity for the country

Bulgaria ranks among the richest in BD among many countries in Europe. Although small in area (110,910 km^2), the country's territory includes parts of three biogeographical regions, viz., Alpine, Black Sea, and Continental regions. The geographical position of the country in the SE part of the continent, the complex palaeogeographic and paleoclimatic history, diverse topography and climate, considerable forest resources, sufficient FW resources, proximity to the Black Sea, etc., are basic and important factors determining the rich diversity of species, communities, and habitats. In fact, BD of Bulgaria is an invaluable asset, which is a significant part of the national identity of the country. BD is an important basis for the performance of ecosystem services (provisioning, regulating, supporting, and cultural), which provides various things for the quality of life of the population and socioeconomic prosperity of the country. The study reflected that rationale and efficient use of BD are among the national priorities and also underpins the development of cognitive and ecological tourism, green energy and transport, modern agriculture, stock breeding, forestry, fisheries, and AQC. Further, there had been expansion of the raw material base for the pharmaceutical, food, and cosmetics industries through traditional herbal works and collection of "gifts of nature."

Further, there is a need for conservation of BD and making favorable conditions for its sustainable use and, in particular, genetic resources in agriculture are vital for the development of agriculture. Incidentally, conservation of genetic resources is the basis for the conservation of traditions in agriculture and provides the opportunities for the development of related industries, contributing to the enrichment of nutritional as well as vocational requirements and acquisition of new positions in the changing market. Endemic and subendemic species and communities together with those having conservation value constitute a significant part of the national BD. Their conservation, especially the endemics, is not only of national importance but also important at European and global levels. To conserve this BD as well as to protect and preserve areas with specific biological, aesthetic, and cultural values, a National Ecological Network (including PAs and Natura 2000 sites) was established.

Notwithstanding the above, the vertebrates are the best studied animals in Bulgaria, of which *c.* 781 species were known till 2009, which included two species of Cyclostomata, four species of Chondrichthyes, 213 species of Osteichthyes, 19 species of Amphibia, 37 species of Reptilia, 409 species of Aves, and 97 species of Mammalia. Moreover, it may be mentioned that of all the European countries only Spain and Greece have some amount of similarity with Bulgaria in having rich herpetofauna. Also, of the 35–36 species of bats identified in Europe, *c.* 33 species are found in Bulgaria. Further, the total number of endemic genera, species, and subspecies of animals in Bulgaria was estimated to be *c.* 1200 until 2009, of which the Bulgarian endemics were *c.* 790 and the Balkan endemics were *c.* 410. Particularly diverse and unique were the cave and subterranean fauna of Bulgaria, mainly because the country is rich in crust formations and groundwater. Incidentally, >780 species of cave animals of different taxonomic groups had been registered, of which >110 species inhabit the deepest parts of the caves and belong to the category of typical troglobionts. Further, Bulgaria is among the first countries in Europe in species richness, diversity, and endemism of the cave fauna.

Further, information about the faunistic diversity of Bulgaria expanded through the application of modern taxonomic and faunistic methods during the period 2009–13. Furthermore, a number of species, new to science or new for the territory of Bulgaria had been discovered and described. According to Red Data Book of Republic of Bulgaria (2011) in Family Acipenseridae:

(a) *Acipenser gueldenstaedtii* has not been categorized as CR.
(b) *A. nudiventris* was threatened by extinction and may be extinct (EX).
(c) *Acipenser ruthenus* is not categorized as endangered (EN).
(d) *Acipenser stellatus* is not categorized as CR.
(e) *Acipenser sturio* is threatened by extinction and may often be considered as extinct (EX).

Further, two species of ZP, one species of zoobenthos, and three species of fishes were recorded along the Bulgarian Black Sea coast for the first time. Moreover, the most vulnerable to the introduction and establishment of alien plant and animal species were the aquatic ecosystems, riparian habitats, and man-made habitats. Further, the invasive alien species *C. fluminea* was established for the first time in Bulgaria in 2001 in the Danube River at Vetren village. The species spread throughout the Bulgarian part of the Danube River within a period of 11 years, reaching a very high abundance at some sites (at Zagrazhden village, 2012) and changing the entire ecosystem. Further, the species spread rapidly also to the tributaries of Danube (possibly at the rate of 8 km/year) and reached *c.* 80 km upstream, e.g., in Iskar River. Incidentally, 22 new localities of *C. fluminea* were registered in the Danube River during the period 2009–13 alone as well as in 24 localities in the inland rivers and 4 localities in standing water basins in Bulgaria. Most probably, the distribution of the species in the inland waters of the country had been supported by human activities, such as fishing, fish stocking, recreation, etc. Proper management of all these require **CBM** strategies.

5.4.4.16 Fish resources

Commercial and recreational fishing

More than 40 species of Black Sea and FW fish as well as other aquatic organisms (mussels, crabs, snails) are used for commercial and recreational fishing in the country. Commercial fishing is done in the Black Sea, the Danube River, and inland waters (reservoirs). The largest share belongs to the catches in the Black Sea, followed by the catches from the Danube River. The catches during the period 2009–12 varied slightly, showing general decrease with regard to marine and anadromous species. Proper management of all these require CBM approach. Further, one of the most valuable species in the Black Sea, an object of commercial fishing, not only for Bulgaria but also for all Black Sea countries is the turbot (*Scophthalmus maximus*). Stocks and catches of this species substantially dwindled after 1980, mainly because of the intensive exploitation of the stocks of the species in the past. Current estimates from Bulgaria portrayed low stocks and poor size/age structure. Further, the biomass index tended to decrease during the period 2008–11. Unlawful fishing and unsustainable use of the common resources by the Black Sea countries are the main reasons for this phenomenon.

Incidentally, the main fish species that are subject to commercial fishing in the **Danube River** are the native species, notably, *Alosa immaculata, Barbus barbus, Abramis brama, C. carpio, Silurus glanis, Vimba vimba, Sander lucioperca, Alburnus alburnus, Chondrostoma nasus, Aspius aspius,* and *Esox lucius,* and from the Asian carps—*Carassius gibelio, Hypophthalmichthys molitrix, Aristhichthys nobilis,* and *Ctenopharingodon idella.* Their sustainable harvest require **CBFM**.

On the other hand in the absence of effective CBFM, the total catch from the River Danube is relatively low. In addition to the absence of effective CBFM mechanisms, it could be because the populations of economically valuable ichthyospecies in the river are very unstable. Hence, their catch could neither be guaranteed nor planned. However, the average annual catch for the period 2009–12 was *c.* 107.5 tons, which is only *c.* 1% of the total catch in the country. Further reason for the low catch could be the poor and even critical state of the populations of the anadromous species, Pontic shad (*Alosa immaculata*) and the sturgeons. Further, a large part of the catch (*c.* 24%) for the period 2010–12 was due to the alien carp species imported from Asia. The largest catches were registered for the *Aristhichthys nobilis, Hypophthalmichthys molitrix, Carassius gibelio,* and *Ctenopharingodon idella. Aristhichthys nobilis* is the only species, the catch of which has substantially increased during 2012. The percentage of the alien species in the catches had been maintained or even slightly increased in comparison with the previous period from 2005 to 2008, when it was *c.* 22%. According to NAFA, no fish stocking to the Danube River had been conducted in recent years. Therefore, it is possible that the maintenance of the populations of *Aristhichthys nobilis, Hypophthalmichthys molitrix,* and *Ctenopharingodon idella* were due to natural reproduction in the river.

Incidentally, the principal fish species important for commercial fishing in the inland waters of Bulgaria are *Aristhichthys nobilis, Hypophthalmichthys molitrix, C. carpio,* and *Carassius gibelio.* Incidentally, the average annual catch for the period 2009–12 was 1292.3 tons, which was *c.* 13% of the total catch. The largest share of the catch falls to *C. carpio* (average annual catch for 2010–12: *c.* 575 tons), *Aristhichthys nobilis* (*c.* 338 tons), *Carassius gibelio* (*c.* 158 tons), *Hypophthalmichthys molitrix* (*c.* 30 tons), *Ctenopharingodon idella* (*c.* 19 tons), *Silurus glanis* (*c.* 17 tons), and *Sander lucioperca* (*c.* 15 tons).

5.4.4.17 *Aquaculture (AQC)*

The constant decline of fish resources in natural waters was mainly due to (a) unsustainable harvest and (b) environmental alteration(s).

These had resulted in increased interest in AQC mainly to fill the void supposedly caused due to overexploitation. Also, the **CBM** operations, notably the CBFM, could play a significant role in AQC aspects too.

Incidentally, AQC production in Bulgaria plays an increasingly important role adding to the sea and river fishing. The total AQC production (for fish stocking and for fish and other aquatic organisms for consumption) was to the tune of *c.* 8427.4 tons per year during the period 2009–12. However, the highest production was recorded in 2010, which decreased in the subsequent years. For example, at the end of 2012, the total number of active fish farms in the country was *c.* 388 compared to *c.* 422 in the previous year. Of these, *c.* 347 were for FW AQC, and 41 for marine AQC. However, the total AQC production, during 2012, from the stocking material was *c.* 7557.1 tons. This was *c.* 2.5% less than the previous year due to a 29.2% decline in the production of stocking material, while the production of fish for consumption had depicted an increase of *c.* 8.2%.

The largest quantities were produced from *Oncorhynchus mykiss, C. carpio,* and *Aristhichthys nobilis* followed by *Ctenopharingodon idella, Hypophthalmichthys molitrix,* and *Silurus glanis.*

5.4.4.18 *Genetic resources*

According to the CBD, "genetic resources" refers to the genetic material of actual or potential value. They are the basis for the development of important economic sectors in Bulgaria, such as agriculture, stock breeding, fisheries, perfumery, cosmetics, pharmaceutical industry, etc. Ex situ stored genetic resources of wild species of conservation value could be used as source for a variety of conservation and restoration activities related to extinct or endangered species.

5.4.4.19 *Threats to biodiversity in Bulgaria*

The threats to biodiversity in Bulgaria were mainly due to the following reasons:

(a) deterioration, fragmentation, and loss of habitats;
(b) pollution in FW and brackish water (BW) habitats;

Concomitant to the above, it may be mentioned that there had, and still have, been major alteration(s) in habitat related to the inland (FW) water bodies of the country mainly due to human activities, thereby leading to loss of BD. Some of the eco-unfriendly measures are construction and operation of hydro-electric (HE) power plants; construction of hydraulic structures and their maintenance; unplanned dredging of the river basin, notably, the Danube River for navigation; sand and gravel extraction in the river beds; withdrawal of often large volume of water from surface of the water bodies; felling of riparian vegetation; and so on. Hydro-morphological alterations and regulation of the water flow are the main reasons for not achieving a good ecological status of surface waters.

Notwithstanding the above, with regard to fish resources, the main reasons for the deterioration of the stocks of valuable fish species and other aquatic organisms in the Black Sea, the Danube River, and the inland waters are as follows:

(a) overexploitation of fish and nonfish resources;
(b) lack of or failure to implement a comprehensive science-based program for regular stocking in order to maintain optimum stocks in water bodies under intensive fishing operations;
(c) unregulated or uncontrolled stocking mostly in inland water bodies, which violate the composition and species structure of the fish populations as well as the balance of ecosystems;
(d) lack of in-depth and comprehensive studies on the ichthyofauna and fish stocks as well as improper management of fish resources. In this connection, it may be emphasized that **CBFM** could be an effective tool for proper management of the fish resources in the aquatic domain;
(e) lack of systematic monitoring of the status of fish and nonfish resources, their exploitation, and environmentally oriented development of fisheries. In fact, when stock breeding and crops are considered, the main causes of the loss of genetic resources are complex but more essential could be the application of intensive farming methods and exploitation; conducting one-sided selection; frequent change(s) in the production relations; changes in tenure and land use and land management; lack of markets for products made from local breeds that leads to spontaneous crossings in search of improved productivity and better commercial properties; introduction of foreign breeds and varieties, by displacing and replacing the indigenous local ones; gaps in legislation related to the protection of local breeds and their improper culture, scarcity of **efficient management** structure of the fish resources, etc.

Notwithstanding the above, during the period 2009–13, water bodies infested with *Dreissena* spp. have portrayed changes in physical and chemical properties of water; changes in the structure and abundance of phytoplankton, ZP; and changes in the trophic structure of benthic macro-invertebrates and fish populations.

5.4.4.20 *Climate change*

Bulgaria is in a geographical region that is among the most vulnerable to climate change. The following facts may be mentioned:

(a) It may be mentioned that there was a trend toward warming in Bulgaria since the late 1970s.
(b) Further, 20 of the last 23 years, since 1989 have depicted positive anomalies of mean annual air temperature compared to the climate norm (1961–90).
(c) In fact, the longest droughts had been observed in the 1940s and the most significant droughts were in 1945 and 2000.
(d) The number of longer periods of drought followed by severe storms and flooding with heavy damage and casualties has increased.

5.4.4.21 *Socioeconomic and cultural implications*

Involvement of the community in CBM operations certainly has socioeconomic and cultural implications. Further, the introduction and wide spreading of a number of invasive alien species have negative socioeconomic consequences. For example, the impact of invasive alien plant species could be considered in competition with cultivated plants and may lead to economic losses to agriculture. Incidentally, many of these plants often become the worst weeds to crops. Further, 11 terrestrial arthropod species have a negative impact on the economy. Their control requires substantive amount of funds.

Notwithstanding the above, the very significant environmental and socioeconomic impacts in inland waters of the country were caused by the invasions of *D. polymorpha* and *D. rostriformis bugensis*.

Incidentally, it may be mentioned that the mussels of the genus *Dreissena* may interfere with vital aquatic facilities, such as reservoirs and water supply systems, pumping stations, measuring equipment, pipe systems, turbines, protective screens, exit tunnels and canals, AQC facilities, etc., as major biofoulers. They may also cause plugging of the ducts, pipes, and fencing, underwater corrosion of metal equipment, impeding the operation of hydro- and thermoelectric power plants, water supply, and drainage hydraulic equipment(s), fish cages, etc. Thus, they were believed to have caused significant amount of economic damage in different regions of the country and different sectors, notably, power generation, drinking and industrial water supply, irrigation and fish farming, etc. during the period 2009–13.

Concomitant to the above, the **wetlands** are inseparably linked with the life and livelihood of the people. As such, maintaining ecosystem functions of wetlands through sustainable use of their abiotic and biotic resources is very much related to the quality of life of local communities and their livelihoods. At the same time, it may also contribute to limiting the adverse effects of climate change and facilitating adaptation to them. Various models and a number of studies have shown that climate change and anthropogenic impacts cause drought and temperature alteration(s) in the country, which led to significant change(s) in the BD, ecosystem function(s), and ecosystem services, including wetlands.

Notwithstanding the above, **community** involvement, mainly through **CBFM,** could play a significant role in the appropriate **management** of the wetlands and their resources. Concomitant to the above, stringent national policies are definitely needed for the conservation of BD. Further, there is to be augmentation in the activities related to the conservation of the near-extinct and endangered breeds and varieties in order to prevent the loss of **traditional knowledge** and practices and for the maintenance of local plant and animal varieties and breeds. This knowledge and practice(s) are part of the culture of local communities, and proper maintenance of their lifestyle and traditions are supposed to be part of our national identity. Loss of entire populations of old, traditionally grown and adapted to local conditions could be regarded as a gross national loss.

5.4.4.22 Policies and goals related to biodiversity in Bulgaria

The conservation of BD is one of the priorities in the National Plan for Development of the Republic of Bulgaria. The National Reform Program is prepared within the European semester, a new tool for better coordination of economic policies within the EU in implementing the strategy for smart, sustainable, and inclusive growth, "Europe 2020." The main purpose of this mechanism is the consistency of actions for sustainable economic growth, financial stability, increased employment, social inclusion, etc. The National BD Conservation Strategy, adopted in 1998, outlines the priorities for immediate action to the following:

(a) To strengthen the scientific basis of nature conservation.
(b) To support legislative initiatives.
(c) To expand and strengthen the network of PAs.
(d) To impart environmental education and supplementary training.
(e) To implement development and implementation of a policy on ecotourism.
(f) To promote the protection of the Black Sea basin.
(g) To promote the conservation of the Balkans.

For the complete and fruitful implementation of all these aspects, involvement of the community is a prerequisite. **CBM** could yield beneficial outcomes.

Notwithstanding the above, "fisheries" sector is of profound significance. The National Program for Fishery and AQC, with the strategic objectives, are as follows:

(i) Sustainable use of fish resources and protection of the FW and marine environments.
(ii) National Strategic Plan for Fishery and AQC had been structured according to the seven areas of the CFP of the EU, including the sustainable exploitation of fish resources, sustainable development of AQC, and conservation of environment including water. Further, National Program for the Sustainable Development of Fish Resources aimed toward ensuring sustainable development of the fish resources, restoration and conservation of biological balance, and enrichment of the diversity of fish resources in aquatic ecosystems.

Notwithstanding the above, the main financial mechanisms to support activities related to BD were the EU Structural Funds. Based on the National Development Plan, National Strategic Reference Framework was developed, which sets general guidelines and strategies as well as OPs for certain key sectors.

Notwithstanding the above, a number of academic forums were organized in order to promote the significance of BD in Bulgaria. Some of these are as follows:

(1) National conference, "BD, and Environment."
(2) International conference, "Wetlands: Functioning, Management, and Restoration."
(3) Seventh National Botanical Conference and "40 Years of the Program on 'Man and the Biosphere (MAB),' UNESCO."
(4) Balkan workshop on networking on invasive alien species.
(5) EEA/Eionet Balkan Meeting on Networking Activities on Invasive Alien Species (2011).
(6) The Sixth European Congress of Hemiptera (2012).
(7) XVIII European Congress of Lepidopterology (2013).
(8) Academic celebrations of the International day for Biological Diversity.

(9) Various international, national, and regional workshops related to the achievement of specific research projects in the areas of BD.
(10) Annual open-door visits in government and scientific institutions.
(11) Further, the Bulgarian NGOs had implemented a large number of projects and information campaigns to promote the values of BD and approaches for its conservation.

In addition to the above, a large number of campaigns were launched in order to promote the conservation of agro-BD. In addition, a number of events had been organized related to the conservation and maintenance of innovations and practices of local communities, e.g., Mildew Honey Festival—Tsarevo, Smilyan Beans Festival, Plum Festival in Troyan, Cherry Festivals—Kyustendil, Harvest Fest—Kyustendil, etc.

Notwithstanding the above, there were some measures of common interest as follows:

(i) Collective activities have improved the management and conservation of resources.
(ii) Promotion of the use of selective fishing equipment and methods.
(iii) Measures aimed at preservation, protection, and propagation of the aquatic flora and fauna, and so on.

Concomitant to the above, there were efforts for sustainable development of the fishery areas through financial support for local strategies, for development, for covering operational expenses of the fishermen groups, etc.

In addition to the above, The National Program for the Sustainable Development of Fish Resources include general measures for the conservation, restoration, and protection of resources as well as could provide basis for developing annual programs to support sustainable fish populations and research activities, thus could ensure effective **management** of fish resources in the long run. Its activities were focused on the following two directions with the corresponding results:

(a) The control on unregulated fishing, ban on the use of unlawful tools of fishing, and promotion of seizure of unlawful harvest of natural resources in water bodies.
(b) Restocking and other activities have to be conducted in order to support and maintain stable populations of fish and other aquatic organisms that are consistent with the results from the research on their status.

Concomitant to the above, in terms of fishing restrictions, each year a ban on catching fish and other aquatic organisms is imposed. The restrictions are year-round or during the breeding season of the species in the Black Sea, the Danube, and the inland waters in order to protect BD and setup conditions for natural reproduction of the populations. A long-term ban on sturgeon fishing was introduced because its populations are in a critical condition. Moreover, the fishing of the following species was forbidden in Bulgarian waters of the Danube and the Black Sea:

A. gueldenstaedtii, *A. ruthenus*, *A. stellatus*, and *Huso huso* for quite a long period. In addition, due to the highly reduced stocks of turbot in the Black Sea, the annual catch quotas for the species identified by the Council of the EU had been reduced from 50 to 43.2 tons. Restocking activities in the Danube and other inland rivers were conducted according to the various programs.

Notwithstanding the above, there had been the National Plan for the Protection of the Most Important **Wetlands** in Bulgaria. There were provisions and horizontal measures that may be implemented at national level. Incidentally, control of poaching is a contentious issue. Initially, stringent measures were considered as a specific measure for certain areas but it was finally formulated on a nationwide level because of the national dimension of the phenomenon. A number of projects had been launched and implemented in order to collect relevant data in support of the conservation and management of fish resources.

Notwithstanding the above, field studies are important tools for the collection of information. The field studies cover the real breeding habitats, providing a minimum population size for the survival of the species and restoration of the natural reproduction, development of tripartite management measures to ensure the reproduction of sturgeon or any other endangered species of fish, and trying to buildup an electronic database and online portal for sturgeon or any other endangered species of fish. Bulgaria participates in the project which include collaborative efforts to raise awareness on the overexploitation of the sturgeon populations (or other endangered population of fish) in the Danube in Romania and Bulgaria financed by the Life+ Program. The project aimed to stop the exploitation of the CR sturgeon species in the Danube River in the region of Bulgaria and Romania and to ensure long-term survival of these species and preserving their natural and high economic value. Bulgaria participates in the Danube Sturgeon Task Force (DSTF) network and in developing programs for the conservation and restoration of the sturgeon in the Danube.

The management plans for river basins include measures for certain water bodies aiming at species adaption and mitigation of the impact of climate change on BD. There were programs for proper land use and using crops requiring less water, application of water-saving technologies, reforestation of watersheds, etc. Further, Bulgaria had participated in the project "Climate Change and Impacts on Water Supply (CC-WATERS, 2009–2012)," financed by the OP for South East Europe in the EU. An assessment was carried out of the impact of forest ecosystems on the quantity and quality of water resources in the watershed of Struma River. This catchment area is among the hardest hit by climate change in Bulgaria. Recommendations were made regarding the application of appropriate silvicultural methods for forest management in the area to achieve adaptation to adverse climate changes. Further, the impact of climate change could be significant for BD mainly due to the closed nature of the Black Sea. Moreover, geographical isolation of the Black Sea makes it impossible or limits the natural migration of species to the north, following suitable climatic conditions. Further, the migration of cold-adapted organisms at depth is hindered by the presence of H_2S in the water of the Black Sea under 150–200 m. This, further, hampers the adaptation of species to climate warming. There seemed to be threat to the endemic species in the Black Sea, which are of limited distribution and hence are particularly vulnerable to climate change(s), mainly because of the narrow range of environmental tolerance.

Concomitant to the above, there had been efforts to improve the status of BD by safeguarding ecosystems, species, and genetic diversity. The National Ecological Network gives priority to CORINE sites, Ramsar sites, important plant areas, and important bird areas. In Bulgaria there are six categories of PAs: reserve, national park, natural landmark, managed reserve, nature park, and protected site. For the successful implementation of all the management and conservation-related activities, involvement of the **community** is a prerequisite; **CBFM** could be an efficient tool for **CBM** of the fish resources.

In fine, the structural policy of EU fisheries most closely reflects the future direction of the CFP and the integration of the common **fisheries policy with other community policies and development priorities**.

Notwithstanding the above, it is known that **Bulgaria** is a SE European country with *c.* 378-km long coastline on the Black Sea. The main cities on the Black Sea coast are Varna and Burgas, which are among the most populous cities in Bulgaria. Bulgaria is a member of the EU since January 1, 2007. The Bulgarian fisheries policy is, thus, covered by the EU CFP. The National Agency for Fisheries and AQC, established within the Ministry of Agriculture and Food, is the executive body of the central administration for implementation of the national policy on fishery and AQC and for the application of the fisheries legislation in **Bulgaria** (Irina, 2011).

Fisheries management includes a TAC system for turbot and sprat as well as other measures, such as effort regulation, closed seasons, restocking, closed areas, and gear restrictions. Further, **CBFM** could be an effective operational and **management** tool. Overall fish production in Bulgaria relies on commercial fishing (*c.* 86% of the total production) and AQC (*c.* 14%). The total fish and AQC production in Bulgaria was 227,788 tons in 2005. Input of commercial catches was from the marine fishing subsector (in the Black Sea) and from the inland fishing subsector (in the River Danube, and in other rivers as well as from lakes, both natural and man-made).

Incidentally, the fish catches from the inland waters account for *c.* 10.3% of the commercial fish production and Danube River catches (*c.* 1.8%) and catches from the inland reservoirs (*c.* 8.5%). The most important species for fishing in the inland lentic water basins (*c.* 1663.86 tons in 2005) are common carp, crucian carp, silver carp and bighead carp, grass carp, black carp, bream, European catfish, and pike-perch. The common carp has, by far, the most significant contribution (*c.* 43% of the volume of the catches).

Notwithstanding the above, the total AQC production increased during the last decade and reached *c.* 3143 tons in 2005. There are two AQC subsectors in Bulgaria, viz., FW fish farming (in warm or cold water) and marine AQC farming. Incidentally, FW fish farming accounts for *c.* 95% of the Bulgarian AQC. Incidentally, the main AQC fish species in Bulgaria were the rainbow trout (*c.* 36.6% of the total AQC production in 2005) and the common carp (*c.* 29.5%). It had been estimated that all fisheries activities (including processing) contribute to an added value of *c.* €14.3 million to the Bulgarian economy, which is *c.* 0.14% of the GDP.

Sometimes, imports may exceed exports and the fish processing industry may have to depend on imported raw materials. However, there was a gradual increase in exports of fisheries and AQC during the recent times.

5.4.4.23 *Fisheries in Bulgaria*

Bulgaria has a rich potential in AQC and fisheries. Incidentally, the total export of fisheries and AQC products amounted to *c.* 6381.5 tons in 2008, with exports to the EU accounting for *c.* 55% of the total exports. It may be noted that the principal characteristics of the Bulgarian fisheries exports is the limited range of products. Incidentally, exports of molluscs occupied the largest share in total exports of fish and fishery products (*c.* 46%). In fact, the principal product was the sea snail "Rapana" caught along the Black Sea coast and usually exported prepared (boiled and frozen). Molluscs were mainly exported to Turkey, Macedonia, Republic of Korea, Japan, and Greece. Frozen fish was the second significant export item (*c.* 33% of total exports of fish). Further, most of the frozen fish exports (especially sprat) go to Romania (*c.* 90% of the total exports of frozen fish), while a limited quantity is exported to the United Kingdom, the Netherlands, and Germany.

On the other hand, total imports of fishery products in Bulgaria reached to *c.* 29,249 tons in 2008. However, trade with EU countries increased significantly over the past few years. In spite of this, however, imports of fish and fishery products from third countries remained essential. It may be mentioned that frozen fish forms the largest share in total imports of fish and fishery products to Bulgaria to the tune of *c.* 67% in 2008. The main suppliers were Canada, Peru, Spain, and the United States.

Concomitant to the above, the Bulgarian fisheries sector, including fish processing, provides a relatively limited contribution to employment at the national level (*c.* 0.38% of the national workforce). Nevertheless, it provides vital employment at regional level, especially in some coastal zones, where it makes a significant contribution to the local economies (e.g., Primorsko in the district of Bourgas: *c.* 16.6% of jobs in fishing and Byala in the district of Varna: *c.* 11.9%). Incidentally, it may be noted that there were *c.* 12,260 employees in the fisheries sector in 2003, of which *c.* 16% were women.

Several institutes and universities were involved with fisheries research in Bulgaria. The Institute of Fish Resources and the Institute of Oceanology of the Bulgarian Academy of Sciences (both located in Varna) undertake scientific research activities and give recommendations for sustainable exploitation of fish and other aquatic resources in the Black Sea. Research on the dynamics of fish populations in inland waters is carried out at the Institute of Zoology of the Bulgarian Academy of Sciences as well as in the Department of Biology of the Sofia University and also in the Central Laboratory of Ecology in Sofia. The Institute of fisheries and Aquaculture in Plovdiv mainly engaged in scientific research activities in AQC and, especially, **in the development of technologies for breeding commercially important species as** well as in the production of raw materials from productive species.

5.4.4.24 *Structural and cohesion policies in fisheries in Bulgaria*

Bulgaria, officially called "the Republic of Bulgaria," is located in Southeastern Europe. Bulgaria borders Turkey and Greece to the south, Macedonia and Serbia to the west, and Romania to the north, and has a coastline on the Black Sea to the east. Bulgaria has a territory of 111,001.9 km^2 and a population of *c.* 7.35 million inhabitant. The population density of Bulgaria was *c.* 68.2 inhabitants per km^2 with *c.* 73% of the population living in urban areas.

Moreover, the coastal region is an important hub of tourism, particularly, during the summer season and constitutes one of the country's most popular tourist destinations. Further, the Bulgarian coastline is *c.* 378 km long. The Balkan Mountains reach the Black Sea at Cape Emine (north of Nessebar) and divides the coastline into a northern part dominated by rocky headlands, and a southern part with wide sandy beaches. The continental shelf adjacent to the Bulgarian coast is *c.* 40 km wide with the shelf break located at *c.* 100 m depth. Further, below *c.* 150 m depth, there is a permanent lack of oxygen (anoxia) which results in deep Black Sea waters and sea bottom is virtually lifeless. In the Black Sea, all the waters are under the jurisdiction of the Coastal States and there are no high seas. The main forms of national jurisdiction of the Black Sea maritime space, as defined by the UN Convention on the Law of the Sea (UNCLOS) in 1982, are as follows: (i) internal waters, (ii) territorial sea, (iii) contiguous waters, and (iv) exclusive economic zone (EEZ).

Notwithstanding the above, Bulgaria had signed and ratified UNCLOS on May 15, 1996. It has jurisdiction of > *c.* 34,288 km^2 of water representing *c.* 6.8% of all maritime jurisdictions in the Black Sea. Bulgaria signed an agreement for the delimitation of its jurisdictional waters with Turkey in 1997 and has a theoretical boundary with Romania.

5.4.4.25 *Fisheries in Bulgaria*

Fisheries management measures

The basic conditions for **fisheries management in Bulgaria** were said to be determined by the CFP. Central in this overall framework are the TACs. In 2008, for the first time, TAC for sprat and turbot in Bulgarian waters were set. In 2008 and 2009, the catch limit for turbot was set at 50 tons, in 2010 at 48 tons (after an initial commission proposal of 38 tons) and in 2011 at 43.2 tons. The TAC for sprat was 15,000 tons in 2008 and 12,750 tons in 2009 and 2010. In 2011, the Bulgarian TAC for sprat was set at 8032.5 tons. The technical measures accompanying the quantitative restrictions for the turbot fishery consisted of a fishing ban from April 15 to June 15, a minimum landing size of 45 cm, and the use of a 400-mm minimum legal mesh size for bottom-set nets. **Management**, involving the community, could be an effective tool in achieving the goal.

Concomitant to the above, various **management strategies had been formulated in Bulgaria** to control fishing efforts and promote rehabilitation and conservation of aquatic resources and ecosystems. These measures include (FAO, 2002), among others, the following:

(a) Direct limitation of fishing effort through registration of fishers and licensing of fishing gear and fishing vessels.
(b) To observe "closed seasons" in order to ensure reproduction and survival of juveniles, i.e., for winter and spring reproduction of trout and other cold-water species.

On the other hand, there was a 30–60-day closure for Danube fisheries, in compliance with the international agreements on sturgeon species and pontic shad.

(c) Restocking of the Danube River, and also be other inland water bodies, with sturgeon and cyprinid juveniles may be thought of.
(d) Control on size and power of fishing vessels (through the FAA, the EU-supported Fishing Vessel Register Data project, the fishing vessel management scheme and control on size and power of fishing vessels in line with the EU, CFP).
(e) Closed fishing areas and restriction of some gears (bottom trawling and dredging).
(f) Establishment of **management zones** through a licensing scheme. In this connection, it may be noted that the two management zones had been attempting to provide equitable allocation of resources and to reduce conflict between traditional and commercial fishermen.

Notwithstanding the above, commercial fishing may be carried out by Bulgarian citizens and legal entities that have been granted a commercial fishing license. Foreign vessels could be allowed to carry out commercial fishing in national waters. Incidentally, however, there was no **ITQ** system in Bulgaria.

5.4.4.26 Catches

It may be mentioned that the overall fish production in Bulgaria relies on commercial fishing (*c.* 86% of the total production) and AQC (*c.* 14%). Incidentally, the total fish and AQC production was *c.* 227,788 tons in 2005 (MAF-NSPFA 2007–13).

Incidentally, the commercial fishing is composed of the following:

(i) marine fishing subsector (in the Black Sea) and
(ii) inland fishing subsector (in the River Danube, in the other rivers, and in the lakes, both natural and artificial).

Further, fish catches from the inland waters accounted for *c.* 10.3% of the commercial fishing operations and included the Danube River catches (*c.* 1.8%) and catches from the reservoirs (*c.* 8.5%).

Notwithstanding the above, AQC production also comprises two distinctive subsectors, viz.,

(a) FW fish farming and
(b) marine farming of mussels and algae.

AQC of FW species (trout, common carp, grass carp, bighead carp, catfish, pike) produce about five times more than FW commercial fishing. On the other hand, marine AQC of mussels (*Mytilus galloprovincialis*) had produced *c.* 3.5 times more than the commercial catch of these molluscs. Conversely, FW farming is traditionally a well-developed sector but restructuring of production process is required. There is also a need for introducing new high value species with good market prospects.

Notwithstanding the above, marine AQC may consist only of mollusc farming and accounts for $<1.0\%$ of the total production. There is no big marine fish farm yet in the country in spite of the good experience from the past. Conversely, marine AQC may have good prospects for future development and so needs special measures of support (MAFNSPFA: 2007–13).

In addition to the above, a small- and medium-sized dammed lake is, at the moment, treated in the legislation as sites for commercial fishing. Incidentally, many of them partly function as system breeding farms for fattening fish for consumption. The fishing boats in these lakes are actually used in the fish breeding technological process. For the big dams, where zones for commercial fishing still exist, these zones tend to be gradually abandoned in favor of angling, AQC, or other activities, such as fishing tourism and ecotourism.

CBFM involving the community could be an effective tool for the efficient management of the fisheries.

5.4.4.27 Danube fisheries

River Danube often depicts rich fish catches. The annual catches from the Danube during the last few years varied between 250 and 500 tons, with 360.82 tons of fish in 2005. The species commonly fished are European catfish (*Sillurus glanis*), Danube bream (*A. brama*), barbel (*B. barbus*), common carp (*C. carpio*), crucian carp (*Carassius* sp.), common rudd (*Scardinius erythrophthalmus*), lots of other herbivorous species as well as anadromous species, notably, sturgeons and Danube shad (*Alosa pontica*), etc.

Concomitant to the above, the populations of one of the most valuable Danube species are not stable. Therefore, the catches cannot be planned beforehand. Some of the sturgeons, such as the ship sturgeon (*A. nudiventris*) and the

Atlantic sturgeon (*Acipenser sturio*) do not exist in the river anymore. The populations of other sturgeon species and especially the populations of beluga are in steady depletion and thus require special protection measures.

5.4.4.28 Aquaculture (AQC)

The total AQC production increased during the last decade and had reached *c.* 3143 tons in 2005. Incidentally, there are two AQC subsectors in Bulgaria, viz.,

(a) FW fish farming (in warm or cold water) and
(b) marine AQC farming.

5.4.4.29 Freshwater aquaculture (FW AQC)

In Bulgaria, FW fish farming accounts for *c.* 95% of the Bulgarian AQC. The total water spread area used for fish farming is *c.* 9178 ha. Production of fish for consumption is not always very good. In fact, they accounted for *c.* 60% of the total number of fish farms. However, traditionally, intensive and semi-intensive production systems are used. Incidentally, the main AQC fish species in Bulgaria is the rainbow trout (*Oncorhynchus mykiss*), which accounted for *c.* 36.6% of the total AQC production in 2005. Likewise, the common carp (*C. carpio*) is another significant AQC ichthyospecies, which accounted for *c.* 29.5% of the total AQC production. Nevertheless, "polyculture" dominate the semi-intensive AQC in Bulgaria, involving mainly the fish species like the common carp, grass carp, silver carp, and bighead carp, On the other hand, the superintensive AQC in net cages is related mainly to rainbow trout monocultures. In this connection, it may be stated that AQC production is directed mainly to the internal market with *c.* 100% of the carp production being marketed in Bulgaria itself. The frozen fish products form the major part of the exported products.

In contrast to the above, the main warmwater AQC species are the common carp (*C. carpio*), bighead carp (*Aristhichthys nobilis*), silver carp (*Hypophthalmichthys molitrix*), grass carp (*Ctenopharingodon idella*), black carp (*Mylopharingodon piceus*), buffalo species (*Ictiobus* spp.), and tench (*Tinca tinca*). In addition, there seemed to be a growing importance of the European catfish (*S. glanis*), American catfish (*Ictalurus punctatus*), pike-perch (*Sander lucioperca*), and pike (*Esox lucius*), as products of warmwater AQC. Moreover, successful tests for introducing and breeding the eel (*Anguilla anguilla*) had been made. Incidentally, the production from warmwater farms was *c.* 1813 tons in 2005, which accounted for *c.* 48% of the total AQC production in Bulgaria.

5.4.4.30 Marine aquaculture (marine AQC)

Marine AQC is quite promising in Bulgaria. Marine AQC facilities, in Bulgaria, include mainly the Black Sea mussel farms and farms for the production of kelp algae (*Cystoseira* sp.). With a total of 18 farms in 2005 (16 mussel farms and 2 algae farms), they represented *c.* 9% of the number of farms in the country. Incidentally, the main species in marine AQC is the Black Sea mussel (*M. galloprovincialis*). Its production in 2005 was *c.* 170.59 tons, i.e., *c.* 5.4% of the total AQC production in Bulgaria. In addition, there was a steady tendency for development of more marine AQC farms along the Bulgarian coast during the last few years.

5.4.4.31 Processing, trade, and employment

It was estimated that all fisheries activities (including processing) contributed an added value of *c.* €14.3 million to the Bulgarian economy, which was *c* 0.14% of the GDP. This, however, was significantly lower than the EU fisheries average of *c.* 1.2%. Incidentally, processing accounted for *c.* 69% of the added value to the sector.

Notwithstanding the above, there are important regional impacts and related activities that are very much dependent on fisheries, which justify the immense significance of fisheries in the national economy. These impacts act with due consideration of "fisheries" sector in the policy framework for agricultural and food sectors.

5.4.4.32 Fish processing

In 2008, the fish and seafood processing industry in Bulgaria consisted of 45 farms. There had been an estimated turnover of *c.* €49 million and a total income of *c.* €52 million. About half of these processing firms (*c.* 49%) are quite large, having between 50 and 249 employees (STECF, 2010). Incidentally, the processing industry is closely related to marine fishing and AQC activities and less linked to FW and AQC activities. Some of the fish and AQC producers have their own processing facilities situated close to the fresh fish production sites, which help to improve the quality of the final products. Nevertheless, there are several fishermen or fish producers' associations with the status of NGOs, which could form the basis for producers' organizations.

5.4.4.33 *Fisheries research*

The principal Bulgarian scientific research and education centers related to the fisheries sector in the country are as follows:

(A) Faculty of Biology, Department of Hydrobiology and Ichthyology within Sofia University, "St. Climent Ohridski," the main structure for higher education in the field of fisheries.
  (a) Institute of Fish Resources in Varna.
  (b) Institute of Oceanology in Varna.
  (c) The above institutes are specialized mainly in Black Sea research.
  (d) Institute for Fisheries and Aquaculture in Plovdiv, specialized in AQC research.
  (e) Institute of Zoology of the Natural Museum in Sofia.
  (f) Central Laboratory of Ecology in Sofia.
  (g) Trakia University in Stara Zagora.
  (h) National Veterinary Institute for Research and Diagnostic Activities within the National Veterinary Service of the MAF.

(B) Agriculture and forestry: Agricultural Academy with 27 institutes, 19 experimental stations, and 2 experimental plots; professional high schools for agriculture, forestry, and food technology; Agrarian University—Plovdiv; Thracian University, Stara Zagora; Forestry University, etc.

(C) BD and biological resources: Bulgarian Academy of Sciences. It is a group of institutes dedicated to BD, bioresources, and ecology such as Institute of Biodiversity and Ecosystem Research; Forestry Institute; Institute of Plant Physiology and Genetics; National Natural History Museum and Botanic Garden.

These research institutes support NAFA activities for organization and management of the fisheries sector in Bulgaria. NAFA also benefits from organizing and/or participating in education and training activities for its administration, notably, the professional organizations and for people working in the sector. Incidentally, the Institute of Fish Resources and the Institute of Oceanology of the Bulgarian Academy of Sciences (both of them located in Varna) are known to undertake scientific research activities and give recommendations for the sustainable exploitation of fish and other aquatic resources in the water bodies.

Notwithstanding the above, research on the dynamics of fish populations in inland waters (in particular, the commercially important and endangered/protected species in rivers and natural or artificial lakes) is carried out at the Institute of Zoology of the Bulgarian Academy of Sciences, in the Department of Biology of the Sofia University, and in the Central Laboratory of Ecology in Sofia.

Concomitant to the above, the Institute of Fisheries and AQC in Plovdiv is mainly engaged in scientific and research activities in AQC and especially in the development of technologies for breeding commercially important species as well as in the production of raw materials from productive species. However, the existing scientific organizations are confronted with significant funding problems. In addition, there seemed to be insufficient communication between scientific institutes, AQC producers, and fishermen organizations. These problems could be solved by using the options of the community financial measures for the implementation of the EC CFP as well as through public and the private partnerships or PPP mode (MAF NSPFA, 2007–13).

5.4.4.34 *Inference*

Notwithstanding the above, for all those activities indicated above, to achieve the goal an effective **management** system is a prerequisite. Also, CBM involving the entire community from the grassroots level is the best strategic means of management. Also, CBFM is an efficient tool for the management of AQC and fisheries.

5.4.5 Denmark

5.4.5.1 *Nordic cooperation*

Cooperation, coordination, collaboration, and comanagement are few key words in fisheries management, notably, in CBFM today. Incidentally, Nordic cooperation is one of the world's most extensive and significant forms of regional collaboration, involving **Denmark**, Finland, Iceland, Norway, Sweden, and three autonomous areas, viz., the Faroe Islands, Greenland, and Åland. Having reference to **Denmark**, Eliasen and Sverdrup (2009) dealt with the Nordic experience of fisheries management in relation to the reform of the EU CFP. Nordic cooperation has firm traditions in politics, economy, and culture. It plays an important role in European and international collaboration, and aims at making a strong Nordic community in a powerful Europe.

Nordic cooperation seeks to safeguard Nordic and regional interests and principles in the global community. Common Nordic values help the region to solidify its position as one of the world's most innovative and competitive one. Nordic experience of fisheries management is often seen in relation to the reform(s) of the EU CFP.

Concomitant to the above, the Nordic countries developed seven distinct types of fisheries management (Søreng, 2008). Unlike the West Nordic countries (Greenland, Iceland, the Faroe Islands, and Norway), which are not members of the EU, Denmark, Sweden, and Finland are subject to the EU's CFP. It may be mentioned that the seven systems operate in societies, which, historically and socially, could be characterized by a close sense of affinity and shared values. However, these systems operate under conditions that differ significantly from one nation to another, especially in terms of the countries' economic dependence on fishery resources, their sociocultural approaches to fisheries, and so on. As a result of these factors, the Nordic countries have a relatively diverse system of fisheries management. International evaluations of global fisheries management takes place at regular intervals. These are carried out by different organizations, which employ different criteria. All these evaluations have ranked the Nordic systems, especially the West Nordic ones, among the best in the world. As such, Nordic diversity in fisheries management had proven successful at an international level. Incidentally, the Nordic region is the home to many highly sustainable solutions.

There had been report(s), some of which form part of the Nordic Council of Ministers' contribution to the debate, about reform of the EU CFP. The Commission's Green Paper on the CFP, which published in 2009 had posed many relevant questions about future fisheries policy and stressed the need for a wide ranging and basic debate on the current situation. The whole basis of EU fisheries policy is in the field for discussion and the Nordic Council of Ministers had decided to make an active contribution to that debate.

Concomitant to the above, fisheries management in **Denmark** is of great significance. There had been report(s) to bring Nordic experience into the discussion in Denmark for the reform of the CFP. The discussion takes, as its point of departure, the so-called Green Paper in which the commission summarized the status of the EU fisheries and also put themes and questions up for discussion. Incidentally, the themes and cases of the report(s) are issues which feature prominently in the Green Paper, notably the following:

(a) reduction in overcapacity by establishing RBM systems;
(b) increase in the responsibility of the industry through development of **comanagement** arrangements; and
(c) reduction in discard of fish.

Incidentally, the waters surrounding the Nordic countries are rich in fish resources. In fact, fisheries had been among the most important productive sectors in the Nordic economies; and in the Faroe Islands and in Greenland, it is by far one of the most important sectors. However, management of marine fisheries has over time had various objectives in the Nordic countries, ranging from fiscal and social purposes to stock conservation and socioeconomic well-being in recent years. Nevertheless, there are strong similarities between the Nordic countries but differences in the ecosystem and the sociocultural settings of the fisheries led to different management systems.

Incidentally, Nordic experience with regard to fleet reduction by using RBM systems are as follows:

(a) The introduction of RBM with transferability of rights has contributed to adjusting fleet capacity with the fish resources available for exploitation.
(b) The economic performance of the remaining active Nordic fishing vessels improved significantly in all fleet segments with the reduction in the fleet capacity to match the TACs and fish quotas, even though debts increased.
(c) RBM systems could be designed to cater for social concerns in relation to small-scale fisheries and coastal communities through special coastal fisheries schemes or restrictions on allocation and transfer between regions and/or geographical areas.
(d) Another concern that had been addressed in the design of systems in some of the Nordic countries is the problem of increasing cost of investments for new fishers.
(e) Special quota funds for new fishers could lower the barrier to entrance, which is formed by the increasing cost of investments in the fishery. This was especially important for the livelihood of coastal communities.

Notwithstanding the above, the introduction of RBM systems is highly sensitive. There had been allays and concerns about the impacts on fishing communities employment, in spite of the advantage in terms of economic efficiency, etc. Concerns existed also about the principles applied for the allocation of fishing rights, and the distribution of the associated wealth. Incidentally, the Nordic experience, in this regard, was that the introduction of RBM could be adaptive and include an open discussion among all stakeholders about the features of the system in relation to policy objectives in fisheries and in society at large.

Incidentally, the Green Paper aimed at generating industry's responsibility for the CFP.

(f) Comanagement or self-management is, generally, viewed as tool for this. Incidentally, there were a number of Nordic instances of groups of fishers or broader groups of stakeholders taking responsibility of parts of the management of specific fisheries under comanagement arrangements with consultative or advisory elements.
(g) Concomitant to the above, the highest legitimacy of the regulations has been found, where the groups also had an advisory role, formal or informal, and their proposals and recommendations were included more or less directly into the management regulations.

Notwithstanding the above, comanagement arrangements brought out potential conflicting interests in the open, which could be between the fishers at the local and national level, and other groups seeking to get influenced and a role in comanagement.

The cases illustrate the following:

(a) High internal conflict levels in the comanagement groups reduce their capacity to influence decision-making, especially, when the mandate for the group is unclear.
(b) However, with a clear mandate or room for maneuvring, the stakeholder groups may take on responsibility and reach compromises or consensus also when conflict levels are high.
(c) Incidentally, the potential for collective action and strong comanagement engagement was considered even stronger within relatively homogenous groups, like RBM groups with clearly delimited membership.
(d) Notwithstanding the above, the governmental agencies easily gets directly involved in such conflicts and the decision to lend authority to stakeholder groups, formally or informally, may be considered carefully.

Notwithstanding the above, "discard" refers to a waste of a precious resource, and prevents stocks from recovering in spite of low quotas. Incidentally, the fisheries systems in the Nordic countries applied a mix of effort and quota regulations as their main principles for regulating fisheries. The effort regulation in the Faroe Islands and the discard ban in the Faroe Islands, Iceland, and Norway seemed to have had some success. It is briefly discussed as follows:

(a) Effort regulation, in combination with effective monitoring of fishing grounds, eliminate quota-related discard. However, incentives to high fishing efficiency during the fishing days available may lead to size-related discard, which still persists under effort regulation regimes without effective systems for monitoring, control, and surveillance.
(b) In addition to the above, discard bans are combined with systems for control and enforcement and with legal mandates to intervene when the rules are violated.
(c) Further, discard bans are, sometimes, supported by technical measures regulating the use of fishing gear(s), and also imposition of closed fishing areas.
(d) Tradition for comanagement and general fisher influence had meant that the management of discard could be considered as reasonable and legitimate by the fishers.
(e) Further, discard bans are supported by incentives for the fishers to avoid discard. These incentives could be in the form of economic benefits to the fishers.
(f) Furthermore, incentives for the fishers to avoid discard(s) were discussed in Denmark. If the fishers through the use of electronic devices could document all catches and accept that they are all counted against the quota, then they could receive a share of what is set aside for discard in the TAC calculations.

In fine, the acceptance of this new approach among fishers and in the political system is yet to be seen at large. However, there are lessons to be learned from the Nordic countries in the discussion of the revision of the CFP. Various examples of solutions and models were seen but a central lesson could be that they may be modified to suit the specific context. With this note, it is hoped to contribute to a widespread discussion of models and solutions and also a discussion on how they could help solve the problems in the CFP as well as in management at local, regional, and national levels.

5.4.5.2 *Iceland*

Prior to the introduction of the ITQ system in the late 1970s, Iceland practiced a wide range of fisheries management systems. These included access licences, fishing effort restrictions, investment controls, and vessel buy-back schemes, all of which were found not satisfactory. The first ITQ systems were introduced in the herring fisheries, following the establishment of the Icelandic EEZ. Since then, the system had been extended in several steps. In 1984, a limited form of ITQs was introduced in the demersal fisheries, which were, by far, Iceland's most important fisheries. In 1991, a uniform Nordic experience of fisheries management with fairly complete ITQ system was adopted in all Icelandic fisheries, applying to all vessels above a certain minimum size. In 2004, the system was expanded to cover all commercial fishing vessels. Currently, the system comprises 25 fish species and about 35 different fisheries.

5.4.5.3 *Norway*

In 1973, when a resource crisis hit the pelagic industry, a license scheme was introduced in order to limit the number and size of the pelagic purse seiners. At the same time, license schemes were also implemented for saithe seiners and for shrimp trawlers. This means that when Norway got its 200 miles EEZ in 1977, it regulated most of the offshore fleets to curb their further expansion in numbers.

Concomitant to the above, the next step in establishing RBM in Norwegian fisheries was a quasi-individual quota regime in the cod fisheries in 1990. Due to a crisis in the cod fisheries, the total quota had to be dramatically reduced, and open access had to be closed, dividing the coastal fishers into two groups:

(i) Group I: With guaranteed individual quotas granted according to vessel size and previous participation.
(ii) Group II: Originally with maximum quotas within a limited group quota.

While the remedy originally was intended as a crisis measure, it soon turned out to be beneficial to the ones with guaranteed quotas. Hence, during the last 10 years, more and more fisheries closed according to the same principle: a participation right based on previous participation and a minimum catch in a qualifying period was considered. As of 2009, most Norwegian fisheries of economic importance were closed either by limiting access (Groups I and II) or by specifying the removal right in terms of quotas. There had been detailed allocation keys for all the major fish stocks, specifying group quotas for the different fleet groups.

5.4.5.4 *The Faroe Islands*

In the beginning of the 1990s, the Faroe Islands experienced catches at a historically low level from the most important demersal fish stocks. This had fostered a severe economic crisis in the fishing industry. As the regulation of the fishery was based on the use of technical regulations only, it was decided to introduce a more effective fisheries regulation.

Incidentally, an ITQ management system was preliminarily introduced in 1994 for vessels of 20 GT and above, whereas the smaller coastal fishing vessels were fishing on a general annual quota. This system was met with much reluctance from both the fishing industry and a number of politicians and caused a substantial revision of the Commercial Fisheries Act.

Notwithstanding the above, the main elements of the RBM management system applied in the Faroe Islands since 1997 are as follows:

(i) A capacity policy which limits the size of the fishing fleet to the existing (1996) level.
(ii) A grouping of the fishing fleet into vessel segments based on size and type.
(iii) Allocation of individual and transferable rights to the industry by means of a dual license system comprising:
  (a) catch permits that follow the individual vessels and outline the capacity for the vessel groups and
  (b) fishing permits that are used for the management of fishing patterns.

Notwithstanding the above, the permits include the number of days allowed to be at sea, which are used to regulate the catch of demersal species such as cod, haddock, and saithe on the Faroe Plateau. A similar regulation was established for the Faroe Bank fishery. The regulation sets the number of days at sea that each individual vessel is allocated for a specified fishery. The number of days is regulated each year in a process including both scientific advice and advice from the industry.

Concomitant to the above, there had been technical regulations, *in vogue*, including gear regulation and minimum size limits, and a system of closed areas regulating the admittance of the different vessel segments. Further, bycatch quotas were used to regulate the fishery in the zone outside the Faroe Plateau.

Notwithstanding the above, in 2003, ITQs were introduced in the **Danish herring fishery** on an experimental basis. The reason for the experiment was, first and foremost, the need for a modernization of the rather old Danish pelagic fishing fleet in Nordic experience of fisheries management in order to meet the quality requirements of the fish processing industry. ITQs were allocated according to vessel track records for the previous 3 years. From 2007, the herring ITQ system was made permanent, and mackerel and industrial species, such as sandeel, sprat, blue whiting, and horse mackerel, were put under ITQ management as well.

In 2007, a quasi-ITQ system (VTQ) was implemented in the Danish demersal fisheries for cod, saithe, plaice, haddock, hake, sole, turbot, monkfish, Norway lobster, and prawns. The major driving force behind this was the need to improve the economy of the demersal fishing fleet and reduce the pressure on demersal fish stocks through a substantial capacity reduction.

Incidentally, the main difference between the ITQ and VTQ systems is that in the former, the quota could be transferred independently of the fishing vessels. In the latter, the fish quotas (allocated on a 3-year historic record) and the vessels to which they are allocated are inseparable and only transferable together. However, a (new) vessel owner can transfer the quota utilization to another fishing vessel in his possession and if there is >1 (new) owner the quota utilization can be split among them and transferred proportionately to other vessels in their possession. Fishers holding VTQs could form "quota pools" and through quota lease or swaps among pool members they can ensure efficient use of the pool's fleet capacity. Concomitantly, the discard(s) related to individual quota limitations is reduced. Quota loans between fishing vessels outside quota pools are also permitted with some limitations.

Notwithstanding the above, since 2009, the tying of the demersal fish quotas to vessels was abolished and the VTQ system turned into a proper ITQ system. In 2009, the Danish blue mussel fishery had also come under ITQ management, meaning that all Danish commercial marine fisheries are now managed through an ITQ system.

5.4.5.5 *Nordic experience of comanagement*

Comanagement is about involvement of the users of management. In this case, it is mostly the fishers' involvement in the management of fisheries. It could be seen as a more democratic way to manage. But the managerial intentions are often to get more trustworthy data on catches and fishing practices so that better decisions about regulations could be made. However, regulations could better be implemented with a novel approach and a high level of compliance could be achieved.

Incidentally, comanagement covers systems with varying degrees of user involvement in the management. With increasing fisher involvement, the comanagement arrangements could be labeled as follows:

(i) **Instructive management**, where decisions are made in the government system and fishers are instructed in decisions. This is not a comanagement arrangement.
(ii) **Consultative comanagement**, where fishers are consulted on proposals before decisions.
(iii) **Cooperative comanagement**, which occurs between fishers and authorities as equal partners in developing proposals for management decisions.
(iv) **Advisory comanagement,** where fishers advise governments on decisions to be made and the government endorses these decisions.
(v) **Informative comanagement**, where governments have delegated the authority to make decisions to user groups, who are responsible for informing government about the decisions made.

Concomitant to the above, self-management in a clear-cut version is hard to imagine in the modern society fisheries. There has to be some governmental involvement to regulate access to the fish resources and to secure sustainable resource use. However, self-management within a general policy framework is possible. As an example, TACs are set at EU level. MSs allocate the quota among the user groups and leave it to the groups to self-manage the further allocation of the quota internally. This is seen in the producer organizations (POs) in the United Kingdom and the quota pools in Denmark. In these cases, the fishers practice self-management within a framework set by a higher level of authority and under the condition that they report and document the quota uptake to the higher level. The delegation can be withdrawn if the self-management is not working as intended within the politically accepted framework.

5.4.5.6 *The green paper on comanagement*

The Green Paper proposes to distinguish between deciding on the principles for the CFP and implementing the CFP. Principles and standards such as fishing within MSY, adjusting fleet capacity to resources, abolishing discards, etc., may still be decided at the EU level. Implementation of decisions could be delegated to the MSs for possible self-management by the industry. In that way, there could be a higher flexibility locally and regionally, and micro-management from Brussels could be avoided. The specific implementation may be in accordance with community standards and under EU control. Therefore, the discussion of comanagement and self-management in the CFP context is under the condition that the overall standards and principles for fisheries management are set at the EU level and that implementation is subject to the community control. This could be a community initiated control or control of documentation received from the MSs or the industry. Comanagement and self-management could be seen by the commission as a tool for encouraging the industry to take on more responsibility in implementing the CFP. It is considered critical to the success of reform and industry is expected to understand the need for it, followed by supporting it and finally to have a genuine stake in its successful outcome.

Concomitant to the above, comanagement is often seen as an alternative to the top-down approach taken up till now in implementing the CFP. In fact, there is a need to involve the fishers in the management of community fisheries through comanagement or self-management arrangements. This could generate incentives to act responsibly with regard to sustainable use of the fish resources as well as other principles in the CFP. But the following question arise:

(i) How could this be done in practice?
(ii) How could more responsibility be given to the fishing industry?
(iii) How could industry be best structured to take on more responsibility?
(iv) Which mechanisms could supervise the self-management and ensure that the industry actually takes on the responsibility and acts in a responsible way?

5.4.5.7 *Comanagement in the EU*

In the present CFP management system, the overall policies and principles of the fisheries policy are decided upon/revised every 10 years by the Council of Ministers based on input from the EU Commission. The details in the implementation of the CFP are decided at the same level. In this top-down approach, the stakeholders are involved in the decision-making process in two consultative comanagement arrangements, which are said to be as follows:

(i) The Advisory Committee on Fisheries and Aquaculture (ACFA) is composed of interest groups representing the fishing industry, the processing sector, traders, and employees within fisheries as well as the consumers and environment. It was established in 1971 for the purpose of establishing a close dialog with the fisheries sector and other groups affected by the CFP. The ACFA comments on proposals from the EU Commission before decisions are made, and ACFA has the possibility of raising questions on almost all other issues. In that way, the ACFA could be seen as an institution for consultative comanagement. It is consultations at a very high level and within a nested system, which tends to establish a long distance between the consultative arrangement and the practical fisher. This reduces the legitimacy effect of the process being consulted. Incidentally, the Regional Advisory Councils (RACs) were introduced with the reform of the CFP in 2002. The intention was to establish a higher involvement of the stakeholders at an early stage in the decision-making process. It was initially considered to delegate some decision competence to the RACs, but so far their role seemed to be only advisory. The RACs comment on proposals formulated by the commission, but during their first year of existence, the RACs have also taken own initiatives in formulating proposals. Such proposals from the RACs may be taken into consideration, although, the Commission and the Council of Ministers are not formally obliged to take the proposals into consideration. Formally, the function of the RACs should be labeled as helping in consultative comanagement arrangement, but depending on the political will, the RACs could function as advisory comanagement partners as well. In the MSs, different types of comanagement arrangements exist for the country-specific implementation of the CFP.

5.4.5.8 *Comanagement in the Nordic countries*

In the Green Paper, the POs, which are central bodies in the CFP market policy, are mentioned as the type of institution that could be turned into bodies through which the industry could take on management responsibilities. In contrast to many EU countries, in the Nordic experience of fisheries management, the role of POs was limited in the Nordic countries, but the Nordic experience presented in the following show that other industrial organizations at national and local levels can take on the role of comanagement partners or even self-managers. The Nordic comanagement experience includes the issue of ensuring stakeholder representation. The following questions may arise:

(a) Who is represented by the organization?
(b) Is the internal authority of the organization strong enough to bear the responsibility of the self-management or comanagement?

The **first case** of comanagement is said to be a formal consultative arrangement in **Norway**. In this arrangement, an informal advisory role is given to the Norwegian Fishermen's Association in the decision-making process on the principles for distributing shares of the TACs to different fleet categories. The formal consultative arrangement and the informal advisory arrangement may coexist with severe consequences for the many groups of stakeholders involved and for the legitimacy of the decisions on quota allocations.

In the **second case,** focus is on comanagement at the local level. In **Sweden,** a government led comanagement initiative supported six local fisheries management groups. The groups are to identify local problems and eventually draft and implement local management plans. The local groups consisted of fishers or a broader range of stakeholders and were formed with a high level of enthusiasm and energy. Having identified the management needs, the groups began to develop the local management plans. During this process, it became obvious that the mandate of the group(s) was not very clear. However, the groups had no formal power to implement the plans. Also, perhaps, no procedures were in place to take the plans to higher management levels for decision-making or even for consideration. Some of the plans were simply overruled by other interests at the national or EU level.

The **third case** focused on the local self-established fisher groups in the mussel fishery in the Limfjord in **Denmark**. These groups managed to get a strong, yet informal, advisory role in the local management by collectively developing fishing plans and other self-management initiatives within the framework of the formal regulations. Incidentally, a conflict between the fishers regarding the enforcement of internal agreements illustrated how public power over enforcement could be "borrowed" by the groups to achieve compliance and curb possible free riding.

5.4.5.9 *Formal and informal comanagement in Norway*

Norwegian fisheries management has a strong tradition of user-group involvement. This tradition stretches back to >100 years ago, when the fishermen became directly involved in the management of the famous Lofoten cod fishery. Since then, several important comanagement arenas were introduced in the Norwegian fisheries, both in the organization of the firsthand fish markets and in the bargaining about state subsidies. Today, the most important area for user-group participation in fisheries management is **quota management**.

Concomitant to the above, most of the commercially important Norwegian fish resources are said to be shared with other countries. Therefore, the TACs are negotiated and decided upon in various international forums. Representatives from the sector, most notably, the Norwegian Fishermen's Association, form part of the Norwegian delegation in these negotiations. This allows the fishermen to be close to and have a say in the most critical management decisions. Incidentally, the two most important arenas for user-group influence are within the national management of the TACs after they had been decided upon.

Firstly, there is a formal system of consultation, the management meeting, which is held every year in October or November. Here, stakeholder groups are invited to present their viewpoints on the government's management plans for different fisheries. The management meeting has a consultative function, but the Minister of Fisheries and the government make the final management decisions. However, firm advice from the core stakeholder groups is most often considered.

Secondly, there is an informal practice by which one of the stakeholder groups, the Norwegian Fishermen's Association, is granted a privileged role in connection with the formulation of the principles by which the Norwegian national quotas are distributed to the fleet categories (segments). This practice developed from a hard-won compromise that was reached in 1989 and often is referred to as the "Trawler Ladder."

In the comanagement context, the Norwegian government had left an important aspect of the fisheries policy, the annual quota allocation, to be decided in an internal process within the Norwegian Fishermen's Association. This was an informal and conditional arrangement. The government remains in charge and is not formally obliged to follow the allocation principles established by the NFA. There has so far been a strong political commitment to this arrangement. The fishermen were generally involved in the decision-making process, through their representatives in the NFA, on an issue that directly affects their employment and income.

Notwithstanding the above, the system of user-group participation in Norwegian quota management has both strengths and weaknesses. It is an obvious strength that the difficult task of quota distribution may be left to the industry itself. This not only relieves the government of a difficult and time-consuming task but also lends legitimacy to the whole management system.

Concomitant to the above, there is increasing criticism of the present allocation system for not taking the interests of the most fisheries-dependent communities into account. In particular, there were criticisms from the northern part of Norway, pointing to the trend that the coastal fisheries in the North were losing quota shares to the more industrial fisheries on the southwestern coast. However, the Norwegian Fishermen's Association quota allocation system could not able to prevent this, despite heavy mobilization from members in the North. A major reason for this could be the power balance within the association, where the North Norwegian representatives are not in a majority position.

5.4.5.10 *Local comanagement with unclear delegation of mandate*

In 2004, the **Swedish** government launched a comanagement project with the intention of gaining experience with different comanagement forms and decision-making processes in local fisheries. The initiative was based on the hypothesis that "common resources are best managed in cooperation." Within this project framework, the Swedish Board of Fisheries (SBF), the government agency responsible for fisheries management in Sweden supported the formation of six comanagement initiatives dealing with different types of fisheries and involving different stakeholder groups. The SBF, which was said to be responsible for fisheries administration, research and development, and enforcement and control, had a tradition of consultative or advisory comanagement with industry organizations at the national level Nordic experience of fisheries management. In the project in question, the SBF supported the development of local management initiatives by supplying expertise and funds for the establishment of local organizations that could take part in fisheries management at the local level.

Notwithstanding the above, the commercial fishers were the key stakeholders in all the six comanagement organizations. Other stakeholders, such as owners of water bodies and land adjacent to lakes and rivers (inland), anglers, fishing tourism, environmentalists, etc., usually had participated as active observers in some of the groups, whereas in others they were partners *at par* with the commercial fishers. The SBF had participated in all subprojects, both as a facilitating partner and as the authority responsible for formulating and implementing fisheries regulation and control.

Under the above condition, the overall conclusion from the project is that a government agency could initiate and facilitate the formation of local organizations/groups that could comanage local resources together with the agency. Such organizations have the possibility of identifying management needs at an early stage, exchanging knowledge among stakeholders internally and with the authorities; and develop proposals and eventually implement them. Within the organization, possible local disagreements about the resource use and other subjects could be articulated and resolved and mutual understanding of positions could be established.

Concomitant to the above, seen from the SBF perspective, the local initiatives could offer factual knowledge, ideas, and proposals, which could not have possibly been obtained otherwise. This also includes the development of new and selective fishing gears, and other self-controlled instruments.

Notwithstanding the above, based on the experience, both the SBF and the various stakeholder participants, fishers, and others realized that the comanagement process is quite consuming, both in terms of time and money, and especially in terms of time. However, the major shortcomings relate to unresolved conflicts between local and national interests and to the lack of legal mandate in the local comanagement groups.

5.4.5.11 *Organization of the industry*

A local organization of the fishers was developed in the late 1970s, the Central Association of Fishers in the Limfjord (CFL). The CFL is a member of the national Danish Fishermen's Group. The industry could generate a lot of funds through trades but everything is suggested to be done in a sustained and eco-friendly way.

5.4.5.12 *Empowering the local organizations*

The CFL has no formal powers to force the fishers to comply with the regulations decided upon by the association regarding quota reduction, area closures, and allocation of fishing days and fishing areas. Fishing plans for each vessel are usually forwarded as advice to the Fisheries Directorate that normally transforms this advice into official licences for the individual vessels.

Notwithstanding the above, the representativeness of the actors in the group is also very important. Incidentally, in the Nordic countries, the national fishermen's organizations have traditionally been the natural partner in comanagement. This is further enhanced with the growing tendency of allocating rights to clearly defined groups or individuals. Such well-defined groups with rights and interests have the possibility of a high capacity for collective action, and thereby promote comanagement or self-management.

In the cases from **Norway and Sweden**, the conflicting interests between fishers at local and national level within the fishermen's associations are illustrative of this. Likewise, other groups tend to get influenced for a role in comanagement arrangements. In fact, the conflicting interests could decrease the capacity for collective action within the comanagement groups. Conflicts are to be taken care of internally within the comanagement groups.

However, the governmental management system could easily become directly involved in such conflicts, if they want to, and government agencies have to decide if they actively or passively would lend authority to the groups recognized formally or informally as comanagement partners as the **Danish** case(s) illustrate.

5.4.5.13 *Nordic experience of reduction of discards*

According to the Food and Agriculture Organization (FAO) of the UN, **discards are the portions of fish catches that are thrown back into the sea**. Discarding may have many reasons, wrong catch, wrong sizes, damaged catch, no quota or high grading. High grading refers to the discard of less valuable catch or sizes in favor of bigger or more valuable fish. Discarding could be a problem, when the organisms returned to the sea are dead or not able to survive because it represents both a biomass removed from the sea that is not usually taken into account in stock estimates and a reduction in potential future value of the discard. Discard of a fish and organisms that are in a healthy condition and that are likely to survive is not regarded as a problem and is either allowed or even mandatory in some cases to ban catching of those (endangered) animals, e.g., sharks, whales, turtles, etc., in many water bodies.

5.4.5.14 *Reduction of discards in the Nordic countries*

The Nordic countries have various kinds of experiences with arrangements to avoid discards. However, the situation in the mixed ground fish fisheries in the North Sea is complex and, hence, experiences from less mixed and complex fisheries in Nordic countries may not be seen as a formula. Nevertheless, some of the experiences from the Nordic countries might be relevant. The Nordic countries, notably, Denmark, the Faroe Islands, Iceland, and Norway, have different systems of fishing regulations as well as various solutions to avoiding discards. In this connection, it may be mentioned that the Faroe Islands, Iceland, and Norway do not have huge mixed fisheries. Thus, the problem to be solved is less complex than in the EU. In addition, the countries are small with, more or less, homogenous populations. The four countries have a combination of effort regulation and output regulation (quota regulations), although effort and output regulations have different effects on discarding.

5.4.5.15 *Bycatch rules*

Bycatch rules is a measure that makes it possible to undertake fishing, even if the targeted species is mixed with other species. In the Nordic countries, the fishers are allowed to have certain rates of species in their catches and to land amounts of fish equivalent to those rates.

Notwithstanding the above, it may be noted that the waters surrounding the Nordic countries are rich in fish resources. Fisheries have been among the most important productive sectors in the Nordic economies till date and are by far the most important sector in the Faroe Islands and Greenland. Management of marine fisheries had various objectives in the Nordic countries, over time, ranging from fiscal and social purposes from stock conservation to socioeconomic well-being in recent years. There are intimate similarities between the Nordic countries but, nevertheless, differences in ecosystem and the sociocultural settings of the fisheries have led to different management systems over time.

In fine, it could be said that there were various kinds of notes in order to focus and portray Nordic experience(s) into the discussion of the reform of the CFP. The discussion often could initiate from its point of departure, the so-called "Green Paper," in which the commission has summarized the status of the EU fisheries and also had put gists and questions for deliberations.

5.4.6 Finland

Finland is rich in AQC and fisheries. There are a number of lentic bodies that harbor a rich diversity of fishes. As an example, works of Salonen and Mutenia (2007) portrayed that the Lake Inari is a large, subarctic, oligotrophic, and regulated lake. The fish community of the lake is composed of 10 indigenous species, mostly salmonids. Three new species, viz., lake trout (*Salvelinus namaycush*), landlocked salmon (*Salmo salar* m. *sebago*), and vendace (*Coregonus albula*) were introduced into the watershed in the 1950s and 1960s. Vendace established a self-reproducing population in the lake that expanded fantastically, thus making a boom in fisheries. However, the stock of this fish collapsed in the early 1990s, which has, however, rebounded in the 2000s. But, the Lake trout and landlocked salmon could not establish self-reproducing populations and the catches of these species were based on stocking.

Concomitant to the above, two large reservoirs, Lokka and Porttipahta, were formed in the region during the late 1960s for the requirements of hydroelectric (HE) power plants. The fish community in the reservoirs composed of 14 native species. On the other hand, an alien species, viz., peled (*Coregonus peled*) was introduced in 1972 in order to utilize the ZP population in the new, extensive open-water areas.

Peled was established a self-reproducing population in the reservoirs in the 1980s and was expanded by leaps and bounds, making a fisheries boom and stimulating new fishing technology around 1990. However, its recruitment and stock, subsequently, declined astonishingly, thus causing problems for fisheries and **management**.

Today, the low peled stock is composed of both a continuously stocked and occasionally naturally reproducing fish population. The four alien species, viz., the lake trout, landlocked salmon, vendace, and peled had not hybridized with the native fishes nor had any serious ecological impacts on native fish communities. They comprise an economically important, additional catch in both Lake Inari and the reservoirs.

Introduction of alien fish species is quite a common practice today and could be regarded as a "bad practice," as such a practice, generally, replaces the coveted native indigenous fish species. However, alien fish species are introduced for a variety of reasons, notably, AQC, sport fisheries, improvement of wild stocks, ornamental fish trade, biological control, and so on. In this connection, it may be noted that during the 1920s and 1930s, a large number of alien fish species were brought into the European FWs for general **fishery management** purposes. The next intensive period was the 1960s and 1970s, the period which saw the introduction of many alien ichthyospecies in the Finnish waters (Lehtonen, 2002).

Incidentally, there are innumerable examples of negative impacts of the introduction of new fish species among the FW communities (Moyle and Light, 1996). Notwithstanding the above, such so-called "biological invasions" also had given rise to a diversity of terms, which are often applied wrongly and could be in a misleading manner (Falk-Petersen et al., 2006). Lake Inari is regulated and the two reservoirs, viz., Lokka and Porttipahta were significant commercial and recreational fishing habitats in the northernmost part of Finland.

Concomitant to the above, it may be noted that three alien fish species, viz., vendace (*Coregonus albula*), landlocked salmon (*Salmo salar* m. *sebago*), and lake trout (*Salvelinus namaycush*) were introduced into Lake Inari area for fishing purposes over the last 50 years. In addition to these, one densely rakered whitefish (*Coregonus lavaretus*) form was also introduced on an experimental basis from 1976 to 1989. Moreover, one alien peled (*Coregonus peled*) was introduced into the reservoirs in 1972. The vendace spread downstream from Lake Inari into the Pasvik watercourse of Norway and Russia (first observation 1989). However, in Norway, in particular, it was considered as an unwanted invasive species (Amundsen et al., 1999). However, the vendace had caused negative impacts on ZP and the native whitefish morphs in the Pasvik river system since the invasion (Bøhn and Amundsen, 2004). Further, peled also migrated a distance of *c.* 100 km downstream from the Porttipahta reservoir into the Kemijoki watershed and no negative effects of the invasion had been documented (Huttula and Autti, 2006). However, the purpose of introducing these new alien species differ considerably in the two study areas. Moreover, this is the first case study in which the background, establishment, sustainable use, and impacts of these new alien resources in both the areas, viz., Lake Inari and the reservoirs, had been explored together.

Lake Inari is a subarctic oligotrophic lake that has been regulated (max. Range 2.36 m) since the 1940s. The lake flows into the Arctic Ocean via the River Pasvik. The surface area of the lake is *c.* 1102 km^2 (Marttunen et al., 1997; Salonen, 1998). The fish community consist of 10 indigenous fish species, most of which were salmonids. Concomitantly, Lokka, the largest reservoir in the EU and Porttipahta were formed around 1970 for the generation of HE from the Kemijoki river system, which flows into the Baltic Sea. The boreal reservoirs are regulated (mean range 2.5–3.0 m) and characterized as mesotrophic to eutrophic. Their maximum combined area was *c.* 630 km^2 (Mutenia, 1985; Virtanen et al., 1993; Sutela et al., 2002). The fish community of the reservoirs was quite different from that of the adjacent Lake Inari. Many cyprinid species were present in the reservoirs, whereas they were absent in Lake Inari, with the exception of the minnow (*Phoxinus phoxinus*). The fish community of the reservoirs was based on naturally occurring fish species in the area. There were four indigenous fish species in the reservoirs (Sundbäck, 1977a, b) and only one alien species, peled.

5.4.6.1 *Statistics of stocking, fisheries, and catches*

There had been statistical data on stocking and catches in connection with the obligatory monitoring program of Lake Inari and its tributaries being ordered by a Finnish court in 1975. The fisheries and catch data were gathered from the annual records that commercial trawl and trap net fisheries were supposed to keep as well as from the yearly surveys of all groups of fishermen (Kaatra and Simola, 1985; Mutenia and Salonen, 1994; Salonen and Mutenia, 2004). In the case of the reservoirs, stocking data were collected in connection with the monitoring studies by Finnish Game and Fisheries Research Institute. The fisheries and catch data were collected from the annual records, which were supposed to be maintained by commercial trawl and trap net fisheries. Concomitantly, the surveys of subsistence fishermen and the related statistics were obtained from the local sources of fishermen, fish traders, fish buyers, etc. (Mutenia, 1995; Salonen and Mutenia, 2004). All these aquatic resources, naturally, had called for efficient **management** of these abiotic (notably, the water bodies) and biotic (notably, the fishes) resources by effective *modus operandi* like the **CBFM** in a **participatory** mode **(PRA)** involving the stakeholders from the grassroots level.

5.4.6.2 *Socioeconomic impacts*

In view of the foregoing account, it could be stated that Vendace and peled became new targets for fisheries and, further, led to the development of new fishing technology in the study areas. Moreover, strong boom-and-bust development in the case of vendace and peled had brought many kinds of problems for commercial fisheries, e.g., difficulties in predicting and managing fisheries and investment, etc. Nevertheless, lake trout could be considered as a new, significant catch for both commercial and local subsistence fishermen. Despite occasional criticism of the introduction of alien invasive species, public opinion in the study areas was mainly positive, the exception being the case of the lake trout, which splits the opinions among fishermen to some extent. Nevertheless, **in fine**, after examining the ecological and socioeconomic impacts, it could be concluded that the overall effects of these four alien fish species on fisheries were mainly positive.

5.4.7 France

France possesses a rich potential for AQC and fisheries. **CBFM** in a **participatory** mode is, generally, considered as one of the efficient *modus operandi* for the effective management and conservation of water bodies and fisheries. Attempts had been made in a number of countries, notably in **France**, etc., for the management of fisheries by CBFM technique.

Of late, France's sea fisheries produced *c.* 655,000 mt (including 24,000 tons from the overseas French departments) of fish, covering a range of > 230 different species including 150 species of fishes. The main species were tuna, sardine, saithe, herring, mackerel, horse mackerel, and anchovy in terms of volume, while in terms of value, the leading species were tuna, anglerfish, sole, scallop, hake, and Norway lobster. On the other hand, the industry had registered a total turnover of €1.12 billion and had employed a total of *c.* 23,500 fishermen (*c.* 20,525 in metropolitan **France** and *c.* 2975 in the overseas departments) (Anon, 2001a, b).

Notwithstanding the above, the French fishing industry had undergone far-reaching change(s) since 1945. It had lost more than half of its vessels and more than two-thirds of its workforce; and the overall tonnage had been in slow decline since 1970. At the same time, however, the combined engine power of its fishing fleet increased considerably and onboard equipment now is far more sophisticated. The extension of fisheries under the jurisdiction of third countries has led to a decline in oceanic fishing, and a sharp rise in small-scale fishing.

Concomitant to the above, another feature of French fisheries is the wide range of harvesting methods in both inland and marine sectors. The bewildering diversity of fish catching devices does not seem to match the usual distinction between "small-scale" (mainly in FW) and "industrial" (in both FW and the sea) fisheries.

Notwithstanding the above, the average consumption of marine products in France is at *c.* 29 kg per capita per year. However, there is slight rise every year across the continent as a whole. This puts France midway between Northern and Southern Europe and therefore means that French fishery output is not high enough for self-sufficiency. A little different scenario exists for the inland FW system.

5.4.7.1 *Institutional framework for fisheries management*

European framework: CFP

France implements CFP in the community waters being a member of the EU. The CFP, which was in force since 1983, aims at reconciling resource conservation with the preservation of income and jobs in the aquatic (FW, marine, and coastal) domain, which offers few alternatives in terms of production or employment. It, therefore, covers not just resources but also markets and structures.

Concomitant to the above, with regard to resource management, at the moment, the CFP regulations comprise the following:

(a) a traditional management tool based on TACs and quotas;
(b) technical measures relating to gear or catch;
(c) effort-related management based on vessel engine power and the number of days in the aquatic regime.

Notwithstanding the above, the CFP also provides for the introduction of measures to rebuild, over a period of several years, stocks that are threatened in terms of sustainable harvesting and, thus, could decide for recourse to effort-related management strategies.

In continuation of the above, the Outline Act on fisheries and AQC states that resources are a public good and that government has a leading role in allocating fishing licences and catch quotas. Moreover, the Act also confirms that the latter are nontransferable in order to prevent them from being considered as assets, thereby maintaining the economic and social equilibrium in fishery-dependent domains. Further, the Decree of January 9, 1852 (as amended) also specifies that resource management may take into account producers' catch histories, market trends, and socio-economic equilibria.

5.4.7.2 *Quota and management of species by the community*

Fishing quotas are distributed to POs for allocation to their members. Incidentally, POs are groups set up by the producers and recognized by the government, on the basis of EU criteria ensuring that they are representatives. Their role is further strengthened by the Outline Act in order to facilitate resource management and promote the sector's organization into a real industry. In return, POs may have to draw up a management plan for the relevant species, i.e., a market-led fisheries management strategy.

Concomitant to the above, the quotas are allocated by the Minister for Fisheries after consulting the National Committee for Fisheries. The latter represents all those in the fishing industry. The Minister splits up the national quota of the main stocks into subquotas for the POs, based mainly on vessel size and catch history.

5.4.7.3 *Species not subject to quota*

For the species not covered by EU management measures, some stocks require regulation at the national or regional level. Further, the Minister for Fisheries could lay down certain broad rules, on the introduction of licences, for instance, and may delegate the management to regional bodies (Regional Fishery Committees).

Notwithstanding the above, the Normandy scallop fishery, for example, is subject to specific management rules to prevent overfishing, including opening and closing dates, mandatory licences, mesh size, and gear restrictions as well as maximum catch quotas per vessel/person/day.

In addition, other species, subject to similar management measures, include the crustaceans and the estuarine fish. Further, with regard to the anadromous and catadromous fishes (e.g., mainly the Atlantic salmon, shad, and eel), France's Migratory Fish Management Committees (COGEPOMI), established under the 1984 Fisheries Act and Decree No. 94-157 of February 16, 1994 had drawn up 5-year management plans for individual river basins, watercourses, or groups of watercourses, setting out fishery management regulations tailored to the type of commercial or recreational fishing practised there as well as conservation measures, wherever necessary. These management committees represent the authorities and the industry.

5.4.7.4 *Reduction in catch capacity*

Fisheries management was based on community quota allocation per species and per country, or "stock management" until 1988. Subsequently, the quota policy continued but the EU had also asked the MSs to reduce their catch capacity.

Notwithstanding the above, France opted to set up a system of fishery exploitation licences (*permis de mise en exploitation*) under the Act of July 3, 1991 via the Multi-Annual Guidance Programs. Perhaps, a subsequent decree had specified the criteria governing their issue. Finally, the licences were allocated to various types of vessels and the regions concerned. Concomitantly, a range of measures were introduced, in 2001, at the national level to reduce fleet capacity and fishing effort in order to meet the objectives of various plans.

5.4.7.5 *Specific case of the Mediterranean*

The Mediterranean fisheries are distinctive in terms of: (a) geography, (b) diversity and density of users of aquatic resources, (c) target species, and (d) absence of any TAC or quota systems (one exception being bluefin tuna).

Concomitant to the above, a forum for international cooperation on the management of fishery resources exists, which was the former General Fisheries Council for the Mediterranean, an advisory body within the FAO. The later had become known as the General Fisheries Commission for the Mediterranean (GFCM) when it took on wider responsibilities.

Concomitantly, France is a member of the GFCM (others include Spain, Italy, and Greece) and is responsible for budget, legal, and procedural issues. The European Community, on the other hand, is a member by virtue of its responsibility for resource management. In this context, France has, at the national level, since long had a special resource management system governed by decree 90–95 of January 25, 1990, which laid down the general requirements for the Mediterranean Sea fisheries. The system is part of a broader set of EU regulations aimed, in particular, at technical measures.

5.4.7.6 *Specific case of the FSAT*

The French Southern and Antarctic Territories (FSAT) were not covered by the EU CFP. The relevant sea fisheries regulations were based on the Act of June 18, 1966 on sea fishing. These instruments lay down the rules for resource management and more specifically TAC and the technical requirements governing fishing. The regulatory system also includes the measures adopted by France as a member of the Commission for the Conservation of Antarctic Living Marine Resources (CCAMLR).

5.4.7.7 *Industry participation*

The industry's participation and involvement in resource management are ensured, in particular, by the National Committee for Sea Fisheries and Aquaculture, a joint trade organization, representing all the stakeholders in the sector. It is mandatory to consult the National Committee over any national or EU measure regarding resource conservation/management, and the conditions are said to be applicable to commercial fisheries. Moreover, the regional and local fishery committees, for their part, provide the industry with technical assistance and information, and play an active role in drawing up measures taken at the national level. In addition, there are 39 local committees, regional committees, and one national committee. Their aim is to maintain a balance between resource management and economic activity in terms of not only profitability but also social, economic, and local development (e.g., jobs, land/sea structures). These industry-generated measures supplement the community standards.

Notwithstanding the above, Olivier and Olivier (2001) opined that RBM regimes are considered by economists as an important solution to the problems of excess capacity and biological overharvesting of fisheries. In practice, adoption of such regimes, particularly those relying on individual quota allocations, has often met with resistance from within the fisheries concerned. A key reason for this resistance appears to be the distributional conflicts which arise in the process of implementing the regimes. An economic analysis of the nature of these conflicts in the different contexts in which they had been observed is often proposed. The approach centers on the way in which distributional conflicts could influence the operation of management systems and their impacts on fisheries. Further, these may trickle down from the country to the individual firm level. As an illustration, an analysis of the economic processes at the firm level is developed based on the simulation of a fishery managed under ITQs.

5.4.7.8 Distributional issues, economic simulation

With reference to the above context, there are works related to EU concerted action on economics and the CFP, presented at various meetings, notably the "Workshop on the definition and allocation of use rights in European fisheries, Brest, France May 5–7, 1999."

There are statements like "if economics is to advice policy, it seems difficult to believe that the distributional issue could simply be swept aside." The issue of who is to receive the wealth associated with the fishery and the impact that this wealth could have on the regional and national economy, is intimately linked to the kind of management system adopted and the way in which it is implemented. Therefore, economics could, preferably, consider the distributional as well as the efficiency implications of policy options while formulating advice on the implementation of management systems' (Cunningham, 1994). Conventional economic approaches to **fisheries management**, built on the well-known model of free- and open-access renewable resources, are available. This model is centered on the identification of a level of efficient operation, defined as that which provides the maximum net returns to inputs in comparison to other economic sectors. Analysts tried to determine the levels of harvest at which economic efficiency will be achieved as well as identify policy measures needed to achieve such levels in the face of market failures. These failures originate from the lack of definition of rights to the resources, the resulting absence or imperfection of markets for them, and the externalities that this entails. Incidentally, a particular class of phenomena, which these studies were allowed to explain, is the race for a share of the catch in fisheries, where only total capture is being restricted. It is now well understood that such management measures result in inefficient outcomes, where "too many vessels chase too few fish." Thus, economists have suggested that fishing rights to resource flows be explicitly allocated to individual harvesters in order to restore efficiency in these open-access managed fisheries (Christy, 1973). Such an allocation tried to avoid the derby effect due to open access by ensuring a fixed share of the catch to harvesters. It could generate efficiency gains, given that harvest activities could be planned temporally to take advantage of the market demand; and in some cases, operating costs could be reduced in the absence of "race for fish" conditions. A further step toward economic rationalization of the fishery could be to allow the subsequent allocations of fishing rights to be determined by bargaining between economic agents on an open market (Moloney and Pearse, 1979; Scott, 1988; Neher et al., 1989; Arnason, 1991; Hannesson, 1991; Bjorndal and Munro, 1997). The implication could be that if agents could freely trade rights under conditions of perfect competition, these could end up in the hands of those who value them most, i.e., those who are most efficient in using them, whatever the initial allocation be (Moloney and Pearse, 1979).

Notwithstanding the above, the economists had been advocating the adoption of such management regimes for some time. However, the cases in which explicit allocation of rights to fish had been established as a basis for the regulation of capture fisheries remained limited in number. In practice, plans to restrict entry and allocate rights of catch to a limited number of participants often met with difficulties. As such, this could not be regarded as a good management strategy.

Concomitant to the above, Scott (1988) underlined the fact that commissions had been setup with powers to coordinate national fishery controls under various international treaties dating back to the beginning of the century. These commissions often implemented national quotas but the MSs had not distributed these as individual quotas but instead distributed to licensed vessels in order to participate in the international fishery (Scott, 1988). In addition, evaluations of the consequences of extended fisheries jurisdiction had emphasized that increased national control over fisheries did not lead to the expected improvements in terms of limited entry (FAO, 1993; Troadec, 1994; Hannesson, 1996; Bjorndal and Munro, 1997). Moreover, where catch allocations had been established, an extensive length of time was needed before the opposition to the allocation schemes could be overcome (McCay et al., 1995). The main reasons for these difficulties appear to be related to wealth distribution issues, as Macinko (1993) opined that any type of access control or alteration of an existing access control regime could entail a specific distribution of benefits advantaging some participants over others. This also pointed toward an **unmeritorious method of**

management. As such the resulting conflicts had been recognized by some authors as a major obstacle in the way of efficient management of fisheries (Johnson and Libecap, 1982; Karpoff, 1987; Scott, 1993; Christy, 1996). However, efforts had been made to interpret the nature of these conflicts in a number of general lines of analysis (Moloney and Pearse, 1979). Moreover, the conclusion rests on a number of analytical and practical assumptions. From an analytical point of view, the idea of a market for fishing rights, identical in principle to the market for a private good, is usually seen by economists as the main alternative to "command and control" regulation of fisheries or to intervention by a public agency based on less tractable taxing schemes.

Notwithstanding the above, there is a need for a number of conditions to be met, for a market to be efficient, including the divisibility and transferability of rights and the presence of a sufficient number of separate owners to ensure competitive behavior. Moreover, there could be no costs linked to the establishment, enforcement, and exchange of rights as well as no income effects related to trade in rights, for the initial allocation to remain neutral. Further, from a practical perspective, it had been shown that the implementation of such rights-based regimes may involve certain difficulties (e.g., monitoring and enforcement problems) and **spillover effects** (e.g., bycatch and discards or redirections of fishing effort to other fisheries), which may undermine the overall benefits expected from these regimes (Copes, 1986, 1997a, b).

Notwithstanding the above, the first line of analysis is in terms of equity and accepted rules of fairness. For instance, the fishery participants preferred regulations in the allocation of access rights (Scott, 1993). One reason, which is usually advanced, to explain this is that participants consider that it gives equal opportunity to every participant to use his skills and this mechanism may reward the most skilful, which they consider is fair (Matthiasson, 1992; Karpoff, 1989).

The second line of analysis is in terms of rent-seeking behavior (Edwards, 1994). For example, authors have shown that the ability to make decisions on wealth distribution could be inversely related to the degree of vested interests in the fishery. Thus, according to Christy (1996), it is not surprising that countries with the most advanced catch allocation regimes are those where the fisheries are the youngest (New Zealand and Australia), while in areas where the history of fisheries is the longest, the catch allocation regime may not be too advanced, but no doubt well established (Northeast Atlantic).

Nevertheless, whatever the interpretation is, it is widely accepted that distributional issues have a strong potential to affect management systems and, in particular, those which setout to resolve the overcapitalization problem. Incidentally, there have been efforts to show how economic analysis could help in understanding the implications of this for the operation of fisheries management systems. The works often consist of different parts, notably,

(a) One part may review the existing literature on distributional issues in fisheries management and develop a framework for the interpretation of these issues from an economic perspective, followed by examination of some of the empirical evidence(s) in support of this framework at the international, national, and fleet and firm levels.

(b) The other part may try to build on the empirical evidence(s) at fleet and firm level, and develop a simple model of quota trading in order to simulate the effects of various distributional scenarios on the management system.

Under the above conditions of complicacies and uncertainties in the management scenario of aquatic regimes and their resources, it had often been contemplated that why not there could be an efficient management system to take care of the management and conservation of the aquatic regime and its resources involving the stakeholders from the grassroot level. Also, while perusing and striving upon such an aspect, **CBFM**, in a **participatory mode**, has been adjudged as one of the most effective *modus operandi* to achieve the goal of efficient and sustained management and conservation of the water bodies and their resources.

5.4.8 Germany

Herring forms a significant fishery in the Baltic Sea. Also, trawls are much operated in the region, thus, posing keen competition among the different groups and types of fishermen. Incidentally, the Griefswalder Bodden at the **German** Baltic Sea coast is one of the principal spawning areas of the western Baltic herring (*Clupea harengus membras* L.) stocks. Prior to 1990, both trawlers and boats with static gear had been fishing there during the spawning season. After Germany's reunion, only the coastal fishermen with static gear continued fishing for herring because for bigger vessels only fishing for cod (*Gadus morhua* L.) was economically feasible. Eventually, the declining cod stocks and a recently built processing plant for herring on Ruegen Island had changed the situation. In 2002, after 10 years of absence, trawl fishermen resumed herring fishing, as revealed by the studies of Döring et al. (2003). This, consequently, led to competition between small-scale gear operators and the trawl operators. In fact, the coastal fishermen depending on local fish stocks are more conversant and aware of the moderate exploitation methods. As such, the position of the coastal fishery with their relatively smaller scale exploitation will be strengthened, mainly because of its importance to the increasing tourist and the tight job market as well as because of less negative impacts on the ecosystem. In addition, a recovery plan for cod had become necessary in order to avoid further competition with the trawl fishermen. Incidentally, there had been efforts to deal with the present situation by means of a **community-based fishing** model. However, there could be various practical problems to be considered, although, in theory, different solutions seem to be easy to realize.

Notwithstanding the above, fishing has a long tradition around Ruegen Island off the **German** Baltic coast. In fact, the wealth of this region depended on the herring fishery during the Middle Ages and was one of the main reasons for the foundation of the famous trade organization in the Baltic, the **Hanseatic League**. Subsequently, after World War II, the herring fishery was shared out between the fleet using pelagic trawl nets and the coastal fishermen who used static gears. Incidentally, both the groups could obtain a very good income in comparison to other jobs mainly because of the subsidized prices of fish. However, the situation changed after the German reunification in 1990. Fishermen with static gear continued with the herring fishery, mainly because of the cancelled subsidies. Also, the fishermen with pelagic trawls switched over to fishing for cod.

Concomitant to the above, a new fish processing plant in Neu-Mukran on Ruegen Island began filleting herring, cod, and flounder (*Platichthys flesus* L.) during the spring of 2003. Thus, fishermen using pelagic trawls were able to earn their livelihood from herring once again and in 2002, they fished for herring for the first time since 1990, but it is doubtful if this could provide a long-term perspective. In fact, lower catch quotas for herring may lead to increasing competition between mobile and static gear fishermen.

Notwithstanding the above, the coastal fishery is considered as very significant for the whole area. On the one hand, the fishery offers jobs, while, on the other hand, the stocks are generally exploited in a moderate way, being derived from centuries of tradition. In addition, the local fishermen with their small cutters and their fresh fish products are a tourist attraction. All these factors, taken together, make it necessary to preserve the small-scale fishing sector. However, one solution might be to allot the fishing rights exclusively to the small-scale fishers so as to improve their socioeconomic situation as well as to secure the future of the moderate exploitation system.

Concomitant to the above, there have been contributions to outline a sustainable approach to the exploitation of fish stocks in the FW, marine, and coastal regions. As an example, the Baltic coastal region could be cited. Such accounts, generally, begin with a historical description, before offering an explanation for the increasing competition, and then look for a possible model for resolving the competition on a sustainable basis. There had been also plans for the application of efficient techniques, like **CBFM** for the effective management of the fisheries and water bodies in a **participatory** mode.

5.4.8.1 The history of the fishery in NE Germany

Herring fishing around the island of Ruegen was mentioned, for the first time, in a chronicle about the Slavs written in the 12th century (Peesch, 1961; Klinkhardt, 1996). The fishery played a significant role in the earlier times. When the **Germans** arrived in the area during the 13th century, the villagers became "vassals." Property, in terms of both land and fishing rights, were in the hands of the landowners. The number of fishermen was also limited, and only when an older fisherman retired could a younger man gain access to the fishery. Thus, the number of fishermen remained constant for centuries. Later, the local people were able to buy their own land as well as fishing rights with the end of bondage in 1806. However, the trap nets had to be leased from the government and were, therefore, limited (Fraude, 1925). Further, the local fishermen formed so-called "Kommünen" in order to cope with the big trap nets during the herring season and it consisted of 4–5 members. Subsequently, the coastal fish stocks were inevitably exploited more intensively, as the population increased at the end of the 19th century, and thus fishing became a full-time occupation for some fishermen.

In continuation of the above, the next social change occurred after the World War II, when the coast between Swinouscje and Lübeck became part of the German Democratic Republic (GDR). Subsequently, nothing changed in the distribution of fishing rights, and the fishermen worked as private businessmen. However, there was a bigger demand for fishing opportunities in view of settlement of many fishermen from former parts of Germany, such as East Pommerania and East Prussia. However, the government decided to allocate new zones for trap nets, as these new immigrants were not admitted to the existing "Kommünen," and in 1954 some of the fishers organized themselves into the first **fishery cooperative society**. Later, more and more fishermen were compelled to join this new type of cooperative society. The public authorities allowed trade only through these cooperative societies, and only then the organizations could continue to fish commercially. In this connection, it may be mentioned that in those days, the fishermen had above-average incomes profiting from guaranteed fish prices. The herring catches have risen to *c.* 40–

50,000 tons per year in 1988–89 (Kruse, 1991), a quantity that was divided approximately 50:50 between the coastal and trawl fishers. However, some larger trawlers with onboard fish processing facilities stayed in coastal waters after the closure of some fishing grounds, but after the establishment of the EEZ in 1977 they were able to transfer the catch from the trap nets and gill nets to the processing chain directly. This situation lasted until Germany's reunion in 1989–90, when landings from coastal waters amounted to *c.* 66,000 tons per year, a figure including both fresh and saltwater fish.

Concomitant to the above, the economic framework was altered substantially due to the political upheaval in 1989. However, the allocation of the fish and fishing rights had remained the same. The maximum number of gill nets was fixed and the sites for the trap nets were leased by the government each year. Further, with the end of the GDR's economic system, guaranteed prices were abandoned and most of the primary processing plants were closed, leaving only some of the canning factories in business. Moreover, the fishermen had to adjust to the new circumstances of free enterprise. Some of the fish species could be sold for high prices like the cod and sole (*Solea solea* L.) but for others such as herring and sprat (*Sprattus sprattus* L.) prices fell sharply. Moreover, for some species, there was no market at all.

With reference to the above, the fishermen were little hesitant about investing in new boats, after the reunion, mainly because of uncertainties over future fishing opportunities (Scharmann, 1994).

Incidentally, the cooperatives survived and remained responsible for setting the catch. However, negotiations over bank credits were difficult because of the short-term perspective with the quota system. None of the fishermen was in a position to really forecast how much he might be able to catch in the following years and how the prices will develop. Incidentally, of late, the cod quota in the Baltic declined. This further depressed the profit margins. Moreover, switching to herring was not an alternative because of the very low prices, especially, for mixed catches. Eventually, the trawl fishermen had to concentrate on fishing for cod and flounder, while the coastal fishermen with their smaller vessels had to continue with the herring fishery, as they had no other alternatives. In addition, the new framework regulations for the CFP, which were adopted in December 2002, and the end of subsidies for investment in 2004 added to the fear of bankruptcy among the fishermen.

5.4.8.2 *Fisheries management*

Fisheries management is an important aspect in the aquatic domain. Different *modus operandi* have to be adopted for fisheries management in different scenarios. Some amount of light could be thrown here on the management of Baltic Sea fisheries as an instance of fisheries management in an overall situation.

With reference to the above, one of the most controversial issues in fisheries management concerns the nature of fishing rights. On the one hand, there are individual exploitation rights (ITQs as in New Zealand and Iceland) and on the other hand, the so-called **CBM**, in which a group of members holds the fishing rights. Both the systems secure an efficient allocation of fishing rights and quotas.

Notwithstanding the above, no official management of fishing practice in the coastal waters of EU MSs was necessary before the introduction of the 200-mile EEZ. But, with the declaration, in the late 1970s, the fishing sector became a major political field for the governments, owing to the responsibility of the Council of Ministers for the whole of EU waters. However, the 12-mile inshore zones are under the jurisdiction of the individual MSs today. The EU is solely responsible for fisheries management outside this zone. Moreover, the EU Commission developed a CFP with the introduction of the EEZ and also had established a special Directorate General (DG) for fisheries. Incidentally, the main instruments for management are the catch quotas and technical conservation measures (Holden, 1994).

Concomitant to the above, based on the advice of the International Council for the Exploration of the Sea (ICES) and Scientific, Technical, and Economic Committee on Fisheries of the EU (STECF), the Commission develops proposals each year for TACs for each major species, to be decided by the Council of Ministers at the Council meeting. Moreover, the agreed TACs are allocated to the MSs by means of fixed percentage shares, agreed in 1982, which formed the basis of the principle of relative stability. Also, different systems of allotment to individual fishermen are *in vogue* within the MSs. In Germany, however, the cooperatives receive a share of the quota and the same are re-allocated to their members. The advantage of such a procedure is that the fishermen receive quotas of species which they can catch cost effectively, taking account of the fact that some fishermen are able to move further offshore with their boats while others are not. These quotas are also transferable among the members of the cooperative society in order to achieve a better balance between fishing opportunities and fishing capacity.

Notwithstanding the above, the problem of present fisheries management lies not in the allocation of catch quotas within a country but the decision taken at a higher level over the size of the TAC. However, an acute example was the cod crisis in the North Sea and ICES' recommendation for a zero quota in 2003, followed by a decision to reduce the quota by *c.* 45% which was not enough to secure the stock's recovery.

In continuation of the above, together with the latest CFP-framework regulations, the Council of Ministers introduced some new instruments that may eventually lead to a better stock management, including multi-annual management plans for all major commercial species.

Notwithstanding the above, studies on CBM **systems** suggest that the coastal fishermen (depending exclusively on local fish stocks) are much more aware of the need not to overfish the stocks. However, it may be noted that the coastal fishermen of Germany, for example, did not exhaust the cod quota in 2002, as they knew that the stock was at low ebb.

5.4.8.3 *The competition between small-scale and trawler fishermen*

It may be noted that most of the fish for German canning factories comes from Scotland. This may change after the new processing plant starts production, when the herring fishery could again be profitable for bigger vessels. Accordingly, a trial fishery was made in 2002 to find out whether the resumption of herring fishing would be economically feasible and viable.

Fishing gears are the fish catching devices which are of variegated types in different parts of the world. Incidentally, the gill nets had bad consequences for most fisheries all over the world. Therefore, it is difficult to demand for the use of gill nets because of ecological reasons (Lien, 1996: 219 pp.; Kar, 2013).

Notwithstanding the above, the greatest advantage of **CBFM** is the conservation of the aquatic domain along with its resources by experienced fishermen who possess considerable knowledge not only of the local circumstances but also of when, where, and how to fish for different species with the best available technique. However, the number of fishermen in a cooperative society is a limiting factor for the fishing capacity in the region, mainly because of their limited fishing range. For any particular area, there are a specific number of fishermen who are able to make their living from the existing fish stocks. We could call such a group as a "production club."

Concomitant to the above, Buchanan (1965) called "goods," which are neither private nor common, as "club goods" and the benefit of these goods is restricted to the club members only. However, all benefits will eventually be dissipated, if more and more users are allowed to join the club. Further, if membership of the group is opened, in such a case, a social planner (which is often assumed in theoretical models) has to reduce the number of members to the number that will take only the MEY. On the other hand, we could theoretically define the optimum number of club members, if we had sufficient information about the extent of the fish stocks. This optimum club size could be automatically achieved by a self-regulated system like the old community-based cooperatives. In addition, the coastal fishermen are not able to augment the fishing effort at will mainly because they use the fishing gears which depend on the quantity of fish, and hence only a certain number of gears could be used on any one day. From these arguments, we may preliminarily conclude that the essential elements of CBM systems may be retained in the best interest of both the fishing communities and the fish stocks.

5.4.9 Hungary

Hungary has a rich potential in fisheries. There are only *c.* 4% of the surface waters borne within the country. The majority of the incoming waters are, however, drained through the three large rivers, viz., the Danube, Tisza, and Drava. However, the per capita amount of total water drained (*c.* 120 billion m^3/year) is the highest in Europe, mainly due to its basin-like property.

5.4.9.1 *General description of the fishery sector in Hungary*

An analysis of the national and local circumstances

Hungary is one of the AQC and fishery-rich countries in **Europe**. However, the most commonly used indicator is the contribution of the fisheries and AQC sector to the GDP of the country in order to estimate the significance of the sector in the national economy (Anon, 2008a, b).

As a gross statistics, Hungarian agriculture sector contributed *c.* 3.0% to GDP, *c.* 6.1% to the export, *c.* 4.6% to the investments, and *c.* 5.0% to the employment in 2005. Hungarian animal husbandry, crop production, agricultural services, and the nonagricultural secondary activities accounted for *c.* 36.0%, 54.2%, 6.9%, and 2.9% of the gross production value of agriculture.

Currently, the gross value of the Hungarian fisheries sector is *c.* 2.5% of the gross production value of animal husbandry and *c.* 1% of the total Hungarian agricultural production. Further, in 2006, the total value of the table size category products in the Hungarian fishery sector was *c.* 9.5 billion HUF (*c.* 37.3M EUR); AQC: *c.* 6 billion HUF (*c.* 23.5M EUR); and natural water fishery: *c.* 3.5 billion HUF (*c.* 13.7M EUR), which contributed *c.* 0.04% to the GDP. However, the importance of the fishery sector was much higher than that shown by the simple data, as the existence and economic performance of several supplying and servicing sectors and also the entire recreational fishery and angling were based on this sector. The average income level of the employees of this sector did not reach *c.* 75% of the national average in 2005. Moreover, both the volume and the value of the Hungarian fish and fish product import were significantly higher than the export (in volume, it was *c.* 14 times; in value, it was *c.* 13 times higher in 2006).

Notwithstanding the above, the climatic and hydrological conditions in **Hungary** are favorable for fish production in fishponds, wherever the classical fishpond technologies are applied. Incidentally, *c.* 68% of the Hungarian fish production come from the fishponds, where mainly common carp, herbivorous fish species, like the bighead carp, silver carp and the grass carp, and some predator fish (wels, pike-perch, pike) were produced. Moreover, the total fish production and catch from natural waters (fishpond production + intensive system production + fishing in natural waters + recreational purposes) in Hungary was *c.* 30,363 tons in 2006, out of which market size products (for human consumption) added up to *c.* 21,859 tons (Anon, 2008a, b).

However, the fishery sector could become less attractive for the financial investors mainly due to the income disparity. Moreover, the fisheries and AQC sector could be a low profitability sector in comparison to other sectors of the national economy, mainly because of the uncertainty of the catches. Nevertheless, the Hungarian fisheries production meets the FW fish demand of its population when the weather conditions permit it. In fact, the common carp, bighead carp, silver carp, and the African catfish all have a significant role. However, the sector definitely needs import of marine fishery products in order to meet the consumers' expectations and to provide a wide range of fish products. Further, in the market of domestic fisheries products, the situation concerning food safety is almost the same as in the case of more developed western nations of the EU.

Concomitantly, due to the liberalization of international trade, the import was increasing; and, at the same time, the related control activities put a burden on the Hungarian food safety system.

5.4.9.2 *Aquaculture (fishpond and intensive production)*

Hungary is a country with much AQC practices. In Hungary, there are two main fields of AQC, viz. (a) pondfish production and (b) intensive fish production. These two farming methods are with each other in a number of ways. Hence, both the activities could be carried out by the same company. Nevertheless, one of the significant differences is that while one of them is exposed to weather conditions, the other enables programmed production to a full extent. On the other hand, if we consider the aspects of environmental sustainability, pondfish production could obviously be a very much environmentally friendly method of production, while in the case of intensive fish farming, the organic load of effluents could be extremely high. Therefore, cleaning of water and repeated use of water is of great importance in the case of intensive fish production. Nevertheless, the utilization of geothermal energy is significant in this area.

5.4.9.3 *Pondfish production*

According to the statistical data, there were *c.* 361 fishpond farms (*c.* 263 enterprises) in operation, during 2006, in Hungary. Based on the reliable sources, the total area used by the sector was *c.* 34,000 ha. According to the data provided, the total fishpond surface was *c.* 26,248 ha, out of which *c.* 23,878 ha was in operation. The difference comes from the area of those fishponds which were kept dry for reconstruction works. The area of the fishponds, which fall under the scope of Natura 2000, is *c.* 15,615 ha and the total area of fishponds included in the **program for extensive fishponds** is *c.* 23,194 ha (578 fishponds).

Concomitantly, *c.* 68% of the fishponds operating in Hungary produced market size fish for human consumption. Moreover, *c.* 17% of the fishponds produce "two-summer-old" fish for restocking, *c.* 10% had produced fingerlings ("one-summer-old" fish), and the remaining *c.* 5% for other purposes. According to the survey of the Hungarian Fish Producers' Association and Product Board (HFFAPB), *c.* 40% of the fishponds are old, and in principle, modernization or reconstruction was required.

Concomitant to the above, the total amount of fish harvested from the fishponds, of late, was *c.* 20,762 tons, of which the amount of fish produced as market size fish was *c.* 12,898 tons. Concomitantly, the amount of fish sold directly from fishpond farms by angling was *c.* 355 tons.

Notwithstanding the above, economically, the vast majority of fish produced as market fish in Hungary is the common carp. It represents *c.* 61% of the total annual catch and production. Its role is dominant not only in pondfish farms but also in the fish farming activities in water reservoirs, oxbow lakes, and other natural waters. Concomitantly, the second most important group of fish, after the common carp, is the group of herbivorous fish species with a proportion of *c.* 15%; the share of predators being *c.* 13% and the remaining *c.* 11% are the breams and other low market value species. Notwithstanding the above, the amount of land areas suitable for pondfish production was quite high mainly due to the favorable geographical and climatic conditions. The investments in the sector increased compared to previous years, when the support possibilities for developments played a very important role. This was due to the favorable geographical and climatic conditions.

5.4.9.4 Intensive production

There were 10 **intensive fish farms**, of late, in Hungary, the majority of which had utilized geothermal energy. Concomitantly, the amount of fish produced in intensive systems was *c.* 2081 tons, out of which *c.* 1789 tons were produced as market size fish for human consumption.

In this regard, it may be mentioned that the total production of the intensive systems had been continuously increasing in the last few years. Thus, when comparing the year 2006 to 2003, the total production volume increased by *c.* 67% in 2006 while the output of these increased by *c.* 22% in 1 year (between 2005 and 2006), which also indicates a fast growth in the demand. Incidentally, the most dominant fish species bred in Hungarian intensive fish production was the African catfish, the market introduction of which was a success story in the Hungarian fishery sector. The producers sold *c.* 75% more in 2006 compared to 2003. Incidentally, the African catfish constituted *c.* 96% of the total volume of fish produced in intensive systems, and this ratio was expected to remain similar because the other species were produced in the fraction of this output.

5.4.9.5 Inland fisheries

Inland fisheries is said to be a potential sector for the production of fishes through the utilization of natural waters. According to the data from 2006, catches were reported from *c.* 133,948 ha of water surface, of which *c.* 62,841 ha were covered by Lake Balaton and small-Balaton reservoir. It is categorized into two types: (i) commercial fisheries and (ii) recreational fisheries.

The bigger segment of recreational fisheries is angling (> 320 thousand persons) and the smaller segment is the so-called "traditional-gear fisheries" (*c.* 3000 persons). Incidentally, in 2006, the total catch of inland fisheries was *c.* 7540 tons, of which *c.* 7172 tons was table-sized fish. On the other hand, commercial fishing captured *c.* 2806 tons of fish in 2006, which could be *c.* 37% of the total catch. Further, *c.* 41% of this was common carp, *c.* 29% herbivorous fish, *c.* 9% predators, and *c.* 21% other species. Moreover, in 2006, *c.* 4744 t, i.e., *c.* 63% of the total catch had originated from recreational fishing (mainly angling).

Notwithstanding the above, in **Hungary**, commercial fishing was done in reservoirs, in addition to other kinds of water bodies. Fishing was done partly by using traditional fishing techniques and partly by applying certain techniques of fishpond farming. In addition, fishing was sometimes also done in the oxbow lakes.

Concomitant to the above, the maintenance of capture fishery in Hungary was well founded, which was based on the traditions unique in Europe. In fact, the capture fishery always had an outstanding role in the utilization of natural resources in the aquatic ecosystems and with its uniqueness and culture, it was able to apply a systematic and the so-called **sound management approach** in the utilization of natural resources. In fact, reasonable natural water fishery management is the common task and responsibility of anglers and the fishery sector.

5.4.9.6 Fish consumption and processing

"Fish" is an item of staple food in many parts of the globe. In Hungary, the supply of fish and fish products per capita displayed a slow but steady increase. In 2006, fish consumption per capita was *c.* 396 kg/year, which was *c.* 5% of the total meat consumption. This amount included both the domestically produced and the imported products. The consumption of carps was *c.* 1.5% of the total annual consumption, and the consumption of African catfish was *c.* 0.17% of the domestic production. These were consumed mainly in the form of fresh fish, while the imported fish was consumed in processed form (canned, frozen ready-to-cook, frozen fillet, pickled, smoked, etc.).

Notwithstanding the above, **Wetlands** are precious abiotic resources that serve as potential habitats for the coveted aquatic biota, notably the fishes, wetland birds, AM, etc. They also play a prominent role in the biogeochemical cycles in nature. Under this significant background of the **wetlands**, their transformation into areas of other agricultural activities as well as misuse of the different kinds of **wetlands** including fish farming activities had been causing significant damages; notably, a depletion of the number of species in its BD resource. On the other hand the ecological value of Hungarian fishponds is outstanding, both on domestic and on European level. In fact, these fishponds have a core position in the European ecological network.

Notwithstanding the above, the population of nesting and migratory bird species is also of great importance on a European scale. In fact, the number of avian species observed in these fishponds sometimes exceed 300, which is said to be > 80% of the bird species occurring in Hungary. Concomitantly, the number of nesting bird species in the fishponds could be > 100. Incidentally, it may be noted that a very significant portion of the population of waders and water-linked bird species nest, feed, or rest in the fishpond areas. Thus, it could be remarkable not only from the point of view of number but also from the point of view of population size.

In addition to birds, population of the otter (*Lutra lutra*) had also expanded. Incidentally, the results of the population survey of the "Association for the Otters" depicted that *c.* 60% of the Hungarian otter population, which was outstanding even on a European scale, live in the fishponds. It could be mainly because they find optimal environmental condition and living situation (feeding, resting, and hiding places), which are inevitable for population establishment and its increase. In addition, the fishponds also play their role as habitats of amphibians and reptiles, both consisting of a number of endangered species.

5.4.9.7 *Ramsar convention and its links with Hungarian fisheries sector*

Hungary is a signatory to the **Ramsar Convention**. Since 2003, Hungary was designated 19 areas in the Ramsar list covering *c.* 154,147 ha.

Notwithstanding the above, there had been some amount of thoughtful exercises done on the aspect of management and conservation of the water bodies and the fishes with the application of efficient *modus operandi*. Also, in this direction, **CBFM** in a **participatory** mode was considered as an effective method for the management of water bodies and their bioresources, notably, the fishes.

5.4.9.8 *EU water framework directive (WFD)*

AQC and, in particular, fishpond farming are closely interconnected with the programs for the development of the environment and rural areas, thus ensuring the maintenance of Natura 2000 network, the fulfilment of BD-related obligations (Gothenburg), and fulfilment of the WFD and the Kyoto Protocol requirements. The measures related to agricultural environment protection and farming-related payments contribute to the achievement of the Gothenburg requirements concerning the reversing of the decline in BD by 2010. With regard to BD, maintenance of the extensive farming systems is very important in the conservation of habitats of the protected or endangered bird and mammal species.

Concomitantly, Hungarian fishery sector interlocks with the WFD in two ways:

(a) through the water quality of rivers;
(b) through the requirements of agricultural water management.

These waters could also be used in fishery and angling while they serve as a habitat for valuable flora and protected fauna. These are the expectations of the WFD, which pays special attention to the ecological state and the potential of these water bodies. Incidentally, significant impacts of human activities on the environment are also to be analyzed in connection with these issues. The legal base for the designation of the Natura 2000 areas was formed by the Hungarian Parliament with the modification of the Law on nature conservation, which came into force on July 16, 2004. This modification entitled the government to issue its decree on the Natura 2000 network.

5.4.9.9 *Gender mainstreaming*

However, it could be pointed out, in connection with this matter, that fishery and fish production in the traditional sense have been based on employment of men for many thousands years. Nowadays there is a possibility of increasing the level of employment of women in intensive fish production and processing beyond fisheries-related administration. This incentive could be made visible even in the application process but extra attention should also to be paid during implementation and monitoring.

In fine, the main strategic vision of the Hungarian fisheries sector is as follows:

Competitive as well as economically, socially, and environmentally sustainable and modern fisheries sector via the modernization of production installations and the construction of new facilities in AQC and processing.

Notwithstanding the above, "sustainability" is an important issue in the management and conservation of the fish resources. Sustainable development is an idea, or perhaps an ideal, as much grounded in hope as in reality (Peter and Arico, 2002).

Fishes are potential sources of nutrition and avocation. Hungary seemed to be doing well in the field of fishery and AQC. Experts including scientists from 19 countries of Central and Eastern Europe, i.e., Albania, Belarus, Bosnia and Herzegovina, Bulgaria, Croatia, the Czech Republic, Estonia, Hungary, Latvia, Lithuania, Moldova, FYR Macedonia, Poland, Romania, Serbia and Montenegro, Slovakia, Slovenia, and Ukraine were invited to provide a survey and analysis to evaluate the major underlining causes of FW fishery-related BD decline in these countries in transition. The studies (Robert et al., 2004) against a background of a rich and diverse FW heritage and sound ecological science showed that, of late, there is a serious situation in relation to the ecological sustainability of FW fisheries in the CEE countries, with similar trends in nearly all of them. FW commercial fishing capacity and the number of recreational fishers in the surveyed countries were making a situation where the pressures on fish stocks and associated BD were excessive. The position is, generally, adversely affected by inefficient controls resulting, at least in some cases, in high illegal and unreported catches, and unlawful trade and movement of fish. In general, BD of FW fishery ecosystems is threatened by overfishing, habitat destruction, pollution, and mismanagement of aquatic resources resulting from poor integration of interests of the stakeholders. Consequently, the most urgent measure for all the surveyed countries may be the significant improvement in the efficiency of control regimes, by using the ecosystem-based approach as defined by the CBD. The education, training, and awareness raising of the FW fishery stakeholders on BD and sustainable use issues could address the generally rather poor compliance of the stakeholders with the relevant conservation measures. Taking into account the trend of a gradual decrease in numbers of professional fishers and increase in numbers of sport and recreational fishers, the latter could be a priority target group for IUCN cooperation. Recreational fishers in effective organizations and with their awareness sufficiently raised could be the force who could change the practices currently leading to overexploitation of FW fish diversity in Central and Eastern Europe. The report suggested that developing a partnership between BD conservationists and fishers could yield tangible results toward effective conservation by opening up constructive contacts between IUCN European Sustainable Use Specialists Group (ESUSG) and the IUCN Program Office for Central Europe on the one hand and the European Anglers Alliance (EAA) and the European Inland Fisheries Advisory Commission of Food and Agriculture Organization (FAO) of the United Nations (UN) (EIFAC of the FAO/UN) on the other.

Notwithstanding the above, the **Hungarian** BD Monitoring System (HBMS) (Anon, 2001a, b) is a national program for monitoring the biological diversity in Hungary, supervised by the Authority for Nature Conservation of the Ministry of Environment. Since 1994, when the CBD was ratified in Hungary, and published in Act no. LXXXI of 1995, it was a legal obligation to monitor biological diversity for nature conservation purposes. Subsequently, data collection started after elaborating the methodology of the monitoring system.

Concomitant to the above, it could be said that the diversity of living organisms is a basic biological feature, which in the long term sustains life. Moreover, biological diversity enables, through natural selection, the adaptation to the continuously changing environment, which results in the relative stability of living systems. Biological diversity varies in time and space. In evolutionary and palaeontological perspective, species have always evolved and become extinct, just as today. The number of species, based on *c.* 1.7 million known today, could be estimated anywhere between 10 and 100 million. As a result of human activities (agriculture, forestry, fisheries, urbanization, industry, transport, tourism, energy production, mining, etc.) and due to the exponential growth and enhancement of human population, wild life have become endangered today. Moreover, the fast extinction of species has drawn attention to the vulnerability of the biosphere. The rate of extinction of species has multiplied in today's scenario as compared to earlier days. Moreover, the injudicious exploitation of natural resources has led to the fragmentation and loss of habitats, expansion of invasive species, overexploitation of animal and plant species, pollution of water, soil, and air, and furthermore, led to global climate change. All these all contribute to the loss of biological diversity. Concomitantly, nature provides those renewable resources and services that are used every day by humans (e.g., soil formation, medicine from medicinal plants, etc.). Further, the damage to nature also threatens our own life support system.

In this connection, an international convention was signed in 1992, called the Convention of Biological Diversity (CBD), at Rio de Janeiro in Brazil. Hungary was among the first signatories to sign the CBD, and also had ratified it in 1994. One of the basic commitments of the parties is that a national strategy and supporting legislation are to be framed, which would enable the conservation of biological diversity and the sustainable use of its certain components. To fulfil these requirements, it is vital to know and monitor the state and changes of the biota and this knowledge could be based on continuous, repeated observations.

Further, biological diversity also includes the infraspecific variability of species and the diversity of associations of living organisms, which are beyond the diversity of species. In addition, BD monitoring implies the long-term observation of certain characteristics of selected species and communities. Further, the aim of monitoring could also be to study the effects of certain known or predictable environmental changes on ecosystem, notably, lowering of the water table level, climate change, etc. Taking into account the extremely large number of species and many habitats, it simply may not be feasible to monitor everything everywhere.

Notwithstanding the above, the following principles were taken into consideration when designing the Hungarian monitoring activities:

(a) Monitoring of the state of protected and threatened natural values.
(b) Observation of the indicators regarding the general state of the ecosystems in the country.
(c) Study of direct or indirect effects of certain human intervention(s) or environmental factor(s).

5.4.9.10 *The Hungarian biodiversity monitoring system*

The Hungarian Biodiversity Monitoring System (HBMS) is a national program for observing biological diversity in Hungary, supervised by the Authority for Nature Conservation of the Ministry of Environment.

A monitoring program was launched by the Authority for Nature Conservation in 1996 with the support of the PHARE program of the EU involving several research institutes. This was done in compliance with the National Nature Conservation Concept, 1994 (Bonnie Sutherland, 1997) and in line with the **National Biodiversity Strategy**.

As a result, a proposal for the **Hungarian National Biodiversity Monitoring Program** was developed, which outlines the theoretical background and the methodology of monitoring of selected species and communities. The results were published in a series of 10 volumes in 1997, and an additional volume was produced in 2001. Incidentally, the system, originally proposed in 1997, was running with an increasing number of component projects; and was named as the HBMS. The service has also welcomed the contribution of institutes, universities, or nongovernmental organizations (NGOs) that could assist the Central Coordination Unit or the regional teams. The program had also proposed to establish a BD Monitoring Service that could provide guidance and administrative management and also carry out sampling within the state nature conservation organizations. The service also played a significant role in raising the awareness of the general public. Moreover, the conservation of the Hungarian natural heritage, which is still outstanding at the European level, needs a concerted response from the society. In fact, conservation goals may not be achieved, without the support of the local community stakeholders and the public administration. Moreover, the monitoring and long-term observation of BD is not only a requirement of the CBD but also a priority task in Europe. In fact, the nature conservation *acquis communautaire* of the EU, including the so-called Wild Birds and the Habitats Directives, could also be implemented in Hungary, upon accession of Hungary to the EU.

Concomitant to the above, it may be noted that the monitoring program provide the theoretical background, the selection process and the sampling methods, being based on the expert advice of many ecologists. As a novelty, even for professionals, the program described all the habitat types of Hungary, and now this classification system (the Hungarian National Habitat Classification System) provide the basic units for habitat mapping.

The projects were formulated by the definition of the objectives and the exact descriptions of the tasks are as follows:

(i) Monitoring of protected and threatened plant and animal species.
(ii) Monitoring of the wetlands and other aquatic habitats and their communities.
(iii) Monitoring of habitat types in Hungary.
(iv) Monitoring of populations of invasive plant and animal species.
(v) Monitoring of selected sites of the Hungarian Forest Reserve Network.
(vi) Regional monitoring of the biota of the Kis-Balaton wetlands.
(vii) Regional monitoring of the Szigetköz wetlands.
(viii) Monitoring of salt-affected habitat types.
(ix) Monitoring of dry grasslands.
(x) Monitoring of mountain hay meadows.

Concomitant to the above, the observation of bird species is one of the most popular nature conservation activities with the longest tradition and the largest scientific databases, especially in Europe. Concomitantly, it is one of the most rapidly expanding nature conservation activities. Further, the monitoring of birds is carried out at the population level of selected species, and also at the community level. Moreover, the Hungarian Ornithological and Nature Conservation Society can also meet the requirements of the HBMS.

Notwithstanding the above, the otter (*Lutra lutra*) is protected almost all over the whole range of its occurrence. In fact, its population became fragmented or extinct almost throughout Europe; although a recent recovery is apparent. Survival of the species is threatened by habitat loss, water pollution, and poaching. The standard sampling methods used by the Hungarian Ornithological and Nature Conservation Society could also meet the requirements of the HBMS. Moreover, the Department of Ethology of the Eötvös Loránd University, Budapest had played an active role in designing methodologies for the studies.

In addition to the above, monitoring had also revealed the occurrence of *c.* 14 species of amphibians and *c.* 12 species of reptiles from Hungary, all of which are protected.

Notwithstanding the above, investigation of the aquatic communities needs special expertise in hydrobiology. In fact, water itself is highly variable as a habitat, the sampling method for aquatic species and communities have to be adapted to its main features as well. For this reason, no single standard sampling methodology could be chosen, given the diverse abiotic and biotic features of the bewildering varieties of lentic and lotic habitats, from small to big size. In fact, the methodologies may have to be frequently edited in monitoring of wetlands according to requirements and experiences. Moreover, macroscopic invertebrates were used in biological water qualification for many years in Hungary, which included the investigation of several taxa, including, perhaps, insects and molluscs.

Concomitant to the above, the most commonly used method is the population size estimation, which could be done by counting or by estimating the cover. If necessary, the population may also be mapped. Further, the native plant associations were first surveyed and described in Hungary during the first half of the 20th century, based on a classification system according to Zürich-Montpellier school of phyto-sociology. Further, the highly developed state of vegetation science was accompanied by an advanced, widespread technique, the so-called **coenological** sampling method. This method is based on estimating the cover of the species in the sampling squares.

In continuation of the above, a considerable part of the monitoring efforts focused on the area of the Danube-Dráva National Park. The project was established by a team of experts in harmony with the methodology of the HBMS, during 1999, and in 2000 the field monitoring had commenced. The habitats were characterized by exceptional diversity and variability, which posed a challenge for the experts taking part in the monitoring. The examined components were plant associations, protected plant species, molluscs, ZP, dragonflies, mayflies, butterflies, large moths, fishes, amphibians, birds, bats, small mammals, and small carnivores. Further, the data generated during the study had been entered into a GIS-based database. Concomitantly, the Szigetköz region of the River Danube impacted by Gabcikovo-Nagymaros water barrage system when the flow of the river was diverted in 1992. Incidentally, a monitoring project was carried out, since 1987, by organizations other than the state nature conservation service. Thus, the data collected so far could be used to follow the increasing desiccation process due to the diversion of the course of the river. This monitoring project included studies of the following components: fishes, birds, macroscopic aquatic invertebrates, algae, vascular plant species, and vegetation. The HBMS contributed to the project by monitoring the dragonflies and small mammals.

Notwithstanding the above, the Kis-Balaton, part of the Balaton Uplands National Park, is a large wetland area with a unique fauna and because of its international significance, it had been designated as a Ramsar site. The Kis-Balaton also had formed part of a water-quality improvement system of the Zala River, in addition to its natural values. A GIS-based database of the results of monitoring exercise compiled covering the following components: Birds (rare and colonially nesting species, reed, and reed-nesting birds), small mammals, fish, molluscs, leeches, biting mosquitoes, dragonflies, rotifers; and, among the plants, the algae. In addition, the vegetation and the population of protected and invasive plant species, bats, amphibians, reptiles, carabid beetles, and spiders that may be collected by soil traps have also been frequently surveyed.

Notwithstanding the above, the most essential parts of the HBMS have been developed by 2001. The staff and management capacity for building the system were ensured and the conditions for the proper functioning were refined during the development.

In continuation of the above, there had been a regular exercise of thoughts with regard to management and conservation of the water bodies and their resources, in view of the enormous diversity of abiotic and biotic resources in **Hungary**. Also, in this direction there had been efforts to implement **CBFM** in a **participatory** mode as an efficient *modus oprandi* for the effective management of the water bodies and their bioresources.

5.4.10 Ireland

5.4.10.1 Integrated water resources management (IWRM)

Kenneth Irvine (2014) dealt with IWRM, from local to global sustainability, having relation to **Ireland**.

Integrated Water Resources Management (IWRM) is a significant program, which draws its inspiration from the Dublin Principles. It is a systematic process for the sustainable development, allocation, and monitoring of water resources use with regard to its both abiotic and biotic components in the context of social, economic, and environmental objectives.

The three pillars of IWRM, as discussed in the Global Water Project's GWP-TAC (2004), are as follows:

(i) Moving toward an *enabling environment* of appropriate policies, strategies, and legislations for sustainable water resources development and management.
(ii) Putting in place the *institutional framework* through which the policies, strategies, and legislation(s) could be implemented.
(iii) Setting up of *management instruments* required by these institutions to do their job.

Incidentally, an extensive suite of tools for IWRM is already available. These include the (a) Global Water Partnership Toolbox for IWRM and the UNEP Handbook on the Development and (b) Implementation of National Programs of Action for the Protection of the Marine Environment from Land-based Activities. Similarly, there could also be programs dealing with FW. Moreover, there is also a body of knowledge that had been generated through the EU Directives emanating from the sixth EU Framework Program.

Notwithstanding the above, a flagship integrated project, titled, the "New Approaches to Adaptive Water Management Under Uncertainty (NeWater)" project, cofunded by the "Global Change and Ecosystem" subpriority of the 6th EU Framework Programe, was conducted between 2005 and 2008, with the major objective to develop a theoretical and practical framework to support the transition of existing water management regimes toward greater adaptability. Moreover, specific approaches and tools were tested and further developed based on this framework for practical applications in various lentic and lotic basins.

In addition, with regard to islands, the whole island could be considered as a system of highly integrated ecosystems and subecosystems such that any intervention within a hydrological system considers the impact of the intervention on the whole hydrological system. Further, there could be the question of whether the governance systems, the management capacities, and the types of financing required for traditional IWRM could be considered as actually culturally and politically sensitive to the needs of the developing countries, particularly the Small Island Developing States (SIDS). Incidentally, the successful implementation of classic IWRM plans require SIDS to perform significant institutional and legislative reforms. This process calls for the establishment of a reformed institutional structure that involves decision makers at the highest political level, from ministries, agencies, local administrations, the civil society, NGOs, private sector, and other actors concerned with the management of water resources.

Incidentally, the EMPOWERS approach to Water Governance partnership is a group of organizations and individuals committed to improving water governance, and, more particularly, the rights and long-term access of disadvantaged social groups to reliable water supplies and management. Further, the EMPOWERS partnership developed a process of *modus operandi*, practical tools, and knowledge for improved participatory water resource management and water service provision at the local level.

Notwithstanding the above, the UNEP, in partnership with the Global Environment Facility (GEF), was eager to develop an IWRM methodology which could portray the unique geological, biogeographical, socioeconomic, and cultural attributes of SIDS as well as take into consideration the need to conceptualize IWRM within the framework of an entire watershed stretching from the ridge to the reef (GWP Technical Advisory Committee, 2000). Incidentally, the aim of the intervention was to develop a globally applicable resource book for developing and implementing IWRM for SIDS. It may, however, be noted that each of the SIDS regions, particularly the Pacific and the Caribbean regions, had also adopted their own tool kit for developing IWRM. The Pacific Islands Applied Geoscience Commission, for example, had been designed to provide important materials on the different aspects of IWRM planning and implementation. Further, "the Integrated Watershed and Coastal Area Management (GEF-IWCAM)" project in the Caribbean developed a checklist of the steps required in the development of a national integrated water resources management plan (Blomquist et al., 2005).

Concomitant to the above, until recently, there had also been a preoccupation with the lotic or lentic basins or large catchment being the most sensible unit for IWRM. In this regard, it may be mentioned that although some functions could be best handled at this scale, the second Dublin principle speaks of acting at the lowest appropriate level, i.e., much of the water management could happen at the local level (Moriarty et al., 2000).

Notwithstanding the above, the experiences gained from the implementation of SIDS indicate that IWRM for SIDS is a pragmatic approach toward better and more sustainable water management because such process promotes the coordinated development and management of water, land, and related resources, without compromising the sustainability of vital ecosystems.

5.4.11 Italy

Italy is a country with high potential in AQC and fisheries. However, there may be various background needs for implementing successful management of natural resources, notably the water bodies and their bioresources, especially, the fisheries (Anon, 2004). Incidentally, **CBFM,** by involving the community stakeholders from the grassroots level in a **participatory mode**, is efficient *modus operandi* for the successful management of the natural resources, notably, the water bodies and the fisheries.

Notwithstanding the above, some of the background materials are as follows:

5.4.11.1 *Background experience in community-based natural resource management*

Some of the countries experienced in different management patterns are as follows:

(i) Philippines: coastal fisheries management.
(ii) Vanuatu and Fiji: fisheries and coral reef management.
(iii) Bangladesh: wetlands fisheries management.
(iv) Indonesia: wetlands (floodplain) fisheries management.
(v) India: wetlands fisheries management; research on poverty in coastal areas.

5.4.11.2 *Conditions for successful community-based natural resource management*

(i) "Homogeneous" communities: common objectives, recognized common interests, and social cohesion.
(ii) Benefits exceed costs.
(iii) Clearly defined boundaries to resources to be managed.
(iv) Limited uses and users.
(v) Decentralized decision-making.
(vi) "Simple" administrative structures.
(vii) "Friendly" or at least neutral power structures.
(viii) An "enabling" environment: institutional, political, social, cultural, and economic.

5.4.11.3 *The "enabling" environment*

(i) Devolution of power: choice including choice about NRM.
(ii) Legal rights to natural resources.
(iii) Devolution of resources (funds).
(iv) Institutionalization/mainstreaming of participatory approaches.
(v) Flexibility: time frames and funding mechanisms.
(vi) Long-term engagement.
(vii) Leadership: knowledgeable persons to lead the process.

5.4.11.4 *Issues*

(i) NRM is generally a priority of all under the present general depleting trend of the resources.
(ii) Power structures in management could be changed through friendly interventions.

5.4.12 Mediterranean and the Black Sea

The **Mediterranean and the Black Sea** region borders a good number of countries and have a rich potential in fisheries. Gwenaël et al. (2008) reported about the artisanal fishermen in the area of Port-Cros National Park in the **Mediterranean**.

Artisanal fisheries involve local fishing units, small amounts of capital, small vessels, and short coastal trips for small commercial or self-consumption purposes.

(i) **Artisanal fisheries involve local fishing units**, small amounts of capital, small vessels, and short coastal trips for small commercial or self-consumption purposes.
(ii) In fact, artisanal fisheries are more sustainable than industrial fisheries in many aspects.
(iii) Incidentally, the Mediterranean artisanal fishers increasingly need to adapt to new business scopes arising mainly from tourism.
(iv) The artisanal fishers are to be included and properly accounted for in MPA designation and for management processes.
(v) The MPA managers as well as the fishers share the same aim, i.e., to sustain healthy and productive seas.

5.4.12.1 *Done at SR on 23 Nov 19 arnd 4–45pm*

Concomitant to the above, Gutierrez (2015) dealt with the management and comanagement options for small-scale fisheries in the **Mediterranean and the Black Sea region**. There had also been a detailed discussion about the history of management of the fisheries. It may be mentioned that the words "countries" and "states" appearing in the text refers to countries, territories, and areas without distinction.

5.4.12.2 *Management and comanagement: actual comanagement versus participatory advisory schemes*

Definition of comanagement

The management of natural resources, notably, the fisheries in particular using the traditional as well as the self-designed *modus operandi* was *in vogue* since time immemorial. Time has tested many methodologies and, incidentally, comanagement, of late, was found to be an approach that has been more recently adopted globally in response to the perceived failure of centralized management of fisheries. At the same time, comanagement was also considered as a strong tool to counter the declining FY. Suspected lacunae in the state management infrastructure also paved the way for comanagement of fishery resources. Incidentally, comanagement systems and processes may vary in terms of the nature of power sharing, composition, and functions but one of the main aims of this *modus operandi* is to bring together the fishers, the government officials, and others operating within the fold of the fisheries sector.

Notwithstanding the above, the term, "**comanagement**" **usually refers to a bunch of arrangements with different degrees of power sharing allowing joint decision-making by the state and the user groups (UGs) about a set of resources or an area. However, there is no single unique globally accepted definition of comanagement** (Armitage et al., 2007; Berkes, 2007a, b). In fact, comanagement shares many features with other kinds of partnerships and cooperative environmental governance arrangements involving multiple actors (Plummer and Fitz Gibbon, 2004). However, a critical feature of comanagement is the presence of at least one strong vertical link between the community or the user group (UG) with the government, including formal arrangements for sharing responsibilities and authority. Otherwise, ad hoc public participation in management decisions or simply consultation may not be regarded as comanagement.

Concomitant to the above, the term comanagement is relatively recent, as its earliest use could be traced back to late 1970s. However, as mentioned earlier, the practice of power sharing in resources management could be traced back to the earlier times (Ostrom, 1990). Nevertheless, most definitions of comanagement entails some **institutionalized arrangements** for user participation in management as well as in the decision-making process, often portraying a dynamic partnership using the capacities and interests of local fishers and communities, complemented by the ability of the state to provide with enabling policies and legislations as well as enforcement and other kinds of assistance.

5.4.12.3 *Types of comanagement*

Notwithstanding the above, the term "**comanagement**" could also be defined in a different way, despite the inclusiveness of the term "comanagement." **It could also be defined as the sharing of responsibility and/or authority between the government and local stakeholders to manage a resource efficiently**. Incidentally, the term "comanagement" covers a broad spectrum of management arrangements, as in the literature, as well as the amount of sharing of the responsibility and/or authority that the government and the local resource users could shoulder or may differ depending on country-specific and site-specific conditions (Pomeroy, 1995). Hence, a certain level of involvement and mode of communication between the government and the fishers is needed for successful implementation. However, some of the salient features are as follows:

- **(i)** Comanagement refers to a host of arrangements with different degrees of power sharing allowing joint decision-making by the state and the UGs about a set of resources or an area.
- **(ii)** Comanagement tries to entail institutionalized arrangements for user participation in management and decision-making process.
- **(iii)** Ad hoc public participation in management decisions or simple consultations may not be regarded as comanagement.
- **(iv)** A fishery could be said to be comanaged by well-established comanagement organizations and/or institutions with decision-making power in local fisheries management. Further, most fisheries management systems, particularly in the developed countries involve the users/stakeholders from the grassroots level in the decision-making process, e.g., involvement in the consultative committees. However, those cases where the importance of the legal and political systems dominate, the comanagement aspects of the fisheries are often excluded from this categorization.

Notwithstanding the above, there are **various degrees of comanagement**. For example:

- **(i)** Within an "instructive type of management arrangement," there is only minimal exchange of information between the government and the users and it only differs from centralized management by the existence of mechanisms for dialog with users.
- **(ii)** A "consultative type of management arrangement" exists when mechanisms for consultation between government and users have been established but where decisions are still taken by the government.

In fact, these two types of management arrangements cannot be considered as true comanagement.

(iii) On the other hand, the "cooperative types" are closer to a true definition of comanagement, which involves arrangements where governments and users cooperate on an equal basis in management and in decision-making.

(iv) Further, "advisory comanagement" arrangements are those where the users advise the government of decisions to be taken and government usually endorses them.

(v) "Informative comanagement" occurs when government has delegated responsibility to user groups who are responsible for informing government of their decisions. Finally, self-governance, such as CBM or traditional marine tenure systems, entail a total power devolution from the central government to the local users.

Notwithstanding the above, some authors try to exclude CBM or self-governance from the concept of comanagement. On the other hand, however, others argue to the contrary, since, these arrangements are recognized in the national legislation or they form part of sectorial development policies.

Concomitant to the above, it is important to bear in mind that this typology is just a simplification of very complex management and governance structures. In fact, there is a multitude of tasks that could be comanaged under a different type of comanagement arrangement at different stages in the process. Therefore, comanagement covers a broad spectrum of possible collaborative decision-making between government and communities or UGs.

5.4.12.4 *Executive role of the stakeholders versus participatory advisory schemes*

It may be noted that comanagement is becoming increasingly important in contemporary fisheries management. Comanagement describes power and responsibility sharing agreements that are made between the government and the UGs. Comanagement that involves true power-sharing and joint responsibilities between UGs and the government is difficult to achieve in a setting where stakeholders are polarized and, generally, not encouraged to participate in the process actively. In fact, participation may happen at many different levels of inclusion. At a basic level, participation could be "passive consultation," where information is disseminated and those impacted are informed of future plans (Pomeroy and Berkes, 1997). However, effective participation of the principal stakeholders requires certain degree of empowerment. In fact, Jentoft (2007) defined "empowerment" as "a process through which people become strong enough to participate,; people also get share in the control, and people are influenced by the events and institutions which affect their lives."

Notwithstanding the above, there are as such three main types of comanagement with different levels of involvement in management and decision-making process, viz. (a) cooperative, (b) advisory, and (c) informative, as mentioned earlier.

In this connection, self-management or CBM could be considered as a form of comanagement, since these arrangements could often be part of sectorial development policies and/or legal frameworks.

Concomitant to the above, a true comanagement requires a departure from thinking about the role of local, regional, and federal governments, but on the other hand it considers the political changes that could be necessary for devolution of power. However, a significant concept behind power devolution and executive role of stakeholders is actually to try to define the identity of the relevant stakeholders. The popular term "stakeholder," encompasses all sorts of professions and groups, in addition to fishers, who may possibly have an economic or cultural interest in fisheries. However, many of them may not reside in geographically defined fishing communities. Environmentalists, processors, and shippers, among others, might, therefore, be considered as stakeholders in the general sense of the term. Nevertheless, the local fisheries management may, however, consider the local fishing communities and the fishers themselves as legitimate stakeholders. In general, a dearth of representation of decentralized bodies and institutions usually precludes accountability goals.

Local engagement in management and fishers' participation in the decision-making process are essential components of successful comanagement. Certainly, there could be a number of possible methodologies that could lead to a successful comanagement outcome and which may incorporate both biological and social concerns and, at the same time, could be considered as legitimate by the UGs. However, an essential component which all of these are to contain is direct local community involvement. Thus, it is not only important to characterize the most relevant stakeholders but also essential to clearly define local communities in geographic terms. However, local ownership and control, when embedded into a system of institutions that delegate power and are truly interactive, may be a key principle in facilitating successful comanagement. For instance, in the United States, most fisheries, under the Fishery Management Council systems, include formal consultations with the stakeholders at various stages of the decision-making process. In fact, such consultation is found in fisheries throughout the developed world. However, the US council system is unique (compared to Europe, Canada, Australia, and New Zealand) in that some stakeholders sit on the Councils. However, these systems remain effectively top-down managed, with the US laws and the courts often being the dominant determinant of fisheries management actions. The US councils also manage many fisheries and it is often the case that only one or two stakeholders from an individual fishery could be represented in the council. Thus, the US council system may not be regarded as a true comanagement system, although some exceptions may prevail.

Notwithstanding the above, meaningful comanagement arrangements are to go beyond consultation by redirecting the flow of social and economic benefits from the fishery back into the communities. Communities are allowed to share power and responsibility with the government fisheries managers. In fact, effective participation of main stakeholders in comanagement requires certain level of empowerment.

5.4.12.5 Overview of traditional fisheries management in the region

Traditional schemes involving self-management

"Fish" was a significant item of food in the diet of human beings since time immemorial in the world. In fact, the Mediaeval Europeans, like many other races in the world used to consume high quantities of fish as a source of protein. In fact, attempts at managing European fisheries are quite ancient. Hence, the need to manage fisheries comes from the ancient times. However, legislators and managers had already acknowledged the problems associated with overfishing, by the 13th century; at the same time, believing that anthropogenic impacts on habitats were also affecting fishery resources and yields. Subsequently, the Europeans had become aware of shifts in the availability and the exhaustibility of fishery resources, between 1200 and 1400, thereby stimulating the evolution of fisheries management. However, by that time, the fishermen were grouped into guilds, which were in charge of exploiting and managing the resources at the local level. Nevertheless, the public authorities, such as the Royal Persons and the lords, had undertaken regulation of fisheries for both consumption and conservation purposes. Incidentally, the first fisheries laws in Europe had come into force during the 13th century, and both the size of the fish caught and gear and temporal restrictions and closures were considered. Incidentally, this simple fisheries management had worked till the Industrial Revolution in the late 1800s.

Concomitant to the above, a study performed in other parts of the world, notably, in Hawaii and in Florida (USA), has portrayed that fisheries management during the four centuries, before the arrival of the Europeans, was characterized by adaptive management with characteristics of common property resource governance systems. These management systems involved PAs of the coral reefs, protection of vulnerable life-history stages of fish, and species with high susceptibility to overfishing. Moreover, the fisheries management and governance included strategies and social institutions to support resources allocations among community members as well as strict enforcement by a local rule-making authority. This form of self-management enabled the coastal communities to manage their resources sustainably for centuries.

Notwithstanding the above, of late, there seemed to be a number of examples of self-governance or sharing mechanisms for the management of natural resources, including government-community partnerships and collaborations for forest management in the Kumaon Himalayas in India around 1900s (Agrawal, 2005). Similar examples were *in vogue* in the council forests of Kirinyaga, Kenya, around 1930s. However, in the fisheries sector, the earliest documented legal arrangement seemed to be the Lofoten Islands cod fishery in Norway in the 1890s (Jentoft and McCay, 1995) and the Japanese inshore fisheries under Japan's Fisheries Act of 1901 and its subsequent amendment(s) and revisions (Lim et al., 1995). Incidentally, the early literature depicted comanagement as a class of relatively simple partnership arrangements, for example, in the implementation of indigenous land and resource claims. However, the wide range of international experience, which were accumulating since the 1980s, indicated that comanagement had become more complex and dynamic than might be concluded from the earlier literature and had evolved in diverse directions (Plummer and Armitage, 2007a, b). For instance, the fishing communities in Kenya used TEK and social norms until the 1920s in order to define the rules for resource management and governance. Traditionally, an elder was the leader of a landing site, who provided the necessary advice about seasonality, issuing permission to fishers from other areas, ensuring social cohesion, and restricting specific fishing gears. Moreover, the local social structures were used to enforce such rules or to impose sanctions when needed. After independence, the state had taken the responsibility of management of fishery resources and the management decisions were subsequently prescribed with little or no input from the fishers and other stakeholders, thus resulting in negative consequences for the sustainability of the resource and the economic viability of the fisheries. Some of the other instances include the Native American communities in the US Pacific Northwest, the Eastern Indonesian "Sasi" system, and the inshore northern cod fisheries of Newfoundland.

Incidentally, they were mostly artisanal, although some amount of commercial exchange was associated with these systems. These so-called subsistence fisheries were either ignored or legislated away by governments with the emergence of industrial fisheries.

Concomitantly, many traditional, preindustrial fisheries were also managed under some kind of self- or cooperative management involving different local actors. On the other hand, fishing communities often controlled access to the fishery resources mostly through the use of some sort of access rights (e.g., territorial use rights or TURFs) in the less developed nations. Incidentally, these CBM systems often involved elaborate social customs and regulations at the local level aimed at controlling effort and/or limit harvests. Under such systems, and given the technological limitations of fishing, control over access to fishery resources was mostly in the hands of those most directly dependent on them. In these communities, however, self-regulatory measures were mostly used to manage their resources sustainably. In fact, many preindustrial fishery-based communities have embraced cooperative management of some kind.

5.4.12.6 *Transfer of management responsibility from primary stakeholders to centralized state agencies in modern times*

It may be mentioned that many of the traditional access systems changed drastically after industrialization in the 1970s, when government policies in most developed countries were directed toward centralization of management functions and scientific advice. In many less-developed nations, governments looking for foreign exchange capital signed joint venture agreements with distant water fishing companies, who had moved their operations inshore, and also forced community-based fishers to move from their traditional grounds or had reduced stocks and obliged them to fish harder and farther offshore to meet their domestic needs. These conditions led to conflict, thus confronting stakeholders in detriment of effective management of fishery resources. In addition, lack of proper management and enforcement of regulations in these centralized systems led to depletion of many coastal resources. For example, decades of top-down management with little or no enforcement led to a virtual collapse of many inland fisheries as well as severely depleted artisanal coral reef fisheries in Kenya.

Similarly, the traditional or local CBM systems have a long history of existence in **Southeast Asia**. However, a majority of these systems weakened or had disappeared due partly to the institutional restructuring under colonial administrations, technological modernization, socioeconomic stratification, and unequal concentration of power and wealth within coastal communities.

Notwithstanding the above, some of the significant and pertinent points in this regard are as follows:

- **(i)** Self-governance and sharing mechanisms for management of natural resources including government-community partnerships in forest and fisheries were common since mediaeval times.
- **(ii)** In Kenya, the fishing communities used the TEK and social norms to define the rules for resource management and governance since 1900.
- **(iii)** By 1970s, many of the traditional access systems changed drastically after industrialization, when government policies in most developed countries were directed toward centralization of management functions and scientific advice.

5.4.12.7 *Contemporary schemes including comanagement elements*

(e.g., Cofradias, Prud'Hommies)

By the late 1980s, there had been a trend of decentralization from centralized governance to plan, administer, and implement development. In fact, community-based cooperative management arose out of crises caused by stock depletion, a growing perception that the central governments are incapable of managing stocks centrally; and political pressure stemming from them led to a so-called increasing sense of alienation among coastal communities toward their governments. As such, decentralization, stakeholder participation, and community involvement had come to be considered as essential components of development and management in order to protect and sustain the fishery resources.

5.4.13 Norway

Norway possesses rich potential in AQC and fisheries. Vera et al. (2012) dealt with the conditions under which CBM may work effectively along with an example from Sa'mi Pastoralists regarding their management of pastures sustainably.

CBM is a good exercise and has been implemented in socioecological systems (SES) globally. CBM has also been the prevailing policy in Sa'mi pastoral SES in Norway, but the outcomes tend to vary extensively among the resource groups.

Further, in **Norway**, Søreng (2008) dealt with fishing rights discourses with a note on indigenous versus non-indigenous voices. In this connection, a pertinent question may often arise: Who is and who should be entitled to fishing rights? This is an important issue in global fisheries management discourses. Privatization of fishing rights together with an increased emphasis on indigenous rights, are among the most important factors that have put rights on the agenda. There had been discussions with regard to a dispute on indigenous rights that had appeared in the Norwegian media in 2006 between "Sami" and "non-Sami" small-scale fishing interests. Both the interests groups were struggling to regain fishing rights lost due to recently introduced resource management schemes, with the probable goal of securing the future of small-scale fisheries and coastal and fjord settlements. Despite sharing some of the same objectives, disputes in the media have driven them apart. Various contributions on the aspect had tried to answer the reason(s) for such happenings as well as the redressal measures which could be adopted to prevent further separation and also to repair the rift.

5.4.14 The Netherlands

Managing water in ways that best meet all our diverse needs for water including its protection in the natural ecosystem is very essential.

The **Netherlands** is a country which is rich in fisheries. Within the Netherlands, as in many parts of the world, the quality of peoples' lives is directly related to the quality of the natural environment, viz., the land, the air, and the water resources. A good quality of the environment is considered as crucial to human health and economic and social development as well as for ecosystem preservation and diversity. The way and the standard of the management of our natural resources done today determine how well these resources could serve our descendants and us. As such, there is an utmost need for caring much about the management of the coveted resources, especially our water resources. **Incidentally**, Stedinger et al. (2005) dealt with water resources systems and their planning and management with an introduction to methods, models, and applications.

It may be noted that it is absolutely essential for managing water in ways that best meet all our diverse needs of water and its conservation has to be endured. But, in spite of knowledge and experience, the Dutch, as in many parts of the world, still experience droughts, floods, and water pollution. However, these adverse impacts are not unique to Europe only. In fact, in many other regions of this world, the need for improved water management is much more critical and urgent. Many people, especially the children, suffer each day because of the lack of it. We take pride in our abilities to manage water and, consequently, we are also take pride in our abilities to help others to manage it.

It is quite understood that much of the world has witnessed ever-greater demands throughout history for reliable high-quality, and inexpensive water supplies for domestic consumption, agriculture, and industry. Moreover, there had also been increasing demands, in recent decades, for hydrological regimes that support healthy and diverse ecosystems; provide for water-based recreational activities; reduce, if not prevent, floods and droughts; and, in some cases, provide for the production of hydropower and ensure WLs adequate for ship navigation. However, the water managers are challenged to meet these multiple and often conflicting demands. Concomitantly, public stakeholder interest groups have depicted, of late, an increasing desire to take part in the water resources development and management as well as in many decision-making processes. Further, added to all these management challenges are the uncertainties of natural water supplies and demands due to changes in our climate, changes in people's standards of living, changes in watershed land uses, and changes in technology. It, thus becomes a challenge for the managers to develop or redevelop and restore, and then manage water resources systems. Incidentally, such systems could range from small watersheds to those encompassing large river basins and coastal zones. Moreover, these are to be taken care in a way that meets society's changing objectives and goals. As such, uncertainties arise with regard to integration and sustainability of water resources. As such, before engineering and implementing the projects, efforts are to be directed toward planning and addressing the water management problems. Incidentally, such efforts could involve identifying various alternatives or scopes for addressing the problems. Further, the various impacts of each proposed alternative need to be estimated and evaluated. In fact, for this purpose, a variety of optimization and simulation models and modeling approaches are developed to help the water planners and managers in identifying and evaluating plans. The Netherlands is having expertise in hydrology since long. The country is making every efforts in developing efficient *modus operandi* in water resources management.

5.4.15 Romania

Romania possesses rich potential in AQC and fisheries. The GEF helps Romania and also other countries through funding for activities that protect the global environment, particularly, the aquatic environment.

In 2002, the GEF started a study to "analyze how attaining global environmental goals could contribute to the generation of local benefits and vice versa in accordance with the GEF mandate."

The final report of one of the 18 GEF projects dealt with the local impacts on the Romanian Danube Delta Biodiversity Project (DDBP). The project (Mirela et al., 2005) intended to protect the Romanian Danube Delta Ecosystem. It contributed to the conservation of BD)in the Delta, strengthening the capacity of the Danube Delta Biodiversity Reserve Authority (DDBRA) and the Danube Delta National Institute (DDNI). The project was very successful in strengthening both the institutions, and this had left them well positioned to contribute to the sustainable management of the Delta, one of Europe's most important wetlands. However, an analysis of the four stages of the project cycle has shown that there was very little conceptualization of the role of local people. The project document had seen the local people as the main threat to BD through their almost unrestricted access to fish, the major natural resource in the Delta. In fact, the study team had agreed that overexploitation of the natural resources, such as fish has to be stopped. Concomitantly, this could have been done differently, viz., by ensuring that the local population was involved in changing management approaches and become part of the solution. However, only a small proportion of the planned area was restored and, in consequence, the global, national, and local benefits were considerably less than planned mainly because the local issues had not been taken into account. The project's impact on the sustainable use of the Danube Delta's fish resources could benefit in the long term from the strengthening of the DDBRA and the DDNI. However, the other components were only marginal impacts. For example, the pilot projects for alternative income generating activities under the project, such as handicrafts and ecotourism were only a very minor impact (Mirela et al., 2005).

Notwithstanding the above, the study (Mirela et al., 2005), further, looked into the various NRM approaches used in the Delta over the last two centuries, focusing particularly on the latest regime that of private concessions. This system was built on and is in line with project arrangements, although introduced only after project completion in 2002. While in paper, the concessions controlled the amount of fish caught, it, however, had remained unclear whether this arrangement had actually done so. What was, perhaps, very clear was that the concessions had reduced the income of professional fishermen by *c.* 30%–50% and made subsistence fishing by local inhabitants more difficult. Moreover, the concessions increased the amount of tax received by the government.

In fine, the analysis of the field data showed that while the project had been somewhat successful in conserving nature and its BD, the livelihoods of many of the inhabitants were negatively impacted. Further, the project had missed many scopes to benefit the local population and, thereby, enhance the sustainability of the global environmental benefits.

Nevertheless, an alternative perspective on the project was provided by Dr. Gonzalo Castro, Head of Biodiversity in the GEF Secretariat.

Concomitant to the above, the report concluded with eight recommendations to the GEF:

(1) To ensure a balance between global-national-local benefits and include a sociological appraisal at the time of project design and appraisal.
(2) to protect resources and ensure the livelihoods of the local communities and proper compensation to the project affected persons.
(3) To study the impact of the so-called currently privatized natural resources management arrangements.
(4) To ensure ongoing and independent monitoring of sensitive biosphere resources.
(5) To strengthen harmonization of the activities of the various in-county administrative levels.
(6) To monitor key indicators and to ensure greater flexibility during implementation to respond to changing circumstances.
(7) To strengthen the socioeconomic capacity of the implementing agencies.
(8) To proactively identify and follow-up local development initiatives for GEF support.

Concomitantly, Giger (2008) dealt with certain aspects of **Romanian** lentic systems and their management.

The Black Sea coastal lakes are important and threatened ecosystems. The Taşaul Lake (area 23.35 km^2, max. depth 4 m) near Constanţa was transformed from an open coastal lagoon (salt water) to a FW lake in the 1920s by technical constructions (shallow, turbulent, and eutrophic), classified as heavily modified water body according to the EU-WFD. Incidentally, there is an unconfirmed history and concepts of sustainable fishery in Taşaul Lake, Romania. Brackish biota was replaced by typical FW species. Incidentally, commercially less valuable fish, such as the giebel carp and roach, became abundant while carps and pike perches drastically dwindled.

The eutrophic Lake Taşaul was used mainly for AQC and fish production. The latter is strongly dependent on stocking and exploitation. Incidentally, the historical fish data had shown a rich BD in the 1970s but the catch records declined drastically in 1994–95, from *c.* 180 tons to *c.* 50 tons, and later on to 5 tons/year, until 2000.

Incidentally, the fish stock in Taşaul Lake was assessed in 2005 from a total of *c.* 581 specimens investigated. However, the fishery was unsustainable, since the fishing mortality coefficient was greater than the natural mortality coefficient. Moreover, there were incidents of poaching, minor restocking, etc., but only infrequently. According to the model based on standard fish catch, the average MSY was *c.* 10.9 tons/year for giebel carp and *c.* 12.2 tons/year for roach. However, the conclusions and recommendations toward a more sustainable fishery management in Lake Taşaul, apart from reducing nutrient input from Casimcea River by fighting point sources, were as follows:

(i) To perform a detailed monitoring of yearly restocking to quantify fish stock input.
(ii) To perform a detailed monitoring of the fishermen active per year, the number of gear nets used per day, and the catch per unit effort (CPUE) in order to quantify the fishing effort.
(iii) To perform detailed statistics about the net catches and angling, and to estimate poaching, in order to quantify fishing output.
(iv) Effective management of the water bodies and fishes by efficient management *modus operandi*, like the **CBFM** could yield fruitful results.

5.4.16 Switzerland

Switzerland is a country with many water bodies and has rich potential in AQC and fisheries. Incidentally, Anon (2013) reviewed the key factors that were believed to support the small-scale coastal fisheries management.

There had been reviews aimed at helping to identify the role that certain foundation(s) could play in managing the small-scale coastal fisheries. Incidentally, there had been expert groups (EG), notably, the Integrated Marine Management (IMM), the leading research and development group with expertise in fisheries and AQC in coastal zone and their management. Such organizations could undertake a rapid review of the key sources of knowledge (evaluations, literature reviews and key informant interviews) and identify key factors, which could influence success in small-scale coastal fisheries management.

Incidentally, there are no easy and simple universal solutions to the problems that the small-scale coastal fisheries face. However, such organization(s) help in identifying a number of key and interconnected success factors. In fact, identification of such factors helps the organizations (like the different Foundations) in three specific ways are as follows:

(i) It could help improve the focus of its fisheries strategies and programed investments. Incidentally, the review provided a broad overview of the different areas of knowledge in the fisheries and marine conservation areas, which allowed the foundation(s) to narrow down the focus of their grant allocation into a subset of extremely complex, globalized, and interconnected systems.
(ii) There were efforts to recognize the contribution of many different factors in successful outcomes for fisheries. These were pointed out to the foundation and the partner(s) toward interventions and approaches that seek to negotiate and resolve the various trade-offs among the success factors. For instance, to determine whether or not fisheries are to be managed for wealth-raising sustainability or for other kinds of social benefit(s), etc. Further, such review(s) help in guiding organizations like the foundation(s) to focus on processes that might bring in various diverse points of views, for example, **comanagement** or a fisheries improvement partnership, which could work toward mechanisms that could help to resolve the differences in viewpoints.
(iii) The review could serve as an entry point for the grantees and the partners to examine their own perspectives on what works and what may not work, along with evidence(s) from other sectors in fisheries.

Further, it may also be noted that 20 success factors could be explored in detail through an interactive data visualization tool online. The focus of such works is to trace the key success factors in achieving a balance of social, economic, and ecological benefits from the management of small-scale coastal fisheries.

5.4.16.1 Background to coastal fisheries

Small-scale coastal fisheries is a significant sector. They contribute very significantly to food and nutritional security, employment, and income generation and in providing livelihoods for many poverty-stricken people. In fact, the diversity of the coastal resources makes scopes for these people, and the ease of becoming a fisher means that many people depend on fisheries for all or part of their household livelihood strategies. However, as people may often try to enter the fishery sector, many fisheries are overcapitalized with too much investment in fishing capacity; some are overexploited and their contribution to wider economic development is poorly used. Today, *c.* 87.3% of fisheries are either overexploited or fully exploited (FAO, 2012). These issues are compounded by a number of other human activities which result in coastal pollution and other types of habitat destruction, thus further depleting resources. However, subsidies to the fisheries sector in the past had aggravated excessive pressure on the resources, and destructive fishing methods were always been threatening wider ecosystem functions in many coastal areas. Further, growing global demand for fish by a variety of users as well as the penetration of global marketing chains to almost all the corners of the world have also been giving incentives to overexploit fisheries. Moreover, climate change is beginning to affect coastal habitats, such as coral reefs, which are under severe threat. These are some of the richest habitats that provide a wide diversity of scopes for the coastal people.

Notwithstanding the above, climate change has other effects on both the ecology and the livelihoods of the people who live near the coast, including increasing their exposure to natural hazards. The climate is only one of a number of factors that are changing. Today's fishers are faced with fuel and food price rises, changes in demand for fish, and changes in market structures, which are forcing fisheries and fisheries management to confront issues in the much broader contexts of the wider ecology and local and national economies.

5.4.16.2 *Fisheries management*

Much of the current situation of poorly managed fisheries results from past policies and practices. The momentum of growth meant that fishing capacity was allowed to expand beyond that required for full and efficient utilization of the fisheries. Many different forms of fisheries management had evolved, depending on the perceived aims of the management process. Early forms of customary marine tenure focused on the locally controlled management of resources and benefits from them. As fisheries developed, yield maximization approaches were adopted which, in turn, gave rise to concerns for the environment and for maximizing sustainable economic returns from fisheries management. Thus, approaches had been added that conferred rights to individuals and community in the sector to achieve both efficiency and equity gains. This recognition of the wider ecological and economic implications of fisheries management had also stimulated an interest in more holistic approaches that go beyond fisheries and include **ecosystem-based approaches and integrated coastal management**. Such approaches were increasingly looking closely again at **community involvement** in **fisheries management** through **comanagement arrangements**, under which the community and the government share roles and responsibilities.

5.4.16.3 *Impacts*

The social, economic, and ecological impacts of fisheries management are difficult to measure. However, in a gross manner, a balance of sustainability, efficiency, and equity goals had not been fully achieved. In fact, with the current state of the world's fisheries, on an overall basis, the management process is not largely successful everywhere.

5.4.16.4 *Key success factors*

In spite of this rather little negative picture of the **management** process, there are indications that some elements of fisheries management, in some locations and under some circumstances had contributed to success. Although, there is no single approach or a management system that could address all the problems facing fisheries management globally, there is a growing belief that if enough of the factors affecting success are brought together, they might have the potential to reverse the fisheries' downward spiral.

5.4.17 United Kingdom (UK)

There have been instances of promotion of **comanagement** between the local communities and the government agencies as a strategy to improve **fisheries management**. Such efforts are quite *in vogue* in the **UK**. Incidentally, Campbell et al. (2008) dealt with studies on comanagement (particularly comamagement of sea turtle fisheries with notes on biogeography versus geopolitics).

There had been focal discussions on fishers' incentives for engaging in comanagement and on the potential to scale up comanagement to a regional level. Such instances often also portray fishers' participation in socioeconomic surveys. Results of the surveys, based on three established criteria for comanagement, viz. (i) perceived crisis in stock, (ii) willingness to participate, and (iii) community cohesion revealed that fishers' support existed for comanagement but the extent of support and the views of specific management interventions could vary. Further, there had been discussions pertaining to the implications of the results of comanagement in each territory. There had also been deliberations on the efforts for establishing coordinated management regimes in the region.

It is quite well known that the aspect of appropriate institutions and actors for managing common pool resources had long been debated in the resource management bodies. In this connection, it may be pointed out that general concerns about the failures of top-down management by the government, during the 1980s, had led to the generation of enhanced interests in communities to serve as resource managers.

Notwithstanding the above, CBNRM emerged as a bottom-up alternative for a number of reasons, including documented examples of success, beliefs about community's interests in resource sustainability, and shifts toward governance models that emphasized both local participation along with the government regulations. This could be the initiation of the emergence of the CBM paradigm. Further, more recently, the abilities of the communities to effectively manage resources critically examined, and the arguments to scale-up conservation and management are resurging. However, many issues were at stake in such debates and underlying it, are the assumed strengths and weaknesses of the actors involved in resource management, the scale at which strengths could be capitalized and weaknesses could be overcome, and how they vary for particular resources. For instance, communities may be seen as capable of managing resources that could be termed as "local" (e.g., a water body) but at the same time, may be less able to manage resources that are shared and also nonlocal, e.g., the migratory species. Concomitantly, the governments may be less willing to delegate management to communities when there are competing and conflicting interests in the resource, it has high market value, or there is limited and incomplete scientific understanding of it regarding its actual potential.

5.4.17.1 Glimpses of management of AQC and fisheries in Great Britain

Further, in Great Britain, as in many other countries in the world, there are many issues with regard to BD management and conservation, in general, and AQC and fisheries management and conservation, in particular. For instance, in January 2005, the WWF-UK reported that *c.* 13 of the *c.* 16 surveyed marine species were on decline, including the commercial fish species. Causes for the decline included, among others, overfishing, dredging, increased shipping traffic, oil and gas exploration, and coastal developments, etc., to cite a few. As a consequence, in December 2004, the British conservationists called for large areas of the North Sea, in particular, to be designated as no fishing areas, mainly because of a perceived decline in stocks. However, the British fishing industry had heaved a sigh of relief after the EU Fishery/Fishing Ministers rejected such a plan preferring to agree to less draconian measures, including tougher surveillance on illegal fishing and the introduction of measures to help endangered species. As a consequence, the days that European vessels could fish was reduced from 15 to 14 each month, although excluded from this constraint were the trawlers, which used mesh sizes large enough to avoid hauling of the juvenile fishes. In addition, the most endangered species of cod was afforded further protection by a ban of 18 weeks duration on fishing in the Baltic, coupled with mandatory inspections of the vessels catching >1.0 tons of Cod in the North Sea.

5.4.18 Macedonia (Yugoslavia)

Fisheries and AQC are potential sources of nutrition and avocation. **Macedonia (Yugoslavia)** is rich in AQC and fisheries (Anon, 2008a, b). Incidentally, EU requirements on resource and fleet management and inspection and control do not seem to apply to inland fishing, except for control of marketing and traceability of fishery products (Anon, 2013). The administrative capacity of the unit for fisheries and AQC in the Ministry of Agriculture, Forestry, and Water Economy and the relevant unit in the Agricultural Inspectorate was insufficient. Preparations in this area were at infancy. The country did not have an operational fisheries program to serve as a basis for structural measures as regards structural action for small-scale commercial fisheries and inland fisheries are concerned. Nevertheless, measures in the field of market policy, databases, and records relating to production and trade in FW fish and fish reproductive materials are regularly updated. Incidentally, the 2013 allocation for financial support for fisheries and AQC, as state aid was enhanced by one-fifth (to *c.* €1.46 million).

Notwithstanding the above, there were no formal international agreements. But, the government was actively seeking informal working arrangements with Albania and Greece regarding the management of resources in Lakes Ohrid, Prespa, and Dojran. Incidentally, there was limited progress in the field of fisheries mainly because administrative capacity had remained insufficient.

Concomitant to the above, management of BD is a significant issue in the management and sustainable harvest of natural resources in **Macedonia**, as in many other countries (Anon, 2008a, b). Against the Report to the SBD COP 9 held in Bonn, Germany in May 2008, as background, the Fourth National Report to the CBD tried to describe the development of the basic components of the BD in the Republic of Macedonia along with plans of achievements and suggestions for the future activities. However, situation in many cases may not be easily implemented in view of competing goals with regard to natural resources. The WG had used some indicators in the fourth national report, which could be considered as important tools for monitoring the status and trends of BD at various levels they serve as communication tools to summarize data on complex BD issues. Also, it could serve as a signal for key issues for addressing through policy or management interventions. Such a basis enabled the **CoP** and each **Contracting Party** (CP) to develop a set of BD indicators as part of their national BD strategies and action plans. Moreover, under the present scenario, attempts had been made to involve the stakeholders from the grassroots level for efficient management, sustainable harvest, and effective conservation of the BD and also the abiotic resources. Efforts had also been made to involve the NGOs, which was an added feature. Such coordination could furthermore enhance opportunities for synergy in the national implementation of effective management, sustained harvest, and conservation of the resources.

In addition to the above, various means of communication had been used, like publicly launching national reports on International BD Day, making national reports accessible to a wider audience through national clearing-house mechanisms or other media, developing and disseminating by-products of national report, and so on.

5.4.18.1 *Overview of biodiversity status, trends, and threats*

Overview

The Republic of Macedonia is located in Europe, in the central part of the Balkan Peninsula. Out of the total land area of the Republic of Macedonia, *c.* 40% is forested and *c.* 51% is under agricultural production. Incidentally, Macedonia has a significant amount of biological diversity, notably, *c.* 1580 species of algae, *c.* 340 species of lichens, *c.* 1250 species of fungi, *c.* 3700 species of plant species, and *c.* 9339 species of animals (consisting of *c.* 8833 invertebrates and *c.* 506 vertebrates). Significantly, there are *c.* 854 endemic species (probably, consisting of *c.* 135 species of algae, *c.* 117 species of plants, *c.* 579 of invertebrates, and *c.* 23 species of vertebrates). Unfortunately, increasing exploitation of wildlife species, during the last two decades for commercial gains, have negative impact on the basic components of biological diversity, especially on the diatoms, medicinal plants, invertebrates. and vertebrates. Macedonian Government adopted a decision for the classification of endangered flora, fungi, and faunal species included in international trade (TARIC, CITES and D4).

5.4.18.2 *Main changes in the status and trends of BD components: Status, trends, and treads of key ecosystems and species*

The enormous floral and faunal diversity at national level could be due to central position of Macedonia and the various influences to which its territory were exposed to. With reference to the structure of the ecosystems and their species, the territory of Macedonia is existed in different biogeographical regions. Some of the principal ones are as follows:

(a) The sub-Mediterranean area of the southern part of the Vardar Valley and the area near Doyran Lake, which are dominated by the Mediterranean and sub-Mediterranean species.
(b) The Middle-European biogeographical region, which include a major part of Macedonia and is dominated by various climate-related broad-leaf forests.
(c) The Steppolic area in the central part of Macedonia, with steppe-like vegetation.
(d) The Boreal biogeographical region includes the biome of the European primarily coniferous forests of the boreal type.
(e) The Middle-south European mountainous biogeographical region includes the alpine and partly subalpine zone of the highest mountains and biome of the arctic-alpine rocky terrains, pastures, snow banks, and screens.
(f) The Oreo-tundral area with ordeal floristic elements, invertebrates (butterflies), and mammals.
(g) The aquatic area (lakes, wetlands, flowing waters) which dominates the Mediterranean and Pontus-Caspian species.

5.4.18.3 *Wetlands ecosystems*

Macedonia is said to be rich in water bodies, AQC and fisheries. Hydrologic status of three big natural lakes, viz., Ohrid Lake, Prespa Lake, and Doyran Lake as well as the glacial mountain lakes are rich in aquatic flora and fauna. Also, the rivers are very important habitats mainly for aquatic fauna. The Ohrid Lake has a specific ecosystem with floating AM. In addition, there are few representatives of threatened fish species. Further, the rivers Sateska and Crni Drim, and the Sent Naum spring are significant habitats for endemic species of trouts, whose populations have dwindled today. Moreover, The Prespa Lake possess critical ecological situation with continuous depletion in WL. As such, many sectors of the lake with floating AM and littoral faunal communities have been affected. In this regard, it may be mentioned that the status of Prespa-carp is endangered, mainly due to uncontrolled fishing.

Concomitant to the above, the establishment of the strictly protected ornithological reserve Ezerani and the trans-boundary Prespa Park could contribute to the improvement in the state of the lake ecosystem. The protection of river Golema Reka would further add to the improvement process. In addition, one of the most significant types of ecosystem is the Island Golem Grad in the Prespa Lake with specific floral and faunal species.

Notwithstanding the above, the status of Doyran Lake ecosystem was very critical since 1988–2002. The accelerated trend of Eutrophication had resulted in intensive sedimentation and a dramatic reduction in the epibenthic communities. However, it is concomitantly believed that the current status of the benthic communities still has enough of genetic potential to completely restore itself. Nevertheless, the status of the benthos could be approximately inferred from the amount of annual fish catch. However, the situation had undergone change(s), since 2006 mainly because the level of water had been rapidly rising, being connected to an irrigating canal system, which drew water from underground. Thus, the disturbed ecological balance of the lake was improved and also led to the improvement of the state of the biological communities within the lake ecosystem.

Concomitant to the above, the status of the riverine ecosystems has also rapidly changed. Almost all the big rivers, viz., Vardar, Treska, Lepenec, etc., were under tremendous amount of anthropogenic pressures. At the same time, reservoirs had also been built on some rivers and these represented a sink for persistent substances. Moreover, the benthic communities in the riverine ecosystems were showing reduced abundance and declining trend, which may ultimately pave the way for depleting fish population especially the six endemic fish species with possible status of globally threatened species.

Incidentally, the Wetland vegetation, which was used to develop over large areas of swamps and marshes within the entire valley of Macedonia, had undergone significant amount of changes, might be due to a drainage regime, which converted most of these ecosystems into arable land. As such, the so-called remnants of the wetland communities exist mainly in a fragmentary state and their floral and faunal species, if any enmeshed in them are highly endangered. Incidentally, the most important wetland communities are located near the natural lakes and rivers. Some of the wetlands, which are still preserved, are important in explaining the genesis of wetland vegetation in Macedonia. Many wetlands in Macedonia are derelicted with depleted biota. However, Belchista Marsh still exist in its natural state having plant and animals in quite stable status.

Very important genetic resources are 20 endemic species of fish and 60 exotic fish species, 15 species of amphibians, 32 species of reptiles, 328 species of international indicators (CBD) birds, and 82 species of mammals. The framework of transboundary Prespa Park, established by littoral countries had been formed between Micra and Macro Prespa Lake and their river network is important for migration of waterbirds, raptors, vultures, and endemic fish species. This ecosystem provides environmental services for the fishery sector. Economic incentives to local stakeholders is a good basis for encouraging them to actively participate in forest and aquatic resource managements. Further, water management and sustainable use of bioresources and BD conservation also provide a multitude of other cultural and socioeconomic benefits. There had been efforts to maintain sociocultural diversity of indigenous and local communities as well as protection of traditional knowledge, innovations and practices, etc.

Concomitant to the above, the biological resources support sustainable livelihoods, local food security, and health care. There had also been efforts to ensure the fair and equitable sharing of benefits arising out of the use of genetic resources. Incidentally, Macedonia had introduced specific regulations on the access to genetic resources and benefit sharing. Incidentally, there had been attempts to promote the sustainable management and utilization of genetic resources, thus contributing to conservation of the biological resources.

Notwithstanding the above, there had been efforts to address challenges to BD from climate change and pollution. In addition, there had also been attempts to reduce pollution and its impacts on BD. Further, as a result of ban on the use of phosphorous in detergents, there were probably efforts to improve the quality of water. The phosphorus load along with the loads of nitrates and heavy metals were depicting a declining trend in the waters and the water quality had improved.

Notwithstanding the above, there had been efforts to maintain capacity of ecosystems to deliver goods and services and support livelihoods. Incidentally, biological resources support sustainable livelihoods, local food security, and health care. Further, there had been efforts for stakeholders' participation in natural forest management and water management as well as for sustainable use of bioresources and BD conservation. All these were aimed at providing with a multitude of cultural and socioeconomic benefits. Nevertheless, efforts were made for fair and equitable sharing of benefits arising out of the use of genetic resources. Moreover, there had been guidelines for all transfers of genetic resources in line with the convention on BD and as per the International Treaty on Plant Genetic Resources for Food and Agriculture and other applicable agreements were taken into consideration.

Notwithstanding the above, ITK of the indigenous local people is a precious asset and are to be protected and propagated with due seriousness. Such ITKs are *in vogue* in almost every discipline. At the same time, it may be remembered that management and conservation of the abiotic and biotic resources are indispensable for sustainable harvest of the resources. In this regard, there may be ITKs available for effective management and conservation of resources. Also, **CBFM** in a **participatory** mode could be regarded as an efficient *modus operandi* for efficient management of the fish resources.

Concomitant to the above, it may be noted that the gene of the local indigenous community is to be maintained, conserved, and propagated in order to keep faith with the traditional profession with precious ITK and in order to continue with the race. In this regard, the community is to be preserved from our side with enough of rehabilitating measures and socioeconomic support; else the coveted gene may be lost from us for ever. For instance, it could be mentioned that many of the traditional indigenous fishing communities with enough of ITK in India, particularly in NE India, not being rehabilitated properly as yet are losing faith in their traditional, age-old, hereditary profession of fishing and, by and large, are getting converted into nonfishermen, having been suffered from socioeconomic hardships and status. As such, an urgent initiative is essential for a letup in the situation in order to not only conserve the fishermen gene but also for effective management and conservation of the fishes and the water bodies (Kar, 1990, 2003, 2007, 2013; Dey and Kar, 1989). Macedonia has done some amount of works in this direction.

However, some efforts were made for protecting traditional knowledge, innovations, and practices.

5.4.18.4 *Millennium traditional fishing*

Probably, as a traditional practice, the indigenous inhabitants adopt measures to protect the waterbirds in Doyran Lake. Also, the nets which the fishermen place in the water bodies in places, generally, allow only big fish to be caught from the fishing areas. Concomitantly, efforts had been made to ensure fair and equitable sharing of benefits arising out of the use of genetic resources. In brief, such an effort promotes the sustainable management and utilization of genetic resources, thereby contributing to conservation of biological resources.

In addition, there had been practice in which benefits arising from the commercial and other utilization of genetic resources are shared among the countries providing with such resources.

In fine, an overall assessment was done on the fact that whether implementation of the convention has had any impact on improving conservation and sustainable use of BD, and the fair and equitable sharing of benefits arising out of the utilization of genetic resources in the country. The convention's first target, viz., conservation of BD is based on traditions and this aspect had been actively dealt with by many workers. There had been good expertise in this field with a number of experienced workers. Nevertheless, there is yet much to be achieved in putting the knowledge into practice. Endangered biological resources had been registered, they are being researched, monitored, and tried to be rehabilitated. CBD and sustainable use of natural resources formed the platform of whole human activity. One of the important needs in fulfilling the convention's requirements on national level lies in more integrity in the efforts and developments. One of the main bottlenecks in development is insufficient exchange of information. Moreover, less communication between the sectors and low awareness about the role of different sectors in fulfilling the CBD had been some of the factors decelerating the progress.

Notwithstanding the above, another important need for progress in the convention's agenda and objectives in the particular country is a gradual transition toward a positive attitude for sustained harvest and conservation of BD. This is true both for individual responsible persons and for the society at large to take acre of such aspects. Awareness development for sustained use of natural resources (if required through nature education) could be high priority issues before proceeding for works based on technological developments. For any action(s) toward enhancements of the objectives of CBD, suggestions may be taken at regional, national, and global levels. However, some of the aspects in this regard are as follows:

(i) Refining existing programs of work or developing new ones to address emerging issues.
(ii) Suggesting goals and objectives to be included in the future Strategic Plan of the Convention. Also very important is to identify effective mechanisms for implementation.
(iii) Sustainable use of biological diversity and fair and equitable sharing of benefits of genetic resources are very impactful for protection of nature and for BD conservation.

Concomitant to the above, at the global level of the convention, measures had already been taken for inclusion of the issues of BD conservation into sectoral policies more effectively. Incidentally, one of the most important needs for development is information exchange and cooperation between sectors. The Macedonian Government and National Committee for CBD fully support the convention's implementation both at regional and global levels. Also, very important are the programs and projects of UN, UNEPP, UNDP, UNESCO, FAO, and other international and regional organizations and institutions. Coordination with the Secretariat of UNEP/CBD could be carried forward for integrated policies toward sustainable use and conservation of Biological Diversity. In this regard, an overall assessment was made whether the implementation of the convention has had an impact. It had shown an improvement in the conservation and sustainable use of BD, and a fair and equitable sharing of benefits arising out of the utilization of genetic resources, in their country.

Incidentally, the convention's first target was the conservation of BD based on traditions, and this aspect is usually actively dealt with by many researchers. In this regard, in Macedonia, endangered biological resources are continuously registered. They are being researched, monitored, and tried to be rehabilitated whenever required. In fine, sustainable use and conservation of the natural resources, notably the fishes, could be achieved through efficient management processes, like the CBFM involving the stakeholders from the grassroot level in a participatory mode.

5.4.19 Summary

(1) **Europe,** in the temperate zone, has different types of water bodies in different parts, with diverse types of aquafauna. There is a need for efficient *modus operandi* for the management of the water bodies and their bioresources in different parts of the continent.

(2) There have been many works on fisheries management in **Europe (in general)**. Anon (2006) dealt with the report of The Bergen Conference on Implementing the EAF. Arlinghaus et al. (2002) dealt with reconciliation of traditional inland fisheries management and sustainability in industrialized countries, with emphasis on **Europe (in general)**. Anon dealt with securing European fisheries, through CBFM, i.e., securing a future for the **European fisheries (in general)**.

(3) In **Austria,** the proposed lake typology could provide a basis for the **effective management** of the fisheries of the Austrian lakes (Gassner et al., 2005). Gassner et al. (2005) generated the basic information for the management of fisheries in **Austria**. Further, Schmutz et al. (2007) dealt with fish-based methods for assessing European running waters along with a synthesis, and a brief note on management.

(4) In **Belgium**, Depestele et al. (2014) worked on the sensitivity assessment as a tool for spatial and temporal gear-based fisheries management. Depestele et al. (2012) reported about an integrated impact assessment of trammel net and beam trawl fisheries. Further, in **Belgium**, Margrethe et al. (2014) dealt with the aspect(s) of 'what does stakeholder involvement mean for fisheries management?' Further, Newman (2014) dealt with structural and cohesion policies with regard to inland fisheries and the CFP in **Belgium**. Incidentally, Bartch et al. (2009) dealt with a handbook reporting,' Voices from the Fisheries for preserving local fisheries knowledge, which help in linking generations, and in improving environmental literacy linking to sustainable fisheries management involving the stakeholders from the grassroots level through the *modus operandi* of CBFM. Further, Anon (2009a, b) analyzed the existing RBM instruments in MSs and setting up the best practices in the EU, through various pilot projects for carrying out the works related to CFP.

(5) In **Bulgaria**, Anon (2009a, b) discussed the Statement of the **Bulgarian** Fisheries Administration and the Stakeholders in the Fisheries and AQC sector in **Bulgaria** in relation to the proposed EC Green Paper on the reform in the CFP. Anon (2013) dealt with the Fifth National Report of the CBD of the Republic of Bulgaria with some aspects of **management** and conservation of the resources. Concomitantly, the **wetlands** in Bulgaria are inseparably linked with the life and livelihood of the people and there had been national plan(s) for the protection of the most significant **Wetlands** in Bulgaria. As such, maintaining ecosystem functions of wetlands through sustainable use of their abiotic and biotic resources by the application of efficient management *operandi*, like the **CBFM** is very much related to the quality of life of local communities and their livelihoods. At the same time, it may also contribute to limiting the adverse effects of climate change and facilitating adaptation to them. Further, **community** involvement, mainly through **CBFM**, could play a significant role in the appropriate **management** of the wetlands and their resources. Concomitant to the above, stringent national policies are definitely needed for the conservation of BD and increase in the activities related to the conservation of the near-extinct and endangered breeds and varieties. There is also a need to prevent the loss of **traditional knowledge** and practices and for the maintenance of local plant and animal varieties and breeds. At the same time, the "fisheries" sector is of profound significance. Incidentally, the National Agency for Fisheries and AQC, established within the Ministry of Agriculture and Food, is the executive body of the central administration for implementation of the national policy on fishery and AQC and for application of the fisheries legislation in **Bulgaria** (Irina, 2011).

(6) In **Denmark**, Nordic cooperation is one of the world's most extensive and significant forms of regional collaboration, involving **Denmark**, Finland, Iceland, Norway, Sweden, and three autonomous areas, viz., the Faroe Islands, Greenland, and Åland. Having reference to **Denmark**, Eliasen and Sverdrup (2009) dealt with the Nordic experience of fisheries management in relation to the reform of the EU CFP. Nordic cooperation has firm traditions in politics, economy, and culture. It plays an important role in European and international collaboration, and aims at making a strong Nordic community in the powerful Europe. Further, according to Søreng (2008), the Nordic countries developed seven distinct types of fisheries management. In 2003, ITQs were introduced in the **Danish herring fishery** on an experimental basis. The reason for the experiment was,

first and foremost, the need for a modernization of the rather old Danish pelagic fishing fleet in order to meet the quality requirements of the fish processing industry. Further, with regard to Nordic experience of comanagement, the comanagement arrangements could be labeled as follows:

(i) **Instructive management**, where decisions are made in the government system and fishers are instructed in decisions. This is not a comanagement arrangement.
(ii) **Consultative comanagement**, where fishers are consulted on proposals before decisions.
(iii) **Cooperative comanagement**, which occurs between fishers and authorities as equal partners in developing proposals and for management decisions.
(iv) **Advisory comanagement**, where fishers advise governments on decisions to be made and the government endorses these decisions.
(v) **Informative comanagement**, where governments have delegated the authority to make decisions to user groups, who are responsible for informing government about the decisions made.

With regard to Nordic experience of reduction of discards, discards are the portions of fish catches that are thrown back into the sea as per FAO. Discards are to be avoided. Concomitantly, bycatch rule is a measure that makes it possible to undertake fishing, even if the targeted species is mixed with other species. In the Nordic countries, the fishers are allowed to have certain rates of species in their catches and to land amounts of fish equivalent to those rates.

Notwithstanding the above, there had been a focus on the local self-established fisher groups in the mussel fishery in the Limfjord in **Denmark**. These groups had managed to get a strong, yet informal, advisory role in the local management by collectively developing fishing plans and other self-management initiatives within the framework of the formal regulations. Incidentally, a conflict between the fishers regarding the enforcement of internal agreements illustrated how public power over enforcement could be "borrowed" by the groups to achieve compliance and curb possible free riding.

(7) In **Finland**, Salonen and Mutenia (2007) portrayed that the Lake Inari is a large, subarctic, oligotrophic, and regulated lake.

(8) In **France**, Anon (2001a, b) dealt with country note on fisheries management systems. CBFM in a participatory mode is applied in a number of countries, notably in **France**, etc., for the management of fisheries. Further, the industry had registered a total turnover of €1.12 billion and employed a total of 23,500 fishermen (20,525 in metropolitan **France** and 2975 in the overseas departments). Incidentally, France implements CFP in the community waters being a member of the EU. Incidentally, Olivier and Olivier (2001) opined that RBM regimes are considered by economists as an important solution to the problems of excess capacity and biological overharvesting of fisheries in many countries, including France. Further, Scott, A.1 (988) underlined the fact that commissions had been set up in certain countries, including France, with powers to coordinate national fishery controls under various international treaties dating back to the beginning of the century.

(9) In **Germany,** Döring (2001) worked on the conservation and sustainable exploitation of fisheries resources under the CFP. There had been efforts to deal with the issue of competition between small-scale gear operators and the trawl operators by means of a **community-based fishing** model. Concomitant to the above, Buchanan (1965) called "goods," which are neither private nor common, as "club goods" and the benefit of these goods is restricted to the club members only.

(10) In **Hungary**, Anon (2008a, b) dealt with fisheries OP of the Republic of Hungary for the period 2007–13 in connection with the renewal of the fishery sector in Hungary. **In Hungary**, the most commonly used indicator is the contribution of the fisheries and AQC sector to the GDP of the country in order to estimate the significance of the sector in the national economy. Further, in **Hungary**, commercial fishing is done in reservoirs, in addition to other kinds of water bodies, partly by using traditional fishing techniques and partly by applying certain techniques of fishpond farming. In fact, the capture fishery always has an outstanding role in the utilization of natural resources in the aquatic ecosystems and with its uniqueness and culture, it was able to apply a systematic and the so-called **sound management approach** in the utilization of natural resources. **Wetlands** are precious abiotic resources that serve as potential habitats for the coveted aquatic biota, notably the fishes, wetland birds, AM, etc. The attempts for the transformation of the **wetlands** into areas of other agricultural activities as well as misuse of the different kinds of **wetlands** have been causing significant damages including a depletion of the number of species in its BD resource. Nevertheless, the ecological value of Hungarian fishponds is also outstanding, both on domestic and European level. In fact, these fishponds have a core position in the European ecological network. Meanwhile, the HBMS (Anon, 2001a, b) was a national program for observing the trend of biological diversity in Hungary, supervised by the Authority for Nature Conservation of the Ministry of

Environment. In Hungary, there had been efforts to implement **CBFM** in a **participatory** mode as an efficient *modus oprandi* for effective management of the water bodies and their bioresources.

(11) In **Ireland**, Kenneth Irvine (2014) dealt with integrated water resources management (IWRM) from local to global sustainability having relation to **Ireland**.

(12) In **Italy**, there may be various background needs for implementing successful management of natural resources, notably the water bodies and their bioresources, especially, the fisheries as opined by Anon (2004). Incidentally, **CBFM** by involving the community stakeholders from the grassroot level in a **participatory mode** is an efficient *modus operandi* for the successful management of the natural resources, notably, the water bodies and the fisheries.

(13) In **Mediterranean and the Black Sea region**, Gwenaël et al. (2008) reported about the artisanal fishermen in the area of Port-Cros National Park in the **Mediterranean**. Incidentally, the **Artisanal fisheries involve local fishing units**, small amounts of capital, small vessels, and short coastal trips for small commercial or self-consumption purposes. Gutierrez (2015) dealt with the management and comanagement options for small-scale fisheries in the **Mediterranean and the Black sea**. There had also been a detailed discussion about the history of management of the fisheries.

(14) In **Norway**, Vera et al. (2012) dealt with the conditions under which CBM may work effectively along with an example from Sa'mi Pastoralists regarding their management of pastures sustainably. Further, in **Norway**, Søreng (2008) dealt with fishing rights discourses with a note on indigenous versus nonindigenous voices. Still further, in **Norway**, in 1973, when a resource crisis hit the pelagic industry, a license scheme was introduced in order to limit the number and size of the pelagic purse seiners. The next step in establishing RBM in Norwegian fisheries was a quasi-Individual Quota (IQ) regime in the cod fisheries in 1990. Further, the **first case** of comanagement is said to be a formal consultative comanagement arrangement in **Norway**. In this arrangement, an informal advisory role is given to the Norwegian Fishermen's Association in the decision-making process on the principles for distributing shares of the TACs to different fleet categories. The formal consultative arrangement and the informal advisory arrangement may coexist with severe consequences for the many groups of stakeholders involved and for the legitimacy of the decisions on quota allocations. Further, the Norwegian fisheries management has a strong tradition of user-group involvement. Incidentally, today, the most important area for user-group participation in fisheries management is **quota management**. Incidentally, the most of the commercially important Norwegian fish resources are shared with other countries. The TACs are negotiated and decided upon in various international forums. The two most important arenas for user-group influence are within the national management of the TACs after they had been decided upon. **Firstly**, there is a formal system of consultation, the management meeting, where the stakeholder groups are invited to present their viewpoints on the government's management plans for the different fisheries. **Secondly**, there is an informal practice by which one of the stakeholder groups, the Norwegian Fishermen's Association, is granted a privileged role in connection with the formulation of the principles by which the Norwegian national quotas are distributed to the fleet categories (segments). This practice often is referred to as the "Trawler Ladder."

(15) In **Sweden**, the focus was on comanagement at the local level. In **Sweden**, a government led comanagement initiative had supported six local fisheries management groups. The groups were to identify local problems and eventually could draft and implement local management plans. The local groups consisted of fishers or a broader range of stakeholders and were formed with a high level of enthusiasm and energy. Having identified the management needs, the groups began to develop the local management plans. During this process, it became obvious that the mandate of the group(s) was not very clear. However, the groups had no formal power to implement the plans. Also no procedures were in place to take the plans to higher management levels for decision-making or even for consideration. Some of the plans were simply overruled by other interests at national or EU level. Further, in 2004, the **Swedish** government had launched a comanagement project to gain experience with different comanagement forms and decision-making processes in the local fisheries. The initiating hypothesis was "common resources are best managed in cooperation." Incidentally, the Swedish Government Board of Fisheries (SBF), responsible for fisheries administration, research, and development, and enforcement and control had suggested six comanagement initiatives to deal with different types of fisheries involving different stakeholder groups and, further, had a tradition of consultative or advisory comanagement with industry organizations at the national Nordic experience of fisheries management. Further, the SBF usually supports local management initiatives by supplying expertise and funds for the making local organizations to work for fisheries management locally. In both **Norway and Sweden**, there is some conflicting interests between fishers at local and national level within the fishermen's associations. Nevertheless, the other groups tend to get influenced for a role in comanagement arrangements. However, the conflicting interests, if any, could decrease the capacity for collective action within the comanagement groups. Conflicts are better to be taken care of internally within the comanagement groups.

(16) In **the Netherlands,** Stedinger et al. (2005) dealt with water resources systems planning and management with an introduction to methods, models, and applications.

(17) In **Romania,** Mirela et al. (2005) dealt with Romania Danube Delta Biodiversity project with reference to local benefits case study report. In 2002, the GEF started a study to analyze how attaining global environmental goals contribute to the generation of local benefits and vice versa in accordance with the GEF mandate. The report also dealt with the local impacts on the Romanian Danube Delta Biodiversity Project (DDBP). The project contributed to the conservation of BD in the delta; strengthening the capacity of the Danube Delta Biodiversity Reserve Authority (DDBRA), and the Danube Delta National Institute (DDNI). The project contributed to the sustainable management of the delta, one of Europe's most important wetlands. However, analyses have shown that there was very little conceptualization of the role of local people. It had been revealed that the local people are the main threat to BD through their almost unrestricted access to fish, the major natural resource in the delta. There had been efforts for the management and conservation of the various natural resources through different means, including the **CBFM**. Concomitantly, Giger (2008) dealt with certain aspects of **Romanian** lentic systems and their management.

(18) In **Switzerland,** Anon (2013) reviewed the key factors which supported the small-scale coastal fisheries management. The recognition of the wider ecological and economic implications of the so-called poor fisheries management had also stimulated an interest in more holistic approaches that go beyond fisheries and include **ecosystem-based approaches and integrated coastal management**. Such approaches were increasingly looking closely again at **community involvement** in **fisheries management** through **comanagement arrangements**, under which the community and the government share roles and responsibilities.

(19) In **United Kingdom,** Campbell et al. (2008) dealt with comanagement of sea turtle fisheries with notes on biogeography versus geopolitics. In Great Britain, as in many other countries in the world, there are many issues with regard to BD management and conservation, in general and AQC and fisheries management and conservation, in particular. There are instances of promotion of **comanagement** between local communities and the government agencies as a strategy to improve **fisheries management**. Incidentally, CBNRM emerged as a bottom-up alternative for a number of reasons, including documented examples of success, beliefs about community's interests in resource sustainability, and shifts toward governance models that emphasized both local participation along with the government regulations.

(20) In the **former Yugoslav Republic of Macedonia,** fisheries and AQC are potential sources of nutrition and avocation. Anon (2008a, b) dealt with Fourth National Report of the implementation of CBD in Republic of Macedonia for the period 2006–08. Further, in connection with the Progress Report for 2013, Anon (2013) dealt with an extract from the communication from the Commission to the European Parliament and the Council with regard to enlargement strategy and main challenges. Moreover, management of BD is a significant issue in the management and sustainable harvest of natural resources in **Macedonia,** as in many other countries (Anon, 2008a, b). Further, there may be ITKs available for effective management and conservation of resources. Also, **CBFM** in a **participatory** mode could be regarded as an effective *modus operandi* for efficient management of the fish resources. Still further, it may be noted that the gene of the local indigenous community is to be maintained, conserved, and propagated in order to keep faith with the traditional profession with precious ITK and in order to continue with the race. In this regard, the community is to be preserved from our side with enough of rehabilitating measures and socioeconomic support; else, the coveted gene may be lost from us for ever. For instance, it could be mentioned that many of the traditional indigenous fishing communities with enough of ITK in India, particularly in NE India, not being rehabilitated properly as yet are said to be losing faith in their traditional, age-old, hereditary profession of fishing, and, by and large, are getting converted into nonfishermen, having been suffered from socioeconomic hardships and status. As such, an urgent initiative is essential for a letup in the situation to conserve the fishermen gene and also for the effective management and conservation of the fishes (Kar, 1990, 2003, 2007, 2013; Dey and Kar, 1989). Macedonia had conducted some amount of works in this direction.

(21) Prior to the introduction of the ITQ system in the late 1970s, Iceland had practiced a wide range of fisheries management systems. These included access licences, fishing effort restrictions, investment controls, and vessel buyback schemes, all of which were found to be not satisfactory. The first ITQ systems were introduced in the herring fisheries, following the establishment of the Icelandic EEZ. Since then, the system had been extended in several steps. In 1984, a limited form of ITQs was introduced in the demersal fisheries, which were, by far, Iceland's most important fisheries. In 1991, a uniform Nordic experience of fisheries management with fairly complete ITQ system was adopted in all Icelandic fisheries, applying to all vessels above a certain minimum size. In 2004, the system was expanded to cover all commercial fishing vessels. Currently, the system comprises 25 fish species and about 35 different fisheries.

(22) In the beginning of the 1990s, **the Faroe Islands** experienced catches at a historically low level from the most important demersal fish stocks. This had fostered a severe economic crisis in the fishing industry. It was decided to introduce a more effective fisheries regulation and an ITQ management system was preliminarily introduced in 1994 for vessels of 20 GT and above, whereas the smaller coastal fishing vessels were fishing on a general annual quota. This system was met with much reluctance from both the fishing industry and a number of politicians, and caused a substantial revision of the Commercial Fisheries Act. Incidentally, there had been certain elements of the RBM management system applied in the Faroe Islands, since 1997.

(23) Some of the other works related to management of water bodies in Europe and also may be elsewhere are those of Tammi et al. (2002), Argillier et al. (2002a, b) Mehner et al. (2005), Tonn and Magnusson (1982), Johnson et al. (1977), Dolman (1990), Schupp (1992), Ruoppa and Karttunen (2002), EU (2000), Gassner et al. (2003a, b, c), Zick et al. (2004), Clarke and Warwick (1997), Clarke and Gorley (2001), Gassner et al. (2005), Haempel (1930), Bauch (1963), Johnson et al. (1977), Tonn and Magnusson (1982), Dolman (1990), Schupp (1992), Spindler (1997), Argillier et al. (2002a, b), Tammi et al. (2002), Mehner et al. (2005), Gassner et al. (2005), Steedman et al. (1996), Kelso et al. (1996), Steedman et al. (1996), Gassner et al. (2005), Tyler-Walters et al. (2009), Stelzenmüller et al. (2010), Stelzenmüller et al. (2010), Hiddink et al. (2007), Moyle and Light (1996), Falk-Petersen et al. (2006), Kaatra and Simola (1985), Mutenia and Salonen (1994), Salonen and Mutenia (2004), Mutenia (1995), Salonen and Mutenia (2004), Cunningham (1994), Moloney and Pearse (1979), Scott (1988), Neher et al. (1989), Arnason (1991), Hannesson (1991), Bjorndal and Munro (1997), FAO (1993), Troadec (1994), Hannesson (1996), Bjorndal and Munro (1997), Copes (1986, 1997a, b), Scharmann (1994), Moriarty et al. (2000), Agrawal (2005), Jentoft and McCay et al. (1995), Lim et al. (1995).

5.5 America

The **continent of America** is bestowed with rich and varied resources of water bodies and BD under a variety of agroclimatic conditions. As such, there is a need for the efficient management of the lentic and lotic water bodies and their bioresources for sustained harvest and conservation. Incidentally, there could be different types of management *modus operandi* in different parts of the continent.

5.5.1 Argentina

Argentina is a country with rich fishery potential. The elements of effective fisheries management are well understood and appreciated. Strong management means enacting measures to prevent overfishing and more importantly implementing measures to reduce fishing pressure, if stocks become depleted. Key practices revealed through new surveys included evaluating the status of fish and shellfish stocks; designing appropriate management measures to limit fish mortality; and enforcing these regulations to prevent or reduce negative impacts of fishing.

5.5.1.1 Governance predicts upside potential

Despite awareness of what makes up strong fisheries management, there seemed to be a lack of publicly available information on the status of fisheries management in Argentina. Likewise, there appeared to be some amount of void in information on fisheries management globally. Thus, in order to address this gap, **Ray Hilborn** and **Michael Melnychuk** from the University of Washington conducted an analysis to further the understanding of the effectiveness of management approaches in regulating fishing mortality. The study included an independent survey done by experts of a country's fisheries discipline, providing an initial rapid assessment of fisheries management and governance systems. A total of 129 surveys across 28 countries were done, including the 20 countries that catch the maximum amount of fish in the world. For each country, respondents were presented with a list of 10 species: four with the highest volume of landings, four with the greatest estimated landed value, and the remaining 2–6 species were sampled randomly in proportion to their landings and landed value. The fisheries management survey was characterized by four dimensions of fisheries governance. Within each dimension, 8–14 questions were asked about the use and effectiveness of specific management strategies and tactics employed.

5.5.1.2 *Effectiveness of fishery governance by country*

Though the countries with the lowest index values, notably, China, Thailand, and Myanmar have low assessments across all dimensions; nevertheless, these results had highlighted the opportunity for comprehensive improvements in fisheries management. Further, countries, which fall toward the bottom of the index, notably, Indonesia and Brazil were introducing large reforms that could help tip the scales toward more positive rankings in the future. In Brazil, there is a commitment to implement an ambitious fisheries management plan, which could position the country to improve its governance system in the areas of management and enforcement. Similarly, Indonesia have introduced measures intended to improve its fisheries governance, such as fighting illegal fishing, particularly targeted at foreign fleets.

The results of the various studies helped in augmenting the ability to identify, specific areas and species, countries which are performing well, and where they stand to improve. For instance, Japan and India had relatively strong research capabilities; nevertheless, there are opportunities for improvements in management and enforcement dimensions. However, one caveat of the methodology was that the results were susceptible to potential biases of self-reporting since individuals most commonly responded to the countries in which they work and reside. Moreover, countries with poor regulation of corruption may rank higher, as the assessors might have rated the systems more in theory than in practice.

Notwithstanding the above, among the emerging markets, four countries that performed well in the survey were South Africa, **Argentina**, Chile, and Peru. South Africa, with an especially high country index and also moderate index value in the socioeconomic dimension had strong regulation of industrial fleets and had, of late, introduced policies to benefit previously disenfranchised coastal communities. Chile is also often cited as a model for its well-crafted artisanal fishing policies, which empower local communities to develop their own management strategies for high-value shellfish. Further, the survey questions did not distinguish CBM from centralized government management systems but, instead, focused on how effective the system was at limiting fishing pressure irrespective of the level of centralization.

Notwithstanding the above, the works of Daniela et al. (2006) related to an estimation of compliance with Article 7 (Fishery Management) of the UN Code of Conduct for Responsible Fishing (FAO, 1995) is a "live document," although it could alter with time. It is one of 53 such country evaluations covering the top *c.* 96% of the world's fish catch. The document represents the best attempt by authors at presenting a fair and objective evaluation of compliance using 44 questions derived from the code, using a wide range of cited source materials. Questions had been divided into six evaluation fields, viz., management objectives; framework (data and procedures); precautionary approach; stocks, fleets and gears; social and economic factors; and monitoring, control, and surveillance. However, the derivation of the 44 questions was described in Pitcher (1999).

Of these, the first three fields covered intentions of a country's legislation to adhere to the code, while the last three evaluation fields are intended to rate actual performance. Full details of the methods are available in Pitcher et al. (2006). Although precautions had been taken, but still there had been sufficient awareness that omissions and errors of interpretation may still remain for some countries.

5.5.1.3 *Management objectives*

Reference information obtained for main commercial fisheries, such as hake, short-fin squid, "abadejo," and "angostino" (FAO, 2003). The management of short-fin squid uses an escapement of *c.* 40% threshold to avoid high probabilities of low recruitment in the following season (Basson et al., 1996). The process of closing the fishery when the threshold is reached may not be immediate, taking from 2 to 3 weeks to be implemented. For this reason, the real level of the escapement is often between *c.* 35% and 20% of the stock (CEDEPESCA, 2003). On the other hand, in the case of the hake fishery, the stock assessment advice is based on a risk assessment that evaluated the consequences of the management strategies to the recovery and maintenance of the spawning stock above-defined targets: 400,000 tons and/or 500,000 tons (INIDEP, 2004). Whenever there was an offer to opt among three possible scenarios of larger to smaller risk, the fishing Federal Council and the application authority opted for the riskiest scenario. But, in 2005, the political authorities chose an option that had surpassed the greatest risk. The assessment scientists strongly recommended the use of the selective fishing gears and increasing no-take areas to protect juveniles. However, there was no clear evidence of acceptance of those recommendations.

5.5.1.4 *Are small-scale fishers considered in the reform plan?*

Notwithstanding the above, there had been no special national regulations for small-scale fisheries and the general regulations were applicable. Fishers, in fact require the national licenses and, preferably, have to be registered with the DNPyA (Dirección *Nacional de Pesca y Acuicultura*). However, since the fish in the areas of so-called provincial jurisdiction, the operation was regulated by each province (FAO, 2003). However, one of the most significant fleet targets was a multi-specific fishery called "coastal varied." There were suggestions for the identification of fishing units related to the spatial distribution patterns. These areas of ichthyologic coastal associations constituted as a useful tool for the design of the fishery management plan (Carozza et al., 2005). The CFP (Resolution 07/2005) established management and administrative measures applied to the "coastal varied" group in response to the recommendations. The measures included the use of a vessel monitoring system and the use of observers on board whenever possible. Part of the offshore fleet (targeted on hake) had fished in the coastal zone in some areas and times. This was an additional impact on the resources, working competitively with the small-scale fleet.

5.5.1.5 *Impacts of fishery on BD*

Some of the impacts of fisheries on marine BD were identified and assessed, such as bycatch and incidental capture of marine fauna. Several species of small cetaceans, such as dusky (*Lagenorhynchus obscurus*), Commerson's dolphin (*Cephalorhynchus commersonii*), and common dolphin (*Delphinus delphis*) were caught as bycatch in trawl fisheries in Patagonian waters including those directed at hake, red shrimp and, more recently, anchovy (Dans et al., 2003). The Franciscana dolphin (*Pontoporia blainvillei*) was also a bycatch of a small-scale artisanal gillnet fleet that supports the livelihoods of many coastal communities. The bycatch rates were almost 800 individuals per year.

Notwithstanding the above, according to FAO (2003), Argentina initiated an evaluation of the status of shark populations and the incidental catch of seabirds in response to International Plans of Action (IPoA) on sharks and seabirds. At the same time, there were also efforts toward the elaboration of management plans of action for mitigating these effects. However, uncertainty existed regarding rays and sharks mainly because of a lack of information. Nevertheless, Argentina is one of the 10 countries that is to land a large quantity of cartilaginous fish. Some species are completely discarded. Therefore, it is very much essential to get detailed information in order to know their status. In fact, Argentina had signed FAO agreements to develop a Plan of Action for Sharks and Rays but that assessment and long-term strategy have not been implemented. Nevertheless, Argentina had carried out an assessment of albatross and petrel bycatch taken by the national long-line fleet, operating in the EEZ waters. The Argentine government had signed an Agreement on Conservation of Albatrosses and Petrels (ACAP) document but, till 2006, the said matter was waiting for legal ratification by the National Congress. However, there was investigative report concerning the interaction of marine turtles (lute turtle, *Dermochelys coriacea*), green turtle (*Chelonia mydas*), stubborn Tortuga (*Caretta caretta*) with fisheries in the Southwest Atlantic. The fisheries and institutions involved had been identified in order to gather information but, unfortunately, there were no policies or conservation measures in place at local or regional levels (Domingo et al., 2005).

5.5.1.6 *Does the management plan aim to restore depleted stocks in this fishery?*

Management measures, such as restricted access, closed areas and seasons, and minimum sizes are applied to fisheries (FAO, 2003). Moreover, an emergency plan was in place, since 1998, to stop the trend toward the collapse of the hake stock. However, the plan was later undermined by different factors, although some recovery was observed in 2000 and 2001. The management set the objective to recover hake spawning biomass and recommended a TAC of 261,000 tons for the southern stock. Further, additional measures had included the effective use of selectivity devices in target and nontarget fisheries (e.g., bycatch in the shrimp fishery) and an increase in the permanently closed area for the protection of hake juveniles. Bottom trawls were forbidden inside the closed area. The authorities had chosen the more-risky catch option with a spawning biomass recovery in just 25 years.

5.5.1.7 *Is fishing gear mandated by the management plan to avoid bycatch of nontarget species, to limit environmental and habitat damage?*

According to the Argentine report to FAO (2003), the use of selective fishing gears was mandated in some fisheries in order to restrict incidental catch and discards of nontargeted species. Moreover, maximum trawling time and speed were also regulated; fishing depth zones were limited; observers were required on board; and monitoring and inspections were also required. In addition, specific mitigation measures had been applied to avoid the incidental catch of seabirds in long-liners including releasing lines at night, the use of tory lines; and avoiding the use of surface longlines. Designed selective devices were included in some regulations. However, the industry had rejected their use and, therefore, the existing regulations were not applied.

5.5.1.8 *Are ecosystem linkages with the fishery made explicit in the management plan?*

It may be mentioned that the management measures are usually based on species rather than on the ecosystem. However, there appeared to be no ambitious integrated management plans in Argentina.

5.5.1.9 *Are all the stakeholders in fishery resource identified and considered?*

Godelman (2004) asserted that the suspension of Resolution 73/2004 (which tried to establish an emergency plan for the hake fishery) had the virtue of triggering a political discussion about the implementation of the Federal Fisheries Code. In this connection, it may be mentioned that discussion tables were established by the national authorities to receive criticisms and proposals from the different sectors and provinces and so this may be considered as a case of identification and consultation with the interested parties in fisheries management. The stakeholders in these fisheries were identified, but there were no clear and transparent systems of participatory management to empower all sectors of the fishing industry in policy formation and management decisions.

5.5.1.10 *Are social, economic, and institutional factors related to sustainability evaluated with data?*

The significance of socioeconomic information in achieving conservation goals is generally appreciated in Argentina but there had not been much progress in this direction. There is a gap of complete information. Filling this gap could help in identifying the complex economic and social relationships between the industry and the communities and this could also assist to work with economic incentives to diminish excess capacity; to promote responsible fishing; to generate market incentives; and to encourage regional developments.

5.5.1.11 *Are management instruments under continuous review?*

In **Argentina**, there are examples of reviews dealing with ways and means to deal with situations during crisis, but this might not have been a regular feature. For instance, the management instruments were adjusted by different resolutions and decrees, since the near-collapse of the hake fishery in the late 1990s. However, the management instruments were not systematically reviewed under an adaptive management scheme. There were no long-term management plans, management procedures, indicators, monitoring, or evaluation systems. However, alteration (s) in management measures were motivated by urgent crises, industry complaints, technical proposals, or politics.

5.5.1.12 *Social and economic issues*

Is the fishery managed so as to minimize conflict among different sectors?

According to the **Argentine** report to FAO (2003), there were conflicts between coastal and industrial fisheries and among various kinds of fishing gears, etc. In addition, many conflict resolution mechanisms had been mentioned. However, the reality appears somewhat different from those engaged in the fisheries. Incidentally, the hake fishery had suffered from a high level of conflict among different sectors, for instance, the industry sectors, like the factory trawlers vs the shoreside delivery fleets, provinces vs nation, etc. Efforts had been made to identify the stakeholders in these fisheries, but, unfortunately, there were no clear and transparent system(s) of **participatory management** to empower all sectors of the fishing industry in policy framing and management decisions. However, there was also no proactive policy to solve the overcapacity problem. The government was working to implement the ITQ system, but there were problems with the first quota allocation. On the other hand, the shrimp fishery showed other conflict scenarios due to a large bycatch of hake. Moreover, there were intentions to set something like a hake bycatch limit to manage the shrimp fishery, especially, inside the hake closed areas. However, political intervention(s) had made it difficult to achieve responsible decisions based on consensus. In fact, there had never been any clear and transparent conflict resolution mechanism(s) to resolve conflicts between coastal and industrial fleets. Evidently, all the fleets seemed to want more resources to fish. Moreover, the coastal and industrial fleet had requested for their fishing zone, limited by 50-m deep isobaths. The government, though had agreed to that, there were implementation problem(s). The coastal and industrial fleets used to compete with the small-scale and artisanal fleet(s) and, the latter had also requested for their own exclusive fishing grounds (at least 3 miles away from the shore). Moreover, there were no clear rules for all of them and, hence, no effective management plans were in place to deal with such competitive schemes.

Are indigenous peoples' rights and needs being met?

Notwithstanding the above, the **Mapuche** and other indigenous coastal groups and communities in the south of Chile were not formally recognized in Chile but might have been recognized in Argentina. As such many people of Mapuche communities were involved as crew and workers in the hake fishery. However, there were no native Argentine groups with special fishing rights related to maritime fisheries.

Are the needs of local fishing communities being met?

There was no clear and transparent participatory mechanism where the artisanal and small-scale fishermen and coastal communities could defend their rights. The Federal Fishery Council was mostly concerned with industrial fleets.

When a change to the management of a fishery is made, is its cost-effectiveness evaluated?

Some of the facts had revealed that there was a lack of consideration for cost-effectiveness, alternatives, etc. Moreover, there was no clear understanding of the ITQ system's requirements and its consequences. Conversely, if market forces alone were to be allowed to manage the overcapacity issue under an ITQ context, there could be a lot of problems for small companies and their employees. The management system may, preferably, understand the impact of its management decisions on the social and economic scenario, which, in turn, could affect most of those involved in the fishery sector and, thus, there was a need to act for reducing the main impacts over the communities.

When a change to the management of a fishery is made, is its social impact evaluated?

With reference to the above, it may be mentioned that there were strikes by the Labor Union workers in two of the main fishing cities (Mar del Plata and Puerto Madryn) in 2005, on demand for better incomes. Unproductive seasons for shrimp and squid as well as low hake quotas had negative economic impacts on Patagonia. Thus, was an increased social tension because of that. Moreover, there were no explicit policies for different catch alternatives, for example, how to deal with an *c.* 30% hake TAC reduction. Thus, there were issues that were to be taken into consideration in a long-term management plan.

Concomitant to the above, the **Argentine** ITQ program is believed to manage four of the country's most commercially important species by allocating quota to individual vessels under a single CSP (Young, 2013). Incidentally, the program goals focus on long-term stock conservation, maximization of domestic employment, and the promotion of social stability. In this regard, it may be mentioned that the managers had incorporated special design features within the program to meet each fishery's requirements. Quota set-aside had given managers the flexibility to address fishery-specific social and biological goals, while the multi-criteria allocation process had incentivized investment in the domestic economy and compliance with fishing regulations. In 1998, the Argentine government enacted its Federal Fisheries Act (*Ley Nacional* 24.922) to achieve more sustainable development of its aquatic (FW and marine) resources. Part of this legislation had stipulated the formation of an ITQ, or CSP. After several years of design and experimentation, managers launched an official CSP in 2010, locally known as *Régimen de Cuotas Individuales Transferibles de Captura*. The CSP managed four species, viz., Argentine hake (*Merluccius hubbsi*), Patagonian grenadier (*Macruronus magellanicus*), Patagonian toothfish (*Dissostichus eleginoides*), and southern blue whiting *(Micromesistius australis)*.

5.5.1.13 *Road to a catch share*

Argentina rapidly developed its fisheries sector over the past few decades. Policies favoring foreign investment and economic development formed thousands of jobs in the processing and harvesting sectors. Increased catches and export-oriented production expanded national earnings (Johns, 1992). However, Argentina was not well equipped to sustainably manage the rapid growth of its fisheries. Incidentally, no comprehensive national fisheries law had existed prior to the Federal Fisheries Act of 1998 (UNEP, 2002).

The **limited entry management** approach, established in 1991 wasnot sufficient in halting the rising capacity in Argentine waters (Schonberger and Agar, 2001). Incidentally, fishing licenses could be easily transferred to new, higher capacity vessels upon decommission of outdated or sunken vessels (Abaza and Jha, 2002; Kalikoski et al., 2006). Meanwhile, European firms had transferred excess capacity from European waters; financed vessel upgradations; and had driven capacity growth through joint ventures (UNEP, 2002). Incidentally, the processing fleet experienced a fivefold increase in aggregate power, between 1989 and 1996; while, the ice trawler fleet was tripled in its capacity (UNEP, 2002). Meanwhile, in the Argentine hake fishery ice trawlers were replaced with much larger processor vessels imported from Europe (Schonberger and Agar, 2001; Abaza and Jha, 2002).

5.5.1.14 *Performance*

The CSP is said to be largely meeting its goals, 3 years after its implementation, and a number of notable improvements had emerged. The program provided a stable platform for management during the recent economic recession and had also provided with flexibility to buffer against economic shock. Incidentally, the fishermen achieved *c.* 100% catch limit compliance for each of the four species (Fish Source, 2013).

5.5.1.15 *Define program goals*

Fisheries became an increasingly significant component of Argentina's export economy and are particularly important to the local economy of the coastal communities, such as *Mar del Plata* and *Puerto Deseado*. Argentina's Federal Fisheries Act had outlined three primary objectives in order to promote stability in the sector:

Notwithstanding the above, the CSP was designed to achieve the objectives outlined in the Federal Fisheries Act. Specifically, the program had sought to maximize fisheries production at sustainable levels, while preserving traditional fleet composition and employing as many national citizens as possible.

5.5.1.16 Define and quantify the available resource

The CSP is to manage the Argentine hake (*Merluccius hubbsi*) stock; the Patagonian grenadier (*Macruronus magellanicus*); Patagonian toothfish (*Dissostichus eleginoides*); and the Southern blue whiting (*Micromesistius australis*). These species represent four of the most significant commercial species in Argentina, as they generate significant employment opportunities and export earnings for the country. Argentine hake, Patagonian toothfish, and southern blue whiting were believed to be the historically overfished species. Thus, in order to preserve these species and their associated economic benefits, CFP had included them as the first species to be managed by the CSP. The program also aimed at preventing the overexploitation of Patagonian grenadier, which is an increasingly **sought-after species**.

5.5.1.17 Define eligible participants

Shares were allocated to the vessels, and eligibility to participate in the program was restricted to the vessels owned by individuals or companies holding fishing licenses for one or more of the four species. In Argentina, fishing licenses were reserved for national citizens and companies registered in the country. This effectively limited participation in the program to Argentine citizens and domestic companies only. Further, the program was specifically designed to promote employment and retain a diverse fishing fleet. For the Argentine hake and Patagonian grenadier fisheries, managers were incorporated additional eligibility provisions to address fleet diversity. For Argentine hake, province-specific "social reserves" were specially designed for smaller-scale vessels, while, a "reallocation fund" for Patagonian grenadier had provided with fishing opportunities under the CSP for vessels with lower catch histories.

5.5.1.18 Develop administrative systems

The CSP was administered by CFP. It was said to oversee quota transfers and trading. CFP had also to approve all trades, whether they are full or partial. In addition, there were arrangements to manage a system of onboard and onshore observers in order to provide with precise information on fishing mortality, size, sex, and age of catch. However, the vessel operators had to pay for their onboard observers.

Further, CFP had collected a small transaction fee from each quota transfer, which went to the National Fisheries Fund (NFF), the so-called main source of funding for the CSP. In addition to the fees from the quota transfers, revenue was generated from the issuance of a fishing permit(s) through entry fees from foreign vessels; penalties for **infractions** of the Federal Fisheries Law and sales from confiscated equipments and vessels.

5.5.1.19 Assessment, performance, and innovations

Three years after its implementation, the CSP met most of its goals. Each of the four fisheries achieved catch limit compliance. The most significant improvement occurred in the Argentine hake fishery, in which, catches had exceeded legal limits on a regular basis. Patagonian toothfish appeared to be slowly rebuilding; while, improved accountability under the CSP was helping stabilize the Patagonian grenadier stock (MSC, 2012; Fish Source, 2013). However, one of the most notable achievements was the certification of the Patagonian grenadier fishery by the Marine Stewardship Council in May 2012 (MSC, 2012). The program was also said to be meeting most of its economic and social goals. Export prices had increased for each of the four species since program implementation and allocation were specifically designed to maximize the utilization of Argentine labor (Villasante and Sumaila, 2009).

Enhancing the welfare of fishing communities, achieving sustainable management of fish resources, and sustainable trade in fish and fish products are critical developmental goals, which had been reaffirmed in the Rio+20 Outcome Document (FAO, 2018). While dealing with related aspects, FAO (2018) opined that by transitioning to a green economy, through measures, such as the elimination of harmful subsidies and the use of greener production and processing methods, fisheries resources could be expected to be managed more sustainably for the benefit of present and future generations in many places, including **Argentina**. In addition, better information management and conservation measures, including monitoring, control and surveillance systems, need to be developed and implemented with a view to changing awareness and incentives for economic operators with regard to the exploitation of fishery resources. In this context, it may be pointed out that trade could both support and benefit from the transition to a green economy by encouraging better long-term management of global fish resources and improved sustainability along the supply chain.

Notwithstanding the above, there had been works, which tried to identify general trends and trade opportunities associated with the transition to a green economy in the fisheries and AQC sectors in different places including

Argentina. It tried to explore ways of responding to the international demand for sustainable fish and fish products and how they could be better integrated into relevant markets. Incidentally, growing international demand for the integration of sustainability considerations offers increased trade opportunities in.

(a) exports of sustainably and legally produced fish and fish products;
(b) exports of certified wild-capture fish and processed fish products;
(c) exports of certified AQC and processed AQC products;
(d) exports of sustainable aquarium fish; and.
(e) promotion of nonconsumptive uses, such as tourism.

5.5.1.20 Environmental and economic context for greening the economy

Current state of world fisheries and aquaculture

It is quite well-known that fish and fish products constitute a major source of avocation, nutrition, and recreation in the global economy. Incidentally, fish products originate from two main modes of production, viz.,

(a) harvesting of wild fish (marine and FW) and.
(b) aquaculture.

The latter is defined by the Food and Agriculture Organization (FAO) of the United Nations (UN) as "the farming of aquatic organisms including fish, molluscs, crustaceans and aquatic plants."

It may be mentioned that, in 2010, wild capture fisheries and AQC supplied the world with *c.* 148 MT of fish, worth *c.* US $217.5 billion. Out of this, *c.* 128 MT (*c.* 86%) had gone for human consumption. Incidentally, preliminary data (2011) indicated that production had increased to *c.* 154 MT, out of which, *c.* 131 MT (*c.* 85%) were consumed as food.

Notwithstanding the above, even though, wild fish capture represented *c.* 60% of the total fish production in 2010, the total catch from the wild fisheries remained relatively stable throughout the last three decades. Incidentally, the source of significant growth in the global production of fish, since the late 1980s, had been AQC. Since then, it increased almost 12-fold, at an average annual rate of *c.* 8.8%. However, AQC today accounted for almost half of the total fish supply for human consumption. On the other hand, in 2010, the global production of farmed fish was *c.* 59.9 MT, compared to *c.* 55.7 MT in 2009 and *c.* 32.4 MT in 2000. When farmed aquatic plants and nonfood products are included, world AQC production in 2010 was *c.* 79 MT, with a market value of *c.* US $125 billion. Incidentally, AQC is also the world's one of the fastest-growing food production activities based on animal protein.

Concomitant to the above, it may be pointed out that "overexploitation" not only causes negative ecological consequences but also reduces fish production, which leads to negative social and economic consequences. Thus, there had been suggestions that `in order to increase the contribution of fisheries (marine + FW) to the food security and the economy, as well as, for the well-being of the coastal communities, effective management plans may, urgently, be put in place to rebuild the overexploited stocks.

Greening the fisheries and aquaculture sectors

The Green Economy Report (GER) stressed the importance of greening the fisheries and AQC sectors, so as to ensure their medium- and long-term profitability. The GER tried to predict that an investment of US$ 100 to 300 billion in strengthening fisheries management and vessel decommissioning could lead to a significant increase in the catch. It would also increase resource rents from a negative US$ 26 billion to a positive US$ 45 billion annually.

Key findings of the green economy report

The GER predicts that a single investment of US$ 100–300 billion would significantly reduce excess capacity. In addition, it should result in an increase in fisheries catch from the current 80 MT a year to 90 MT by 2050, following the recovery of fish stocks. The present value of benefits from greening the fishing sector is about 3–5 times the necessary additional costs. In a scenario of a bigger amount of spending of 0.1%–0.16% of GDP over the period 2010–50, the plans could be, to reduce the vessel fleet; to relocate employment; and, to better manage the stocks in order to increase the catch in the medium and longer term. Incidentally, *c.* 27%–59% more employment could be achieved, relative to the baseline by 2050. In this same scenario, >70% of the fish resources than in 1970 could be available by 2050 (between 50 MT and 90 MT per year). This could be against a mere small amount under a BAU scenario, where no additional stock management activities are practiced.

Notwithstanding the above, the value of fisheries ecosystems was assessed by The Economics of Ecosystems and Biodiversity (TEEB).

5.5.1.21 *The economics of ecosystems and biodiversity*

The TEEB is a global initiative focused on drawing attention to the economic benefits of BD. Its objective is to highlight the growing cost of BD loss and ecosystem degradation. TEEB presents an approach that may help the decision-makers to recognize, demonstrate, and capture the values of ecosystems and BD, including how to incorporate these values into decision-making. TEEB highlights the economic importance of fisheries by considering the impacts of overfishing, fish stock collapse, and environmental damage on food and job security and economic efficiency. One example in **Argentina** illustrated that continued overfishing of Argentinean hake threatens their long-term ecological and economic viability. Moreover, ecological models had projected that if the existing quota were met, no-fishing zones in the nursery grounds were respected, the *c.* 120% overcapacity of fishing vessels (which are in vogue) were reduced by *c.* 25%–50%, and the stock of hake would recover leading to significant economic gains increasing the net present value from US $65.7 to US $118.5 million for fresh fish and from US $263 to > US $460 million for frozen fish over the period 2010–30.

Concomitant to the above, greening the world's fisheries could help to restore damaged marine ecosystems. Further, the adoption of green technologies, such as low-impact fuel-efficient fishing methods, and investments in alternatives to the use of fossil fuels would help to reduce the carbon footprint of the fisheries sector. In fact, when managed sustainably, the fisheries sector could support a great number of communities and enterprises. They are said to be potential in generating employment opportunities and raise household income, particularly for those engaged in artisanal fisheries. The fisheries sector is also equally the potential to provide people with rich nutrition. In fact, sustainably managed fisheries could generate more income than unsustainably managed fisheries, mainly due to more secured and larger fish stocks.

5.5.1.22 *Trends, challenges, and opportunities*

Fisheries

AQC production is said to have the potential to take some of the pressure off from wild fish stocks while supporting livelihoods and food production. This is particularly the case in developing countries, which, of late present the largest share of world AQC production by volume. Sustainably managed coastal, FW, and marine AQC (mariculture) offer significant scope for green growth and employment opportunities. Further green technologies for AQC are in the process of development in order to harness this green growth potential, including innovative production systems that use environmentally friendly feeds. The involvement of local communities is a key tenet of the transition to a greener economy. In this respect, the sustainable management of artisanal fisheries may also yield significant benefits, particularly, where local communities form part of integrated approaches to fisheries management. In the community-based EAFM, the local community is said to manage the fish resources in close cooperation with the government and the other partners. This applies, for example, to the "Piaba" (Brazil) and "Seahorse" (The Philippines) aquarium fish projects. The marine aquarium trade, in fact, supports jobs in predominantly rural, low-income coastal communities. Further, where resources are managed sustainably, aquarium trade provides with strong economic incentives for coral reef conservation in regions where other options for generating revenue are limited, as the relevant products are among the highest value-added products originating in the reef.

Fisheries subsidies and marine resources management: Lessons learned from studies in Argentina and Senegal

Notwithstanding the above, much amount of information had been collected about **fisheries Subsidies and marine resources management** from **Argentine and Senegalese** studies discussed in the following account (Anon, 2002). **These** were conducted within the framework of two broader UNEP Country Projects that involved the collaboration of a number of individuals and organizations. Each of these projects aimed to assess the environmental and related social and economic effects of trade liberalization and trade-related policies as well as to develop policy response packages that contribute to mitigating the negative impacts and enhancing the positive impacts identified through the assessments.

United nations environment program

The United Nations Environment Program (UNEP) is the overall coordinating environmental organization of the United Nations system. Its mission is to provide leadership and encourage partnerships in caring for the environment by inspiring, informing, and enabling nations and people to improve their quality of life without compromising that for the future generations. In accordance with its mandate, UNEP works to observe, monitor, and assess the state of the global environment and improve our scientific understanding of how environmental changes occur; and, in turn, how such changes could be managed by action-oriented national policies and international agreements. UNEP, which is headquartered at Nairobi is the apex body to deal with such matters.

Fish is the primary source of protein for *c.* 950 million people worldwide and represents an important part of the diet of many more. Fisheries are also a source of employment for *c.* 200 million people directly depending on ocean fishing for their livelihoods. Approximately 40% of the world fishery production enters international trade with nearly half of fishery exports deriving from developing countries, and, further, some of the developing countries may represent even up to *c.* 80% of the total exports (Dommen and Deere, 1999).

Division of technology, industry, and economics

The mission of the Division of Technology, Industry, and Economics (DTIE) is to encourage decision makers in government, industry, and business to develop and adopt policies, strategies, and practices that are cleaner and safer, use natural resources more efficiently, and reduce pollution risks to both human beings and the environment. The approach of DTIE is to raise awareness by fostering international consensus on policies, codes of practice, and economic instruments through capacity-building and information exchange and by means of pilot projects.

Economics and trade branch

The Economics and Trade Branch (ETB) is one of the units of the Division of Technology, Industry, and Economics (DTIE). The work program of the branch consists of three main components, viz., economics, trade, and financial services. Its mission is to enhance the capacities of the countries, particularly the developing countries and countries with economy in transition, to integrate environmental considerations in development planning and macroeconomic policies, including trade policies. UNEP's mission in this field is also to address the linkages between environment and financial performance and the potential role of the financial services sector in promoting sustainable development. The trade component of the program focuses on improving countries' understanding of the linkages between trade and environment and enhancing their capacities in developing mutually supportive trade and environment policies, and providing with technical inputs to the trade and environment debate through a transparent and a broad-based consultative process.

In recent years, after four decades of steadily expanding catches, there had been significant declines in fish stocks, especially of preferred species for human consumption, such as cod, haddock, and plaice. Incidentally, it may be mentioned that fisheries production increased by *c.* 6% per year for the two decades following 1950 **(trebling** from 18 to 56 MT), but the average rate of increase declined to *c.* 2% between 1970 and 1980, and fallen to almost zero in the 1990s. Major ecological, economic, and social damages were evident. In particular, declining catches had cost >100,000 jobs in the last few years among the world's *c.* 15–21 million fishers. Concomitantly, the cost of fish in some local market places had risen by leaps and bounds, placing fish out of reach for many low-income consumers (Weber, 1994).

Notwithstanding the above, overfishing of the world's marine resources was the main cause for the decline of fisheries productivity. According to the FAO, *c.* 50% of all fishery resources were fully utilized, and, of late, *c.* 25% still holds potential to be further exploited, and the remaining *c.* 25% are in severe danger of depletion and require major interventions to restore sustainable yields.

Concomitant to the above, widespread overfishing is widely recognized as a growing threat to the sustainable management of the world's fisheries. In major fishing states, investment in new capacity (vessels, equipment, and labor force) supported fishing efforts at levels significantly exceeding the reproductive capacity of fishery grounds. Excessive government support policies and especially subsidies to the fishing industry were suspected by many experts to have a direct causal relationship to recent trends in overfishing.

However, opinions still differ as to the relative importance of fisheries subsidies as a factor affecting the stability of fisheries resources. More work is required to address the linkages between fisheries subsidies and fisheries resource sustainability to guide progress toward a potential reform of fisheries policies worldwide. Policy reforms are, preferably, to integrate environmental, social, economic, and trade objectives to ensure the long-term sustainability of entire fishery ecosystems while minimizing any negative social and economic impacts on segments of the population relying on fishing for employment and/or for food. In particular, more empirical studies at the country level are needed to define and categorize current forms of government support and to assess their environmental, social, and economic effects.

Concomitant to the above, the UNEP supported these two studies on fisheries subsidies in Argentina and Senegal in 2001 to help to meet the need for additional study. Examining the main types of support and subsidies provided to the fishery sector over the last decades in Argentina and Senegal, these studies could provide further detailed assessment of their environmental and related socioeconomic impacts. They would offer valuable insights to trade and environment officials who would like to increase their understanding of the intricate relationship between subsidies and the environment in the fishery sector. They could also provide inputs to the national policy-makers seeking to promote productive and sustainable fishing industries in their countries.

5.5.1.23 *Historical background: Fisheries subsidies and world trade negotiations*

Subsidies in argentine fisheries

Notwithstanding the above, **Argentine** fisheries faced crisis situation(s) in some part of its history, as revealed from the studies on **subsidies in Argentine fisheries** (Maria Onestini, 2002). In Argentina, fisheries exploitation subsidized from both international and to some extent national sources, making this an intricate part of the issue. Fisheries exploitation in Argentina grew at unprecedented rates in recent years and this had been one of the country's most dynamic economic sectors in the past few years. Value addition was grown steadily and exports had also grown *c.* 478% between 1985 and 1995 while, in comparison, the total exports increased *c.* 159% during the same period.

Environment and trade

Concomitant to the above, a significant shift from overall underutilization to overexploitation of some fishing resources occurred in the 1990s. A clear connection with international markets is presented as a result of:

(a) The trans-nationalization of capital in the fisheries industry.
(b) The issue of permits to foreign boats to operate in Argentine jurisdiction.
(c) The fact that up to *c.* 90% of production is exported during the same periods.

In addition, a notable alteration had also occurred internally, given the high degree of growth of fisheries exploitation from Patagonian ports.

Notwithstanding the above, charter agreements with fleets from Asian countries were signed, during the early part of the decade mainly for the capture of squid. Another major change experienced in relation to Argentine fisheries, was the agreement with the EU, negotiated and ratified in the early 1990s, which came into place until the late 1990s. This accord the first of the so-called "second-generation" agreements on the subject signed by the European trade bloc, involved the participation of joint ventures and temporary associations of companies of Argentine and European capital in the fisheries industries. These alteration(s) as well as the global transformations in the fisheries sector have radically modified the way that fishing activities are conducted in the country. The above processes had taken place in the context of much economic change. Although the opening up of the economy had begun in the 1980s, the major transformations of the regulatory context took place in the 1990s with a new set of policies that had combined stabilization programs with structural reforms and liberalization of trade. However, its main components are as follows:

(a) Stabilization program: fixed foreign exchange rate and tight monetary policy.
(b) Commercial openness (trade liberalization)
(c) State reform (privatization of public utilities)
(d) Deregulation of markets and economic activities.

Concomitant to the above, the opening up of trade occurred within this structural reform context, thus paving the way for a series of effects. Among the effects, some were believed to be more salient, such as.

(i) an initial increase of foreign direct investment (FDI);
(ii) increases in domestic consumption, production, productivity, investments, and exports;
(iii) access to new technologies on process and products, logistics, and communication;
(iv) organizational innovations; and.
(v) modernization of infrastructure and services.

Concomitantly, however, these profound transformations resulted in an economic system that had remained increasingly dependent on the supply of foreign capital inputs and capital goods. In addition, some negative trends could be pointed out, such as.

(a) higher rates of concentration and trans-nationalization of the economy, with the crowding of many small- and medium-sized firms;
(b) "de-clustering" processes in many industrial chains, as a consequence of the substitution of national productions by imports;
(c) high rates of unemployment and underemployment and deterioration of labor conditions. Etc.

Fisheries and the environment

The WTO's Agreement on Subsidies and Countervailing Measures ("SCM Agreement") comprises more concise and circumscribed definitions. The **definition of a subsidy** found in the SCM Agreement is now the only legal definition of a subsidy in international trade law. This agreement **defines a subsidy** in specific terms as a "financial contribution provided by, or at the direction of a government" that confers a "benefit" (WTO, 1999). It further specifies that in order to be considered as a subsidy, a contribution has to involve one of the following:

(i) direct transfer of funds or liabilities;
(ii) forgone or uncollected revenues that would otherwise be owned by the government;
(iii) provision of a good service to a firm or industry other than general infrastructure; and.
(iv) any type of income or price support.

Notwithstanding the above, analysis of fisheries subsidies and their relation to sustainable development had lacked in many aspects, not the least, being working definitions accepted by all parties involved. It was revealed that:

(a) a thorough analysis was missing on what is the amount of subsidies involved today in fisheries exploitation;
(b) a whole set of non-budgeted or non-evident subsidies that is to be taken into account.

The so-called failure by the states to recuperate the full economic rent in the trade of publicly owned resource rights is one of the most pervasive issues in natural resource exploitation and subsidies as well as one of the most difficult ones to unravel at this stage. As it had been pointed out that subsidizing natural resource production through the sale of access at such a low price that the rent is transferred from the state to the producer is one of the most ubiquitous forms of subsidies in natural resource exploitation, including fisheries exploitation. This had been the case in Argentina. Further, the impacts of subsidies on sustainable development appeared to be analyzed differently from "where one stands" and even from the confined or local-type analysis. Further global analysis on this issue is needed to be done, especially on subsidies which are classified as "good" subsidies when perceived as having positive environmental impacts. However, two examples of unsound classification of "good" subsidies could be found in the Argentine case. Firstly, the European subsidies employed for access to distant waters were categorized as positive subsidies; given that they were believed to have reduced pressure on natural resources in European waters. Nevertheless, as can be seen in this case, the outcome had been a transfer of the problem of overcapacity to distant waters. Another subsidy, cataloged as positive is the use of government funds for vessel buy-back schemes in order to reduce capacity, an instance of which is the Canadian buy-back scheme for the closed cod fishery. Nevertheless, this program, probably, had shifted excess capacity from one region to another because vessels retired from this type of exploitation were sold to other countries. These were mainly done in the developing countries, among which Argentina could also figure.

In fine, and more generally speaking, from an analytical as well as from a policy-setting point of view, more work still needs to be done regarding the impacts of subsidies on the use of marine fisheries. Nevertheless, it is now clear that subsidies play a negative role in overfishing practices in Argentina as well as in many other countries.

Notwithstanding the above, Orensanz and Seijo (2013) dealt with rights-based management in **Latin American** fisheries. For benthic fisheries, there had been a documentation of management experiences. For finfish fisheries, there were documentations of rights-based management studies covering various aspects. Some of the salient aspects are as follows:

(i) Limited entry or moratoria combined with a total allowable catch (TAC) in the Galapagos Islands sea cucumber fishery as well as in the sea urchin and Juliana clam fisheries of Chile, moratoria on catch shares in the diving fisheries for scallops from San José Gulf (Argentina), and so on.

(ii) Territorial use privileges based on seabed tracts in Chilean territorial use rights in fisheries (TURFs, "AMERBs").

(iii) Territorial community rights (traditional and indigenous users) in the Seri Indian fishery of Mexico.

On the other hand, for finfish fisheries, there were documentations of rights-based management studies for aspects like individual vessel quota (IVQ); management of the anchovy (*Engraulis ringens*) fishery of Peru; individual stakeholder quota; management of the hake (*Merluccius gayi gayi*) fishery of Chile; Community territorial use rights in the Gulf; weakfish (*Cynoscion othonopterus*) fishery of the Gulf of California, Mexico; individual effort quotas (IEQs) for artisanal communities in the multispecies fishery at Coiba National Park, Panama, and so on.

Each case had specified the main attributes of the access rights (in a broad sense, including privileges), whether formal or informal, with regard to:

(i) how the rights are conferred and upheld,
(ii) exclusivity of participation in the fishery,
(iii) duration of the rights conferred,
(iv) security or quality of the title conferred by the rights,
(v) transferability, divisibility, and flexibility in the use of the rights, and.
(vi) actual rights enforceability and corresponding compliance with use rights limitations.

Notwithstanding the above, the studies also reported on aspects of the harvest strategies in place, including the following:

(i) fishing methods and gears,
(ii) when fishing is authorized to take place,
(iii) harvest controls and monitoring, and so on.

Concomitant to the above, Anon (2005a, b) reported about sustainable fisheries and livelihoods in **Latin America** with reference to the imperative of recognizing the resource access rights of artisanal fishermen, coastal communities, and the indigenous people.

The rationale of artisanal fisheries is to provide benefits for both current and future generations, and its intrinsic characteristics make it the most effective way to exploit aquatic ecosystems on a sustainable basis.

Providing the basis for both economic and social activities, artisanal fisheries also have important cultural dimensions. The rationale of artisanal fisheries is to provide with benefits for both current and future generations, and its intrinsic characteristics make it the most effective way to exploit aquatic ecosystems on a sustainable basis.

There had been a congregation of *c.* 50 participants, belonging to various organizations including artisanal fish worker organizations, indigenous people's organizations, NGOs, research institutions and universities, from the Southern Region of Latin America, notably, from Ecuador, Venezuela, Peru, Chile, Brazil, Uruguay, and Argentina along with delegates from UK, India, and Norway, from 1 to 4 March 2005 to discuss issues and consider measures to guarantee fishing access rights for artisanal fish workers, coastal communities, and indigenous people in the Southern Latin American Region. The meeting was organized by the International Collective in Support of Fish workers (ICSF) and CeDePesca and supported by the FAO Fish Code Program.

The understanding of how globalization processes in fisheries impacting at the regional level was increased, and, amidst considerable diversity among the participants, there was sharing of many common characteristics and concerns.

Concomitant to the above, it was affirmed that the artisanal fishing plays a vital role in providing with food and livelihoods to a big part of the globe, including the Southern Latin American region, where artisanal fishing is characteristically a small scale, extensive activity, carried out directly by fish workers using selective fishing gears. Providing the basis for both economic and social activities, artisanal fisheries also have important cultural dimensions. The rationale of artisanal fisheries is to provide with benefits for both current and future generations, and its intrinsic characteristics make it the most effective way to exploit aquatic ecosystems on a sustainable basis. Concomitantly, the management of the water bodies and their resources through CBM could further improve the situation.

Concomitant to the above, it is known about the important roles played by both men and women in the harvesting, processing, and trading with fish and fishery products, and in sustaining coastal communities and indigenous people whose livelihoods, quality of life, and culture depend on aquatic ecosystems. In view of the above, there are high concerns about the negative impacts of globalization and liberalization in the region. These are believed to be due to unregulated development and expansion of economic activities in the coastal areas, including intensive AQC, industrial development of fisheries and other sectors, and luxury tourism. The result is that the coastal ecosystems are being degraded, aquatic resources depleted, with artisanal fish workers, coastal communities, and indigenous people being displaced from their homes with their livelihoods disrupted. Further, privatization of the natural resources is leading to the concentration of ownership, further enhancing these disturbing trends.

Concomitant to the above, in addition to effective management mechanisms involving the community, there are urgent needs on other aspects, like the following:

(i) To recognize the priority access rights of artisanal fish workers, coastal communities, and indigenous people to coastal lands and near-shore coastal waters.
(ii) To establish legal frameworks and institutional arrangements for applying an integrated approach to granting access and use rights in coastal ecosystems, in ways that optimize equity and sustainability.
(iii) To institutionalize systems of comanagement for artisanal fisheries, with participatory and transparent mechanisms for decision-making, and with provisions for strengthening the capacity of organizations of artisanal fish workers, coastal communities, and indigenous people in order to enable them to participate fully in the management decisions.
(iv) To establish effective regulation for the prevention of potentially destructive fishing practices, such as trawling as well as to halt the development and expansion of intensive and other unsustainable AQC activities.
(v) To support coastal communities to develop community tourism, ecological agriculture, and other activities which may complement local economic development.

(vi) To design international and national level fishery research programs with the active participation of organizations of artisanal fish workers, coastal communities, and indigenous people in ways that may incorporate their traditional knowledge, in order to meet their specific technical, economic, and social requirements.

(vii) To motivate the artisanal fish workers, coastal communities, and indigenous people for the appreciation of the importance of the FAO Code of Conduct for Responsible Fisheries in promoting sustainable fisheries and food security, and in addressing the specific needs of artisanal fisheries.

(viii) To persuade the different national governments to take the necessary measures to implement the code and related international instruments for effective fisheries management.

(ix) To try for demarcation of fishing zone exclusively for the artisanal fishers.

(x) To support the calls of artisanal fish workers' organizations, coastal communities, and indigenous people to ban the use of potentially destructive fishing techniques, like trawling in the zone reserved exclusively for artisanal fishing, etc.

(xi) To try to reject the use of ITQs as a management tool for artisanal fisheries, as the use of ITQs could jeopardize the legitimate rights of artisanal fish workers, coastal communities, and indigenous people to secure their livelihood activities.

(xii) To recognize the significant role played by women in artisanal fisheries, who in addition to undertaking household tasks and struggling with their children's education and welfare work as fishers, shellfish and seaweed gatherers, traders, fish processors, and factory workers. The fisherwomen are to be guaranteed about the recognition of their resource access rights as well as make room for women to participate in the organizations of artisanal fish workers, coastal communities, and indigenous people.

(xiii) To support the Fortaleza Declaration of Red Manglar International of September 4, 2004, which calls for states to fulfill their obligations under the 1971 Ramsar Convention, particularly with regard to para 15 of Resolution VII.21 on Intertidal Wetlands, signed in Costa Rica in 1999 at COPS 7, which "urges all Contracting Parties to suspend the promotion, making of new facilities, and expansion of unsustainable aquaculture activities harmful to coastal wetlands until assessments of the environmental and social impact of such activities, together with appropriate studies, identify measures aimed at establishing a sustainable system of aquaculture that is in harmony both with the environment and with local communities."

(xiv) To recognize the rights of artisanal fish workers, both as food producers and as workers. They are to benefit from labor laws that respect traditional systems of wage distribution based on catch shares. The artisanal fishermen are also to get social security benefits that include provisions for death and injury, maternity, medical costs, and retirement pension. The artisanal fishermen are also to get training and education in line with their needs.

(xv) To help the artisanal fishermen, coastal communities, and indigenous people at the time of natural calamities, like the tsunami, famine, etc.

5.5.2 Canada

There was a famous Bruegel engraving, stating "Big Fish Eat Little Fish," which illustrates a marine food web by showing how the stomach of each size of fish contains fish of the smaller size class. This is used to dramatize trophic relationships (Pitcher and Hart, 1982). Similarly, there are other Bruegel paintings depicting various types of instructions. The Ecopath Models, like the Bruegel engravings, are take all trophic relationships within an aquatic ecosystem into account.

Notwithstanding the above, the application of various types of models in fisheries and AQC seems to be quite common today. There have been contributions related to the application of **Ecosystem models to investigate multispecies management strategies for capture fisheries** (Tony and Kevern, 2002). "Ecosim," the dynamic version of "Ecopath" (Walters et al., 1997, 2000), has recently been endowed with routines that not only simulate the consequences of changes in fisheries for all elements in the ecosystem but also look for fishing rates that could maximize ecological, social, or economic goals. This new facility had interested FAO, because it could lead to a way of managing multi-species fisheries that takes into consideration all elements of the ecosystem and not just those that are subjected to fishing. As such, the fisheries center and the FAO decided to hold a workshop to explore the potential of the new software.

Back to that Bruegel, the "Land of Cockayne" is said to be a derisive medieval comment on those who imagine themselves in a paradise, where food and luxuries are so easily obtainable that life is comprised of little more than a happy indolence or boast that they are in such a place. Bruegel's painting depict Cockayne is evidently such a fool's paradise. The term "Cockayne" had been derived from "cake," and there are similar terms in medieval French. It had given rise to "Cockneys," inhabitants of London, noted for boasting of the miraculous nature of their city, and a similar idea had also become the "Big Rock Candy Mountain" of US hobo lore.

It could all be too easy to be mesmerized by these elegant simulations. However, the "Land of Cockayne" is intended to provide with a warning that we are to continually check simulation results with the real world. Also, not be tempted to boast that Ecosim optima place us in a far better world. It is said to be a reality check.

Concomitant to the above, there had been reports comprising the edited proceedings of the workshop held at the Fisheries Centre, the University of British Columbia in July 2000, jointly sponsored by the FAO, Rome, and the Government of Japan. This was the first published account of the new Ecosim policy search software that was said to aim at finding fishing rates which maximize objective functions for economic, ecological, employment, or mixed goals.

In addition to the above, there have been literature, which examines fisheries and their ecosystems in the Strait of Georgia (BC); the Bali Strait (Indonesia); the North Sea; The Faroe Islands, Hong Kong; Lake Malawi; Newfoundland; Port Philip Bay (Tasmania); Prince William Sound (Alaska); the San Matias Gulf (Argentina); the Scotian Shelf (Canada); Southern Benguela (South Africa); Campeche Sound (Mexico); Gulf of California (Mexico); the Caribbean (Columbia); and so on.

5.5.2.1 *Providing with scientific information for management*

The single-species or single-resource approach, which is still prevalent in the vast majority of fisheries in the world, is incomplete and inadequate. This is known to the people who are involved in the utilization and management of aquatic resources in recent years. Being faced with mounting evidence of failures in peoples' ongoing attempts to use the living aquatic resources sustainably and responsibly, there had been efforts to reexamine the science, regarding the management methods applied in fisheries. One of the most basic deficiencies, which is now widely considered, is the focus on fishery resources as essentially independent stocks, driven entirely by their inherent population dynamics, as had been the practice during the dramatic expansion in fishing power and catches that characterized fisheries in the 20th century. Incidentally, as a response to this assumption, global attention within fisheries is believed to be turning toward recognizing that biological populations and communities function within and are ultimately regulated by the ecosystems in which they occur. The challenge now is, thus, to translate this intellectual acceptance of the need to manage fisheries as an integral part of the ecosystem into an effective methodology that allows us to make optimal use of our diverse aquatic ecosystems in a responsible and sustained manner. In fact, the principles of ecosystem-based management of fisheries are to incorporate and extend the conventional principles for sustained fisheries development. The EBFM strives to consider the capacity of the ecosystem as a whole, instead of focusing on a single species, to produce food, revenues, employment, and other essential and desirable services for humankind. From that starting point, it is necessary to devise utilization and management strategies that enable us to optimize that capacity, taking into account variability in the system and uncertainty in our knowledge.

Concomitant to the above, these strategies may also need to refer to limits and targets related to conservation of ecosystem components, structures, processes, and interactions, instead of setting only relatively simple reference points related to single population. An immediate implication of this is that the complexity of the system being considered increases, and the number of objectives and the conflicts between those objectives increases substantially. This expansion in uncertainty and complexity is not, however, a consequence of adopting ecosystem-based management, rather it is said to be the result of recognizing and attempting to consider the full complexity and uncertainty that had always been there; but, which, previously, perhaps, had been ignored. Fishery scientists, managers, and interest groups are generally aware of the need to consider this full range of complexity, but there is still the prevalence of ignorance as to how to implement an effective ecosystem-based management system in practice and, also, the practical problems raised by this recognition are said to be considerable. Uncertainties and conflicting objectives have severely hindered successful application of efforts to implement single-species management, and could even be harder to deal with, as ecosystem interactions are recognized and incorporated. Both the need for and difficulty in EBFM were recognized by the 95 states which had met at Kyoto in Japan, from December 4 to 9, 1995 on the occasion of the International Conference on the Sustainable Contribution of Fisheries to Food Security. They had formulated the Kyoto Declaration which proposed, among other important principles, that the states could base their fisheries policies, strategies, and resource management and utilization for sustainable development of the fisheries sector on the following aspects:

- **(i)** maintenance of ecological systems;
- **(ii)** use of the best scientific evidence(s) available;

(iii) improvement in economic and social well-being; and.
(iv) inter and intragenerational equity.

Notwithstanding the above, it may be noted that subsequent to the Kyoto Conference, the Government of Japan had provided with financial support to the FAO for a program to assist countries in implementing the Kyoto Declaration. One of the projects undertaken under this program was titled "Multispecies Fisheries Management." This aimed at providing with methods and scientific advice for improved multispecies and EBFM. Under this project, a workshop was held at the Fisheries Centre, UBC, in March 1998 to consider the "Use of Ecopath with Ecosim to Evaluate Strategies for Sustainable Exploitation of Multispecies." The workshop did not come up with formal conclusions and recommendations, but there was widespread agreement that Ecopath with Ecosim (EwE), as a well-developed and generic ecosystem-modeling tool, could play a useful role in providing with important information to decision makers on fisheries policy and strategies from an ecosystem perspective, complementary to that available from conventional single-species assessments.

It was designed to follow-up from the preliminary meeting and to look into more details, using specific case studies. There had been attempts to look at the type of information, including its limitations, which could be sorted out with our existing ability and could be assessed and forecast at the scale of ecosystems.

5.5.2.2 The objective(s) of the workshop

A key requirement for effective management is good information on which good decisions could be based. Incidentally, it is a key role for science and scientists to provide the best available information on the state of the resources and their likely response, or responses, to any planned fisheries or management action. The workshop was intended primarily to address this task, and, particularly, to examine the nature of the information at the scale of the ecosystem which scientists could provide for EBFM.

The objective of the workshop was, therefore, to use quantitative ecosystem models to investigate the impact of different multispecies harvesting strategies on the community structure and fishery yields of different ecosystem types with a view to identifying preferred harvesting strategies. Different models and information were used by each participant to consider different sets of objectives for their ecosystem. Further, the objectives had also included the following:

(i) To use the models for identifying the management strategies which could come closest to achieving those objectives.
(ii) To estimate the ecosystem consequences of each management strategy.

Notwithstanding the above, attempts were made to consider the key sources of uncertainty in different models and the possible implications of these for their results and conclusions.

This approach was effectively used in the ecosystem models as operating models (e.g., Hilborn and Walters, 1992; Cochrane et al., 1998). Such an application was one of the primary goals of the developers of Ecosim, who suggested that it could be used to "conduct fisheries policy analyses that explicitly accounted for ecosystem trophic interactions" (Walters et al., 1997), and routine (basics) had been provided within Ecosim to assist the user in such exploration of fisheries strategies or policies. A full description of these routines could be found in the Help System of EwE.

Concomitant to the above, the first of the routines is the "open-loop" policy search, which estimates the time series of relative fleet sizes that would maximize a multi-criteria objective function which included the net economic value, social employment value, and ecological stability criteria. The relative fleet sizes were used to calculate the relative fishing mortality rates by each fleet type on the affected stocks. The user could, therefore, specify the relative priority of economic value, social employment value, and ecosystem stability, where ecosystem stability could be defined in terms of relative abundance of the different biological groups included in the model. These relative priorities represent the management goals. The open-loop search then tries to identify the strategy, in terms of relative fleet sizes that comes closest to meeting those goals.

In addition, the Ecosim also includes the "closed-loop" policy simulations, which allow the user to examine the consequences of a given management strategy, taking into account the dynamics and uncertainties of the stock assessment and regulatory processes.

Notwithstanding the above, the "open-loop" routine uses a nonlinear search procedure to determine the optimum fishing rates over time across the fleet, according to the objective function specified by the user in the open-loop input screen. On the other hand, the closed-loop routine is believed to use the fishing pattern identified as optimum for a specified management goal using the open loop, and, then runs a series of forwarding projections using that fishing pattern but maintaining the target fishing rates of the different fleets in accordance with the annual observations, with specified errors, on the status of the stock. This routine enables the user to estimate the actual performance of the selected management strategy (i.e., the optimum fishing pattern over time estimated by the open-loop routine) given the observation error inevitably encountered by the fisheries manager striving to maintain the fishing rates specified by the policy in a real fishery. The information generated by these routines could be used, in combination with other sources of information, including the results of single-species assessments and forecasts, to inform the manager on the strategy or strategies that would best achieve the agreed objectives.

5.5.2.3 *Procedure for identifying the strategies*

The following specifications and procedures could be followed to ensure reliability and comparability in the results of this kind of study. All strategies are to be tested over 20 years of simulation, unless the results had indicated that the ecosystem had not stabilized over that period, in which case the simulation period could be extended until stability was achieved.

Further, with any nonlinear search procedure, there could be a danger that the procedure may converge on a local minimum, and not on the global minimum for the given objective function. In fact, testing the performance of the current fishing strategy in each ecosystem required running of the model over the 20-year simulation period with the fishing rates of each fleet maintained at the rates used in the underlying Ecopath model.

5.5.2.4 *Definition of ecosystem stability*

It refers to the capability of a natural system to apply self-regulating mechanisms to return to a steady state after an outside disturbance. Once the optimum management strategy had been identified for a given set of objectives, it could, then, be tested for performance in simulations of 20 years in the presence of observation error using the closed-loop routine. Under this, the strategy could be run for 100 times with randomly generated observation error in the index of abundance of each biomass group used each year by the "manager" in order to maintain the fishing rates at those specified in the strategy. These indices could be specified as either estimated catch or biomass, or the directly estimated exploitation rate of the previous year arising from, for example, a tagging program. The closed-loop routine, therefore allows the user to estimate the effects of observation and implementation error on the performance of the strategy identified by the open-loop search.

5.5.2.5 *The wise use of policy search routines*

The results of various studies, perhaps, had clearly illustrated the significant role of the open- and closed-loop policy search routines in exploring ecosystem-based management strategies. These routines provide efficient and objective means of identifying strategies that could best achieve clearly specified and precise objectives.

Notwithstanding the above, the sensitivity of the model behavior to the assumed values of the vulnerability settings was also emphasized in various studies. The overall conclusion for anyone considering using the EwE policy search routine to explore possible management strategies seriously is that it is a powerful and sophisticated tool. The user, however, could better be conversant with the principles of nonlinear optimization and must follow the practices and guidelines that apply to such statistical tests. Incidentally, there are more chances of obtaining misleading results and conclusions without the cautious, thoughtful, and correct use of the software.

Notwithstanding the above, in **Canada**, there had been works on the evaluation of the policy optimization routine of Ecopath/Ecosim, using the model of the Newfoundland-Labrador shelf as a case study (Marcelo et al., 2002). The routine was used to calculate the best combination of harvesting strategies for multiple fleets and to illustrate the type of trade-offs expected when fisheries management aims at maximizing economic, social, and ecological goals. In order to maximize social and economic returns from the fishery, the optimization routine was driven the ecosystem to a very simplified state, where high trophic level species have depleted and low trophic level species had a net increase in biomass after 20 years of simulation. Further, the model predicts, almost in all cases, a substantial decrease in all fishing fleets, in order to maximize ecological stability. The performance of calculated optimal harvesting strategies is generally affected by increasing errors in stock assessment procedures. However, more practical analysis of future fisheries management policies for the Newfoundland-Labrador shelf may require more accurate data for prices per species, fleet operational costs, employment indicators for each fishery, and also ecological parameters that better represent the current status of the ecosystem.

5.5.2.6 *Simulation results*

There had been results of studies about the policy optimization for various contrasting management goals. The model is to calculate a reduction in inshore and offshore trawlers and an increase in mixed gears, foreign trawlers, and the harvesting of seals and sea birds in order to meet the economic goal of maximizing the net economic return taken from the system. However, in this scenario, the best solution seemed to be to increase the relative fishing effort of the fleets with highest yields (mixed gears and foreign trawlers), which also target low trophic level species and to less consider the top predators in the system (seals and birds), which are competitors to fisheries.

Notwithstanding the above, the optimal policy involves an increase in all fishing fleets, particularly mixed gears that have a higher jobs/landings ratio in order to meet the social goal of maximizing employment. Further, on the other hand, the model also predicts, in almost all cases, a substantial decrease in all fishing fleets in order to maximize ecological stability. Probably, a balance has to be stuck in order to determine the optimum status keeping in view various aspects. Incidentally, the predictions of optimal fishing rates and performance values are sensitive to the value of the vulnerability parameter, although the general qualitative pattern of predictions is similar under both trophic control scenarios. Moreover, the hypothetical manager drives the system to a very simplified state, where high trophic level species are depleted and low trophic level species have a net increase in biomass in order to maximize economic return and employment (measured as a variable dependent on landings). Thus, there seemed to be two apparent factors contributing to increase in landings:

(a) the increase in fishing pressure on the groundfish stocks and.
(b) the increase in biomass and hence the productivity of lower trophic level species also targeted by fisheries (e.g., shrimps, capelin, planktivorous fish, etc.). A more balanced distribution of biomass across the food web is obtained when the goal is to maximize ecological stability. Releasing fishing pressure could lead to a system that presents, at the end of the simulation, similar characteristics encountered at the starting condition. Further, there are believed to be examples, which are illustrative of the type of trade-offs encountered when managers aim at balancing ecological and socioeconomic goals. Moreover, using the simple goal functions, the model(s) could predict that improvements in ecological performance could occur at the expense of substantial losses in socioeconomic performance. Interestingly, the optimal policy calculated by the model in the scenario could be the one that closes fisheries with low jobs/catch ratio (offshore and foreign trawlers) and also the direct harvesting of charismatic fauna. In addition to the above, the effect of the lack of perfect information about stock sizes and the actual fishing mortality may hinder obtaining a definite inference and could hamper the simulation exercises. Further, Ecosim evaluates the performance of fishing strategies in a closed-loop scheme where managers have to rely on stock assessment to determine the annual optimal policies. However, typically, errors in stock assessment could cause variability in the implemented fishing rates, and hence, could affect the performance of fishing strategies. Moreover, there seemed to be examples depicting the performance of a policy that optimizes the balance among economic, social, and ecological goals and are usually evaluated under three levels of observation errors. The results had portrayed that there was a general decrease of performance for economic, social, and overall indicators, but a slight improvement in ecological performance with increasing error, meaning that, the variability imposed by stock assessment error is predicted to attenuate the impact of fishing on ecological stability. The results also, further, indicated that ecological stability either improved or remained relatively unchanged with the increasing assessment.

In inference, the various analyses may not always provide a realistic policy evaluation for the model area. Instead, the aim often could be to test the overall behavior of the model under different parameter settings as well as to check the consistency of the results according to the different policy scenarios. However, a more realistic analysis includes better economic data (prices per species, fleet operational costs), better social indicators (such as employment for each fishery), and also ecological parameters that better represent the current status of the system and its ecosystem dynamics. Moreover, of particular importance, the vulnerability parameters are included among the species, which seemed to influence the magnitude of performance values and optimal fishing rates. Further, when running the policy optimization routine, attention is to be paid to long-term changes in productivity, or oceanic regimes, which might influence the overall performance of policy options considerably.

Incidentally, Wibera et al. (2004) worked on participatory research supporting community-based fishery management. They emphasized the engagement of researchers and fishers together in adopting social science approaches to the development of community-based fishers' organizations. The work was carried out in the Scotia-Fundy Region of Atlantic Canada (the Bay of Fundy and Scotian Shelf). Its rationale reflected the following arguments:

(a) Effective CBM requires that managers can pose and address social science questions.
(b) Participatory research, involving true cooperation at all stages, could support this process.
(c) There is a need to overcome practical and methodological barriers faced in developing participatory research protocols, which could, subsequently, serve the needs of the CBM; but, at the same time, it is not to demand high transaction costs.

There have been attempts to report on the works done with the fishers' organizations, both aboriginal and non-aboriginal, in which, social science priorities were set by each organization and small-scale research projects were designed and carried out to meet these requirements.

Notwithstanding the above, there had also been works which had identified interests among fishers in research on three different levels of meaning as follows:

(a) Practical livelihood concerns, including what, when, and where to fish, and with what intensity of effort.
(b) Social, economic, and political issues, e.g., on institutional structures, policies of access and allocation, overlap and conflict between regulatory regimes, etc.; and,
(c) Values and ethics that implicitly or explicitly guide policy development and implementation.

Concomitant to the above, various research themes was found to be crucial, including those of power-sharing, defining boundaries of a community-based group, access and equity, designing effective management plans, enforcement, and scaling-up for effective regional and ecosystem-based management. However, the research results have demonstrated the effectiveness of extending participatory methods, and this could serve as an alternative to the traditional scientific notions of the research process.

5.5.3 The Caribbean

The Caribbean islands possess a rich diversity and potential for fish and fisheries. Anon (2005a, b) dealt with the Fishing Community initiatives in CBFM and The Sitka Declaration as a proposal for Forum of Fishermen, NGO's, and the Government Agencies in order to explore expansion of the use of sectors, catch share(s), and CBFM in the NE and the **Caribbean.** The engaged and informed fishermen of today now believe that the communities are in a position to responsibly manage resources and regain access to diverse fisheries as those stocks were rebuild. In fact, they were, perhaps, working toward this goal with an efficient *modus operandi* or tool called the "CBFM," an approach where communities are said to directly manage their resources.

5.5.3.1 Community-based fisheries management

It could be said that our fishing communities were once sustained by many different species, notably, the featherbacks, shads, carps, catfishes, trouts, loaches, perches, murrels, glass fishes, puffer fishes, lobsters, the codfish, herring, winter flounder, and even tuna. Now, mainly due to overfishing, pollution, etc. (due to the application of advanced technology in coastal development, etc.), many of these historic fisheries are, perhaps, unavailable today to the local fishermen. Hence, stricter regulations are said to be in place today for dealing with such situations. Incidentally, our coastal communities are precariously dependent on few commercial species. Incidentally, the marine ecosystem is said to be complex and interdependent. The loss of one species may certainly affect the others. In fact, all the fishermen, along with other stakeholders are to be involved in managing the resources together, if they want to ensure a bright future for the fishing communities and the world. In fact, the aware fishermen, now believe that, the communities could responsibly manage resources and regain access to diverse fisheries as those stocks were rebuild. In fact, they were, perhaps, working toward this goal with an efficient *modus operandi* or tool called the CBFM, an approach where communities are directly manage their resources. This management style is to protect and sustain communities and resources alike.

Concomitant to the above, CBFM is practiced worldwide, from Alaska to Thailand, from Mexico to New England. In addition, some organizations, like The Stonington Fisheries Alliance, The Cape Cod Commercial Hook Fishermen, etc., also practice CBFM. The alliance, which was started in 1999 to introduce the concept of managing fisheries based on principles that lead to sustainability, had gone a long way in this direction.

5.5.3.2 Why community based fisheries management?

It may be noted that the community-based groups have more in common with community development/social action and environmental groups, unlike the traditional fishermen's organizations that have historically been formed to protect the interests that most business lobbyists (who deal with allocation, keeping regulations at a minimum, avoiding taxes, getting government price supports or loans), etc. Incidentally, all the stakeholders have an economic interest and they also have an interest in protecting the environment that supports and surrounds their hometowns.

Contrary to the above, today, the coastal fishing communities are in trouble. The sense of the traditional connection of community residents to marine resources is gradually vanishing. Current management systems are undermining the ecological, social, and economic basis of community sustainability. In many cases, this, perhaps, had contributed to the exhaustion of resources, decreased biological diversity, and the loss of a way of life that formed the basis of much of our history and culture. Moreover, increased absentee ownership and capital costs deny the traditional entry of future generations into the livelihood avocations of the coastal communities. On the other hand, the community's knowledge, economic dependence, and attachment to fisheries resources are essential in promoting a strong sense of stewardship. There has to be a national and an international interest in maintaining healthy, dynamic, and diverse coastal communities.

5.5.3.3 *Fishing community initiatives in CBFM and the Sitka Declaration: Proposal for forum of fishermen, NGO's, and government agencies to explore expansion of the use of sectors, Catch Share, and CBFM in the Northeast and Caribbean*

A common saying is said to be that "wealth, which is free for all is valued by none" It could be mainly because, he who is, perhaps, little foolish, to wait for its proper time of use, could only find that it had been taken by another (H. Scott Gordon: The Economic Theory of a Common Property Resource: The Fishery, 1954).

Concomitant to the above, the world's fisheries today also offer a contemporary example of such a tragedy of the commons, because oceans are un-owned, and each fishing fleet has no incentive to conserve or replenish the fish it takes; rather, it has every incentive to take as many fish as possible lest the benefits of a larger catch go to someone else. This was **commented by Kent Jeffreys in "Rescuing the Oceans," in The True State of the Planet, Ron Bailey, ed. (New York: The Free Press, 1995).**

Notwithstanding the above, "CBM" had been advocated worldwide as a means to overcome the overexploitation of fisheries. Yet, the researchers and the managers are believed to be divided in their opinions regarding the effectiveness of these measures. In fact, the "tragedy of the commons" model is often accepted as a universal paradigm, which assumes that unless managed by the state or privatized, common-pool resources are inevitably overexploited due to conflicts between the self-interest of individuals and the goals of a group as a whole. Under this situation, the emergence and maintenance of effective community-based efforts (that may include cooperative risky decisions) leading to the establishment of marine reserves, might have come into existence. However, it could often be found that locally crafted and enforced harvesting rules, may lead to a fast augmentation in resource abundance. Nevertheless, news about this increase may spread quickly at a regional scale, resulting in poaching from outsiders, and a subsequent rapid cascading deleterious effect on fishing resources, and there could also be noncompliance of locally designed rules. Nevertheless, cooperation for management of common-pool fisheries could emerge and evolve rapidly and could become effective at a local scale even in recently organized fisheries. Moreover, stakeholder participation in monitoring could play a key role in reinforcing cooperation. Further, the increase of local fishery stocks may attract outsiders who, if not restricted, could overharvest and threaten local governance. In fact, the fishers and the fishing communities are indeed supposed to require incentives to maintain their management efforts. Rewarding the local effective management with the support and recognition of formal governance could generate these incentives.

5.5.3.4 *Local management by the commons*

Concomitant to the above, CBM of the marine reserve network relied primarily on the following:

(a) a suite of simple rules and means of their enforcement;
(b) leadership of key individuals;
(c) meeting venues that allowed for social and ecological feedbacks; and.
(d) capitalizing on the region's physical and environmental characteristics.

Incidentally, the Fishers had designed, monitored, and enforced three main types of formal and informal rules and sanctions:

(i) resource-based rules (fishing banned June and July; fishing banned within reserves)
(ii) monitoring rules (mandatory participation in financial contributions for monitoring)
(iii) administrative rules (mandatory: participation in cooperative meetings, monthly financial contributions to the cooperative, and timely provision of paperwork for cooperative).

These rules and their sanctions were built primarily on foundations of trust and reciprocity and indicating the concerns for the group's well-being. Hence, the most effective and usual form of enforcement had relied on peer pressure and public shame. This, in essence, could finally threaten the rule-breaker's reputation (also known as social capital), and his social bonds with the rest of the members of the group. These were de facto sanctions with no legal standing under the statutes of the cooperative. Further, the local government officials, sometimes, provided enforcement support on occasions that was based entirely on the rapport built between the fishers and the officials. Further, formal and informal sanctions by rules were, perhaps, devised by the local fishers.

Concomitant to the above, *c.* 100% of the fishers were interviewed, in order to exemplify this. Many of them trusted that other fellow fishers, for the most part, respected the reserves. Likewise, on certain occasions, when asked about the open-ended question, regarding "in what way would violating cooperative rules could affect all involved?" In many instances, the answers were spontaneous under three categories as follows:

(i) Personal guilt and a sense of betrayal to the group.
(ii) Concern of the rest of the group's opinion about one's actions.
(iii) Concern over the possibility of losing trust and friendship.

Open access to fisheries could lead to overuse and decline in the fish stock, and in the catch. As such, generally, the response by the resource managers was to limit new access, but to permit those already permitted to continue to use the fishery. If this does not work, then the next step would be to impose additional technical constraints, with the aim to limit the catch of those who had still been fishing.

Concomitant to the above, it is necessary to ascertain what the economists had thought what was to be done, i.e., either to eliminate the "commons" of the fishery and make property rights (or "quasi-property rights") in order to make incentives for stewardship. But, in such cases, what could be the *modus operandi*? One, perhaps, may not be able to fence the water. But, one could, probably, make ITQs and allocate them to the cooperatives. To limit the number of fishes to be caught, people are allowed to choose how and when to catch them. This could let the fishermen be flexible and efficient about when and how to fish. Incidentally, "sectors" are said to be groups of permit holders (fishermen), voluntarily joining together to manage a combined allocation of fish. To be in a sector, in fact, one has to sign a binding contract, and, as a group, one could not exceed the allocated limit. Moreover, sectors may decide how to ration the allowable fish catch among the members. Fishermen could believe that they may not be brought to book by the bad actions of fishermen outside the sector, particularly, those who, sometimes, may misbehave to have their rights and keep their allocation. In addition, the sectors might also help with marketing, bargaining, and group buying.

Notwithstanding the above, as an instance, for the first time in decades alteration(s) occurred for New England's groundfish fleet. The New England Fishery Management Council approved an amendment to the current management plan for groundfish stocks (including cod, haddock, and flounder) that, perhaps, had offered fishermen a better way to do business. Moreover, the fishermen had got a scope to work in community-based harvesting cooperatives, called harvesting sectors. Meanwhile, implementation of a model, which could help to rebuild fish stocks, conducting safer fishing trips, reduce fish waste, and enhance profits for fishermen by allocating each sector a certain amount of fish and allowing more flexibility in how those fish are harvested.

Notwithstanding the above, fisheries management regimes may take many forms, but, some of them may fail to designate shares of the catch. This failure may make strong incentives for some individuals to maximize their share without regard to long-term sustainability. This could be mainly because the benefits of conservation actions may not accrue to individuals. Incidentally, the competition to maximize catch may usually entail excessive capital investments in fishing vessels and gears; thus, resulting in intense fishing pressure, leading to overfishing, high by-catch rates, and the use of large, efficient types of gears, which could harm the habitat. However, the managers may have to respond by increasing regulations; but, this, at the same time, often may exacerbate perverse incentives. In addition, many fisheries could be producing more value than the current system permits, i.e., enormous quantities of fish could be landed during short seasons; thus, they may force the fishermen to sell at low prices. Meanwhile, the conservation and economic problems facing fisheries could be addressed in an integrated way by designating access privileges (specifying shares of the catch) to individuals; by initiating harvest cooperatives; or other appropriate entities. Incidentally, it may be noted that designated access privilege (DAP) systems could demonstratively end the competition to maximize catch as well as often could result in better conservation and financial performances. However, the cost of implementing these systems could be comparatively high and was a hindrance for better management. In fact, this may not always happen to be so. Fisheries could accept investments from a variety of sources and use a portion of the increased financial performance to repay outstanding grants and loans.

Notwithstanding the above, the key to protecting fish stocks, habitats, and the communities, which depend on them, will be to implement DAPs that are appropriate for each fishery or community, making investments on sustainability and making financing mechanisms that, perhaps, are themselves sustainable.

5.5.3.5 Catch shares

In order to respond to mounting threats to fish populations, fishermen, and their communities, in 2008, The Cape Cod Commercial Hook Fishermen's Alliance (CCCHFA) launched the Cape Cod Fisheries Trust (CCFT). In its efforts, CCCHFA have been raising funds and buying cod, scallop and lobster permits to make sure that there is the next generation of fishermen on the Cape. CCCHF's work with local fishermen had forged a new and promising model for other fishing communities that, perhaps, were working to save their fish, their fleets, and preserve a way of life. The CCFT was said to be a home-grown idea to buy fishing permits and keep them in the community. Local fishermen were supposed to sign a binding lease agreement with the trust, and, fish with traditional gears and small boats. In fact, the trust exhibited market savvy, management excellence, creativity, and business acumen for a year under its belt. There is no doubt that this approach had helped to stem the tide of dwindling fishermen in the community. There had been a threat to the fishing community, for few years, as permit prices had gone up as well as the bigger fishing farms had acquired small boat fishing permits. As such, the effect seemed to be threatening to wipe out the fishing heritage for ever.

Concomitant to the above, of late, there had been material developments with regard to commercial fishing permits that required us to buildup the CCFT program as quickly as possible. In fact, the CCFT was a solution that could protect our fish, the fishermen, and, most importantly, our fishing community. There could be an expectation to ensure the long-term survival of several commercial fishing enterprises through the success of such a program.

5.5.3.6 *Can CBFM work for fishermen in places as diverse as Maine, Massachusetts, New Jersey, and Puerto Rico?*

It could be quite known that Belford, New Jersey and Vieques, Puerto Rico appear to be as different as any two places could be. However, a second look might reveal that the communities and their fishermen have more in common with each other and with the communities and fishermen in Maine and Massachusetts. Belford, however, was an anomaly in New York Harbor. It was the last commercial fishing port, and the oldest fishing cooperative in the region. However, the fishermen were disillusioned being encroached upon by development and hampered by pollution and a complex array of regulatory systems. It was believed that several generations of families had fished from Belford but a new condominium development, a commuter ferry terminal, and last but not the least, an inadequate community development plan had increased the number of fishermen, who do not want to see their children follow them into the fishery.

Notwithstanding the above, the fishermen of Vieques had formed the Vieques Fishermen's Association, in 1975. Further, many fishermen in Vieques had used fish traps as a mainstay of their art. These fish traps, locally called "nasas," were, in those days, made of chicken wire and reinforced with saltwater-resistant wooden sticks cut from the local trees. However, a major problem for the Viequense fishermen was that the large naval ships using Vieques for target practice and mock battles were running over the fish trap buoys and cutting the lines leading to the traps. It was at this juncture that the Vieques Fishermen's Association had entered the political arena. The association brought a lawsuit against the United States Navy for the loss of these traps. At the hearing, the judge was naturally sympathetic to the fishermen. He admonished the navy lawyers for their bad works. The judge had delivered the verdict in favor of the fishermen, and the navy had to reimburse the funds for the lost fish traps.

Notwithstanding the above, even today, the fishermen are struggling, but this time because of increased tourism and shoreline development. In recent years, the fishermen believed that a sustainable fishery, which includes lobster AQC, could be maintained by having an alliance with the Vieques Conservation and Historic Trust, and this is the only way to preserve their culture, community, and livelihood.

5.5.4 Cuba

Cuba possesses rich potential in AQC and fisheries. The Republic of Cuba is a multi-island country comprising the island of Cuba, which is the largest island in the Caribbean, together with many other islands and bays. The Republic of Cuba has a total land area of *c.* 114, 525 km^2. There had been attempts (Julio, 2006) to portray the level of resource exploitation and the present fisheries management regime in **Cuba**, including an overview of the Cuban fisheries; some theoretical background on management regimes; a critical review of the fisheries regulations through time; and, an account of the legal and institutional framework of the fisheries sector as well as an evaluation of the current status of the most important fisheries based on catch and effort data using the **Gordon-Schaefer model.**

Analyses of four fisheries, viz., lobster, shrimp, tuna, and other fishes revealed that at the current effort levels, the catches are sustainable both in ecological and economic terms, not surpassing reference points, like MSY and MEY. The management regime was successful, even though there seemed to be other factors influencing the outcome. The results showed a relation for the lobster, shrimp, and other fishes; thus, statistically indicating a change after 1996 in these fisheries. Nevertheless, this alteration could also be influenced by external factors, like economic constraints. Attempts had also been made to include some recommendations for future sustainable management of fisheries in Cuba.

The Cuban archipelago is situated in the Caribbean Sea, nearly south of the Tropic of Cancer. The subtropical climate is moderated by the Trade Winds and two seasons, rainy and dry. The insular shelf is *c.* 61,525 km^2 and the EEZ is *c.* 350,751 km^2. More than75% of the population live in urban areas. Among others, the agricultural products, forests, fishes, and shellfishes in the territorial waters form the most important natural resources of Cuba. Fishing has been an important source of nutrition, avocation, and a potential means in revenue generation in Cuba. Sustainable and efficient fishing is supposed to be significant for the Cuban economy. The Cuban fisheries management regime had undergone some significant changes during the last decades in response to suspected stock depletion, which had indicated non-sustainable utilization during the 1970s–80s. The significant changes to the fisheries management were with regard to increased emphasis on sustainability as well as a greater focus on the value of the catch rather than the quantity. It is, perhaps, advisable to revise management strategies periodically to adapt to the dynamic conditions of the stock and resource users as well as to changes in the inter-temporal preferences of the fishing sector (FAO, 1997a, b; Seijo et al., 1998).

Notwithstanding the above, a resource management regime may have to often allow for several different and often conflicting interests. Fishermen and the fish processing industry have an interest in how much could be caught as well as when and how. However, their short-run interests may conflict with society's best and greater interests, notably, the interests for the posterity, As such, a study which attempts to evaluate a management regime constructively, may, invariably include a description of the current management regime and its historical resume' to lay the foundations for evaluation and future policy recommendations. However, there is no well-documented study on the current Cuban fisheries management regimes on which such an evaluation could be based.

Notwithstanding the above, it is quite well accepted that fishing is done for both nutrition and avocation. There are the publication(s) which depict the fishing regulations in Cuba and also describe the way they are grouped (Sánchez, 1997). Some documents vividly present the previously quoted works, but lack an analysis of the current management regime in the terms of effectiveness. Thus, to fill the void, there are contributions, which attempt to provide a more detailed description of Cuban fisheries management that could be used as a basis for the evaluation of the management regime. Incidentally, a considerable amount of expectations was associated with the recent changes in the Cuban fisheries management regime. This can clearly be seen from the following quotation:

In the case of Cuba, a redesigning of the monitoring system has occurred recently, giving coverage in a systematic and extensive way, all fishing areas. Further, Bio-economic models are applied, to help and guide in resource management activities (Carcamo et al., 2003). Incidentally, some of the available information may indicate that *c* 40% of the Cuban fisheries were believed to be in the senescence phase and *c* 50% in the mature phase during 1995 (Baisre, 2000a). The latest milestone in Cuban fisheries management, the Decree-Law 164 was issued during 1996. This new law constituted a significant shift in emphasis on fisheries management, focusing on sustainability and profitability. Now, a long time has elapsed; and, it is, perhaps, the right time to assess whether it has achieved the goals that were set out initially of sustainable utilization and improved catch value. Moreover, there is contribution(s) which dealt with case studies on Cuban fisheries management, their historical background, current status, and future prospects. The main objective of such case studies could be to provide with an overview and evaluation of the Cuban fisheries management regime (FMR). Nevertheless, the evaluation of the current policy could, perhaps, be based on one main aspect: sustainability of a given fishery measured both in ecological and economic dimensions.

5.5.4.1 *Cuban fisheries*

A brief introduction

During the 19th century and the first half of the 20th century, the commercial fishing industry in Cuba was characterized by a fleet composed of small boats and vessels plying in the coastal waters. These crafts, which were typically low capacity and technically unsophisticated, have primarily targeted a combination of reef fish: spiny lobster, sponge, and a few pelagic finfish species, such as mackerels, tunas, and billfish. The landings were handled by small-scale processing facilities, and the products were then mostly directed into the local domestic markets and the tourism industry (Adams et al., 2000; Claro et al., 2001). Catches were around 5000–10,000 mt/year during the early 1900s and around 20,000 mt/year during the 1950s. After the Cuban Revolution of 1959, the growth of the fisheries industry was an important objective of the government. Given the cited changes in the early 1990s, Cuba's commercial fishing industry was changed dramatically. As a result, production emphasis shifted from high-volume, low-valued, pelagic stocks toward high-valued, nearshore fisheries. Moreover, Cuba more recently played an increasingly significant role in the global market for these high-valued finfish and shellfish seafood products harvested primarily within Cuba's nearshore waters (Adams et al., 2000).

History of fisheries management in Cuba

The commercial fisheries in Cuba became a part of the economy only after the early 19th century, during the colonial era. Nevertheless, it was still an artisanal, small-scale economic sector. However, the number of fishing communities grew, and, consequently, the exploitation of the fishing resources had picked up at that time. The first fishing regulations recorded in Cuban history were issued by the Temporary Commission of Fisheries (*Comisión Temporal de Pesca*). This commission was established in 1880 and included a number of prominent scientists (Baisre, 1987a). Unfortunately, there is not much information on those regulations.

Concomitant to the above, the country went under the US military intervention between the end of the Spanish-American War and the proclamation of the Republic of Cuba. During this period (1898–1902), the US occupied army issued a number of rules for fisheries in the form of military orders. These regulations included general rules for fisheries and fishing gears, a closed season and minimum legal sizes for the sponge fisheries, closed season, and fishing gear restrictions for turtle fisheries, fishing restrictions, and spawning zones for lane snapper and the appointment of the Coast Guard Service to control the fulfillment of the issued rules. Even though those rules were not issued in view of the concern of the fishing resources, they could be considered as positive because they had formed a legal control system for the fisheries in Cuba (González, 2004).

Notwithstanding the above, the Republic of Cuba was proclaimed on May 1902, 20 and The National Board of Fisheries *(Junta Nacional de Pesca)* was formed in its early years. This body was established in 1911 and was attached to the MOA. It had started issuing the fishing regulations; thus, continuing the work of the previous Temporary Commission of Fisheries. However, the regulations(s) of this period did not differ much from the previous ones, focusing on minimum legal size and season closures for the few commercial species (reef fishes, crustaceans, molluscs, sponges, and turtles).

Concomitant to the above, the next milestone in the Cuban Fisheries Management Regime was the issuing of the Decree-Law 704 "General Law of Fisheries" in 1936. It compiled and updated the previous fisheries legislation that was either scattered as agreements, enactments, and military orders or outdated. This law contained several regulatory measures that were empirical, supported by fishermen and fishery officers, and were not based on intensive research works. Some other rules were copied from similar ones existing in the United States, and were generally aimed to protect species by season and zone closures (Baisre, 1987b). Besides, the season closures for all commercial species (thus, allowing the authorized fishing gears only) and pollution prevention; the ban on the use of specific gear(s) during the breeding zones were also contained in that Decree-Law.

Concomitantly, the National Institute of Fisheries *(Instituto Nacional de la Pesca)* was formed in 1955 to study and research different aspects of fisheries. The new Decree 2724, "Rules for the General Law of Fisheries" was formed by this institution and issued a year later (1956). But the scarce budget and the government's so-called lack of interest in the fishing industry were damaging this institution and its efforts.

Notwithstanding the above, The Cuban Revolution of 1959 "revolutionized" the fishing industry. The two main objectives of the new government for this sector of the Cuban economy were.

(a) Increasing the catches as well as expanding the fishing zones.
(b) Developing the processing industry.

In order to accomplish such objectives, some previously issued regulations were inconsistently eliminated by the end of the 1960s. There also had begun the use of set gillnets and even pair trawling for some fishes. Moreover, the use of trawl nets for shrimp and drift nets for lane snapper were generalized during this period, consequently increasing the catches (Baisre, 1987c; Claro et al., 2001).

Notwithstanding the above, the overfishing generated by the misuse of fishing gears, the increased efforts, and the lack of proper regulations caused an alert on the perils of overexploitation of fish resources. All this, and the fact that regulations for the lobster fishery were significantly improved in 1978 with optimistic results, showed the effectiveness of an appropriate management system. Such improvements included an extended season closure to 90 days (previously reduced to 45 days) and more control on minimum legal size (Baisre, 2004). Thus, the management rules were changed and season closures for several species were implemented again. Also, the restriction and/or ban of certain types of fishing gears were applied. Moreover, there, perhaps, were also limits on effort (Baisre, 1987c).

All these changes were completely put into practice in 1981, so that the year could be regarded as setting the beginning of a new period of properly managed fisheries in the annals of Cuban ichthyology. The historical series of Cuban catches as well as the milestones in fisheries management is evident from the different literature (Baisre, 2000a). Concomitantly, the effects of the increased emphasis on catch quantity from the mid-1960s onward were also evident from various sources. The signs of overharvesting are also clearly evident, along with a period of stagnation and a trend of decline in catches. Thus, there was obviously a need for a revision of the fisheries policy. This revision came with the Decree-Law 164 "Rules of Fisheries," which was approved in 1996, and represent the latest milestone in the Cuban fisheries. There is also a management history of Cuban fisheries.

Notwithstanding the above, there were efforts in the management of the water bodies and fisheries through **CBFM** involving the stakeholders from the grassroots level.

Legal and institutional framework

Today's main rules for fisheries management could be found in the Decree-Law 164 "Rules of Fisheries." This was approved by the Council of State of the Republic of Cuba on May 26, 1996 (GOO, 1996). The most relevant aspects of this legislation are as follows:

(a) The formation of a system of fishing licenses by which any fisherman or fishing boat must have a proper license or permit. This is issued and/or canceled on an annual basis.
(b) A system of penalties to be applied to the persons who violate the law(s).
(c) A monitoring, control, and surveillance system by means of an enforcement agency (National Office for Fishery Inspection) could be initiated with its offices in almost every province.
(d) The establishment of an advisory and consulting body (Fisheries Consulting Commission) involving all the stakeholders in the use and exploitation of fishery resources in FW, and sea.
(e) Watching over the conservation of the FW and marine environment, including penalties for spills and other forms of pollution.

The regulatory measures either approved by the Consulting Commission or directly released by the Fisheries Regulations Division become legal through resolutions. Then, these resolutions are signed by the Minister of Fishing Industry. Concomitantly, Sánchez (1997), opined that the resolutions could be grouped according to their purpose into the following:

(a) Areas under special regime of use and protection (FW/Marine Reserves or FW/MPAs.
(b) Permanent or temporary closures for some species.
(c) Regulatory measures to marine species entering the FW at random.
(d) Minimum legal size.
(e) Prohibition on catching and marketing of potentially toxic species.
(f) (ciguatera)
(g) **(f)**Quotas for sport fisheries.
(h) Regulations on construction and use of fishing gear(s)
(i) Methodology for the concession of fishing licenses and authorizations.

In addition to the above groups of regulations, there is also Article 51 of Law-Decree No.164, which establishes the penalties for the ones who catch, land, transport, process, market, or consume any of the endangered species reported by CITES without the proper authorization by the Ministry of Fishing Industry. There are also other pieces of legislation related to FW and marine fisheries resources.

Notwithstanding the above, even though some of the documents do not regulate the fisheries directly, they are useful tools for the fisheries policy, e.g., Law 81 of Environment considers the endangered species, the environmental impact of fisheries projects and the marine pollution. The same applies to the Decree-Law on PAs, some of which are FW and MPAs. There was also joint resolution reflecting the importance of coral reefs as the main habitat of demersal fishes.

Notwithstanding the above, Cuban legislation on fisheries is supported by international law to which Cuba is a signatory, such as:

(i) FAO Code of Conduct for Responsible Fisheries (CCRF)
(ii) United Nations Convention on the Law of the Sea (UNCLOS)
(iii) Convention on Biological Diversity (CBD)
(iv) Convention on International Trade in Endangered Species (CITES)
(v) Convention on Wetlands (RAMSAR)

5.5.4.2 Current fisheries

The Cuban fisheries combine elements from both artisanal and industrial fisheries and their most outstanding features are as follows:

(a) A complete production system that includes a wide network of facilities for the landing, storing, processing, transporting, and marketing of fish products.

(b) The use of small boats (rarely exceeding 20 m in length), typical of short-range fisheries.
(c) The wide range of fishing implements and methods used, mainly due to the diversity of species and the bottom topography.
(d) The seasonality of catches determined by the behavior patterns of some species. These animals are grouped only in specific seasons, either for spawning or in response to weather conditions.

Notwithstanding the above, the fishing industry products in 2005 had amounted to *c.* 72.4 millions Cuban Pesos (*c.* 66.6 millions US$), i.e., *c.* 3.6% of the total Cuban exports (ONE, 2005). This sector employs *c.* 34,000 people including fishery, industrial process, shipyards, marketing, and research workers.

Concomitant to the above, there is a regional fishing division for statistical purposes that began to be used, in a natural way, by the fishermen themselves, i.e., the fishermen from different ports were established their fishing zones without government intervention, and, such de facto division later became official.

5.5.4.3 *National fisheries strategy*

The present Cuban strategy for the development of the fishing industry, until 2010 (MIP, 2005), was based on some important elements, which are briefly as follows:

The large diversity of the fisheries resources and its moderate abundance (common to tropical waters) provide highly valued products. It makes the marine fisheries sector a source of capital for the rest of the fishing and AQC sector and the country's economy. However, almost every fish resource in Cuban waters is being gradually exploited, and, therefore, a big rise in the catches may not be very much expected (Baisre, 2000a). Incidentally, the MPAs are a useful and effective tool for fisheries management and recovery of some resources. In that case, the strategic objective is to use the Cuban fisheries resources in a sustainable manner in order to ensure an efficient, high-quality production by reduced costs and added values (MIP, 2005).

5.5.4.4 *Modeling approach*

The evaluation of the Cuban fisheries policy was based on a surplus production model, which, in this case, could be the Schaefer model, also known as **Gordon-Schaefer model**. The choice is based on the fact that such models are holistic, i.e., they deal with the entire effort and total yield obtained from the stock, without entering into details like growth and mortality or the age and size (Hannesson, 1993; Sparre and Venema, 1998; Seijo et al., 1998). Later, a curve is formed, which plots revenues and costs against the effort in order to determine the reference points: MSY and MEY. This analysis was carried out on the four most important fisheries: lobster, shrimp, tuna, and other finfishes (collectively known in Cuba as *escama*), and, the data source for all fisheries is the statistical database from MIP.

5.5.4.5 *Recommendations*

In fine, the studies on the status of Cuban fishing resources by Baisre (2000a) and Claro et al. (2001) have influenced the management regime, particularly, in view of an attempt toward the management and conservation of the water bodies and fisheries through the **participatory** mode of **CBFM** involving the stakeholders from the grassroots level. The latest changes included licensing of the fishing boats, increasing the closed season for several species, increasing mesh size in trawl nets, declaring FW and MPAs, and the elimination of unselective gears like set nets as well as the strengthening the property rights in the existing cases. The few analyzed fisheries showed levels below the MEY and general trends to decreased effort levels in recent years. Moreover, the test of the hypothesis that the same Schaefer model applies to the period before the implementation of Decree-Law 164 is rejected for all the species except tuna. Although these tests lend support to the hypothesis that the Decree-Law 164 has had a significant effect on stocks, it is not conclusive since other factors, such as general economic conditions may also contribute to the changes by restricting effort. However, tuna, a migratory species is not greatly affected by the management rules. It seemed to further lend support to the theory that the regulatory changes, rather than the economic conditions, have had a significant effect on stocks. In this connection, it may be pointed out that the current institutional structure could jeopardize the ecological objectives favoring the economic one because conservation and production are managed by the same entity (Claro et al., 2001). However, some studies are believed to opine to the contrary. The two functions in the Ministry of Fishing Industry are working separately and the fisheries policy is proving to be successful. Therefore, the current management rules are recommended to be maintained and a study to implement TURFs is also recommended for demersal fishes and other fisheries as an added regulatory measure. Concomitantly, a more detailed study using biological indicators like age, size, catch composition, and spawning stock biomass could be recommended. Such studies could be carried out using analytical tools like the **Beverton and Holt model**.

Incidentally, Anon (2009) threw some amount of light on sustainable economic and social development (with some amount of reference to Cuba) in which fisheries play a significant role.

5.5.4.6 *The environment*

Cuba is making great efforts to deepen its approach toward environmental management in all aspects (including fisheries management), but particularly, in the context of the relationship between the economic, social, and environmental components (Anon, 2009).

5.5.4.7 *Biological diversity*

Cuba adopted a National BD Strategy and Action Plan which is the most outstanding in the Caribbean Region by reference to the degree of national participation involved in its development and execution as well as the territorial scope of its implementation. The main features of this strategy are territorial adjustment, and the development of research projects that respond to the main questions raised in the National BD Study. The advances recorded in this area may be illustrated, inter alia, by the activities that are currently being implemented at the national level. Among these activities are the development of programs for the management and restoration of hydrographic basins, the strengthening of the National System of Protected Areas, the rehabilitation of BD (including fishes) in the degraded areas, the strengthening of air, water (FW, marine and coastal) and soil pollution controls, the execution of projects to manage biological diversity in hydrographic basins, integrated studies of coastal ecosystems and their bioresources including fisheries; the conduct of enabling activities for the monitoring of BD including agro-BD and fisheries, and economic incentives for the conservation of BD and related enabling mechanisms. These activities continue to enjoy the support of the Global Environmental Facility (GEF), the United Nations Environmental Program (UNEP), and the UNDP.

5.5.4.8 *Fisheries resources*

Fisheries and AQC are the potential sectors for the nutrition and avocation of the country. However, it may be mentioned that trends in fisheries resources statistics try to indicate a historical loss of *c.* 20,000 mt, due to the combined action of overfishing and changes in the aquatic (FW and marine) ecosystem. Although some of these losses could be recovered through the regulation of fishing activities, some changes, arising from impacts on the coastal zones and the alteration(s) that, perhaps, had taken place due to unfavorable climatic changes could be potentially irreversible. Moreover, one of the main reasons for the decrease in capture is, without a doubt, the indiscriminate harvest of several species during their breeding period; although, however, it could be incorrect to attribute the total decline to this factor. An analysis reflected in a study of 1995, showed that of the 21 most important fish species or group of species of the Cuban continental shelf, *c.* 38.9% were at the *senile* stage; *c.* 48.7% were in the *developed* stage with a high level of exploitation; and, only *c.* 12.4% were in the *developing* stage with the possibility of increasing catches. This corroborates the fact that *c.* 87.6% of this resource was in a state of crisis and demanded urgent implementation of existing regulatory measures, such as the prohibition of the capture of certain species and the enforcement of regulations in respect of the size of catches; fishing techniques; the issue of fishing permits; and the demarcation of permitted fisheries zones. Incidentally, there had been attempts in Cuba through the **participatory** model of **CBFM** involving the stakeholders from the grassroot level for efficient management of the water bodies and fisheries.

In addition to the above, there have been many publications on fish, fisheries, fishing efforts, and fisher folk. However, there had been only a limited number of studies from a social perspective. The main aim of certain significant communications, like Doyon (2007), was to comprehend and appreciate the development and transformation(s) in **Cuban** fisheries. Such publications tried to portray, how the national fisheries had been managed since the 1959 revolution, and how the resulting processes of decentralization had occurred in the wake of the 1990s' social, economic and political crisis. Attempts had been made to examine how large-scale fisheries had been reorganized both ideologically and practically, as a consequence of these events. Efforts had also been made to explore how large-scale fisheries in the region are connected to an emerging small-scale fishery. Cuban fisheries demonstrate the complexities that arise from both the local context and its links to regional, national, and international processes.

Cuban revolution is a significant event in Caribbean history. One of the principal goals of the Cuban revolution was to transform the environment in order to serve the agenda of nation's economic development. Cuba's leaders sought to extricate the country from its underdeveloped status, which was said to be characterized by a blend of the local subsistence economy and foreign (mainly US) domination of Cuba's main productive sectors and exports. As such, the socialist State "toolbox," including collectivization, centralization, and gigantism used with a Cuban twist in order to achieve the country's new objectives through the sustained harvest of its natural resources (Díaz-Briquets and Perez Lopez, 2000).

Concomitant to the above, "Fisheries" might not have been as important as agriculture in the government's development plans, but it did serve two principals with different objectives:

(a) "Fisheries" were to be part of the government's development strategy, most probably and notably, via industrialization. Therefore, they had to shed their small-scale, artisanal characteristics; and, tried to turn into a single large-scale, well-equipped fleet, which was supposed to be better suited to deep-sea fishing.

As Castro had stated "We will no longer be speaking of small fishers with a dug-out canoe or sailboat; we will be speaking of fish harvesters with increasingly modern means of production and larger boats; they will not be fishing only on the continental shelf but will be going out into the oceans" (translated: Fidel Castro, 1962 in Morales, 1972, 29). Meanwhile, the innumerable small fisher organizations and unions were to be united into one large centralized State organization, which was an onerous task. This first approach to fisheries led directly to the second.

(i) Not only would the fisheries be the object of development, but also they could represent Cuba's high level of development. Thus, fisheries automatically had been embodied in the revolutionary project in view of its high potential for nutrition and avocation. Concomitantly, there were also thoughts and efforts for **CBFM** through the sustainable harvest of the resources, involving the stakeholders from the grassroot level.

Notwithstanding the above, it may be mentioned that the socialist ideal had remained as a central element of the government's ideology and discourse, since the beginning of the revolution in 1959, although there is often a gap between the socialist ideal and peoples' everyday practices on the one hand, and the government's policy on the other hand. All these were said to be linked to the economic challenges the country had faced. Indeed, the island nation faced major crises and setbacks throughout the years, particularly in the early 1990s, with the collapse of the socialist block, leading to what had been named the "special period in time of peace" (hereinafter referred to as "special period"). Using various techniques (Brotherton, 2005), the state had maintained its ideological stance, despite the many profound transformations in the organization of power and the daily lives of the people. The state began decentralizing, leaving more responsibilities to provincial, regional, and local entities as well as government-owned enterprises and newly formed cooperatives. Meanwhile, the population has had to develop new social practices in order to survive this decentralization process. Moreover, the citizens were also to subtly interpret some revolutionary principles while negotiating their way through the changing system of the State institutions; thereby, reinventing the revolution on their terms (Doyon, 2005a, b). These alteration(s) affected the fishing sector and coastal populations. It is evident that the fisheries expanded rapidly, but later experienced a dramatic decline in the 1990s, from which they are slowly recovering. In the process, Cuban fisheries went from a centrally organized large-scale, deep-sea, industrial fishing fleet to a small-scale, low-tech, coastal, and regionally organized fishery. The state, however had never discursively acknowledged these transformations, playing with semantics. In doing so, it downplayed the downscaling of its fisheries and attempted to make the "effect of truth" via discourse (Escobar, 1995).

Attempts had been made on various occasions to examine the changes that have affected the fisheries in Cuba since the revolution, using an anthropocentric and ethnographic approach. More specifically, the analyses were done mainly from a qualitative perspective, what is at stake in the changes in the fisheries following the "special period." This could lead us to consider the case of the reorganization of the fisheries in the formal and informal sectors. Publications also contribute to the understanding of Cuban fisheries, which received very little research attention, particularly from a social science perspective. Such a perspective could be explored, at first, by studying how the government fisheries were developed in the wake of the 1959 revolution and how they changed during the "special period" in the 1990s as the state decentralized. Then, there could be efforts to examine how large-scale fisheries (also named industrial deep-sea fisheries by the State) had been organized in La Coloma, in the province of Pinar del Río. This coastal town (population *c.* 5000) is where one of island's most important fishing ports is located. It is known mainly for lobster fishing, and it experienced major transformations in production and in labor organization during the crisis of the "special period." Finally, efforts would be made to explore, how the fishing industry of la Coloma is related to the newly emerging small-scale subsistence informal and illegal fishing activity in the neighboring village of Las Canas.

The present communication is based on 14 months' of fieldwork in Cuba during the period 2001–02. During the period of stay in Las Canas village in Cuba, there were genealogical works, participants' observation, and informal and semi-directed interviews. Among other things, the interviews focused on how people spoke about their environmental practices, such as fishing. The interviews had also focused on the perceptions of the people with regard to the environment and its management. Further, peoples' response was also observed on sustainable development projects put in place by outside aid agencies and the State. There had been acquaintances with *c* 50 fishermen working in the *Combinado pesquero de la Coloma* (the fishing plant and port, hereafter referred to as the *Combinado*) as well as with the subsistence artisanal fishermen.

The interviewees also had included University Professors, Scientists, and officials of the Forest department. Participation in community-based sustainable development project, notably, of the mangroves in the region of La Coloma-Las Canas, was also *in vogue*. Moreover, the interviewees also had included the government representatives working in the provincial Ministry of Science, Technology, and Environment (CITMA) and some related agencies. Approximately 90 formal semi-directed interviews were conducted in the field.

5.5.4.9 *Cuban fisheries and the revolution*

It is believed that the Cuban revolution had resulted in significant changes to technical organization in the fisheries as well as to the volume of production. The sector consisted mainly of subsistence activities, before 1959, with a small commercial sector. Moreover, the Cuban fisheries barely had met domestic and tourist needs. The revolutionary government promoted industrialization throughout the Cuban economy, including fisheries. A deep-sea fleet was launched, which could eventually reach into international waters from the Atlantic to the Pacific. The government's aim was to supply the local population with more sources of protein; to increase the influx of foreign currency; to better protect the national territory; and to establish bilateral agreements with other Latin American countries (Adams, 1998). A modernization program had led to the construction of harbors, fish plants, and new ships during the 1960–70s. At that time, with the strong support of the Soviet Union through preferential rates for fuel and mechanical parts, the Cuban fisheries were making some of the largest catches in Latin America and the Caribbean. Production had gone up from 20,000 mt in 1959 to >100,000 at the beginning of the 1970s, reaching 200,000 in 1976 (Adams et al., 2000).

This expanded fishery was managed within a socialist economy framework for >30 years (Verdery, 1996). Management activities were centralized and the 63 government fishing enterprises reported directly to the Ministry of Fishing Industry. All administrative decisions related to production quotas and the adoption or prohibition of fishing techniques were made directly by the Ministry. The fishing enterprises had to conform to these decisions or risk facing sanctions.

Notwithstanding the above, the centralization and accumulation of the means of production, typical of socialist regimes, stimulated the construction of several fish plants and shipyards. Further, the collectivization of the means of production also had resulted in the formation of several fishing cooperatives and state enterprises (similar to state farms), which later played a significant role in **CBFM.** Workers were paid not according to their volume of production, but in set monthly wages determined by the government through a common national policy. This often had a negative impact on production and conservation, since (as in the agricultural sector) employees had little scope for changes and the state ultimately made all decisions (Doyon, 2005a).

Concomitant to the above, the management style and change(s) in scale conferred a strong political-governmental dimension to fisheries. They were held up as a model of Cuba's progress, modernism, and development, which could no longer be compared to other Latin American countries but rather to Western standards. Technologies were developed, larger marine zones were fished, and a greater number of species were caught; all of which were aimed to lift the country out of its Third World condition (Eckstein, 1993). As might be imagined, these accomplishments and their associated processes had also influenced relations between fish harvesters and the government. Fish harvesters became representatives of Cuban modernity as well as primary protectors of Cuban sovereignty. Moreover, they were innovative producers, having abandoned outmoded local practices and subsistence activities that had kept them in poverty and to the contrary adopted modern well-equipped boats. The collectivization of the means of production, the new wage labor system, the socialist moral model (according to which **each worker was encouraged to constantly improve himself or herself)**, the stabilization of working conditions, and the overall socialist orientation; all played key roles in transforming pisciculture and fishing practices in Cuba. Further, these transformations also influenced fish harvesters' relations to space, their social organization, and their revolutionary subjectivities.

Notwithstanding the above, industrial fishing was a profitable sector until the end of the 1970s, when different countries, such as Mexico (1976), the USA (1977), and Haiti (1977), established their EEZs; thus, depriving Cuba of access to rich and relatively close fishing zones. Thereafter, the Cuban fleet was turned to more distant international waters as well as to Soviet fishing zones, and, thus, became more dependent on the USSR. In this reorientation of Cuban fisheries, new species were fished in quite different conditions than those prevailing in the Caribbean. The new situation required new equipments and entailed higher costs for traveling to more distant waters and for extra maintenance, which was supported by financial and technical assistance from the former USSR. However, the situation for Cuban fisheries deteriorated with the economic crisis of 1990, following the collapse of the erstwhile Soviet Union (USSR) and the Eastern Block. The Cuban government became unable to maintain its fleet due to high costs and had to drastically reduce its fishing activities. The FCP fleet was condemned (it was eventually used for maritime transportation), while the other fleets were considerably reduced with many employees losing their jobs. Catches of *c.* 232,000 mt in 1988 dropped to *c.* 88,000 mt in 1994 and have stabilized at around 60,000 mt since 2002. Fisheries, which were earlier based on large catches of low-value pelagic species, were reoriented toward high-value species from the island's coastal zones, mainly for external markets (Japan, France, Spain, Italy, and Canada). This had generated *c.* 100 million US $ per year, *c.* 60% of which is from the lobster (*Panulirus argus*) industry. In this connection, it may be mentioned that, throughout this upscaling and downscaling, the government had argued that, its fleet had maintained an industrial status, although the fishermen and researchers were well aware of the true nature of the transformations.

Concomitant to the above, Cuban fisheries returned almost to their prerevolutionary characteristics as a consequence of these changes, mainly, perhaps, in terms of equipment(s) and scope. Significant changes had taken place in economic organization and socialist ideologies, during this period. Castro declared: "capitalism, yes; capitalists, no!" and the country had proceeded to "dollarize" its economy. Meanwhile, the country, during this period had opened up to foreign trade, promoted joint ventures, and concentrated its investments in high-value production for export; all of which have had an impact on fisheries. Moreover, efforts were made to redirect the industry, after many fish harvesters were laid off. As a measure to effect implementation, at first, changes were made in the management scenario (Joyce, 1999). In 1996 (through a decree-law 164 for the management of fisheries), the government had proposed a **decentralized approach**, establishing a licensing system and protection zones for certain species at risk. In this new context, the daily management of fishing operations and production was not in the hands of the Ministry but was entrusted with producer associations (one for each province) that manage fisheries and fish plants with a preestablished budget and quota system. When an association exceeds its quota, the fish harvesters and the enterprises they run, receive a monthly bonus. Paradoxically, however, these associations are also responsible for protecting the environment. Observations suggested, however, that protection efforts are generally limited to a ban on fishing within 10 m of submerged coral reefs, as identified by the CITMA. Such a narrow definition of environmental protection raised doubts about the effectiveness of these measures since the country's priority was to secure hard currency; as the fish harvesters had been paid in dollars according to their level of production.

Notwithstanding the above, structural modifications in the Ministry of Fisheries Industry also affected the internal division of labor among the associations. Firstly, each association classified its fishing territory into production zones where different groups could fish (Joyce, 1997). Officially, this decision was made for conservation purposes. In principle, it decreases competition for the resource, since, each is to remain, preferably, in a specific zone, and is subjected to penalties, if they do not comply with. Since they have access to US dollars, fishers have higher incomes than most Cubans, but they also have to be more entrepreneurial, which entails the risk of showing "individualist" behaviors that were often discouraged and criticized by the State. Nonetheless, most of the fish harvesters had preferred this system. Thus, although they have more responsibilities and are at greater economic risk, they also have access to higher incomes.

Concomitant to the above, several economic and political national and international drivers have affected the development of Cuban fisheries in the recent years. The United States, for example, is currently taking an interest in the sector and is concerned with how Cuban fisheries might affect its foreign trade, a situation that generates speculation about the post-Castro period (Adams, 1998). **These government-driven economic and political changes were justified by economic necessities and the government's desire to maintain a "renewed" socialism with a stronger accent on decentralization**. The transition from an open-sea large-scale industrial fishing sector to a focus on more restricted small-scale FW and coastal zone production is motivated by the need for hard currency, making it a "new industrial" fishing sector. Moreover, the fish harvesters have become "responsible" workers, and a "responsible citizen" through the process of decentralization. This is seen as a "good comrade" in this late socialist period (Doyon, 2003; Premat, 2004). There had been efforts directed toward the inception of management and conservation of water bodies and fisheries through **CBFM,** as a modern tool for the purpose.

In fine, the **Cuban** case presented here had exposed the complex process of fisheries transformations in relation to changing national and international economic and political scenarios. Based on ethnographic research, the present communication tried to document the case of fisheries since the revolution, revealing the recent changes in the sector, in particular since the "special period." The fisheries model, which was promoted in the first 30 years of the revolution, was based on an industrial model, which was intended as proof of the modernist, progressive, and scientific trends of the Cuban revolution.

However, the **Cuban** fisheries underwent major changes following the economic crisis of the 1990s. In a context, where the government no longer had the resources to provide with even the most basic needs of its population, the industrial regime that prevailed during the revolution's years of plenty could no longer be maintained. Investments in the fisheries sector were abandoned and production had to be reoriented. The large scale, deep-sea focus of fisheries had been shifted; thus, changing the fishing practices of the *Combinado* fish harvesters. The new model that was adopted included the processed products sold to mass international markets. Incidentally, there had been a high division and specialization of labor; strong reliance on science and modernization; changes in the equipment patterns; some amount of technological sophistication; small crafts with small engines with generally locally handmade gears with medium catch capacity that is highly diversified to adapt to the fishing season; and an organization based sufficiently on close family and friendship ties. These alteration(s) were supported by State discourse. The government, however, had continued to assert that the gains and principles of the revolution remained almost intact in its fisheries, in spite of the extent of the transformations, which it continued to characterize as industrial. On the other hand, it sought that, the workers adopt new practices and attitudes, such as entrepreneurship, which contrasts with socialist ideology, but which enables the government to deal with its economic challenges while preserving the revolution.

Notwithstanding the above, these changes in the fisheries tried to highlight certain connections between large- and small-scale sectors, specifically through the many family ties between large- and small-scale fishers. As a result, informal fishing practices in Las Canas were nourished and maintained, to some extent, by the fishing activities of the *Combinado*. Equipment, techniques, and ecological knowledge are exchanged through various connections rooted in family ties and local markets. However, the links between the different types of fisheries may sometimes be overlooked. This communication, thus, tried to suggest the significance of focusing attention on them, as they provided information on important ecological and political knowledge (Butler, 2005), on how *savoir-faire* and techniques are handed down from one generation to another and, moreover, also between individuals within a single generation. For example, in Las Canas, ecological and technical knowledge that was neglected for >30 years were passed on in only one generation.

Concomitant to the above, additionally, in **Cuba**, the distinctions and boundaries between the formal and informal sectors; between what is prohibited and tolerated; and between the margins and the mainstream are variable. This could be blurred and subject to changes with different interpretations. However, these aspects are crucial for natural resource management. Incidentally, it may be stated here, that, taking systematically into account, it may contribute to a better understanding of some of the more hidden, but no less influential dimensions of the dynamics that play out within and between different types of fisheries.

5.5.5 Mexico

Mexico possess a rich potential in AQC and fisheries. However, the progress toward better management of fisheries in Mexico had been slow, despite increasing concerns. The main political, social, and economic forces behind overfishing in the country remain largely at the place. Incidentally, demographic pressure, open access or ill-defined property rights, excessive centralization of management decisions, and a wide array of market and regulatory disincentives for conservation were the major problems. There have been papers (Enriquez-Andrade and Batalla-Camargo, 2000) presenting an overview of the current situation of fisheries management in **Mexico** and discussions on the potential benefits and limitations of the implementation of rights-based fisheries management with the expectation to reverse current overexploitation trends. The discussions were also believed to consider the likely management scenarios in the country. The papers often tried to conclude that property rights-based management has an important potential to promote efficiency, profitability, and increased stewardship of fishery resources; particularly, if implemented at the local human community level. However, an effective implementation may have to surmount substantial constitutional, political, economic, and cultural barriers.

Notwithstanding the wealth of its fishery resources, and the rich potential in AQC and fisheries, Mexico's fishing industry seemed to present a low-performance level, caused by poor management and fishing practices, environmental degradation, and fleet overcapacity.

Concomitant to the above, the total fishery landings in Mexico reached a maximum of *c.* 1,391,282 mt in 1997, up from *c.* 524,689 mt in 1976 when the EEZ was declared. It was said to be *c.* 165% increase (Anon, 1998). Incidentally, the growth of production was characterized by three periods, viz.,

(i) a development period with moderate growth (1950–72);
(ii) a period of rapid growth associated with the extension of the nation's jurisdiction over fishery resources (1973–81); and.
(iii) a period of output stagnation and variability (1981–99).

Incidentally, average landings, perhaps, had not increased, during the last few decades due to a combination of poor management practices and environmental variability. From the total production of *c.* 71.5% corresponding to the Pacific Ocean, *c.* 26% correspond to the Gulf of Mexico and the Caribbean Sea and *c.* 2.5% correspond to the FW production. Concomitantly, the total value of Mexican production, of late, was said to be *c.* US $9600 million.

Notwithstanding the above, the country declared a 200-mile EEZ, in 1976, comprising an extension of *c.* 3,149,920 km^2 of rich oceanic ecosystems. The explicit goal was to utilize the contiguous fishery resources by developing a domestic fleet and to exclude all foreign fishing activities. The National Fishery and AQC policies, at this time, had a profound effect, many of them negative, on the coastal zone demographic trends and the use of natural resources.

Notwithstanding the above, it was believed to have been estimated that under ideal management and environmental conditions, MSY in Mexican waters could be *c.* 1.65 mmt, up, perhaps, *c.* 19% from the *c.* 1.39 million estimated in 1997. This could require significant hike in the production of currently underutilized species the sardines, anchovies, squids, and certain species of tunas, and through the practice of AQC in the total production (Anon, 1996). Further, a study by the National Fishery and AQC Chamber (Anon, 1999) conceived a scenario for the year 2020, in which, Mexican production from conventional methods and AQC is to reach *c.* 4.8 mmt, of which *c.* 2.2 mmt could be from capture and *c.* 2.6 mmt from AQC. Hike in production could also lead to an increase in exports from an estimated *c.* 780 million US $ in 1999 to *c.* 2340 million US $ in 2020.

Concomitant to the above, the total Mexican AQC production in 1997 had reached *c.* 173.4 thousand mt (Anon, 1999) representing *c.* 11% of the total fish and shellfish production in volume and *c.* 15.65% in value. The Pacific Coast accounted for *c.* 57,660 mt or *c.* 33.6%. On the other hand, AQC production in Mexico had not grown for > 10 years in spite of official support. The probable causes could include environmental degradation, the growing scarcity of suitable land, diminishing supply of FW, low productivity, lack of funds, inadequate technologies, and conflict(s) with other users. With regard to price, the actual prices increased in comparison with beef, pork, and chicken and are expected to grow even more in the future (FAO, 1997a, b).

5.5.5.1 *Research activities and procedures*

Research activities consisted of extensive (directed) field interviews with a wide array of stakeholders as well as a multi-sectoral workshop in order to.

(a) investigate the institutional and market failures that were said to favor fishery overexploitation in Mexico, particularly, in the Baja California Peninsula; and.
(b) assess the potential for property-rights-based management instruments to address the problems detected by means of qualitative multi-criteria evaluation methods (Nijkamp et al., 1990).

5.5.5.2 *The possible causes of management failure*

There may be a number of reasons for management failure, some of which may be the root causes. Conclusions from the multi-sectoral workshop had corroborated the previous findings related to the use of the marine-coastal resources in the Baja California Peninsula (Enríquez-Andrade and Danemann, 1998), which found widespread institutional failures that were common to all coastal States in Mexico. The main problems, in addition to demographic pressure, could include the following:

(i) Open access.
(ii) Undefined or unenforceable property rights.
(iii) Excessive centralization of decision-making and market power.
(iv) Administrative inefficiency, corruption, and weak enforcement of regulations.

5.5.5.3 *Rationing of scarce fishery resources*

Rationing of access to scarce fishery resources is essential to protect the depleted fishery resources. However, the poorly defined property rights to land, natural resources, and the environment continue to be the major obstacles for the development of the coastal zone in Mexico. Concomitantly, it is widely documented in fisheries worldwide that lack of clearly defined property rights are said to promote overexploitation of many fish stocks and wasteful use of scarce capital assets. However, in Mexico, the public access to the marine commons is granted by the Constitution. Thus, in spite of current fishing regulations, a de facto open access to fishery resources existed due to this mandate and cultural tradition.

Further, for most of the second part of the last century, the fishing sector received widespread support from the Federal Government in the form of several types of subsidies; thus to make the situation even worse. In fact, the federal government regulates all forms of fishing operations in Mexico. Incidentally, the current version of the Mexican fishery law was passed in 1992. In 1999, the rules and regulations emanating from the fishery law had come into force.

Notwithstanding the above, both concessions and permits constituted highly attenuated forms of access rights to fish stocks. However, neither permits nor concessions are divisible, nor could be owned by the foreigners. Moreover, these instruments cannot be traded, and, also, cannot be leased nor subleased. However, they were not exclusive, and, perhaps, could be transferred only with the permission of the Federal Fishing Authority. In fact, ineffective enforcement made these already weak property rights instruments even more attenuated.

In fine, in certain management scenarios, efforts were made to address the problems raised by the current method of managing coastal and marine fisheries in Mexico, by working in collaboration with the stakeholders. Incidentally, from *c.* 20 m inland of the mean high tide to the extent of the EEZ, *c.* 200 nautical miles out to sea, is constitutionally defined as a public property, and is de facto an open-access area without effective restrictions to utilization. However, the current federal system of management is too centralized and, perhaps, not working. Entrusting control of the resources with the state and the municipality, or, even better at the community level, and allowing them to develop access rules could be an effective solution. However, in addition to an extensive decentralization of the administration and their efforts, the widespread implementation of rights-based management in Mexico could need major constitutional amendments.

Notwithstanding the above, McCay et al. (2014) dealt with fishing cooperatives, concessions, and comanagement on the Pacific coast of **Mexico**. Studies revealed that the Mexican case(s) support arguments for greater community-level engagement in catch share and territorial management throughout the Pacific.

Incidentally, 10 fishery cooperatives of the Pacific coast of Mexico were studied to examine the aspects for successful CBM of the fishery commons. The cooperatives held the exclusive rights to "concession" territories for major fisheries and were linked by geographic adjacency and through a Federation. Some of the case studies underscored the role of factors, such as the smallness of scale; the productivity, visibility, and the so-called legitimacy of the fisheries resources involved, etc. Moreover, certain aspects, like the so-called clarity of social and territorial boundaries, adjacency and linkages among territorial units developed a moderate sense of community management. In addition, the cooperatives had also made considerable investments in attaining high levels of knowledge, leadership, transparent, and democratic decision-making process and, more importantly, "vigilance," and enforcement of the rules for running of the organization. The study also showed the opportunities and experiences with environmental change(s) in the development of strong and adaptive capacities for **comanagement** between local organizations and government agencies. However, particular histories and larger legal, political, and cultural contexts may matter, but the Mexican case(s) support arguments for greater community-level engagement in the catch, share, and territorial management throughout the Pacific.

Catch shares and TURFs community-oriented fishery management is recognized in the United States in the framework of a national policy advocating the management of fisheries through "catch shares." Incidentally, "catch share" is a general term for several fishery management strategies that allocate a specific portion of the total allowable fishery catch to individuals, cooperatives, communities, or other entities. Each recipient of a catch share is directly accountable to stop fishing when its specific quota is reached. Catch share-based management is often interpreted as a **euphemism** for using ITQs. Incidentally, a recent flurry of research papers on catch shares had tried to depict their effectiveness in achieving the biological and ecological goals of fisheries management. However, the policy has opened the door to broader interpretations. As such, allocations could be to cooperatives, communities, and other entities besides individuals. In fact, concerns about the often negative effects of ITQ-based catch shares on communities had led to efforts to make allocations of shares of a fishing quota directly to community-based organizations and cooperatives. In this connection, it may be noted that the cooperative-like sectors, of late had implemented such aspects in New England's groundfish fishery. But, they had little explicit reference to the community. In fact, some of the sectors had developed in ways intended to reflect and protect community resources and values. Conversely, on the US Pacific coast, plans had appeared for Community Fishing Associations to hold shares of quota, as was believed to be already existing in certain Alaskan fisheries.

Concomitant to the above, it is a common view that good management and conservation practices coupled with the sustained harvest of resources are the keys to achieve success in the sectors of AQC and fisheries. However, of late, the fishery management scholars have suggested the need to develop a better-suited small-scale fisheries management approach for the developing countries than the one offered by conventional Western Fishery Science. This alternative approach is based on the development of better resource access controls, CBM, and increased use of local traditional fishery knowledge. In response, there were works (Guillermo and Basurto, 2002), which aimed to understand what are the most important social and ecological elements that could contribute to the successful CBM of the Seri Indians' (**Comcaac Indians)** *callo de hacha* (pen shell scallops) fishery **in Sonora, Mexico**. With this broad aim in view, there had been studies to document and analyze the Seri controls over access to the fishery as well as the integration of TEK into local fishing practices. Results of such studies were depicted that the success of this locally managed fishery, which had originated from a good fit between well-defined property rights, locally designed institutions, and the natural system. Incidentally, the outside fishers were allowed to fish in the Seri waters regularly, in exchange for benefits to the Seri. In fact, the integration of Seri community's fishing norms and beliefs, into local management rules had allowed them to achieve low-cost monitoring and successful exclusion of the outsiders from their fishing grounds whenever necessary. Therefore, certain case studies suggested that absolute exclusion may not be necessary to avoid overexploitation and the attainment of successful local management of fish resources could be gradually possible.

Notwithstanding the above, some of the important Seri fishing practices that were responsible for promoting resilience and sustainable use of the *callo de hacha* fishery are as follows:

(i) multispecies management,
(ii) existence of no-take fishing areas,
(iii) rotation of fishing grounds, etc.

5.5.5.4 Introduction to the project

Justification

Fisheries and AQC are the potential sources of nutrition and avocation of the people. Today >1.0 billion people rely on fish as a major source of their food, income, and livelihood. More than 99% of the world's innumerable fishers participate in small-scale fisheries and practically all of them pertain to developing countries. Together they produce *c.* 58% of the world's annual marine catch consisting of enormous amounts and varieties of fishes as well as they supply most of the fish consumed in the developing world. Concomitantly, the importance of small-scale fisheries cannot be ignored (Berkes et al., 2001, b).

Notwithstanding the above, current fishery science is largely devoted to stock assessment with a geographical focus on industrialized countries and a disciplinary focus on biology, and to some extent to economics. Such an approach may not always suit the fishery-management needs of the developing world and overexploitation and habitat degradation scenarios, perhaps, continue to threaten the livelihood of millions of people in the inland and coastal areas. In fact, the management and governance of small-scale fisheries are in urgent need of reform(s) (Berkes et al., 2001, b). However, research works, of late, focused on small-scale fisheries management, and the results of such studies in the developing world portrayed that managers need to be able to exert more control over access to their resources, if an improvement in the inland and coastal fisheries resources conservation is contemplated. In fact, CBM and increased use of local traditional fishery knowledge have emerged as promising alternatives for fisheries, and resource management, in general (Christy, 1982; Cordell, 1989; McGoodwin, 1990; Ostrom, 1990; Dyer and McGoodwin, 1994; Pomeroy, 1994; Berkes and Folke, 1998; Kothari et al., 1998; Newell and Ommer, 1999; National Research Council, 1999; Cudney-Bueno, 2000; Durenberger and King, 2000; Berkes et al., 2001, b; Bourillon, 2002).

Notwithstanding the above, The Seri *callo de hach* (CDH) fishery provides an excellent opportunity to study these promising alternatives. This fishery is one of many **Mexican** small-scale fisheries unregulated by the Mexican Government. There is no federal or state mandate to regulate fishing efforts in the form of season closures, restrictions on fishing gear, or minimum harvestable size. The Seri CDH fishery is managed entirely by local users. The federal government granted the Seri Indians a portion of their historic coastal territory and a fishing concession, in 1975, as a *modus operandi* to guarantee the survival of the tribal culture and reduce conflicts with other local fishing communities. The CDH fishery was practiced entirely in the *Canal del Infiernillo* (Infiernillo Channel), a section of the Seri fishing concession where the Seri were able to control access to non-Seri fishermen. This area acted as an EFZ for the Seri, and it was evident that within it a Seri common property regime existed. This contrasted with the de facto open-access regime that prevailed in the rest of the fishing concession and the coastal waters nearby. Locally and regionally, it is *vox populi* among the fishermen from nearby towns (i.e., *Bahia de Kino*) that the Seri fishing zone was rich in CDH beds, compared to the open-access fishing grounds nearby. This belief was supported by the fact that the Seri CDH beds were able to sustain the fishery all the year round. Elsewhere, the fishery takes place only during part of the year, since, it was hard to find an abundant supply of CDH the year around. Nevertheless, the preliminary data suggested that the CDH fishery was almost 30 years old and for the last few years had supported good fishing. The seemingly successful CDH community-based fishery allowed us to look to the inner workings of the access control mechanisms and fishing practices. It was based mainly on ITK, which the Seri people had in place and in this way could contribute to the development of successful fisheries management approaches that could be better suited for the small-scale fisheries in the developing countries.

Goals and objectives

The success of the *callo de hacha* (*CDH*) fishery had variously been suggested to be due to certain social and ecological mechanisms in place. The principal among them could be the following:

(i) controlled access and fishing.
(ii) management practices based on ITK/TEK.

Therefore, the aim of various research works in the field was to answer to the following question:
Which could be regarded as the most significant social and ecological elements that might contribute to the successful local management of the *callo de hacha* fishery?
Incidentally, the following main objectives were chalked out to answer this question:

(a) To analyze why and how Seri control of access of the outside fishers becomes effective in their channel.
(b) To document and validate how ITK/TEK was integrated into local fishing practices.
(c) To ascertain which fishing practices were significant for the success of this community-based and managed fishery.

Notwithstanding the above, specific objectives are also developed, so as to complement and help accomplish these main objectives:

(i) To describe the Seri physical and historical background as it pertains to the CDH fishery.
(ii) To revise the literature which is believed to address the theoretical foundations of common property resources and ITK/TEK.
(iii) To describe the historical and current Seri *callo de hacha* fishing practices.

5.5.5.5 General procedure for the project

The general procedure for the project could be listed as follows.

To establish rapport and trust with the Seri community

It is essential to be able to conduct participants' observations, which, in turn, is said to characterize the most ethnographic research. It is crucial to effective fieldworks on this topic. The initial effort to establish rapport and trust with the Seri community (or, for that matter with any other community) and local fishermen, is by living in the area for quite a long time. This could be started much before the actual research data collection starts. Association with the previous workers could be an advantage.

Subsequently, the researcher could start identifying his own data informants, contacts, and friends among the people of the community, say "Seri" community. It may be important to mention here, that, some community, notably, the Seri, for various reasons, may not like the presence of wildlife researchers in their community, and hence entry could be generally difficult.

In-depth literature search

In such studies, a thorough literature review on the following aspects is conducted:

(i) The region's physical environment, history of the local community, say the Seri Indians' in this case.
(ii) History of specific type(s) of fisheries of interest to the researcher, say, early diving fisheries in the Gulf of California.
(iii) Biology and ecology of the fish of interest, say, *callo de hacha.*
(iv) Common property theory.
(v) The intellectual roots of ITK/TEK.
(vi) Presentation in the proper platform.
(vii) Conduct of extensive Ethnographic and Ecological Field Research.

Notwithstanding the above, ethnographic and ecological research in the village (e.g., in the Seri village) is conducted in order to accomplish the goals of this project. Moreover, the same ethnographic methods could be used for collecting all the types of sociocultural data. Conduct of detailed research in such fields, including under different scenarios, could lead to useful results for the management studies.

Notwithstanding the above, there have been publications dealing with the **agricultural and fisheries policies in Mexico and the concomitant recent achievements and continuing with the reform agenda** (Anon, 2006). Such communications tried to analyze the effects of Mexico's ambitious reforms in the fields of agricultural and fisheries policies since 1990 as well as attempts to make recommendations for further reforms. Such evaluation(s) are based on the criteria for good agricultural and fisheries policy as agreed upon by the OECD countries. Such criteria, if implemented, could support various economic sectors, which contributed to the aspects like, wider economy; sustainability of natural resources; etc.

Agricultural policy

Mexico's efforts were commendable for the direction it could give to its agricultural policy reform, coupled with achievements over the last few decades. The reforms were reduced the degree of commodity market distortion; improved the effectiveness of income transfers to producers; reduced consumer food costs; reduced the bias against rural areas in social programs, such as food aid; reduced the extreme regressivity of agricultural transfers; improved transparency of policies through decentralization; recognition of the misalignment of public and private incentives as regard use of land, water, and other resources; reduced the cost of agricultural policy in terms of total support in GDP; and the share of producer support in farm receipts. Concomitantly, there had been efforts to identify the remaining obstacles, which lay between Mexico's current agricultural policy and a policy regime that is fully consistent with the principles defined by the OECD Ministers.

Notwithstanding the above, the extensive stretches of the community land restrict development and resource management. Use of natural resources, such as water remains mostly unmeasured and property rights are infrequently enforced. Hence, the degree of exploitation is often not known and private decision-making is, perhaps, not aligned with public requirements. Moreover, most agricultural policy transfers are distributed regressively. Some of the highest obstacles are the remaining barriers to trade, payments tied to outputs (target income), and new energy subsidies. These policies do not provide with much benefit to the farmers for all their costs to consumers and taxpayers.

Fisheries policy

Mexico possesses a rich potential in AQC and fisheries. The country had made significant amount of progress, since 1990, in reforming the policies governing the fisheries sector. However, the regulatory environment in the early 1990s was not very conducive to developing a sustainable fishery sector and, consequently, hampered the sector's longer-term economic prospects. However, after experiencing multiple shifts in policy direction during the 1990s, the current policy framework is now more appropriate in helping the sector move toward a more sustainable and profitable future. In fact, the high-value commercial tuna and shrimp fisheries are sustainably managed today, and thus, the AQC sector is said to be flourishing. Moreover, there was remarkable success in reducing bycatch in the tuna and shrimp fisheries. Also, the negative environmental impacts of AQC had significantly reduced. Further, the transparency of stock assessments, resource status, and management measures improved continuously, together with the extent of fisheries surveillance. However, it is clear that, more reforms, perhaps, are needed to be undertaken, if the fisheries sector is to remain on a sustainable path and generate long-term net economic and social benefits for Mexico.

To develop a vision for the sector in order to provide with institutional stability over time

A major priority for Mexico is to develop a higher-level, long-term vision for the future of the fisheries and AQC sector in order to provide with a sound basis for achieving a balance between industrial development, resource sustainability, and poverty alleviation.

To try for controls on fishing effort

Management reforms are required to regulate the fishing efforts that could be restricted from overfishing. Market-based mechanisms, including transferable effort quotas and CBM, may, preferably, be introduced for those fisheries where it could be feasible.

To go for fisheries management plans and stock recovery programs

In Mexico, only a few fisheries have management plans in place. However, these had generally been ad hoc, fragmented and little effective in only some cases. Mexico, could, preferably, introduce a system of legally enforceable, integrated fisheries management plan(s), building on the current system of *Normas Oficiales Mexicanas* and the National Fisheries Chart, and, the consultative framework that is already established and functioning for a number of fisheries. Moreover, stock recovery programs for overexploited species could be an integral component of management plans and could include measurable goals for stock recovery based on scientifically derived reference points. Stronger enforcement of existing regulations for the unlawful fishing practices could be a part and parcel of stock recovery plans.

To include the artisanal sector under stronger management

It may be opined that the continued use of artisanal fisheries as a social safety net could not only exacerbate existing problems of resource conflict(s), stock degradation and rural poverty, but could also deplete the stock. It could be said to be a both poor fisheries policy and poor social policy. Fisheries and rural development policies need to be mutually supportive. While broader social policies could be necessary to address rural poverty and development concerns, fisheries policies could be a part of the solution. However, a necessary first step could be to determine the size of the artisanal fleet. The use of local **comanagement** initiatives may be effective for localized stocks and area-based fisheries.

To go on with the reform(s) of the institutional arrangements in order to provide more effective management of the sector

Mexico could examine the establishment of different organizations as functionally independent agencies having independent budget lines, devolved decision-making powers on fisheries issues, and a direct line of reporting to the Secretary for Agriculture. The decentralization and regionalization of the institutional arrangements may continue. At the moment, this process is, perhaps, haphazard with no clear guidance as to the government's intention. In fact, a clear plan and appropriate commitment to funding is required with the aims of increasing transparency, accountability, stakeholder participation, and empowerment. However, a key component in any change in the institutional arrangements is the pressing need to unify fisheries enforcement functions. However, enforcement efforts, of late, are hampered by a poor institutional structure and a lack of resources. The split in enforcement functions between the government agencies and the inability of different other organizations(s) to enforce its own regulations without referring them to the National Prosecutor significantly reduces the effectiveness of fisheries enforcement. Enforcement of regulations within the artisanal fisheries and many of the small-scale fleets is virtually impossible with current levels of resources and without having a significant impact on the livelihoods of the poor fishers.

To augment agency coordination for AQC projects

Of late, there are too many overlapping and contradictory regulations governing AQC developments. However, a "one-stop shop" could be a useful innovation to reduce red-tape culture and streamline the approval process. Further works are also required to develop a coherent, transparent set of environmental parameters for AQC operations in order to reduce the costs and uncertainty that currently exists in relation to environmental compliance.

In fine, it is, perhaps, widely recognized and appreciated that agricultural and fisheries policy reforms are difficult to be undertaken in isolation. Reforms are probably, more readily launched in the context of multilateral agreements. More importantly, overall status of economic management, including public services, such as education, safety net provisions and health care, etc., are to be improved in parallel with further reforms to agricultural and fisheries policies. It may appear premature to restrain agricultural and fishery policies that serve functions relating to rural poverty alleviation or which offset high costs imposed by finance or energy policy, until and unless, these objectives are met by other policies. However, the reforms are critical steps toward a regime that could support agricultural and fisheries sectors which are significantly contributed to overall, rural, economic development of the nation.

In addition to the above, Saenz-Arroyo et al. (2005) dealt with community-based marine reserve network in North-Western **Mexico**. There is seemed to be an increasingly broad recognition and appreciation that fully protected FW and marine reserves could be some of the most powerful tools to overcome globally spread overfishing problems (Halpern, 2003; Gell and Roberts, 2003; Pikitch et al., 2004). Workers all over the globe have made a clear and open call to implement networks of fully protected aquatic (FW and marine) reserves that could represent the world biogeographic regions and could help in recovering vulnerable and depleted species (Pikitch et al., 2004; Pauly et al., 2002). In fact, this seemed to be a clear policy statement with little controversy among the workers and scientists. However, the practical design of these networks are sometimes, misunderstood; thus, said to be encouraging the academicians to propose for large (aquatic) reserves, perhaps, with little consideration of socioeconomic constrains (Christie et al., 2003). A significant example was the model developed for the Gulf of California and published in 2002 in the journal "Science". Incidentally, this model brings a novel mathematical approach to incorporate ecological, social, and economic concerns in the design of fully protected reserve networks. However, it was tested on the wrong scale. As a result, the model, perhaps, had ended up after proposing large areas as "no-take zones," fully fishing villages were proposed to be excluded from their fishing rights (i.e., the Seri Indian marine territory). In spite of the fact that making such large "no-take," "no-fishing" PAs could make such restricted areas appear very promising to the conservationists, it may better be good to remember the lesson provided by Icarus, the Greek son from Architect Daedalus. According to the Greek mythology, Daedalus built wings with feathers and wax to allow them to escape from Crete (Wilson, 1998). Despite his father's caveats on flying close to the sun, Icarus approached the aster and his wings melted and he fell into the sea without accomplishing his venture. Proposing such large fully protected reserves (whether in the FW bodies or in the sea) ignoring the dependence of coastal economies on the aquatic (FW/marine) resources, is just as dangerous as Icarus' approach to the sun. They may risk losing the fishers' support the best potential allies for reserve's stewardship.

Concomitant to the above, there have been information available in the literature, proposing alternative approach(es) to establishing large FW/marine reserves networks, but one step at a time, and, in partnership with the fishing communities, who are said to be genuinely worried about the FW/marine species depletion, on which, their subsistence largely depends (Saenz-Arroyo et al., 2005). Such an approach includes the direct capacity building of local fishing communities and NGOs in the design, establishment, and monitoring of FW/marine reserves, funding for the initiatives, and the design for a multi-criteria, decision-making tool.

5.5.5.6 *The initial seed (in Mexico)*

In 2000, "Comunidad y Biodiversidad A.C. (COBI)," a Mexican conservation organization started a pilot project to develop a community-based methodology to implement networks of fully protected marine reserves. The first networks were planned within the boundaries of two natural PAs in the Gulf of California (San Pedro Mártir Island Biosphere Reserve and Loreto Bay National Park); and in full agreement with the local communities. The idea, perhaps, was to encourage setting these reserves, as community experiments, to recover (or, at least to monitor, how they recuperate) the depleted species. Incidentally, the fisher's concerns on depleted species matched the ones identified by the literature as very vulnerable to fishing, such as the large predatory fish, molluscs, sea turtles, etc., that are said to be easy to catch (Tegner et al., 1996; Musick, 1999).

Concomitant to the above, in San Pedro Mártir Island, almost one-third of the coastal ecosystem was declared as "no-take zones" in 2002 through an official decree of protection. Incidentally, the main objective of the reserve was to recover some depleted species identified by the fishers, such as turtles, groupers and molluscs, etc. On the other hand, in Loreto Bay, two small seamounts were established in 2003, as no-take zones mainly for scientific experimentations.

N.B.: It may be noted that, the term "marine reserve" is ongoing, and is used as fully protected. However, the term "natural protected area" could be used for official decrees of protection.

Notwithstanding the above, governmental allocation of financial resources and Park staff in Loreto Bay made the experiment successful and 3 years after the establishment of Loreto's no-take reserves, interesting results were gathered, which was of a broad regional relevance. For example, in order to evaluate the success of the communication program directed to the stakeholders of the park, a survey had been conducted to assess how the management rules were understood by the users before the program had taken place and how the changed rules were understood and complied with after our program had started. Further, in 2002, the first series of surveys were applied to elucidate awareness of users on the management rules. Subsequently, in 2003, a communication program was started that consisted of producing and distributing widely the pamphlets, posters, etc., in public places as well as giving talks to the community for disseminating precise information of Loreto park boundaries and rules. The percentage of users who could localize the park boundaries, correctly, was increased from *c.* 13% to *c.* 73%, and the percentage that knew correctly where the no-fishing zones were located increased from *c.* 13% to *c.* 50% after the program was initiated. This, in turn had a positive effect on the compliance of the rules while in 2002–03, there was no fisher within the no-take zones boundaries while during the monitoring visits by 2003–04, *c.* 67% of the team's workers had increased to *c.* 88%.

Concomitant to the above, another significant result of these no-take reserves was new insight on the impact of artisanal fisheries on the reef ecosystems. The results of 4 years of monitoring by underwater censuses depicted that populations of some of the most vulnerable reef fishes (viz., the groupers, snappers, and parrot fishes) significantly decreased outside the reserve's boundaries during the last few years but remained almost stable within the boundaries of the no-take zones. Further, parallel to these species' depletions, reef fish diversity had also decreased significantly outside the reserve boundaries but not inside the reserves.

5.5.5.7 *Multi-criteria decision supporting tools: From simple to sophisticated reserve designs and frameworks*

Several criteria are often involved in dealing with different aspects of management of resources. As such, supporting tools are also evolved to take care of the multi-criteria decision-making process. Incidentally, the formation of network of marine/FW reserves could be one of the greatest examples to illustrate the complexity of an environmental decision. In fact, its assessment requires information from multiple points of views and could involve wide groups of stakeholders. Moreover, there could be demands to compromise with objectives, and, further, with relatively rapid solutions. Hence, multi-criteria evaluation techniques are some of the most suitable tools to deal with such problems (Munda et al., 1994; Van Pelt, 1993).

Concomitant to the above, as implied in its name, **multi-criteria evaluation** is a method, which allows decision-makers to incorporate different points of views when evaluating a set of possible decisions.

Notwithstanding the above, applying this technique also offers an opportunity to incorporate ITK and local stakeholders' insight in describing the ecological features of sites of particular significance, such as breeding aggregations or nursery areas. Further, to deal with the problem of how different FW/MPAs' networks could be more suitable to different localities, a technique, called EVAMIX (Voogd, 1983) could be used and the algorithms could be programed in visual basic to make it user friendly.

Another example regarding the use of multi-criteria to design marine reserves is inducted from a fishing cooperative that has exclusive fishing rights for almost all benthic species (except the kelp *Macrosystis* spp.) in the waters adjacent to Natividad Island. This small populated island (< 10 km^2 and *c.* 500 inhabitants) is located right off the central west coast of the Baja California Peninsula. Concerned with the historical depletion of abalone species (*Haliotis* spp) and its fragility to environmental fluctuations, fishing cooperative members acceded to evaluate the possibility of establishing a small network of reserves, mainly to recover abalone species, as a scientific experiment. In order to present to the cooperative assembly (made of all 85 members) about the cost and possible benefits of a network of reserves, the assembly approved to constitute a team to evaluate different possibilities of networking the reserves. This team of experts was made of fisheries technicians from the cooperative, expert fishers who are locally recognized by their fishing and diving experiences, and staff from, COBI, as the conservation and scientific counterpart. Criteria and alternatives were designed by the team. However, in this case, the process was very much different from San Pedro Martír, due to the large databases of information available on resource distribution, abundance, fishing statistics, characterization of habitats, etc.

5.5.5.8 Fisher fund

It may be mentioned that the grassroot level conservation groups and the local fisher's organizations from various fishing communities in the Mexican Northwest region called for peoples' support to plan and implement parallel projects in their own communities, after 2 years of implementing the pilot project in San Pedro Mártir Island Biosphere Reserve and Loreto Bay National Park. Subsequently, COBI with the support of The Nature Conservancy decided to organize a 1-week hands-on workshop entitled "Design, establishment and evaluation of no-take fishing zones as a tool for conservation and management of marine ecosystems in Northwestern México." This was planned after recognizing and appreciating the difficulties of the local organization(s) to attend to many projects in various communities simultaneously. Incidentally, the main objectives of the workshop were to build the capacity among the workers.

In fine, restricted areas protect the endangered and less abundant biota in order to save them from extinction. The community-based reserves are made for protection of the less abundant biota from endangered sea turtles to variously depleted abalone species, or to monitor some of the northern limits of globally important coral reefs. In addition to ecological and biological reasons, they could also be conserved for the purpose of scientific studies with them. They are promoted through the Fondo Pescador and could represent a robust network of marine reserves, mainly because they are supported by the local fishers. From 2000, till today, little by little, almost *c.* 10 km^2 of the fully protected reserves have been formed in Mexico. However, even such a vast area have been far less from the target to be achieved toward the protection of species at a regional scale. Nevertheless, it offers a step-by-step process in the formation of safe havens for the endangered species. Moreover, during the period of the brief time that Fondo Pescador had operated, it could, perhaps, be learnt that the fishers are fully aware that, totally protected marine reserves, perhaps, offer benefits as **management tools**. However, the lack of technical, financial, and legal support for these community efforts had diluted them in the daily activities of the fishers. Incidentally, the role of the conservationist, then, is to encourage local willingness to implement and support such strategies by implementing "tenure system" on resources and the fishing areas as exclusive fishing rights. The conservationists could also financially support such initiatives which could make marine reserves. They could also assist by giving technical advice during all steps of the process, including monitoring.

Notwithstanding the above, the term "community," is often interpreted as a society that belongs to an isolated, rural, and poor village. However, this concept could be applied at different scales to different fisheries. To coastal fisheries (and even in FW fisheries), a community is that body or aggregation of people, who are usually involved in the decision-making process, like, how to protect certain parts of the fishing grounds; how to protect some of the endangered species of fishes; and, so on. However, for the migratory species, like the sharks or tunas, "the community" might be represented by fishers of several villages, along large territories, some of them may be involving many countries.

Similar exercises of trying to make a community-based network of reserves could also be done in the FW sector. The methodology that would be applied and the software, which could be produced, may be applied to any of the similar situations. Incidentally, some of the significant community-based models, which are framed using a bottom-up approach have the potential to be emulated and implemented in other places of the world, and are also said to be strong instruments to eventually develop large-scale networks of FW or marine reserves with local participation of stakeholders.

In addition to the above, in the **Gulf of Mexico**, the catch shares for red snapper had cut the accidental killing of fish by *c* 70%, and, perhaps, had led to a *c* 60% increase in red snapper (as per Environmental Defence Fund). Other ecological benefits might also arise when secured access to a catch gives incentives to fishers for reducing overfishing and avoid sensitive spawning areas to promote stock recovery.

5.5.6 Peru

Peru, as a nation, is rich in water bodies and fisheries. Pinedo et al. (2000) reported about CBNRM as a nonlinear process with regard to a case study in the **Peruvian** Amazon *Varzea*. Their study revealed that small communities along the waterways in the northern Peruvian Amazonia depend on a wide variety of natural resources. Their long-term survival and development require management systems that could assure the future availability and sustainability of these resources.

The *ribereño* peasants, who live in small communities along the waterways in the northern Peruvian Amazonia, depend on a wide variety of natural resources. Their long-term progress could require such a type of management system that could ensure the availability and sustainability of these resources in the future. In this context, the management of natural resources by the local communities could be an important alternative to the prevailing private and State models for the sustainable use of the resources. However, there may be many factors which could contribute to the success or failure of CBNRM. In fact, there could better be a long-term and dynamic process that may include performance peaks, valleys, and interruptions. Moreover, there had been propositions that, certain factors, essential for the establishment and continuity of a CBNRM initiative, might not remain constant through time. This is the result of the dynamic nature of the social and natural setting, in which, these riverine communities live. For this reason, a CBM system may, preferably, be viewed as a nonlinear process that may not be understood without analyzing its historical development and the dynamic nature of the factors. These factors regulate the use of resources. As such, management systems that are flexible in their response to these factors are better adapted to the existing conditions in the Amazon Basin.

5.5.7 Trinidad

Trinidad and Tobago are trying to develop themselves in AQC and fisheries, and there have been works related to this field. Shing (2001) dealt with a case study of the Integrated Coastal Fisheries Management (ICFM) Project, which was a pilot project for the Gulf of Paria, **Trinidad** under the UNDP ICFM Project, funded by the UNDP Division for Global and Interregional Programs and executed by the FAO of the UN.

The Gulf of Paria was selected as one of three sites for the implementation of the project. Criteria for the site selection have been based on the importance of coastal fisheries to the country; management issues to be addressed; government policies; institutional framework; and the existence of programs to complement project activities both nationally and regionally. Incidentally, the project was implemented in March 1994 for 12 months in Trinidad and Tobago.

The Gulf of Paria is a shallow semi-enclosed basin bounded by the west coast of Trinidad and the east coast of Venezuela. The Atlantic Ocean water enters the Gulf, in the south, via the constricted Serpent's Mouth and exits through the similarly constricted Dragon's Mouth in the North. The average depth is *c.* 20 m. A clockwise circulation undergoes in the Gulf water. Further, the salinity and circulation are influenced by the flow of the rivers, such as the Amazon and Orinoco as well as the local rivers such as the Caroni and the Oropuche. Concomitantly, the volume of sediments brought-in by these rivers gives rise to soft, muddy, sandy substratum, while the nutrient-rich waters contribute to the high productivity of the area, supporting rich and diverse fisheries resources. Further, the fringing mangroves and the wetlands along the Gulf provide shelter for the juvenile fish as well as support populations of shellfish such as oysters, mussels, snails, crabs, shrimps, etc. In addition, the avifauna of the wetlands is also rich and diverse. This had been encouraging > 16,000 ecotourists to visit the Caroni Swamp every year.

Concomitant to the above, > 2500 fishers operate in communities and recreational sites along the coast of the Gulf of Paria. The estimated landed catch from the Gulf of Paria was *c.* 7800 mt, in 1991, at an ex-vessel value of *c.* TT $70.5 million. Incidentally, a variety of fishing methods are used in the marine capture fishery, which include, among others, the gill nets, bottom longlines, fish pots, hand lines, and trawl nets.

The Gulf is the most populous and active coast of Trinidad. More than 90% of the population inhabit along the Gulf coast. The capital city, Port of Spain as well as other major cities and towns are located along this coast. The watershed and hinterland of the Gulf coast support the majority of the country's agriculture, ranging from small farms to large estates supporting sugar cane, rice, and citrus cultivation. Major industries are also located along this coast. These include the manufacture of ammonia, methanol, and chemical fertilizers. Incidentally, the Agro-industries include the processing of sugar cane and the manufacture of food products. There is also exploration and extraction of oil and natural gas both offshore and onshore, and many of the major oil refineries are located in the area.

5.5.7.1 *Project details*

In 1992, the Fisheries Division of the Ministry of Agriculture, Land, and Marine Resources commissioned a bibliographic survey to review the possible impacts of human activities in the coastal waters and on the fisheries resources of Trinidad with emphasis on the Gulf of Paria (Boodoosingh, 1992). It was made quite clear that fisheries management, particularly for the Gulf of Paria, required an expansive and extensive approach. The marine fisheries resources were stressed not only by the fishing activities, but also by the range of activities which result in the alteration of the habitat and deterioration of the environmental quality. The report had also identified activities, such as coastal reclamation and port construction; poor watershed management practices; and quarrying, contributing to increased sediment loads. Moreover, offshore drilling during oil and gas exploration; improper solid and liquid waste disposal (both domestic and industrial as well as from the ships in the sea); and the widespread use of chemicals in agriculture and industry had been further deteriorating the situation.

Notwithstanding the above, it became obvious that the management of the fisheries of the Gulf of Paria involved both the regulation of the activities of fishers as well as greater involvement of the fisheries sector in the decision-making process to regulate the development of the coastal zone of the Gulf. The Gulf of Paria coastal zone or management area was defined as the marine areas up to the watershed of the Gulf.

Notwithstanding the above, the Fisheries Division demonstrated a keen interest in implementing the ICFM Project to portray its strategy for the management of the fisheries resources of the Gulf of Paria. In fact, the ICFM approach was receiving global attention at that time. The Global ICFM Project was viewed as a scope to secure the necessary support to establish a framework for the integration of the concerns of the fisheries into coastal area planning. Incidentally, the objectives of the ICFM project are as follows:

(a) to improve methods and coordinating mechanisms for integrated coastal fisheries management and.
(b) to develop an enlarged assistance program for ICFM of the marine fisheries resource of the Gulf of Paria.

Concomitant to the above, the project had sought to strengthen the capability of the Fisheries Division of the Ministry of Agriculture, Land and Marine Resources specifically, in order to provide the necessary state leadership to have the concerns of the fisheries sector incorporated into the framework of development planning for the coastal zone, which was specific to the Gulf of Paria. However, the strategies for achieving this involve the following:

(a) to improve the information base on the fisheries sector and the impact of other sectors on it;
(b) to enhance the general awareness of the socioeconomic contribution of the fisheries sector and its role as a natural indicator of environmental quality;
(c) to strengthen representation and associations within the fishing and related concerns; and.
(d) to promote the supportive role of the Fisheries Division in matters related to coastal zone management.

Notwithstanding the above, in effect, the aim was to use this project to identify the prerequisites for successful ICFM of the Gulf of Paria and to initiate a suitable framework for this purpose.

5.5.7.2 *Issues and conflicts*

Multi-gear, multispecies fisheries

Fisheries resources of the Gulf of Paria have a multispecies composition and are harvested by a heterogeneous, multi-gear fleet. Thus, among the fishery groups, there could obviously be conflicts between them, e.g., between trawlers and fishermen using other methods, who blame the trawlers for the marked depletion of fisheries resources and destruction of the habitat. Further, also, within the trawl group, there could be conflicts between the artisanal trawler owners and the semi-industrial and industrial trawl fleet. The artisanal trawler owners, generally, accuse the "bigger" trawlers, both local and foreign fleets, for contributing the most to the problems. The industrial trawlers, on the other hand, usually, blame the inshore trawlers and foreign fleets, as the number of industrial vessels permitted to fish has been fixed since 1988. Some fishermen, however, attribute reduction in the catches due to industrial development along the gulf coast, which has displaced fish landing sites in some areas.

Notwithstanding the above, an "open-access" situation exists with regard to the management framework for fisheries resources.

Multi-sectoral use of the Gulf of Paria coastal zone and associated watershed

A number of sectors are involved in the development and use of the coastal zone, often with conflicting and overlapping jurisdictions. In general, the legally established procedures for the development of the coastal zone did not adequately provide for:

(a) conflict resolution among the users of coastal resources and between resource users and other sectors;
(b) networking among the sectors to ensure that, most of the relevant information was available to decision makers;
(c) decision-making based on a consideration of the effects of development on the entire coastal system, considering land/sea continuity; and.
(d) consistency in decision-making within sectors with regard to coastal development and exploitation of resources.

5.5.7.3 *Weak consultative process in decision-making with regard to coastal area development*

There have been few avenues for the fisheries sector's involvement in the wider context of coastal area planning, prior to the project. The Fisheries Division is the state agency responsible for the development, management, and conservation of fishery resources in the territorial waters, EEZ and in inland waters of Trinidad and Tobago; fisheries policy formulation; administration of fisheries regulations; and the provision of extension services. Nevertheless, the associated Caribbean Fisheries Training and Development Institute provide training in fishing gear technology, fish handling and processing, and in other aspects. However, the division collaborates with the fishing industry and a number of different agencies on an ad hoc basis, in carrying out its responsibilities.

5.5.7.4 *Weak organization of the fisheries resource users*

There were few fishery organizations capable of representing the sector in environmental matters. In general, however, the development of fishery organizations chequered history. There were many registered organizations for the Gulf of Paria but only a few were functional. Moreover, there were, perhaps, *c.* 39 fish/shrimp landing sites accounting for *c.* 60% of the landing sites around Trinidad. However, the communities associated with these sites, along the Gulf coast, are varied. They ranged from interests dispersed within large cities (such as Port of Spain and San Fernando, where no physical community exists) to towns and villages, like Cedros, Fullerton, and Icacos (where fishing is said to be a major occupation and livelihood alternative). Moreover, there were three fishing cooperatives (notably, Icacos, Cedros and Almoorings) and one fishing association (notably, San Fernando). Incidentally, the inshore artisanal fishermen formed the National Organization of Fishing and Allied Cooperative Societies Limited (NOFACS) in 1988 in order to coordinate the activities of its members; represent their interests to the Government; promote the development of fishing cooperatives; and seek to revive inactive ones. There were seven cooperatives comprising NOFACS including the three from the Gulf of Paria. The activities of the cooperatives as well as the NOFACS continued to be constrained by the independent nature of the individual fisherman; consequently, limiting the effectiveness of the group in representing the interests of its members.

Notwithstanding the above, there was one fishing interest group, viz., the Trinidad and Tobago Trawler Owners' Association, representing the interests of the industrial trawl fleet. This association was highly active, as an interested group, and effectively represented the concerns of its members. In addition, the Orange Valley Association had been formed, at Orange Valley, to represent the interests of the semi-industrial trawlers.

5.5.7.5 *Fisheries versus other sectors*

It is assumed that the Fisheries Division may have to compete with two highly visible sectors, within the MOA, Land and Marine Resources, viz., forestry and agriculture, which contribute more to the GDP than fisheries. Moreover, the ministry may have to compete, in terms of national accounting, with other more prominent ministries, such as energy, trade and tourism, and Finance. Further, heterogeneity in the scale of activities within the sector and across the sectors generate an interesting wave of power relationships, for example, artisanal and industrial fishing concerns; cash crop agriculture; etc. The classical perception of fisheries as the employer of last resort may be a factor in relegating the sector to a comparatively low status.

5.5.7.6 *Geographic extent of the area under consideration*

The boundaries of the management area included the watershed, coastal zone, and marine areas of the Gulf of Paria. It was further felt that for a program in ICFM to be effective, it could seek to include all the sectors likely to have an impact on the fisheries resources.

5.5.7.7 *Types of stakeholder groups*

(a) organizations with management authority.
(b) organizations representing users and communities.
(c) users and nonorganized stakeholders.
(d) National Organization of Fishing and Allied Cooperatives Society Ltd.
(e) fish vendors.
(f) other fishers.

5.5.7.8 *Stakeholder details*

To initiate ICFM activities focused on improving the profile of the fisheries sector through enhancing the capability of the Fisheries Division and to establish a framework for ICFM.

5.5.7.9 *The fisheries division of the Ministry of Agriculture, land, and marine resources*

There are a number of agencies involved in various activities related to fisheries development and assessment. However, this division is vested with the responsibility of management and policy formulation. In terms of the development of the coastal zone and influencing the approval process for such developments, it was determined that, management issues, involving other sectors, could require state leadership. It is, perhaps, only a state agency that could approach other state agencies for obtaining relevant data and information to add to the process of decision-making.

The following points could be useful in this regard:

(a) The donor agencies (FAO, UNDP) provide the financial support for project implementation.
(b) The fishers of the Gulf of Paria are those who fish in the Gulf of Paria as well as use fish landing sites along its coast. In general, the fisheries resources are spatially dispersed and are said to be exploited by a diversity of resource users in spatially scattered communities. It was felt that fishermen who live, fish, and market their catch in communities along the Gulf could have local knowledge of the resource as well as be more intensely aware of the changes in the availability of the resource and the environment. As such, they would be more active in conserving the resources for the posterity. In addition, they could be more alert to any changes locally that may be a consequence of developments by other sectors.

5.5.7.10 *Other fishermen*

There had been plans to be extended to other areas, notably to the other coasts. It was, however, felt that if other fishermen were aware of the activities, implementation of ICFM for other coasts may not be too difficult.

5.5.7.11 *Strategies applied for stakeholder participation*

Involvement of all stakeholders in the management process and not just the fishing communities

All the stakeholders including those at the grassroot levels are to be involved in the management process of the resources. For this purpose, a final list of stakeholders could be prepared through public meetings. The efficient management of the resources could be assured through ensured participation of all the stakeholders associated with peoples' support.

The need for awareness generation on environmental and resource management issues among all stakeholders (resource users and the general public)

Awareness building among all the stakeholders as well as among the people is a significant component of any effort dealing with management and conservation of resources. In this exercise, there could be a need to host a number of community meetings in order to promote **community-based** and integrated management as well as a range of related, interactive, promotional activities. These could include the launching of an annual "Clean Coast Day";

production of a video program on ICFM as a resource management strategy for the locality; an ICFM alphabet poster and booklet; and ICFM curriculum teaching materials for secondary schools; and, so on. Almost similar programs could also be chalked out for the FW sector.

Identification of critical areas, particular resources, and user groups along the Gulf of Paria on which to try to focus on project activities

Two fishing communities were identified in order to highlight some of the major management issues and conflicts within the fishing community as well as those associated with human activities in the coastal zone. Incidentally, three types of surveys were conducted, which are given below:

(a) **A community survey**, which could be conducted by a local consultant using participatory observation methods (he may have to live for about 6 months in one community) collecting data by direct interviews as well as through general interaction with the community. This could identify some of the major problems which could arise in implementing ICFM. More importantly, it could provide with information on community structure to define strategies for the implementation of ICFM as well as could identify the major problems as the community, perhaps, witness them.
(b) A **household survey** in communities focusing on the female head of the household. This could try to identify gender perspectives related to NRM and the impact of human activities on natural resources.
(c) **Generation of secondary data profiles**, taking help from published, census-type information in order to obtain background demographic, socioeconomic, and geographic information for each area for having a general understanding of the lifestyles and standard of living of both the fishing and interrelated non-fishing communities which might have to be considered in ICFM.
(d) In addition, a **"local knowledge" survey** could be conducted with fishing communities in order to obtain information on the fishers' knowledge of the resources, the environment, and environmental changes. There could be large-scale participation of the fishers as well as consensus with the conclusions of the survey, which could be presented to them at several community meetings after the survey.

Consolidation of all available information on the Gulf of Paria to avoid duplication of works

There had been works done on related aspects. A review of all organizations and institutions likely to have relevant information on the Gulf of Paria, which indicated that *c.* 51 organizations were likely to have information related to aquatic resources of the Gulf coastal zone relevant for decision-making with regard to planning and development. All the stakeholders identified were contributed toward the compilation and publication of The Gulf of Paria Bibliographic Database (GULP). Moreover, > 25 organizations, some representing several agencies, contributed to the consolidation of geographic information on the Gulf of Paria. The resulting integrated GIS-based information could be used as a planning and management tool for the development of the coastal zone.

Consideration of alternatives to the prevailing "open-access" approach to harvesting of fisheries resources

Suggestions for approaching the abovementioned aspect were obtained through the local knowledge survey. Fishers provided their views on the fisheries factors that are said to contribute to the abundance of fisheries resources. These were considered in the later development of a draft Fisheries Policy as well as a draft Fisheries Management Bill, which have been presented to the fishing communities.

Consultations at the community and national levels to focus project objectives and activities and thereby enhance institutional linkages

It may be mentioned that the community surveys may support consultation at the community level. On the other hand, results of the surveys of specific case studies seemed to had revealed that the compilation and publication of GULP as well as the GIS-based database on the Gulf have contributed to the enhancement of institutional linkages and encouraged widespread involvement in two spheres of consultations as follows:

(a) National Conference on Information Networking on Fisheries and the Coastal Zone (February 1995). The electronic bibliographic database developed under the project was formally handed over to the National Library and Information Services for national use and continual updating at the national level.
(b) Interagency Seminar on Integrated Planning Coordination and Management for the Coastal Zone (April 1995). This was resulted in an arrangement; whereby, the Fisheries Division became a part of the review process for Environmental Impact Assessments for developments within the coastal zone.

5.5.7.12 *Lessons learnt*

Lessons were learnt in a good way through various types of experiences. At the same time, it could be important to consider prevailing perceptions and culture. There could exist perceptions (which may be erroneous) about the fisheries sector, and such (so-called erroneous!!) perceptions could contribute to its low profile as well as may constrain its ability to participate in the wider context of the coastal zone or FW management planning. Concomitantly, with regard to the fishers, the independent nature of the fishers and the history of their associations could affect the extent to which they could be organized to be effective.

5.5.7.13 *Stakeholder identification*

Identification of the stakeholders is an important aspect of effective management. In planning the project and seeking confirmation of the initial list of stakeholders, it may be done by identifying all the relevant so-called interested stakeholders. However, this may be insufficient for the identification of the actually interested stakeholders who could ensure their participation. Thus, in order to identify the actual stakeholder, there is, perhaps, a need for going further than simply identifying who should be involved in the process. The methodology employed may also be able to identify the extent of participation possible and how this participation could be effective.

5.5.7.14 *Stakeholder identification and capacity building assessment*

The process is pointed toward the disorganized or weakly organized groups, which may be difficult to include in a collaborative framework effectively. Incidentally, some groups could be excluded, mainly because they are not organized and in a number of ways the nature of their activities did not facilitate bonding.

Notwithstanding the above, the Convenor's bias could lead to inadvertent marginalization of some stakeholders. Further, relevant concerns of other groups may also be marginalized due to the convenor's focus on its own general goals. In an instance, for example, for the Gulf of Paria project, the Fisheries Division was the convenor. Therefore, while seeking to include the fisheries sector's concerns in the development planning process, the Division adopted the role of representing the views of the entire fisheries sector at the expense of other groups. In addition, donor support may not be easily available, if the project attempts for a narrow focus, rather than a holistic approach.

5.5.7.15 *Understanding power relations*

In continuation of the foregoing account, stakeholder identification could help in understanding the power relations among sectors which may affect the consultative process. It is evident that economic hegemony among the various sectoral interests in the Gulf had determined the ratio of power relations.

Notwithstanding the above, research and networking are the key elements for capacity need assessment. **Sometimes, the entire project or a scheme could be a learning experience which may facilitate extensive public and private sector collaboration, and may also identify the skills and structures that need to be considered for the participatory process.** It may, therefore, be important to identify the groups and areas which need to be enhanced, if the process is to be effective. Nevertheless, within the project cycle of the present example, it was, perhaps, possible to enhance the capacity of the Fisheries Division to ensure that the Division was equipped with the necessary skills to foster participatory management of the fishery resources in the Gulf of Paria.

5.5.7.16 *Rationalizing urgency with practicability*

The urgency to implement approaches could affect implementation strategies and outcomes. In the present example, the Gulf of Paria ICFM Project was said to be very ambitious given its 1-year duration. Perhaps, it could have been limited in terms of geographic coverage or sectoral interests, as a pilot project. The project was achieved its objectives in enhancing the image of the fisheries sector, and had also laid a foundation for interagency collaboration, but, perhaps, the time did not allow for thorough the examination of options for establishing a legal framework with appropriate stakeholder representation.

5.5.7.17 *Feasibility in determining the size of the area to be managed*

Ferreira (1998) listed a number of criteria for successful **fisheries comanagement,** which are applicable in the case of ICFM. Among them, it was pointed out that the area to be managed may, preferably, be not too large. In fact, for greater convenience, it could be confined to the local waters of a particular community. Moreover, in applying formal techniques for comanagement, in the context of a project, the implementation strategy could, preferably, have relevance to the practicability.

5.5.8 USA

The USA has been known for possessing rich resources in AQC and fisheries. In fact, there have been large-scale research works done in the field of fisheries and AQC including management with a focus on CBFM involving the stakeholders from the grassroot level in a participatory mode.

Incidentally, LFK is an essential area of application. There have been several works done on the LFK in different parts of the world (Bartch et al., 2009). LFK is significantly contributed to the knowledge in Fisheries Management, particularly, CBFM. Such a project was piloted in Maine, USA. It was a learning journey for everyone involved. Designing an educational program may be quite difficult, particularly, where knowledge and participation intersect to engage young people and the stakeholders as vital community members, i.e., in brief, the "community" as a whole. Such an exercise is done in order to address issues which impact the future vitality and sustainability of the small fishing communities. Some of the outcomes which sustained us through the many challenges of implementing the LFK project include, among others, the following:

(a) Receptivity from the community

Generating partnerships and common agenda between and among the diverse stakeholders, viz., the regulatory agencies, schools, students, and local community members, while, concomitantly, a challenging task had been one of the greatest strengths of this project. These partnerships resulted in such outcomes, which might not be able to be predicted at the project's inception.

5.5.8.1 Why is it important to document and preserve marine-related oral history?

Since the dawn of civilization, there had been a dependency on nature and natural resources, e.g., on FW, sea, living beings, etc., i.e., on both biotic and abiotic resources of nature, in order to sustain ourselves by keeping the body and soul together and making a living. However, that way of life is changing fast. The rivers, lakes, wetlands, oceans, and even the groundwater are threatened, both domestically and commercially around the world; thereby, resulting in uncertainty for the future of our ecosystem and for the denizens which are present in the ecosystem.

Under the present practice of so-called assault to the nature, which is *in vogue*, ITK in different aspects including those in the fisheries arena are deteriorating and gradually vanishing from among us leading to a so-called void in the field.

Concomitant to the above, fishery and fishing economies are being transformed or replaced around the nations. In some places, there seemed to be both emigrations and immigrations in the look-out for avocation in view of resource depletion in one place or the other. All these economic and social forces contribute to increased environmental pressures, and cultural and social transformations.

Concomitant to the above, a rich body of knowledge about the history and culture of community fishing and CBFM are disappearing. In fact, preserving the stories and experiences in management and harvest by the local fisherfolk are especially important in those regions, which are gifted with myriads of bewildering diversity of biotic and abiotic forms of aquatic resources. Incidentally, the coastal States, like Maine historically had depended on the marine or Great Lakes environment for their economic and cultural base. However, the demography of the lake or wetland shoreline, the river bank region, or the coastal region undergoes change(s) in course of time. But, it is important that the posterity is to know and also should have minimum knowledge about the glorious history and culture of their province's and nation's people.

This information is also useful to social scientists interested in coastal societies and their economies whose research is drawn on by marine fisheries managers when anticipating the impacts of regulations on local fishing communities. Further, fishermen's detailed knowledge about the local FW and marine environment as well as the characteristics and the past distribution of the local fish stocks are certainly useful to fishery biologists who try to reconstruct, what FW or marine fish stocks were, probably like, before scientists had begun collecting this information. Today, it is important that we better try to understand complex relationships among diverse species in local and regional environments, in view of the fact that the NOAA Fisheries Service management adopted an ecosystems-based approach. However, properly focused, oral histories, related to the ecology and management of the water bodies, collected from experienced fishermen in specified localities could further contribute to this effort.

Notwithstanding the above, a strategy for documenting and preserving the cultural, environmental, and historical knowledge of fishing communities, particularly with special emphasis on CBFM may come out through engaging schools and students. Through personal interviews with the community residents, who have been associated with different aspects of FW fisheries as well as interactions with other individuals with long experience in the FW environment, students may be able to explore the connections between fisheries, the FW environment, their community, and the CBM of the water bodies and their resources. For many students, integrating this kind of learning experience into the school curriculum could be the catalyst for engagement, as they see the connections between abstract academic concepts and their own experiences.

In addition, a successful project embedded in a school not only preserves local knowledge and local voices of a community, but it also makes place-based knowledge accessible to a larger national and international community. Incidentally, all ages could improve their awareness and knowledge possibly, through participating in well-designed, well-executed projects. Such activities could help them to know about how their FW environment influences human behavior and beliefs. Moreover, participants could also gain an increased appreciation of how such patterned human behavior and beliefs impact FW regions and their riparian area.

5.5.8.2 *Overview of the LFK through pilot project*

The idea for a LFK pilot project usually evolves from a series of conversations, e.g., it involved a lot of conversations in 2002 at NOAA's National Marine Fisheries Service (NMFS), being heard by Senior Social Scientists. The first conversation concerned the need for improving relationships with fishing communities affected by agency regulatory actions. Certain agencies may regulate FW fisheries resources; may administer the Marine Mammal Protection Act; and, may oversee the fates of other important biotas, notably, the FW terrapins; fish; and, some FW AM, which are said to be protected under the Endangered Species Act. Their actions, often involving rules and regulations, may profoundly affect fishermen's ability to make a decent living, and, thus, the welfare of fishermen's families could be jeopardized. Also, the fishing communities and the fish landing centers and ports, which are involved in fishing, could also be impacted. Thus, the regulatory power over fisheries could generate antagonism toward both Federal and State regulators from those relying on fisheries for their livelihood.

Another angle of conversation, which had concerned the anticipated loss was that, if *c* 50% of the agency's senior workforce plan to retire in near future, this could be a problem common to, perhaps, all federal agencies. Aware of the problem, NOAA is looking for ways and means to introduce youth to the wide array of career possibilities available throughout the different avenues, including the water bodies-related occupations.

Concomitant to the above, the third aspect of the conversation had concerned many commercial fishermen's belief that fisheries scientists and managers may not accept their views and knowledge about the fish and other FW resources that they use, and the FW environment that they inhabit. While some scientists and managers may disdain LFK; many people face difficulty just, in systematically collecting this knowledge followed by its transformation into a usable format for scientific and management workers.

Notwithstanding the above, the fourth pattern of conversation involved NOAA's expanding education role to promote an environmentally literate public. In addition to its scientific mandate, NOAA play a leading role in the conservation, management, and restoration of FW bodies, oceans, coastal, and Great Lakes resources. The stewardship of these resources for current and future generations is critical to the long-term sustainability of our society and the planet. Monitoring the health of these ecosystems and building an understanding of the relationships between the water bodies and other Earth systems is one of the core missions of the organization. Over the last century, human actions greatly altered these natural systems; thus, seriously threatening the sustainability of the resources. All of NOAA's scientists, including atmospheric, oceanic, biological, and social, recognize that the pressure on the environment could increase with the globalization of world markets, population shifts, and the race for economic growth. NOAA or, for that matter, any other organization, may not be able to manage these burgeoning issues. As such, collaborating with the public to share stewardship responsibilities is, perhaps, a prerequisite. Environmental literacy is one step forward in that direction and process.

Concomitant to the above, the possibility for a project to exist in the intersection of these four conversations is, perhaps, not very bleak. The new project could amalgamate general fishing community outreach and could make efforts to introduce young people to FW-related careers with the formation of a school-based, FW-focused oral history project supported by an agency research tool that could archive and preserve LFK including CBM. The LFK Pilot Project is, perhaps, trying to make an initiation in this direction (Abbott-Jamieson, 2007).

Notwithstanding the above, Anon (2007) dealt with Natural Resources (**in USA**) with regard to their Conflicts and Conflict Resolution. The World Bank **defined natural resources** as "materials that occur in nature and are essential or useful to humans, such as water, air, land, forests, fish and wildlife, topsoil, and minerals, etc." There have been various types of conflicts with the use of natural resources since the dawn of human civilization.

5.5.8.3 *Types of natural resources*

The natural resources may be classified as renewable or nonrenewable. In most cases, renewable resources such as cropland, forests, and water could be replenished over time by natural processes, and, if not overused, are indefinitely sustainable. Nonrenewable resources, such as diamonds, minerals, and oil are found in finite quantities, and their value increases as supplies dwindle. Further, **a nation's access to natural resources often determines its wealth and status in the world economic system.**

Agriculture, though strongly linked to natural resources, is generally not thought of as a natural resource as it depends largely on cultivation. However, economies that are heavily dependent on agriculture certainly depend heavily on other natural resources, such as water and land. Conflicts often may arise with regard to the distribution and use of natural resources. In addition, they are resolved using various *modus operandi*. Incidentally, CBM, involving the community stakeholders from the grassroot level could be a potential method of resolving various types of conflicts with resources.

Notwithstanding the above, Soeftestad (1999) dealt with the report from the **CBNRM** International Workshop held at Washington D.C. in May 1998.

This workshop was organized because of the growing interest in CBNRM as a third alternative to command and control market-based solutions to NRM.

CBNRM is an approach to managing renewable natural resources. Most policy-makers and practitioners would have regarded CBNRM as an oxymoron (Elinor Ostrom, 1998). But, today, there is said to be a growing consensus in favor of CBNRM. There had, perhaps, been efforts to assist people who are interested in pursuing this approach. Different people on various occasions and at different times, perhaps, had given rise to different understandings of the term and its approach to sustainable development. Thus, CBNRM has variously been described, inter alia, as a tool (or a set of tools), a checklist, a method, a means, a set of activities, a model, a process, and an approach.

Notwithstanding the above, CBNRM encompasses a large number of experimentations and regional variations. Incidentally, there is a large number of experimentation going on worldwide in the area of local NRM, and the practitioners involved may or may not use the label CBNRM. Such differences in understanding and terminology are caused by different factors including culture, history, and language.

Notwithstanding the above, the following core aspects of CBNRM could be identified:

(i) There is a "community" aspect of CBNRM. CBNRM, obviously, starts with communities as a foundation, and it also ends with communities as a focus. In-between is the strategic process of identifying needs and local capacities and involving and aligning stakeholders, both within and outside the local community. In this regard, CBNRM could be understood as a "system" of NRM.

The essential feature of CBNRM is "starting" with communities, taking them into confidence and having confidence in them. The process engages their ideas, experiences, values, and capabilities on behalf of resource conservation objectives. Concomitantly, it also seeks ways and means for communities to become better remunerated and also better served. It is prepared to accommodate local interests' needs and norms, as well, that are compatible with long-term preservation of ecosystems and their biological resources. There is a burden of proof on outsiders for proceeding contrary to these interests.

(ii) There is a "natural resource management" aspect of CBNRM.

CBNRM is to address the management of locally available and renewable natural resources. More specifically, it focuses primarily on natural resources that are under some form of community or collective management. Incidentally, the natural resources that are managed in this way whether traditional or recent is often referred to as "common property resources" (CPRs).

(iii) There is a "comanagement" aspect of CBNRM.

As a strategy, CBNRM parallels in social and policy terms and the parallel nested in the natural world, comprising organisms, species, associations, and ecosystems, i.e., CBNRM activities take place at different levels. This speaks of the significance of establishing relations between stakeholders, who are located on different levels, based on comparative advantages. Furthermore, this underlines the argument that successful comanagement, preferably, give parallel and strategic emphasis on both community-based groups (the horizontal axis) and to operational linkages (the vertical axis).

Concomitant to the above, Peter F. Sale et al. (2005) dealt with critical science gaps, which impede the use of no-take fishery reserves. **No-take fishery reserves, in addition to serving as valuable BD conservation roles also protect a portion of the fishery stock as insurance against future overfishing.** So long as there is adequate compliance by the fishing community, it is likely that, they may also sustain and even enhance fishery yields in the surrounding water bodies.

Incidentally, there are believed to be significant gaps in scientific knowledge that are to be filled, if no-take reserves are to be used effectively as fishery management tools. However, these gaps are glossed over by some uncritical advocacy. Hence, there had been attempts to review the plan; identify the most crucial gaps; and, to offer suggestion(s) of ways and means to fill them; so that, a promising and efficient management tool could help to meet the growing challenges faced by the management bodies of aquatic resource management.

Globally, the aquatic resources; and, notably, the fishery, in different domains, are, believed to be under ever-increasing threats of collapse. It is quite a common experience today that **efforts increase, but yields rarely remain stable** but, generally, decline as well as other anthropogenic impacts degrade habitats and ecological systems on which the fisheries depend.

Notwithstanding the above, the MPAs of various types could be a form of resource management method, that regulates the human activities in particular locations. Although there are many types of MPAs, the concern often is **no-take fishery reserves**, which could also be called "closed areas" or "harvest refugia" and, in this regard, the focus is generally laid on fishery benefits, while recognizing that no-take reserves have significant BD conservation benefits. Such reserves **commonly increase the density, biomass, and average size of target species within their borders**. Further, there are a few well-documented cases, where they supplement the fished-out stocks in the surrounding areas. However, it is little premature to assume that no-take reserves are invariably effective in efficient fisheries management. It is because, there had been relatively few empirical studies, many of which are, perhaps, poorly documented.

Notwithstanding the above, there have been discussions on the opportunities and challenges, with regard to the management of **US Regional Fishery**.

Yellow goatfish (*Mulloidichthys martinicus*) school over a coral reef in the bright clear waters of the Caribbean. Goatfish also disperse at night in search of food, using their chemosensory barbels ("chin whiskers") to detect worms and other small invertebrates in the sediments. In fact, the **yellow goatfish** are not a target of any fishery. However, the Caribbean Fishery Management Council works to protect the reef habitat for these and other reef dwellers. Moreover, a similar species, **yellowfin goatfish** (*M. vanicolensis*) is a significant commercial fish that is highly esteemed as food in the US Pacific Islands. It is known in Hawaii as weke "ula," in American Samoa as "vete," and in Guam and the Commonwealth of the Northern Mariana Islands as "satmoneti" (Chamorro language) and as "wichigh" (Refaluwasch language). Incidentally, the fisheries for yellow fin goatfish are regulated by the Western Pacific Fishery Management Council as a component of the Fishery Management Plan for Coral Reef Ecosystems.

Fisheries and AQC are the potential resources for human beings in terms of both nutrition and avocation. However, this is a challenging time for fisheries. Incidentally, the present depleting trend of fish production is not in a position to keep pace with the growing trend of human population by leaps and bounds at an exponential rate. Moreover, the developmental measures implemented in the aquatic domain coupled with the unrestricted and unsustainable harvest of the fishes have been exerting pressure on the habitat and fish populations alike. Climate change is building new, unpredictable problems, which may not be possible to be solved by a single agency or the government. In addition, it may be noted that energy generation, the Federal and State MPAs, and other uses of the ocean are competing spatially with fisheries. Concomitantly, the Americans are exploring innovative ways to achieve conservation goals, including new approaches to fisheries management, harvesting, marketing, and consumption. In fact, these ideas and approaches are discussed by the fishermen, fishing and marketing organizations, conservation groups, public and scientific advisory groups, fishery managers, etc.

Notwithstanding the above, eight Regional Fishery Management Councils have recognized current challenges in fisheries management.

Notwithstanding the above, fisheries cannot be effectively managed by a single State or Federal entity, unlike forestry, mining, and many other natural resource industries. Further, fisheries differ a great deal across the country. For instance, Western Pacific fisheries focus on pelagic and migratory fish stocks, such as tuna, which require international management. Further, the South Atlantic fisheries provide recreational opportunities for millions of people. On the other hand, in the North Pacific, high-volume commercial fisheries target crabs, salmon, and groundfish, such as the pollock.

Concomitantly, with the passage of the **Magnuson-Stevens Fishery Conservation and Management Act,** the Regional Fishery Management Council system was formed in 1976 in order to address the regional differences. In fact, the unique council system is designed to allow regional, participatory governance by knowledgeable people with a stake in fishery management. Moreover, each council's voting members were believed to include one National Marine Fisheries Service representative, a representative of each state fishery agency in the council area, several private citizens nominated by the state governors and approved by the Secretary of Commerce because of their specific qualifications, and in some regions a representative from local tribal or territorial governments. Further, the nonvoting members include regional representatives from the US Coast Guard, the Department of State, the US Fish and Wildlife Service, and Marine Fisheries Commission.

Notwithstanding the above, the eight Regional Fishery Management Councils serve as the frontline of fisheries management, where, regionally specific management measures (such as fishing seasons, quotas, and closed areas) are initiated, developed, and ultimately adopted in a fully transparent public process. After adoption by council vote, these measures are subjected to approval by the Secretary of Commerce, the regulatory implementation by the National Marine Fisheries Service, and enforcement by the US Coast Guard and other authorities. Further, the councils develop fishery output in the offshore area cooperatively with the State authorities.

Notwithstanding the above, the council decisions are subjected to rigorous scientific analyses. Scientists and policy analysts evaluate potential fishery regulations for both environmental and socioeconomic impacts. Moreover, the proposed regulations are vetted by expert panels of scientists, stakeholders, and by the public, before a council takes a final decision. Thus, the open process provided by the council system allows everyone to have a say in the stewardship of our marine resources and how the fisheries are to be managed sustainably. In fact, the commercial and recreational fisheries have a major economic impact in the United States, both nationally and in the communities where fishing takes place. Further, according to the National Marine Fisheries Service, commercial and recreational saltwater fishing generated >US $185 billion in sales as well as supported >2.0 million jobs in 2006. On the other hand, the commercial fishing industry, consisting of harvesters, fish food processors and dealers, fish food whole sellers and retailers, generated *c.* US $103 billion in sales and *c.* US $44 billion in income and supported *c.* 1.5 million jobs in 2006. Conversely, the recreational fishing (probably, alone) generated *c.* US $82 billion in sales, *c.* US $24 billion in income, and supported *c.* 534,000 jobs in the same year. Further, the efficient management of the resources provide stability in fisheries employment for the nation, while protecting marine BD and, in some cases helping in rebuilding depleted fish stocks.

Concomitantly, Donald et al. (2006) worked on the Fisheries Management Plan for the Minnesota waters of Lake Superior. The Lake Superior fish community had undergone dramatic changes since the mid-1900s due to overfishing, the introduction of nonnative species, pollution, and land-use changes in the watershed. The Lake Superior fish community had become much more complex, since the 1950s and, is now composed of both native and exotic species. The most devastating introduction to the Lake Superior community was the sea lamprey, which virtually eliminated the lake trout in all but a few isolated areas of Lake Superior. Incidentally, rehabilitation efforts, since the 1960s, including sea lamprey control, harvest regulations, and stocking programs, along with stricter pollution standards and best management practices for land use have led to the restoration of healthy fish stocks throughout much of Lake Superior.

Incidentally, there had been studies serving as comprehensive guides on how to best manage Minnesota's portion of the Lake Superior fishery. There had been plans based on a fish community approach to fisheries management. The strategies and actions in this plan have focused on the work of the MNDNR (Minnesota Department of Natural Resources) Fisheries Management Section over the next decade.

5.5.8.4 *An overview of the general plan*

Fisheries management in Minnesota was the responsibility of the Fisheries Management Section of the Minnesota Department of Natural Resources (MNDNR).

Notwithstanding the above, the long-term goal for fisheries management in the Minnesota waters of Lake Superior is as follows:

- **(i)** To protect the Lake Superior ecosystem, restore its watershed, and manage for a diverse, stable, self-sustaining fish community that could provide with recreational, commercial, and fishing opportunities to the people in a sustained manner. The MNDNR had used an ecosystem-based management approach. In its mission, it aimed to protect the Lake Superior ecosystem and manage the fishery based on ecological principles. Incidentally, the plan had recognized the following aspects:
- **(ii)** Fish production in Lake Superior is finite. However, the users may like to have more fish from its waters, but the lake simply may not have the capacity to produce higher levels. Moreover, additional stocking of trout and salmon cannot take place without considering impacts on the forage base.
- **(iii)** Lake Superior is the least productive, but one of the most pristine of the Great Lakes was demonstrated for the capacity to support self-sustaining fish populations through natural reproduction. The plan had emphasized the continued need for habitat protection, and the desire for managing self-sustaining fish populations that were the best suited to the lake's environment.

Notwithstanding the above, the user groups (UG) in the Lake Superior possess diverse types of interests, and the Lake Superior Management Plan (LSMP) attempted to balance resource protection, recreational opportunities, cultural beliefs, and economic development for the benefit of both present and future generations. Moreover, citizen participation was the cornerstone of the planning process and could be an ongoing process.

Notwithstanding the above, the Lake Superior fishery management was an expensive program when compared to other fisheries in the state. The lake is accounting for *c.* 4.5% of the total fisheries budget. Although Lake Superior is a unique resource and offers diverse recreational opportunities, UGs recognized that increased expenditure for Lake Superior fishery programs could be difficult to justify when viewed from a State-wide perspective.

Concomitant to the above, the effectiveness of Lake Superior management programs is to be continually evaluated. If established criteria indicate program changes are required, interested citizens could be consulted and the necessary action(s) could be taken.

Notwithstanding the above, *c.* 7% of Lake Superior is included within the State of Minnesota. In fact, The Great Lakes Fishery Commission provides the structure for cooperative management among the various jurisdictions around the lake and continued involvement with the commission is required to address the ever-increasing complexity of issues that arise. Further, citizen participation could be crucial, and there were said to be efforts to implement this plan. Moreover, it could be expected that the plan could focus and stimulate ongoing conversations about future fisheries management for the Minnesota waters of Lake Superior. Incidentally, it is believed that this plan may be required for sound management of the Lake Superior fishery, its success could ultimately be determined by citizens' support, and this could lead to the long-term benefits to the resources and its users.

Notwithstanding the above, Granados-Dieseldorff and Montaña (2012) reported about advancing conservation with innovative ideas for the future generations as a community-based approach to the management of an aggregating reef fishery in the **Meso-American reef.** Conservation scientists advocate for the establishment of MPAs as a global precautionary approach to the sustainability of world fisheries.

Small-scale fisheries supply most of the seafood the entire world eats (FAO, 2010). They employ >98% of the fishers of the world, most of them from developing countries and are said to be in the grip of poverty (Berkes et al., 2001, b; FAO, 2010). Faced with degrading marine ecosystems (Jackson et al., 2001; Pauly et al., 2003; Worm et al., 2006) and in the midst of the controversy around the legitimate decline rates of fishery stocks (Branch et al., 2011; Worm et al., 2009), conservation scientists continue to advocate for the establishment of MPAs as a global precautionary approach to the sustainability of world fisheries (Foley et al., 2010; Lubchenco et al., 2003). However, the effectiveness of the MPAs in benefitting both the protected resource and its users remains largely untested (McCay and Jones, 2011; Mascia et al., 2010). This dilemma had frequently led to discussions about whether fisheries conservation and management have actually yielded the expected benefits to fish, fishers, and the society at large. Concomitantly, rather, it had often ignited social and political conflicts at the community, national, and regional levels (Berkes, 2006; Mascia and Claus, 2009). Moreover, the frequent piecemeal governance of the commons' management (Armitage et al., 2009; Berkes, 2006; Hardin, 1968) are often geographically misplaced fishery legislation(s) (St. Martin, 2001) and, thus, the underrepresentation of key stakeholders in management strategies had continued to jeopardize the fate of fisher-fish relationships within socio-ecological systems (Johannes, 1998; Johannes et al., 2000).

5.5.8.5 Epilogue

The continent of America possesses large number of water bodies having a rich diversity of fishes. Also, various types of *modus operandi* for fisheries management are *in vogue.* Incidentally, fisheries are globally acclaimed resources which play a significant role in the nutrition and avocation of the people. In fact, we need to manage fish in the context of the environment in which they live, that is, an ecosystem. Such an effort could help us to manage the fisheries in an effective way. In other words, it means that, there is a need for understanding the complex ecological and socioeconomic environments in which fish and fisheries exist. It also means that anticipating the effects that fishery management will have on the ecosystem and vice versa. This could be regarded as an important aspect, and is *in vogue* in America, which is also followed in other parts of the world.

Notwithstanding the above, practical application of a comprehensive **EBFM** approach is still in its rudimentary stage of development in the FW, marine, and coastal environments. Also, the regional organizations concerned with fisheries are yet to give a comprehensive shape to EBFM. However, it is time that these organizations grasp the potential of CBFM for efficient management of the water bodies and the fisheries therein. This awareness may be expanded and actions to be taken to implement this approach.

Incidentally, there have been efforts toward ecosystem-based management of fisheries with regard to opportunities and challenges based on coordination between marine Regional Fishery Bodies and Regional Seas Conventions, which are under the fold of UNEP Regional Seas Reports and Studies (Anon, 2001). This is an aspect, which is *in vogue* in different parts of the globe including America.

In such exercises, joint effort between UNEP and FAO is developed with the technical assistance of the Advisory Committee on Problems of the Sea (ACOPS) and with inputs coming from Regional Seas Conventions (RSCs) and Regional Fisheries Bodies (RFBs).

Concomitant to the above, it may be mentioned that there are publications, which benefited from the inputs of the participants to various meetings. Incidentally, the main aim of many such publications is to present considerations which could serve as the basis for potential cooperation between RFBs and RSCs. They try to discuss the concept of EBFM, the relevant mandates and activities of RFBs and RFCs, and the relationship and the mutual relevance of their work. Possible mechanisms for cooperation, and issues for future consideration(s), are often identified. It may be anticipated that such cooperation could best be implemented on a site-specific or regional basis, after initial consideration at the global level by RFBs and RSCs. It is contemplated that UNEP will embark upon a process of collaboration with FAO, during the years to come, to explore the opportunities and challenges for coordinated activities on EBFM.

The ecosystem approach to the management of oceans and their resources was consolidated in Agenda 21. The practical application of this approach is still in the early stages of development by regional organizations who are related to fisheries, in FW, marine, and coastal environments. It was seen as a platform for potential cooperation between these bodies and the Subcommittee on Ocean and Coastal Areas of the UN Administrative Committee on Coordination. To this end, paper(s) was developed jointly by FAO and UNEP to be deliberated at the third Global Meeting of the RSCs in November 2000 and the second Meeting of FAO and non-FAO RFBs meeting in February 2001. Incidentally, a brief description of marine RFBs and RSCs have been presented, noting the origins, status, geographic coverage, and basis for the consideration of EBFM approach by each. In fact, the establishment of RFBs had taken place throughout the past century; many had been formed since the adoption of the United Nations Convention on the Law of the Sea (UNCLOS) in 1982. They all have clear management functions, although only a small proportion has specific ecosystem-related mandates. However, all had the scope to consider this approach in some form or the other. Current challenges include the development of the concept, objectives, indicators, reference points, and implementation mechanisms, as appropriate in a climate of cooperation with other international organizations and each other.

On the other hand, the establishment of RSCs began in 1972. The 11 major RSCs in force were designed for the protection of the marine environment is mostly in the form of "comprehensive framework conventions." Their protocols and annexes had specified the concrete measures expected to be implemented by the parties, and "Action Plans" were related to all issues relevant to the development and protection of the marine environment and their resources. However, periodic revisions of the Action Plans broadened their scope to emphasize issues contained in Agenda 21, such as integrated management.

Incidentally, various communications had tried to develop the concept and rationale behind EBFM. In fact, the broad ecosystem objectives appeared in international instruments. Moreover, the approach often had formed part of some national laws/strategies, and had been considered at global conferences. However, the practical application of the concept had emerged only recently as a subject of international attention. In this regard, two leading examples are the March 1999 symposium on "Ecosystem Effects of Fishing" convened by the ICES and the September 2001 "Reykjavik Conference on Responsible Fisheries in the Marine Ecosystem." It was considered that the implementation of the ecosystem approach could prompt changes to institutional, training, capacity, information, monitoring, evaluation, governance, and regulatory requirements. Incidentally, general provisions of the RSCs, and structure and strategies of the action plans were covered; noting that, the latter usually includes five main components, viz., environmental assessment; environmental management; environmental legislation; institutional arrangements; and financial arrangements.

Incidentally, five RSCs had been described which contributed to the goals of fisheries management by dealing with the control of land-based resources and activities that may have deleterious effects on the marine environment, including its living resources. In almost all regions, intensive monitoring of the quality of the marine environment was carried out, and the data provided good background information for fisheries management. Further, activities of RFBs relating to eecosystem-based management had been described. Moreover, the work of three leading RFBs was profiled, and a synthesis of RFBs' action(s) presented. In general, the RFBs' activities are related to the following:

(i) the impact of fisheries on the ecosystem,
(ii) the impact of other sectors on fisheries,
(iii) the impact of climate and ozone depletion on fisheries, and.
(iv) ecosystem monitoring.

Actions for future consideration by RFBs included defining ecosystem objectives in parallel with the current conservation objectives of fisheries management. It was suggested that the new objectives are to address BD, habitat productivity, and aquatic (FW and marine) environmental quality. Some additional needs included the definition of indicators and reference points, and new monitoring activities, and data products for the indicators.

Notwithstanding the above, the relationship and mutual relevance of the works carried out by RFBs and RSCs reviewed, especially in areas relating to BD of species, habitat, marine environmental quality, climate change, and land-based pollution of the marine environment. In addition, monitoring and assessment, which applied to these areas, had also been done. Ecosystem consideration and indicator framework were increasingly factored into the fisheries management. As such, the functionalities of RFBs and RSCs were adapted in a practical, cost-effective way to meet the future needs. This could be done in a way that would not overburden either RFBs or RSCs and could be built on current programs. Some examples of activities which could form a basis for practical cooperation are often suggested in the current scenario.

Concomitant to the above, considering the current activities of RFBs and RSCs as well as the experiences gained through the cooperation already established between RFBs and RSCs, concrete suggestions were made for options that may lead to enhanced cooperation on ecosystem-based fishery management.

5.5.8.6 Rationale

The ecosystem-based approach to management of the water bodies (FW and marine) and their resources was consolidated in Agenda 21. Review and coordination of implementation of this area among United Nations (UN) agencies were facilitated by the Subcommittee on Ocean and Coastal Areas (SOCA) of the Interagency Committee on Sustainable Development. They had operated under the umbrella of the UN Administrative Committee on Coordination (ACC).

In the meantime, SOCA considered the relevance of the regional setting for improving coordination among different UN programs, at its ninth session, in July 2000, addressing different aspects of Coastal and Ocean Management. Its main purpose was the exploration of new ways to integrate the work of the agencies and, in particular, to look for synergies between regional organizations, respectively, competent for fisheries and the FW, marine, and coastal environment. It was felt that the challenge(s) posed by the development of ecosystem approaches to fisheries management and integrated FW and coastal management could be considered by both types of bodies as a potential platform for practical cooperation. However, the need for such a type of coordination had also been recognized by FW and Marine RFBs and the UNEP.

In view of the above, there was a plan for a paper, centered around ecosystem-based management in fisheries, which could be jointly developed by FAO and UNEP. The purpose of the paper is to present considerations which could serve as the basis for potential cooperation between RFBs and RSCs. It could try to describe the concept of EBFM, the relevant mandates and activities of RFBs and RSCs, and the relationship and mutual relevance of their work. Further, possible mechanisms for cooperation, and issues for future consideration, could be identified. It was further believed that such a type of cooperation could be best carried out on a site-specific or on a regional basis, after the initial consideration at the global level by RFBs and RSCs.

In this connection, there had been contributions presented at the Third Global Meeting of the RSCs, organized by UNEP (at Monaco, November 6–9, 2000) and at the Second Meeting of FAO and Non-FAO RFBs or Arrangements organized by FAO (Rome, February 20–21, 2001).

Notwithstanding the above, RFBs are, in general, concerned with the impacts from other sectors affecting fisheries, including land-based pollution and habitat as well as the impacts of non-sustainable fishing or fishing techniques and climate.

Concomitantly, the representative(s) of the Alliance of Small Island States (SIDS) emphasized that a major challenge for SIDS is the need for development and management programs aimed at achieving ecological and economic sustainable use of FW, coastal, and marine resources in various regions, including sustainable fisheries.

5.5.8.7 Brief description of RFBs and RSCs

Regional fishery bodies

Origins There is said to be a saying "The tragedy of the commons," which may mean the overuse of common fishery resources because of the absence of a sufficiently strong system of cooperative and rational management. In addition, this is said to be a classic challenge to fisheries governance. The phenomenon may often result from massive overinvestment and subsidies in the fisheries sector. However, it could not only be a tragedy of the commons, but could also be an opportunity for the commons.

Incidentally, the National Authorities alone may not be able to protect areas that do not fall within their jurisdiction. International cooperation and the legal regimes are the only ways by which global fisheries governance could be achieved effectively, although the problems obviously differ greatly from one region to another.

In view of the above, the need for such international cooperation contributed to the establishment of the first recorded marine RFB, the ICES, in 1902. ICES had, as its mandate, the development of cooperation in scientific studies of the North Atlantic marine environment and its resources through cooperation among scientists. It was an advisory body, with an ecosystem approach.

Concomitant to the above, a number of other RFBs were established, following the establishment of ICES, as follows:

(i) The International Commission for Scientific Exploration of the Mediterranean Sea in 1919 and.
(ii) The North American Council for Fishery Investigations in 1920. The findings of these bodies had led to regulatory action through the conclusion of a number of conventions.

Notwithstanding the above, ICES and some of the other bodies, which were established in the first half of the century, particularly before 1945 had concentrated their efforts on the generation of scientific information and the promotion of scientific cooperation with emphasis on the conservation of fishery resources.

Concomitant to the above, the process of further development had begun with the establishment of the Food and Agriculture Organization (FAO) as a specialized Agency of the United Nations in 1945 having a clear mandate to contribute to the development and management of fisheries at the international, regional, and national levels. This was followed by the First UN Conference on the Law of the Sea (UNCLOS) in Geneva in 1958.

Concomitant to the above, 40 RFBs were established during the past 100 years approximately. The evolution of these bodies corresponded to the three distinct phases of the evolution of international fisheries in the 20th century, viz.,

(i) the pre-UNCLOS period 1902–50;
(ii) the Law of the Sea negotiating period 1951–82; and.
(iii) the post-Third UNCLOS period.

5.5.8.8 Ecosystem-based management of fisheries

Concept and rationale

Background The concept and rationale of ecosystem-based management of fisheries (EBMF) take into account the interrelationships between the planet's web of life and ongoing human action. It had emerged in the 1982 based on the United Nations Convention on the Law of the Sea (UNCLOS). Although specific reference to "ecosystem" management only may appear in relation to rare or fragile ecosystems, the provisions on fisheries management, nevertheless, described that, ecosystem-like considerations are to be taken into account, such as associated and dependent species, the interdependence of stocks, and minimum standards at all levels. Instruments agreed at the 1992 UNCED were the Agenda 21, and the Convention on Biological Diversity (CBD), etc. There were further, efforts to refer more specifically to the ecosystem approach, and this was carried forward by the 1995 United Nations Fish Stocks Agreement and the 1995 Code of Conduct for Responsible Fisheries. The latter called for the integration of fisheries into FW and coastal area management, including the fragility of FW and coastal ecosystems.

Three leading countries engaged in developing and implementing the EAFM are Australia, Canada, and the United States.

The concept of ecosystem-based management of fisheries was adopted by some RFBs, but their approaches might have varied. Incidentally, considerations related to BD and the effect of climate on the fisheries were also recognized as important. Incidentally, the fisheries ecosystem could, sometimes, be a little complicated. Most ecosystems contain a great number of species, and the number of potential biological and economic interactions could increase exponentially with the number of species.

Notwithstanding the above, at present, there is no consensus on criteria for defining ecosystem overfishing, nor on a hierarchy of biological attributes for which ecosystems are to be managed. However, it was suggested that ecosystem considerations may be incorporated into fisheries management by modifying existing overfishing paradigms or by developing new approaches to account for ecosystem structure and function in relation to harvesting. In particular, existing concepts of overfishing do not seem to provide direct guidance on issues, such as BD depletion, habitat degradation, and changes in the food web caused by fishing. However, it is believed that ecosystem considerations do not need to substitute for existing overfishing concepts. Instead, they could be used to evaluate and modify primary management guidance for important fisheries and species. In practice, they are emphasized on the need to manage fishing capacity, supported by broader use of technical measures, such as aquatic (FW and marine) PAs and gear restrictions.

Finally, the concept of ecosystem management is related to the question of the geographical boundaries of the ecosystem. Their definition could be a somewhat arbitrary process, depending on the particular interests of the ecologists and the details of the issue being addressed. Thus, an appropriate approach is to be adopted for consideration of specific ecosystem features and associated human activities that would accommodate areas which may be larger or smaller than needed for ecosystem and fisheries considerations.

Objectives Recent international instruments have described broad ecosystem objectives for the aquatic domain with particular emphasis on the marine environment. Some of these are as follows:

(a) to manage aquatic (say, marine) living resources sustainably for human nutritional, economic, and social goals;
(b) to protect and conserve the aquatic (say, marine) environment;
(c) to use preventive, precautionary, and anticipatory planning and management implementation;
(d) to protect and maintain the relationships and dependencies among species; and.
(e) to conserve genetic, species, and ecosystem BD.

These needs were to be translated into practical objectives, which could be defined parallel with presently used conservation objectives of fishery management plans. The objectives may, preferably, focus on the maintenance of BD, habitat productivity, and FW/marine environmental quality as conditions for the long-term sustainable development of fisheries.

Notwithstanding the above, it was suggested that the BD objective would need to include several components, such as ecosystem and species diversity, genetic variability within species, and species at risk. The habitat productivity objective would need to address directly impacted species, ecologically dependent species, and trophic level considerations. Once the objectives were set, the next steps were to provide with the respective performance measures and reference points. Implementation of strong objectives could require effective monitoring systems, evaluation processes, and governance.

Advantages The advantages of an ecosystem-based approach to fisheries management extend beyond the realms of sustainable fisheries, BD, habitat protection, and so on. Some of the aspects related to advantages are briefly stated below:

(a) The ecosystem approach acknowledges the paradigm shift in the management of all-natural resources, and the increasing need to integrate fisheries and environment management objectives as well as human and ecosystem well-being.
(b) Movement toward such integration, coupled with a precautionary approach, could allow managers to distance themselves from the present so-called "firefighting approach" in favor of a more holistic approach.
(c) Assessment of the potential effects of other human activities on the fisheries resource could also be encouraged.

Notwithstanding the above, the EAF and environmental management could accumulate energy, inspiring the development of appropriate scientific and related yardsticks. It could also prompt alteration(s) to institutional, training, capacity, information, monitoring, evaluation, governance, and regulatory requirements. Moreover, it could serve as a model for environmental management of other sectors whose activities affect, and are affected by, the aquatic (notably, marine) environment.

In addition to the above, the ecosystem management is enhanced by the Millennium Ecosystem Assessment, which is, further, designed to determine the condition of global ecosystems and to analyze the capacity of an ecosystem to provide goods and services which are said to be significant for human development. However, the fisheries and those dependent on them are affected, when the capacity is diminished.

One benefit of this integrated ecosystem assessment program is that it provides the information necessary to weigh trade-offs among various goods and services and to identify opportunities for benefits. Information and data collected in EBFM could be input into this assessment, and the relative importance of fisheries strengthened.

5.5.8.9 Impact of non-fishery activities on the ecosystem

Non-fishery activities may exert a profound influence on the ecosystem. Thus, assessment and management of the impact of non-fishery activities on the ecosystem are important to EBFM. Such problems often center around pollution of the water bodies.

Notwithstanding the above, different measures of coastal developments could also affect the ecosystem. It aggravates pollution and could also alter the coastlines radically. Notably, clearing, land reclammation, and channelization of flood and tidal waters are believed to destroy coastal wetlands. Port development, road construction, coastal building, and the mining of beach sand could not only cause pollution but also, could obliterate shoreline habitats.

5.5.8.10 Provisions of the RSCs and the activities of associated action plans relevant to ecosystem-based management of fishery resources

General provisions of the conventions

The general obligations specified in almost all conventions are to prevent, reduce, abate, combat, and control pollution in the convention area. Some conventions also include obligations to preserve rare and fragile ecosystems as well as habitats of depleted, threatened, and endangered species. The specific obligations common to most conventions include the following:

(i) Control of pollution caused by dumping discharges from ships, exploration, and exploitation of continental shelf and land-based sources.
(ii) Scientific and technical cooperation, including joint monitoring and research programs, data and information exchange, and technical assistance.
(iii) Reporting on the measures adopted in the implementation of the conventions at national levels.

5.5.8.11 *Activities relevant to ecosystem-based management of fisheries*

Majority of the RSCs are to specify their concern for the status of marine living resources and emphasize the need for their protection. But none of the conventions had dealt, in a major way, with the management of fishery resources. Nevertheless, some of the convention provisions, and particularly, some of the protocols and action plans associated with the conventions could be seen as contributing directly or indirectly to these goals.

Concomitant to the above, it may be mentioned that five conventions (Barcelona, Bucharest, Cartagena, Kuwait, Lima) were associated with protocols dealing with the control of land-based resources and activities that may have deleterious effects on the aquatic (notably marine) environment, including its living resources. Protocols or annexes dealing specifically with PAs, endangered species, and BD were attached to six conventions (Barcelona, Cartagena, Kuwait, Lima, Nairobi, OSPAR). Moreover, some of the BD-related protocols were included specific reference to fisheries and to activities relevant to the management of fisheries resources.

5.5.8.12 *Activities of RFBs related to the ecosystem approach to fisheries management (continued)*

It may be mentioned that "Environment" was first considered seriously in connection with fisheries management by an RFB in the 1970s. ICES responded to public concerns about the environmental threats from chemical contamination. One obvious concern was whether the fish harvested for human consumption might present a health risk for the consumer. ICES responded by establishing the Advisory Committee on Aquatic (notably, Marine) Pollution (ACMP) in 1972.

At that time, "environment" was used almost exclusively to mean contaminant chemistry and biology.

Notwithstanding the above, the concept of environment had expanded to merge with the concept of ecosystems, in more recent years. However, much work still needs to be done to identify the objectives, indicators, and reference points of ecosystem management, including fisheries. Efforts are also to be made to establish the programs, monitoring, training, capacity, and other elements needed for implementation. Attempts are required to identify the ecosystems themselves, and to establish an appropriate system of governance. The instruments establishing the RFBs do not, generally, try to accord a mandate for EBFM. Those which do either refer to:

(i) Ecosystem-based management, including CCAMLR and two initiatives, establishing new RFBs in the South-East Atlantic and Western Central Pacific oceans.
(ii) Management of ecologically related species, including the NPAFC Convention, which mandates scientific research, cooperation, information requirements and consideration by the Commission of "ecologically related species."

5.5.8.13 *International Council for the exploration of the sea*

The leading advisory of RFB in this field is the ICES. It is an international scientific organization studying and helping to safeguard North Atlantic marine ecosystems and the living resources they sustain. However, a major area of its work is to provide with information and advice to its 19 member countries and international regulatory commissions. Being an international scientific community, it considered its vision as the one, which was relevant, responsive, sound, and credible with respect to marine ecosystems and their relation to humanity. Its scientific goals, among others had included the following:

(i) quantification of aquatic (notably, marine) environmental processes and their variability;
(ii) determination of implications of climate variability on ecosystem dynamics;
(iii) characterization of biological diversity and its role in ecosystem functioning;
(iv) enhancing knowledge on populations of living resources focusing on the scientific understanding, which is said to be a prerequisite for a well-founded ecosystem approach to fisheries; and.
(v) to gradually emphasize on **CBFM** as an effective tool for the efficient management of water bodies and the fisheries.

5.5.8.14 *Inferences and possible options*

There had, of late, been 11 major RSCs in force and two additional conventions were, probably being negotiated. They had been designed for the protection of the marine environment, including its living resources, under the jurisdiction of the contracting parties to these conventions. A total of 99 sovereign States and the EU were the parties to one or several of the conventions in force. Geographically, the conventions covered large parts of the most productive and most threatened areas of the oceans. However, none of the conventions had dealt with the management of fishery resources effectively. However, a number of activities, associated with the conventions have been carried out in the framework of programs, and these were directly or indirectly contributed to the improvement in the management of fisheries.

Notwithstanding the above, of late, there were *c.* 27 active marine RFBs, 7 of which had been established under the FAO Constitution. Moreover, FAO also had facilitated the establishment of other bodies. Negotiations were going-on for the establishment of a Regional Fishery Commission under the FAO Constitution in the SW Indian Ocean.

Concomitant to the above, the convergence of fisheries management and ecosystem management paradigms and requirements were a reality. Its steadily growing acceptance and implementation were reflected in the ongoing activities and in the various initiatives, such as the 2001 Reykjavik Conference on Responsible Fisheries in the Ecosystem.

In fine, with the increasing establishment of new information technology and systems, which offered integrated ecosystem assessment, it was imperative that regional bodies built on their strengths and successes to date, evaluate their future needs, and begin to work on a mutually beneficial framework of cooperation.

5.5.8.15 *Activities of RSCs and programs relevant to fishery resources (continued)*

Following are some of the examples of past, ongoing, or planned activities of RSCs and programs that are directly or indirectly relevant to the management of fishery resources.

5.5.8.16 *Abidjan convention*

The biennial work program adopted in March 2000 by the Parties to the Convention included a section on "management of fisheries, mangroves and associated ecosystems." Some of the envisaged main fisheries-related activities included the following:

(i) Reviewing the status and effectiveness of management structures for fisheries, mangroves, and associated ecosystems.
(ii) Implementing effective management structures where nonexistent.
(iii) Monitoring and surveying of organic and inorganic pollutants and their effects on fisheries.
(iv) Building capacities and raising public awareness on the impact of mangrove deforestation on the fish spawning grounds and the loss of productivity.

5.5.8.17 *Barcelona convention*

The revised action plan (MAP Phase II) identified five objectives: the first one of which was "to ensure sustainable management of natural aquatic (FW and marine) and land resources and to integrate the environment with social and economic developments, and land-use policies."

Further, the revised action plan on living aquatic resources had recognized that "a number of fish stocks were under heavy pressure," and, thus, called for "urgent" action at various levels, notably, the fishing communities, fisheries organizations, and the government.

5.5.8.18 *Bucharest convention*

Article XIII of the Convention (Protection of the aquatic, notably, marine living resources) reads "The Contracting Parties, when taking measures in accordance with this Convention …., shall pay particular attention to avoiding harm to aquatic (notably marine) life and living resources; in particular, by changing their *modus operandi* of fishing practices and by other legitimate uses of the Sea (notably, the Black Sea) and, in this respect, shall give due regard to the recommendations of the competent international organizations."

5.5.8.19 *Cartagena convention*

The Caribbean Environment Program (CEP) and, particularly, the protocol concerning specifically PAs and Wildlife have been instrumental in raising awareness about the importance of aquatic (FW and marine) PAs (MPAs) and their values in the recovery of fisheries as well as for their role in the management of the aquatic resources in the region. In this connection, a network of MPA managers established to facilitate the exchange of information and expertise and to promote, among others, the development of fishery reserves and zoning regulations. Further, the region's capacity to manage MPAs was considerably enhanced through a comprehensive training of trainers program for MPA managers which had addressed all aspects of MPA management, including zoning and applications to fisheries management.

5.5.8.20 *Caspian convention*

A proposal was developed for a large-scale (*c.* US $35 million) project that was expected to be supported by the GEF. The project entitled "Conservation of Biological Diversity of the Caspian Sea and its Coastal Zone" had envisaged the evaluation and development of a Convention/agreement on Conservation and Utilization of Bioresources of the Caspian Sea; the establishment of a "commission" to guide and coordinate the implementation of the convention and a Secretariat to implement a large number of activities related to fishery management. Subsequently, a considerably scaled-down project was approved by the GEF, which had served as the basis for the Caspian Environment Program (CEP) that was supported and implemented jointly by GEF(UNDP) and the EU (TACIS). In the framework of the CEP, a Regional Strategic Action Program (SAP) was planned to be developed. Fisheries-related issues were among those that were planned to be treated in the framework of the SAP.

5.5.8.21 *Helsinki convention*

The Baltic Prime Ministerial Conference had adopted the Presidency Declaration, followed by the Kalmar Communique' by the foreign ministers (July 1996) that had requested updating and strengthening HELCOM. The Action Programs adopted at Kalmar requested, inter alia, "the development of a coherent policy for sustainable fishing in the Baltic Sea based on a comprehensive plan to be elaborated by the International Baltic Sea Fishery Commission in consultation with HELCOM and ICES." Subsequently, the 1999 Extraordinary Meeting of HELCOM had identified the "environmental impacts of fishery management and practices" among the priorities within the framework of HELCOM.

5.5.8.22 *Jeddah convention*

The Convention recognized "the need to develop management approach to the use of the FW, marine, and the coastal areas." Management of fisheries were said to be among the issues addressed through the action plan associated with the Conventions. A report on the Status of the Living Marine Resources in the Red Sea and Gulf of Aden region and associated FW habitats, and their Management was prepared and issued in May 2000. The report was a comprehensive overview of the status and development of fisheries in the region and their effect on the environment, including an overview of the current national legislation and the dominant socioeconomic and cultural conditions among the fishing communities.

5.5.8.23 *Kuwait convention*

The preamble paragraphs of the convention recognized, inter alia, that pollution of the convention area was to "present a growing threat to aquatic (FW and marine) life, fisheries, etc." and the region's "special hydrographic and ecological characteristics and its particular vulnerability to pollution." As such, there was a need to prevent the large-scale industrial development because these are, obviously, to "jeopardize the elimination of living resources." Concomitantly, there was an urgent need "to develop an integrated management approach for the use of the aquatic environment."

5.5.8.24 *Nairobi convention*

The Eastern African Coastal and Marine Environment Resources Database and Atlas were prepared as a georeferenced information base and planning tool. Inter alia, it was extensively covered information on fisheries (e.g., commercial, subsistence, and big-game fisheries; aquatic farming; location and extent of fishing grounds; fishing methods; and catch statistics) and fishery-related topics.

5.5.8.25 *Noumea convention*

SPREP was involved in the preparation of documents for presentation at the World Summit on Sustainable Development (South Africa, September 2002). The Council of Regional Organizations of the Pacific (CROP) helped in the process. The management and sustainability of inshore and oceanic fisheries were one of the major issues, which was raised in the submission. It may be mentioned that a Strategic Action Program (SAP) for the International Waters of the Pacific Small Island Developing States was implemented by SPREP, with the assistance of UNDP and financial support of GEF. The objective of the program's oceanic component was "to enable conservation and sustainable yield of living resources of the ocean" through a coordinated program of the South Pacific Commission (SPC) and the Forum Fisheries Agency (FFA).

5.5.8.26 *OSPAR convention*

The preamble part of the convention was recognized that "questions relating to the management of fisheries were appropriately regulated under international and regional agreements dealing specifically with such questions." However, where the commission had considered that action was desirable in relation to a question, it could try to draw the question to the attention of the authority or international body competent for that question. Where actions within the competence of the "commission was desirable to deal with, the commission could endeavour to cooperate with them."

5.5.8.27 *Definition of sustainable fisheries*

Sustainable, productive fisheries are achieved when appropriate management ensures a high probability of stocks being able to replenish themselves over a long period within a sound ecosystem while offering stable economic and social conditions for all those involved in the fishing activity.

5.5.8.28 *Goals*

Development of economically and socially sustainable, environmentally safe, and responsible fisheries by:

(i) maintaining biologically viable fish stocks, by taking care of the marine and aquatic environment and associated BD,
(ii) establishing maximum fishing possibilities by appropriate selective fishing techniques for harvesting stocks within these limits, and.
(iii) distributing the direct and indirect benefits of the open sea and coastal fishery resources between local communities in an equitable manner.

5.5.8.29 *Indicators*

There had been some indicators intended to highlight the trends in biological systems and the economy of the fishery-dependent communities around the water bodies.

5.5.8.30 *Biological indicators*

(i) Spawning stock biomass (SSB): The part of the biomass of cod, herring, and sprat, etc., taking part in the reproduction process, in tons. This is an important indicator of the biological health of a given stock.
(ii) Fishing mortality: the proportion of the average population removed annually by fishing.
(iii) **Recruitment**: Recruitment may be defined as the number of fish surviving to enter the fishery or to some life-history stage, such as settlement for maturity. In other words, it may mean that the number of fish reaching the water body followed by their entry into fisheries. In fact, it is a well-known fact that changes in fish populations are primarily driven by variability in recruitment.

5.5.8.31 *Economic indicators*

(i) Landings per country: the total amount of landings in tons of cod, salmon, herring, sprat.
(ii) Number of fishing vessels per country operating in the sea say the Baltic Sea.
(iii) Fish consumption per capita per country.

5.5.8.32 *Social indicators*

(i) Number of full-time fishermen engaged in the sea, e.g., in the Baltic Sea Region, per country.

5.5.8.33 *Impact of eutrophication*

Eutrophication is the highly nutritious status of a water body caused by anthropogenic inputs of nutrients. This has had extensive effects on the water bodies, including Baltic marine ecosystem, etc. Anoxia, caused by a combination of eutrophication and hydrographic/climatic conditions, occurs frequently in the western Baltic and in the deep waters of the central and northern Baltic. This was wiped out the benthic communities over large areas of the seabed. In the case of cod, DO concentrations $>2\,mL/L$ at the depths, where eggs are neutrally buoyant, are a necessary prerequisite for obtaining good recruitment, and anoxia in the bottom layer combined with low salinity at the spawning grounds reduces recruitment. All of these could affect the management process.

5.5.8.34 *Impact of pollutants*

Toxic pollutants are to exert a great impact on animals at the top of food webs, such as fish, seabirds, seals, etc. Incidentally, the population of seals was crashed to very low levels due to combined effects of persistent organic pollutants and hunting. Effluents from pulp mills had been harming the fish populations, thus, affecting their management as well as altered patterns of genetic variation in mussel populations. However, levels of toxic pollutants are now generally believed to be declining in marine biota, but levels are still high in sediments and Baltic seals still show signs of chronic disease. Changes in the food web due to toxic pollutants contribute to the extensive mortality of salmon fry caused by the M74 syndrome in the Baltic Sea.

5.5.8.35 *Impact of human activities on fish stocks*

Human interventions in the terrestrial, aquatic, and aerial domains have been said to be affecting the lives of all biota in these domains. The utilization of coastal and offshore waters for activities such as fishing, AQC, shipping, recreation, electric or engineering projects, dredging, dumping, extraction of gravel and sand, and oil prospecting had been having a large combined effect on the BD of species, including the BD of the fishes and their concomitant management aspects. Moreover, the habitats, notably, the shallow-water sites, which are of importance as spawning and nursery grounds, are at risk from a combination of factors such as FW, terrestrial and coastal developments, tourism and eutrophication, etc. Some waters of the Baltic Sea are of international importance as overwintering areas of seabirds. Hydroelectric schemes built on the rivers flowing into the Baltic reduced the number of rivers in which salmons spawned in quite large numbers. The genetic diversity of wild salmon (which according to ICES now constitute <10% of total salmon populations) was eroded by a combination of damming and environment unfriendly activities. Further, *c.* 30 nonnative organisms had become established, and most of the alien species were found close to the shore but a few became common in deep waters as well.

5.5.8.36 *Impact of fisheries on the ecosystem; the need to protect species and habitats*

Fisheries have both direct and indirect effects on the ecosystem. Foremost are the direct effects caused by the removal of fish and shellfish for landing, which may lead to difficulties in the management of the fisheries. Other direct effects include, inter alia, the mortality of nontarget fish and shellfish, and other animals. There is a threat to the BD and a loss of habitats and biotopes in different parts of the ecosystem. The protection of habitats and species in FW, marine, and coastal areas are essential for the functioning of the ecosystems. The ecosystems are to be used in a manner consistent with sustainable development, ensuring the conservation of biological diversity and the sustainable use of its components.

Further, the impact of fisheries on the ecosystem is viewed from different perspectives by RFBs. Many link this to the protection of species and habitat, the impact of gears on nontarget, associated and dependent species (NADS), such as the seabirds and marine mammals. The impact of AQC on the marine environment is also considered in this context. On the whole, consideration of such impacts, especially those relating to gears and NADS, tends to focus on that part of the ecosystem, which is related to the fisheries resource.

5.5.8.37 *The impact of climate, ozone depletion on fisheries*

Climate change is an ecological consideration. There could be an impact of climate change on the aquatic biota including the fishes; thereby, it could hamper in their efficient management process. In particular, **global warming** (GW) due to the **greenhouse effect** (GHE) could alter inputs of saltwater, freshwater (FW), DO, nutrients, and pollutants with potentially large consequences for aquatic (FW and marine) ecosystems and species. Change(s) in currents would also influence the recruitment of organisms, including fishes to coastal and offshore waters. Changes in the input of FW from rivers would have particularly large effects on the different bays.

5.5.8.38 *Toward a global Federation of Regional Observing Systems*

It could be opined that the continued development of the aquatic (FW and marine) RFBs and RSCs could be critical to the successful design and implementation of the coastal management issues. However, the advantages and mutual benefits of collaboration fall into three important categories:

(a) the requirements for managing the environment and living aquatic resources in an ecosystem context,
(b) regional requirements for designing and implementation, and.
(c) making more cost-effective use of existing infrastructure, expertise, and knowledge.

In fine, Ecosystem-Based Management is an efficient method of management of aquatic bioresources, including the fishes. The data requirements of FW, coastal, and marine services are, for the most part, common to all the themes. Safe and good FW, coastal, and marine operations require accurate and timely forecasts of storms, floods, and precipitation; data regarding wind and water current, waves, water depth, temperature, and visibility. The set of variables that are to be measured and assimilated in near real time are barometric pressure, winds, air and water temperature, sea level, stream flows, and surface currents and waves. In addition to these variables, protecting, sustaining, and restoring aquatic ecosystems require timely data on environmental variables needed to detect and predict changes in biological, chemical, and geological phenomena, e.g., nutrient concentrations, attenuation of visible light, suspended sediments, biomass, species composition, concentrations of contaminants and pathogens, etc.

5.5.9 Summary

(1) The **continent of America** is bestowed with rich and varied resources of water bodies and BD under a variety of agroclimatic conditions. As such, there is a need for the efficient management of the lentic and lotic water bodies and their bioresources for sustained harvest and conservation.

(2) In **Argentina,** the works of Daniela et al. (2006) related to an estimation of compliance with Article 7 (Fishery Management) of the UN Code of Conduct for Responsible Fishing (FAO, 1995). It is a "live document," although it could alter with time. Moreover, Argentina initiated an evaluation of the status of shark populations and the incidental catch of seabirds in response to IPoA on sharks and seabirds (FAO, 2003). Godelman (2004) asserted that the suspension of Resolution 73/2004 (which tried to establish an emergency plan for the hake fishery) had the virtue of triggering a political discussion about the implementation of the Federal Fisheries Code. Concomitant to the above, the **Argentine** ITQ program is to manage four of the country's most commercially important species by allocating quota to individual vessels under a single CSP (Young, 2013). Further, while dealing with related aspects, FAO (2018) opined that by transitioning to a green economy, through measures, such as the elimination of harmful subsidies and the use of greener production and processing methods, fisheries resources could be expected to be managed more sustainably for the benefit of present and future generations in many places, including **Argentina**. Incidentally, Anon (2002) dealt with Fisheries Subsidies and Marine Resources Management in connection with lessons learned from studies in Argentina and Senegal. Dommen and Deere (1999) discussed Fisheries, International Trade, and Sustainable Development in connection with Natural Resources International Trade and Sustainable Development having some relation with Argentina. Incidentally, Weber Peter (1994) cautioned about major ecological, economic, and social damages that were evident. In particular, declining catches had cost > 100,000 jobs in the last few years among the world's *c.* 15–21 million fishers. Concomitantly, **Argentine** fisheries have faced crisis situation(s) in some part of its history as revealed from the studies by Maria Onestini (2002) on **subsidies in Argentine fisheries.** Incidentally, Orensanz and Seijo (2013) dealt with rights-based management in **Latin American** fisheries. Concomitant to the above, Anon (2005a, b) reported about sustainable fisheries and livelihoods in **Latin America** with reference to the imperative of recognizing the resource access rights of artisanal fishermen, coastal communities, and the indigenous people.

Some of the other works related to Management of resources, including fish resources, in America and could be elsewhere are Johns (1992), UNEP (2002), Schonberger and Agar (2001), Abaza and Jha (2002), Kalikoski et al. (2006).

(3) In **Canada,** Tony and Kevern (2002) dealt with the use of ecosystem models to investigate multispecies management strategies for capture fisheries. In fact, the application of various types of models in fisheries and AQC seems to be quite common today. Walters et al. (1997) dealt with structuring dynamic models of exploited ecosystems from trophic mass-balance assessments. Further, Walters et al. (2000) worked on representing density dependent consequences of life-history strategies in aquatic ecosystems. In fact, "Ecosim," the dynamic version of "Ecopath" had, of late, been endowed with routines that not only simulate the consequences of changes in fisheries for all elements in the ecosystem, but could also look for fishing rates that could maximize ecological, social, or economic goals. This had encouraged the FAO mainly for its versatility in managing multispecies fisheries with regard to almost all elements of the ecosystem. As such, the Fisheries Centre and the FAO had decided to hold a workshop to explore the potential of the new software. This approach was effectively used in the ecosystem models as operating models (e.g., Hilborn and Walters, 1992; Cochrane et al., 1998). Such an application was one of the primary goals of the developers of Ecosim, who had suggested that it could be used to "conduct fisheries policy analyses that explicitly accounted for ecosystem trophic interactions" (Walters et al., 1997). Incidentally, ecosystem stability refers to the capability of a natural system to apply self-regulating mechanisms, to return to a steady state after an outside disturbance. Notwithstanding the above, in **Canada**, there had been works on the evaluation of the policy optimization routine of Ecopath/Ecosim, using the model of the Newfoundland-Labrador shelf as a case study (Marcelo et al., 2002). In other words, there had been the use of Ecosim to investigate multispecies harvesting strategies for capture fisheries of the Newfoundland-Labrador shelf. The routine was used to calculate the best combination of harvesting strategies for multiple fleets and to illustrate the type of trade-offs expected when fisheries management aims at maximizing economic, social, and ecological goals. Incidentally, Wibera et al. (2004) worked on participatory research supporting CBFM. They had emphasized the engagement of researchers and fishers together in adopting social science approaches to the development of community-based fishers' organizations. Incidentally, there have been attempts to report on the works done with the fishers' organizations, both aboriginal and non-aboriginal, in which, social science priorities were set by each organization; and, small-scale research projects were designed and carried out to meet these requirements.

(4) In **The Caribbean islands,** Anon (2005a, b) dealt with the Fishing Community initiatives in CBFM and The Sitka Declaration as a proposal for Forum of Fishermen, NGO's, and the Government Agencies in order to explore expansion of the use of sectors, catch share(s), and CBFM in the **Caribbean.** The aware and active fishermen seem to be confident about responsibly managing the resources with efficient *modus operandi* of "CBFM," an approach where communities directly manage their resources as well as regain access to diverse fisheries as those stocks were rebuild. In fact, CBFM is practiced worldwide, from Alaska to Thailand, from Mexico to New England. Incidentally, it is believed that "a wealth, which is free for all is valued by none." It could be mainly because he who is little foolish, to wait for its proper time of use, could only find that it had been taken by another (H. Scott Gordon: The Economic Theory of a Common Property Resource: The Fishery, 1954). In fact, the world's fisheries today also offer a contemporary example of such a tragedy of the commons because oceans are un-owned, and each fishing fleet has no incentive to conserve or replenish the fish it takes, rather, it has every incentive to take as many fish as possible lest the benefits of a larger catch go to someone else (**Kent Jeffreys in "Rescuing the Oceans," in The True State of the Planet, Ron Bailey, ed. (New York: The Free Press, 1995)**. Nevertheless, "CBM" is advocated worldwide as a means to overcome overexploitation of fisheries. However, there is a conflict in belief between researchers and the managers regarding the effectiveness of these measures. In fact, the "tragedy of the commons" model is often accepted as a universal paradigm, which assumes that, unless managed by the State or privatized, common-pool resources are inevitably overexploited due to competition between the self-interest of individuals and the goals of a group as a whole. Under this scenario, the emergence and maintenance of effective community-based efforts could lead to the establishment of marine reserves. In fact, it could often be found that locally crafted and enforced harvesting rules, may lead to a fast augmentation in resource abundance. Incidentally, the Fishers had designed, monitored, and enforced three main types of formal and informal rules and sanctions: (i) resource-based rules (fishing banned June and July; fishing banned within reserves); (ii) monitoring rules (mandatory participation and financial contributions for monitoring); and (iii) administrative rules (mandatory: participation in cooperative meetings, monthly financial contributions to the cooperative, and timely provision of paperwork for cooperative). In addition, there had been deliberations with regard to effective working of CBFM.

(5) In **Cuba,** Julio (2006) attempted to portray the level of resource exploitation and the present fisheries management regime in **Cuba**, including an overview of the Cuban fisheries; some theoretical background on management regimes; a critical review of the fisheries regulations through time; and an account of the legal and institutional framework of the fisheries sector as well as an evaluation of the current status of the most important fisheries based on catch and effort data using the **Gordon-Schaefer model.** Attempts had also been made to include some recommendations for the future sustainable management of fisheries in Cuba. The Cuban fisheries management regime switched over to increased emphasis on sustainability and a greater focus on the value of the catch, rather than on the quantity. There had been deliberations on the "current management regime and its historical resume" to help in evaluations of future policy recommendations. However, a redesigning of the monitoring system in Cuba, of late had covered all fishing areas; including application of bio-economic models (which are components used in resource management) in a systematic and extensive way. During the 19th century and the first half of the 20th, the commercial fishing industry in Cuba consisted mainly of low capacity and technically unsophisticated, small boats and vessels plying in the coastal waters, which, primarily targeted a combination of reef fish; spiny lobster, sponge, and a few pelagic finfish species. The landings were handled by small-scale processing facilities and the products were, then, mostly directed into the local domestic markets.

Apropos to the Cuban Revolution of 1959, Cuba's commercial fishing industry underwent dramatic changes with production emphasis shifted from high-volume, low-valued, pelagic stocks toward high-valued, nearshore fisheries. Moreover, Cuba played a significant role in the global market for these high-valued finfish and shellfish seafood products harvested primarily within Cuba's nearshore waters. Incidentally, the commercial fisheries in Cuba became a part of the economy only after the early 19th century (during the colonial era). However, it was still an artisanal, small-scale economic sector, with an occasional increase in the number of fishing communities.

The first fishing regulations recorded in Cuban history were issued by the Temporary Commission of Fisheries (*Comisión Temporal de Pesca*, 1880). The US occupied army, during the period of its military intervention between the end of the Spanish-American War and the proclamation of the Republic of Cuba, in 1898–1902, had issued a number of rules for fisheries in the form of military orders, e.g., general rules for fisheries and fishing gears, a closed season, minimum legal sizes. There were also Coast Guard Services to control the fulfillment of the issued rules. However, such rules could be considered positive because they had formed a legal control system for the fisheries in Cuba (González, 2004). The Republic of Cuba was proclaimed in May 1902 and The National Board of Fisheries *(Junta Nacional de Pesca)* was formed 1911, being attached to the MOA and issued fishing regulations, thereby, continuing the work of the previous Temporary Commission of Fisheries. The regulations(s), as usual, had focused on minimum legal size and season closures, etc.

The next milestone in the Cuban Fisheries Management Regime was the issuing of the Decree-Law 704 "General Law of Fisheries" in 1936, as an updated version of the previous fisheries legislation, which included season and zone closures (Baisre, 1987b) for commercial species; pollution prevention; and, the ban on the use specific gear(s) in the breeding zones.

Incidentally, the National Institute of Fisheries *(Instituto Nacional de la Pesca)* was formed in 1955 with the objective of studying and researching on fisheries. The new Decree 2724 "Rules for the General Law of Fisheries" was formed by this institution and issued in 1956, but budget constraints, etc. could not make much headway.

The Cuban Revolution of 1959 "revolutionized" the fishing industry. The two main objectives of the new government for this sector of the Cuban economy were mainly: (a) increasing the catches as well as expanding the fishing zones and (b) developing the processing industry. It is that the Cuban revolution had resulted in significant changes to technical organization in the fisheries as well as to the volume of production. The sector had consisted mainly of subsistence activities, before 1959, with a small commercial sector. Moreover, the Cuban fisheries barely had met domestic and tourist needs. The revolutionary government promoted industrialization throughout the Cuban economy, including fisheries. A deep-sea fleet was launched, which could eventually reach into international waters from the Atlantic to the Pacific. The government's aim was to supply the local population with more sources of protein; to increase the influx of foreign currency; to better protect the national territory; and to establish bilateral agreements with other Latin American countries (Adams, 1998).

Technologies were developed, larger marine zones were fished, and a greater number of species were caught; all of which were aimed to lift the country out of its Third World condition (Eckstein, 1993).

Further, some of the previously issued regulations were eliminated by the end of the 1960s. Set gillnets and pair trawling were initiated. The use of trawl nets for shrimp and drift nets for lane snapper were generalized for increasing the catches. Consequently, overfishing caused due to misuse of the fishing gears increased efforts, and the lack of proper regulations had imperiled the fish population. Later, the management rules were changed and season closures for several species were implemented again including restriction of certain types of fishing gears and efforts. All these changes were completely put into practice in 1981, so that, the year could be regarded as setting the beginning of a new period of properly managed fisheries in the annals of Cuban ichthyology. The historical series of Cuban catches as well as the milestones in fisheries management are evident from different literature. The increased emphasis on catch quantity from the mid-1960s resulted in overharvesting, followed by stagnation and decline in catches. Thus, there was a revision of the fisheries policy with the Decree-Law 164 "Rules of Fisheries" being approved in 1996 to represent the latest milestone in the Cuban fisheries.

Nevertheless, there were efforts in the management of the water bodies and fisheries through **CBFM** involving the stakeholders from the grassroot level. Today's main rules for fisheries management could be found in the Decree-Law 164 "Rules of Fisheries," being approved by the Council of State of the Republic of Cuba on 26 May 1996. In addition, Law 81 of Environment considers the endangered species, the environmental impact of fisheries projects, and the marine pollution. The same applies to the Decree-Law on PAs, some of which are FW and MPAs and the Joint Resolution reflecting the importance of coral reefs as the main habitat of demersal fishes.

In fine, the studies on the status of Cuban fishing resources by Baisre (2000a) and Claro et al. (2001) influenced the management regime, particularly, in view of an attempt toward management and conservation of the water bodies and fisheries through the **participatory** mode of **CBFM** involving the stakeholders from the grassroot level. The latest changes included licensing of the fishing boats, increasing the closed season for several species, increasing mesh size in trawl nets, declaring FW and MPAs, and the elimination of unselective gears like set nets as well as strengthening the property rights issue.

Concomitantly, **the Schaefer model** might not be very suitable for all cases. However, a more detailed study using biological indicators, like age, size, catch composition, and spawning stock biomass could be done using analytical tools, like the **Beverton and Holt model**.

Incidentally, Anon (2009) threw some amount of light toward sustainable economic and social development (with some amount of reference to Cuba) in which, fisheries could play a significant role.

Further, **Cuba** is making great efforts to deepen its approach toward environmental management in all aspects (including fisheries management), but, particularly, in the context of the relationship between the economic, social, and environmental components (Anon, 2009).

In addition, **Cuba** adopted a National BD Strategy and Action Plan which is the most outstanding in the Caribbean Region with regard to the degree of national participation involved in its development and execution.

Moreover, Fisheries and AQC are the potential sectors for nutrition and avocation of the country. However, there seemed to be a decreasing trend and a historical loss of *c.* 20,000 mt, due to the combined action of overfishing and changes in the aquatic (FW and marine) ecosystem. Although some of these losses could be recovered through the regulation of fishing activities, some changes arising from impacts on the coastal zones and the alteration(s) that had taken place due to unfavorable climatic changes could be potentially irreversible. Moreover, the indiscriminate harvest of several species during their breeding period could be other causes of the depleting trend.

Incidentally, there had been attempts in Cuba through **participatory** model of **CBFM** for efficient management of the water bodies and fisheries.

In addition to the above, in view of the limited number of studies from a social perspective, the main aim of Doyon (2007) was to comprehend and appreciate the development and transformation(s) in **Cuban** fisheries. Such publications tried to portray, how the national fisheries had been managed since the 1959 revolution and how the resulting processes of decentralization had occurred in the wake of the 1990s' social, economic, and political crisis.

Concomitantly, as already stated there had been thoughts and efforts for **CBFM** by **PRA** through sustainable harvest of the resources.

Indeed, the island nation faced major crises and setbacks throughout the years, particularly in the early 1990s, with the collapse of the socialist block, leading to what had been named the "special period in time of peace."

Attempt had been made on various occasions to examine the changes that have affected the fisheries in Cuba since the revolution using an anthropocentric and ethnographic approach.

Concomitant to the above, Cuban fisheries returned almost to their prerevolutionary characteristics as a consequence of these changes, mainly in terms of equipment(s) and scope. Significant changes have taken place in economic organization and socialist ideologies, during this period.

The government-driven economic and political changes were justified by economic necessities and the government's desire to maintain a "renewed" socialism with a stronger accent on decentralization.

Nevertheless, the **Cuban** fisheries underwent major changes following the economic crisis of the 1990s. In a context, where the government no longer had the resources to provide even the most basic needs of its population, the industrial regime that prevailed during the revolution's years of plenty could no longer be maintained. Investments in the fisheries sector were abandoned and production had to be reoriented. The large-scale, deep sea focus of fisheries had been shifted; thus, changing the fishing practices of the *Combinado* fish harvesters. The new model that was adopted, included the processed products sold to mass international markets. Incidentally, there had been a high division and specialization of labor; strong reliance on science and modernization; changes in the equipment patterns; some amount of technological sophistication; small crafts with small engines with generally locally handmade gears with medium catch capacity that is highly diversified to adapt to the fishing season; and, an organization based sufficiently on close family and friendship ties. These alteration(s) were supported by state discourse. The government, however, had continued to assert that the gains and principles of the revolution remained almost intact in its fisheries, in spite of the extent of the transformations, which it continued to characterize as industrial. On the other hand, it sought that the workers adopt new practices and attitudes, such as entrepreneurship, which contrast with socialist ideology, but which enabled the government to deal with its economic challenges while preserving the revolution.

(6) In **Mexico,** there had been papers (Enriquez-Andrade and Batalla-Camargo, 2000) presenting an overview of the current situation of fisheries management in **Mexico** and discussions on the potential benefits and limitations of the implementation of rights-based fisheries management with the expectation to reverse current overexploitation trends. Incidentally, the progress toward better management of fisheries in Mexico was slow, despite increasing concerns. The main political, social, and economic forces behind overfishing in the country were largely at place. Incidentally, demographic pressure, open access or ill-defined property rights, excessive centralization of management decisions, and a wide array of market and regulatory disincentives for conservation were the major problems. The discussions were also considered the likely management scenarios in the country. The papers often tried to conclude that property rights-based management has an important potential to promote efficiency, profitability, and increased stewardship of fishery resources, particularly, if implemented at the local human community level. However, an effective implementation may have to surmount substantial constitutional, political, economic, and cultural barriers.

Incidentally, the growth of production were characterized by three periods, viz., (i) a development period with moderate growth (1950–72); (ii) a period of rapid growth associated with the extension of the nation's jurisdiction over fishery resources (1973–81); and, finally, (iii) a period of output stagnation and variability (1981–99).

Incidentally, average landings had not increased, during the last few decades, due to a combination of poor management practices and environmental variability.

Mexico had declared a 200-mile EEZ in 1976, comprising an extension of *c.* 3,149,920 km^2 of rich oceanic ecosystems. The main aim was to use the contiguous fishery resources by developing a domestic fleet and to exclude all foreign fishing activities.

It was estimated that under ideal management and environmental conditions using the currently underutilized species, like, the sardines, anchovies, squids, and the tunas; and through the practice of AQC in the total production, the MSY in Mexican waters could be *c.* 1.65 mmt up by *c.* 19% from the *c.* 1.39 million estimated in 1997. Further, by the year 2020, the Mexican production from conventional methods and AQC could reach *c.* 4.8 mmt; out of which, *c.* 2.2 mmt could be from capture and *c.* 2.6 mmt from AQC. Hike in production could also lead to an increase in exports from an estimated *c.* 780 million US $ in 1999 to *c,* 2340 million US $ in 2020 (Proceedings of IIFET, 2000).

Incidentally, the research activities (through extensive field interviews with the stakeholders) were directed mainly to ascertain the institutional and market failures as well as to assess the potential for property-rights based on management instruments (Nijkamp et al., 1990).

Incidentally (i) open access; (ii) undefined or unenforceable property rights; (iii) excessive centralization of decision-making and market power; (iv) administrative inefficiency, corruption, and weak enforcement of regulations were some of the possible causes of management failure(s) in Mexico.

Moreover, rationing of access to scarce fishery resources were essential to protect the depleted fishery resources. In fact, the poorly defined property rights to land, natural resources, and the environment were among the major obstacles for the development of the coastal zone in Mexico. Concomitantly, it is globally acclaimed that a lack of clearly defined property rights promote overexploitation of many fish stocks.

However, in Mexico, the public access to the marine commons is granted by the Constitution.

Further, for quite a long period, the fishing sector had received much support from the federal government as various types of subsidies, thus, worsening with regard to unsustained harvest.

Incidentally, indivisible and nontransferable, non-tradable and non-leasable concessions and permits had led to attenuated forms of access rights to fish stocks. Further, there was ineffective enforcement of these management tools.

The Federal system of management is too centralized and unworkable. Thus, entrusting control of the resources with the State and the Municipality or with the Community and allowing them to develop access rules could be an effective solution. However, widespread implementation of rights-based management in Mexico could need major constitutional amendments, in spite of much decentralization of the administration.

Incidentally, McCay et al. (2014) dealt with fishing cooperatives, concessions, and comanagement on the Pacific coast of **Mexico**. Studies revealed that the Mexican case(s) support arguments for greater community-level engagement in catch share and territorial management throughout the Pacific. Incidentally, catch share-based management is often interpreted as a **euphemism** for using ITQs.

Probably, good management and conservation practices coupled with sustained harvest of resources could help to achieve success in the sectors of AQC and fisheries.

In fact, of late, the fishery management scholars opined for a better-suited alternative small-scale fisheries management approach for the developing countries, rather than the one offered by conventional Western Fishery Science; the former being based on the development of better resource access controls, CBM, and an increased use of local traditional fishery knowledge. In this regard, Guillermo and Basurto (2002) worked with the aim of understanding, what are the most important social and ecological elements that could contribute to the successful CBM of the Seri Indians' **(Comcaac Indians)** *callo de hacha* (pen shell scallops) fishery **in Sonora, Mexico**. This fishery is one of many **Mexican** small-scale fisheries unregulated by the Mexican Government. There is no Federal or State mandate to regulate fishing efforts in the form of season closures, restrictions on fishing gear, or minimum harvestable size. The Seri CDH fishery is managed entirely by local users. The study has analyzed the Seri controls over access to the fishery and integration of TEK into local fishing practices. The federal government granted the Seri Indians a portion of their historic coastal territory and a fishing concession, in 1975, as a *modus operandi* to guarantee the survival of the tribal culture and reduce conflicts with other local fishing communities. Results had portrayed the success of this locally managed fishery, which had originated from a good fit between well-defined property rights, locally designed institutions, and the natural system.

The seemingly successful CDH community-based fishery allowed us to look to the inner workings of the access control mechanisms and fishing practices, based mainly on ITK, which the Seri people had in place and in this way could contribute to the development of successful fisheries management approaches that could be better suited for the small-scale fisheries in the developing countries.

In fact, the success of the *callo de hacha (CDH)* fishery has variously been suggested due to certain social and ecological mechanisms in place. The principal among them could be the following:

(i) controlled access and fishing; (ii) management practices based on ITK/TEK.

Further, Anon (2006) dealt with the **Agricultural and Fisheries Policies in Mexico and the concomitant recent achievements and continuing with the reform agenda**. Such communications tried to analyze the effects of Mexico's ambitious reforms in the fields of agricultural and fisheries policies since 1990 as well as attempts to make recommendations for further reforms. Such evaluation(s) were based on the criteria for good agricultural and fisheries policy as agreed upon by the OECD countries. Such criteria, if implemented, could support various economic sectors, which contribute to the aspects like wider economy; sustainability of natural resources, etc.

Incidentally, the extensive stretches of the community land restrict development and resource management. Water remains mostly unmeasured in its use with an unknown amount of exploitation and property rights are often enforced. But private decision-making is not aligned with *the* public requirements. Incidentally, some of the highest obstacles are the remaining barriers to trade, payments tied to outputs (target income) and new energy subsidies. These policies do not provide much benefit to the farmers for all their costs to consumers and taxpayers.

The Fisheries policy of Mexico made significant amount of progress in policy reform(s), with multiple shifts, since 1990, which were not very conducive, before 1990, for developing a sustainable fishery sector; thus hampering the sector's longer-term economic prospects. The current policy is more appropriate for more sustainable and profitable fisheries, with the high-value commercial tuna and shrimp fisheries being sustainably managed today, and associated with a flourishing AQC sector; accompanied with a reduction in bycatch in the tuna and shrimp fisheries; fall in negative environmental impacts of AQC; increased transparency in stock assessments, resource status; and improved management measures including fisheries surveillance.

However, more reforms are needed, with a clear vision; regulation on fishing effort; efforts for reform(s) of the institutional arrangements for efficient fisheries management plans and stock recovery programs; inclusion of the artisanal sector under stronger management framework; augmentation agency coordination for AQC projects; and so on. Moreover, the use of local **comanagement** initiatives may be effective for localized stocks and area-based fisheries. All these are essential for continued sustainable fisheries with long-term economic and social benefits for Mexico.

In addition to the above, Saenz-Arroyo et al. (2005) dealt with community-based marine reserve network in Northwestern **Mexico**. There were broad appreciations that fully protected FW and marine reserves, could be very powerful tools to overcome globally spread overfishing problems (Halpern, 2003; Gell and Roberts, 2003; Pikitch et al., 2004). Workers worldwide had called for a clear policy statement with little controversy, to implement networks of fully protected aquatic (FW and marine) reserves representing the world biogeographic regions, and this could help in recovering vulnerable and depleted species (Pikitch et al., 2004; Pauly et al., 2002). However, due to practical difficulties with such schemes, there had been suggestions to propose for large (aquatic) reserves with little consideration of socioeconomic constrains (Christie et al., 2003). A significant example is the model developed for the Gulf of California (Sala et al., 2002). However, the model, having been tested on a wrong scale, had ended-up after proposing large areas as "no-take zones"; although this model was believed to bring a novel mathematical approach to incorporate ecological, social, and economic concerns in the design of fully protected reserve networks. Further, the fully fishing villages were proposed to be excluded from their fishing rights in the said model (i.e., the Seri Indian marine territory) (Sala et al., 2002). No doubt, making such large "no-take," no-fishing' PAs could make such restricted areas appear very promising to the conservationists. Proposing such large fully protected reserves (whether in the FW bodies or in the sea) ignoring the dependence of coastal economies on the aquatic (FW/marine) resources is just as dangerous as Icarus' approach to the sun. They may risk losing the fishers' support, the best potential allies for reserve's stewardship.

Further, the application of **multi-criteria evaluation** technique also offers a scope to incorporate ITK and local stakeholders' insight in describing the ecological features of sites of particular significance, such as breeding aggregations or nursery areas.

It may be mentioned that the grassroot level conservation groups and the local fisher's organizations from various fishing communities in the Mexican Northwest region were called for peoples' support to plan and implement parallel projects in their communities, which could help in generating a **Fisher Fund.** Such efforts could be exemplified in the implementation of the pilot project in San Pedro Mártir Island Biosphere Reserve and Loreto Bay National Park.

In fine, restricted areas protect the endangered and less abundant biota in order to save them from extinction. The community-based reserves are made for the protection of the less abundant and endangered biota from overexploitation and extinction.

Incidentally, similar exercises of trying to make a community-based network of reserves could also be done in the FW sector.

(7) In **Peru**, Pinedo et al. (2000) reported about CBNRM as a nonlinear process with regard to a case study in the **Peruvian** Amazon *Varzea.* Their study had revealed that small communities along the waterways in the northern Peruvian Amazonia depend on a wide variety of natural resources. Their long-term survival and development require management systems that could assure the future availability and sustainability of these resources. The *ribereño* peasants live in small communities along the waterways in the northern Peruvian Amazonia. They depend on a wide variety of natural resources. For the sustainable use and management of natural resources, local CBM could be a more viable alternative than the prevailing private and State models of management. In fact, certain factors, essential for the establishment and continuity of a CBNRM initiative, may occasionally undergo changes, because of the dynamic nature of the social and natural setting, in which, these riverine communities live. Thus, a CBM system may, preferably, be viewed as a nonlinear process that may not be understood without analyzing its historical development and the dynamic nature of the factors. Incidentally, the management systems that are flexible in their response to these factors are better adapted to the existing conditions in the Amazon Basin.

(8) In **Trinidad and Tobago**, Shing (2001) dealt with a case study of the ICFM Project, which was a pilot project for the Gulf of Paria, **Trinidad** under the UNDP ICFM Project, funded by the UNDP Division for Global and Interregional Programs and executed by the FAO of the UN.

The Gulf of Paria was selected based on the importance of coastal fisheries to the country; management issues to be addressed; government policies; institutional framework; and the existence of programs to complement project activities both nationally and regionally.

The Gulf of Paria is a shallow semi-enclosed basin having an average depth of *c.* 20 m. A clockwise circulation occurs in the Gulf water, with the salinity and circulation being influenced by the flow of the rivers, like, the Amazon and Orinoco, etc. The substratum is soft, muddy, and sandy due to the sediments brought-in by these rivers. The nutrient-rich waters support rich and diverse fisheries resources contributing to the high productivity of the area. Moreover, the fringing mangroves and the wetlands along the Gulf give shelter to the juvenile fish, shellfish, such as oysters, mussels, snails, crabs, shrimps, etc. The avifauna of the wetlands is also said to be rich and diverse.

More than 2500 fishers operate in communities and recreational sites along the coast of the Gulf of Paria. They fish with a variety of fishing methods, like the gill nets, bottom longlines, fish pots, hand lines, trawl nets, etc., revealing an estimated landed catch of *c.* 7800 mt in 1991.

Incidentally, >90% of the population inhabit along the Gulf coast. Many industries include ammonia, methanol, and chemical fertilizers, oil and natural gas extraction, etc. The Agro-industries include the processing of sugar cane and the manufacture of food products. The fisheries management for the Gulf of Paria required an expansive and extensive approach. There had been certain activities which could alter the habitat and deteriorate the environmental quality, like, coastal reclamation and port construction; poor watershed management practices and quarrying, leading to increased sediment loads; pollution due to offshore drilling and improper solid and liquid waste disposal and the widespread use of chemicals in agriculture and industry, and so on.

Management of the fisheries of the Gulf of Paria included more involvement of the fisheries sector in the decision-making process to regulate the development of the coastal zone of the Gulf. The Gulf of Paria coastal zone or management area was also formed.

Further, the Fisheries Division showed keen interest in implementing the ICFM Project with the following main objectives: (a) to improve methods and coordinating mechanisms for integrated coastal fisheries management and (b) to develop an enlarged assistance program for ICFM of the marine fisheries resource of the Gulf of Paria.

Moreover, the project had sought to strengthen the capability of the Fisheries Division of the Ministry of Agriculture, Land, and Marine Resources specifically, in order to provide with the necessary state leadership to have the concerns of the fisheries sector incorporated into the framework of development planning for the coastal zone, which was said to be specific to the Gulf of Paria, and the following strategies were formed:

(a) to improve the information base on the fisheries sector and the impact of other sectors on it; (b) to enhance the general awareness of the socioeconomic contribution of the fisheries sector and its role as a natural indicator of environmental quality; (c) to strengthen representation and associations within the fishing and related concerns; and (d) to promote the supportive role of the Fisheries Division in matters related to coastal zone management.

Fisheries resources of the Gulf of Paria have a multispecies composition. Hence, they are harvested by a heterogeneous, multi-gear fleet. As such, there could be **conflicts** between and among the fishery groups, like, between trawlers and fishermen using other methods, who blame the trawlers for the marked depletion of fisheries resources and destruction of the habitat.

Further, a number of sectors are involved in the development and use of the coastal zone, often with conflicting and overlapping jurisdictions. In general, the legally established procedures for the development of the coastal zone did not adequately provide for: (a) conflict resolution among the users of coastal resources and between resource users and other sectors; (b) networking among the sectors to ensure that, most of the relevant information was available to the decision-makers. The second aspect is to be taken care of.

Incidentally, some of the salient points with regard to general management aspects of natural resources in Trinidad and Tobago are given as follows:

(i) Weak consultative process in decision-making with regard to the coastal area development.

(ii) Weak organization of the fisheries resource users. There were few functional fishery organizations capable of representing the sector in environmental matters. In general, however, the development of fishery organizations have a chequered history. Moreover, there were three fishing cooperatives (notably, Icacos, Cedros, and Almoorings) and one fishing association (notably, San Fernando). Incidentally, the inshore artisanal fishermen formed the National Organization of Fishing and Allied Cooperative Societies Limited (NOFACS) in 1988 in order to coordinate the activities of its members; represent their interests to the Government; promote the development of fishing cooperatives; and seek to revive inactive ones. Nevertheless, there was one fishing interest group, viz., the Trinidad and Tobago Trawler Owners' Association, representing the interests of the industrial trawl fleet. This Association was very active, as an interested group and effectively represented the concerns of its members.

(iii) Issues of competition between Fisheries versus other sectors: The Fisheries Division may have to compete with two other sectors, viz., (a) with the forestry and agriculture, within the MOA, Land, and Marine Resources (which contribute more to the GDP than fisheries) and (b) Fisheries Division may have to compete with the departments of Energy, Trade and Tourism, Finance, etc.

(iv) Geographic extent of the area under consideration to be taken care of.

(v) Different types of stakeholder groups are to be identified, notably,

  (a) organizations with management authority, (b) organizations representing users and communities, (c) users and nonorganized stakeholders, (d) National Organization of Fishing and Allied Cooperatives Society Ltd., (e) fish vendors, and (f) other fishers.

(vi) Detailed Records of Stakeholders are to be maintained.

(vii) To initiate ICFM activities focused on improving the profile of the fisheries sector through enhancing the capability of the Fisheries Division and to establish a framework for ICFM.

(viii) Records of Other fishermen are also to be maintained.

(ix) Strategies applied for stakeholder participation are to be chalked out. In fact, there should be involvement of all stakeholders in the management process and not just the fishing communities.

(x) There is a need for awareness generation on environmental and resource management issues among all stakeholders including the resource users and the general public.

(xi) There could be an exercise of identification of critical areas including particular resources and user groups along the Gulf of Paria, on which, there is need to focus different types of management activities, like community surveys; household surveys; generation of secondary data profiles including demographic, socioeconomic, and geographic information; "local knowledge" surveys including information on the fishers' knowledge of the resources, the environment, and environmental changes.

(xii) Efforts could be made for consolidation of all available information on the Gulf of Paria to avoid duplication of works.

(xiii) Some amount of considerations is made with regard to the alternatives to the prevailing "open-access" approach to the harvesting of fisheries resources.

(xiv) There could be consultations at the community and national levels to focus project objectives and activities, and, thereby, to enhance institutional linkages.

(xv) There is a need for proper identification of the Stakeholders accompanied by capacity building assessment.
(xvi) There is a need for understanding power relations.
(xvii) Sometimes, the entire project or a scheme could be a learning experience which may facilitate extensive public and private sector collaboration and may also identify the skills and structures that need to be considered for the participatory process.
(xviii) There is a need for rationalizing urgency with practicability.
(xix) There should, preferably, be feasibility in determining the size of the area to be managed.

(9) In **United States.**
Bartch et al. (2009) discussed the voices from the fisheries domain with regard to the preservation of LEK for linking generations and thereby improving environmental literacy.
LFK contributes significantly to the knowledge in Fisheries Management, particularly, CBFM. Such a project was piloted in Maine, USA. Designing an educational program may pose a difficulty, particularly, where knowledge and participation intersect to engage young people and the stakeholders as vital community members. Such efforts are done for addressing issues which impact the future vitality and sustainability of the small fishing communities. However, some outcomes which had sustained us through the many challenges of implementing the LFK project include, among others, the following:

(i) Receptivity from the community.
The question may arise, why is it important to document and preserve marine-related oral history? A simple answer could be that under the present practice of so-called assault to nature, which is *in vogue,* ITK in different aspects, including those in the Fisheries arena, are be deteriorating and gradually vanishing from among us leading to a so-called void in the field. Further, fishery and fishing economies are being transformed mainly in view of emigrations and immigrations in the look-out for avocation because of resource depletion in one place or the other.
Concomitantly, a rich body of knowledge about the history and culture of community fishing and CBFM are disappearing. It could be possible for the local young people to preserve and conserve the coveted indigenous traditional ecological and fisheries knowledge and the rich history and heritage of their province's and nation's people and culture; particularly, in those regions, which are gifted with myriads of bewildering diversity of biotic and abiotic forms of aquatic resources. Incidentally, the coastal States, like Maine historically had depended on the marine or Great Lakes environment for their economic and cultural base. The young people in these regions having access and minimum knowledge of the rich history of their province, nation, and the people could accomplish this task.
Such information could also be useful to the social scientists who deal with socioeconomics of the coastal societies.
Further, fishermen's detailed knowledge about the local FW and marine environment as well as the past distribution and characteristics of local fish stocks are certainly useful to fishery biologists who try to reconstruct, what FW or marine fish stocks were like before scientists had begun collecting this information.
Today, it could be better to understand complex relationships among diverse species in local and regional environments, as the NOAA Fisheries Service management adopted an ecosystems-based approach. Concomitantly, authentic oral histories on the ecology and management of the water bodies could further contribute to this effort.
Moreover, detailed information about the cultural, environmental, and historical knowledge of fishing communities, their management based on CBFM, may be collected by engaging school students.
Notwithstanding the above, Anon (2007) dealt with Natural Resources (**in USA**) with regard to their conflicts and conflict resolution. Incidentally, natural resources may be renewable or nonrenewable. Further, a nation's access to natural resources often determines its wealth and status in the world economic system. Incidentally, agriculture, though strongly linked to natural resources, is generally not thought of as a natural resource as it depends largely on cultivation, although economies that are heavily dependent on agriculture certainly depend heavily on other natural resources, such as water and land. Conflicts often may arise with regard to the distribution and use of natural resources. In addition, they are resolved using various *modus operandi.* Incidentally, CBM, involving the community stakeholders from the grassroot level, could be a potential method of resolving various types of conflicts with resources.
Further, Soeftestad (1999) dealt with the report from the **CBNRM** International Workshop held at Washington D.C. in May 1998.
This workshop was organized because of the growing interest in CBNRM as a third alternative to command and control market-based solutions to NRM.
CBNRM is an approach to managing renewable natural resources. Today, there is said to be a growing consensus in favor of CBNRM, which has variously been described, inter alia, as a tool (or a set of tools), a checklist, a method, a means, a set of activities, a model, a process, and an approach.

However, CBNRM have the following core aspects: (i) there is a "community" aspect of CBNRM; (ii) there is a "natural resource management" aspect of CBNRM; and (iii) there is a "comanagement" aspect of CBNRM.

Incidentally, Peter F. Sale et al. (2005) dealt with critical science gaps, which impede the use of no-take fishery reserves. No-take fishery reserves, in addition to serving as valuable BD conservation roles, protect a portion of the fishery stock as insurance against future overfishing.

Incidentally, there are significant gaps in scientific knowledge that are to be filled, if no-take reserves are to be used effectively as fishery management tools. However, these gaps are glossed over by some uncritical advocacy. Hence, there had been attempts to review the plan; identify the most crucial gaps; and to offer suggestion(s) of ways and means to fill them,; so that a promising and efficient management tool could help to meet the growing challenges faced by the management bodies of aquatic resource management.

Globally, the aquatic resources and the fishery, in different domains, are under ever-increasing threats of collapse. It is quite a common experience today that **efforts increase, and yields, rarely, remain stable** but, generally, decline as well as other anthropogenic impacts degrade habitats and ecological systems on which the fisheries depend. Incidentally, the MPAs could be said to be a form of resource management method, but the **no-take fishery reserves** could also be called "closed areas" or "harvest refugia" and, in this regard, the focus is generally laid on fishery benefits, while recognizing that no-take reserves have significant BD conservation benefits. Such reserves commonly increase the density, biomass, and average size of target species within their borders.

Notwithstanding the above, Iverson (2013) discussed the opportunities and challenges, with regard to the management of **US Regional Fishery**.

Fisheries and AQC are the potential avenues of both nutrition and avocation. However, this is a challenging time for fisheries. Incidentally, the present depleting trend of fish production is not in a position to keep pace with the increasing trend of the human population by leaps and bounds at an exponential rate. Moreover, the developmental measures implemented in the aquatic domain coupled with the unrestricted and unsustainable harvest of the fishes have been exerting pressure on the habitat and fish populations alike.

Moreover, climate change is building new and unpredictable problems. Nevertheless, the Americans are exploring innovative ways to achieve conservation goals, including new approaches to fisheries management, harvesting, marketing, and consumption. Incidentally, eight Regional Fishery Management Councils were recognized current challenges in fisheries management.

Concomitantly, with the passage of the **Magnuson-Stevens Fishery Conservation and Management Act**, the Regional Fishery Management Council system was formed in 1976 in order to address the regional differences. The unique Council system was designed to allow regional, participatory governance by knowledgeable people with a stake in fishery management.

Eight Regional Fishery Management Councils had served as the frontline of fisheries management, where regionally specific management measures (such as fishing seasons, quotas, and closed areas) are initiated, developed, and ultimately adopted in a fully transparent public process.

Concomitantly, Donald et al. (2006) worked on the Fisheries Management Plan for the Minnesota waters of Lake Superior. The Lake Superior fish community had undergone dramatic changes since the mid-1900s due to overfishing, the introduction of nonnative species, pollution, and land-use changes in the watershed. The Lake Superior fish community, since 1950s, have composed of both native and exotic species, the latter included the sea lamprey, which had caused devastating effects by eliminating the lake trout, in all but a few isolated areas of Lake Superior. However, rehabilitation efforts, since the 1960s, including sea lamprey control, harvest regulations, and stocking programs, along with stricter pollution standards and best management practices for land use, have led to the restoration of healthy fish stocks throughout much of Lake Superior.

Moreover, there had been plans based on a fish community approach to fisheries management of Lake Superior fishery. The Lake Superior today is the least productive, but one of the most pristine of the Great Lakes, and have demonstrated the capacity to support self-sustaining fish populations through natural reproduction. The plan had emphasized the continued need for habitat protection, and the desire for managing self-sustaining fish populations that were the best suited to the lake's environment.

Further, the user groups (UG) in the Lake Superior possess diverse types of interests, and the LSMP attempted to balance resource protection, recreational opportunities, cultural beliefs, and economic development for the benefit of both present and future generations. Moreover, citizen participation was the cornerstone of the planning process and could be an ongoing process.

Concomitant to the above, the effectiveness of Lake Superior management programs is to be continually evaluated. If established criteria indicate program changes are required, interested citizens could be consulted and the necessary action(s) could be taken.

Notwithstanding the above, Granados-Dieseldorff and Montaña (2012) reported the advancing conservation with innovative ideas for the future generations as a community-based approach to the management of an aggregating reef fishery in the **Meso-American reef**. In fact, the Conservation scientists advocate for the establishment of MPAs as a global precautionary approach to the sustainability of global fisheries.

Other workers, who had worked on **CBM** and associated use of local traditional fishery knowledge in management, in general, were Christy (1982), Cordell (1989), McGoodwin (1990), Ostrom (1990), Dyer and McGoodwin (1994), Pomeroy (1994), Berkes and Folke (1998), Kothari et al. (1998), Newell and Ommer (1999), National Research Council (1999), Cudney-Bueno (2000), Durenberger and King (2000), Berkes et al. (2001, b), Bourillon (2002), Tegner et al. (1996), Musick (1999), Munda et al. (1994), Van Pelt (1993), Díaz-Briquets and Perez Lopez (2000), and so on.

Suggested reading

Asia

Acharjee, P.R., n.d. G.C. College Credit and Thrift Co-operative Society Ltd., Silchar, Assam, pp. xiv + 124.

Acharya, S., Barman, R., Bhattacharjee, P.C., Botre, S., Chhatre, A., Gandhi, K., Ganguli, A., Ghate, U., Kar, D., Mishra, M., Panigrahi, R., Pramod, P., Seshagiri Rao, P.R., Srivastava, D.S., Vardhan, M., 1998. Conservation. pp. 107–137, In: Gadgil, M. (Ed.), The Hindu Survey of the Environment. Kasturi and Sons, Chennai, p. 216.

Adams, T., 1993. Forthcoming Changes in the Legal Status of Traditional Fishing Rights in Fiji. SPC Tradit. Mar. Resour. Manage. Knowl. Inf. Bull. 2, 21–22.

ADB, 1999. Annual Report. pp. 1–322.

ADB, 2007. Outlook 2007: Growth amid change. Asian Development Bank, pp. 1–388.

Adger, W.N., 2000. Social and ecological resilience: are they related? Progr. Hum. Geogr. 24, 347–364.

Agardy, T., 2000. Effects of fisheries on marine ecosystems: a conservationist's perspective. ICES J. Marine Sci. 57 (3), 761–765.

Agrawal, A., 2002. Common resources and institutional sustainability. In: Ostrom, E., Dietz, T., Dolsak, N., Stern, P.C., Stonich, S., Weber, E.U. (Eds.), Drama of the Commons. National Academy Press, Washington, DC, pp. 41–85.

Agriculture, Fisheries and Conservation Department, 2005. Hong Kong Eco-map. In: Friends of the Country Parks and Cosmos Book Limited (Hong Kong).

Ahmed, M., Navy, H., Vuthy, L., Tiongco, M., 1998. Socio-Economic Assessment of Freshwater capture Fisheries of Cambodia. Report on a Household Survey, p. 185. MRC/DOF/DANIDA.

Ahmed, M., Capistrano, A.D., Hossain, M., 1997. Experience of partnership models for the co-management of Bangladesh fisheries. Fish. Manage. Ecol. 4, 233–248.

Ahmed, M.M., Dickson, M., 2007. Institutional Issues in the CBFM-2 Project, Conference Paper 02. The WorldFish Centre, Bangladesh and South Asia Office, Banani, Dhaka, Bangladesh, pp. 2–12. 1 Pathfoot Drive, Bridge of Allan, Stirling FK9 4SD, UK: pp. 1-345.

Aizaki, M., 1986. Trophic status and water quality of high altitude lakes in Mount Annapurna region. In: Ishida, Y. (Ed.), Studies on Distribution, Adaptation and Evolution of Micro-organisms in Nepal Himalaya. Second Report. Tokyo University Press, Tokyo.

Akimichi, T., Ruddle, K., 1984. The Historical Development of Territorial Rights and Fishery Regulations in Okinawan Inshore Waters. SENRI Ethnol. Stud. 17, 37–88.

Alexander, P., 1995. Sri Lankan Fishermen: Rural Capitalism and Peasant Society. Sterling Publishers, New Delhi.

Amarasinghe, U.S., 1985. Studies on the exploitation of minor cyprinids in Parakrama Samudra, a man-made lake in Sri Lanka, using gillnets. J. Natl. Aquat. Resour. Agency (Sri Lanka) 32, 11–23.

Amarasinghe, U.S., Pitcher, T.J., 1986. Assessment of fishing effort in Parakrama Samudra, an ancient man-made lake in Sri Lanka. Fish. Res. 4, 271–282.

Amarasinghe, U.S., 1987. Status of the fishery of Pimburettewa wewa, a man-made lake in Sri Lanka. Aquacult. Fish. Manag. 18, 375–385.

Amarasinghe, U.S., Samarakoon, J.I., 1988. Some factors affecting contribution of the cichlid species, *Etroplus suratensis* and *Tilapia rendalli* to the gillnet catches in a man-made lake in Sri Lanka. Asian Fish. Sci. 2, 17–26.

Amarasinghe, U.S., 1988. Empirical determination of a desirable mesh size for the gillnet fishery of *Oreochromis mossambicus* (Peters) in a man-made lake in Sri Lanka. Asian Fish. Sci. 2, 59–69.

Amarasinghe, U.S., De Silva, S.S., Moreau, J., 1989. Spatial changes in growth and mortality and effects on the fishery of *Oreochromis mossambicus* (Pisces, Cichlidae) in a man-made lake in Sri Lanka. Asian Fish. Sci. 3, 57–68.

Amarasinghe, U.S., 1990. Minor cyprinid resources in a man-made lake in Sri Lanka: a potential supplementary source of income for fishermen. Fish. Res. 9, 81–89.

Amarasinghe, U.S., 1992. Recent trends in the inland fishery of Sri Lanka. In: Baluyut, E.A. (Ed.), FAO Fisheries Report No. 458, Supplement. Food and Agriculture Organization of the United Nations, Rome.

Amarasinghe, U.S., De Silva, S.S., 1992a. Population dynamics of *Oreochromis mossambicus* and *O. niloticus* (Cichlidae) in two reservoirs of Sri Lanka. Asian Fish. Sci. 5, 37–61.

Amarasinghe, U.S., De Silva, S.S., 1992b. Empirical approaches for evaluating the efficiencies of different fishing methods in tropical, shallow reservoirs: a Sri Lankan case study. In: De Silva, S.S. (Ed.), Reservoir Fisheries of Asia. International Development Research Centre, Ottawa, Canada.

Amarasinghe, U.S., De Silva, S.S., 1994. In: Chou, L.M., Munro, A.D., Lam, T.J., Chen, T.W., Cheong, L.K.K., Ding, J.K., Tan, C.H. (Eds.), Selectivity Patterns in the Multi-mesh Gillnet Fishery for Cichlid Species in a Sri Lankan Reservoir. The Third Asian Fisheries Forum, Asian Fisheries Society (Manila).

Amarasinghe, U.S., Chandrasekara, W.U., Kithsiri, H.M.P., 1997. Traditional practices for resource sharing in an artisanal fishery of a Sri Lankan Estuary. Asian Fish. Sci. 9, 311–323.
Amarasinghe, U.S., 1997. Observations on the trends in the cichlid fishery in an ancient reservoir in Sri Lanka. J. Natl. Aquat. Resour. Agency (Sri Lanka) 35, 57–68.
Amarasinghe, U.S., 1998a. Reservoir fisheries management in Sri Lanka: achievements, mistakes and lessons for future. Int. Rev. Hydrobiol. 83 (Special Issue), 523–530.
Amarasinghe, U.S., 1998b. How effective are the stocking strategies for the management of reservoir fisheries in Sri Lanka? In: Cowx, I.G. (Ed.), Stocking and Introductions of Fish. Fishing News Books, Blackwell Science Ltd., Oxford.
Amarasinghe, U.S., De Silva, S.S., 1999. The Sri Lankan reservoir fi -shery: a case for introduction of a co-management strategy. Fish. Manag. Ecol. 6, 387–399.
Amarasinghe, U.S., 2002. The fishery and population dynamics of *Oreochromis mossambicus* and *Oreochromis niloticus* (Osteichthyes, Cichlidae) in a shallow irrigation reservoir in Sri Lanka. Asian Fish. Sci. 15 (1), 7–20.
Amarasinghe, U.S., Ajith Kumara, P.A.D., Ariyaratne, M.H.S., 2002a. Role of non-exploited fishery resources in Sri Lankan reservoirs as a source of food for cage aquaculture. In: Cowx, I.G. (Ed.), Management and Ecology of Lake and Reservoir Fisheries. Fishing News Books, Blackwell Science Ltd., Oxford (U.K.).
Amarasinghe, U.S., De Silva, S.S., Nissanka, C., 2002b. Evaluation of the robustness of predictive yield models based on catchment characteristics using GIS for reservoir fisheries in Sri Lanka. Fish. Manag. Ecol. 9, 293–302.
Amarasinghe, U.S., 2003. New approaches to inland capture fisheries statistics in Sri Lanka. In: New approaches for the improvement of inland capture fishery statistics in the Mekong Basin. Ad-hoc expert consultation, 2–5 September, 2002, Udon Thani, Thailand. RAP Publication 2003/01. Food and Agriculture Organization of the United Nations, Bangkok.
Amarasinghe, U.S., 2006. Reservoirs of Sri Lanka: A major source of animal protein for rural communities. In: Odada, E.O., Olago, D.O., Ochola, W., Ntiba, M., Wandiga, S., Gichuki, N., Oyieke, H. (Eds.), Proceedings of the 11th World Lake Conference, Nairobi, Kenya, 31 October–04 November 2005. International Lake Environment Committee, Japan.
Amarasinghe, U.S., Shirantha, R.R.A.R., Wijeyaratne, M.J.S., 2006. Some aspects of ecology of endemic freshwater fishes of Sri Lanka. In: Bambaradeniya, C.N.B. (Ed.), The Fauna of Sri Lanka: Status, Taxonomy, Research and Conservation. The World Conservation Union (IUCN), Colombo, Sri Lanka and Government of Sri Lanka.
Amarasinghe, U.S., Wijenayake, W.M.H.K., 2015. Results of a decade of R and D efforts on culture-based fisheries in Sri Lanka. pp. 59–71, In: De Silva, S.S., Ingram, B.A., Wilkinson, S. (Eds.), Perspectives on culture-based fisheries developments in Asia. Network of Aquaculture Centres in Asia-Pacific, Bangkok, Thailand. pp. 110.
Amos, M., 1993. Traditional-based marine management systems in Vanuatu. South Pacific Commission; Traditional marine resource management and knowledge. Inf. Bull. (2), 14–17.
Anon, 1997. In: Reservoir Fisheries Management in Savannakhet Province, Lao PDR. Final Report; Fisheries Management Science Programme; Overseas Development Administration; MRAG Ltd. August 1997, pp. 1–46.
Anon, 1999. Study on Fisheries Resources and Fishing Operations in Hong Kong Waters. Legislative Council Panel on Economic Services Consultancy, pp. 1–5.
Anon, 2004. India: Good Management practice in sustainable fisheries: Policy briefs. Published by World Bank, pp. 1–3.
Anon, 2005. Wetland Conservation in Myanmar. S:WorkingDocs\Species\CentralAsianFlyway\Delhi2005\Docs_CAF\inf_04_14_Myanmar.doc, pp. 1–4.
Anon, 2006a. Report of the Working Group of Fisheries for the Eleventh Five Year Plan, 2007-2012. Government of India; Planning Commission, pp. 1–259.
Anon, 2006b. XI Five Year Plan for Fisheries, Report of the Working Group. Planning Commission. Government of India, New Delhi.
Anon, 2010a. Liquid Assets: Community-Based Fisheries Management in Bangladesh. World Fish Centre, UKAid, DFID (Department for International Development), pp. 1–5.
Anon, 2010b. India Marine Fisheries: Issues, Opportunities and Transitions for Sustainable Development. The World Bank, ARD; Agriculture and Rural Development; Report No. 54259-IN, Agriculture and Rural Development Sector Unit South Asia Region, pp. 1–124.
Anon, 2011a. Possible Approaches in Community Based Fisheries Management. FAO, pp. 1–2.
Anon (2011b). Review of National Laws and Legislations in the Philippines in Relation to Ecosystem Approach to Fisheries Management (EAFM). Palawan State University; Centre for Strategic Policy and Governance; The USAID Coral Triangle Support Partnership, Honolulu, Hawaii, pp. 1-94.
Anon, 2012. National Fisheries Strategy (2012-2025). Ministry of Fish Wealth, Republic of Yemen. UNDP and IFAD, pp. 1–42.
Anon, 2015. PaFF Programme: Support to Community Forestry and Fisheries in Kratie and Stung Treng Provinces. In: Scaling up under a new Partnership Programme with Switzerland; Support to Community Forestry and Fisheries in Kratie and Stung Treng Provinces; Organised by: The Non-Timber Forest Products Exchange Programme Asia (NTFP-EP), in association with Representatives from the Partners for Forestry and Fisheries Cambodia (PaFF) Programme together with delegates from the provincial governments of Kratie, Stung Treng and of the Forestry and Fishery Administrations and the Swiss Agency Cooperation Office in Cambodia.
Anon, 2016. Blue Peace in the Middle East: Mekong learning journey in Cambodia and Lao PDR. Strategic Foresight Group and Mekong River Commission: Supported by Swedish International Development Cooperation Agency (Sida), pp. 1–60.
Anon, 2018. Hong Kong: The facts in Agriculture and Fisheries. pp. 1–4.
Arnason, R., 2000. Economic Instruments to Achieve aquaculture field and cultural management in South Ecosystem Objectives in Fisheries Management, Korea. ICES J. Mar. Sci. 57, 742–751.
Andreasson, A., 2007. In: Dickson, M., Brooks, A. (Eds.), Community management as part of the inland capture fisheries strategy in Bangladesh. The World Fish Centre Bangladesh and South Asia Office, Dhaka, Bangladesh, pp. 24-25. CBFM-International Conference on Community-Based Approaches to Fisheries Management. The World Fish Centre, Conference Proceedings, 75 Print: pp. 38; CD-ROM, p. 337.
Atapattu, A.R., 1987. Territorial use rights in fisheries in Sri Lanka. In: Symposium on the Exploitation and Management of Marine Fishery Resources in Southeast Asia. Indo-Pacific Fisheries Commission, RAPA Report 10, pp. 379–401.
Ayyappan, S., Diwan, A.D., 2004. Fisheries potential of India: challenges ahead. Fish. Chimes 20 (11), 32–36.

Bailey, C., Zerner, C., 1992. Community-based Fisheries Management Institutions in Indonesia. In: Pollnac, R.B., Bailey, C., Poernomo, A. (Eds.), Paper presented at the 1992 Meetings of the Society for Applied Anthropology, Memphis, 25-29 March, 1992. Local Management of Fisheries Resources in Indonesia: Opportunities and Constraints. Contributions to Fishery Development Policy in Indonesia; Central Research Institute for Fisheries; Agency for Agricultural Research and Development; Ministry of Agriculture, Jakarta, pp. 38–56.
Baird, I., 2000. Integrating Community-Based Fisheries Co-Management and Protected Areas Management in Lao PDR: Opportunities for Advancement and Obstacles to Implementation. IIED, Evaluating Eden Series Discussion Paper No. 14, pp. 1–22.
Balasuriya, L.K.S.W., Tilak, C.R., Kumarasiri, W.S.A.A.L., Sirisena, H.K.G., 1983. Induced breeding of Indian carps in Sri Lanka. J. Inland Fish. 2, 63–67.
Banerjea, S.M., 1967. Water quality and soil condition of the fish ponds in some States of India in relation to fish production. Indian J. Fish 14 (1–2), 115–144.
Barber, C.V., Vaughan, R.P., 1997. Policy reform and community-based programmes to combat cyanide fishing in Philippines. World Resources Institute and International Marinelife Alliance, The Philippines, SPC Live Reef Fish Inform. Bull. 3, 26–35.
Barbhuiya, A.H., Das, B., Darlong, L., Tarafdar, R.G., Sharma, R., Sharma, S.S., Datta, S., Das, B., Barman, R., Deb, S., Saha, B., Kar, D., 2009. Fish Biodiversity in certain rivers of Tripura. Environ. Ecol. 27 (1), 222–227.
Barbhuiya, A.H., Kar, D., 2009. Status of *Tor tor* mahseer in river Barak in North-East India with a note on their morphometry and meristics. J. Natcon 21 (2), 239–246.
Barbhuiya, A.H., Kar, D., 2012. Treatise on Mahseer Fishes. pp. vii + 293, Dominant Publishers, New Delhi.
Barbhuiya, M.H., Kar, D., 2002. Biology of rice weevil, *Sitophilus oryzae* L. (Coleoptera: Curculionidae) in stored wheat, *Triticum vulgare* and its control. Environ. Ecol. 20 (3), 700–702.
Barman, R.C., Kar, D., Dana, S.S., Majumdar, D., 2012. Role of availability of critical aquaculture inputs and community-managed markets on community-based floodplain wetland management. J. Appl. Nat. Sci. 4 (2), 164–166.
Barman, R.C. (2004). Community-based fisheries management with peoples participation for sustainable development of Beel fisheries. M.F.Sc Thesis, West Bengal University of Animal and Fisheries Sciences, Mohanpur, West Bengal (Nadia).
Barman, R.C., Dana, S.S., Bhattacharjya, B.K., Kar, D., Barbhuyan, A.H., 2008. Effect of situational variables on knowledge level of fishermen for sustainable development of Beel Fisheries. Environ. Ecol. 26 (4B), 2092–2094.
Barman, R.C., Kar, D., 2014. Concept definition and approaches of community based Wetlands Fishers and management. In: Kar, D. (Ed.), Research Frontiers in Wetlands, Fishery and Aquaculture. Dominant Publishers and Distributors Pvt. Ltd., New Delhi, pp. 291–310. xi + 320.
Barman, R.C., Kar, D., 2015. Role of Education and Social Participation on community-based Floodplain Wetland Management. ESTIA 1, 16–22. ISSN: 2454-1451.
Bartlett, M.S., 1954. A note on the multiplying factors for various chi-square approximations. J. R. I. State Dent. Soc. 16 (series B), 296–298.
Baticados, D., Agbayani, R., 1998. Case study of institutional arrangements in the fisheries co- management of Malalison, Island, Central Philippines, 22-94. Southeast Asian Fisheries Development Center, Iloilo.
Baugh, 1965. (Clinostomatidae) and one new species of genus *Allocreadium* Looss, 1900 (Allocreadiidae). Biospectra, 1–6.
Baylon, C.C., 2001. Micro-enterprise Development in Selected Fishing Communities in the Province of Iloilo, Philippines. Conference Paper 17; College of Fisheries and Ocean Sciences, University of the Philippines in the Visayas, Miagao, Iloilo, Philippines, pp. 1–18.
Benchakarn, M., 1972. Fish stocking in Ubolratana Reservoir. Annual Report of Ubolratana Fishery Development Unit. National Inland Fisheries Institute, Bangkok (In Thai).
Benchakarn, M., Nookuor, A., 1988. Increase efficiency of stocking fi-sh in the fingerling of rohu by pen culture in Ubolratana Reservoir. Annual Report. Ubolratana Reservoir Fisheries Development Station, Department of Fisheries, Bangkok.
Bennet, G.W., 1962. Management of Artificial Lakes and Ponds. Reinhold, New York, pp. 1–283.
Berkes, F., 1986. Local-level management and the commons problem: A comparative study of Turkish coastal fisheries. Mar. Policy 10, 215–229.
Berkes, F. (Ed.), 1989. Common Property Resources: Ecology and Community-Based Sustainable Development. Belhaven Press, London.
Berkes, F., 1994. Co-management: Bridging the two solitudes. North. Perspect. 22, 18–20.
Berkes, F.; Mahon, R.; McConney, P.; Pollnac, R.; Pomeroy, R. (2001). Managing small-scale fisheries: Alternative directions and methods. International Development Research Centre (Ottawa).
Berkes, F., 2002. Cross-scale institutional linkages: Perspectives from the bottom-up. In: Ostrom, E., Dietz, T., Dolsak, N., Stern, P.C., Stonich, S., Weber, E.U. (Eds.), Drama of the Commons. National Academy Press, Washington, DC, pp. 293–321.
Bernacsek, G.M., 1997. Large Dam Fisheries of the Lower Mekong Countries. Report No. 97023, Mekong River Commission, Vientiane.
Berry, S., 1989. Social institutions and access to resources. Africa 59 (1), 41–55.
Bhandari, B., 1998. An inventory of Nepal's wetland. pp. xvi + 309, Final Report, IUCN, Nepal, Wetlands and Heritage Unit, Kathmandu (Nepal).
Bhandari, B., 2006. Wetlands and culture. In: The Proceedings of the Seminar on Cultural aspects of Nepal's Wetlands, Kathmandu; Forum for Ecosystem Management (FEM), p. 120.
Bhatta, R., Bhat, M., 1998. Impacts of aquaculture on the management of estuaries in India. Environ. Conserv. 25 (2), 109–121.
Bhukaswan, T., 1985. The Nam Pong Basin (Thailand). In: FAO Fishery Technical Paper, 265. United Nations (Rome), Food and Agriculture Organization of the.
Bhukaswan, T., 1987. Inland fisheries. In: Thai fisheries in the future. SEAFDEC, Samut Prakran (In Thai).
Bhukaswan, T., Chookajorn, T., 1988. Present status of the reservoir fi- sheries in Thailand. In: De Silva, S.S. (Ed.), Reservoir Fisheries Management and Development in Asia. Interntional Development Research Center, Ottawa.
Binky, K., Kar, D., 2010. Intensity of Parasitisation among Fishes of Karbhala wetland in Cachar district of Assam. Life Sci. Bull. 7 (2), 233–234.
Binky, K., Kar, D., 2011. Ichthyospecies Diversity of Karbhala Wetland in Cachar district of Assam. Environ. Ecol. 29 (1), 17–19.
Binky, K., Shomorendra, M., Kar, D., 2011. Nematode Parasites of Karbhala Wetland in Silchar, Assam. Biol. Forum 3 (2), 18–21.
Binky, K., Shomorendra, M., Kar, D., 2012. Redescription of two known species of genus *Metaclinostomum* Pandey andRedescription of two known species of genus Metaclinostomum Pandey and Baugh, 1965 (Clinostomatidae) and one new species of genus Allocreadium Looss, 1900 (Allocreadiidae). Biospectra, 1–6.
Binky, K., Singh, M.S., Kar, D., 2015. A Treatise on the Helminth Fauna of Fishes. pp. xxvi + 262, Asiatic Publishing House, New Delhi.

Blaikie, P., 1989. Environment and access to resources in Africa. Africa 59 (1), 18–40.
Blair, E.H., Robertson, J.A., 1903-09. The Philippine Islands. vol. 55 A.H. Clark Co., Ohio, pp. 1493–1898.
Bradshaw, A.D., 1996. Underlying principles of restoration. Can. J. Fish. Aquat. Sci. 53 (Suppl. 1), 3–9.
Brandt, S.B., Hartman, K.J., 1993. Innovative approaches with bioenergetics models: Future applications to fish ecology and management. Trans. Am. Fish. Soc. 122, 731–735.
Brehm, 1953. Some aquatic fauna from Kalipokhari. Eastern Nepal.
Buyali, M.E., 2001. Community-based cooperative fisheries management for Lake Victoria in Tanzania. Ministry of Natural Resources and Tourism; Fisheries Division; Dar-es-salaam, Tanzania, p. 46, University of Iceland, Skulagata 4, 120 Reykjavik (Iceland).
Cairn Jr., J., 2003. Ethical issues in ecological restoration. Ethics Sci. Environ. Polit., 50–61.
Carpenter, S.R., Kitchell, J.F., 1988. Consumer control of lake productivity. BioScience 38, 764–769.
Chabjinda, K., 1996. Cage culture of channel catfish (Ictalurus punctatus) in reservoir. Technical Paper No. 40/1996. Inland Fisheries Division, Department of Fisheries (Bangkok.) (In Thai).
Chabjinda, K., Teanchareon, P., 1991. Preliminary study on cage culture of *Clarias gariepinus* in reservoir. In: Proceeding of the seminar on fi sheries in 1991. National Inland Fisheries Institute, Bangkok (In Thai).
Chakraborty, M., 1998. Conflict and participation in community-based fresh water resource management: The case of Chandrabhaga stream in Garhwal Himalaya. The World Bank/WBI's CBNRM Initiative, Uttar Pradesh.
Chakrabarty, R.D., Samaranayake, R.A.D.B., 1983. Fish culture in seasonal tanks in Sri Lanka. J. Inland Fish. 2, 125–140.
Chambers, R., 1997. Whose Reality Counts: Putting the First Last. Intermediate Technology Publications, London.
Chan, K.L., Tseng, W.Y., 1982. A comparative study of the benthic fishes of Tolo Harbour and the northeastern waters of Hong Kong. In: Morton, B., Tseng, C.K. (Eds.), The Marine Flora and Fauna of Hong Kong and Southern China. Hong Kong Univ. Press, pp. 801–814.
Chan, L.K (Compiler) (1984). Hong Kong Fish List. Hong Kong Agriculture Fisheries Department.
Chan, L.K., Wong, S.W., 1989. Identification guide to fishes of Hong Kong. Hong Kong Agriculture and Fisheries Department, Fisheries Branch, Capture Fisheries Division, pp. 55.
Chan, W.L., 1965. A new anacanthobatid skate of genus *Springeria* from the South China Sea. Jpn. J. Ichthyol. 13, 40–45.
Chan, W.L., 1966a. *Oxymetopon compressus*, a new eleotrid fish from Hong Kong. Jpn. J. Ichthyol. 14, 1–3.
Chan, W.L., 1966b. Notes on opisthognathid jaw fishes from Hong Kong, with the description of a new species. Jpn. J. Ichthyol. 14, 9–11.
Chan, W.L., 1966c. *Cocotropus hongkongiensis*, a new species of the Aploactidae from Hong Kong. Jpn. J. Ichthyol. 14, 12–16.
Chan, W.L., 1968. Marine fishes of Hong Kong, Part I. Hong Kong Govt. Press, Hong Kong, p. 128.
Chan, W.L., 1970. A new genus and two new species of commercial snappers from Hong Kong. Hong Kong Fish. Bull. 17, 19–38.
Chandrasoma, J., 1986. Trends in reservoir fisheries in Sri Lanka with special reference to *Etroplus suratensis*, *Tilapia rendalli* and *Oreochromis niloticus*. In: Maclean, J.L., Dizon, L.B., Hosillos, L.V. (Eds.), The First Asian Fisheries Forum. Asian Fisheries Society, Manila (Philippines).
Chandrasoma, J., 1992. Performance of rohu, *Labeo rohita* (Ham.), in some perennial and seasonal reservoirs of Sri Lanka: an evaluation. J. Appl. Ichthyol. 8, 228–233.
Chandrasoma, J., Kumarasiri, W.S.A.A.L., 1986. Observations on polyculture of fish in seasonal tanks in Ratnapura and Moneragala districts in Sri Lanka. J. Inland Fish. 3, 49–55.
Chang, X., 1995. Preliminary studies on particulate organic detritus in Heilongtan Reservoir. Reserv. Fish. 3, 24–27 (in Chinese).
Chang, I.K.Z., Lee, J.H., Lee, J.B., 2011. An Integrated Fisheries Risk Analysis Method for Ecosystems (IFRAME) approach for assessing, forecasting and managing Korean fisheries under a changing climate. In: Proc. Cook Island Workshop, pp. 1–48.
Chen, Q., Zheng, B. (Eds.), 1987. Systematic Synopsis of Chinese fishes. vols. 1–2. Science Press, Beijing, p. 1458 (in Chinese).
Chen, X., 2006. Practice and discussion on investigation and development of fisheries resources in You Xi Jie Mian Reservoir in Fujian province. J. Fujian Fish. 1, 84–86.
Chong, C.K., 1984. The demersal fishery resources in Hong Kong waters. Agriculture and Fisheries Department, Fisheries Branch, Capture Fisheries Division.
Choudhury, A.B., 1998. Wetland Ecology Resources, Research and Conservation: An Indian Scenario. MEPS Publishers, Kalyani, Nadia (West Bengal).
Chris, F. (1991). Notes on Chaos in Fisheries Management: by: Estellie Smith. Memorial University of Newfoundland, MAST, Vol. 3(2).
Christopherm, D., Harley, G., Hughes, A.R., Hultgren, K.M., Miner, B.G., Sorte, C.J.B., Thornber, C.S., Rodriguez, L.F., Tomanek, L., Williams, S.L., 2006. Ecol. Lett. 9, 228–241.
CIFRI, 1986. Annual Report. Central Inland Fisheries Research Institute, Barrackpore (West Bengal).
CIFRI, 2000. Ecology and Fisheries in Beels of Assam. Bulletin No. 104, Central Inland Fisheries Research Institute, Barrackpore (India), p. 48.
CIFRI, 2006. Annual Report. Central Inland Fisheries Research Institute, Barrackpore (West Bengal).
Claridge, G.F., Sorangkhoum, T., Baird, I.G., 1997. Community Fisheries in Lao PDR: A survey of Techniques and Issues. IUCN-The World Conservation Union, Vientiane (Lao PDR).
Clarke, S., Wai-yin, A.L., Mak, Y.M., Kennish, R., Nigel Haggan Clarke, S., Leung, A.W.-Y., Mak, Y.M., Kennish, R., Haggan, N., 2002. Consultation with local fishers on the Hong Kong artificial reefs initiative. ICES J. Marine Sci. 59, 171–177.
Coates, D., 2002. Inland capture fishery statistics of Southeast Asia: Current status and information needs. FAO RAP Publication 2002/11, Food and Agriculture Organization of the United Nations, Bangkok.
Collette, B.B., Su, J., 1986. The halfbeaks (Pisces: Beloniformes, Hemiramphidae) of the Far East. Proc. Acad. Nat. Sci. Philadelphia 138, 250–301.
Cooper, B.L., 1980. Fisheries management in streams. In: Lackey, R.T., Nielsen, L.A. (Eds.), Fisheries Management. Blackwell Scientific, Boston, pp. 297–322.
Costa, H.H., Abayasiri, R., 1978. The limnology of Colombo (Beira) lake 7. Food and feeding ecology of the fish, *Tilapia mossambica*. Spolia Zeylanica 32, 94–110.
Costa, H.H., de Silva, P.K., 1995. Limnological research and training in Sri Lanka: State of the art and future needs. In: Gopal, B., Wetzel, R.G. (Eds.), Limnology in Developing Countries. vol. 1. International Association for Limnology Publication, International Scientific Publication, New Delhi, (India).

Costa-Pierce, B.A., 1998. Constraints to the sustainability of cage aquaculture for resettlement from hydropower dams in Asia: An Indonesian case study. J. Environ. Dev. 7, 333–368.

Cowx, I.G., 1994. Stocking strategies. Fish. Manage. Ecol. 1, 15–30.

Crocombe, R.G., 1964. Land Tenure in the Cook Islands. Melbourne University Press, Melbourne.

Cycon, D.E., 1986. Managing fisheries in developing nations: A plea for appropriate development. Nat. Resour. J. 26 (1), 14.

Dai, Z., Xu, X., 1996. An approach to eutrophication classification standard in reservoirs. Reserv. Fish. 2, 3–6 (in Chinese).

DaN, A.L., 1989. Traditional Environmental Knowledge and Resource Management in New Caledonia. In: Johannes, R.E. (Ed.), Traditional Ecological Knowledge: A Collection of Essays. International Union for the Conservation of Nature, Gland, pp. 45–53.

Dana, S.S., 1987. A study of factor affecting the attitude of livestock owners towards artificial insemination in cattle. Indian Veterinary Research Institute, Izatnagar (Uttar Pradesh), p. 127. M.V.Sc.Thesis.

Darlington, P.J. (Jr.), 1957. Zoogeography: The Geographical Distribution of Animals. John Wiley and Sons, New York, p. 675.

Das, B., Kar, D., 2011a. Essentials of Limnology and Fishery Science. (New Delhi), Mangalam Publishers, p. 190.

Das, B.K., Kar, D., 2011b. Habitat mapping, spatial analysis of fish diversity of River Subansiri during winter season in Assam and Arunachal Pradesh (India). Environ. Ecol. 29 (4A), 1948–1951.

Das, B.K., Singh, N.R., Dutta, B., Kar, D., 2013. Length-weight relationship of *Labeo rohita* (Hamilton-Buchanan) and *Labeo gonius* (Hamilton-Buchanan) from Sone Beel, the biggest wetland of Assam, India. J. Environ. Res. Dev. 8 (4), 6613–6618.

Das, B.K., Boruah, P., Kar, D., 2014a. Potential Application of Geographic Information System (GIS) and Remote Sensing (RS) in River Analysis. Int. J. Recent Sci. Res. 5 (4), 828–830.

Das, B.K., Boruah, P., Kar, D., 2015e. Fish diversity and drainage analysis of River Siang, East Siang District of Arunachal Pradesh. Proceeding of the National Conference on Conservation of Natural Resources and Biodiversity for Sustainable Development-2014. J. Biosci. Discov. 6 (1 - I Special), 16–20.

Das, B.K., Boruah, P., Kar, D., 2014b. Fish diversity and Habitat Mapping of River Siang in Arunachal Pradesh using Remote Sensing and GIS. pp. 13-2, In: Mishra, G.C. (Ed.), Innovative Energy Technology Systems and Environmental Concerns: A Sustainable Approach. Research India Publication, New Delhi.

Das, B.K., Ghosh, A., Boruah, P., Kar, D., 2015b. Studies on the Physico-chemical parameters of water and Drainage analysis using RS and GIS Technology of River Siang in Arunachal Pradesh. In: Biodiversity Conservation and Sustainable Use. Print World Print and Publication, Jorhat, Assam (India), pp. 197–204.

Das, B.K., Kar, D., 2015a. Drainage Pattern and their Geological Significance of River Siang in Arunachal Pradesh, India. ESTIA 1, 32–38.

Das, B.K., Kar, D., 2015b. Habitat Inventory Parameters and Drainage Analysis of River Siang in Arunachal Pradesh. Int. J. Mar. Sci. 5 (48), 1–7 (Canada).

Das, B.K., Singh, N.R., Boruah, P., Kar, D., 2015c. Fishing devices of the River Siang in Arunachal Pradesh, India. J. BD Fish 4 (31), 571–576.

Das, B.K., Kar, D., 2015c. Digital Elevation Model (DEM) and the Drainage Analysis of River Siang in Arunachal Pradesh, India. J. Remote Sens. (In press).

Das, B.K., Singh, N.G., Das, S., Kar, D., 2015f. Length-weight relationship (LWR) of *Channa striata* (Bloch, 1793) of River Siang, Arunachal Pradesh. Int. J. Env. Nat. Sci. 7, 1–6 (ISSN 2349-3763).

Das, B.K., Kar, D., 2015d. Physico-chemical parameters and drainage types of River Siang in Arunachal Pradesh, India. pp. 53–56, In: Mishra, G.C. (Ed.), Conceptual Framework and Innovations in Agroecology and Food Sciences, pp. iii + 82. Krishi Sanskriti Publications, New Delhi.

Das, B.K., Singh, N.G., Das, S., Kar, D., 2015g. Length-weight relationship (LWR) of *Channa striata* (Bloch, 1793) of River Siang, Arunachal Pradesh. Int. J. Env. Nat. Sci. 7, 1–6.

Das, B.K.; Boruah, P.; and, Kar, D. (2015). Ichthyofaunal diversity of Siang River in Arunachal Pradesh, India. Proc. Zool. Soc., https://doi.org/10.1007/s12595-015-0155-6.

Das, B.K., Boruah, P., Kar, D., 2016b. Habitat Mapping and Fish Diversity of River Siang in Arunachal Pradesh. Glob. J. Environ. Sci. Res. 3 (1), 9–18. Taiwan.

Das, B.K., Boruah, P., Kar, D., 2014c. Icthyo-Faunal Diversity and Conservation Strategies of Siang River in Arunachal Pradesh. In: Singha, D.N.G.C. (Ed.), Natural Resources: Utilizing and Conserving for Human Welfare and Sustenance. Global Publishing House, Visakhapatnam, (India), pp. 139–151.

Das, B.K., Boruah, P., Kar, D., 2014d. Physico-Chemical Parameters of River Siang in Arunachal Pradesh, India. In: Banerjee, P., Rajguru, U., Sarkar, C.R. (Eds.), Proceedings of the National Seminar on Advances in Physical and Biological Sciences; Jagiroad College, Morigaon, Assam, (India), pp. 78–88.

Das, B.K., Kar, D., 2016a. Diversity of zooplankton of River Siang of Arunachal Pradesh, India. Oceanogr. Fish. J. California, (USA).

Das, N., Das, B.K., Das, S., Singh, N.R., Kar, D., 2015a. Preliminary studies on Ichthyofauna diversity and Physico-Chemical Parameters of River Bhogdoi during winter and Pre-Monsoon Season in Assam. IOSR J. Agric. Vet. Sci. 8 (6), 35–41. Ver. I.

Das, N.G., Hossain, M.S., 2005. Livelihood and Resource Assessment for Aquaculture Development in Waterlogged Paddy Lands: Remote Sensing, GIS and Participatory Approaches. Greater Noakhali Aquaculture Eextension Project and Institute of Marine Sciences; University of Chittagong, Chittagong.

Das, N.G.D., Hossain, M.S., Islam, M.S., 2009. Waterlogged Area as New Horizon for Aquaculture Development: A Golden Dream to the Rural Communities of Begumgonj, Noakhali. Bang. J. Mar. Sci. Fish. 1 (1), 47–62.

Das, P., Kar, D., 2016b. Composition, abundance and diversity of zooplankton population from three different wetlands from Barak valley, Assam. Int. J. Appl. Nat. Sci. 5 (5), 41–46.

Das, P., Kar, D., 2016c. Study of piscine and aquatic macrophytic biodiversity in wetlands of Barak valley, Assam. A study on aquatic biodiversity of wetlands of Barak valley, Assam. J. Appl. Life Sci. Int.

Das, S., Das, T., Das, B.K., Kar, D., 2015d. Study of fish diversity and physico-chemical characteristics of River Singla. Int. J. Env. Sci. 5 (6), 881–887.

Das, S., Das, T., Das, B.K., Kar, D., 2014. Length-Weight Relationship and Condition Factor in *Barilius sacra* of River Torsa in West Bengal. Int. J. Sci. Res. Rev. 3 (3), 65–69.

Das, S., Das, B.K., Kar, D., 2015h. Study of fish diversity and physico-chemical parameters of River Singla, Assam, India. Int. J. Environ. Sci. 5 (6), 881–887.

Das, S., Das, B.K., Kar, D., 2015i. Length-weight relationship and relative condition factor in *Xenentodon cancila* of River Mat in Mizoram. Int. J. Biol. Biol. Sci. 4 (2) (ISSN 2327-3062). 027-031 (USA).
Das, S., Kar, D., 2016d. A glimpse into the fish diversity of River Tuirial in Mizoram. Indian J. Appl. Res.
Das, S., Purkayastha, D., Kar, D., 2016a. Length-weight relationship and Relative Condition factor in *Puntius ticto* of River Singla in the Karimganj district of Assam, India. Int. J. Appl. Res.
Das, S., Kar, D., 2016e. A preliminary study on the ichthyospecies diversity of River Tlawng in Mizoram, India. Int. J. Appl. Res. ISSN 2394-7500.
Datta, S.K., Pal Singh, S., Chakrabarti, M., Biswas, S., Sah, B., 2010. A Perspective on Fisheries Sector Interventions for Livelihood Promotion. Centre for Management in Agriculture, Indian Institute of Management, Ahmedabad, pp. 1–32. W.P. No. 2010-02-03.
Department of Fisheries, 2007. Fisheries statistics of Thailand. Available at http://www.fisheries.go.th/it-stat/.
Desai, V.R., Singh, R.K., 1979. Occurrence of thermocline in Rihand Reservoir of U.P. and its depth criterion in relation of reservoir level. J. Inland Fish. Soc. India 11 (2), 69–72.
De Silva, S.S., 1985. Body condition and nutritional ecology of *Oreochromis mossambicus* (Pisces: Cichlidae) populations in man-made lakes in Sri Lanka. J. Fish Biol. 27, 621–633.
De Silva, S.S., 1987. Stocking practices in Sri Lanka. In: Petr, T. (Ed.), Reports and papers presented at the Indo-Pacific Fisheries Commission Expert Consultation on the Inland Fisheries of Larger Indo-Pacific Countries. Food and Agriculture Organization of the United Nations, Rome, pp. 223–232. FAO Fisheries Report, 371 (Supplement).
De Silva, S.S., Sirisena, H.K.G., 1987. New fish resources of reservoirs in Sri Lanka: Feasibility of introduction of a subsidiary gillnet fishery for minor cyprinids. Fish. Res. 6, 17–34.
De Silva, S.S., 1988. Reservoirs of Sri Lanka and their fisheries. In: FAO Fisheries Technical Paper, 298. Food and Agriculture Organization of the United Nations, Rome.
De Silva, S.S., Sirisena, H.K.G., 1989. New fish resources of reservoirs in Sri Lanka. 3. Results of commercial-scale trials and yield estimates of a gillnet fishery for minor cyprinids. Fish. Res. 7, 279–287.
De Silva, S.S., Moreau, J., Amarasinghe, U.S., Chookajorn, T., Gurrero, R.D., 1991. A comparative assessment of the fisheries in the lacustrine inland waters in three Asian countries based on catch and effort data. Fish. Res. 11, 177–189.
De Silva, S.S., Lin, Y., Tan, G., 1992. Possible yield-predictive models based on morphometric characteristics and stocking rates for three groups of Chinese reservoirs. Fish. Res. 13, 369–390.
De Silva, S.S., Amarasinghe, U.S., 1996. Reservoir fisheries in Asia. In: De Silva, S.S. (Ed.), Perspectives in Asian fisheries: A volume to commemorate the 10th anniversary of the Asian Fisheries Society. Asian Fisheries Society, Manila, pp. 189–216.
De Silva, S.S., Wijenayake, W.M.H.K., Gunaratne, A.B.A.K., Amarasinghe, U.S., 2005. Use of GIS tools to develop a scale for the selection of non-perennial reservoirs for culture-based fisheries activities. In: Nishida, T., Kailola, P.J., Hollingworth, C.E. (Eds.), GIS Spatial Analyses in Fishery and Aquatic Sciences. Vol. 2. Fishery and Aquatic GIS Research Group, Japan.
Devi, M.B., Das, P., Das, U., Kar, S., Aditya, G., Kar, D., 2016. Rotifer Species Assemblages in Three Freshwater Habitats of Manipur, India. In: Proceedings of the Zoological Society. Springer, pp. 1–9, https://doi.org/10.1007/s12595-016-0201-z.
Dey, S.C., 1973. Studies on the Distribution and Taxonomy of the Ichthyofauna of the hill streams of Kamrup-Khasi-Garo Regions of Assam with special reference to the Functional morphology of some rheophillic Fishes. In: D.Sc. Thesis, xi + 299. University of Calcutta, India.
Dey, M.M., Kanagaratnam, U., 2007. In: Dickson, M., Brooks, A. (Eds.), Community-Based Management of Small Scale Fisheries in Asia: Bridging the Gap Between Fish Supply and Demand. CBFM-International Conference on Community Based Approaches to Fisheries Management, The World Fish Center, Conference Proceedings, 75 Print: pp. 38; CD-ROM: pp. 337. The World Fish Centre, Bangladesh and South Asia Office, Dhaka, Bangladesh, p. 24.
Dey, S.C., 1981. Studies on the Hydrobiological Condition of Some Beels of Kamrup District of Assam and Their Bearing on Fish Production. North-Eastern Council (Shillong), p. 177. Final Technical Report.
Dey, S.C., Kar, D., 1987. Physico-chemical complexes of water and soil in Sone, an ichthyologically potential tectonic lake of Assam. J. Assam Sci. Soc. 30 (1), 1–11.
Dey, S.C., Kar, D., 1989a. Aquatic macrophytes of Lake Sone in Assam. Environ. Ecol. 7 (1), 253–254.
Dey, S.C., Kar, D., 1989b. Fishermen of Lake Sone in Assam: their socio-economic status. Sci. Cult. 55, 395–398.
Dey, S.C., Kar, D., 1989c. An account of *Hilsa ilisha* (Hamilton) of Lake Sone in the Karimganj district of Assam. Bangladesh J. Zool. 17 (1), 69–73.
Dey, S.C., Kar, D., 1990. Fish yield trend in Sone: a tectonic lake of Assam. Matsya 15–16, 39–43.
Dey, S.C., Kar, D., 1994. Phytoplankton dynamics of Lake Sone in Assam. Bull. Life Sci., IV, pp. 49–54.
DFID, 1999. Sustainable Livelihoods Guidance Sheets. Department for International Development, London, SW1E SJL (UK).
Dhar, N., Goswami, M.M., Kar, D., 2004a. Aquatic macrophytes in Baskandi Lake, an Oxbow Lake of Barak basin, Assam. Environ. Ecol. 22 (Spl-4), 606–608.
Dhar, N., Goswami, M.M., Kar, D., 2004b. Co-operative Movement: Need of the hour in Capture Fishery Sector. In: Goswami, P.J. (Ed.), Cooperative Movement in Assam: Problems and Prospects, pp. 113–116.
DoF, 2012. Jatiya Matshya Soptaho, 2012: Department of Fisheries. Ministry of Fisheries and Livestock, The Government of Peoples Republic of Bangladesh, Dhaka.
DOFD, 2006. Country Profile–Nepal 2005/2006: Fisheries Sub-sector. Directorate of Fisheries Development (DOFD), Kathmandu (Nepal).
Duangswasd, I.M., Janesirisak, S., 1973. Fish stocking in Ubolratana Reservoir. In: Annual Report of Ubolratana Fishery Development Unit. Bangkok, National Inland Fisheries Institute (In Thai).
Dubey, G.P., Chatterjee, S.N., 1977. Case study of Gandhisagar reservoir, Madhya Pradesh, India. Proc. Indo-Pacific Fish. Council 17 (3), 232–246.
Dutta, B., Kar, D., 2011. Aquaculture in Dibrugarh district of Assam. Environ. Ecol. 27 (2), 1520–1523.
Dutta, B., Biswas, S.P., Kar, D., 2013. Hydrobiological parameters of ponds of NR Hatchery at Lahowal at Dibrugarh in Assam. Environ. Ecol. 31 (2B), 844–846.
Dutta, B., Boruah, P., Kar, D., 2015b. Fishes of the Towkak River: With special emphasis on the habitat requirements. Int. J. Manage. Social Sci. 3 (3), 38–44.
Dutta, B., Baruah, L., Baruah, P., Kar, D., 2015a. Studies on the potential Habitat Parameters for the fishes of the Towkak River in Assam and Nagaland, India. In: Biodiversity Conservation, Crisis and Sustainable Use. Print World Print and Publication, Jorhat, Assam, India, pp. 52–57.

Dutta, B., Das, N., Kar, D., 2016. The Fish Catching Devices with their Efficacy and Cost-benefit Analysis in the Towkak River in Assam and Nagaland, India. Fish. Aquacult. J. 07 (1), 1000156. https://doi.org/10.4172/2150-3508.1000156 (USA).
Elizabeth, A.F., Anthony, D.M.S., Craig, R.J., 2003. Mortality and predation in ecosystem models: Is it important how these are expressed? Ecol. Model. 169, 157–178.
Emmerson, D.K., 1975. Orders of Meaning: Understanding Political Change in a Fishing Community in Indonesia. In: Paper presented at the Annual Meeting of the American Political Science Association, 2-5 September.
ERM, 1998. Fisheries Resources and Fishing Operations in Hong Kong Waters. In: Final Report to the Hong Kong SAR Government, Agriculture and Fisheries Department. Environmental Resources Management – Hong Kong Ltd.
Eschmeyer, W.N., 1990. Catalog of the genera of recent fishes. Academy of Sciences (San Francisco), California, p. 697.
Eschmeyer, W.N. (Ed.), 1998. Catalog of Fishes. Vols. 1-3. California Academy of Sciences, (San Francisco).
FAO, 1983. The state of Food and Agriculture. In: World review: The situation in Sub-Saharan Africa. Women in developing agriculture, pp. 1–240.
FAO, 1999. Food Balance Sheet, Data and Statistics Service (1997 data). United Nations Food and Agriculture Organization, Rome.
FAO, 2003. The ecosystem approach to fisheries: Issues, terminology, principles, institutional foundations, implementation and outlook. FAO Fish. Techn. Pap. 443, 1–76.
FAO, 2010. The State of Food Insecurity in the World: Addressing Food Insecurity in Protracted Crises. Food and Agriculture Organization of the United Nations, Rome, pp. 1–62.
FAO, 2011. The State of Food and Agriculture. Food and Agriculture Organization of the United Nations, Rome, pp. 1–160.
FAO/UNDP, 1980. In: Strategy for Development of Aquaculture in Sri Lanka. Report of a Planning Mission, September-October 1980. ADCP/MR/81/15.
FBMA, 2006. China Fisheries Yearbook. Agricultural Press of China, Beijing (in Chinese).
Feeny, D., Berkes, F., McCay, B.J., Acheson, J.M., 1990. The tragedy of the commons: twenty-two years later. Hum. Ecol. 18, 1–19.
Fernando, C.H., 1967. The beach seine fishery in Ceylon's freshwaters. Indo-Pacific Fisheries Council Current Affairs Bulletin. No. 50.
Fernando, C.H., 1984. Reservoirs and lakes of Southeast Asia (Oriental region). In: Taub, F.B. (Ed.), Lake and Reservoir Ecosystems. Elsevier, Amsterdam (The Netherlands).
Fernando, C.H., 1993. Impact of Sri Lankan reservoirs, their fisheries, management and conservation. In: Erdelen, W., Preu, C., Ishwaran, N., Bandara, C.M.M. (Eds.), Ecology and Landscape Management in Sri Lanka. Proceedings of the International and Interdisciplinary Symposium 12-16 March 1990, Colombo, Sri Lanka. Margraf Scientific Books, Weikersheim.
Fernando, C.H., 2000. A view of the inland fisheries of Sri Lanka: Past, present and future. Sri Lanka J. Aquat. Sci. 5, 1–26.
Fernando, C.H., Ellepola, W.B., 1969. A preliminary study of two village tanks (reservoirs) in the Polonnaruwa area with biological notes on these reservoirs in Ceylon. Bull. Fish. Res. Station Ceylon 20, 3–13.
Fernando, C.H., Furtado, J.J., 1975. Reservoir fishery resources of South-East Asia. Bull. Fish. Res. Stn. Sri Lanka 26 (1–2), 83–95.
Fernando, C.H., Holcik, J., 1991. Fish in reservoirs. Int. Rev. Ges. Hydrobiol. 76, 149–167.
Fernando, C.H., Indrasena, H.H.A., 1969. The freshwater fisheries of Ceylon. Bull. Fish. Res. Station Ceylon 20, 101–134.
Fernando, C.H., Weerawardhena, S.R., 2002. Sri Lanka Freshwater Fauna and Fisheries. Volume Publishing, Kitchener, Ontario (Canada).
Ferrow, W., Swar, D.B., 1978. Bathymetric maps from three lakes in Pokkhara valley (Nepal). J. Inst. Sci. Technol. 1, 177–188.
Fisheries Administration (FiA), 2012. Cambodia Fisheries: Annual Report 2012., p. 16.
Fleischer, A., Pizam, A., 1997. Rural tourism in limitations of the conventional fisheries management Israel. Tour. Manag. 18 (6), 367–372.
Flewwelling, P., 2001. Fisheries Management and MCS in South Asia. FAO, Rome, p. 56. FAO/FISHCODE Project, GCP/INT/648/NOR: Field Report C-6 (En).
Foltz, C., Pomeroy, R.S., Barber, C. (Eds.), 1996. Proceedings of the Visayas-wide conference on community-based coastal resources management and co-management. In: Fisheries Co-management Research Project Working Paper No. 4. ICLARM, Manila, pp. 1–115.
Fowler, H.W., 1930a. Notes on Japanese and Chinese fishes. Proc. Acad. Nat. Sci. Philadelphia 81, 597–616. for 1929.
Fowler, H.W., 1930b. A synopsis of the fishes of China. Part I. The sharks, rays and related fishes. Hong Kong Nat. 1, 24–34.
Fowler, H.W., 1930c. A synopsis of the fishes of China. Part I (continued). The sharks, rays and related fishes. Hong Kong Nat. 1, 79–88.
Fowler, H.W., 1930d. A synopsis of the fishes of China. Part I (continued). The sharks, rays and related fishes. Hong Kong Nat. 1, 129–138.
Fowler, H.W., 1930e. A synopsis of the fishes of China. Part I (concluded). The sharks, rays and related fishes. Hong Kong Nat. 1, 177–189.
Fowler, H.W., 1931a. A synopsis of the fishes of China. Part II. The herrings and related fishes. Hong Kong Nat. 2, 69–79.
Fowler, H.W., 1931b. A synopsis of the fishes of China. Part II (continued). The herrings and related fishes. Hong Kong Nat. 2, 111–123.
Fowler, H.W., 1931c. A synopsis of the fishes of China. Part II (concluded). The herrings and related fishes. Hong Kong Nat. 2, 198–208.
Fowler, H.W., 1931d. Studies of Hong Kong fishes, no. 2. Hong Kong Nat. 2, 287–317.
Fowler, H.W., 1932a. A synopsis of the fishes of China. Part III. The eels. Hong Kong Nat. 3, 46–63.
Fowler, H.W., 1932b. A synopsis of the fishes of China. Part III (concluded). The eels. Hong Kong Nat. 3, 126–144.
Fowler, H.W., 1932c. A synopsis of the fishes of China. Part IV. The cats, lizard fishes, green gars, half beaks and flying fishes. Hong Kong Nat. 3, 247–279.
Fowler, H.W., 1933. A synopsis of the fishes of China. Part V. The cods, opahs, flounders, soles, John dories, berycoids, pipe fishes, silversides, mullets, barracudas and thread fishes. Hong Kong Nat. 4, 156–175.
Fowler, H.W., 1934a. A synopsis of the fishes of China. Part V (continued). The cods, opahs, flounders, soles, John dories, berycoids, pipe fishes, silversides, mullets, barracudas and thread fishes. Hong Kong Nat. 5, 54–67.
Fowler, H.W., 1934b. A synopsis of the fishes of China. Part V (continued). The cods, opahs, flounders, soles, John dories, berycoids, pipe fishes, silversides, mullets, barracudas and thread fishes. Hong Kong Nat. 5, 146–155.
Fowler, H.W., 1934c. A synopsis of the fishes of China. Part V (continued). The cods, opahs, flounders, soles, John dories, berycoids, pipe fishes, silversides, mullets, barracudas and thread fishes. Hong Kong Nat. 5, 210–225.
Fowler, H.W., 1934d. A synopsis of the fishes of China. Part V (continued). The cods, opahs, flounders, soles, John dories, berycoids, pipe fishes, silversides, mullets, barracudas and thread fishes. Hong Kong Nat. 5, 304–319.
Fowler, H.W., 1935a. A synopsis of the fishes of China. Part V (continued). The cods, opahs, flounders, soles, John dories, berycoids, pipe fishes, silversides, mullets, barracudas and thread fishes. Hong Kong Nat. 6, 62–77.

Fowler, H.W., 1935b. A synopsis of the fishes of China. Part V (continued). The cods, opahs, flounders, soles, John dories, berycoids, pipe fishes, silversides, mullets, barracudas and thread fishes. Hong Kong Nat. 6, 132–147.
Fowler, H.W., 1935c. A synopsis of the fishes of China. Part V (continued). The cods, opahs, flounders, soles, John dories, berycoids, pipe fishes, silversides, mullets, barracudas and thread fishes. Hong Kong Nat. 6, 276–284.
Fowler, H.W., 1936a. A synopsis of the fishes of China. Part VI. The mackerels and related fishes. Hong Kong Nat. 7, 61–80.
Fowler, H.W., 1936b. A synopsis of the fishes of China. Part VI (continued). The mackerels and related fishes. Hong Kong Nat. 7, 186–202.
Fowler, H.W., 1936c. A synopsis of the fishes of China. Part VI (continued). The mackerels and related fishes. Hong Kong Nat. 7, 271–315.
Fowler, H.W., 1937a. A synopsis of the fishes of China. Part VII. The perch-like fishes. Hong Kong Nat. 8, 124–145.
Fowler, H.W., 1937b. A synopsis of the fishes of China. Part VII (continued). The perch-like fishes. Hong Kong Nat. 8, 249–289.
Fowler, H.W., 1938a. A synopsis of the fishes of China. Part VII (continued). The perch-like fishes. Hong Kong Nat. 9, 58–86.
Fowler, H.W., 1938b. Studies of Hong Kong fishes, no. 3. Hong Kong Nat. 6, 1–52.
Fowler, H.W., 1939a. A synopsis of the fishes of China. Part VII (continued). The perch-like fishes. Hong Kong Nat. 9, 141–164.
Fowler, H.W., 1939b. A synopsis of the fishes of China. Part VII (continued). The perch-like fishes. Hong Kong Nat. 9, 203–220.
Fowler, H.W., 1940a. A synopsis of the fishes of China. Part VII (continued). The perch-like fishes. Hong Kong Nat. 10, 39–62.
Fowler, H.W., 1940b. A synopsis of the fishes of China. Part VII (continued). The perch-like fishes. Hong Kong Nat. 10, 109–121.
Fowler, H.W., 1941. A synopsis of the fishes of China. Part VII (continued). The perch-like fishes. Hong Kong Nat. 10, 205–222.
Fowler, H.W., 1949. A synopsis of the fishes of China. Part VII. The perch-like fishes (continued). J. Hong Kong Fish. Res. Station 2, 3–65.
Fowler, H.W., 1953. A synopsis of the fishes of China. Part VII. The perch-like fishes (continued). Quart. J. Taiwan Museum, 1–77.
Fowler, H.W., 1954. A synopsis of the fishes of China. Part VII. The perch-like fishes (continued). Quart. J. Taiwan Museum, 1–110.
Fowler, H.W., 1956. A synopsis of the fishes of China. Part VII. The perch-like fishes (completed). Quart. J. Taiwan Museum 9, 161–354.
Fowler, H.W., 1958. A synopsis of the fishes of China. Part VIII. The Blenniioid and related fishes. Quart. J. Taiwan Museum 11, 147–339.
Fowler, H.W., 1959. A synopsis of the fishes of China. Part VIII. The Blenniioid and related fishes. Quart. J. Taiwan Museum 12, 67–97.
Fowler, H.W., 1960. A synopsis of the fishes of China. Part IX. The Gobioid fishes. Quart. J. Taiwan Museum 13, 91–161.
Fowler, H.W., 1961a. A synopsis of the fishes of China. Part IX. The Gobioid fishes. Quart. J. Taiwan Museum 14, 49–87.
Fowler, H.W., 1961b. A synopsis of the fishes of China. Part IX. The Gobioid fishes. Quart. J. Taiwan Museum 14, 203–250.
Fowler, H.W., 1962. A synopsis of the fishes of China. Part X. The Gobioid fishes (concluded). Quart. J. Taiwan Museum 15, 1–77.
Franks, T., Cleaver, F., 2007. Water governance and poverty: A framework for analysis. Prog. Dev. Stud. 7 (4), 291–306.
Frid, C.L.J., Clark, R.A., Hall, J.A., 1999. Long-term changes in the benthos on a heavily fished ground off the NE coast of England. Mar. Ecol. Prog. Ser. 188, 13–20.
Froese, R., Pauly, D. (Eds.), 1997. FishBase—a biological database on fish (Software). ICLARM.
Gadgil, M., Iyer, P., 1989. On the diversification of common-property use by Indian society. In: Berkes, F. (Ed.), Common Property Resources: Ecology and Community-Based Sustainable Development. Belhaven Press, London, pp. 240–255.
Gadgil, M., Seshagiri Rao, P.R., Utkarsh, G., Pramod, P., Chatre, A., 2000. New meanings for old knowledge: The people's biodiversity registers programme. Ecol. Appl. 10, 1251–1262.
Gaiger, P.J., 1974. A Provisional List of Inshore Marine Fishes. Agriculture and Fisheries Department. Fisheries Occasional Paper No. 2.
Chandra, G., 2011. Management Regimes and Institutional Arrangement in Floodplain Wetlands Fisheries of Assam: An Evaluation. Indian J. Ext. Educ. 47 (1 and 2), 27–33.
Gao, H., Chen, C., Lin, Y., 1997. Epidemiological investigation of outbreak disease of fishes and its prevention treatment in some reservoirs. Reserv. Fish. 1, 8–10 (in Chinese).
Gao, Y., Xu, Q., et al., 2006. Technical measures for continuous high production and efficiency of fisheries in Huanyuan Reservoir. Reserv. Fish. 26, 44–45.
Garcia, S.M., Charles, A.T., 2007. Fisheries systems and linkages: from clockwork to soft watches. ICES J. Mar. Sci. 64, 580–587.
Garcia, S.M., Zerbi, A., Aliaume, C., Do Chi, T., Lasserre, G., 2003. The Ecosystem Approach to Fisheries: Issues, Terminology, Principles, Institutional Foundation, Implementation and Outlook. FAO Fisheries Technical Paper No. 443, Food and Agriculture Organization of the United Nations, Rome.
Gautam, A.P., Shivakoti, G.P., 2002. Evolution and impacts of community-based forest management in the hills of Nepal. Doctoral thesis, School of Environment, Resources and Development; Asian Institute of Technology, Pathum Thani, Thailand, p. 30.
Gautam, G., 2003. Study on the fish diversity and aquatic life resources of Rupa Lake, Kaski. Dissertation submitted to the Central Department of Zoology, T.U. (Kathmandu), In: Stock Assessment Tools (FiSAT) User's Guide. FAO Computerized Information Series (Fisheries), No. 8. Food and Agriculture Organization of the United Nations, Rome.
Geetarani, B., Shomorendra, M., Vishweshwari, A., Kar, D., Jha, A.N., 2010. Helminthic infection in relation to the host length and sex in some freshwater fishes, *Anabas testudineus* (Bl.) and *Channa gachua* (Ham.). Biospectra 5 (2), 233–236.
Geetarani, B., Shomorendra, M., Kar, D., 2011. Helminth-fauna of Air-breathing Fishes of Utra Lake in Manipur. Environ. Ecol. 29 (1), 176–178.
Gomes, A., 2007. Empowerment of women through fisheries and aquaculture. In: Dickson, M., Brooks, A. (Eds.), Proceedings of the CBFM-2 International Conference on Community Based Approaches to Fisheries Management, Dhaka, Bangladesh, 6–7 March 2007. The WorldFish Centre – Bangladesh Office, pp. 1–4.
Gopal, B., 2005. Does inland aquatic biodiversity have a future in Asian developing countries? Hydrobiologia 542, 69–75.
Government of Kerala, 1993. Inland Fisheries at a Glance—1993. (Trivandrum).
Green, L.W., George, M.A., Daniel, M., Frankish, C.J., Herbert, C.P., Bowie, W.R., et al., 2003. In: Minkler, M., Wallerstein, N. (Eds.), Guidelines for participatory research in health promotion. Community-based participatory research for health, Jossey-Bass (San Francisco), pp. 419–428.
Groombridge, B., Jenkins, M., 1998. Freshwater Biodiversity: A Preliminary Global Assessment. World Conservation Monitoring Centre; WCMC Biodiversity Series No. 8, World Conservation Press, Cambridge, UK.
Gunderson, L.H., Holling, C.S. (Eds.), 2002. Panarchy: Understanding Transformations in Human and Natural Systems. Island Press, Washington, DC.
Gurung, T.B., 2007. Restoration of small lakes through cooperative management: A suitable strategy for poverty-laden areas in developing countries? Lakes Reserv. Res. Manag. 12, 237–246.

Gurung, T.B., Bista, J.D., 2003. Livelihood improvements through fisheries in the Pode Community in Pokhara, Nepal. Stream J. 2, 1–2.
Gurung, T.B., Wagle, S.K., Bista, J.D., et al., 2005. Participatory fisheries management for livelihood improvement of fishers in Phewa Lake, Pokhara, Nepal. Himalayan J. Sci. 3, 47–52.
Gurung, T.B., Wagle, S.K., 2002. Importance of Water Quality for Optimizing in Aquaculture, October 18-19, 2000. Kathmandu. Fisheries Training Centre, Janakpur, Nepal.
Hamilton, F.B., 1822. An Account of the Fishes found in the River Ganges and its Branches. Edinburgh and London, London.
Hayes, F.R., 1957. On the variation in bottom fauna and fish yield in relation to trophic level and lake dimensions. J. Fish. Res. Board Can. 14, 1–32.
Hayes, F.R., Phillips, J.E., 1958. Lake water and sediment, IV. Radiophosphorus equilibrium with mud, plants, and bacteria under oxidised and reduced conditions. Limnol. Oceanogr. 3, 459–475.
Heerdt, G., Hootsmans, M., 2007. Why biomanipulation can be effective in peaty lakes. Hydrobiologia 584, 305–316.
Henderson, H.F., Ryder, R.A., Kudhongania, A.W., 1973. Assessing fishery potentials of lakes and reservoirs. J. Fish. Res. Board Can. 30, 2000–2009.
Henderson, H.F., Welcomme, R.L., 1974. The Relationship of Yield to Morphoedaphic Index and Numbers of Fisherman in African Inland Fisheries. CIFA Occasional Paper, No. 1, Central Institute of Freshwater Aquaculture, India.
Henderson, S., Markland, H.R., 1987. Decaying Lakes: The Origins and Control of Cultural Eutrophication: Detailed Analysis. John Wiley and Sons, New York.
Henocque, Y., Tandavanitj, S., 2011. Fisheries and co-management in practice, Thailand. In: Climate of Coastal Cooperation. Coastal and Marine Union, EUCC, The Netherlands, pp. 102–103.
Hewett, S.W., Johnson, B.L., 1992. Fish Bioenergetics Model 2. Sea Grant Institute; University of Wisconsin, Madison. WIS-SG-92-250.
Hickel, B., 1973. Limnological investigation in lakes of Pokhara Valley, Nepal. Int. Rev. Ges. Hydrobiol. 58, 659–672.
Himalayan Power Consultant, 1889. Feasibility Report on Karnali (Chisapani) Multipurpose Project. 28 Vol. Kathmandu, Nepal.
Hjalager, A. (1996). Agricultural diversification into objective of fisheries management policy from the tourism: evidence of a European community maintenance of fishery order or the fishery adjustment to development program. Tourism Management, fish stock recovery. Also establishment of FSRP that can 17(2): 103-11.
Hodgkiss, I.J., 1988. Fifty common marine food fishes of Hong Kong. Mem. Hong Kong Nat. Hist. Soc. 18, 19–33.
Hodgkiss, I.J., 1996. Species new to Hong Kong. Mem. Hong Kong Nat. Hist. Soc. 20, 237–238.
Hogan, Z.S., 2004. Threatened fishes of the world: *Pangasianodon gigas* Chevey, 1931 (Pangasiidae). Environ. Biol. Fishes 92, 120.
Hollowed, A.B., Bond, N.A., Wilderbuer, T.K., Stockhausen, W.T., A'mar, Z.T., Beamish, R.J., Overland, J.E., Schirripa, M.J., 2009. A framework for modelling fish and shellfish responses to future climate change. ICES J. Mar. Sci. 66, 1584–1594.
Hong, S.K., Koh, C.H., Harris, R.R., Kim, J.E., Lee, J.S., Ihm, B.S., 2010. Land use in Korean tidal wetlands: Impacts and management strategies. Environ. Manag. 45 (5), 1014–1026.
Hong Kong Agriculture and Fisheries Department, 1968. 32 Important Food Fish and the Fishing Industry in Hong Kong.
Hong Kong Agriculture and Fisheries Department, 1972. 50 Important Food Fish and the Fishing Industry in Hong Kong.
Hong Kong Agriculture and Fisheries Department, 1996. Agriculture and Fisheries Annual Departmental Report, 1994-1995. Hong Kong Government.
Hong Kong Agriculture and Fisheries Department, 1998. Hong Kong Fishery Resources Survey: 1996–1997. Agriculture and Fisheries Department; Fisheries Branch; Capture Fisheries Division.
Hong Kong Environmental Protection Department, 1996. Hong Kong—The Environmental Challenge. Hong Kong Environmental Protection Department, pp. 1986–1996.
Hong Kong Government, 1996. Fishing Industry: Hong Kong. Information Services Department of Hong Kong Government, p. 2.
Hossain, M.S., 2009. Floodplain aquaculture in Begumgonj: New Horizon for rural livelihoods in Bangladesh. Aquacult. Asia 14 (3), 7–10.
Hu, C., Huang, D., et al., 2005. Study on coordination of fisheries development and multi-functions of large and medium reservoir in China. Reserv. Fish. 25 (5), 1–4.
Hu, C., Lin, T., Wu, H., Gao, H., Chang, X., 1998. Studies on comprehensive fishery exploitation of Heilongtan Reservoir in Sichuan Province, China. Int. Rev. Hydrobiol. 83 (special issue), 535–538.
Huang, D., Liu, J., Hu, C., 2001. Fish Resources in Chinese Reservoirs and their utilisation. In: De Silva, S.S. (Ed.), Reservoir and Culture Based Fisheries: Biology and Management. Proceedings of an International workshop held in Bangkok, Thailand. IDRC; Ottawa, pp. 16–21.
Huang, Z.G. (Ed.), 1994. Marine Species and Their Distributions in China Seas. Ocean Press (Beijing), China, p. 730 (in Chinese).
Hutchinson, G.E., 1967. A Treatise on Limnology, vol. II: Introduction to Lake Biology and the Limnoplankton. John Wiley and Sons Inc., New York, p. 1115.
Hwang, G.H., Kim, S., Lee, Y., 1998. Marine Tourism.
ICES, 1998. Report of the Working Group on the Ecosystem effects of Fishing Activities. ICES, Copenhagen, p. 263. ICES CM 1998/ACFM/ACME: 01 Ref.: E.
ICES, 2000. Ecosystem effects of fishing. In: ICES Marine Science Symposia. 210. ICES J. Mar. Sci. 57, 791.
IHHP (Institute of Hydrobiology of Hubei Province), 1976. Fishes of the Yangtze River. Science Press, Beijing (in Chinese).
IISD, 2001. In: Creech, H., Willard, T. (Eds.), Strategic Intentions: Managing Knowledge Networks for Sustainable Development. International Institute for Sustainable Development, pp. 1–158.
Immanuel, S., Kurup, K.N., 2003. Marine fisheries along the southwest coast of India. In: Silvestre, G., et al. (Eds.), Assessment, Management and Future Directions for Coastal Fisheries in Asian Countries, pp. 757–792. World Fish Centre Conference Proceedings, 67: pp. 1120.
Immanuel, S., Vivekanandan, E., Pillai, V.N., Kurup, K.N., Srinath, M., 2003. A preliminary assessment of the coastal fishery resources in India—socioeconomic and bioeconomic perspectives. In: Silvestre, G., et al. (Eds.), Assessment, Management and Future Directions for coastal fisheries in Asian Countries; Conference Proceedings; World Fish Centre, 67, p. 1120.
Indrasena, H.H.A., 1965. The development of fresh water fisheries in Ceylon. In: Proceedings of the Symposium on the Development of Ceylon's Fisheries. Bulletin of Fisheries Research Station, Ceylon, 17, pp. 287–289.
Institute of Oceanology Academia Sinica, 1991. Atlas of the Sea Fishes of China in Live Colour. Vols. 1-2 Shanghai Scientific and Technical Publ (in Chinese).

Institute of Oceanology Academia Sinica, 1995. Yellow Sea and East China Sea Continental Shelf and Coastal Water Living Resources Preliminary Assessment (1991-1995). pp. 172–202 (in Chinese).
Institute of Zoology Academia Sinica (China), 1962. Fishes of the South China Sea. Vols. 1-3 Science Press, Beijing, p. 1184 (in Chinese).
Islam, G.M.N., Dickson, M., 2007. Turning Social Capital Into Natural Capital: Changing Livelihoods of Fishers Through CBFM. Conference Paper 04, World Fish Centre, Bangladesh and South Asia Office, Dhaka, Bangladesh, pp. 1–21.
Islam, A., Rahman, M.M., 2007. Cluster-based management approaches for wetlands: CNRS's experience in the CBFM-2 project. In: Dickson, M., Brooks, A. (Eds.), 2007 CBFM-International Conference on Community Based Approaches to Fisheries Management; The World Fish Centre; Conference Proceedings; 75 Print: pp. 38; CD-ROM: p. 337. Centre for Natural Resource Studies; The World Fish Centre; Bangladesh and South Asia Office, Dhaka, Bangladesh, p. 13.
IUCN, 1996. Environmental Studies of Nepal's Begnas and Rupa Lakes: A Collaboration Between the National Planning Commission and IUCN., p. 156.
IUCN, 1999. IUCN Sri Lanka. The 1999 List of Threatened Fauna and Flora of Sri Lanka. prepared with financial support from The worldwide Fund for Nature (WWF) and The German Federal Ministry for Economic cooperation and Development (BMZ) through the IUCN Asia Regional Biodiversity Programme, IUCN-The World Conservation union, pp. 1–14.
IUCN, 2004. In: Karki, S., Thomas, S. (Eds.), A Review of the Status and Threats to Wetland in Nepal. Jagdamba Press, Hattiban, Lalitpur, (India). pp. v + 78.
Izzat, H.F., 1998. Fisheries Development in the Arab World. Middle Eastern Natural Environments, Bulletin 103. Food and Agriculture Organization of the United Nations (FAO), Rome (Italy), pp. 388–406.
Janet, A.N., Jason, S.L., Jonathan, A.H., William, J.O., 2009. Changing spatial distribution of fish stocks in relation to climate and population size on the Northeast United States continental shelf. Mar. Ecol. Prog. Ser. 393, 111–129.
Janssen, M.A., Carpenter, S.R., 1999. Managing the resilience of lakes: a multi-agent modeling approach. Conserv. Ecol. 3, 15.
Jennings, S., 2005. Indicators to support an ecosystem approach to fisheries. Fish Fish. 6 (3), 212–232.
Jentoft, S., 2000. The community: a missing link of fisheries management. Mar. Policy 24, 53–59.
Jeppesen, E., Jensen, J.P., Sondergaard, M., Lauridsen, T., Pedersen, L.J., Jensen, L., 1997. Top-down control in freshwater lakes: the role of nutrient state, submerged macrophytes and water depth. Hydrobiologia 342, 151–164.
Jhingran, A.G., 1986. Reservoir Fisheries Management in India. Bulletin No. 45, Central Inland Fisheries Research Institute, Barrackpore.
Jhingran, V.G., 1991. Fish and Fisheries of India. Hindustan Publishing Corporation, New Delhi. xxiii + 727.
Jim, C. Y. and Corlett, R. T. (eds.); Friends of the Country Parks; The World Conservation Union; World Commission of Protected Areas and Cosmos Book Limited.
Jiwyam, W., Chareontesprasit, N., 2001. Cage culture of Nile tilapia and its loadings in a freshwater reservoir in Northeast Thailand. Pak. J. Biol. Sci. 4 (5), 614–617.
Johannes, R.E., 1988. The Role of Marine Resource Tenure Systems (TURFS) in Sustainable Nearshore Marine Resource Development and Management in U.S. - Affiliated Tropical Pacific Islands. In: Smith, B.D. (Ed.), Topic Reviews in insular resource Developement and Management in the Pacijic U.S.-Afiliated Islands. University of Guam Marine Laboratory. Technical Report 88.
Johnson, C., 2001. Community formation and fisheries conservation in southern Thailand. Dev. Change 32, 951–974.
Johnson, D., 2002. Fishery comparisons or valid comparisons? Reflections on a comparative approach to the current global fisheries malaise with reference to Indian and Canadian cases. Marit. Stud. 1 (1), 103–121.
Jonathan, A.H., Michael, A.A., Michael, J.F., Erik, H.W., James, D.S., 2010. Forecasting the dynamics of a coastal fishery species using a coupled climate–population model. Ecol. Appl. 20 (2), 452–464.
Juatagate, T., De Silva, S.S., Mattson, N.S., 2001. Socio-economic status of river sprat (*Clupeichthys aesarnensis* Wongratana 1983): Lift net fishers in Sirinthorn Reservoir, Thailand. In: De Silva, S.S. (Ed.), Reservoir and Culture-Based Fisheries: Biology and Management. ACIAR Proceedings No. 98. Australian Centre for International Agricultural Research, Canberra.
Jutagate, T., Lamkom, T., Satapornwanit, K., Naiwinit, W., Petchuay, C., 2001. Fish species diversity and ichthyomass in Pak Mun Reservoir, five years after impoundment. Asian Fish. Sci. 14, 417–425.
Jutagate, T., Mattson, N.S., 2003. Optimizing fishing gear operation in Sirinthorn Reservoir, Thailand. Nat. Hist. Bull. Siam Soc. 51 (1), 105–122.
Jutagate, T., De Silva, S.S., Mattson, N.S., 2003a. Yield growth and mortality of Thai river sprat *Clupeichthys aesarnensis* in Sirinthorn Reservoir, Thailand. Fish. Manag. Ecol. 10, 221–231.
Jutagate, T., Krudpan, C., Ngamsnae, P., Payooha, K., Lamkom, T., 2003b. Fisheries in the Mun River: A one-year trial of opening the sluice gates of the Pak Mun Dam, Thailand. Kasetsart J. Sci. Technol. 37, 101–116.
Jutagate, T., Krudpan, C., Ngamsnae, P., Lamkom, T., Payooha, K., 2005a. Changes in the fish catches during the trial of opening the sluice gates of a run of-the river reservoir in Thailand. Fish. Manag. Ecol. 12, 57–62.
Jutagate, T., Lamkom, T., Grudphan, C., Ngamsnae, P., Payooha, K., 2005b. Multivariate analysis in upstream migration of pangasiids and silurids from the Mekong River into the Mun River, Thailand. J. Ubon Rajathanee Univ. 7, 45–68.
Kabir, I., Hassan, S.R., 2007. Legal issues pertaining to community based fisheries management. Bangladesh Environmental Lawyers' Association, In: Dickson, M., Brooks, A. (Eds.), CBFM-International Conference on Community Based Approaches to Fisheries Management, The World Fish Centre, Conference Proceedings, 75 Print, pp. 38; CD-ROM: pp. 337, p. 12.
Kalikoski, D.C., Vasconcellos, M., Lavkulich, L., 2002. Fitting institutions and ecosystems: the case of artisanal fisheries management in the Patos Lagoon. Mar. Policy 26, 179–196.
Kar, D., 1984. Limnology and Fisheries of Lake Sone in the Cachar district of Assam (India). viii + 201, PhD Thesis, University of Gauhati (Assam).
Kar, D., 1990. Limnology and Fisheries of Lake Sone in the Cachar district of Assam (India). viii + 201, PhD Thesis (Published), University of Gauhati, Assam (1984): Matsya, 15–16: 209–213.
Kar, D., 1996. Biodiversity Conservation Prioritisation Project (BCPP) in India. In: Proc. International Project Formulation Workshop of BCPP, World Wide Fund (WWF) for Nature-India, 1 (New Delhi).
Kar, D., 1998. Biodiversity Conservation Prioritisation Project (BCPP) in India. In: Proc. International Project Finalisation Symposium of BCPP, World Wide Fund (WWF) for Nature-India, 1 (New Delhi).
Kar, D., 1999a. Preliminary Study of Limnology and Aquatic biota of Rudra Sagar wetland in South Tripura. In: Proc. Regional Seminar on Biodiversity, Guwahati, Assam, 1.

Kar, D., 1999b. In: Microbiological and Environmental Studies in relation to Fish Disease in India. Gordon Research Conference, Connecticut (USA).
Kar, D., 2000a. Present status of Fish Biodiversity in South Assam and Tripura. In: Ponniah, A.G., Sarkar, U.K. (Eds.), Fish Biodiversity of North-East India. NBFGR-NATP Publication No. 2, Lucknow, pp. 80–82. pp. 228.
Kar, D., 2000b. Socio-economic Development of the Fisherwomen through Aquaculture with emphasis on Integrated Farming in the villages around Chatla Haor Wetland in Silchar, Assam. In: Seminar presented on DBT-sponsored Awarness Workshop on Biotechnology-based Programmes for Women and Rural Development, NEHU, Shillong, 1, pp. 13–14.
Kar, D., 2001. Species composition and distribution of Fishes in the rivers in Barak valley region of Assam and the Principal rivers in Mizoram and Tripura in relation to their Habitat parameters. In: Proc. National Workshop, NATP-ICAR Project Mid-term Review, 1: Central Marine Fisheries Research Institute, Cochin. 25 p.
Kar, D., 2002a. In: Bhattacharjee, M.K., Dattachoudhury, M., Mazumder, P.B. (Eds.), Present status of Biodiversity of Fishes of Barak valley region of Assam with a note on their Management and Conservation. Proc. UGC-Sponsored State-level Seminar on Biodiversity of Assam, Session Chairman's Lecture, 30 Jan 2000. Karimganj College, Assam University, Assam, pp. 3–10.
Kar, D., 2002b. Water bodies and fishes: resources and development. Conserv̄. Forum J. 1. 15 p.
Kar, D., 2003a. Fishes of Barak drainage, Mizoram and Tripura. In: Kumar, A., Bohra, C., Singh, L.K. (Eds.), Environment, Pollution and Management. APH Publishing Corporation, New Delhi, pp. 203–211. pp. xii + 604.
Kar, D., 2003b. An Account of the Fish Biodiversity in South Assam, Mizoram and Tripura Along With a Brief Account of Epizootic Ulcerative Fish Disease Syndrome in Freshwater Fishes. UGC-sponsored Invited Lecture in Dept. of Environmental Engg, Guru Jambeswar University, Hissar (Haryana).
Kar, D., 2003c. Peoples' Perspective on Fish Conservation in the Water bodies of South Assam, Mizoram and Tripura. In: Mahanta, P.C., Tyagi, L.K. (Eds.), Participatory Approach for Fish Biodiversity Conservation in North-East India. National Bureau of Fish Genetic Resources (ICAR), Lucknow, pp. 325–328. v + 412.
Kar, D., 2004a. A Glimpse into the Fish Bioresources of North-East India with a note on their Management, Conservation and Biotechnological Potential. In: Proc. (Invited Lecture) DBT-sponsored National Symposium on Biodiversity Conservation and Sustainable Utilisation of Environmental Resources; Tripura University, 10-11 Jan., 2004. 9 p.
Kar, D., 2004b. Inventorying of Fish Biodiversity in North-East India with a note on their Conservation. In: Proc. National Conference on Fish and their Environment, 9-11 Feb 2004, B.A. Marathwada University, Aurangabad, 1.
Kar, D., 2004c. Conservation and Ichthyodiversity in the River Systems of Barak drainage, Mizoram and Tripura. In: Proc. National Symp. Management of Aquatic Resources for Biodiversity Maintenance and Conservation, 2-4 Feb 2004, J.N. Vyas University, Jodhpur, 1.
Kar, D., 2004d. Fish Fauna of Barak Drainage in Assam, of Mizoram and of Tripura with a note on Conservation. J. Freshwater Biol. 16 (1-4), 31–39.
Kar, D., 2005a. An enigmatic fish disease called Epizootic Ulcerative Syndrome. J. Freshwater Biol. 17 (1-4), 83–84.
Kar, D., 2005b. Fish Genetic Resources and Habitat Diversity of the Barak drainage. In: Ramachandra, T.V., Ahalya, N., Rajsekara Murthy, C. (Eds.), Aquatic Ecosystems, Conservation, Restoration and Management. Capital Publishing Company, Bangalore, pp. 68–76. xiii + 396.
Kar, D., 2005c. Fish Diversity in the Major Rivers in Southern Assam, Mizoram and Tripura. In: Nishida, T., Kailola, P.J., Hollingworth, C.E. (Eds.), Proc. 2nd International Symposium on GIS and Spatial Analyses in Fisheries and Aquatic Sciences, 2-6 Sep 2002, University of Sussex, Brighton (UK). Vol. 2. Fisheries and Aquatic GIS Research Group, Kawagoe, Saitama (Japan), pp. 679–691.
Kar, D., 2005d. Fish Biodiversity and Habitat Parameters of rivers in Barak drainage (Assam), in Mizoram and in Tripura. Himalayan J. Environ. Zool. 19 (1), 41–45.
Kar, D., 2006a. Panorama of wetlands in North-East India with special emphasis on their Eutrophication status. Sci. Soc. 4 (2), 167–180.
Kar, D., 2006b. Study of Fish Diversity in a Wetland. Activity Guide `Biodiversity'. National Childrens' Science Congress, NCSTC Network and Dept. of Science and Technology, Govt. of India, pp. 37–38.
Kar, D., 2007a. Fundamentals of Limnology and Aquaculture Biotechnology. xiv + 609, Daya Publishing House, New Delhi.
Kar, D., 2007b. Sustainability issues of Inland Fish Biodiversity and Fisheries in Barak drainage (Assam), in Mizoram and Tripura. In: Kurup, M.B., Ravindran, K. (Eds.), Sustain Fish. School of Industrial Fisheries, Cochin University of Science and Technology (CUSAT): Proceedings of International Symposium on `Improved sustainability of Fish Production Systems and Appropriate Technologies for Utilisation', 16-18 March, 2005 (Cochin), pp. 555–560. xii + 863.
Kar, D., 2007c. Lentic Fishery: fishery of a tropical wetland (Beel) in Assam. Sci. Soc. 5 (1), 53–72.
Kar, D., 2010a. Biodiversity Conservation Prioritisation. pp. xi + 167, Swastik Publications, New Delhi, ISBN: 978-93-80138-26-8.
Kar, D., 2010b. Fish diversity in rivers of Mizoram. Fish. Chimes 30 (7), 18–22.
Kar, D., 2010c. Water bodies, Fish diversity and sustainable aquaculture. Ann. J. Womens' College (Silchar), 168–194.
Kar, D., 2012a. Wetlands, Rivers, Fish diversity, Fish disease and Aquaculture in North-East India. Fish. Chimes 31 (12), 35–37.
Kar, D., 2012b. Essentials of Fish Biology. pp. vii + 244, Dominant Publishers, New Delhi.
Kar, D., 2012c. Taxonomy, BSc, MSc, NET, SLET. APH Publications, New Delhi, p. 102.
Kar, D., 2013. Wetlands and Lakes of the World. pp. xxx + 687, Springer, London.
Kar, D., 2014. Research Frontiers in Wetlands, Fishery and Aquaculture. xi + 320, Dominant Publishers and Distributors Pvt. Ltd, New Delhi.
Kar, D., 2015. Epizootic Ulcerative Fish Disease Syndrome. xix + 293, Elsevier (Academic Press), USA.
Kar, D., 2016. Wetlands, Rivers, Fish, Plankton resources and Fish disease and Aquaculture in North-East India: An Overview. In: Proc. International Symposium, Lake 2016. Indian Institute of Science, Bengaluru, and Alva's Education Foundation, Mengaluru (India). pp. 37.
Kar, D., Dey, S.C., 1982a. *Hilsa ilisha* (Hamilton) from Lake Sone in Cachar, Assam. Proc. Indian Sci. Congr. 69. 77 p.
Kar, D., Dey, S.C., 1982b. An account of *Hilsa ilisha* (Hamilton) of Sone Beel (Cachar district, Assam, India). Proc. All-India Sem. Ichthyol. 2. 3 p.
Kar, D., Dey, S.C., 1986. An account of ichthyospecies of Lake Sone in Barak valley of Assam. Proc. All-India Sem. Ichthyol. 2. 3 p.
Kar, D., Dey, S.C., 1987. An account of the Fish and Fisheries of Lake Sone in the Barak valley of Assam (India). In: Proc. Workshop Development of Beel Fisheries in Assam, vol. 1. 13 p.
Kar, D., Dey, S.C., 1988a. An account of Fish yield from Lake Sone in the Barak valley of Assam. Proc. Indian Sci. Congr. 75. 49 p.
Kar, D., Dey, S., 1988b. Preliminary Electron Microscopic studies on Diseased fish tissues from Barak valley of Assam. In: Proc. Annual Conference of Electron Microscopic Society of India, vol. 18. 88 p.
Kar, D., Dey, S.C., 1988c. A critical account of the Recent Fish Disease in the Barak valley of Assam. In: Proc. Regional Symp. on Recent outbreak of Fish Diseases in North-East India, vol. 1. 8 p.

Kar, D., Dey, S.C., 1988d. Impact of Recent Fish Epidemics on the Fishing Communities of Cachar district of Assam. In: Proc. Regional Symp. on Recent outbreak of Fish Diseases in North-East India, vol. 1. 8 p.
Kar, D., Dey, S.C., 1990a. Fish Disease Syndrome: A Preliminary Study from Assam. Bangladesh J. Zool. 18, 115–118.
Kar, D., Dey, S.C., 1990b. A Preliminary study of Diseased Fishes from Cachar district of Assam. Matsya 15–16, 155–161.
Kar, D., Dey, S.C., 1990c. Epizootic Ulcerative Syndrome in Fishes of Assam. J. Assam Sci. Soc. 32 (2), 29–31.
Kar, D., Dey, S.C., 1991. Gill nets in Lake Sone of Assam with their Economics and Impact on Fishery. J. Appl. Zool. Res. 2 (2), 76–79.
Kar, D., 1991a. Fish Disease Syndrome affects the people of Barak valley. Fish. Chimes 10 (11). 63 p.
Kar, D., 1991b. EUS still prevails in South Assam. Fish. Chimes 10 (11). 63 p.
Kalita, M.N., Kar, D., Singh, H.T., 1993. Studies on toxicity of certain piscicidal plants of Assam to fishes. J. Assam Sci. Soc. 35 (2), 125–129.
Kar, D., Dey, S.C., 1993a. Variegated Encircling gears in Lake Sone of Assam. J. Appl. Zool. Res. 4 (2), 171–175.
Kar, D., Dey, S.C., 1993b. Interrelationship and Dynamics of Fish Population of Lake Sone in Assam. Environ. Ecol. 11 (3), 718–719.
Kar, D., Dey, S.C., 1996. Scooping Gears of Lake Sone in Assam. J. Appl. Zool. Res. 7 (1), 65–68.
Kar, D., Dey, S.C., 2000. Yield and Conservation of Indian major carps of Lake Sone in Assam. Environ. Ecol. 18 (4), 1036–1038.
Kar, D., Dey, S.C., 2002. On the occurrence of advanced fry of *Hilsa (Tenualosa) ilisha* (Hamilton-Buchanan) in Chatla Haor Seasonal Wetland of Assam. Proc. Zool. Soc. (Calcutta) (Springer) 55 (2), 15–19.
Kar, D., Barbhuiya, M.H., 2000a. Length-weight Relationship and Condition Factor in *Gudusia chapra* (Ham-Buch) and *Botia dario* (Ham-Buch) from Chatla Haor (flood plain wetland) in Cachar district of Assam. Environ. Ecol. 18 (1), 227–229.
Kar, D., Barbhuiya, M.H., 2000b. Ichthyodiversity of Chatla Haor: A Floodplain wetland in Barak valley Region of Assam. pp. 3–6, In: Pandey, B.N., Singh, B.K. (Eds.), Advances in Zoology and Environmental Degradation and Biodiversity, pp. 279. Daya Publishing House, New Delhi.
Kar, D., Barbhuiya, M.H., 2001. Ecology of aquatic macrophytes of Chatla Haor, a floodplain wetland in Cachar district of Assam. Environ. Ecol. 19 (1), 231–233.
Kar, D., Barbhuiya, M.H., 2002a. Macrophytic diversity in certain Wetlands of Barak valley region of Assam. In: Ramachandra, T.V., Rajasekara Murthy, N., Ahalya, N. (Eds.), Restoration of Lakes and Wetlands. Allied Publishers (P) Ltd., Bangalore, pp. 86–89. xxii + 400.
Kar, D., Barbhuiya, M.H., 2002b. Waterfowls in Chatla Haor, a seasonal floodplain in Cachar district of Assam with a note on waterfowl and wetland conservation. Environ. Ecol. 20 (3), 689–691.
Kar, D., Barbhuiya, M.H., 2002c. Status Survey of Gangetic Dolphin in Barak drainage of Assam with a note on its Management and Conservation. Environ. Ecol. 20 (1), 249–250.
Kar, D., Barbhuiya, M.H., 2002d. Aquatic macro-invertebrates of Baraknadi Salchapra Anua in Barak valley region of Assam. Environ. Ecol. 20 (2), 696–699.
Kar, D., Barbhuiya, M.H., 2004. Abundance and Diversity of Zooplankton in Chatla Haor, a floodplain wetland in Cachar district of Assam. Environ. Ecol. 22 (1), 247–248.
Kar, D., Barbhuiya, M.H., 2005. Length-weight relationship and relative condition factoring *Hilsa (Tenualosa) ilisha* (Hamiilton) of Barak drainage of Assam. Indian J. Environ. Ecoplan. 10 (1), 265–267.
Kar, D., Barbhuiya, A.H., 2013. Frontiers of Wetlands, Fishers and Aquaculture Research. Mangalam Publications, New Delhi, p. 263.
Kar, D., Mazumdar, J., 2004. In: Biodiversity, Disease and Conservation of Fish Bioresources: A perspective. Proc. National Workshop on Sci. and Tech. for Regional Development: Case for North-East India, 3-6 Feb 2004, Tezpur University and I.I.T. Guwahati. 34 p.
Kar, D., Das, M., 2004a. Preliminary Histochemical studies with diseased fishes suffering from Epizootic Ulcerative Syndrome in Assam. In: Proc. National Symposium for Management of Aquatic Resources for Biodiversity Maintenance and Conservation, vol. 1: 27 p.
Kar, D., Das, M., 2004b. Preliminary Enzymological studies with diseased fishes suffering from Epizootic Ulcerative Syndrome in Fishes. In: Proc. National Conference on Fish and their Environment, vol. 1. 25 p.
Kar, D., Sen, N., 2007. Systematic List and Distribution of Fish Biodiversity in Mizoram, Tripura and Barak drainage in North-East India. ZOOs' Print J. 22 (3), 2599–2607.
Kar, D., Barbhuiya, A.H., 2009. Mahseer Fishes of Barak Drainage, Mizoram and Tripura. Souvenir. In: National Symposium on Coldwater Fisheries Management: New Strategies and Approaches, 24 Oct 2009, pp. 77-80, Directorate of Coldwater Fisheries Research (ICAR), Bhimtal (Uttarakhand, India).
Kar, D., Barbhuiya, A.H., 2011a. Length-weight relationship and condition factor in *Puntius amphibious* (Valenciennes) and *Puntius vittatus* Day from Barak valley region of Assam. J. Inland Fish. Soc. India 42 (1).
Kar, D., Barbhuiya, A.H., 2011b. Status of Chocolate Mahseer of River Barak with a note on its morphometry and meristics. J. Inland Fish. Soc. India 42 (1).
Kar, S., Kar, D., 2016a. Zooplankton diversity of a freshwater pond in Cachar district of Assam, India. Int. J. Life Sci. 4 (1), 125–128.
Kar, S., Kar, D., 2016b. Zooplankton diversity of a freshwater perennial pond in Silchar city of Assam, India. Am. Int. J. Res. Formal Appl. Nat. Sci. 14 (1), 10–14.
Kar, S., Das, P., Das, U., Bimola, M., Kar, D., Aditya, G., 2018. Culture of the zooplankton as fish food: Observations on three freshwater species from Assam, India. AACL Bioflux 10 (5), 1210–1220.
Kar, D., Dey, S.C., Michael, R.G., Kar, S., Changkija, S., 1990b. Studies on Fish Epidemics from Assam, India. J. Indian Fish. Assoc. 20, 73–75.
Kar, D., Bhattacharjee, S., Kar, S., Dey, S.C., 1990a. An account of `EUS' in Fishes of Cachar district of Assam with special emphasis on Microbiological studies. In: Souv. Annual Conference of Indian Association of Pathologists and Microbiologists of India (NE Chapter), 8, pp. 9–11.
Kar, D., Dey, S.C., Kar, S., Bhattacharjee, N., Roy, A., 1993. In: Virus-like particles in Epizootic Ulcerative Syndrome of Fish. Proc. International Symp. on Virus-Cell Interaction: Cellular and Molecular Responses, p. 1. 34 p.
Kar, D., Dey, S.C., Kar, S., Roy, A., Michael, R.G., Bhattacharjee, S., Changkija, S., 1994. A Candidate virus in Epizootic Ulcerative Syndrome of Fish. In: Proc. National Symp. of the Indian Virological Society, vol. 1. 27 p.
Kar, D., Roy, A., Dey, S.C., Menon, A.G.K., Kar, S., 1995a. Epizootic Ulcerative Syndrome in Fishes of India. World Congress of In Vitro Biology. In Vitro 31 (3). 7 p.
Kar, D., Kar, S., Roy, A., Dey, S.C., 1995c. Viral Disease Syndrome in Fishes of North-East India. In: Proc. International Symp. of International Centre for Genetic Engg. And Biotechnology (ICGEB) and the Univ. of California at Irvine, vol. 1. 14 p.

Kar, D., Dey, S.C., Kar, S., 1995d. A Viral Disease among the Fishes of North-East India. Annual Congress on Man and Environment. In: National Environment Sci. Acad. And National Institute of Oceanography, vol. 10. 62 p.
Kar, D., Dey, S.C., Kar, S., Roy, A., 1995b. Epizootic Ulcerative Disease Syndrome among the Fishes of North-East India. J. North-Eastern Coun. (Govt. India) 14 (4), 21–24.
Kar, D., Dey, S.C., Kar, S., 1995e. Fish and Fisheries of Lake Sone in Assam. J. North-Eastern Council 15 (2-3), 49–57.
Kar, D., Dey, S.C., Kar, S., Michael, R.G., Gadgil, M., 1996c. Ichthyoecology, Management and Conservation of. Fish Resources of Lake Sone in Assam (India). In: Tiger Paper (Food and Agricultural Organisation of the United Nations) (FAO), Vol. XXIII, No. 3, pp. 27–32.
Kar, D., Dey, S.C., Purkayastha, M., Kar, S., 1996a. An overview of the Impediments in Conservation of Biodiversity of Lake Sone in Assam. In: Proc. Seminar on Conservation of Biodiversity in Indian Aquatic Ecosystems, Jawahar Lal Nehru University, New Delhi, vol. 1.
Kar, D., Dey, S.C., Roy, A., Kar, S., 1996d. Viral Disease Syndrome in Fishes of India. In: Proc. International Congress of Virology, vol. 10.
Kar, D., Purkayastha, M., Kar, S., 1996b. Biodiversity Conservation Prioritisation Project: A case study from Sone Beel in Assam. In: Proc. National Workshop on Biodiversity Conservation Prioritisation Project (BCPP), World Wide Fund (WWF) for Nature-India and Centre for Ecological Sciences, Indian Institute of Science (Bangalore).
Kar, D., Saha, D., Laskar, R., Barbhuiya, M.H., 1997. Biodiversity Conservation Prioritisation Project (BCPP) in Barak valley region of Assam. In: Proc. National Project Evaluation Workshop on BCPP, Betla Tiger Reserve and National Park, 1 (Palamu).
Kar, D., Saha, D., Dey, S.C., 1998b. Epizootic ulcerative syndrome in Barak valley of Assam. In: Project Report submitted and presented at the National Symposium of Biodiversity Conservation Prioritisation Project (BCPP) held at WWF-India, 18–19 Jan, 1998 (New Delhi), pp. 2–4.
Kar, D., Dey, S.C., Roy, A., 1998c. Present status of Epizootic Ulcerative Syndrome (EUS) in Southern Assam. In: Proc. Regional Project Initiation Workshop for NATP-ICAR-NBFGR Project, vol. 1. 9 p.
Kar, D., Dey, S.C., Kar, S., Roy, A., 1998a. An account of Epizootic Ulcerative Syndrome in Assam. In: Final Project Report of Biodiversity Conservation Prioritisation Project (BCPP). Submitted and presented at the International Project Finalisation Workshop of BCPP, 1, 22-28 April 1998 (New Delhi), pp. 1–3.
Kar, D., Dey, S.C., Roy, A., 1998d. Fish Disease Diagnosis and Fish Quarantine problems in India and South-East Asia with particular emphasis on Epizootic Ulcerative Syndrome (EUS) fish disease problems. In: Proc. International Workshop on Fish Disease Diagnosis and Quarantine, Indian Council of Agricultural Research (ICAR)-Ministry of Agriculture (MoA), Govt. of India (GoI)-Network of Aquaculture Centres in Asia (NACA)-Food and Agricultural Organisation (FAO) of the United Nations (UN)-OIE, France: Held at the Central Institute of Freshwater Aquaculture (CIFA), vol. 1.
Kar, D., Upadhyaya, T., 1998. Histopathological Studies of Fish tissues affected by Epizootic Ulcerative Syndrome in Assam. In: Technical Bulletin, XIII; Convention and National Symposium of Indian Association of Veterinary Anatomists (IAAM), 11-13 Dec 1998. College of Veterinary Sciences, Assam Agricultural University, Guwahati.
Kar, D., Dey, S.C., Kar, S., Ramachandra, T.V., 1999c. Trawls of Lake Sone in Assam. J. Appl. Zool. Res. 10 (2), 170–172.
Kar, D., Rahaman, H., Barnman, N.N., Kar, S., Dey, S.C., Ramachandra, T.V., 1999b. Bacterial pathogens associated with epizootic ulcerative syndrome in freshwater fishes of India. Environ. Ecol. 17 (4), 1025–1027.
Kar, D., Mandal, M., Bhattacharjee, S., 1999a. Fungal Pathogens associated with Epizootic Ulcerative Syndrome in Fishes of Barak valley region of Assam. In: Proc. 1st National Conference on Fisheries Biotechnology, CIFE, vol. 1. 34 p.
Kar, D., Dey, S.C., Roy, A., 2000d. Fish Genetic Resources in the Principal rivers and wetlands in North-East India with special emphasis on Barak valley (Assam), in Mizoram and in Tripura, with a note on Epizootic Ulcerative Syndrome fish disease. In: Proc. National Project Initiation Workshop of the NATP-ICAR World Bank- aided Project on `Germplasm Inventory, Evaluation and Gene Banking of Freshwater Fishes', National Bureau of Fish Genetic Resources (NBFGR), Lucknow, 1. 12 p.
Kar, D., Mandal, M., Laskar, B.A., Dhar, N., Barbhuiya, M.H., 2000a. Ichthyofauna of some of the oxbow lakes in Barak valley region of Assam. In: Proceedings of the National Symposium on Wetlands and Fisheries Research in the New Millennium, 1. 16 p.
Kar, D.; Dey, S.C.; Roy, A.; and, Mandal, M. (2000c). Epizootic Ulcerative Syndrome Fish Disease in Barak valley region of Assam, India, Proc. Nat. Symp. Current Trends in Wetlands and Fisheries Research in the New Millennium, vol. 1: 2 p.
Kar, D., Dey, S.C., Mandal, M., Lalsiamliana, 2000b. Epizootic Ulcerative Syndrome among the fishes of Assam. In: Proc. National Workshop of NATP-ICAR-NEC North-East Programme, Shillong, vol. 1. 24 p.
Kar, D., Mandal, M., Lalsiamliana, 2001b. Species composition and distribution of Fishes in the rivers in Barak valley region of Assam and in the Principal rivers of Mizoram and Tripura in relation to their habitat parameters. In: Proc. National Workshop, NATP-ICAR Project, Mid-term Review, 1: Central Marine Fisheries Research Institute, Cochin. 25 p.
Kar, D., Laskar, B.A., Lalsiamliana, 2001a. Further studies on the Ichthyospecies composition and distribution of Freshwater Fishes in Barak drainage and in the Principal rivers in Mizoram and in Tripura with a note on their Feeding and Breeding biology. In: Proc. National Project Monitoring Workshop of NATP-ICAR Project, National Bureau of Fish Genetic Resources, Lucknow, vol. 1. 22 p.
Kar, D., Laskar, B.A., Mandal, M., Lalsiamliana, Nath, D., 2002a. Fish Genetic Diversity and Habitat parameters in Barak drainage, Mizoram and Tripura. Indian J. Environ. Ecoplan. 6 (3), 473–480.
Kar, D.; Laskar, B.A.; Nath, D.; Mandal, M.; and, Lalsiamliana (2002d). *Tor progenius* (McClelland) under threat in River Jatinga, Assam. Sci. Cult., 68 (7-8): 211.
Kar, D., Laskar, B.A., Nath, D., 2002f. *Tor* sp. (Mahseer fish) in river Mat in Mizoram. Aquaculture 3 (2), 229–234.
Kar, D., Dey, S.C., Roy, A., 2002h. In: Prevalence of Epizootic Ulcerative Syndrome among fishes of Assam. Proc. Regional Symp. on Biodiversity, S.S. College, Assam (Central) University, Hailakandi (Assam).
Kar, D., Dey, S.C., Roy, A., 2002i. On the diseased fishes suffering from hitherto unknown Epizootic Ulcerative Syndrome in Fishes in India. In: Proc. All-India Congress of Zoology, Dec., 2002 (Bangalore).
Kar, D., Dey, S.C., Roy, A., 2002j. In: Prevalence of Epizootic Ulcerative Syndrome (EUS) among fishes of Mizoram. Proc. Regional Symp. on Aquaculture, p. 1.
Kar, D., Laskar, B.A., Lalsiamliana, 2002c. Germplasm Inventory, Evaluation and Gene Banking of Freshwater Fishes. 3rd Annual Technical Report. pp. 57, In: National Project Evaluation Workshop, 3: NATP-ICAR World Bank-aided Project: National Bureau of Fish Genetic Resources (Lucknow).

Kar, D., Dey, S.C., Mandal, M., Laskar, B.A., Lalsiamliana, 2002b. Preliminary survey of the fish genetic resources of the Rivers in Barak Drainae, Mizoram and Tripura. In: Ramachandra, T.V., Rajasekahra, M.C., Ahalya, N. (Eds.), Restoration of Lakes and Wetlands. Allied Publishers (P) Ltd., Bangalore, pp. 73–81. pp. xxii + 400.

Kar, D., Dey, S.C., Roy, A., Mandal, M., 2002e. Epizootic ulcerative syndrome in Fishes of Barak valley of Assam, India. In: Ramachandra, T.V., Murthy Rajasekhara, S., Ahalya, N. (Eds.), Restoration of Lakes and Wetlands. Allied Publishers (P) Ltd, Bangalore, pp. 303–307. pp. xii + 400.

Kar, D., Dey, S.C., Datta, N.C., 2003e. Welfare Biology in the New Millennium. xx + 97, Allied Publishers Pvt. Ltd., Bangalore.

Kar, D., Dey, S.C., Roy, A., 2003g. Capture Fishery in Lentic Systems with some light on Sone Beel in Assam with special reference to prevalence of EUS in the Barak valley region of Assam. In: Proc. Lecture Series in UGC-sponsored Vocational Course in Industrial Fish and Fisheries, Cachar College, Assam University, Silchar, Abstracts, vol. 1.

Kar, D., Roy, A., Dey, S.C., 2003f. Epizootic Ulcerative Disease Syndrome in Freshwater Fishes with a note on its Management for Sustainable Fisheries. In: Proc. International Conference on Disease Management for Sustainable Fisheries, 26-29 May, 2003, Department of Aquatic Biology and Fisheries, University of Kerala, Trivandrum.

Kar, D., Lohar, M., Ngassepam, Z., Sinom, S., Tiwary, B., Bawari, M., 2003b. Fish Bioresources in certain rivers of Assam and Manipur with a note on their assessment, management and conservation. In: Proc. Nat. Symp. Assessment and Management of Bioresources, North Bengal University and The Zoological Society, Calcutta, 28-30 May, 2003. 56 p.

Kar, D., Dey, S.C., Roy, A., 2003h. Epizootic Ulcerative Syndrome: A virulent disease among the freshwater fishes of India. In: Proc. Guest Lecture delivered at the Dept. of Zoology, Arunachal University, Arunachal Pradesh, vol. 1.

Kar, D., Dey, S.C., Roy, A., 2003i. Biodiversity of freshwater fishes of North-East India with a note on their health status. In: Proc. DBT-sponsored Guest Lecture Series delivered at the Dept. of Molecular Biology and Biotechnology, Tezpur (Central) University, Tezpur, vol. 1.

Kar, D., Dey, S.C., Roy, A., 2003j. Epidemiology of Epizootic Ulcerative Syndrome disease among the freshwater fishes of India. In: Proc. Regional Symp. on World Bank-aided programme, entitled, `Assam Rural Infrastructure and Agricultura Services Project (ARIASP), Silchar, vol. 1.

Kar, D., Dey, S.C., Roy, A., 2003k. Review of Epizootic Ulcerative Syndrome disease among the freshwater fishes of India. In: Proc. UGC-sponsored lecture series delivered at the Department of Environmental Engineering and Biotechnology. Guru Jambeswar University, Hisar (Haryana).

Kar, D., Laskar, B.A., Nath, D., 2003d. Length-weight relationship of *Salmostoma phulo phulo*(Hamilton Buchanan) collected from River Karnafuli along Indo-Bangladesh border (in Mizoram). J. Appl. Zool. Res. 14 (2), 188–190.

Kar, D., Dey, S.C., Roy, A., Mandal, M., 2003c. Epizootic ulcerative syndrome fish disease in Barak Valley Region of Assam. In: Kar, D., Dey, S.C., Datta, N.C. (Eds.), Welfare Biology in the New Millennium. Allied Publishers Pvt. Ltd., Bangalore, pp. 73–78. pp. xx + 97.

Kar, D., Baishya, A.K., Mandal, M., Arjun, J., Laskar, B.A., 2003a. A Preliminary account of the Vegetations in Dargakona locality of Silchar in Assam. Environ. Ecol. 21 (2), 254–258.

Kar, D., Roy, A., Dey, S.C., 2004c. An overview of Fish Genetic Diversity of North-East India. In: Garg, S.K., Jain, K.L. (Eds.), Proceedings of National Workshop on Rational use of Water Resources for Aquaculture, 18–19 March 2004, CCS Haryana Agricultural University, vol. 1, pp. 164–171.

Kar, D., Roy, A., Mazumder, J., Patil, P., 2004a. Biotechnological approach for defining Epizootic Ulcerative Syndrome Disease in Freshwater Fishes. In: Proc. Mid-term Review: DBT-sponsored Project, New Delhi, vol. 1.

Kar, D., Dey, S.C., Roy, A., Mazumder, J., Patil, P., Kohlapure, R.M., 2004b. Fish Disease Prevalence in Assam with particular reference to Epizootic Ulcerative Syndrome and its pathogens. In: Proc. International Conference `Bioconvergence', 18-20 Nov 2004, Thapar Institute of Engineering and Technology, Patiala, vol. 1. 220 p.

Kar, D., Mazumdar, J., Devi, P., Devi, B.R., Devi, V., 2006b. Isolation of *Aeromonas hydrophila* from fishes affected by Epizootic Ulcerative Syndrome as well as from corresponding healthy fish species and from their habitat. J. Curr. Sci. 9 (1), 323–327.

Kar, D., Laskar, B.A., Nath, D., 2006c. An account of Fecundity of *Eutropiichthys vacha* (Hamilton-Buchanan): a commercially important fish in Assam. Environ. Ecol. 24 (Spl (3)), 726–727.

Kar, D., Nagarathna, A.V., Ramachandra, T.V., Dey, S.C., 2006a. Fish diversity and conservation aspects in an aquatic ecosystem in North-East India. ZOOS' Print J. 21 (7), 2308–2315.

Kar, D., Mazumdar, J., Halder, I., Dey, M., 2007b. Dynamics of initiation of disease in fishes through interaction of microbes and the environment. Curr. Sci. (Bangalore) 92 (2), 177–179. 25 January, 2007.

Kar, D., Mazumdar, J., Barbhuiya, M.A., 2007c. Isolation of Mycotic flora from fishes affected by Epizootic Ulcerative Syndrome in Assam, India. Asian J. Microbiol. Biotechnol. 9 (1), 37–39.

Kar, D., Barbhuiya, A.H., Arifuddidn, Barbhuiya, M.A., Chetia, P., Islam, R., Rahman, S., 2007a. Traditional riverine fish catching devices of Assam. Fish. Technol. 44 (2), 137–146.

Kar, D., Barbhuiya, A.H., Thangjam, G., Devi, S.M., Deb, S., Das, B., Chanu, H., Nishima, 2008. Panorama of Fish Biodiversity in certain rivers and wetlands in Manipur. Proc. Zool. Soc. India 7 (2), 123–134.

Kar, D., Barbhuiya, A.H., Baruah, A.R., Choudhury, C., Banerjee, P., Pal, R., Bhattacharjee, A., Saikia, R., Das, B., Barman, R., Saha, B., 2009. Panorama of Fish Diversity in certain rivers, wetlands and protected areas in Assam. Geobios 36, 57–64.

Kar, D., Barbhuiya, A.H., Das, B., Purkayastha, M., Barbhuiya, M.H., Saha, D., Laskar, R.A., Arjun, J., 2010b. Medicinal plants in certain regions of Assam, India. Environ. Ecol. 28 (3B), 1965–1968.

Kar, D., Barbhuiya, A.H., Saha, B., Devi, S., Medeline, G.M., Das, B., 2010a. Protection of fish diversity in Meghalaya needs conservation measures: in the context of fishing devices used. Fish. Chimes 30 (1), 124–127.

Kar, D., Singha, R., Das, B.K., John, R., George, R., Mansoor, M., Selva, M., 2015. Preliminary Epidemiological Studies on Epizootic ulcerative Syndrome (EUS) in Freshwater Fishes of Assam. Int. J. Sci. Res. Sci. Eng. Technol. 1 (5), 256–263.

Kar, D., Laskar, B.A., Nath, D., 2006d. Length-weight relationship of *Neolissochilus hexagonolepis* (McClelland) and *Garra lissorhynchus* (McClelland) from River Jatinga in Cachar district of Assam. Indian J. Fish. 52 (4), 495–496.

Kar, D., Shomorendra, M., Singha, R., Puinyabati, H., Geetarani, B., Binky, K., Sangeeta, O., Ranibala, T., 2011. Fish diversity and Helminth fauna in the fishes of Assam and Manipur, India. Fish. Chimes, 55–65.

Kar, S., Kar, D., 1989. Food habit and malaria among the Ao Nagas of Nagaland: a brief report. Soc. Sci. 2 (2), 129–132.

Kar, S., Kar, D., 1994. Health and Disease pattern among the Ao Nagas. Soc. Sci. VI (7), 417–422.

Kar, S., Kar, D., Arjun, J., 2002g. Genetic Survey among the Kuki Tribal Populations in Loharbond area of Cachar, Assam. Environ. Ecol. 20 (1), 82–85.

Kar, S., Kar, D., 2004a. Osmotic fragility and malaria among the Ao Nagas of Nagaland. Environ. Ecol. 22 (1), 243–244.

Kar, S., Kar, D., 2004b. Genetics, Culture, Environment and Health among the Kuki Hill Tribe in Cachar District of Assam, India. Sci. Cult. 70 (11–12), 357–359.

Karr, J.R., 1991. Biological Integrity, a long neglected aspect of Water Resource management. Ecol. Appl. 1, 66–87.

Katiha, P.K., 1994. Natural Resource Management: A Case Study of Reservoir Fisheries in Himachal Pradesh. PhD. Thesis, Department of Social Sciences, Dr. Y.S. Parmar University of Horticulture and Forestry, Nauni, Solan (Himachal Pradesh).

Katiha, P.K., Vass, K.K., Shrivastava, N.P., Sharma, V.K., Sarkar, S., 2007. Inventory of Reservoirs in Indian Indo-Gangetic Basin for fishery enhancements. Abstr. No. P 184, In: `Taal', 2007, 12th World Lake Conference. Organised by MoEF, GoI and ILEC, Japan.

Katon, B., Pomeroy, R., Salamanca, A., 1997. The Marine Conservation Project for San Salvador: A Case Study of Fisheries co-management in the Philippines. ICLARM, Manila, pp. 10–95. Fisheries Co-management Research Project Working Paper No. 23.

Katon, B., Pomeroy, R., Ring, M., Garces, L., 1998. Mangrove Rehabilitation and Coastal Resource Management of Mabini-Candijay: A Case Study of Fisheries Co-management Arrangements in Cogtong Bay, Philippines. ICLARM, Manila, pp. 20–147. Fisheries Co-management Research Project Working Paper No. 33.

Kennish, R., Wilson, K.D.P., Lo, J., Clarke, S.C., Laister, S., 2002. Selecting sites for large-scale deployment of artificial reefs in Hong Kong: Constraint mapping and prioritization techniques. ICES J. Mar. Sci. 59, S164–S170.

Khan, M.S., Apu, N.A., 1998. Fisheries Co-Management in the Oxbow Lakes of Bangladesh. Chittagong University, Chittagong, pp. 1–61.

Kim, S.G., 1999. The Evaluation of Fishing Village Through Controlling of Individual Resources Based on Revitalization Projects (1994-1998) and Effective Creation of Ecological System Through Traditional Stock Measures in the Future. Ministry of Maritime Affairs Enhancement Programmes.

Kimberly, Z., Robert, C., 2004. An introduction to cooperatives. In: Cooperatives: Principle and Practices in the 21st Century. University of Wisconsin, Extension–Madison, Wisconsin (USA), p. 89.

Kirkpatrick, W., 1793. An Account of the Kingdom of Nepaul, Being the Substance of Observations Made During a Mission to That Country, in the Year 1793. William Miller, 1811 (London). Reprinted: Asian Educational Services, 1996 (New Delhi).

Klee, G.A., 1991. Conservation of Natural Resources. xvi + 410, Prentice Hall, Englewood Cliffs, NJ.

Klessig, L.L., Sorge, B., Robert, K., Michael, D., Bode, J., 2004. A Model Lake Plan for a Local Community. Produced by Cooperative Extension Publishing, University of Extension–Madison, Wisconsin. RP/05/2004.

Klessig, L.L., 2001. Lakes and society: the contribution of lakes to sustainable societies. Lakes Reserv. Res. Manag. 6, 95–101.

Krishnan, M., Birthal, P.S., Ponnusamy, K., Kumaran, M., Singh, H., 2000. Aquaculture Development in India: Problems and Prospects. In: Workshop Proceedings held at National Centre for Agricultural Economics and Policy Research, September 6-7, 1999 (New Delhi).

Kumar, Anjani; Joshi, P.K.; and, Birthal, Pratap S. (2003). Fisheries Sector in India: An Overview of Performance, Policies and Programmes, pp. 1-16. In: A Profile of People, Technologies and Policies in Fisheries Sector in India, (eds.), Kumar, Anjani; Katiha, Pradeep K.; and, Joshi, P K. Workshop Proceedings Series 10, National Centre for Agricultural Economics and Policy Research; Indian Council of Agricultural Research, pp. 1-171.

Kurien, J., 1978. Towards an Understanding of the Fish Economy of Kerala State. Centre for Development Studies, Trivandrum. Working Paper 68.

Kurien, J., Achari, T., 1990. In: Overfishing along Kerala Coast: Causes and Consequences. Economic and Political weekly, XXV, Sep-1-4 (Mumbai).

Kurien, J., Achari, T., 1998. Overfishing the Coastal Commons: Causes and Consequences. In: Guha, R. (Ed.), Social Ecology. Oxford University Press, New Delhi, p. 389.

Kurien, J., 1992. Ruining the commons and responses of the commoners: Coastal overfishing and fishermen's actions in Kerala State, India. In: Ghai, D., Vivian, J. (Eds.), Grassroots Environmental Action: Peoples' Participation in Sustainable Development. Routledge, London.

Kurien, J., 1993. Ruining the commons—coastal overfishing and fishworkers, action in South India. Ecologist 23 (1), 5–11.

Kurien, J., 1995a. Collective action for common property resource rejuvenation: The case of peoples artificial reefs in Kerala State, India. Hum. Organ. 54, 160–168.

Kurien, J., 1995b. Impact of Joint Ventures on Fish economy in Economic and Political Weekly. XXX (6) (Mumbai).

Kurien, J., 1995c. The Kerala Model: Its Central Tendency and the Outlier. Soc. Sci. 23.

Kurien, J., Paul, A., 2000. Nets for social safety: An analysis of the growth and changing composition of social security programmes in the fisheries sector of Kerala State, India. In: Samudra Monograph. International Collective in Support of Fishworkers, Chennai, India.

Kurup, B.M., Sebastian, M.J., Sankaran, T.M., Ravindranath, P., 1993. An account of inland fishing gear and fishing methods of Kerala. Fishery Technology (Special Issue), Society of Fisheries Technologists, India.

Kwok, K.Y., Ni, I.H., 1999. Reproduction of cutlass fishes, *Trichiurus* spp., from South China Sea. Mar. Ecol. Prog. Ser. 176, 39–47.

Lam, C., 1986. A new species of *Bathygobius* (Pisces: Gobiidae) from Hong Kong. Asian Mar. Biol. 3, 75–87.

Lam, C. (1990a). Intertidal gobies (Pisces: Gobiidae) from Hong Kong. In: Morton, B. (ed.), The marine flora and fauna of Hong Kong and Southern China II, Hong Kong. Hong Kong Univ. Press: pp. 673-690.

Lam, C., 1990b. Hong Kong mudskippers (Pisces: Periophthalmidae). In: Morton, B. (Ed.), The Marine Flora and Fauna of Hong Kong and Southern China II. Hong Kong Univ. Press, Hong Kong, pp. 691–697.

Lam, C., 1990c. Benthic ichthyofauna of Tolo Harbour and the entrance to Tolo Channel, Mirs Bay. In: Morton, B. (Ed.), The marine flora and fauna of Hong Kong and southern China II. Hong Kong Univ. Press, Hong Kong, pp. 899–905.

LaRoe, E.T., 1986. Instream impacts of soil erosion on Fish and Wildlife. In: Waddell, T.E. (Ed.), The Offsite Cost of Soil Erosion. Washington, DC.

Laskar, B.A., Das, S., Nath, D., Kar, D., 2002. Ecological studies in Puneer Haor Wetland in Cachar district of Assam with special emphasis on Aquatic macrophytes, ichthyofauna and wetland birds. Conserv. Forum J. 1. 25 p.

Leal, D.R., 1996. Community-Run Fisheries: Avoiding the `Tragedy of the Commons'. PERC Policy Series Issue No. PS–7, In: Improving Environmental Quality Through Markets.

Lee, H.K., 1997. Night fishing survey with purse-seine at Long Harbour, Double Haven and Hoi Ha Wan (January 1995 to December 1996). Agriculture and Fisheries Department; Fisheries Branch; Capture Fisheries Division.

Lee, S.-G., Rahimi, M.A., 2014. Creative economy potentials and fisheries development in South Korea. World J. Fish Mar. Sci. 6 (1), 49–56.

Leigh, P., 2005. The ecological crisis: the human condition, and community-based restoration as an instrument for its cure. Ethics Sci. Environ. Polit., 3–15.

Leung, A.W., Leung, Y.K.F., Lam, K.K.Y., Morton, B., 1997. The Evaluation of an Experimental Artificial Reef for Marine Habitat Rehabilitation and the Feasibility of Using Stabilized Coal Combustion Solid Residues for Reef Construction. Final report.

Leung, A.W., 1991. Abundance and diversity of benthic fishes: Tolo and Mirs Bay, Hong Kong. In: Morton, B. (Ed.), The Marine Flora and Fauna of Hong Kong and Southern China, III. vol. 1. Hong Kong Univ. Press, Hong Kong, pp. 459–473.

Leung, A.W., 1994. The Fish Fauna of Lobster Bay, Cape D'Aguilar. Master's Thesis, Hong Kong Univ. (Hong Kong).

Leung, A.W., 1997a. The impacts of dredging and fishing. In: Morton, B. (Ed.), The Marine Flora and Fauna of Hong Kong and Southern China IV. Hong Kong Univ. Press, Hong Kong, pp. 437–461.

Leung, A.W., 1997b. The epibenthic ichthyofauna of Tolo Harbour and Hong Kong's north-eastern waters: A long term record of change. In: Morton, B. (Ed.), The Marine Flora and Fauna of Hong Kong and Southern China IV. Hong Kong Univ. Press, Hong Kong, pp. 463–487.

LGSP, 2011. The World Bank, Report No: 63726, BD, project appraisal document to the people's republic of Bangladesh, for a second local governance support project, October. 31, 2011. Sustainable Development Unit. Urban and Water Unit, South Asia Region, pp. 1–90.

Limsong, S., Sitha, H., 2015. Cambodian attempts to develop culture-based fisheries. 121, In: De Silva, S.S., Ingram, B.A., Wilkinson, S. (Eds.), Perspectives on Culture-Based Fisheries Developments in Asia. NACA Monograph Series No. 3, p. 126.

Li, S., 1988. The principles and strategies of fish culture in Chinese reservoirs. In: De Silva, S.S. (Ed.), Reservoir Fishery Management and Development in Asia. International Development Research Centre, Ottawa.

Li, S., Xu, S., 1995. Culture and Capture of Fish in Chinese Reservoirs. International Development Research Centre, South-bound Sdn. Bhd, Penang (Malaysia).

Lin, C.K., Kaewpaitoon, K., 2000. An overview of freshwater cage culture in Thailand. In: Lin, C.K., Liao, I.C. (Eds.), Proceedings of the First International Symposium on Cage Aquaculture in Asia. Asian Fisheries Society, Manila (the Philippines).

Liu, J., Li, Z., Xie, S., 2008. Achievements, challenges and strategies for reservoir fisheries development in China. In: De Silva, S.S., Amarasinghe, U.S. (Eds.), Status of Reservoir Fisheries in Five Asian Countries. NACA Monograph No. 2. Network of Aquaculture Centres in Asia-Pacific, Bangkok, Thailand, pp. 17–29.

Liu, Z., Tang, Z., et al., 2003. Baseline survey of fisheries resources in Beishi Reservoir. Reserv. Fish. 23, 43–46.

Liu, J., He, B., 1992. Cultivation of the Chinese Freshwater Fishes, third ed. Science Press, Beijing, p. 750 (in Chinese).

Liu, J., Huang, Y., 1998. Fisheries and fish culture practices in Fuqiaohe Reservoir, China. Int. Rev. Hydrobiol. 83 (special issue), 569–576.

Liu, J., Peng, J., Yu, F., Jiang, J., Yao, X., Yi, Y., 2001. Estimation of the new ice fish, *Neosalanx taihuensis*, yield in Zhanghe Reservoir, China. In: DeSilva, S.S. (Ed.), Reservoir Culture-based Fisheries: Biology and Management. Proceedings of an International Workshop held in Bangkok, Thailand from 15th-18th February 2000. Australian Centre for International Agricultural Research, Canberra.

Liu, J., Cui, Y., Liu, J., 1998. Food consumption and growth of two piscivorous fishes, the mandarin fish and the Chinese snakehead. J. Fish Biol. 53, 1071–1083.

Liu, J., Hu, C., 2000. Sturgeons and their status and function in reservoir fisheries. Reserv. Fish. 20, 1–4.

Liu, J., Yu, Z., 1990. The paddle fish (*Polyodon spathula*) and its fisheries in the USA. Acta Hydrobiol. Sin. 14 (1), 75–83 (in Chinese).

Liu, J.K., 1984. Lakes of the middle and lower basins of the Chang Jiang (China). In: Taub, F.B. (Ed.), Lakes and Reservoirs. Elsevier Publication, Amsterdam, pp. 331–355.

Lobe, K., Berkes, F., 2003. The padu system of community-based fisheries management: Change and local institutional innovation in south India. Mar. Policy 28, 271–281.

Loffler, H., 1969. High altitude lakes in Mt. Everest region. Verh. Int. Ver. Theor. Angew. Limnol. 17, 373–385.

Lopez, M.D.G., 1985. Notes on Traditional Fisheries in the Philippines. In: Ruddle, K., Johannes, R.E. (Eds.), The Traditional Knowledge and Management of Coastal Systems in Asia and the Pacific. UNESCO-ROSTSEA, Jakarta, pp. 190–206.

Lorenzen, C.J., 1967. Determination of chlorophyll and pheopigments: spectrophotometric equations. Limnol. Oceanogr. 12, 343–346.

Lorenzen, K., 2001. Using population models to assess culture-based fisheries: a brief review with an application to the analysis of stocking experiments. In: De Silva, S.S. (Ed.), Reservoir and Culture-Based Fisheries: Biology and Management. ACIAR Proceedings No. 98, Australian Centre for International Agricultural Research (Canberra).

Manosroi, A., Meng-Umphan, K., Manosroi, J., 2003. Annual sex hormonal profiles, gonad development and age determination of the Mekong giant catfish (*Pangasianodon gigas*, Chevey). Aquacult. Res. 34, 1379–1385.

Marafa, L.M., Chau, K.C., 2006a. Ecotourism and nature-based recreation in the protected areas. In: Jim, C.Y., Corlett, R.T. (Eds.), Proceedings of the 5th Conference on Protected Areas of East Asia-Sustainable Management of Protected Areas for Future Generations. Friends of the Country Parks, The World Conservation Union, World Commission of Protected Areas and Cosmos Book Limited.

Marafa, L.M., Chau, K.C., 2006b. Ecotourism and nature-based recreation in the protected areas. In: Proceedings of the 5th Conference on Protected Areas of East Asia: Sustainable Management of Protected Areas for Future Generations.

Masae, A., 1998. In: An analysis of fisheries co-management arrangements: the case of Ban Laem Makham, Sikao District, Trang Province, South Thailand. Paper presented at the Asian Regional Workshop on Fisheries Co-management, Phuket (Thailand).

Mathew, S., 1991. Study of territorial use rights in small-scale fisheries: Traditional systems of fisheries management in Pulicat Lake, Tamil Nadu, India. In: FAO Fisheries Circular 839 (Rome).

Maude, H.E., 1982. Slavers in Paradise. Institute of Pacific Studies, University of the South Pacific, Suva.

McConney, P., 1998. In: Creating Conditions for Community-Based Small Scale Fisheries Management in Caribbean. International Workshop on Natural Resources Management (CBNRM), 10–14 May, 1998 (Washington D.C.).

McCutcheon, M.M., 1980. In: Changing Systems of Resource Procurement from the Lagoon and Reef in Palau. Paper presented at the Annual Meeting of the American Anthropological Association (Washington D.C.).

McGregor, J.A., 1995. The assessment of policy and Mangement options in inland capture fisheries: Summary guidelines. In: Presented at the seminars on Fisheries Researces Dissemination, University of Bath (U.K.) and Proshika (Bangladesh).

McQueen, D.J., 1990. Manipulating lake community structure: Where do we go from here. Freshw. Biol. 23, 613–620.

Meaden, G.J., Kapetsky, J.M., 1991. Geographical information systems and remote sensing in inland fisheries and aquaculture. In: FAO Fisheries Technical Paper, 318. Food and Agriculture Organization of the United Nations, Rome.

Mendis, A.S., 1965. A preliminary survey of 21 Ceylon lakes. 2. Limnology and fish production potential. Bull. Fish. Res. Station Ceylon 16, 7–16.

Mendis, A.S., 1977. The role of man-made lakes in the development of fisheries in Sri Lanka. Proc. Indo-Pacific Fish. Council 17 (3), 247–254.

Michael, R.G., 1969. Seasonal trends in Physico-chemical factors and Plankton of freshwater fish pond and their role in fish culture. Hydrobiologia 33 (1), 145–160.
Middendrop, H.A.J., Balarin, J.D., 1999. Fisheries enhancement and participatory resource management: Two types of management in the oxbow lakes projects in Bangladesh. In: Middendrop, H.A.J., Thompson, P.M., Pomeroy, R.S. (Eds.), Sustainable Inland Fisheries Management in Bangladesh. ICLARM Conference Proceedings Vol. 58, ICLARM, (Manila).
Ministry of Forest and Soil Conservation, 2003. National Wetland Policy—2059. His Majesty's Government, Ministry of Forest and Soil Conservation, Nepal, p. 16 (Approved on March 2003 (2059/11/22).
Minkin, S.F., Rahman, M.M., Halder, S., 1997. Fish biodiversity, human nutrition and environmental restoration in Bangladesh. In: Tsai, C., Ali, M.Y. (Eds.), Open water fisheries of Bangladesh. The University Press Limited, Dhaka, pp. 183–198.
Mironga, J.M., 2005. The effect of water hyacinth, *Eichornia crassipes*, infestation on phytoplankton productivity in Lake Naivasha and the status of control. In: 11th World Lake Conference, Nairobi, Kenya, Proceedings, Volume 2, pp. 573–579.
Mitaura, H., Mitsunaga, Y., Arai, N., Yamagichi, Y., Khachaphichat, M., Viputhanumas, T., 2005. A review of the Mekong giant catfish tracking project (MCTP) from 2002 to 2004. In: Proceedings of the 2nd International Symposium on SEASTAR 2000 and Asian Bio-logging Science. Graduate School of Informatics, Kyoto University.
Mitchell, C.P., 1980. Control of water weeds by grass carp in two small lakes. N. Z. J. Mar. Freshw. Res. 14, 381–390.
Mohamed, K.S., 2004. Towards Ecosystem Based Management of Marine Fisheries: Building Mass Balance Trophic and Simulation Models. Winter School, Central Marine Fisheries Research Institute, Cochin (Kerala), pp. 125–129. Technical Notes.
Mohsin, S.M., 2007. Sunamganj Community-Based Resource Management Project (SCBRMP): An approach to sustainable Beel fisheries management. In: Dickson, M., Brooks, A. (Eds.), CBFM-International Conference on Community-Based Approaches to Fisheries Management, The World Fish Centre, Conference Proceedings. The World Fish Centre, Bangladesh and South Asia Office, Dhaka, Bangladesh, pp. 15–16. pp. 38; CD-ROM: pp. 337.
MOMAF, 1999. A collection of successful fishing and Fishing Ports Division. Fishing Ports Division (MOMAF). Village Revitalization Projects. Ministry of Maritime Affairs and Fisheries, and Fishing Ports Division, Seoul.
Moreau, J., Villanueva, M.C., Amarasinghe, U.S., Schiemer, F., 2001. Trophic relationships and possible evolution of the production under various fisheries management strategies in a Sri Lankan reservoir. In: De Silva, S.S. (Ed.), Reservoir and Culture-based Fisheries: Biology and Management. ACIAR, Canberra, (Australia).
Moreau, J., Lek, S., Leelaparata, W., Srichareondham, B., Villaneuva, M.C., 2004. A comparison of various fitting techniques for predicting fish yield in Ubolratana Reservoir (Thailand) from a time series data. In: Lek, S., Scardi, M., Verdonchot, P.F.M., Descy, J.P., Park, Y.S. (Eds.), Modelling Community Structure in Freshwater Ecosystems. Springer, Berlin.
Moreau, J., De Silva, S.S., 1991. Predictive fish yield models for lakes and reservoirs of the Philippines, Sri Lanka and Thailand. In: FAO Fisheries Technical Paper, 319. Food and Agriculture Organization of the United Nations, (Rome).
Morton, B., 2005. Over Fishing: Hong Kong's Fishing Crisis Finally Arrives. Department of Zoology, The Natural History Museum, Cromwell Road, London SW7 5BD, UK, pp. 1–5. Marine Pollution Bulletin xxx.
Morton, B., Wu, S.S., 1974. The hydrology of the coastal waters of Hong Kong. Environ. Res. 10, 319–347.
Morton, B., Tseng, C.K. (Eds.), 1982. The Marine Flora and Fauna of Hong Kong and Southern China. Proceedings of the First International Marine Biological Workshop, Hong Kong, 18 April-10 May, 1980. Hong Kong Univ. Press, Hong Kong.
Morton, B. (Ed.), 1990. The marine flora and fauna of Hong Kong and southern China, II. Proceedings of the Second International Marine Biological Workshop, Hong Kong, 2-24 April 1986. Hong Kong Univ. Press, (Hong Kong).
Morton, B. (Ed.), 1992. The marine flora and fauna of Hong Kong and southern China, III. In: Proceedings of the Fourth International Marine Biological Workshop, Hong Kong, 11-29 April, 1989. Hong Kong Univ. Press, (Hong Kong).
Morton, B., 1996. Artificial reefs, fish and cynanide. Mar. Pollut. Bull. 32 (7), 522–523.
Morton, B. (Ed.), 1997. The marine flora and fauna of Hong Kong and southern China, IV. Proceedings of the Eighth International Marine Biological Workshop, Hong Kong, 2–20 April 1995. Hong Kong Univ. Press, (Hong Kong).
Moser, C., 1998. The asset vulnerability framework: Reassessing urban poverty reduction strategies. World Dev. 26 (1), 1–19.
Motwani, M.P., Jayaram, K.C., Sehgal, K.L., 1962. Fish and Fisheries of Brahmaputra River System, Assam, I. Fish Fauna with observation on their zoogeographical significance. Trop. Ecol. 3, 17–43.
Moyle, P.B., 1976. Fish Introduction in California. History and impact on Native Fishes. Biol. Conserv. 9, 101–118.
Murray, F., 2002. In: When Co-management Fails: A Review of Theory and Lessons Learned from Reservoir Fisheries in the Dry-Zone of Sri Lanka. Institute of Aquaculture, University of Stirling, Stirling FK9 4LA, U.K., Conference Paper 11, pp. 1–22.
Mustafa, M.G., Halls, A.S., 2007. Impact of the Community-Based Fisheries Management on sustainable use of inland fisheries in Bangladesh. World Fish Centre, Bangladesh and South Asia; Mekong River Commission Secretariat (MRCS); Inland Fisheries Research and Development; Institute (IFReDI), Department of Fisheries; Phnom Penh, Cambodia, Conference Paper 03: pp, 1–19; and Proceedings of the CBFM-2 International Conference on Community Based Approaches to Fisheries Management, Dhaka: p. 11, (Bangladesh).
Naqi, S.A., 1989. Licensing *vs* leasing system of fishing access. In: Aguero, M., Huq, S., Rahman, A., Ahmed, M. (Eds.), Inland Fisheries Management in Bangladesh. Department of Fisheries, Bangladesh; Bangladesh Centre for Advanced Studies; and, International Centre for Living Aquatic Resources Managenement, Manila, (Philippines), pp. 83–92.
NARC, 1991. Nepal Agriculture Research Study (TA No. 1182 NEP), ANNEX No. 4-Fisheries and Aquaculture in Nepal, (M 1975-2005). A study Team Report Prepared for the National Agriculture Research Centre and Asian Development Bank, Lalitpur (Nepal).
Narzary, A., Das, S., Das, B.K., Singh, N.R., Kar, S., Das, P., Dutta, B., Kar, D., 2015. A preliminary study on zooplankton diversity of Ramnagar Anua, Srikona Beel and Tapang Hoar of Cachar district, Assam, India: a Project Report. J. Chem. Biol. Phys. Sci. 5 (3), 2809–2817.
Nasuchon, N., 2009. Coastal Management and Community Management in Malaysia, Vietnam, Cambodia and Thailand, With a Case Study of Thai Fisheries Management. Division for Ocean Affairs and the Law of the Sea; Office of Legal Affairs, the United Nations, New York, pp. 1–91.
Natarajan, A., 2005. Enabling Better Management of Fisheries Conflicts: A Case Study in India. In: Paper presented at in the Regional Consolidation Workshop on Fish Fights over Fish Rights: Managing Conflicts and Exit from the Fisheries and Security Implications for South and Southeast Asia, 17-20 May 2005, International Rice Research Institute Complex (IRRI), Los Baños, Laguna, The Philippines. In: Enabling

Better Management of Fisheries Conflicts; NRSP Production System; Land Water Interface. Final Technical Report (R8294), Revised Version, November 2005; Submitted by: The World Fish Centre; in collaboration with research partners: The World Fish Center, Bangladesh; Fisheries Action Coalition Team, Cambodia; Mitraniketan, India; and, The University of Reading, United Kingdom: pp. 1-91.

Natarajan, A.V., 1976. Ecology and state of fishery development in some of the man made reservoirs of India. In: Symposium on the Development and Utilization of Inland Fishery Resources, Colombo, 27–29 October 1976.

Natarajan, A.V., 1977. In: Fifth Workshop of AICP on Ecology and Fisheries of Freshwater Reservoirs. Ukai, 7–8 September, 1977.

Natarajan, A.V., 1979. Ecosystem oriented approach for reservoir fisheries development in India. In: Souvenir on the ICAR Golden Jubilee Year, CIFRI, Barrackpore, West Bengal, Part I, pp. 107–111.

Natarajan, A.V., Pathak, V., 1983. Pattern of Energy Flow in Tropical and Sub-Tropical Impoundments. Bulletin No. 36, Central Inland Fisheries Research Institute, Barrackpore.

Nath, M., Singh, N.R., Das, B.K., Dutta, B., Das, U., Das, P., Kar, S., Kar, D., 2015. A preliminary study on Fish Diversity of Kakri and Deo rivers around Dharmanagar in Tripura. Int. J. Theor. Appl. Sci. 7 (2), 6–13.

Nathan, D., Apu, N.A., 2007. In: Dickson, M., Brooks, A. (Eds.), Access to Common Property Resource and Poverty Reduction: Inland Open-Water Fisheries in Bangladesh. CBFM-International Conference on Community-Based Approaches to Fisheries Management. The World Fish Centre, Conference Proceedings, 75 Print: pp. 38; CD-ROM: pp. 337. The World Fish Centre, Bangladesh and South Asia Office, Dhaka, Bangladesh, p. 24.

Nathanael, S., Edirisinghe, U., 2002. Developing co-management in an artisanal gill net fishery of a deep hydro-electric reservoir in Sri Lanka. Fish. Manag. Ecol. 9, 267–276.

Naylor, R.L., Goldburg, R.J., Primavera, J., Kautsky, N., Beveridge, M.C.M., Clay, J., Folkes, J., Lubchence, M.H., Troell, M., 2000. Effects of aquaculture on world food supplies. Nature 405, 1017–1024.

NEC, 1991. Summary Report on Integrated Inland Fisheries Development Project in North-Eastern States. Eastern States, p. 176. North-Eastern Council, Govt. of India, Shillong.

Neis, B., Schneider, D.C., Felt, L., Haedrich, R.L., Fischer, J., Hutchings, J.A., 1999. Fisheries assessment: What can be learned from interviewing resource users? Can. J. Fish. Aquat. Sci. 56, 1949–1963.

Nelson, J.S., 1994. Fishes of the World, third ed. J. Wiley, USA, p. 600.

Newrkla, P., Duncan, A., 1984. The biology and density of *Ehirava fluviatilis (clupeiod) in Parakrama Samudra, Sri Lanka*. Verh. Int. Ver. Theor. Angew. Limnol. 22, 1572–1578.

Nguyen, T.K.A., 2012. Community-Based Co-Management in Vietnamese Fisheries: The Case of the Fisheries Associations in Tam Giang-Cau Hai Lagoon. Master's Thesis in Fisheries and Aquaculture Management and Economics, FSK-3911 (30 ECTS), The Norwegian College of Fishery Science, University of Tromso, Norway and, Nha Trang University, Vietnam, pp. 1–105.

Nguyen, H.S., Bui, A.T., Le, L.T., Nguyen, T.T.T., De Silva, S.S., 2001. The culture-based fisheries in small, farmer-managed reservoirs in two provinces of northern Vietnam: an evaluation based on three production cycles. Aquacult. Res. 32, 975–990.

Niering, W.A., 1987. Vegetation dynamics (succession and climax) in relation to plant community management. Conserv. Biol. 1, 287–295.

Ni, I.H., 1994a. Environmental Impact Assessment for Borrow Area South of Ninepin Island (Fisheries Survey). Contract Report Submitted to Binnie Consultants, Ltd. For Civil Engineering Department of Hong Kong Government, p. 44.

Ni, I.H., 1994b. Independent Environmental Monitoring and Audit of Dredging in the East Lamma Channel (Fish Trawl Survey). Contract Report Submitted to Binnie Consultants Ltd. for Civil Engineering Department of Hong Kong Government, p. 35.

Ni, I.H., 1995a. The Impact of Dredging on Fisheries Sensitive Receiver (Hong Kong's Territorial Baseline Survey for Marine Fisheries). Contract Report Submitted to Binnie Consultants Ltd. for Civil Engineering Department of Hong Kong Government, p. 87.

Ni, I.H., 1995b. EIA on Sand Dredging and Backfill of Borrow Pits at Potential Eastern Water Marine Borrow Area: Impact on Fisheries. Contract report submitted to AXIS Environmental Consultants Ltd. for Civil Engineering Department of Hong Kong Government, p. 9.

Ni, I.H., 1995c. The Demersal Fisheries Resources in Eastern Waters: Comparative Study of Fishing Surveys Conducted in 1982/83. Contract Report Submitted to AXIS Environ-mental Consultants Ltd. for Civil Engineering Department of Hong Kong Government, p. 31.

Ni, I.H., 1997. Seasonal variability of fisheries resources in the east of Sha Chau. In: Proceedings of a Colloquium on the Development of Management Strategy for Chinese White Dolphins, pp. 69–83.

Ni, I.-H., Kwok, K.-Y., 1999. Marine fish fauna in Hong Kong Waters. Zool. Stud. 38 (2), 130–152.

Nichols, J.T., 1928. Fishes of the White Nile (with Table of World's Freshwater Fish Faunae). American Mus. Novitates No. 319.

Nissanka, C., 2001. Empirical Models for Fish Yield Prediction in Selecte Reservoirs in the Dry Zone of Sri Lanka. Ph.D. Thesis, University of Kelaniya, Sri Lanka.

Niumnampeth, W., 1981. Fisheries Law and Regulations. Department of Fisheries, Bangkok.

Noble, R.L., 1980. Management of lake, reservoirs and ponds. In: Robert, T., Larkey, Larry, A. (Eds.), Fisheries Management. Niellsen Blackwell Scientific, (Boston), pp. 265–295.

NRC, 1999. Sustaining Marine Fisheries. National Research Council, National Academy Press, Washington, D.C., p. 164.

Nuov, S. (2015). Culture-based fisheries in Cambodia: pp. 124. In: De Silva, S.S.; Ingram, B.A.; and, Wilkinson, S. (eds.), Perspectives on Culture-Based Fisheries Developments in Asia. NACA Monograph Series No. 3,: pp. 126.

Oglesby, R.T., 1981. A Synthesis of the Reservoir Fisheries in Sri Lanka. FI: TCP/SRL/8804, Field Document 2, Food and Agriculture Organization of the United Nations, (Rome).

O'Riordan, B., 2013. The global fisheries crisis: dispelling myths and misconceptions, and fishing for solutions. Pract. Action Publ. 3 (1-2), 1–10.

Ostrom, E., 1990. Governing The Commons: The Evolution of Institutions for Collective Action. pp. ix + 271, Cambridge University Press.

Ostrom, E., 1992. Crafting Institutions for Self-Governing Irrigation Systems. Institute for Contemporary Studies Press, San Francisco.

Ostrom, E., Burger, J., Field, C.B., Norgaard, R.B., Policansky, D., 1999. Revisiting the commons: local lessons, global challenges. Science 284, 278–282.

Owen, W., 1985. Natural Resources Conservation: An Ecological Approach, fourth ed. Macmillan, New York.

Padisák, J. (Ed.), 2002. Shallow Lakes 2002. International Conference on Limnology of Shallow Lakes. Veszprém University Press, Veszprém (Hungary).

Pal, J.N., 1979. In: Physico-Chemical Features of Getalsud Reservoir. Proceedings of the Summer Institute on Capture and Culture Fisheries of the Man-made Lakes in India. Central Inland Fisheries Research Institute, Barrackpore.

Panabokke, C.R., 2001. The nature and properties of small tank systems of the dry zone and their sustainable production thresholds. In: Gunasena, H.P.M. (Ed.), Food Security and Small Tank Systems in Sri Lanka. Proceedings of the Workshop Organized by the Working Committee on Agricultural Science and Forestry, 9 September 2000. National Science Foundation (Colombo).

Panayotou, T., 1980. Economic conditions and prospects of small-scale fishermen in Thailand. Mar. Policy 4 (2), 142–146.

Panini, D., 2002. Addressing livelihood issues in conservation-oriented projects: A case study of Pulicat Lake, Tamil Nadu, India. In: Roger, J., Bhaskar, V. (Eds.), Conflict and Cooperation in Natural Resource Management. Palgrave, New York, pp. 63–74.

Pan, S., Ni, I.H., 1996. Genetic Variation and Identification of Butter-Fish (*Pampus* sp.) in the Chinese Waters. MPhil thesis, Hong Kong Univ. Science and Technology.

Parameswaran, S., Murugesan, V.K., 1984. A note on an inexpensive coracele. Indian J. Fish. 31 (3), 387–389.

Paterson, J., Pernetta, C., 2015. Integrating Fisheries and Habitat Management: Fisheries *Refugia* in the South China Sea. In: De Silva, S.S., Amarasinghe, U.S. (Eds.), Status of Reservoir Fisheries in Five Asian Countries. Network of Aquaculture Centres in Asia-Pacific, Bangkok (Thailand), p. 113. NACA Monograph No. 2.

Patgiri, A., Goswami, M.M., Kar, D., Barbhuiya, M.H., 2001. Comparative Study of Length-Weight Relationship and Relative Condition Factors in Major carps in ponds in Guwahati. Indian J. Environ. Ecoplan. 5 (1), 179–180.

Patgiri, A., Goswami, M.M., Kar, D., Barbhuiya, M.H., 2002. Quality assessment of biological feed in four carp species of Assam. Environ. Ecol. 20 (1), 86–88.

Paul, S., 1990. Inland Fisheries marketing and marketing intervention. In: Sugunan, V.V., Bhawmik, U. (Eds.), Technologies for Inland Fisheries Development. Central Inland Capture Fisheries Research Institute, Barrackpore (India), p. 219.

Paul, S., Sugunan, V.V., 1990. Development of reservoir fisheries: areas of hopes and despair. In: Jhingran, A.G., Unnithan, V.K. (Eds.), Reservoir Fisheries in India. Proceedings of the National Workshop on Reservoir Fisheries, 3–4 January, 1990. Asian Fisheries Society Indian Branch, Mangalore (India). Special Publication No. 3.

Paul, Thompson and Chowdhury, S.N. (2007). Experiences in Wetland Co-Management: The MACH Project. The World Fish Centre Bangladesh and South Asia Office, Dhaka, Bangladesh: pp. 14-15. In: Dickson, M. and Brooks, A. (eds.) CBFM-International Conference on Community Based Approaches to Fisheries Management; The World Fish Centre, Conference Proceedings, 75 Print: pp. 38; CD-ROM: pp. 337.

Paul, T., Nishida, J., 2008. South Asia Region, Environment Matters (Annual Review 2008). In: Traditional Management Systems in Two Fishing Villages in East Godavori District, Andra Pradesh, India. Business and Economics, pp. 44–47. By: Huggins, Laura E. (2013): pp. 1–215.

Paul, P.K., Rahman, A.F.M., Arifur, Nazrul, K.M.S., 2014. Successful Practice of Community-Based Fisheries Management in seasonal water logged paddy lands of Begumganj, Noakhali, Bangladesh. Int. J. Res. Agric. Food Sci. 2 (2), 29–36.

Pauly, D., Morgan, G.R. (Eds.), 1987. Length-based methods in fisheries research. ICLARM Conference Proceedings, No. 13; International Centre for Living Aquatic Resources Management, Manila, Philippines; and, Kuwait Institute for Scientific Research, Safat, (Kuwait).

Pawaputanon, O., 1988. Study to Determine the Most Effective Size of Seed Fish for Stocking in Public Water. Technology and Production Plan No. 2/1986, Northeast Fishery Project, Department of Fisheries, Bangkok (In Thai).

Pawaputanon, O., 1992. Inland Capture Fisheries in Thailand. FAO Fisheries Report No. 458, Supplement, Food and Agriculture Organization of the United Nations, Rome.

Pearsall, W.H., 1938. The soil complex in relation to Plant communities. I. Oxidation-reduction potentials in soils. J. Ecol. 26, 180–193.

Peng, J., Liu, J., Xiong, B., 2004. Carrying capacity of water body for cage culture. Acta Ecol. 1, 28–34.

Perrow, M.R., Meijer, M.-L., Dawidowcz, P., Coops, H., 1997. Biomanipulation in shallow lakes: state of the art. Hydrobiologia, 342–343.

Pet, J.S., Piet, G.J., 1993. The consequences of habitat occupation and habitat overlap of the introduced *Oreochromis mossambicus* and indigenous fish species for fishery management in a Sri Lankan reservoir. J. Fish Biol. 43 (Supplement), 193–208.

Pet, J.S., Gevers, G.J.M., van Densen, W.L.T., Vijverberg, J., 1996. Management options for a more complete utilization of the biological fish production in Sri Lankan reservoirs. Ecol. Freshw. Fish 5, 1–14.

Pethiyagoda, R., 2006. Conservation of Sri Lankan freshwater fishes. In: Bambaradeniya, C.N.B. (Ed.), The Fauna of Sri Lanka: Status, Taxonomy, Research and Conservation. The World Conservation Union (IUCN), Colombo, Sri Lanka and Government of Sri Lanka.

Petr, T., 2007. Intensification of reservoir fisheries in tropical and subtropical countries. Int. Rev. Hydrobiol. 79, 131–138.

Petruny-Parker, M.E., Castro, K.M., Schwartz, M.L., Skrobe, L.G., Somers, B. (Eds.), 2004. Who Gets the Fish? Proceedings of the New England Workshops on Rights-Based Fisheries Management Approaches. Rhode Island Sea Grant, Narragansett, R.I, p. 28.

Pholprasit, S., Waiwuttho, N., 1970. Stocking of Chinese carps in Ubolratana Reservoir. In: Annual Report of Ubolratana Fishery Development Unit. National Inland Fisheries Institute, Bangkok (In Thai).

Phounsavath, S., 1998. Community-Based Fisheries Management as an Approach to Participatory Management of Reservoir Fisheries: A Case Study of Two Fishing Villages at the Nam Ngum Reservoir, LAO, P.D.R. A thesis submitted in partial fulfillment of the requirements for the degree of Master of Science, Asian Institute of Technology; School of Environment, Resources and Development, Bangkok (Thailand), pp. 1–119.

Phukan, G.C., 2006. The Assam Fishery Rules, 1953 as amended upto the Assam Fishery (Amendment) Rules, 2005, second ed. Assam Law House, Guwahati, p. 180.

Pido, M., 1996. A Handbook for Rapid Appraisal of Fisheries Management Systems. ICLARM, Manila.

Pido, M.D., Pomeroy, R.S., Carlos, M.B., Garces, L.R., 1996. A handbook for rapid appraisal of fisheries management systems (version 1). ICLARM, Educ. Ser. 16, 85.

Piet, G.J., Vijverberg, J., 1998. An ecosystem perspective for the management of a tropical reservoir fishery. Int. Rev. Hydrobiol. 83 (Special Issue), 103–112.

Pillai, T.G., Swallows, J.D., 1980. Cage Culture of Fish in Nepal: A Report Prepared for the Integrated Fisheries and Fish Culture Development Project, Nepal. FAO, Rome, p. 28. F1:DP/NEP/73/025, Field Document, 8.

Pitcher, T.J., Watson, R., Haggan, N., Guénette, S., Kennish, R., Sumaila, R., Cook, D., Wilson, K., Leung, A., 2000. Marine Reserves and the Restoration of Fisheries and Marine Ecosystems in the South China Sea. Bull. Mar. Sci. 66 (3), 543–566.

Polovina, J.J., 1991. Fisheries applications and biological impacts of artificial habitats. pp. 153–176, In: Seamen, W., Sprague, L.M. (Eds.), Artificial Habitats for Marine and Freshwater Fishes. Academic Press, London, p. 285.

Pomeroy, R.S., Williams, M., 1994. Fisheries Co-Management and Small-Scale Fisheries: A Policy Brief. ICLARM, Manila, pp. 1–15.

Pomeroy, R.S., 1995. Community-based and co-management institutions for sustainable coastal fisheries management in Southeast Asia. Ocean Coast. Manag. 27, 143–162.
Pomeroy, R.S., Pido, M., 1995. Initiatives towards fisheries co management in the Philippines: the case of San Miguel Bay. Mar. Policy 19 (3), 199–211.
Pomeroy, R., Pollnac, R., Predo, C., Katon, B., 1996. Impact evaluation of community-based coastal resource management projects in the Philippines. In: Fisheries Co-management Research Project Research Report No. 3. ICLARM, Manila, pp. 38–282.
Pomeroy, R., Berkes, F., 1997. Two to tango: the role of government in fisheries co-management. Mar. Policy 21 (5), 465–480.
Pomeroy, R.S., Katon, B.M., Harkes, I., 1998. In: Fisheries Co-management: Key Conditions and Principles Drawn from Asian Experiences. International Centre for Living Aquatic Resources Management (ICLARM), MC PO Box 2631, Makati City 0718 (Philippines), pp. 1–23.
Pomeroy, R., Len, G., Pido, M., Silvestre, G., 2010. Ecosystem-based fisheries management in small-scale tropical marine fisheries: emerging models of governance arrangements in the Philippines. Mar. Policy 34, 298–308.
Prabhu, R.R., 2011. Institutional Dynamics in Community-Based Fisheries Resource Management for Sustainable Development of Marine Fisheries in Kerala. PhD thesis, Department of Applied Economics; Cochin University of Science and Technology (CUSAT), Kochi-22 (Kerala), p. 477.
Pradhan, B.R., 1982. In: Preliminary Studies of Syarpu Daha, A Mid Hill Lake in Rukum District, Nepal. Proceedings of the National Science and Technology Congress. National Council of Science and Technology, Kathmandu (Nepal).
Pradhan, B.R., Swar, D.B., 1988. Limnology and Fishery Potential of the Indrasarovar Reservoir, Kulekhani, Nepal. In: De Silva, S.S. (Ed.), Proceedings of a Workshop on Reservoir Fishery Management and Development in Asia. International Development Research Centre, Ottawa (Canada).
Pradhan, G.B.N., Pantha, M.B., 1995. In: Nepal Country Report (Annex.11-12). Report on a Regional Study and Workshop on the Environmental Assessment and Management of Aquaculture Development. Food and Agriculture Organization of the United Nations and Network of Aquaculture Centres in Asia-Pacific, Bangkok (Thailand).
Pradhan, Gagan, B.N. (2009). Reservoir Fisheries in Nepal, pp.: 55-67. In: De Silva, S.S. and Amarasinghe, U.S. (eds.) (2009). Status of Reservoir Fisheries in Five Asian Countries, pp. 116, NACA Monograph No. 2. Network of Aquaculture Centres in Asia-Pacific, Bangkok (Thailand).
Prajoo, N., 2007. Study on Fish Diversity of Harpan Khola and Socio economics of the Jalari Women of Phewa Lake. An M.Sc. Thesis submitted to Tribhuvan University (Nepal).
Prchalová, M., Draštík, V., Kubecka, J., Sricharoendham, B., Schiemer, F., Vijverberg, J., 2003. Acoustic study of fish and invertebrate behaviour in a tropical reservoir. Aquat. Living Resour. 16, 325–331.
Prosser, J., 1978. Environmental Report in Gandaki River Basin Power Study. Nepal Electricity Authority, Kathmandu (Nepal).
Puinyabati, H., Shomorendra, M., Kar, D., 2010a. Helminth parasites of fishes of Chatla Haor, Silchar, Assam. Environ. Ecol. 28 (3A), 1852–1854.
Puinyabati, H., Shomorendra, M., Kar, D., 2010b. Infection of Helminth parasites in relation to the length of fish host. Natl. J. Life Sci. 7 (1), 49–50.
Puinyabati, H., Shomorendra, M., Kar, D., 2013a. *Senga silcharensis* n. sp., a new cestode from the intestine of the fresh water fish, *Channa punctatus* (Bloch) from Chatla Haor, Silchar, Assam. Sci. Cult. 79 (5-6), 245–247.
Puinyabati, H., Shomorendra, M., Kar, D., 2012. Trematode parasites of some freshwater fishes of Manipur and Assam. Biospectra 7 (4), 61–68.
Puinyabati, H., Shomorendra, M., Kar, D., 2013b. Basic Parasitology. pp. xxi + 201, Asiatic Publishing House, New Delhi.
Pushpalatha, K.B.C., 2001. Community-based freshwater fish culture in Sri Lanka. In: De Silva, S.S. (Ed.), Reservoir and Culture-Based Fisheries: Biology and Management. ACIAR Proceedings No. 98; Australian Centre for International Agricultural Research, Canberra.
Pushpalatha, K.B.C., Amarasinghe, U.S., 2007. Co-management strategies for the development of culture-based fi sheries in village reservoirs of Sri Lanka. In: Paper presented at the Eighth Asian Fisheries Forum, 21-23 November 2007. Kochi (India).
Rai, A.K., 2000a. Limnological characteristics of subtropical Lakes, Phewa, Begnas and Rupa in Pokhara valley, Nepal. Limnology 1, 1–13.
Rai, A.K., 2000b. Evaluation of natural food for planktivorous fish in Lakes Phewa, Begnas, and Rupa in Pokhara Valley, Nepal. Limnology 1, 81–89.
Rai, A.K., Shrestha, B.C., Joshi, P.L., Gurung, T.B., Nakanishi, M., 1995. In: Bathymetric maps of Lakes Phewa, Begnas and Rupa, Pokhara Valley, Nepal. Mem. Fac. Sci. Kyoto Univ. Ser. Biol., XVI (No. 1), pp. 49–54.
Rajan, K.V., 1993. Fish trapping devices and methods of southern India. Fish. Technol. 30, 85–93.
Ramachandran, A., 2005. Development of model fisherman villages. In: Community Based Integrated Coastal Management Overview, pp. 1–107. India.
Ramakrishniah, M., 1990. Morpho-drainage Index: A fish yield predictor for Indian reservoirs. In: Paper presented at the 2nd Asian Reservoir Workshop, 15–19 October, Hangzhou (P.R. China).
Ramsar Convention Secretariat (Ed.), 2004. Ramsar Handbooks for Wide Use of Wetlands, second ed. Ramsar Convention Secretariat, Gland, (Switzerland).
Ranibala, T., Vishweshwari, A., Kar, D., Jha, A.N., Puiyabati, H., Shomorendra, M., 2010. Seasonal occurrence and infestation intensity of a digenetic trematode and two cestodes in fishes of Awangsoi fishery, Bishnupur, Manipur. Biospectra 5 (2), 229–232.
Ranibala, T., Shomorendra, M., Kar, D., 2011a. Prevalence and intensity of Nematode parasites infecting the fish, *Anabas testudineus* in Loktak Lake. Environ. Ecol. 29 (3B), 1611–1612.
Ranibala, T., Shomorendra, M., Kar, D., 2011b. Intensity of Helminth infection with special reference to Cestodes in the fish, *Clarias batrachus* (Linn.) of Loktak lake, Manipur. Conserv. Forum J. 3, 52–55.
Ranibala, T., Singh, M.S., Kar, D., 2015. Essentials of Fish Helminthology. pp. xx + 207, Asiatic Publishing House, New Delhi.
Rao, G.S., 2011. Marine fisheries in India: The path ahead. In: Ninth Indian Fisheries Forum, 19–23 December, 2011 (Chennai).
Rawson, D.S., 1952. Mean depth and fish production in large lakes. Ecology 33, 513–521.
Rawson, D.S., 1955. Morphometry as a dominant factor in the productivity of large lakes. Verhandl. Int. Verein Limnol. 12, 164–175.
Reis, A.C.F., 2008. Creative Economy as economy path. a development strategy: a view of developing CBFM introduced in order to bring together the countries. URL: http://www.garimpodesolucoes.com.fishermen and culture shifts them towards a better br/downloads/ebook_en.pdf.
Richards, J., Wu, R.S.S., 1985. Inshore fish community structure in a subtropical estuary. Asian Mar. Biol. 2, 57–68.
Richter, A.F., 1986. Biomanipulation and its feasibility for water quality management in shallow eutrophic water bodies in the Netherlands. Aquat. Ecol. 20, 165–172.

Rick Gregory; Kazi Ali Toufique; Md. Nuruzzaman (2007). Common interests, private gains: A study of co-operative floodplain aquaculture, pp. 16-17. The World Fish Centre; Bangladesh and South Asia Office, Dhaka, Bangladesh. In: Dickson M. and Brooks, A. (eds.) (2007). CBFM-International Conference on Community Based Approaches to Fisheries Management; The World Fish Centre, Conference Proceedings; pp. 38; 75 Print; CD-ROM: pp. 337.

Riji, J.K., Rosalind, M., Kar, D., Mansoor, M., Kumar, M., Singha, R., Waikhom, G.Z., 2016. Detection of Ranavirus infection in cultivated carps of North-East India. Fish Pathol. 51 (Spl. issue), 66–74.

Roberts, T.R., 2001. On the river of no returns: Thailand's Pak Mun Dam and its fish ladder. Natural History Bull. Siam Soc. 49, 189–230.

Rosenthal, H., 1979. Preliminary report and recommendations on reservoir and tank stocking activities in inland fisheries in Sri Lanka. Interim Report to FAO for Reservoir Fisheries in Sri Lanka. Development of fisheries in the man-made lakes and reservoirs. Food and Agriculture Organization of the United Nations (Rome). (97 the Project TCP/SRL/8804).

Rowe, D.K., 2007. Exotic fish introductions and the decline of water clarity in small North Island, New Zealand lakes: A multi-species problem. Hydrobiologia 583, 345–358.

Ruddle, K., 1993. In: van Girzkel, R., Verrips, J. (Eds.), External Forces and Change in Traditional Community-Based Fishery Management Systems in the Asia-Pacific Region. Maritime Anthropological Studies (MAST), 6: (1/2), pp. 1–37.

Ruddle, K., Akimichi, T. (Eds.), 1984. Maritime institutions in the Western Pacific. National Museum of Ethnology, Senri Ethnological Studies, Number 17 (Osaka).

Ruddle, K., 1985. The continuity of traditional management practices: the case of Japanese coastal fisheries. In: Ruddle, K., Johannes, R.E. (Eds.), The Traditional Knowledge and Management of Coastal Systems in Asia and the Pacific. UNESCO-ROSTSEA, Jakarta, pp. 157–179.

Ruddle, K., 1990. The continuity of traditional management practices: the case of Japanese coastal fisheries. In: Ruddle, K., Johannes, R.E. (Eds.), The Traditional Knowledge and Management of Coastal Systems in Asia and the Pacific. UNESCO-ROSTSEA, Jakarta, pp. 263–285.

Saaty, T.L., 1977. A scaling method for priorities in hierarchical structures. J. Math. Psychol. 15, 234–281.

Sadovy, Y., 1999. Artificial reefs and marine fishery reserves as tools in the management of reef fisheries. In: Porcupine: Newsletter of the Department of Ecology and Biodiversity, University of Hong Kong, vol. 19, pp. 22–24.

Sampath, V., 2005. An Overview of the International and National Fishery Policy: Related Issues: A Desk Research. Desk Study on Fisheries Policy, pp. 1–134.

Santos, A.B., 1980. Relevant Laws on Municipal Fisheries. Bureau of Fisheries and Aquatic Resources, Manila (Philippines).

Sangeeta, O., Shomorendra, M., Kar, D., 2010a. Study of seasonal occurrence of fish nematodes of Oinam Lake in Bishnupur District of Manipur. Life Sci. Bull. 7 (2), 259–261.

Sangeeta, O., Shomorendra, M., Kar, D., 2010b. Study of Seasonal occ + urrence of Fish Nematodes of Oinam Lake in Bishnupur district of Manipur. Life Sci. Bull. 7 (2), 259–261.

Saowakoon, S., Boonjung, H., Ngamsnae, P., Jutagate, T., 2008. Financial analysis of the professional fisheries in the man-made lakes in Surin Province, Thailand. Paper presented at the 49th Kasetsart University Annual Conference, February 2008, Bangkok (Thailand).

Sarker, A.K.M.R.A., 2005. Community-Based Aquaculture Development in the Waterlogged Area for the Sustainable Rural Livelihoods. Remote Sensing, GIS and Participatory Approach, Institute of Marine Science and Fisheries, University of Chittagong, Bangladesh. MSc thesis.

Satapornwanit, K., Wongrat, P., 2000. Inland fisheries resources of Thailand. In: Proceedings of the 12th Asian Agriculture Symposium: Agriculture and Water. Kyushu Tokai University, Kyushu.

Sathiadhas, R., Ramachandran, C., Vipinkumar, V.P., Sangeetha, K.P., 2007. Status of fisheries economics and extension research in India: Opportunities and challenges. In: Modayil, M.J., Pillai, N.G.K. (Eds.), Status and Perspectives in Marine Fisheries Research in India. Central Marine Fisheries Research Institute, Kochi. pp. 404.

Schiemer, F., Hofer, R., 1983. A contribution to the ecology of the fish fauna of the Parakrama Samudra reservoir. In: Schiemer, F. (Ed.), Limnology of Parakrama Samudra–Sri Lanka. Dr. W. Junk Publishers, The Hague (The Netherlands).

Schiemer, F., Amarasinghe, U., Frouzova, J., Sricharoendham, B., Silva, E.I.L., 2001. Ecosystem structure and dynamics: A management basis for Asian reservoirs and lakes. In: De Silva, S.S. (Ed.), Reservoir and Culture-based Fisheries: Biology And Management. ACIAR Proceedings No. 98. Australian Centre for International Agricultural Research, Canberra.

Schmidt, C.-C., 2003. Fisheries and Japan: A case of multiple roles? In: OECD (Paris), Proc. International Symposium on Multiple Roles and Functions of Fisheries and Fishing Communities, 13 February 2003, Aomori (Japan), pp. 1–18.

Schouten, R., Sripatraprasit, P., Amornsakchai, S., Vidthayanon, C., 2000. Fish and fisheries up- and down-stream of the Pak Mun Dam. World Commission on Dams (Cape Town).

Seixas, C.S., Berkes, F., 2003. Dynamics of social-ecological changes in a lagoon fishery in southern Brazil. In: Berkes, F., Colding, J., Folke, C. (Eds.), Navigating Social-Ecological Systems. Cambridge University Press, Cambridge, pp. 271–298.

Selby, I., Evans, N.C., 1997. Origins of mud clasts and suspensions on the seabed in Hong Kong. Cont. Shelf Res. 17 (1), 57–78.

Sen, A., 1997. Editorial: Human capital and human capability. World Dev. 25 (12), 1959–1961.

Sen, A., 1981. Poverty and Famines: An Essay on Entitlement and Deprivation. Oxford University Press, Oxford.

Sen, S., Nielsen, J.R., 1996. Fisheries co-management: A comparative analysis. Mar. Policy 20, 405–418.

Shajahan, K.M., 1996. Deep sea fishing policy: a critique. In: Economic and Political Weekly, XXXI (5) (Mumbai).

Shapiro, J., 1990. Biomanipulation in next phase-making it stable. Hydrobiologia 200–201, 13–27.

Sharma, B.P., 1990. Economics of cage culture of carps in the lakes of Pokhara valley, Nepal. Report submitted to Interntional Development Research Centre (Singapore).

Sharma, C.K., 1997. A Treatise on Water Resources of Nepal. Mass Printing Press, Kathmandu (Nepal).

Sharma, R.N., 1979. Cage fish culture in Nepal. In: International workshop on pen and cage culture of fish (11–12 Feb 1979) Tigbaun, Iloilo, Philippines. International Development Research Center and Aquaculture Department. SEAFDEC, Philippines.

Shawna, L.M., Lawrence, W.G., Cargo, M., Margaret, A.P., Daniel, M., Olds, R.S., Reed-Gross, E., 2003. Reliability-Tested Guidelines for Assessing Participatory Research Projects. pp. 1–13.

Sheikh, A.A., Arya, H.P.S., Nikam, T.R., 1993. Knowledge level of farmers on poultry production technology. Ind. J. Extens. Educ. 29 (3 and 4), 92–93.

Shen, S.C., Shao, K.T., Chen, C.T., Chen, C.H., Lee, S.C., Mok, H., 1993. Fishes of Taiwan. Department of Zoology, National Taiwan Univ., Taipei. pp. 960 (in Chinese).
Shen, S.C., 1964. A list of fishes from Hong Kong. Quart. J. Taiwan Mus. 17 (3–4), 193–208.
Shen, S.C., 1967a. Study of flat fishes (Pleuronectiforms or Heterosomata) in the adjacent water of Hong Kong. Quart. J. Taiwan Mus. 20 (1–2), 149–281.
Shen, S.C., 1967b. Additions to the study of flat fishes in the adjacent water of Hong Kong. Rept. Inst. Fishery Biol. 2, 19–27.
Shen, S.C., 1984a. Synopsis of Fishes of Taiwan. Southern Materials Centre, Taipei. pp. 553 (in Chinese).
Shen, S.C., 1984b. Coastal Fishes of Taiwan. Dept. of Zoology, National Taiwan Univ., Taipei. pp. 160 (in Chinese).
Shi, W., 1996. Fish Enhancement and Aquaculture in Inland Open Waters. China Agricultural Press, Beijing (China).
Shrestha, S.B., Shrestha, T.K., Pradhan, B.R., 1979. Reports on the study of aquatic ecology and the potential of fisheries development in Bagmati River Project. A Report prepared for GTZ, Kathmandu (Nepal).
Shrivastava, N.P., Bhattacharya, B.K., 2003. Status of fisheries of floodplain wetlands of Assam. In: Vinci, G.K., Jha, B.C., Bhaumik, U., Mitra, K. (Eds.), Fisheries management of floodplain wetlands in India, Bull. No. 125. Central Inland Fishery Research Institute (CIFRI), ICAR, Barrackpore, pp. 22–29.
Shrivastava, N.P., Das, A.K., 2003. Ecology and Fisheries of selected reservoirs of Madhya Pradesh. CIFRI Bulletin No. 118, Central Inland Fishery Research Institute (CIFRI), ICAR, Barrackpore.
Shirantha, R.R.A.R., Amarasinghe, U.S., 2006. Alien ichthyofauna and aquatic biodiversity in torrential riverine habitats of Sri Lanka. In: Proceedings of the International Conference on Humid Tropical Ecosystems: Changes, Challenges and Opportunities, 4–9 December 2006. Kandy (Sri Lanka).
Sidayat, M., 2008. Community-Based Fishery Management in the Post-Conflict situation: A Case Study in Coastal Villages North Malukud. In partial fulfilment of the requirements for obtaining the degree of Masters of Arts in Development studies, Specialisation: Rural Livelihoods and Global Change (RLGC).
Sihapitukgiat, P., Junsawang, B., Srichareontham, B., Ingthamchit, S., Chittapalapong, T., Su-a-run, K., Ratanachumnong, D., Kakkaew, M., Boonchuwong, P., Sidthimunka, C., Noradee, P., 2002. Limnological, fisheries and socio-economical characteristics of Ubonratana Reservior, Khon Kaen, Thailand. National Inland Fisheries Institute, Bangkok.
Silva, E.I.L., Davies, R.W., 1988. Notes on the biology of *Hemiramphus limbatus* (Hemiramphidae: Pisces) in Sri Lanka. Trop.l Freshwater Biol. 1, 42–49.
Silva, D., Sena, S., Song, S.L., 2015. Challenges and constraints for developing CBF in Cambodia and a possible strategy for success, pp. 51–57. In: De Silva, S.S., Ingram, B.A., Wilkinson, S. (Eds.), Perspectives on culture-based fisheries developments in Asia. NACA Monograph Series No. 3, p. 126.
Sims, N.A., 1990. Adapting Traditional Marine Tenure and Management Practices to Modem Fisheries Framework in the Cook Islands. In: Ruddle, K., Johannes, R.E. (Eds.), Traditional Marine Resource Management in the Pacific Basin. UNESCO-ROSTSEA, Jakarta, pp. 323–358.
Singh, R.K., Shrivastava, N.P., Desai, V.R., 1980. J. Inland Fish. Soc. India 12 (1), 100–111.Singha, R., Puinyabati, H., Shomorendra, M., Kar, D., 2010. Helminth Parasites in relation to length of three Freshwater Fishes of Dolu Lake of Silchar, Assam, India. Environ. Ecol. 28 (4A), 2506–2508.
Singha, R., Singh, M.S., Kar, D., 2014. Parasite Fauna in the Wetland Fishes of India. Today and Tomorrow's Printers and Publishers, New Delhi, p. 209.
Singh, N.R., Kar, D., 2014. Abundance and Distribution of Helminth Parasites in the Fishes of Sone Beel, the biggest wetland in Assam. Int. J. Scient. Res. 3 (12), 67–69.
Singh, N.R., Singh, M.S., Kar, D., 2015a. A check list of Fish species diversity of Sone Beel, the biggest wetland in Assam. NeBIO (Int. J. Environ. Biodiv.) 6 (1), 21–26.
Singh, N.R., Singh, M.S., Kar, D., 2015b. Role of Fishing gears and their significance in Fishery technology of Pumlen Lake. Life Sci. Leaflets 65, 13–25.
Singh, M.S., Singh, N.R., Devi, I.N., Devi, B.D., Ranibala, T., Kar, D., 2015. Check List of Fish species of Loktak Lake, Bishnupur District. Biol. Forum: An Int. J. 7 (1), 171–179.
Sinha, M., Katiha, P.K., 2002. Management of inland fisheries resources under different property regimes. In: Marothia, D.K. (Ed.), Institutionalizing Common Pool Resources. Concept Publishing Company, New Delhi.
Siriweera, W.I., 1986. The inland fishery in Sri Lanka: A historical perspective. Occasional Publication No. 33, Agrarian Services and Training Institute, Colombo (Sri Lanka).
Siriweera, W.I., 1994. A Study of the Economic History of pre-Modern Sri Lanka. Vikas Publishing House Pvt. Ltd., Jangpura, New Delhi (India).
Sirichaipun, N., Manee, M., Janesirisak, S., 2001. Production of Nile tilapia rearing difference feeding regimes in Sirindhorn Reservoir. Technical paper No. 17/2001. Inland Fisheries Division, Department of Fisheries, Bangkok. (In Thai).
Slkkerveen, L.J., 1989. Changing Values and Attitudes of Social and natural Scientists towards Indigenous Peoples and their Knowledge Systems. In: Warren, D.M., Slkkerveen, L.J., Titilola, S.O. (Eds.), Indigenous Knowledge Systems; Implications for Agriculture and International Developement. Studies in Technology and Social Change No. 11. Ames Technology and Social Change Programme, Iowa State University, USA, pp. 121–137.
Smith, E., 1991. Chaos in fisheries management. Maritime Anthropol. Stud. 3 (2), 1.
Soede, L.P., Merkl, A., Claussen, J., Thompson, H., Wheeles, D., 2001. Integrated Marine Management Concessions: A New Approach to an Old Problem. The Conservation and Community Investment Forum (CCIF), Sanur, Bali. pp. 1–15.
Sonowal, M., Kar, S., Das, B.K., Das, S., Singh, N.R., Das, P., Dutta, B., Kar, D., 2015. Preliminary studies on Zooplankton diversity of certain wetlands of Southern Assam. Life Sci. Leaflets 67, 68–78.
South China Sea Fisheries Institute, 1979. The fishes of the islands in the South China Sea. Science Press, Beijing, p. 613 (in Chinese).
Sreenivasan, A., 1989. Principles of ecology and fisheries management in man-made lakes. In: Jhingran, A.G., Sugunan, V.V. (Eds.), Conservation and Management of Inland Capture Fisheries Resources of India. Inland Fisheries Society of India, CIFRI, Barrackpore.

Sricharoendham, B., Amarasinghe, U.S., Jayasinghe, R.P.P., Kaypa, S.M., 2008. Status of the capture fisheries in four Asian reservoirs and a volcanic lake. In: Schiemer, F., Simon, D., Amarasinghe, U.S., Moreau, J. (Eds.), Aquatic Ecosystem and Development: Comparative Asian Perspectives. Biology of Inland Waters Series. Backhuys Publishers, Leiden (The Netherlands).
Srivastava, U.K. (Ed.), 1985. Inland Fish Marketing in India. Vol. 4A: Reservoir Fisheries. Concept Publishing, New Delhi.
Srichareondham, B., Leelapatra, W., Rattanachamnong, D., Kaewjaroon, P., Aimsab, M., 2000. Variation on fish community and catch of Siridhorn Reservoir. Ubon Ratchathani Province. Technical paper No. 206. National Inland Fisheries Institute, Bangkok. (In Thai).
Sriwatana, W., Kasisuwan, S., 1996. Effects of stocking densities on growth and yield of the green catfi sh Mystus nemurus (Cul. & Val.) cultured in fl oating cages. Technical Paper No. 30/1996. Pattani Inland Fisheries Development Centre, Department of Fisheries, Thailand. (In Thai).
Stevenson, G.G., 1991. Common Property Economics: A General Theory and Land Use Application. Cambridge University Press.
Sugunan, V.V., Bhattacharjya, B.K., 2000. Ecology and Fisheries of Beels in Assam. Bull. No. 104. Central Inland Fishery Research Institute (CIFRI), ICAR, Barrackpore.
Sugunan, V.V., Sinha, M., 2001. Sustainable Capture and culture based fisheries in fresh waters of India. In: Pandian, T.J. (Ed.), Sustainable Indian Fisheries. National Academy of Agricultural Sciences, New Delhi, pp. 43–70.
Sugunan, V.V., 1995. Reservoir Fisheries in India. FAO Fisheries Technical Paper No. 345, Food and Agriculture Organization of the United Nations, Rome.
Swar, D.B., Gurung, T.B., 1988. Introduction and cage culture of exotic carps and their impact on fish harvested in Lake Begnas, Nepal. Hydrobiologia 166, 277–283.
Swar, D.B., Pradhan, B.R., 1992. Cage fish culture in the lakes of Pokhara valley, Nepal and its impact on lake fisheries. Asian Fish. Sci. 5, 1–13.
Swingle, H.S., 1950. Relationship and dynamics of balanced and unbalanced fish populations. Bull. Alabama Agric. Exp. Sta. 264, 73.
Tabachnick, B., Fidell, L., 1996. Using Multivariate Statistics, third ed. Harper Collins, New York.
Takamura, N., Kadono, Y., Fukushima, M., Nakagawa, M., Kim, B.-H.O., 2003. Effects of aquatic macrophytes on water quality and phytoplankton communities in shallow lakes. Ecol. Res. 18, 381–395. Publishers (Oxford).
Tang, D.L., Ni, I.H., 1996. Remote sensing of Hong Kong waters: spatial and temporal changes of sea surface temperature. Acta Oceanogr. Taiwanica 35, 173–186.
Tasker, M.L., Knapman, P.A., Laffoley, D., 2000. Effects of fishing on non-target species and habitats: Identifying key nature conservation issues. In: Kaiser, M.J., de Groot, S.J. (Eds.), Effects of Fishing on Non-target Species and Habitats. Blackwell Scientific, pp. 281–289.
Teiwaki, R., 1988. Management of Marine Resources in Kiriati. Atoll Research Unit and Institute of Pacific Studies, University of the South Pacific, Suva.
Terashima, A., 1984. Three new species of the Cyprinoid genus *Schizothorax* from Lake Rara, North-Western Nepal. Jpn. J. Ichthyol. 31, 122–135.
Thanasomwang, W., Tewaratmaneekul, P., Viputhanumas, T., Kongphol, C., 1983. Cage Culture in Thailand. Extension Paper No. 1/1983, Department of Fisheries, Bangkok (In Thai).
Thawornnan, W., Sorasit, S., 1997. Experiment on cage culture of giant gorami *Osphronemus gorami* Lacepede in Bung Sifai, Pichit. Technical Paper No. 8/1997, Inland Fisheries Division, Department of Fisheries (In Thai).
Thayaparan, K., 1982. The role of seasonal tanks in the development of freshwater fisheries in Sri Lanka. J. Inland Fish. 1, 133–167.
The Work Foundation (UK) Staying Ahead, 2011. Korean fisheries: the economic performance of the UK's creative ITQ model has been recently added in Korean industries. URL: http://www.theworkfoundation.com/fisheries only for one species and still in the primary level. assets/docs/publications/176_stayingahead.pdf.
Thilsted, S.H., Wahab, M.A. (Eds.), 2014. Production and conservation of nutrient-rich small fish (SIS) in ponds and wetlands for nutritional security and livelihoods in South Asia. Proceedings of a World Bank/SAFANSI-Funded Regional Workshop on Small Fish and Nutrition; Dhaka, Bangladesh, 1–2 March 2014: pp. 47. The World Bank/South Asia Food and Nutrition Security Initiative (SAFANSI).
Thompson, P.M., Sultana, P., Islam, M.N., Kabir, M.M., Hussain, M.M., Kabir, M.S., 1999. An assessment of co management arrangements developed by the community Based Fisheries Management project in Bangladesh. Paper presented at the International Workshop on Fisheries Co management, 23–27 August, 1999, Penag, (Malaysia).
Thompson, G.B., Horikoshi, M., 1982. Distribution of subtidal benthos collected by trawling in Tolo Harbour and Tolo Channel, Hong Kong. In: Morton, B., Tseng, C.K. (Eds.), The marine flora and fauna of Hong Kong and southern China. Hong Kong Univ. Press, Hong Kong, pp. 733–743.
Thompson, G.B., Wu, R.S.S., Philips, D.J.H., 1982. A trawl survey of the benthos of Tolo Harbour and Tolo Channel in 1978. In: Morton, B., Tseng, C.K. (Eds.), The marine flora and fauna of Hong Kong and Southern China. Hong Kong Univ. Press, Hong Kong, pp. 733–743. pp. 745–760.
Ticheler, H.J., Kolding, J., Chanda, B., 1998. Participation of loçal fishermen in scientific fisheries data collection: A case study from Bangweulu swamps, Zambia. Fish. Manage. Ecol. 5, 81–92.
Toloa, F., Gillett, R., Pelasio, M., 1991. Traditional Marine Conservation in Tokelau. Paper presented to the South Pacific Commission; 23rd Regional Technical Meeting on Fisheries. Noumea, New Caledonia, 5–9 August.
Tom'tavala, D.Y., 1990. National Law, International Law and Traditional Marine Claims: A Case Study of the Trobriand Islands, Papua New Guinea. Dalhousie University, Department of Law, Halifax, Nova Scotia. pp 228.
Tuantong, J., 2015. Reservoir Fisheries of Thailand. In: De Silva, S.S., Amarasinghe, U.S. (Eds.), Status of reservoir fisheries in five Asian countries. Network of Aquaculture Centres in Asia-Pacific, Bangkok (Thailand). pp. 113, NACA Monograph No. 2.
Tuyen, T.V., 2008. Evaluation of co-management issues/community-based coastal resource management. Hue University of agriculture and forestry, 5.
Tuyen, T.V., 2010. Property rights and rights allocation for resource co- management in Tam Giang lagoon, Vietnam. Hue University of agriculture and forestry.
Ullah, M., 1985. Fishing rights, production relations and profitability: A case study of Jamuna River fishermen in Bangladesh. In: Panayotou, T. (Ed.), Small Scale Fisheries in Asia: Socio-economic Analysis and policy. International Development Research Center, Ottawa (Canada).
Ulluwishewa, R., 1995. Traditional practices of inland fishery resources management in the dry zone of Sri Lanka: Implications for sustainability. Environ. Conserv. 22, 127–132.

Umi, N.D., Singh, N.R., Kar, D., 2015. Studies on Fish affected with Epizootic Ulcerative Syndrome with special emphasis on Parasitic infestation: A Project Report. Biol. Forum: An Int. J. 7 (1), 155–164.
UNCTAD, Creative Economy Report 2008, 2009. ITQ systems, fishing Villages are only some unique. URL:http://www.unctad.org/en/docs/ditc20082cer_cultural effects that Korean government made in this en.pdf. sector in order to develop fisheries sector in creative mode.
UNCTAD, 2010. The contribution 6 of tourism to trade and development.
UNCTAD, Creative Economy Report 2010, 2013. Well as delicious Korean dishes. URL: http://www.unctad.org/en/docs/ditc20082cer_.
United Nations Educational, Scientific and Cultural Organization (UNESCO), 1969. Determination of Photosynthetic Pigments in Seawater. UNESCO Monographs on Oceanographic Methodology, UNESCO (Paris).
Unni, K.S., 1985. Comparative limnology of several reservoirs in central India. Int. Revue gesamten Hydrobiol 70 (6), 845–856.
Unni, K.S., 1993. Limnology of Central Indian lakes, reservoirs and ponds: A review. In: Mishra, P.C., Trivedy, R.K. (Eds.), Ecology and Pollution of Indian Lakes and Reservoirs. Ashish Publishing House, New Delhi.
Upreti, B., Upreti, Y.G., 2002. Factors leading to agrobiodiversity loss in developing countries: The case of Nepal. Biodivers. Conserv. 11, 1607–1621.
USAID, 2011. USAI D Evaluation Policy. pp. 1–20 (Washington DC).
Valente, R.M., Evans, N.C., Whiteside, P.G.D., 1996. Environmental monitoring in Hong Kong using the REMOTS seabed camera. Coast. Manage. Trop.l Asia 6, 26–29.
van Chung, N., 2014. Impact of fishery association management on fishery community. Stud. Mundi, Econom. 1 (1), 28–37.
van Mulekom, L., 1998. Capacity and Opportunity Building in a Small-Scale Fisherfolk Community for Community-Based Coastal Resources Management: A Community Development Approach to CB-CRM.
Vass, K.K., 1997. Floodplain wetlands: An important inland fishery resource of India. In: Fisheries Enhancement of Small Reservoir and Floodplain Lakes in India. CIFRI, ICAR, Barrackpore, West Bengal (India), pp. 23–30.
Vaas, K.F., Schurman, J.J., 1949. On the Ecology and Fisheries of some Javanese Freshwaters. Common Gen. Agric. Res. Sta. Buitenzorg Java 97, 1–60.
Vass, K.K., Sugunan, V.V., 2009. Status of Reservoir Fisheries in India. pp. 31–54, In: De Silva, S.S., Amarasinghe, U.S. (Eds.), Status of Reservoir Fisheries in Five Asian countries: China-India-Nepal-Sri Lanka-Thailand. Network of Aquaculture Centres in Asia-Pacific, Bangkok (Thailand), p. 116. NACA Monograph No. 2.
Vass, K.K., 2007a. Indian Reservoir Fisheries Development: Constraints and Solutions. Fishing Chimes 27 (1), 76–81.
Vass, K.K., 2007b. Comparative limnology of lakes in Indus, Gangetic and Brahamaputra basins in India. Abstr. SIL Congress, Canada.
Vass, K.K., 2007c. Conservation and restoration of Lakes and Wetlands. Keynote Abstr. No. KN13, pp. 12, In: Taal, 2007, 12th World Lake Conference. Organised by MoEF GoI and ILEC, Japan.
Verdegem, M.C.J., 1999. Lakes and reservoirs in Indonesia, Malaysia, the Philippines, Singapore and Thailand. In: van Densen, W.L.T., Morris, M.J. (Eds.), Fish and Fisheries of Lakes and Reservoirs in Southeast Asia and Africa. Westbury Publishing, Otely.
Vidthayanon, C., Karnasutra, J., Nabhitabhata, J., 1997. Diversity of Freshwater Fishes in Thailand. Office of Environmental Policy and Planning, Bangkok.
Vipinkumar, V. P. ; Ramachandran, C.; and, Shyam, S. Salim (2013). Co-management Paradigm and Sociological issues in RFM in Indian Marine Fisheries Sector. Winter School on ICT-oriented Strategic Extension for Responsible Fisheries Management; Socio-Economic Evaluation and Technology Transfer Division; Central Marine Fisheries Research Institute: pp. 253-268.
Virapat, C., Mattson, N.S., 2001. Inventory of reservoir fishery resources in Thailand. In: De Silva, S.S. (Ed.), Reservoir and Culture-based Fisheries: Biology and Management. Australian Centre for International Agricultural Research, Canberra. ACIAR Proceeding No. 98.
Virapat, C., Phimonbutra, U., Chantarawaratid, C., 2000. Fishery and Fisheries Management in Thai Reservoirs: Review and Assessment. In: Management of Reservoir Fisheries in the Mekong Basin. Mekong River Commission, Vientiane.
Virapat, C. (1993). Bionomics of fish stocking in two Thai reservoirs: biological, management and development perspectives. Ph.D. thesis, Dalhousie University, Nova Scotia (Canada).
Viswanathan, K.; Garforth, C.; Salayo, N. D.; Ananth, P. N.; and Sithrith, M. (2004). Communication planning for participatory Management of fisheries conflicts in Cambodia and India. Abstract of a presentation at the 7th Asian Fisheries Forum, Penang (Malaysia). Final Technical Report (R8294) Annex 7.5: 8.
Vivekananda Chandrika Dheevara Sabha Household Survey, 2001. Vallarpadam Island, Kerala (India). productivity in aquatic environment. pp. 213, IBP Hand Book No. 12., Blackwell Scientific Pub.(U.K.).
Wikipedia (2011). URL:http://en.wikipedia.org/wiki/Creative Maritime Affairs and Fisheries, Fishing Communities industry.
Wilson, J.A., Kleban, P., 1992. Practical implications of chaos in Fisheries: Ecologically-Adapted Management. University of Maine, pp. 67–75.
WSSD, 2002. Report of the World Summit on Sustainable Development. Johannesburg, South Africa, 26 August-4 September 2002; A/CONF.199/20, United Nations Publication.
Wright, A., 1985. Marine Resource Use in Papua New Guinea: Can Traditional Concepts and Contemporary Development be Integrated. In: Ruddle, K., Johannes, R.E. (Eds.), The Traditional Knowledge and Management of Coastal Systems in Asia and the Pacific, Jakarta. UNESCO-ROSTSEA, Jakarta, pp. 79–99.
Wright, A., 1990. Marine Resource Use in Papua New Guinea: Can Traditional Concepts and Contemporary Development be Integrated. In: Ruddle, K., Johannes, R.E. (Eds.), The Traditional Knowledge and Management of Coastal Systems in Asia and the Pacific, Jakarta. UNESCO-ROSTSEA, Jakarta, pp. 301–321.
Warren, D.M., 1989. The Impact of Nineteenth century Social Science in Establishing Negative Values and Attitudes towards Indigenous Knowledge Systems. In: Warren, D.M., Slikkerveen, L.J., Titilola, S.O. (Eds.), Indigenous Knowledge Systems: Implications for Agriculture and International Development. Iowa State University, USA, pp. 171–183. Studies in Technology and Social Change No. 11. Ames; Technology and Social Change Programme.
Wu, J., 1998. Reservoir fisheries in China. Int. Rev. Hydrobiol. 83 (special issue), 611–618.
Wang, Q., Liu, J., Li, Z., Zhang, T., 2015. Culture-based fisheries in lakes of the Yangtze River basin, China, with special reference to stocking of Mandarin fish and Chinese mitten crab. In: De Silva, S.S., Ingram, B.A., Wilkinson, S. (Eds.), Perspective on culture-based fisheries development in Asia. Network of Aquaculture Centers in Asia-Pacific (NACA). pp. 126, Monograph series No. 3.

Wang, K.L., Liu, L.Y., You, F., Xu, C., 1992. Studies on the genetic variation and systematics of the hairtails fishes from South China Sea. Mar. Sci. 2, 69–71 (in Chinese).
Wang, K.L., Zhang, P.J., Liu, L.Y., You, F., XU, C.; Wang, J.F., 1994. Bio-chemical assessment of the taxonomic status of Trichiuridae population from China coastal water. Acta Oceanol. Sin. 16, 93–104 (in Chinese).
Wang, K.L., Zhang, P.J., Liu, L.Y., You, F., XU, C., 1993. Studies on Trichiuridae from China coastal water. Acta Oceanol. Sin. 15, 77–85 (in Chinese).
Wheeler, A., 1979. Fish in an urban environment. In: Laurie, I.C. (Ed.), Nature in Cities. John Wiley, New York.
Williamson, G.R., 1970. The hydrography and weather of the Hong Kong fishing grounds. Hong Kong Fisheries Bull. 1, 43–49.
Wilson, D.C., Ahmed, M., Susanna, V.S., Kanagaratnam, U., 2006. Cross-scale linkages and adaptive management: fisheries co-management in Asia. Mar. Policy 30, 523–533.
Wong, K.C., 1982. A preliminary survey of the benthic fishes of Tolo Harbour. In: Morton, B., Tseng, C.K. (Eds.), The marine flora and fauna of Hong Kong and southern China. Hong Kong Univ. Press, Hong Kong, pp. 921–933.
Wu, R.S.S., 1984. The feeding habits of seven demersal fish species in a sub-tropical estuary. Asian Mar. Biol. 1, 17–26.
Wu, R.S.S., Cheung, R.Y.H., Shin, P.K.S., 1998. The 'Beneficial Uses' approach in coastal management in Hong Kong: A compromise between rapid urban development and sustainable development. Ocean Coast. Manag. 41 (1), 89–102.
Water Resources Strategy of Nepal, 2002. Hydro Power Commission; Ministry of Water Resources. Government of Nepal, Kathmandu, (Nepal).
Welcomme, R.L., 1979. Fisheries Ecology of the Floodplain Rivers, viii + 317. Longmans, London.
Welcomme, R.L., Bartley, D.M., 1998. Current approaches to the enhancement of fisheries. Fish. Manage. Ecol. 5, 351–382.
Weliange, W.S., Amarasinghe, U.S., Kakkaeo, M., Sricharoendham, B., Schiemer, F., Lopez, N.C., 2008. Trophic patterns of fish assemblages in five Asian lacustrine water bodies. In: Schiemer, F., Simon, D., Amarasinghe, U.S., Moreau, J. (Eds.), Aquatic Ecosystem and Development: Comparative Asian Perspectives; Biology of Inland Waters Series. Backhuys Publishers, Leiden (The Netherlands).
Weerakoon, D.E.M., 1979. Induced breeding of two major species of Chinese carps, *Ctenopharyngodon idellus* and *Aristichthys nobilis* in Sri Lanka. Bull. Fish. Res. Station, Sri Lanka 29, 55–62.
Weimin, M., 2008. Development of Reservoir Fisheries in China. In: De Silva, S.S., Amarasinghe, U.S. (Eds.), Status of Reservoir Fisheries in Five Asian Countries. NACA Monograph No. 2. Network of Aquaculture Centres in Asia-Pacific, Bangkok (Thailand), pp. 1–15. p. 113.
Wijenayake, W.M.H.K., Jayasinghe, U.A.D., Amarasinghe, U.S., Gunaratne, A.B.A.K., De Silva, S.S., 2005. Quantification of morphometric characteristics and aquatic vegetation cover in non-perennial reservoirs using remote sensing techniques for planning culture-based fisheries. In: Proceedings of the International Workshop on Fisheries and Aquatic Research in Sri Lanka. 29–31 March 2005, NARA, Colombo (Sri Lanka).
Wijeyaratne, M.J.S., Amarasinghe, U.S., 1987. Estimations of maximum sustainable fish yield and stocking densities of fish fingerlings in freshwater lakes and reservoirs. Archiv. Hydrobiol. Beiheft Ergeb. Limnol. 28, 305–308.
Xing, Y., Lin, Z., et al., 2007. Investigation to plankton in Kunlong Reservoir and analysis on the fish productivity. Reservoir Fisheries 27, 61–62.
Xiong, B., 1996. A principal component analysis of ecological factors affecting fish yield in reservoirs. Reservoir Fisheries, 3–6 (in Chinese).
Yeung, K.M., 2008. Geodiversity: A new challenge for nature conservation in Hong Kong. In: Agriculture, Fisheries and Conservation Department; The Government of the Hong Kong Special Administrative Region. Proceedings of Asia-Pacific Seminar on Geological Conservation and Sustainable Development (Hong Kong), pp. 47–58.
Yung, Y.K., Yau, K., Wong, C.K., Chan, K.K., Yeung, I., Kueh, C.S.W., Broom, M.J., 1999. Some observations on the changes of physico-chemical and biological factors in Victoria harbour and vicinity, Hong Kong, 1988–1996. Mar. Pollut. Bull. 39, 315–325. Consultation with local fishers on the Hong Kong artificial reefs initiative S177.
Yadava, Y.S., Goswami, M.M., Kar, D., Choudhuri, M., 1989. Present status of hilsa resources in Assam. In: Das, P., Jhingran, A.G. (Eds.), Fish Genetics in India. Today and Tomorrow's Printers and Publishers, New Delhi, pp. 251–256.
Yamamoto, T., 1996. Community-Based Fisheries Management. Nihon University, CBFM in Asia and the Pacific 227, Tokyo, Japan, pp. 209–227.
Yadav, U.K.R., 1994. Studies on the water quality and benthic fauna of the feeding river, Palung and Streams, Charkhu, Thudo, Chitlang of the Kulekhani reservoir. In: Proceedings of Second National Congress on Science and Technology (June 8-11, 1994). RONAST, Kathmandu, Nepal.
Yamagichi, Y., Mitaura, H., Arai, N., Yasushi, M., Kawabata, Y., Khachaphichat, M., Viputhanumas, T., 2005b. Feeding habits of hatcheryreared young Mekong giant catfi sh in a fi sh pond and in Mae Peum Reservoir. In: Proceedings of the 2nd International symposium on SEASTAR2000 and Asian Biologging Science. Graduate School of Informatics, Kyoto University.
Yamagichi, Y., Mitaura, H., Mitsunaga, Y., Arai, N., Khachaphichat, M., Viputhanumas, T., 2005a. Study on feeding habits of Mekong giant catfish in Mae peum Reservoir, Thailand. In: Proceedings of the 2nd International symposium on SEASTAR2000 and Asian Bio-logging Science. Graduate School of Informatics, Kyoto University (Japan).
Yashmin, S., Romita, D.P., Devi, B.R., Kar, D., 2007. Panorama of Fish catching devices in Manipur State of India. Sci. Soc. 5 (2), 181–188.
Zerner, C., 1991. Sharing the Catch in Mandar: Changes in an Indonesian Raft Fishery (1970–1989). In: Poggie, J.J., Pollnac, R.B. (Eds.), Small-scale Fishery Development: Socio cultural Perspectives, Kingston. University of Rhode Island, Kingston, R.I, R.I. International Centre for Marine Resource Development, pp. 42–70.
Zhang, J., 1986. Fisheries Resources in Amur River System. China Agricultural Press, Beijing (China).
Zhang, C.L., 1955. Yellow Sea and East China Sea fishes Assessment Report. Science Press, Beijing (in Chinese).
Zhang, C.I., Lim, J.-H., Park, H.W., Kim, D.-H., Kwon, H.C., 2010. A study on indicators and reference points for the ecosystem-based resource assessment. Bull. Korean Soc. Fish. Technol. 46 (1), 32–49.
Zhu, Y.D., Zhang, C.L., Cheng, Q.T. (Eds.), 1963. Fishes of East Chinese Sea. Science Press, Beijing (in Chinese).

Africa

Adams, W.M., Brockington, D., Dixon, J., Vira, B., 2000. Common choices: policy options for common pool resources. DFID natural resources systems programme semi-arid production. In: Project R7973, policy implications of CPR knowledge in India, Zimbabwe and Tanzania, London, Mimeo. 40 pp.
African Economic Digest, 1993. Namibia: Fishing for Growth. African Economic Digest, p. 4. May 17.
Anderson, T.L., Leal, D., 2001. Free Market Environmentalism. Palgrave, New York.

Anon, 2008. Community-Based Natural Resource Management: How Knowledge is Managed, Disseminated and Used. International Fund for Agricultural Development (IFAD), pp. 1–51.
Anon, 2009. FAO and the World Bank: 'Sunken Billions' Report: Progress and Challenges in Global Marine Fisheries: The Sunken Billions Revisited: Overview. International Bank for Reconstruction and Development/The World Bank.
Anon, 2012. Princes charities' international sustainability unit: international sustainability unit: interim REDD + finance. In: Current Status and Ways Forward for 2013–2020, pp. 1–33.
Anon, 2013a. Protecting and promoting marine biodiversity through small-scale, time-sensitive, community-based programmes. The MCAF (Marine Conservation Action Fund). Newsletter 2 (1), 1–4.
Anon, 2013b. Case study on Concerted Management of the Sea bream Fisheries in Kayar, Senegal. pp. 98–106.
Anon, 2013c. Community-Based Natural Resource Management: Stock-Taking Exercise in Tanzania. Tanzania Natural Resource Forum, pp. 1–84.
Anon, 2015. Community-Based Aquaculture: A Pioneering Viable Alternatives to Fishing. Blue Ventures, London, N7 9DP, United Kingdom, pp. 1–9.
Atta-Mills, J., Alder, J., Sumaila, R.U., 2004. The unmaking of a regional fishing nation: the case of Ghana and West Africa. Nat. Resour. Forum 28, 13–21.
Bavnick, M., 2001. Marine resources management: conflict and regulation in the fisheries of the Coromandel cost. Sage Publications, New Delhi.
BBC, 2010. Namibia's first female trawler captain. BBC World News (August 6). Available at: http://news.bbc.co.uk/2/hi/africa/889336.stm.
Beaudry, F.H., William, F.B., David, R.J., 1993. World fishing fleets: an analysis of distant-water Fleet operations: past–present–future, volume II: Africa and the Middle East. National Marine Fisheries Service, National Oceanic and Atmospheric Administration, Silver Spring, MD, p. 51.
Bergh, P.E., Davies, S., 2005. Against all odds: Taking control of Namibian fisheries. In: Sumailia, U.R., Boyer, D., Skogen, M., Stenshamn, S. (Eds.), Namibia's Fisheries: Ecological, Economic and Social Aspects. Eburon Academic Publishers, Delft, The Netherlands.
Blaikie, P.M., 2000. Development, post-anti- and populist: a critical review. Environ. Plan. A 32, 1033–1050.
Bulayi, M.E., 2001. Community Based Cooperative Fisheries Management for Lake Victoria Fisheries in Tanzania. Final Project, UNU Fisheries Training Programme. The United Nations University, Reykjavik, Iceland.
Butterworth, D.S., Rademeyer, R.A., 2005. Sustainable management initiatives for the southern African hake fisheries over recent years. Bull. Mar. Sci. 76 (2), 287.
Cheung, W.W.L., Vicky, L.W.L., Jorge, S.L., Kelly, K., Reg, W., Zeller, D., Daniel, P., 2009. Redistribution of Fish Catch by Climate Change: A Summary of a New Scientific Analysis. Ocean Science Series; PEW Environment Group of the PEW Charitable Trusts, Washington, DC.
CIA, 2011. The World Fact Book. Central Intelligence Agency, Washington, DC. Available at: www.cia.gov/library/publications/the-world-fact-book/geos/30PERCPolicySerieswa.html.
Costello, C., Gaines, S.D., Lynham, J., 2008. Can catch-shares prevent fisheries collapse? Science 321, 1678–1681.
Costello, C., Lynham, J., Sarah Lester, E., Steven, G.D., 2010. Economic incentives and global fisheries sustainability. Ann. Rev. Resour. Econ. 2, 299–318. Available at: www.annualreviews.org.
Crawford, B., Olsen, S., Mills, D., Tsamenyi, M., Agbogah, K., Hutchful, G., 2013. Solving the Fisheries Crisis in Ghana: A Fresh Approach to Collaborative Fisheries Management, 2013. USAID-URI Integrated Coastal and Fisheries Governance (ICFG) Initiative. Coastal Resources Center, Graduate School of Oceanography, University of Rhode Island. 20 p.
Creel, L., 2003. Ripple effects: population and coastal regions: population reference bureau. Washington, DC. Available at: www.prb.org/Publications/PolicybBriefs/Ripple EffectsPopulationandCoastalRegions.aspx.
Cunningham, S., Neiland, A.E., 2010. African fisheries development aid. In: Leal, D. (Ed.), The Political Economy of Natural Resource Use: Lessons for Fisheries Reform. World Bank, Washington, DC, pp. 21–44.
Das, B., Kar, D., 2011. Essentials of Limnology and Fishery Science. Mangalam Publishers, New Delhi, p. 190.
De Soto, H., 2000. The Mystery of Capital: Why Capitalism Triumphs in the West and Fails Everywhere Else. Basic Books, New York.
Deacon, R.T., 2009. Creating marine assets: property right in ocean fisheries. In: PERC Policy Series, vol. 43. PERC, Bozeman, MT.
Deacon, R.T., 2010. Pathways to fishery reform: accounting for political economy. In: The Political Economy of Natural Resources Use: Lessons for Fisheries Reform.
Deme, M., Failler, P., 2013. Case Study on Concerted Management of Seabream Fisheries in Kayar, Senegal. pp. 98–106.
Demsetz, H., 1967. Toward a theory of property rights. In: papers and proceedings of the seventy-ninth annual meeting of the American economic association. Am. Econ. Rev. 57 (2), 347–359.
Dey, S.C., Kar, D., 1990. Fish yield trend in Sone, a tectonic lake of Assam. Matsya 15–16, 39–43.
Dhar, N., Goswami, M.M., Kar, D., 2004. Co-operative movement: need of the hour in capture fishery sector. In: Goswami, P.J., Acharjee, P.R. (Eds.), Cooperative Movement in Assam: Problems and Prospects. GC College Credit and Thrift Co-operative Society Ltd., Silchar (Assam), pp. 113–116. pp. xiv + 124.
Economist, 2008. A Rising Tide: Scientists find proof that Privatising Fishing stocks can avert a disaster. September 18. Available at: www.economist.com/node/12253181.
Economist, 2011. Happy but Unequal Namibia: On We Go: Swapo Rules and Harmony Prevails. March 10. Available at: www.economist.com/node/18340509.
Eide, A., Peter, M., Steinshamn, S.I., 2003. Assessment of the Economic Benefits African Countries Received from their Marine Resources: Three Case Studies. SNF-Report No. 05/03. Institute for Fencing Fisheries in Namibia and beyond; Research in Economics and Business Administration, Bergen, Norway.
Encyclopedia Britannica, 2010. Britannica World Data-Namibia. Available at: www.britannica.com/EBchecked/topic/402283/Namibia/44014/Independencebefore-the-conquest.
Environmental Defense Fund, 2010. How Catch Shares Work: A Promising Solution, Reviving Fisheries and Fishing Communities. EDF, Washington, DC. Available at: www.edf.org/page.cfm?tagID=3332.
Food and Agriculture Organization, 2008. The state of world fisheries and aquaculture. Fisheries and Aquaculture Department, Food and Agricultural Organization of the United Nations (FAO), Rome. Available at: www.fao.org/docrep/011/i0250e/i0250e00.htm.
Food and Agriculture Organization, 2010. Fishery and Aquaculture Country Profiles: Namibia (Washington DC). Available at: http://www.fao.org/fishery/countrysector/FI-CP_NA/en.

Food and Agriculture Organization, 2011. Fishery and Aquaculture Country Profiles: India, Washington, DC. Available at: www.fao.org/fishery/countrysector/FI-CP_IN/en.
FOROYAA Newspaper, 2010. Africa: Losing U.S. $2–5 Billion Fisheries Revenue Due to Mismanagement, Says DFID Fisheries Advisor. September 22 (Serrekunda, Gambia). Available at: http://allafrica.com/stories/201009220814.html.
Griffith, D.R., 2008. The ecological implications of individual fishing quotas and harvest cooperatives. Front. Ecol. Environ. 6, 191–198.
Gwartney, J.D., Robert, L., Walter, B., 1990. Economic Freedom of the World. 32 PERC Policy Series. Fraser Institute, Vancouver, BC, Canada.
Gwartney, J.D., Joshua, H.C., Robert, L., 2010. Economic Freedom of the World. Annual Report. Fraser Institute, Vancouver, BC, Canada. Available at: www.freetheworld.com/reports.html.
Hartman, A., 2010. Higher Quotas for Hake, Monkfish. The Namibian, Windhoek, Namibia. April 30. Available at: stories/201005030147.html.
Heal, G., Wolfram, S., 2008. Sustainable fisheries. Nature 455, 1044–1045.
Heita, D., 2010. Unemployment the Big Worry: New Era (Namibia). December 17. Available at: www.newera.com.na/article.php?articleid=14629&title=Unemployment%20the%20big%20worry.
Institute for Security Studies, 2007. The Crisis of Marine Plunder in Africa. ISS, Tshwane (Pretoria), South Africa. Available at: www.iss.co.za/pgcontent.php?UID=22399.
Ithindi, A.P., 2003. Rent Capture in the Namibian Fisheries: The Case of Hake. Final Project, UNU-Fisheries Training Programme; the United Nations University. Reykjavik, Iceland.
iWatch News, 2011. Spain's Hake Appetite Threatens Namibia's Most Valuable Fish. The Center for Public Integrity. iWatch News. Available at: http://www.iwatchnews.org.
Jurgen, J.D., 1998. A Tongue in Cheek Assessment: Namibia Brief: Focus on Fisheries and Research. UNU-Fisheries Training Programme, No. 20; the United Nations University. Reykjavik, Iceland.
Kar, D., 1990. Limnology and Fisheries of Lake Sone in the Cachar District of Assam (India). viii + 201, PhD thesis (Published), 15–16 University of Gauhati, Assam (1984), Matsya, pp. 209–213.
Kar, D., 1996. Biodiversity Conservation Prioritisation Project. WWF (India) and WWF (USA).
Kar, D., 1997. Biodiversity Conservation Prioritisation Project (BCPP). Mid-term Review Workshop, WWF (India)-WWF (USA).
Kar, D., 1998. Fish diversity in North-East India with particular reference to Barak drainage. In: Proceedings of the Regional Symposium, Department of Zoology. Cotton College, Guwahati, Assam. vol. 1, 5p.
Kar, D., 1999. An account of the limnology and fisheries of Lake Sone in Assam. In: Ahmad, S.H. (Ed.), Advances in Fish Biology. Hindustan Publishing Corporation, New Delhi.
Kar, D., 2000. Socio-economic development of the fisherwomen through aquaculture with emphasis on integrated farming in the villages around Chatla wetland in Silchar, Assam. In: Report on DBT-Sponsored 'Awareness Workshop on Biotechnology-Based Programmes for Women and Rural Development', Shillong. vol. 1, pp. 13–14.
Kar, D., 2002. Germplasm inventory. evaluation and gene banking of freshwater fishes. In: Proceedings of the NATP-ICAR World Bank-Aided Project Monitoring Workshop, National Bureau of Fish Genetic Resources, vol. 5, 10–11.
Kar, D., 2003. Peoples perspective on fish conservation in the water bodies of South Assam, Mizoram and Tripura. In: Mahanta, P.C., Tyagi, L.K. (Eds.), Participatory Approach for Fish Biodiversity Conservation in North-East India. National Bureau of Fish Genetic Resources, Lucknow, pp. 325–328. pp. v + 412.
Kar, D., 2004. Fish biodiveristy in wetlands and rivers of north-east India with a note on epizootic ulcerative fish disease syndrome. In: Proceedings of the International Symposium. Thapar Institute of Engineering and Technology. vol. 1, 25 p.
Kar, D., 2005a. Fish biodiversity in river Luva in Meghalaya along indo-Bangladesh border. In: Proc. National Symposium on Re-assessment of Fish Genetic Resources in India and Need to Evolve Sustainable Methodology for Conservation. National Bureau of Fish Genetic Resources, ICAR, Lucknow.
Kar, D., 2005b. Fish diversity in the major rivers in Southern Assam, Mizoram and Tripura. In: Nishida, T., Kailola, P.J., Hollingworth, C.E. (Eds.), Proc. 2nd International Symposium on GIS in Fisheries and Aquatic Sciences. vol. 2. University of Sussex, UK, pp. 679–691 (Japan).
Kar, D., 2005c. Fish genetic resources and habitat diversity of the Barak drainage. In: Ramachandra, T.V., Ahalya, N., Rajasekara Murthy, C. (Eds.), Aquatic Ecosystems (Conservation, Restoration and Management). Capital Publishing Company, Bangalore, pp. 68–76. pp. xiii + 396.
Kar, D., 2007a. Fundamentals of Limnology and Aquaculture Biotechnology. Daya Publishing House, New Delhi. pp. vi + 609.
Kar, D., 2007b. Sustainability issues of inland fish biodiversity and fisheries in Barak drainage (Assam), in Mizoram and Tripura. In: Kurup, M.B., Ravindran, K. (Eds.), Sustain Fish. School of Industrial Fisheries, Cochin University of Science and Technology (CUSAT): Proceedings of International Symposium on 'Improved sustainability of Fish Production Systems and Appropriate Technologies for Utilisation', pp xii + 863, 16–18 March, 2005 (Cochin), pp. 555–560.
Kar, D., 2010. Biodiversity Conservation Prioritisation. Swastik Publications, New Delhi. pp. xi + 167.
Kar, D., 2012. Essentials of Fish Biology. Dominant Publishers, New Delhi. pp. vii + 244.
Kar, D., 2013. Wetlands and Lakes of the World. Springer, London. pp. xxx + 687.
Kar, D., 2014. Research Frontiers in Wetlands, Fishery and Aquaculture. Dominant Publishers and Distributors Pvt. Ltd., New Delhi. pp. xi + 320.
Kar, D., 2015. Epizootic Ulcerative Fish Disease Syndrome, Elsevier. Academic Press, USA. pp. xix + 293.
Kar, D., Barbhuiya, A.H., 2011a. Length-weight relationship and condition factor in Puntius amphibious (Valenciennes) and Puntius vittatus Day from Barak valley region of Assam. J. Inland Fish. Soc. India. 42 (1), 15–20.
Kar, D., Barbhuiya, A.H., 2011b. Status of Chocolate Mahseer of River Barak with a note on its morphometry and meristics. J. Inland Fish. Soc. India 42 (1), 32–36.
Kar, D., Dey, S.C., 1982a. Hilsa ilisha (Hamilton) from Lake Sone in Cachar. Assam. Proc. Indian Sci. Congr. 69. 77 p.
Kar, D., Dey, S.C., 1982b. An account of Hilsa ilisha (Hamilton) of Sone Beel (Cachar district, Assam, India). Proc. All-India Sem. Ichthyol. 2. 3 p.
Kar, D., Dey, S.C., 1986. An account of Ichthyospecies of Lake Sone in Barak valley of Assam. Proc. All-India Sem. Ichthyol. 6. 29 p.
Kar, D., Dey, S.C., 1987. In: An account of Fish and Fisheries of Lake Sone in the Barak valley of Assam (India). Proceedings of the Workshop Development of Beel Fisheries in Assam, vol. 1, 13 p.
Kar, D., Dey, S.C., 2002. On the occurrence of advanced fry of Hilsa (Tenualosa) ilisha (Hamilton-Buchanan) in Chatla Haor Seasonal Wetland of Assam. Proc. Zool. Soc. (Calcutta) 55 (2), 15–19.

Kar, D., Barbhuiya, A.H., 2013. Frontiers of Wetlands, Fishers and Aquaculture Research. Mangalam Publications, New Delhi, p. 263.
Kar, D., Barbhuiya, A.H., Thangjam, G., Devi, S.M., Deb, S., Das, B., Chanu, H., Nishima, 2008. Panorama of Fish Biodiversity in certain rivers and wetlands in Manipur. Proc. Zool. Soc. India 7(2), 123–134.
Lake Victoria Fisheries Organization, 2010. Lake Victoria Fisheries Organization: An Institution of the East African Community. LVFO, Jinja, Uganda. Available at: www.lvfo.org/.
Lange, G.-M., 2003. The Value of Namibia's Commercial Fisheries. Ministry of Environment and Tourism, Directorate of Environmental Affairs. February. DEA Research Discussion Paper No. 55. Windhoek, Namibia. Available at: http://mdgs.un.org/unsd/envaccounting/ceea/archive/Fish/Namibia_Value_Commercial_Fisheries.PDF.
Laura, E.H., 2011. In: Meiners, R. (Ed.), Fencing Fisheries in Namibia and Beyond: Lessons from the Developing World, p. 49. PERC Policy Series No. 49.
Leal, D., 2002. Fencing the Fishery: A Primer for Ending the Race to Fish. PERC, Bozeman, MT. Available at: www.perc.org/articles/article120.php.
Leal, D., 2005. Evolving Property Rights in Marine Fisheries. Rowman and Littlefield. Fencing Fisheries in Namibia and beyond, Lanham, MD.
Leal, D., 2010. The Political Economy of Natural Resource Use: Lessons for Fisheries Reform. World Bank, Washington, DC.
Leal, D., Michael, D.A., Baker, P., 2008. Beyond IFQs in Marine Fisheries: A Guide for Federal Policy Makers. PERC, Bozeman, MT.
Lévêque, C., Guégan, J.F., 1990. Les grands Barbus (Teleostei, Cyprinidae) d'Afrique de l'Ouest: révision systématique et parasitofaune branchiale. Revue d'Hydrobiologie Tropicale 23, 41–65.
Lévêque, C., Paugy, D., Teugels, G.G., 1990. In: Orstom-Mrac (Ed.), Faune des Poissons d'Eaux Douce et Saumdtrede I'Afiiqzre de l'Ouest. Coll. Faune tropicale XXVIII, Tome I. Paris: Tervuren.
Libecap, G., 1989. Contracting for Property Rights. Cambridge University Press, New York.
Lopes, S., Pinto, M.A., 2006. Illegal Fishing: The Case of Mozambique. In: Paper Presented at the 'Forging Unity: Coastal Communities and the Indian Ocean's Future': Conference Organized at IIT Madras, Chennai, India, 9–13 October 2001. Available at: www.icsf.net/icsf2006/uploads/publications/proceeding/pdf/english/issue_11/chapter868.pdf. (Accessed 28 October 2010).
Lynch, J., 1996. Lake Victoria and Invasive Species and Trade. TED Case Studies No. 388. Available at: www1.american.edu/ted/victoria.htm.
Manning, P., 2005. The Namibian hake fishery. In: Cunningham, S., Bostock, T. (Eds.), Successful Fisheries Management: Issues, Case Studies and Perspectives. Eburon Academic Publishers, Delft, Netherlands, pp. 169–196.
McClanahan, T.R., Cinner, J.E., Abunge, C., Rabearisoa, A., Paubert, M., Ramahatratra, F., Andrianarivelo, N., 2014. Perceived benefits of fisheries management restrictions in Madagascar. Ecol. Soc. 19 (1), 5.
Meese III, E., 2005. Ed Meese: Reagan Would Still Oppose Law of the Sea Treaty. Human Events. April 25. Available at: www.humanevents.com/article.php?id=7249.
Melber, H., 2003. Of big fish and small fry: the fishing industry in Namibia. Rev. Afr. Polit. Econ. 30 (95), 142–149.
Melcher, A., Ouedraogo, R., Savadogo, M., Sendzimir, J. (Eds.), 2013. Current questions in sustainable fisheries, water management and higher education. Book of Abstracts, Susfish Workshop and West Africa symposium 2013 Vienna, Austria, 23–28 of June 2013. Sustainable Management of Water and Fish Resources in Burkina Faso, Project 56, pp. 1–39.
Menges, W., 2004. PG to Decide Fate of Pirate-Fishing Suspect. The Namibian, Windhoek, Namibia. July, 5.
Meshack, C.K., 2005. Impact of Participatory Forest Management on Local Forest Based Livelihoods: Experience from Eastern Arc Mountains Forests of Tanzania., p. 62.
Ministry of Fisheries and Marine Resources (MFMR), 2003. Annual Report. Windhoek, Namibia. Available at: http://209.88.21.36/opencms/export/sites/default/grnnet/MFMR/downloads/docs/annual_report2003.pdf.
Ministry of Fisheries and Marine Resources (MFMR), 2007. Annual Report. MFMR, Windhoek, Namibia. Available at: http://209.88.21.36/opencms/export/sites/default/grnnet/MFMR/downloads/docs/2007_Report.pdf. 34 PERC Policy Series.
Mukherjee, K., Upadhyay, S., Sane, A., 2008. Policy Brief on Governance and Integrated Coastal Management: India. Mangroves for the Future, Bangkok, Thailand.
Namibia, 2010. Namibia Government—Economy. Available at: www.namibiagovernment.com/economy.htm.
Nichols, P., 2005. Marine fisheries management in Namibia: has it worked? In: Sumailia, U.R., Boyer, D., Skogen, M., Stenshamn, S. (Eds.), Namibia's Fisheries: Ecological, Economic and Social Aspects. Eburon Academic Publishers, Delft, Netherlands, pp. 319–332.
North, D., 1990. Institutions, Institutional Change and Economic Performance. Cambridge University Press, Cambridge, UK.
Omar, M.I.V., 2006. In: Overview of Fisheries Resources: Mozambique. Mozambique Ministry of Fisheries, Department of Aquaculture. Presented at the Workshop on Fisheries and Aquaculture in Southern Africa: Development and Management, Windhoek, Namibia, 21–24 August, 2006.
Ostrom, E., 1990. Governing the Commons: The Evolution of Institutions for Collective Action. Cambridge University Press, Cambridge, UK.
Panduleni, N.E., 2004. Duration of fishing rights and investment: an empirical study of investment in Namibian fisheries. In: Final Project, UNU Fisheries Training Programme. The United Nations University, Reykjavik, Iceland. Available at: http://innri.unuftp.is/Proj04/PanduPRF04.pdf.
Paul, T., Nishida, J., 2008. South Asia Region. Environment Matters. Annual Review. pp. 44–47. Available at: http://siteresources.worldbank.org/EXTENVMAT/Resources/EnvironmentMatters2008-a.pdf.
Pfliegner, K., 2007. Is joint Forest management viable in protection of Forest reserves? Experience from Morogoro region. Arc J. 21, 17–20.
PFM, 2009, June. Proceedings of the First Participatory Forest Management (PFM) Research Workshop: Participatory Forest Management for Improved Forest Quality, Livelihood and Governance.
Piers, B., 2006. Is small really beautiful? Community-based natural resource management in Malawi and Botswana. World Dev. 34 (11), 1942–1957.
Pomeroy, R., 1998. A Process for Community-Based Fisheries Co-Management. NAGA (No. 1448).
Pomeroy, R., Berkes, F., 1997. Two to tango: the role of government in fisheries co-management. Mar. Policy 21 (5), 465–480.
Puddington, A., 2011. Freedom in the World 2011: The Authoritarian Challenge to Democracy. Freedom House, Washington, DC.
Ragnar, A., 2001. Alternative Fisheries Management System: The Icelandic Experience. In: Paper presented at the seminar on `The Future of Commons: Fencing Fisheries in Namibia and beyond: Fishery Policy: The Voice of Europe's Fishing Regions. European Centre for the Regions, Santiago de Comostela, Spain, November 29–30.
Ragnar, A., Kelleher, K., Willmann, R., 2009. The Sunken Billions: The Economic Justification for Fisheries Reform. Agriculture and Rural Development Series. World Bank and the Food and Agriculture Organization of the United Nations, Washington DC.

Roe, D., Nelson, F., Sandbrook, C. (Eds.), 2009. Community Management of Natural Resources in Africa: Impacts, Experiences and Future Directions. International Institute for Environment and Development, London, UK, pp. 1–207. Natural Resource Issues No. 18.
Saito, K., Hailu, M., Spurling, D., 2005. Raising the Productivity of Women Farmers in Sub-Saharan Africa. World Bank, Washington, DC.
Salagrama, V., 2005. Traditional community based management systems in two fishing villages in East Godavari District, Andhra Pradesh, India. In: Cunningham, S., Bostock, T. (Eds.), Successful Fisheries Management Issues: Case Studies and Perspectives. Eburon Academic Publishers, Delft, Netherlands, pp. 111–148.
Salehe, M.A., 2008. Capability and Legitimacy of Beach Management Units (BMUs) to Improve Fisher's Income Through Management of First-Hand Sales System in Lake Victoria, Tanzania. Department of Social and Marketing Studies, Norwegian College of Fishery Science, University of Tromsø, pp. 1–107. Master's thesis in International Fisheries Management.
Scott, A., 2010. New Directions in Fisheries Management. In: Leal, D. (Ed.), The Political Fencing Fisheries in Namibia and beyond; Economy of Natural Resource Use: Lessons for Fisheries Reform. World Bank, Washington, DC, pp. 1–20.
Shyamsundar, P., Araral, E., Suranjan, W., 2005. Devolution of Resource Rights; Poverty, and Natural Resource Management: A Review. Environmental Economic Series No. 104. The World Bank, Washington, DC.
Srinvasan, U., Thara, W., Cheung, W.L., Watson, R., Rashid, U.S., 2010. Overfishing Trends and the Global Food Crisis: A Summary of a New Scientific Analysis. Ocean Science Series: PEW Environment Group of the PEW Charitable Trusts, Washington, DC.
Sumailia, U.R., David, B., Morten, S., Stein, S., 2005. Namibia's Fisheries: Ecological, Economic and Social Aspects. Eburon Academic Publishers, Delft, Netherlands.
Taylor, M., 2001. Whose Agenda? Reassessing the Role of Community-Based Natural Resources Management in Botswana, Mimeo.
Taylor, H., 2002. Insights into participation from critical management and labour process perspectives. In: Cooke, B., Kothari, U. (Eds.), Participation: The New Tyranny. Zed Books, London, pp. 122–138.
The World Bank, 2009. Annual Report: Year in Review. pp. 1–68.
Thurston, E., Rangachari, K., 1909. Castes and Tribes of South India. Government Press, Madras, India.
United Nations, 2011. United Nations Convention on the Law of the Sea, December, 10, 1982. Available at: www.un.org/depts/los/convention_agreements/convention_overview_convention.htm.
Van Zyl, B.J., 2001. Managing Namibia's Marine Fisheries: A Decade of Rebuilding. United Nations Environment Programme, Nairobi, Kenya. Available at: www.unep.org/bpsp/Fisheries/Fisheries%20Case%20Study%20Summaries/VanZyl(Summary).pdf.
Wilson, D.C., 2001. Lake Victoria fishers' attitudes towards management and co-management. In: Geheb, K., Sarch, M.-T. (Eds.), Broaching from the Inland Waters of Africa the Management Impasse: Perspectives on Fisheries and Their Management. Innovative Fisheries Management, Aalborg, Denmark.
Windhoek Observer (Namibia), 2011. Spain's Economic Woes May Hit Hake Sector. August 17. Available at: www.observer.com.na/59-spains-economic-woesmay-hit-hake-sector.
World Bank, 1999. World Bank to Help Bangladesh Boost Environmentally Friendly and Sustainable Fish and Shrimp Production. World Bank, Washington, DC.
World Bank, 2004. Some Key Considerations for Improving Fisheries Management Performance. Policy Brief 15. World Bank, Washington, DC.
World Bank, 2010. Integrated Coastal Zone Management—India. World Bank, Washington DC. Available at: http://web.worldbank.org/external/projects/main?pagePK=64283627&piPK=73230&theSitePK=40941&menuPK=228424&Projectid=P097985.
World Bank, 2011, July 20. Available at: http://web.worldbank.org/external/default/main?pagePK=34370&piPK=34424&theSitePK=4607&menuPK=36PERCPolicySeries34463&contentMDK=20015453.
World Bank, 2011b. Fisheries and Aquaculture. World Bank, Washington, DC. Available at: http://go.worldbank.org/DM59TK1531. (Accessed 1 September 2011).
World Resources Institute (WRI) in collaboration with United Nations Development Programme, United Nations Environment Programme, World Bank, 2005. World Resources: The Wealth of the Poor—Managing Ecosystems to Fight Poverty. Washington, DC. Available at: http://pdf.wri.org/wrr05_lores.pdf.
Yadava, Y.S., Goswami, M.M., Kar, D., Chaudhuri, M., 1989. Present status of hilsa resources in Assam. In: Das, P., Jhingran, A.G. (Eds.), Fish Genetics in India. Today and Tomorrow's Printers and Publishers, New Delhi, pp. 251–256.

Web Links

http://susfish.boku.ac.at/index.html.
http://www.appear.at/.
http://www.entwicklung.at/en/.
http://www.oead.at/.
http://www.boku.ac.at/home.html.
http://www.wau.boku.ac.at/al.html.
http://www.ird.fr/.
http://www.seeterrasse.at/.
http://hydropeaking.boku.ac.at/hytec.htm.
http://www.lod.sk.
http://www.iucn.org/fr/propos/union/secretariat/bureaux/paco/paco_burkinafaso.

Australasia

Adams, T., 1998. The interface between traditional and modern methods of fishery management in the Pacific Islands. Ocean Coast. Manag. 40, 127–142.
Adams, V.M., Mills, M., Jupiter, S.D., Pressey, R.L., 2010. Improving social acceptability of marine protected area networks: a method for estimating opportunity costs to multiple gear types in both fished and currently unfished areas. Biol. Conserv. 144, 350–361.

Agrawal, A., Gibson, C.C., 1999. Enchantment and disenchantment: the role of community in natural resource conservation. World Dev. 27, 629–649.
Ainsworth, C.H., Varkey, D.A., Pitcher, T.J., 2008. Ecosystem simulations supporting ecosystem-based fisheries management in the Coral Triangle, Indonesia. Ecol. Model. 214, 361–374.
Alcala, A.C., 1998. Community based coastal resource management in the Philippines: a case study. Ocean Coast. Manag. 38, 179–186.
Allen, M.S., 2007. Three millennia of human and sea turtle interactions in Remote Oceania. Coral Reefs 26, 959–970.
Amos, M., 1993. Traditionally based marine management systems in Vanuatu. Tradit. Mar. Resour. Manag. Knowl. Inf. Bull. 2, 14–17.
Amos, M.J., 1995. Combination of different management principles for *Trochus niloticus* resources in Vanuatu. SPC Trochus Inf. Bull. 4, 35–41.
Anders, K., Chris, R., Mitchell, L., Stuart, P., Stacy, J., 2011. Mapping fish community variables by integrating field and satellite data; object-based image analysis and modeling in a traditional fijian fisheries management area. Remote Sens. (Basel) 3, 460–483.
Andréfouët, S., Guzman, H.M., 2005. Coral reef distribution: status and geomorphology-biodiversity relationship in Kuna Yala (San Blas) archipelago, Caribbean Panama. Coral Reefs 24, 31–42.
Andréfouët, S., Kramer, P., Torres-Pulliza, D., Joyce, K.E., Hochberg, E.J., Garza-Perez, R., Mumby, P.J., Riegl, B., Yamano, H., White, W.H., Zubia, M., Brock, J.C., Phinn, S.R., Naseer, A., Hatcher, B.G., Muller-Karger, F.E., 2003. Multi-site evaluation of IKONOS data for classification of tropical coral reef environments. Remote Sens. Environ. 88, 128–143.
Andréfouët, S., Gilbert, A., Yan, L., Remoissenet, G., Payri, C., Chancerelle, Y., 2005. The remarkable population size of the endangered clam *Tridacna maxima* assessed in Fangatau Atoll (Eastern Tuamotu, French Polynesia) using *in situ* and remote sensing data. ICES J. Mar. Sci. 62, 1037–1048.
Andréfouët, S., Muller-Krager, F.E., Robinson, J.A., Kranenburg, C.J., Torres-Pulliza, D., Spraggins, S.A., Murch, B., 2006. Global assessment of modern coral reef extent and diversity for regional science and management applications: a view from space. In: Proceedings of 10th International Coral Reef Symposium, 28 June–2 July 2006, Okinawa, Japan, pp. 1732–1745.
Andréfouët, S., Muller-Krager, F.E., Robinson, J.A., Millennium Coral Reef Landsat Archive. Available online http://oceancolor.gsfc.nasa.gov/cgi/landsat.pl.
Andrew, N.L., Bene, C., Hall, S.J., Allison, E.H., Heck, S., Ratner, B.D., 2007. Diagnosis and management of small-scale fisheries in developing countries. Fish Fish. 8, 227–240.
Annette, M.-H., Joeli, V., Nicholas, V.C.P., Selina, S., Nicholas, A.J.G., 2006. Community-based marine resource management in Fiji: from yesterday to tomorrow. In: Conference Paper: International Coral Reef Symposium, pp. 1–8.
Anon, 2004. Transforming the live reef food fish trade towards sustainability in the Republic of the Marshall Islands through community-based coral reef conservation and fisheries management: International Marinelife Alliance, Republic of the Marshall Islands, phase 1 final report, October 2002—March 2004, NOAA coral reef conservation grant program FY 2003. IMA, pp. 1–29.
Anon, 2014. SPC Activities: Community-Based Ecosystem Approach to Fisheries Management (CEAFM) and Climate Change Adaptation in the State of Yap, FSM: Current Status of Marine Resources in Federated States of Micronesia (FSM). Department of Resources and Development, Government of the Federated States of Micronesia, pp. 1–2.
Asafu-Adjaye, J., Tapsuwan, S., 2008. A contingent valuation study of scuba diving benefits: case study in Mu Ko Similan Marine National Park, Thailand. Tour. Manag. 29, 1122–1130.
Aswani, S., 2002. Assessing the effects of changing demographic and consumption patterns on sea tenure regimes in the Roviana Lagoon, Solomon Islands. Ambio 31, 272–284.
Aswani, S., 2005. Customary sea tenure in Oceania as a case of rights-based fishery management: does it work? Rev. Fish Biol. Fish. 15, 285–307.
Aswani, S., Hamilton, R., 2004. The value of many small *vs.* few large marine protected areas in the Western Solomon Islands. SPC Tradit. Mar. Resour. Manage. Knowl. Inf. Bull. 16, 3–14.
Aswani, S., Albert, S., Sabetian, A., Furusawa, T., 2007. Customary management as precautionary and adaptive principles for protecting coral reefs in Oceania. Coral Reefs 26, 1009–1021.
Barker, N., Roberts, C., 2008. Attitudes to and preferences of divers toward regulation. In: Garrod, B., Gössling, S. (Eds.), New Frontiers in Marine Tourism: Diving Experiences, Sustainability, Management. Elsevier, Oxford, pp. 172–187.
Bartlett, C.Y., Pakoa, K., Manua, C., 2009. Marine reserve phenomenon in the Pacific Islands. Mar. Policy 33, 673–678.
Bascompte, J., Melián, C.J., Sala, E., 2005. Interaction strength combinations and the overfishing of a marine food web. Proc. Natl. Acad. Sci. U. S. A. 102, 5443–5447.
Bataille-Benguigui, M.-C., 1997. Men/fish relationships in traditional societies, their effects on human behaviour. In: Séret, B., Sire, J.Y. (Eds.), Proceedings of the 5th Indo-Pacific Fish Conference. Soci´et´e Franc¸aise d'Ichtyologie & Institute de Recherche pour le D´eveloppement, Paris, pp. 849–860.
Bell, J.D., Ratner, B.D., Stobutzki, I., Oliver, J., 2006. Addressing the coral reef crisis in developing countries. Ocean Coast. Manag. 49, 976–985.
Berkes, F., 1989. Common Property Resources: Ecology and Community-Based Sustainable Development. Belhaven, London, UK.
Berkes, F., 2008. Sacred Ecology, second ed. Routledge, New York, USA.
Berkes, F., Colding, J., Folke, C., 2000. Rediscovery of traditional ecological knowledge as adaptive management. Ecol. Appl. 10, 1251–1262.
Blaikie, P., 2006. Is small really beautiful ? Community-based natural resource management in Malawi and Botswana. World Dev. 34, 1942–1957.
Blaschke, T., 2010. Object based image analysis for remote sensing. ISPRS J. Photogramm. Remote Sens. 65, 2–16.
Breiman, L., 2001. Random forests. Mach. Learn. 45, 5–32.
Brundtland, G., 1987. Our Common Future: The World Commission on Environment and Development. OUP.
Brunnschweiler, J.M., 2009. Tracking free-ranging sharks with hand-fed intragastric acoustic transmitters. Mar. Freshw. Behav. Physiol. 42, 201–209.
Brunnschweiler, J.M., Earle, J.L., 2006. A contribution to marine life conservation efforts in the South Pacific: the Shark Reef Marine Reserve, Fiji. Cybium 30 (Suppl), 133–139.
Brunnschweiler, J.M., Sazima, I., 2008. A new and unexpected host for the shark-sucker (*Echeneis naucrates*) with a brief review of the echeneid–host interactions. Mar. Biodivers. Rec. 1, e41.
Bryant, D., Burke, L., McManus, J., Spalding, M., 1998. Reefs at Risk: A Map-Based Indicator of Potential Threats to the World's Coral Reefs. World Resources Institute, Washington, DC, USA, p. 56.
Buckles, D. (Ed.), 1999. Cultivating Peace: Conflicts and Collaboration in Natural Resource Management. International Development Research Centre, London.

Bush, K., 1998. A measure of peace: peace and conflict impact assessment of development projects in conflict zone. In: Working Paper, 1, The Peacebuilding and Reconstruction Program Initiative. IDRC, Ottawa, Canada.
Cadoret, A., 2009. Conflict dynamics in coastal zones: a perspective using the example of Languedoc-Roussillon (France). J. Coast. Conserv. Plann. Manage. 13, 151–163.
Caillaud, A., Boengkih, S., Evans-Illidge, E., Genolagani, J., Havemann, P., Henao, D., et al., 2004. Tabus or not taboos? How to use traditional environmental knowledge to support sustainable development of marine resources in Melanesia. SPC Tradit. Mar. Resour. Manage. Knowl. Inf. Bull. 17, 14–35.
Calamia, M.A., 1996. Traditional ecological knowledge and geographic information systems in the use and management of Hawaii's coral reefs and fish ponds. High Plains. Appl. Anthropol. 16, 144–164.
Carney, D., 1999. Approaches to sustainable livelihood for the rural poor. In: ODI Poverty Briefing Paper 2. Overseas Development Institute, London.
Carr, M.H., Anderson, T.W., Hixon, M.A., 2002. Biodiversity, population regulation, and the stability of coral-reef fish communities. Proc. Natl. Acad. Sci. U. S. A. 99, 11241–11245.
Cash, D.W., Moser, S.C., 2000. Linking global and local scales: designing dynamic assessment and management processes. Glob. Environ. Chang. 10, 109–120.
Cater, C., 2008. Perceptions of and interactions with marine environments: diving attractions from great whites to pygmy seahorses. In: Garrod, B., Gössling, S. (Eds.), New Frontiers in Marine tourism: Diving Experiences, Sustainability, Management. Elsevier, Oxford, pp. 49–64.
Chapman, L.B., Lewis, A.D., 1982. UNDP/MAF Survey of Walu and Other Large Coastal Pelagics in Fiji Waters. Suva, Fiji., p. 36.
Charles, E., Boude, J., Murray, A., Paquotte, P., 2003. Coastal fishing: resource's enhancement and preservation. Ocean Coast. Manag. 46, 421–437.
Christensen, J.D., Jeffrey, C.F., Caldow, C., Monaco, M.E., Kendall, M.S., Appeldoorn, R.S., 2003. Cross shelf habitat utilization patterns of reef fishes in southwestern Puerto Rico. Gulf Caribb. Res. 14, 9–27.
Christie, P., White, A.T., 2007. Best practices for improved governance of coral reef marine protected areas. Coral Reefs 26, 1047–1056.
Chupp, M., 1991. When Mediation is Not Enough. MCS Conciliation Quarterly.
Cillaurren, E., David, G., Grandperrin, R., 2001. Coastal Fisheries Atlas of Vanuatu: A 10-Year Development Assessment. IRD, Paris, France.
Cinner, J., 2005. Socio-economic factors influencing customary marine tenure in the Indo-Pacific. Ecol. Soc. 10, 36.
Cinner, J.E., 2007. Designing marine reserves to reflect local socioeconomic conditions: lessons from long-enduring customary management systems. Coral Reefs 26, 1035–1045.
Cinner, J.E., Aswani, S., 2007. Integrating customary management into marine conservation. Biol. Conserv. 140, 201–206.
Cinner, J.E., McClanahan, T.R., 2006. Socio-economic factors that lead to overfishing in small-scale coral reef fisheries of Papua New Guinea. Environ. Conserv. 33, 73–80.
Cinner, J.E., Marnane, M.J., McClanahan, T.R., Almany, G.R., 2006. Periodic closures as adaptive coral reef management in the Indo-Pacific. Ecol. Soc. 11, 31–65.
Cinner, J.E., Sutton, S.G., Bond, T.G., 2007. Socio-economic thresholds that affect use of customary fisheries management tools. Conserv. Biol. 21, 1603–1611.
Clarke, K.R., Gorley, R.N., 2001. PRIMER v5: User Manual/Tutorial. PRIMER E.Ltd, Plymouth, UK.
Clarke, P., Jupiter, S.D., 2010. Law, custom and community-based natural resource management in Kubulau District (Fiji). Environ. Conserv. 37, 98–106.
Cohen, P.J., Foale, S.J., 2013. Sustaining small-scale fisheries with periodically harvested marine reserves. Mar. Policy 37, 278–287.
Colding, J., Folke, C., 2001. Social taboos: `invisible' systems of local resource management and biological conservation. Ecol. Appl. 11, 584–600.
Colin, P., Arneson, C., 1995. Tropical Pacific Invertebrates. Publication of The Coral Reef Research Foundation, Coral Reef Press, CA, USA, p. 296.
Congalton, R., Green, K., 1999. Assessing the Accuracy of Remotely Sensed Data: Principles and Practices. Lewis Publishers, Boca Raton, FL (USA), p. 137.
Connell, J.H., 1978. Diversity in tropical rain forests and coral reefs—high diversity of trees and corals is maintained only in a non-equilibrium state. Science 199, 1302–1310.
Conroy, C.R., Singh, A.N., Chan, M.K., 1998. Conflicts affecting participatory forest management: some experiences from Orissa. In: Revised Version of a Paper Presented at the Workshop on Participatory Natural Resource Management in Developing Countries: Mansfield College, 6–7th April, 1998 (Oxford).
Cooke, A.J., Polunin, N.V.C., Moce, K., 2000. Comparative assessment of stakeholder management in traditional Fijian fishing-grounds. Environ. Conserv. 27, 291–299.
Costello, C., et al., 2008. Can catch shares prevent fisheries collapse? Science 321, 1678. registered trademark of AAAS.
Craighill Handy, E.S., 1932. Houses, Boats and Fishing in the Society Islands. vol. 90 Bernice P. Bishop Museum, Bulletin, Honolulu, Hawaii, p. 111.
Dalzell, P., Adams, T.J.H., Polunin, N.V.C., 1996. Coastal fisheries in the Pacific Islands. Oceanogr. Mar. Biol. Annu. Rev. 34, 395–531.
Davenport, J., Davenport, J.L., 2006. The impact of tourism and personal leisure transport on coastal environment: a review. Estuar. Coast. Shelf Sci. 67, 280–292.
David, G., Cillaurren, E., 1992. National fisheries development policies for coastal waters: small-scale village fishing, and food self-reliance in Vanuatu. Man Cult. Oceania 8, 35–58.
Davis, D., Tisdell, C.A., 1998. Tourist levies and willingness to pay for a whale shark experience. Tour. Econ. 5, 161–174.
Depondt, F., Green, E., 2006. Diving user fees and the financial sustainability of marine protected areas: opportunities and impediments. Ocean Coast. Manag. 49, 188–202.
di Sciara, G.N., Hanafy, M.H., Fouda, M.M., Afifi, A., Costa, M., 2009. Spinner dolphin (*Stenella longirostris*) resting habitat in Samadai Reef (Egypt, Red Sea) protected through tourism management. J. Mar. Biol. Assoc. U. K. 89, 211–216.
Diedrich, A., 2007. The impacts of tourism on coral reef conservation awareness and support in coastal communities in Belize. Coral Reefs 26, 985–996.

Dorenbosch, M., Grol, M.G.G., Christianen, M.J.A., Nagelkerken, I., van der Velde, G., 2005. Indo-Pacific sea-grass beds and mangroves contribute to fish density coral and diversity on adjacent reefs. Mar. Ecol. Prog. Ser. 302, 63–76.

Dressler, W.B., Usher, B., Schoon, M., Brockington, D., Hayes, T., Kull, C.A., et al., 2010. From hope to crisis and back again? A critical history of the global CBNRM narrative. Environ. Conserv. 37, 5–15.

Drew, J.A., 2005. Use of traditional ecological knowledge in marine conservation. Conserv. Biol. 19, 1286–1293.

Dulvy, N.K., Freckleton, R.P., Polunin, N.V.C., 2004. Coral reef cascades and the indirect effects of predator removal by exploitation. Ecol. Lett. 7, 410–416.

Elizabeth, H.F., 2005. Sustainability certification in community-based fisheries. In: International Conference of The Coastal Society, pp. 92–96.

English, S., Wilkinson, C., Baker, V., 2011. Survey Manual for Tropical Marine Resources. Australian Institute of Marine Science, Townsville, QLD, Australia, p. 390. 1997: Remote Sens., 3: 482.

Faasen, H., Watts, S., 2007. Local community reaction to the 'no-take' policy on fishing in the Tsitsikamma National Park, South Africa. Ecol. Econ. 64, 36–46.

Faasili, U., 1997. The Use of Village By-Laws in Marine Conservation and Fisheries Management. Pacific Science Association Inter-Congress. July 1997, Fiji.

Falanruw, M.V.C., 1994. Traditional fishing on yap. In: Science of Pacific Island Peoples: Ocean and Coastal Studies. vol. 1. Institute of Pacific Studies, University of the South Pacific.

Ferretti, F., Myers, R.A., Serena, F., Lotze, H.K., 2008. Loss of large predatory sharks from the Mediterranean Sea. Conserv. Biol. 22, 952–964.

Foale, S., Manele, B., 2004. Social and political barriers to the use of Marine Protected Areas for conservation and fishery management in Melanesia. Asia Pac. Viewp. 45, 373–386.

Friedlander, A.M., Parrish, J.D., 1998. Habitat characteristics affecting fish assemblages on a Hawaiian coral reef. J. Exp. Mar. Biol. Ecol. 224, 1–30.

Friedlander, A.M., Brown, E.K., Monaco, M.E., 2007. Coupling ecology and GIS to evaluate efficacy of marine protected areas in Hawaii. Ecol. Appl. 17, 715–730.

Froese, R., Pauly, D., Fishbase. Available online: www.fishbase.org.

Gaichas, S.K., 2008. A context for ecosystem-based fishery management: developing concepts of ecosystems and sustainability. Mar. Policy 32, 393–401.

Garla, R.C., Chapman, D.D., Shivji, M.S., Wetherbee, B.M., Amorim, A.F., 2006. Habitat of juvenile Caribbean reef sharks, *Carcharhinus perezi*, at two oceanic insular marine protected areas in the south-western Atlantic Ocean: Fernando de Noronha Archipelago and Atol das Rocas, Brazil. Fish. Res. 81, 236–241.

Garrod, B., 2003. Local participation in the planning and management of ecotourism: a revised model approach. J. Ecotour. 2, 33–53.

Garrod, B., Wilson, J.C., 2004. Nature on the edge? Marine ecotourism in peripheral coastal areas. J. Ecotour. 12, 95–120.

Gillet, R., Lam, M., 1999. The *Beche-de-mer* divers of Ontong Java Atoll. SPC Women Fish. Inf. Bull. 5.

Govan, H., 2009. Achieving the potential of locally managed marine areas in the South Pacific. SPC Tradit. Mar. Resour. Manage. Knowl. Inf. Bull. 25, 16–25.

Grafton, R.Q., Kompas, T., Schneider, V., 2005. The bio-economics of marine reserves: a selected review with policy implications. J. Bioecon. 7, 161–178.

Gratwicke, B., Speight, M.R., 2005. The relationship between fish species richness, abundance and habitat complexity in a range of shallow tropical marine habitats. J. Fish Biol. 66, 650–667.

Grimble, R., Wellard, K., 1997. Stakeholder methodologies in natural resource management: a review of principles, contexts, experience and opportunities. Agr. Syst. 55, 173–193.

Grober-Dunsmore, R., Frazer, T.K., Lindberg, W.J., Beets, J.P., 2007. Reef fish and habitat relationships in a Caribbean seascape: the importance of reef context. Coral Reefs 26, 201–216.

Grober-Dunsmore, R., Frazer, T.K., Beets, J.P., Lindberg, W.J., Zwick, P., Funicelli, N.A., 2008. Influence of landscape structure, on reef fish assemblages. Landsc. Ecol. 23, 37–53.

Gutiérrez, N.L., Hilborn, R., Defeo, O., 2011. Leadership, social capital and incentives promote successful fisheries. Nature 470, 386–389.

Halpern, B.S., 2003. The impacts of marine reserves: do reserves work and does size matter? Ecol. Appl. 13, 117–137.

Halpern, B.S., Warner, R.R., 2002. Marine reserves have rapid and lasting effects. Ecol. Lett. 5, 361–366.

Hamel, M.A., Réfouët, S., 2010. Using very high resolution remote sensing for the management of coral reef fisheries: review and perspectives. Mar. Pollut. Bull. 60, 1397–1405.

Harriss, J., de Renzio, P., 1998. Missing link or analytically missing? The concept of social capital. In: Harriss, J. (Ed.), Policy Arena: Social Capital. Development Studies Institute, London School of Economics, London (1997).

Hasler, H., Ott, J.A., 2008. Diving down the reefs? Intensive diving tourism threatens the reefs of the northern Red Sea. Mar. Pollut. Bull. 56, 1788–1794.

Hastie, T., Tibshirani, R., Friedman, J.H., 2001. The Elements of Statistical Learning: Data Mining, Inference, and Prediction. Springer-Verlag, New York, NY, USA, p. 763.

Hawkins, J.P., Roberts, C.M., 2004. Effects of artisanal fishing on Caribbean coral reefs. Conserv. Biol. 18, 215–226.

Heupel, M.R., Williams, A.J., Welch, D.J., Ballagh, A., Mapstone, B.D., Carlos, G., et al., 2009. Effects of fishing on tropical reef associated shark populations on the Great Barrier Reef. Fish. Res. 95, 350–361.

Hickey, F.R., 2006. Traditional marine resource management in Vanuatu: acknowledging, supporting, and strengthening indigenous management systems. SPC Tradit. Mar. Resour. Manage. Knowl. Inf. Bull. 20, 11–23.

Hickey, F.R., 2007. Traditional marine resource management in Vanuatu: world views in transformation. In: Haggan, N., Neis, B., Baird, I.G. (Eds.), Fishers' Knowledge in Fisheries Science and Management. UNESCO, Paris, France, pp. 147–168. Coastal management source books, 4.

Hicks, C.C., McClanahan, T.R., 2012. Assessing gear modifications needed to optimize yields in a heavily-exploited, multi species, sea-grass and coral reef fishery. PLoS One 7, 1–12.

Hilborn, R., 2007. Managing fisheries is managing people: what has been learned? Fish Fish. 2007 (8), 285–286.

Hill, J., Wilkinson, C., 2004. Methods for Ecological Monitoring of Coral Reefs. Version 1: A Resource for Managers. Australian Institute of Marine Science and Reef Check, Townsville, QLD, Australia, p. 117.
Hirth, H.F., 1993. Marine turtles. In: Wright, A., Hill, L. (Eds.), Nearshore Marine Resources of the South Pacific. Forum Fisheries Agency, Honiara (Solomon Islands), pp. 329–370.
Hobley, M., Shah, K., 1996. What makes a local organisation robust? Evidence from India and Nepal. In: Natural Resources Perspectives, Paper 11. Overseas Development Institute, London.
Hoffmann, T.C., 2002. Coral reef health and effects of socio-economic factors in Fiji and Cook Islands. Mar. Pollut. Bull. 44, 1281–1293.
Hooker, S.K., Gerber, L.R., 2004. Marine reserves as a tool for ecosystem-based management: the potential importance of mega-fauna. Bioscience 54, 27–39.
Horsman, N., Mulipola, A., 1995. Catch data and collection from market surveys in Western Samoa. In: South Pacific Commission and Forum Fisheries Agency Workshop on the Management of South Pacific Inshore Fisheries. Integrated Coastal Fisheries Management Project Technical Document. South Pacific Commission, New Caledonia, p. 17.
Hunt, C., 2003. Economic globalization impacts on Pacific marine resources. Mar. Policy 27, 79–85.
Hviding, E., 1991. Traditional Institutions and their Role in the Contemporary Coastal Resource Management in the Pacific Islands. Naga, ICLARM Quarterly, October 1991, pp. 3–6.
Hviding, E., 2003. Contested rain forests, NGOs, and projects of desire in Solomon Islands. Int. Soc. Sci. J. 55, 539–554.
ICIMOD, 1996. Seminar on Conflict Resolution in Natural Resources. International Centre for Integrated Mountain Development, Kathmandu.
IIED, 1994. Whose Eden: An Overview to Wildlife Management. International Institute for Environment and Development, London.
International Alert, 1996. Resource Pack. International Alert, London.
Isoun, E., Fletcher, C., Frazer, N., Gradie, J., 2003. Multi-spectral mapping of reef bathymetry and coral cover, Kailua Bay, Hawaii. Coral Reefs 22, 68–82.
IUCN, 1994. Guidelines for Protected Area Management Categories. IUCN, Gland, Switzerland and Cambridge.
Jameson, S.C., Tupper, M.H., Ridley, J.M., 2002. The three screen doors: can marine `protected' areas be effective? Mar. Pollut. Bull. 44, 1177–1183.
Jennifer, J., Dirk, Z., Daniel, P., 2010. Counting fish: a typology for fisheries catch data. J. Integr. Environ. Sci. 7 (2), 135–144. Taylor and Francis.
Jennings, S., Polunin, N.V.C., 1996. Fishing strategies, fishery development and socio-economics in traditionally managed Fijian fishing grounds. Fish. Manag. Ecol. 3, 335–347.
Jiménez-Badillo, L., 2008. Management challenges of small-scale fishing communities in a protected reef system of Veracruz, Gulf of Mexico. Fish. Manag. Ecol. 15, 19–26.
Johannes, R.E., 1981. Words of the Lagoon: Fishing and Marine Lore in the Palau District of Micronesia. University of California Press, USA, p. 245.
Johannes, R.E., 1982. Traditional conservation methods and protected marine areas in Oceania. Ambio 11 (5), 258–261.
Johannes, R.E., 1998. Government-supported, village-based management of marine resources in Vanuatu. Ocean Coast. Manag. 40, 165–186.
Johannes, R.E., 2002. The renaissance of community-based marine resource management in Oceania. Annu. Rev. Ecol. Syst. 33, 317–340.
Johannes, R.E., Hickey, F.R., 2001. Evolution of Village-Based Marine Resource Management in Vanuatu Between 1993 and 2001. Coastal Region and Small Island: UNESCO, 2004, Paris, France Papers., p. 15.
Jones, G.P., McCormick, M.I., Srinivasan, M., Eagle, J.V., 2004. Coral decline threatens fish biodiversity in marine reserves. Proc. Natl. Acad. Sci. U. S. A. 101, 8251–8253.
Juerg, M.B., 2009. The Shark Reef Marine reserve: a marine tourism project in Fiji involving local communities. J. Sustain. Tour. 18, 29–42. iFirst.
Jupiter, S.D., Egli, D.P., 2011. Ecosystem-based management in Fiji: successes and challenges after five years of implementation. J. Mar. Biol. https://doi.org/10.1155/2011/940765. Remote Sens. 2011, 3: 480.
Jupiter, S.D., Weeks, R.P., Jenkins, A., Egli, D.P., Cakacaka, A., 2012. Effects of a single intensive harvest event on fish populations inside a customary marine closure. Coral Reefs 31, 321–334.
Jusoff, K., Majid, N.M., 1995. Integrating needs of the local community to conserve forest biodiversity in the State of Kelantan. Biodivers. Conserv. 4 (1), 108–115.
Kallie, J., Taua, T., Faasili, U., 1999. An assessment of community-based management of subsistence fisheries in Samoa. In: MRAG Workshop on Aspects of Coastal Fisheries Resource Management, Fiji, p. 12.
Kelleher, G., Kenchington, R., 1992. Guidelines for establishing marine protected areas. A marine conservation and development report. IUCN, Gland, Switzerland.
Kenchington, R.A., Hudson, B.E.T., 1988. Coral Reef Management Handbook. UNESCO, p. 321.
King, M.G., 1988. Coral reefs in the South Pacific. In: South Pacific Regional Environment Programme Handbook. South Pacific Commission, New Caledonia, p. 40.
King, M.G., 1991. Understanding Fisheries in the South Pacific. Forum Fisheries Agency; Institute of Pacific Studies of the University of the South Pacific, Fiji, p. 42.
King, M.G., 1995. Fisheries biology: assessment and management. In: Fishing News Books. Blackwell Science, Oxford, England, p. 341.
King, M.G., Faasili, U., 1998a. A network of small, community-owned fish reserves in Samoa. PARKS 8, 11–16.
King, M.G., Faasili, U., 1998b. Village fisheries management and community-owned marine protected areas. Naga, ICLARM Q. 21 (2), 34–37.
King, M.G., Faasili, U., 1999. Community-based management of subsistence fisheries in Samoa. Fish. Manag. Ecol. 6, 133–144. UK.
King, M.G., King, S.A., 1995. The oceans and coastal areas and their resources. In: UNESCO-UNEP International Environmental Education Programme; Environmental Education Module, p. 99.
King, M., Lambeth, L., 2000. Fisheries Management by Communities: A Manual on Promoting the Management of Subsistence Fisheries by Pacific Island Communities. Community Fisheries Section; Secretariat of the Pacific Community, Noumea, New Caledonia; New Zealand Official Development Assistance; Australian Agency for International Development, pp. 1–95.
Knudby, A., LeDrew, E., Brenning, A., 2010a. Predictive mapping of reef fish species richness, diversity and biomass in Zanzibar using IKONOS imagery and machine-learning techniques. Remote Sens. Environ. 114, 1230–1241.

Knudby, A., Brenning, A., LeDrew, E., 2010b. New approaches to modelling fish-habitat relationship. Ecol. Model. 221, 503–511.
Kohler, K.E., Gill, S.M., 2006. Coral Point Count with Excel extensions (CPCe): a Visual Basic programme for the determination of coral and substrate coverage using random point count methodology. Comput. Geosci. 32, 1259–1269.
Kronen, M., 2004. Fishing for fortunes? A socio-economic assessment of Tonga's artisanal fisheries. Fish. Res. 70, 121–134.
Krueger, R.A., Casey, M.A., 2000. Focus Groups: A Practical Guide for Applied Research. Sage Publications, Thousand Oaks, USA.
Kuffner, I.B., Brock, J.C., Grober-Dunsmore, R., Bonito, V.E., Hickey, T.D., Wright, C.W., 2007. Relationships between reef fish communities and remotely-sensed rugosity measurements in Biscayne National Park, Florida, USA. Environ. Biol. Fishes 78, 71–82.
Lambeth, L., 1999. An Assessment of the Role of Women Within Fishing Communities in the Republic of Palau. Community Fisheries Section; Secretariat of the Pacific Community, New Caledonia, p. 32.
Lane, M.B., 2008. Strategic coastal governance issues in Fiji: the challenges of integration. Mar. Policy 32, 856–866.
Lane, M.B., McDonald, G., 2005. Community-based environmental planning: operational dilemmas, planning principles and possible remedies. J. Environ. Plan. Manag. 48, 709–731.
Lara, E.N., González, E.A., 1998. The relationship between reef fish community structure and environmental variables in the southern Mexican Caribbean. J. Fish Biol. 53, 209–221.
Lederach, J.P., 1996. The Mediator's Cultural Assumptions. MCS Conciliation Quarterly.
LeFur, J., Guilavogui, A., Teitelbaum, A., 2011. Contribution of local fishermen to improving knowledge of the marine ecosystem and resources in the Republic of Guinea, West Africa. Can. J. Fish. Aquat. Sci. 68, 1454–1469.
Lester, S.E., Halpern, B.S., 2008. Biological responses in marine no-take reserves *versus* partially protected areas. Mar. Ecol. Prog. Ser. 367, 49–56.
Lester, S.E., Halpern, B.S., Grorud-Colvert, K., Lubchenco, J., Ruttenberg, B.I., Gaines, S.D., Airame, S., Warner, R.R., 2009. Biological effects within no-take marine reserves: a global synthesis. Mar. Ecol. Prog. Ser. 384, 33–46.
Lobe, K., Berkes, F., 2004. The padu system of community-based fisheries management: change and local institutional innovation in south India. Mar. Policy 28, 271–281.
Lyzenga, D.R., 1978. Passive remote-sensing techniques for mapping water depth and bottom features. Appl. Optics 17, 379–383.
Lyzenga, D.R., Malinas, N.R., Tanis, F.J., 2006. Multispectral bathymetry using a simple physically based algorithm. IEEE Trans. Geosci. Remote Sens. 44, 2251–2259.
Machlis, G.E., Tichnell, D.L., 1985. The State of the World's Parks: An International Assessment for Resource Management, Policy and Research. Westview Press, Boulder, Colorado.
Mael, J., 2007. Census of Agriculture 2007: Thematic Studies in Fisheries and Forestry. Vanuatu National Statistical Census, Port-Vila, Vanuatu, p. 2007.
Maliao, R.J., Pomeroy, R.S., Turingan, R.G., 2009. Performance of community-based coastal resource management programmes in the Philippines: a meta-analysis. Mar. Policy 33, 818–825.
Mangi, S.C., Roberts, C.M., 2007. Factors influencing fish catch levels on Kenya's coral reefs. Fish. Manag. Ecol. 14, 245–253.
Marc, L., Jennifer, B., Jeremie, K., Jason, R., Caillon, S., 2013. Community-based management of near-shore fisheries in Vanuatu: what works? Mar. Policy 42, 167–176.
Maritorena, S., Morel, A., Gentili, B., 1994. Diffuse-reflectance of oceanic shallow waters—influence of water depth and bottom albedo. Limnol. Oceanogr. 39, 1689–1703.
McClanahan, T., Glaesel, H., Rubens, J., Kiambo, R., 1997. The effects of traditional fisheries management on fisheries yields and the coral-reef ecosystems of southern Kenya. Environ. Conserv. 24, 105–120.
McClanahan, T.R., Marnane, M.J., Cinner, J.E., Kiene, W.E., 2006. A comparison of marine protected areas and alternative approaches to coral-reef management. Curr. Biol. 16, 1408–1413.
McClanahan, T.R., Graham, N.A.J., Maina, J., Chabanet, P., Bruggemann, J.H., Polunin, N.V.C., 2007. Influence of instantaneous variation on estimates of coral reef fish populations and communities. Mar. Ecol. Prog. Ser. 340, 221–234.
McGoodwin, J.R., 1990. Crisis in the world's fisheries: people, problems, and policies. In: Crean, K., Symes, D. (Eds.), Fisheries Management in Crisis. Blackwell Scientific, Cambridge, MA.
McIlwaine, C., 1998. Contesting civil society: reflections from El Salvador. Third World Q. 9 (4), 651–672.
Mellin, C., Andréfouët, S., Ponton, D., 2007. Spatial predictability of juvenile fish species richness and abundance in a coral reef environment. Coral Reefs 26, 895–907.
Mellin, C., Bradshaw, C.J.A., Meekan, M.G., Caley, M.J., 2010. Environmental and spatial predictors of species richness and abundance in coral reef fishes. Glob. Ecol. Biogeogr. 19, 212–222.
Metcalf, S.C., 1995. Communities, parks, and regional planning: a co-management strategy based on the Zimbabwe experience, (Chapter 1). In: Saunier, R.E., Meganck, A. (Eds.), Conservation of Biodiversity and the New Regional Planning. International Union for the Conservation of Nature and Natural Resources, Switzerland.
Metcalf, S.J., Gaughan, D.J., Shaw, J., 2009. Conceptual models for ecosystem based fisheries management (EBFM) in Western Australia. Fisheries research report no. 194, Fisheries Research Division; Western Australian Fisheries and Marine Research Laboratories, pp. 1–42.
Mollica, M., 1999. Village-based marine resource management in Samoa. SPC Women Fish. Inf. Bull. 5.
Moore, C., 1996. The Mediation Process: Practical Strategies for Resolving Conflict. Jossey-Bass.
Mora, C., Andréfoüet, S., Costello, M.J., Kranenburg, C., Rollo, A., Veron, J., et al., 2006. Coral reefs and the global network of marine protected areas. Science 312, 1750–1751.
MSC, 2001. MSC Certification Methodology, Issue 3, March 2001, p. 40.
Muehlig-Hofmann, A., 2007. Traditional authority and community leadership: key factors in community-based marine resource management and conservation. SPC Tradit. Mar. Resour. Manage. Knowl. Inf. Bull. 21, 31–44.
Mumby, P.J., 2001. Beta and habitat diversity in marine systems: a new approach to measurement, scaling and interpretation. Oecologia 128, 274–280.
Mumby, P.J., Edwards, A.J., 2002. Mapping marine environments with IKONOS imagery: enhanced spatial resolution can deliver greater thematic accuracy. Remote Sens. Environ. 82, 248–257.

Mumby, P.J., Harborne, A.R., 1999. Development of a systematic classification scheme of marine habitats to facilitate regional management and mapping of Caribbean coral reefs. Biol. Conserv. 88, 155–163.
Mumby, P.J., Green, E.P., Edwards, A.J., Clark, C.D., 1997. Coral reef habitat-mapping: how much detail can remote sensing provide with ? Mar. Biol. 130, 193–202.
Mumby, P.J., Edwards, A.J., Arias-Gonzalez, J.E., Lindeman, K.C., Blackwell, P.G., Gall, A., Gorczynska, M.I., Harborne, A.R., Pescod, C.L., Renken, H., Wabnitz, C.C.C., Llewellyn, G., 2004a. Mangroves enhance the biomass of coral reef fish communities in the Caribbean. Nature 427, 533–536.
Mumby, P.J., Hedley, J.D., Chisholm, J.R.M., Clark, C.D., Ripley, H., Jaubert, J., 2004b. The cover of living and dead corals from airborne remote sensing. Coral Reefs 23, 171–183.
Munasinghe, M.; Dyson, M.; Egli, N.; Peuker, A.; and, Wencelius, F. (Eds.) n.d. Conservation of West and Central African Rainforests World Bank Environment Paper 1, pp. 222–228, World Bank (Washington, D.C.).
Murray, K., Conner, M.M., 2009. Methods to quantify variable importance: implications for the analysis of noisy ecological data. Ecology 90, 348–355.
Muslim, A.M., Foody, G.M., 2008. DEM and bathymetry estimation for mapping a tide-coordinated shoreline from fine spatial resolution satellite sensor imagery. Int. J. Remote Sens. 29, 4515–4536.
Myers, R.A., Baum, J.K., Shepherd, T.D., Powers, S.P., Peterson, C.H., 2007. Cascading effects of the loss of apex predatory sharks from a coastal ocean. Science 315, 1846–1850.
Nash, W.J., 1990. Spawning and Hatchery Rearing of Trochus (*Trochus niloticus*) in Vanuatu, With a Discussion of Management of the Vanuatu Trochus Fishery. FAO Report, Suva, Fiji.
Ndelu, T., 1998. Conflict management and peace building through community development. Community Dev. J. 33 (2), 109–116.
Neis, B., Schneider, D.C., Felt, L., Haedrich, R.L., Fischer, J., Hutchings, J.A., 1999. Fisheries assessment: what can be learned from interviewing resource users? Can. J. Fish. Aquat. Sci. 56, 1949–1963.
Newman, C., Knudby, A., LeDrew, E., 2007. Assessing the effect of management zonation on live coral cover using multi-date IKONOS satellite imagery. J. Appl. Remote. Sens. 1, 011504.
O'Reilly, S., 1997. The Contribution of Community Development to Peace-Building. World Vision's Area Development Programmes. World Vision, London.
Ochiewo, J., 2004. Changing fisheries practices and their socio-economic implications in South Coast Kenya. Ocean Coast. Manag. 47, 389–408.
OECD, 1998. Conflict, peace and development co-operation on the threshold of the 21st century. In: Development Cooperation Guidelines Series. Organisation for Economic Development and Cooperation, Paris.
Ostrom, E., 1990. Governing the Commons: The Evolution of Institutions for Collective Action. Cambridge University Press, Cambridge, UK.
Ostrom, E., Walker, J., 2003. Trust and Reciprocity: Interdisciplinary Lessons from Experimental Research. Russell Sage Foundation, New York, USA.
Ostrom, E., Burger, J., Field, C.B., Norgaard, R.B., Policansky, D., 1999. Revisiting the commons: local lessons, global challenges. Science 284, 278–282.
Palumbi, S.R., 2004. Marine reserves and ocean neighbourhoods: the spatial scale of marine populations and their management. Annu. Rev. Env. Resour. 29, 31–68.
Phinn, S., Roelfsema, C., Mumby, P., 2011. Multi-scale image segmentation for mapping coral reef geomorphic and benthic community zone. Int. J. Remote Sens. 2011.
Pitcher, T.J., Kalikoski, D., Short, K., Varkey, D., Pramod, G., 2009. An evaluation of progress in implementing ecosystem-based management of fisheries in 33 countries. Mar. Policy 33, 223–232.
Pittman, S.J., McAlpine, C.A., Pittman, K.M., 2004. Linking fish and prawns to their environment: a hierarchical landscape approach. Mar. Ecol. Prog. Ser. 283, 233–254.
Pittman, S.J., Christensen, J.D., Caldow, C., Menza, C., Monaco, M.E., 2007. Predictive mapping of fish species richness across shallow-water seascapes in the Caribbean. Ecol. Model. 204, 9–21.
Pittman, S.J., Costa, M.B., Battista, T.A., 2009. Using lidar bathymetry and boosted regression trees to predict the diversity and abundance of fish and corals. J. Coast. Res. S1, 27–38.
Pomeroy, R.S. (Ed.), 1994. Community management and common property of coastal fisheries in Asia and the Pacific: concepts, methods and experiences. ICLARM Conf. Proc. 45, 124–144.
Pomeroy, R.S., 1998. A process for community-based fisheries co-management. Naga, ICLARM Q. 21 (1), 71–75.
Preston, G.L., Chapman, L.B., Mead, P.D., Taumaia, P., 1987. Trolling Techniques for the Pacific Islands. SPC handbook no. 28, South Pacific Commission, Noumea.
Pretty, J., 2003. Social capital and the collective management of resources. Science 302, 1912–1914.
Purkis, S.J., Graham, N.A.J., Riegl, B.M., 2008. Predictability of reef fish diversity and abundance using remote sensing data in Diego Garcia (Chagos Archipelago). Coral Reefs 27, 167–178.
Putman, R., 1993. Making Democracy Work: Civic Traditions in Modern Italy. Princeton University Press, Princeton, NJ.
R Core Development Team, 2008. A Language and Environment for Statistical Computing: R Foundation for Statistical Computing; Vienna, Austria. Available online http://www.R-project.org/.
Rawlinson, N.J.F., Milton, D.A., Blaber, S.J.M., Sesewa, A., Sharma, S.P., 1995. A Survey of the Subsistence and Artisanal Fisheries in Rural Areas of Viti Levu, Fiji. ACIAR monograph no. 35, p. 138.
Reid-Grant, K., Bhat, M.G., 2009. Financing marine protected areas in Jamaica: an exploratory study. Mar. Pollut. Bull. 33, 128–136.
Rhodes, K.L., Warren-Rhodes, K., Houk, P., Cuetos-Bueno, J., Fong, Q., Hoot, W., 2011. An interdisciplinary study of market forces and nearshore fisheries management in Micronesia. In: Coastal fisheries management programme; Asia Pacific Resource Centre; a report of the marine programme of the Asia Pacific conservation region. The nature conservancy, pp. 1–141. Report no. 6/11.
Robbins, W.D., Hisano, M., Connolly, S.R., Choat, J.H., 2006. Ongoing collapse of coral-reef shark populations. Curr. Biol. 16, 2314–2319.
Roberts, C.M., Polunin, N.V.C., 1991. Are marine reserves effective in management of reef fisheries? Rev. Fish Biol. Fish. 1, 65–91.

Rodríguez-Martínez, R.E., 2008. Community involvement in marine protected areas: the case of Puerto Morelos reef, México. J. Environ. Manage. 88, 1151–1160.
Roelfsema, C., Phinn, S., 2010. Integrating field data with high spatial resolution multispectral satellite imagery for calibration and validation of coral reef benthic community maps. J. Appl. Remote. Sens. 2010, 4.
Roelfsema, C., Phinn, S., Jupiter, S., Comley, M., Beger, M., Peterson, E., 2010. Object-based analysis of high spatial resolution imagery for mapping large coral reef systems in the West Pacific at geomorphic and benthic community scales. In: Proceedings of 2010 IEEE International Geoscience and Remote Sensing Symposium, Honolulu, HI, USA, 25–30 July 2010.
Ruddle, K., 1994. A Guide to the Literature on Traditional Community-Based Fishery Management in the Asia Pacific Tropics, Rome, Italy. FAO fisheries circular no. 869.
Ruddle, K., 1995. A guide to the literature on traditional community-based fishery management in Fiji. SPC Tradit. Mar. Resour. Manage. Knowl. Inf. Bull. 5, 7–15.
Ruddle, K., 1998. The context of policy design for existing community-based fisheries management systems in the Pacific islands. Ocean Coast. Manag. 40, 105–126.
Ruddle, K., Hviding, E., Johannes, R.E., 1992. Marine resources management in the context of customary tenure. Mar. Resour. Econ. 7, 249–273.
Russ, G.R., Alcala, A.C., 1999. Management histories of Sumilon and Apo Marine Reserves, Philippines, and their influence on the national marine resource policy. Coral Reefs 18, 307–319.
Russ, G.R., Alcala, A.C., 2003. Marine reserves: rates and patterns of recovery and decline of predatory fish, 1983–2000. Ecol. Appl. 13, 1553–1565.
Ruttan, L.M., 1998. Closing the commons: cooperation for gain or restraint ? Hum. Ecol. 26, 43–66.
Sadovy, Y., 2005. Trouble on the reef: the imperative for managing vulnerable and valuable fisheries. Fish Fish. 2005 (6), 167–185.
Sale, P.F., Cowen, R.K., Danilowicz, B.S., Jones, G.P., Kritzer, J.P., Lindeman, K.C., et al., 2005. Critical science gaps impede use of no-take fishery reserves. Trends Ecol. Evol. 20, 74–80.
Sano, Y., 2008. The role of social capital in a common property resource system in coastal areas: a case study of community-based coastal resource management in Fiji. SPC Tradit. Mar. Resour. Manage. Knowl. Inf. Bull. 24, 19–32.
Saucerman, S., Kinsolving, A., 1995. Fisheries management and conservation in American Samoa. In: Country Paper 6. Joint FFA/SPC Workshop on the Management of South Pacific Inshore Fisheries. South Pacific Commission, Noumea, p. 7.
Schafer, C.L., 1999. National park and reserve planning to protect biological diversity: some basic elements. Landsc. Urban Plan. 44, 123–153.
Schärer, R., 2001. Community-Based Certification: A Road Map for Sustainable Fisheries., p. 12.
Schoeffel, P., 1997. Myths of Community Management: Sustainability, The State and Rural Development in Papua New Guinea, Solomon Islands and Vanuatu. State, Society and Governance in Melanesia. Discussion paper no. 2007/5, Research School of Pacific and Asian Studies, Australian National University, Canberra, Australia.
Sen, S., Raakkaer-Nielsen, 1996. Fisheries co-management: a comparative analysis. Mar. Policy 20 (5), 405–418.
Shackleton, C.M., Willis, T.J., Brown, K., Polunin, N.V.C., 2010. Reflecting on the next generation of models for community based natural resources management. Environ. Conserv. 37, 1–4.
Shannon, C.E., Weaver, W., 1963. The Mathematical Theory of Communication. University of Illinois Press, Urbana, IL (USA), p. 144.
Souter, D.W., Linden, O., 2000. The health and future of coral reef systems. Ocean Coast. Manag. 43, 657–688.
South, G.R., Goulet, D., Tuqiri, S., Church, M., 1994. Traditional marine tenure and sustainable management of marine resources in Asia and the Pacific. In: Proceedings of an International Workshop Held at the University of the South Pacific, Fiji, p. 312.
SPC, 1998. Pacific Island populations. In: Report prepared by the South Pacific Commission for the international conference on population and development, 5–13 September 1994, Cairo. Revised edition (1998), p. 68.
Stefansson, G., Rosenberg, A.A., 2005. Combining control measures for more effective management of fisheries under uncertainty: quotas, effort limitation and protected areas. Philos. Trans. R. Soc. Lond. B Biol. Sci. 360, 133–146.
Strobl, C., Boulesteix, A.-L., Zeileis, A., Hothorn, T., 2007. Bias in random forest variable importance measures: illustrations, sources and a solution. BMC Bioinf. 8. https://doi.org/10.1186/1471-2105-8-25.
Strobl, C., Boulesteix, A.L., Kneib, T., Augustin, T., Zeileis, A., 2008. Conditional variable importance for random forests. BMC Bioinf. 9. https://doi.org/10.1186/1471-2105-9-307.
Stumpf, R.P., Holderied, K., Sinclair, M., 2003. Determination of water depth with high-resolution satellite imagery over variable bottom types. Limnol. Oceanogr. 48, 547–556.
Su, H.B., Liu, H.X., Heyman, W.D., 2008. Automated derivation of bathymetric information from multi-spectral satellite imagery using a Non-Linear Inversion Model. Mar. Geod. 31, 281–298.
Syms, C., Jones, G.P., 2000. Disturbance, habitat structure, and the dynamics of a coral-reef fish community. Ecology 81, 2714–2729.
Thompson, M.H., Dumont, C.P., Gaymer, C.F., 2008. ISO 14001: towards international quality environmental management standards for marine protected areas. Ocean Coast. Manag. 51, 727–739.
Timothy, E.E., Michael, C.M., Trevor, A.B., Selina, S.H., Olaf, P.J., Jason, S.L., Steven, J., Martell, D., Ana, M.P., John, G.P., Anthony, D., Smith, M., 2012. Catch shares, fisheries, and ecological stewardship: a comparative analysis of resource responses to a rights-based policy instrument. Conserv. Lett. 5 (3), 1–12.
Tongilava, S.L., 1994. Traditional practices and their implications for sustainable development in Tonga. South Pacific regional environment programme. SPREP reports and studies no. 71, p. 56.
Topelko, K.N., Dearden, P., 2005. The shark watching industry and its potential contribution to shark conservation. J. Ecotour. 4, 108–128.
Townsend, C., 2008. Dive tourism, sustainable tourism, and social responsibility: a growing agenda. In: Garrod, B., Gössling, S. (Eds.), New Frontiers in Marine Tourism: Diving Experiences, Sustainability, Management. Elsevier, Oxford, pp. 140–152.
Turner, M.G., Gardner, R.H., O'Neill, R.V., 2001. Landscape Ecology in Theory and Practice: Pattern and Process. Springer Science + Business Media, New York, NY, (USA), p. 404.
Turner, R.A., Cakacaka, A., Graham, N.A.J., Polunin, N.V.C., Pratchett, M.S., Stead, S.M., et al., 2007. Declining reliance on marine resources in remote South Pacific societies: ecological *versus* socio-economic drivers. Coral Reefs 26, 997–1008.

UKFSP/ODI, 1998. Manual on Alternative Conflict Management in Community-Based Natural Resource Projects. Overseas Development Institute, London.
UNCED, 1992. Agenda 21: United Nations Conference on Environment and Development (UNCED). United Nations General Assembly, New York.
Veitayaki, J., Ram-Bidesi, V., Matthews, E., Gibson, L., Vuki, V., 1995. Overview of destructive fishing practices in the Pacific Islands region. South Pacific Regional Environment Programme (SPREP). Reports and studies no. 93, p. 32.
Wagner, J., 2007. Conservation as development in Papua New Guinea: the view from Blue Mountain. Hum. Organ. 66, 28–37.
Warner, M., 1999. Conflict Management and Consensus-Building in Community-Based Natural Resource Projects: Context, Principles, Tools and Training. Overseas Development Institute, London.
Warner, M., 2000. Conflict management in community-based natural resource projects: experiences from Fiji and Papua New Guinea. In: Working Paper 135; Results of ODI Research Presented in Preliminary Form for Discussion and Critical Comment. Overseas Development Institute. ISBN 0 85003 484 1.
Warner, M., Jones, P., 1999. Assessing the need to manage conflict in community-based natural resource projects. In: Natural Resources Perspective, Paper 35. Overseas Development Institute, London.
WCS, 2009. Ecosystem-Based Management Plan. Wildlife Conservation Society, Suva, Fiji, Kubulau District, Vanua Levu, Fiji, p. 121.
Webb, E.L., Maliao, R.J., Siar, S.V., 2004. Using local user perceptions to evaluate outcomes of protected area management in the Sagay Marine Reserve, Philippines. Environ. Conserv. 31, 138–148.
Wedding, L.M., Friedlander, A.M., 2008. Determining the influence of seascape structure on coral reef fishes in Hawaii using a geospatial approach. Mar. Geod. 31, 246–266.
Wedding, L., Friedlander, A.M., McGranaghan, M., Yost, R., Monaco, M., 2008. Using bathymetric LIDAR to define nearshore benthic habitat complexity: implications for management of reef fish assemblages in Hawaii. Remote Sens. Environ. 112, 4159–4165.
Weiant, P., Aswani, S., 2006. Early effects of a community-based marine protected area on the food security of participating households. SPC Tradit. Mar. Resour. Manage. Knowl. Inf. Bull. 19, 16–31.
Wells, M., Brandon, K., 1992. Parks and People: Linking Protected Area Management With Local Communities. World Bank, Washington, D.C.
Wells, M., Brandon, K., 1993. The principles and practice of buffer zones and local participation in biodiversity conservation. Ambio 22 (2–3), 157–162.
Wescott, G., 2006. The long and winding road: the development of a comprehensive, adequate, and representative system of highly protected marine protected areas in Victoria, Australia. Ocean Coast. Manag. 49, 905–922.
White, G., 2007. Indigenous governance in Melanesia. In: State, Society and Governance in Melanesia Discussion Paper No. 2007/5. Research School of Pacific and Asian Studies, Australian National University., Canberra, Australia.
White, A.T., Vogt, H.P., Arin, T., 2000. Philippine coral reefs under threat: the economic losses caused by reef destruction. Mar. Pollut. Bull. 40, 598–605.
Wilson, D.C., Ahmed, M., Siarb, S.V., Kanagaratnamb, U., 2013. Cross-scale linkages and adaptive management: fisheries co-management in Asia. Mar. Policy 30, 523–533.
Winterbottom, R., 1992. In: Cleaver, K. (Ed.), Tropical Forest Action Plans and Indigenous People: The Case of Cameroon.
World Bank, 1995. World Bank Participation Source Book. Environment Department, World Bank.
World Bank, 1998. Post-Conflict Reconstruction: The Role of the World Bank. World Bank, Washington, D.C.
Wright, A., Hill, L. (Eds.), 1991. Inshore Marine Resources of the South Pacific: Information for Fishery Development and Management. Forum Fisheries Agency, Honiara, Solomon Islands.
Zann, L.P., 1980. Tuvalu's Subsistence Fisheries. Institute of Marine Resources, University of the South Pacific, Suva, Fiji, p. 77.
Zeller, D., Booth, S., Craig, P., Pauly, D., 2006. Reconstruction of coral reef fisheries catches in American Samoa: 1950–2002. Coral Reefs 25, 144–152.

Europe

Agrawal, A., 2005. Environmentality: Technologies of Government and the Making of Subjects. Duke University Press, Durham and London.
Alverson, D.L., 1997. Global assessment of fisheries by-catch and discards: a summary overview. In: Pikitch, E.K., et al. (Eds.), Global Trends in Fisheries Management, pp. 115–125. Bethesda.
Amundsen, P.A., Staldvik, F.J., Reshetnikov, Y.S., Kashulin, N., Lukin, A., Bøhn, T., Sandlund, O.T., Popova, O.A., 1999. Invasion of vendace *Coregonus albula* in a subarctic watercourse. Biol. Conserv. 88, 405–413.
Anon, 1994. Information Paper on the southern Bluefin Tuna Fishery. The Australian Fisheries Management Authority, p. 6.
Anon, 2001a. Country note on fisheries management systems: France. Mar. Policy 25 (2), 103–112.
Anon, 2001b. Hungarian Biodiversity Monitoring System, 1998–2001. Ministry of Environment Authority for Nature Conservation, pp. 1–13.
Anon, 2004. Community-driven development and natural resource management. In: IFAD Workshop on Community-Driven Development, 4–5 June, 2004, Casa San Bernardo (Rome), pp. 1–246.
Anon, 2006. Report of The Bergen Conference on Implementing the Ecosystem Approach to Fisheries. pp. 1–28. 26–28 September 2006, (Bergen, Norway).
Anon, 2008a. Fisheries Operational Programme of the Republic of Hungary: 2007–2013: For the Renewal of the Fishery Sector in Hungary. Ministry of Agriculture and Rural Development of the Republic of Hungary, pp. 1–79.
Anon, 2008b. Fourth National Report of the Implementation of CBD in Republic of Macedonia, (2006–2008). pp. 1–32.
Anon, 2009a. Studies and pilot projects for carrying out the common fisheries policy no FISH/2007/03. In: European Commission: An Analysis of Existing Rights-Based Management (RBM) Instruments in Member States and on Setting-Up the Best Practices in the EU, Final Report; Part I, pp. 1–135.
Anon, 2009b. Green Paper: Reform of the Common Fisheries Policy and A Sustainable Future for Aquaculture: 82nd Plenary Session; Committee of the Regions; DEVE-IV-045; European Union; Opinion of the Committee of the Regions. pp. 1–16.
Anon, 2013. Fisheries: 2013 Progress Report: Extract From the Communication From the Commission to the European Parliament and the Council: Enlargement Strategy and Main Challenges 2013–2014: COM (2013)700 Final: The Former Yugoslav Republic of Macedonia 2013 Progress Report. pp. 1–67.

Anon, 2013a. Convention on Biological Diversity, 5th National Report, 2009–2013. Ministry of Environment and Water, Republic of Bulgaria, pp. 1–129.
Anon, 2013b. Key Factors Supporting Small-Scale Coastal Fisheries Management: Synthesis Review. Evaluation Office, Rockefeller Foundation, IMM Ltd, pp. 1–122.
Argillier, C., Pronier, O., Irz, P., 2002a. Approche typologique des peuplements piscicoles lacustres Francais. 1. Les coummunautés des plans d' eau d' altitude supérieure á 1.500 m. Bull. Fr. Pêch Piscic. 365/366, 373–387.
Argillier, C., Pronier, O., Irz, P., Molinier, O., 2002b. Approche typologique des peuplements piscicoles lacustres Francais. 2. Structuration des communautés dans les plans d' eau d' altitude inférieure à 1.500 m. Bull. Fr. Pêch Piscic. 365/366, 389–404.
Arlinghaus, R., Mehner, T., Cowx, I.G., 2002. Reconciling traditional inland fisheries management and sustainability in industrialized countries, with emphasis on Europe. Fish Fish. 3, 261–316.
Armitage, D., Berkes, F., Doubleday, N. (Eds.), 2007. Adaptive Co-Management: Collaboration, Learning, and Multi-Level Governance. University of British Columbia Press, Vancouver.
Armstrong, C.W., Clark, D.J., 1997. Just fishing? Equity and efficiency in fisheries management regimes. Mar. Resour. Econ. 12, 203–220.
Arnason, R., 1991. Efficient management of ocean fisheries. Eur. Econ. Rev. 35, 408–417.
Balon, E.K., 1975. Ecological guilds of fishes: a short summary of the concept and its application. Verh. Int. Ver. Limnol. 19, 2430–2439.
Bartch, J., Abbott-Jamieson, S., Joyce, W., 2009. Voices From the Fisheries: Preserving Local Fisheries Knowledge, Linking Generations, and Improving Environmental Literacy: Handbook. U.S. Dept. of Commerce, NOAA, Natl. Mar. Fish. Serv., Silver Spring, MD, p. 132.
Bauch, G., 1963. Die einheimischen Süsswasserfische, 4. Auflage. Neumann Verlag, Neudamm, Melsungen.
Berkes, F., 2002. Cross-scale institutional linkages: perspectives from the bottom up. In: Ostrom, E., Dietz, T., Dolsak, N., Stern, P.C., Stonich, S., Weber, E.U. (Eds.), The Drama of the Commons. National Academy Press, Washington DC, pp. 293–321.
Berkes, F., 2007a. Adaptive co-management and complexity: exploring the many faces of co-management. In: Armitage, D., Berkes, F., Doubleday, N. (Eds.), Adaptive Co-Management. University of British Columbia Press, Vancouver, pp. 19–37.
Berkes, F., 2007b. Community-based conservation in a globalized world. Proc. Natl. Acad. Sci. U. S. A. 104, 15188–15193.
Berkes, F., 2008. Sacred Ecology, second ed. Routledge, New York.
Berkes, F., George, P., Preston, R., 1991. Co-management. Alternatives 18 (2), 12–18.
Berkes, F., Colding, J., Folke, F. (Eds.), 2003. Navigating Social-Ecological Systems: Building Resilience for Complexity and Change. Cambridge University Press, Cambridge (UK).
Berkes, F., Bankes, N., Marschke, M., Armitage, D., Clark, D., 2005. Cross-scale institutions and building resilience in the Canadian North. In: Berkes, F., Huebert, R., Fast, H., Manseau, M., Diduck, A. (Eds.), Breaking Ice: Renewable Resource and Ocean Management in the Canadian North. University of Calgary Press, Calgary, pp. 225–247.
Berkes, F., Armitage, D., Doubleday, N., 2007. Synthesis: adapting, innovating, evolving. In: Armitage, D., Berkes, F., Doubleday, N. (Eds.), Adaptive Co-Management. University of British Columbia Press, Vancouver, pp. 308–327.
Binkley, M., 1989. Nova Scotian offshore groundfish fishmen—effects of the enterprise allocation and the drive for quality. Mar. Policy 13, 274–284.
Bjorndal, T., Munro, G., 1997. The economics of fisheries management: a survey. In: Papers on Fisheries Economics, p. 41.
BLE, 2002. Monatsbericht Dezember 2002. Bundesanstalt für Landwirtschaft, Hamburg.
BLE (n.d.) (Different Years): Der Markt für Fischereierzeugnisse in der Bundesrepublik Deutschland, Hamburg (Bundesanstalt fur Landwirtschaft).
Blomquist, W., Dinar, A., Kemper, K., 2005. Comparison of institutional arrangements for river basin management in eight basins. In: World Bank Policy Research Working Paper 3636.
BMLFUW, 2002. Gewässerschutzbericht, Bundesministerium für Land- und Forstwirtschaft. Umwelt und Wasserwirtschaft, Wien, p. 213.
Bøhn, T., Amundsen, P.A., 2004. Invasion-mediated changes in the population biology of a dimorphic whitefish, *Coregonus lavaretus* population. Ann. Zool. Fennici 41, 125–136.
Boncoeur, J., Mesnil, B., 1997. Surexploitation des stocks et conflits dans le secteur despêches. Une discussion du "triangle des paradigmes" d'Anthony Charles dans le contexte européen. Association Française d'Halieumétrie—3ème forum halieumétrique, Montpellier.
Bonnie Sutherland, 1997. Nature conservation on private land in Nova Scotia. Proc. N.S. Inst. Sci. 41 (3), 77–89.
Borne, V.D.M., 1880. Die Fischerei Verhältnisse des Deutschen Reiches, Oestereich-Ungarns der Schweiz und Luxenburgs. W. Moeser Hofbuchdruckerei, Berlin.
Bromley, D.W., 1977. Distributional implications of the extended economic zone: some policy and research issues in the fishery. Am. J. Agric. Econ. 59, 887–892.
Bromley, D.W., Bishop, R.C., 1977. From Economic Theory to Fisheries Policy: Conceptual Problems and Management Prescriptions. Economic Aspects of Extended Fisheries Jurisdiction. An Arbor Science Publishers Incorporated, Mishigan, pp. 281–301.
Buchanan, J.M., 1965. An economic theory of clubs. Economica 32, 1–14.
Busch, W.D.N., Sly, P.G., 1992. The Development of an Aquatic Habitat Classification System for Lakes. CRC Press, Boca Raton FL.
Campbell, L.M., Silver, J.J., Gray, N.J., Ranger, S., Broderick, A., Fisher, T., Godfrey, M.H., Gore, S., Jeffers, J., Martin, C., Andrew, M., Richardson, P., Sasso, C., Slade, L., Godley, B., 2008. Co-management of sea turtle fisheries: biogeography *versus* geopolitics. Mar. Policy 33, 137–145.
Casey, K., Dewees, E., et al., 1995. The effects of individual vessel quotas in the British Columbia halibut fishery. Mar. Resour. Econ. 10, 211–230.
Charles, A.T., 1988. Fishery socioeconomics: a survey. Land Econ. 64, 276–295.
Charles, A.T., 1994. Towards sustainability: the fishery experience. Ecol. Econ. 11, 201–211.
Christy, F.T., 1996. Paradigms lost: the death rattle of open access and the advent of property right regimes in fisheries. In: Communication présentée à la huitième conférence bi-annuelle de l'Institut International d'Economie des Pêches, Marrakech.
Christy Jr., F.T., 1973. Fishermen Quotas: A Tentative Suggestion for Domestic Management, Law of the Sea Institute, University of Rhode Island, Occasional Paper No. 19. Kingston, Rhode Island.
Clarke, K.R., Gorley, R.N., 2001. Primer v5. In: User Manual/Tutorial. Primer-E Ltd., Plymouth.
Clarke, K.R., Warwick, R.M., 1997. Change in Marine Communities: An Approach to Statistical Analysis and Interpretations. Marine Laboratory, Plymouth.
Copes, P., 1986. A critical review of the individual quota as a device in fisheries management. Land Econ. 62, 278–291.

Copes, P., 1997a. Adverse impacts of individual quota systems on conservation and fish harvest productivity. In: Discussion Paper 96-1 Presented at the 10th EAFE, Quimper.
Copes, P., 1997b. Social impacts of fisheries management regimes based on individual quotas. Social implications of quota systems in fisheries. In: Proceedings of a Seminar Held in the Vestman Islands, Iceland, pp. 61–90.
Copes, P., 1998. The extended economics of an innate common use resource: the fishery. In: Paper Presented at the Ninth Biennial Conference of the International Institute of Fisheries Economics and Trade (Tromso).
Cunningham, S., 1994. Fishermen's incomes and fisheries management. Mar. Resour. Econ. 9 (3), 241–252.
Davidse, W.E., 1997. Property Rights in Fishing: Effects on the Industry and Effectiveness for Fishery Management Policy. LEI-DLO.
Depestele, J., Courtens, W., Degraer, S., Haelters, J., Hostens, K., Houziaux, J.-S., Merckx, B., Polet, H., Rabaut, M., Stienen, E.W.M., Vandendriessche, S., Verfaillie, E., Vincx, M., 2012. An integrated impact assessment of trammel net and beam trawl fisheries. In: Science for a Sustainable Development, SD/NS/08A. Belgian Science Policy Office (SSD), Brussels (Belgium), p. 233.
Depestele, J., Wouter, C., Steven, D., Jan, H., Kris, H., Mardik, L., Eunice, P., Bea, M., Hans, P., Marijn, R., Henning, R., Eric, S., Sofie, V., Filip, A.M.V., Magda, V., 2014. Sensitivity assessment as a tool for spatial and temporal gear-based fisheries management. Ocean Coast. Manag. 102, 149–160.
Dey, S.C., Kar, D., 1989. Fishermen of Lake Sone in Assam: their socio-economic status. Sci. Cult. 55, 395–398.
Dolman, W.B., 1990. Classification of Texas reservoirs in relation to limnology and fish community associations. Trans. Am. Fish. Sci. 119, 511–520.
Döring, R., 2001. Die Zukunft der Fischerei im Biosphärenreservat Südost-Rügen. Lang-Verlag, Frankfurt a. M. European Fisheries Council; Regulation 2371/2002: On the conservation and sustainable exploitation of fisheries resources under the common fisheries policy. Off. J. Eur. Commun., 31.12.2002—L358/59.
Döring, R., Hart, P., Johnson, M., 2003. Small scale *versus* trawl fisheries in coastal waters: the case of the German Baltic Herring fishery. In: Proceedings of the Workshop, 'Who Owns the Sea', Tjärno, Sweden, pp. 113–140.
Eckmann, R., 1995. Fish species richness in lakes of the northeastern lowlands in Germany. Ecol. Freshw. Fish 4, 62–69.
Edwards, S.F., 1994. Ownership of renewable ocean resources. Mar. Resour. Econ. 9, 253–273.
Eliasen, S., Sverdrup, S., 2009. Nordic Experience of Fisheries Management: Seen in Relation to the Reform of the EU Common Fisheries Policy. TemaNord, 579, pp. 1–93.
EU, 2000. Directive 2000/60/EC of the European Parliament and the Council of 23 October 2000 Establishing a Framework for Community Action in the Field of Water Policy. The European Parliament and the Council of the European Union, p. 72.
Eythorsson, E., 1996. Theory and practice of ITQs in Iceland: privatization of common fishing rights. Mar. Policy 20, 269–281.
Falk-Petersen, J., Bøhn, T., Sandlund, O.T., 2006. On the numerous concepts in invasion biology. Biol. Invasions 8 (6), 1409–1424.
FAO, 1993. Marine Fisheries and the Law of the Sea: A Decade of Change. FAO Fisheries Circular, Rome.
FAO, 2002. Information on Fisheries Management in the Republic of Bulgaria.
FAO, 2012. The State of Food and Agriculture. Investing in Agriculture for a Better Future. Food and Agriculture Organization of the United Nations, pp. 1–182.
Fisherman Vancouver B.C., 1987. February 23, p. 8.
Fraude, H., 1925. Gewässer und Fischfang um Rügen, Bergen auf Rügen. Verlag von Walter Krohß.
Garthe, S., Hüppop, O., 2004. Scaling possible adverse effects of marine wind farms on seabirds: developing and applying a vulnerability index. J. Appl. Ecol. 41, 724–734.
Gassner, H., Zick, D., Wanzenböck, J., Lahnsteiner, B., Tischler, G., 2003a. Die Fischartengemeinschaften der grossen Österreichischen Seen – Vergleich zwischen historischer und aktueller Situation, fischökologische Seentypen – Schriftenreihe des BAW, Band 18, Wien, 163 Seiten, Bundesamt für Wasserwirtschaft.
Gassner, H., Zick, D., Wanzenböck, J., Lahnsteiner, B., Tischler, G., 2003b. Development of a Fish Based Lake Typology for Austrian Lakes >50 ha. Nordic Councel of Ministers, Copenhagen.—Tema Nord 2003/547: 47–49 (Fish Based Lake Typology in Austria 43; © 2005b). WILEY-VCH Verlag Gmb H & Co. KGaA, Weinheim.
Gassner, H., Wanzenböck, J., Tischler, G., 2003c. Ecological integrity assessment of lakes using fish communities—suggestions of new metrics developed in two Austrian prealpine lakes. Int. Rev. Hydrobiol. 88, 635–652.
Gassner, H., Wanzenböck, J., Zick, D., Tischler, G., Pamminger-Lahnsteiner, B., 2005. Fish based lake typology in Austria. Int. Rev. Hydrobiol. 90 (4), 422–432.
Geen, G., Nayar, M., 1989. Individual transferable quotas in the southern bluefin tuna fishery: an economic appraisal. In: Neher, P.A., et al. (Eds.), Rights Based Fishing. NATO ASI Series, pp. 355–381.
Giger, W. (Ed.), 2008. Proceedings of the Swiss—Romanian Research Programme on Environmental Science and Technology (ESTROM), 61. vol. 14. National Institute of Marine Geology and Geo-Ecology; Geo-Eco-Marina, pp. 1–95.
Gutierrez, N.L., 2015. Management and Co-management options for small-scale fisheries in the Mediterranean and the Black sea. Updating national biodiversity strategies and action plans. In: Srour, A., Ferri, N., Bourdenet, D., Fezzardi, D., Aurora, N. (Eds.), First Regional Symposium on Sustainable Small-Scale Fisheries in the Mediterranean and Black Sea. Food and Agricultural Organization of the United Nations, pp. 147–184.
Guyader, O., 1998a. Evaluation économique et simulation des effets des régulations sur les comportements des agents: le cas des pêcheries de thon germon de l'Atlantique Nord Est. Thèse de doctorat Université de Rennes 1, p. 420.
Guyader, O., 1998b. Simulation de la mise en oeuvre de quotas individuels transférables et application à la pêche germonière. In: La dynamique des systèmes—Complexité et chaos. Hermes, Paris, pp. 147–169.
Gwenaël, C., Charles, F.B., Patrick, B., Le Diréach, L., 2008. The management of artisanal fishing within the Marine Protected Area of the Port-Cros National Park (North-West Mediterranean Sea): a success *story?*. ICES J. Mar. Sci. 66 (1), 41–49.
GWP Technical Advisory Committee, 2000. TAC Background Paper Number 4. Integrated Water Resources Management.
Haempel, O., 1930. Biologische und fischereiliche Typen der Alpenseen. Verh. Int. Ver. Limnol. 5, 337–344.
Hanna, S., Munasinghe, M., 1995. Property Rights and the Environment: Social and Ecological Issues. The Beijer International Institute of Ecological Economics and the World Bank, Washington.
Hannesson, R., 1985. Inefficiency through government regulations: the case of Norway's fishery policy. Mar. Resour. Econ. 2, 115–141.

Hannesson, R., 1991. From common fish to rights-based fishing. Eur. Econ. Rev. 35, 397–407.
Hannesson, R., 1996. Fisheries Mismanagement: The Case of the North Atlantic Cod. Fishing News Books, Oxford, p. 160.
Hiddink, J.G., Jennings, S., Kaiser, M.J., 2007. Assessing and predicting the relative ecological impacts of disturbance on habitats with different sensitivities. J. Appl. Ecol. 44, 405–413.
Holden, M., 1994. The Common Fisheries Policy. Blackwell Science, Oxford.
Holland, D., 2000. Fencing the fisheries commons: regulatory barbed wire in the Alaskan ground fish fisheries. Mar. Resour. Econ. 15, 141–149.
Huttula, E., Autti, J., 2006. Kemijoen jokialueen kalatalousvelvoitteen tarkkailutulokset vuosina 2000-2004. Tutkimusraportti 8. Kemijoki Oy. Rovaniemi. (in Finnish). 174 s.
Illies, J., Botosaneanu, L., 1963. Probèmes et méthodes de la classification et de la zonation écologique des eaux courantes, considerées surtout du point de vue faunistique. Verh. Int. Ver. Limnol. 12, 1–57.
Irina, P., 2011. Fisheries in Bulgaria: Structural and Cohesion Policies in Fisheries. Directorate General for Internal Policies; Policy Department B, Bulgaria, pp. 1–48.
Irvine, K., 2014. Integrated Water Resources Management (IWRM): From Local to Global Sustainability: Towards Integrated Water Management, TIMe Project 2014-W-DS-20. pp. 1–57.
Jentoft, S., 2000. Legitimacy and disappointment in fisheries management—prospects of user participation. Mar. Policy 24, 141–148.
Jentoft, S., 2007. Limits to governability: institutional implications for fisheries and coastal governance. Mar. Policy 31, 360–370.
Jentoft, S., McCay, B.J., 1995. User participation in fisheries management: lessons drawn from international experiences. Mar. Policy 19, 227–246.
Johnson, R.N., Libecap, G.D., 1982. Contracting problems and regulation: the case of the fishery. Am. Econ. Rev. 72, 1005–1022.
Johnson, M.G., Leach, J.H., Minns, C.K., Olver, C.H., 1977. Limnological characteristics of Ontario lakes in relation to associations of walley (*Stizostedion vitreum vitreum*); northern pike (*Esox lucius*); lake trout (*Salvelinus namaycush*), and small mouth bass (*Micropterus dolomieui*). J. Fish. Res. Board Can. 34, 1592–1601.
Kaatra, K., Simola, O., 1985. Water Level Regulation of Lake Inari: Impacts and Compensation Measures.
Kar, D., 1990. Limnology and Fisheries of Lake Sone in the Cachar district of Assam (India), viii + 201. PhD thesis (Published), University of Gauhati, Assam. (1984): Matsya, 15–16: 209–213.
Kar, D., 2003. Fishes of Barak drainage, Mizoram and Tripura. In: Kumar, A., Bohra, C., Singh, L.K. (Eds.), Environment, Pollution and Management. A P H Publishing Corporation, New Delhi, pp. 203–211. p. xii + 604.
Kar, D., 2007. Fundamentals of Limnology and Aquaculture Biotechnology, vi + 609. Daya Publishing House, New Delhi.
Kar, D., 2013. Wetlands and Lakes of the World. Springer, London, p. xxx + 687.
Karpoff, J.M., 1987. Sub-optimal controls in common resource management: the case of the fishery. J. Polit. Econ. 95, 179–194.
Karpoff, J.M., 1989. Characteristics of limited entry fisheries and the option component of entry licenses. Land Econ. 65, 386–393.
Kelso, J.R.M., Steedman, R.J., Stoddart, S.J., 1996. Historical causes of change in Great Lakes fish stocks and the implications for ecosystem rehabilitation. Can. J. Fish. Aquat. Sci. 53 (Suppl. 1), 10–19.
Klee, O., 1991. Angewandte Hydrobiologie, 2. Aufl. Georg Thieme Verlag Stuttgart, New York.
Klinkhardt, M., 1996. Der Hering. Westarp Wissenschaften, Magdeburg.
Krebs, C.J., 1989. Ecological Methodology. Harper and Row Publishers, New York, p. 654.
Kruse, I., 1991. Gesamtfänge der Hochsee-, See- und Küstenfischerei der DDR in den Jahren 1978–1990. In: Fischerei-Forschung, Jahrgang 29, pp. 3–9.
Lehtonen, H., 2002. Alien freshwater fishes of Europe. In: Leppäkoski, E., Gollasch, S., Olenin, S. (Eds.), Invasive Aquatic Species of Europe. Kluwer Academic Publisher, pp. 153–161.
Lien, J., 1996. Conservation aspects of fishing gear: Cetaceans and gillnets. In: University of Alaska (Ed.), Solving Bycatch. Fairbanks, pp. 219–224.
Lim, C.P., Matsuda, Y., Shigemi, Y., 1995. Co-management in marine fisheries: the Japanese experience. Coast. Manag. 23, 195–221.
Macinko, S., 1993. Public or private?: United States commercial fisheries management and the public trust doctrine: reciprocal challenges. Nat. Resour. J. 33, 919–955.
MAF NSPFA, 2007-2013. National Strategic Plan for Fisheries and Aquaculture, 2007–2013. Ministry of Agriculture and Forestry, National Agency for Fisheries and Aquaculture.
Margrethe, A., Claire, W.A., Helen, J.B., Christine, R., 2014. What does stakeholder involvement mean for fisheries management? Ecol. Soc. 19 (4), 35.
Marttunen, M., Hellsten, S., Puro, A., Huttula, E., Nenonen, M.-L., Järvinen, E., Salonen, E., Palomäki, R., Huru, H., Bergman, T., 1997. Inarijärven tila, käyttö ja niihin vaikuttavat tekijät. Suomen ympäristö. 58, 1–197 (in Finnish with English summary).
Matthiasson, T., 1992. Principles for distribution of rent from a 'commons'. Mar. Policy 16, 210–231.
Matulich, S.C., Mittelhammer, R.C., et al., 1996. Toward a more complete model of individual transferable fishing quotas: implications of incorporating the process sector. J. Environ. Econ. Manag. 31, 112–128.
McCay, B.J., 1995. Social and ecological implication of ITQs: an overview. Ocean Coast. Manag. 28, 3–22.
McCay, B.J., Creed, C.F., et al., 1995. Individual transferable quotas (ITQs) in Canadian and US fisheries. Ocean Coast. Manag. 28, 85–115.
Mehner, T., Diekmann, M., Brämick, U., Lemcke, R., 2005. Composition of fish communities in German lakes as related to lake morphology, trophic state, shore structure and human-use intensity. Freshw. Biol. 50, 70–85.
Memon, P., Cullen, A., et al., 1992. Fishery policies and their impact on the New Zealand Maori. Mar. Resour. Econ. 7, 153–167.
Mirela, A., Michael, M.C., Alexandra, C., Nicoleta, D., Dirk, R.F., Veronica, M., Emil, P., Cosima, R., 2005. Romania Danube Delta Biodiversity project: Local Benefits Case Study Report. Stockholm Environment Institute; Global Environment Facility; Romania DDBP Local Benefits Case Study Report. pp. 1–143.
Moloney, D.G., Pearse, P.H., 1979. Quantitative rights: an instrument for regulating commercial fisheries. J. Fish. Res. Board Can. 36, 859–866.
Moriarty, P.B., Visscher, J.T., Bury, P., Postma, L., 2000. The Dublin principles revisited for WSS. In: 26th WEDC Conference: Water, Sanitation and Hygiene: Challenges of the Millennium; Dhaka (Bangladesh).
Moyle, P.B., Light, T., 1996. Biological invasions of fresh water: empirical rules and assembly theory. Biol. Conserv. 78, 49–161.
Mutenia, A., 1985. Fish stock and fishing in the Lokka and Porttipahta reservoirs, Northern Finland. In: Alabaster, J.S. (Ed.), Habitat Modification and Freshwater Fisheries. Food and Agriculture Organisation of United Nations, London, pp. 195–201.

Mutenia, A., 1995. Lokka ja Porttipahta. In: Railo, E., Tuunainen, A.-L., Ahvonen, A., Lappalainen, A., Leinonen, K., Nylander, E. (Eds.), Fish and Game by Region: Regional Statistics on Fisheries and Hunting in Finland, Ympäristö-Miljö (SVT) 12/1995, pp. 50–53.
Mutenia, A., Salonen, E., 1994. Rehabilitation of Fisheries of Regulated Lake Inari, Northern Finland: Rehabilitation of Freshwater Fisheries. Fishing News Books, London, pp. 280–288.
Neher, P.A.E., Arnason, R.E., et al., 1989. Rights-based fishing. In: Proceedings of a Workshop on the Scientific Foundations for Right-Based Fishing, Reykjavik, June 27-July 1, 1988. Workshop on the Scientific Foundations for Rights Based Fishing, 1988; Reykjavik, p. XIV-541.
Newman, S., 2014. Structural and Cohesion Policies: Inland Fisheries and the Common Fisheries Policy. Institute for European Environmental Policy; European Union, pp. 1–46.
OCDE, 1997. Towards Sustainable Fisheries: Economic Aspects of the Management of Living Marine Resources. OCDE Publications, Paris.
Olivier, G., Olivier, T., 2001. Distributional issues in the operation of rights-based fisheries management systems. Mar. Policy 25 (2), 103–112.
Ostrom, E., 1990. Governing the Commons: The Evolution of Institutions for Collective Action. Cambridge University Press, Cambridge.
Ostrom, E., 2005. Understanding Institutional Diversity. Princeton University Press, Princeton.
Ostrom, E., 2007. A diagnostic approach for going beyond panaceas. Proc. Natl. Acad. Sci. U. S. A. 104, 15181–15187.
Palsson, G., Helgason, A., 1995. Figuring fish and measuring men: the individual transferable quota system in the Icelandic cod fishery. Ocean Coast. Manag. 28, 117–146.
Palsson, G., Petursdottir, G.E., 1997. Social Implications of Quota Systems in Fisheries. TemaNord, Copenhague.
Peesch, R., 1961. Die Fischerkommünen auf Rügen und Hiddensee. Akademie-Verlag, Berlin.
Peter, B., Arico, S., 2002. Conserving and managing biodiversity sustainably: the role of science and society. Nat. Resour. Forum 26, 245–248.
Platteau, J.P., Nugent, J., 1989. Contractual Relationships and Their Rationale in Marine Fishing. Cahiers de la Faculté des Sciences Economiques et Sociales de Namur, May. 26.
Plummer, R., Armitage, D., 2007a. Crossing boundaries, crossing scales: the evolution of environment and resource co-management. Geogr. Compass 1 (4), 834–849.
Plummer, R., Armitage, D., 2007b. A resilience-based framework for evaluating adaptive co-management: linking ecology, economics and society in a complex world. Ecol. Econ. 61, 62–74.
Plummer, R., Fennell, D., 2007. Exploring co-management theory: prospects for socio-biology and reciprocal altruism. J. Environ. Manag. 85, 944–955.
Plummer, R., Fitz Gibbon, J., 2004. Some observations on the terminology in co-operative environmental management. J. Environ. Manag. 70, 63–72.
Plummer, R., Fitz Gibbon, J., 2007. Connecting adaptive co-management, social learning, and social capital through theory and practice. In: Armitage, D., Berkes, F., Doubleday, N. (Eds.), Adaptive Co-Management. University of British Columbia Press, Vancouver, pp. 38–61.
Pomeroy, S.R., 1995. Community-based and co-management institutions for sustainable coastal fisheries management in Southeast Asia. Ocean Coast. Manag. 27 (3), 143–165.
Pomeroy, R., Berkes, F., 1997. Two to tango: the role of government in fisheries co-management. Mar. Policy 21 (5), 465–480.
Premazzi, G., Chiaudani, C., 1992. Ecological Quality of Surface Waters: Quality Assessment Schemes for European Community Lakes. Commission of the European Communities, Brussels, p. 123.
Robert, A., Robin, S., Tamara, K., 2004. Freshwater Fisheries in Central and Eastern Europe: The Challenge of Sustainability. Overview Report. IUCN, Warsaw, Poland. English-language edition by: Robin Sharp, Project Coordinators: Zenon Tederko Tamara Kutonova.
Ruoppa, M., Karttunen, K., 2002. Typology and Ecological Classification of Lakes and Rivers. TemaNord 2002/566.
Salonen, E., 1998. The vendace stock and fisheries in Lake Inari. Boreal Environ. Res. 3, 307–319.
Salonen, E., Mutenia, A., 2004. The commercial coregonid fishery in northernmost Finland: a review. Ann. Zool. Fennici 41, 351–355.
Salonen, E., Mutenia, A., 2007. Alien Fish Species in Northernmost Finland. Finnish Game and Fisheries Research Institute, Helsinki, pp. 1–17. TUTKIMUKSIA, 2/2007, Finland.
Scharmann, L., 1994. Die Fischwirtschaft in den norddeutschen Küstenländern. In: Europa Regional. vol. 2, Nr. 4, pp. 1–9.
Schmutz, S., Cowx, I.G., Haidvogl, G., Pont, D., 2007. Fish-based methods for assessing European running waters: a synthesis. Fisheries Manag. Ecol. 14, 369–380. Blackwell Publishing Ltd.
Schupp, D.H., 1992. A Classification of Minnesota Lakes With Associated Fish Communities. Minnesota Department of Natural Resources, p. 27. Investigational Report, 417.
Scott, A., 1988. Development of property in the fishery. Mar. Resour. Econ. 5, 289–311.
Scott, A., 1993. Obstacles to fishery self-government. Mar. Resour. Econ. 8, 187–199.
Sharp, B.M., 1998. Integrating recreational fisheries into rights-based management systems. In: Paper Presented at the First World Congress of Environmental and Resource Economics, Venice (Italy).
Smith, W., 1996. A case study: the Dutch co-management system. In: European Association of Fisheries Economist Conference, 1–3 April (Barcelona).
Søreng, S.U., 2008. Fishing rights discourses in Norway: indigenous versus non-indigenous voices. Marit. Stud. 6 (2), 77–99.
Spindler, T., 1997. In: Gassner, et al. (Eds.), Fischfauna in Österreich. Ökologie-Gefährdung-Bioindikation-Fischerei-Gesetzgebung. Bundesministerium für Umwelt. Monographien 23. 432 H. Wiley-VCH Verlag, GmbH and Co. KGaA, Weinheim. 2005.
Squires, D., Kirkley, J., Tisdell, C., 1995. Individual transferable quota as a management tool. Rev. Fish. Sci. 3, 141–169.
STECF, 2010. In: Doring, R., Guillen, J. (Eds.), Report of the Working Group on the Evaluation of Data Collected on the Fish Processing Sector (SGECA 10-04). Joint Working Group on Economic Affairs (SGECA) of the Scientific, Technical and Economic Committee for Fisheries (STECF), p. 309.
Stedinger, J.R., Jozef, P.M., Monique, D., Villars, T., 2005. Water resources systems planning and management: an introduction to methods, models and applications. In: Loucks, D.P., Van Beek, E. (Eds.), Studies and Reports in Hydrology. UNESCO Publishing, pp. 1–690.
Steedman, R.J., Whillans, A.P., Behm, K.E., Bray, T.H., Cullis, K.I., Holland, M.M., Stoddart, S.J., White, R.J., 1996. Use of historical information for conservation and restoration of Great Lakes aquatic habitat. Can. J. Fish. Aquat. Sci. 53 (Suppl. 1), 415–423.
Stelzenmüller, V., Ellis, J.R., Rogers, S.I., 2010. Towards a spatially explicit risk assessment for marine management: assessing the vulnerability of fish to aggregate extraction. Biol. Conserv. 143, 230–238.

Sundbäck, K., 1977a. Results of research on the fisheries of lokka reservoir and plan for fishing and management of the fish stocks. Finnish Game and Fisheries Research Institute. Tiedonantoja 9, 68–89 (in Finnish).
Sundbäck, K., 1977b. Results of research on the fisheries of Porttipahta reservoir and plan for fishing and management of the fish stocks. Finnish Game and Fisheries Research Institute. Tiedonantoja 9, 90–105 (in Finnish).
Sutela, T., Mutenia, A., Salonen, E., 2002. Relationship between annual variation in reservoir conditions and year-class strength of peled (*Coregonus peled*) and whitefish (*C. lavaretus*). Hydrobiologia 485 (1–3), 213–221.
Sutinen, J.-G., 1979. Fishermen remuneration systems and implications for fisheries development. Scott. J. Political Econ. 26 (2), 147–162.
Suuronen, P., 1995. Conservation of young fish by management of trawl selectivity. Finn. Fish. Res. 15, 97–116.
Tammi, J., Lappalainen, A., Rask, M., 2002. An evaluation of the preliminary typology of Finnish lakes by using fish community data. In: Rouppa, M., Karttunen, K. (Eds.), Typology and Ecological Classification of Lakes and Rivers, pp. 50–53. TemaNord 2002/566.
Terry, J.M., 1993. Quotas individuels transférables pour la pêche à la morue charbonnière et au flétan avec des engins fixes au large de l'Alaska. In: L'utilisation des contingents dans la gestion des pêches. OCDE, pp. 103–130.
Thébaud, O., 1998. Les choix collectifs de conservation des ressources marines vivantes: analyse économique appliquée aux cas des pêcheries coralliennes et de l'exploitation internationale des grands cétacés. Thèse pour le Doctorat de l'EHESS, Paris, p. 470.
Tonn, W.M., Magnusson, J.J., 1982. Patterns in the species composition and richness of fish assemblages in northern Wisconsin lakes. Ecology 63, 1149–1166.
Troadec, J.P., 1994. Le nouvel enjeu de la pêche: l'ajustement des institutions aux nouvelles conditions de pleines exploitation des stocks. C. R. Acad. Agric. France 80, 41–60.
Tyler-Walters, H., Rogers, S.I., Marshall, C.E., Hiscock, K., 2009. A method to assess the sensitivity of sedimentary communities to fishing activities: aquatic conservation. Mar. Freshwat. Ecosyst. 19, 285–300.
van den Bergh, J.C.J.M., Carbonell, A.F., Munda, G., 1998. Models of Individual Behaviour and Implications for Environmental Policy. Working Paper. Free University, Amsterdam.
Vera, H., Fauchald, P., Jernsletten, J.-L., 2012. Community-based management: under what conditions do sami pastoralists manage pastures sustainably? PLoS One 7 (12), 1–9.
Virtanen, M., Hellsten, S., Koponen, J., Riihimäki, S., Nenonen, O., 1993. Water quality model of northern reservoirs and its validation with field measurements. VTT Tiedotteita 1525, 1–205 (In Finnish with English abstract).
Walter, M.E., 1903. Die Fischerei Als Nebenbetrieb Des Landwirtes und Forstmannes. Verl. J. Neumann, Neudamm, p. 801.
Wang, S., 1995. The surf clam ITQ management: an evaluation—Thalassorama. Mar. Resour. Econ. 10, 93–98.
Wilen, J.E., Casey, K., 1997. In: Palsson, G. (Ed.), Employment and Remuneration Effects: Social Implications of Quota Systems in Fisheries. Petursdottir Gudrun, Copenhagen, pp. 315–333.
Zick, D., Gassner, H., Wanzenböck, J., Lahnsteiner, B., Tischler, G., Patzner, R., 2004. Die Veränderungen der Fischartengemeinschaften der großen österreichischen Seen im Laufe des 19. Jahrhunderts. Österreichs Fischerei 57, 20–27.

America

Abaza, H., Jha, V., 2002. Integrated assessment of trade and trade-related policies. In: UNEP Country Projects, Round II: A Synthesis Report. Nations Environmental Programme United, New York/Geneva.
Abbott-Jamieson, S., 2007. Using Oral History Techniques in a NOAA Fisheries Service (NMFS) Education and Outreach Project: Preserving Local Fisheries Knowledge; Linking Generations; and, Improving Environmental Literacy. NAPA Bulletin, No. 28, American Anthropological Association, Arlington, VA, pp. 134–142.
Adams, C., 1998. An Overview of the Cuban Fishing Industry and Implications to the Florida Seafood Industry If Renewed Trade. International Working Paper IW93-3, Institute of Food and Agricultural Sciences, University of Florida, Gainsville.
Adams, C., et al., 2000. An overview of the Cuban commercial fishing industry and recent changes in management structure and objectives. Edis 106 MAST, 2007, vol. 6(1): 83–108, In: Document FE218, International Working Paper IW93-3. Institute of Food and Agricultural Sciences, University of Florida, Gainsville.
Anon, 2000. Programa de Pesca y Acuacultura 1995–2000. Secretaría de Medio Ambiente, Recursos Naturales y Pesca, Mexico, DF, p. 96.
Anon, 2001. Ecosystem-Based Management of Fisheries: Opportunities and Challenges for Coordination Between Marine Regional Fishery Bodies and Regional Seas Conventions. UNEP Regional Seas Reports and Studies No. 175, UNEP, p. 52.
Anon, 2002. Fisheries Subsidies and Marine Resources Management: Lessons learned from Studies in Argentina and Senegal. United Nations Environment Programme, pp. 1–64.
Anon, 2005a. Sustainable fisheries and livelihoods in Latin America: the imperative of recognising the resource access rights of artisanal fishermen. In: Coastal Communities and Indigenous Peoples, Santa Clara, Argentina, 1–4 March: Proc. Workshop, pp. 1–4.
Anon, 2005b. Fishing Community Initiatives in Community Based Fisheries Management and the Sitka Declaration: Proposal for Forum of Fishermen, NGO's, and the Government Agencies to Explore Expansion of the Use of Sectors, Catch Share, and Community Based Fisheries Management in the Northeast and Caribbean. pp. 1–2.
Anon, 2006. Agricultural and Fisheries Policies in Mexico: Recent Achievements and Continuing With the Reform Agenda. OECD-Multi-lingual Summaries, pp. 1–36.
Anon, 2007. A study guide series on peace and conflict for independent learners and classroom instructors. In: Natural Resources: Conflict, and Conflict Resolution. United States Institute of Peace, Washington, DC, pp. 1–36.
Anon, 2009. Towards Sustainable Economic and Social Development. pp. 163–180.
Anon. n.d. Governance and Marine Fisheries: Comparing Across Countries and Stocks, pp. 1–2.
Anonymous, Anuario Estadístico de Pesca, 1997. Secretaría de Medio Ambiente Recursos Naturales y Pesca. Dirección General de Comunicación Social, México, D.F., p. 241.
Anonymous, Desarrollo de los Sectores Pesqueros y acuícola de México, 1999. Prospectiva y estrategias. Camara Naciona de la Industria Pesquera, México, DF, p. 41.

Anonymous, Anuario Estadístico de Pesca, 1998. Secretaría de Medio Ambiente Recursos Naturales y Pesca. Dirección General de Comunicación Social, México, DF, p. 241.
Anonymous, Anuario Estadístico de Pesca, 1999. Secretaría de Medio Ambiente Recursos Naturales y Pesca. Dirección General de Comunicación Social, México, DF, p. 241.
Armitage, D.R., Plummer, R., Berkes, F., Arthur, R.I., Charles, A.T., Davidson-Hunt, I.J., Diduck, A.P., Doubleday, N.C., Johnson, S., Marschke, M., McConney, P., Pinkerton, E.W., Wollenberg, E.K., 2009. Adaptive co management for social-ecological complexity. Front. Ecol. Environ. 7, 95–102.
Arnason, R., 1994. Theoretical and practical fishery management. In: Loayza, E. (Ed.), Managing Fishery Resources. World Bank, Washington, pp. 3–10. World Bank Discusión Papers, 217.
Arnason, R., 2001. Modern fisheries management. In: Lecture Presented at EFTA Seminar on Fisheries Management (Brussels).
Baisre, J., 1987a. La pesca en Cuba: Apuntes para su historia (II) La pesca desde 1492 hasta 1902. Etapa colonial. Mar y Pesca 263, 38–41.
Baisre, J., 1987b. La pesca en Cuba: Apuntes para su historia (IV) La pesca desde 1902 hasta 1952. Etapa de la seudo república. Mar y Pesca 265, 34–39.
Baisre, J., 1987c. La pesca en Cuba: Apuntes para su historia (V) La pesca después de 1959. Etapa revolucionaria. Mar y Pesca 266, 37–41.
Baisre, J., 2000a. Chronicle of Cuban marine fisheries (1935–1995). In: Trend Analysis and Fisheries Potential. FAO. Fish. Tech. Paper, 394.
Baisre, J., 2000b. Concesiones para la pesca de la langosta en Cuba: Experiencias y resultados. In: Informe del Taller sobre manejo y asignación de recursos pesqueros a los pescadores artesanales en América Latina, Valparaíso, Chile, 25–28 de abril del 2000. Oficina Regional de la FAO para América Latina y el Caribe, Santiago, Chile.
Baisre, J., 2004. La Pesca Marítima en Cuba. Científico-Técnica, La Habana.
Baisre, J., Booth, S., Zeller, D., 2003. Cuban fisheries catches within FAO area 31 (Western Central Atlantic): 1950–1999. In: Zeller, D., Booth, S., Mohammed, E., Pauly, D. (Eds.), From Mexico to Brazil: Central Atlantic Fisheries Catch Trends and Ecosystem Models. vol. 11 (6). Fisheries Centre Research Reports, pp. 133–139.
Bartch, J., Abbott-Jamieson, S., Whitmore, J., 2009. Voices From the Fisheries Handbook: Preserving Local Fisheries Knowledge; Linking Generations; and, Improving Environmental Literacy. U.S. Dep. Commerce; National Oceanic and Atmospheric Administration (NOAA), (National Marine Fisheries Service), Natl. Mar. Fish. Serv., Silver Spring, MD. 132 pp.
Barton, D., Agnew, D., Purchase, L., 2002. The Southwest Atlantic: achievements of bilateral management and the case for multilateral arrangement. In: CEFAS Fisheries Symposium, 2002.
Basson, M., Beddington, J.R., Crombie, J.A., Holden, S.J., Purchase, L.V., Tingley, G.A., 1996. Assessment and management techniques for migratory annual squid stocks: the *Illex argentines* fishery in the Southwest Atlantic as an example. Fish. Res. 28, 3–27.
Berkes, F., et al., 2001. Managing Small-Scale Fisheries: Alternative Directions and Methods. International Development Research Centre, Ottawa.
Berkes, F., Folke, C. (Eds.), 1998. Linking Social and Ecological Systems: Management Practices and Social Mechanisms for Building Resilience. Cambridge University Press, UK.
Berkes, F., 2006. From community—based resource management to complex systems: the scale issue and marine commons. Ecol. Soc. 11, 45.
Bertolotti, M.I., Verazay, G., Akselman, R. (Eds.), 2001. El Mar Argentino y sus recursos pesqueros. Tomo 3. Evolución de la flota pesquera argentina, artes de pesca y dispositivos de selectividad. Publicaciones especiales INIDEP, Mar del Plata, p. 165.
Boodoosingh, M., 1992. A Review of Possible Impacts of Human Activities on Fisheries in the Coastal Waters of Trinidad with Special Emphasis on the Gulf of Paria. Fisheries Occasional Paper Series No. 2, Ministry of Agriculture Land and Marine Resources, Trinidad, p. 126.
Bourillon, L., 2002. Exclusive Fishing Zone as a Strategy for Managing Fishery Resources by the Seri Indians, Gulf of California, Mexico. Ph.D. dissertation, School of Renewable Natural Resources. University of Arizona at Tucson.
Branch, T.A., Jensen, O.P., Ricard, D.M., Hilborn, R., 2011. Contrasting global trends in marine fishery status obtained from catches and from stock assessments. Conserv. Biol. 25, 777–786.
Breton, Y., Brown, D., Davy, B., Haughton, M., Ovares, L. (Eds.), 2006. Coastal Management in the Wider Caribbean: Resilience, Adaptation and Community Diversity. International Development Research Centre; IAN Randle Publishers, Ottawa.
Breton, Y., Doyon, S., 1999. La noción de 'familia' en la antropología marítima: del parentesco al manejo costero. In: Perspectivas Rurales 6, Numero especial: Sustenabilidad ambiental y economica de las poblaciones costeras. vol. 3 (2), pp. 37–45.
Brotherton, P.S., 2005. Macro-economic change and the biopolitics of health in Cuba's special period. J. Lat. Am. Anthropol. 10 (2), 339–369.
Bundy, A., Lilly, G.R., Shelton, P.A., 2000. A Mass Balance Model of the Newfoundland-Labrador Shelf. Canadian Technical Report of Fisheries and Aquatic Sciences.
Burawoy, M., Verdery, K. (Eds.), 1998. Uncertain Transition: Ethnographies of Change in the Postsocialist World. Rowman and Littlefield, Lanham, MD.
Butler, C., 2005. Understanding the Coho crisis: political knowledge in a fractured salmon industry. In: Proceedings of the Mare Conference, July 7–9, (Amsterdam).
Caddy, J.F., Mahon, R., 1995. Reference Points for Fisheries Management. FAO Fisheries Technical Paper No. 347. FAO, Rome.
Carcamo, R., de León, M.E., Wakidi, A., Medina, A., 2003. Appendix F. Sub-regional reports (Group North Central: Belize, South Cuba and Mexico). In: Report of Second Workshop on the Management of Caribbean Spiny Lobster Fisheries in the WECAFC Area. FAO Fisheries Report No. 715, pp. 211–214.
Carozza, C., Fernandez, A.N., Lagos, N., 2005. Associación íctica costera bonaerense "variado costero" evolución de las capturas desembarcadas por la flota comercial argentina (1990–2003). In: Informe Tecnico Interno, INIDEP N° 04, p. 21.
Cauhepe, M., 1999. Management of the Argentine Hake. Final Project. Fisheries Training Programme. The United Nations University, Iceland.
CEDEPESCA, 2003. La pesqueria de calamar en Argentina. Comunidad Pesquera No 9, July 2003. CEDEPESCA.
CEDEPESCA, 2004. El fracaso total de la zafra 2004 obliga a repensar todo. CEDEPESCA.
Christensen, V., Walters, C.J., Pauly, D., 2000. Ecopath With Ecosim Version 4: Help System. University of British Columbia, Fisheries Centre, Vancouver, Canada and ICLARM, Penang, Malaysia. Fisheries Centre/FAO Workshop, p. 73.
Christie, P., McCay, B.J., Miller, M.L., Lowe, C., White, A.T., Stoffle, R., Fluharty, D.L., McManus, L.T., Chuenpagdee, R., Pomeroy, C., Suman, D.O., Blount, B.G., Huppert, D., Eisma, R.-L.V., Oracion, E., Lowry, K., Pollinac, R.B., 2003. Towards developing a complete understanding: a social science research agenda for marine protected areas. Fisheries 28, 22–26.

Christy, F.T., 1982. Territorial Use Rights in Marine Fisheries: Definitions and Conditions. FAO Fisheries Technical Paper No. 227, Food and Agriculture Organization of the United Nations, Rome.

Claro, R., Baisre, J., Lindeman, K., García-Arteaga, J., 2001. Cuban fisheries: historical trends and current status. In: Claro, R., Lindeman, K., Parenti, L. (Eds.), Ecology of the Marine Fishes of Cuba. Smithsonian Institution Press, Washington, pp. 194–219.

Cochrane, K.L., Butterworth, D.S., De Oliveira, J.A.A., Roel, B.A., 1998. Management procedures in a fishery based on highly variable stocks and with conflicting objectives: experiences in the South African pelagic fishery. Rev. Fish Biol. Fish. 8, 177–214.

Colburn, L.L., 2006. Research and resource management in North American fisheries. Hum. Organ. Special Iss. 65 (3).

Consejo Federal Pesquero (CFP), 2009. Regimen general de administracion por cuotas individuales transferibles de captura (CITC). Resolucion 10, 2009. Consejo Federal Pesquero (CFP).

Consejo Federal Pesquero (CFP), 2012. In: Consejo Federal Pesquero (CFP) (Ed.), Individual Transferable Quota Catch Management Regime (CITC).

Constable, A.J., de la Mare, W.K., Agnew, D.J., Everson, I., Miller, D., 1996. Managing fisheries to conserve the Antarctic marine ecosystem: practical implementation of the convention on the conservation of Antarctic marine living resources (CCAMLR). ICES J. Mar. Sci. 27, 1054.

Cordell, J. (Ed.), 1989. A Sea of Small Boats. Cultural Survival, Inc., Cambridge, MA. Cultural Survival Report, 26.

Cordo, H., 2004. Evaluación del estado del efectivo norte de 41°S de la merluza (Merluccius hubbsi) ystimación de la captura biológicamente aceptable correspondiente al año 2004. In: Informe Tecnico Interno; INIDEP N° 17, p. 35.

Cordo, H., 2005. Evaluación del estado del efectivo sur de 41°S de la merluza (Merluccius hubbsi) ystimación de la captura biológicamente aceptable correspondiente al año 2005. In: Informe Tecnico Interno; INIDEP N° 48: pp. 27; INIDEP N° 37: pp. 29.

Cudney-Bueno, R., 2000. Management and Conservation of Benthic Resources Harvested by Small-Scale Hookah Divers in the Northern Gulf of California, Mexico: The Black Murex Snail Fishery. MS thesis, School of Renewable Natural Resources; The University of Arizona, Tucson.

Daniela, K., Marcelo, V., Pitcher, T.J., 2006. An Estimation of Compliance of the Fisheries of Argentina With Article 7 (Fisheries Management) of the UN Code of Conduct for Responsible Fishing. pp. 1–19.

Dans, S., Koen-Alonso, M., Pedraza, S.N., Crespo, E.A., 2003. Incidental catch of dolphins in trawling fisheries off Patagonia, Argentina: can populations persist? Ecol. Appl. 13 (3), 754–762.

Davis, A., Wagner, J., 2006. A right to fish for a living? The case for coastal fishing people's determination of access and participation. Ocean Coast. Manag. 49, 476–497.

Decoster, J., 2001. Challenges Facing Artisanal Fisheries in the 21st Century. ICSF, Bruxelles.

Díaz-Briquets, S., Perez Lopez, J., 2000. Conquering Nature: The Environmental Legacy of Socialism in Cuba. University of Pittsburgh Press, Pittsburgh.

Domingo, A., Bugoni, L., Prosdocimi, L., 2005. Diagnóstico de Situación: Las Tortugas Marinas ysu Interacción con las Pesquerías en el Océano Atlántico Sur Occidental (ASO), vol. 22, pp. 67–78.

Dommen, C., Deere, C., 1999. Fish for Thought: Fisheries, International Trade and Sustainable Development. Natural Resources International Trade and Sustainable Development Series No. 1. ICTSD and IUCN, Geneva.

Donald, R.S., Joseph, J.O., Theodore, N.H., Steven, A., *2006*. Fisheries Management Plan for the Minnesota Waters of Lake Superior. Fisheries Management Section; Division of Fish and Wildlife; Minnesota Department of Natural Resources in Cooperation With the Lake Superior Advisory Group; Lake Superior Fishery Area Special Publication, 163, pp. 1–100.

Doyon, S., 2003. L'environnement 'révolutionnaire': pratiques, discours et dynamique socio-environnementale de la gestion des resources côtières à Cuba (Doctoral dissertation). Department of Anthropology (McGill University), Montréal.

Doyon, S., 2005a. Une révolution à sa mesure: Diversité des pratiques environnementales communautaires à Cuba. Anthropol. Soc. 29 (1), 121–144.

Doyon, S., 2005b. Savoir environnemental local et scientifique: entre la révolution, la gouvernance nationale et le développement durable à Cuba. VertigO 6 (2), 4–11.

Doyon, S., 2007. Fishing for the revolution: transformations and adaptations in Cuban fisheries. MAST 6 (1), 83–108.

Doyon, S., Brotherton, P.S., 2008. Forthcoming Capitaliser la révolution cubaine. Pratiques et politiques dans les secteurs de la santé et de l'environnement. Anthropol. Soc. 32 (1), 193–216.

Durenberger, E.P., King, T.D. (Eds.), 2000. State and Community in Fisheries Management: Power, Policy, and Practice. Bergin and Garvey, Westport, CT.

Dyer, C.L., McGoodwin, J.R. (Eds.), 1994. Folk Management in the World's Fisheries: Lessons for Modern Fisheries Management. University Press of Colorado, Niwot Colorado.

Eckstein, S., 1993. Back From the Future: Cuba Under Castro. Princetown University Press, New Jersey.

Enriquez-Andrade, R., Batalla-Camargo, G., 2000. Perspectives of rights based fishery management in Mexico, pp. 1–4. In: Perspectives for Rights Based Fishery Management in Mexico: Micro-Behavior and Macro-Results: Proceedings of the Tenth Biennial Conference of the International Institute of Fisheries Economics and Trade Presentations. IIFET 2000 Proceedings; Perspectives of Rights Based Fishery Management in Mexico, pp. 1–27.

Enríquez-Andrade, R.R., Danemann, G.D., 1998. Identificación yestablecimiento de prioridades para las acciones de conservación y oportunidades de uso sustentable de los recursos marinos de la Península de Baja California. In: Reporte presentado al Fondo Mexicano para la Conservación IIFET 2000 Proceedings de la Naturaleza A.C.; Pronatura Península de Baja California: pp. 6; 8 annex and a CD., 1998.

Escobar, A., 1995. Encountering Development. The Making and Unmaking of the Third World. Princeton University Press, Princeton.

European Commission (EC), 2002. Communication From the Commission on the Reform of the Common Fisheries Policy 'Roadmap'. Office for Official Publications of the European Communities, Luxembourg.

FAO, 1995. Code of Conduct for Responsible Fisheries. FAO, Rome, p. 41.

FAO, 1997a. Review of the state of world fishery resources: marine fisheries. FAO Fish. Circ. 920.

FAO, 1997b. El estado mundial de la pesca y la acuicultura 1996. Organización de las Naciones Unidas para la Agricultura y la Alimentación, Departamento de Pesca, Roma, p. 125.

FAO, 2003. Summarized Extracts on the Status of Implementation of the Code of Conduct for Responsible Fisheries. 25th Session, 24–28 February, 2003. FAO.
FAO, 2018. National Aquaculture Sector Overview. Fisheries and Aquaculture Department, Argentina, pp. 1–15.
Ferreira, L., 1998. Barriers to fisheries co-management in the Gulf of Paria, Trinidad. In: Research Paper, Contemporary Issues in Ocean Development and Management, Marine Affairs Programme, Dalhousie University, p. 45.
Filippo, P., 2005. Análisis de la legislacion argentina en materia de ordenamiento y operaciones pesqueras a la luz del Código de Conducta para la Pesca Responsable de la Organizacion de las Nacionecs Unidas para la Agricultura y la Alimentación FAO. Fundación Vida Silvestre Argentina. Publicación especial. en prensa.
Fish Source, 2013. Status and Environmental Performance of Fisheries Worldwide. Database and reports http://www.fishsource.com/site/index.
Fishery and Aquaculture Country Profiles, n.d. In: FAO Fisheries and Aquaculture Department, Rome (Italy).
Foley, M.M., Halpern, B.S., Micheli, F., Armsby, M.H., Caldwell, M.R., Crain, C.M., Prahler, E., Rohr, N., Sivas, D., Beck, M.W., Carr, M.H., Crowder, L.B., Emmett, D.J., Hacker, S.D., McLeod, K.L., Palumbi, S.R., Peterson, C.H., Regan, H.M., Ruckelshaus, M.H., Sandifer, P.A., Steneck, R.S., 2010. Guiding ecological principles for marine spatial planning. Mar. Policy 34, 955–966.
Food and Agricultural Organization (FAO), 1997. Fisheries Management. FAO Technical Guidelines for Responsible Fisheries No. 4. FAO, Rome.
Food and Agricultural Organization (FAO), 1999a. Guidelines for the routine collection of capture fishery data. In: Prepared at the FAO/DANIDA Expert Consultation. Bangkok, Thailand, 18–30 May, 1998. FAO, Rome. FAO Fisheries Technical Paper No. 382.
Food and Agricultural Organization (FAO), 1999b. Indicators for Sustainable Development of Marine Capture Fisheries. FAO Technical Guidelines for Responsible Fisheries, No. 8. FAO, Rome.
Food and Agriculture Organization of the United Nations (FAO), 2001. Fishery and Aquaculture Country profiles. FAO, Argentina.
Food and Agriculture Organization of the United Nations (FAO), 2009. Fish. Stat. Fishery Statistical Collections Global Database. FAO.
Food and Agricultural Organization of the United Nations (FAO), 2010. The State of World Fisheries and Aquaculture, 2010. FAO, Rome.
Gaceta Oficial Ordinaria (GOO), 1996. Decreto-Ley 164 "Reglamento de Pesca", Edición Ordinaria. Gaceta Oficial de La República de Cuba, pp. 405–410. Año XCIV, No. 26.
Gell, F., Roberts, C.M., 2003. Benefits beyond boundaries: the fisheries effect of marine reserves. Trends Ecol. Evol. 18 (9), 448–455.
Gianni, M., Simpson, W., 2005. The Changing Nature of High Seas Fishing: How Flags of Convenience Provide Cover for Illegal, Unreported and Unregulated Fishing. Australian Department of Agriculture, Fisheries and Forestry, International Transport Workers' Federation, and WWF International, p. 83.
Godelman, E., 2004. Propuesta para el ordenamiento de la pesquería de merluza en Argentina. CeDePesca, Argentina.
Goedefroit, S., Chaboud, C., Breton, Y., 2002. La ruée vers l'or rose: regards croisés sur la pêche crevettière traditionnelle à Madagascar. Editions IRD, Paris.
González, E., 2004. Tradiciones Pesqueras Cubanas (1). Reglamentos durante la primera intervención norteamericana. Mar y Pesca 343, 40–42.
Grainger, R.J.R., Garcia, S.M., 1996. Chronicles of Marine Fishery Landings (1950–1994): Trend Analysis and Fisheries Potential. FAO Fisheries Technical Paper No. 359, FAO, Rome.
Granados-Dieseldorff, P., Montaña, C.G., 2012. A community-based approach to the management of an aggregating reef fishery in the Meso-american Reef. In: Applied Biodiversity Sciences Perspectives, pp. 1–11.
Guillermo, Basurto, X., 2002. Community-Based Conservation of the *callo dehacha* Fishery by the Comcaac Indians, Sonora, Mexico. A Thesis submitted to the Faculty of the School of Renewable Natural Resources in partial fulfillment of the requirements for the Degree of Master of Science with a major in Renewable Natural Resources; Studies in the Graduate College the University of Arizona, pp. 1–187.
Halpern, B.S., 2003. The impact of marine reserves: do reserves work and does reserve size matter? Ecol. Appl. S(13) (1), S117–S137.
Hannesson, R., 1993. Bioeconomic Analysis of Fisheries. Fishing News Books, Oxford (By arrangement with FAO).
Hardin, G., 1968. Tragedy of commons. Science 162, 1243–1248.
Hilborn, R., Walters, C.J., 1992. Quantitative Fisheries Stock Assessment: Choice, Dynamics and Uncertainty. Chapman and Hall, New York, p. 570.
INIDEP, 2004. Evaluación del estado del efectivo sur de 41oS de la merluza (Merluccius hubbsi) y estimación de la captura biológicamente aceptable correspondiente al año 2004. Informe Tecnico Interno INIDEP No. 17, p. 35.
International Conference on Responsible Fishing, Cancun, Mexico (Cancun Declaration on Responsible Fishing), 1992. Ministerial Conference on Fisheries (Rome Consensus on World Fisheries), 1995; Kyoto Conference on Sustainable Contribution of Fisheries to Food Security (The Kyoto Declaration and Plan of Action), 1995; World Food Summit (Rome Declaration on World Food Security and World Food Summit Plan of Action), 1996., p. 470.
Jackson, J.B.C., Kirby, M.X., Berger, W.H., Bjorndal, K.A., Botsford, L.W., Bourque, B.J., Warner, R.R., 2001. Historical overfeishing and the recent collapse of coastal ecosystems. Science 293, 629–638.
Johannes, R.E., 1998. The case for data—less marine resource management: examples from tropical near shore fin fisheries. Trends Ecol. 13, 243–246.
Johannes, R.E., Freeman, M.M.R., Hamilton, R.J., 2000. Ignore fishers' knowledge and miss the boat. Fish Fish. 1, 257–271.
Johns, M., 1992. Industrial capital and economic development in turn of the century Argentina. Econ. Geogr. 68 (2), 188–204.
Johnson, D., 2006. Category, narrative, and value in the governance of small-scale fisheries. Mar. Policy 30, 747–756.
Johnson, D., et al., 2005. Fish capture. In: Kooiman, J., et al. (Eds.), Fish for Life. Interactive Governance for Fisheries. Amsterdam University Press, Amsterdam, pp. 71–92.
Joyce, I.T., 1997. The spiny-lobster fishery in Cuba. Geogr. Rev. 87 (4), 484–503.
Joyce, I.T., 1999. Socialism and the fishery: a case study of the Cuban shrimp fishery. Can. Geogr. 43 (1), 53–69.
Julio, A.B.H., 2006. Cuban Fisheries Management Regime: Current State and Future Prospects. Final Project. Institute of Economic Studies, University of Iceland and Dirección de Regulaciones Pesqueras, Ministerio de la Industria Pesquera, 5ª Ave. y 246, Barlovento. Playa, Ciudad de La Habana, Cuba, pp. 1–33.
Kalikoski, D., Vasconcellos, M., Pitcher, T.J., 2006. An estimation of compliance of the fisheries of Argentina with article 7 (fisheries management) of the UN code of conduct for responsible fishing. In: Pitcher, T.J., Kalikoski, D., Pramod, G. (Eds.), Evaluations of compliance with the UN Code of Conduct for Responsible Fisheries. vol. 14. Fisheries Centre Research Reports. 2.

Kelleher, K., 2004. Discards in the World's Marine Fisheries: An Update. Food and Agriculture.
Kooiman, J., Bavinck, M., Jentoft, S., Pullin, R. (Eds.), 2005. Fish for Life: Interactive Governance for Fisheries. Amsterdam University Press, Amsterdam.
Kooiman, J., Bavinck, M., 2005. The governance perspective. In: Kooiman, J., et al. (Eds.), Fish for Life; Interactive Governance for Fisheries. Amsterdam University Press, Amsterdam, pp. 11–25.
Kothari, A., Pathak, N., Anuradha, R.V., Taneja, B. (Eds.), 1998. Communities and Conservation: Natural Resource Management in South and Central Asia. Sage Publications, New Delhi, India.
Le Gallic, B., 2002. Fisheries sustainability indicators: the OECD experience. In: Tools for Measuring (Integrated) Fisheries Policy Aiming at Sustainable Ecosystem; Joint Workshop EEA-EC DG Fisheries-DG Environment (Brussels).
León, F., 1997. Socialism and sociolismo: social actors and economic change in 1990s Cuba. In: Miguel, A. (Ed.), Centeno ant Mauricio Font Toward a New Cuba? Legacies of a Revolution. Boulder, London, pp. 39–52.
Lubchenco, J., Palumbi, S.R., Gaines, D., Andelman, S., 2003. Plugging a hole in the ocean: the emerging science of marine reserves. Ecol. Appl. 13, S3–S7.
Marcelo, V., Johanna, H., Alida, B., 2002. The use of Ecosim to investigate multispecies harvesting strategies for capture fisheries of the Newfoundland-Labrador shelf, pp. 68–72. In: Pitcher, T., Cochrane, K. (Eds.), The Use of Ecosystem Models to Investigate Multispecies Management Strategies for Capture Fisheries, *Labrador Shelf*. Fisheries Centre, University of British Columbia, Canada, pp. 1–159. A report co-published as an FAO Fisheries Report: Using Ecosim for Fisheries Management: FAO/Fisheries Centre Workshop, Page 3; A Research Report from The UBC Fisheries Centre and the Food and Agriculture Organisation of the United Nations: Fisheries Centre Research Reports, 10 (2). 1–159.
Marine Stewardship Council, 2012. Public Certification Report: Assessment Against MSC Principles and Criteria for Argentine Hoki. May 24, 2012.
Mascia, M.B., Claus, C.A., Naidoo, R., 2010. Impacts of marine protected areas on fishing communities. Conserv. Biol. 24, 1424–1429.
Mascia, M.B., Claus, C.A., 2009. A property rights approach to understanding human displacement from protected areas: the case of marine protected areas. Conserv. Biol. 23, 16–23.
McCay, B.J., Micheli, F., Ponce-Díaz, G., Murray, G., Shester, G., Ramirez-Sanchez, S., Weisman, W., 2014. Cooperatives, concessions, and co-management on the Pacific coast of Mexico. Mar. Policy 44, 49–59.
McCay, B.J., Jones, P.J.S., 2011. Marine protected areas: the governance of marine ecosystems and fisheries. Conserv. Biol. 25, 1130–1133.
McGoodwin, J., 1990. Crisis in the World's Fisheries: People, Problems, and Policies. Stanford University Press, Stanford, California.
Ministry of Agriculture of the Republic of Argentina (Min. Agri.), 2012. Landings Data (Disembarques). Regimen Federal de la Pesca. Ley Nacional 24: 922.
Ministerio de la Industria Pesquera (MIP), 2005. Proyección Estratégica del Ministerio de la Industria Pesquera 2005–2010.
Ministerio de la Industria Pesquera (MIP), 2006. Informe de Cuba. In: Taller Regional sobre Mejoramiento de los Sistemas de Información y Recolección de Datos Pesqueros para América Central y el Caribe. FAO-OSPESCA. San Salvador, 23–26 Enero.
Morales, P., 1972. Cinco anos de Cuba en el mundo pesquero. Mar y Pesca, 26–35.
Munda, G., Nijkamp, P., Rietveld, P., 1994. Qualitative multi-criteria evaluation for environmental management. Ecol. Econ. 10 (2), 97–112.
Musick, J.A., 1999. Ecology and conservation of long lived marine animals. In: Life in the Slow Lane: Ecology and Conservation of Long Lived Marine Animals. Proceedings of the Conference on Conservation of Long Lived Marine Animals; American Fishery Society (Monterrey, California).
National Research Council, 1999. Sharing the Fish: Toward a National Policy on Individual Fishing Quotas. National Academy Press, Washington, DC.
Newell, D., Ommer, R.E. (Eds.), 1999. Fishing Places, Fishing People. Traditions and Issues in Canadian Small-Scale Fisheries. University of Toronto Press, Toronto.
Nijkamp, P., Rietveld, P., Voogd, H., 1990. Multicriteria Evaluation in Physical Planning. North Holland, Amsterdam, p. 219.
Oficina Nacional de Estadísticas (ONE), 2005. Anuario Estadístico de Cuba 2005. Oficina Nacional de Estadísticas, La Habana.
Onestini, M., 2002. Subsidies in Argentine fisheries, pp. 11–24. In: Anon (Ed.), Fisheries Subsidies and Marine Resources Management. United Nations Environment Programme, pp. 1–64. (2002), Lessons learned from Studies in Argentina and Senegal.
Orensanz, J.M., Seijo, J.C., 2013. Rights-Based Management in Latin American Fisheries. FAO, Rome, pp. 1–155. Fisheries and Aquaculture Technical Paper No. 582.
Organization of the United Nations, Rome, Italy, 1994. FAO Fisheries Technical Paper 470. New York Times, p. 131. (1994) Argentines Sink Vessel. Wednesday, May 22, 1994.
Organisation for Economic Co-operation and Development (OECD), 2000. Transition to Responsible Fisheries. Economic and Policy Implications. OECD, Paris.
Ostrom, E., 1990. Governing the Commons: The Evolution of Institutions for Collective Action. Cambridge University Press, New York, NY.
Ostrom, E., 1998. How Communities Beat the Tragedy of the Commons (Slide Document); and, `Self-Governance and Forest Resources' (text document).
Panayotou, T., 1982. Management Concepts for Small Scale Fisheries: Economic and Social Aspects. Fisheries Technical Paper 228, FAO, Rome.
Pauly, D., Christensen, T.A., Guenette, S., Pitcher, T., Sumaila, R., Walkers, C.J., Watson, R., Zeller, D., 2002. Towards sustainability in fisheries. Nature 418, 689–695.
Pauly, D., Alder, J., Bennett, E., Christensen, V., Tyedmers, P., Watson, R., 2003. The future for fisheries. Science 302, 1359–1361.
Pérez-López, J.F., 1997. Cuba's second economy and the market transition. In: Centeno, M.A., Font, M. (Eds.), Toward a New Cuba? Legacies of a Revolution. Boulder, London, pp. 171–186.
Peter, W., 1994. Net loss: fish, jobs and the marine environment. Worldwatch 1994.
Pikitch, E.K., Santora, C., Babcock, E.A., Bakun, A., Bonfil, R., Conover, E.O., Dayton, P., Doukakis, P., Fluharty, D., Heneman, B., Houde, E.D., Link, J., Livingston, P.A., Mangel, M., McAllister, M.K., Pope, J., Sainsbury, K.J., 2004. Ecosystem-based fishery management. Science 305, 346–347.
Pinedo, D., Summers, P.M., Smith, R.C., Saavedra, J., Zumaeta, R., Almeyda, A.M., 2000. Community-based natural resource management as a non-linear process: a case study in the Peruvian Amazon *varzea*. In: Eighth IASCP Conference, Bloomington, Indiana, 30 May–4 June, 2000, pp. 1–28.

Pitcher, T.J., Hart, P.J.B., 1982. Fisheries Ecology. Chapman and Hall, London, p. 414.
Pitcher, T.J., 1999. Rapfish: A Rapid Appraisal Technique for Fisheries, and Its Application to the Code of Conduct for Responsible Fisheries. FAO Fisheries Circular No. 947, p. 47.
Pitcher, T.J., Kalikoski, D., Pramod, G. (Eds.), 2006. Evaluations of Compliance With the UN Code of Conduct for Responsible Fisheries. vol. 14. Fisheries Centre Research Reports. 2.
Pomeroy, R.S. (Ed.), 1994. Community Management and Common Property of Coastal Fisheries in Asia and the Pacific: Concepts, Methods and Experiences. International Center for Living Aquatic Resources Management (ICLARM), Manila Philippines.
Poon, S.E., Bonzon, K., 2013. Catch share design manual. In: Territorial Use Rights for Fishing. vol. 3. Environmental Defense Fund, pp. 1–161.
Prado, J., Drew, S., 1999. Research and development in fishing technology in Latin America. FAO Fish. Circ. 944, 31.
Premat, A., 2004. Cuban Counterpoint of the Public and the Private: Reflections on the Making of Urban Agriculture Sites in Havana, Cuba. Doctoral dissertation, Graduate Programme in Social Anthropology; York University.
Puga, R., de León, M.E., 2003. La pesquería de la langosta en Cuba. In: Report of Second Workshop on the Management of Caribbean Spiny Lobster Fisheries in the WECAFC Area. FAO Fisheries Report No. 715, pp. 85–91.
Rodrígez, G., 2003. Base para el Manejo Sostenible de un Bosque de Manglar en Estado de Deterioro. Tesis para el grado de Doctor en Ciencias, Universidad de Pinar del Rio, Cuba, Universidad Alicante, España.
Rosendahl, M., 1997. Inside the Revolution: Everyday Life in Socialist Cuba. Cornell University Press, Ithaca.
Roux, A., Piñero, R., 2004. Estado de los fondos de distribución de langostino patagónico (Pleoticus muelleri). Informe Técnico Interno. INIDEP No. 50, p. 8.
Saenz-Arroyo, A., Torre, J., Bourillon, L., Kleiberg, M., 2005. A community-based marine reserve network in North-Western Mexico. In: Proc. Symposium and Workshop of the North American Marine Protected Areas Network; North American Commission for Environmental Co-operation, p. 19.
Sainsbury, K., Sumaila, U.R., 2003. Incorporating ecosystem objectives into management of sustainable marine fisheries, including 'best practise' reference points and use of marine protected areas. In: Sinclair, M., Valdimarson, G. (Eds.), Responsible Fisheries in the Marine Ecosystem. FAO, Rome, pp. 343–361.
Sala, E., Aburto-Oropeza, O., Paredes, G., Parra, I., Barrera, J.C., Dayton, P.K., 1993. A general model for designing networks of marine reserves. Science 298, 1991–1993.
Sale, P.F., Cowen, R.K., Danilowicz, B.S., Jones, G.P., Kritzer, J.P., Lindeman, K.C., SergePlanes, Polunin, N.V.C., Russ, G.R., Sadovy, Y.J., Steneck, R.S., *2005*. Critical science gaps impede use *of* no-take fishery reserves. Trends Ecol. Evol. *20* (*2*), 74–80.
Sánchez, P., 1997. In: Las regulaciones pesqueras en Cuba. Proceedings From Pesca, 97. CIP, La Habana.
Schonberger, S., Agar, J., 2001. Argentina: Towards Rights-Based Fisheries Management. Rural Development Working Paper. The World Bank, Washington, D.C.
Seijo, J.C., Defeo, O., Salas, S., 1998. Fisheries Bioeconomics: Theory, Modelling and Management. FAO Fisheries Technical Paper No. 368, FAO, Rome, p. 108.
Shing, C.C.A., 2001. Case Study of the Integrated Coastal Fisheries Management Project: A Pilot Project for the Gulf of Paria, Trinidad. CANARI Technical Report No. 280, Division of Fisheries; Caribbean Natural Resources Institute (Canari), Trinidad and Tobago, pp. 1–13.
Soeftestad, L.T., 1999. The International Workshop on Community-Based Natural Resource Management (CBNRM), Workshop Report; Washington D.C., 10–14 May 1998; World Bank (Washington D.C.). pp. 1–36.
Sparre, P., Venema, S.C., 1998. Introduction to Tropical Fish Stock Assessment. Manual, Part 1. FAO Fisheries Technical Paper No. 306.1, Rev. 2, FAO, Rome.
SSPyA, 2004. Condiciones generals para que los insepctores que se embarquen en los buques desarrolen su tarea de control delcumplimiento de las normas que regulan la actividad pesquera. Disposicion 424, Subsecretaria de Pesca yAcuicultura (SSPyA).
St. Martin, K., 2001. Making space for community resource management in fisheries. Ann. Assoc. Am. Geogr. 91, 122–142.
Tegner, M.J., Basch, L.V., Dayton, P.K., 1996. Near extinction of an exploited marine invertebrate. Trends Ecol. Evol. 11 (7), 278–280.
Tony, P., Kevern, C. (Eds.), 2002. The Use of Ecosystem Models to Investigate Multispecies Management Strategies for Capture Fisheries. vol. 10. Fisheries Centre Research Reports; University of British Columbia, p. 156. Vancouver and FAO Fisheries Report (2).
United Nations Environmental Programme (UNEP), 2002. Fisheries Subsidies and Marine Resource Management: Lessons Learned From Argentina and Senegal. United Nations Environmental Programme, New York and Geneva.
Van Pelt, M., 1993. Ecologically sustainable development and project appraisal in developing countries. Ecol. Econ. 7 (1), 19–42.
Vasconcellos, M., Cochrane, K., 2005. Overview of world status of data-limited fisheries: inferences from landings statistics, pp. 1–20. In: Kruse, G.H., Gallucci, V.F., Hay, D.E., Perry, R.I., Peterman, R.M., Shirley, T.C., Woodby, D. (Eds.), Fisheries Assessment and Management in Data-Limited Situations. Alaska Sea Grant College Programme, University of Alaska Fairbanks, p. 958.
Verdery, K., 1996. What was socialism and what comes next? Princeton University Press (Princeton); 1991 theorizing socialism. Am. Ethnol. 18 (3), 419–439.
Villasante, S., Sumaila, R., 2009. Economics of Fisheries Management of Straddling Fish Stocks in the Patagonian Marine Ecosystem. Working Paper Series No. 2009-WP15, Latin American and Caribbean Environmental Economics Programme.
Voogd, H., 1983. Multicriteria Evaluation for Urban and Regional Planning. Pion, London.
Walters, C., Christensen, V., Pauly, D., 1997. Structuring dynamic models of exploited ecosystems from trophic mass-balance assessments. Rev. Fish Biol. Fish. 7, 139–172. FAO/Fisheries Centre Workshop: Page 11.
Walters, C., Pauly, D., Christensen, V., Kitchell, J.F., 2000. Representing density dependent consequences of life history strategies in aquatic eco-systems: EcoSim II. Ecosystems 3 (1), 70–83.
Wibera, M., Berkesb, F., Anthony, C., John, K., 2004. Participatory research supporting community-based fishery management. Mar. Policy 28, 459–468.
Wilson, E.O., 1998. Consilience. Vintage Books, p. 332.
World Trade Organization, 1999. Trade and Environment Special Studies, 4 (Geneva).
Worm, B., Barbier, E.B., Beaumont, N., Duffy, J.E., Folke, C., Halpern, B.S., Jackson, J.B.C., Lotze, H.K., Micheli, F., Palumbi, S., Sala, E.K.A., Stachowicz, J.J., Watson, R., 2006. Impacts of biodiversity loss on ocean ecosystem services. Science 314, 787–790.

Worm, B., Hilborn, R., Baum, J.K., Branch, T.A., Collie, J.S., Costello, C., Fogarty, M.J., Fulton, E.A., Hutchings, J.A., Jennings, S., Jensen, O.P., Lotze, H.K., Mace, P.M., McClanahan, T.R., Minto, C., Palumbi, S.R., Parma, A.M., Ricard, D., Rosenberg, A.A., Watson, R., Zeller, D., 2009. Re-building global fisheries. Science 325, 578–585.

WT/CTE/W/167, 2000. Environmental Benefits of Removing Trade Restrictions and Distortions: The Fisheries Sector. Note by the Secretariat, WTO Committee on Trade and Environment.

Young, J., 2013. Catch Shares in Action: Argentine Individual Transferable Quota Programme. Multi-Species, Individually-Allocated, Quota-Based, Transferable. Environmental Defense Fund, pp. 1–10.

Zapata, F., Fazio, A., Aspalanato, N., Bernisone, L., Caride, A., Echezarreta, A.F., Gonzalez, V., Oberti, A., 2005. Reducing incidental mortality of Franciscana dolphin *Pontoporia blainvillei* with acoustic warning devices attached to fishing nets. Mar. Mamm. Sci. 18 (4), 833–842. Bordino, P., Kraus, S., Albareda, D., Fazio, A., Palmerio, A., Mendez, M., Bottac, S. (2002).

CHAPTER

6

Climate change, fisheries management, and economics

OUTLINE

Community Based Fisheries Management. https://doi.org/10.1016/B978-0-12-821723-8.00006-3

Abstract

Our lives are inseparably connected to the climate of the **Mother Earth**. Similarly, other living beings are also connected. Human societies have adapted to the relatively stable climate since the last ice age which ended several thousands of years ago. Incidentally, a warming climate could bring the changes that can affect our water supplies, agriculture, power and transportation systems, the natural environment, and even our own health and safety.

Keywords: Natural environment, Climate change, Human population, Aquaculture, Fisheries management, Economics

6.1 Aspects of climate change

6.1.1 Introduction

Our lives are inseparably connected to the climate of the **Mother Earth**; and, similarly other living beings are also connected. Human societies have adapted to the relatively stable climate since the last ice age which ended several thousands of years ago. Incidentally, a warming climate could bring the changes that can affect our water supplies, agriculture, power and transportation systems, the natural environment, and even our own health and safety.

The warming of the Earth is the subject of intense debate and concern for many scientists, policy-makers, and citizens in the past few decades. Several researches characterized the global warming trend over the last century; and, these efforts tried to portray, what lay in store for the 21st century; and, the extent to which warming could be attributable to human activities.

Incidentally, the global climate varies naturally over time scales from decades to thousands of years or even longer. These natural variations may originate in the following two ways:

(a) from internal influences on the climate system including exchange of energy, water, and carbon between the atmosphere, oceans, land and ice, and
(b) from external influences on the climate system, including variations in the energy received from the Sun and by the effects of volcanic eruptions.

At this stage, it could be useful to differentiate between the terms "weather" and "climate." "Weather" is the state of the atmosphere—its temperature, humidity, wind, rainfall, and so on—over a short period ranging from hours to weeks. It has influenced by the oceans, land surfaces, and ice sheets, which together with the atmosphere form what is called the "climate system." "Climate," in its broadest sense, is the statistical description of the state of the climate system over quite a long period.

Climate change refers to any significant **change** in the measures of **climate** lasting for an extended period. In other words, **climate change** portrays major **changes** in **temperature**, precipitation, or wind patterns, among other **effects**, that occur over several decades or longer, or as long as thousands of years. Alternatively, climate change is a change in the statistical properties of the climate system that lingers for several decades or longer—usually at least 30 years. These statistical properties include averages, variability, and extremes. Climate change may be due to natural processes, such as changes in the Sun's radiation, volcanoes, or internal variability in the climate system, or due to human influences such as changes in the composition of the atmosphere or land use. Incidentally, it is possible to see and measure **climate changes** around the world, for example global temperature has increased, weather patterns have changed, polar ice caps have melted, and sea level has risen. In addition, it is mentioned that human activities have influenced climate by changing concentrations of CO_2, methane, and nitrous oxide in the atmosphere as well as by some chemically manufactured greenhouse gases, such as halocarbons, etc. Climate change has further influenced the alterations in the concentration of aerosols and changes in the land cover that affects the reflectivity of the earth's surface. In fact, the aerosols have reflected some amount of incoming sunlight. Incidentally, the human-generated gases enhance the natural greenhouse effect and further warm the surface. In addition to the direct effect, the warming that results from increased concentrations of long-lived greenhouse gases can be amplified by other processes, for example, water vapor amplification.

The planet's oceans and glaciers have also experienced some major changes—oceans are warming and becoming more acidic, ice caps are melting, and sea levels are rising. As these and other changes become more pronounced in the coming decades, they will likely present challenges to our society and environment.

Climate change, of late, seems to be a menacing issue in the globe affecting both the abiotic and the biotic domains. The increasing trend of climate change seemed to have been looming large as an added potential threats to both biotic and abiotic resources. Extensive burning of fossil fuel coupled with extensive deforestation over the last few centuries had enhanced the level of atmospheric CO_2 by *c.* 40% more than the preindustrial values, which were higher than at any time over the past 800,000 years or longer (Doney and Schimel, 2007). Further, future projections will forecast even faster accumulation of CO_2, unless meaningful and effective steps are taken to reduce the level of CO_2 emissions by humankind.

Notwithstanding the above, average prediction(s) may somewhat be further conservative; in that, the estimated CO_2 emissions from fossil fuel in 2005 had exceeded those forecasted by the most extreme scenario from the 1990s (Raupach et al., 2007); thus, implying that future atmospheric CO_2 levels may exceed current forecast(s) by the model(s), and the oceans may acidify faster than predicted today. It may be noted that the oceans of the earth absorb now *c.* 30% of the released anthropogenic CO_2 (Sabine et al., 2004). This, basically, have been changing the chemistry of the ocean water by acidifying surface waters (Caldeira and Wickett, 2003). Organisms' net responses to rising CO_2 may, however, vary depending on sensitivities to decreased seawater pH, carbonate concentration, etc.

It became a surprising event when a climate change was visibly evident in America in 2012. A record heat wave affected many regions of the United States in 2012. Moreover, the so-called superstorm sandy had pounded the Northeastern (NE) region of the United States. The sea surface temperature (SST) anomaly might have influenced the trajectory of superstorm sandy (Greene et al., 2013). Incidentally, the 2012 heat wave was remarkable in its intensity on the continental shelf, in addition to its large size and persistence, and warming was especially pronounced in the Gulf of Maine, where temperature had shot by 0.026°C $year^{-1}$ on the average since 1982, which accelerated by 0.26°C $year^{-1}$ beyond 2004. In fact, the intense warming in the Gulf of Maine triggered rapid and unprecedented changes in the ocean ecosystem; thus, leading to complex interactions among physics, ecology, economics, and policy matters. Moreover, the same climatic conditions that had led to the marine heat wave had also contributed to an extreme melting event in Greenland (Nghiem et al., 2012).

Incidentally, it may be noted here that the temperature of the ocean waters in the world increases fast, with some of the fastest-warming events occurring in the NW Atlantic, particularly in the Labrador Sea (Belkin, 2009; Taboada and Anadón, 2012). In other words, a less visible heat wave was occurring over a large portion of the Northwest (NW) the Atlantic Ocean. This ocean heat wave of 2012 was one of the largest and most intense warming events in the NW Atlantic in the last 30 years. Incidentally, the SST was 1–3°C warmer than the average of 1982–2011; particularly, over the region, from Cape Hatteras to Iceland and northward into the Labrador Sea. In fact, there is a need for a thorough analysis of the atmosphere-ocean interactions that produced this anomaly. However, the level of warming was *at par* with the mean SST change projected for the end of the century (Meehl et al., 2007).

Such a marine heat wave had affected the coastal ecosystems and the associated economies. Naturally, many marine species responded to warmer temperatures in different ways including shifting their geographic distribution and seasonal cycles. Incidentally, the warmwater species had moved northward, while some species undertook local migrations earlier in the season. Nevertheless, both these had affected fisheries targeting different species.

Notwithstanding the above, the shell-forming marine organisms build up carbonate structures using certain approaches. Detailed reviews on such aspects are found in Fabry et al. (2008), Doney et al. (2009), and others. Briefly, it is stated that organisms which exert low biological control on calcification, deposit $CaCO_3$ directly in their inner shell walls, and consequently, they depend on a sufficient ambient carbonate concentration to accumulate shells successfully. Notably, the commercially valuable molluscs, such as the bivalves (e.g., the scallops, oysters, etc.) and some gastropods (e.g., the conchs) use this method to build shells.

Extreme phenomena may become more common as climate change progresses (Tebaldi et al., 2006; Hansen et al., 2012). Incidentally, the NW Atlantic heat wave of 2012 had provided valuable insights into the generation of scientific database related to various aspects of the fishery, including fishery management frameworks, particularly, with regard to the need of adopting effective measures for different aspects including variability of marine temperatures. In the event of failure to anticipate the climatic alterations and the consequent non-incorporation of these aspects into the management *modus operandi* could exacerbate economic and social problems. Incidentally, there is a growing trend of underperformance of global fisheries economically, both in the freshwater (FW) and marine domains, mainly because of overharvesting of resources (particularly the fishes), pollution, climate change effects, and habitat degradation. However, if lessons were drawn from events like the 2012 thermal wave situation, it could help in developing adaptation strategies that could enhance existing capabilities to sustain marine ecosystems and fisheries in the event of climate change.

Notwithstanding the above, climate change is evident in other parts of the globe, notably, in Asia and, particularly, in the Indian subcontinent, due to shifting of seasonal durations. Incidentally, there were several periods of persistent atmospheric blocking during both winter and summer seasons, which had brought warm air into the region and might have contributed to ocean warming (Greene and Monger, 2012).

In fine, the planet's oceans and glaciers have also experienced some major changes. As mentioned earlier, oceans are warming and becoming more acidic, ice caps are melting, and sea levels are rising. As these and other changes become more pronounced in the coming decades, they will likely present challenges to our society and environment. As such, an effective management system based on the ecosystem approach and involving the stakeholders from the grassroots level is a prerequisite for efficient and sustained utilization of the natural resources, including fishes.

6.2 The growing human population and impact on climate

An enormous threat to the ecology and biodiversity of the planet in the coming decades will lead to the global **climate** disruption due to the buildup of human-generated greenhouse gases (GHGs) in the atmosphere, with the burgeoning population.

There is no doubt that the growth of human population is a major contributor to global warming, given that humans use fossil fuels to power their increasingly mechanized lifestyles. More people means more demand for oil, gas, coal, and other fuels mined or drilled from below the Earth's surface, that, when burned, spew enough carbon dioxide (CO_2) into the atmosphere to trap warm air inside like a greenhouse.

According to the United Nations Population Fund (UNPF), the human population grew from 1.6 billion to 6.1 billion people in the 20th century. During that time, emissions of CO_2, the leading GHG grew 12-fold. In addition, with a worldwide population, it is expected to surpass 9 (nine) billion over the next 50 years. Thus, the environmentalists and others are worried about the ability of the planet to withstand the added load of GHG entering the atmosphere and wreaking havoc on ecosystems down below.

Lately, a study performed on the relationship between population growth and global warming determined that the "carbon legacy" of just one child can produce 20 times more GHG than a person could save by driving a high-mileage car, doing recycling works, or using energy-efficient appliances, etc. Each child born in the United States could add *c.* 9441 metric tons (mt) of CO_2 to the carbon legacy of an average parent. As such the potential savings from reduced reproduction are found huge as compared to the savings that can be achieved by changes in lifestyle.

Among the western countries, the United States have the largest population in the developed world and is the only developed nation experiencing significant population growth. Incidentally, the population of the United States may double before the end of the century. A total of 300 million inhabitants of the United States produce greenhouse gases at a *per capita* rate, which are more than double that of Europe. Incidentally, the US greenhouse gas contribution is driven by a disastrous combination of high population, significant industrial growth, and huge consumption levels. Incidentally, the transportation sector accounts for one-third of all US carbon emissions.

Notwithstanding the above, another one-fifth of US carbon emissions come from the residential sector. Average home sizes have increased dramatically in recent decades, as accompanying with the footprint of each home.

Incidentally, suburban sprawl contributes significantly to deforestation, reducing the capacity of the planet to absorb the increased amount of CO_2 that we emit. Concomitantly, house building outpaces the population growth due to a dramatic decrease in household size, from 3.1 persons per home in 1970 to 2.6 in 2000. Incidentally, more Americans are driving farther to reach bigger homes with higher heating and cooling demands and fewer people per household than ever before. All these trends exacerbate the carbon footprint inherent in the basic energy needs of a burgeoning US population. Incidentally, the availability and practice of family planning methods have declined the fertility rates. If this has a bearing on decreased population growth, it could mean less *per capita* emission of GHGs into the atmosphere. This, in turn, could mitigate global warming. Population reduction to ecologically sustainable levels could solve the global warming crisis and move us toward a healthier and more stable society.

People around the globe today, are trying to address the problem by reducing their carbon footprint through less consumption and better technology. However, unsustainable human population growth may deter these efforts. This leads us to conclude that we not only need smaller footprints but also fewer feet.

In the United States, Portland and Oregon, for instance, had reduced their combined *per capita* residential energy and car driving carbon footprint by *c.* 5% between 2000 and 2005. During the same period, however, their population was grown by *c.* 8%.

According to the Worldwatch Institute, a nonprofit environmental think tank, the overriding challenges facing our global civilization are to curtail climate change and slow down the rate of population growth. The population is growing rapidly, far outpacing the ability of our planet to support it, given current practices. Overpopulation is associated with negative environmental and economic outcomes ranging from the impacts of over-farming, deforestation, and water **pollution** to eutrophication and global warming.

Population, global warming, and consumption patterns are inextricably linked in their collective global environmental impact. As the contribution of the developing countries to global emissions grows, population size and growth rates could become significant factors in magnifying the impacts of global warming.

There are suggestions from certain quarters that continuing population growth and increasing global consumption of resources are considered along with carbon emissions of an individual over his or her lifetime. It is, sometimes opined that future growth amplifies the consequences of people's reproductive choices in the same way, that compound interest amplifies a bank balance.

Notwithstanding the above, with regard to the growing human population vis-à-vis climate change and aquaculture, even with the right management changes, there could be winners and losers, and humanity has to tackle all these. In fact, reduction in emissions coupled with multilateral cooperation and real changes in fisheries management could pave the way to success. The burgeoning growth of the global population coupled with the growing needs of fishes has led to increased aquaculture to find an answer to the diminishing protein supply in our nutrition as well as for an avocation. Hence, alterations in the management process could be critical for meeting the United Nations Sustainable Development Goals. Concomitantly, the impacts of inaction are also clear. Innumerable people depend on fish as their primary source of protein. At the same time, many fishing nations are not responding fast enough to make alteration(s). Moreover, successful transboundary management programs are comparatively rare. But proper action does not take long time to have a good impact on some species. Studies have revealed that many fisheries could bounce back from overfishing in as little as 10 years' under the correct management policies.

6.3 The impact on water bodies, fisheries, and aquaculture

6.3.1 Introduction

Fish is an essential and significant entity of nutrition and avocation for the people of the globe. Even a small quantity of fish could bring the improvement(s) in peoples' diets, as it provides vital nutrients that are absent in typical starchy staples that dominate poor people's diets (FAO, 2005). Further, fish provides *c.* 20% of animal protein intake in 127 developing countries (Thorpe et al., 2006), which could reach up to *c.* 90% in small island developing states (SIDS) or coastal areas (FAO, 2007). In fact, aquaculture contributes to an increasingly significant proportion of fish production over the recent decades.

Climate change is an issue of grave concern today. Climate changes have a wide-ranging impact on fisheries and those who depend on them. In fact, a significant amount of information on the biophysical impacts of climate change on aquatic ecosystems is available, but lacks knowledge on how these influences could be taken care of in terms of the socioeconomic context of fisheries as well as how these adaptation(s) could proceed. Incidentally, the vulnerability of fisheries to climate change is determined not only by the degree of change or impact, but also by the sensitivity of individuals or fisheries systems and their adaptive capacity. Adaptive capacity relies on various assets and may be constrained by factors including culture or marginalization. Incidentally, the priority and responsibility for governments, the civil society, and the international organizations as well as people, in general, are to take care of the excess GHG emissions, global warming, etc. It is because the long-term consequences of climate change are highly complex, unknown, and potentially irreversible. Incidentally, fisheries make a moderate contribution to GHG emissions through fossil-fuel-based catching operations and transportation, which may be reduced with improved technology and management of stocks by eco-friendly methods. Different global emissions indicate that climate change could affect both marine and FW systems and fishing communities as well. The civic society, thus, have a responsibility to facilitate adaptation, especially for vulnerable groups because of their exposure, sensitivity, or lack of adaptive capacity.

A management plan is given in the following aspects:

(i) to identify the most vulnerable individuals and communities;
(ii) to examine the possible government-assisted adaptation(s);
(iii) to consider the difficulties with regard to private adaptations; and
(iv) to look for desirable adaptations which may help in the long-term reductions in vulnerabilities, rather than short-term coping strategies which may enhance vulnerability.

Notwithstanding the above, there were attempts for reviewing the potential impacts of climate change on fisheries, which have suggested a role of public policy in adaptation; possibility to reduce vulnerability and, further, to provide with information for planning and stimulating adaptation as well as to ensure that adaptation actions may not negatively affect other ecosystem services and the viability of fisheries in the long run.

In this connection, the **first rationale** for promoting adaptation is to protect those parts of the fishing sector and communities in FW, marine, and coastal areas that have the least ability to cope. In the case of developing countries, there are additional embargos, like, lack of public infrastructure, high disease burden, and many other factors, which restrict the ability to adapt.

Concomitant to the above, the **second public policy** response is the provision of high-grade information on the risks, vulnerability, and threats posed by climate change. Such information includes the scenarios of climate change at the global scale. Moreover, it also involves significant investment in the collection of a huge amount of climate-related information; land use planning; and, other forms of regulations. Hence, there could be a need for policy integration across various sectors, including the aquatic domain (FW, marine, and coastal). A significant amount of planning may be required for agriculture, fishery, and health services, where climate change risks could play a significant role.

The **third domain** of public policy response is the provision and enhancement of the public good aspects of fishery-related biodiversity and ecosystem services. Incidentally, the Millennium Ecosystem Assessment highlighted the significance of ecosystem services for the well-being of humanity. In fact, climate change impacts represent enhanced reasons for sustainable fisheries management and incentives to promote aquatic biodiversity conservation in FW, marine, and coastal regions, given the potential threat of habitat decline and species extinction throughout the globe. Incidentally, there are suggestions to implement the ecosystem approach to fisheries.

Concomitant to the above, pro-poor governance of small-scale fisheries is now promoted by international organizations in order to address the marginalization of fishers and equity (FAO, 2005). These formidable challenges for governance continue and become more imperative in the face of climate change. Incidentally, variability, vulnerability, and uncertainty, which had historically been significant factors that managers were struggling to take account of, could become more prevalent under climate change situation(s). Incidentally, poverty in small-scale fisheries and marginalization of fishers reduce their adaptive capacity. Moreover, the wider context of fisheries is almost equally significant because of the ways in which policies, socioeconomics, demographics, ecology, and markets could influence fisheries. Moreover, there could be other significant pathways to depict climate change impacts, because they evolve rapidly with the process of globalization.

Notwithstanding the above, it may be noted that future climate change may not interact with fisheries in the way it is doing today because it is supposed to affect future fisheries in a futuristic context. This naturally gives rise to additional uncertainty, and also, emphasizes the need for adaptive governance as well as the integration of fisheries with other allied and linked sectors, particularly agriculture, which may itself affect fisheries due to climate impacts and adaptation. In fact, Current problems with fisheries management call for strong and reliable institutions, which could lacks govern resource use. However, paradoxically, top-down, or rigid approaches, which may seem attractive, may not offer the flexibility to ensure resilient fisheries systems under the scenario of climate change. Conversely, approaches, such as adaptive comanagement are proposed to address uncertainty and harness the knowledge and commitment of resource users at multiple scales (Armitage et al., 2009). Such approaches may offer the best hopes for resilient fisheries. In fact, the governance systems with a focus on continual learning from experience could be more likely to address new challenges as they arise. However, policies, which place too much emphasis on stability, certainty, and top-down control may lead to unexpected and unregulated consequences and may end in a "lock in" condition of fisheries; thus, preventing desirable and sustainable adaptation. However, abandoning fisheries as a livelihood may become a necessary reality in some fisheries, particularly, where the output is very poor. Incidentally, the social and value-based adaptation emphasizes the need for equitable and just processes.

The global population could be *c.* 9 billion by 2050, resulting in increased food needs (McMichael, 2001). As such, food production must be augmented through various means including AQC after taking into account the impact of climate change scenario as well as the capacity, efficiency, and sustainability of the production methods.

Notwithstanding the above, declines in terrestrial food production could be supplemented by augmentation through AQC. In fact, humans and fish are inextricably linked for millennia not only from the nutritional and avocational point, but also from an evolutionary point of view. Incidentally, the development of the human brain, and hence, what humans are today is linked to food sources rich in n-3 (DHA, EPA) and n-6 (AA) PUFAs because, fish had constituted a major part of the diet of our ancestors. In this connection, evidence revealed that *Homo sapiens* might have evolved not only in a savannah habitat, but also in a habitat that was rich in fish and shellfish resources (Crawford et al., 1999). Many health reports recommend fish in a healthy diet, physical growth, and general well-being. Concomitantly, some of the ailments may be alleviated by supplementing with PUFAs. Thus, there seemed to be a growing trend in increased fish consumption appreciating the importance of omega-3 fatty acids in the human diet. In fact, two crucial ones, viz., EPA and DHA are primarily found in certain fish. Moreover, food fish provides essential micronutrients in the form of vitamins and minerals, e.g., best sources of iodine and selenium, and also some of the coenzymes (Q 10). Incidentally, climate change may impact on the globe in many ways hike in global mean temperatures (planet may be heated up by *c.* 1.1°C this century, and, up to 3°C elevation if the GHGs continue to rise at the current rate) (Zwiers and Weaver, 2000; Kerr, 2006); increase in the atmospheric GHG (Brook et al., 1996); devastating cyclones, blizzards; fragmented monsoon; erratic floods, droughts, etc. All these impacts on the globe could also affect humans. Moreover, humans may also be affected by food scarcity, diseases, etc. Nevertheless, AQC and fisheries could help us in mitigating the impacts.

Incidentally, food fish production is classified into two subsectors, viz.:

(i) the capture fisheries, which overtly include the naturally occurring and recruited wild populations, the great bulk of which (*c.* 85%–90%) are in the FW and the seas and
(ii) the culture fisheries, which include the farmed food fish subsector, achieve relatively growing significance, and are popularly called "aquaculture."

Concomitant to the above, various kinds of research works are found addressing the potential impacts of climatic change on aquaculture. Concomitantly, efforts are being made to reveal the contribution of aquaculture (AQC) to the total food fish supply, and to address the potential adaptations and mitigation.

In some of the main fish producing countries, the AQC sector contributes significantly to the gross domestic product (GDP) accounting for *c.* 45% of total food fish consumption. The burgeoning growth of the human population, vis-à-vis stagnation in capture fishery, warrants the augmentation of AQC production.

Notwithstanding the above, today, it is essential to evaluate the potential effects of aquaculture on climate change and vice versa. Efforts are made to analyze current aquaculture practices in relation to tropical, subtropical, and temperate climates as well as in relation to marine, FW, and brackish water domains across different continents. Notably, AQC practices are found predominantly in tropical and subtropical climatic regions, and geographically, in the Asian region, with most commonly cultured commodities being the finfish, molluscs, crustaceans, and seaweeds. However, under the prevailing conditions, it could be necessary to focus on the development of adaptive strategies for addressing and mitigating climate change impacts on aquaculture, which could also help in minimizing the predicted gap between demand and supply for food fish requirement. Incidentally, elements like sea level and temperature rise; change in monsoonal rainfall patterns; and extreme climatic events and water stress could cause impacts on AQC.

Notwithstanding the above, climate change may show different types of impacts in different culture systems in FW, marine, brackish water, etc. Incidentally, *c.* 65% of the aquaculture production is inland; and, also, concentrated mostly in the tropical and subtropical regions of Asia. Further, it is likely that diseases affecting aquaculture could be aggravated both in incidence and in impact due to climate change. In fact, the impact could be small, if eco-friendly practices are followed, while, it could be big, if unusually aggravated artificial means are followed to raise the growth of the cultured stocks. In addition, climate change could impact on water availability, weather patterns, such as extreme rain events and augmentation of eutrophication and stratification in the lentic water bodies, and so on. Further, local adaptive measures are essential, particularly, for finfish AQC, which is usually and predominantly based on species feeding low in the food chain. As such, there is often a need for the greater availability of phytoplankton and zooplankton in the water body.

In marine cage culture, the introduction of improved technologies to withstand extreme weather events could be a significant adaptive measure. Sea-level rise and consequently increased saltwater intrusion are some of the impacts in the deltaic areas in the tropics, and, hence, some of the salt-tolerant species are introduced. In fact, the negative impacts are likely to be felt mostly in the temperate regions where the finfish aquaculture is based mainly on carnivorous species. Further, in the event of calamities, AQC could provide alternative livelihoods to the agriculturists. While considering the contribution of AQC to carbon emission, on the whole, AQC is often said to be less energy costly and could rather contribute to carbon sequestration more than other terrestrial farming systems.

6.3.2 A general overview of climate change

After bringing-out of the fourth Assessment Report of the Intergovernmental Panel on Climate Change (IPCC) in 2007, the perils of climate change to mankind and the natural ecosystems have become a very important item of discussion in different forums for taking effective steps to mitigate the ill effects. The issue became an urgent one for the aquatic domain, particularly, with regard to the deleterious impacts of warming on the rise of water level and the impacts on fisheries and aquaculture. In fact, the coastal and riparian communities, in general, are immediately and very badly affected. Hence, such issues are difficult to ignore, and, rather need urgent attention. Incidentally, it may be emphatically stated that fisheries and aquaculture certainly have a significant role in the nutrition and avocation of the people from local to global levels; besides, "fish" is a precious entity in the garland of the food chain. In addition, the fisherfolk happened to be a significant component in the domain of fisheries and aquaculture. In fact, *c.* 43.5 million people work directly in this sector, having a majority in the developing countries. Further, in a holistic view, the sector is to support *c.* 200 million livelihoods, adding those who work with processing, marketing, distribution, and supply network. Incidentally, it may be noted that aquatic foods have a high nutritional quality, contributing to *c.* 20% or more of average *per capita* animal protein intake for >1.5 billion people, mostly from developing countries. They are also the most widely traded foodstuffs as well as are essential components of export earnings for many less affluent nations. Moreover, the sector is of particular significance to the small island states which depend on fisheries and aquaculture for *c.* 50% of their animal protein. In fact, climate change affects broadly across ecosystems, societies, and economies, enhancing pressure on all kinds of livelihoods.

FAO may take a lead in informing fishers and policy-makers about the likely consequences of climate change for fisheries and aquaculture. Thus, having been convened to identify and review the key issues of climate change in relation to fisheries and aquaculture, the Expert Workshop of FAO discussed threadbare the impacts of climate change on aquatic resources and ecosystems as well as how these ecological impacts could be mitigated through adaptive measures within fisheries and aquaculture. The Workshop developed three comprehensive background documents, which were also published. The three documents are as follows:

(i) physical and ecological impacts of climate change relevant to marine and inland capture fisheries and aquaculture;
(ii) climate change and capture fisheries: potential impacts, adaptation, and mitigation; and
(iii) climate change and aquaculture: potential impacts, adaptation, and mitigation.

The FAO Workshop analyzed the three documents, and one of the key messages was that climate change is a persisting and compounding threat to the sustainability of capture fisheries and aquaculture development. Impacts may occur to both in the form of gradual warming and associated physical changes. The Workshop summarized the main impacts on the ecosystem and livelihoods and their implications for food security. Urgent adaptation measures are needed in response to threats to food and livelihood provision due to climatic variations. Incidentally, the increasing trend of climate change have loomed large as a potential threat to fisheries and aquaculture. This threat has been aggravated due to habitat degradation and overharvests of the resources. Studies revealed that climate change could alter primary productivity; shifts in distribution; and changes in the potential yield of exploited aquatic species; thus, impacting the economics of fisheries globally. However, the poorly managed fisheries would be the worst affected due to the impacts of climate change. In fact, there is a clear roadmap to construct a potential future with more fish, more food, and more prosperity; provided, the people act in the right direction, with the right *modus operandi* and right now.

Notwithstanding the above, climate change becomes more evident, and as it increases, it could alter the productivity of fisheries and the distribution of fish stocks. Further, **climate change could have a cascading effect on the global fisheries and their management**. Thus, from an economic point of view, the economic and policy issues related to the impact of climate change on fisheries are to be dealt with urgently. In addition, the changes could have impacts on the livelihood of the fishing communities in different ways. Concomitantly, climate change may cause both physical and biological impacts on the fishes, which, in turn, may affect fishery economic performance. This, further, may influence management effectiveness and policy issues.

The emission of carbon-di-oxide (CO_2) is constantly augmented, mainly due to human-made interferences. Concomitantly, global warming is on the rise. At the same time, ocean acidification is a consequence of rising anthropogenic CO_2 emissions. This is poised to alter marine ecosystems markedly by enhancing dissolved CO_2 and diminishing ocean pH almost worldwide (Doney et al., 2009). Such situations deter the growth of calcium carbonate shells and skeletons of a large number of marine animals. In fact, the ongoing ocean acidification may harm a wide range of marine organisms and the food webs that depend on them. Incidentally, laboratory studies indicated that the molluscs, including species that support valuable marine fisheries, such as the mussels and oysters (Gazeau et al., 2007) and especially their juveniles (Kurihara et al., 2007, 2009) are particularly sensitive to the alterations due to climate change. Reduction in calcification decreases the fitness of commercially valuable molluscs by directly damaging shells (Kurihara et al., 2007, 2009; Gazeau et al., 2007).

However, the first direct impact of climate change on humans is through the decreasing harvests of fish and fishery-related revenues. Incidentally, the societies, which are dependent on marine calcifiers, could consequently experience significant economic losses and even social disruptions over the next several decades. In fact, the fishermen are supposed to be one of the worst affected people due to the impact(s) of climate change.

6.3.3 Acidification of the ocean

The oceanic uptake of anthropogenic CO_2 occurs through a series of chemical reactions that enhance the level of aqueous CO_2, and decrease the pH and carbonate ion level of seawater. Incidentally, the anthropogenic CO_2 have lowered the average surface ocean pH from a preindustrial value of 8.2 to 8.1 recently (Caldeira and Wickett, 2003). Further, according to prediction(s) by model(s), atmospheric CO_2 trajectories may rise from *c.* 385 ppm in 2008 to 450–650 ppm by 2060. This could lower mean ocean surface pH by an additional 0.2–0.3 units. Incidentally, further, the expected hike in the emissions of anthropogenic CO_2 emissions over the next half a century is mainly due to fast industrial growth in developing nations. This could accelerate alteration(s) in the ocean chemistry to rates which could be unprecedented in the recent annals of atmospheric and geological records (Doney et al., 2009; Doney and Schimel, 2007). In fact, seasonal acidification events are already in the offing. Water with poor physicochemical characteristics upwells along with the California coastline during summer (Feely et al., 2008). Also, some high-latitude polar and subpolar waters may witness poor physicochemical characteristics by mid-century or even earlier (Orr et al., 2005; Steinacher et al., 2009). Further, acidic conditions could damage the corals, and may also prevent their growth by destroying crucial benthic habitats (Kleypas et al., 2006; Lumsden et al., 2007). Incidentally, ecological shifts to macro-algal overgrowth and decreased species diversity may follow after coral disturbances leading to often stable new ecosystem states (Scheffer et al., 2001) dominated by herbivores and less commercially valuable species.

6.3.4 Impacts of climate change on the ecosystem

Climate change modifies the distribution of marine and FW species when considered in terms of physical and biological impacts. In general, the warmwater species are being shifted toward the poles, where they experience changes in the size and productivity of their habitats. However, ecosystem productivity could be reduced in most tropical and subtropical oceans, lakes, etc., and may also affect the physiological processes of the fish; thus, exerting both positive and negative impacts on fisheries and aquaculture systems based on the region and latitude. Moreover, climate change affects the seasonality of particular biological processes, which could change marine and FW food webs, and there could be unpredictable consequences for adequate fish production. There could also be enhanced threats of exotic species invasions and the spread of vector-borne diseases, which could be an additional concern. Further, differential warming between land and oceans, and between polar and tropical regions affects the intensity, frequency, and seasonality of climate patterns (e.g., *El Niño*); and extreme weather events (like floods, droughts, and storms, etc.) may occur. These events could also affect the stability of related marine and FW resources. Concomitantly, rise in sea level; melting of the glaciers; acidification of oceans and alteration(s) in precipitation pattern and intensity; change(s) in the groundwater and river flows, etc.; could occur due to the impacts of climate change, and, these could significantly impact the coral reefs, wetlands, rivers, lakes, and estuaries. Thus, adaptive measures are required for their mitigation and to minimize their impacts on fisheries and aquaculture systems.

6.3.5 Impacts of climate change on livelihoods

Due to the impacts of climate change, there could be a need for alteration(s) in the practice of fishing and aquaculture; as well, in the location of landing, farming, and processing facilities.

Change(s) in the distribution, species composition, and habitats, due to climate change, could lead to alteration(s) in the fishing practices and aquaculture operations. Moreover, extreme events of climate change could exert impact(s) on infrastructure, ranging from landing and farming sites to postharvest facilities and transport paths. Further, climate change may affect safety at sea as well as the habitations of the fishermen, with increased risks of low-lying areas being more affected. Still, further, water stress (caused due to climate change) and competition for water resources could affect aquaculture practices and inland fisheries production and are likely to enhance conflicts among water-dependent activities. Further, livelihood strategies may need to be changed. There could also be changes in fishers' migration patterns due to changes in the timing of fishing activities. Declined livelihood options inside and outside the fishery sector may compel occupational alterations. Livelihood diversifications could be considered as an established means of risk transfer; otherwise, they could negatively affect livelihood outcomes. There could also be risk(s) from extreme events and occupational change(s).

Notwithstanding the above, climate change affects the four dimensions of food security, which are as follows:

(i) availability of aquatic foods may vary with the change(s) in habitats, stocks, and species distribution;
(ii) stability of supply is affected by the change(s) in seasonality and increased variability in ecosystem productivity;
(iii) access to aquatic foods is affected by changes in livelihoods and catching or farming opportunities; and,
(iv) utilization of aquatic products could also be impacted; for instance, some communities may need to adjust to species not traditionally consumed.

6.3.6 Carbon footprints of fisheries and aquaculture

It is said that fisheries and aquaculture activities make a minor but still significant contribution to GHG emissions during production, operations, storage, and transport of fish. However, there could be significant differences in the emissions for different species. Moreover, the eco-friendly use of fossil fuels could mitigate the carbon footprints of fisheries and the aquaculture sector. Further, the GHG contributions of this sector are considered as small compared to other sectors; nevertheless, should be improved. Furthermore, there should be reduction in energy expenses in fishing practices and aquaculture production and more energy-efficient post-harvest and distribution systems through technological innovations. Moreover, the sector may have fruitful interactions with environmental services, e.g., maintaining the quality and function of the coral reefs, coastal margins, inland watersheds, potential carbon sequestration, and other nutrient management systems.

Concomitant to the above, by sustainable use of genetic diversity, including by sustainable application of biotechnological tools, there could be particular efficiency impacts, e.g., for widening production scope of low-impact aquaculture species, or for making agricultural crop materials or waste products usable for growing carnivorous aquatic species, and so on. However, such aspects are to be evaluated on wider social, ecological, and political criteria. Policies may, occasionally, support the use of a more efficient and effective approach(es) and practice(s), but possible negative impacts of mitigation on food security and livelihoods are taken into consideration along with efforts for their minimization.

6.3.7 Adaptation to change

It is a common notion that the resource-dependent communities are required to adapt themselves to different aspects of climate change. This could be seen throughout the history, but the projected climate change poses multiple additional risks to fishery-dependent communities that might limit the effectiveness of past adaptive strategies. Nevertheless, the adaptation strategies should be context and location specific. They are also to take care of the impacts at both (i) short-term (e.g., increased frequency of severe events) and (ii) long-term (e.g., reduced productivity of aquatic ecosystems) aspects. Moreover, all the levels of adaptation (community, national, and regional) may clearly require stronger capacity building, which might also give them more benefits through augmentation of the level of their awareness about the impacts of climate change on fisheries and aquaculture. Moreover, the standing practice of management is to be strengthened for increasing resilience and adaptability. Further, ecosystem approaches to fisheries (EAF) and aquaculture (EAA) may, preferably, be adopted to increase the resilience of aquatic resources in the domain of fisheries and aquaculture.

Notwithstanding the above, adaptation options may also encompass diversification of livelihoods and promotion of aquaculture crop insurance in the face of potentially reduced or more variable yields. Further, in the event of inclement and severe weather episodes (may be under the influence of climate change), there are strategies in place for minimizing the vulnerabilities to fishing and fish farming communities. Such strategies could include the augmentation of capacity building for weather forecasting; early warning systems; safer harbors and landings; and safety at sea. In addition, the adaptation strategies, in general, should also promote disaster preparedness and disaster risk management. Concomitantly, the National climate change adaptation and food security policies and programs could also be integrated with the fisheries and aquaculture sector. This could help to ensure that potential climate change impacts will be integrated into broader national development and planning. Incidentally, in the event of salinity increase in the deltaic areas leading to loss of agricultural land, aquaculture-based livelihoods as alternative livelihood options could be promoted.

6.3.8 Technical and organizational structures: Capacity building

In the event of climate change being *in vogue*, there is a need for policy preparation and action plans in response to climate change. It involves the technical agencies, like the departments responsible for fisheries and aquaculture, science and education, planning and finance, etc., who could play significant roles in national development. Also, the community and the political representatives, the public, the civil society, and the nongovernmental organizations (NGO) at the local, regional, and national levels are to be involved in a collaborative and participatory mode for capacity building and getting targeted information.

Notwithstanding the above, some of such works attempt to review the physical and ecological impacts of climate change relevant to marine and inland capture fisheries and aquaculture. There have been several effects of climate change on ecosystem and fish production processes.

6.3.9 Marine water

The oceans are said to be warming, which may not be geographically homogeneous. The accumulated impacts of thermal and salinity regimes reduce the density of the ocean and increase vertical stratification in deep lake surface, which may perturb water column stability and alter surface mixing along with concomitant possible reduction in nutrient availability. Further, ocean acidification have lowered seawater pH by 0.1 units in the last two centuries, which could further decrease by 0.3–0.5 pH units over the next century. However, the impacts of ocean acidification could be particularly severe for shell-borne organisms, tropical coral reefs, and cold-water corals. In addition, there are some increased intensities of coastal upwelling in recent decades, but the global circulation models do not show a clear pattern of upwelling response to earth's warming at the global scale. Incidentally, the sea levels are rising globally at an enhanced rate, risking particularly the Atlantic coasts and the Gulf of Mexico, the Mediterranean, the Baltic, Asian mega-deltas, various small-island regions, and other low-lying coastal areas. In fact, there could be lots of human settlements in the submerging-prone regions. Impacts of climate change could also be evident in the form of alteration in the ecology and biology of the dominant phytoplankton group.

Notwithstanding the above, in most of the Pacific ocean, surface warming and freshening act in the same direction and contribute to reduced mixing. In the Atlantic and Indian Oceans, on the other hand, temperature and salinity trends act in the opposite directions, and, changes in mixing not adequately quantified. Large salinity changes were observed in the North Atlantic along with sporadic changes in freshwater inputs. These Great Salinity Anomalies (Dickson et al., 1988) had resulted from the strengthening of the subpolar gyre. Three such anomalies were documented from 1968 to 1978 and in the 1980s and 1990s (Houghton and Visbeck, 2002).

6.3.10 Inland water

Concomitant to the above, there are evidence that inland waters are also warming, with differential impacts on the river runoff affecting the primary and secondary productions. Incidentally, the inland water temperatures are strongly linked to the dynamics of the hydrological cycle. Many 20th-century studies have recorded trends in river flows and the changes in lake levels at scales ranging from catchment to global. Significant trends in rising water levels (WL), in response to increased melting of snow and ice or declines in WL due to the combined effects of drought, warming, and human activities were recorded. However, overall, no globally homogeneous trend was reported (Rosenzweig et al., 2007).

Notwithstanding the above, year-to-year variation in river flows is influenced, in some regions by large-scale atmospheric circulation patterns associated with *El Niño* Southern Oscillation (ENSO), North Atlantic Oscillation (NAO), and other decadal variability systems. In addition, there is a broad coherent pattern of changes in annual runoff, portraying an increase at higher latitudes and a decrease in parts of West Africa, southern Europe, and southern Latin America (Milly et al., 2005). Probably, there was an approximately 4% enhancement in global total runoff per 1°C rise in temperature in the 20th century, with regional variation(s) around this trend (Labat et al., 2004) However, there are views contrary to this (Legates et al., 2005).

Concomitant to the above, worldwide, in some lakes, notably, the Lake Chad, there was a diminution of the water spread area due to declining precipitation and human interventions (Coe and Foley, 2001). Atmospheric warming, reduces rainfall in the subtropics and increases at higher latitudes and in parts of the tropics. Also, human interventions are some of the significant causes of inland water shrinkages (Christensen et al., 2007).

In addition, climate change affects the marine and inland ecosystems with regard to changes in land-use pattern, including alteration(s) in sediment loads, fluvial dynamics, and associated physicochemical parameters, like hypoxia, stratification, changes in salinity, etc. However, their complex consequences could certainly impact the community composition, production, and seasonality processes in plankton and fish populations. Further, under climate change, the high latitude/altitude lakes, in general, could experience reduced ice cover, warmer water temperatures, and a longer growing season, as a consequence of enhanced algal abundance and productivity. On the other hand, some of the deep tropical lakes could experience decreased algal abundance and a fall in productivity. Further, differences could be observed among different forms of plankton with regard to responses to temperature and light intensity. This could suggest that the marine and FW trophodynamics may be altered due to disturbance in the predator-prey relationship caused by warming of the ocean or FW.

According to the IPCC (2007), the FW resources are prone and vulnerable to climate change (Bates et al., 2008), which may cause a reduction of *c*. 10%–30% of average river runoff at mid-latitudes in the dry tropics by mid-century (Kundzewicz et al., 2008), and, an increase of *c*. 10%–40% at high latitudes in the wet tropics (Milly et al., 2005). Concomitantly, there could be shifts from snow to rain and a consequent change in the timing of peak river flows with concomitant alteration(s) in flood and drought frequency and intensity. Further, according to the IPCC (*op. cit*), impacts of climate change and effective adaptations could depend on local conditions, including socioeconomic conditions and other kinds of pressures on water resources (Kundzewicz et al., 2008).

Concomitant to the above, surface water temperatures had increased by 0.2–2°C in lakes and rivers in Europe, North America, and Asia Since the 1960s (Rosenzweig et al., 2007). This situation, along with longer ice-free seasons had influenced thermal stratification in various lentic systems in Europe and North America, and its period, had advanced by up to 20 days and lengthened by 2–3 weeks, because of increased thermal stability (O'Reilly et al., 2003).

6.3.11 Fishes

Incidentally, climate change may expand the range of many warm water ichthyospecies toward the poles, whereas it may contract the range of cold-water species. However, the pelagic species may exhibit fast changes, and which may also portray vertical movements to counteract surface warming. Some of the migratory fishes follow the trends in ocean water temperature and their process of migration could be preponed or postponed during summer or winter, respectively, depending on the thermal regime of water. The migration of a bigger school of the fish population toward the poles during summer could render aquatic desertification of the equatorial region. Thus, there were changes in the phenology of many FW and marine fishes due to climate change.

Notwithstanding the above, consequences of climate change are evident along with temporal scales. Augmented temperatures may negatively impact the physiology of fish, affecting life history events, recruitments, abundance, distributions, and also aquaculture processes. Concomitantly, adaptation to climate change events is a continuous process.

A brief summary of the likely physical impacts of climate change is given as follows:

6.3.12 Physical features

Warm African continent has more than the global annual mean warming, in all seasons, with the drier, warmer subtropical region, and wetter tropics. There is decreased annual rainfall in many regions of the Mediterranean Africa, southern Africa, and northern Sahara, while increased rainfall in East Africa (Christensen et al., 2007). Warming may be above the global mean in central Asia, the Tibetan Plateau, and in northern, eastern, and South Asia, while it is similar to the global mean in Southeast (SE) Asia. Precipitation in boreal winter may increase in northern Asia and the Tibetan Plateau, in eastern Asia, and in the southern parts of SE Asia. Concomitantly, summer rainfall may increase in northern Asia, East Asia, South Asia, and in many parts of SE Asia. However, it may decrease in central Asia. The hot spells in summer are of longer duration, more intense, and more frequent in East Asia. Incidentally, cold days are likely to be in East and South Asia during winter. Conversely, an increase in the frequency of intense precipitation events are likely in parts of South and East Asia during the monsoon period. Concomitantly, severe rainfall with tropical cyclones may be aggravated in East, SE, and South Asia. Hiked moisture convergence could increase monsoonal circulations with profuse rainfall in the regions (Christensen et al., 2007).

Further, significant negative impacts are felt across *c.* 25% of Africa's inland aquatic ecosystems by 2100 (De Wit and Stankiewicz, 2006), and both water quality and ecosystem goods and services are deteriorated because it is generally difficult and costly to regulate hydrological regimes.

6.3.13 Meridional overturning circulation in the sea

The Atlantic meridional overturning circulation (MOC) carries warm upper waters into far-northern latitudes. In the process, it cools, sinks, and returns southwards at depth. Alteration(s) in the hydrological cycle (including sea ice dynamics, as freezing water releases salt), due to climate change have the potential to influence the strength of the MOC. The heat transport of the MOC, substantially contributes to the climate of continental Europe, and any slowdown may have important atmospheric climate consequences (Vellinga and Wood, 2002). Incidentally, model predictions suggest increased FW input in the Arctic and sub-Arctic both through precipitation and reduced sea ice formation (Schrank, 2007); thus, potentially increasing stratification, with increased stability of the surface mixed layer, reduction in salt flux, reduced ocean convection, and less deepwater formation (Stenevik and Sundby, 2007). These could lead to a prolonged reduction in thermohaline circulation in the Atlantic.

According to Schmittner (2005), a disruption of the thermohaline circulation (THC) may collapse North Atlantic zooplankton stocks to less than half of their original biomass. Further, Kuhlbrodt et al. (2005) studied the physical, biological, and economic consequences of a THC change in northern Europe. They had opined that a major THC change might increase sea level by >50 cm and suggested possible strong impacts on the whole marine food web (from algae to plankton, shrimp, and fish) in the northern North Atlantic.

Incidentally, the El Niño events have three major impacts in coastal upwelling systems, which are given as follows:

(i) they increase coastal temperatures;
(ii) they reduce plankton production by lowering the thermocline (which inhibits upwelling of nutrients); and
(iii) they alter trophodynamic relationships (Lehodey et al., 2006).

In non-upwelling areas, they may not always change the vertical structure of the water column; and there could be either increasing or decreasing available habitats (Lehodey, 2004). The strong 1997 ENSO caused bleaching in every ocean (up to *c.* 95% of corals in the Indian Ocean), ultimately, resulting in *c.* 16% of the corals destroyed globally (Hoegh-Guldberg, 1999, 2005; Wilkinson, 2000). However, some studies are expected stronger and more frequent El Niño as a result of global warming (Timmermann et al., 1999; Hansen et al., 2006).

Further, positive correlations were found between upwelling and atmospheric temperature in paleo records in the California Current (Pisias et al., 2001). Moreover, sea surface temperature (SST) records obtained from sediment cores off Morocco, have indicated the anomalous and unprecedented change in temperature in the 20th century, which could be consistent with increased climate change-driven upwelling (McGregor et al., 2007). Inferences could be arrived at preliminarily through paleo records linking declining winter and spring snow cover over Eurasia with stronger southwest (summer) monsoon winds.

6.3.14 Sea-level rise

Sea level, globally, has been rising at an average rate of 1.8mm per year since 1961 (Miller and Douglas, 2004; Church et al., 2004), threatening many low altitude regions. This rate has increased from 1993 to about 3.1mm per year mainly due to the melting of mountain glaciers and declines in snow cover in both the hemispheres as well as losses from the ice sheets of Greenland and Antarctica (Bindoff et al., 2007). Increased variability in sea level, which is evident in recent decades, is consistent with more frequent and intense El Niños (Folland et al., 2001). However, the global average sea-level rise will range between 0.18 and 0.59 m by the end of the 21st century (2090–2099; relative to 1980–1999) based on model-based projections (Meehl et al., 2007).

All coastal ecosystems are vulnerable to sea-level rise. Also, there may be more direct anthropogenic impacts on coral reefs and coastal wetlands including salt marshes and mangroves. Further, there is a global loss of *c.* 33% of coastal wetland areas and a projected *c.* 36 cm rise in sea level between 2000 and 2080. The largest losses may be on the Atlantic and the Gulf of Mexico coasts of America, the Mediterranean, the Baltic, and small island regions (Nicholls et al., 2007b). Sea-level rise may, consequently, reduce the intertidal habitat area in ecologically important North American bays by *c.* 20%–70% over the next 100 years (Galbraith et al., 2002). Key human vulnerabilities to climate change and sea-level rise exist, where the stresses on natural low-lying coastal systems coincide with low human adaptive capacity, e.g., the deltas, especially Asian mega-deltas (for instance, the deltaic region of the River Ganges in the Ganges-Brahmaputra-Meghna basin in India and Bangladesh); low-lying coastal urban areas, especially areas prone to natural or human-induced subsidence and tropical storm landfall (e.g., New Orleans, Shanghai); and small islands, especially low-lying atolls (e.g., the Maldives) (Nicholls et al., 2007); etc.

6.3.15 Acidification and other chemical properties

It is approximately half of the CO_2 released by human activities between 1800 and 1994 have been stored in the ocean (Sabine et al., 2004), and *c.* 30% of modern CO_2 emissions are taken up by oceans today (Feely et al., 2008). Incidentally, continued uptake of atmospheric CO_2, had lowered the pH of surface seawater by 0.1 units in the last 200 years, with further possible pH reduction in the range from 0.3 to 0.5 units over the next 100 years, and from 0.3 to 1.4 units over the next 300 years, depending on the CO_2 emission scenario (Caldeira and Wickett, 2003). In fact, there could be greater impacts of these changes in some regions and severe for shell-borne organisms (Orr et al., 2005).

6.3.16 Fisheries' contribution to GHG emissions

Some amount of GHGs, which are emitted through fisheries activities also contribute to climate change both, during capture operations and subsequently during the transport, processing, and storage of fish. In fact, various works on fisheries' contribution to climate change have concluded that the minimal contribution of the sector to climate change does not warrant much focus on mitigation (Troadec, 2000). However, Tyedmers et al. (2005), calculated that fishing fleets consume the large quantity of fuel for their various kinds of operations.

6.3.17 Emissions from trade

According to an FAO estimate (FAO, 2007), *c.* 53 million tons of fish were internationally traded in 2004 including products of both fisheries and aquaculture. The transport of this fish have resulted in the emissions of the huge amount of GHGs. Moreover, high-value tuna transport by air freight from Japan, etc., could add significantly to GHG emissions. Incidentally, air freight exports of fish to the United States, Europe, and different parts of Asia are estimated at 200 000, 100 000, and 135 000 tons, respectively (Conway, 2007).

6.3.18 Contributions from fisheries toward mitigation

There have been efforts for the utilization of waste products from fish processing installations for producing biodiesel. This may give an alternative option(s) to fossil fuels in specific instances where large quantities of fish fats are available.

6.3.19 Impacts of climate change on fisheries

Increased CO_2 and ocean acidification could cause effects on calciferous animals, e.g., molluscs, crustaceans, corals, and echinoderms, etc. and potentially reduce the production of calciferous marine resources. It leads to decline in yields; warming of upper layers of the ocean; and warmwater species replacing cold-water species.

Moreover, there could be the shift in the distribution of plankton, invertebrates, fishes, and birds toward the north or south poles. The plankton species are moving to tropical waters, that is higher latitudes, where the timing of phytoplankton blooms may change and there may also found changes in zooplankton composition.

However, it leads to potential mismatch between prey (plankton) and predator (fish populations); reduced production and biodiversity and increased variability in yield; sea-level rise; loss of coastal fish breeding and nursery habitats, e.g., mangroves, coral reefs, etc.; and reduced production and yield of coastal and related fisheries.

6.3.20 Impact of climate change on Inland fishing operations and livelihood

Climate change impacts may cause the altered levels of precipitation; decrease in the volume of rainfall; and reduced scopes for fishing, fish farming, and aquaculture; however, fisheries and aquaculture are the potential sources of rural livelihood incomes. Incidentally, thus, there could be **reduced diversity of rural livelihoods** and more reliance on nonfarming income(s). Further, population **displacement** from the coastal areas could cause influx of new fishers, while, more droughts and floods damage their productive assets, like, fish ponds, weirs, rice fields, etc. as well as homes (Allison et al., 2005).

Notwithstanding the above, coastal zones throughout the world experience erosion (Nicholls et al., 2007a); thereby, threatening the coastal communities with flooding and loss of coastal ecosystems. However, a variety of processes are responsible for this, including changes in land use. Also, erosion may be exacerbated by climate-mediated sea-level rise (Nicholls et al., 2007a).

Notwithstanding the above, innumerable people, particularly the small-scale fishers, in SE Asia, East Africa, and throughout the Pacific, depend on the coral reefs, the near-shore habitats, the mangroves, and sea-grass beds, etc., for their livelihoods (Whittingham et al., 2003). Incidentally, the people living in these areas are impacted by sea-level rise and other climate change impacts (Nicholls et al., 2007b). These people with less mobility and adaptability are affected by the change in species distributions in response to climate change, which may also cause the relocation of species with the disappearance and enhancement of landings of target species in different regions. For instance, the landings of shrimp and octopus increase in northern Peru in the Humboldt Current system during El Niño years, whereas, in the south, the landings of scallops increase under tropical warm water conditions.

Conversely, freshwater input in estuaries may favor the appearance of brackish water species. For instance, during the El Niño of 1997–1998, increased precipitation in northern Peru had altered the salinity patterns in estuaries, favoring the mullet fishery (Badjeck, 2008); and, in Columbia, during the La Niña event of 1999–2000, a tilapia fishery boom was observed due to salinity changes (Blanco et al., 2007).

6.3.21 Adaptation of fisheries to climate change

Adaptation to climate change may be defined as an adjustment to ecological, social, or economic systems, in response to observed or expected alteration(s) in climate and their effects, in order to alleviate adverse impacts of climate change. However, various dimensions of adaptation may be implemented in response to impacts generated by a changing climate.

Incidentally, there are many classifications of adaptation options summarized in Smit et al. (2000) based on their purpose, mode of implementation, or on the institutional form they take. Incidentally, Coulthard (2009) had highlighted resource fluctuations in fisheries and diversification of livelihoods.

6.3.22 Adaptation of fisheries management

Fisheries management was unknown among the humankind at the dawn of civilization, when the natural resources, along with the fishes were plentifully available. Incidentally, many fisheries managements are still, loosely based on maximum sustainable yields or similar fixed ideas of the potential productivity of a stock. There have, been not much effort for the integration of environmental processes into management targets and measures, although climatic influences on the productivity of many fish species are often much recognized (Anon, 2007). In fact, more fishery managers explicitly consider different kinds of variations and move beyond static management parameters, as management processes become more complicated. Such alteration(s)in the management *modus operandi* may pave the way to implement the ecosystem approach to fisheries (EAF) management, with a holistic, integrated, and participatory approach to obtain sustainable fisheries (FAO, 2006).

6.3.23 Climate variability patterns

There is an atmospheric circulation pattern, which originates primarily due to heating contrasts between the poles and the equator, modulated by seasonality. Incidentally, **El Niño events** are associated with many atmospheric and oceanic patterns, including abnormal patterns of rainfall over the tropics, Australia, southern Africa and India, and parts of the Americas. Further, coincident **ecological alteration(s)** are observed in both abiotic and biotic aspects; and, include influences on plankton (MacLean, 1989), macrophytes (Murray and Horn, 1989), crustaceans (Childers et al., 1990), fish (Sharp and McLain, 1993), marine mammals (Vergani et al., 2004), seabirds (Testa et al., 1991), and marine reptiles (Molles Jr. and Dahm, 1990).

6.3.24 Observed effects of climate variability and climate change on ecosystem and fish production processes

Direct effects of climate change impact the performance of individual organisms (notably, the fishes) at different stages in their life history via alterations(s) in physiology, morphology, and behavior. Climatic impacts also occur at the population level *via* various kinds of alterations, which may influence dispersal and recruitment, distributions, biodiversity, productivity, and microevolutionary processes (Harley et al., 2006, b).

There are, changes in physical and biological systems in every continent, including Antarctica as well as in most of the oceans of the world, in response to climate change. Most such studies are from the mid and high latitudes in the Northern Hemisphere and less from the tropical regions and Southern Hemisphere (Parry et al., 2007). Incidentally, the marine and FW systems, generally, respond to the combined and synergistic effects of physical and chemical changes acting directly and indirectly on all biological processes. Concomitantly, the marine and FW populations respond to environmental and climate variability through various physiological, spawning, and recruitment processes.

Incidentally, it may be noted that the term "bioclimate envelope" was used to define the interacting effects and limits of temperature, salinity, oxygen, etc. on the performance and survival of species (Pearson and Dawson, 2003). Such bioclimate envelopes could be used to model changes in species distributions and abundance patterns as a result of climate change.

6.3.25 Climate change and fish recruitment processes

It may be pertinent to note that one of the most important difficulties in fisheries science over the past century was the issue of recruitment variability and its causes and consequences to commercial fish populations. Incidentally, many hypotheses were proposed to explain the alarming level of depletion in the number of many marine and FW species as they develop from egg to adult passing through larval and juvenile stages (Ottersen et al., 2008). These hypotheses may be grouped into three general categories, which are as follows: (i) starvation and predation; (ii) physical dispersal; and (iii) synthesis processes.

The match-mismatch hypothesis of Cushing (Cushing, 1969, 1990; Durant et al., 2007) relates the impact of starvation on recruitment and is clearly connected with climate variability and change(s).

Concomitant to the above, three hypotheses relate climate effects directly to the recruitment and abundance of marine fish populations, which are given as follows:

(i) The optimal environmental window hypothesis of Cury and Roy (1989)
(ii) The triad hypothesis of Bakun (1996)
(iii) The oscillating control hypothesis of Hunt Jr. et al. (2002)

According to the **optimal environmental window hypothesis** (Cury and Roy, 1989), species are adapted to the typical ("optimal") conditions within their preferred habitats. This could imply that better recruitment success may be expected with "mean," rather than, with "extreme" (either high or low) conditions.

According to **Triad's hypothesis** (Bakun, 1996), optimal conditions of enrichment processes (upwelling, mixing, etc.); concentration processes (convergences, fronts; water column stability); and retention within appropriate habitats are necessary for good recruitment. Locations in which these three elements exist to support favorable fish habitats are called "**ocean triads.**" However, they are likely to be sensitive to disruptions or systematic alterations due to climate change.

According to the **oscillating control hypothesis** (Hunt et al., 2002), the pelagic ecosystem is driven by low plankton production in the cold days, but is predominately driven by predation in warm periods. In the cold days, production of walleye pollock (*Theragra chalcogramma*) is limited by cold temperatures and low food reserves, whereas in the early warm period, strong plankton production promotes good fish recruitment. This hypothesis was developed for the southern Bering Sea.

6.3.26 Primary production

There may be a reduction in **nutrient availability to the euphotic zone due to climate change impacts**, followed by a concomitant reduction in **primary** (Behrenfeld et al., 2006) and **secondary** (Roemmich and McGowan, 1995) production. This was revealed from general observations and model(s), which suggested that climate change is likely to lead to increased vertical stratification and water column stability in oceans and lakes.

6.3.27 Secondary production

Among the scanty global scale works on the impacts of climate change on secondary productivity, Richardson (2008) provided with a general review of the potential climate warming impacts on zooplankton. In fact, demographic features of marine zooplankton, make them potential candidates for assessing the rapid impacts of climate change because of their life span (Mackas and Beaugrand, 2008). Incidentally, shifts and trends in plankton biomass have been observed in the North Atlantic (Beaugrand and Reid, 2003), the North Pacific (Karl, 1999; Chavez et al., 2003), and in the southern Indian Ocean (Hirawake et al., 2005). Incidentally, the more eurythermal and specifically heat tolerant, mid-to-high-intertidal species might actually be more vulnerable to climate change than the less heat-tolerant species, because they may live closer to their physiological limits (Harley et al., 2006, b). Concomitantly, one of the better-studied impacts of climate variability and change on marine zooplankton is the North Atlantic copepod community, which contributes up to *c.* 90% of the zooplankton biomass in the region.

6.3.28 Inland waters

As in the case of marine environments, global warming also influences the biological production in the inland waters, which, depends on the combination of processes, such as ice cover, water flows, stratification, and nitrification, along with the additional impact of human beings on water and land use. Incidentally, atmospheric warming has already reduced ice cover, caused warmer water temperatures, longer growing seasons, and increased algal abundance and productivity (Karst-Riddoch et al., 2005) in high-latitude or high-altitude lakes. There are similar instances of increases in the abundance of zooplankton, correlated with warmer water temperatures and longer growing seasons (Schindler et al., 2005). Further, rapid increases in water temperature after ice breakup had enhanced fish recruitment in oligotrophic lakes (Nyberg et al., 2001). However, some deep tropical lakes may depict reduced algal abundance and declines in productivity, because stronger stratification, reduces upwelling of nutrient-rich deep water (Hecky et al., 2006). Primary productivity in Lake Tanganyika decreased by up to *c.* 20% over the past 200 years (O'Reilly et al., 2003; Lambert et al., 2020). Vollmer et al. (2005) documented rising temperatures in Lake Malawi, over the last 60 years, leading to reduced nutrient loading and, presumably, reduced productivity.

6.3.29 FW systems and palaeontological records

FW systems, particularly the lentic ones, are highly vulnerable to climate change. In terms of time scale, which is time immemorial, paleontological records seem to portray that the lentic bodies (notably, the lakes, wetlands, etc.) have altered their shapes and distributions. Ultimately, the prehistoric ones had disappeared entirely, mainly due to climate change (Poff et al., 2002).

Concomitant to the above, FW ecosystems are also highly biodiverse. They support *c.* 40% of all fish species despite accounting for only a small proportion (*c.* 0.01% by volume) of aquatic habitats (Arthington et al., 2003). Further, *c.* 20% of FW species are threatened, endangered, or extinct, in which, climate change might have an impact (Revenga et al., 2000). As such, the protection of FW biodiversity is increasingly recognized as a major conservation priority in the globe (Abell et al., 2002).

6.3.30 Aquaculture systems

Aquaculture systems are affected by climate change impacts, either directly or indirectly. According to Handisyde et al. (2006) and De Silva and Soto (2009), climate change effects on aquaculture have both direct (e.g., through physical and physiological processes) and indirect (e.g., through variations in fishmeal supplies and trade issues) impacts. With regard to direct issues, Handisyde et al. (2006), noted that, the physical changes related to climate change (viz., alteration(s) in temperature, solar radiation, current and wave actions, sea-level rise, water stress, and the frequency of extreme weather events, etc.) may have significant impacts on the physiological, ecological, and operational (e.g., species and site selection, culture technologies, etc.) aspects. The Third (McLean and Tsyban, 2001) and the Fourth (Easterling et al., 2007) Assessment Reports of the IPCC identified the impacts of climate change on aquaculture (De Silva and Soto, 2009), which are given in the following:

(i) stress due to enhanced temperature and oxygen demands;
(ii) uncertain supplies of FW;
(iii) extreme weather events;

(iv) water level rise;
(v) increased frequency of diseases and toxic events; and
(vi) uncertain supplies of fishmeal from capture fisheries.

On the other hand, the positive impacts of climate change on aquaculture are given in the following:

(i) increased food conversion efficiencies and growth rates in warmer waters;
(ii) increased length of the growing season and range expansions poleward due to decreases in ice (Easterling et al., 2007); etc.

Further, aquaculture in temperate regions may be adversely impacted by the increased prevalence of pathogens (Handisyde et al., 2006).

6.3.31 Food web impacts from plankton to fish

Plankton is considered as the natural food organisms of the fishes. Climatologically driven alteration(s) in species composition and abundance may alter species diversity, which, in turn, may have implications on ecosystem functions, such as productivity (Duffy, 2003).

6.3.32 Alteration(s) in distribution

Climate change plays a significant role in the habitat and distribution of both marine and FW fishes through its impact(s) on the physical properties of marine and FW environments. These include temperature, salinity, vertical mixing rates, and thermohaline and wind-driven circulations. Populations have evolved with some amount of environmental tolerances (so-called bioclimate envelopes). Subsequently, they interact with the climate-regulated environmental conditions to determine the preferred or suitable habitats. In fact, climate change drives most terrestrial and marine species ranges toward the poles (Parmesan and Yohe, 2003), as it was the case in the Pleistocene-Holocene transition (Fields et al., 1993) although the amplitude might be different.

6.3.33 Changes in abundance

Changes in the recruitment and growth rates may lead to alteration(s) in the abundance and biomass of marine populations. Further, changes in temperature have direct impacts on fish abundance and biomass.

Further, recent declines in fish abundance in the East African Rift Valley lakes, were linked with climate impacts on lake ecosystems (O'Reilly et al., 2003). Lake Tanganyika had historically supported one of the world's most productive pelagic fisheries. However, there was, possibly a decline in clupeid catch by *c.* 30%–50% in the late 1970s, due to climate change impacts.

6.3.34 Phenological changes

Parmesan and Yohe (2003) estimated that more than half (*c.* 59%) of 1598 terrestrial, FW, or marine species had exhibited measurable changes in their phenology and/or distributions over the past 20–140 years, mainly due to the impacts of climate change (Root et al., 2003).

6.3.35 Change of phenology in the marine environment

According to the match-mismatch hypothesis by Cushing (1969), shifts in the timing of blooms of primary or secondary producers may cause a mismatch with their predators. Ocean warming may radically alter the marine trophodynamics through predator-prey mismatch (Visser and Both, 2005). For instance, dinoflagellates in the North Sea had advanced their seasonal peak by nearly 1 month (Edwards and Richardson, 2004). On the other hand, the responses of copepods were more variable (Edwards and Richardson, 2004). Incidentally, Beaugrand et al. (2003) and Reid et al. (2003) had shown fluctuations in plankton abundance in the North Sea, which were due to climate change impacts. Concomitantly, the zooplankton and high trophic-level organisms (fish, birds, etc.) in the North Pacific depicted consistent patterns, being linked to year-to-year and decade-to-decade ocean climate fluctuations.

6.3.36 Change of phenology in the FW environment

Changes seem to be also evident in the inland water scenario. Perhaps under the influence of climate change, many lakes are now responding with phenological adaptations with warmer water temperatures. The spring algal blooms occurring about 4 weeks earlier in various large European lakes (Gerten and Adrian, 2000). Also, spring phytoplankton bloom has advanced in many cases. Further, phenological shifts had also been demonstrated for some wild and farmed fish species (Ahas, 1999).

6.3.37 Species invasions and diseases

Outbreaks of diseases have increased on a global scale over the last few decades among many animals including corals, echinoderms, fishes, mammals, molluscs, and turtles (Ward and Lafferty, 2004). Causes for increases in diseases of many animals, remained largely uncertain, although the temperature is one significant factor that has been implicated (Harvell et al., 2002). In fact, earlier, unseen diseases had emerged in new areas through shifts in the distribution of the hosts or pathogens; many of which were said to be in response to climate change (Harvell et al., 1999). Incidentally, the ecology of some of the human pathogenic microbes was associated with the aquatic environment, and efforts were made to link them to alteration(s) in temperature.

Vibrio parahaemolyticus causes gastroenteritis associated with the consumption of raw oysters. This cosmopolitan pathogen was rarely isolated from water which is $<15°C$ (ICMSF, 1996). Further, *Vibrio cholerae* is transmitted through water and fish, and its outbreak is, linked with climate change (Lipp et al., 2002). Furthermore, storms and flash floods due to climate change might lead to the transport of viruses (noroviruses, hepatitis A virus) from wastewater sources to shellfish growing areas. In view of this, fish safety management programs are needed to be considered in order to combat these hazards. Moreover, latitudinal shifts of diseases in terrestrial and aquatic environments are also to be taken care of.

Concomitant to the above, climate warming may increase pathogen development and survival rates, disease transmission, and host vulnerability (Harvell et al., 2002). Incidentally, the growth rates of marine bacteria and fungi in the coral ecosystems may be positively correlated with temperature (Harvell et al., 1999). In addition, winter temperatures consistently <3 °C limit the development of the multinucleated spore unknown (MSX) disease caused by the protozoan pathogen, *Haplosporidium nelsoni* (Hofmann et al., 2001), and the poleward spread of this and other pathogens could be expected to continue.

This example also indicates the relevance of seasonal information when considering the effects of climate change. Further, massive mortalities among the Moroccan sardines in 1997, seemed to be more related to abrupt environmental changes. Moreover, Marcogliese (2001) opined that parasites of FW and marine organisms are directly affected by climate change, and also indirectly by the effects of climate change on their hosts. In addition, climate change may also influence the selection of different modes of transmission and virulence (Marcogliese, 2001). Incidentally, warming temperatures may facilitate the establishment and spread of deliberately or accidentally introduced species (Carlton, 2000).

Notwithstanding the above, harmful algal blooms (HABs) increase globally due to anthropogenic influences (Smayda, 1990). Incidentally, HABs are increasing in some areas of the Northeast Atlantic. According to geological records taken from Atlantic and Pacific Canada, it is believed that when summer SST was much warmer during the late glacial-early Holocene, there was a period of sustained high production of red-tide blooms (Mudie et al., 2002). This led the authors to suggest that global warming is strongly implicated in the historical increase in the frequency of red tides and HABs (Dale, 2001). In fine, a final impact involves the depletion of the ozone layer (Austin et al., 1992). Incidentally, marine phytoplankton was protected from UVB damage when cocultured with marine viruses (Jacquet and Bratbak, 2003).

Concomitant to the above, the FW fishes in a major portion of the world have been affected by a *hitherto* unknown virulent, enigmatic disease, variously called epizootic ulcerative (Fish) disease syndrome or EUS, which has been sweeping the FW fishes, in epidemic dimension, on a semi-global scale, causing large-scale mortality among them, rendering many of them endangered, devastating the economy of certain countries, and causing fear psychosis among the consumers and untold misery to the poor fishermen.

6.3.38 Summary of EUS

EUS has been causing large-scale mortality among FW fishes since 1988, initially affecting four species of fishes very widely. Our study revealed fluctuation in the intensity of the disease in relation to species affected. Large hemorrhagic cutaneous ulcers, epidermal degeneration, and necrosis followed by sloughing of scales are the principal symptoms of EUS. Low total alkalinity (TA) could be a predisposing "Stress factor." Sick fishes show low hemoglobin and polymorphs but high ESR and lymphocytes. The communicative nature of EUS revealed variation in the time gap between fish and infection in different species. Inoculation of microbes in the test animals did not reveal any sign of ulcerations for 2 years. Bacterial culture revealed the occurrence of hemolytic *E. coli*, *Aeromonas hydrophila*, *Pseudomonas aeruginosa*, *Klebsiella* sp., *Staphylococcus epidermidis* in the surface lesions as well as in the gut, liver, gills, heart, kidney, and gonads of sick fishes, all of which are sensitive to Chloramphenicol, Septran, Gentamycin, etc. Fungal isolation revealed the occurrence of *Aphanomyces* sp. with the concomitant occurrence of the same fungal genus in histological sections of EUS-affected fishes. Histopathological (HP) studies showed the focal areas of increased fibrosis and chronic inflammatory cell infiltration in muscles; focal areas of fatty degeneration of hepatocytes surrounding the portal triads in the liver. Inoculation of 10% tissue homogenate of EUS-affected *Clarias batrachus* into 80% confluent monolayer form BF2, RTG, and EPC fish cell lines in Leibovitz L-15 medium, revealed progressive CPE, which was passable in subsequent cultures; thus, indicating the "isolation" of the virus.

Electron microscopic studies with the ultrathin sections of still-occurring EUS-affected fish tissues revealed the presence of virus-like particles (inclusion bodies), and, preliminarily, virus belonging to birna group has been electron microscopically identified as the primary etiological agent of EUS. Further studies in this regard are being conducted (Kar, 2015).

Incidentally, the outbreak COVID19 (Coronavirus) global pandemic in 2019–2020 may be linked with some kind of climate change impacts as in the case of outbreak of epizootic ulcerative (fish) disease syndrome (EUS) among the fishes, where temperature and total alkalinity (TA) of water had been found to act as predisposing stress factors in the outbreak of EUS.

Climate change have a range of direct and indirect impacts on marine and FW capture fisheries, with implications for fisheries-dependent economies, coastal communities, and fisherfolk.

Notwithstanding the above, fisheries and fisherfolk are impacted in a wide range of ways due to climate change. These may include biophysical impacts on the spatial distribution or productivity of marine and FW fish stocks through processes, such as ocean acidification; habitat alteration(s) and damage; fragmentation and disruption in precipitation vis-à-vis FW availability; and so on.

6.3.39 Fisheries' contribution to livelihoods and economic development

As indicated earlier, fisheries and aquaculture are potential sources of nutrition and avocation. The number of people directly employed in fisheries and aquaculture is conservatively estimated at 43.5 million, out of which >90% are small-scale fishers (FAO, 2005). Moreover, there are "forward linkages" to other economic activities generated by the supply of fish (trade, processing, transport, retail, etc.) in addition to those directly employed in fishing as well as "backward linkages" to supporting activities (boat building, net making, engine manufacture and repair, the supply of services to fishermen and fuel to fishing boats, etc.). As such, considering these other activities, >200 million people are dependent on small-scale fishing in developing countries, in addition to millions for whom fisheries provide supplementary income (FAO, 2003a,b, 2004). In fact, fisheries are often available in remote and rural areas where other economic activities are generally limited, and thus, fisheries are important engines for economic growth and livelihoods in rural areas (FAO, 2005). Incidentally, some fishers are specialized and rely entirely on fisheries for their livelihood, whereas for many others, especially in inland fisheries in developing countries, fisheries usually form part of a diversified livelihood strategy (Smith et al., 2005). Further, fisheries may serve as a "safety net" to landless poor or in the event of other livelihoods failing to support them economically (FAO 2005, 2006). In this connection, it may be noted that many small-scale fisherfolk live in poverty, mainly because of various kinds of economic constraints and oppressions. This is either due to the degradation of resources or due to the failure of the safety net function of fisheries' not being able to support the poor fisherfolk economically. In addition, this generalized understanding of the economic poverty of fishers mainly in the developing world is due to multidimensional factors including, health issues as well as sometimes, their political marginalization (Bene, 2003; FAO, 2005), being poorly recognized in terms of their contribution to food security and poverty alleviation. Incidentally, fisheries and fishermen are often subjected to various kinds of climate change-related issues.

6.3.40 Current trends and status of fisheries

As indicated above, there are impacts on fisheries due to climate change, and these could interact with the existing drivers, trends, and status of fisheries. Further, there seems to be a growing awareness of the significance of understanding human aspects of fisheries and focusing on fisheries governance rather than purely management. More attention is now being paid to incentives framed through management measures and institutional arrangements around fisheries, including the incorporation of local fishers and their knowledge through comanagement and community-based management initiatives (Jentoft, 2006). This trend was, accompanied by a greater awareness of the significance of taking account of ecosystems within which fisheries are embedded. Both the aspects, notably, (i) the involvement of the stakeholders from the grassroots level and (ii) the need to consider the wider ecosystem approach are incorporated in the ecosystem approach to fisheries (FAO, 2003a).

In addition to the above, fisheries could also be exposed to a diverse range of direct and indirect climate change impacts, leading to possible displacement and migration of human populations, followed by further impacts on coastal communities and infrastructure due to sea-level rise as well as alteration(s) in the frequency, distribution, or intensity of tropical storms. In fact, fisheries are dynamic socioecological systems which experience fast change(s) in markets, exploitation, and governance; thus ensuring a constantly developing context for future climate-related impacts. These existing socioeconomic trends and the indirect effects of climate change may interact, amplify, or even overwhelm biophysical impacts on fish ecology.

In continuation of the above, in fact, it could be difficult to predict the impact of the variety of different impacting mechanisms on fisheries, and complex interactions between social, ecological, and economic systems as well as the possibility of sudden and surprising changes, which may occur due to climate change impacts on fisheries. Further, the vulnerability of fisheries and fishing communities depends on their extent of exposure and sensitivity to change(s), and on the ability of individuals or systems to anticipate and adapt. Further, this adaptive capacity relies on various assets and is constrained by culture or marginalization. Moreover, vulnerability may vary between countries and communities, and, between demographic groups within the society. Usually, poorer and less empowered countries and individuals are more vulnerable to climate change impacts, whereas the vulnerability of fisheries is likely to be higher where they have already suffered from overexploitation or overcapacity. Incidentally, adaptation(s) to climate impact(s) includes reactive or anticipatory actions by individuals or public institutions. These range from abandoning fisheries altogether for alternative occupation(s), to developing insurance and warning systems and changing fishing operations. However, the governance of fisheries affects the range of adaptation options available, and may need to be flexible enough to take care of changes in stock distribution and abundance. In fact, accepting inherent uncertainty, governance aimed toward equitable and sustainable fisheries and ecosystem-based approach is one of the best types of governance systems.

In fact, the ecological systems which support fisheries are already known to be sensitive to climate variability. For instance, in 2007, the Intergovernmental Panel on Climate Change (IPCC) had highlighted the various kinds of risks to aquatic systems from climate change, including loss of coastal wetlands, coral bleaching, and changes in the distribution and timing of FW flows, etc.

Further, regarding seas, oceanic water acidification may profoundly impact on marine ecosystems (Orr et al., 2005). Concomitantly, the fishing communities and the related industries in the coastal or low-lying zones could be increasingly at risk from sea-level rise, extreme weather events, and a wide range of human pressures (Nicholls et al., 2007b). Moreover, poverty and marginalization among fishing communities could decrease their ability to adapt and respond to the impacts of climate change(s). Highly productive seasonal floodplain fisheries, permit the professional, part-time and occasional fishermen (Dey and Kar, 1989; Kar, 2007, 2013; Smith et al., 2005) in many SE and south Asian nations, like India, Bangladesh, etc. to have lucrative earnings during certain intense fishing seasons, which could help them to mitigate the adverse impacts of climate change. This is a significant feature of the socioeconomics of inland fisheries in these countries. Further, increasingly globalized fish markets may cause new vulnerabilities to market disruptions.

Different discussions lead to a more holistic analysis of fisheries and climate change, because fisheries are looked at not as a simple relationship between a community and an aquatic system, but as a potential sector-related nutrition and avocation of the people. Incidentally, some of the relevant aspects of fisheries with regard to climate change are as follows:

Exposure (E): Nature and degree to which fisheries production systems are exposed to climate change.
Sensitivity (S): Degree to which national economies are dependent on fisheries.
Potential impacts (PI): All impacts that may occur without taking into account the planned adaptation.
Adaptive capacity (AC): Ability or capacity of a system to modify or change to cope with alteration(s) in actual or expected climate-related stress(es).

6.3.41 Importance of socioeconomic drivers in Fijian fishing communities

In Fiji, the Lau islands in the remote South-East province have limited land and transport networks. The majority of the households participate in subsistence fishing. Incidentally, there had been a temperature-induced mass coral mortality event in 2000. There was, also quite a large-scale damage to corals from the crown of thorns of starfish in 1999. Thus, it might be expected that fisheries and local communities, who used those reefs, would be directly impacted. However, a socioeconomic survey conducted in the area in 2006 found that while some fishers were aware of the bleaching and starfish phenomena, and few had identified them as a threat to fish populations.

6.3.42 Impact(s) of climate variability on Peruvian Anchoveta fisheries

The dominating pelagic anchovies (*Engraulis ringens*) form a potential marine fishery around Peru, forms >95% of Peruvian fisheries catches, and is landed by the industrial sector, which is, the second-highest generator of foreign currency; and, accounted for *c.* US $1124 million through exports in 2001 (FAO, 2003b). However, under the influence of the warm modes of the El Niño-Southern Oscillation (ENSO; or, commonly known as El Niño), the harvest of anchovies was highly variable with much population fluctuations. Incidentally, El Niño events reduce upwelling along the Peruvian coast, and thereby, impacting on the natural process. There is a significant decrease in anchovy biomass due to a possible reduction in nutrient supply. The anchovy biomass was, *c.* 1.2 million tons, which was the lowest in the 1990s (Ñiquen and Bouchon, 2004). Further, the total volumes of fisheries landings had dwindled by *c.* 55% from 1997 to 1998 El Niño, as compared to 1996 (CAF, 2000).

Incidentally, Peru is, considered as the main producer of fishmeal and fish oil in the world, and fluctuations of anchovy stocks, impact on both national and global aquaculture feed market. Moreover, the industrial fishery sector was, adversely affected by the reduced stock of anchovies and sardines in the eastern Pacific upwelling areas, while the fisherfolk in Denmark have received near-record prices for Baltic sprat, a competing species for fishmeal production (MacKenzie and Visser, 2001). Nevertheless, the later La Niña events (cooling of sea surface temperatures) had increased catches of anchovies and revenues for the industrial sector (Ordinola, 2002).

6.3.43 Rainfall and inland African fisheries

The continent of Africa possesses a rich potential in aquaculture and fisheries. Incidentally, the shallow, highly productive Lake Chilwa in Malawi supports trade with fish worth *c.* US $10 million a year. However less rainfall and its variations (may be due to climate change impacts) are drying out of the entire lake, with a decline in fish productivity which is strongly tied to the amount of water in the lake. However, during the dry periods, some fishers diversify their livelihoods to farming, pastoralism, etc., while other wealthier and more specialized fishers, try to migrate to fisheries in other lakes in the region (Allison et al., 2007).

6.3.44 Multiple impacts on Mekong Delta fisheries

The River Mekong is one of the biggest rivers in SE Asia, and the Mekong Delta is also one of the largest Deltas supporting innumerable fishermen with a livelihood. Incidentally, the lower Mekong Delta supports >1000 fish species; a capture fishery of *c.* 1.5 million tons; and, fishery-based livelihoods for *c.* 40 million people. Incidentally, climate-mediated threats to these fisheries include, altered precipitation rates and rising sea level, which, impacts on various aspects of the delta's ecology and on human settlements. Moreover, the overwhelming impacts on fisheries in the delta are from human activities including overfishing, land-use changes, and hydrological disruptions, in addition to the climate change impacts. However, enhanced rainfall in the future may possess the potential to increase fishery yields (Easterling et al., 2007).

6.3.45 Vulnerability of communities

Vulnerability is the extent to which a thing is prone to change(s) due to some impact. The vulnerability may also be analyzed based on statistics at the subnational level. There have been derivation of an index of adaptive capacity with respect to fishing livelihoods of 29 coastal communities in five nations in the western Indian Ocean, viz., Kenya, Madagascar, Mauritius, Seychelles, and Tanzania. The index had combined eight variables proposed to be important for adaptive capacity which was, weighted according to their relative significance. The resultant ranking of communities had been broadly predicted from the national level development statistics.

6.3.46 Food fish production and basis of the need

A review of climate change and its impacts on fish production considers the needs of the food fish for human consumption and the amounts available for the production of by-products, like fishmeal and fish oil, which are used in the production of feeds for the cultured aquatic organisms, like the shrimps and the carnivorous finfish.

6.3.47 Food fish needs

Food fish from aquaculture accounts for *c.* 50% of the total amount of fish consumed globally; out of which, freshwater production contributes to *c.* 30% (FAO, 2009). In fact, it aims to enhance the contribution from AQC to *c.* 60% by the first part of the 21st century (FAO, 2008). Incidentally, the capture fisheries component of fish supplies probably have reached almost a saturation level of *c.* 100 million tons per year and accounts for *c.* 25% of the food-fish supply.

Fishes are obtained from the wild (capture fisheries) and also from aquaculture (culture fisheries). However, a big portion of the cultured product is channeled to the processing industries and is, therefore, unavailable for direct human consumption (Jackson, 2006). Incidentally, the inland fisheries sector is gaining sufficient momentum in the tropics. The concept of good health through fish consumption generated among the burgeoning human population, along with increased *per capita* consumption of food fish, have accelerated the demand for inland fisheries and AQC. Thus, there is an increase in the demand for food fish in various continents. Hence, AQC have to provide *c.* 60 million tons of food fish in order to meet the increasing food fish needs, and this aspect has to be considered while evaluating the impacts of climate change on aquaculture.

6.3.48 Food security and fish

"Food security" is the existence of that situation, when all people, at all times, have physical and economic access to sufficient, safe, and nutritious food to meet their dietary needs and food preferences for an active and healthy life (FAO, 2003a,b). It may be mentioned that according to Sen (1981), food security may be achieved directly or indirectly through the following means:

(i) production-based entitlements: producing food for self
(ii) trade-based entitlements: selling/bartering goods or other assets
(iii) labour-based entitlements: selling own labor
(iv) transfer-based entitlements: receiving gifts or transfers of food

Food security is a universal human right. Kurien (2005) summarized various issues related to food security. The accepted recommended minimal daily calorie intake for a healthy person is 1800. This was increased to 2100 per day by the United States Department of Agriculture (USDA). Food fish provides essential micronutrients in the form of vitamins and minerals. Incidentally, climate change may impact on food security issues in the form of less production through AQC.

6.3.49 Aquaculture production

It may be important to base the evaluations on the past trends of aquaculture production on the basis of approximate climate regimes, viz., tropical, subtropical, and temperate regions, in order to assess the potential changes in aquaculture under different climate change scenarios.

6.3.50 Environmental-climatic distribution of aquaculture

Activities related to AQC occur in all **three basic environments, viz., freshwater, marine, and brackish water** and in all the **three climatic regimes, viz., tropical, subtropical, and temperate**. The climate change impacts, if any, are likely to produce more significant effects on the FW aquaculture in tropical and subtropical regions than elsewhere because production is more concentrated there.

6.3.51 Climatic-national-regional distribution of aquaculture

The production of the four major commodities, viz., **finfish, molluscs, crustaceans, and seaweeds** have been analyzed, in order to obtain a view on the distribution of aquaculture products in relation to climatic regimes, viz., tropical, subtropical, and temperate. Incidentally, it was revealed that the production of all four major cultured commodities in the three climatic regimes under consideration occurs predominantly in Asia. The differences in production between Asia and the other continents are extremely large and approximate to >90% in all instances. Therefore, the adaptive strategies needed to prevent and counteract the potential impacts of climatic change on aquaculture could preferably be initiated and targeted in Asian countries.

6.3.52 Growth trends in aquaculture

Aquaculture is often regarded as one of the fastest-growing production sectors in the world in the past few decades. It has witnessed an annual rate of growth of *c.* 10%. However, aquaculture growth may not be able to proceed at the same rate in all regions of the world (De Silva, 2001). Incidentally, it is important to remember these aspects, when considering the climate change impacts on aquaculture.

6.3.53 Brief synthesis of previous studies on climate change effects on aquaculture and fisheries

Apart from a few dedicated studies, fisheries and aquaculture have not received a large amount of attention with regard to the impact of climate change on them.

Nevertheless, the first notable attention to climate change issues in relation to fisheries was made a few decades ago (Wood and McDonald, 1997). Further such issue(s) are, dealt with giving more emphasis on a physiological point of view, e.g., with regard to effects of temperature on growth (Jobling, 1997), larval development (Rombough, 1997), reproductive performance (Van der Kraak and Pankhurst, 1997), etc. In addition, certain policy briefs pertaining to threats to fisheries and AQC from climate change were dealt with, as they relate to communities depending on these sectors for livelihoods. In addition, efforts were also made to find ways, means, and strategies to combat/adapt to climate change-induced effects (WFC, 2006). Incidentally, two sectors could provide with opportunities to adapt to climate change through:

- **(i)** integrating aquaculture with agriculture; and
- **(ii)** fisheries management, which could augment efforts to do more adaptive research; rather than, seeking to only maximize yield (WFC, 2006).

Furthermore, there were suggestions for further research to find innovative ways to improve the existing adaptability of fishers and aquaculture farmers. Incidentally, Sharp (2003), considered future climate change impacts on regional fisheries by assessing historical climate changes and evaluating the consequences of climate-related dynamics on the evolution of species, society, and fisheries. Further, efforts have been taken to rank the impacts of climate changes on regional fisheries, and the following types of fisheries have been recognized as most responsive to climatic variables (in descending order of sensitivity):

- **(i)** FW fisheries in small rivers and lakes, in regions with larger temperature and precipitation change;
- **(ii)** fisheries within exclusive economic zones (EEZ), where access-regulation mechanisms automatically and artificially reduce the mobility of the fishing groups and fishing fleets;
- **(iii)** fisheries in large lakes and rivers;
- **(iv)** fisheries in the estuaries, particularly, the regions in the estuaries which are impacted by sea-level rise or decreased river flow; and
- **(v)** high sea fisheries.

Concomitant to the above, the marine fisheries with large-scale production could also be under the impacts of climate change (Sharp, 2003). Incidentally, Handisyde et al. (2006) did comprehensive study on aquaculture and climate change. They tried to reflect the influence of climate change parameters, such as temperature, precipitation, sea-level rise, ocean currents, other extreme events, etc., on global aquaculture, aquaculture production, aquaculture-dependent livelihoods, indirect livelihood through fishmeal and fish oil availability, etc. Their models and submodels tried to cover the exposure to extreme climatic events, adaptive capacity, and vulnerability. The study was exemplified by a case study from Bangladesh, which has extensive deltaic areas in the world and one of the most sensitive areas to sea-level rise and severe weather-related damage.

Concomitant to the above, 2WE Associate Consulting (2000) dealt with the influence of climate change on Canadian aquaculture and portrayed the physical and ecological impacts of climate change(s) on fisheries and aquaculture. In addition, the effects of synergistic interactions between current stressors, including fishing and ecosystem resilience and the abilities of marine and other aquatic organisms to adapt and evolve according to climatic changes were discussed.

Under the backdrop of more emphasis to be laid on FW, marine, and estuarine researches, Roessig et al. (2004) worked on the physiology and ecology of marine and estuarine fishes in the tropics. On the other hand, Ficke et al. (2007) opined that climate change impacts on FW are more reflected through increased water temperature and toxicity of pollutants and decreased oxygen levels. Further, lotic fishes are impacted through altered hydrological regimes and enhanced groundwater temperature. Moreover, eutrophication is exacerbated in the lentic systems. Stratification become more pronounced with a consequent impact on food webs and habitat availability and quality. In this connection, Nunn et al. (2007) dwelt upon the recruitment success of cyprinid fish in low-lying rivers, in relation to the potential changes induced in the Gulf Stream by climatic change.

Concomitant to the above, climate change impacts on fisheries could indirectly affect aquaculture, such as the decline in ocean productivity (Schmittner, 2005). In this connection, Atkinson et al. (2004) noted a decrease in Antarctic krill density (*Euphausia superba*) and a corresponding increase in salps (mainly *Salpa thompsoni* Foxton, 1961), one of the main grazers of krill exacerbated by sea temperature rises and a decrease in polar ice. Krill once served as a major protein source for the replacement of fishmeal in aquaculture feeds (De Silva and Turchini, 2008). It is also a major food item of baleen whales and many wild fish species.

6.3.54 Impacts of climate change on aquaculture

Climate change may cause a differential impact on the various aspects of AQC practiced in different regions and continents as well as on various sectoral practices. Such practices may be driven by cultural attributes such as "living with water" and the associated historical trends to farm fish by certain ethnic groups. Moreover, aquaculture growth in different regions may undergo alteration(s) due to climatic change(s), particularly, in regions where aquaculture, by itself, may be able to provide adaptation possibilities for other sectors (e.g., coastal agriculture).

6.3.55 Major climatic changes that could potentially impact on aquaculture

Climate change may alter different parameters of the environment, like temperature, humidity, rainfall, etc., due to various kinds of impacts. However, all the climatic parameters may not exert an equal magnitude of impact upon fisheries and aquaculture. Moreover, the potential impacts on farming activities may not be attributed to one single recognizable factor of climatic change, rather, it could be a chain of compounded effects that become causative in most instances. Nevertheless, based on the IPCC forecast (2007), those elements of climatic change, which are likely to impact on aquaculture may be summarized as follows:

(i) *Global warming*: There is the consensus that our planet earth will be heated by *c.* 1.1°C in this century and this increase could be up to *c.* 3°C.

(ii) *Rise in Sea level*: Sea-level rise, associated with global warming, could go to the range of 10–100 cm over this century of which the thermal increase could contribute to *c.* 10–43 cm, and melting glaciers may contribute to *c.* 23 cm rise. Incidentally, the sea level increases could profoundly affect the deltaic regions, including an increase in the saline water intrusion vis-à-vis bringing about major biotic changes.

(iii) *Ocean productivity and changes in circulation patterns*: Major changes could occur in ocean productivity and circulation patterns; the most impacted one is the North Atlantic (Schmittner, 2005) and Indian oceans (Gianni et al., 2003). These changes could impact on the individual fisheries as well as the biomasses of phyto and zooplankton groups due to changes in the food webs.

(iv) *Changes in monsoons and occurrence of extreme climatic events*: Increase in the frequency of occurrence of extreme climatic events, like devastating floods, changes in monsoonal rainfall patterns and storms, etc. could generally occur.

(v) *Water stress*: According to IPCC (2007), by the early part of 21st century, *c.* 75–250 million people in Africa could be under water stress, and freshwater availability in Central, South, East, and SE Asia, particularly in the larger river basins, is projected to decrease.

(vi) *Alteration(s) in the hydrological regimes in the inland waters*: Global warming may alter aquaculture activities in both lentic and lotic water bodies. Eutrophication could be exacerbated. Stratification could be more pronounced with impacts on food webs and habitat quality and availability (Ficke et al., 2007). Both these aspects, in turn, could have a bearing on aquaculture activities, in particular, inland cage and pen aquaculture.

6.3.56 Direct impacts on aquaculture

The major issues are impacts on water, land, seed, feed, and energy. An example of the direct impact of climate change on AQC relates to a smog cloud generated over SE Asia during the 2002 El Niño. Human activities *per se* might not be the only reason but it cuts sunlight and heat to the lower atmosphere and the ocean by *c.* 10%, and such a situation had contributed to significant dinoflagellate blooms that had impacted on aquaculture in the coastal areas from Indonesia to the Republic of Korea; thus, causing damage to aquaculture worth millions of US dollars (Swing, 2003). Further, climatic disasters, such as the 2008 cyclone in Myanmar (with relevant impacts on coastal communities) or the repetitive hurricanes in the Caribbean, are usually connected with climate change.

6.3.57 Potential impacts

Separate treatments are preferred to deal with climate change impacts in FW, marine, and brackish water environments. There is also a need to address appropriate adaptive measures. Global warming enhances temperature, which could bring associated change(s) in the hydrology and hydrography of water bodies and could exacerbate the occurrence of algal blooms and red tides, etc.; thus, significantly impacting on aquaculture.

6.3.58 Inland aquaculture

6.3.58.1 *Pond aquaculture*

Finfish culture is the account for the great bulk of aquaculture in the tropical and subtropical ponds, whose water temperature is determined by solar radiation, air temperature, wind velocity, humidity, etc. Incidentally, the predicted hike in air temperature could increase evaporation and cloud cover (IPCC, 2007) and could reduce solar radiation reaching the ponds. However, overall, there may not be any major increase in water temperature in all inland aquaculture ponds. Thus, immediate need for species change(s) or in the *modus operandi* of the current aquaculture practices may not be required. Conversely, in the temperate FW ponds, the cultured salmonids, viz., the rainbow trouts (*Oncorhynchus mykiss*) and the brown trouts (*Salmo trutta*) have a very narrow optimal range of temperature and a relatively low upper lethal limit. Incidentally, temperature hike in these ponds could impact on productivity and may also cause mortality. However, high water exchange may soften the potential impact(s). Moreover, less availability of water due to climate-change-driven droughts could become an issue for these systems.

6.3.58.2 *Integrated aquaculture*

Integrated aquaculture, encompassing, paddy-*cum*-fish culture, fish culture integrating with animal husbandry, etc. is an age-old traditional single-unit family practice mostly in rural India, China, etc. as well as in a few temperate regions in eastern Europe (Deb and Kar, 2007; Nath and Kar, 2016; Singha and Kar, 2016; Suklabaidya and Kar, 2016). Fish species, feeding low in the trophic chain, are generally cultured. External feeds are generally not added. The cultured fish stock usually feeds on the phyto- and zooplankton and the benthos which grow naturally in the cultured water bodies and their productions are also to be promoted. As such, these practices essentially act as **carbon sinks**.

Notwithstanding the above, such practices of integrated poultry/duck aquaculture may harbor and spread the highly pathogenic avian influenza, commonly known as "bird flu," and is mostly caused by the H5N1 strain of type A virus from the Orthomyxoviridae family of the viruses (Feare, 2006). As such, integrated aquaculture falls within sectors 3 and 44 of the FAO (2004) farm management guidelines for biosecurity. These practices may imperil certification of the aquaculture produce *vis-a- vis* their marketability, and there is a need to adopt further precautionary approaches. Nevertheless, these farming practices are efficient means of recycling biological wastes; and these practices act as **carbon sinks** and they also help in **carbon sequestration**.

6.3.58.3 *Cage culture*

Cage aquaculture becomes an increasingly important facet of aquaculture development in the globe and could continue to do so for the foreseeable future (Halwart et al., 2007). This trend is being driven by the following aspects:

(i) A realization of the need to utilize existing inland waters for food fish production, in the wake of possibility of limitations on land and water availability (De Silva and Phillips, 2007) in the future (which may be due to climate change).
(ii) To fulfill the increasing demands for higher quality/high-valued food fish in the marine environment.

(iii) Inland cage culture, in addition to the supply of food fish, could also be an important means of providing with alternative livelihoods for people displaced from reservoir impoundment (Abery et al., 2005), a situation occurs at the highest rate in Asia (Nguyen and De Silva, 2006).

Inland cage culture practices may be very much variable in intensity and mode of operation, and, in the species to be cultured. However, the unregulated proliferation of cage culture practices in many water bodies in the tropical region could result in the regular occurrence of fish kills, disease transmission, and, consequently, lower profits, mostly as a result of a high density of cages in a single water body without much consideration of the carrying capacity of the ecosystem (Abery et al., 2005).

Incidentally, Ficke et al. (2007) opined that the climatic change could exert some impacts, like exacerbation of eutrophication; more pronounced stratification in the lentic systems with concomitant depletion of dissolved oxygen (DO) levels during the dawn hours; abrupt alteration(s) in the wind patterns and rainfall; and, upwelling, bringing deep DO-depleted waters to the surface. This could adversely affect both cultured naturally recruited fish stocks in the water body. Further, deoxygenation problems due to upwelling restricted cage culture to one crop per year as opposed to two crops per year previously. Also, better planning of cage culture activities is suggested to avoid the impacts of climate change.

Concomitant to the above, traditional, river-based cage culture also occurs in most tropical regions in Asia. Climate change may also impact on this practice to a big or small extent. In fact, improved farming practices, including more efficient feeding, could be an adaptation which may help the sector to face the potential climate change impacts.

6.3.59 Mariculture

Mariculture, a significant type of aquaculture is practiced in inshore and offshore areas of tropical, subtropical, and temperate regimes of some countries and involves sand and bottom culture and raft and cage culture. However, in temperate regions, it includes mainly the culture of salmonids in cages (Halwart et al., 2007). Mariculture also involves the culture of finfish, like groupers (e.g., *Epinephalus* spp.), snappers (e.g., *Sparus* spp.), cobia (*Rachycentron canadum*), etc., in the tropics and subtropics; molluscs such as *Ruditapes* sp., *Mytilus* sp.; and seaweeds, such as *Gracilaria* sp., etc. The culture of some of these imply minimal energy consumption; and, concomitantly, they are essentially **carbon sequestering**. Incidentally, the main energy costs are related to the transportation of the products to the consumers. These cultures are carbon friendly and could cause minor environmental perturbation(s) associated with changes in the hydrographical conditions in the culture area and sedimentation of feces and pseudo-feces on the bottom. Incidentally, global warming can exert an impact on mariculture in temperate regions. In salmon farms, the rise of temperature > 17°C could be detrimental, when feed intake drops, and feed utilization efficacy is reduced. Thus, possible adaptive measures include enhancements in feed utilization efficacy as well as protein and lipid usages for growth. Incidentally, low-fat feeds could give better results at higher temperatures (Bendiksen et al., 2002).

Notwithstanding the above, the increased temperature may augment hatchery production of seeds, if proper feeds are given and diseases are controlled (Benetti et al., 2008).

6.3.60 Saltwater intrusion

Saltwater intrusion generally occurs in the deltaic regions, where a significant amount of aquaculture activities occur in tropical Asia in the middle to upper levels of the tidal ranges. Incidentally, in addition to estuarine shrimp farming activities in Asia, South America, and the Caribbean, other AQC activities include recently emerged catfish culture (*Pangasianodon hypophthalmus*) and rohu fish (*Labeo rohita*) culture in the River Mekong in Viet Nam and in the River Irrawaddy in Myanmar (Aye et al., 2007), respectively. Incidentally, rohu fish culture is *in vogue* in different water bodies in the Indian subcontinent.

Notwithstanding the above, sea-level rise, in addition to causing ecological and habitat changes (including mangroves, which serve as nursery grounds for many euryhaline species), could lead to increase in the salinity level with the intrusion saltwater further upstream of rivers beyond the estuary region and consequently, shifting the AQC operations further upstream as an adaptive measure or shifting to more salinity tolerant strains of these species. However, such shifts could be cost intensive and may impact on the socioeconomic status of the communities involved. Further, saltwater intrusion may render agriculture unsuitable and thus, enabling the agriculture farmers to switch over to AQC as an alternative means of livelihood. Hence, this could be a positive aspect of AQC. However, developing adaptive measures are of paramount importance to combat climate change impacts.

Notwithstanding the above, specific predictions of approximately 1 m rise in sea level (due to climate change impacts) is available for the Mekong Delta, a site for promising aquaculture and literally the food basket of Vietnam, which accounts for *c.* 46% of the nation's agricultural production and *c.* 80% of the rice exports. Incidentally, there could be an inundation of *c.* 15–20,000 km^2 of the land, with a loss of *c.* 76% of arable land.

Further, there could be devastations in different parts of the world due to increased frequency of severe cyclones, typhoons, hurricanes, floods, droughts, etc., because of climate change impacts and the colossal losses could be to the tune of US$4 and US$38 billion, reaching as high as US$58 billion in a single year (IPCC, 2007). Concomitantly, El Niño and La Niña events cause extreme weather conditions in temperate regions. The El Niño of 1994–1995 devastated the salmon industry in southern Chile due to severe storms and also caused a large number of escapes from sea cages (Soto et al., 2001). Further, the El Nino event hindered the development of offshore aquaculture in the different parts of the temperate region, including North America, by severe winter storms. Moreover, the **intensity of climatic change is likely to increase further**. Moreover, extreme weather conditions due to climate change impact on aquaculture activities in tropical and subtropical regions in Asia and elsewhere. The recent unusual blizzard in southern China, destroying the AQC facilities points to the impact of climate change. The range of devastation could be from the physical destruction of aquaculture facilities to loss of stock and spread of disease.

Further, due to climate change, the central part of Viet Nam experienced the worst flood in 50 years in 2007 with devastating damages to AQC (Nguyen, 2008). Notwithstanding the above, the supply of **fishmeal and fish oil** is also impacted by climate change.

Notwithstanding the above, **destructive effects of extreme climatic events** cause a big school of cultured fishes to enter natural waterways, which, in turn, cause **environmental disturbances**, and the potential for impacting biodiversity becomes considerably higher.

In addition to the above, **climate change** in some regions of the world, is known to cause inclement weather conditions; water quality changes (e.g., plankton blooms); and increased pollutants and other damaging runoff from land-based sources. Such weather conditions may increase the **vulnerability of sea-based aquaculture**, particularly cage aquaculture and also seaweed farming in the coastal bays in Asia.

Climate change may also increase the vulnerability of near-shore land-based coastal aquaculture to **all forms of severe weather conditions** causing erosion, infrastructural damage(s), escapes, loss of livelihoods of aquaculture farmers, etc., particularly, in the large coastal deltas of Asia, which are the hub of AQC activities. Nevertheless, few **adaptive measures** are available for such impacts.

Notwithstanding the above, **Hurricanes**, caused due to climate change impacts have devastated the coastal rural aquaculture in Central America. In fact, hurricanes with different names have damaged the AQC installations in different regions, for instance, hurricane Mitch devastated the shrimp farms in Nicaragua, which flourished from 1990 to 1998, and hurricane Felix wiped out many rural areas in Nicaragua, some of them, in fact, with incipient aquaculture activities. Some of the other potential hurricanes were hurricanes, Dennis and Emily, in Jamaica, and hurricane Stan in El Salvador and Guatemala.

Incidentally, **adaptive measures** could include the availability of weather forecasting infrastructure, provision of alternative livelihood, etc.

6.3.61 Water stress

Water stress, due to climate change, exerts a significant impact on aquaculture. The predicted water stress could decrease water availability in the major rivers in Central, South, East and SE Asia as well as in Africa (IPCC, 2007), and thus, it could affect the sustainability of intense aquaculture activities in the deltaic regions of the Mekong, the Ganges-Brahmaputra-Barak-Meghna drainage (Kar, 2007, 2013), and the Chindwin-Irrawardy drainage system in Asia. However, reduction in aquaculture water, as an adaptive measure, in general, is achieved through:

(a) selection of feed ingredients that need little water to be produced;
(b) enhancement of within-aquaculture-system feed production through periphyton-based technologies (Van Dam et al., 2002); and
(c) integration of AQC with agriculture (Verdegem et al., 2006).

Further, the issue of water stress have been considered along with saltwater intrusion, and monsoon disruption, wherever required (Hughes et al., 2003).

Concomitant to the above, climate change in some regions in Asia and Africa, is predicted to increase drought periods (IPCC, 2007); thus, resulting in less water retention time in non-perennial water bodies. This may render such water bodies relatively unsuitable for aquaculture purposes because a minimum period of 6 months of water retention is needed for most fish to attain marketable size.

As an adaptive measure against water stress, land-based AQC could be popularized as a viable alternative. Cost-intensive, energy-intensive, and skill-intensive recirculation technology may be remotely considered in this case. Moreover, species feeding high in the food chain and having high market value are used in this technology. It entails the requirement of the good and rich quality of feed. It may not be possible for an ordinary fish farmer to maintain all these aspects. Incidentally, energy costs of maintaining recirculation systems are rather high, with the concomitant emission of GHG (primary causative factor for climate change), which is contrary to the main aims of adaptive measures to minimize climate change impacts, i.e., the system should rather revolve around "energy savings" (De Ionno et al., 2006).

In fact, the development of offshore mariculture was pleaded as a plausible means of augmenting food fish production over the last few decades. Concomitantly, there are minimal immediate environmental perturbations in doing so (Grøttum and Beveridge, 2008).

6.3.62 Indirect impacts of climate change on aquaculture

Handisyde et al. (2006) considered indirect impacts that climate change have on aquaculture *vis-à-vis* possible influences on price fluctuations of capture fishery produces. Incidentally, capture fisheries are a major source of inputs for aquaculture, providing with feed, in particular, and seed to some degree. As such, alteration(s) in capture fisheries caused by global climate change(s) could hamper the aquaculture industry.

6.3.63 Impacts of climate change on diseases

Climate changes have impacted the incidence of various kinds of diseases on the earth, including associated risks for human health (Epstein, 2005). Incidentally, the terrestrial vector-borne and diarrhoeal diseases increase under the influence of climate change. However, other than documenting few large-scale disease episodes, impacts of climatic change on aquatic organisms have been mainly restricted to coral bleaching and associated changes (Harvell et al., 1999; Kar, 2015).

Incidentally, an enhancement in the incidence of disease outbreaks in corals and marine mammals, together with the incidence of new diseases was reported (Harvell et al., 1999; Kar, 2015). Coral bleaching was linked with the high El Niño temperatures in 1997–1998, probably, associated with the global transport of fish species, bringing together pathogens, which were previously unexposed in the populations. Climatic alteration(s) also have some amount of influence in this regard (Hughes et al., 2003).

Under the impact of global warming, increased agricultural intensification and associated translocations not only caused adverse effects on biodiversity but also exacerbated emerging infectious diseases in free-living wild animals and fishes (Daszak et al., 2000). Nevertheless, there have been efforts to reveal the changes in the parasite populations in temperate and boreal regions of eastern North America in the wake of the associated effects of climate change. Based on the simulations and predictions from a general circulation model, an overall inference portrayed a significant influence of climate change on different life-history traits affecting parasite transmission and virulence (Marcogliese, 2001). However, it may be difficult to predict the exact consequences of climate change on aquaculture *per se*, doubtlessly, the aquaculture sector needs to be aware of the potential and new threats from parasitism.

Concomitant to the above, there was a profuse proliferation of **harmful algal blooms (HAB)**, etc., particularly, in the North Atlantic and the North Sea (Edwards et al., 2006) due to anthropogenic influences (Smayda, 1990) associated with **climatic alteration(s)**. Incidentally, the HABs affect human health through the consumption of affected filter-feeding molluscs, commonly referred to as "shellfish poisoning." Apart from this impact, the HABs could also bring about harmful effects on cage culture operations of salmon. Accordingly, adaptive measures are to be set up in place(s) for regular monitoring and vigilance of the aquaculture facilities in areas with potential vulnerability to eutrophication and HABs.

Notwithstanding the above, the probability of climate change enabling the Pacific oyster (*Crassostrea gigas*) and the associated pathogens to spread into new areas, was highlighted in different communications. Concomitantly, the spread of two protozoan parasites (*Perkinsus marinus* and *Haplosporidium nelsoni*) northward from the Gulf of Mexico to Delaware Bay, under the influence of higher winter temperatures (in fact, the pathogens, otherwise, are kept in check by temperatures $<3°C$), resulted in mass mortalities in the Eastern oyster (*Crassostrea virginica*) (Hofmann et al., 2001). Hence, due to climate change, there seemed to be an upsurge of pathogens impacting on cultured organisms, such as the molluscs. Further, the outbreak of *Vibrio parahaemolyticus* in oysters in Alaska as well as in many seafood products in Chile has been related to the warmer seawater temperatures during summer (Paris-Mancilla, 2005). Incidentally, the main adaptation measures were: (i) to avoid the edible organisms (especially the bivalves) reaching high temperatures while in transit or in storage (since the multiplication of the pathogen takes place at an optimum temperature of 37°C). The pathogen could also remain encysted to tide over the unfavorable condition(s). Hence, practices of eating raw seafood in Chile was banned during summer; (ii) eatables should be eaten only after boiling.

In continuation of the above, it could is possible to predict a general impact of water warming on the spread of diseases, notably, the bacterial diseases in aquaculture because, in most cases, these diseases persist under the stressed condition of the fish. Abrupt changes in water temperatures and other physicochemical parameters of water, causing a stressed situation, could lead to the initiation of diseases in fish (Snieszko, 1974). In this regard, changes in the limnological parameters of water, notably, the temperature and total alkalinity (TA), were acted as the predisposing stress factors in the initiation of the *hitherto* unknown virulent enigmatic epizootic ulcerative fish disease syndrome (EUS), which turned out to be a semi-global fish pandemic causing large-scale mortality among the fishes (rendering many of them endangered); causing fear psychosis among the fish consumers; and, had devastated the economy of the poor fisherfolk (Kar, 2015). Concomitantly, ocean acidification could impact on the immune response of common AQC candidates, blue mussels (*Mytilus edulis*) (Bibby et al., 2008). Further, increased uptake of toxicants and heavy metals by cultured, FW filter-feeding molluscs through accelerated metabolic rates in warm water condition was reported (Ficke et al., 2007). This, in turn, could require food safety certification.

Nevertheless, essential adaptive measures include regular monitoring of the water quality and the cultured product(s) for human health risks. The spread of diseases is the most feared threat in aquaculture. Examples of disease-related catastrophes in the aquaculture industry include the spread of the white spot disease in shrimp farming in different parts of the globe (Morales and Morales, 2006); the EUS disease among the fishes (Kar, 2015); and more recently, the case of infectious salmon anemia (ISA), which had been seriously impacting Chile's Atlantic salmon industry.

In view of the above, given that, the issue of the spread of pests, parasites, pathogens, and diseases are considered as the major threats under the climate change scenario, appropriate biosafety and biosecurity steps are to be adopted as main adaptation measures, considering aquaculture as one of the potential industries guaranteeing, to a large extent, the nutrition and avocation of the people.

6.3.64 Impacts of climate change on biodiversity

Climate change has impacted various aspects of the globe, including biotic and abiotic. Incidentally, IPCC (2002) stressed upon issues related to impacts of climate change on biodiversity *vis-a-vis* their remedies. Some amount of works was done on coral bleaching and subsequent loss of coral habitats, which are one of the most biodiversity-rich habitats on the earth. Further, **many species** are either extinct or critically endangered today due to various types of diseases and other impacts of **climate change** in contrast to the extinction of only the **golden toad** (*Bufo periglenes*) from Costa Rica during the earlier days (Crump, 1998). Further, large-scale mortality of fishes (due to **EUS disease**), shrimps (due to white spot disease), etc., are due to impacts of climate change at least to some extent. However, it could be little difficult to predict the **overall loss of biodiversity** due to **climate change**. In this connection, Thomas et al. (2004) opined probable extinction of at least, one out of five living species on this planet due to the current levels of emissions of **GHGs**.

The introduction of alien species in AQC is common today (Turchini and De Silva, 2008). However, their impacts on local biodiversity (BD) *per se* probably have not been well evaluated. But, certain evidence point toward the devastating impact(s) of one such translocation, associated with the introduction of a fungal plague and the consequent dissemination among the native European freshwater crayfish (Edgerton et al., 2004). Further, Gienapp et al. (2008) addressed the issue on possible relatedness between **climate change and evolution** and concluded that:

(i) many alterations perceived as an adaptation to changing climate could be environmentally induced plastic responses; rather than, microevolutionary adaptations and
(ii) clear cut evidence is lacking to indicate a significant role for evolutionary adaptation to ongoing climate warming.

Incidentally, the balance of evidence suggested that **global climate change** may not always enhance impacts on **biodiversity through aquaculture** *per se*. However, in view of the alteration(s) in the **temperature regimes**, etc., particularly in the temperate regions, the probabilities of diseases occurring among the filter-feeding molluscs and fishes, for instance, could be higher. Furthermore, any new introductions for aquaculture purposes, one should have to take into consideration including different factors in the initial risk assessments and decision-making.

Notwithstanding the above, there have been translocations of salmons, tilapias, and the white-legged shrimp in cool temperate waters; in warm tropical waters; and in brackish water in order to assess their production under different climate conditions *vis-à-vis* impact of climate change on them. The first two groups have accounted for >1.0 million tons of production, closely followed by the white-legged shrimp (*Penaeus vannamei*). Incidentally, climate change could impact on the culture of all three species groups. However, warming in the temperate regions could narrow the distribution range of salmonids. On the other hand, the opposite could be true for tilapia and shrimp and their distribution could be extended well into the subtropics.

Notwithstanding the above, **coral bleaching and coral reef deterioration** are not only due to the **climate change impacts,** but the **destructive fishing methods** for trades, etc. also significantly cause the devastation (Scales et al., 2007). However, recent efforts in meeting the trade supply through AQC has ameliorated the situation, leading to some amount of conservation of the coral reef biodiversity.

Mass escapes, due to adverse weather conditions, are common in AQC operations, which could influence the genetic makeup of native stocks. Thus, the design of aquaculture facilities, particularly those located in areas vulnerable to unusual climatic events needs to consider measures that could minimize mass-scale escapes.

6.3.65 Social impacts of climate change on aquaculture

Climate changes, among others, have impacted the fisheries, the aquaculture, and the society at large (Allison et al., 2005). However, the **potential social impacts of climate change** on fisheries are manifold and primarily arise from the following:

- **(i)** diminished revenues to fishers resulting from declines in catch and stock abundance (Luam Kong, 2002; Mahon, 2002);
- **(ii)** changes in migratory routes and biogeography of the stocks affecting fishing efforts, an example of having increased travel time to fishing grounds (Dalton, 2001; Mahon, 2002);
- **(iii)** changes in harvest technologies and processing costs brought about by the need to capture new species (Broad et al., 1999);
- **(iv)** damage to physical capital from severe weather events (Jallow et al., 1996);
- **(v)** impacts on transportation and marketing chains/systems (Catto, 2004); and
- **(vi)** reduced human capital from severe weather events, increased incidence of red tides, and associated shellfish poisoning (Patz, 2000).

Importantly, some of the above, for instance, may cause damage to physical capital in various forms due to climate change impacts, like infrastructure, transport, etc. Further, the majority of the small-scale AQC enterprises is located in the tropical and subtropical regions; are often **farmer owned and farmer managed** and are also usually **clustered** together. As such, **extreme climatic events could destroy the** entire **cluster as a whole** affecting many **poor** and vulnerable farming communities. As an **adaptive measure**, a form of **cluster insurance** scheme (with governmental policy backup) **may be instituted to help** such clusters to spring back to their livelihoods.

Concomitantly, marine tropical AQC are usually confined in enclosed coastal bays and generally consist of smallholdings **clustered together**. They could become potentially vulnerable to extreme climatic events (due to climate change), such as sea storms, wave surges, etc. The marine farming communities, as usual, are **vulnerable to adverse weather** events.

On the other hand, FW finfish aquaculture, in the tropics, is usually done in the ponds, tanks, and in the various kinds of other lentic water bodies, locally called "*Beel, Haor,* **and** *Anua*" (Kar, 2007, 2013). These water bodies could also be influenced by extreme climatic events (due to climate change), such as cyclones, floods, etc. Similarly, the farming communities around them could become potentially vulnerable to inclement weather due to climate change. Incidentally, the **aquaculture** sector, at least in Asia, is serviced to a significant extent by small-scale **artisanal** fishers providing the required trash fish or low-valued fish to feed the cultured stock. Moreover, these artisanal farmers could also be impacted by climate change issues.

Notwithstanding the above, in the high climate change impacted deltaic region (with issues like sea-level rise, water stress, etc.), land-based **agriculture** has **abandoned** and replaced by aquaculture as means of alternative livelihoods. The possible major social upheavals in the lifestyles have carefully tailored with initial capacity building needed to adjust to an altered livelihood pattern. Guidance has been obtained from almost a similar type of change in livelihood pattern from agriculture to aquaculture with respect to communities displaced by **reservoir impounding**. Incidentally, **adoption of aquaculture have** led to the **socioeconomic improvement** of the incumbents (Wagle et al., 2007).

Incidentally, some people opined that **aquaculture** is not always ecologically sustainable in a world that is becoming increasingly conscious of **carbon emission, including those resulting from food production.** It is possible that, shortly, consumers could make a demand for carbon-emission labeling. Concomitantly, the shrimp farming process may return to **indigenous species,** such as *P. monodon*, which could be cultured using better management practices (BMPs) that are **less energy demanding.**

In continuation of the above, climate change enhance the **intensity of diseases affecting aquaculture,** which, in turn, affects the farming communities and associated workers for examples, 'ISA Virus 6' affecting Chile's salmon farming industry and the local economy.

Almost similar kinds of issues have also been observed during the outbreak of the EUS global pandemic among FW fishes (Kar, 2015).

6.3.66 Potential impacts of aquaculture on climate change

AQC is the global practice today. It often helps in carbon sequestration by acting as a carbon sink. In integrated AQC, the stocked fish usually feed on the phyto and zooplankton and benthos, which grow naturally in the pond, and their production is also promoted. As such, the pond with phytoplankton acts a **carbon sink**, and also helps in **carbon sequestration**. In other words, the AQC-based farming practices are efficient and effective means of recycling biological wastes, and, they act as **carbon sinks** and help in **carbon sequestration**.

In addition, enhancement of within-aquaculture system of fish food production through periphyton-based technology (Wahab et al., 1999; Van Dam et al., 2002) is often recommended to get rid of externally obtained fish meal and fish oil, whose production involves the spending of much energy and evolution of a lot of GHGs.

Concomitant to the above, fisheries and aquaculture make a moderate contribution to GHG emissions through fossil-fuel-based catching operations and transportation, which are reduced by improved technology and management of stocks using eco-friendly methods. Different global emissions mean that climate change could affect both marine and FW systems and fishing communities as well.

6.3.67 Carbon sequestration

One of the major causative factors of climate change is the accumulation of GHGs in the atmosphere, irrespective of the source(s) of emission (Brook et al., 1996; Flannery, 2005; Kerr, 2006; IPCC, 2007). Incidentally, **carbon sequestration is the process through which agriculture and forestry practices remove atmospheric carbon dioxide**. In fact, forestation, reforestation, and forest preservation are considered to be highly favorable nature friendly practices that sequester and/or preserve carbon; and, all these assist in alleviating climate change by enhancing carbon storage (Miller, 2008).

6.3.68 Institutional, policy, and planning measures

In the event of climate change, the following are the priority areas for the development of the aquaculture sector, in terms of institutional and policy measures:

(i) to implement an ecosystem approach to aquaculture (EAA) as a global strategy;
(ii) to prioritize and enhance mariculture and especially non-fed aquaculture (filter feeders, algae);
(iii) to accelerate the use of suitable inland water bodies through culture-based fisheries with appropriate stock augmentation practices;
(iv) the EAA tries to integrate aquaculture within the wider ecosystem in such a way that it could promote sustainability of interlinked socio-ecological systems (Soto et al., 2008);
(v) the EAA is stressed upon the need to integrate aquaculture with other sectors (e.g., fisheries, agriculture, urban development, etc.), which share and utilize the common resources (land, water, feeds, etc.).
(vi) the EAA tries to focus on different spatial scales; notably, (i) the farm, (ii) the aquaculture zone, water body or watershed, where the activity takes place, and (iii) the global scale (Soto et al., 2008). Thus, the implementation of EAA at the water body scale is one of the most relevant adaptations to climate change.

Other relevant aspects to be considered in the policies and planning are described in the following.

6.3.69 Aquaculture insurance

There may be losses caused by various reasons, like, climatic events; disease outbreaks; escapes from the farms, etc. As such, an economic adaptive measure, that could help in limiting bankruptcies in aquaculture business as a result of disasters, is to be adopted by aquaculture farmers in the form of an insurance cover against damage to stock and property from disasters. Adequate insurance cover could help, particularly, the small-scale farmers with finance for recommencing the business operations.

6.3.70 Research and technology transfer

Technology transfer from "lab to land" is an important aspect of aquaculture research, particularly to tackle the climate change impacts. Hence, there is possible need for streamlining the research works in the fields of aquatic ecology, limnology, fishery, aquatic animal physiology, etc. encompassing significant issues, such as prophylactic and therapeutic measures against new types of diseases, including their antidotes; finding new and better-adapted species to climate change situation(s); formulation of better feeds and feeding practices that are more ecosystem friendly; and so on.

6.3.71 Zooplankton culture

The culture of zooplankton under seminatural condition(s) could throw a significant amount of light on feeding the ichthyospecies in culture. The zooplankton is considered as the potential natural fish food organisms for the fishes. A significant amount of attempts have been made to culture zooplankton in the laboratory under seminatural conditions (Kar et al., 2018a,b; Das et al., 2018a,b, 2020; Devi et al., 2016; Lambert et al., 2020; Kar, 2019).

6.3.72 Conclusion

Aquaculture is one of the potential sectors for nutrition and avocation guaranteeing answers to our diminishing protein supply across all communities irrespective of living standards. It is quite likely that aquaculture, in view of its resilience and adaptability, should be able to respond positively to climate change impacts. However, there are needs for related policies to be in place, in order to respond to climate alteration impact(s). There could also be needs for institutional and socioeconomic changes, backedup, and supplemented by relevant technical developments. In fact, there can preferably be a holistic approach, which should work from the bottom-up rather than top-down direction. The matter is important because the great bulk of aquaculture is usually owned by the small-scale farmer, particularly in Asia, being operated by themselves or by the farming cooperative societies.

In this regard, it is possible for adaptive changes to be implemented effectively and in a timely manner only by incorporating indigenous knowledge and by obtaining cooperation from the grassroot level. Incidentally, many climate-related changes have occurred on our planet, over many thousands of millennia, bringing about major changes in flora and fauna. However, the reason(s) for such changes are not always obvious and/or universally accepted. However, we now believe that the climate change now facing the earth has been primarily brought about by anthropogenic activities, which, might have started since the beginning of the last industrial revolution. In fact, the causative agents of the changes and, therefore, the required mitigating measures are understood and have been subjected to rigorous scientific scrutiny (IPCC, 2007).

Human food needs and food production are impacted by climate change, and such changes in the climate during the coming decades are the major concern, particularly for the developing nations. However, it is predicted that climate change has many negative impacts for the poor developing countries and, also, hit them the hardest. The assumed belief is based on the exponential growth rate of the human population coupled with the fact that food production is not evenly distributed throughout the globe and global total food security is still not fully attained (Kerr, 2006). In addition, the plants and animals, which are sometimes, less capable of adapting to the changes, could be another casualty of the climate change impacts.

It is believed that even a modest climate change in the next few decades could begin to decrease both aquacultural and agricultural crop production in low latitudes (Kerr, 2006). These include those regions, where aquaculture is very much predominant. It is, however, encouraging to note that a significant proportion of innovations regarding aquaculture have originated from the grassroots initiatives. As such, the rural, small-scale aquaculture farmers could be expected to be alert to the climate change impacts, and they could make the necessary adaptations.

Notwithstanding the above, there are often discussions regarding capture and culture fisheries. It is often believed that the principal difference between the two sectors is that the food fish supplies are still predominantly maintained by the former; while the future enhancements in demand are supposed to be met mostly by the latter. The significance of capture fisheries could at best be static and there is a high probability that climate change may cause it to decline. Consequently, aquaculture will try to fill-in the gap between demand and supply as well as could also meet the growing human fish food needs. In fact, aquaculture is a significant contributor to the animal protein component of the food basket, although it is only a relatively small food production sector. Incidentally, aquaculture had augmented its contribution from 0.7 kg per capita in 1970 to 6.4 kg per capita in 2002, with *c.* 10 million people active in the production sector. This increase is higher than that is known for terrestrial livestock farming, which was grown only @ *c.* 2.8% per year during the same period (Bunting and Pretty, 2007).

Moreover, this reflects the late emergence of aquaculture as a significant contributor to the human food basket. In this connection, it may be pertinent to emphasize that aquaculture, of late, has been much scrutinized from an environmental impact point of view; presumably; because this sector had gained prominence only in the last few decades coinciding with a surge of global awareness about sustainable development and environmental integrity (CBD, 1994). Incidentally, unlike many other animal meat production sectors, aquaculture, which farms the poikilothermic animals, is, infrequently distributed with more concentrations in tropical and subtropical regions of Asia, and to a lesser extent on the temperate coasts of Europe and South America. Given this distribution, it could be expected that the major climate change impacts on aquaculture could be through global warming and the consequent temperature enhancements in water.

Sea-level rise and associated saltwater intrusion, compounded by monsoonal weather pattern changes are a concern in the tropical and subtropical regions, where the bulk of aquaculture activities take place. Further, the impact is likely to be more profound in the major deltaic areas in the tropics. Incidentally, adaptive measures are feasible, such as changing a species or moving the main current aquaculture operations away from the shore. In fact, seawater intrusion could disable land-based agriculture and enable and promote AQC practices as alternative means of livelihood. Incidentally, this process might be fuelled, in part, by the fact that financial returns from aquaculture production tend to be usually significantly higher than those from the traditional agricultural practices on a unit area basis and is also relatively less energy demanding than terrestrial animal husbandry.

In the tropics and subtropics, inland aquaculture is predominantly practiced. It is often opined that inland water aquaculture in the existing water bodies (both lentic and lotic) are being augmented, through various types of fish culture, operations, notably, cage culture, polyculture, integrated fish farming, and so on. Incidentally, expected climatic changes could have a profound impact on the lentic water bodies through various kinds of processes, like enhanced eutrophication and stratification, and thus, it brings the mortality of cultured stocks through upwelling, oxygen depletion, etc. However, there are many adaptive measures available to avoid such calamities, foremost being the development of aquaculture activities in accordance with the potential carrying capacities of the water bodies, and there should be continuous monitoring of the environmental variables in relation to nutrient loading, etc.

Concomitant to the above, the impacts of climate change on wild fish populations are likely to have a significant impact on aquaculture, particularly, with regard to the availability of raw materials for the production of fishmeal and fish oil. Feeds for the farmed animals bear a very high ecological cost (Bartley et al., 2007). Incidentally, aquaculture of carnivorous species (which constitute only a small proportion of all cultured commodities) could require much amount of animal protein-rich feed. Such carnivorous fish usually have much high value in the market. But, such fishes are better to be adapted to different other types of feeds, so that, there is not much harvest of wild fish from the capture fisheries in nature, and thus, capture fisheries are not overharvested. Hence the correct way to address such issues will be the development of suitable diets that use less amount of fishmeal and fish oil. This process had got underway a few years ago, with the development of high energy diets for the salmons. It is also important to curtail the use of diets containing fish oil through the growth phase of the fish, and, instead, adopt "finishing diets" (Jobling, 2004; Turchini et al., 2007) prior to harvesting in order to satisfy consumer demands and maintain the fish quality (Mourente et al., 2005). Incidentally, aquaculture, unlike terrestrial animal husbandry, relies on a wide range of species, which could be around 300 today (FAO, 2007).

Notwithstanding the above, coral bleaching exacerbated by climatic changes and its effects on biodiversity is a major and a growing concern. It is important to consider the process in conjunction with coral destruction caused by destructive fishing methods undertaken mainly to meet the demands of the live fish restaurant trade, which is a growing luxury and lucrative trade in limited locations in some countries. However, the dependence on wild-caught fish for this trade was markedly declined because of the growing public concern, and this market is increasingly making use of cultured fish. This indicates that aquaculture seems to be capable of helping in lessening the exacerbation of coral reef destruction and, thereby, assisting in enhancing the preservation of biodiversity.

Incidentally, aquaculture is criticized as ecologically cost-intensive and environmentally degrading. However, such conclusions are almost always based on aquaculture of high-value commodities, such as shrimps and carnivorous finfish species, such as the salmons. Nevertheless, such perceptions have generated erroneous misconceptions among the public, planners, developers, and investors. In this connection, the fact is that the great bulk of aquaculture is still dependent on fish and molluscs feeding low in the food chain as well as on the seaweed commodities that essentially act as carbon sinks and help in carbon sequestration.

Concomitant to the above, aquaculture plays an increasingly significant role (in the wake of climate change) by fostering carbon sequestration. In addition, aquaculture offers a high degree of elasticity and resilience to adapt to changes in the climate. AQC tries to rely on the provision of natural fish food through energy-saving periphyton generation technology. This could, even further, reduce the sector's contribution to climatic change (Van Dam et al., 2002).

Notwithstanding the above, the production of fish meal and fish oil for shrimp and carnivorous fish culture is energy expensive and contributes to carbon emissions. People may be encouraged to give up the consumption of such food fishes.

Ultimately, it might have to be conceded that there is a need to collate a robust quantity of information to address issues regarding the role of aquaculture in relation to climatic change. Incidentally, the efforts of the globe are directed toward reducing all forms of carbon emissions that are from food production processes or transport. However, with regard to food production, there is a need for the estimation of energy requirements. For instance, **carp aquaculture uses minimal energy but have a potential significance in the carbon cycle, fixing CO_2 through phytoplankton;** some of which are end as fish by the way of the food web. Further, fertilization and phytoplankton-based aquaculture systems are equally more climate/carbon friendly **than the more intensive forms which utilize considerable amounts of external energy inputs.** However, all the issues mentioned above have to be balanced against the food and development needs, and in order to arrive at tangible decisions, a large amount of data will be required as well as people's active participation.

In fine, it may not be possible to end any discussion simply by considering the relations between aquaculture and climatic change *per se*. However, aquaculture cannot be considered to occur in a vacuum. In fact, the world has accepted that there should be unified efforts to reduce GHG emissions in order to mitigate further exacerbation of global climate change. In this regard, one option could be to reduce the dependence on fossil fuels as an energy source, and this could be done by increasing dependence on biofuels. The first-generation production of biofuels is from the conversion of plant starch, sugars, oils, and animal fats into an energy source that could be combusted to replace fossil fuels. In fact, among biofuels, the most popular, at present, is bioethanol, which is produced by fermentation of many food crops, such as maize, cassava, and sugar cane (Worldwatch Institute, 2006). At present, the Brazilian sugarcane bioethanol has the highest net GHG mitigating potential (Macedo et al., 2004). The biofuels, as an alternative, have a ripple effect on food crops, their prices, availability, access, food security, and poverty, and, an overall impact on sustainable development (Naylor et al., 2006). It may be mentioned that aquaculture of carnivorous fishes and most forms of animal husbandry operations depend, to varying extents, for feeds obtained from different bio-resources, and some of the same bio-resources are also used for biofuels production. In fact, the equation of relation between aquaculture and climatic change seems to be not a straight forward one rather, many other factors have to be built into this complex equation in order to bring about adaptive measures with collective efforts and responsibility with an ecosystem perspective.

6.4 Management of fisheries and fish farming

Management is a significant aspect of fisheries and aquaculture. There may not be any long-term pragmatic solution in the offing with regard to a true solution or mitigation option for ocean acidification except limiting the CO_2 emissions to the atmosphere generated from the burning of fossil fuel. In fact, the management of ocean acidification is a long-term goal, which requires a fundamental reorganization of energy and transportation infrastructures globally. However, climate geo-engineering approaches, which do not regulate atmospheric CO_2, may not be able to address acidification. In fact, seawater chemistry changes due to ocean acidification are already apparent and are expected to grow over the next few decades (Feely et al., 2008). Hence, short-term responses intended to conserve sustainable marine environmental resources may also focus on adaptation to the inevitable near-future CO_2 increases. On the other hand, addressing the global problem of ocean acidification to preserve commercially valuable fisheries resources could require regional solutions. Incidentally, some of the local-scale strategies, notably, enhanced electrochemical weathering, directly combat seawater acidification by increasing alkalinity. However, such methods could be cost and energy intensive for a small benefit. As such, other strategies, like updating fishery management plans to include acidification, are less expensive as well as could be regionally tailored as needed to accommodate biological, economic, and social variations. In fact, designing new policies could begin with comprehensive research targeted toward regional needs (Doney et al. 2009). Some of these aspects could be as follows:

- **(i)** Expanded time-series studies of coastal and open-ocean seawater chemistry are needed to monitor ocean acidification's progress and to place it in context with historical data.
- **(ii)** Basic studies at the organism level are needed to enhance our currently limited knowledge of responses of commercial and keystone species to lowered pH and elevated CO_2. In this connection, topics of particular interest could include aspects like:
  - **(a)** the role that life history and population variability could play in shaping acidification responses;
  - **(b)** the sensitivity responses of molluscs, crustaceans, and fin-fish juveniles and adults to changing ocean water chemistry;
- **(iii)** Ecosystem-wise and ecosystem-wide studies are needed to throw light on secondary effects from habitat and prey losses. Such kind of information could be particularly useful for fisheries dominated by predatory finfish, as in the US Pacific regions, where the relative effects of prey switching; keystone species change(s); benthic and habitat degradations; and, overall biomass reduction are to be understood for long-term planning.
  In fact, the biological research results could enable the managers to identify and help the regionally significant species better. For instance, research suggests adjustments to fishery quotas or marine protected areas, to portray that aquaculture of juvenile molluscs is warranted along with some coastlines, or, that, preservation of a particular keystone predator could keep some crustacean fisheries robust.
- **(iv)** Economic and social science studies are required to understand in a better way, how markets, prices, and communities will respond to declining fishery harvests, and how best to mitigate potential socioeconomic impacts.

Incidentally, it may be noted that, the quantitative assessments of marine organisms' responses to ocean acidification and climate change must be explicitly incorporated into fishery management plans for improved long-range planning activities. Further, mathematical fisheries models are to be enhanced with chemistry and temperature-driven climate change and acidification terms, based on species-specific observational studies, in order to assist in determining appropriate harvest levels for many fisheries. Such model refinement could help in ensuring that catch levels remain sustainable, despite ongoing environmental changes. However, the likelihood of complex secondary effects resulting from ocean acidification emphasizes on the need for developing and using ecosystem-based management models. Nevertheless, more accurate predictions of ocean acidification's regional economic effects could arise from bio-economic models, which are adjusted for ocean acidification and climate change; thus, enabling the timely implementation of fiscally sound responses.

In continuation of the above, it may be noted that increasing fishery capacity could encourage multispecies fishing; could develop new markets; could minimize wastes; could increase aquaculture; and, last but not the least, could support research for selecting low pH-tolerant species or strains. However, initiating fishing activities *via* these methods could avoid wide-spread unemployment, but, at the same time, it could also require labor market adjustments, such as retraining fishers, etc. In addition, social measures are to be pursued to support the marine resource-dependent communities; as they may experience alteration(s) in avocation, demography, livelihoods, etc. Further, there may also be change(s) in local economies, community reorganization(s), through the involvement of government machinery, etc.

Incidentally, different aspects may hinder the planning process, like the uncertainty of whether ocean acidification's effects will appear incrementally or after profound ecosystem reorganizations, etc. Nevertheless, regional-scale marine resource management plans may begin soon, despite the various bottlenecks to estimate the outcomes of ocean acidification. Concomitantly, these short-term efforts are to be followed by long-term CO_2 mitigation plans. Moreover, the scope of the studies is to be wider. Instead of focusing only on one country, like the USA, the ambit of the study could be wider covering many countries in various geographical zones. The studies could also include economic consequences of ocean water acidification on coral ecosystems, and so on. Incidentally, it may be noted that marine resources are significant nutritional resources, which provide *c.* 20% of the global population with protein (FAO 2007, 2008, 2009). In fact, some of the developing island and coastal nations, which depend very much on marine and coral ecosystems for food, tourism, and exportable natural resources suffer the most economically from the consequences of ocean acidification and climate change. Ocean acidification may decrease their food supplies; and, rising sea levels physically endanger these communities. Further, the loss of the mangrove ecosystem may expose low-lying coastline communities to storms and wave damage(s); thus, causing economic and social disruption following harsh and inclement weather events. Further, damages to corals could also affect the ecosystem. Fortunately, the chemistry of ocean acidification is predictable, which often may allow us to anticipate its effects, and, accordingly, planning of management programs could be done. These may protect the country's economic interests as well as may provide strategies, which could be helpful for other nations of the world.

Concomitant to the above, fishery management and conservation studies should also enable sufficient proportions of noncommercial species to survive in the changing ocean chemistry and any ensuing ecological shifts, so that, fundamental ecosystem functions and services are preserved. Thus, by following a careful approach to management, fishing pressure reduction and environmental stress mitigation should begin before the effects of ocean acidification on marine resources become obvious, effective, and irreversible. Incidentally, the consequences of a careful approach may temporarily reduce fishery revenues in the short term; nevertheless, such a conservation action plan may, in fact, result in bigger fish stocks and increased revenues in the long run. Hence, adjustments of fishery management plans are, preferably, to take into account not only the economic considerations, but also the biological conservation goals and social outcomes, like community participation, preservation, etc. In fact, an "objectives-based" approach to address and take care of ocean acidification could help in balancing both ecosystem and social objectives along with proper management and conservation of the resources.

The ecological and economic impacts of the 2012 marine thermal wave in the United States revealed the need for planning for adaptation to (altered) climatic conditions within fisheries management. In fact, it was possible to identify many directions based on the 2012 experience, which could assist in sustaining the fisheries and fishing communities in a warming and more variable climate. Incidentally, both, improved capacities to detect and predict ecosystem impacts of climate change as well as nimble fishery management systems that may adapt to ecosystem changes, could be important.

Consequent upon generous funding availability from different national and international organizations, a considerable amount of progress has been made over recent decades in refining models of the climate system. However, there had not been much satisfactory progress in modeling the impacts of climate change (including gradual warming of the aquatic domain and the associated events thereof) on the ocean ecosystems, fishery resources, and human communities. Further, such works are needed to be done on the coupled models that link physical alterations to biological outcomes and economic impacts at a variety of spatial and temporal scales. Such contributions could, no doubt, help the fishery managers in identifying and evaluating the adaptation strategies due to climate change.

The targeted forecasting models could be valuable for supporting fishery management decisions, in addition to the general models, in the context of issues related to climate change. In fact, several biological responses to the 2012 warm conditions could have been predicted using data that are routinely collected by ocean observing systems. For instance, the timing of the lobster inshore migration and molting is closely related to temperature, and observing systems are supposed to routinely monitor the coastal temperatures. Incidentally, the models which integrate multiple data systems to anticipate species distributions, migrations, or other biological conditions in near-real time are few and far between.

Concomitant to the above, a cellular automata model was generated for mimicking possible distributional patterns and different management alternatives under varying assumptions on the fleets' fishing aptitude. In the study, it was revealed that fisheries management and fishing aptitude, including technological development(s) and local knowledge, turned out to have the greatest impact on the spatial distribution of the fishing effort when comparing the IPCC's SRES A1B scenario with repeated sequences of the current environmental situation over 45 years. In both cases, the highest profits in the simulation period of 45 years were obtained at low exploitation levels and moderate fishing aptitude.

Further, researches conducted with regard to climate change and fisheries management often tried to infer the expected economic consequences of climate change in different regions, notably, the Baltic Sea using different types of models, for example, the dynamic age-structured bio-economic model that takes species interactions into account. In such exercises, various management objectives are analyzed under different management scenarios, mainly to see how a manager would try to react to such changes. However, other models, viz., the baseline model, global climate model, etc., are used for various purposes (Nakićenović, 2000).

Several fishery management frameworks are often designed based on previously observed biological patterns. In fact, the status of commercially fished stocks is usually assessed against biomass levels over the past 20–30 years. In addition, stock boundaries and protected areas are established based on the distribution of species when they were originally designated. Further, temporal spawning closures are timed to historical spawning periods, and fishing seasons are set to open after a certain reproductive threshold is likely to be met or when product quality is expected to be superior. However, these basic biological patterns and events may not be stationary and are expected to alter in response to climatic conditions. Further, in the event of adaptation of fisheries management aspects to climate change could need that management tools may suitably incorporate the aspect of nonstationarity due to climate-driven physical variability and biological changes.

Further, the start-up of a new fishery could pose challenges to management agencies, which may not immediately have the authority over or may not be prepared to monitor or regulate a stock in a new area. There could be need for amending the existing rules and policies and governance arrangements (including both domestic jurisdictions and international agreements on transboundary stocks) in order to facilitate access, monitoring, and regulation of fisheries as distribution patterns of species shift.

Notwithstanding the above, studies had tried to peruse the potential future outcomes for *c.* 915 fish stocks across the globe under alternative management and climate scenarios. There had been attempts to model the impact(s) of climate change(s) on fish productivity and the range of geographical distribution, which could portray the availability of the school of fishes and where they could be caught, under four climatic projections. These climatic projections range from a global temperature rise of 1°C (strong climate mitigation) to a rise of 4°C by 2100 (disarrayed business-as-usual). For each of these simulated climatic scenarios, in the bio-economic modeling approach, there have been efforts to examine the future biomass, harvest, and profit(s) under alternative management approaches using the climatic projections.

Studies depicted that *c.* 50% of species examined shifted across national boundaries, and almost all species are expected to experience changes in productivity in response to rising ocean temperatures. These alterations, no doubt, present new challenges for fishing nations. Researchers found that the implementation of management practices that take care of alteration(s) in productivity and range of geographic distribution, may lead to gains on a global scale, as compared to the harvest of the available biomass of today. These practices range from flexible management strategies (which include responsible harvest policies and take care of changing stock productivity) to the development and improvement of existing governance institutions to deal with various kinds of *modus operandi* including multilateral fishery agreements, and so on.

Notwithstanding the above, various outcomes differ locally and regionally, although improved management yield augmented and better global outcomes. The results further indicated that future profits in the fishery dwindle in tropical latitudes even with the management that fully adapts to the challenges of climate change. This means that the nations on and around the equator face the brunt of the impacts, and many of these nations have developing economies, and are highly dependent on seafood as a source of nutrition and avocation. In addition, harvests and profit(s) from fisheries depend on how much planetary warming occurs, which lead to a significant difference in the abundance.

In this connection, it may be noted that there is many scientific databases to highlight the necessity for implementing climate change mitigations and adaptation policies to lessen its impact(s) on aquaculture and fisheries.

Notwithstanding the above, a new study by Environmental Defence Fund (EDF) is portrayed that there could be significant increases in food, fish, and economic activities by tackling sustainable fisheries management and climate change together. However, the nations are to act promptly in order to materialize these gains. Further, the works revealed that the globe's seas have the potential to be significantly more abundant with resources than it is today, although climate change(s) are *in vogue*. However, good management practices are to be put in place and warming is restricted to not more than 2°C.

Better fisheries management helps to offset the negative effects of climate change. If we, with our own initiative, try to limit global temperature rise *vis-a-vis* implement effective fisheries management, the world's oceans still have the potential to be significantly more plentiful in the future than they are today.

Notwithstanding the above, it is opined that the expected effects of climate change on our oceans are broadly negative; nevertheless, still, there are enough opportunities to turn the tide and make a more bountiful future. In fact, with concerted and adaptive responses to climate change, the world's oceans could make more abundant fish populations, more food for human consumption, and more profit for fishermen, despite the negative impacts of climate change. Conversely, studies cautioned that inaction with regard to fisheries management in order to tackle climate change could mean even more significant losses of fish and the benefits that they provide to people. Scientists from the United States and Japan claimed that there have been studies which had tried to examine future fishery outcomes under both climate change projections and alternative human responses. Studies indicated that our oceans could be highly productive for decades to come if actions are taken now to put effective and forward-looking management practices in place.

Concomitant to the above, it is asserted that cooperation among the nations is important for ensuring future fisheries benefits as stocks could shift across management boundaries. It is because rising ocean temperatures may send fish stocks beyond their traditional geographical ranges as they track their preferred and optimum thermal habitats.

Thus, such shifts could undermine previously effective and well-designed management approaches. Incidentally, there could be incentivization of overfishing in the nations which have access to the shifted fish stocks. This, consequently, may weaken existing fishing agreements.

Moreover, even with the right management changes, there could be winners and losers and, humanity has to tackle all these. In fact, reduction in emissions coupled with multilateral cooperation and real changes in fisheries management could pave the way to success. The burgeoning growth of the global population coupled with the growing needs of fishes has led to increased aquaculture to find an answer to the diminishing protein supply in our nutrition as well as for an avocation. Hence, alterations in the management process could be critical for meeting the United Nations Sustainable Development Goals. Concomitantly, the impacts of inaction are also clear. Innumerable people depend on fish as their primary source of protein. At the same time, many fishing nations are not responding fast enough to make alteration(s). Moreover, successful transboundary management programs are also comparatively rare. But proper action does not take a long time to have a good impact on some species. Studies revealed that many fisheries could bounce back from overfishing in 10 years under the correct management policies.

Notwithstanding the above, climate change is expected to have brunt in many places where fisheries are already poorly managed, and matters are likely to get a lot worse if we do not act properly. Incidentally, inaction could bring about increased conflicts as fish move into new waters, along with dangers to food security in some of the world's most vulnerable places. In fact, the fishermen would be among the most affected groups due to climate change. No doubt, the window is narrow, but we have the tools and a clear road map to build a future with more fish, more food, and more prosperity; provided if we act in the right direction now.

6.5 The economic aspects

6.5.1 Introduction

The crucial role which natural systems play in underpinning economic activity and anthropocentric well-being is of growing concern as evidence mounts of the increasing pressures being placed on such systems by human activities (CBD, 1994). One reflection of that concern is the recent undertaking of major global assessments of the status of the services provided by ecosystems. Economic analysis is an increasing feature of such undertakings and has prompted a rapidly expanding literature regarding the implementation of such analyses. Various kinds of works contribute to this literature in many ways. Incidentally, there have been general frameworks and nomenclature for integrating economic analyses within ecosystem service assessments. Given the inherently interdisciplinary nature of such undertakings, there have been reviews of some of the basic principles of economic analysis so as to introduce these to a natural science audience. Concomitantly, attempts have been made to address certain key challenges that economists may have to face in order to adequately represent the complex nature of ecosystem services within economic analyses.

There have been works contributing to the expanding literature on ecosystem service assessment by considering its integration with economic analyses of such services. The *modus operandi*, in some cases, included an initial consideration of a single period during which ecological stocks are maintained at sustainable levels, followed by focussing upon analyses for future-orientated policy and decision-making. Later, the flow of ecosystems services and their contribution to welfare-bearing goods (notably, the fishes) is considered.

There had been methods for valuing resultant benefits through reviews followed by illustrations by case studies of land-use change. Further, there had been efforts to broaden the time horizon to discuss the aspects of future costs and benefits. Lastly, the sustainability assumption of the researchers had been relaxed for the incorporation of depleting ecological assets with a particular focus upon stocks which exhibit thresholds, below which, restoration is uncompromisable.

Notwithstanding the above, certain types of communications have a direct empirical context, as it provides much of the economic methodology underpinning the ongoing UK National Ecosystem Assessment (UK NEA). This exercise, coordinated by UNEP-WCMC, is a direct successor to the UN Millennium Ecosystem Assessment, adapted for the specific context of the United Kingdom, and taking on a number of conceptual and scientific advances which have arisen since 2005. The MA conceptual framework linked primary "supporting services" (e.g., soil formation) to "provisioning" (e.g., food production), "regulating" (e.g., climate), and "cultural" (e.g., recreation) services. Its findings not only demonstrated the importance of ecosystem services to human well-being, but also shown that, at global scales, many key services are being degraded and used unsustainably. The MA challenge was taken up by the UK House of Commons Environmental Audit Committee (2007) which recommended that "ultimately the Government should conduct a full MA-type assessment for the UK to enable the identification and development of effective policy responses to ecosystem service degradation."

Ecosystem service assessments and accompanying economic analyses can be roughly divided into following two types:

(i) "Sustainability analyses" are typically focused on changes occurring up to the present day and involve assessing whether or not the observed development path of an economy has been progressively running down the natural asset base which underpins its viability. Such analyses provide important early warning signals for non-sustainable growth patterns.
(ii) "Programme evaluation" analyses provide forward-looking, policy-relevant assessments, and comparisons of available alternative future development strategies. Such analyses are typically implemented through assessments of one or more future scenarios, the drivers of which include forecasts of environmental change, trends in domestic and world markets, and potential (and possibly dynamically compensating) policy shifts.

Notwithstanding the above, many assumptions are generally made. The initial assumptions hold constant two complicating factors, viz.,

(i) the passage of time and
(ii) the sustainability of stocks of ecosystem assets.

The **former** issue is neutralized by initially focussing on a given scenario operating in a single period, across which it is assumed that individuals are indifferent as to when flows are received. While considering the **latter** issue, it is often noted that the ecosystem services are flows generated by the stocks of ecosystem assets. Various methods for isolating the ecosystem service elements within the production of such goods are generally considered before judging the various methods used to value ecosystem services. These instances serve to underline the need to ensure that economic analyses recognize the inherently spatial nature of most ecosystem services.

The Millennium Ecosystem Assessment defines "ecosystem services" as "the benefits accrued by the people from the ecosystems. There have been efforts to add to this definition to propose that "ecosystem services are the aspects of ecosystems utilized (actively or passively) to produce well-being for the humanity." Incidentally, it is the human focus that necessitates the integration of economic analysis within such assessments, so that we can quantify and value ecosystem services ensuring that their significance and worth could be incorporated within the decision-making process. In fact, decision-making through economic valuation could be a reflection of human preferences. It may be noted that the economists may refer to flows as income and stocks as wealth, the true intergenerational value of which is referred to as "inclusive" or "comprehensive" wealth. Thus, from an economic perspective, ecosystem services could be regarded as those contributions of the natural world, which generates goods which people value.

Notwithstanding the above, the following three forms of ecosystem service benefit have been recognized:

(i) "goods" (e.g., products obtained from ecosystems, such as resource harvests, water, and genetic material, etc.);
(ii) "services" (e.g., recreational and tourism benefits or certain ecological regulatory functions, such as water purification, climate regulation, and erosion control); and
(iii) "cultural benefits" (e.g., spiritual and religious beliefs, heritage values).

Concomitant to the above, at this stage, certain terminologies are to be clarified. These are given as follows:

(i) "Ecological assets" are the stocks of potential services that the ecosystem provides being conditioned by structure and processes. In economic terms, we can think of these as the "wealth" of the ecosystem.
(ii) "Ecosystem services" refer to the flow of services (outcomes of structure and processes) provided by ecological assets in some assessment period.

Incidentally, accurate quantification of natural assets and services is a prerequisite for any credible ecosystem assessment. Further, there have been attempts to provide value estimates for the total stock of all ecosystem services globally. Concomitantly, a safe minimum standard (SMS) can be thought of as a precautionary approach to the management of natural asset.

Concomitant to the above, as compared to today's scenario, the estimated future global outcomes include a *c.* $14 billion USD increase in profits, *c.* 25 billion additional servings of seafood, and *c.* 217 million more metric tons of fish in the sea, which could be approximately a third more fish than which exist today, if the imperative of the Paris Climate Accord is fulfilled. At the same time, it ensures that global temperatures do not rise >2°C. However, these expected results depend on implementing fisheries management practices that are take care of climate-driven alterations in species productivity as well as the range of geographical distribution. Also, the warmings from the emissions are to be limited to an optimum level. Inaction on such grave issues on fisheries management and climate change could mean net losses of fish as the planet's population grows.

Concomitant to the above, in a case study of US commercial fishery revenues, there have been efforts to constrain the economic effects of ocean acidification over the next half a century. Incidentally, it may be noted that the $3.8 billion US annual domestic ex-vessel commercial harvest finally contributed $34 billion to the US gross national product during 2007. In this connection, it may be noted that the molluscs had contributed to *c.* 19%, or $748 million, of the ex-vessel revenues in a year. Incidentally, it may be important to state that there could be sufficient amount of revenue declines, layoffs in employments, and, at the same time, indirect economic costs may occur, if ocean acidification broadly damages marine habitats, alters marine resource availability, and disrupt other ecosystem services. Efforts had been made to review the implications for marine resource management and propose possible adaptation strategies designed to support fisheries and marine-resource-dependent communities, many of which have already possessed little economic resilience.

Notwithstanding the above, nationwide, income, and jobs generated by **US fisheries** were multiplied amazingly from catch to retail sale. For instance, in 2007, the domestic commercial fisheries' harvest from outside US territories and aquaculture were provided a primary sale value of *c.* US $5.1 billion. On the other hand, processing, wholesale and retail activities had resulted in the sales of *c.* US $68.3 billion, contributing to *c.* US $34.2 billion in value added to the US gross national product in 2007. However, the number of individuals employed directly and indirectly by commercial fishing bodies is difficult to quantify, because fishermen are frequently self-employed. Further, the middlemen who may not always handle the ocean products solely are not counted in the surveys of the industries. Nevertheless, commercial fish processing and whole selling together had supported *c.* 63 000 jobs in the United States in 2007. Further, in 1999, commercial fishing was employed *c.* 10, 500 people in the New York State; wholesale and processing had supported *c.* 5060 jobs and retail sales were supported an additional *c.* 10,100 jobs. Concomitantly, the seafood sales at New York restaurants and cafeterias had supported an equivalent of *c.* 70, 000 full-time jobs. Hence, in total, the seafood industry was supported *c.* 100, 000 jobs in the New York State alone (New York Sea Grant 2001).

Concomitant to the above, supplementing the economic benefit from commercial fishing, the US **recreational fishing** encourages expenditure on permits, equipment, and travel, and also funds to support industries, and thereby generating jobs, profits, tax revenues, and trade to trade revenue. For instance, the recreational saltwater fishing had generated *c.* US $12 billion of income in the United States during 2000, and, this was supported *c.* 350,000 jobs, for a total economic benefit of *c.* US $43 billion that year. However, the total impact of ocean acidification is not yet known for every commercially and recreationally valuable species; nevertheless, emerging data suggest that the number or quality of many high-value, aragonite-forming molluscs decreases, and this is followed by declining economic revenues in that fishery sector. This was supported by various works, notably, on dwindled molluscan populations in acidified ecosystems. There have been further works with regard to malformation on juvenile oyster shells in aragonite-undersaturated laboratory studies, and decreased survival of oyster larvae in upwelling Oregon seawater with decreased pH and altered biogeochemistry. In contrast to the above, the molluscs and the crustaceans comprise the bottom or middle trophic levels of many ecosystems. This could imply that acidification-related damage to either of these groups may impact their primary and secondary predators negatively.

A fish lives in water, whether FW or marine. As such, any change in the physicochemical and biological characteristics of the aquatic domain is destined to affect these coveted aquatic denizens. Further, both FW or marine acidification may affect mankind through a variety of socioeconomic linkages, potentially starting with reduced harvests of commercially important ichthyospecies. In this connection, it may be noted that the total ex-vessel or basic primary value of US commercial harvests from US waters and at-sea processing was *c* US $4 billion in 2007. Out of the total, the molluscs were accounted for *c.* 19%, crustaceans *c.* 30%, and finfish *c.* 50%, and, *c.* 24% of total US ex-vessel revenue was harvesting fish that prey directly on calcifiers. However, different groups seemed to dominate regional revenues, e.g., the molluscs were more important in the New England and mid-to-south Atlantic regions; crustaceans had contributed greatly to New England and Gulf of Mexico fisheries; and, the predators had dominated the Alaskan, Hawaiian, and Pacific-territory fisheries.

6.5.2 Concluding comments

There are an expanding literature and interest in the concept of ecosystem services as an underpinning guide for decision-making. Various attempts were made to contribute to this literature by providing an overview of issues arising from the economic analysis of such ecosystem service assessments. The initiation is by considering a single assessment period, during which, ecosystem assets are used sustainably. With these assumptions in place, discussions could continue by developing a terminology for future applications and employ this to focus on the flow of ecosystem services and their role in generating welfare-bearing goods and resultant benefits. There could be distinction(s) between sustainability analyses of historic trends and assessments of alternative policy options through scenario analyses. Economic analysis of the role and value of ecosystem services starts through isolating their contribution to welfare-bearing goods. This contribution is then valued through the application of a range of methods and techniques ranging from adjustments of market prices to the measurement and valuation of preferences for nonmarket goods. The preceding principles are illustrated through a case study focusing on the economic value of the ecosystem services and associated goods arising from land-use change. As part of this illustration, the central concepts of marginal analysis and substitution effects are considered within economic assessments, and the single period assumption(s) are relaxed to discuss the incorporation of multiple period costs and benefits within the assessment of shadow values. This process introduces the concept of discounting. Ultimately, the assumptions are relaxed, considering that assets are necessarily being managed sustainably. Attempts are made to incorporate such problems within the economic analysis.

Notwithstanding the above, there have been discussions at various platforms regarding the **economic and policy issues related to the impact of climate change on fisheries.** In this connection, it may be noted that climate change may cause physical and biological impacts on the fishes, which, in turn, may affect fishery economic performance. This, further, in turn, may influence management effectiveness and policy issues. There have been discourses on the linkages between these issues.

Concomitant to the above, there have been works reflecting significant community shift, which is sometimes observed was likely to be a function of multiple ecosystem factors and not just declining calcification or organisms' health. Furthermore, biological studies have not yet quantitatively and fully identified ameliorative long-term processes that could offset losses, like the natural selection of resistant species or strains, or initiation of self-defensive strategies, and so on. In addition, for the economic projections, there have been no assumptions about changes in fishing intensity or the effects of supply and demand on marine resource prices. Rather, the potential revenue losses from decreased molluscan harvests in the future are calculated, and efforts are made to adjust to recent values. However, anticipated future revenue losses may worth less than losses today because of the compounding effects of interest and capital return rates.

Notwithstanding the above, ecosystem processes, such as productivity and decomposition rates are always show a positive relationship with biodiversity measured as functional trait diversity, although the relationship generally saturates at some point.

However, empirical and theoretical studies indicate that, in the case of long-term environmental change or a constantly fluctuating environment and to realize the full insurance value of biodiversity, much less loss of biodiversity can be tolerated. Finally, in the case of ecosystem service, such as the nonuse values held for wild species diversity and potential benefits from bioprospecting, any loss of diversity in species or genetic terms represents a significant loss.

Given the wealth of issues which have to be embraced within an economic analysis of ecosystem services discussions could be deliberately uncritical. However, it is concluded by briefly covering some of the major problems facing different issues. A major issue concerns the substantial gaps in natural science knowledge regarding the processes and interactions through which ecosystem services, are provided and maintained. Incidentally, one of the most significant problems facing ecosystem service assessments is the ignorance and uncertainty concerning ecosystem functions.

Notwithstanding the above, the past century of natural science research had witnessed the development of disciplinary expertise, which, although vital in facilitating a deep understanding of individual systems had inadvertently discouraged the cross-disciplinary cooperation and holism necessary to comprehend the interplay of ecological systems. However, a marked reversal of this trend was evident, of late, with interdisciplinary research being brought to the fore. However, still, the interactions between systems are relatively poorly understood. Nevertheless, this knowledge is stretched to its limits when we consider the system interactions within the context of a general setup and a stressor, such as global climate change.

Notwithstanding the above, concerns and gaps exist within the economic literature. Incidentally, interdisciplinary research, collaborating economics with other social sciences has advanced significantly over recent decades.

Incidentally, it may not be surprising to assume that certain natural scientists may not have much experience as well as may not have the informational requirements for economic analysis (e.g., economic valuation process; damage and threshold analyses for resilience measures; etc.).

In continuation of the above, the issues related to various aspects of the environment and climate change are not fully settled. Nevertheless, the importance of incorporating ecosystem services within policy appraisal and decision-making techniques with regard to the management of the environment and its resources (notably, fishes) may be underestimated but are to be dwelt upon within a community management perspective.

6.6 Thoughts on preventive actions

Fisheries provide a vital source of food for more than half of the world's population and support the livelihoods of > 56 million people globally. Although climate change alters the productivity and distribution of marine fisheries, climate-adaptive fisheries management could mitigate many of the negative impacts on human society.

Beyond improving information about how climate-driven physical changes could impact the ocean ecosystems, it may be also necessary to (i) willingly identify ways in which fishery management infrastructures and processes are required to adapt to future climatic conditions and (ii) confirm that fishery policies facilitate these alterations.

Climate change has been altering the productivity and distribution of FW and marine fisheries. However, climate-adaptive fisheries management could mitigate many of the negative impacts on human society. There have been forecasts of global fisheries biomass, catch, and profits up to 2100 AD under three climate scenarios and five levels of management reforms to:

(i) determine the impact of climate change on national fisheries and
(ii) quantify the national-scale benefits of implementing climate-adaptive fisheries reforms.

Management reforms dealing with shifting productivity and shifting distributions could yield higher catch and profits in the future relative to today for *c.* 60%–65% of countries under the two least severe climate scenarios, but for only *c.* 35% of countries under the most severe scenario. Furthermore, these management reforms could yield higher cumulative catch and profits than business-as-usual management for approximately all countries under the two least severe climate scenarios but yield lower cumulative catch for *c.* 40% of the countries under the most severe scenario. Fortunately, perfect fisheries management may not require achieving these benefits. Transboundary cooperation, with 5-year intervals between adaptive interventions could result in comparable outcomes.

However, the ability for realistic management reforms to offset the negative impacts of climate change is bounded by changes in underlying biological productivity. Although realistic reforms could generate higher catch and profits for *c.* 23%–50% of the countries experiencing reductions in productivity, the remaining countries could need to develop, expand, and reform aquaculture and other food production sectors to offset losses in capture fisheries. Still, climate-adaptive management is more profitable than business-as-usual management in all countries. Incidentally, guidance may be provided on implementing and achieving the benefits of climate-adaptive fisheries reform along with a gradient of scientific, management, and enforcement capacities.

Notwithstanding the above, the harvest rate is based on the following five management scenarios:

(i) business-as-usual (i.e., no adaptation),
(ii) productivity shift adaptation,
(iii) range shift adaptation,
(iv) full adaptation, and
(v) "realistic" adaptation.

As mentioned, **Business-as-usual** could be regarded as no adaptation. **Productivity shift adaptations** improve fisheries management by implementing a dynamic, economically optimal harvest policy under the given current biological conditions, which optimally adjusts harvest mortality on the basis of available biomass and is, therefore, naturally adaptive to climate-driven productivity changes. **Range shift adaptations** assume that transboundary cooperation results in the maintenance of management, rather than the degradation of management to open access, as stocks shift across boundaries. **Business-as-usual management** fails to implement either adaptation. It maintains current harvest rates for species that do not shift spatially, while management degrades to open access for stocks that shift across boundaries. Concomitantly, **full adaptation** assumes that both challenges are addressed: the dynamic economically optimal harvest policy is implemented and maintained even as stocks shift across boundaries. Realistic adaptation refines the full adaptation scenario by implementing productivity shift adaptations at possible management intervals. It determines the economically optimal harvest rates on 5, 10, or 20-year intervals and tries to maintain these rates until the next management intervention.

Furthermore, acquiring the benefits of climate-adaptive fisheries reform may require accounting for shifting productivity and distributions along a gradient of scientific, management, and enforcement capacities. However, some countries may lack the monitoring programs required to detect and describe shifts in distribution and productivity; the scientific capability for conducting either climate-agnostic or climate-adaptive stock assessments; and the management capacity for setting and enforcing the fisheries regulations [25,33,34]. This is frequently the case for the tropical developing countries, which have to forecast the weather conditions in the event of experiences of the greatest losses in fisheries catch and profits under climate change. They also exhibit the greatest vulnerability to these reductions in food and income [35]. The tools for enacting climate-adaptive fisheries reforms and acquiring biological and socioeconomic resilience to climate change may have to cover this gradient of capacity.

Adapting to climate change is supposed to require dynamic, flexible, and forward-looking management. This may be achieved by aligning management policies with the spatiotemporal scales of climate change, ecosystem change, and socioeconomic responses. In systems of higher capacity, this could involve four broad strategies, which are given as follows:

- **(i)** Managers could envision and may prepare for alternative futures using tools, such as forecasts, structured scenario planning, holistic ecosystem models, risk assessments, and climate vulnerability analyses.
- **(ii)** The collection of near-real-time biological, oceanographic, social, and/or economic data may be done for proactive and dynamic adjustments in spatial and temporal management actions.
- **(iii)** Developing harvest control regulations, which may account for or are competent in changing environmental conditions affecting productivity, may help to increase catch while also reducing the probability of overfishing.
- **(iv)** All the abovementioned management *modus operandi* may be simulation tested through management strategy evaluations (MSE) in order to measure the efficacy of alternative strategies and their competence under different climate scenarios. Incidentally, in the lower capacity systems, forward-looking fisheries management could include precautionary management to buffer against uncertainties as well as management strategies that try to conserve population resilience, age structure, and genetic diversity.

In fine, it is opined that there is need for adopting the following measures for developing management-related forecasting capability: (i) maintenance and augmentation of observing programs which track changes in physical conditions, biological attributes, and fisheries activities and (ii) integrating these individual databases into operational models of biological and fishery responses to physical changes. Such steps could lead to long-term sustained fisheries in the world.

6.7 Summary

(1) Our lives are inseparably connected to the climate of the **Mother Earth**.

(2) The warming of the Earth has been the subject of intense debate and concern for many scientists, policymakers, and citizens for the past few decades. In the 21st century, warming of the earth may be attributable largely to human activities.

(3) "Weather" is the state of the atmosphere: its temperature, humidity, wind, rainfall, and so on; usually, over a short period ranging from hours to weeks. "Climate," in its broadest sense, is the statistical description of the state of the climate system over quite a long period.

(4) **Climate change** refers to any significant **change** in the measures of **climate** lasting for an extended period.

(5) It may be mentioned that human activities are influenced climate by changing concentrations of CO_2, methane, and nitrous oxide in the atmosphere as well as by some chemically manufactured greenhouse gases, such as halocarbons, etc. Climate change is further influenced by altering the concentrations of aerosols and by altering the reflectivity of Earth's surface by changing land cover. Climate change, of late, seems to be a menacing issue in the globe affecting both the abiotic and the biotic domains. Extensive burning of fossil-fuel coupled with extensive deforestation over the last few centuries had been enhancing the level of atmospheric CO_2

(6) It became a surprising event when climate change was visibly evident in America in 2012. A record heatwave affected many regions of the United States in 2012. Moreover, the so-called superstorm sandy had pounded the North-Eastern (NE) region of the United States.

(7) Climate change was also seemed to be evident in other parts of the globe, notably, in Asia and, particularly, in the Indian subcontinent, where the shifting of seasonal durations seemed to be evident, at least to some extent.

(8) An enormous threat to the ecology and biodiversity of the planet in the coming decades will be the global **climate** disruption due to the buildup of human-generated GHGs in the atmosphere, with the burgeoning population. The expected population increase from approximately 1.6 billion to 6.1 billion could increase GHG emission 12-fold. Further, there may be impacts of over-farming, deforestation, and water **pollution** to eutrophication and global warming.

(9) Fish is an essential and significant entity of nutrition and avocation for the people of the globe. Even a small quantity of fish could bring about improvement(s) in peoples' diets, as they could provide vital nutrients which may be absent in typical starchy staples that dominate poor people's diets.

(10) Notwithstanding the above, there were attempts for reviewing the potential impacts of climate change on fisheries, which had suggested a role of public policy in adaptation; possibility to reduce vulnerability; and, further, to provide information for planning and stimulating adaptation as well as to ensure that, adaptation actions may not negatively affect other ecosystem services and the viability of fisheries in the long run.

(11) Incidentally, food fish production is classified into two subsectors, viz.,

  (i) the capture fisheries, which overtly include the naturally occurring and recruited wild populations, the great bulk of which (*c.* 85%–90%) are in the FW and the seas; and

  (ii) the culture fisheries, which include the farmed food fish subsector, which is achieved relatively growing significance and is popularly called "aquaculture."

  Concomitant to the above, various kinds of research works are found addressing the potential impacts of climatic change on aquaculture. Concomitantly, efforts are being made to reveal the contribution of aquaculture (AQC) to the total food fish supply, and efforts to address the potential adaptations and mitigation.

  In some of the main fish producing countries, the AQC sector contributes significantly to the gross domestic product (GDP) accounting for *c.* 45% of total food fish consumption. The burgeoning growth of the human population, vis-a-vis stagnation in capture fishery, warrants the augmentation of AQC production.

(12) Climate change may show different types of impacts in different culture systems in FW, marine, brackish water, etc. Incidentally, *c.* 65% of the aquaculture production is inland, and, also, concentrated mostly in the tropical and subtropical regions of Asia. Further, it is likely that diseases affecting aquaculture could be aggravated both in incidence and in impact due to climate change. In marine cage culture, the introduction of improved technologies to withstand extreme weather events could be a significant adaptive measure. Further, in the event of calamities, AQC could provide alternative livelihoods to the agriculturists. While considering the contribution of AQC to carbon emission, on the whole, AQC is often less energy costly, and could rather contribute to carbon sequestration more than other terrestrial farming systems.

(13) FAO may take a lead in informing fishers and policy-makers about the likely consequences of climate change for fisheries and aquaculture. Thus, having been convened to identify and review the key issues of climate change in relation to fisheries and aquaculture, the Expert Workshop of FAO discussed threadbare the impacts of climate change on aquatic resources and ecosystems.

(14) Further, **climate change could have a cascading effect on global fisheries and their management**. Thus, from an economic point of view, the economic and policy issues related to the impact of climate change on fisheries are to be dealt with urgently.

(15) The oceanic uptake of anthropogenic CO_2 occurs through a series of chemical reactions that enhance the level of aqueous CO_2, decrease the pH, and carbonate ion level of seawater. Incidentally, the anthropogenic CO_2 have lowered the average surface ocean pH from a preindustrial value of 8.2–8.1 till recently. There could be a further reduction in mean ocean surface pH by an additional 0.2–0.3 units.

(16) Climate change may expand the range of many warm water ichthyospecies toward the poles, while it may contract the range of cold-water species. However, the pelagic species may exhibit fast changes and also portray vertical movements to counteract surface warming. Some of the migratory fishes follow the trends in ocean water temperature and their process of migration could be preponed or postponed during summer or winter, respectively, depending on the thermal regime of water.

(17) The El Niño events have three major impacts in coastal upwelling systems, which are given as follows:

  (i) they increase coastal temperatures;

  (ii) they reduce plankton production by lowering the thermocline (which inhibits upwelling of nutrients); and

  (iii) they alter trophodynamic relationships (Lehodey et al., 2006).

(18) Sea level, globally, has been rising at an average rate of 1.8 mm per year since 1961, threatening many low altitude regions. This rate has increased from 1993 to about 3.1 mm per year due to the melting of mountain glaciers and declines in snow cover in both the hemispheres as well as losses from the ice sheets of Greenland and Antarctica. Increased variability in sea level, which is evident in recent decades, could be consistent with more frequent and intense El Niños. However, the global average sea-level rise could range between 0.18 m and 0.59 m by the end of the 21st century (2090–99; relative to 1980–99) based on model-based projections.

(19) All coastal ecosystems are vulnerable to sea-level rise. Also, there may be more direct anthropogenic impacts on coral reefs and coastal wetlands including salt marshes and mangroves. Further, key human vulnerabilities to climate change and sea-level rise exist, where the stresses on natural low-lying coastal systems coincide with low human adaptive capacity, e.g., the deltas, especially Asian mega-deltas (for instance, the deltaic region of the River Ganges in the Ganges-Brahmaputra-Meghna basin in India and Bangladesh); low-lying coastal urban areas, especially areas prone to natural or human-induced subsidence and tropical storm landfall (e.g., New Orleans, Shanghai); small islands, especially low-lying atolls (e.g., the Maldives) (Nicholls et al., 2007a,b); etc.

(20) It is believed that approximately half of the CO_2, released by human activities between 1800 and 1994, had been stored in the ocean (Sabine et al., 2004), and *c.* 30% of modern CO_2 emissions are taken up by oceans today (Feely et al., 2008). Incidentally, continued uptake of atmospheric CO_2 had lowered the pH of surface seawater by 0.1 units during the last 200 years, with further possible pH reduction in the range from 0.3 to 0.5 units over the next 100 years and from 0.3 to 1.4 units over the next 300 years, depending on the CO_2 emission scenario (Caldeira and Wickett, 2003). In fact, there could be greater impacts of these changes in some regions and severe for shell-borne organisms (Orr et al., 2005).

(21) Some amount of GHGs, which are emitted through fisheries activities also contribute to climate change both during capture operations and subsequently during the transport, processing, and storage of fish. Various works on fisheries' contribution to climate change have concluded that the minimal contribution of the sector to climate change does not warrant much focus on mitigation.

(22) Climate change impacts may cause altered levels of precipitation; decrease in the volume of rainfall; and, reduced scopes for fishing, fish farming, and aquaculture, whereas fisheries and aquaculture could have been potential sources of rural livelihood incomes. Incidentally, thus, there could be **reduced diversity of rural livelihoods**, and more reliance on nonfarming income(s). Further, population **displacement** from the coastal areas could cause an influx of new fishers, while more droughts and floods may cause damage to productive assets, like, fishponds, weirs, rice fields, etc. as well as homes.

(23) The term "**bioclimate envelope**" is used to define the interacting effects and limits of temperature, salinity, oxygen, etc. on the performance and survival of species. Such bioclimate envelopes could be used to model changes in species distributions and abundance patterns as a result of climate change.

(24) There may be enhancements in **Species invasions and diseases** due to the impacts of climate change.

In addition to incidences of *Vibrio parahaemolyticus*, there could be the development of the multinucleated spore unknown (MSX) disease caused by the protozoan pathogen, *Haplosporidium nelsoni* (Hofmann et al., 2001) and there could be increased in the occurrences of harmful algal blooms (HABs) globally due to anthropogenic influences (Smayda, 1990). Incidentally, climate change may have some amount of role in the outbreak of the *hitherto* unknown virulent, enigmatic disease, variously called epizootic ulcerative (fish) disease syndrome or EUS pandemic, among the FW fishes in a large portion of the world, which has been sweeping the FW fishes, in epidemic dimension, on a semi-global scale, causing large-scale mortality among them, rendering many of them endangered, devastating the economy of certain countries, causing fear psychosis among the consumers and causing untold misery to the poor fishermen, since the 1980s.

Incidentally, the outbreak COVID19 (Coronavirus) global pandemic during 2019–20 may be linked with some kind of climate change impacts as in the case of outbreak of epizootic ulcerative (fish) disease syndrome (EUS) disease among the fishes, where temperature and total alkalinity (TA) of water had been found to act as predisposing stress factors in the outbreak of EUS.

(25) Climate changes have a range of direct and indirect impacts on marine and FW capture fisheries, with the implications for fisheries-dependent economies, coastal communities, and fisherfolk.

(26) Climate change might have some role in the importance of socioeconomic drivers in Fijian fishing communities due to temperature-induced mass coral mortality events in 2000. Moreover, there were some impact(s) of climate variability on Peruvian Anchoveta fisheries as well as on the fisheries of the Gangetic, Irrawaddy and Mekong Delta in Asia.

(27) AQC is the global practice today. It often helps in carbon sequestration by acting as a carbon sink. In integrated AQC, the stocked fish usually feed on the phyto and zooplankton and benthos, which grow naturally in the pond and their production is also promoted. As such, the pond with phytoplankton acts a **carbon sink** and also helps in **carbon sequestration**. In other words, the AQC-based farming practices are efficient and effective means for recycling biological wastes and they act as **carbon sinks** and help in **carbon sequestration**.

(28) **The culture of zooplankton** under seminatural condition(s) could throw a significant amount of light on feeding the ichthyospecies in culture. The zooplankton acts as the potential natural fish food organisms for the fishes. A significant amount of attempts have been made to culture zooplankton in the laboratory under seminatural conditions (Kar et al., 2018a,b; Das et al., 2018a,b, 2020; Devi et al., 2016; Lambert et al., 2020; Kar, 2019).

(29) Aquaculture is the one of the potential sectors for nutrition and avocation guaranteeing answers to our diminishing protein supply across all communities irrespective of living standards. It is quite likely that aquaculture, in view of its resilience and adaptability, should be able to respond positively to climate change impacts. However, there are needs for related policies to be in place in order to respond to climate alteration impact(s). Aquaculture seems to be capable of helping in lessening the exacerbation of coral reef destruction, and, thereby, also assisting in enhancing the preservation of biodiversity.

(30) Management is a significant aspect of fisheries and aquaculture. There may not be any long-term pragmatic solution in the offing with regard to a true solution or mitigation option for ocean acidification except limiting the CO_2 emissions to the atmosphere generated from the burning of fossil fuel. Management of ocean acidification is a long-term goal, which requires a fundamental reorganization of energy and transportation infrastructures globally.

(31) The crucial role which natural systems plays in underpinning economic activity and anthropocentric well-being is of growing concern as evidence mounts of the increasing pressures being placed on such systems by human activities. Attempts were made to address certain key challenges that economist may have to face in order to adequately represent the complex nature of ecosystem services within economic analyses.

(32) In fine, it is opined that there is need for adopting the following measures for developing management-related forecasting capability: (i) maintenance and augmentation of observing programs which track changes in physical conditions, biological attributes, and fisheries activities and (ii) integrating these individual databases into operational models of biological and fishery responses to physical changes. Such steps could lead to long-term sustained fisheries in the world.

Suggested reading

2WE Associate Consulting, 2000. Aquaculture and climate change in Canada. A discussion paper. Canada, Victoria, BC. 11 pp. Also available at www.cisc.univ.ca/workshop/Aquaculture&climate-in-Cda.htm.

Abell, R., Thieme, M., Lehner, B., 2002. Ecoregion Conservation for Freshwater Systems, With a Focus on Large Rivers. World Wildlife Fund, Washington, DC.

Abery, N.W., Sukadi, F., Budhiman, A.A., Kartamihardja, E.S., Koeshendrajana, S., Buddhiman, A.A., De Silva, S.S., 2005. Fisheries and cage culture of three reservoirs in west Java, Indonesia; a case study of ambitious developments and resulting interactions. Fish. Manage. Ecol. 12, 315–330.

Ahas, R., 1999. Long-term phyto-, ornitho- and ichthyophenological time-series analyses in Estonia. Int. J. Biometeorol. 42, 119–123.

Allison, E.H., Adger, W.N., Badjeck, M.-C., Brown, K., Conway, D., Dulvy, N.K., Halls, A., Perry, A., Reynolds, J.D. (2005). Effects of climate change on the sustainability of capture and enhancement fisheries important to the poor: analysis of the vulnerability and adaptability of fisher folk living in poverty. London, Fisheries Management Science Programme MRAG/DFID, Project no. R4778J. Final technical report. 164 pp.

Allison, E.H., Andrew, N.L., Oliver, J., 2007. Enhancing the resilience of inland fisheries and aquaculture systems to climate change. J. Semi-Arid Tropic. Agricult. Res. 4 (1). Also available at e-journal: www.icrisat.org/Journal/SpecialProject/sp15.pdf.

Anonymous, 2007. Symposium on cod recovery. Friday, 9 March and Saturday, 10 March 2007, Edinburgh, Scotland. (also available at www.nsrac.org/cod-symposium/docs/Cod_Recovery_Symposium_Report.pdf.

Armitage, D.R., Plummer, R., Berkes, F., Arthur, R.I., Charles, A.T., Davidson-Hunt, I.J., Diduck, A.P., Doubleday, N.C., Johnson, D.S., Marschke, M., McConney, P., Pinkerton, E.W., Wollenberg, E.K., 2009. Adaptive co-management for socialecological complexity. Front. Ecol. Environ. 7 (2), 95–102.

Arthington, A.H., Lorenzen, K., Pusey, B.J., Abell, R., Halls, A., Winemiller, K.O., Arrington, D.A., Baran, E., 2003. River fisheries: ecological basis for management and conservation. In: Welcomme, R.L. (Ed.), Proceedings of the 2nd International Large Rivers Symposium. FAO, Rome.

Atkinson, A., Slegel, V., Pakjomov, E., Rothery, P., 2004. Long-term decline in krill stocks and increase in salps within the Southern Ocean. Nature 432, 100–103.

Austin, J., Butchart, N., Shine, K.P., 1992. Possibility of an Arctic ozone hole in a doubled- CO2 climate. Nature 360, 221–225.

Aye, K.M., Ko, L.K., Win, H., De Silva, S.S., 2007. A new fresh water aquaculture practice that has successfully targeted a niche export market with major positive societal impacts: Myanmar. Aquacult. Asia XII (4).

Badjeck, M.-C. (2008). Vulnerability of coastal fishing communities to climate variability and change: implications for fisheries livelihoods and management in Peru. University of Bremen, Bremen, Germany. (Also available at http://elib.suub.uni-bremen.de/diss/docs/00011064.pdf. Ph.D thesis.).

Bakun, A., 1996. Patterns in the Ocean. Ocean Processes and Marine Population Dynamics. California Sea Grant College System, California. p. 323.

Bartley, D.M., Brugère, C., Soto, D., Gerber, P., Harvey, B., 2007. Comparative assessment of the environment costs of aquaculture and other food production sectors: methods of meaningful comparisons. In: Rome. FAO Fisheries Proceedings, No 10, 240 pp.

Bates, B.C., Kundzewicz, Z.W., Wu, S., Palutikof, J.P. (Eds.), 2008. Climate Change and Water. IPCC Secretariat, Geneva. Technical paper of the IPCC. 210 pp.

Beaugrand, G., Reid, P.C., 2003. Long-term changes in phytoplankton, zooplankton and salmon related to climate. Glob. Chang. Biol. 9 (6), 801–817.

Beaugrand, G., Brander, K.M., Lindley, J.A., Souissi, S., Reid, P.C., 2003. Plankton effect on cod recruitment in the North Sea. Nature 426, 661–664.

Behrenfeld, M.J., O'Malley, R.T., Siegel, D.A., McClain, C.R., Sarmiento, J.L., Feldman, G.C., Milligan, A.J., Falkowski, P.G., Letelier, R.M., Boss, E.S., 2006. Climate-driven trends in contemporary ocean productivity. Nature 444, 752–755.

Belkin, I.M., 2009. Rapid warming of large marine ecosystems. Prog. Oceanogr. 81, 207–213. https://doi.org/10.1016/j.pocean.2009.04.011.

Bendiksen, E.A., Jobling, M., Arnesen, A.M., 2002. Feed intake of Atlantic salmon parr Salmo salar L. in relation to temperature and feed composition. Aquacult. Res. 33, 525–532.

Bene, C., 2003. When fishery rhymes with poverty: a first step beyond the old paradigm on poverty in small-scale fisheries. World Dev. 31 (6), 949–975.

Benetti, D.D., Orhun, M.R., Sardenberg, B., O'Hanlon, B., Welch, A., Hoenig, R., Zink, I., Rivera, J., Denlinger, B., Bacoat, D., Palmer, K., Cavalin, F., 2008. Advances in hatchery and grow-out technology of Cobia, Rachycentron canadum. Aquacult. Res. 39, 701–711.

Bibby, R., Widdicombe, S., Parry, H., Spicer, J., Pipe, R., 2008. Effect of ocean acidification on the immune response of the blue mussel, Mytilus edulis. Aquat. Biol. 2, 67–74.

Bindoff, N.L., Willebrand, J., Artale, V., Cazenave, A., Gregory, J., Gulev, S., Hanawa, K., Le Quéré, C., Levitus, S., Nojiri, Y., Shum, C.K., Talley, L.D., Unnikrishnan, A., 2007. Observations: Oceanic climate change and sea level. In: Solomon, S., Qin, D., Manning, M., Chen, Z., Marquis, M., Averyt, K.B., Tignor, M., Miller, H.L. (Eds.), Climate Change 2007: The physical science basis. Cambridge University Press, Cambridge, UK and New York, USA. Contribution of Working Group I to the Fourth Assessment Report of the Intergovernmental Panel on Climate Change.

Blanco, J.A., Narváez Barandica, J.C., Viloria, E.A., 2007. ENSO and the rise and fall of a tilapia fishery in northern Colombia. Fish. Res. 88, 100–108.

Broad, K., Pfaff, A.S.P., Glantz, M.H., 1999. Climate information and conflicting goals: El Niño 1997 to 1998 and the Peruvian fishery. In: New York, USA. Public philosophy, environment and social justice. Carnegie Council on Ethics and International Affairs.

Brook, E.J., Sowers, T., Orchardo, J., 1996. Rapid variation in atmospheric methane concentration during past 110 000 years. Science 273, 1087–1990.

Bunting, S., Pretty, J., 2007. Aquaculture development and global carbon budgets: emissions, sequestrations and management options. In: UK, University of Essex, Centre for Environment and Society, Occasional Paper 2007-1, 39 pp.

CAF, 2000. Las lecciones del El Niño: Memorias del fenómeno El Niño 1997 to 1998: retos y propuestas para la región Andina, volumen V: Perú. Corporación Andina de Fomento.

Caldeira, K., Wickett, M.E., 2003. Anthropogenic carbon and ocean pH. Nature 425, 365.

Carlton, J.T., 2000. Global change and biological invasions in the oceans. In: Mooney, H.A., Hobbs, R.J. (Eds.), Invasive Species in a Changing World. Covelo, CA, Island Press, pp. 31–53.

Catto, N.R., 2004. Impacts of climate variation and adaptation for Atlantic Canadian fisheries communities. In: Moncton, Canada, Canadian Association of Geographers, Annual Meeting, May 2004.

CBD (Convention on Biological Diversity), 1994. Text and Annexes. Interim Secretariat for the Convention on Biological Diversity, Switzerland. 34 pp.

Chavez, F.P., Ryan, J., Lluch-Cota, S.E., Niquen, M., 2003. From anchovies to sardines and back: multidecadal change in the Pacific Ocean. Science 299, 217–221.

Childers, D.L., Day, J.W., Muller, R.A., 1990. Relating climatological forcing to coastal water levels in Louisiana estuaries and the potential importance of El Niño Southern Oscillation events. Climate Res. 1, 31–42.

Christensen, J.H., Hewitson, B., Busuioc, A., Chen, A., Gao, X., Held, I., Jones, R., Kolli, R.K., Kwon, W.T., Laprise, R., Magaña Rueda, V., Mearns, L., Menéndez, C.G., Räisänen, J., Rinke, A., Sarr, A., Whetton, P., 2007. Regional climate projections. In: Solomon, S., Qin, D., Manning, M., Chen, Z., Marquis, M., Averyt, K.B., Tignor, M., Miller, H.L. (Eds.), Climate Change 2007: The Physical Science Basis. Contribution of Working Group I to the Fourth Assessment Report of the Intergovernmental Panel on Climate Change. Cambridge University Press, Cambridge, UK and New York, USA.

Church, J.A., White, N.J., Coleman, R., Lambeck, K., Mitrovica, J.X., 2004. Estimates of the regional distribution of sea-level rise over the 1950 to 2000 period. J. Climate 17, 2609–2625.

Coe, M.T., Foley, J.A., 2001. Human and natural impacts on the water resources of the Lake Chad basin. J. Geophys. Res. 106, 3349–3356.

Conway, P., 2007. Tipping the scales. If the world is eating more seafood, how come air cargo isn't seeing a corresponding benefit? Air Cargo World. February 2007. Also available at www.aircargoworld.com/archives/features/3_feb07.htm.

Coulthard, S., 2009. Adaptation and conflict within fisheries: insights for living with climate change. 2009 in press, In: Adger, W.N., Lorenzoni, I., O'Brien, K. (Eds.), Adapting to Climate Change: Thresholds, Values, Governance. Cambridge University Press, Cambridge, UK, pp. 255–268.

Crawford, M.A., Bloom, M., Broadhurst, C.L., Schmidt, W.F., Cunnane, S.C., Galli, C., Gehbremeskel, K., Linseisen, F., Lloyd-Smith, J., Parkington, J., 1999. Evidence for the unique function of docosahexaenoic acid during the evolution of the modern hominid brain. Lipids 34, S39–S47.

Crump, M., 1998. In Search of the Golden Frog. University of Chicago Press, Chicago, USA.

Cury, P., Roy, C., 1989. Optimal environmental window and pelagic fish recruitment success in upwelling areas. Can. J. Fish. Aquat. Sci. 46, 670–680.

Cushing, D.H., 1969. The regularity of the spawning season of some fishes. ICES J. Mar. Sci. 33, 81–92.

Cushing, D.H., 1990. Plankton production and yearclass strength in fish populations- an update of the match-mismatch hypothesis. Adv. Mar. Biol. 26, 249–293.
Dale, B., 2001. The sedimentary record of dinoflagellate cysts: Looking back into the future of phytoplankton blooms. Sci. Mar. 65, 257–272.
Dalton, M.G., 2001. El Niño expectations and fishing effort in Monterey Bay, California. Environ. Econ. Manage. 42, 336–359.
Das, P., Mukherjee, S., Kar, S., Das, U., Kar, D., Aditya, G., 2018a. Zooplnakton as dietary component of selected freshwater fish: Network Analysis based on gut content. AACL Bioflux 11 (1), 232–244.
Das, U., Das, P., Kar, S., Jayapalan, C., Kar, D., 2018b. An assessment on zooplankton diversity between three freshwater habitats of Barak valley, Assam. Uttar Pradesh J. Zool. 38 (1), 26–34.
Das, P., Kar, S., Das, U., Bimola, M., Kar, D., Aditya, G., 2020. Day time variations of Zooplankton species composition obdervations from the wetlands of Assam, India. Acta Limnol. Bras. (in press).
Daszak, P., Cunningham, A.A., Hyatt, A.D., 2000. Emerging infectious diseases of wildlife threats to biodiversity and human health. Science 287, 443–448.
De Ionno, P., Wines, G., Jones, P., Collins, R., 2006. A bioeconomic evaluation of a commercial scale recirculating finfish growout system—An Australian perspective. Aquaculture 259, 315–327.
De Silva, S.S., 2001. A global perspective of aquaculture in the new millennium. In: Subasinghe, R.P., Bueno, P., Phillips, M.J., Hough, C., McGladdery, S.E., Arthur, J.R. (Eds.), Aquaculture in the Third Millennium. Bangkok, NACA, pp. 431–459.
De Silva, S.S. and Phillips, M. (2007). A review of cage culture: Asia (excluding China). In M. Halwart, D. Soto and J.R. Arthur (eds). 2007. Cage Aquaculture: Regional Reviews and Global Overview. FAO Fisheries Technical Paper, No. 498: 21–47.
De Silva, S.S., Soto, D., 2009. Climate change and aquaculture: potential impacts, adaptation and mitigation. In: Cochrane, K., De Young, C., Soto, D., Bahri, T. (Eds.), Climate Change Implications for Fisheries and Aquaculture. Overview of Current Scientific Knowledge. FAO, Rome, pp. 151–212. FAO Fisheries and Aquaculture Technical Paper. No. 530.
De Silva, S.S., Turchini, G.M., 2008. Towards understanding the impacts of the pet food industry on world fish and seafood supplies. J. Agricult. Environ. Ethics 21, 459–467.
De Wit, M., Stankiewicz, J., 2006. Changes in surface water supply across Africa with predicted climate change. Science 311, 1917–1921.
Deb, S., Kar, D., 2007. Integrated Fish Farming in Silcoorie and Irongmara villages in Cachar district of Assam. M. Phil Dissertation, Assam University, Silchar (India).
Devi, Maibang, B., Das, P., Das, U., Kar, S., Aditya, G., Kar, D., 2016. Rotifer Species Assemblages in Three Freshwater Habitats of Manipur, India. In: Proceedings of the Zoological Society, Springer, 1–9. https://doi.org/10.1007/s12595-016-0201-z.
Dey, S.C., Kar, D., 1989. Fishermen of Lake Sone in Assam: their socio-economic status. Sci. Cult. 55, 395–398.
Dickson, R.R., Meincke, J., Malmberg, S.-A., Lee, A.J., 1988. The "Great Salinity Anomaly" in the North Atlantic, 1968–1982. Prog. Oceanogr. 20, 103–151.
Doney, S.C., Schimel, D.S., 2007. Carbon and climate system coupling on timescales from the Precambrian to the Anthropocene. Annu. Rev. Environ. Res. 32, 31–66.
Doney, S.C., Fabry, V.J., Feely, R.A., Kleypas, J.A., 2009. Ocean acidification: the other CO2 problem. Ann. Rev. Mar. Sci. 1, 169–192.
Duffy, J.E., 2003. Biodiversity loss, trophic skew and ecosystem functioning. Ecol. Lett. 6, 680–687.
Durant, J.M., Hjermann, D.Ø., Anker-Nilssen, T., Beaugrand, G., Mysterud, A., Pettorelli, N., Stenseth, N., 2007. Climate and the match or mismatch between predator and prey requirements and resource availability. Climate Res. 33, 271–283.
Easterling, W.E., Aggarwal, P.K., Batima, P., Brander, K.M., Erda, L., Howden, S.M., Kirilenko, A., Morton, J., Soussana, J.-F., Schmidhuber, J., Tubiello, F.N., Parry, M.L., Canziani, O.F., Palutikof, J.P., 2007. Food, fibre and forest products. In: Linden, P.J.v.d., Hanson, C. (Eds.), Climate Change 2007: Impacts, Adaptation and Vulnerability. Cambridge University Press, Cambridge, UK, pp. 273–313. Contribution of working group II to the Fourth Assessment Report of the Intergovernmental Panel on Climate Change.
Edgerton, B.F., Henttonent, P., Jussila, J., Mannonent, A.R., Paasonens, P., Taugbøl, T., Edsman, L., Souty-Grosset, C., 2004. Understanding the causes of disease in European fresh water crayfish. Conserv. Biol. 18, 1466–1473.
Edwards, M., Richardson, A.J., 2004. Impact of climate change on marine pelagic phenology and trophic mismatch. Nature 430, 881–884. https://doi.org/10.1038/nature02808.
Edwards, M.J., Johns, D.G., Leterme, S.C., Svendsen, E., Richardson, A.J., 2006. Regional climate change and harmful algal blooms in the North East Atlantic. Limnol. Oceanogr. 51 (2), 820–829.
Epstein, P.R., 2005. Climate change and human health. N. Engl. J. Med. 353, 1433–1436.
Fabry, V.J., Seibel, B.A., Feely, R.A., Orr, J.C., 2008. Impacts of ocean acidification on marine fauna and ecosystem processes. ICES J. Mar. Sci. 65, 414–432.
FAO, 2003a. Fisheries Management. 2. The Ecosystem Approach to Fisheries. FAO Technical Guidelines for Responsible Fisheries. No. 4, Suppl. 2, FAO, Rome. 112 p.
FAO, 2003b. Fishery Country Profile: Peru. FAO, FID/CP/PER Rev. 2, Rome, FAO.
FAO, 2004. FAO Recommendations on the Prevention, Control and Eradication of Highly Pathogenic Avian Influenza (HPA1) in Asia. FAO Position Paper, 49 pp FAO, Rome.
FAO, 2005. Increasing the Contribution of Small-Scale Fisheries to Poverty Alleviation and Food Security. FAO Technical Guidelines for Responsible Fisheries. No. 10, FAO, Rome. 79 p.
FAO, 2006. State of World Aquaculture: 2006. FAO Fisheries Technical Paper. No. 500, FAO, Rome. 134 pp. Also available at www.fao.org/docrep/009/a0874e/a0874e00.htm.
FAO, 2007. The State of World Fisheries and Aquaculture—2006. FAO, Rome. 162 p.
FAO, 2008. Fisheries and Aquaculture Information and Statistics Service. Total fisheries production 1950 to 2006. FishStat Plus—Universal software for fishery statistical time series (online or CD-ROM), FAO, Rome. Also available at www.fao.org/fi/statist/FISOFT/FISHPLUS.asp.
FAO, 2009. FishStat Plus. Universal software for fishery statistical time series. Available at www.fao.org/fishery/statistics/software/fishstat/en.
Feare, C.J., 2006. Fish Farming and the Risk of Spread of Avian Influenza. Wild Wings Bird Management, Bird Life International, Cambridge, UK. 11 pp. Also available at www.birdlife.org/action/science/species/avian.flu/index.html.

Feely, R.A., Sabine, C.L., Hernandez-Ayon, J.M., Ianson, D., Hales, B., 2008. Evidence for upwelling of corrosive 'acidified' water onto the continental shelf. Science 320, 1490–1492.

Ficke, A.D., Myrick, C.A., Hansen, L.J., 2007. Potential impacts of global change on freshwater fisheries. Rev. Fish Biol. Fisheries 17, 581–613.

Fields, P.A., Graham, J.B., Rosenblatt, R.H., Somero, G.N., 1993. Effects of expected global climate change on marine faunas. Trends Ecol. Evol. 8, 361–367.

Flannery, T., 2005. The Weather Makers. Text Publishing, Melbourne, Australia.

Folland, C.K., Rayner, N.A., Brown, S.J., Smith, T.M., Shen, S.S., Parker, D.E., Macadam, I., Jones, P.D., Jones, R.N., Nicholls, N., Sexton, D.M.H., 2001. Global temperature change and its major uncertainties since 1861. Geophys. Res. Lett. 28, 2621–2624.

Galbraith, H., Jones, R., Park, R., Clough, J., Herrod-Julius, S., Harrington, B., Page, G., 2002. Global climate change and sea level rise: potential losses of intertidal habitat for shorebirds. Water birds 25, 173–183.

Gazeau, F., et al., 2007. Impact of elevated CO_2 on shellfish calcification. Geophys. Res. Lett. 34, L07603.

Gerten, D., Adrian, R., 2000. Climate-driven changes in spring plankton dynamics and the sensitivity of shallow polymictic lakes to the North Atlantic oscillation. Limnol. Oceanogr. 45, 1058–1066.

Gianni, A., Saravanan, R., Chang, P., 2003. Oceanic forcing of Sahel rainfall on interannual to interdecadal time scales. Science 320, 1027–1030.

Gienapp, P., Teplitsky, C., Alho, J.S., Mills, J.A., Merila, J., 2008. Climate change and evolution: disentangling environmental and genetic responses. Mol. Ecol. 17, 162–176.

Greene, C.H., Monger, B.C., 2012. An Arctic wild card in the weather. Oceanography 25 (2), 7–9. https://doi.org/10.5670/oceanog.2012.58.

Greene, C.H., Francis, J.A., Monger, B.C., 2013. Superstorm Sandy: a series of unfortunate events? Oceanography 26 (1), 8–9. https://doi.org/10.5670/oceanog.2013.11.

Grøttum, J.A., Beveridge, M., 2008. A review of cage aquaculture: northern Europe. In: Halwart, M., Soto, D., Arthur, J.R. (Eds.), Cage aquaculture. Regional Reviews and Global Overviews, Rome, FAO, pp. 126–154. FAO Fisheries Technical Paper No. 498.

Halwart, M., Soto, D., Arthur, J.R., 2007. Cage Aquaculture. Regional Reviews and Global Overviews. FAO, Rome. FAO Fisheries Technical Paper. No. 498. Rome, FAO. 240 pp.

Handisyde, N.T., Ross, L.G., Badjeck, M.-C., Allison, E.H., 2006. The effects of climate change on world aquaculture: a global perspective. Final Technical Report, DFID Aquaculture and Fish Genetics Research Programme, Stirling Institute of Aquaculture, Stirling, U.K. 151 pp. Available at www.aqua.stir.ac.uk/GISAP/pdfs/Climate_full.pdf.

Hansen, J., Sato, M., Ruedy, R., Lo, K., Lea, D.W., Medina-Elizade, M., 2006. Global temperature change. Proc. Natl. Acad. Sci. U. S. A. 103, 14288–14293.

Hansen, J., Sato, M., Ruedy, R., 2012. Perception of climate change. Proc. Natl. Acad. Sci. U. S. A. 109, E2415–E2423. https://doi.org/10.1073/pnas.1205276109.

Harley, C.D.G., Hughes, R.A., Hultgren, K.M., Miner, B.G., Sorte, C.J.B., Thornber, C.S., Rodriguez, L.F., Tomanek, L., Williams, S.L., 2006. The impacts of climate change in coastal marine systems. Ecol. Lett. 9, 228–241.

Harvell, C.D., Kim, K., Burkholder, J.M., Colwell, R.R., Epstein, P.R., Grimes, D.J., Hofmann, E.E., Lipp, E.K., Osterhaus, A.D.M.E., Overstreet, R.M., Porter, J.W., Smith, G.W., Vasta, G.R., 1999. Emerging marine diseases: climate links and anthropogenic factors. Science 285, 1505–1510.

Harvell, C.D., Mitchell, C.E., Ward, J.R., Altizer, S., Dobson, A.P., Ostfeld, R.S., Samuel, M.D., 2002. Climate warming and disease risks for terrestrial and marine biota. Science 296, 2158–2162.

Hecky, R.E., Bootsma, H.A., Odada, E., 2006. African lake management initiatives: the global connection. Lakes Reserv.: Res. Manage. 11, 203–213.

Hirawake, T., Odate, T., Fukuchi, M., 2005. Long-term variation of chl.a in the Southern Ocean during 1965–2002. Geophys. Res. Lett. 32. https://doi.org/10.1029/2004GL021394.

Hoegh-Guldberg, O., 1999. Climate change, coral bleaching and the future of the world's coral reefs. Mar. Freshw. Res. 50, 839–866.

Hoegh-Guldberg, O., 2005. Low coral cover in a high-CO2 world. J. Geophys. Res. 110. https://doi.org/10.1029/2004JC002528.

Hofmann, E., Ford, S., Powell, E., Klinck, J., 2001. Modelling studies of the effect of climate variability on MSX disease in eastern oyster (*Crassostrea virginica*) populations. Hydrobiologia 460, 195–212.

Houghton, R.W., Visbeck, M.H., 2002. Quasi-decadal salinity fluctuations in the Labrador Sea. J. Phys. Oceanogr. 32, 687–701.

Hughes, T.P., Baird, A.H., Bellwood, D.R., Card, M., Connolly, S.R., Folke, C., Grosberg, R., Hoegh-Guldberg, O., Jackson, J.B.C., Kleypas, J., Lough, J.M., Marshall, P., Nyström, M., Palumbi, S.R., Pandolfi, J.M., Rosen, B., Roughgarden, J., 2003. Climate change, human impacts and the resilience of coral reefs. Science 301, 929–934.

Hunt Jr., G.L., Stabeno, P., Walters, G., Sinclair, E., Brodeur, R.D., Napp, J.M., Bond, N.A., 2002. Climate change and control of the southeastern Bering Sea pelagic ecosystem. Deep-Sea Res. II 49, 5821–5853.

International Commission on Microbiological Specifications for Foods (ICMSF), 1996. Characteristics of Microbial Pathogens in Microorganism in Food 5. Blackie Academic and Professional.

IPCC, Gitay, H., Suarez, A., Watson, R.T., Dokken, D.J. (Eds.), 2002. Climate change and biodiversity. In: IPCC Technical Paper V. Inter-Governmental Panel on Climate Change, WMO and UNEP. 76 pp.

IPCC, 2007. Summary for policymakers. In: Parry, M.L., Canziani, O.F., Palutikof, J.P., Linden, P.J.v.d., Hanson, C.E. (Eds.), Climate change 2007: impacts, adaptation and vulnerability. Contribution of Working Group II to the Fourth Assessment Report of the Intergovernmental Panel on Climate Change. Cambridge University Press, Cambridge, UK, pp. 7–22.

Jackson, A., 2006. The importance of fishmeal and fish oil in aquaculture. Int. Aquafeed, 16–19. November–December 2006.

Jacquet, S., Bratbak, G., 2003. Effects of ultraviolet radiation on marine virus phytoplankton interactions. FEMS Microbiol. Ecol. 44, 279–289.

Jallow, B.P., Barrow, M.K.A., Leatherman, S.P., 1996. Vulnerability of the coastal zone of The Gambia to sea level rise and development of response strategies and adaptation options. Climate Res. 6, 165–177.

Jentoft, S., 2006. Beyond fisheries management: the phronetic dimension. Mar. Policy 30 (6), 671–680.

Jobling, M., 1997. Temperature and growth: modulation of growth rate via temperature change. In: Wood, C.M., McDonald, D.G. (Eds.), Global Warming: Implications for Fresh Water and Marine Fish. Cambridge University Press, Cambridge, UK, pp. 225–253.

Jobling, M., 2004. "Finishing" feeds for carnivorous fish and the fatty acid dilution model. Aquacult. Res. 35, 706–709.

Kar, D., 2007. Fundamentals of Limnology and Aquaculture Biotechnology. vi + 609, Daya Publishing House, New Delhi. ISBN: 81-7035-455-2.

Kar, D., 2013. Wetlands and Lakes of the World. pp. xxx + 687, Springer, London. Print ISBN 978-81-322-1022-1.
Kar, D., 2015. Epizootic Ulcerative Fish Disease Syndrome. xix + 293, Elsevier, Academic Press, USA. ISBN: 9780128025048.
Kar, Devashish (2019). Wetlands diversity and their fishes in Assam, India. Transyl. Rev. Syst Ecol. Res. 21.3 (2019), The Wetlands Diversity, pp. 1-94 (Romania).
Kar, S., Das, P., Das, U., Bimola, M., Kar, D., Aditya, G., 2018a. Culture of the zooplankton as fish food: observations on three freshwater species from Assam, India. AACL Bioflux 10 (5), 1210–1220.
Kar, S., Das, P., Das, U., Bimola, M., Kar, D., Aditya, G., 2018b. Correspondence of zooplankton assemblage and water quality in wetlands of Cachar, Assam, India: Implications for environmental management. Limnolog. Rev. 18 (1), 9–19.
Karl, D.M., 1999. A sea of change: biogeochemical variability in the north Pacific subtropical gyre. Ecosystems 2, 181–214.
Karst-Riddoch, T.L., Pisaric, M.F.J., Smol, J.P., 2005. Diatom responses to twentieth century climate-related environmental changes in high-elevation mountain lakes of the northern Canadian Cordillera. J. Paleo. 33, 265–282.
Kerr, R., 2006. Global warming is changing the world. Science 316, 188–192.
Kleypas, J.A., et al., 2006. Impacts of Ocean Acidification on Coral Reefs and Other Marine Calcifiers. In: A Guide for Future Research NSF, NOAA. US Geological Survey, St Petersburg, FL. 18–20 April.
Kundzewicz, Z.W., Mata, L.J., Arnell, N.W., Döll, P., Jimenez, B., Miller, K., Oki, T., Şen, Z., Shiklomanov, I., 2008. The implications of projected climate change for freshwater resources and their management. Hydrol. Sci. I 53, 3–10.
Kurien, J., 2005. Responsible Fish Trade and Food Security. FAO, Rome. FAO Fisheries Technical Paper No. 456, 102 pp.
Kurihara, H., Kato, S., Ishimatsu, A., 2007. Effects of increased seawater *p*CO2 on early development of the oyster *Crassostrea gigas*. Aquat. Biol. 1, 91–98.
Kurihara, H., Asai, T., Kato, S., Ishimatsu, A., 2009. Effects of elevated *p*CO2 on early development in the mussel *Mytilus galloprovincialis*. Aquat. Biol. 4, 225–233.
Labat, D., Godderis, Y., Probst, J.L., Guyot, J.L., 2004. Evidence for global runoff increases related to climate warming. Adv. Water Resour. 27, 631–642.
Lambert, N., Giri, A., Mishra, B.P., Kar, D., 2020. The impact of abiotic factors on the occurrence, assemblages and diversity of freshwater zooplanktonin lake Tanganyika, Burundian littoral. Int. J. Entomol. Res. 5 (2), 74–82.
Legates, D.R., Lins, H.F., McCabe, G.J., 2005. Comments on "Evidence for global runoff increase related to climate warming" by Labat et al. Adv. Water Resour. 28, 1310–1315.
Lehodey, P., 2004. Climate and fisheries: an insight from the central Pacific Ocean. In: Stenseth, N.C., Ottersen, G., Hurrell, J., Belgrano, A. (Eds.), Marine Ecosystems and Climate Variation: The North Atlantic. Oxford. Oxford University Press, pp. 137–146.
Lehodey, P., Alheit, J., Barange, M., Baumgartner, T., Beaugrand, G., Drinkwater, K., Fromentin, J.-M., Hare, S.R., Ottersen, G., Perry, R.I., Roy, C., van der Lingen, C.D., Werner, F., 2006. Climate variability, fish and fisheries. J. Climate 19, 5009–5030.
Lipp, E.K., Huq, A., Colwell, R.R., 2002. Effects of global climate on infectious disease: the cholera model. Clin. Microbiol. Rev. 15, 757–770.
Luam Kong, A., 2002. Impact of climate change on Caribbean fisheries resources: research needs. Caribbean food systems: Developing a research agenda, 19–20 September 2002, St. Augustine, Trinidad. Global Environmental Change and Food Systems (GECAFS).
Lumsden, S.E., Hourigan, T.F., Bruckner, A.W., Dorr, G. (Eds.), 2007. The state of deep coral ecosystems of the United States NOAA Technical Memorandum CRCP-3 Silver Spring, Maryland New York Sea Grant 2001 The Economic Contribution of the Sport Fishing Commercial Fishing and Seafood Industries to New York State NYSGI-T-01-001.
Macedo, I., Verde, M.L., Azevedo, J., 2004. Assessment of Greenhouse Gas Emissions in the Production and Use of Fuel Ethanol in Brazil. Government of the State of Săn Paulo and Secretariat of the Environment, Brazil.
Mackas, D.L., Beaugrand, G., 2008. Comparisons of zooplankton time series. J. Mar. Syst., 25–45.
MacKenzie, B.R., Visser, A.W., 2001. Fisheries and climate change: the Danish perspective. In: Jørgensen, A.M.K., Fenger, J., Halsnaes, K. (Eds.), Climate Change Research—Danish Contributions. Copenhagen, Danish Climate Centre.
MacLean, J.L., 1989. Indo-Pacific red tides, 1985–1988. Mar. Pollut. Bull. 20, 304–310.
Mahon, R., 2002. Adaptation of fisheries and fishing communities to the impacts of climate change in the CARICOM region. In: Mainstreaming Adaptation to Climate Change (MACC). The Caribbean Centre for Climate Change, Washington, DC. p. 33.
Marcogliese, D.J., 2001. Implications of climate change for parasitism of animals in the aquatic environment. Can. J. Zool. 79, 1331–1352.
McGregor, H.V., Dima, M., Fischer, H.W., Mulitza, S., 2007. Rapid twentieth century increase in coastal upwelling off North West Africa. Science 315, 637–639.
McLaughlin, J.B., DePaola, A., Bopp, C.A., Martinek, K.A., Napolilli, N.P., Allison, C.G., Murray, S.L., Thompson, E.C., Bird, M.M., Middaugh, J.P., 2005. Outbreak of *Vibrio parahaemolyticus* gastroenteritis associated with Alaskan oysters. N. Engl. J. Med. 353 (14), 1463–1470.
McLean, R.F., Tsyban, A., 2001. Coastal zones and marine ecosystems. In: McCarthy, J.J., Canziani, O.F., Leary, N.A., Dokken, D.J., White, K.S. (Eds.), Climate Change 2001: Impacts, Adaptation, and Vulnerability. Cambridge University Press, Cambridge, UK, pp. 343–380. Contribution of working group II to the Third Assessment Report of the Intergovernmental Panel on Climate Change.
McMichael, A.J., 2001. Impact of climatic and other environmental changes on food production and population health in the coming decades. Proc. Nutr. Soc. 60, 195–201.
Meehl, G.A., T.F. Stocker, W.D. Collins, P. Friedlingstein, A.T. Gaye, J.M. Gregory, A. Kitoh, R. Knutti, J.M. Murphy, A. Noda, and others. (2007). Global climate projections. Pp. 747–845 in Climate Change 2007: The Physical Science Basis. Contribution of Working Group I to the Fourth Assessment Report of the Intergovernmental Panel on Climate Change. S. Solomon, S.D. Qin, M. Manning, Z. Chen, M. Marquis, K.B. Averyt, M. Tignor, and H.L. Miller, eds, Cambridge University Press, Cambridge, UK.
Miller, J.B., 2008. Sources, sinks and seasons. Nature 451, 26–27.
Miller, L., Douglas, B.C., 2004. Mass and volume contributions to twentieth century global sea level rise. Nature 248, 407–409.
Milly, P.C.D., Dunne, K.A., Vecchia, A.V., 2005. Global pattern of trends in streamflow and water availability in a changing climate. Nature 438, 347–350.
Molles Jr., M.C., Dahm, C.N., 1990. A perspective on El Niño and La Niña: global implications for stream ecology. J. N. Amer. Benth. Soc 9, 68–76.
Morales, V.V.Q., Morales, R.R., 2006. Regional review on aquaculture development 1. Latin America and the Caribbean - 2005. FAO Fisheries Circular, Rome. F1017/1. 177 pp.

Mourente, G., Good, J.E., Bell, J.G., 2005. Partial substitution of fish oil with rapeseed, linseed and olive oils in diets for European sea bass (*Dicentrarchus labrax* L.): effects on flesh fatty acid composition, plasma prostaglandins E2 and F2, immune function and effectiveness of a fish oil finishing diet. Aquacult. Nutr. 11, 25–40.

Mudie, P.J., Rochon, A., Levac, E., 2002. Palynological records of red tide-producing species in Canada: past trends and implications for the future. Palaeogeogr. Palaeoclimatol. Palaeoecol. 180, 159–186.

Murray, S.N., Horn, M.H., 1989. Variations in standing stocks of central California macrophytes from a rocky intertidal habitat before and during the 1982–1983 El Niño. Mar. Ecol. Prog. Ser. 58, 113–122.

Nakićenović, N., 2000. Greenhouse gas emissions scenarios. Technol. Forecast. Soc. Chang. 65, 149.

Nath, S., Kar, D., 2016. Integrated Fish Farming in Village Irongmara, Cachar, Assam: An intricate analysis. MSc Dissertation, Assam University, Silchar (Inda).

Naylor, R.L., Liska, A.J., Burke, M.B., Falcon, W.P., Gaskell, J.C., Rozelle, S.D., Cassman, K.G., 2006. The ripple effect. Biofuels, food security, and the environment. Environ. 49, 32–43.

Nghiem, S.V., Hall, D.K., Mote, T.L., Tedesco, M., Albert, M.R., Keegan, K., Shuman, C.A., DiGirolamo, N.E., Neumann, G., 2012. The extreme melt across the Greenland ice sheet in 2012. Geophys. Res. Lett. 39. https://doi.org/10.1029/2012GL053611.

Nguyen, K.C., 2008. Surviving the flood. Outlook, Viet Nam News Mon. Publ. 51 (March 2008), 14–16.

Nguyen, T.T.T., De Silva, S.S., 2006. Fresh water finfish biodiversity and conservation: an Asian perspective. Biodivers. Conserv. 15, 3543–3568.

Nicholls, R.J., Hanson, S., Herweijer, C., Patmore, N., Hallegatte, S., Corfee-Morlot, J., Chateau, J., Muir-Wood, R., 2007a. Ranking Port Cities with High Exposure and Vulnerability to Climate Extremes: Exposure Estimates. OECD, Paris. ENV/WKP (2007)1: 64 pp. Also available at www.oecd.org/dataoecd/16/58/39720578.pdf.

Nicholls, R.J., Wong, P.P., Burkett, V.R., Codignotto, J.O., Hay, J.E., McLean, R.F., Ragoonaden, S., Woodroffe, C.D., 2007b. Coastal systems and low-lying areas. In: Parry, M.L., Canziani, O.F., Palutikof, J.P., Linden, P.J.v.d., Hanson, C.E. (Eds.), Climate Change 2007: Impacts, Adaptation and Vulnerability. Cambridge University Press, Cambridge, UK, pp. 315–356. Contribution of working group II to the Fourth Assessment Report of the Intergovernmental Panel on Climate Change.

Ñiquen, M., Bouchon, M., 2004. Impact of El Niño events on pelagic fisheries in Peruvian waters. Deep-Sea Res. II Top. Stud. Oceanogr. 51, 563–574.

Nunn, A.D., Harvey, J.P., Britton, J.R., Frear, P.A., Cowx, I.G., 2007. Fish, climate and Gulf Stream: the influence of abiotic factors on the recruitment success of cyprinid fishes in lowland rivers. Fresh Water Biol. 52, 1576–1586.

Nyberg, P., Bergstand, E., Degerman, E., Enderlein, O., 2001. Recruitment of pelagic fish in an unstable climate: studies in Sweden's four largest lakes. Ambio 30, 559–564.

O'Reilly, C.M., Alin, S.R., Pilsnier, P.D., Cohen, A.S., McKee, B.A., 2003. Climate change decreases aquatic ecosystem productivity of Lake Tanganyika, Africa. Nature 424, 766–768.

Ordinola, N. (2002). The consequences of cold events for Peru. In. M.H. Glantz, (Ed.), La Niña and its Impacts: Facts and Speculation, pp. 146–150. Toronto, Canada, New York, USA, Paris, United Nations University Press.

Orr, J.C., Fabry, V.J., Aumont, O., Bopp, L., Doney, S.C., Feely, R.A., Gnanadesikan, A., Gruber, N., Ishida, A., Joos, F., Key, R.M., Lindsay, K., Maier-Reimer, E., Matear, R., Monfray, P., Mouchet, A., Najjar, R.G., Plattner, G.-K., Rodgers, K.B., Sabine, C.L., Sarmiento, J.L., Schlitzer, R., Slater, R.D., Totterdell, I.J., Weirig, M.-F., Yamanaka, Y., Yool, A., 2005. Anthropogenic ocean acidification over the twenty-first century and its impact on calcifying organisms. Nature 437, 681–686.

Ottersen, G., Kim, S., Huse, G., Polovina, J.J., Stenseth, N.C., 2008. Major routes by which climate signals force marine populations. J. Mar. Syst., 24–42.

Paris-Mancilla, E., 2005. Intoxicación Por *Vibrio parahaemolyticus*. Cuad Méd Soc (Chile) 45, 43–47.

Parmesan, C., Yohe, G., 2003. A globally coherent fingerprint of climate change impacts across natural systems. Nature 421, 37–42.

Parry, M.L., Canziani, O.F., Palutikof, J.P., van der Linden, P.J., Hanson, C.E. (Eds.), 2007. Cross-chapter case study. In: Climate Change 2007: Impacts, Adaptation and Vulnerability. Cambridge University Press, Cambridge, UK, pp. 843–868. Contribution of working group II to the Fourth Assessment Report of the Intergovernmental Panel on Climate Change.

Patz, J.A., 2000. Climate change and health: new research challenges. Ecosyst. Health 6 (1), 52–58.

Pearson, R.G., Dawson, T.P., 2003. Predicting the impacts of climate change on the distribution of species: are bioclimate envelope models useful? Glob. Ecol. Biogeogr. 12, 361–371.

Pisias, N.G., Mix, A.C., Heusser, L., 2001. Millennial scale climate variability of the North East Pacific Ocean and Northwest North America based on radioloaria and pollen. Q. Sci. Rev. 20, 1561–1576.

Poff, N.L., Brinson, M.M., Day Jr., J.W., 2002. Aquatic ecosystems and global climate change. Potential impacts on inland freshwater and coastal wetland ecosystems in the United States. In: Report prepared for the Pew center on Global Climate Change. Also available at www.pewclimate.org.

Raupach, M.R., et al., 2007. Global and regional drivers of accelerating CO_2 emissions. Proc. Natl. Acad. Sci. U. S. A. 104, 10288–10293.

Reid, P.C., Edwards, M., Beaugrand, G., Skogen, M., Stevens, D., 2003. Periodic changes in the zooplankton of the North Sea during the twentieth century linked to oceanic inflow. Fish. Oceanogr. 12, 260–269.

Revenga, C., Brunner, J., Henninger, N., Kassem, K., Payne, R., 2000. Pilot Analysis of Global Ecosystems: Freshwater Systems. World Resources Institute, Washington, DC.

Richardson, A.J., 2008. In hot water: zooplankton and climate change. ICES J. Mar. Sci. 65, 279–295.

Roemmich, D., McGowan, J.A., 1995. Climatic warming and the decline of zooplankton in the California Current. Science 267, 1324–1326.

Roessig, J.M., Woodley, C.M., Cech Jr., J.J., Hansen, L.J., 2004. Effects of global change on marine and estuarine fishes and fisheries. Rev. Fish Biol. Fisheries 14, 251–275.

Rombough, P.T., 1997. The effects of temperature on embryonic and larval development. In: Wood, C.M., McDonald, D.G. (Eds.), Global Warming: Implications for Fresh Water and Marine Fish. Cambridge University Press, Cambridge, UK, pp. 177–223.

Root, T.L., Price, J.T., Hall, K.R., Schneider, S.H., Rosenzweig, C., Pounds, J.A., 2003. Fingerprints of global warming on wild animals and plants. Nature 421, 57–60.

Rosenzweig, C., Casassa, G., Karoly, D.J., Imeson, A., Liu, C., Menzel, A., Rawlins, S., Root, T.L., Seguin, B., Tryjanowski, P., 2007. Assessment of observed changes and responses in natural and managed systems. In: Parry, M.L., Canziani, O.F., Palutikof, J.P., van der Linden, P.J., Hanson, C.E. (Eds.), Climate Change 2007: Impacts, Adaptation and Vulnerability. Cambridge University Press, Cambridge, UK, pp. 79–131. Contribution of working group II to the Fourth Assessment Report of the Intergovernmental Panel on Climate Change.

Sabine, C.S., Feely, R.A., Gruber, N., Key, R.M., Lee, K., Bullister, J.L., Wanninkhof, R., Wong, C.S., Wallace, D.W.R., Tillbrook, B., Millero, F.J., Peng, T.-H., Kozyr, A., Ono, T., Rios, A.F., 2004. The oceanic sink for anthropogenic CO_2. Science 305, 367–371.
Scales, H., Balmford, A., Manica, A., 2007. Monitoring the live reef food fish trade: lessons learned from local and global perspectives. SPC Live Reef Fish Inform. Bull. 17, 36–44.
Scheffer, M., Carpenter, S., Foley, J.A., Folke, C., Walker, B., 2001. Catastrophic shifts in ecosystems. Nature 413, 591–596.
Schindler, D.E., Rogers, D.E., Scheuerell, M.D., Abrey, C.A., 2005. Effects of changing climate on zooplankton and juvenile sockeye salmon growth in southwestern Alaska. Ecology 86, 198–209.
Schmittner, A., 2005. Decline of the marine ecosystem caused by a reduction in the Atlantic overturning circulation. Nature 434, 628–633.
Schrank, W.E., 2007. The ACIA, climate change and fisheries. Mar. Policy 31, 5–18.
Sen, A., 1981. Poverty and Famines: An Essay on Entitlement and Deprivation. Clarendon, Oxford, UK.
Sharp, G.D., 2003. Future Climatic Change and Regional Fisheries: a Collaborative Analysis. FAO, Rome. FAO Fisheries Technical Paper, No. 452: 75 pp.
Sharp, G.D., McLain, D.R., 1993. Fisheries, El Niño Southern Oscillation and upperocean temperature records: an eastern Pacific example. Oceanography 6, 13–21.
Singha, S., Kar, D., 2016. Integrated Fish Farming in Village Irongmara, Assam: A field study. M.Sc. Dissertation, Assam University, Silchar (India).
Smayda, T.J., 1990. Novel and nuisance phytoplankton blooms in the sea: evidence for a global epidemic. In: Graneli, E., Sunderstroem, B., Edler, L., Anderson, D.M. (Eds.), Toxic Marine Phytoplankton. Elsevier, pp. 29–40.
Smit, B., Burton, I., Klein, R.J.T., Wandel, J., 2000. An anatomy of adaptation to climate change and variability. Clim. Change 45, 223–251.
Smith, L.E.D., Nguyen Khoa, S., Lorenzen, K., 2005. Livelihood functions of inland fisheries: policy implications for developing countries. Water Policy 7, 359–383.
Snieszko, S.F., 1974. The effects of environmental stress on outbreaks of infectious diseases of fishes. J. Fish Biol. 6 (2), 197–208.
Soto, D., Jara, F., Moreno, C., 2001. Escaped salmon in the inner seas, southern Chile: facing ecological and social conflicts. Ecol. Appl. 11 (6), 1750–1762.
Soto, D., Aguilar-Manjarrez, J., Brugère, C., Angel, D., Bailey, C., Black, K., Edwards, P., Costa Pierce, B., Chopin, T., Deudero, S., Freeman, S., Hambrey, J., Hishamunda, N., Knowler, D., Silver, W., Marba, N., Mathe, S., Norambuena, R., Simard, F., Tett, P., Troell, M., Wainberg, A., 2008. Applying an ecosystem-based approach to aquaculture: principles, scales and some management measures. In: Soto, D., Aguilar-Manjarrez, J., Hishamunda, N. (Eds.), Building an Ecosystem Approach to Aquaculture. FAO, Rome, pp. 15–35. FAO/Universitat de les Illes Balears Expert Workshop. 7–11 May 2007, Spain, Mallorca. FAO Fisheries Proceedings. No. 14.
Steinacher, M., Joos, F., Frölicher, T.L., Plattner, G.-K., Doney, S.C., 2009. Imminent ocean acidification in the Arctic projected with the NCAR global coupled carbon cycle-climate model. Biogeosciences 6, 515–533.
Stenevik, E., Sundby, S., 2007. Impacts of climate change on commercial fish stocks in Norwegian waters. Mar. Policy 31, 19–31.
Suklabaidya, M., Kar, D., 2016. Studies on Integrated Fish Farming in certain locations in village Irongmara, Cachar. Assam. M.Sc. Disseration, Assam University, Silchar (India).
Swing, T.G., 2003. What future for the oceans? New York, USA. Foreign Aff., 139–152. September–October.
Taboada, F.G., Anadón, R., 2012. Patterns of change in sea surface temperature in the North Atlantic during the last three decades: Beyond mean trends. Clim. Change 115, 419–431. https://doi.org/10.1007/s10584-012-0485-6.
Tebaldi, C., Hayhoe, K., Arblaster, J.M., Meehl, G.A., 2006. Going to the extremes: An intercomparison of model-simulated historical and future changes in extreme events. Clim. Change 79, 185–211. s10584-006-9051-4.
Testa, J.W., Oehlert, G., Ainley, D.G., Bengtson, J.L., Siniff, D.B., Laws, R.M., Rounsevelle, D., 1991. Temporal variability in Antarctic marine ecosystems: periodic fluctuations in the phocid seals. Can. J. Fish. Aquat. Sci. 48, 631–639.
Thomas, C.D., Cameron, A., Green, R.E., Bakkenes, M., Beaumont, L.J., Collingham, Y.C., Erasmus, B.F.N., Siqueira, M.F.D., Grainger, A., Hannah, L., 2004. Extinction risks from climate. Nature 427, 145–148.
Thorpe, A., Reid, C., Anrooy, R.V., Brugere, C., Becker, D., 2006. Poverty reduction strategy papers and the fisheries sector: an opportunity forgone? J. Int. Dev. 18 (4), 489–517.
Timmermann, A., Ohberhuber, J., Bacher, A., Esch, M., Latif, M., Roeckner, E., 1999. Increased El Niño frequency in a climate model forced by future greenhouse warming. Nature 398, 694–696.
Troadec, J.P., 2000. Adaptation opportunities to climate variability and change in the exploitation and utilisation of marine living resources. Environ. Monit. Assess. 61, 101–112.
Turchini, G.M., De Silva, S.S., 2008. Bio-economical and ethical impacts of alien finfish culture in European inland waters. Aquacult. Int. https://doi.org/10.1007/s10499-007-9141-y. available online DOI. 33 pp.
Turchini, G.M., Francis, D.S., De Silva, S.S., 2007. Finishing diets stimulate compensatory growth: results of a study on Murray cod, *Maccullochella peelii peelii*. Aquacult. Nutr. 13, 351–360.
Tyedmers, P.H., Watson, R., Pauly, D., 2005. Fuelling global fishing fleets. Ambio 34 (8), 635–638.
Van Dam, A.A., Beveridge, M.C.M., Azim, M.C., Verdegem, M.C.J., 2002. The potential of fish production based on periphyton. Rev. Fish Biol. Fish. 12, 1–31.
Van der Kraak, G., Pankhurst, N.W., 1997. Temperature effects on the reproductive performance of fish. In: Wood, C.M., McDonald, D.G. (Eds.), Global Warming: Implications for Fresh Water and Marine fish. Cambridge University Press, Cambridge, UK, pp. 159–176.
Vellinga, M., Wood, R.A., 2002. Global climatic impacts of a collapse of the Atlantic thermohaline circulation. Clim. Change 54, 251–267.
Verdegem, M.C.J., Bosma, R.H., Verreth, J.A.V., 2006. Reducing water use for animal production through aquaculture. Water Resour. Devel. 22, 101–113.
Vergani, D.F., Stanganelli, Z.B., Bilenca, D., 2004. Effects of El Niño and La Niña events on the sex ratio of southern elephant seals at King George Island. Mar. Ecol. Prog. Ser. 268, 293–300.
Visser, M.E., Both, C., 2005. Shifts in phenology due to global climate change: the need for a yardstick. Proc. R. Soc. B 272, 256–2569.
Vollmer, M.K., Bootsma, H.A., Hecky, R.E., Patterson, G., Halfman, G.D., Edmond, J.M., Eccles, D.H., Weiss, R.F., 2005. Deep-water warming trend in Lake Malawi, east Africa. Limnol. Oceanogr. 50, 727–732.
Wagle, S.K., Gurung, T.B., Bista, J.D., Rai, A.K., 2007. Cage fish culture and fisheries for food security and livelihoods in mid hill lakes of Pokhara Valley, Nepal: post community based management adoption. Aquacult. Asia XII (3), 21–27.

Wahab, M.A., Azim, M.E., Ali, M.H., Beveridge, M.C.M., Kahn, S., 1999. The potential of periphyton based culture of the native major carp calbaush, *Labeo clabasu* (Hamilton). Aquacult. Res. 30, 409–419.
Ward, J.R., Lafferty, K.D., 2004. The elusive baseline of marine disease: are diseases in ocean ecosystems increasing? PLoS Biol. 2, 542–547.
WFC, 2006. The Threat to Fisheries and Aquaculture from Climate Change. WFC, Penang. World Fish Center Policy Brief: 8 pp.
Whittingham, E., Campbell, J., Townsley, P., 2003. Poverty and reefs. In: A global overview. vol 1. Exeter, UK. DFID-IMM-IOC-UNESCO, 260 pp.
Wilkinson, C. (Ed.), 2000. Status of coral Reefs of the World: 2000. Australian Institute of Marine Science. Townsville, Australia.
Wood, C.M., McDonald, D.G., 1997. Global Warming: Implications for Fresh Water and Marine Fish. Cambridge University Press, Cambridge, UK. 425 pp.
Worldwatch Institute, 2006. Biofuels for transportation: global potential and implications for sustainable agriculture and energy in the twenty-first century. Washington, DC.
Zwiers, F.W., Weaver, A.J., 2000. The causes of the twentieth century warming. Science 290, 2081–2083.

CHAPTER

7

Discussion

OUTLINE

Abstract

There have been detailed deliberations, in the present treatise, on different aspects of community-based management of resources, on a global scenario, including different management options, development aspirations, conflicts and consensus, etc. Incidentally, an attempt has been made in the following paragraphs to deal with little different but related aspects.

Keywords: Harvesters, Fisheries, Outlets, Global scenario, Fishing vessels, Shell middens

There have been detailed deliberations, in the present treatise, on different aspects of community-based management of resources, on a global scenario, including different management options, development aspirations, conflicts and consensus, etc. Incidentally, an attempt has been made in the following paragraphs to deal with little different but related aspects.

7.1 Fishery management before 1900

The natural resources are God-gifted assets to the living world including humanity. The bountiful resources at the dawn of civilization vis-à-vis less number of harvesters and consumers, probably, could not generate any sense of sustainability and rational management. The idea came into reality with the growing number of consumers. And the concept of sustainability, management, and conservation reached its zenith in a compelling way with the burgeoning growth of consumers; in particular, the exponential population explosion of human beings by leaps and bounds, especially, in the recent era. Incidentally, a common type of dupe or fallacy of the late 20th century is that the first generation has encountered the difficult problems of natural resource management, and the rapid advances in technology have eased many of our problems through different types of technological *modus operandi.* Certainly the wide spread application of science through human endeavors, since the beginning of the century, had, probably, focused more on methodologies of control; rather than improving harvesting efficiency. There is, probably, much to be learned from the past. However, it is now quite difficult to find citations of early works and experiences in fisheries management in publications issued over the last few decades. The current view tends to assume that the use of radar and electronic communications for monitoring control and surveillance of fishing fleets, multi-criteria decision making, etc., were not much prevalent before the advent of computers. Nevertheless, the reality of course seems to be that fisheries management has a rich and important history and glorious heritage, some of whose lessons have been rediscovered only during the past few decades. New *modus operandi* could often emerge from methodologies discarded earlier; they could be renovated through application of new knowledge and technology, as well as by bringing in novelties therein. The imposition of "closed areas" is one example, which could be cited here. Incidentally, the ancient methodologies, being reemphasized in coastal areas after decades of neglect and nonuse, are difficult or expensive to enforce the regulation in the mainland, being distant from the coast. On the other hand, satellite monitoring of fishing vessels equipped with transducers as a requirement for licensing, probably, not only makes such measures feasible in the open seas and oceans, but, at the same time, opens the door to the possible leases or exclusive access rights for resources of specific ocean areas. However, the prediction of possible new approaches, which could be made feasible by future technological change(s), is, of course, largely difficult. Only a range of options could be suggested and in looking for these options, one needs to look not only forwards but also backwards. Incidentally, it may be interesting to note here that fishing itself is, perhaps, as old as mankind. Extensive shell middens seem to reflect the early concern with marine resources of prehistoric coastal settlements. One of the oldest fishing implements known is said to be the copper and bronze fishing hook from Mesopotamia, which is dated back to *c.* 3500 BCE, and also by that time, salted fish became an important commodity in Mesopotamian and Mediterranean trade.

Concomitant to the above, in the 10th century, in **Sicily**, according to Arab sources quoted by Spallanzani in writings in the 18th century, precious red coral was fished by rotation over a 10-year cycle between 10 zones established near Messina, which corresponded fairly well to the time needed for the colonies to reach commercial size. With the increasing participation in fisheries on the high seas during the early part of the second millennium came in the conflicts, crises, and the forerunners of management decisions. Nicolson wrote of the petition to Edward III in 1376, expressing his concern about a new type of fishing gear, which appeared to be an early type of small mesh beam trawl. The petitioning fishermen were concerned about the damage done to the substrate (it was believed to have destroyed the flowers of the land below the water), damage to spat of oysters and mussels, and were concerned about the effect of excessive catches of small fish. A commission investigated the matter and advised that the gear could, preferably, be used only in deep water but unhappily legislation to this effect was, probably, never enacted. Despite this failure, fisheries management organizations appeared to be *in vogue* in many fisheries in the past.

Notwithstanding the above, in **France**, the local regulatory boards associated with fishing harbors known as prud'homies (or "wise men" of the port) predate the French Revolution. The functions of these bodies, and still today, ensure fair allocation of resources within the local jurisdiction of a prud'homie, protecting the territory from outside intrusions and protecting the resources through the regulation of, for instance, mesh sizes, fishing seasons, and so on.

The cofrad'ias of **Spain** were similar in purpose and many coastal communities worldwide were believed to have systems of taboos, access rights, and traditional practices which effectively protect the *status quo*.

In **Japan**, fisheries management also has a centuries-old history. Kalland has reported that during the Tokugawa period, from 1603 to 1868, fishing territories were established, and coastal waters were considered to be extensions of the land, thus they were part of the feudal domain. The feudal lords, probably, have partitioned the coastal waters and allocated areas of fishing ground to local fishing communities under the control of the village heads. This system remain largely intact today, although during late last century, authority for managing the coastal territories, probably, has shifted from village heads to fishing cooperatives. Such community approaches to management are widespread and researchers referred to similar practices in the **Philippines, Oceania, the Pacific Coast of North America, and Mexico**. Other researchers have also suggested the existence of controls which indicate that the communities knew the value of these resources and the importance of conserving them. Further, there are records of the existence of explicit fishing rights in the medieval times.

Notwithstanding the above, the local fishing rights, probably, were not developed further within the common law of property and in many cases have been subsequently lost or superseded. However, there is danger in overromanticizing the "equilibrium" between pretechnological cultures and their resources. Nevertheless, it is true to say that before industrial-scale exploitation, fisheries were, probably, managed by using traditional knowledge. In fact, the early technology and practices were, perhaps, largely sustainable, even without scientific information about the marine ecosystem. Nevertheless, a major challenge to "traditional" small-scale fishers and their rights was probably inevitable with the development of a market economy for fish, such that the price of a commodity, and hence, the incentive to harvest, probably, has remained almost constant, even when the local demand was fulfilled. Subsequently, the development of large motorized fishing vessels, that had followed the industrial revolution, was said to be an inevitable result not only of technological advance but also of the wider market including the means of storage, processing, and transportation of fish that had developed concomitantly to meet the needs. Consequently, the center of decision-making automatically shifted from coastal communities to the business community, banks, cities, and the central government, since these new developments required substantial investments.

Notwithstanding the above, on a global scale, beyond coastal waters, in terms of national economies, fisheries are often considered of secondary importance to other national interests in many countries, although, it should not have been, because the coveted fishery resources are, no doubt, highly potential sources of nutrition and avocation. As such, the fisheries domain, being a potential arena for providing with nutrition, seems to be capable of handling strong strategies and efficient modus operandi for effective and sustained production, propagation, management, and conservation of the fish resources. In fact, the fisheries, by virtue of their objective, are devoted to maintain a sustained and continuous supply of protein; and, thus, help in getting an answer to the diminishing protein supply in protein-deficient areas.

A brief glimpse into the historical resume' reveals that the early role of shipping for trade and territorial protection and expansion has had enormous implications for fisheries. Incidentally, an early manifestation of the overriding of local rights in favor of national policy was said to be the declaration by Queen Elizabeth I of England to the Spanish ambassador, and she claimed the freedom of the seas as being "common to all." In fact, the concept of the seas and oceans as common property became firmly entrenched with the study of Hugo Grotius, an attorney working for the Dutch East India Company. His contribution took the form of a pamphlet titled "*Mare Liberum,*" published in 1609 in which he claimed that "the sea cannot be attached to the possessions of any particular nation." This is an aspect of the legal context that European countries needed to justify their colonial expansions. The consequence, as believed, was that the indigenous people, who had survived and adapted to the new colonial regimes, were acculturated to the ways and mentalities of the conquerors. As a result, a wealth of native knowledge about the management and conservation of natural resources was, probably, forgotten. Earlier, the fishery regulations applied in the aquatic environment, perhaps, had tended to reflect the original practices and established rights in the husbandry of the fisheries, for example, by the monasteries, where they were believed to have imposed ban on the consumption of meat on Fridays and this, probably, had encouraged the need for conservation of fishery resources. Concomitantly, the growing demand has also encouraged the need for aquaculture. Such concepts and activities had, probably, automatically led to the generation of the concept and impelling need of sustained and rational harvest of the fishery resources; efficient management through different kinds of *modus operandi,* notably, the community-based fisheries management (CBFM), comanagement, etc.; and conservation of the fishery resources through pragmatic and effective measures. Concomitantly, regulations, such as "closed seasons," etc., to protect the spawners (brooders); maintenance of minimum mesh sizes; and bans on different types of gears and fish catching devices, which were believed to be destructive, had, probably, all been around for a long time; even though, the origin of these measures, probably, were not well documented, they were almost certainly reinvented in different areas. Further, a total allowable catch (TAC), based on scientific analysis, is a concept that dates back to the 1960s but measures such as the use of bag limits and limited access, probably, have a much longer history. In this connection, it could be believed that, many earlier size limits had, probably, reflected such aspects, which might had been related to aspects other than the issues of biological concerns. Thus, the strict size regulations for controlling the Dutch herring fishery were, probably, initially based on product requirements of a curing method developed in the 14th century. These regulations were designed, perhaps, to ensure high quality of the product, also probably, had specified mesh sizes and seasons.

Concomitant to the above, Territorial Use Rights in Fisheries (TURFs) were widespread and controlled the number of entrants to a fishery. There had been discussions on the traditional access rights as they applied to inshore fisheries. Incidentally, the 20th century has been an age of discovery and tremendous amount of progress for fisheries management; the world overwhelmingly witnessed a revolutionary change in the overall management scenario of the aquatic resources! Nevertheless, efforts are going on.

7.2 Financial support to fisheries: Implications for sustainable development

7.2.1 Summary profiles of GFTs in selected OECD countries

Notwithstanding the above, there has been survey(s) performed in the OECD member countries to obtain an inventory of GFT (**Government Financial Transfers)** programs in their countries.

An account of the GFT programs in different countries is briefly given in the following paragraphs:

7.2.1.1 Canada

Canada provided with funding for a total of 32 programs of financial support for the sector at the Federal level over the period 2001–2003. The major single element of financial support is the employment insurance (EI) scheme for fishers which totaled CAD 312.7 million in 2002–2003.

There were programs to provide short-term income support for self-employed fishers (primarily in inshore commercial fisheries) during the non-fishing season and it is administered by Human Resources and Skills Development Canada (HRSDC). Fishers' EI is designed to protect self-employed fishers from the uncertainties of the fishing industry due to high seasonality and fluctuating catch rates. Two benefit periods of 26 weeks are available to accommodate the summer and winter fisheries, respectively. Incidentally, expenditure for research, management, and enforcement amounted to CAD 300 million in 2002–2003 with research accounting for *c*. 39% of the total. Infrastructure expenditure is directed to the development and maintenance of harbors for the commercial and recreational sector and amounted to CAD 92.9 million in 2003.

Funding for the development of aboriginal commercial fisheries is the other major element of financial support in Canada. Following the *Sparrow* decision by the Supreme Court of Canada in 1990, DFO launched the Aboriginal Fisheries Strategy (AFS) in 1992. The objective of this program was to provide a framework for the regulation of aboriginal peoples' right to fish for food, social, and ceremonial purposes and to contribute to the economic self-sufficiency of aboriginal communities. CAD 10.9 million was expended under the allocation. Transfer programs were, probably, in place to assist aboriginal communities to obtain access to commercial fisheries and/or other economic development opportunities. The program, probably, had facilitated the voluntary retirement of commercial licenses and the issuance of licenses to eligible aboriginal groups in a manner that does not add to the existing effort on the resources. Under the Fisheries Access Program established in 1999, assistance is provided to eligible aboriginal groups in the form of nonrepayable contributions to facilitate the voluntary retirement of commercial licenses, vessels, and gear(s), construction of new vessels and gear(s), and to provide training and other skill development activities. Around CAD 68.3 million was provided under this program in 2003.

7.2.1.2 Czech Republic

In the Czech Republic, financial support is provided to the aquaculture and fish processing industries under the Special Accession Program for Agriculture and Rural Development (SAPARD), the European Agricultural Guidance and Guarantee Fund (EAGGF), and the Financial Instrument for Fisheries Guidance (FIFG). The structural subsidies provided under SAPARD had begun in May 2002 and grants for 15 fish-processing projects had been made by the end of 2003 worth a total of EUR 1,533,804 in public co-funding by the European Union and the state budget.

The supports were meant for the following:

(i) Support aimed at maintaining and improving the genetic potential of fish, including monitoring of the fish yield. Incidentally, subsidy granted in 2003 was CZK 4,179,700.
(ii) Support aimed at conserving and promoting the genetic resources of fish. Subsidy granted in 2003 was CZK 3,000,000.
(iii) Support for broader and easier access to information and key concepts from the scientific sphere and research. Subsidy granted in 2003 was CZK 105,000.
(iv) Support for education programs to improve conditions in organizations providing practical training to pupils. Subsidy granted in 2003 was CZK 1,140,000.
(v) Support for pond functions not related to fish production, such as flood prevention, environmental protection, aquacultural functions, recreation, protection of vegetation, game and birds, etc. This support is granted as part of general support provided to Czech regions. The main functions of the pond-based aquaculture industry are water storage, a means of keeping water in regions improvement of water quality, etc. Subsidy granted in 2003 was said to be CZK 71,000,000.

7.2.1.3 Iceland

In Iceland, limited transfers are provided to their fishing industry. Research and management services are provided through the Directorate of Fisheries, the Ministry of Fisheries, the Marine Research Institute and the Icelandic Fisheries Laboratories, and the Coast Guard provides fisheries surveillance. The Ministry of Fisheries is responsible for management of the Icelandic fisheries and the implementation of legislation to this effect. The Marine Research Institute is the Center of Scientific Research for marine resources and responsible for the recommendation of the annual TAC for the stocks subject to catch restrictions. The Directorate of Fisheries and the Coast Guard are responsible for ensuring compliance with the Fisheries Management Act.

Wage costs for the fishing industry are subsidized through an income tax deduction for fishers. The deduction is generally available to the fisheries sector and is also available to seamen on coast guard vessels, research vessels, dredging vessels, ferries, cruise ships, rescue vessels, and harbor pilot boats. All persons employed as fishers on fishing vessels are entitled to the income tax deduction for fishers.

Concomitant to the above, the Icelandic parliament "Althingi" decided to allow fishing vessel owners a transitional period, phasing in the fishing fee in equal steps from 2004 to 2009 (the fee was set at *c*. 6%). A fisheries surveillance fee is also levied and includes the fee for onboard surveillance and processing permit, operating license, payments for export certificates, and catch quota transfers. The fisheries surveillance fee covers the costs of the Directorate of Fisheries for fisheries surveillance, but part of the fee has been abolished and replaced by the fishing fee.

In addition to the above, the Fisheries' Development Fund was intended to encourage increased profitability in the fisheries sector and has been in operation since 1994. The fund made grants for retirement of vessels and purchase of infrastructure used for fish processing and their accompanying processing equipment(s); and was also entrusted with the financing of a major share in a new research vessel for the Marine Research Institute. To cover the cost of these extensive tasks and financial obligations, fees were levied on vessel owners and, for a time, on owners of processing plants. The levy was abolished when a fishing fee was introduced.

7.2.1.4 *Italy*

Italy provided with *c.* EUR 190 million in financial transfers to the fishing industry in 2004. This support was provided through the EU structural fund Financial Instrument for Fisheries Guidance (FIFG) and through the National Triennial Plans overseen by the Directorate General for Fishing and Aquaculture in the Ministry for Agricultural and Forestry Policies. The National Triennial Plans call for the rationalization and development of the whole fishing industry, where not only short- and medium-term targets but also long-term considerations are included.

The program financing provided by the FIFG covers 11 main areas:

(a) decommissioning schemes
(b) fleet renewal and modernization of fishing vessels
(c) protection of marine resources in coastal waters
(d) aquaculture
(e) fishing port facilities
(f) processing and marketing of fishery and aquaculture products
(g) small-scale coastal fishing
(h) socioeconomic measures
(i) measures to find and promote new market outlets
(j) operations by members of the trade
(k) innovative actions and technical assistance

Concomitant to the above, financial support provided through the National Triennial Plans is directed toward:

(a) compensation for closed seasons
(b) compensation for damage to the fishery sector
(c) support to promotion and marketing
(d) interest rebates and loan guarantees
(e) management expenditure
(f) research expenditure

Expenditures on management, research, and enforcement and infrastructure accounted for *c.* 65% of total transfers in 2004.

7.2.1.5 *Japan*

The main items of financial support provided by Japan to the fishing industry are direct payments for fishery restructuring, interest subsidies, infrastructure expenditure, and general services expenditure. The "Fishery Restructuring Program for Fishery Resource Rehabilitation" provides direct payment assistance for fishing fleet reduction. The program was started in 1981 following the reduction in access to distant water fishing grounds after the introduction of the 200-mile EEZ zones by the coastal countries. The fleet reduction program was established to avoid overcrowding in Japanese fishing grounds.

In Japan, the annual amount of the transfer budget had been rather stable in the range of JPY 2.0–3.8 billion in most of the years. Further, the major form of cost reducing transfers in Japan is an interest subsidy. The interest subsidy program is designed to assist structural adjustment of coastal fisheries under certain conditions. The main purpose of the program is to contribute to the introduction of advanced fisheries' management for structural adjustment of small-scale coastal fisheries. The actual differences between the commercial and subsidized interest rates were, probably, within a range of *c.* 1.25%–0.01%. The renewal of small fishing boats and equipments are said to be supported under the program in an effort to facilitate improvement in worker's safety on family-owned coastal boats. The annual amount of the budget for cost reducing transfers was stable around JPY 2.5–4.1 billion since mid-1997. However, the largest type of transfer to the Japanese fishing sector was allocated to the construction of coastal infrastructure. This transfer is said to be the government payment for the construction of new, or, the extension of the existing, fishing ports and other coastal public facilities including breakwaters, public wharves, navigation routes, roads, water supply and sewerage systems, and park facilities. This is a government program meant for establishing regional public infrastructures, in an effort to improve maritime transportation bases, to ensure safety navigations and to enhance quality of peoples' living environment of the regional communities, as well as disaster prevention. This transfer is, probably, not paid directly to the industry, but is used to provide infrastructure support. The annual amount of the transfer for this category was within the range of *c.* JPY 190–336 billion since the 1980s and totaled *c.* JPY 203 billion in 2003. Further, the provision of general services, such as management, research, and enforcement expenditures, amounted to around JPY 62 billion in 2003.

7.2.1.6 *Norway*

The financial support provided by Norway to its fishing sector has declined from NOK *c.* 153 million in 2001 to a budgeted NOK *c.* 121 million in 2004. This represents a significant decline from a peak in 1991 when total support amounted to NOK *c.* 1100 million. Payments for vessel decommissioning and license retirement represent the major item of expenditure, totaling NOK *c.* 35 million in 2004. The decommissioning scheme for vessels that are <15 m, was said to be established in July 2003, and was partly industry-financed (*c.* 50% of the total payments). Incidentally, *c.* 197 applications were received for decommissioning and *c.* 113 vessels were decommissioned.

Notwithstanding the above, "The Fisheries General Agreement Reserve Fund" was established to serve as a buffer for the schemes covered by the General Agreement and to provide financial help to fisheries-related projects. The reserve funds allow for strict control of spending; and, do not seem to permit spending in the other schemes covered by the agreement. It, thereby, contributes to the reduction in the need for government funding in the fisheries sector. Some transfers were, probably, related to research activities. There were also cost reducing transfers; and when there is a deficit on the minimum wage scheme, it is said to be covered by the reserve fund.

7.2.1.7 *Slovak Republic*

The Slovakian fisheries sector is a part of the program for sustainable rural development, under the general Sectoral Operational Program. This involves investments for the period of 2004–2006, with two measures being available to the fisheries sector relating to (a) fish processing and the promotion of fisheries products, and (b) aquaculture. Incidentally, the Slovak Republic does not seem to have any marine capture fisheries. However, their efforts seem to demonstrate the particular experience of an aquaculture-oriented country.

The broad objective of the program on fish processing and the promotion of fisheries products is to increase the competitiveness and quality of processed fishery products while reducing the adverse impacts on the environment. Incidentally, more specific objectives are to.

(i) decrease production costs and improve working conditions;
(ii) increase the quality of products;
(iii) improve hygienic conditions;
(iv) reduce negative impacts on the environment.

The support is provided as a nonrepayable financial contribution, with the processing and aquaculture sectors both being eligible for the aid; *c.* EUR 2.1 million is available under the program.

The objectives of the financial support provided to the aquaculture sector are to increase the effectiveness, competitiveness, and quality of fish-farming facilities while adhering to the environmental protection standards. This is said to be achieved by helping to:

(i) reduce the production costs and increase the labor productivity;
(ii) introduce new environmentally friendly technologies;
(iii) preserve specific genetic attributes of the local farmed lines and original lines and species;
(iv) expand fish breeding in freshwater aquaculture, and
(v) offer new employment opportunities.

Incidentally, *c.* EUR 3.1 million is available under the aquaculture program over the period 2004–2006.

7.2.1.8 *Sweden*

Sweden provides financial support to the fisheries sector under the EU's Financial Instrument for Fisheries Guidance (FIFG). The objectives of the program are to

(i) Contribute to achieving a sustainable balance between fishery resources and their exploitation.
(ii) Strengthen competitiveness of the sector and develop economically viable enterprises in the sector by generating a favorable structure.
(iii) Improve market supply and value added to fishery and aquaculture products.

(iv) Contribute to revitalizing areas dependent on fisheries and aquaculture.

The support was provided in the form of direct payments, cost reducing transfers, and as a type of loan. Under the loan conditions, beneficiaries have no obligation to pay back the loan as long as the stipulated conditions are met. The repayment of the loan amount was said to be interest-free, hence, might be called a "conditioned subsidy." In normal cases, the loan is written off during a 10-year period. A security has to be given if the aid is in the form of a loan and if the value of the aid (loan) amounts to > SEK *c.* 400,000. Further, support for the renewal of the fishing fleet is limited to vessels which are < 18 m. Further, for modernization of the fishing fleet, the vessels have to be listed in the EU's register of fishing vessels for at least 5 years. Incidentally, there is a higher grant for projects of collective interest within the small-scale coastal fisheries. Moreover, aids are granted only on condition that the fishing effort is not increased. Priority is given to the projects aimed at increasing the fish quality, improving the selectivity of the fishing gears, and improving working conditions and safety onboard. Further, the Swedish Government has appointed and authorized the Swedish Board of Fisheries for managing and paying various types of fisheries programs. Incidentally, the Swedish Board of Fisheries possesses the competence to issue national instructions concerning the FIFG. Moreover, the Industry Development unit of the Swedish Board of Fisheries, probably, has the practical function as the managing authority, and the head of the unit decides to approve or reject aid to individual project. The unit coordinates with the administration of the FIFG and, probably, also prepares a manual for the Swedish Board of Fisheries and the County Boards on handling the payments, etc. Incidentally, the approved beneficiaries of the support scheme are said to be specified for each measure, and include the fishers, authorities, institutes, organizations, aquaculture companies, public owners of fishing harbors, companies active in fishing harbors, fishery associations, processing industries, trade associations, producer's organizations, associations of companies and nonprofit associations, county boards, the Swedish Board of Fisheries, external consultants, and, so on. In this connection, it may be mentioned here that *c.* SEK 173 million was provided to the sector over the period 2001–2003 (half the period of the FIFG cycle). The estimated number of recipients is 1000, comprising 500 fishers, 200 processing industries, 100 aquaculture companies, and 200 other recipients.

7.2.1.9 United Kingdom

The primary method of direct support for the fishing industry is through the Financial Instrument for Fisheries Guidance, as with the other EU countries. The objectives of financial assistance measures are outlined in detail in the European Council Regulation (EC) No. 2792/1999. The objectives are to

(i) contribute to achieving a sustainable balance between fishery resources and their exploitation;
(ii) strengthen the competitiveness of the structures and the development of economically viable enterprise in the sector;
(iii) improve market supply and the value added to fishery and aquaculture products, and
(iv) contribute to revitalizing areas dependent on fishing and aquaculture.

Concomitantly, the objectives of the accompanying UK strategic program are:

(i) sustainable exploitation of resources;
(ii) adoption of an inclusive approach to fisheries management, i.e., managers, scientists, and industries working collaboratively;
(iii) to support fisheries-dependent communities;
(iv) to ensure an economically viable fishing industry; and
(v) to develop a culture of quality by facilitating traceability and quality control.

It may be mentioned here that with the exception of decommissioning schemes, grants are generally provided in the form of matching contributions. Aid to the fishing industry under FIFG is covered in the certain program(s), and is subject to a number of rules and conditions. There was, probably, a single program of fisheries aid for the whole of the UK with subprograms for England, Scotland, and Wales.

Notwithstanding the above, time and space, probably, did not permit detailed account in this regard about the other countries in the world and a compromise has to be made between length and depth.

7.3 Summary

(1) There have been detailed deliberations, in the present treatise, on different aspects of community-based management of resources, on a global scenario, including different management options, development aspirations, conflicts and consensus, etc., to name a few. Incidentally, an attempt has been made in this chapter to deal with little different but related aspects.

(2) With regard to fishery management before 1900, the bountiful resources at the dawn of civilization vis-à-vis less number of harvesters and consumers, probably, could not generate any sense of sustainability and rational management. The idea has come into reality with the growing number of consumers.

(3) Technological advancements have eased the management *modus operandi* of the natural resources.

(4) Incidentally, it may be interesting to note here that fishing itself is, perhaps, as old as mankind. Extensive shell middens seem to reflect the early concern with marine resources of prehistoric coastal settlements.

(5) With the increasing participation in fisheries on the high seas during the early part of the second millennium came in the conflicts, crises, and the forerunners of management decisions.

(6) In **France**, since the prerevolution period, the local regulatory boards (prud'homies) are regulating fair allocation of resources within the local jurisdiction of a prud'homie, protecting the territory from outside intrusions, and protecting the resources through various kinds of regulations, like mesh sizes, fishing seasons, etc.

(7) The cofrad'ias of **Spain**, probably, has similar types of regulations. Also, many coastal communities globally still have various systems of taboos, access rights, and traditional practices, which effectively protect the overharvest of the resources.

(8) In **Japan,** between 1603 and 1868, fishing territories were established, and coastal waters were considered as extensions of the land; the feudal lords partitioned the coastal waters and allocated areas of fishing ground to local fishing communities under the control of the village heads, subsequently shifted to the responsibility of the cooperatives. Such community approaches are also *in vogue* in the **Philippines, Oceania, the Pacific Coast of North America, and Mexico**.

(9) A brief glimpse into the historical resume′ reveals that the early role of shipping for trade and territorial protection and expansion has had enormous implications for fisheries. Incidentally, there, probably, had also been early manifestations of the overriding of local rights in favor of national policies.

(10) The growing demand for fish as food and for avocation has encouraged the need for aquaculture. Such concepts and activities have automatically led to the generation of the concept and impelling need of sustained and rational harvest of the fishery resources; efficient management through different kinds of *modus operandi,* notably the community-based fisheries management (CBFM), comanagement, etc., and conservation of the fishery resources through pragmatic and effective measures. Concomitantly, regulations, such as "closed seasons," etc., to protect the spawners (brooders), maintenance of minimum mesh sizes, and bans on different types of gears and fish catching devices, which are believed to be destructive, had, probably, all been around.

(11) Further, the Territorial Use Rights in Fisheries (TURFs) gradually become effective, widespread, and controlled a number of entrants to a fishery.

(12) Further, survey(s) were performed in the OECD member countries to obtain an inventory of GFT (**Government Financial Transfers)** programs in their countries, with regard to financial support to fisheries and their implications for sustainable development

(13) As a little detailed account, **Canada** has provided funding for a total of 32 programs of financial support for the sector at the Federal level over the period 2001–2003. **Czech Republic** has supported aquaculture and fish processing industries with an amount of EUR 1,533,804. Incidentally, **Iceland** was believed to have made limited transfers to their fishing industry. **Italy** provided *c.* EUR 190 million as financial transfers to the fishing industry in 2004. In **Japan,** the annual amount of the transfer budget has been rather stable with a range of JPY 2.0–3.8 billion in most of the years. However, the financial support provided by **Norway** to its fishing sector has declined from NOK *c.* 153 million in 2001 to a budgeted NOK *c.* 121 million in 2004. In **Slovak Republic,** investments for the period 2004–2006 include two measures being available to the fisheries sector relating to fish processing and the promotion of fisheries products, and aquaculture. Moreover, an amount of *c.* EUR 3.1 million was provided under the aquaculture program over the period 2004–2006. In **Sweden,** *c.* SEK 173 million was provided to the sector over the period 2001–2003 (half the period of the FIFG cycle). In **UK,** with the exception of decommissioning schemes, grants are generally provided in the form of matching contributions.

Glossary

Fisheries management may be defines as the pursuit of certain objectives through the direct or indirect control of effective fishing effort or some of its components.

Comanagement It usually refers to a bunch of arrangements with different degrees of power sharing allowing joint decision-making by the state and the user groups (UG) about a set of resources or an area. However, there is, perhaps, no single unique globally accepted definition of co-management (Armitage et al., 2007; Berkes, 2007). In other words, it brings together the fishers, the government officials, and others operating within the fold of the fisheries sector.

Different types of comanagement:

(i) *Instructive management*, where decisions are made in the government system, and fishers are instructed in decisions. This is not a comanagement arrangement.
(ii) *Consultative comanagement*, where fishers are consulted on proposals before decisions.
(iii) *Cooperative comanagement*, between fishers and authorities as equal partners in developing proposals for management decisions.
(iv) *Advisory comanagement*, where fishers advise governments on decisions to be made, and the government endorses these decisions.
(v) *Informative comanagement*, where governments have delegated the authority to make decisions to user groups, who are responsible for informing government about the decisions made.

Sustainable development According to the World Conference on Environment and Development (1987), sustainable development could simply be defined as "Development that meets the needs of the present generation without compromising the ability of future generations to meet their own needs."

Natural resources The World Bank has *defined natural resources* as `materials that occur in nature and are essential or useful to humans, such as water, air, land, forests, fish and wildlife, topsoil, and minerals, etc. There have been various types of conflicts with the use of natural resources, probably, since the dawn of human civilization.

Sustainable fisheries Sustainable, productive fisheries are achieved when appropriate management ensures a high probability of stocks being able to replenish themselves over a long period of time within a sound ecosystem, while offering stable economic and social conditions for all those involved in the fishing activity.

Abbreviations

ACOPS	Advisory Committee on Problems of the Sea
AFS	Aboriginal Fisheries Strategy
AQC	Aquaculture
ACC	(UN) Administrative Committee on Coordination
CEP	Caspian Environment Program
CROP	The Council of Regional Organizations of the Pacific
EAGGF	European Agricultural Guidance and Guarantee Fund
EBFM	Ecosystem-Based Fisheries Management
FAO	Food and Agricultural Organisation (of the UN)
FFA	Forum Fisheries Agency
FIFG	Financial Instrument for Fisheries Guidance
FW	Freshwater
GEF	Global Environment Facility
GFT	Government Financial Transfers
ICES	International Council for the Exploration of the Sea
MPA	Marine Protected Areas
NADS	nontarget, associated and dependent species
RFB	Regional Fishery Bodies
RSC	Regional Seas Conventions
SAP	Strategic Action Program
SOCA	Subcommittee on Ocean and Coastal Areas
SAPARD	Special Accession Program for Agriculture and Rural Development
SIDS	Alliance of Small Island States
SPC	South Pacific Commission
TAC	Total Allowable Catch
TURF	Territorial User Rights in Fisheries
UNEP	United Nations Environment Programme
UNCLOS	United Nations Convention on the Law of the Sea

Index

Note: Page numbers followed by *f* indicate figures.

B

C

F

G

J

K

L

T